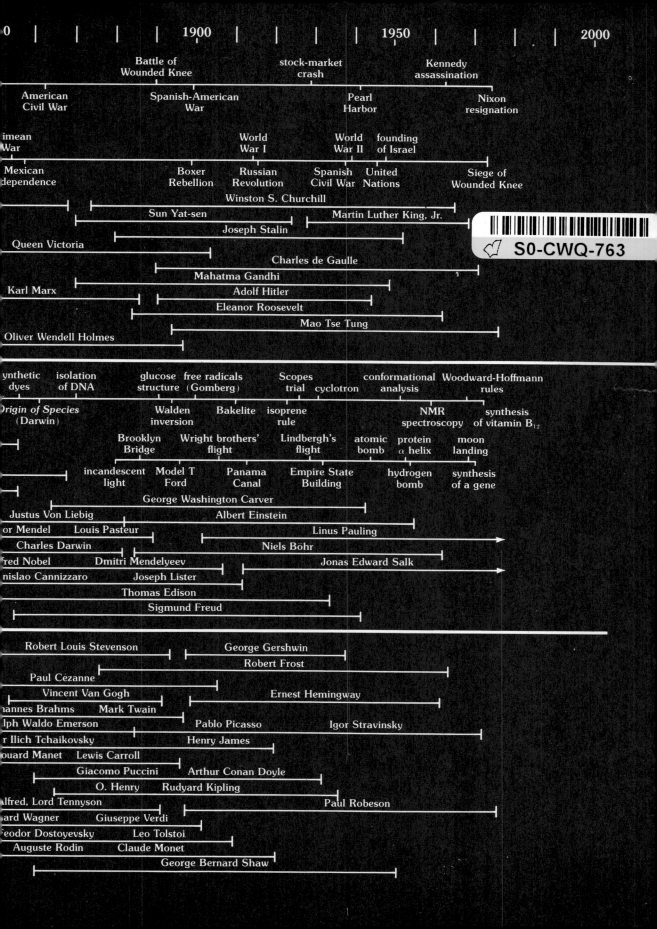

AN INTRODUCTION TO
ORGANIC CHEMISTRY

AN INTRODUCTION TO ORGANIC CHEMISTRY

WILLIAM H. REUSCH

Michigan State University

HOLDEN-DAY, INC.
San Francisco
Düsseldorf   Johannesburg   London
Panama   Singapore   Sydney

AN INTRODUCTION TO ORGANIC CHEMISTRY

Copyright © 1977 by Holden-Day, Inc.,
500 Sansome Street, San Francisco, Ca.
All rights reserved. No part of this
book may be reproduced by mimeograph,
or any other means, without the permission
in writing of the publisher.

Library of Congress Catalog Card Number: 76–50855
ISBN: 0–8162–7161–5

Printed in the United States of America

34567890   0987

Permission for the publication of the Sadtler Standard Spectra® listed below has been granted, and all rights are reserved by Sadtler Research Laboratories, Inc.

figure number

3.1	Infrared spectrum of 1,2-dichloroethane
P3.3	Infrared spectrum of carbon tetrachloride
4.1	Infrared spectrum of limonene
4.2	Pmr spectrum of limonene
8.2	Infrared spectrum of 3-methyl-2-buten-1-ol
8.3	Pmr spectra of methanol
13.3	Pmr spectra of *N,N*-dimethylformamide

The following spectra are reproduced by courtesy of Varian Associates:

figure number

2.7	Pmr spectrum of *n*-octane
3.2	Pmr spectra of 1,2,2-trichloropropane and 1,2-dibromo-2-methylpropane
P4.1	Pmr spectra of α- and β-pinene
8.5	Pmr spectrum of trimethylene oxide
8.14	Pmr spectrum of isopropyl mercaptan
9.11	Pmr spectra of *p*-cymene and *p*-methylanisole
11.4	Pmr spectra of α-methylbenzylamine and morpholine
12.3	Pmr spectra of *n*-butyraldehyde and *p*-chloroacetophenone
13.3	Pmr spectra of *p*-ethoxyacetonilide and methyl methacrylate

to my family

PREFACE

IN CHOOSING an organic chemistry text from the assortment of introductory books now available, teachers of this subject must evaluate the relative importance of the many features—some common, some unique—that each book offers their instructional program. This is by no means a trivial task, and I hope my comments here will prove helpful to those facing such a decision. In the long run, of course, it will be the student users of a text who will determine its instructional value.

My purpose in writing this book has been to provide a thorough but concise introduction to the most important and fundamental principles of organic chemistry. In this effort I have tried wherever possible to show the importance of the experimental evidence on which these principles are founded. Innovations in presentation have been made where a clear improvement in understanding results. However, traditional approaches of proven effectiveness have been retained, and no changes were made just for the sake of being different. I believe that careful inspection—or better yet, use—of this text will disclose a freshness of approach, a clarity of exposition, and a variety of content which make it unique. In the following paragraphs I have outlined some important features that make this book an ideal text for a beginning course in organic chemistry.

By a judicious selection of topics, this book has been kept to a reasonable size for use in a one-year course serving both chemistry majors and nonmajors. To this end, I have tried to emphasize those aspects of organic chemistry that are truly central and necessary if understanding is to be achieved, and in cutting topics I have chosen to sacrifice breadth rather than depth of coverage. As a consequence of this decision, instructors may find some old friends missing (the benzoin condensation) or treated only in the supplemental problems (the haloform reaction). I apologize if I have ruffled any feathers with these omissions, but I firmly believe that it is unrealistic to ask beginning students to familiarize themselves with all important reactions and mechanisms in one year.

I designed and wrote this book to serve as a teaching text, and not as an encyclopedia. Since it is good pedagogy to develop complex subjects gradually, this results in some important topics being discussed progressively in different sections of the book. For example, Brønsted acid-base concepts are first treated in conjunction with $E2$ elimination reactions (Chapter 3), placed in a quantitative framework during the development of aromaticity (Chapter 9), and then reviewed and further refined in Chapters 12 and 14. A similar sequential development is used for spectroscopy, bonding, and electrophilic aromatic substitution. My guideline here has been to carry the exposition of a subject only to that stage actually used in accompanying discussions. If a later topic requires a more sophisticated treatment, the necessary elaboration is provided at that time.

I have chosen a functional-group organization for the first fourteen chapters of the text. The effectiveness of this approach has been established many times, and within this context particular emphasis is placed on the structure and reactivity of common classes of organic substances. The order of presentation is arranged so that the characteristic reactions of one functional group illustrate preparations of subsequently discussed groups. Most commonly used organic compounds come from petroleum or coal; the book therefore begins with a discussion of alkane and cycloalkane chemistry (mainly combustion and halogenation), followed immediately by a chapter on alkyl halides. This allows a very early treatment of substitution and elimination reactions, which in turn lead naturally to chapters on alkenes, alkynes, and alcohols.

The characteristic reactions of each functional-group class are organized according to reaction types and simple mechanistic concepts. In this way key similarities and differences between functional classes are underlined, making it easier for students to manage the large number of reactions they are encountering for the first time.

The importance of chemical reactions in effecting specific molecular transformations (that is, as tools for synthesis) is clearly described, and a summary of key reactions, organized according to use, is given in a separate section following the last chapter. However, the degree to which synthesis should be emphasized has been left up to the instructor, since in my opinion nonmajors need not be well versed in this skill. For those wishing to stress synthesis, brief reviews of important preparation methods are given in turn for alkenes, alkynes, alcohols, amines, and carbonyl compounds. Furthermore, the ability to devise two- or three-step reaction sequences leading to specific compounds is gradually nurtured by carefully chosen problems, and more complex synthesis pathways are discussed in the chapters on carbonyl compounds and carboxylic acid derivatives. Finally, a comprehensive treatment of the strategy and methodology of synthesis is provided in Chapter 19. The chemistry major should find this a stimulating and useful review of the entire subject.

A clear distinction between fact and theory is maintained throughout. Reaction mechanisms are useful in correlating, organizing, and remembering experimental observations. However, if mechanisms are presented in a manner that overemphasizes their importance, we risk establishing a situation in which students memorize mechanisms rather than reactions, often with a minimum of understanding. In accordance with this principle, I have also chosen to discuss theoretical bonding models at several different points, increasing the sophistication of the model as required by the case under consideration. A simple covalent electron-pair model is used in the introductory chapters; then a σ,π bonding model is presented for use with alkenes and alkynes. This model is subsequently elaborated for aromatic systems, a subject which also lends itself to a comparison of resonance and molecular-orbital models. Orbital-symmetry considerations are treated in still a later chapter (16).

In accordance with my effort to show the experimental foundation of organic chemistry, important spectroscopic tools (notable pmr and infrared spectroscopy) are introduced in the first chapter and are used over and over again in subsequent chapters. In my opinion pmr spectroscopy is particularly suitable for beginning students because the chemical-shift concept serves to illustrate and reinforce the recognition of structurally equivalent groups of atoms, a facet of organic chemistry that is often troublesome for the novice. Furthermore, pmr is easier to understand and yields more immediately useful information than infrared spectroscopy. The treatment of pmr in Chapter 1 is very elementary and dwells only on the chemical-shift concept. The student is made aware of spin-spin splitting effects, but a full exposition of this aspect of nmr spectra is deferred to Chapter 8.

Naturally occurring compounds representative of each functional group are presented and discussed in turn in each chapter. An artificial distinction between synthetic compounds and natural products is thus avoided. This incorporation of biologically related substances and topics begins in Chapter 1 with the subject of isomerism (progesterone and tetrahydrocannabinol are isomers) and continues in later chapters with topics as diverse as insect pheromones, nerve-signal transmission, transformations of vitamin-D precursors, and biosynthetic pathways. Because I consider proteins and carbohydrates to be special cases, involving relatively subtle variations on given structural themes, I have chosen to develop these topics separately, in Chapters 17 and 18.

Within the limitations implied by an arrangement of topics according to functional groups, this book is organized in a way that allows for flexibility of use to meet a variety of instructional needs. Each important topic and subtopic is identified by a heading, so that the instructor can easily rearrange or select material for reading assignments. In teaching chemistry majors, for example, many instructors may wish to consolidate the aromatic-substitution discussion in Section 15.2 with that in Chapter 10. For a short nonmajors course, the treatment in Chapter 9 (and 10) is adequate.

Elaboration of the fundamental facts and principles developed in the first fourteen chapters may proceed in any of several directions with the final group of special-topic chapters. One of these, Chapter 15, provides a unique review of earlier material in the context of interacting functional groups. Another, Chapter 16, discusses two important classes of nonionic reactions. Chapters dealing with carbohydrates, proteins, and the strategy of organic synthesis complete this group. Although these special chapters may be taken up in any order, the chapter on synthesis is more effective if it is preceded by the material on interacting functional groups.

As an aid to the beginning student, important terms and concepts are summarized at the end of each functional-group chapter. In addition, problems designed to probe the reader's understanding of important topics and to extend a particular line of reasoning are inserted throughout the text. Complete answers to all these problems are provided at the end of the book. These text problems are buttressed by additional exercise sets at the end of each chapter. Answers to the exercise sets are given in a separate study guide and solutions manual, prepared by Prof. Ronald Starkey of the University of Wisconsin, Green Bay. Students undertaking a self-instruction program in organic chemistry will find this study guide especially helpful.

Many people have helped produce this book, and any success it may enjoy is due in part to their efforts. Professors Peter Yates and Frank Lambert provided helpful commentary as the manuscript was being written. This "feedback" helped avoid a major rewriting of the first draft. Other valuable reviews of part or all of the text were conducted by

Carl Djerassi	Peter Yates	Roland Flynn
Thomas Bond	Addison Ault	Arthur Schultz
James Coke	Samuel Danishefsky	Joseph Ciabattoni
Jeremiah P. Freeman	Kenneth L. Marsi	Robert O. Hutchins
Weston Thatcher Borden	Lawrence J. Altman	Frank L. Lambert
David Todd	D. M. S. Wheeler	Margaret J. Jorgenson

Ronald Starkey not only assembled the problem sets at the end of each chapter, but also read the entire manuscript and offered many constructive suggestions. His study guide and solutions manual reflects both his years of teaching experience and his intimate association with this book. The artwork which enhances and clarifies many technical discussions was prepared by Basil Wood; Rosetta Reusch did most of the indexing.

Finally, special thanks are due two persons. Nancy Clark guided the manuscript and its author through the labyrinth of events leading to a published book. Her professionalism and attention to detail are evident throughout this text. Fred Murphy, President of Holden-Day, demonstrated unflagging enthusiasm and patience during the long years I spent writing this book. His role in encouraging me to undertake the project and in assembling a superb production team is deeply appreciated.

East Lansing *William H. Reusch*
December 1976

CONTENTS

preface *vii*
to the student *xxvii*

1 FUNDAMENTAL FACTS AND PRINCIPLES 1

1·1 DETERMINATION OF MOLECULAR FORMULA 2
purification methods 2
combustion analysis 4
  composition 4
  empirical formula 5
  molecular formula 5
mass spectroscopy 6

1·2 DETERMINATION OF STRUCTURAL FORMULA 8
the problem of isomerism 8
diffraction analysis 8
molecular spectroscopy 10
  infrared spectroscopy 12
  electronic and rotational spectroscopy 14
  nuclear-magnetic-resonance spectroscopy 15

1·3 THEORETICAL MODELS FOR MOLECULAR STRUCTURE 19
the ionic bond 19
the covalent bond 20
molecular geometry 22
charge distribution 23

1·4 CHEMICAL REACTIVITY 25
types of organic reactions 25
reaction mechanisms 26
reactive intermediates 28
functional groups 29

SUMMARY 31
EXERCISES 32

2 ALKANES AND CYCLOALKANES 35

2·1 NOMENCLATURE AND REPRESENTATION 35
nomenclature of simple alkanes 35
representation of molecular structure 36
conformations 39
the IUPAC nomenclature system 40

2·2 PROPERTIES OF ALKANES 43
physical properties 43
spectroscopic properties 46

2·3 CHEMICAL REACTIVITY OF ALKANES 46
combustion 46
  heat of reaction 47
  bond energies 47
halogenation 51
  chlorination of methane 51
  other alkanes 54

2·4 REACTION-RATE THEORY 54
energy of activation 55
the probability factor 57

2·5 CYCLOALKANES 59
stereoisomerism of substituted cycloalkanes 60
ring strain 61
conformations of cycloalkanes 63
chemical reactivity 68

SUMMARY 68

EXERCISES 69

3 ALKYL HALIDES 73

3·1 NOMENCLATURE OF ALKYL HALIDES 74

3·2 PROPERTIES OF ALKYL HALIDES 74
physical properties 74
infrared spectroscopy 75
pmr spectroscopy 76
conformational effects 78

3·3 NUCLEOPHILIC SUBSTITUTION REACTIONS AT CARBON 79
structure and reactivity 79
reaction-rate studies 80

3·4 THE S_N2 MECHANISM   81
    transition-state configuration   82
    nucleophilicity   83

3·5 THE S_N1 MECHANISM   84
    solvent effects   85
    the rate-determining step   86
    stereochemistry of S_N1 reactions   87

3·6 CARBONIUM-ION INTERMEDIATES   88
    carbonium-ion stability   88
    the Hammond postulate   89
    molecular rearrangement   90

3·7 ELIMINATION REACTIONS   91
    the E2 mechanism   92
        the Saytzeff rule   92
        transition-state configuration   93
    basicity and acidity   94
    the E1 mechanism   96

3·8 ORGANOMETALLIC COMPOUNDS   97
    formation of organometallic compounds   97
    reactions of organometallic compounds   98

    SUMMARY   100

    EXERCISES   101

4 ALKENES   107

4·1 NOMENCLATURE OF ALKENES AND CYCLOALKENES   108

4·2 STRUCTURE AND PROPERTIES OF ALKENES   110
    spectroscopic properties   110
    stereoisomerism of alkenes   112

4·3 THE LCAO BONDING MODEL   115
    sigma bonding   116
    pi bonding   118
    tau bonding   120

4·4 STRUCTURE AND STABILITY IN ALKENES   121
    heats of hydrogenation   121
    thermodynamic stability   122

    SUMMARY   124

    EXERCISES   125

5 REACTIONS OF ALKENES 127

5·1 ADDITION OF BRØNSTED ACIDS TO ALKENES 127
Markovnikov's rule 128
mechanism for addition reactions 129
testing the mechanism 130

5·2 ADDITION OF LEWIS ACIDS TO ALKENES 132
halogen addition 132
hypohalous acid addition 133
diborane addition 133

5·3 ADDITION MECHANISMS AND STEREOCHEMISTRY 135

5·4 COORDINATION OF ALKENES WITH TRANSITION METALS 138

5·5 OXIDATION REACTIONS OF ALKENES 140
oxidation-reduction terminology 140
hydroxylation 140
ozonolysis 142
  ozone reactions 142
  mechanism of ozonolysis 144

5·6 ADDITION OF RADICALS TO ALKENES 144

5·7 DIMERIZATION AND POLYMERIZATION 145

5·8 ADDITION OF CARBENES TO ALKENES 148

5·9 ADDITION REACTIONS OF CYCLOPROPANE 149

SUMMARY 151

EXERCISES 152

6 ALKYNES 155

6·1 NOMENCLATURE OF ALKYNES 156

6·2 STRUCTURE AND PROPERTIES OF ALKYNES 157
structure-composition relationship 157
structural and bonding characteristics 158
spectroscopic properties 159
acidity of ethynyl hydrogen atoms 160

6·3 ADDITION REACTIONS OF ALKYNES 162
hydrogen addition 162

   reactions with electrophilic reagents   163
    halogen   163
    Brønsted acids   163
   hydration and tautomerism   164
   the nucleophilicity of alkynes   166
   reactions with nucleophilic reagents   167

6·4 ADDITION REACTIONS WITH DIBORANE   168

6·5 OXIDATION REACTIONS OF ALKYNES   169

6·6 DIMERIZATION, TRIMERIZATION, AND POLYMERIZATION   169

 SUMMARY   172

 EXERCISES   173

7 STEREOCHEMISTRY   177

7·1 OPTICAL ACTIVITY   177
 plane-polarized light   177
 optical isomerism   179

7·2 ENANTIOMERISM   180
 elements of symmetry   181
  axis of symmetry   181
  mirror plane of symmetry   182
 center of symmetry   182
 molecular chirality   183
 racemic modifications   186

7·3 DIASTEREOISOMERISM   187

7·4 RESOLUTION OF ENANTIOMERS   189

7·5 CONFIGURATION   191
 the structural dilemma   191
 configurational notation   192
 the Cahn-Ingold-Prelog nomenclature rules   193

7·6 CONFORMATIONAL ANALYSIS   196

 SUMMARY   199

 REFERENCES   200

 EXERCISES   200

8 ALCOHOLS AND ETHERS 205

8·1 NOMENCLATURE OF ALCOHOLS AND ETHERS 206

8·2 PROPERTIES OF ALCOHOLS AND ETHERS 208
hydrogen-bonding effects 208
spectroscopic properties 209

8·3 SPIN-SPIN INTERACTIONS 211
predicting spin-spin splitting patterns 213
the spin-spin coupling constant 214
proton exchange and spin-spin coupling 215

8·4 CHEMICAL REACTIVITY OF ALCOHOLS AND ETHERS 217
acidity and basicity of alcohols and ethers 219

8·5 ELECTROPHILIC SUBSTITUTION AT OXYGEN 221
ester formation 221
Williamson ether synthesis 222
oxirane formation 224

8·6 NUCLEOPHILIC SUBSTITUTION REACTIONS 225
sulfonate esters as leaving groups 225
sulfite and phosphite ester intermediates as leaving groups 227
conjugate acids as leaving groups 229
ring-opening reactions of oxiranes 230

8·7 ELIMINATION REACTIONS 231
dehydration 231
oxidation and dehydrogenation 233

8·8 THIOLS, SULFIDES, AND PEROXIDES 235

SUMMARY 237

EXERCISES 239

9 AROMATIC COMPOUNDS 245

9·1 THE BENZENE PUZZLE 246

9·2 CONJUGATED DIENES AND POLYENES 248
thermodynamic stability 248
ultraviolet-visible spectroscopy 250
conjugation in benzene 253

9·3 THEORETICAL MODELS FOR BENZENE 254
the molecular-orbital model 254
the resonance model 256

9·4 NOMENCLATURE AND PROPERTIES OF AROMATIC COMPOUNDS   259
    compounds related to benzene   259
    properties of aromatic hydrocarbons   261
        ultraviolet-visible spectroscopy   261
        infrared spectroscopy   262
        pmr spectroscopy   262

9·5 CONFORMATIONAL ENANTIOMERISM IN BIPHENYLS   265

9·6 ELECTROPHILIC SUBSTITUTION REACTIONS OF BENZENE   268
    electrophilic attack on the benzene ring   268
        halogenation   268
        nitration   268
        sulfonation   269
        Friedel-Crafts alkylation   269
    a mechanism for electrophilic aromatic substitution   270
    oxidation of alkyl side chain   272

9·7 CRITERIA OF AROMATICITY   273
    minor changes in the benzene ring   273
    other cyclic conjugated polyenes   274
    the Hückel rule   276
        six-π-electron systems   276
        two-π-electron systems   280
        ten-π-electron systems   280
        fourteen-π-electron systems   281
        eighteen-π-electron systems   282

SUMMARY   282

EXERCISES   284

10 ARYL HALIDES AND PHENOLS   289

10·1 ARYL HALIDES   289

10·2 REACTIONS OF ARYL HALIDES: THE HALIDE GROUP   291
    nucleophilic displacement of halide   291
    elimination reactions   295
    formation of organometallic compounds   296

10·3 REACTIONS OF ARYL HALIDES AT THE AROMATIC RING   297

10·4 PHENOLS AND THEIR DERIVATIVES   299

10·5 REACTIONS OF PHENOLS   302
    acidity of phenols   302
    substitution reactions at oxygen   304

        substitution at carbon   305
        reactions at the aromatic ring   305

10·6 QUINONES   307

10·7 NONBENZENOID COMPOUNDS RELATED TO PHENOLS   309

10·8 BENZYL AND ALLYL HALIDES AND ALCOHOLS   310
        S_N2 reactions of benzyl halides   310
        S_N2 reactions of allyl and propargyl halides   312
        S_N1 reactions of benzyl derivatives   313
        S_N1 reactions of allyl derivatives   315
        stability of benzyl and allyl intermediates   316

SUMMARY   317

EXERCISES   319

11 AMINES   323

11·1 NOMENCLATURE OF AMINES   323

11·2 NATURALLY OCCURRING AMINES   325
        alkaloids   325
        hormones and drugs   327

11·3 STRUCTURE AND PROPERTIES OF NITROGEN COMPOUNDS   328
        bonding and stereochemistry   328
        a modified structure-composition relationship   329
        hydrogen-bonding effects in amines   330
        spectroscopic properties   330
        basicity of amines   333

11·4 ELECTROPHILIC SUBSTITUTION AT NITROGEN   335
        N-alkylation   335
        N-acylation and sulfonation   337
        N-nitrosation   338
        N-oxidation   339

11·5 DISPLACEMENT AND ELIMINATION OF NITROGEN FROM ALIPHATIC AMINES   340
        quaternary salts as leaving groups   340
        amine oxides as leaving groups   342

11·6 RING-SUBSTITUTION REACTIONS OF AROMATIC AMINES   343

11·7 REACTIONS OF ARYL DIAZONIUM SALTS   344
        displacement of nitrogen   344
           the S_N1 mechanism   345
           radical decomposition   346

   nucleophilic attack at nitrogen   347
   diazo coupling reactions   348
   reduction of the diazonium cation   350

11·8   **METHODS OF PREPARING AMINES**   350

11·9   **PHOSPHINES**   352

  SUMMARY   353

  EXERCISES   354

12   ALDEHYDES AND KETONES   361

12·1   **NOMENCLATURE OF ALDEHYDES AND KETONES**   363

12·2   **PROPERTIES OF ALDEHYDES AND KETONES**   364
   structure and bonding of the carbonyl group   364
   physical properties   365
   spectroscopic properties   366
    ultraviolet-visible spectroscopy   366
    infrared spectroscopy   367
    pmr spectroscopy   368

12·3   **REVERSIBLE ADDITION REACTIONS**   369
   formation of hydrates, hemiacetals, and hemiketals   370
   formation of acetals and ketals   371
   cyanohydrin formation   373
   formation of imines and related compounds   374
    oximes, hydrazones, and semicarbazones   375
    reduction of imines to amines   378

12·4   **IRREVERSIBLE ADDITION TO THE CARBONYL GROUP**   379
   reduction by complex metal hydrides   379
   reduction by diborane   380
   reactions with organometallic reagents   380
   the Wittig reaction   381

12·5   **REDUCTIVE DEOXYGENATION OF CARBONYL COMPOUNDS**   383
   oxidation reactions   384

12·6   **REACTIONS VIA ENOL INTERMEDIATES**   385
   conjugate acids and bases of carbonyl compounds   387

12·7   **REACTIONS OF ENOLATE ANIONS**   391
   alkylation reactions   392
   the aldol condensation   394

  SUMMARY   398

  EXERCISES   432

13 CARBOXYLIC ACIDS AND THEIR DERIVATIVES 407

13·1 NOMENCLATURE OF CARBOXYLIC ACIDS 407

13·2 SOURCES OF CARBOXYLIC ACIDS AND THEIR DERIVATIVES 411
methods of preparing carboxylic acids 411
oxidation states of carbon 412
organic acid derivatives 413

13·3 PROPERTIES OF CARBOXYLIC ACIDS AND THEIR DERIVATIVES 416
hydrogen-bonding effects 416
spectroscopic properties 416
infrared spectroscopy 416
pmr spectroscopy 419

13·4 NATURALLY OCCURRING CARBOXYL DERIVATIVES 421
waxes, fats, oils, and phospholipids 421
soaps and detergents 423
flavor and perfume components 423
proteins 424
alkaloids and other physiologically active compounds 425
insect hormones and pheromones 430

SUMMARY 432

EXERCISES 432

14 REACTIONS OF CARBOXYLIC ACIDS AND THEIR DERIVATIVES 437

14·1 SUBSTITUTION OF HYDROGEN IN THE CARBOXYL GROUP 437
acidity and salt formation 437
carboxylate oxygen as a nucleophile 440

14·2 SUBSTITUTION OF HYDROXYL IN THE CARBOXYL GROUP 441
esterification 441
acyl halide formation 443
anhydride formation 443

14·3 DECARBOXYLATION REACTIONS 444

14·4 REDUCTION OF THE CARBOXYL GROUP 445
lithium aluminum hydride 445
diborane 445

14·5 ACYLATION REACTIONS OF CARBOXYLIC ACID DERIVATIVES 446
reactivity to nucleophilic substitution 446
reactions of acyl halides 447
amines, alcohols, water, and hydrogen peroxide 447
Friedel-Crafts acylation of aromatic rings and alkenes 448

   formyl chloride   449
   phosgene   449
   ketenes   450
  reactions of anhydrides   451
  reactions of esters   452
  reactions of amides   454
  reactions of nitriles   456

14·6 BIOLOGICAL ACYLATION REACTIONS   457
  enzyme catalysis   457
  nerve-impulse transmission   458

14·7 SULFONYLATION AND PHOSPHORYLATION   459

14·8 REDUCTION OF CARBOXYLIC ACID DERIVATIVES   463
  reactions with metal hydrides   463
  reactions with organometallic reagents   464

14·9 REACTIONS AT THE α CARBON ATOM   465
  halogenation   465
  the Claisen ester condensation   466
  β-dicarbonyl compounds in synthesis   468

14·10 BIOSYNTHESIS OF COMPLEX MOLECULES   470
  acetate condensation reactions   470
  the isoprene rule   472

14·11 PEROXY ACIDS AND THEIR DERIVATIVES   474

 SUMMARY   476

 EXERCISES   478

15 INTERACTING FUNCTIONAL GROUPS   485

15·1 INTERACTIONS INVOLVING THE HYDROXYL GROUP   487
  two hydroxyl functions   487
  hydroxyl and carbonyl functions   487
  hydroxyl and double-bond functions   489
  aromatic rings and the hydroxyl function   489

15·2 SUBSTITUENT EFFECTS ON ELECTROPHILIC AROMATIC SUBSTITUTION   490
  changes in reactivity   490
  orientation of incoming groups   491
  steric effects   495

15·3 ADDITION OF ELECTROPHILIC REAGENTS TO CONJUGATED DIENES   495

15·4 THE DIELS-ALDER REACTION   497

diene conformations   497
substituent effects   499
stereochemistry   500
dienelike systems   502

15·5 THE MICHAEL CONDENSATION   503

15·6 NEIGHBORING-GROUP PARTICIPATION IN DISPLACEMENT REACTIONS   507

neighboring halogen   508
neighboring oxygen and sulfur   511
neighboring nitrogen   512
neighboring carboxylate derivatives   513
neighboring carbon-carbon double bonds   514

15·7 CARBONIUM-ION-INDUCED REARRANGEMENTS   517

pinacol rearrangement   517
bridged ions involving π-electron systems   519
bridged ions involving σ-electron systems   521

15·8 REARRANGEMENTS TO ELECTRON-DEFICIENT NITROGEN ATOMS   523

Beckmann rearrangement   523
the Schmidt reaction   524
reactions of amide derivatives   525

15·9 REARRANGEMENTS TO ELECTRON-DEFICIENT OXYGEN ATOMS   526

REFERENCES   527

EXERCISES   528

16 NONIONIC ORGANIC REACTIONS   535

16·1 THE DISCOVERY OF STABLE ORGANIC RADICALS   535

detection and observation of free radicals   537

16·2 METHODS OF GENERATING FREE RADICALS   539

covalent-bond homolysis   539
   cracking   539
   peroxides   539
   azo compounds   539
   photolytic bond homolysis   540
electron transfer   541
hydrogen atom abstraction   542

16·3 REACTIONS OF FREE RADICALS   543
   abstraction and fragmentation   543
   addition to multiple bonds   543
   radical coupling or recombination   545
     quenching   546
     solvent-cage effects   546
   some useful radical-recombination reactions   547

16·4 THE STEREOCHEMISTRY OF RADICAL REACTIONS   550

16·5 PERICYCLIC REACTIONS   551
   types of pericyclic reactions   551
     cycloaddition reactions   551
     electrocyclic reactions   553
     sigmatropic reactions   553
   energy requirements of pericyclic reactions   554
   concerted versus nonconcerted reactions   554
   some perplexing aspects of pericyclic reactions   555
   stereochemical nomenclature for pericyclic reactions   558

16·6 THEORETICAL MODELS FOR PERICYCLIC REACTIONS   560
   the Woodward-Hoffmann contribution   560
   the frontier-orbital method: cycloaddition reactions   561
   electrocyclic reactions   564
   sigmatropic reactions   566
   a simple empirical rule for pericyclic reactions   568

16·7 PERICYCLIC REACTIONS OF VITAMIN D   569

REFERENCES   572

EXERCISES   572

17 PROTEINS, PEPTIDES, AND AMINO ACIDS   577

17·1 THE COMPOSITION OF PROTEINS   578

17·2 CONFIGURATION AND PROPERTIES OF α-AMINO ACIDS   581

17·3 THE PRIMARY STRUCTURE OF PROTEINS   582

17·4 PEPTIDES
   naturally occurring peptides   584
   peptide synthesis   586
   properties of simple peptides   587

17·5 SECONDARY AND TERTIARY STRUCTURES OF PROTEINS 588
   coiling and folding of peptide chains 588
   denaturation 594

17·6 CONJUGATED PROTEINS 594

   REFERENCES 596

   EXERCISES 597

18 CARBOHYDRATES 601

18·1 CONSTITUTION AND CONFIGURATION OF SIMPLE SUGARS 602

18·2 CHEMICAL REACTIONS OF MONOSACCHARIDES 604
   osazone formation 604
   oxidation 605
   selective chain shortening and lengthening 606

18·3 THE CONFIGURATION OF (+)-GLUCOSE 607

18·4 CYCLIC ANOMERS OF GLUCOSE 610

18·5 DISACCHARIDES AND POLYSACCHARIDES 614
   (+)-maltose 614
   (+)-cellobiose 615
   (+)-lactose 616
   (+)-sucrose 617
   (+)-raffinose 618

18·6 POLYSACCHARIDES 618
   cellulose 618
   starch 619

18·7 VITAMIN C 621

   REFERENCES 622

   EXERCISES 623

19 THE STRATEGY OF ORGANIC SYNTHESIS   625

19·1 PLANNING A SYNTHESIS   627

19·2 SELECTIVE INTRODUCTION AND MANIPULATION OF FUNCTIONAL GROUPS   631
selectivity in functional-group reactions   632
activating, deactivating, and blocking groups   632
   activating groups   632
   protective or blocking groups   634

19·3 CARBON-CARBON BOND-FORMING REACTIONS   637

19·4 CONCEPTION AND DESIGN OF A SYNTHESIS   640
synthesis strategies   640
judging a synthesis   644

19·5 EXAMPLES OF SYNTHESIS PROBLEMS   645

19·6 PROSPECTS IN SYNTHESIS   657

REFERENCES   658

EXERCISES   659

TABLES   669

ANSWERS TO TEXT PROBLEMS   685

index   789

TO THE STUDENT

WHETHER you approach the study of organic chemistry enthusiastically or with reluctance and trepidation, you may find the following guidelines helpful in achieving mastery of the subject.

First, the study of organic chemistry is in many respects similar to the study of a foreign language. There is, for example, a new vocabulary of terms and symbols. Indeed, reaction equations are essentially sentences describing the consequences of certain experimental operations. As with a language, most of what you learn will be used constantly thereafter. You must therefore try to acquire a firm foundation of fact and principle through a regular program of study and practice (problem solving).

Second, the mastery of organic chemistry will require a significant memorization effort on your part. Some students tend to regard memorizing as an inferior alternative to understanding; however, these two important qualities are so complexly intertwined they cannot be artificially separated in this fashion. One of the best methods of learning organic chemistry is to organize its multitude of facts according to certain well-established principles of structure and reactivity. In this way understanding and memorization can be achieved together. The major difficulty you may encounter in this approach is that many interesting and thought-provoking questions of "how" and "why" are not easily answered. Ironically, a beginning student's desire for full and straightforward explanations sometimes exceeds the professional chemist's knowledge and ability to provide answers. For this reason organic chemistry is still a lively and exciting subject.

FUNDAMENTAL FACTS AND PRINCIPLES

ORGANIC CHEMISTRY is most commonly and simply defined as the *chemistry of carbon compounds*. One way to find out what organic chemistry is all about is to visit and talk to its practitioners. If we were to do this in earnest, we would find ourselves in large industrial research laboratories, in universities and colleges, in oil refineries, in pharmaceutical laboratories, in medical schools, and in government laboratories devoted to such diverse subjects as agriculture, drug evaluation, weapon development, and forensic science. On questioning the organic chemists at work in these facilities, we would find that most of their effort is directed to answering one or more of the following basic questions about a carbon compound:

What is it?
What does it do?
How can it be made?

These are, of course, the very questions that chemists have grappled with for over a century, and we may well ask why an entire field of chemistry is devoted to the element carbon. There is no single clearcut answer to this question; each reader will have to judge the merits of this division for himself after he has become acquainted with the nature of organic chemistry.

Scientists of the early nineteenth century believed that living organisms—plants and animals—contained a "vital force" that permeated all parts and substance of the organism. Compounds derived from such sources were consequently considered to be fundamentally different from mineral substances. This view was gradually discredited as an increasing number of organic compounds were synthesized from inorganic materials. The distinction between the chemistry of organic and inorganic substances has been retained, however, mainly because of the very large number of known carbon compounds (well over 2 million) and their characteristic properties. Certainly we cannot fail to be impressed by the enormous variety of organic substances, ranging from foods such as sucrose and drugs such as strychnine and penicillin to synthetic polymers such as nylon.

During the past century answers to the first two of our basic questions have become increasingly precise and sophisticated. In the early 1800s an elemental analysis of a substance and a crude measurement of its molecular weight were considered an adequate answer to the question "What is it?" Today we can locate each of the atoms within a molecule with considerable accuracy, and even measure the energy required to rotate, stretch, and bend different parts of a molecule. The second question, "What does it do?" concerns the properties of different compounds. Early chemists were often content to classify a new substance as an acid or a base and to describe the colors and/or precipitates it formed with certain reagents. Today we can establish not only the structures of all significant products of a reaction, but also the existence of short-lived intermediates (some surviving for only a fraction of a second), the timing of different steps in a reaction, and the geometric relationships among reacting species. The third question, "How can it be made?" is, of course, the foundation of chemical synthesis.

It may at first seem unbelievable that such detailed knowledge of molecular systems is possible; indeed, this accomplishment deserves wider recognition as one of mankind's great achievements. In this book we shall see how these facts are determined and some of the uses to which they are put.

1·1 DETERMINATION OF MOLECULAR FORMULA

PURIFICATION METHODS

Modern chemists probe the structure of molecules with the aid of many powerful and highly selective instruments. However, before a new or unknown substance can be examined by these techniques, it must first be carefully purified. This is usually accomplished by processes that distribute or partition the components of a mixture between two phases. Since the equilibrium distribution between phases varies from compound to compound, separation of the phases effects a partial separation of the original mixture. Ideally, any degree of separation or purification desired may be achieved by successive repetitions of the procedure. Several such purification procedures are listed in Table 1.1, and two of these are illustrated in Figures 1.1 and 1.2.

A closely related problem is that of determining the purity of a sample and thus establishing the extent to which the purification procedures have been effective. The *melting points* of crystalline substances and the *boiling points* of liquids usually change when small amounts of

1·1 DETERMINATION OF MOLECULAR FORMULA

TABLE 1·1 *Purification methods*

name of process	nature of distribution
crystallization	crystalline solid $\rightleftharpoons$ solution or melt
distillation	liquid $\rightleftharpoons$ vapor
sublimation	solid $\rightleftharpoons$ vapor
extraction	solution A $\rightleftharpoons$ solution B (immiscible A and B)
adsorption chromatography	solid adsorbent $\rightleftharpoons$ vapor *or* liquid
partition chromatography	solution A $\rightleftharpoons$ vapor *or* solution B (immiscible A and B)

impurities are present; consequently, these properties often serve as criteria of purity. Chromatography may also be used for this purpose, provided sensitive detection devices are used. Other properties which may be correlated with the purity of a sample are *refractive index* for liquids and *molar absorptivity*, as determined in ultraviolet, visible, and infrared spectroscopy.

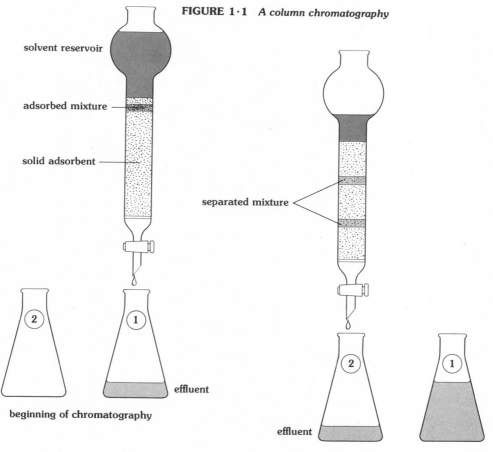

FIGURE 1·1 *A column chromatography*

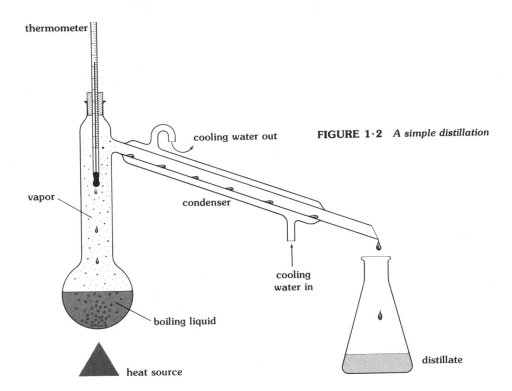

FIGURE 1·2 *A simple distillation*

COMBUSTION ANALYSIS

COMPOSITION Most organic compounds are composed largely of carbon and hydrogen. If necessary, the presence of these elements can be confirmed by detecting carbon dioxide and water among the products of combustion. Other elements, such as nitrogen, phosphorus, sulfur, or chlorine, may also be present, and qualitative tests for these usually involve destructive reduction with molten sodium, followed by analysis of the resulting inorganic ions.

To determine the amounts of carbon and hydrogen in an organic substance we can simply burn a weighed sample and measure the CO_2 and H_2O produced:

$$C_xH_yO_z + \underset{\text{(in excess)}}{O_2} \xrightarrow{\text{combustion}} xCO_2 + \frac{y}{2}H_2O \tag{1.1}$$

Suppose we find that complete combustion of a 4.0 mg sample of this compound yields 8.8 mg of CO_2 and 4.8 mg of H_2O. A simple calculation then shows the compound to be 60.0% carbon and 13.3% hydrogen. The weight fraction of carbon in CO_2 is $12/44$; hence

Weight of carbon in the 4.0-mg sample $= 8.8 \times {}^{12}/_{44} = $ **2.40 mg**
Percentage of carbon $= {}^{2.40}/_{4.0} \times 100 = $ **60.0%**

The weight fraction of hydrogen in H_2O is $2/18$, so

Weight of hydrogen in the 4.0-mg sample $= 4.8 \times {}^{2}/_{18} = $ **0.53 mg**
Percentage of hydrogen $= {}^{0.53}/_{4.0} \times 100 = $ **13.3%**

1·1 DETERMINATION OF MOLECULAR FORMULA

Reliable procedures for the quantitative analysis of other elements allow us to establish the complete composition of an unknown substance. Since the amount of oxygen in a compound is difficult to determine directly, it is often calculated by difference, after all the other components have been measured. In the example above the oxygen content would be 26.7%, provided no other elements were present.

EMPIRICAL FORMULA It is a relatively simple matter to convert the percentage composition of a compound into a formula that shows the relative proportions of different kinds of atoms in the molecule. Such an expression is termed an *empirical formula*. Let us examine this conversion procedure for the versatile solvent, reagent, and drug dimethyl sulfoxide (DMSO). The composition of DMSO has been determined as 30.72% carbon, 7.70% hydrogen, 41.02% sulfur, and 20.56% oxygen. The relative numbers of these atoms in a molecule of DMSO can be obtained by dividing each percentage value by the corresponding atomic weight. For example, the atomic weight of carbon is 12.01, so the proportion of carbon in **DMSO** is $^{30.72}/_{12.01} = 2.56$. However, since fractional atomic quantities are impossible, the resulting formula, $C_{2.56}H_{7.64}S_{1.28}O_{1.29}$, is not an empirical formula. To convert to the empirical formula for DMSO we must multiply each subscript by the smallest factor that will yield a whole-number ratio, in this case $^1/_{1.28}$. In this manner we arrive at the empirical formula C_2H_6SO:

carbon: $\dfrac{30.72}{12.01} = 2.56$ $\quad = 2$

hydrogen: $\dfrac{7.70}{1.008} = 7.64$ $\quad = 6$

$\times \dfrac{1}{1.28}$

sulfur: $\dfrac{41.02}{32.06} = 1.28$ $\quad = 1$

oxygen: $\dfrac{20.56}{15.99} = 1.29$ $\quad = 1$

PROBLEM 1·1
Calculate the empirical formula for an unknown substance analyzed as 60.0% carbon and 13.3% hydrogen. Assume that tests for other elements (except oxygen) are negative.

MOLECULAR FORMULA To convert an empirical formula to a *molecular formula* we must know the molecular weight of the compound in question. For example, the empirical formula of the antimalarial drug quinine is $C_{10}H_{12}ON$, so quinine must have a molecular formula which is an integral multiple of this value: $C_{10}H_{12}ON$, or $C_{20}H_{24}O_2N_2$, or $C_{30}H_{36}O_3N_3$. Since the molecular weights of these formulas vary by increments of 162, even a rough molecular-weight measurement will help to establish the correct formula. Techniques for determining molecular weights include measurement of freezing-point depression and boiling-point elevation, vapor-pressure osmometry, and mass spectrometry. Application of the first method to quinine gives a value of 320 ± 10 for the molecular weight; the molecular formula is thus $C_{20}H_{24}O_2N_2$.

MASS SPECTROSCOPY

Accurate molecular weights can often be obtained by *mass spectroscopy*. In this technique a small sample of a compound at very low pressure (about 10^{-5} torr.) is ionized by bombardment with a high-energy beam of electrons. The positive ions are separated from the negative ions, accelerated, and then sorted according to charge and velocity by being passed through a magnetic field. Figure 1.3 shows the essential features of a simple mass spectrometer.

The distribution of positive ions thus obtained is observed by a suitable detector (a photographic plate or an electron multiplier) and displayed as a series of peaks arranged in increasing ratio of ion mass to ion charge. Since most of the ions have unit charge, the major peaks in such a *mass spectrum* are effectively ordered according to mass, and the heaviest significant ion is usually the *molecular ion* (a positively charged sample molecule). In the mass spectrum of DMSO (Figure 1.4) we find that the molecular ion has a mass-to-charge ratio of 78. This fixes the molecular formula of DMSO as C_2H_6SO, which happens to be the same as its empirical formula.

The many smaller ions observed in such mass spectra are formed by fragmentation of the molecular ion into a variety of positively charged and neutral pieces. This fragmentation is thought to be due to the excess energy acquired by the molecular ion during ionization. The complex pattern of ions that results can serve as a "fingerprint" of the compound being studied. Most commercial single-focusing mass spectrometers

FIGURE 1·3 *A simple single-focusing mass spectrometer*

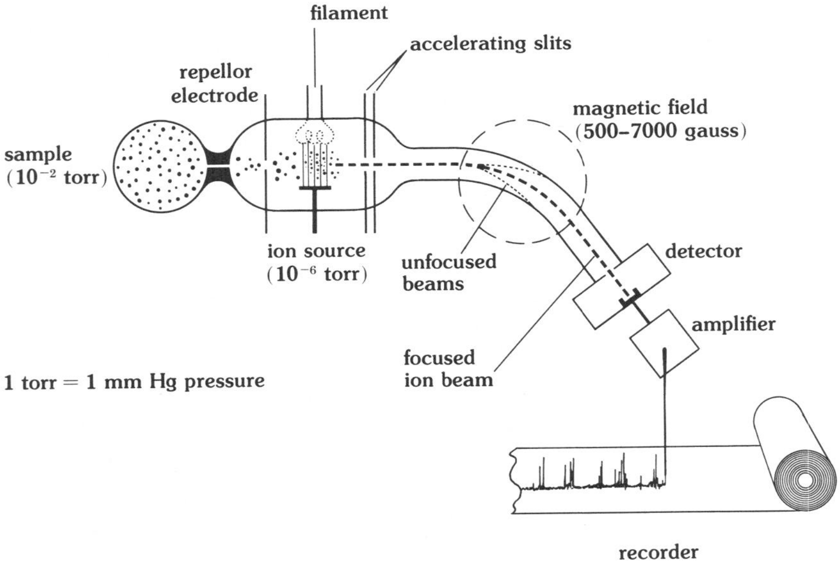

1·1 DETERMINATION OF MOLECULAR FORMULA

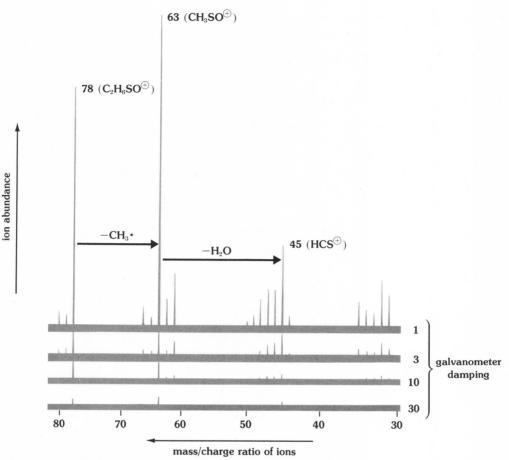

FIGURE 1·4 *Mass spectrum of dimethyl sulfoxide (four-element galvanometer)*

have sufficient resolving power to distinguish peaks separated by one atomic mass unit, even at masses as large as several thousand. Exact molecular weights of even high-boiling substances are thus obtained directly from the mass-to-charge ratio of the molecular ion.

Double-focusing mass spectrometers, with a resolving power over 100 times greater than the single-focusing instruments, are now widely used.† These spectrometers can determine the mass-to-charge ratio of an ion accurately to a hundred-thousandth of an atomic mass unit. Since the atomic weight of each elemental isotope (relative to $^{12}C = 12.00000$) deviates from an integral value by a small but characteristic amount, different molecular compositions will differ in precise molecular weight even though their "gross" molecular weights may be the same. For example, since $^{1}H = 1.007825$, $^{16}O = 15.994915$, and $^{14}N = 14.003074$, the following compounds have different precise molecular masses:

C_6H_{12}	C_5H_8O	$C_4H_8N_2$
84.0938	84.0574	84.0686

†These instruments, which are available commercially, cost more than $100,000.

Clearly, accurate mass measurements of this kind can lead directly to a molecular formula at the expense of only a small sample of material.

1·2 DETERMINATION OF STRUCTURAL FORMULA

THE PROBLEM OF ISOMERISM

In answering the question "What is it?" the organic chemist tries to write a *structural formula*, essentially a map showing the specific relationship of atoms to each other in a particular compound. Such a detailed description is necessary because there are many compounds which have identical molecular formulas but different internal arrangements of their atoms. We call such compounds *isomers*. For example, the liquid tetrahydrocannabinol, the active principle in marijuana, has the molecular formula $C_{21}H_{30}O_2$. So does the crystalline hormone progesterone, which controls the menstrual cycle and other functions of the female reproductive system. More than 100 other compounds having this composition are known, and over 500,000 are theoretically possible. The existence of all these isomers, each having distinct and characteristic properties, provides compelling evidence that these 53 atoms can be connected or bonded together in many different ways.

The different atomic constitutions of progesterone and tetrahydrocannabinol (THC) molecules can be seen in Figure 1.5. The two molecular models show the structure of these molecules according to our present knowledge. Note that the oxygen atoms in THC are fairly close together, whereas in progesterone they lie at opposite ends of the molecule. THC also has a rodlike appendage of five carbon atoms which is not present in progesterone, and although the hydrogen atoms in progesterone cover the carbon framework rather evenly, one of the six-membered rings in THC is relatively exposed (free of hydrogen). Since many atoms are obscured in pictures such as these, chemists usually prefer to represent the unique constitution of each pure compound by a structural formula, which shows individual atoms and the bonds that connect them. The structural formulas in Figure 1.5, for example, show many differences between progesterone and THC molecules which are not apparent in the pictures of these molecules.

DIFFRACTION ANALYSIS

Once we have established the molecular formula of a compound, we face the problem of determining the location of each atom within the molecule. Direct observation with an optical microscope is not possible because the wavelength of visible light is 4000 to 8000 Å (1 Å = 10^{-8} cm or 10^{-1} nm), whereas most molecules are only about 10^{-15} cm² in cross section. Only by using short-wavelength x-rays (0.5 to 2.5 Å) or beams of subatomic particles can we expect to obtain information about the fine structure of molecules. As Nobel Prize winner W. H. Bragg commented many years ago, "Broadly speaking, the discovery of X-rays has increased the keenness of our vision over ten thousand times and we can now 'see' the individual atoms and molecules." During the past 50 years a variety of substances, both simple and complex, have been studied by tech-

1·2 DETERMINATION OF STRUCTURAL FORMULA 9

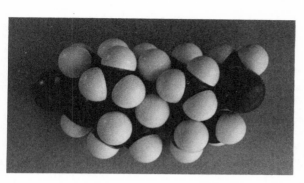

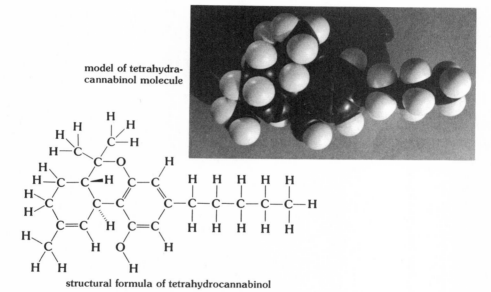

FIGURE 1·5 *Representations of progesterone and tetrahydrocannabinol (photographs by W. H. Reusch)*

niques such as x-ray and electron and neutron diffraction, and the dimensions of their component molecules have been determined to an accuracy of ±0.0005 Å.†

It is clear that a molecule of any substance must be held together by attractive forces or bonds between neighboring atoms. In structural formulas these bonds are represented by lines of arbitrary length. The bonds in actual molecules generally range from 1 to 2 Å, and angles between bonds are found to vary from 60° to 180°.

---

†In the early days of chemical investigation diffraction analysis of complex molecules was not possible; nevertheless, the chemists of that period correctly deduced the structures of many compounds by interpreting the results of a large variety of degradative chemical transformations. Our admiration for the insight and skill of these chemical Sherlock Holmeses must increase when we recognize that this task was roughly equivalent to that faced by a blind person in trying to describe an elephant from the bones lying about an "elephant's graveyard."

1·2 DETERMINATION OF STRUCTURAL FORMULA

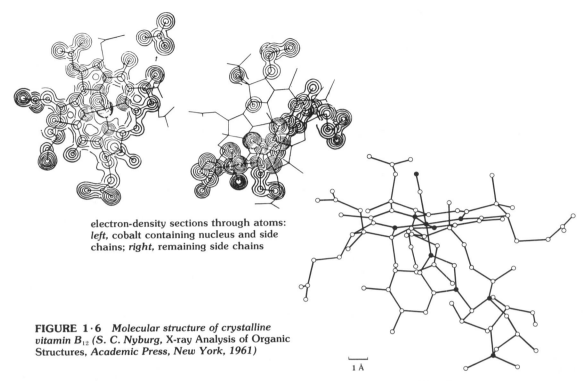

electron-density sections through atoms: *left*, cobalt containing nucleus and side chains; *right*, remaining side chains

FIGURE 1·6 *Molecular structure of crystalline vitamin B_{12}* (S. C. Nyburg, X-ray Analysis of Organic Structures, *Academic Press, New York, 1961*)

atom positions in space

structural formula

MOLECULAR SPECTROSCOPY

Despite the authority and precision of an x-ray structure determination, this technique is not yet routinely used. High-speed computers have helped to shorten considerably the time required for mathematical analysis of x-ray diffraction patterns, but the requirements of crystalline samples, skilled personnel, and computer time have often restricted use of this method to high-priority problems such as the structural analysis of vitamin B_{12}, nucleic acids, and the proteins myoglobin and lysozyme. Fortunately, less demanding spectroscopic methods have been developed to a point where they can provide reliable and useful structural information far more quickly than most chemical tests.

A molecule is not a rigid, precisely defined framework of atoms. Rather, it undergoes a continuous and complex internal motion resulting from the combined vibrations and rotations of individual atoms or groups of atoms. The bond distances and angles obtained by diffraction measurements are thus *average values* which reflect the most probable configuration of the molecule at any given time. The internal (or potential) energy of a molecule can be expressed as the sum of these electronic, vibrational, and rotational contributions:

$$E_{\text{total}} = E_{\text{electronic}} + E_{\text{vibrational}} + E_{\text{rotational}} \tag{1.2}$$

The highly structured nature of molecular spectra indicates that these energies are quantized. This means that a molecule can assume only particular values (or occupy discrete levels) of electronic, vibrational, or rotational energy. The energy in any of these categories will increase from some lower energy level E_1 to a higher level E_2 if the molecule absorbs light (electromagnetic radiation) of the proper wavelength. The *energy difference* ΔE for a transition from one energy level to another is related to the wavelength λ or frequency ν of the absorbed radiation in the following manner:

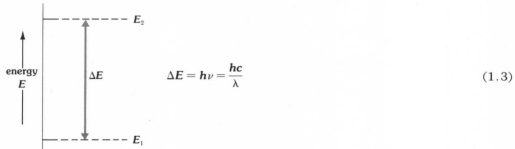

$$\Delta E = h\nu = \frac{hc}{\lambda} \tag{1.3}$$

where h = Planck's constant (6.624×10^{-27} erg-sec) and c = velocity of light (2.998×10^{10} cm sec^{-1}).

Organic chemists generally prefer to measure energy changes in kilocalories per mole. Using this frame of reference, we find that electronic energy transitions require 35 to 150 kcal/mole, vibrational transitions require 1 to 15 kcal/mole, and pure rotational transitions require about 0.1 kcal/mole. The energies associated with different parts of the electromagnetic spectrum are shown in Table 1.2.

1·2 DETERMINATION OF STRUCTURAL FORMULA

TABLE 1·2 *The electromagnetic spectrum*

wavelength (Å)	3×10^{10}	3×10^{8}	3×10^{6}	3×10^{4}	3×10^{2}	3	3×10^{-2}
frequency (Hz)	10^{8}	10^{10}	10^{12}	10^{14}	10^{16}	10^{18}	10^{20}
energy (kcal)	10^{-5}	10^{-3}	10^{-1}	10	10^{3}	10^{5}	10^{7}

gamma rays

x-rays

ultraviolet light

visible light

infrared light

microwaves

television waves

radio waves

INFRARED SPECTROSCOPY One of the most widely used spectroscopic methods for investigating molecular structure is infrared spectroscopy. Fortunately the electromagnetic radiation required to effect transitions between vibrational energy levels usually falls in the relatively accessible infrared region ($\lambda = 2 \times 10^4$ to 16×10^4 Å). All organic compounds absorb radiation in this region, and present-day recording spectrophotometers enable the chemist to obtain an infrared spectrum within a few minutes.† Such spectra consist of absorption peaks or bands appearing at frequencies which are characteristic of various groupings of atoms found in the sample molecules.

The positions of absorption maxima are given in terms of *wavelength*, usually expressed in microns (1 $\mu = 10^{-4}$ cm) or *wavenumber* $\tilde{\nu}$, where $\tilde{\nu} = \nu/c$ and is usually expressed in reciprocal centimeters, cm^{-1}. To simplify our discussion we shall limit our infrared notation to the *wavenumber scale*. A wavelength (in microns) is converted to a wavenumber (cm^{-1}) by use of the relationship $\tilde{\nu} = 10^4/\lambda$.

To a first approximation, the absorption frequency of a fundamental vibration—say, the stretching or bending of a bond between two atoms or groups of atoms—is determined primarily by the masses of the vibrating groups and the resistance to this vibration exerted by the bonding forces. In fact, much of the value of infrared spectroscopy lies in the observation that many stretching and bending motions are essentially independent of remote changes in the molecular structure. Thus certain prominent and characteristic absorption bands may be assigned to particular groupings of atoms and serve to identify these groups when they occur in unknown compounds. Aspirin, for example, a widely used drug which acts as an analgesic (pain-moderating) and antipyretic (fever-reducing) agent, displays an infrared spectrum with the characteristic absorption bands for O—H, C—H, and C=O (two kinds) shown in Figure 1.7.

†Commercially available instruments range in price from $4000 to $25,000.

1·2 DETERMINATION OF STRUCTURAL FORMULA   13

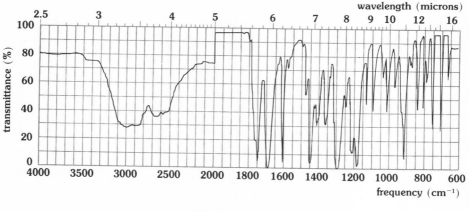

FIGURE 1·7 *Infrared spectrum of aspirin*

Most common atom groupings show characteristic absorption bands between 4000 and 1000 cm^{-1}. Some typical examples are listed in Table 1.3. The 1200- to 600-cm^{-1} region of an infrared spectrum is usually rather complex and is considered to be essentially unique for a particular compound. This part of the spectrum is therefore very useful for establishing the identity or nonidentity of chemical samples and is often called the "fingerprint" region. A noteworthy example of this application was the demonstration by Food and Drug Administration

TABLE 1·3 *Typical infrared absorption bands*

bond	absorption frequency (cm^{-1})	intensity
C—H	2800–3300	strong
C—C	1000–1300	variable
C=C	1620–1680	variable
C≡C	2100–2300	variable
C=O	1680–1780	very strong
O—H	3200–3650	strong
O—H in $\text{C}\begin{smallmatrix}\text{O}\\\text{OH}\end{smallmatrix}$	2500–2700	
N—H	3300–3500	medium

14   1·2 DETERMINATION OF STRUCTURAL FORMULA

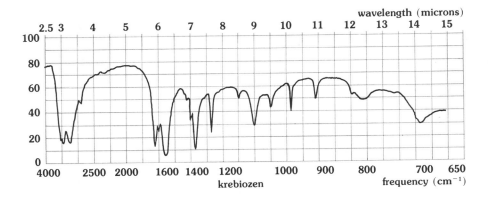

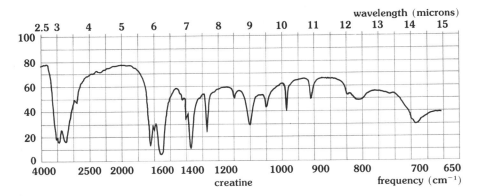

FIGURE 1·8 *Infrared spectra of krebiozen and creatine
(U.S. Food and Drug Administration)*

chemists that the controversial drug krebiozen exhibited an infrared spectrum identical with that of the well-known substance creatine (Figure 1.8).†

ELECTRONIC AND ROTATIONAL SPECTROSCOPY Transitions between electronic energy levels result from the absorption of light in the visible range (4000 to 8000 Å) or the ultraviolet range (2000 to 4000 Å) of the spectrum. Absorption bands in these regions are characteristic of molecules having polarizable electrons. We shall consider this type of spectroscopy in detail when we discuss such compounds in later chapters.

Transitions between the closely spaced rotational energy levels of a molecule require radiation in the microwave region of the spectrum. Since microwave spectrometers are quite complex and the spectra are difficult to interpret, this technique has not yet led to a routine analytic method.

NUCLEAR-MAGNETIC-RESONANCE SPECTROSCOPY During the past 20 years nuclear-magnetic-resonance (nmr) spectroscopy has been applied with increasing frequency to problems of molecular structure,

†The fascinating story of the hopes and disappointments surrounding the anti-cancer drug krebiozen was summarized in *Life* magazine, October 4, 1963.

1·2 DETERMINATION OF STRUCTURAL FORMULA

and it has become as useful an analytic tool as infrared spectroscopy. This technique is based on the fact that many atomic nuclei (^{1}H, ^{19}F, ^{13}C, and ^{31}P, for example), besides having mass and charge, also have *spin angular momentum*. Not all nuclei have spin, but those that do behave like tiny magnets and assume different orientations or energy levels in an applied magnetic field. Transitions between these energy levels are associated with the absorption or emission of rf radiation and may be detected by a suitable spectrometer. As illustrated in Figure 1.9, a sample is rotated in a strong magnetic field to average out the effects of nonuniformity. At the same time it is irradiated by the rf transmitter while the magnetic field is slowly increased (or decreased) by the sweep coils. The *resonance*, or spin flipping, of the different nuclei is detected by the voltage induced in the rf receiver coil and is recorded as a series of peaks on a calibrated chart.

The relationship of the resonance or absorption frequency ν to the magnetic field H is described by the following equations:

$$\nu = \frac{\mu H}{Ih}$$

$$\Delta E = h\nu = \frac{\mu H}{I} \tag{1.4}$$

where μ is the magnetic moment of the spinning nucleus, I is the nuclear spin ($I = \pm 1/2$ for protons), and h is Planck's constant. Nmr spectrometers

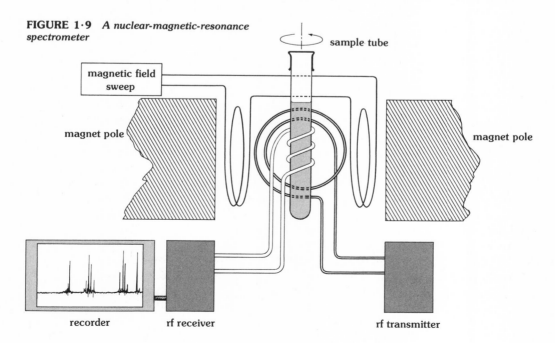

FIGURE 1·9 *A nuclear-magnetic-resonance spectrometer*

must therefore be tuned to the particular kind of nucleus being studied. For example, at a constant magnetic field strength of 14.1 kgauss the nucleus of a hydrogen atom will resonate at a frequency of $\nu = 6.00 \times 10^7$ Hz and the nucleus of fluorine atom will resonate at a frequency of $\nu = 5.64 \times 10^7$ Hz.

Four basic kinds of information about spinning nuclei are obtained from nmr measurements:

1 The presence of a specific kind of nmr-active nucleus in a sample can be demonstrated by tuning the spectrometer to the appropriate resonance frequency.

2 The applied magnetic field H_0 is modified by small local fields generated by electron movements induced in the framework of the molecule. Consequently, the magnetic field H experienced by each nucleus in the molecule differs by a small but characteristic increment from the applied field. These very small differences can be detected and measured by sensitive nmr spectrometers, so that the resulting spectrum of resonance signals contains information about the molecular environment around each nucleus.

3 The intensity of an nmr signal is proportional to the number of nuclei in the sample which absorb at that specific frequency or field strength.

4 Certain orientations of spinning nuclei may interact with each other in a fashion that causes a splitting of their resonance signals. This aspect of nmr spectroscopy and the detection of signal splitting will be discussed further in Chapter 8.

Since hydrogen, with a nuclear spin of $\pm 1/2$, is almost always present in organic compounds, *proton-magnetic-resonance (pmr)* measurements provide us with a valuable means of probing the molecular structure in the vicinity of the hydrogen atoms. As an illustration, consider the pmr spectra of the C_2H_6O isomers dimethyl ether and ethyl alcohol shown in Figure 1.10. The six hydrogen atoms in dimethyl ether (CH_3-O-CH_3) experience identical average environments and therefore exhibit a single sharp resonance signal. However, hydrogen atoms are found in three different locations in the ethyl alcohol molecule (CH_3-CH_2-OH), and there are consequently three rather closely spaced absorption peaks. The intensity of a pmr signal is measured by the area under the absorption peak and is directly proportional to the number of protons giving rise to the absorption. In the case of ethyl alcohol, the ratio of the three peak areas is 1:2:3, and they are therefore assigned to OH, CH_2, and CH_3, respectively.

Most pmr spectrometers are designed to function by varying the magnetic field intensity while the frequency is held constant. A frequency of 6.0×10^7 Hz, or 60 MHz, is commonly used for routine measurements of hydrogen nuclei. This mode of operation has given rise to the relative terms *high field* and *low field* in discussions of pmr spectra. Thus the CH_2 protons in ethyl alcohol exhibit a resonance signal at lower magnetic field than the CH_3 protons in the same compound. Protons having high-field resonance signals are said to be *shielded*, while those appearing at low field are called *deshielded*.

1·2 DETERMINATION OF STRUCTURAL FORMULA 17

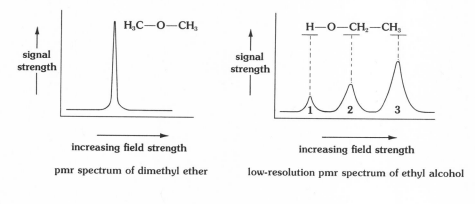

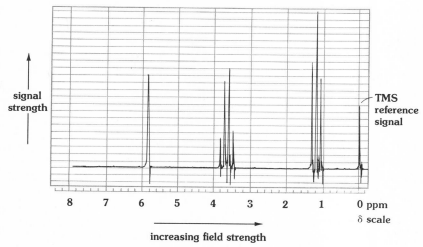

FIGURE 1·10 *Pmr spectra of dimethyl ether and ethyl alcohol*

PROBLEM 1·2
Show how the terms *shielded* and *deshielded* correlate with the modification of applied magnetic fields by local proton environments, as discussed in the preceding paragraphs.

The pmr spectrometers in use today are capable of much better resolution than that shown in the two upper spectra in Figure 1.10.† Under high-resolution conditions ethyl alcohol exhibits a clear fine structure, or splitting, of the basic resonance signals. This splitting is due to the interaction of neighboring protons, a matter we shall discuss further in connection with alcohols and ethers.

In contrast to optical spectroscopy, pmr spectroscopy has no fundamental scale units and no natural zero reference. However, we can locate

†Commercial spectrometers range in price from $10,000 to more than $100,000, depending on their field strength and versatility.

1·2 DETERMINATION OF STRUCTURAL FORMULA

TABLE 1·4 *Typical proton chemical-shift values*

hydrogen type	δ value (ppm)	hydrogen type	δ value (ppm)
$(CH_3)_4Si$	0.0	$C-CH_2-Cl$ $C-CH_2-Br$ $C-CH_2-I$ $C-CH_2-O$	3.0–5.0
$C-CH_3$	0.85–0.95		
$C-CH_2-C$	1.20–1.40		
$\begin{array}{c} C \\ \diagdown \\ C-CH \\ \diagup \\ C \end{array}$	1.40–1.60	$C-OH$ $C-NH_2$	1.0–5.5
		$C-CO_2H$	10.5–12.0
$\begin{array}{c} C \\ \diagdown \\ C=CH_2 \\ \diagup \\ C \end{array}$	4.0–5.5	(benzene ring with H)	6.5–8.0
$C-C\equiv CH$	2.0–3.0		
$O=C-CH_3$	2.0–2.8		
$O=C-H$	9.0–10.0		

each peak, or resonance signal, in a spectrum by measuring its displacement from the signal generated by a suitable reference compound included with the sample. Tetramethylsilane (TMS), with the molecular formula $(CH_3)_4Si$, is currently the reference compound of choice. It is chemically inert, gives a single sharp resonance signal at a somewhat higher field than most common organic protons, and is miscible with (soluble in) most organic solvents. In measuring the displacement, or *chemical shift*, of a resonance signal a dimensionless scale is preferred, so that the values will be independent of the field and frequency characteristics of a particular spectrometer. For this reason the *delta* (δ) *scale* is widely used. The values for this scale are given by

$$\delta = \frac{H_r - H_s}{H_r} \times 10^6 \tag{1.5}$$

where H_r and H_s are the measured field strengths corresponding to resonance (absorption) by the reference protons and the protons in the sample substance, respectively. Some typical δ values, expressed in parts per million, are listed in Table 1.4.†

In summary, the role of pmr spectroscopy in elucidating molecular structures might be likened to intelligence gathering by a network of spies (the hydrogen atoms in a molecule), each broadcasting information about its immediate surroundings (the local molecular environment). Although this information may not always lead to a uniquely defined molecular structure, it will usually enable us to eliminate many alternative isomers from further consideration.

†Another common scale for pmr spectra is the *tau* (τ) *scale*, where $\tau = 10 - \delta$. Since $\delta = 0$ for TMS, the position of this reference compound on the τ scale is 10.00. We shall use the δ scale in this book.

1·3 THEORETICAL MODELS FOR MOLECULAR STRUCTURE

The three-dimensional structures of many molecules have been accurately established from well-documented experimental results. In order to consolidate and extend our understanding of molecular structure, we need theoretical models that will not only account for all the existing facts, but also permit an informed guess as to the structure of molecules that have not been measured. In evaluating such models one criterion outweighs all others in importance: *how well the model correlates with known facts*. The foundation of all theoretical models for molecular structure is the assumption that a molecule is a three-dimensional assemblage of atomic nuclei held together by attractive forces, or bonds, which result from extranuclear electron interactions.† Clearly, the success of a model depends critically on how well it can approximate these electron interactions.

THE IONIC BOND

When electrons are removed from or added to a neutral atom, the result is a charged species called an *ion*. Studies of such ionization processes show that those ions with electron configurations that correspond to the inert-gas atoms helium ($1s^2$), neon ($1s^2$, $2s^2$, $2p_x^2$, $2p_y^2$, $2p_z^2$), and argon ($1s^2$, $2s^2$, $2p_x^2$, $2p_y^2$, $2p_z^2$, $3s^2$, $3p_x^2$, $3p_y^2$, $3p_z^2$) are characterized by exceptional stability. A sodium atom, for example, has one electron in excess of the neon configuration, and consequently it has a relatively low ionization potential (it is easily ionized):

$$\text{Na·} \longrightarrow e^{\ominus} + \underset{\substack{\text{neon}\\ \text{configuration}}}{\text{Na}^{\oplus}} \tag{1.6}$$

In contrast, a chlorine atom is only one electron short of achieving an argon configuration and hence has a high electron affinity:

$$:\ddot{\text{Cl}}\text{·} + e^{\ominus} \longrightarrow \underset{\substack{\text{argon}\\ \text{configuration}}}{:\ddot{\text{Cl}}:^{\ominus}} \tag{1.7}$$

Transfer of an electron from a sodium atom to a chlorine atom should therefore be highly favored. Indeed, the structure of sodium chloride is found to be a three-dimensional lattice of alternating sodium and chloride ions held together by electrostatic forces. This is termed an *ionic bond*.

Unlike inorganic salts, most organic compounds do not conduct electricity and have relatively low melting and boiling points. Consequently, ionic bonding is not believed to be significant in organic molecules. In order to achieve an inert-gas electron configuration, a carbon

---

†This discussion assumes a basic knowledge of the electron structure of atoms. For a review of this subject consult a modern introductory chemistry text such as *Chemistry: A Conceptual Approach* by Mortimer, *Understanding Chemistry* by Pimentel and Spratley, and *Chemistry* by Sienko and Plane.

atom would have to form either C^{+4} or C^{-4} ions. These small ions necessarily have a very high charge density and therefore would be extremely unstable. Even O^{-2}, which has a neon configuration, is not found in any known compounds. A similar problem arises with respect to the hydrogen atoms that are present in almost all organic compounds. Although the $:H^{\ominus}$ ion has a helium configuration, its charge density is also very high, and $H^{\oplus}$ (a free proton) is formed only in atomic accelerators. We therefore conclude that a different kind of bonding-electron interaction must prevail in organic molecules.

THE COVALENT BOND

A second way in which atoms may achieve stable electron states is by a mutual sharing of electrons. The shared electron pairs serve, in a sense, as a negatively charged glue which binds the positive nuclei together. This is termed *covalent bonding*, found in its simplest form in H_2:

$$H\cdot + H\cdot \longrightarrow H:H \tag{1.8}$$

Most stable compounds can be represented by electron-dot formulas, in which all atoms have inert-gas valence-shell configurations. The few exceptions to this rule, such as boron trifluoride in Table 1.5, tend to exhibit exceptional reactivity. The shared electron pair is usually represented in structural formulas by a straight line connecting the appropriate atoms. Nonbonding valence-shell electrons may sometimes be omitted, as in the structural formula for BF_3, but they should not be forgotten or ignored.

We must be careful not to base our model for molecular structure exclusively on analogies with atomic electron configurations. In contrast

TABLE 1·5 *Covalent bond representations*

molecular formula	name	electron-dot formula	structural formula
CH_4	methane	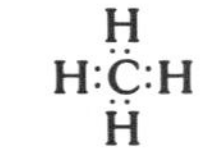 H:C̈:H with H above and H below	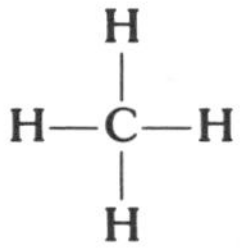 H—C—H with H above and H below
NH_3	ammonia	 H:N̈:H with H below	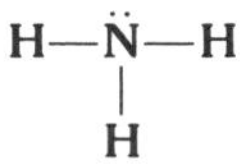 H—N̈—H with H below
C_2H_6O (two isomers)	dimethyl ether	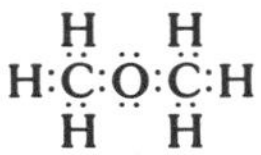 H:C̈:Ö:C̈:H with H's	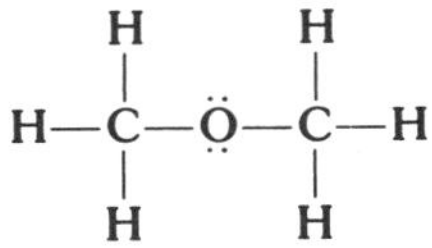 H—C—Ö—C—H with H's
	ethyl alcohol	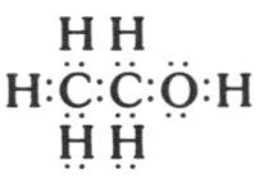 H:C̈:C̈:Ö:H with H's	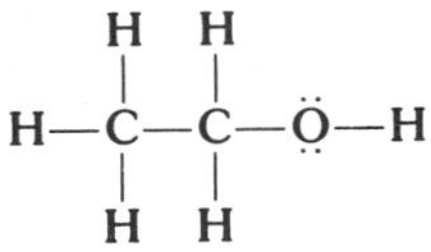 H—C—C—Ö—H with H's
BF_3	boron trifluoride	 :F̈:B:F̈: with :F̈: below	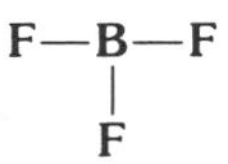 F—B—F with F below

1·3 THEORETICAL MODELS FOR MOLECULAR STRUCTURE

TABLE 1·6 *Multiple covalent bonding in carbon compounds*

molecular formula	name	electron-dot formula	structural formula	bonding partners per carbon	bonding terminology
C_2H_6	ethane	H:C:C:H (with H above and below each C)	H—C—C—H (with H above and below each C)	4	single bond
C_2H_4	ethylene	H:C::C:H (with H's)	H₂C=CH₂	3	double bond
C_2H_2	acetylene	H:C:::C:H	H—C≡C—H	2	triple bond
H_2CO	formaldehyde	H:C::O: (with H)	H₂C=O	3	double bond
HCN	hydrogen cyanide	H:C:::N:	H—C≡N:	2	triple bond

to the spherical positive field that surrounds a single atomic nucleus, the many nuclei in a molecule generate a complex field which perturbs the energy and motion of the valence electrons. As a result of this interaction, valence electrons may occupy *molecular orbitals* encompassing two or more nuclei. This is, in fact, what we ordinarily mean by a covalent bond. For example, the electron configuration of a carbon atom ($1s^2$, $2s^2$, $2p_x^1$, $2p_y^1$, $2p_z^0$) might lead us to predict that it will participate in only two covalent bonds (sharing of the $2p_x$ and $2p_y$ electrons). Such bonding, however, does not provide the carbon atom with a neonlike valence-shell configuration, and derived molecules such as CH_2 are found to be very unstable. Apparently the atoms surrounding a carbon atom in a molecule perturb and interact with its atomic orbitals so as to generate a number of low-energy molecular orbitals sufficient to form a stable molecule.

In a sense, then, carbon can, within limits, adjust its electron sharing (covalent bonding) to accommodate different ratios of bonding partners, as demonstrated by the compounds listed in Table 1.6. It is customary to refer to the multiple bonds formed by the sharing of two or three electrons from each constituent atom (as in C_2H_4 and C_2H_2) as *double* and *triple bonds*, respectively. Once again, note that as a result of this multiple sharing of electrons all the atoms in such molecules have stable inert-gas configurations.

PROBLEM 1·3

a A problem appearing in the *2nd Scientific American Book of Mathematical Puzzles and Diversions* asks the reader to prove that at a recent scientific convention the number of scientists who shook hands an odd number of times is even. On the basis of this puzzle, would you say that the number of hydrogen

22   1·3 THEORETICAL MODELS FOR MOLECULAR STRUCTURE

atoms in stable compounds of C, H, O, and S is odd or even? How would the presence of halogen or nitrogen affect your answer?

b Which of the following molecular formulas represent compounds that are probably stable? Which would you expect to be relatively unstable?

$C_3H_5N_2O$ $C_5H_8S_2$ $C_4H_7NO_2$ $C_{15}H_{25}O_2Cl_3$ $C_{12}H_{20}NS$ $C_{11}H_{19}N_2Cl$

PROBLEM 1·4

Dimethyl sulfoxide (C_2H_6SO) shows a single sharp resonance peak at $\delta = 2.62$ in its pmr spectrum. Write an electron-dot formula which is consistent with this fact. Write formulas for two other isomers having this molecular formula.

MOLECULAR GEOMETRY

In addition to describing the number and kinds of bonds in a molecule, a successful model for molecular structure must also predict the approximate arrangement of the atoms in space. A simple extension of the covalent-bond model accomplishes this: *The valence-shell electrons of any atom in the molecule will occupy either nonbonding atomic orbitals or bonding molecular orbitals, and the electrostatic repulsion between regions of high electron density will keep these orbitals as far apart as possible.*

In applying this rule to specific molecules we find that atoms of carbon and other light elements usually assume one of three different orbital configurations:

1 *Tetrahedral configuration:* The atom in question has four regions of high electron density in its valence shell, and these are directed toward the corners of a tetrahedron centered at its nucleus. Bond angles are found to be about 109.5° (±4°).

compound	structural formula	tetrahedral representation†
CH_4	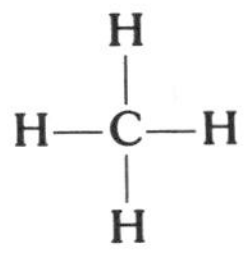 H—C—H with H above and H below	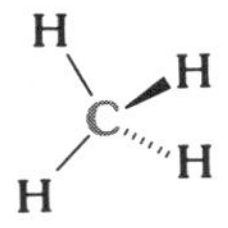 tetrahedral C with 4 H, 109.5°
:NH_3	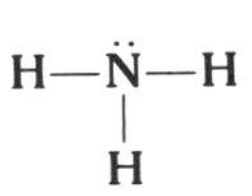 H—N̈—H with H below	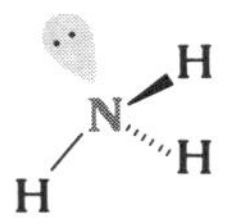 tetrahedral N with 3 H and lone pair
$H_2\ddot{\underset{..}{O}}$	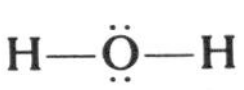 H—Ö—H	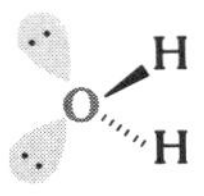 tetrahedral O with 2 H and 2 lone pairs

†The shaded atom and those atoms bonded to it by simple lines are considered to lie in the plane of the paper. The wedge-shaped lines represent bonds to atoms located in front of the paper, and the broken lines represent bonds to atoms lying behind the plane of the paper.

1·3 THEORETICAL MODELS FOR MOLECULAR STRUCTURE

2 *Trigonal configuration:* The atom in question has three regions of high electron density in its valence shell, and these are directed toward the corners of an equilateral triangle. Bond angles are found to be about 120° (±4°).

compound	structural formula	trigonal representation
BF_3	F–B(–F)–F	
C_2H_4	$H_2C{=}CH_2$	
CH_2NH	$H_2C{=}\ddot{N}{-}H$	

3 *Linear configuration:* Two regions of high electron density extend in opposite directions from the atomic nucleus in question. Bond angles are about 180°.

compound	structural formula	linear representation
CO_2	$\ddot{O}{=}C{=}\ddot{O}$	$\ddot{O}{=}C{=}\ddot{O}$
HCN	$HC{\equiv}N{:}$	$H{-}C{\equiv}N{:}$
H_2C_2		$H{-}C{\equiv}C{-}H$

CHARGE DISTRIBUTION

The electron pair in a covalent bond between different atoms is not necessarily shared equally. If one of the atoms has greater *electronegativity*, or affinity for electrons, it will exert a greater attraction on the bonding electrons. This produces a *polar bond*, in which one end is more negative than the other end. The partial charge separation in a polar bond produces a *dipole*, which can be indicated by a charge notation ($\delta\oplus$ or $\delta\ominus$) or by a dipole arrow ($\leftrightarrow$):

$$\overset{\delta\ominus}{\ddot{O}}{=}C{=}\overset{\delta\ominus}{\ddot{O}} \quad \overset{\delta\oplus}{}$$

$$\begin{array}{c}\overset{\delta\ominus}{Cl}\quad\overset{\delta\ominus}{Cl}\\\overset{\delta\oplus}{}C\\Cl\quad Cl\\\delta\ominus\quad\delta\ominus\end{array}$$

1·3 THEORETICAL MODELS FOR MOLECULAR STRUCTURE

The electronegativity of an atom depends on its nuclear charge and the screening effect of the inner electron shells. Thus electronegativity generally increases from left to right across a period of the periodic table, or from bottom to top in one of the groups. The electronegativity order of some commonly encountered atoms is

$$F > O > Cl \approx N > Br > S \approx C > H$$

If several polar bonds occur in a molecule, their dipoles combine to yield a *molecular dipole* whose magnitude can be measured by special techniques and is called a *dipole moment* μ. Molecular dipole moments provide additional evidence of the structure and geometry of certain molecules. For example, water exhibits a strong dipole, while CO_2 and CCl_4 have no measurable dipoles, indicating that their bond dipoles are directed in such a manner that they cancel each other out.

molecular dipole moment
$\mu = 1.84D$

Certain covalent bonds have an unusually large degree of charge separation, which appears to be due to a one-sided sharing of electron pairs (in contrast to the mutual sharing described above). In such cases, if we count the electrons for each atom by dividing shared electrons equally, we find that adjacent atoms each have a *formal charge*. These charges are denoted by an appropriate $\oplus$ or $\ominus$ symbol on each charged atom.

For example, azide ion ($N_3^{\ominus}$) has an electron structure in which all atoms have a complete valence-shell octet, $\ddot{N}=N=\ddot{N}$. An electron count for this structure gives the terminal nitrogen atoms *six* valence electrons each, and the central nitrogen atom *four* valence electrons. Since a neutral nitrogen atom has *five* valence-shell electrons (the atomic number of nitrogen is 7), the terminal atoms in the azide ion must each have a -1 formal charge and the central nitrogen atom a $+1$ formal charge.

TABLE 1·7 *Formal-charge notation*

molecular formula	structural formula	valence-electron count		formal charge	formal-charge formula

BF_3NH_3	:F̈: H
	  | |
	:F̈—B—N—H
	  | |
	:F̈: H
		F
		B
		N
		H
			7e
			4e
			4e
			1e
				0
				-1
				$+1$
				0
					F H
					| |
					F—B$^{\ominus}$—N$^{\oplus}$—H
					| |
					F H

CNO^{-1}	:C≡N—Ö:	C N O	5e 4e 7e	-1 $+1$ -1	$^{\ominus}$:C≡N$^{\oplus}$—O$^{\ominus}$
SO_3^{-2}	:Ö: | :Ö—S—Ö:	S O	5e 7e	$+1$ -1	O$^{\ominus}$ | $^{\ominus}$O—S$^{\oplus}$—O$^{\ominus}$

These charges (two negatives and a positive) add up to −1, the charge on the ion itself. Table 1.7 shows further examples of formal charges in molecules and ions.

PROBLEM 1·5
Write electron-dot formulas for each of the following compounds and calculate the formal charge on each atom other than hydrogen:

HNO_3 H_2SO_4 O_3 CH_2N_2 (two isomers, one linear one cyclic)

1·4 CHEMICAL REACTIVITY

TYPES OF ORGANIC REACTIONS

A major portion of this book is devoted to the study of organic reactions, that is, the manner in which different kinds of organic compounds react with various reagents. Although a casual inspection of the remaining chapters may give the impression of an overwhelming variety of chemical transformations, these transformations actually represent only a few basic reaction types. The most common of these are outlined in Table 1.8.

In an *addition reaction* there is an increase in the number of bonding partners, often called *ligands*, about certain carbon atoms. This is normally accomplished by changing double- or triple-bonded systems into single-bonded ones. An *elimination reaction* is the reverse of this process. *Substitution reactions and rearrangements* do not involve

TABLE 1·8 *Classes of organic reactions*

reaction type	overall transformation†
addition	$\underset{R}{\overset{R}{>}}C=C\underset{R}{\overset{R}{<}} + A-B \longrightarrow R-\underset{R}{\overset{A}{\underset{\|}{C}}}-\underset{R}{\overset{B}{\underset{\|}{C}}}-R$
elimination	$R-\underset{R}{\overset{Y}{\underset{\|}{C}}}-\underset{R}{\overset{Z}{\underset{\|}{C}}}-R \longrightarrow \underset{R}{\overset{R}{>}}C=C\underset{R}{\overset{R}{<}} + Y-Z$
substitution	$R-\underset{R}{\overset{R}{\underset{\|}{C}}}-Y + Z \longrightarrow R-\underset{R}{\overset{R}{\underset{\|}{C}}}-Z + Y$
rearrangement	$R-\underset{R}{\overset{R}{\underset{\|}{C}}}-\underset{H}{\overset{H}{\underset{\|}{C}}}-Y \longrightarrow Y-\underset{R}{\overset{R}{\underset{\|}{C}}}-\underset{H}{\overset{H}{\underset{\|}{C}}}-R$

†R refers to an inert substituent, usually a hydrogen atom or an alkyl group. A, B, Y, and Z refer to atoms or groups of atoms that are chemically reactive.

1·4 CHEMICAL REACTIVITY

any change in the number of ligands, only a change in their nature and arrangement.

Although *proton transfer* is not listed as a basic reaction in Table 1.8, it often plays a major role in organic reactions. For example, the substitution reaction

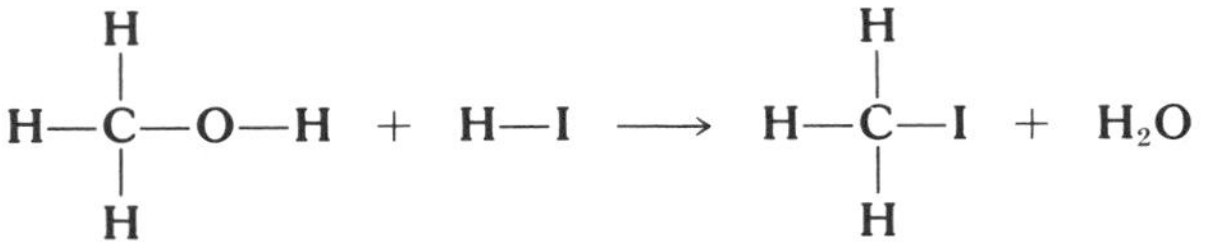

$$\text{CH}_3\text{-O-H} + \text{H-I} \longrightarrow \text{CH}_3\text{-I} + \text{H}_2\text{O} \quad (1.9)$$

actually proceeds in two steps. The first step is a proton transfer and step 2 is a substitution at carbon:

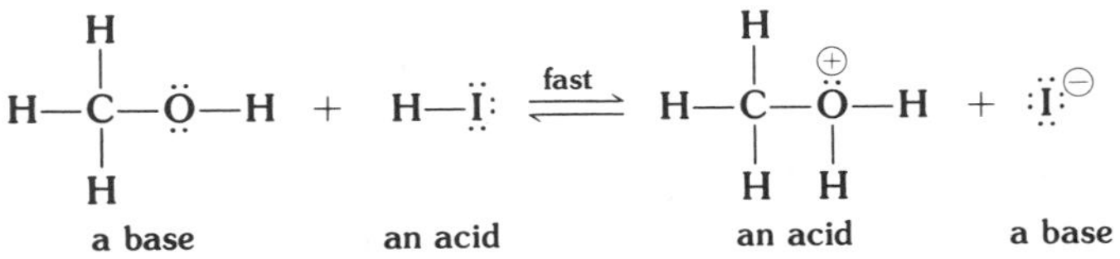

$$\text{CH}_3\text{-}\ddot{\text{O}}\text{-H} + \text{H-}\ddot{\text{I}}\text{:} \xrightleftharpoons{\text{fast}} \text{CH}_3\text{-}\overset{\oplus}{\ddot{\text{O}}}\text{H}_2 + \text{:}\ddot{\text{I}}\text{:}^{\ominus} \quad (1)$$

a base   an acid   an acid   a base

$$\text{CH}_3\text{-}\overset{\oplus}{\ddot{\text{O}}}\text{H}_2 + \text{:}\ddot{\text{I}}\text{:}^{\ominus} \xrightarrow{\text{slow}} \text{:}\ddot{\text{I}}\text{-CH}_3 + \text{H}_2\ddot{\text{O}} \quad (2)$$

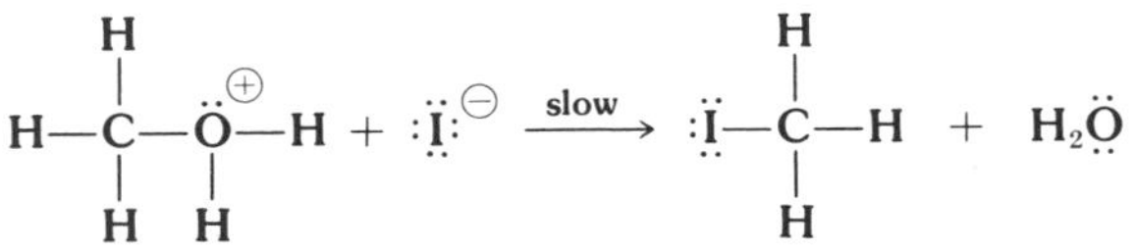

Such proton transfers are best discussed in terms of the *Brønsted-Lowry model*, according to which an *acid* is a proton donor and a *base* is a proton acceptor. We shall examine this acid-base concept in greater detail in Chapter 3.

REACTION MECHANISMS

Today we are not satisfied with merely classifying a reaction. The modern chemist wishes to understand how and why reactions proceed as they do, for the benefits of such understanding reach far beyond organic chemistry. As only one example, the marvelously complex and efficient processes by which living cells transform energy and store information are molecular in nature and must therefore be fundamentally related to the properties and reactions of simpler systems. In at least that sense, then, this text presents the ABC's of molecular biology.

A detailed description and explanation of a reaction is termed a *reaction mechanism*. Some of the conventions and terms used in discussing mechanisms are outlined in Table 1.9. For example, covalent bond breaking may occur with a division of the bonding electrons among the fragments, called *homolysis*, or both bonding electrons may remain with one of the fragments, called *heterolysis*. Curved arrows are used to show the reorganization of electrons in bonds being broken or formed in the reaction. A full arrowhead (⤻) refers to the shift of an electron pair, while a fishhook (⤻) denotes a single electron shift. It is important to remember that these arrows show the directions of *electron* shifts and do not necessarily correspond to the motions of the atoms themselves.

1·4 CHEMICAL REACTIVITY

TABLE 1·9 *Mechanism terminology*

process	terminology
bond breaking	
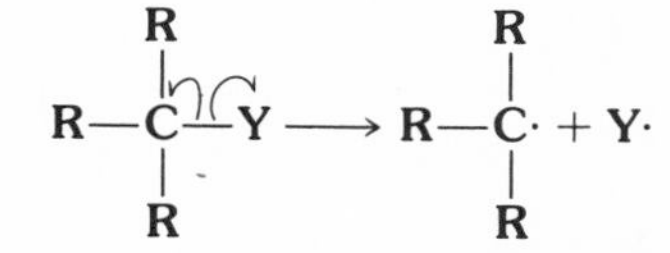 R—C(R)(R)—Y → R—C(R)(R)· + Y· (a carbon radical)	homolysis
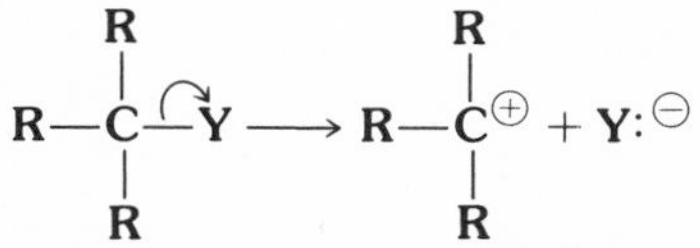 R—C(R)(R)—Y → R—C⁺(R)(R) + Y:⁻ (a carbonium ion)	heterolysis
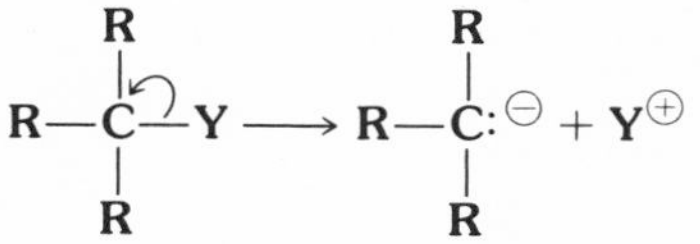 R—C(R)(R)—Y → R—C:⁻(R)(R) + Y⁺ (a carbanion)	heterolysis
bond making	
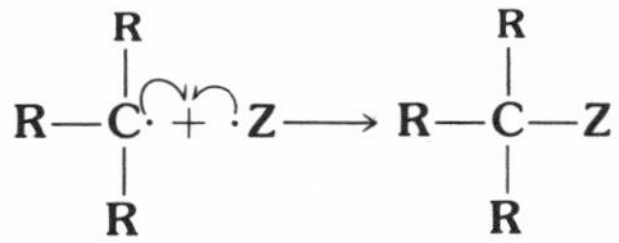 R—C(R)(R)· + ·Z → R—C(R)(R)—Z	radical combination (coupling)
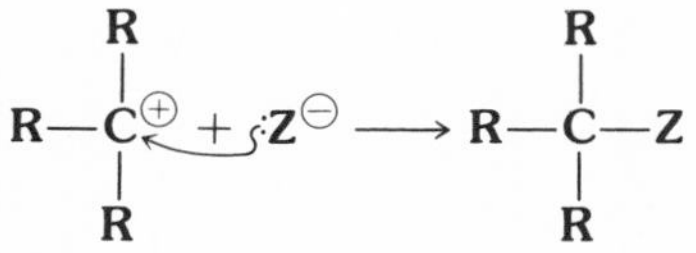 R—C⁺(R)(R) + :Z⁻ → R—C(R)(R)—Z	combination of a carbonium ion with a nucleophile
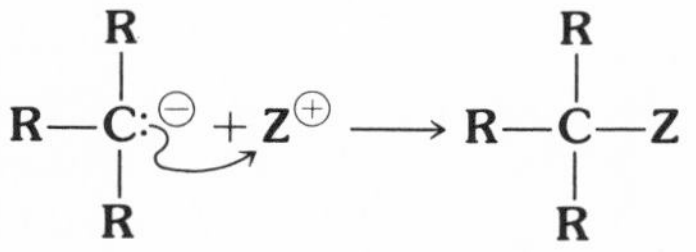 R—C:⁻(R)(R) + Z⁺ → R—C(R)(R)—Z	combination of a carbanion with an electrophile

PROBLEM 1·6
Draw curved arrows in appropriate positions to describe the following four transformations:

a H—CH₂· + :Cl̈—Cl̈: ⟶ H—CH₂—Cl̈: + :Cl̈·

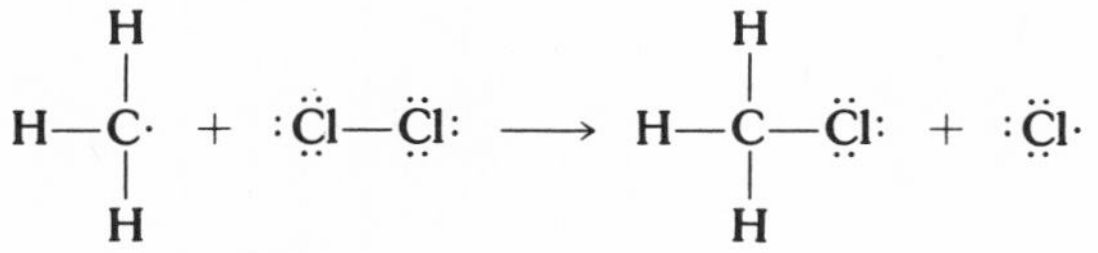

b (CH₃)₃C—Ö· ⟶ CH₃· + (CH₃)₂C=Ö

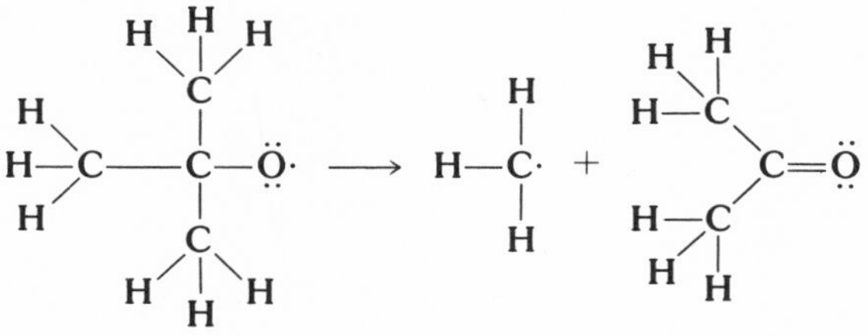

$$c \quad :\ddot{\text{Cl}}:^{\ominus} + \underset{H}{\overset{H}{>}}C=\overset{\oplus}{O}\underset{H}{\overset{H}{<}}C\underset{H}{\overset{H}{<}}H \longrightarrow :\ddot{\text{Cl}}-\underset{H}{\overset{H}{\underset{|}{C}}}-\ddot{O}-\underset{H}{\overset{H}{\underset{|}{C}}}-H$$

$$d \quad :\ddot{\text{Br}}-\underset{H}{\overset{H}{\underset{|}{C}}}-\underset{H}{\overset{H}{\underset{\ominus}{C}}}-\overset{\overset{\ddot{O}}{||}}{C}-\underset{H}{\overset{H}{\underset{|}{C}}}-H \longrightarrow \underset{H}{\overset{H}{>}}C=\underset{H}{\overset{H}{\underset{|}{C}}}-\overset{\overset{\ddot{O}}{||}}{C}-\underset{H}{\overset{H}{\underset{|}{C}}}-H + ?$$

REACTIVE INTERMEDIATES

Breaking a carbon bond produces a reactive species having a carbon atom in an abnormal valence state. There is convincing evidence that these species play an important role as *reactive intermediates* in many organic reactions. Depending on the charge carried by a reactive trivalent carbon atom, these intermediates are called *radicals* (no charge), *carbonium ions* (positive charge), or *carbanions* (negative charge). Subsequent reactions of intermediates primarily reflect their tendency to achieve a stable electron configuration through covalent bonding. Thus an odd-electron intermediate — say, a radical — rapidly combines with other odd-electron species and may even acquire a bonding partner by homolytic attack on an existing weak bond:

$$H-\underset{H}{\overset{H}{\underset{|}{C}}}\cdot + Br-Br \longrightarrow H-\underset{H}{\overset{H}{\underset{|}{C}}}-Br + :\ddot{\text{Br}}\cdot \qquad (1.10)$$

A corresponding reaction tendency exists with systems that are rich or deficient in electron pairs, as shown in Table 1.9. Such systems are termed, respectively, *nucleophiles* and *electrophiles*. A nucleophile is an atom or molecule that can offer a pair of electrons for bond formation; an electrophile is an atom or molecule capable of forming a new bond by accepting this pair of electrons. According to these definitions, which are also the definitions of a *Lewis base* and a *Lewis acid*, a carbanion is a *nucleophile* and a carbonium ion is an *electrophile*. Such ionic intermediates are often involved in reactions occurring in the liquid phase, where the powerful electrostatic forces between ions are attenuated by solvent effects.

Divalent carbon intermediates called *carbenes* have been identified in certain reactions. These highly reactive species are usually generated by elimination reactions (equation 1.11) and react rapidly by addition reactions (equation 1.12):

$$H-\underset{Cl}{\overset{Cl}{\underset{|}{C}}}-Cl + KOH \longrightarrow \underset{Cl}{\overset{Cl}{>}}C: + KCl + H_2O \qquad (1.11)$$

dichlorocarbene

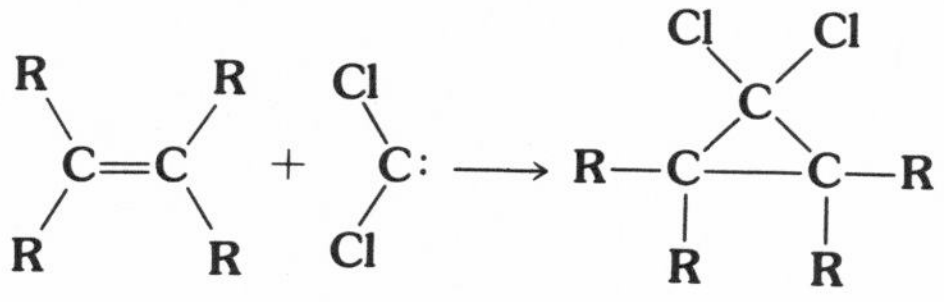

$$\begin{array}{c}R\\R\end{array}C=C\begin{array}{c}R\\R\end{array} + \begin{array}{c}Cl\\Cl\end{array}C: \longrightarrow R-\underset{R}{\overset{R}{C}}-\overset{\overset{ClCl}{\diagdown\diagup}}{\underset{\diagup\diagdown}{C}}-\underset{R}{\overset{R}{C}}-R \qquad (1.12)$$

FUNCTIONAL GROUPS

Most of the chemical transformations exhibited by organic compounds appear to involve particular reactive sites in their molecules. The nature of the reactions which occur at these sites depends to a large extent on the groupings of atoms found there. More than a century of experience has taught us that certain chemical properties are associated with certain characteristic groups of atoms, known as *functional groups,* a fact which greatly simplifies the study of organic chemistry. For example, many of the known reactions of ethyl alcohol (CH_3—CH_2—OH) involve the —O—H atomic grouping, and it is not surprising that *n*-butyl alcohol (CH_3—CH_2—CH_2—CH_2—OH) exhibits a rather similar chemical behavior. Some common functional groups and their characteristic properties are listed in Table 1.10. Each of these functional groups will be discussed in detail in the following chapters.

TABLE 1·10 *A survey of functional groups*

functional group	name of class	characteristic reactions	examples
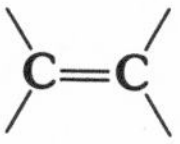 $\diagdown\diagup$ C=C $\diagup\diagdown$	alkenes (olefins)	addition	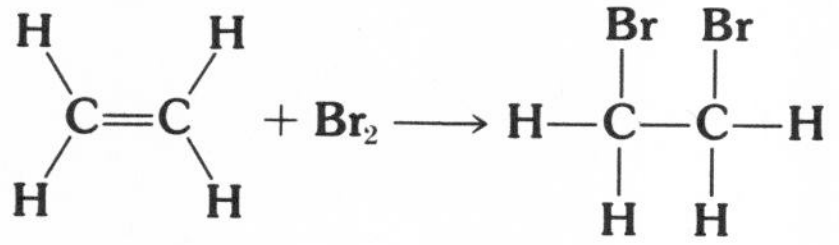 $H_2C=CH_2 + Br_2 \longrightarrow H-CHBr-CHBr-H$
—C≡C—	alkynes (acetylenes)	addition	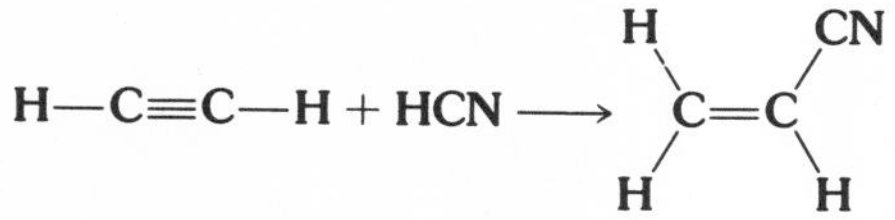 H—C≡C—H + HCN ⟶ H₂C=CHCN
		proton transfer	H—C≡C—H + NaNH₂ ⟶ HC≡CNa + NH₃
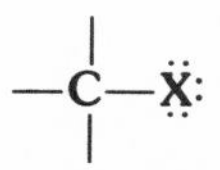 —C—X: (X=Cl, Br, I)	alkyl halides	substitution	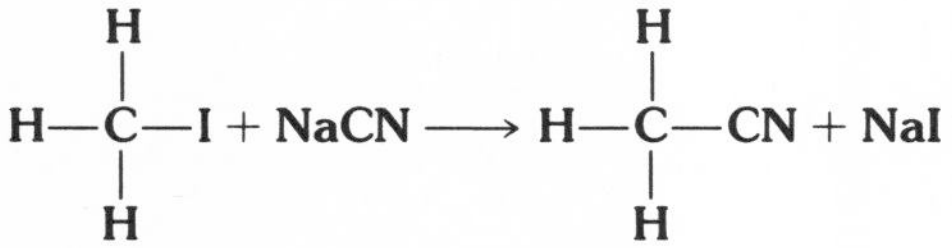 H—CH₂—I + NaCN ⟶ H—CH₂—CN + NaI
		elimination	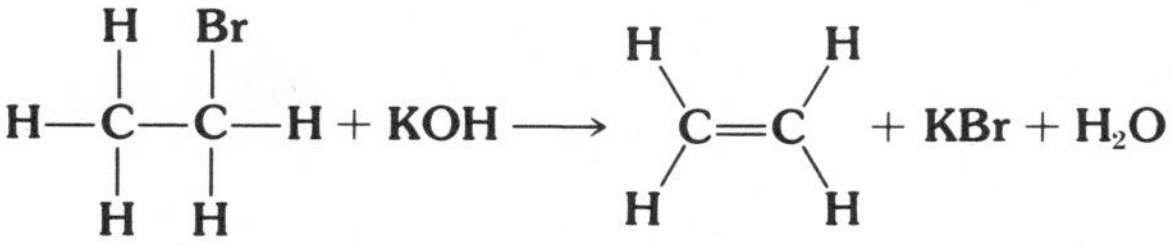 H—CH₂—CHBr—H + KOH ⟶ H₂C=CH₂ + KBr + H₂O

(continued)

TABLE 1·10 *A survey of functional groups (continued)*

functional group	name of class	characteristic reactions	examples
$-\underset{\|}{\overset{\|}{C}}-\ddot{O}-H$	alcohols	substitution	$H-\underset{H}{\overset{H}{\underset{\|}{C}}}-OH + SOCl_2 \longrightarrow H-\underset{H}{\overset{H}{\underset{\|}{C}}}-Cl + SO_2 + HCl$
		proton transfer	$CH_3-OH + NaOH \rightleftharpoons CH_3-ONa + H_2O$
$-\underset{\|}{\overset{\|}{C}}-\ddot{O}-\underset{\|}{\overset{\|}{C}}-$	ethers	substitution	$H-\underset{H}{\overset{H}{\underset{\|}{C}}}-O-\underset{H}{\overset{H}{\underset{\|}{C}}}-H + 2HI \longrightarrow 2H-\underset{H}{\overset{H}{\underset{\|}{C}}}-I + H_2O$
$\overset{}{\underset{}{{>}}}C=\ddot{O}$	carbonyl compounds	addition	$H-\underset{H}{\overset{H}{\underset{\|}{C}}}-\overset{O}{\overset{\|}{C}}-\underset{H}{\overset{H}{\underset{\|}{C}}}-H + HCN \longrightarrow CH_3-\underset{CN}{\overset{O-H}{\underset{\|}{C}}}-CH_3$
$-\underset{:\ddot{O}-H}{\overset{\ddot{O}:}{\overset{\|}{C}}}$	carboxylic acids	substitution	$H-\underset{H}{\overset{H}{\underset{\|}{C}}}-\underset{O-H}{\overset{O}{\overset{\|\|}{C}}} + SOCl_2 \longrightarrow H-\underset{H}{\overset{H}{\underset{\|}{C}}}-\underset{Cl}{\overset{O}{\overset{\|\|}{C}}} + SO_2 + HCl$
		proton transfer	$CH_3-\underset{O-H}{\overset{O}{\overset{\|\|}{C}}} + NaOH \longrightarrow CH_3-\underset{O^{\ominus} Na^{\oplus}}{\overset{O}{\overset{\|\|}{C}}} + H_2O$
$-\ddot{N}H_2$ $\underset{/}{\overset{\backslash}{N}}-H$ $\underset{/\backslash}{\overset{\backslash}{\ddot{N}}}$	amines	proton transfer	$H-\underset{H}{\overset{H}{\underset{\|}{C}}}-\underset{H}{\overset{H}{\underset{\|}{\ddot{N}}}}-H + HCl \longrightarrow H-\underset{H}{\overset{H}{\underset{\|}{C}}}-\underset{H}{\overset{H}{\underset{\|}{\overset{\oplus}{N}}}}-H + Cl^{\ominus}$
		addition†	$CH_3-\ddot{N}H_2 + CH_3-I \longrightarrow CH_3-\underset{H}{\overset{H}{\underset{\|}{\overset{\oplus}{N}}}}-CH_3 + I^{\ominus}$

†We regard this transformation as an addition to nitrogen because the number of bonding partners with the nitrogen has increased in the product.

SUMMARY

organic chemistry   The chemistry of carbon compounds.

empirical formula   A formula showing the relative proportions of different atoms in a molecule.

molecular formula   A formula showing the actual numbers of different atoms in a molecule.

structural formula   A formula which shows the bonding arrangement (constitution) of all the atoms in a molecule.

isomer   One of two or more different compounds having the same molecular formula.

mass spectroscopy   The measurement of mass-to-charge ratios of an assortment of ions by their acceleration and deflection in electrostatic and magnetic fields.

diffraction analysis   The use of very short wavelength radiation as a probe for the study of molecular structure.

infrared spectroscopy   A method of analysis in which vibrational transitions, resulting from the absorption of 4000- to 600-cm^{-1} light, produce a "fingerprint" of the absorbing substance and indicate the kinds of covalent bonds which are present (see Table 1.3).

nuclear-magnetic-resonance (nmr) spectroscopy   A method of analysis in which spinning nuclei are examined by measuring the absorption of rf radiation that can be induced when a sample is exposed to a strong magnetic field. Hydrogen is the most commonly studied nucleus, the procedure being known as *proton-magnetic-resonance (pmr) spectroscopy*. In pmr spectroscopy the positions of the resonance signals are measured with respect to a signal from the reference compound tetramethylsilane (TMS) and are given in parts per million (δ values).

covalent bonding   The attraction between atoms which share pairs of electrons (resulting from electrons occupying molecular orbitals rather than atomic orbitals). A *single bond* is the sharing of one pair of electrons (Y—Z), a *double bond* the sharing of two pairs of electrons (Y=Z), and a *triple bond* the sharing of three pairs of electrons (Y≡Z).

electronegativity   The power of an atom in a molecule to attract electrons to itself.

polar bonds   Covalent bonds which have a bond dipole as a consequence of uneven electron sharing between atoms of different electronegativities.

dipole moment   A measure of the molecular dipole resulting from the addition of all the bond dipoles in a molecule.

formal charge   Charge separation in a molecule resulting from unequal sharing of electrons in one or more covalent bonds. The magnitude and distribution of these charges is equivalent to that expected from a transfer of one or more electrons between atoms.

reaction classes   Addition, elimination, substitution, rearrangement, and proton transfer between acids and bases (see Table 1.8).

reaction mechanism   A detailed description of each step in a chemical reaction (see Table 1.9).

reactive intermediate A short-lived molecule, atom, or ion formed during the course of a chemical reaction.

nucleophile An atom or molecule that can offer a pair of electrons for covalent-bond formation; also called a *Lewis base*.

electrophile An atom or molecule capable of forming a new bond by accepting a pair of electrons from a nucleophile; also called a *Lewis acid*.

functional group A reactive grouping of atoms which shows a characteristic chemical behavior (see Table 1.10).

EXERCISES

1·1 Draw structural formulas showing all valence electrons for the following compounds. Assume that all atoms (except hydrogen) will have a valence-shell electron octet.

a NOCl
b H_2SO_4
c C_3H_6 (two isomers)
d CH_5ON (three isomers)
e CH_3OH
f CH_3ONa
g C_2H_4

1·2 What is the lowest possible molecular weight of a compound that contains 73.39% bromine? If this compound is composed only of carbon, hydrogen, and bromine, suggest a structural formula commensurate with your answer.

1·3 Write structural formulas for the two isomers having the molecular formula $C_2H_4Cl_2$. What spectroscopic tool would be ideal for distinguishing these isomers?

1·4 The combustion analysis used to determine carbon and hydrogen is not useful for oxygen. Explain why.

1·5 What inert-gas atoms are isoelectronic (have identical electronic configuration) with the ions in potassium fluoride?

1·6 A solution of NaCl in water gives a white precipitate immediately on treatment with aqueous silver nitrate reagent. Why is carbon tetrachloride (CCl_4) unaffected by similar treatment?

1·7 For the following compounds, indicate which groups of hydrogen atoms are equivalent (have identical average environments) and should therefore show identical chemical shifts in their pmr spectra:

a $CH_3CH_2OCH_2CH_3$

b $CH_3-\underset{\underset{CH_3}{|}}{\overset{\overset{CH_3}{|}}{C}}-Br$

c $ClCH_2CH_2CH_2Cl$

d $CH_3-\underset{\underset{Cl}{|}}{\overset{\overset{CH_3}{|}}{C}}-O-CH_3$

e $ClCH_2CH_2CH_2OH$

f cyclic structure with $CH_2-CH_2-CH_2-CH_2-C(CH_3)_2$

1·8 Using electron-dot formulas and appropriate curved arrows, write equations for the formation of $H_3C:^\ominus$ and $H_3C^\oplus$ from H_3CBr. Classify each reaction as either a homolysis or a heterolysis.

1·9 Indicate the geometry and bond polarity for H_2CO, CH_3OH, CH_3Cl, and CH_2Cl_2. What is the direction of the resultant molecular dipole for CH_2Cl_2?

1·10 Classify the following reactions according to the terminology given in Table 1.9:

a Br—CH₂—CH₂—Br + Zn ⟶ CH₂=CH₂ + ZnBr₂

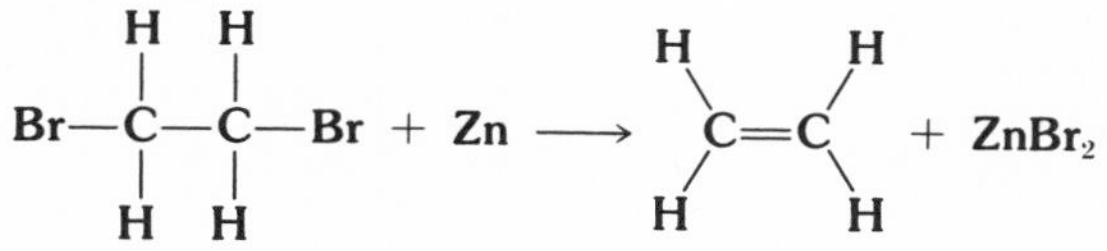

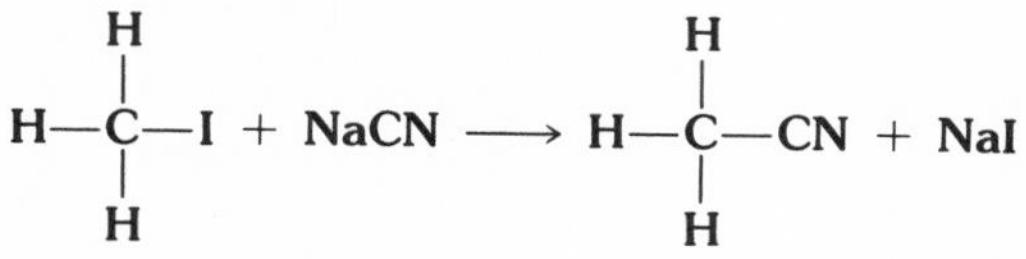

b CH₃—I + NaCN ⟶ CH₃—CN + NaI

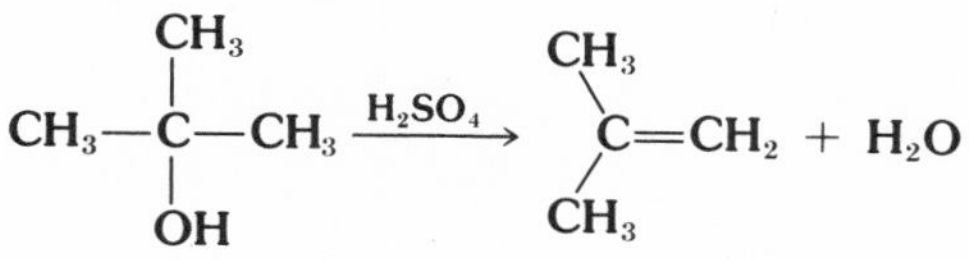

c (CH₃)₃C—OH $\xrightarrow{H_2SO_4}$ (CH₃)₂C=CH₂ + H₂O

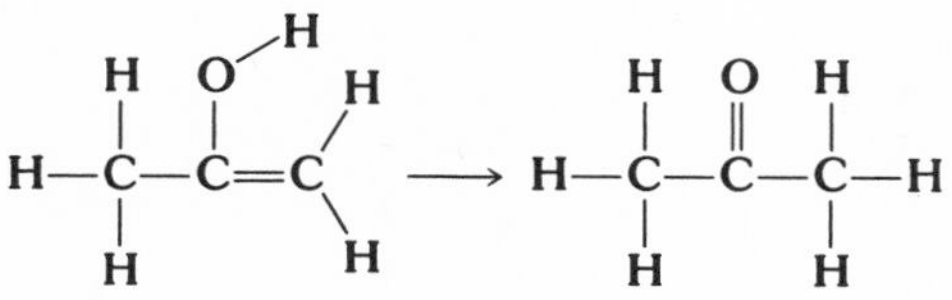

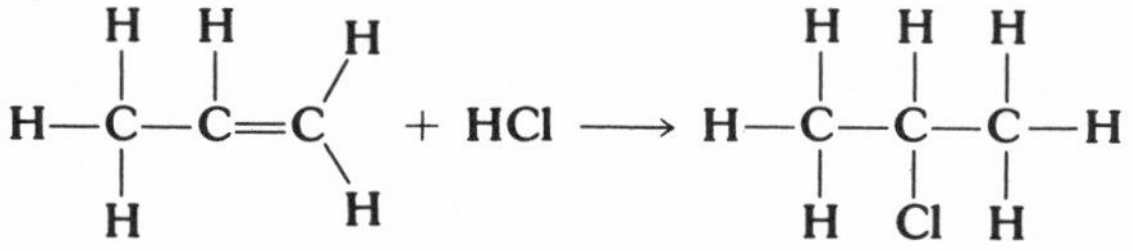

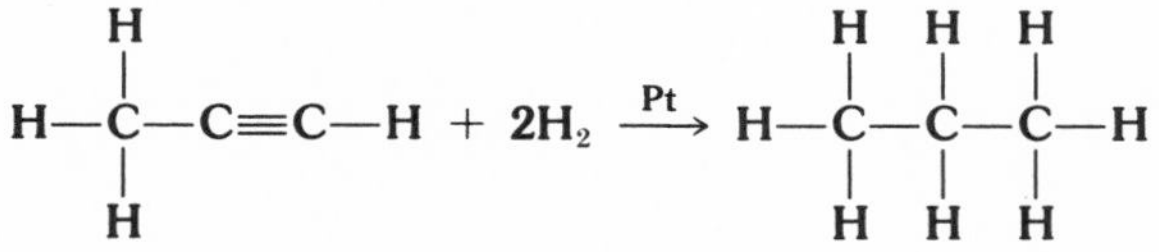

d CH₃—CH=CH—OH ⟶ CH₃—CO—CH₃

e CH₃—CH=CH₂ + HCl ⟶ CH₃—CHCl—CH₃

f CH₃—C≡C—H + 2H₂ $\xrightarrow{Pt}$ CH₃—CH₂—CH₃

g CH₃—N=O ⟶ CH₂=N—OH

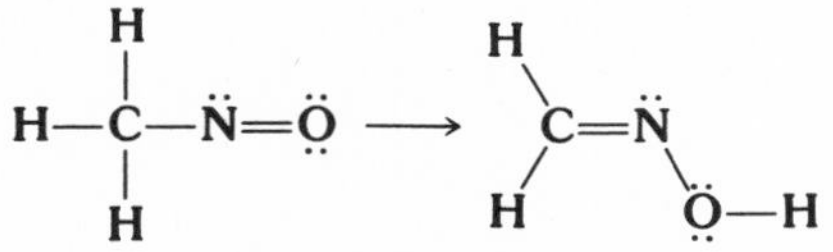

1·11 In Table 1.10 what similarities can be found in the reactions of alkenes and carbonyl compounds? In the reactions of alcohols and carboxylic acids?

1·12 Which of the following species are relatively unstable and could be called reactive intermediates?

a CH₃—Ö—CH₂·

b :Cl—C≡N:

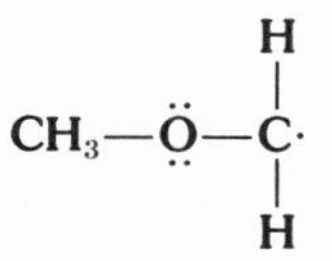

c ⁻:C≡O:⁺

d H—C≡C:⁻

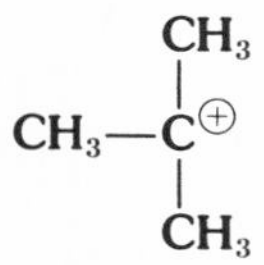

e (CH₃)₃C⁺

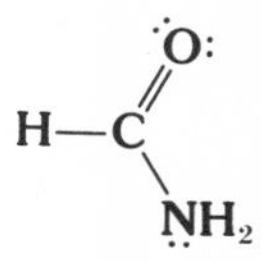

f CH₃—C≡O:⁺

g H—C(=O)—NH₂

1·13 Classify the following as radicals, carbonium ions, carbanions, or carbenes:

a HC≡C:⁻

b H—CH₂—Ö—CH₂·

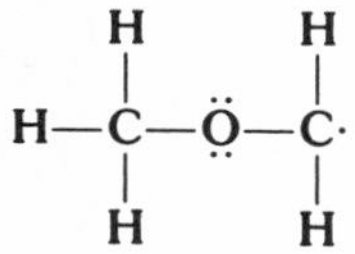

c cyclopropenyl cation CH=CH—CH⁺ · HBF₄⁻

d CH₂:

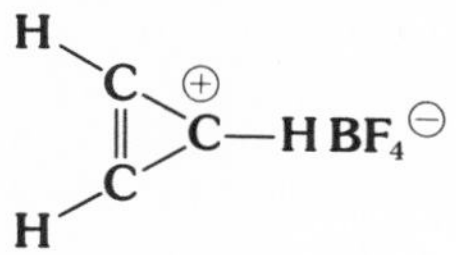

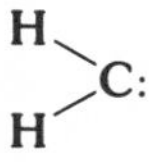

34 EXERCISES

1·14 Calculate τ values for signals which appear at δ = 0.0, 0.9, 5.0, −3.5, and 12.0 ppm.

1·15 Can a relationship between chemical shift and electronegativity of the substituent be discerned from the pmr chemical shifts of the following methyl compounds? If so, what is it?

	δ value
CH_3—F	4.3
CH_3—Cl	3.1
CH_3—Br	2.7
CH_3—I	2.2
CH_3—OH	3.47

1·16 An unknown organic compound displays strong infrared absorption at 2950 cm^{-1} and 1715 cm^{-1}. What types of bonding are probably present in this structure?

1·17 Analysis of pure quinine gives the following composition: 74.1% carbon, 7.41% hydrogen, and 8.62% nitrogen. The molecular weight of quinine is determined to be 320 ± 10.
a What is the empirical formula of quinine?
b What is the molecular formula of quinine?
c The infrared spectrum of quinine shows a strong broad absorption at 3400 cm^{-1}. What functional group is probably present in quinine?

1·18 Write an equation showing the relationship of wavenumbers $\tilde{\nu}$ to frequency ν. Suggest a reason why infrared frequencies are customarily expressed in wavenumbers rather than hertz (cps).

1·19 Write structural formulas for the products expected from the following electron reorganizations:

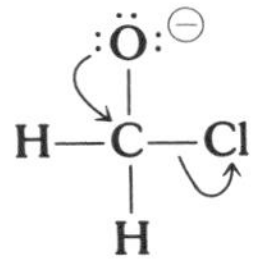

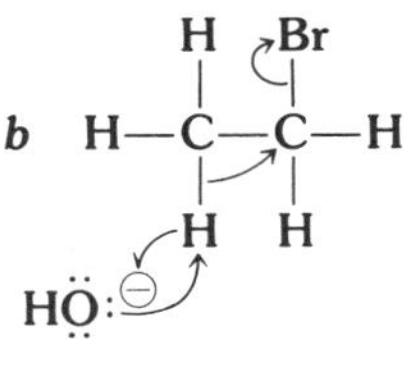

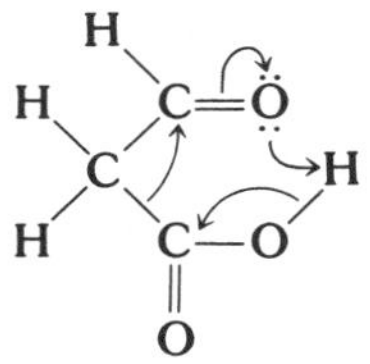

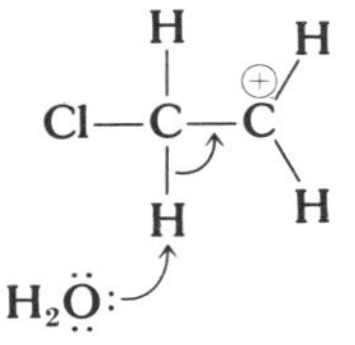

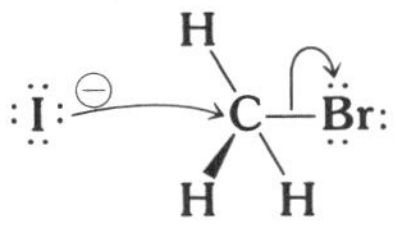

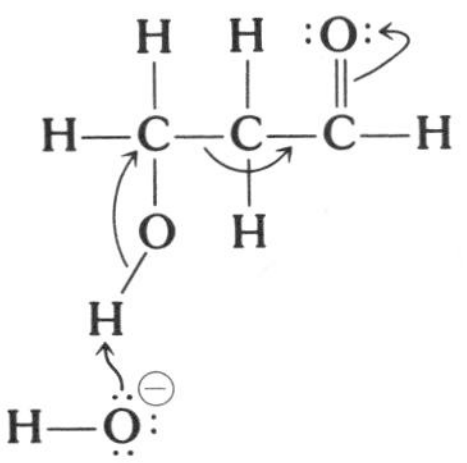

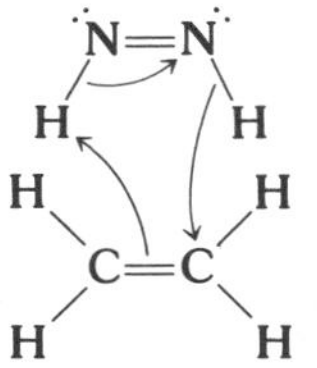

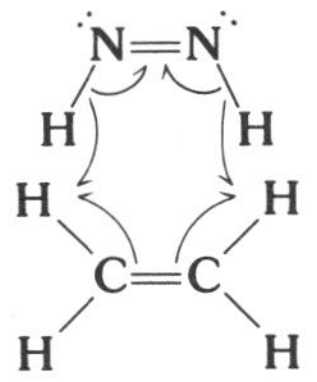

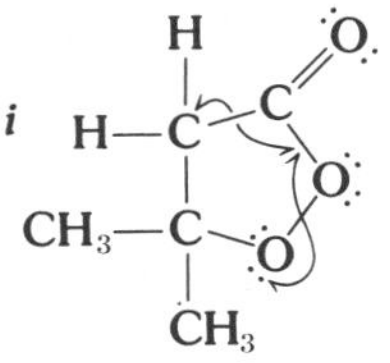

ALKANES AND CYCLOALKANES

A LARGE NUMBER of organic compounds composed only of carbon and hydrogen are obtained from natural gas, petroleum, and coal. These *hydrocarbons* are the basic raw materials for many of the chemical products that have so profoundly affected the life of twentieth-century man. Despite their simple elemental composition, the hydrocarbons display a remarkable variation in structure and behavior, as suggested by the classification scheme in Table 2.1. The *aliphatic hydrocarbons* are composed of continuous or branched chains of carbon atoms combined with sufficient hydrogen atoms to satisfy the tetravalency of carbon. In *alicyclic hydrocarbons* the carbon atoms form rings, and the *aromatic hydrocarbons* are characterized by a unique kind of six-membered ring. It is this exceptional ability of carbon atoms to bond to each other in a variety of different ways that accounts for the enormous numbers of known organic substances.

2·1 NOMENCLATURE AND REPRESENTATION

NOMENCLATURE OF SIMPLE ALKANES

The alkanes are used as the foundation for a systematic nomenclature of organic compounds. Table 2.2 lists the first six continuous-chain alkanes, usually called *normal alkanes* and written *n*-alkanes. All alkane names have the characteristic ending *-ane*. Beginning with pentane, the higher members of this class are named with Greek or Latin prefixes indicating the number of carbon atoms in the chain: *pent*ane for five carbons, *hex*ane for six carbons, *hept*ane for seven carbons, *oct*ane for eight carbons, *non*ane for nine carbons, and so on. The physical properties of these compounds vary in a regular manner as the number of

2·1 NOMENCLATURE AND REPRESENTATION

TABLE 2·1 *Classification of hydrocarbons* ($C_n H_m$)

aliphatic	alicyclic	aromatic
alkanes	cycloalkanes	benzene derivatives
alkenes	cycloalkenes	fused-ring systems
alkynes	cycloalkynes	nonbenzenoid systems

carbon atoms in the chain increases, and the chemical properties show essentially no change. This regularity is understandable, since each member of the group differs from the preceding or following member by only a CH_2 unit. A graded series of chemically related compounds such as this is referred to as a *homologous series*, and the members are called *homologs*.

REPRESENTATION OF MOLECULAR STRUCTURE

Methane, the simplest of the alkanes, consists of a tetrahedral grouping of four hydrogen atoms about a central carbon atom, as shown in Figure 2.1. The C—H bond length is 1.09 Å, and the H—C—H bond angle is 109.5°. X-ray diffraction studies of several higher alkanes show that these molecules exist in the solid state as a zigzag orientation of carbon atoms with C—C bond lengths of 1.54 Å and bond angles ranging from 106° to 112° (slightly distorted tetrahedral bond angles).

Since it is time consuming to draw three-dimensional structures like those in Figure 2.1, structural formulas are ordinarily written to show the bonding of the atoms, but not necessarily their orientation in space. The tetrahedral bonding geometry of tetravalent carbon atoms is then assumed by analogy with a large number of experimentally determined examples and from theoretical models of molecular structure. This being the case, we should attach little significance to the particular form in which a structural formula happens to be written. Because of the tetrahedral bonding geometry of their carbon atoms, the continuous-chain or *n*-alkanes cannot adopt a linear arrangement of carbon atoms. However, structural formulas of such compounds are written in linear form much of the time for convenience. Accordingly, pentane may be represented in any of the following ways:

$$CH_3-CH_2-CH_2-CH_2-CH_3 \equiv \begin{array}{c} CH_3 \\ | \\ CH_2-CH_2 \end{array} \begin{array}{c} CH_2-CH_3 \\ | \\ \end{array} \equiv \begin{array}{c} CH_3-CH_2 \\ \\ CH_3-CH_2 \end{array} CH_2$$

Condensed structural formulas, such as those listed in Table 2.2, are generally preferred because they are the simplest version to write. However, some practice is required in making the translation from expanded to condensed formulas. In cases where the three-dimensional structure of a molecule must be considered, a reasonably accurate model may be constructed from one of a variety of molecular-model kits (Figure 2.2). These models also illustrate the flexibility of single-bonded carbon chains and the numerous orientations they may assume.

2·1 NOMENCLATURE AND REPRESENTATION 37

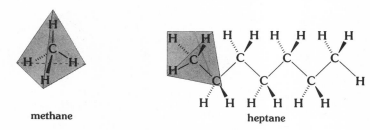

methane   heptane

FIGURE 2·1 *Hydrocarbon structures*

TABLE 2·2 *Continuous-chain alkanes*

molecular formula	name	structural formula	condensed structural formula
CH_4	methane	H—CH₂—H (with H above and below C)	CH_4
C_2H_6	ethane	H—C—C—H	CH_3CH_3
C_3H_8	propane	H—C—C—C—H	$CH_3CH_2CH_3$
C_4H_{10}	butane	H—C—C—C—C—H	$CH_3(CH_2)_2CH_3$
C_5H_{12}	pentane	H—C—C—C—C—C—H	$CH_3(CH_2)_3CH_3$
C_6H_{14}	hexane	H—C—C—C—C—C—C—H	$CH_3(CH_2)_4CH_3$
...		...	...
C_nH_{2n+2}	alkane	H—C—C—∼—C—C—H	$CH_3(CH_2)_{n-2}CH_3$

2·1 NOMENCLATURE AND REPRESENTATION

a Corey-Pauling-Koltun model of methane (Ealing Corporation)

a Brode-Boord ball-and-stick model of methane (Sargent-Welch Scientific Company)

Buchi-Brinkmann-Dreiding models of pentane: *left*, extended conformation; *right*, coiled conformation (Brinkmann Instruments, Inc.)

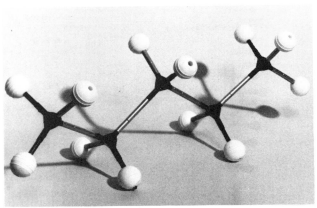

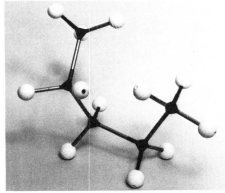

FIGURE 2·2 *Molecular models of methane and pentane (Michigan State University Photographic Laboratory Information Services)*

PROBLEM 2·1
Which of the following formulas represent the same alkane?

a
$$\begin{array}{c} H H \\ H \backslash | \diagup H \\ C \\ H-C-C-H \\ | | \\ C H \\ H \diagup | \backslash H \\ H \end{array}$$

b $CH_3-\underset{\underset{H}{|}}{\overset{\overset{CH_3}{|}}{C}}-\underset{}{\overset{CH_3}{CH_2}}$

c $CH_3CH_2CH(CH_3)CH(CH_3)_2$

d $CH_3-\underset{\underset{CH_3}{|}}{\overset{\overset{CH_3}{|}}{C}}-CH_2-CH_2-CH_3$

e $CH_3(CH_2)_2CH_3$

f $CH_3(CH_2)_2C(CH_3)_3$

g $CH_3CH_2CH(CH_3)_2$

h $CH_3-\underset{\underset{CH_3}{|}}{\overset{\overset{H}{|}}{C}}-\underset{}{\overset{CH_3}{\underset{}{CH-CH_2-CH_3}}}$

FIGURE 2·3 *Conventions for showing the staggered and eclipsed conformations of ethane*

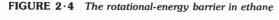

CONFORMATIONS

Despite their tetrahedral bonding geometry, we are still faced with an uncertainty in the structure of all alkanes higher than methane because the relationship between the hydrogens attached to different carbon atoms may vary. For example, ethane (C_2H_6) could be described by a structure in which the hydrogen atoms are staggered, by a structure in which they are opposed, or by an infinite number of intermediate arrangements, all of which are related by rotation about the carbon-carbon bond axis. Figure 2.3 shows a staggered arrangement and an opposed, or eclipsed, arrangement, in which the hydrogen atoms are directly aligned with each other. The *Newman projection* of these structures is a view along the C—C bond axis, with the near carbon atom represented by a dot and the far one represented by a circle. The different spatial orientations that may be assumed by the atoms of a molecule are called its *conformations*.

Spectroscopic studies of ethane have shown that a small energy barrier (2.8 kcal/mole) favors the staggered conformation over the eclipsed conformation (see Figure 2.4). Since the thermal energy of the ethane molecules in a sample at room temperature (25°C) is sufficient to permit rapid interconversion of these conformations, equilibrium between them is quickly established, with the staggered conformation predominating by more than 100:1.

FIGURE 2·4 *The rotational-energy barrier in ethane*

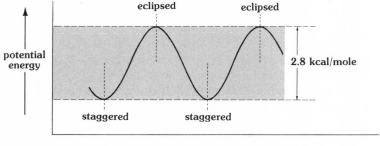

2·1 NOMENCLATURE AND REPRESENTATION

The higher alkanes can assume a multitude of different conformations which are readily transformed from one to another by rotations about carbon-carbon bonds. The zigzag conformations, as shown for heptane in Figure 2.1, are favored in the crystalline state because of the ease with which the atoms pack together in a closely spaced lattice arrangement. In liquid and gaseous states other conformations are present in significant amounts. Again, remember that ordinary structural formulas have no conformational significance. Indeed, the unique representation of a specific compound by a condensed structural formula would not be possible if essentially free rotation about carbon-carbon single bonds did not exist.

> **PROBLEM 2·2**
> Draw eclipsed and staggered conformations about the *central* C—C bond of *n*-butane. Estimate the approximate order of increasing potential energy (decreasing stability) for these conformations. *Hint:* CH_3 is much larger than H.

THE IUPAC NOMENCLATURE SYSTEM

Beginning with C_4H_{10}, the molecular formulas of the alkanes do not invariably represent a continuous chain of carbon atoms. Branching of the carbon chain results in isomeric compounds. For example, butane and pentane have the following isomers:

C_4H_{10} butane

$CH_3—CH_2—CH_2—CH_3$
or
$CH_3(CH_2)_2CH_3$
n-butane

$CH_3—CH(CH_3)—CH_3$
or
$(CH_3)_3CH$
isobutane

C_5H_{12} pentane

$CH_3—CH_2—CH_2—CH_2—CH_3$
or
$CH_3(CH_2)_3CH_3$
n-pentane

$CH_3—CH(CH_3)—CH_2—CH_3$
or
$(CH_3)_2CHCH_2CH_3$
isopentane

$CH_3—C(CH_3)_2—CH_3$
or
$(CH_3)_4C$
neopentane

The number of theoretically possible isomers of a given formula increases rapidly with the number of carbon atoms; thus there are 3 isomeric pentanes, 5 hexanes, 9 heptanes, 18 octanes, 35 nonanes, and 75 decanes.

To provide rational names for the many complex structures that can arise from chain branching, a standard nomenclature system has been adopted by the International Union of Pure and Applied Chemistry (IUPAC). According to this system, a branched hydrocarbon is regarded as a substitution product of a simpler straight-chain compound. On this basis, the various hydrocarbon groups attached to the chain are designated as substituents and are called *alkyl groups*. The alkyl groups are named as shown in Table 2.3. It is important to note that these alkyl groups have no independent existence; they are merely a nomenclature device and have no connection with the radicals described as reactive intermediates in Chapter 1. The terms *primary*, *secondary*, and *tertiary* (abbreviated *p*, *s*, and *t*) refer to the number of carbon atoms attached

2·1 NOMENCLATURE AND REPRESENTATION

TABLE 2·3 *Typical alkyl groups*

structural formula	common name	classification
CH_3-	methyl	primary
CH_3CH_2- or C_2H_5-	ethyl	primary
$CH_3CH_2CH_2-$ or $n\text{-}C_3H_7-$	n-propyl	primary
$\begin{array}{c}CH_3\\ \diagdown\\ CH-\\ \diagup\\ CH_3\end{array}$ or $(CH_3)_2CH-$	isopropyl	secondary
$(CH_3)_2CHCH_2-$	isobutyl	primary
$\begin{array}{c}CH_3\\ \diagdown\\ CH-\\ \diagup\\ C_2H_5\end{array}$ or $C_2H_5(CH_3)CH-$	s-butyl	secondary
$CH_3-\underset{\underset{CH_3}{\mid}}{\overset{\overset{CH_3}{\mid}}{C}}-$ or $(CH_3)_3C-$	t-butyl	tertiary
$(CH_3)_3CCH_2-$	neopentyl	primary
$R-$	alkyl	—

directly to the functional carbon of the group—that carbon atom bonded to the chain being discussed. The methyl group is a special case, since it consists of a single carbon atom (it is sometimes considered a "super primary" group).

Let us now proceed to the *IUPAC nomenclature rules:*

1 The longest continuous chain of carbon atoms is located and named according to the number of carbon atoms in the chain.

2 The groups attached to this chain are identified and named according to Table 2.3:

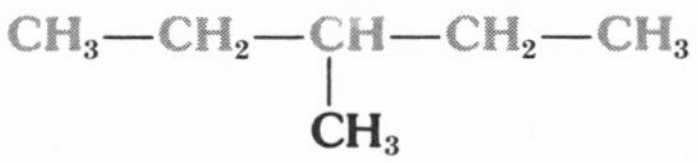

$CH_3-CH_2-\underset{\underset{CH_3}{\mid}}{CH}-CH_2-CH_3$ the longest chain is a *pentane unit*

the substituent is a *methyl group*

3 If more than one longest chain can be found, the one bearing the greatest number of substituent groups is chosen. Thus the *first* of the following two chains is chosen as the longest chain:

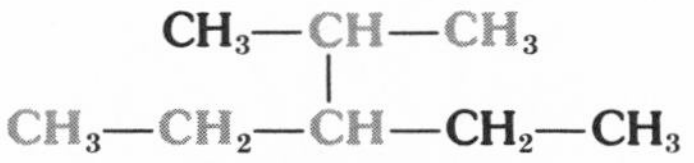

$\begin{array}{c}CH_3-CH-CH_3\\ \mid\\ CH_3-CH_2-CH-CH_2-CH_3\end{array}$ $\begin{array}{c}CH_3-CH-CH_3\\ \mid\\ CH_3-CH_2-CH-CH_2-CH_3\end{array}$

two substituents 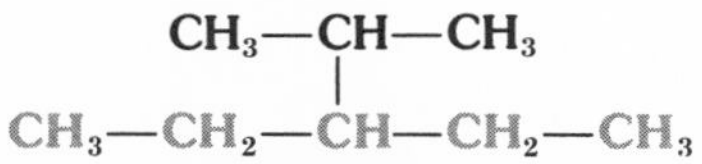   **one substituent**

4 If more than one alkyl group of the same kind is attached to the chain, the name of this group bears a prefix indicating how many such groups are present. Thus the following compound is a dimethylbutane:

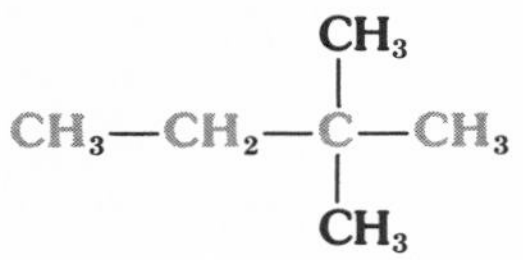

$CH_3-CH_2-\underset{\underset{CH_3}{\mid}}{\overset{\overset{CH_3}{\mid}}{C}}-CH_3$

5 The carbon atoms of the longest continuous chain (found by rule 3) are numbered consecutively from one end to the other, beginning at that end which gives the *lowest number to the site of first difference*. The position of each substituent group on the chain is then denoted by the corresponding number. For example,

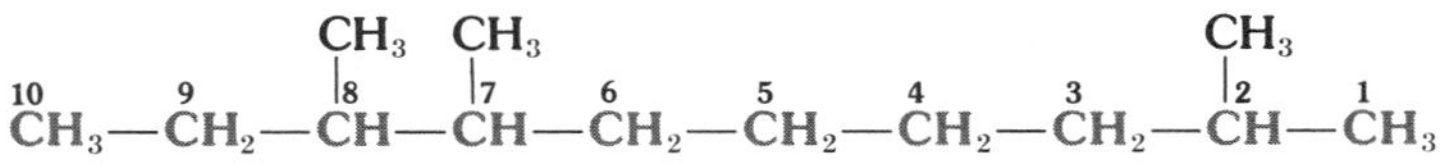

$$\overset{10}{C}H_3-\overset{9}{C}H_2-\overset{8}{\underset{|}{C}}H-\overset{7}{\underset{|}{C}}H-\overset{6}{C}H_2-\overset{5}{C}H_2-\overset{4}{C}H_2-\overset{3}{C}H_2-\overset{2}{\underset{|}{C}}H-\overset{1}{C}H_3$$
with CH$_3$ groups on carbons 8, 7, and 2

is properly named 2,7,8-trimethyldecane (rather than 3,4,9-trimethyldecane) because the site of first difference is closer to the right-hand end of the chain.

6 When there are two or more different substituents attached to the chain, they may be listed in alphabetical order. For example, the following branched octane,

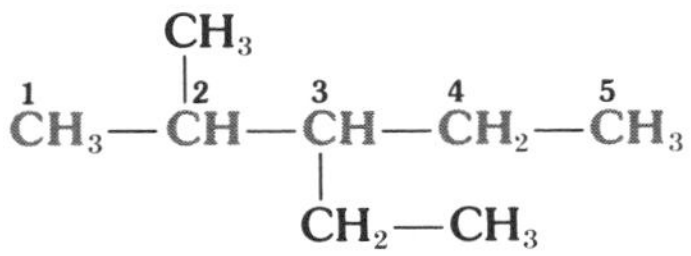

$$\overset{1}{C}H_3-\overset{2}{\underset{|}{C}}H-\overset{3}{\underset{|}{C}}H-\overset{4}{C}H_2-\overset{5}{C}H_3$$
with CH$_3$ on C2 and CH$_2$—CH$_3$ on C3

is 3-ethyl-2-methylpentane. Note that if the seven-carbon chain in the following compound were numbered from the right (rather than from the left, as shown), the resulting name (3-ethyl-3,5,6-trimethylheptane) would not satisfy rule 5:

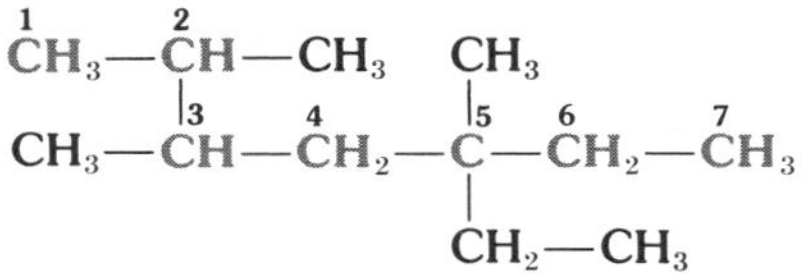

$$\overset{1}{C}H_3-\overset{2}{C}H-CH_3 \quad CH_3$$
$$\overset{3}{C}H_3-\overset{4}{C}H-\overset{5}{C}H_2-\overset{6}{C}-\overset{7}{C}H_2-CH_3$$
$$\overset{|}{C}H_2-CH_3$$

5-ethyl-2,3,5-trimethylheptane

Two techniques may be used to check IUPAC names:

1 The sum of the carbon atoms indicated by the basic name and the carbon atoms of all the substituent groups should equal the total carbon content of the molecule. For 5-ethyl-2,3,5-trimethylheptane, the last example above, we have:

base name:	heptane	C_7
substituent groups:	ethyl	C_2
	trimethyl	C_3
	total	C_{12}

2 A correct IUPAC name will always enable us to reconstruct *unambiguously* the molecular constitution of the compound to which it refers:

5-ethyl-2,3,5-trimethylheptane
⇓
heptane $\underset{1}{C}-\underset{2}{C}-\underset{3}{C}-\underset{4}{C}-\underset{5}{C}-\underset{6}{C}-\underset{7}{C}$
⇓

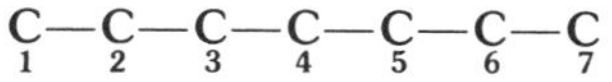

2·2 PROPERTIES OF ALKANES

$$\Downarrow$$

5-ethyl $\underset{1}{C}-\underset{2}{C}-\underset{3}{C}-\underset{4}{C}-\underset{5}{\overset{\overset{\displaystyle CH_2-CH_3}{|}}{C}}-\underset{6}{C}-\underset{7}{C}$

$$\Downarrow$$

2,3,5-trimethyl $\underset{1}{C}-\underset{2}{\overset{\overset{\displaystyle CH_3}{|}}{C}}-\underset{3}{\overset{\overset{\displaystyle CH_3}{|}}{C}}-\underset{4}{C}-\underset{5}{\overset{\overset{\displaystyle CH_2-CH_3}{|}}{\underset{\underset{\displaystyle CH_3}{|}}{C}}}-\underset{6}{C}-\underset{7}{C}$

$$\Downarrow$$

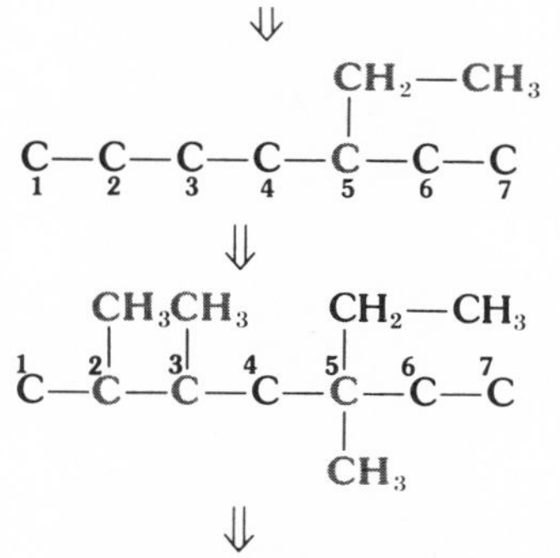

$(CH_3)_2CHCH(CH_3)CH_2C(C_2H_5)_2CH_3$

PROBLEM 2·3
Which of the following drawings represent identical conformations of the same compound? Which represent different conformations of the same compound? Which represent constitutional isomers?

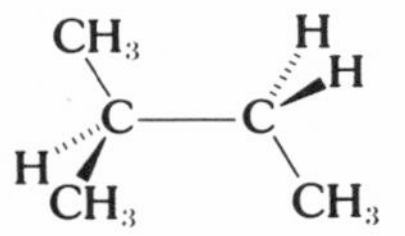

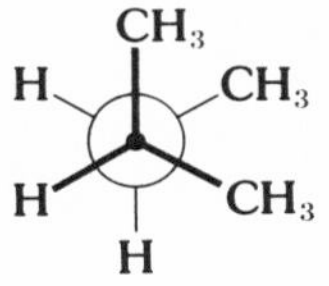

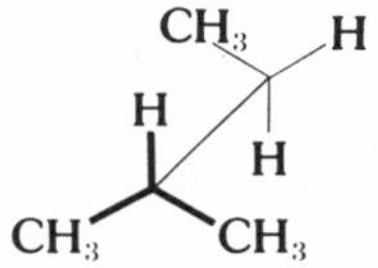

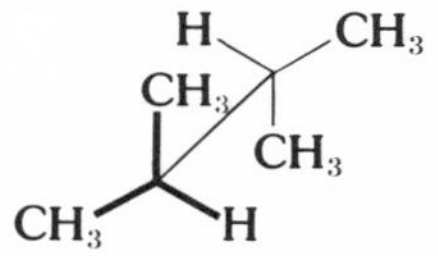

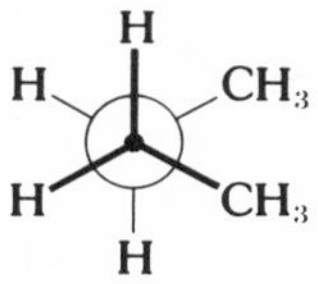

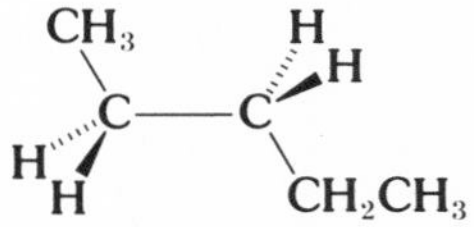

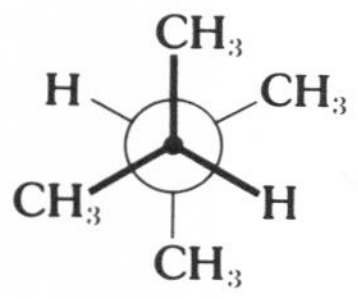

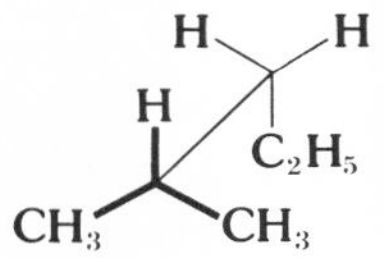

2·2 PROPERTIES OF ALKANES

PHYSICAL PROPERTIES

The alkanes range from flammable gases and liquids to viscous oils and waxlike solids, depending on their molecular weight. Those composed of one to four carbon atoms are gases, the C_5 to C_{17} alkanes are liquids, and the higher members of the series are solids. These generalizations apply, of course, for room temperature (25°) and standard atmospheric pressure (760 torr). At sufficiently high temperatures and low pressures, all alkanes will become gases. The boiling points and melting points of the *n*-alkanes generally increase with the number of carbon atoms in the chain, although the melting points show some alternation (Figure 2.5). The alkanes are all *immiscible* with (insoluble in) and less dense than water.

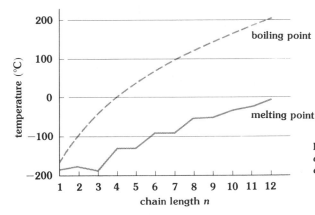

FIGURE 2·5 *Dependence of boiling points and melting points on chain length in continuous-chain alkanes,* $CH_3(CH_2)_{n-2}CH_3$

To understand these physical properties we must consider the *intermolecular forces* (the forces between molecules) in a compound. The fact that all known gases can be liquified, and in most cases solidified, demonstrates the existence of attractive forces among molecules. These forces are quite weak in comparison to the bonding forces which hold the atoms of a molecule together, but it is these weak interactions that determine the three-dimensional configurations of such biologically active substances as enzymes, proteins, and nucleic acids.

The strongest of these intermolecular forces (2 to 10 kcal/mole) is the *hydrogen bond*, a largely electrostatic attraction between a highly electronegative atom such as oxygen, nitrogen, or fluorine and a hydrogen atom bonded covalently to another electronegative atom. Hydrogen bonds have direction (indicated by the location of the component parts) and are usually denoted by dashed lines:

Y—H---Y—R
(Y = N, O, or F)

Water, for example, has an exceptionally high boiling point and viscosity for a substance composed of small, light molecules (molecular weight 18, bp 100°C). This is because hydrogen-bonded aggregations of water molecules, as shown in Figure 2.6, raise the effective molecular weight of this system substantially; it has been estimated that in water at 0°C each molecule is hydrogen bonded to an average of 3.4 other water molecules. Since the boiling point (at constant pressure) may be regarded as a measure of the kinetic energy required to tear a molecule from the liquid aggregate, the fact that methane (CH_4, molecular weight 16, bp $-161.5°C$) and fluorine oxide (F_2O, molecular weight 54, bp $-144.7°C$) boil at temperatures more than 200° below water testifies to the strength of hydrogen bonding in comparison with other intermolecular interactions. The strong hydrogen-bonded lattice structure of water is also resistant to the incorporation or intrusion of nonpolar alkane molecules, and hydrocarbon and water phases are therefore normally immiscible.

Not only are hydrogen bonds much weaker than covalent bonds, they are almost twice as long. In large molecules the cumulative effect of such bonds between repeating units within the molecule can influence

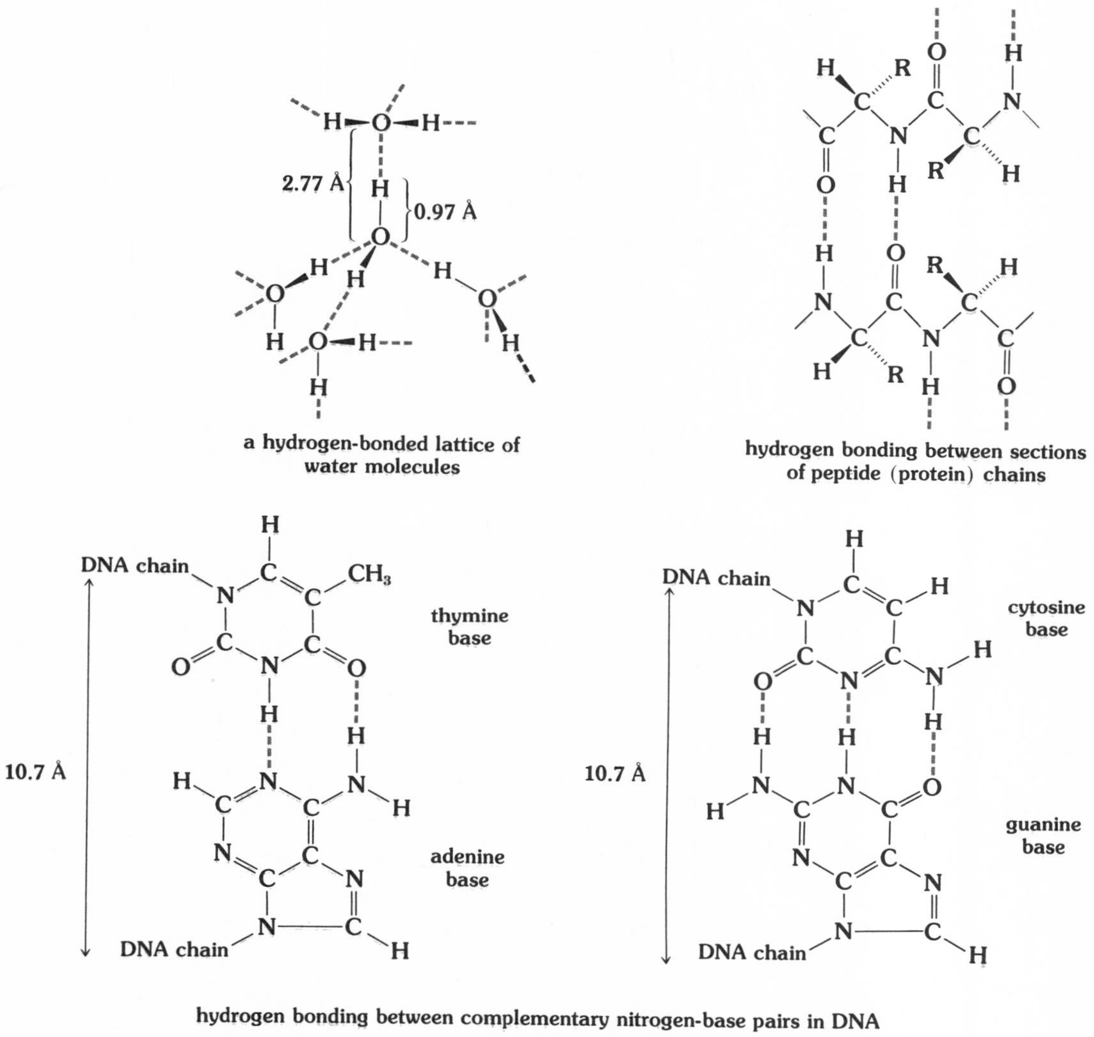

FIGURE 2·6 *Examples of hydrogen bonding*

the three-dimensional structure in dramatic ways. The kinds of hydrogen bonds that control the conformations and higher-ordered structures of proteins and nucleic acids are illustrated in Figure 2.6.†

Dipole-dipole intermolecular attractions are also important in most polar molecules (for example, HCN and H_2CO), but these forces are weaker than the hydrogen bond. The weakest intermolecular attractive forces are the *van der Waals–London forces* that exist between all molecules, including nonpolar substances such as methane. These forces increase with molecular size, or chain length, as indicated by the increasing boiling and melting points in Figure 2.5.

†In his account of the events that led to elucidation of the structure of DNA (*The Double Helix*, Atheneum Press, 1968), Nobel laureate James D. Watson writes: "The crux of the matter was a rule governing hydrogen bonding between bases."

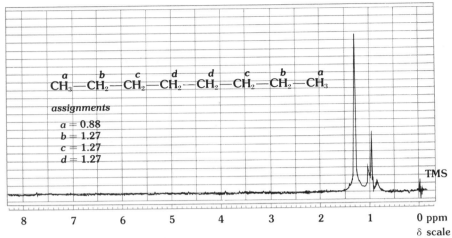

FIGURE 2·7 *Pmr spectrum of* n-*octane*

PROBLEM 2·4
Predict the relative boiling points and water solubility of propane (C_3H_8), ethyl alcohol (C_2H_5OH), and dimethyl ether (CH_3OCH_3).

SPECTROSCOPIC PROPERTIES

Alkanes exhibit characteristic C—H stretching frequencies at 2850 to 3000 cm^{-1} in the infrared spectrum, but since most organic compounds have C—H bonds, this absorption band is of limited value in identifying a particular compound.

The pmr spectra of continuous-chain alkanes show the effects of spin-spin interactions between neighboring hydrogen atoms. Thus the methylene peak at about $\delta = 1.3$ is broadened and the methyl signal at about $\delta = 0.9$ is split into three discernible peaks, as seen in the spectrum of *n*-octane (Figure 2.7). Note that *n*-octane has three pairs of *structurally* nonequivalent methylene groups (CH_2). However, most pmr spectrometers are not sensitive enough to distinguish these different sets of protons, and all 12 secondary hydrogen atoms are observed to resonate at $\delta = 1.27 \pm 0.03$.

2·3 CHEMICAL REACTIVITY OF ALKANES

The alkanes are a relatively unreactive class of compounds, sometimes referred to as the *paraffins* (from the Latin *parum affinis*, little affinity). These hydrocarbons are, for example, inert to treatment with strong acids and bases or with metallic sodium. Two characteristic reactions of the alkanes are *combustion* and *halogenation*.

COMBUSTION

The combustion of natural-gas and petroleum components is a major source of heat and power. Natural gas is mainly methane, while petroleum is a complex mixture of gaseous, liquid, and solid substances which must be refined before it is used. Over 100 different hydrocarbons, many of them alkanes, have been isolated by fractional distillation of a single petroleum sample. The overall chemical transformation that occurs in the complete combustion of an alkane is

$$C_nH_{2n+2} + \frac{3n+1}{2} O_2 \longrightarrow nCO_2 + (n+1)H_2O \tag{2.1}$$

It is clear from this equation that the combustion process is extremely complex. It would, in fact, be difficult to imagine a more extensive reorganization of atoms than has occurred here. Many steps are apparently required for this transformation, which has, therefore, a very complicated reaction mechanism. Indeed, careful studies of hydrocarbon flames have shown the presence of a variety of radical intermediates.

HEAT OF REACTION Since heat is evolved in the combustion of hydrocarbons—that is, the reaction is *exothermic*—we consider the products to have a lower heat content, or potential energy, than the reactants. The difference in heat content between the reactants and products is the *heat of reaction* ΔH. By convention this is defined as

$$\text{heat of reaction} = \text{heat content of products} - \text{heat content of reactants}$$
$$\Delta H = H_{\text{products}} - H_{\text{reactants}} \tag{2.2}$$

Thus ΔH in exothermic reactions will always have a negative value.

Although combustion is always exothermic, other reactions may yield products having a higher heat content than the reactants. In such a transformation heat would be absorbed (ΔH would be positive), and the reaction would be termed *endothermic*. These definitions are summarized in Figure 2.8.

BOND ENERGIES Heat-of-reaction measurements are important to chemists because they can lead to a determination of average bond strengths, or bond energies. In any chemical reaction the kinds and numbers of atoms do not change. We can therefore define a hypothetical high-energy *atomic state* which is the same for both reactants and products. The energy or heat released when the atoms combine to form reactants is equal to Σ_r, the sum of all the bond energies in the reactants. Similarly, if the atoms combine to give the products, then energy is liberated in an amount equal to Σ_p, the sum of all the bond energies in the products. It must be emphasized that these are *imaginary experiments* that only serve to illustrate the concept of bond energies. The difference

FIGURE 2·8 *Exothermic and endothermic reactions*

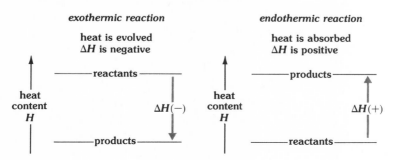

48 2·3 CHEMICAL REACTIVITY OF ALKANES

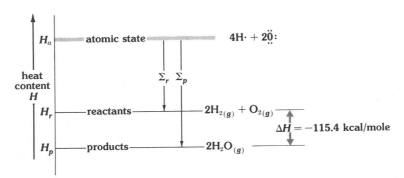

FIGURE 2·9 *Bond-energy relationships in the combustion of hydrogen*

between the total bond energies of the reactants, Σ_r, and the total bond energies of the products, Σ_p, is therefore the heat of reaction ΔH, usually reported for constant-pressure reactions having an initial and final temperature of 25°C:

$$\Delta H^{25°} = \Sigma_r - \Sigma_p \tag{2.3}$$

Figure 2.9 illustrates this relationship for the combustion of hydrogen,

$$2H_{2(g)} + O_{2(g)} \xrightarrow{\text{combustion}} 2H_2O_{(g)} \tag{2.4}$$

where the subscript g refers to a gaseous state.† The fact that this relationship is the reverse of the corresponding heat-content expression (equation 2.2) emphasizes that the bond energies are not an energy possessed by the molecular system, but the energy required to break apart all the covalent bonds to the component atoms. It is important to recognize that the system having the largest total bond energy is *lowest* on the heat-content scale. It must also be remembered that this heat-content, or potential-energy, scale has no fixed or precisely known points. Consequently, we can determine only the *difference in energy* between systems of identical elemental composition.

For the combustion of hydrogen, then, equation 2.3 becomes

$$\Delta H^{25°} = \Sigma_r - \Sigma_p$$
$$-115.4 \text{ kcal/mole} = 2(H\!-\!H) + (O\!=\!O) - 4(O\!-\!H) \tag{2.5}$$

There are too many unknowns for us to solve this equation as it stands; however, bond energies for diatomic molecules such as H_2 and O_2 can be obtained from spectroscopic measurements. Values for H—H and O=O have been found from such measurements to be 104 kcal/mole and 119 kcal/mole, respectively, so that the unknown O—H bond energy can easily be calculated:

$$-115.4 = 208 + 119 - 4(O\!-\!H)$$
$$4(O\!-\!H) = 115.4 + 208 + 119$$
$$O\!-\!H = 110.6 \text{ kcal/mole}$$

†The method of computation described here requires that all experimental data be referred to a completely gaseous reaction system.

2·3 CHEMICAL REACTIVITY OF ALKANES

TABLE 2·4 *Average bond energies*

type of bond	heat content H (kcal/mole at 25°C)	type of bond	heat content H (kcal/mole at 25°C)
H—H	104.2	C—Br	68
F—F	36.6	C—I	51
Cl—Cl	58.0	C—O	85.5
Br—Br	46.1	C=O	166–192†
I—I	36.1	C—N	72.8
O=O	119.1	C=N	147
N≡N	225.8	C≡N	212.6
O—O	35	C—S	65
S—S	54	H—F	134.6
C—C	82.6	H—Cl	103.2
C=C	145.8	H—Br	87.5
C≡C	199.6	H—I	71.4
C—H	98.7	H—O	110.6
C—F	166	H—N	43.4
C—Cl	81	H—S	83

†C=O for CO_2 is 192.0, C=O for aldehydes is 176, and C=O for ketones is 179.

Other average bond energies can be similarly computed, and a collection of such values is presented in Table 2.4. These average bond energies can be used to estimate the heats of reaction for cases that have not been directly measured. For example, the combustion of methane,

$$CH_{4(g)} + 2O_{2(g)} \longrightarrow CO_{2(g)} + 2H_2O_{(g)} \tag{2.6}$$

is calculated from equation 2.3 to have a reaction heat of $\Delta H^{25°} = -194.4$:

$$\Sigma_r = 4(C—H) + 2(O=O) = 4 \times 99 + 2 \times 119 = 634 \text{ kcal/mole}$$
$$\Sigma_p = 2(C=O) + 4(H—O) = 2 \times 192 + 4 \times 111 = 828 \text{ kcal/mole}$$
$$\Delta H^{25°} = 634 - 828 = -194 \text{ kcal/mole}$$

This agrees reasonably well with the *observed value* of −192 kcal/mole.

PROBLEM 2·5
Write a balanced equation and calculate the heat of combustion for the complete combustion of acetylene (H—C≡C—H).

PROBLEM 2·6
At one time a considerable effort was directed (with no success) to achieving the reaction

$$CH_4 + CO_2 \longrightarrow CH_3OH + CO$$

Discuss the feasibility of this transformation at room temperature (note that the C—O bond energy of carbon monoxide is 255.8 kcal/mole).

50 2·3 CHEMICAL REACTIVITY OF ALKANES

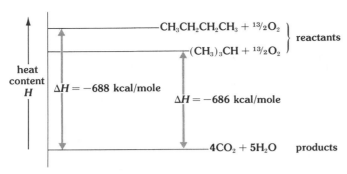

FIGURE 2·10 *Energy changes in the combustion of* n-*butane and isobutane*

In the preceding calculations of bond energies and heats of reaction we have assumed that bonds of a given kind (C—C, C—H, etc.) always have the same energy. However, the variable C=O bond energy noted in Table 2.4 indicates that this assumption is not strictly true. Further evidence is disclosed by a careful measurement of the heats of combustion for the isomers *n*-butane and isobutane. The combustion of both compounds is described by the following reaction:

$$C_4H_{10(g)} + {}^{13}/_2 O_{2(g)} \longrightarrow 4CO_{2(g)} + 5H_2O_{(g)} \tag{2.7}$$

However, as illustrated in Figure 2.10, there is a difference of 2 kcal/mole in the observed heats of combustion for the two isomers. For *n*-butane $\Delta H^{25°} = -688$ kcal/mole, and for isobutane $\Delta H^{25°} = -686$ kcal/mole.

Since identical products are obtained in each case, Σ_p (*n*-butane) = Σ_p (isobutane) and the 2 kcal difference must reside in the reactant bond energies of the two isomers. Both have ten C—H bonds and three C—C bonds, so obviously all C—H bonds and/or all C—C bonds do *not* have the same bond strength. We must draw a careful distinction between the average bond energies listed in Table 2.4 and the *bond-dissociation energies* of specific bonds. The variation in bond-dissociation energies of different C—H bonds is shown in Table 2.5. Diatomic molecules are exceptional in that the average bond energies for H_2, Cl_2, HBr, etc., are also their bond-dissociation energies.

TABLE 2·5 *Bond-dissociation energies of* C—H *bonds*

C—H bond	dissociation energy H (kcal/mole at 25°C)
CH_3—H	102
CH_3CH_2—H	97
$(CH_3)_2CH$—H	95
$(CH_3)_3C$—H	90
CH_2=$CHCH_2$—H	88
CH_2=CH—H	105

Biochemists often speak of "high-energy bonds" in "energy-rich compounds" such as adenosine triphosphate (ATP) and creatine phosphate. These terms do not refer to bond energies as they are defined here, but apply instead to compounds that react readily under physiological conditions and exhibit large negative heats of reaction.

PROBLEM 2·7

a The CH_3—H bond-dissociation energy of 102 kcal/mole in Table 2.5 differs from the C—H bond energy of 98.7 kcal/mole given in Table 2.4. How is it, then, that the heat of combustion of methane can be calculated so closely from equation 2.3?

b The energy required to remove successive hydrogen atoms from methane has been determined as follows. Calculate the bond-dissociation energy of the fourth (and last) hydrogen atom:

$CH_4 \longrightarrow CH_3\cdot + H\cdot \quad \Delta H = 102$ kcal/mole
$CH_3\cdot \longrightarrow CH_2{:} + H\cdot \quad \Delta H = 87$ kcal/mole
$CH_2{:} \longrightarrow \cdot\dot{C}H + H\cdot \quad \Delta H = 125$ kcal/mole
$\cdot CH\cdot \longrightarrow {:}C{:} + H\cdot \quad \Delta H = ?$

PROBLEM 2·8

Using the bond-dissociation energies in Table 2.5 and the average C—C energy value in Table 2.4, calculate the difference in heat content between 2,2-dimethylpropane (neopentane) and *n*-pentane. How will this difference affect the heats of combustion of these isomers?

HALOGENATION

Halogenation is the replacement of one or more of the hydrogen atoms in an organic compound by a halogen (fluorine, chlorine, bromine, or iodine) to form the corresponding fluoride, chloride, bromide, or iodide. Next to combustion, this transformation of alkanes to *alkyl halides* is the most important reaction of the alkanes.

CHLORINATION OF METHANE A mixture of methane and chlorine can be kept in the dark at ordinary temperatures for long periods without showing any signs of change. However, if this mixture is ignited by a spark or exposed to sunlight, an explosive reaction occurs. The overall transformation can be described by a series of *substitution reactions*:

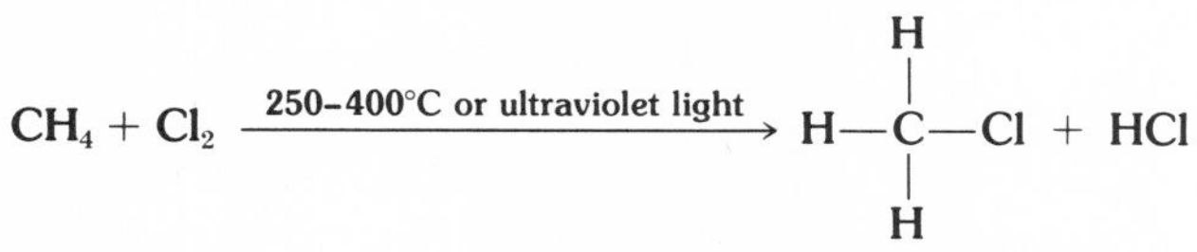

$$CH_4 + Cl_2 \xrightarrow{\text{250--400°C or ultraviolet light}} H-\underset{\underset{H}{|}}{\overset{\overset{H}{|}}{C}}-Cl + HCl \qquad (2.8)$$

chloromethane
or methyl chloride

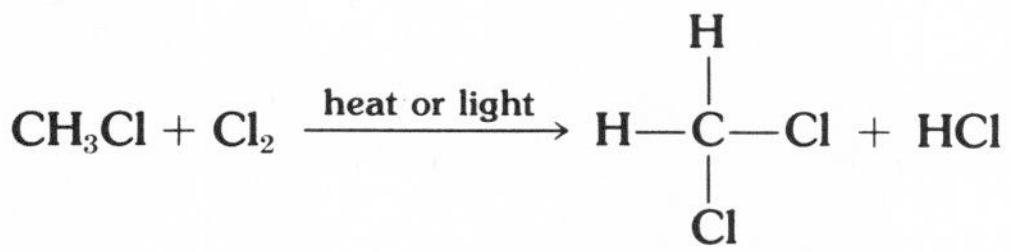

$$CH_3Cl + Cl_2 \xrightarrow{\text{heat or light}} H-\underset{\underset{Cl}{|}}{\overset{\overset{H}{|}}{C}}-Cl + HCl \qquad (2.9)$$

dichloromethane
or methylene chloride

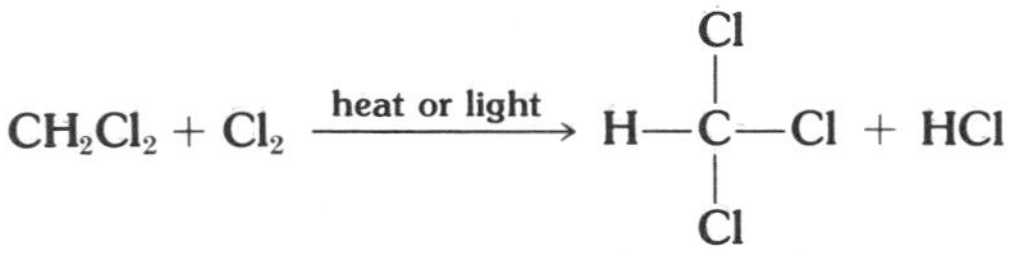

52   2·3 CHEMICAL REACTIVITY OF ALKANES

$$CH_2Cl_2 + Cl_2 \xrightarrow{\text{heat or light}} H-\underset{\underset{Cl}{|}}{\overset{\overset{Cl}{|}}{C}}-Cl + HCl \tag{2.10}$$

trichloromethane
or chloroform

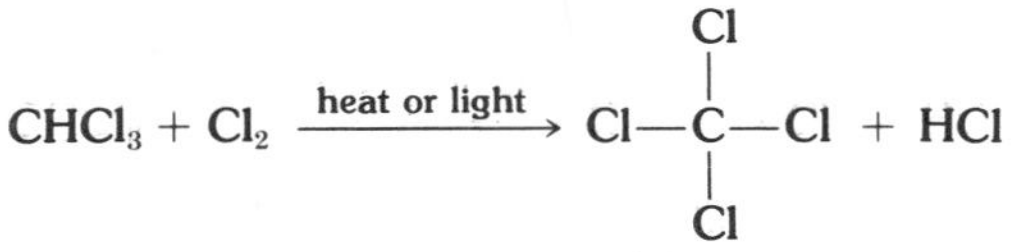

$$CHCl_3 + Cl_2 \xrightarrow{\text{heat or light}} Cl-\underset{\underset{Cl}{|}}{\overset{\overset{Cl}{|}}{C}}-Cl + HCl \tag{2.11}$$

tetrachloromethane
or carbon tetrachloride

If the reaction mixture contains an excess of methane, the most significant reaction is 2.8 and the major product is chloromethane. A similar but slower reaction occurs with bromine. Direct fluorination and iodination are not practical, since fluorine is uncontrollably reactive and iodine does not react.

PROBLEM 2·9
What experimental conditions would be expected to give $CHCl_3$ and CCl_4 as the major products from the chlorination of methane?

PROBLEM 2·10
Consider the alternate *square planar* and *pyramidal* configurations for methane:

 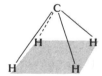

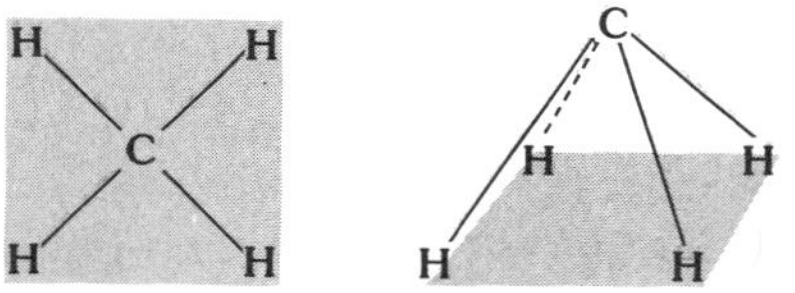

a How many monochloro- (CH_3Cl), dichloro- (CH_2Cl_2), and trichloro- ($CHCl_3$) isomers are possible for the square planar, pyramidal, and tetrahedral configurations of methane?
b Only one compound having the formula CH_2Cl_2 is known. What does this fact imply about the configuration of methane?

PROBLEM 2·11
Calculate $\Delta H^{25°}$ for the following reaction, where X = F, Br, and I in turn:

$$CH_{4(g)} + X_{2(g)} \longrightarrow CH_3-X_{(g)} + HX_{(g)}$$

The chlorination of methane is exothermic ($\Delta H^{25°} = -24$ kcal/mole) and essentially irreversible at temperatures below 500°C. Unlike the complex transformations occurring during combustion, this relatively simple substitution reaction appears to provide an ideal system for speculation and investigation of mechanism. A satisfactory reaction mechanism for the chlorination of methane must accommodate two key facts. First, strong heat or light is necessary for the reaction to proceed at a reasonable rate. Second, if light is used, thousands of molecules react for each photon of light absorbed. This large efficiency factor suggests that a

2·3 CHEMICAL REACTIVITY OF ALKANES

chain reaction is occurring, and since ions are seldom formed in gas-phase reactions, this chlorination chain reaction probably involves neutral intermediates such as radicals and atoms. The following sequence of steps defines a reaction path or mechanism that agrees with these facts:

$Cl_2 \xrightarrow{energy} 2Cl\cdot$ initiation   (2.12)

$\left. \begin{array}{l} Cl\cdot + CH_4 \longrightarrow H-Cl + CH_3\cdot \\ CH_3\cdot + Cl_2 \longrightarrow CH_3-Cl + Cl\cdot \end{array} \right\}$ chain propagation   (2.13)

$\left. \begin{array}{l} Cl\cdot + Cl_2 \longrightarrow Cl_2 + Cl\cdot \\ CH_3\cdot + CH_4 \longrightarrow CH_4 + CH_3\cdot \end{array} \right\}$ not productive   (2.14)

$\left. \begin{array}{l} Cl\cdot + Cl\cdot \longrightarrow Cl_2 \\ Cl\cdot + CH_3\cdot \longrightarrow CH_3-Cl \\ CH_3\cdot + CH_3\cdot \longrightarrow CH_3-CH_3 \end{array} \right\}$ chain-termination reactions   (2.15)

Only two kinds of covalent bonds, C—H and Cl—Cl, are present in the starting mixture of methane and chlorine. At temperatures below 100°C the thermal energy of the reaction mixture cannot break a significant number of these bonds. As the temperature is raised, however, homolysis of the relatively weak Cl—Cl bond takes place preferentially over cleavage of the C—H bonds of methane. Furthermore, since chlorine is the only reactant that absorbs light in the 3000 to 6000-Å region, it is likely that Cl—Cl bond rupture is also the initiating step for the light-induced chlorination. Once the existence of a strongly endothermic initiating step is recognized, it is clear why a methane-chlorine mixture is stable in the dark below 100°.

The chlorine atoms generated in the initiation step react with methane to form HCl and methyl radicals ($CH_3\cdot$). A chain reaction is thus begun (equations 2.13), wherein the reactive chlorine atom is regenerated by attack of methyl radicals on molecular chlorine. Theoretically, a single chlorine atom could completely convert an equimolar mixture of methane and chlorine to chloromethane and hydrogen chloride. In practice, however, the chain reaction is eventually terminated by the disappearance of radicals and atoms through combination with each other (equations 2.15). Although these termination reactions are very rapid and exothermic, chain propagation is favored over termination because the concentration of radicals and atoms in the reaction system is very small and a direct encounter between such species is rare. Chain lengths of many hundreds are common in these reactions.

PROBLEM 2·12

Calculate $\Delta H^{25°}$ for each step of the chain reaction (equations 2.13) operating in the chlorination of methane. Determine the result of *adding* these two equations (include your ΔH calculations).

PROBLEM 2·13

Calculate $\Delta H^{25°}$ for each step in the following *suggested* chain reaction. How does this reaction path compare with that described by equations 2.13?

$Cl\cdot + CH_4 \longrightarrow CH_3-Cl + H\cdot$
$H\cdot + Cl_2 \longrightarrow H-Cl + Cl\cdot$

OTHER ALKANES Chlorination of the higher alkanes takes place in a similar manner, but the formation of isomeric products complicates the analysis of these systems. For example, chlorination of propane gives *both* 1-chloropropane (bp = 47°C) and 2-chloropropane (bp = 36°C):

$$CH_3CH_2CH_3 \xrightarrow[300°]{Cl_2} \underset{\text{48\% 1-chloropropane}}{CH_3-CH_2-CH_2-Cl} + \underset{\text{52\% 2-chloropropane}}{CH_3-\underset{\underset{H}{|}}{\overset{\overset{Cl}{|}}{C}}-CH_3} \quad (2.16)$$

Propane has six equivalent hydrogen atoms attached to primary carbon atoms (that is, six primary hydrogens) and two equivalent hydrogen atoms attached to the secondary carbon (two secondary hydrogens). Thus a probability factor favors formation of 1-chloropropane over 2-chloropropane by 3:1. However, almost equimolar amounts of the two isomers are actually formed at 300°C, which means that the secondary hydrogen atoms must react approximately *three times as rapidly* as the primary hydrogen atoms. To understand the reasons for this difference in reactivity, we must consider some fundamental principles of reaction-rate theory.

PROBLEM 2·14
The isomers 1-chloropropane and 2-chloropropane can be separated by fractional distillation (note the difference in boiling points). Show how pmr spectroscopy can be used to determine the structure of each.

PROBLEM 2·15
Draw structural formulas for the isomeric dichloropropanes. How can these isomers be distinguished?

PROBLEM 2·16
Write a sequence of equations illustrating the mechanism for the chlorination of propane shown in reaction 2.16.

2·4 REACTION-RATE THEORY

A close approach (or collision) between reacting molecules is generally considered to be a critical and necessary event in the course of a reaction. From a knowledge of the concentration of molecules per unit volume, their velocity, and their diameter, it is a relatively straightforward matter to calculate the collision frequency between molecules in a homogeneous system. However, these collision frequencies are so large (about 10^{29} collisions for 1 mole of a gas at standard temperature and pressure) that it becomes obvious that not all encounters between reactant molecules lead to reaction. Apparently only a fraction of the collisions are effective in producing chemical change. This inefficiency has led chemists to propose that the rate of a chemical reaction is a function of characteristic *energy* and *probability factors:*

reaction rate = collision frequency × energy factor × probability factor (2.17)

2·4 REACTION-RATE THEORY

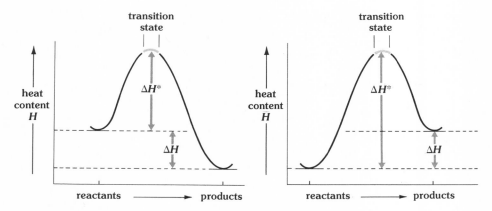

FIGURE 2·11 *Energy diagrams for exothermic and endothermic reactions*

ENERGY OF ACTIVATION

The energy factor that influences reaction rates is essentially the heat content H required to effect the conversion of reactants to products. This energy term should not be confused with the heat of reaction ΔH, even though strongly exothermic reactions are often fast and endothermic reactions are usually slow. Nearly all reactions, particularly those involving some bond breaking, must pass through an unstable high-energy *transition state* in proceeding from reactants to products (Figure 2.11). The difference between the heat content of this transition state and the heat content of the reactants is called the *heat of activation* ΔH^*. This value represents the minimum energy the colliding reactant molecules must possess before they can pass to products For example, the first energy diagram of Figure 2.11 might represent the following reaction, for which the heat of reaction is calculated to be $\Delta H = -1.2$ kcal/mole:

$$Cl\cdot + CH_4 \longrightarrow \begin{bmatrix} \overset{H}{\underset{\delta\cdot}{Cl}} \text{---} H \text{---} \overset{H}{\underset{\delta\cdot}{\underset{H}{C}}} \text{---} H \\ \text{transition state} \end{bmatrix} \longrightarrow HCl + CH_3\cdot \quad (2.18)$$

The dashed lines in the transition state represent *partial bonds* (the C—H bond of methane is partly broken and the H—Cl bond is partly formed). The activation energy in this case has been determined to be $\Delta H^* = 4$ kcal/mole. The reverse of this reaction is, of course, endothermic and would necessarily have the same transition state as the forward reaction. This is best represented by the second energy diagram in Figure 2.11, for which $\Delta H = +1.2$ kcal/mole and $\Delta H^* = 5.2$ kcal/mole.

PROBLEM 2·17

a Calculate the heat of reaction ΔH for the following reaction:

$$CH_4 + Br\cdot \longrightarrow CH_3\cdot + HBr$$

Which energy diagram in Figure 2.11 applies best to this reaction?

b What is the lowest possible ΔH^* for this reaction?

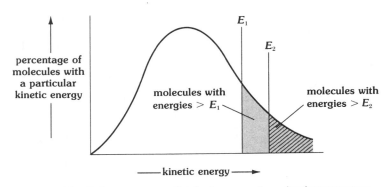

FIGURE 2·12 *Boltzman energy-distribution curve for a fixed temperature*

At a fixed temperature, and in the absence of a significant probability factor, reactions having a *large* heat of activation are slower than those with a small activation energy. This follows from the distribution of kinetic energy within the reaction system and from the law of mass action. Since the number of molecules having a kinetic energy greater than E_1 (or E_2) is proportional to the area under that part of the distribution curve (Figure 2.12), it is clear that this number is larger for the lower energy E_1 than for the higher energy E_2. If we regard E_1 and E_2 as the energies of activation for competing reactions, that reaction with the lower ΔH^* (E_1) will have a greater concentration of sufficiently energetic reactant molecules and can therefore be expected to proceed more rapidly.

The application of these principles to the chlorination of propane (equation 2.16) should now be evident. In the following chain reaction, the first step is the *product-determining step:*

$$\begin{array}{c}CH_3\\ \diagdown\\ CH\cdot + HCl \\ \diagup\\ CH_3\end{array} \longleftarrow \begin{array}{c}CH_3CH_2CH_3\\ +\\ Cl\cdot\end{array} \longrightarrow CH_3CH_2CH_2\cdot + HCl \qquad (2.19)$$

$$C_3H_7\cdot + Cl_2 \longrightarrow C_3H_7-Cl + Cl\cdot$$

FIGURE 2·13 *Product-determining steps in the chlorination of propane*

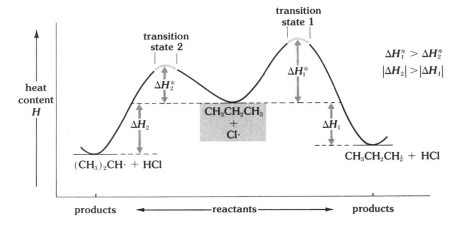

The transition states for the two courses of this step probably resemble the corresponding transition state for methane chlorination in equation 2.18 and will therefore have a partially broken C—H bond. The bond-dissociation energies of C—H bonds vary in the order primary C—H > secondary C—H > tertiary C—H (Table 2.5). Therefore, as Figure 2.13 indicates, the heat of activation $\Delta H_1^{\ddagger}$ for attack of Cl· at one of the six methyl hydrogen atoms in propane should be greater than the corresponding heat of activation $\Delta H_2^{\ddagger}$ for attack at the central methylene group. As a result, the energy factor in these reactions favors the formation of 2-chloropropane.

The energy diagrams in Figures 2.11 and 2.13 are intended as visual aids, rather than as precise representations of a reaction. Instead of a single reaction coordinate, we should imagine many coordinates representing all the various motions displayed by the reacting system. Figure 2.14 shows a front view of a simplified potential-energy surface. The z dimension represents the potential energy, the y dimension represents the reaction coordinate, and the x dimension represents the other degrees of freedom. In such a multidimensional model the transition state appears as a saddle point, or pass, and not as a peak, since the reaction will naturally follow the path of least resistance. With this qualification in mind, the two-dimensional approximations in the energy diagrams can be quite helpful in visualizing the energy changes that take place during reactions.

THE PROBABILITY FACTOR

The probability factor in reaction rates reflects *a change in the order (or disorder)* of the reaction system. Most of us would agree that a disordered or random assemblage has a higher probability of existing than an ordered grouping. Certainly attics, basements, and desk drawers seem to have a perverse tendency toward disorder, and a well-shuffled deck of cards will not arrange itself in separate suits (become more ordered) upon further shuffling. Since propane has six primary hydrogen atoms and only two secondary hydrogen atoms, a transition state for attack at the primary hydrogens (transition state 1 in Figure 2.13) has a greater

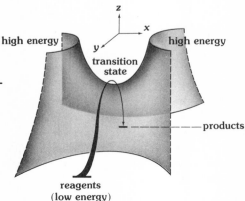

FIGURE 2·14 *Front view of a simplified potential-energy surface (after J. Leffler and E. Grunwald,* Rates and Equilibria of Organic Reactions, *John Wiley & Sons, Inc., New York, 1963)*

statistical probability and is less strictly ordered than a transition state involving the secondary hydrogens (transition state 2). The roughly equal amounts of 1-chloropropane and 2-chloropropane obtained from the chlorination of propane are thus due to the opposition of nearly equal energy and probability factors.

A more precise and quantitative treatment of probability and disorder is provided by the thermodynamic function *entropy*, symbolized by S. Entropy describes the disorder of a system: the greater the disorder, the higher the entropy. Spontaneous processes cause an increase in entropy, and to decrease the entropy of a system, work must be done or heat must be removed.

In the reaction of propane with chlorine atoms (Figure 2.13) the reactants have a higher entropy (are more disordered) than the transition states. For example, visualize a box containing a propane molecule and a chlorine atom. These particles wander independently about the box in a random manner. If the particles collide and a transition state (either 1 or 2) is formed, the orientation of the chlorine atom with respect to the propane molecule is severely restricted, and a lower-entropy (more ordered) system results. This change is described by the *entropy of activation* ΔS^*, which is defined by the relationship

$$\Delta S^* = S^* - S_r \qquad (2.20)$$

where S^* is the entropy of the transition state and S_r is the entropy of the reactants. A negative value of ΔS^* indicates that the transition state is more ordered than the reactants, while a positive ΔS^* points to a less ordered transition state.

We have discussed the influence of the energy factor (heat of activation) on the product-determining step in the chlorination of propane. Now we must include an entropy-of-activation term, which represents the small but significant resistance to organizing the reaction system into the more ordered configuration of the transition state. This entropy factor is temperature dependent and reflects the increase in randomness which accompanies a rise in temperature, thus making it more difficult to align the reacting species. The heat or enthalpy of activation is therefore modified by the entropy of activation, and the resulting thermodynamic function is called the *free energy of activation* ΔG^*:

free energy of activation = heat of activation + entropy factor
$$\Delta G^* = \Delta H^* - T \Delta S^* \qquad (2.21)$$

PROBLEM 2·18
Why is the entropy term in equation 2.21 preceded by a minus sign?

Recall that the chlorination of propane can proceed in two directions (equation 2.19). We must now ascertain whether transition state 1 or 2 in Figure 2.13 has the larger entropy of activation—that is, which transition state requires a greater reorganization of the reactants. Let us again consider our propane molecule and chlorine atom in a box. From Figure 2.15 we see that transition state 1 can be reached by the approach of a chlorine atom from one of six different directions, whereas only two possible orientations can lead to transition state 2. Consequently, ΔS_2^*, the

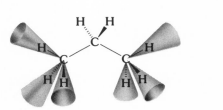

access to transition state 1   access to transition state 2

FIGURE 2·15 *Chlorine atom access corridors*

entropy term for transition state 2, has a larger negative value than ΔS_1^*. At 300°C the difference between the entropies of activation is approximately counterbalanced by the difference between ΔH_1^* and ΔH_2^* (Figure 2.13); thus the free energies of activation ΔG^* for the two reactions are almost equal.

PROBLEM 2·19

Chlorination of 2-methylpropane(isobutane) at 300°C gives 67% 1-chloro-2-methyl-propane and 33% 2-chloro-2-methylpropane. From this fact calculate the discrimination of the energy factor toward nonequivalent sites of reaction.

PROBLEM 2·20

The following equation describes a single-step chain process that could account for the chlorination of methane:

$$Cl\cdot + CH_3-H + Cl-Cl \longrightarrow Cl-CH_3 + H-Cl + Cl\cdot$$

Calculate $\Delta H^{25°}$ for this reaction. What conclusions can you draw regarding the rate of this process? Consider both the energy and probability (or entropy) factors.

2·5 CYCLOALKANES

As noted at the beginning of the chapter, the large number and variety of hydrocarbon compounds is due in part to the almost unique ability of carbon atoms to form chains and rings by bonding to each other. Like the continuous-chain or n-alkanes, the cyclic hydrocarbons, referred to as *cycloalkanes*, form a homologous series with physical properties that vary, like those of the alkanes, according to the number of carbon atoms. Some simple monocyclic (one-ring) compounds are listed in Table 2.6.

Cycloalkanes bearing substituents are referred to in general as *substituted cycloalkanes*. Such compounds are numbered so that the carbon atoms bearing the substituent groups or atoms are assigned the lowest possible numbers:

1,1-dimethylcyclohexane   1-ethyl-2-methyl-cyclopentane   1,1-dichloro-4-methyl-cyclodecane   bromocyclopropane *or* cyclopropyl bromide

TABLE 2·6 *The cycloalkanes*

molecular formula	name	structural formula	abbreviated notation
C_3H_6	cyclopropane	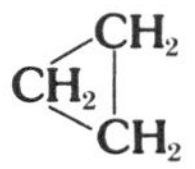 $CH_2{-}CH_2{-}CH_2$ (ring)	 ▷
C_4H_8	cyclobutane	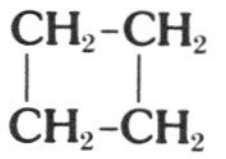 $\begin{array}{c}CH_2-CH_2\\ \mid\quad\ \mid\\ CH_2-CH_2\end{array}$	□
C_5H_{10}	cyclopentane	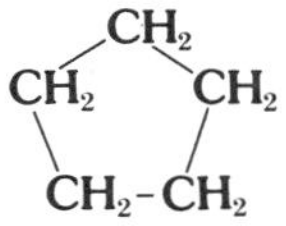 (ring of 5 CH_2)	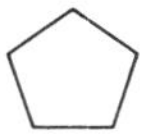 ⬠
C_6H_{12}	cyclohexane	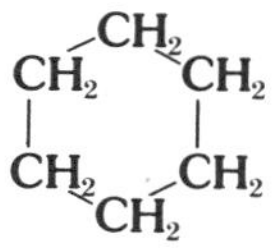 (ring of 6 CH_2)	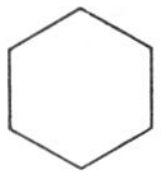 ⬡
C_7H_{14}	cycloheptane	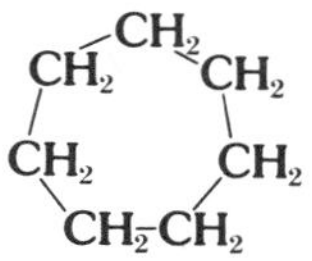 (ring of 7 CH_2)	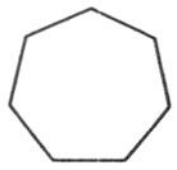 ⬣
. . .	. . .	. . .	. . .
C_nH_{2n}	cycloalkane	$(CH_2)_n$	$(CH_2)_n$

STEREOISOMERISM OF SUBSTITUTED CYCLOALKANES

A novel kind of isomerism occurs in ring systems bearing two or more substituents. For example, in the compound 1,2-dimethylcyclopropane, shown in Figure 2.16, the methyl groups may both occupy the same side of the ring, called a *cis* configuration, or they may lie on opposite sides, called a *trans* configuration. This type of isomerism is called *stereoisomerism*, since the difference in molecular structure is due only to a different arrangement of atoms in space. The particular spatial orientation of atoms that characterizes a given stereoisomer is termed its *configuration*.

Of course, isomers such as *n*-butane and isobutane also have different arrangements of atoms in space, but in addition, the atoms composing these isomers are bonded together in two fundamentally different ways. Thus *n*-butane has two methyl groups with a total of six primary hydrogen

FIGURE 2·16 *Stereoisomers of 1,2-dimethylcyclopropane*

cis configuration   trans configuration

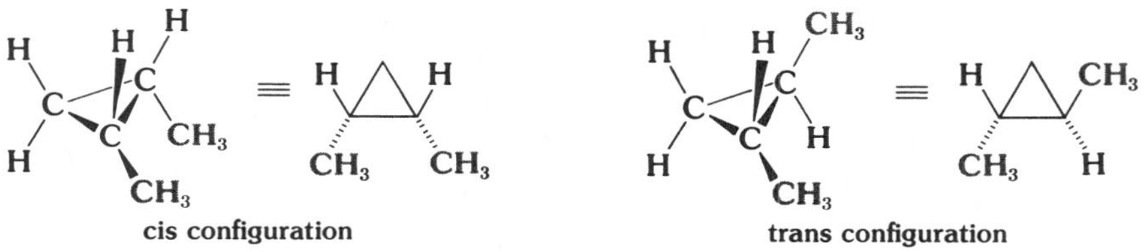

atoms and two methylene groups with four secondary hydrogens. In contrast, isobutane has three methyl groups with nine primary hydrogens and one tertiary carbon bearing a single hydrogen atom.

$CH_3(CH_2)_2CH_3$ $(CH_3)_3CH$
n-butane   isobutane

Isomers differing in the manner in which atoms are connected or bonded together are called *constitutional isomers* to distinguish them from stereoisomers. Constitutional isomers will always have different IUPAC names, whereas stereoisomers must be identified by an additional nomenclature term, such as the *cis-* and *trans-* prefixes. The distinction between stereoisomers and conformational isomers (different conformations of the same molecule) is not so well defined, and some overlap exists. However, interconversion of conformational isomers is often rapid at room temperature, while stereoisomers are configurationally stable under normal conditions.

PROBLEM 2·21
Which, if any, of the four cycloalkanes on page 59 can exhibit stereoisomerism?

PROBLEM 2·22
Which of the following are constitutional isomers? Which are stereoisomers? Name each compound.

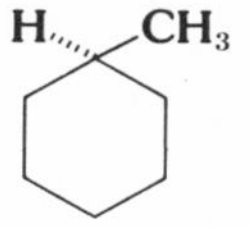

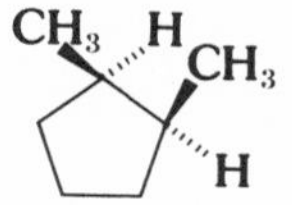

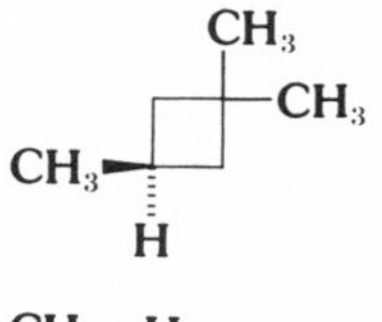

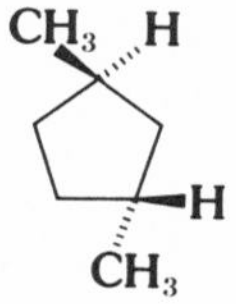

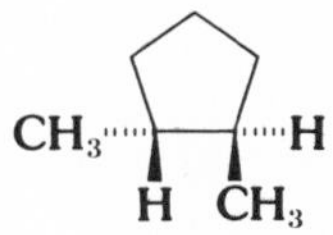

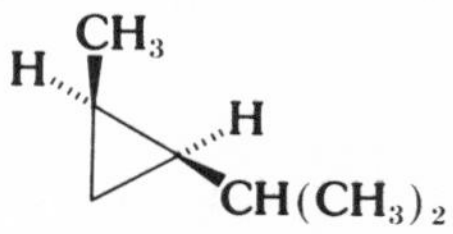

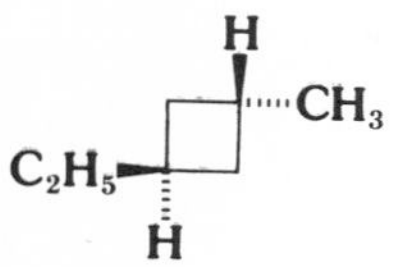

RING STRAIN

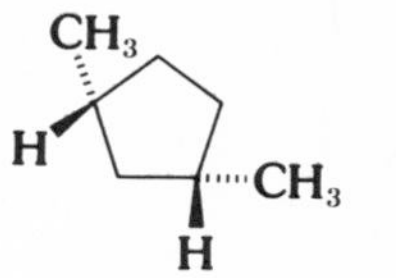

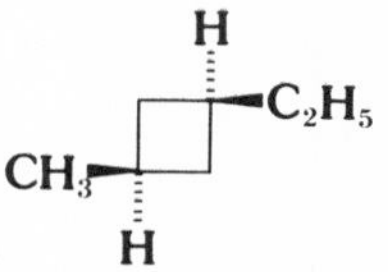

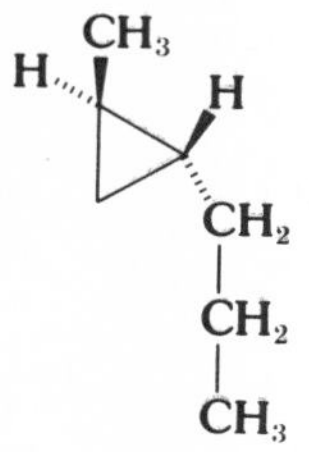

It is clear from their geometry that the C—C—C bond angles in small rings such as cyclopropane and cyclobutane cannot have the normal tetrahedral value of 109.5°. The resulting angle distortion or strain (49.5° for each of the three angles in cyclopropane and 19.5° for each angle in cyclobutane) naturally influences the chemistry and stability of these systems. A quantitative measure of the total strain in the different-size cycloalkanes can be obtained from the heat-of-combustion data in Table

2·5 CYCLOALKANES

TABLE 2·7 *Heats of combustion and strain of cycloalkanes*

cycloalkane $(CH_2)_n$	number of CH_2 units	$\Delta H^{25°}$ (kcal/mole)	$\Delta H^{25°}$ per CH_2	total strain (kcal/mole)
cyclopropane	3	468.6	156.2	27.6
cyclobutane	4	614.3	153.6	26.4
cyclopentane	5	741.5	148.3	6.5
cyclohexane	6	882.1	147.0	0.0
cycloheptane	7	1035.4	147.9	6.3
cyclooctane	8	1186.0	148.2	9.6
cyclononane	9	1336.0	148.4	11.2
cyclodecane	10	1482.0	148.2	12.0
cyclopentadecane	15	2206.5	147.1	1.5
$CH_3(CH_2)_mCH_3$	large	—	147.0	0.0

2.7. To compensate for the changing number of carbon atoms in the different rings, we divide these heats of combustion by the number of carbons n to obtain a $\Delta H^{25°}$ value that is characteristic of a single methylene group. The strain-free continuous-chain alkanes provide a reference for the heat of combustion per methylene group against which the ring strain in the cycloalkanes can be judged. Thus in Figure 2.17 the strain per CH_2 unit is the difference between the ΔH_{ring} value per CH_2 unit in the ring and the ΔH_{chain} value of 147.0 kcal/mole for a CH_2 unit in a continuous chain. The total strain is then obtained by multiplying this difference by n.

The considerable angle strain in cyclopropane and cyclobutane is evident from the data given in Table 2.7. Although early chemists once thought the larger rings (larger than cycloheptane) would also exhibit angle strain, the facts in Table 2.7 do not support this view, and we now recognize that these rings can assume a variety of puckered conformations, having essentially tetrahedral carbon-bond angles.

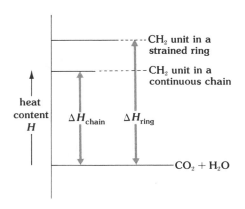

FIGURE 2·17 *Energy relationship for the combustion of methylene groups in alkanes and cycloalkanes*

CONFORMATIONS OF CYCLOALKANES

The conformations of the cycloalkanes can be analyzed in terms of three different strain factors. The first is *angle strain*, the deviation of carbon bond angles from the tetrahedral value of 109.5°. The second is *eclipsing strain*, the rotational preference for a staggered conformation about C—C bonds (see Figure 2.4). The third involves *nonbonded repulsive interactions*, a crowding together of two or more atoms not bonded to each other.

A *planar conformation* of any cycloalkane (see Table 2.8) will necessarily be fully eclipsed, an arrangement we know to be energetically unfavorable. The cycloalkanes therefore tend to adopt puckered conformations in which the eclipsing strain is significantly reduced, even when this may involve a small increase in angle strain (thus cyclobutane is not planar). As an illustration, consider the conformations available to cyclohexane.

The strain-free low-energy chair conformation is so much more stable than the other conformations that 99.9% of the molecules in a sample at 25°C will assume this conformation. Cyclohexane has two distinct but equivalent chair conformations, which are rapidly interconverted by rotations about bonds (molecular models are helpful in visualizing this

TABLE 2·8 *Conformations of cyclohexane*

conformation	name	angle strain	eclipsing strain	nonbonded interactions
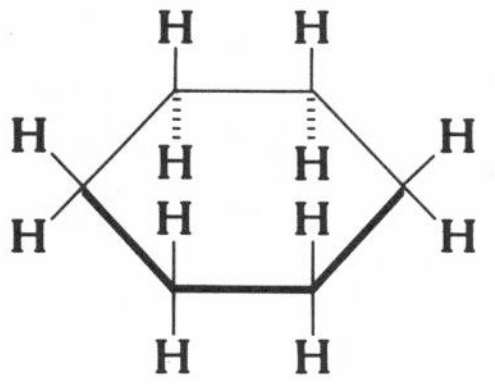	planar	severe (+10.5°)	all bonds	small
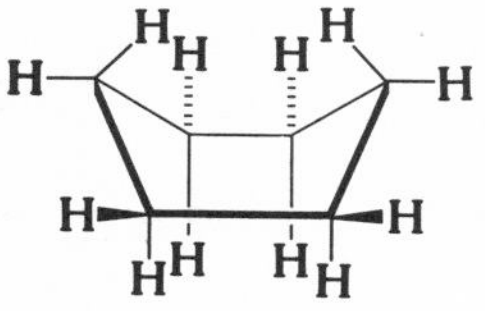	boat	very small	two bonds	moderate
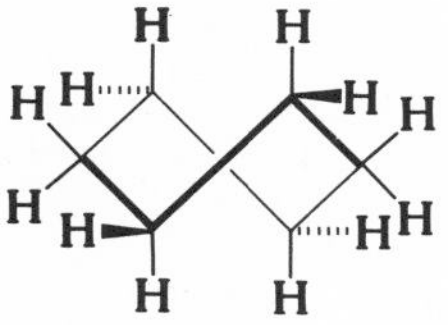	twist	very small	slight eclipsing	small
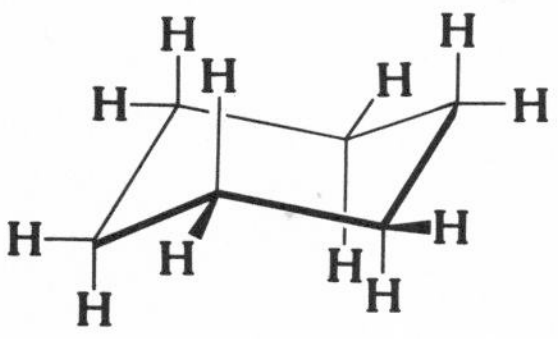	chair	very small	none	small

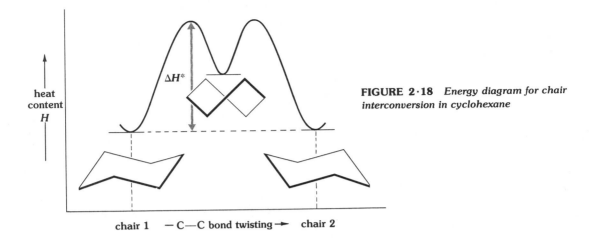

FIGURE 2·18 *Energy diagram for chair interconversion in cyclohexane*

motion). One or more twist conformations are believed to be intermediates in this interconversion, as shown in Figure 2.18.

The twelve hydrogens attached to a chair conformer of cyclohexane fall into two categories. Six of them are oriented above or below the approximate plane of the ring (three on each side) and are called *axial*, because the C—H bonds are parallel to a symmetry axis passing through the center of the ring. The other six hydrogen atoms extend outward from the periphery of the ring and are called *equatorial*. In flipping or inverting one chair conformation into the other, all equatorial bonds become axial and all axial bonds become equatorial.

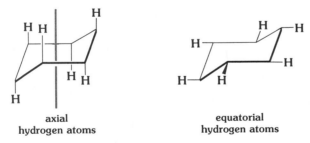

axial hydrogen atoms

equatorial hydrogen atoms

Since axial and equatorial hydrogens have different environments, they should in theory give different pmr signals. However, a sample of cyclohexane at 25°C gives a single sharp resonance signal, indicating that the two chair conformations are in rapid equilibrium. When cyclohexane is cooled, the rate of chair interconversion is reduced, and at low temperature (about −100°C) two groups of resonance signals are in fact observed in the pmr spectrum. Experiments of this kind have led to a determination of the energy barrier for chair inversion (in Figure 2.18 $\Delta H^* = 10.8$ kcal/mole). As a rule, molecular species that can interconvert by pathways having energy barriers (activation energies) less than 18 kcal/mole cannot be obtained as discrete substances at room temperature. This is the case for most conformational isomers. In contrast, constitutional isomers can interconvert only by breaking and reforming covalent bonds, and the very high activation energies required for such processes assure the structural integrity of the isomers.

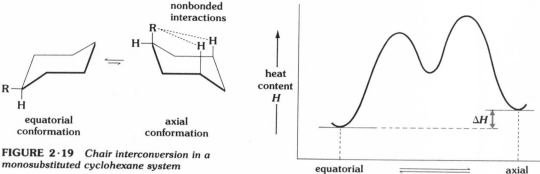

FIGURE 2·19 *Chair interconversion in a monosubstituted cyclohexane system*

The two chair conformations of a cyclohexane ring bearing a substituent are not energetically equivalent. If the substituent **R** occupies an equatorial position, it is subject to less severe nonbonded interactions than it experiences in an axial orientation (Figure 2.19). The energy difference ΔH between these conformations causes the equatorial conformer to predominate at equilibrium. The nonbonded interactions suffered by an axial **R** group are primarily due to crowding by the two axial hydrogens on the same side of the ring as **R**. As **R** becomes larger, these repulsive interactions increase; a very bulky *t*-butyl substituent will effectively freeze the ring system in that chair conformation in which it assumes an equatorial orientation:

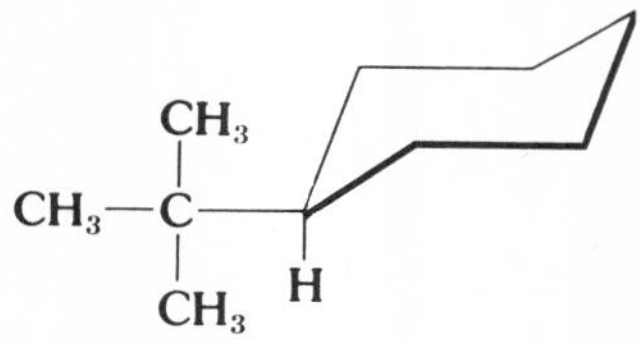

Another way of confining a ring system in a particular conformation is to lock it in place by means of suitable bridging carbon chains. Examples of constrained six-membered rings are shown in Figure 2.20.

FIGURE 2·20 *Bridged boat, twist, and chair ring systems*

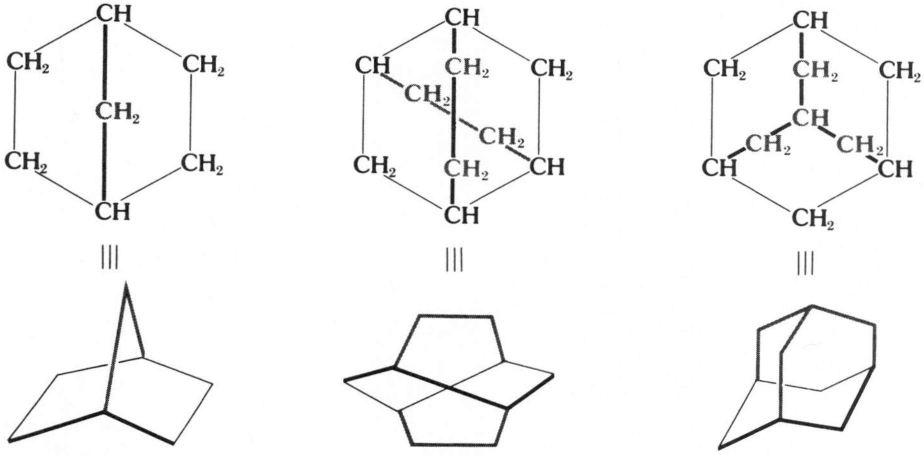

norbornane   twistane   adamantane

66 2·5 CYCLOALKANES

PROBLEM 2·23
According to our nomenclature conventions, norbornane in Figure 2.20 is a bicycloheptane (it has two rings) and the other two compounds are isomeric tricyclodecanes (each has three rings). Devise a simple method of determining the number of rings in a given structure, and use it to determine the number of rings in the following compounds:

Most disubstituted cyclohexanes (systems bearing two substituents) also assume chair conformations, and in these cases the existence of cis-trans configurational isomers must be carefully distinguished from the conformational isomers. The stereoisomeric 1,4-dimethylcyclohexanes in Figure 2.21 illustrate this point. The cis isomer exists as a 50:50 mixture of *equivalent* chair conformations, each having one axial and one equatorial methyl group. The trans isomer has two *nonequivalent* chair conformations, of which the one with two equatorial methyl groups is favored.

PROBLEM 2·24
cis-1,4-Dimethylcyclohexane has a slightly larger heat of combustion (about 1.5 kcal/mole) than the trans isomer. Draw an energy diagram illustrating the origin of this difference and suggest an explanation for it.

Molecules incorporating fused six-membered ring systems are widely distributed in nature. They occur, for example, in such substances as cholesterol and cortisone. The simplest compound of this kind, decalin, may exist in trans or cis fused configurations. The cis isomer consists of two equivalent all-chair conformations, which rapidly interconvert by simultaneous flipping of both six-membered rings. In contrast, *trans*-

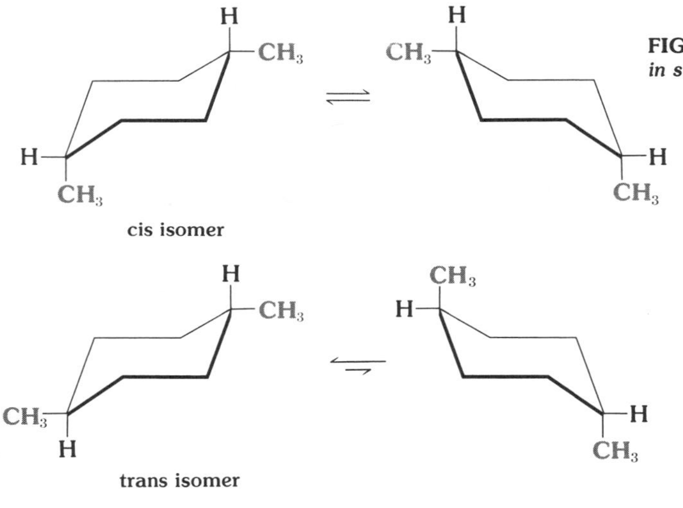

FIGURE 2·21 *Chair conformations in stereoisomers of 1,4-dimethylcyclohexane*

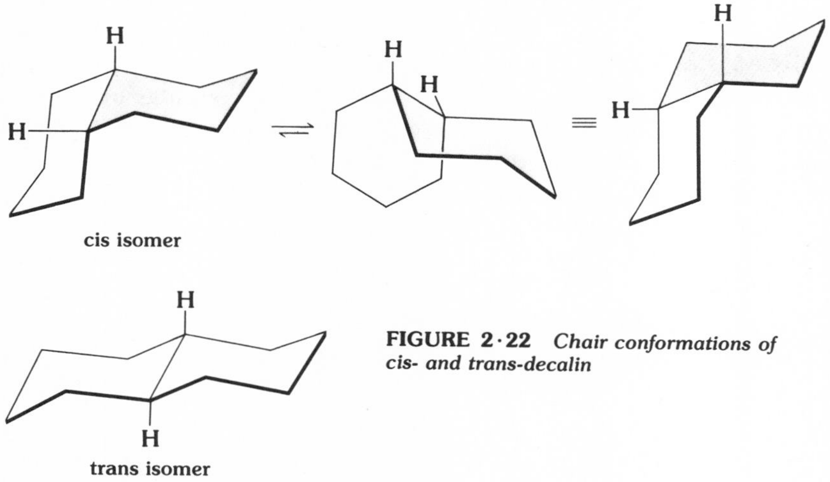

2·5 CYCLOALKANES

FIGURE 2·22 *Chair conformations of cis- and trans-decalin*

decalin has a fairly rigid configuration which resists chair-chair flipping (see Figure 2.22).

Past studies of ring systems have focused greatest attention on six-membered rings because of their wide distribution and importance in natural products. However, conformational equilibria are known to exist in all cycloalkanes larger than cyclopropane. A few examples of such equilibria are illustrated in Figure 2.23.

PROBLEM 2·25
Suggest a reason for the rigidity of *trans*-decalin.

FIGURE 2·23 *Conformational equilibria in some cycloalkanes*

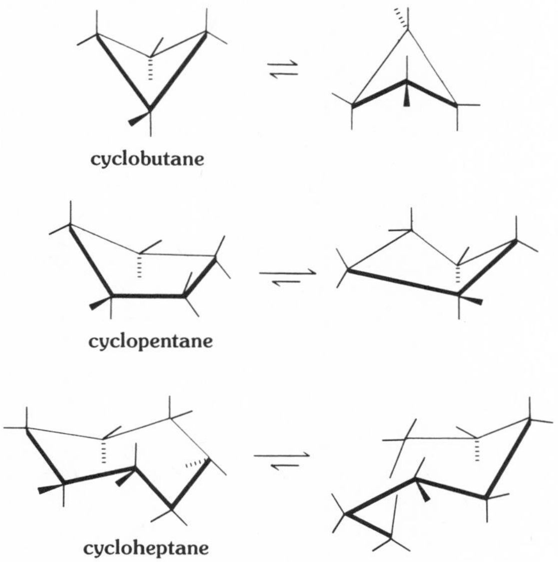

cyclobutane

cyclopentane

cycloheptane

CHEMICAL REACTIVITY

The cycloalkanes exhibit a pattern of reactivity very similar to that of the alkanes. Some typical examples are the following substitution reactions:

$$\triangle + Cl_2 \xrightarrow{light} \triangle\!-\!Cl + HCl \tag{2.22}$$

$$\bigcirc + 9O_2 \longrightarrow 6CO_2 + 6H_2O \tag{2.23}$$

$$\pentagon + Br_2 \xrightarrow{300°} \pentagon\!-\!Br + HBr \tag{2.24}$$

Cyclopropane is atypical in that it also undergoes a number of *addition reactions*. This behavior will be discussed in Chapter 5.

SUMMARY

alkanes A class of relatively unreactive hydrocarbons having the general formula C_nH_{2n+2}. The carbon atoms in normal or *n*-alkanes all have a tetrahedral bonding geometry and are bonded together in chains.

cycloalkanes A class of hydrocarbons similar to the alkanes, but having carbon atoms joined as rings. Cycloalkanes with a single ring of carbon atoms have the general formula C_nH_{2n}.

IUPAC nomenclature system A system of naming organic compounds, adopted as a standard by the International Union of Pure and Applied Chemistry. Its application to alkanes and cycloalkanes is summarized in Tables 2.2, 2.3, and 2.6.

constitutional isomers Isomers differing in the manner in which atoms are connected or bonded together.

conformations The different spatial orientations that may be assumed by the atoms of a molecule, as a result of rotation about single bonds (see Table 2.8); also called *conformers, conformational isomers,* or *rotamers.*

stereoisomers Isomers having the same constitution but differing in the spatial arrangement of their atoms.

configuration The relative arrangement of atoms in space that characterizes a particular stereoisomer. Stereoisomers of disubstituted cycloalkanes have characteristic configurations that may be designated by the prefixes *cis* and *trans.*

heat content (H) A thermodynamic property related to the potential energy of a system; also called *enthalpy.*

heat of reaction (ΔH) The difference in heat content between the reactants and products ($\Delta H = H_p - H_r$). For an *exothermic reaction* (one in which heat is evolved) ΔH is negative; for an *endothermic reaction* (one in which heat is absorbed) ΔH is positive.

bond-dissociation energy The energy required to separate two covalently bonded atoms or groups.

average bond energies Bond energies chosen so that the sum of energies for all the bonds of a molecule is equal to the energy (heat) released when the molecule is formed from its component atoms ($\Delta H = \Sigma_r - \Sigma_p$).

hydrogen bonding A bonding force between a polar Z—H function (Z = N, O, F, Cl, Br, I) and a nucleophilic or basic group Y: (Y = N, O, F, Cl, Br, I, C=C, etc.) often represented by a dashed line, Z—H---:Y.

strain Any structural factor (such as angle distortion, bond eclipsing, or crowding) that raises the potential energy of a molecule constrained in an unfavorable configuration. The chemical reactivity of strained molecules is usually greater than that of similar unstrained molecules.

chain reaction A reaction sequence in which the reactive species required to initiate the transformation is re-formed at the end, allowing the reaction to be repeated.

reaction rate The rate at which reactants are converted to products, defined as a function of collision frequency × probability factor × energy factor.

entropy A thermodynamic property related to the disorder of a system; the greater the disorder, the higher the entropy.

transition state The highest potential-energy state a reacting system must achieve in order to pass from reactants to products.

heat of activation (ΔH^*) The heat content (potential energy) required by the reactant molecules to pass over the transition state, computed as the difference in heat content between the transition state and the reactants ($\Delta H^* = H^* - H_r$); also called *energy of activation* (ΔE^*).

EXERCISES

2·1 Write condensed structural formulas and provide unambiguous names for (*a*) all the isomeric hexanes and (*b*) all the isomeric heptanes.

2·2 Name the following alkanes and cycloalkanes:

a $CH_3CH_2CH(CH_3)_2$

b $CH_3(CH_2)_8CH_3$

c $(CH_3)_3CCH_2CH(CH_3)_2$

d $CH_3CH(CH_2CH_3)_2$

e cyclopropane with two CH_3 substituents

f cyclopentane with H, CH_2CH_3, H, CH_2CH_3 substituents

g cyclohexane with $CH_2CH_2CH_2CH_3$ substituent

h bicyclohexyl (two cyclohexanes connected)

i cyclobutane with $CH(CH_3)_2$ and H substituents

70 EXERCISES

2·3 Indicate the error in each of the following names and write the correct name:
a 3-methylbutane
b 2-ethylpropane
c 3-*iso*propyl-2-methylpentane
d 1,1,3-trimethylbutane
e 2,3-dimethylcyclopentane

2·4 The general molecular formula for alkanes is C_nH_{2n+2}. What is n for the following compounds?
a methane
b ethane
c butane
d *n*-pentane
e isopentane

2·5 Why does the difference in boiling points (ΔT) between members of the alkane homologous series decrease with increasing molecular weight? For example, the difference in boiling point between propane and butane is 42°C, while between nonane and decane it is only 23°C.

2·6 Write structural formulas and names for all the isomeric cyclic hydrocarbons of formula C_5H_{10}. Which of these isomers gives only a single monochloro derivative? Which isomer gives only two monochloro derivatives?

2·7 The three pentane isomers C_5H_{12} are separately chlorinated at 300°C. Isomer A gives three different monochloropentanes, B gives one monochloropentane, and C gives four different monochloropentanes.
a Assign structures to A, B and C.
b Predict the pmr spectrum of B and the monochloropentane obtained from B.
c How many dichloropentane isomers are possible from B?

2·8 Draw the conformations about the 2-3 bond of 2-methylbutane. Which conformation or conformations are preferred (have lowest potential energy)?

2·9 The heat of combustion of propylene ($CH_3CH=CH_2$) is 492.0 kcal/mole and that of cyclopropane is 499.8 kcal/mole. Write balanced equations for the complete combustion of both compounds. Which of these C_3H_6 isomers is thermodynamically more stable?

2·10 Draw structural formulas for all possible cis-trans isomers of 1,3-dimethyl-5-ethylcyclohexane. Draw a favorable chair conformation for each of these isomers and indicate which isomer you would expect to have the lowest potential energy.

2·11 Draw an energy diagram illustrating the chair-chair conformational inversion in (*a*) *cis*-1,2-dimethylcyclohexane and (*b*) *trans*-1,2-dimethylcyclohexane.

2·12 Decalin is an example of a fused-ring cycloalkane (two rings which have two atoms in common) and is found as a mixture of two stereoisomers. Draw the preferred conformations of these isomers and indicate which of the two you would expect to be more stable on the basis of 1,3-diaxial interactions.

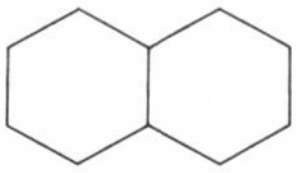

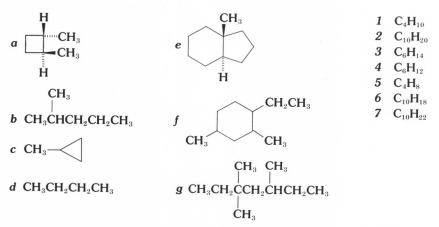

decalin

2·13 Match each of the following structures with the appropriate molecular formula:

a [cyclobutane with H, CH₃, CH₃, H substituents]

b CH₃CHCH₂CH₂CH₃
       |
       CH₃

c CH₃—△ (methylcyclopropane)

d CH₃CH₂CH₂CH₃

e [bicyclic structure with CH₃ and H]

f [cyclohexane with CH₃, CH₃, CH₂CH₃, CH₃ substituents]

g CH₃CH₂CCH₂CHCH₂CH₃
           | |
          CH₃   CH₃
           |
          CH₃

1 C_4H_{10}
2 $C_{10}H_{20}$
3 C_6H_{14}
4 C_6H_{12}
5 C_4H_8
6 $C_{10}H_{18}$
7 $C_{10}H_{22}$

2·14 Given the following reactions, draw the potential-energy diagram for each and calculate ΔH and ΔH^* for the *reverse* reaction.

		ΔH(kcal/mole)	ΔH^*(kcal/mole)
a	A + B ⟶ AB	−14	8
b	EX + Y ⟶ WY + X	4	16
c	E + E ⟶ E₂	−20	0
d	P ⟶ Q	0	6

2·15 The table below shows the relationship between the free-energy change associated with a reaction A ⇌ B and the equilibrium position of that reaction expressed either as the equilibrium constant K_{eq} or the percentage of product B at 25°C.

A ⇌ B, % B	K_{eq}	$\Delta G^{25°}$ (kcal/mole)
50	1.00	0.0
75	3.00	0.65
80	4.00	0.82
85	5.67	1.03
90	9.00	1.30
95	19.00	1.75
98	49.00	2.31
99	99.00	2.73
99.9	999.00	4.09
99.99	9999	5.46

a What is K_{eq} for the conversion of the eclipsed to the staggered conformation of ethane?

b If the equilibrium position for the interconversion of cyclohexane is 99.99% chair conformation, what is ΔH^* for the conversion of chair to twist conformations?

Hint: Assume in both (*a*) and (*b*) that $\Delta G^* = \Delta H^*$ (that is, $\Delta S^* = 0$).

2·16 Oxygen is found to inhibit the vapor-phase chlorination of alkanes. Suggest a reason for this. *Hint:* Oxygen is a diradical, $\cdot O{-}O\cdot$.

ALKYL HALIDES

VERY FEW organic compounds containing halogens have been isolated from terrestrial natural sources, and it is interesting to note that these often have antibiotic properties. Chloromycetin (chloramphenicol), aureomycin (chlorotetracycline), and a recently discovered marine antibiotic which is over 70% bromine by weight are examples of this curious relationship.

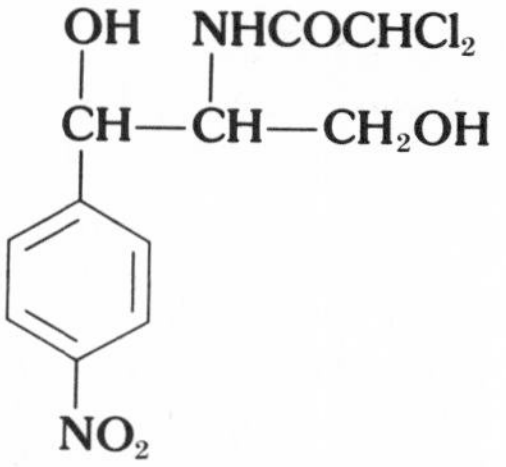

chloromycetin

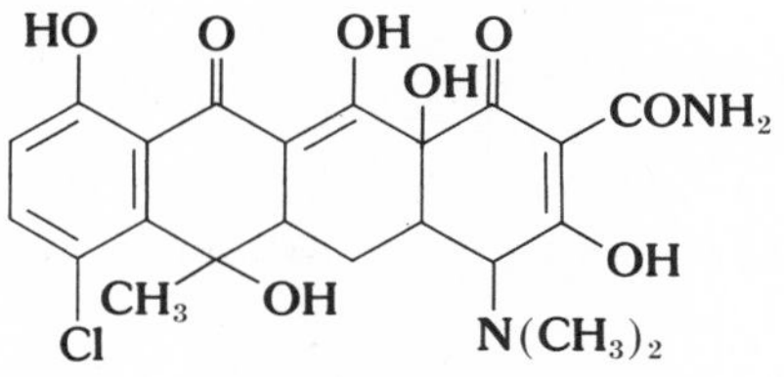

aureomycin

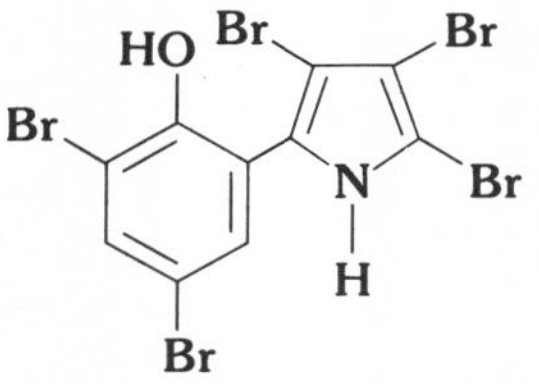

Although halogen-bearing alkanes and cycloalkanes are not abundant in nature, large quantities of these alkyl halides are manufactured every year because of their importance in modern technology. For example, certain relatively simple chlorides, such as methylene chloride (CH_2Cl_2), chloroform ($CHCl_3$), and carbon tetrachloride (CCl_4) are widely used as solvents; and a group of low-boiling, inert fluorine-containing substances called freons (CF_2Cl_2, $CFCl_3$, CF_4, etc.) serve as aerosol propellants and refrigeration fluids.

A number of polychlorinated substances, such as DDT, chlordane, and lindane, are powerful insecticides. However, many of these compounds

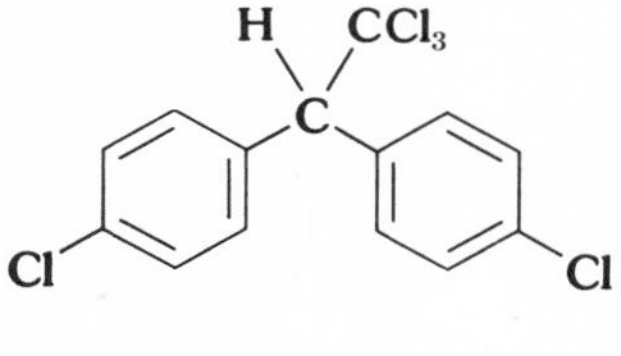

DDT

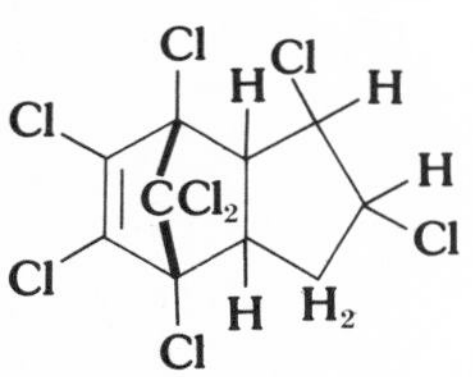

chlordane

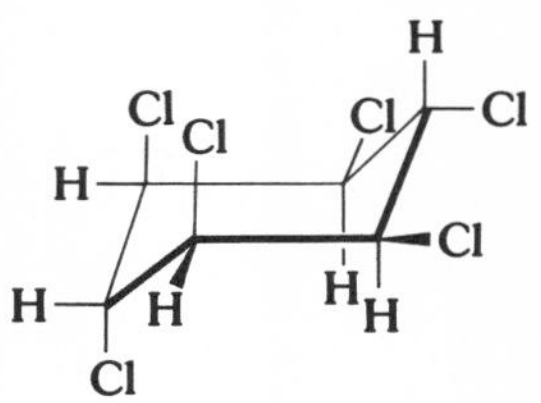

lindane

also act as cumulative poisons to a variety of wild and domestic animals, and their resistance to biological and environmental degradation poses serious pollution problems when massive doses are applied.

Even if the alkyl and cycloalkyl halides had no direct utilitarian value, they would still play a vital role as intermediates and precursors to a multitude of useful substances. This versatility of the organic halides as chemical intermediates will become increasingly apparent in subsequent chapters.

3·1 NOMENCLATURE OF ALKYL HALIDES

The nomenclature of alkyl halides is straightforward; we simply regard the halogen—a fluorine, chlorine, bromine, or iodine atom—as an additional substituent on the hydrocarbon chain or ring. The halogen group, denoted in general by X—, is then referred to as *fluoro-, chloro-, bromo-,* or *iodo-* and is preceded by the appropriate location number. Alternatively, when the organic residue is one of the simple groups listed in Table 2.3, the halogen compound may be named as the corresponding *fluoride, chloride, bromide,* or *iodide*:

$$\overset{1}{C}H_3-\overset{2}{\underset{\underset{Cl}{|}}{\overset{\overset{CH_3}{|}}{C}}}-\overset{3}{C}H_3$$

2-chloro-2-methylpropane
or t-butyl chloride

$$\overset{1}{C}H_3\overset{2}{C}F_2\overset{3}{C}H_2\overset{4}{C}H_3$$

2,2-difluorobutane

$$(CH_3)_3CCH_2I$$

1-iodo-2,2-dimethylpropane
or neopentyl iodide

trans-1-bromo-3-chlorocyclohexane

In discussing alkyl halides we shall find it helpful to extend the designations *primary, secondary,* and *tertiary* to the alkyl group bearing the halogen atom. Thus 1-chloropropane is a primary chloride, bromocyclobutane is a secondary bromide, and 2-fluoro-2-methylpropane is a tertiary fluoride.

bromocyclobutane
or cyclobutyl bromide

3·2 PROPERTIES OF ALKYL HALIDES

PHYSICAL PROPERTIES

Halogenated alkanes have properties similar to those of alkanes having the same molecular weight. Some of these compounds are gases (the freons, for example), but most are water-insoluble liquids and solids. In general, the boiling points of halogenated alkanes rise with increasing molecular weight, as shown in Table 3.1. However, as we saw in Chapter 2, factors other than molecular weight may play an important role, as evidenced by the 70° difference in boiling point between CH_2Cl_2 (bp = 40°C) and CF_2Cl_2 (bp = −30°C).

Halogen-free organic liquids are usually less dense than water (that is, their densities are lower than 1.0 g/ml). As halogen atoms are introduced into the molecular structure, the density increases, and in general, the incorporation of a single bromine or iodine atom or several chlorine

3·2 PROPERTIES OF ALKYL HALIDES

TABLE 3·1 *Typical boiling points and densities of alkyl halides*

structural formula	molecular weight	boiling point (°C)	density (g/ml)
C_2H_5F	48	−37.7	0.82
C_2H_5Cl	64.5	12.2	0.92
C_2H_5Br	109	38.0	1.43
C_2H_5I	156	72.2	1.93
$ClCH_2CH_2Cl$	99	83.5	1.26
$BrCH_2CH_2Br$	188	131.6	2.18
ICH_2CH_2I	282	decomposes	≈2.6

atoms in a molecule will cause the density to become greater than that of water (Table 3.1). Thus CH_2Cl_2, $CHCl_3$, and CCl_4 are liquids having densities greater than 1.3, and the perhalogenated compounds (*per* means thoroughly) CBr_4 and C_2Cl_6 are high-density crystalline solids.

PROBLEM 3·1

n-Pentane and methylene chloride are colorless liquids with similar boiling points. Suggest a simple physical test to distinguish these common solvents.

INFRARED SPECTROSCOPY

The presence of halogen can usually be deduced from the spectroscopic fingerprint of a molecule. In the infrared region, for example, characteristic C—X stretching vibrations are observed at 950 to 1350 cm^{-1} for C—F, at 510 to 775 cm^{-1} for C—Cl, at 490 to 650 cm^{-1} for C—Br, and at 465 to 600 cm^{-1} for C—I. Thus the infrared spectrum of 1,2-dichloroethane (Figure 3.1) clearly shows the presence of C—H and C—Cl bonds.

At first it may seem surprising that more than one distinct C—Cl stretching absorption band is present in this spectrum. Although the two C—Cl bonds in dichloroethane are equivalent, the molecules can assume

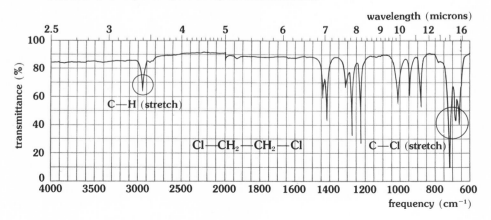

FIGURE 3·1 *Infrared spectrum of 1,2-dichloroethane*

two different staggered conformations, *anti* and *gauche*, and each of these conformational isomers should give rise to characteristic C—Cl vibrational absorptions.† The lower-energy anti conformer predominates in the equilibrium mixture and is responsible for the very strong C—Cl absorption at 710 cm^{-1}.

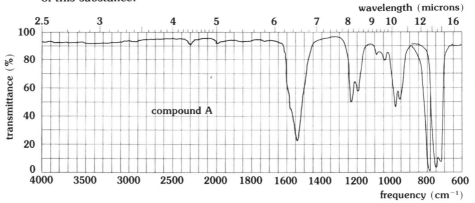

anti conformation   gauche conformation

It is also evident from the spectrum in Figure 3.1 that the absorption intensities of different vibrations are not the same, inasmuch as the four C—H bonds in ClCH$_2$CH$_2$Cl show a relatively weak stretching absorption at 2940 cm^{-1}, compared with the intense C—Cl peaks.

PROBLEM 3·2
The decrease in stretching frequencies in the order

$$C-F \gg C-Cl > C-Br > C-I$$

seems to parallel the decrease in bond energies. Does an examination of other data from Tables 1.3 and 2.4 support this relationship?

PROBLEM 3·3
A colorless oily liquid gives the following infrared spectrum. On the basis of this information, what can you conclude about the composition and structure of this substance?

PMR SPECTROSCOPY

The introduction of halogen atoms into organic molecules usually causes a shift to lower field by some or all of the proton resonance signals in the pmr spectrum. Since halogen atoms are electron-withdrawing, the valence electron clouds around neighboring hydrogen atoms suffer a decrease in density and their nuclei are deshielded. Indeed, the order

†The gauche conformer gives two stretching absorption bands (655 and 675 cm^{-1}) for reasons too complicated to discuss here.

3·2 PROPERTIES OF ALKYL HALIDES

of this effect, F > Cl > Br > I, parallels the change in electronegativity of the halogens. These characteristic chemical shifts are illustrated by the spectra in Figure 3.2, and δ values are summarized in Table 3.2. The presence of more than one halogen atom has a cumulative deshielding effect on neighboring hydrogen atoms, as indicated by the chemical shifts of CH_2Cl_2 ($\delta = 5.30$) and $CHCl_3$ ($\delta = 7.28$).

TABLE 3·2 *Chemical-shift δ values for halogenated alkanes*

hydrogen type	X=F	X=Cl	X=Br	X=I
CH_3—X	4.3	3.1	2.7	2.2
R—CH_2—X	4.3–4.8	3.3–3.8	3.0–3.6	2.5–3.3
R_2—CH—X	←	3.8–4.5		→

FIGURE 3·2 *Pmr spectra of 1,2,2-trichloropropane and 1,2-dibromo-2-methylpropane*

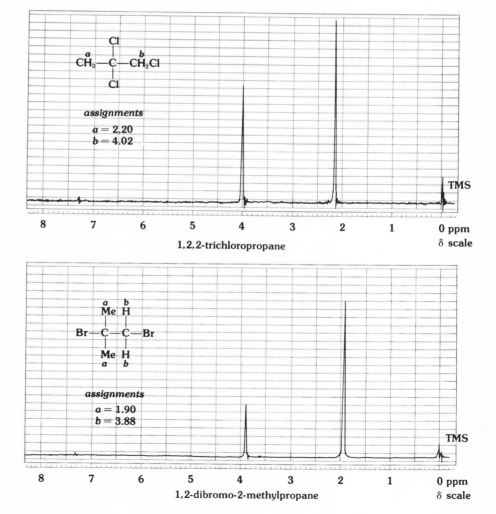

3·2 PROPERTIES OF ALKYL HALIDES

CONFORMATIONAL EFFECTS

An apparent inconsistency is observed when the infrared and pmr spectra of 1,2-dichloroethane are compared. The infrared spectrum in Figure 3.1 discloses the existence of two conformational isomers (anti and gauche), and we would therefore expect to find two or three different resonance signals in the pmr spectrum, depending on whether the H_a hydrogen atoms in the anti conformer have the same chemical shift as the H'_a atoms in the gauche conformer:

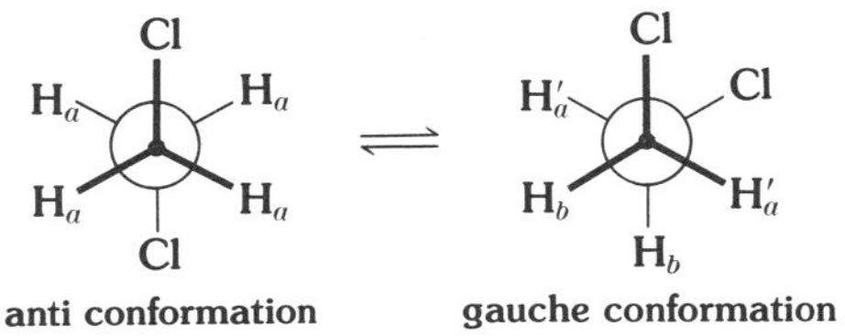

anti conformation ⇌ gauche conformation

This conclusion is, however, at odds with the experimentally determined pmr spectrum of 1,2-dichloroethane, which exhibits only a single sharp resonance signal at $\delta = 4.0$.

This inconsistency is explained when we consider the vastly different time scale of these two spectroscopies. If we regard absorption spectroscopy as a camera which provides us with "pictures" at the molecular level, then visible and infrared spectroscopy have very fast shutter speeds (10^{-14} sec), while pmr spectroscopy has a relatively slow shutter (10^{-2} sec). Since rotations about single bonds are normally rapid at room temperature, the pmr "camera" will see only a time-averaged group of hydrogen atoms which have experienced identical molecular environments. Thus a single pmr signal is entirely consistent with the conformational equilibrium involving the anti and gauche conformers.

PROBLEM 3·4

a A compound with the molecular formula C_3H_7Br exhibits two resonance bands in the pmr spectrum, at $\delta = 1.71$ and 4.32. The relative intensity of these signals is 6:1. Draw a structural formula for this compound.

b A compound with the molecular formula C_3H_6BrCl shows three fine-structured resonance bands, at $\delta = 2.28$, 3.55, and 3.70. The relative intensities of these bands are all the same. Draw a structural formula for this compound.

c A compound with the molecular formula $C_3H_3Cl_5$ shows two resonance bands, at $\delta = 4.52$ and 6.07, with relative intensities of 2:1. Draw a structural formula for this compound.

PROBLEM 3·5

Sketch the kind of pmr spectrum you would expect to find for neopentyl bromide, $(CH_3)_3CCH_2Br$. Indicate chemical shifts and relative intensities.

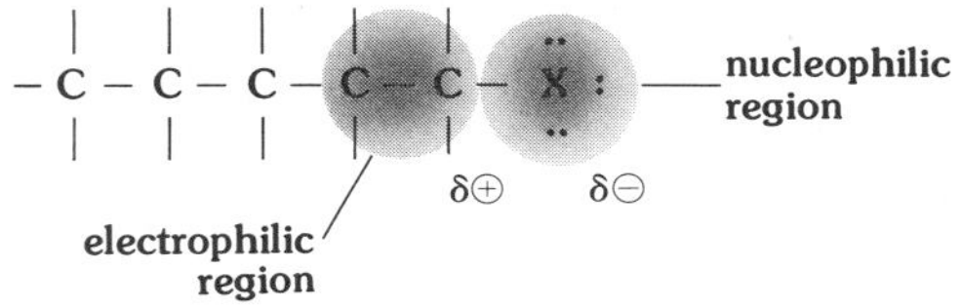

FIGURE 3·3 *The polar character of an alkyl halide*

3·3 NUCLEOPHILIC SUBSTITUTION REACTIONS AT CARBON

The alkyl halides undergo a wider variety of reactions than do the alkanes, and this increased reactivity can be traced to the polar carbon-halogen bond in these molecules. For example, substitution and elimination reactions occur when electron-rich species (nucleophiles and bases) attack the electrophilic region of the alkyl group (Figure 3.3). Of course, the hydrocarbon part of most alkyl halides is still subject to free-radical halogenation as described in equations 2.12 to 2.15; however, halogenation is best accomplished in the gas phase, whereas the reactions of alkyl halides with nucleophilic reagents occur almost exclusively *in solution*.

Combustion appears to be an exception to the superior reactivity of alkyl halides, since these compounds usually do not burn as readily as the corresponding alkanes. Indeed, carbon tetrachloride is sometimes used as a fire-extinguishing agent because its heavy vapors effectively exclude oxygen from the combustion area. Unfortunately, the toxic nature of this compound makes it unsuitable as a general-purpose fire extinguisher.

Alkyl chlorides, bromides, and iodides undergo substitution reactions with a variety of nucleophilic reagents; some typical examples are shown in Figure 3.4. The reactivity of these halides increases in the order Cl < Br < I (alkyl fluorides are essentially unreactive). These reactions are particularly useful for mechanism studies, since moderate rates of conversion are generally observed at ordinary temperatures and common reactant concentrations. In fact, the manner in which the reaction rates change as the concentration of the reactants is varied has provided important clues to the reaction mechanism.

STRUCTURE AND REACTIVITY

Some of the earliest investigations of substitution reactions like those in Figure 3.4 examined the manner in which the reaction rates were affected by changes in the structure of the alkyl group (**R**). In the case of bromide displacement by iodide, reactivity was observed to decrease as the degree of substitution at the carbon bonded to bromine increased:

$$R—Br + NaI \xrightarrow{\text{acetone}} R—I + NaBr$$

reaction rate: $R = CH_3 > C_2H_5 > (CH_3)_2CH > (CH_3)_3C$

(3.1)

There is roughly a thousandfold decrease in reaction rate from methyl bromide to isopropyl bromide, and *t*-butyl bromide reacts so slowly the rate cannot be measured under the same conditions.

FIGURE 3·4 *Some nucleophilic substitution reactions of alkyl halides*

$$R—X' + X^\ominus \xleftarrow{X'^\ominus} R—X \xrightarrow{SH^\ominus} R—SH + X^\ominus$$
$$(R = \text{an alkyl group})$$
$$X = Cl, Br, \text{ or } I)$$
$$R—OH + X^\ominus \xleftarrow{OH^\ominus} \qquad \xrightarrow{CN^\ominus} R—CN + X^\ominus$$

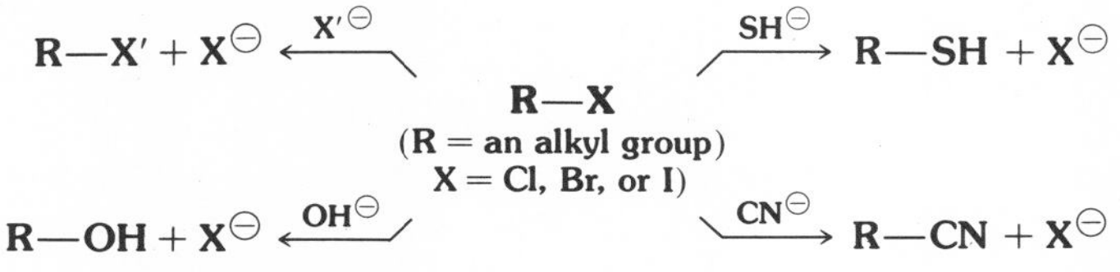

A reasonable interpretation of this substitution effect could be based on either the lower electronegativity of carbon substituents in comparison with hydrogen or the greater size of the groups attached to the functional carbon. Since neopentyl bromide [$(CH_3)_3CCH_2Br$] is essentially unreactive under these conditions, the size or bulk effect, usually referred to as *steric hindrance*, best accounts for the order of reactivity described above. We shall examine this steric effect more closely when we discuss substitution mechanisms.

The hydrolysis of alkyl bromides in 90% aqueous acetone exhibits an unexpected change in the structure-reactivity relationship from that observed above:

$$R{-}Br + H_2O \xrightarrow{\text{90\% aqueous acetone}} R{-}OH + HBr \quad (3.2)$$

reaction rate: $R = (CH_3)_3C \gg (CH_3)_2CH \approx C_2H_5$

The tertiary bromide is more reactive than either the secondary or primary bromides by a factor of roughly 4000 for the reaction in aqueous acetone. Clearly, a simple bulk effect is not operating in this case; in fact, no single variable can reasonably account for the relative reaction rates observed in these two examples. Since complex rate relationships are often the result of two or more *competing reactions*, we conclude that nucleophilic displacement reactions of alkyl halides proceed by at least two different reaction paths.

PROBLEM 3·6
If reaction 3.2 were to proceed solely by the same kind of reaction path as reaction 3.1, what order of reactivity would be observed for the compounds $R = n$-butyl, $R = s$-butyl, and $R = t$-butyl?

PROBLEM 3·7
A study of the pH dependence of reaction 3.2 has shown that an increase in hydroxide-ion concentration causes a corresponding increase in the rate of reaction for $R = C_2H_5$, but not for $R = (CH_3)_3C$. What do these facts suggest about the mechanism of this reaction?

REACTION-RATE STUDIES

If the reaction rate of ethyl bromide with sodium iodide is measured at constant temperature, it is found to be dependent on the concentrations of *both* reactants, and a quantitative relationship is readily formulated:

$$C_2H_5Br + NaI \xrightarrow{\text{acetone}} C_2H_5I + NaBr \quad (3.3)$$

reaction rate $= k_2[C_2H_4Br][NaI]$

where k_2 is the rate constant and the brackets signify molar concentration. This rate equation describes the *kinetic order* of the reaction. By kinetic order we mean the experimentally observed dependence of the reaction rate on reactant concentrations. Since the concentrations of ethyl bromide and sodium iodide each appear to the first power in the rate equation, the reaction is said to be *first order* in each reactant and *second order* overall. Thus doubling the concentration of either C_2H_5Br

or **NaI** would double the rate of substitution, and doubling the concentration of both reactants would quadruple the rate. This variation of reaction rate with concentration is due essentially to changes in the collision frequency of the reactant molecules. The *rate constant* k_2 includes the characteristic energy and probability factors for reaction 3.3 and is constant as long as the reactants and the reaction conditions (temperature, solvent, etc.) are not changed.

A surprisingly different rate relationship is observed for the hydrolysis of *t*-butyl bromide, in that the rate of this reaction is dependent *only* on the alkyl bromide concentration:

$$CH_3-\underset{\underset{CH_3}{|}}{\overset{\overset{CH_3}{|}}{C}}-Br + H_2O \longrightarrow CH_3-\underset{\underset{CH_3}{|}}{\overset{\overset{CH_3}{|}}{C}}-OH + HBr \quad (3.4)$$

reaction rate $= k_1[(CH_3)_3CBr]$

It is evident from the rate equation that this reaction is kinetically *first order*. The first-order rate constant k_1 is larger than the second-order constant k_2, and k_1 is unique to reaction 3.4 (just as k_2 is unique to reaction 3.3).

PROBLEM 3·8

a If the rate of reaction 3.3 is measured by following the loss of NaI (or the formation of NaBr) and is reported in moles per liter per second (moles liter^{-1} sec^{-1}), what units will k_2 assume?

b If the rate of reaction 3.4 is reported in moles per liter per second, what units will k_1 assume?

c The following reaction has been studied:

$$n\text{-}C_3H_7Br + NaI \xrightarrow[25°]{\text{acetone}} n\text{-}C_3H_7I + NaBr$$

The rate of reaction of a solution with 0.01 molarity in NaI and 0.01 molarity in C_3H_7Br is 1.1×10^{-7} mole liter^{-1} sec^{-1}. What is k_2 for this reaction? What will be the rate of reaction of a solution 0.2 molarity in C_3H_7Br and 0.1 molarity in NaI?

3·4 THE $S_N 2$ MECHANISM

Two mechanism models have been devised which account for the facts discussed above. If the reactant molecules interact and proceed directly to products in a single step, as in Figure 3.5, the reaction is termed *bimolecular* and will usually exhibit a second-order rate equation. The simple single-step pathway for this reaction, and in the general case for reaction 3.1, falls into a category known as the $S_N 2$ *mechanism* (substitution nucleophilic bimolecular). In such a process bond making and bond breaking would have to be synchronous (simultaneous), and the transition state would necessarily resemble one of the forms in Figure 3.5. Thus, in reaction 3.3 the attacking nucleophile must approach the

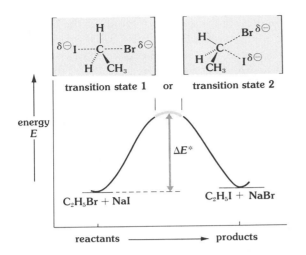

FIGURE 3·5 *Energy diagram for an S_N2 reaction*†

alkyl bromide from either the opposite side or the same side as the leaving bromide ion.

TRANSITION-STATE CONFIGURATION

If S_N2 reactions proceed by way of transition state 1, an *inversion of configuration* should occur at the carbon atom bonded to the halogen. If displacement involves transition state 2, the configuration at this carbon atom will be unchanged. Evidence concerning this aspect of the reaction can be obtained if we examine alkyl halides capable of stereoisomerism. For example, the isomers *cis*- and *trans*-1-bromo-3-methylcyclopentane react with iodide ion in a *stereospecific* manner to yield configurationally inverted products:

cis isomer + I⁻ ⟶ product + Br⁻ (3.5)

trans isomer + I⁻ ⟶ product + Br⁻ (3.6)

Hence an inversion transition state similar to 1 is required for this reaction. Related stereochemical investigations of a wide variety of S_N2 reactions have shown, in fact, that inversion of configuration is a distinguishing feature in all cases.

†The general term *energy* is used here and in subsequent diagrams to avoid the limitations of a more precise term. Most reactions would show similar energy curves regardless of whether enthalpy or free-energy units were used for the ordinate.

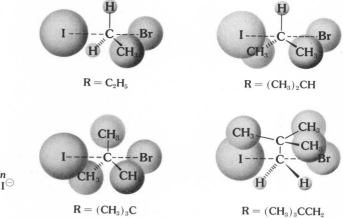

FIGURE 3·6 Steric hindrance in S_N2 transition states for $R-Br + I^\ominus$

As we saw in equation 3.1, sodium iodide reacts more rapidly with ethyl bromide than with isopropyl bromide, and t-butyl bromide and neopentyl bromide are essentially unreactive under similar conditions. This phenomenon is best explained by a *steric*, or bulk, factor. The transition states for this series of reactions show increasing crowding, as illustrated in Figure 3.6. Consequently, nonbonded repulsive interactions are expected to increase along with the crowding, and the activation energies for these reactions (ΔE^* in Figure 3.5) will increase in the same order. The relatively high reactivity of nucleophilic reagents to methyl bromide (about thirty times faster than ethyl bromide for reaction with iodide ion) bears out this interpretation.

NUCLEOPHILICITY

The rates of S_N2 reactions are strongly dependent on the kind of nucleophilic reagents used. This sensitivity to changes in the nucleophile is not unexpected, in view of the partial bonding that exists between the reactants in the transition state (see Figure 3.6). The values in Table 3.3 come from comparative measurements with methyl iodide in methyl alcohol solutions at 25°C; however, this order of nucleophile reactivity, or *nucleophilicity*, probably holds true for other alkyl halides in water or alcohol solvents.

Other factors that we might expect to find correlated with the nucleophilicity of these anionic species are basicity, polarizability, and solvent effects. Although the nucleophilicities of the two oxygen bases in

TABLE 3·3 *Average relative substitution rates for nucleophilic agents*

nucleophile	relative rate	nucleophile	relative rate
$CH_3CO_2^\ominus$	1.0	$SCN^\ominus$	210
$Cl^\ominus$	1.1	$CN^\ominus$	240
$Br^\ominus$	30	$I^\ominus$	1270
$CH_3O^\ominus$	93	$SH^\ominus$	4800

Table 3.3 ($CH_3CO_2^{\ominus}$ and $CH_2O^{\ominus}$) parallel their relative base strengths, the reverse is true for the halide ions. In fact, the strongest halide base, fluoride ion, is much less reactive than chloride ion toward methyl iodide in methanol solution.

The polarizability of a nucleophile is related to the ease with which its valence electrons interact with a nearby electrophile. As a general rule, large atoms or ions are more polarizable than small ones ($S > O$ and $I > Br > Cl > F$) because the nuclear attraction for the outermost electrons is weakened by the shielding effect of the inner filled shells.

The solvation of anions by water or other hydroxylic solvents helps to stabilize these charged particles and also lowers their nucleophilicity. This effect will be discussed in detail in the following section. In general, however, solvation is most pronounced for small anions and will therefore selectively decrease the reactivity of the smaller nucleophiles. The importance of this factor is evident if we compare the order of nucleophilicity for S_N2 reactions of methyl iodide in methanol (see Table 3.3) with the reactivity order for methyl iodide in a poorly solvating non-hydroxylic solvent such as dimethyl sulfoxide:

$$CN^{\ominus} > CH_3CO_2^{\ominus} > Cl^{\ominus} > Br^{\ominus} > I^{\ominus} > SCN^{\ominus}$$

The order of reactivity of the halide *leaving groups* in S_N2 reactions ($X = I > Br > Cl$) parallels the C—X bond energies given in Table 2.3, but the stability of $X^{\ominus}$ (same order) is probably a more important factor in determining this reactivity.

> **PROBLEM 3·9**
> Assuming that ethyl bromide is about forty times more reactive than isopropyl bromide in S_N2 reactions, show how the relative nucleophilicities given in Table 3.3 can be used to estimate the *relative rates* of the following reactions:
>
> a $C_2H_5Br + NaOCH_3 \longrightarrow C_2H_5OCH_3 + NaBr$
>
> b $(CH_3)_2CHBr + NaSH \longrightarrow (CH_3)_2CHSH + NaBr$

3·5 THE S_N1 MECHANISM

To explain the many nucleophilic substitution reactions that are not in accord with the single-step S_N2 mechanism, we must assume a competing *two-step mechanism*. For the timing of events in such a process there are two possible alternatives: either bond breaking at carbon precedes bond making, or bond making is accomplished before bond breaking occurs. The latter mode of reaction is unlikely, since a pentavalent carbon intermediate would be formed. Consequently, the first step in this mechanism must involve bond breaking. Furthermore, the very nature of these reactions demands that the bond cleavage be *heterolytic*, with carbon the cation, or electrophilic atom, and halogen the anion, or nucleophilic atom:

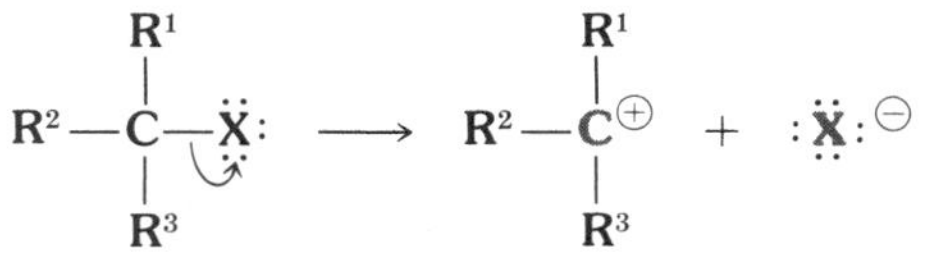

$$R^2-\underset{R^3}{\overset{R^1}{\underset{|}{\overset{|}{C}}}}-\ddot{\underset{..}{X}}: \longrightarrow R^2-\underset{R^3}{\overset{R^1}{\underset{|}{\overset{|}{C}}}}{}^{\oplus} + :\ddot{\underset{..}{X}}:^{\ominus} \tag{3.7}$$

3·5 THE S_N1 MECHANISM

In the second step, then, the intermediate carbonium ion combines with a nucleophilic species to give the product of the reaction:

$$R^2-\overset{\overset{\displaystyle R^1}{|}}{\underset{\underset{\displaystyle R^3}{|}}{C^{\oplus}}} + Y:^{\ominus} \longrightarrow R^2-\overset{\overset{\displaystyle R^1}{|}}{\underset{\underset{\displaystyle R^3}{|}}{C}}-Y \quad (3.8)$$

Equations 3.7 and 3.8 thus define a pathway for nucleophilic substitution which is called the S_N1 *mechanism* (substitution nucleophilic unimolecular). The significance of the term *unimolecular* will become clear as we consider some characteristic features of this mechanism.

PROBLEM 3·10

If we rule out a pentavalent carbon intermediate in the two-step displacement mechanism, how is it we can accept the pentavalent transition state in the S_N2 mechanism?

SOLVENT EFFECTS

In the absence of other factors, the formation and separation of oppositely charged particles (as in equation 3.7) would be a highly endothermic process. In fact, the heat of reaction for the ionic dissociation of methyl chloride,

$$CH_3-Cl \longrightarrow H-\overset{\overset{\displaystyle H}{|}}{\underset{\underset{\displaystyle H}{|}}{C^{\oplus}}} + :Cl:^{\ominus}$$

has been calculated to be $\Delta H = +227$ kcal/mole. Comparison of this value with the C—Cl average bond energy of 81 kcal/mole for homolytic cleavage (see Table 2.4) provides a convincing argument against ionic intermediates in gas-phase reactions. For reactions in solution, however, this calculation does not take into account the interaction of the ions with solvent molecules.

These solvent effects are primarily of two kinds. First, the solvent molecules tend to orient themselves in such a way as to decrease the electrostatic forces between the ions. The degree to which different substances act in this manner is indicated by the *dielectric constant* ϵ. Water ($\epsilon = 81$), methyl alcohol ($\epsilon = 33$), dimethyl sulfoxide ($\epsilon = 45$), formic acid ($\epsilon = 58$), and dimethyl formamide ($\epsilon = 38$) all have relatively large dielectric constants (compare with acetone, $\epsilon = 21$, and chloroform, $\epsilon = 5$) and are widely used as solvents for ionic compounds.

The second solvent effect is *solvation*, in which both anions and cations are stabilized by a surrounding sheath of weakly bonded solvent molecules, as illustrated in Figure 3.7. Water is probably the most generally effective solvation medium known. In contrast to the ΔH value of +227 kcal/mole for ionic dissociation, the heat of ionization for methyl chloride in water is calculated to be $\Delta H = +63$ kcal/mole, indicating a stabilizing solvent effect of about 164 kcal/mole. In a suitable solvent,

3·5 THE S_N1 MECHANISM

carbonium ion   chloride ion

FIGURE 3·7 *Solvation of carbonium and chloride ions by water*

therefore, the ionization of an alkyl halide may be energetically more favorable than the corresponding homolysis.

THE RATE-DETERMINING STEP

Even with solvent stabilization, the ionization step in an S_N1 reaction is endothermic and is undoubtedly much slower than the exothermic neutralization of a carbonium ion by a nucleophile. This relationship is illustrated in Figure 3.8 for the S_N1 reaction sequence in equations 3.7 and 3.8. Since the carbonium ion produced in the ionization reaction is very rapidly converted to neutral products (ΔE_2^* is very small), the overall rate of product formation is virtually identical to the rate of the slow ionization step. For all practical purposes, the slow step in a multistep reaction sequence of this kind determines the rate of the entire sequence and is therefore referred to as the *rate-determining step*.

An important consequence of these rate relationships is that the molecularity of the rate-determining step will usually correspond to the

FIGURE 3·8 *Energy diagram for an S_N1 reaction*

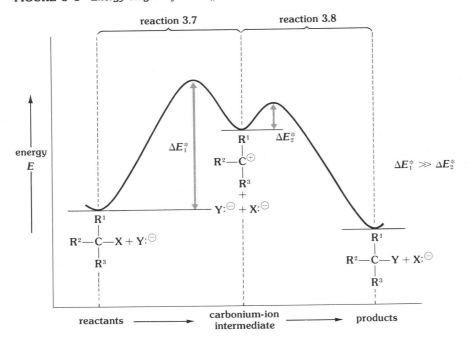

kinetic order of the complete reaction. The hydrolysis of *t*-butyl chloride, for example, proceeds by an S_N1 mechanism, with the rate-determining step being the *unimolecular* ionization of the tertiary chloride (step 1):

$$(CH_3)_3CCl \longrightarrow (CH_3)_3C^{\oplus} + Cl^{\ominus}$$
$$(CH_3)_3C^{\oplus} + 2H_2O \longrightarrow (CH_3)_3COH + H_3O^{\oplus} \tag{3.9}$$

As a result, the overall rate of reaction is not significantly affected by changes in the concentrations or kinds of nucleophilic reagents present in the reaction mixture, although competing nucleophiles may lead to a mixture of products:

$$(CH_3)_3C^{\oplus} + Y{:}^{\ominus} \longrightarrow (CH_3)_3C{-}Y \tag{3.10}$$

The experimentally determined rate equation for reaction 3.4 confirms the first-order kinetics expected in this case.

Carbonium-ion formation can often be accelerated by the addition of an electrophilic cation such as $Ag^{\oplus}$, which will strongly coordinate with the halogen anion. We can refer to this action as a "pulling" effect, in contrast to the "pushing" provided by the interaction of a nucleophilic reagent or solvent at the electrophilic carbon atom. The precipitate of silver halide that forms when most alkyl halides are treated with an alcohol solution of silver nitrate also provides a useful test for the presence of chlorine, bromine, or iodine.

STEREOCHEMISTRY OF S_N1 REACTIONS

In our discussion of the S_N2 mechanism we observed that important geometric characteristics of reactions can be discovered by examining stereoisomeric reactants. A similar study of the S_N1 mechanism is complicated by the elusive nature of the unstable carbonium-ion intermediates formed in the rate-determining step of these reactions. By applying the theoretical model described in Chapter 1 to these intermediates, we predict a *planar trigonal geometry* about the positively charged carbon atom. This configuration is confirmed by other calculations and by recent experimental measurements.

carbonium-ion configuration

Since the intermediate carbonium ion is flat, configurational differences between stereoisomeric alkyl halides (I and II in Figure 3.9) will vanish once this intermediate is formed. The same product mixture is therefore expected from each of the reactant stereoisomers; and in practice this is usually found to be true, although some differences may result from uneven or unsymmetrical solvation of the intermediate carbonium ion.

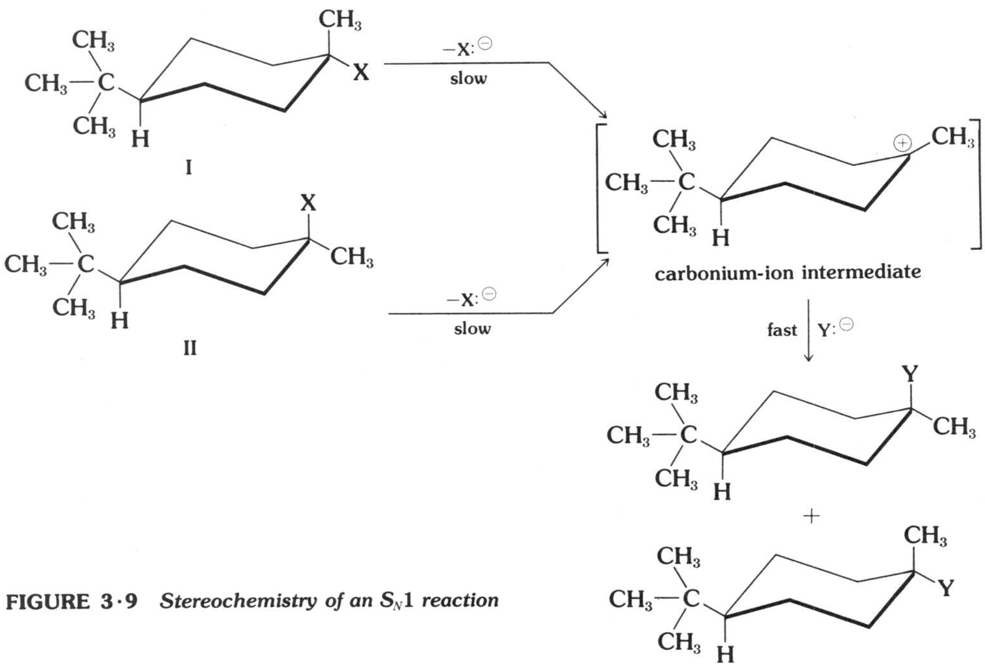

3·6 CARBONIUM-ION INTERMEDIATES

FIGURE 3·9 *Stereochemistry of an S_N1 reaction*

PROBLEM 3·11
If a carbonium-ion intermediate is assumed to be flat, one might expect attack of a nucleophile to occur equally well from either side of the ion. However, in reactions like that outlined in Figure 3.9 a 50:50 mixture of products is seldom obtained. Instead, one product stereoisomer is generally formed in higher yield than the other. Explain.

3·6 CARBONIUM-ION INTERMEDIATES

CARBONIUM-ION STABILITY

The isomeric butyl bromides show an S_N1 reactivity that increases in the following manner:

$CH_3(CH_2)_3Br < CH_3CH_2CHBrCH_3 \ll (CH_3)_3CBr$

Depending on the solvent used, the tertiary halide is several thousand to over a million times more reactive than the secondary and primary isomers. This difference in reactivity has been traced to the relative stabilities of the carbonium-ion intermediates formed in the rate-determining ionization step:

$$C_4H_9\text{—Br} \longrightarrow C_4H_9^{\oplus} + Br^{\ominus} \tag{3.11}$$

Indeed, as a result of many careful and imaginative investigations probing the structure and reactions of carbonium ions, the order of stability shown in Figure 3.10 has been firmly established. When using the term "stability" it is important to distinguish between *chemical stability*, or resistance to chemical transformations, and *thermodynamic stability*,

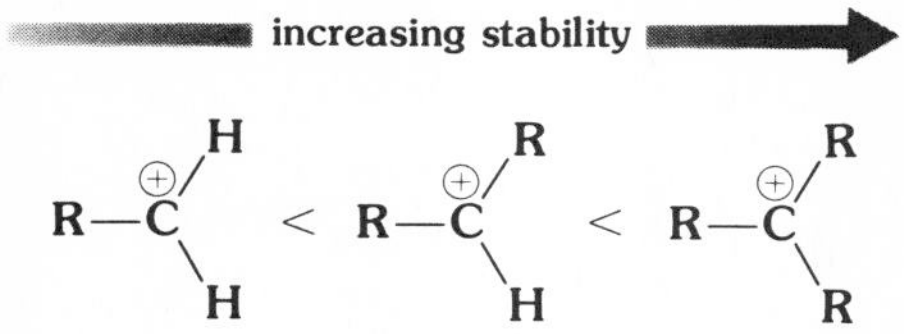

3·6 CARBONIUM-ION INTERMEDIATES

FIGURE 3·10 *Increasing carbonium-ion stability*

◀━━ increasing stability ━━▶

$$R-\overset{H}{\underset{H}{C}}{}^{\oplus} \quad < \quad R-\overset{R}{\underset{H}{C}}{}^{\oplus} \quad < \quad R-\overset{R}{\underset{R}{C}}{}^{\oplus}$$

which refers to the relative heats of formation of two or more molecular species (see Section 2.3). It is the thermodynamic stabilities, or relative heat contents, of the ions that are given in Figure 3.10.

The introduction of a methyl group in place of a hydrogen atom attached to the positive carbon of a carbonium ion appears to stabilize the ion by 15 to 30 kcal/mole. Although it is recognized that ion stabilization requires a spreading out, or *delocalization*, of electrical charge, the mechanism by which this is accomplished is not yet well understood. Two effects seem to be operating in most cases. One of these, an *inductive effect*, is a consequence of unequal sharing of electrons in bonds between atoms of different electronegativity. Since hydrogen is slightly more electronegative than methyl, a $CH_3^{\oplus}$ ion will have a higher concentration of charge on the trivalent carbon atom than will a $(CH_3)_3C^{\oplus}$ ion. The latter ion is therefore more stable. The second effect involves a longer-range delocalization of charge by a more complex mechanism; this mode of stabilization will be discussed in Chapter 10.

Carbonium-ion stability is also affected by changes in the configuration about the positive carbon atom, which shows a preference for the coplanar trigonal geometry described in Section 3.5. For example, the geometric restrictions imposed by the bicyclic ring configuration shown below prevent this tertiary chloride from ionizing to a flat carbonium-ion intermediate. Consequently, S_N1 reactions of this compound are extremely slow (it is unreactive, for example, in a boiling alcohol–silver nitrate solution).

1-chlorobiocyclo[2,2,1]heptane

THE HAMMOND POSTULATE

Since the relative carbonium-ion stabilities given in Figure 3.10 parallel the S_N1 reactivities of the corresponding alkyl halides (tertiary > secondary > primary), it is clear that the activation energy ΔE^* for the rate-determining ionization step is proportional to the potential energy of the carbonium ion being formed. This simple relationship illustrates a very important general principle: *The transition state of a strongly endothermic reaction will reflect the energy and structure of the products to a greater degree than it resembles the reactants. Therefore any substituent which acts to stabilize the product of such an endothermic transformation will also increase the rate at which it is formed.* Furthermore, if we accept the thesis that the products are a reasonable model for the transition state of a strongly endothermic reaction, it follows that the reactants should be a suitable model for the transition state of a strongly exothermic reaction. This principle is known as the *Hammond postulate*.

PROBLEM 3·12
1-Chlorobicyclo[2,2,1]heptane does *not* react by an S_N2 mechanism. Suggest a reason for this inert behavior.

MOLECULAR REARRANGEMENT

The slow S_N1 hydrolysis of neopentyl bromide in aqueous ethyl alcohol gives products (an alcohol and an ether) which are *isomeric* with those expected from a straightforward substitution reaction:

$$\text{CH}_3-\underset{\underset{\text{CH}_3}{|}}{\overset{\overset{\text{CH}_3}{|}}{\text{C}}}-\text{CH}_2-\text{Br} \xrightarrow[\Delta]{\text{C}_2\text{H}_5\text{OH, H}_2\text{O}} \begin{cases} \text{CH}_3-\underset{\underset{\text{CH}_3}{|}}{\overset{\overset{\text{CH}_3}{|}}{\text{C}}}-\text{CH}_2-\text{OR} \\ \text{(R = H or R = C}_2\text{H}_5\text{)} \\ \text{unrearranged products} \\ \\ \text{CH}_3-\underset{\underset{\text{OR}}{|}}{\overset{\overset{\text{CH}_3}{|}}{\text{C}}}-\text{CH}_2-\text{CH}_3 \\ \text{(R = H or R = C}_2\text{H}_5\text{)} \\ \text{rearranged products} \end{cases} \quad (3.12)$$

For example, the pmr spectrum of the resulting alcohol product (R = H) shows resonance signals at $\delta = 0.9$, 1.2, 1.7, and 5.0, having an intensity ratio of 3:6:2:1, whereas the expected product, neopentyl alcohol, exhibits only three resonance peaks: one associated with the nine methyl protons $(\text{CH}_3)_3$, one with the methylene protons CH_2 and the last with the hydroxyl proton OH. The products of this S_N1 reaction are thus 2-methyl-2-butanol (R = H) and the corresponding ethyl ether (R = C_2H_5), indicating that the reaction has involved a *rearrangement* of the carbon backbone or skeleton of the molecule. This remarkable transformation apparently proceeds by a rapid conversion of the initially formed primary carbonium ion to a more stable tertiary carbonium ion (Figure 3.10), as the result of a 1,2 shift of a methyl group with its covalent bonding electrons:

$$(\text{CH}_3)_3\text{CCH}_2\text{Br} \xrightarrow{\text{slow}} \left[\text{CH}_3-\overset{\overset{\text{CH}_3}{\curvearrowright}}{\underset{\underset{\text{CH}_3}{|}}{\text{C}}}-\overset{\text{H}}{\underset{\text{H}}{\text{C}}}\oplus \xrightarrow{\text{fast}} \text{CH}_3-\overset{\overset{\text{CH}_3}{|}}{\underset{\underset{\text{CH}_3}{|}}{\text{C}}}\oplus-\text{CH}_2 \right] \xrightarrow{\text{fast}} \begin{array}{c} S_N1 \\ \text{products} \end{array} \quad (3.13)$$

less stable   more stable

PROBLEM 3·13
a Make chemical-shift assignments for the pmr spectrum of 2-methyl-2-butanol from the data given above.
b Predict the chemical-shift δ values for the pmr spectrum of neopentyl alcohol.

PROBLEM 3·14
a Why are both alcohol and ether products formed in reaction 3.12?
b By what reasoning can we assume that the rearrangement step in equation 3.13 is fast?

Molecular rearrangement should not be considered a rare phenomenon. Indeed, the number and variety of such reactions merited the recent publication of a two-volume review of this subject. This facet of chemical reactivity might seem to pose an almost insurmountable obstacle to our understanding of chemical processes. However, the problem has been largely overcome by the development of rapid spectroscopic techniques for structure determination, and the study of molecular rearrangements has become one of the most fascinating aspects of organic chemistry.

3·7 ELIMINATION REACTIONS

When alkyl halides are treated with a base, there is often an elimination of **HX** from adjacent carbon atoms (referred to as β *elimination* or 1,2 *elimination*. The following equation illustrates the gross structural change in the organic group, but does not imply any particular mechanism or stereochemistry:

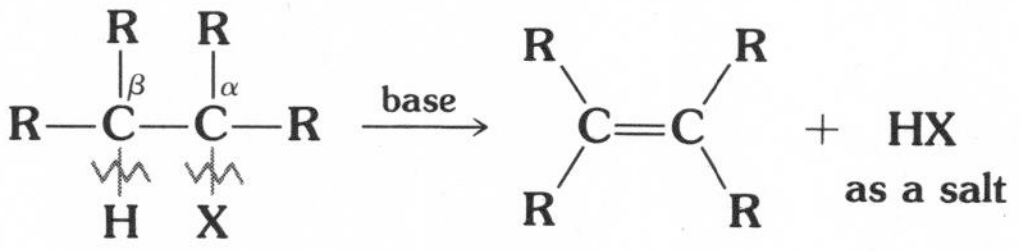

$$\underset{\underset{H\ \ X}{}}{\overset{\overset{R\ \ \ R}{}}{R-\underset{\beta}{C}-\underset{\alpha}{C}-R}} \xrightarrow{\text{base}} \underset{R\ \ \ \ R}{\overset{R\ \ \ \ R}{C=C}} + \text{HX as a salt} \tag{3.14}$$

The ease with which these elimination reactions take place varies markedly with the structure of the alkyl halide substrate, as shown in Figure 3.11, and is often sensitive as well to changes in the base and solvent being used.

Since nucleophilic reactants can also function as bases, a competition between substitution and elimination reactions will occur with some alkyl halides (as in the reaction of 1-bromopropane with potassium hydroxide in Figure 3.11). In such cases the respective yields of substitution and elimination products will be determined by the relative rates of these two processes. The factors which affect the rates of nucleophilic substitution reactions were discussed in Section 3.3. If we can achieve a similar rationalization of elimination reactions, we can then hope to explain and predict the outcome of this competition in specific cases.

FIGURE 3·11 β *elimination in alkyl halides*

$$\text{CH}_3\text{CH}_2\text{CH}_2\text{Br} + \text{KOH} \xrightarrow{\text{C}_2\text{H}_5\text{OH}} \begin{array}{c} \text{CH}_3\text{CH}_2\text{CH}_2\text{OR} \\ 70\%\ (R = H\ \text{and}\ \text{C}_2\text{H}_5) \\ + \\ \text{CH}_3\text{CH}=\text{CH}_2 \\ 30\% \end{array} + \text{KBr}$$

$$(\text{CH}_3)_2\text{CHBr} + \text{KOH} \xrightarrow[\text{slow}]{\text{C}_2\text{H}_5\text{OH}} \text{CH}_3\text{CH}=\text{CH}_2 + \text{KBr}$$

$$(\text{CH}_3)_3\text{CBr} + \text{KOH} \xrightarrow[\text{fast}]{\text{C}_2\text{H}_5\text{OH}} (\text{CH}_3)_2\text{C}=\text{CH}_2 + \text{KBr}$$

PROBLEM 3·15
Equation 3.14 might be interpreted as an equilibrium reaction lying far to the left (that is, K_{eq} is very small). The action of the base would be viewed as a simple neutralization of the HX group, forcing the reaction to the right. Explain why this interpretation is incorrect and suggest an experiment that would prove your point.

THE E2 MECHANISM

The rates of many elimination reactions involving strong bases are found to fit a second-order kinetic relationship:

elimination rate $= k_2$ [alkyl halide][base] (3.15)

This fact, together with the previously noted sensitivity to a change in the nature of the base $B:^\ominus$, suggests a *single-step synchronous mechanism* proceeding through a transition state like the following:

$$B:^\ominus + H-\overset{|}{\underset{|}{C}}-\overset{|}{\underset{|}{C}}-X \longrightarrow \left[\overset{\delta\ominus}{B}\cdots H\cdots \overset{\delta\ominus}{C}=\!=\!=\!\overset{\delta\ominus}{C}\cdots X \right] \longrightarrow B-H + \overset{}{\underset{}{C}}=\overset{}{\underset{}{C}} + X^\ominus \quad (3.16)$$

E2 transition state

It is important to recognize that the transition states of these *E2 reactions* (elimination bimolecular) do not all have the same location on the reaction coordinate, and the extent of C—H and C—X bond breaking and C=C bond making in the transition state will change according to the steric and electronic characteristics of the alkyl halide and the base. We should anticipate, therefore, a less stereotyped group of reactions in the E2 class than in the S_N2 class.

THE SAYTZEFF RULE As a general rule, known as the *Saytzeff rule*, compounds which can give more than one elimination product will yield that isomer having the most highly substituted double bond as the major product:

$$\underset{\underset{CH_3}{|}}{\overset{\overset{CH_3}{|}}{CH_3CH_2\overset{}{C}-Br}} + KOH \xrightarrow{C_2H_5OH} \underset{75\%}{CH_3CH=C\overset{CH_3}{\underset{CH_3}{\diagdown}}} + \underset{25\%}{\overset{CH_3}{\underset{C_2H_5}{\diagup}}C=CH_2} \quad (3.17)$$

Since the thermodynamic stability of isomeric *alkenes* (hydrocarbons with a carbon-carbon double bond) increases with the number of alkyl substituents on the double bond, it is likely that the Saytzeff rule simply reflects the extent of the C=C bond development in the E2 transition state.

PROBLEM 3·16
An alternative explanation of the Saytzeff rule assumes that the order of C—H reactivity to base attack is tertiary > secondary > primary. Discuss the merits and defects of this suggestion.

3·7 ELIMINATION REACTIONS

PROBLEM 3·17

Assuming only 1,2 elimination, write structures for all possible elimination products from each of the following alkyl halides. In each case list the alkenes in order of decreasing double-bond substitution.

a $(C_2H_5)_2CBrCH_3$

b CH₃, Cl on cyclopentane with CH₃

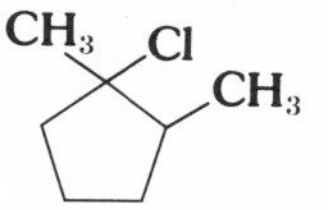

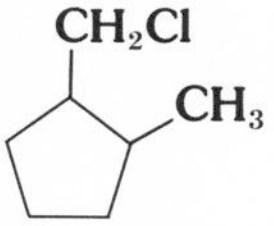

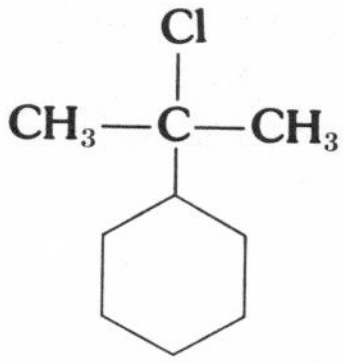

c CH₂Cl on cyclopentane with CH₃

d $CH_3-\underset{\underset{\text{(cyclohexyl)}}{|}}{\overset{\overset{Cl}{|}}{C}}-CH_3$

e Br on cyclohexane with CH₃

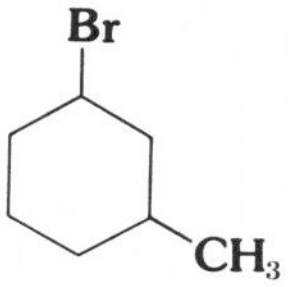

TRANSITION-STATE CONFIGURATION Although not all $E2$ elimination reactions follow the Saytzeff rule, it is often convenient to discuss exceptions in terms of steric or electronic deformations of the common $E2$ transition state in equation 3.16. For example, the different courses taken in the elimination reactions of the stereoisomeric 2-chloro-1-methylcyclohexanes suggests a preferred *trans-coplanar (anti)* orientation in the $E2$ transition state:

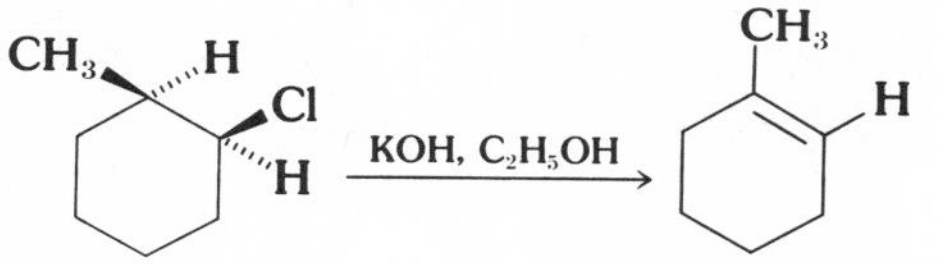

cis isomer $\xrightarrow{\text{KOH, } C_2H_5OH}$ major product   (3.18)

trans isomer $\xrightarrow{\text{KOH, } C_2H_5OH}$ major product   (3.19)

Thus, as illustrated in Figure 3.12, anti elimination of the trans chloride can occur only between C-2 and C-3, whereas the cis isomer is capable of anti elimination between C-1 and C-2 as well as C-2 and C-3. Recent

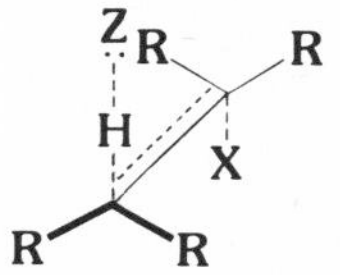

generalized anti geometry

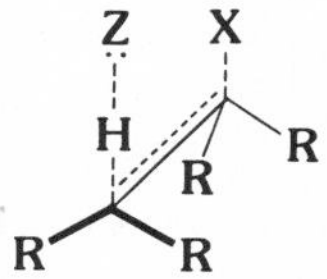

generalized syn geometry

FIGURE 3·12 *E2 transition states*

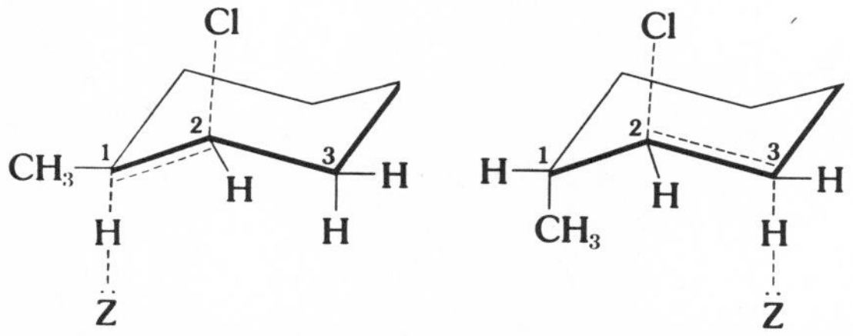

anti elimination of stereoisomeric 2-chloro-1-methylcyclohexanes studies with larger ring systems show that *syn elimination* can also take place, and further investigations are needed to clarify this situation.

> **PROBLEM 3·18**
> In discussing the possible anti orientations for E2 elimination of 2-chloro-1-methylcyclohexane (Figure 3.12), why do we ignore those conformations of the six-membered ring having an equatorial chlorine substituent?

BASICITY AND ACIDITY

The competition between bimolecular substitution and elimination reactions is dependent on the relative reactivities of nucleophilic reagents (bases) toward *carbon* and *hydrogen* atoms. Studies of S_N2 reactions have provided nucleophilicity values (relative reactivities toward electrophilic carbon atoms) for a series of nucleophilic reagents, as listed in Table 3.3. It seems reasonable to assume that the rate of nucleophilic attack at hydrogen will parallel the *basicity* of these reagents.

According to the widely used *Brønsted-Lowry* terminology, an acid is a *proton donor* and a base is a *proton acceptor*. Thus in any acid-base equilibrium there are always two acids and two bases, which are referred to as *conjugate-acid–base pairs*. Note in Figure 3.13 that water is *amphoteric* and can act as either a Brønsted acid or a Brønsted base.

> **PROBLEM 3·19**
> *a* Write formulas for the conjugate acids of the Brønsted bases NH_3, H_2S, C_2H_5OH, and H_2SO_4.
> *b* Write formulas for the conjugate bases of the Brønsted acids H_2SO_4, $HSO_4^\ominus$, NH_3, and CH_4.

We find that the *weakest acid* and *weakest base* in an acid-base reaction always predominate at equilibrium. Consequently, the relative acidities of a pair of acids (or the basicities of a pair of bases) can be estimated from the equilibrium constant of the corresponding acid-base equilibrium. In the reaction of HCl and water, for example, the equilibrium lies far to the right, indicating that in water solution $Cl^\ominus$ is a weaker base than H_2O and $H_3O^\oplus$ is a weaker acid than HCl. Of course, since water is the solvent as well as a base, its high concentration in the reaction mixture also pushes the equilibrium to the right. However, correcting for this concentration factor does not change the relative order of $Cl^\ominus$ and H_2O basicity. Solvents other than water may be used for acid-base equilibrium measurements; in fact, this is necessary if very strong or very weak acids are being studied.

FIGURE 3·13 *Conjugate-acid–base pairs*

		conjugate acid	conjugate base
$HCl + H_2O \rightleftharpoons H_3O^\oplus + Cl^\ominus$ $K_{eq} \gg 1$		HCl	$Cl^\ominus$
acid base acid base		$H_3O^\oplus$	H_2O
$KOH + C_2H_5OH \rightleftharpoons C_2H_5OK + H_2O$ $K_{eq} < 1$		H_2O	$OH^\ominus$
base acid base acid		C_2H_5OH	$C_2H_5O^\ominus$

PROBLEM 3·20

a The ionization constants K_a of weak to moderately strong acids (HA) in water solution are used as a measure of relative acidities:

$$HA + H_2O \rightleftharpoons H_3O^{\oplus} + A^{\ominus} \qquad K_a = \frac{[H_3O^{\oplus}][A^{\ominus}]}{[HA]}$$

If acid HA_1 has a $K_a = 10^{-10}$ and another acid HA_2 has a $K_a = 10^{-5}$, what will be the *relative* base strengths of the conjugate bases $A_1^{\ominus}$ and $A_2^{\ominus}$?

b Why is water a poor solvent to use for determining the acidity of very strong or very weak acids?

Figure 3.14 indicates the relative base strengths of some common nucleophilic ions, as determined from the acidities of the corresponding conjugate acids. Clearly these variations in base strength do not correspond at all with the relative nucleophilicities in Table 3.3, and we now find it entirely reasonable that 2-bromobutane should give a substitution reaction with sodium iodide and an elimination reaction with potassium hydroxide:

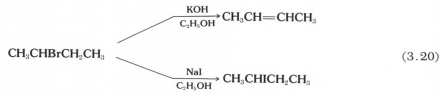

$$CH_3CHBrCH_2CH_3 \xrightarrow[C_2H_5OH]{KOH} CH_3CH{=}CHCH_3$$

$$CH_3CHBrCH_2CH_3 \xrightarrow[C_2H_5OH]{NaI} CH_3CHICH_2CH_3 \qquad (3.20)$$

Table 3.4 summarizes the important rate relationships in substitution and elimination reactions.

TABLE 3·4 ***Relative rates of bimolecular substitution and elimination reactions of R—X + Y:⊖***

variable	*S_N2*	*E2*
X	I > Br > Cl	I > Br > Cl
R	primary > secondary > tertiary	tertiary > secondary > primary
Y:⊖	rate increases with increasing nucleophilicity (Table 3.3)	rate increases with increasing base strength (Figure 3.14)

FIGURE 3·14 ***Relative base strengths of nucleophiles***

increasing base strength →

$I^{\ominus} < Br^{\ominus} < Cl^{\ominus}$ (very weak) $<< F^{\ominus} < CH_3CO_2^{\ominus} < SH^{\ominus} < CN^{\ominus}$ (weak) $<< OH^{\ominus} < OR^{\ominus}$ (strong) $<< NH_2^{\ominus} < CH_3^{\ominus}$ (very strong)

3·7 ELIMINATION REACTIONS

THE E1 MECHANISM

Strong bases are not always necessary in elimination reactions of tertiary alkyl halides; for example,

$$(CH_3)_3CBr \xrightarrow{H_2O, \, DMSO} (CH_3)_2C=CH_2 + HBr \qquad (3.21)$$

In fact, solutions of these compounds in weakly basic high-dielectric solvents yield elimination products at a rate which is *independent* of added nucleophiles or bases. These observations point to a two-step elimination pathway analogous to the S_N1 mechanism discussed in Section 3.6. This E1 *mechanism* (elimination unimolecular) is described by the following reaction sequence:

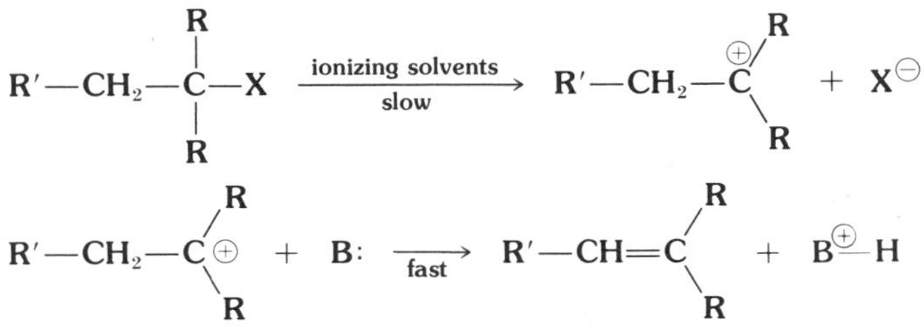

$$R'-CH_2-\underset{R}{\overset{R}{\underset{|}{C}}}-X \xrightarrow[\text{slow}]{\text{ionizing solvents}} R'-CH_2-\underset{R}{\overset{R}{C^{\oplus}}} + X^{\ominus}$$

$$R'-CH_2-\underset{R}{\overset{R}{C^{\oplus}}} + B: \xrightarrow{\text{fast}} R'-CH=\underset{R}{\overset{R}{C}} + B^{\oplus}-H \qquad (3.22)$$

In E1 reactions the base B: is usually the solvent.

Since the E1 mechanism involves the same rate-determining ionization step as the S_N1 mechanism, both reactions show the same substrate reactivity order:

tertiary R—X > secondary R—X > primary R—X > CH$_3$—X

The relative amounts of elimination and substitution products formed in these reactions are therefore determined by the fate of the common carbonium-ion intermediates.

Three modes of reaction are open to a carbonium ion:

1 It may act as a Lewis acid (an electrophile) and combine with a nucleophile to give a substitution product—the S_N1 path.

2 It may act as a Brønsted acid and transfer a proton to a base to give an alkene—the E1 path.

3 It may rearrange to a more stable carbonium ion, which then reacts according to mode 1 or 2.

The competition between mode 1 and mode 2 will be affected by slight changes in the nucleophilicity and basicity of the reaction medium, and it is difficult to go beyond a very general estimate of the relative importance of these factors. For example, the hydrolysis of t-butyl chloride in a water–ethyl alcohol mixture gives a greater proportion of substitution products (compared with elimination products) than a similar reaction in a water-DMSO mixture.

PROBLEM 3·21

If $(CH_3)_3C^{\oplus}$ is a Brønsted acid, what is the corresponding conjugate base?

3·8 ORGANOMETALLIC COMPOUNDS

PROBLEM 3·22

The products of the molecular rearrangement described in equation 3.12 include a small amount of an alkene. What structure would you expect this minor product to have?

PROBLEM 3·23

Would you expect a strong stereochemical bias in $E1$ elimination—that is, a preference for anti or syn elimination? Explain.

3·8 ORGANOMETALLIC COMPOUNDS

FORMATION OF ORGANOMETALLIC COMPOUNDS

Certain active metals such as potassium, sodium, lithium, zinc, and magnesium react readily with alkyl chlorides, bromides, and iodides (alkyl fluorides are relatively inert). Although the vigorous reactions involving sodium and potassium are fairly complex and therefore have only limited use, lithium and magnesium react cleanly to give high yields of alkyl metal derivatives:

$$R-X + 2Li \xrightarrow{\text{pentane}} R-Li + LiX \tag{3.23}$$

$$R-X + Mg \xrightarrow{\text{ether}} R-Mg-X \tag{3.24}$$

Reactive compounds such as **RLi** and **RMgX** are widely used by organic chemists and have become important and versatile reagents. Indeed, the French chemist Victor Grignard was awarded the 1912 Nobel Prize in Chemistry for his discovery and development of organomagnesium compounds (**RMgX**), and these substances are now known as *Grignard reagents*.

Some specific examples of the preparation of organolithium and organomagnesium reagents are given in Figure 3.15. The formulas shown for these reagents are incomplete, inasmuch as recent studies have shown that the metal atoms are also bonded to solvent molecules and/or other solute molecules. Nevertheless, these formulas serve quite well to illustrate the chemical transformations of organometallic compounds.

Dihalides may react with active metals to give *bis*-organometallic

FIGURE 3·15 *Formation of active metal compounds*

$$CH_3CH_2CH_2CH_2Br + 2Li \xrightarrow{\text{pentane or ether}} CH_3CH_2CH_2CH_2Li + LiBr$$
$$\text{n-butyl lithium}$$

$$CH_3I + Mg \xrightarrow{\text{ether}} CH_3MgI$$
$$\text{methyl magnesium iodide}$$

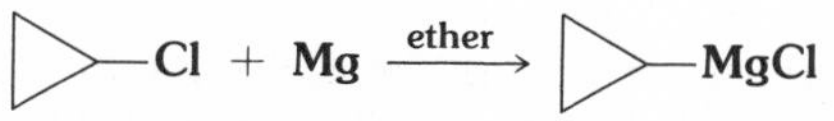

$$\triangleright\!\!-Cl + Mg \xrightarrow{\text{ether}} \triangleright\!\!-MgCl$$
$$\text{cyclopropyl magnesium chloride}$$

reagents, provided the halogen atoms are separated by at least three carbon atoms:

$$Br(CH_2)_4Br \xrightarrow{2Mg, \text{ ether}} BrMg(CH_2)_4MgBr \quad (3.25)$$

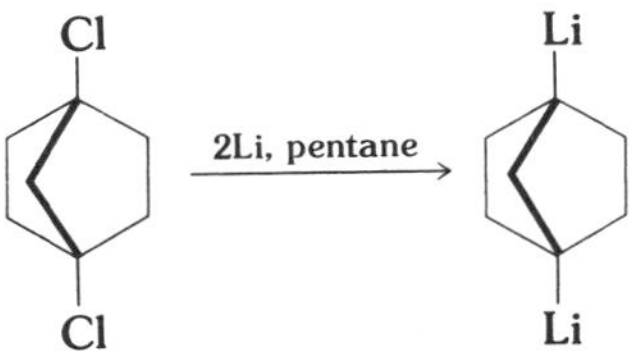

1,4-dichlorocyclohexane $\xrightarrow{2Li, \text{ pentane}}$ 1,4-dilithiocyclohexane (3.26)

Halogen atoms on adjacent carbon atoms normally undergo an elimination reaction when treated with active metals such as zinc or magnesium. A similar transformation of 1,3-dihalides to cyclopropane derivatives has also been observed:

$$CH_3CHBrCH_2Br \xrightarrow{Zn \text{ (or Mg)}} CH_3CH=CH_2 + ZnBr_2 \text{ (or } MgBr_2) \quad (3.27)$$

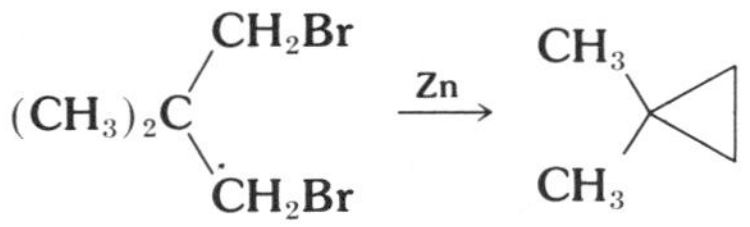

$$(CH_3)_2C\begin{matrix}CH_2Br\\CH_2Br\end{matrix} \xrightarrow{Zn} \begin{matrix}CH_3\\CH_3\end{matrix}\!\!\triangleright \quad (3.28)$$

REACTIONS OF ORGANOMETALLIC COMPOUNDS

The highly reactive organometallic reagents are seldom isolated because they are sensitive to oxygen and may burn spontaneously when exposed to air. In a nonaqueous solution, however, where they are protected and stabilized by solvation, they may be used with little difficulty, provided moisture is excluded. The reaction of these compounds with water is, in fact, characteristic of a general mode of reaction that takes place with all **OH** and **NH** functions:

$$n\text{-}C_4H_9Li + H_2O \longrightarrow n\text{-}C_4H_{10} + LiOH \quad (3.29)$$

$$CH_3MgI + C_2H_5OH \longrightarrow CH_4 + C_2H_5O\text{—}Mg\text{—}I \quad (3.30)$$

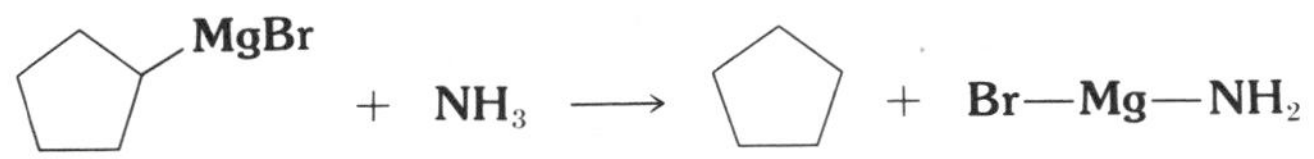

cyclopentyl-MgBr + $NH_3 \longrightarrow$ cyclopentane + Br—Mg—NH_2 (3.31)

PROBLEM 3·24

Chlorine-containing isomers A and B are converted to Grignard reagents by reaction with magnesium turnings in ether. The Grignard reagents from A and B are separately decomposed with water, and in each case the only organic product is 2,2-dimethylbutane. Compound A shows three groups of resonance signals in the pmr spectrum, at $\delta = 0.9$, 1.5, and 4.0, with relative areas of 9:3:1. Compound B also shows three groups of resonance signals, at $\delta = 0.9$, 1.8, and 3.5, with relative areas of 9:2:2. Write structural formulas for compounds A and B.

3·8 ORGANOMETALLIC COMPOUNDS

PROBLEM 3·25

Suggest a chemical procedure for preparing 1-deutero-propane ($CH_3CH_2CH_2D$) and 2-deuteropropane (CH_3CHDCH_3) from propane, assuming that heavy water (D_2O) is available.

It is helpful to consider the substitution reactions of organometallic compounds in terms of the mechanisms outlined in Table 1.9. One important characteristic is the very high selectivity exhibited by these reagents for hydrogen atoms bonded to strongly electronegative atoms such as oxygen and nitrogen. Thus in reaction 3.30 the methyl Grignard reagent attacks *only* the **OH** group in ethyl alcohol, and the remaining five hydrogen atoms are unaffected despite the fact that C—H has a lower bond energy than O—H. Furthermore, the reactions with **OH** (and **NH**) appear to have very small activation energies, since they proceed rapidly even at low temperatures. These facts argue against a mechanism involving radical intermediates and suggest that the polar nature of the O—H and C—M bonds (**M** = Li or Mg) is a key factor here.

Polar bonds may be regarded as partially ionic bonds; in this sense the hydrogen atom in $^{\delta\ominus}O-H^{\delta\oplus}$ and the metal atom in $^{\delta\ominus}C-M^{\delta\oplus}$ are electrophilic, whereas their respective oxygen and carbon atoms are nucleophilic. We see, then, that reactions 3.29 to 3.31 all involve replacement of an electrophilic metal atom by an electrophilic hydrogen atom, as illustrated by the general notation in Figure 3.16. Such reactions are customarily referred to as *electrophilic substitutions at carbon*. Of course, if attention were fixed on the metal atom instead, they could equally well be called nucleophilic substitutions at lithium or magnesium.

PROBLEM 3·26

If attention were focused on the oxygen atom in reaction 3.29, how would this reaction be classified?

Not all organometallic compounds are as reactive as the lithium and magnesium reagents described above. Organomercurials can be prepared by the reaction of Grignard reagents with mercuric bromide and are found to be relatively unreactive with water and alcohols. Electrophilic substitution of these compounds can be achieved, however, in refluxing acid:

$$2RMgBr + HgBr_2 \xrightarrow{ether} R_2Hg + 2MgBr_2 \qquad (3.32)$$

$$R_2Hg + 2HCl \xrightarrow[\Delta]{H_2O} 2R-H + HgCl_2 \qquad (3.33)$$

FIGURE 3·16 *Electrophilic substitution at carbon*

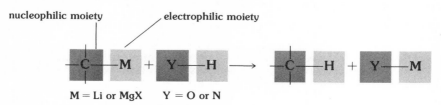

SUMMARY

alkyl halide A derivative of an alkane (or cycloalkane) in which one or more hydrogen atoms are replaced by a halogen (F, Cl, Br, or I). Alkyl halides having two or more Cl atoms or at least one Br or I atom are usually more dense than water (>1.0 g/ml). The presence of a halogen atom also causes neighboring hydrogen atoms to have larger pmr chemical shifts than expected (see Table 3.2).

reactions of alkyl halides Characteristic chemical transformations involving the polar carbon-halogen bond. The halogen atom is nucleophilic and reacts with electrophiles such as $Ag^{\oplus}$, while the neighboring C and H atoms are electrophilic and react with nucleophiles and bases. The most important alkyl halide reactions are *nucleophilic substitution, elimination reactions*, and the formation of *organometallic reagents*:

$$R-X + Y:^{\ominus} \longrightarrow R-Y + X:^{\ominus}$$

$$Z:^{\ominus} + H-\overset{|}{\underset{|}{C}}-\overset{|}{\underset{|}{C}}-X \longrightarrow \underset{\text{an alkene}}{\overset{\diagdown}{\diagup}C=C\overset{\diagup}{\diagdown}} + Z-H + X:^{\ominus}$$

$$R-X + Mg \longrightarrow \underset{\text{Grignard reagent}}{R-Mg-X}$$

$$R-X + 2Li \longrightarrow R-Li + LiX$$

stereospecificity The formation of different products (often stereoisomers) from reaction of stereoisomeric starting materials.

kinetic order The experimentally observed dependence of the rate of a reaction on the concentrations of the reactants. For a reaction A + B $\longrightarrow$ products, the reaction rate is equal to $k[A]^n[B]^m$ (where k is the rate constant) and the kinetic order is $n + m$.

rate-determining step The slowest step in a multistep reaction mechanism.

molecularity The number of reactant molecules taking part in the rate-determining step of a mechanism.

S_N2 (substitution nucleophilic bimolecular) mechanism A one-step process, kinetically second order, in which the reaction rate (1) varies with R according to steric hindrance (1° > 2° ≫ 3°), (2) increases with the nucleophilicity of $Y:^{\ominus}$ ($SH^{\ominus} > I^{\ominus} > CN^{\ominus} > OR^{\ominus} > Br^{\ominus} > Cl^{\ominus} > F^{\ominus}$), (3) varies with X (F ≪ Cl < Br < I), and (4) is not greatly affected by changes in the solvent. Stereoisomeric halides react stereospecifically with inversion of configuration.

S_N1 (substitution nucleophilic unimolecular) mechanism A two-step process, kinetically first order, which proceeds via a carbonium-ion intermediate. The reaction rate (1) varies with R according to carbonium-ion stability (3° ≫ 2° > 1° > CH_3), (2) varies greatly with the solvent (since ions are stabilized by solvation), and (3) is unaffected by changes in $Y:^{\ominus}$ as long as the solvent remains unchanged. All nucleophiles $Y:^{\ominus}$ may appear in the products. The reaction is not stereospecific, and molecular rearrangements may occur.

carbonium ion A reactive intermediate having a positively charged trivalent carbon atom.

solvation The stabilizing interaction of solute molecules with solvent molecules, particularly important for ions in solution. When the solvent is water, this process is termed *hydration*.

thermodynamic stability A qualitative term referring to the relative potential energies of molecules or other systems; the lower the potential energy (or enthalpy), the greater the thermodynamic stability.

rearrangement of carbonium ions The transformation of a primary or secondary carbonium-ion to a more stable carbonium ion by means of a 1,2 shift of a neighboring substituent (usually an alkyl group or hydrogen atom).

Hammond postulate The principle that in a strongly endothermic reaction the products are a reasonable model for the transition state and in a strongly exothermic reaction the reactants may serve as a transition-state model.

Brønsted-Lowry acid-base concept The definition of an *acid* as a substance capable of transferring a proton to a base and a *base* as a substance possessing an unshared pair of electrons that can be used to bond a proton.

E_2 (elimination bimolecular) mechanism A single-step process, kinetically second order, in which the reaction rate varies with R (3° > 2° > 1°) and increases with the strength of the base Z:⊖ (OR⊖ > OH⊖ ≫ CN⊖ > SH⊖ > F⊖ ≫ Cl⊖ > Br⊖, I⊖). The more highly substituted alkene usually predominates among the possible elimination products (the Saytzeff rule), and anti geometry of the leaving groups is usually favored.

E_1 (unimolecular elimination) mechanism A two-step process, kinetically first order, which proceeds via a carbonium-ion intermediate. Rearrangement may occur, and product formation usually follows the Saytzeff rule.

Saytzeff rule The principle that compounds which can give more than one elimination product will generally yield the most highly substituted isomer as the major product.

EXERCISES

3·1 Give two names for each of the following compounds:
a CH_3CH_2Cl
b $(CH_3)_3CBr$
c $(CH_3)_2CHI$

d ⬡—Br

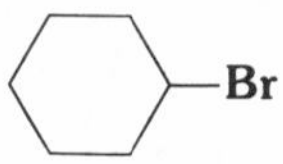

3·2 Name each of the compounds listed:
a $CH_3CHBrCH_2CH_3$
b $(CH_3)_2CHCHBrCH_2CH_2CH_3$

c (cyclopentane with Cl, H, H, CH$_3$ substituents)

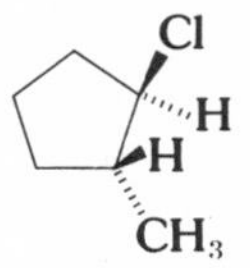

d (cyclohexane with CH$_3$, CH$_3$, Br, H substituents)

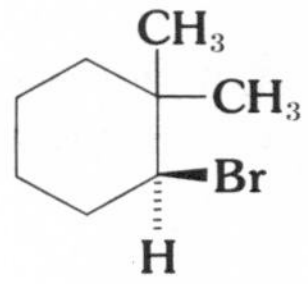

3·3 Draw all the C_4H_9Br isomers and arrange them in order of reactivity toward sodium iodide in acetone.

3·4 Draw structures of all the monochloropentanes, $C_5H_{11}Cl$ and name each according to the IUPAC system.

3·5 For each of the following reactions indicate whether K_{eq} is greater than or less than 1:

a $HCl + NaOH \rightleftharpoons NaCl + H_2O$

b $HCN + NaNH_2 \rightleftharpoons NaCN + NH_3$

c $CH_3CO_2{}^{\ominus}Na^{\oplus} + H_2O \rightleftharpoons CH_3CO_2H + NaOH$

d $CH_3CH_2O^{\ominus}Na^{\oplus} + H_2O \rightleftharpoons CH_3CH_2OH + NaOH$

3·6 Write structural formulas for the major product(s) expected from the following reactions:

a $(CH_3)_2CHCH_2CH_2Cl + KCN \xrightarrow{acetone}$

b $CH_2=CHCH_2Br + NaOCH_3 \xrightarrow{CH_3OH}$

c $CH_3CH_2CH_2Cl + NaI \xrightarrow{acetone}$

d $\triangleright\!\!-CH_2CH_2Br + Mg \xrightarrow{ether} X \xrightarrow{D_2O} Y$

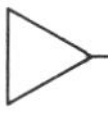

e ⬡$-Br + NaSH \xrightarrow{alcohol, H_2O}$

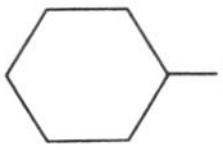

f $ClCH_2CH_2CH_2Br + NaCN \xrightarrow{alcohol}$

g $(CH_3)_3CCl + NaOC_2H_5 \xrightarrow{C_2H_5OH}$

h $CH_3CH_2CH_2Br + NH_3 \longrightarrow$

i $CH_3CH_2I + CH_3CO_2Na \longrightarrow$

3·7 Arrange each of the following groups first in order of reactivity to S_N2 reaction, and then in order of reactivity to S_N1 reaction.

a 1-bromopentane, 2-bromopentane, 2-bromo-2-methylbutane

b 1-chloro-3-methylbutane, 2-chloro-2-methylbutane, 3-chloro-2-methylbutane

3·8 Which of the following pairs of compounds would react more rapidly?

a $(CH_3)_3CCl$ or $CH_3(CH_2)_3Cl$ with NaI in acetone

b $(CH_3)_3CCl$ or $CH_3(CH_2)_3Cl$ with H_2O

c $(CH_3)_3CCl$ or $CH_3(CH_2)_3Cl$ with KOH in alcohol

d NaI or $NaOCH_3$ in CH_3OH with $(CH_3)_3CBr$

e $CH_3CH_2CH_2Cl$ or $(CH_3CH_2)_2CHCl$ with $AgNO_3$ in alcohol

f $CH_3CH_2CH_2Br$ or $CH_3CH_2CH_2Cl$ with NaI in acetone

3·9 Write an example of each of the following reactions. For each reaction indicate the rate-determining step (the slow step) and the product-determining step:

a S_N2

b E2

c S_N1

d E1

3·10 Outline mechanisms which account for the following observed reactions:

a CH$_3$—C(CH$_3$)(CH$_3$)—CHBrCH$_3$ $\xrightarrow[(CH_3)_3COH]{(CH_3)_3COK}$ CH$_3$—C(CH$_3$)(CH$_3$)—CH=CH$_2$

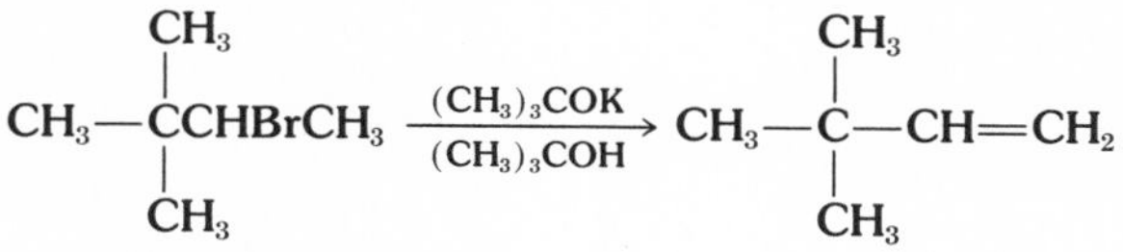

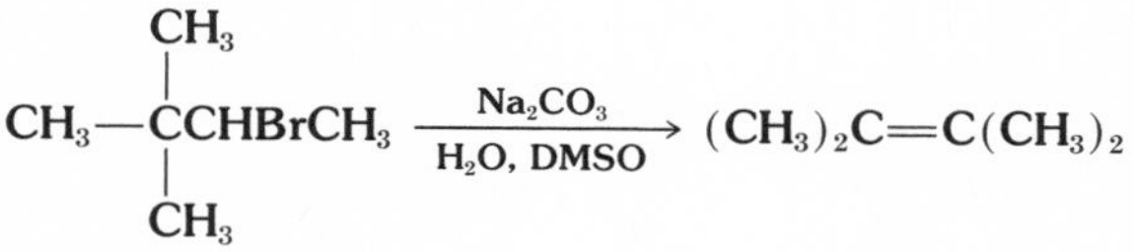

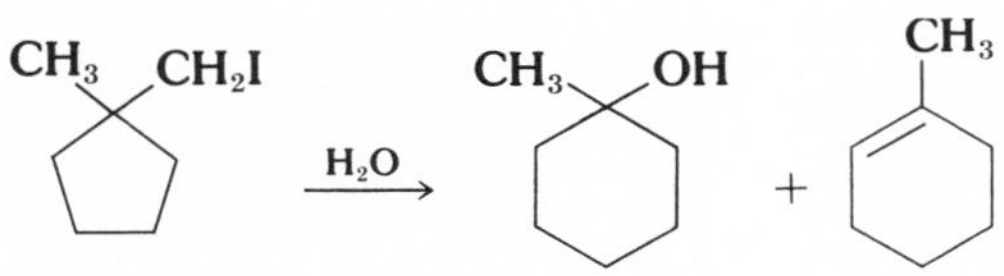

b CH$_3$—C(CH$_3$)(CH$_3$)—CHBrCH$_3$ $\xrightarrow[H_2O, DMSO]{Na_2CO_3}$ (CH$_3$)$_2$C=C(CH$_3$)$_2$

c 1-methyl-1-(iodomethyl)cyclopentane $\xrightarrow{H_2O}$ 1-methylcyclohexanol + 1-methylcyclohexene

d CH$_3$—C(Cl)(CH$_3$)—CH=CHCH$_2$Br $\xrightarrow{NaN_3,\ alcohol}$ CH$_3$—CH(Cl)—C(CH$_3$)(CH)—CH=CH—CH$_2$N$_3$

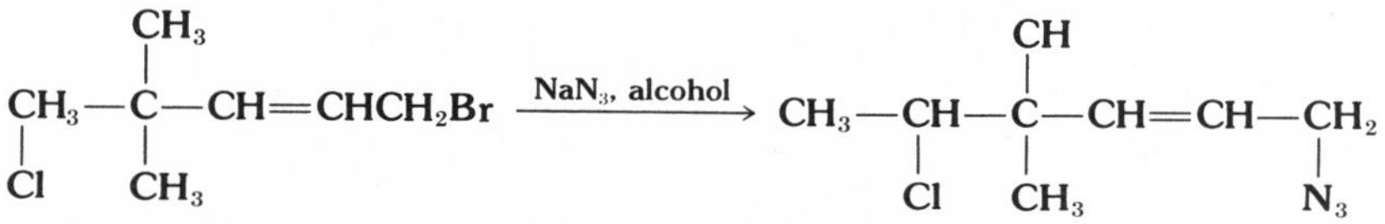

3·11 Draw an energy diagram for reaction 3.9.

3·12 Isopropyl bromide yields two organic products when warmed in an aqueous solution. Draw the products and the mechanism leading to each. Draw a single potential-energy diagram which shows the progress of both the reactions leading to products. What factor will determine the relative abundance of the two products?

3·13 Account for the behavior indicated:

a (CH$_3$)$_3$CCH$_2$Br: extremely slow reaction with either NaI in acetone or AgNO$_3$ in H$_2$O:

b CH$_3$OCHBrCH$_3$: very rapid reaction with H$_2$O

3·14 Indicate reagents and conditions which would accomplish the reactions shown:

a CH$_3$CHBrCH$_2$CH$_3$ ⟶ CH$_3$CHSHCH$_2$CH$_3$

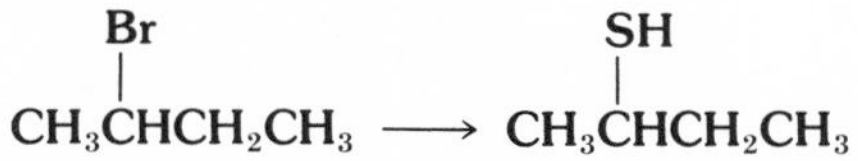

b chlorocyclopentane ⟶ cyanocyclopentane

c 1-methyl-4-chlorocyclohexane ⟶ 1-methylcyclohexene

d CH$_3$CH(CH$_3$)—CHBrCH$_3$ ⟶ CH$_3$C(CH$_3$)(OH)—CH$_2$CH$_3$

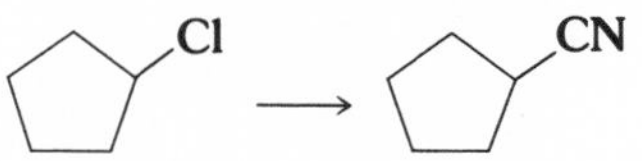

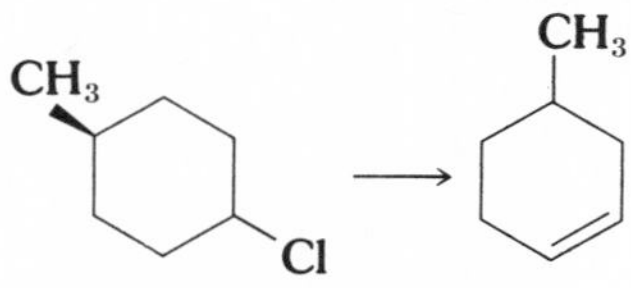

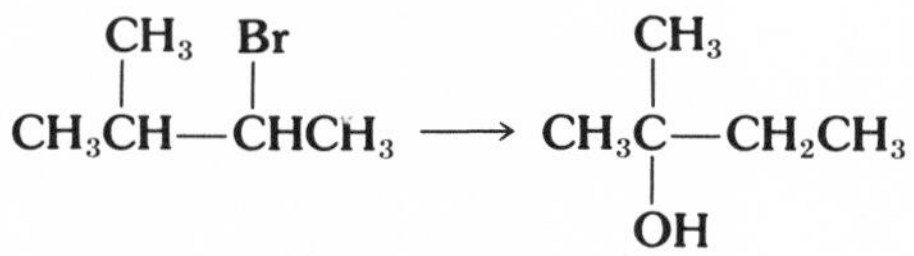

e

f

g

h

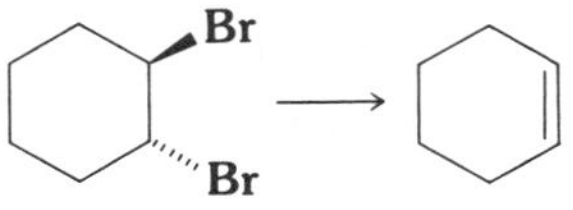

3·15 The strong base potassium *t*-butoxide is frequently used to promote an *E*2 reaction without a competing S_N2 reaction, since *t*-butoxide is a rather poor nucleophile. How do you explain the fact that potassium *t*-butoxide is a much poorer nucleophile than potassium methoxide in an S_N2 reaction, when both anions are equally polarizable and the *t*-butoxide is a stronger base than methoxide?

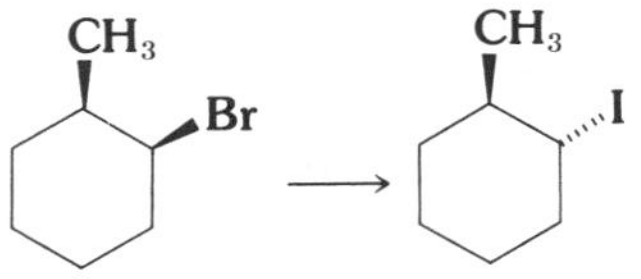

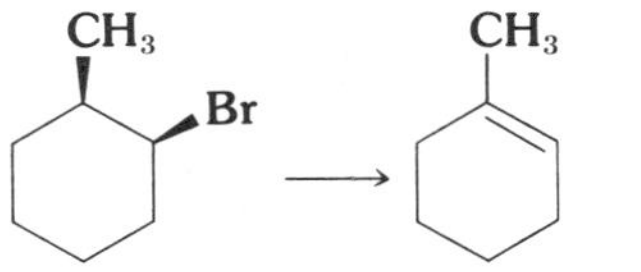

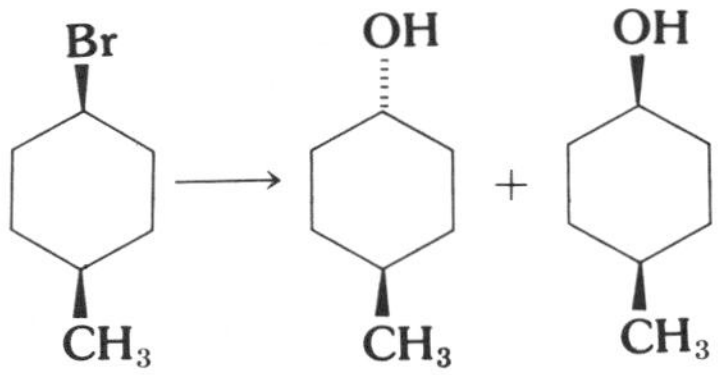

3·16 Some organisms have enzymes that act to convert DDT to DDE, a compound which is even less reactive toward biodegradation than DDT and more toxic to some animals. What type of reaction is the transformation of DDT to DDE? What reagent would you use to accomplish this conversion in the laboratory?

DDE

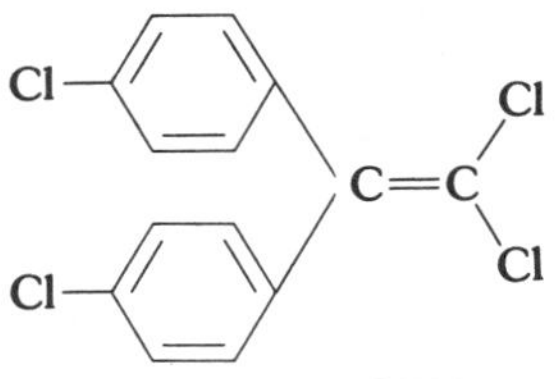

3·17 Explain the difference in the proportion of products for the reaction of neomenthyl chloride and menthyl chloride with base.

neomenthyl chloride 25% 75%

menthyl chloride 100%

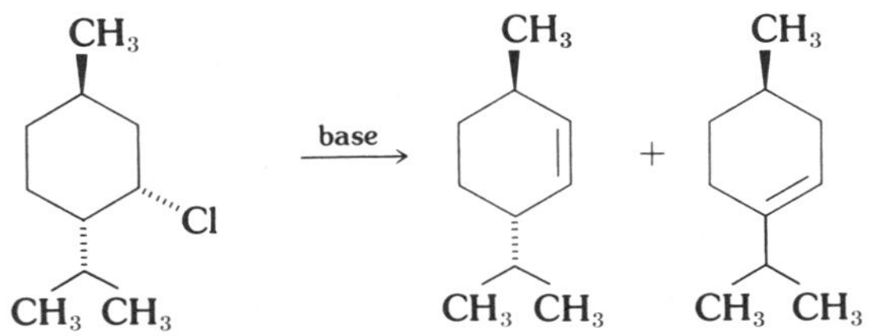

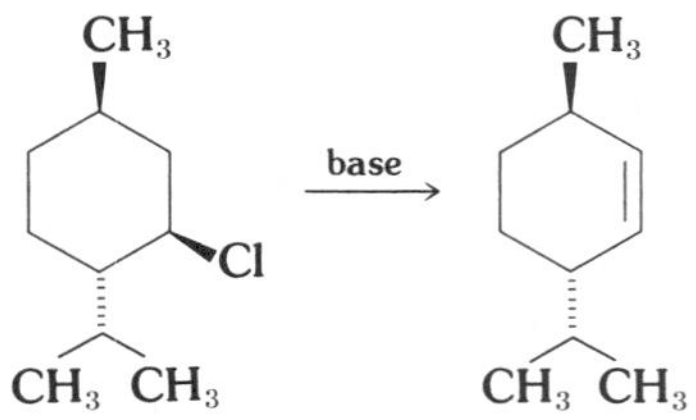

3·18 Using Figure 3.14 as a reference, arrange the following compounds in order of decreasing acid strength:

NH_3 CH_3CO_2H HCl CH_4 H_2O

3·19 Potassium iodide catalyzes the hydrolysis of primary alkyl chlorides:

$$RCH_2Cl + H_2O \xrightarrow[NaHCO_3]{[KI]} RCH_2OH + HCl \text{ neutralized by NaHCO}_3$$

Explain the nature of this action, using equations to illustrate your mechanism.

3·20 Given the reaction

$$CH_3Br + I^\ominus \xrightarrow{acetone} CH_3I + Br^\ominus$$

indicate the effect of each of the following modifications on the rate of the reaction (compared to an original rate).
a Double the $I^\ominus$ concentration
b Double the CH_3Br concentration
c Triple both the $I^\ominus$ concentration and the CH_3Br concentration
d Reduce the $I^\ominus$ by one-half
e Reduce the $I^\ominus$ by one-half and also double the CH_3Br concentration
f Double the volume of solution by the addition of acetone

3·21 Given the reaction

$$CH_3CHBrCH_3 + H_2O \xrightarrow{formic\ acid} CH_3CH(OH)CH_3 + HBr$$

determine the effect of the following changes on the rate of the reaction:
a Double the $(CH_3)_2CHBr$ concentration
b Double the H_2O concentration
c Double both the $(CH_3)_2CHBr$ and the H_2O concentration
d Double the volume of solution by the addition of more formic acid

3·22 We might expect an S_N1 reaction to have a large positive entropy of activation (ΔS^*) because the rate-determining step involves a fragmentation process (the disorder of the system should increase). Experiments show, however, that S_N1 reactions usually have negative entropies of activation (that is, the system becomes more ordered). Explain.

3·23 An alternative mechanism (two steps) for the base-catalyzed elimination reactions of alkyl halides is

$$\underset{\underset{R}{|}}{\overset{\overset{H}{|}}{R-C}} - \underset{\underset{R}{|}}{\overset{\overset{X}{|}}{C}} - R + B{:}^\ominus \underset{}{\overset{fast}{\rightleftarrows}} \underset{\underset{R}{|}}{\overset{}{R-\overset{\ominus}{C}}} - \underset{\underset{R}{|}}{\overset{\overset{X}{|}}{C}} - R + B-H$$

$$\underset{\underset{R}{|}}{\overset{}{R-\overset{\ominus}{C}}} - \underset{\underset{R}{|}}{\overset{\overset{X}{|}}{C}} - R \xrightarrow{slow} \underset{R}{\overset{R}{\diagdown}}C=C\underset{R}{\overset{R}{\diagup}} + X{:}^\ominus$$

Draw an energy diagram for this mechanism and suggest experiments that would test its feasibility.

3·24 The following reaction gives both substitution and elimination products. The rate of formation of both products is proportional to the NaOC$_2$H$_5$ concentration. Write formulas for both products (show configuration).

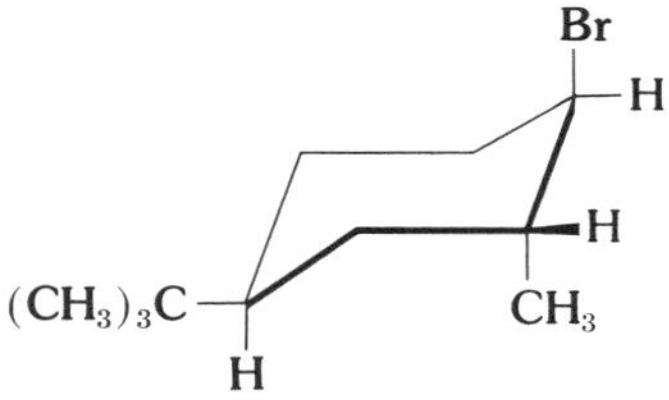

(CH$_3$)$_3$C–[cyclohexane with Br axial up, H; CH$_3$ and H] + NaOC$_2$H$_5$ $\xrightarrow{C_2H_5OH}$ A (substitution product) + B (elimination product)

3·25 The following transformations require more than one step (reaction). Write equations showing how each can be achieved.

a cyclohexane ⟶ cyclohexene

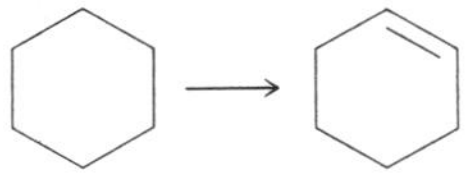

b (CH$_3$)$_4$C ⟶ (CH$_3$)$_3$CCH$_2$D

c cyclopentane ⟶ cyclopentane with I and H (iodocyclopentane)

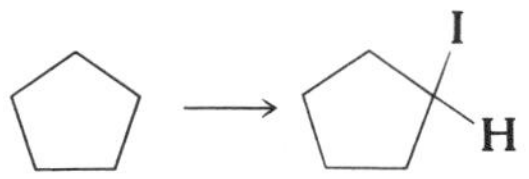

ALKENES 4

THE ALKENES and cycloalkenes are hydrocarbons containing one or more carbon-carbon double bonds. These compounds include gases, liquids, and solids and have melting and boiling points that increase with molecular weight in much the same way as do the alkanes (see Figure 2.5). Carbon-carbon double bonds constitute a particularly important class among the functional groups because they undergo a variety of addition reactions that enable chemists to transform them in many useful ways.

The simpler alkenes such as ethylene (C_2H_4) and propylene (C_3H_6) can be obtained from petroleum and are important industrial raw materials. More complex compounds containing carbon-carbon double bonds are widely distributed in nature. The names commonly used for these compounds are not logically derived from their structural formulas, but were inherited from early chemists, who often based the name of a new substance on its botanical source.

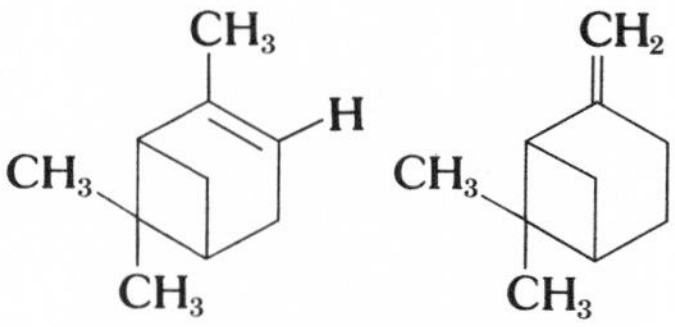

α-pinene β-pinene
(isolated from turpentine)

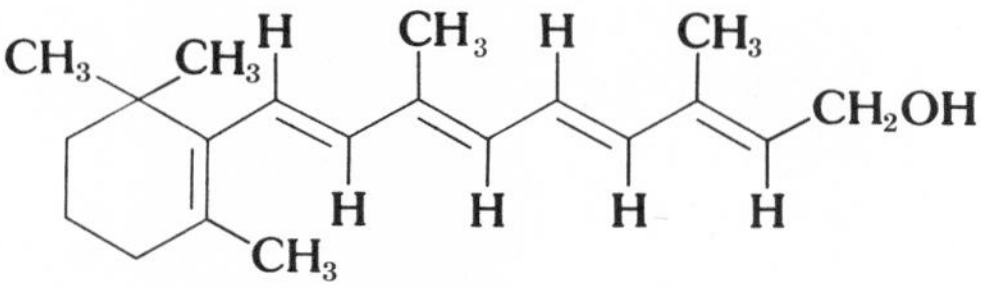

vitamin A

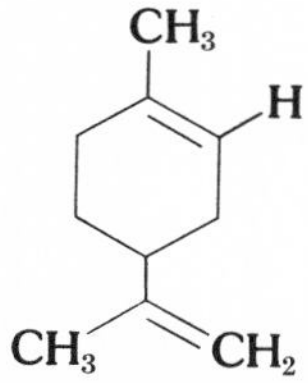

limonene
(from citrus oils)

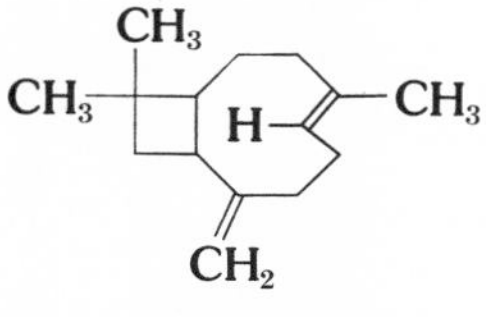

caryophyllene
(from oil of cloves)

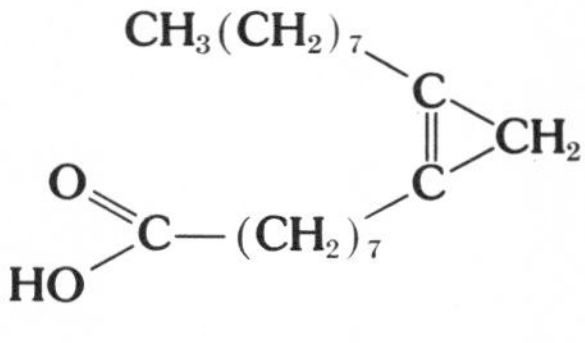

sterculic acid
(from sterculia foetida)

Alkenes that are not available from natural sources can often be synthesized in the laboratory. The most common way to accomplish this is by means of an elimination reaction. Recall from Chapter 3 that dehydrohalogenation of alkyl halides gave alkenes. Other elimination reactions are:

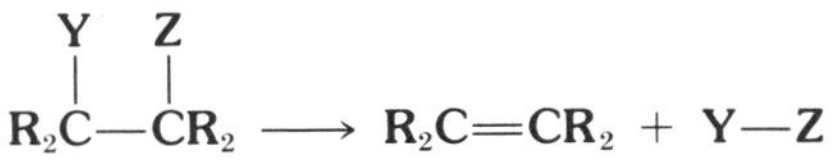

$$R_2C(Y)-C(Z)R_2 \longrightarrow R_2C=CR_2 + Y-Z \tag{4.1}$$

reaction type	Y substituents	Z substituents
dehydrohalogenation	H	Cl, Br, or I
dehalogenation	Cl, Br, or I	Cl, Br, or I
dehydration	H	OH

Another useful method of making disubstituted alkenes is by a controlled addition of hydrogen to a carbon-carbon triple bond, as discussed in Chapter 6:

$$RC\equiv CR + H_2 \text{ (or 2H)} \longrightarrow RCH=CHR \tag{4.2}$$

4·1 NOMENCLATURE OF ALKENES AND CYCLOALKENES

Simple alkenes can be assigned systematic IUPAC names by using a slightly modified form of the general nomenclature rules given for alkanes in Chapter 2.

1 The basic name refers to the longest carbon-atom chain incorporating both carbon atoms of the double bond. The ending *-ene* classifies the compound as an alkene (see Table 4.1).

2 This longest chain is numbered from the end that will assign the lowest possible numbers to the double-bonded carbon atoms. The position of the double bond is then denoted by the lower of the two double-bonded carbon numbers:

$$\overset{1}{C}H_3-\overset{2}{C}H=\overset{3}{C}H-\overset{4}{C}H_2-\overset{5}{C}H_3$$
2-pentene
(*not* 3-pentene)

3 Substituents attached to the longest chain (rule 1) are identified, numbered, and listed in the usual way and precede the double-bond notation in the name:

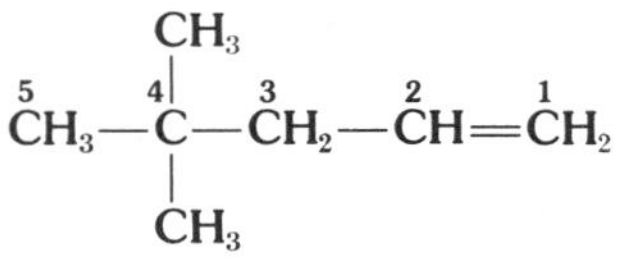

$$\overset{5}{C}H_3-\underset{\underset{CH_3}{|}}{\overset{\overset{CH_3}{|}}{\overset{4}{C}}}-\overset{3}{C}H_2-\overset{2}{C}H=\overset{1}{C}H_2$$
4,4-dimethyl-1-pentene
(*not* 2,2-dimethyl-4-pentene)

4·1 NOMENCLATURE OF ALKENES AND CYCLOALKENES 109

4   In cyclic alkenes the carbon atoms of the double bond are numbered 1 and 2 by definition; hence the double-bond position need not be cited. For example,

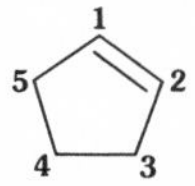

cyclopentene
(*not* 1-cyclopentene)

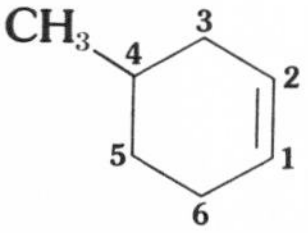

4-methylcyclohexene
(*not* 5-methylcyclohexene)

5   Additional flexibility in naming compounds of this kind is achieved by giving common names to a few simple double-bond groups:

H$_2$C=CH—
vinyl group

H$_2$C=CHCl
vinyl chloride
or 1-chloroethene

H$_2$C=CH—CH$_2$—
allyl group

H$_2$C=CHCH$_2$Br
allyl bromide
or 3-bromo-1-propene

The following examples illustrate IUPAC nomenclature for a variety of alkenes and cycloalkenes:

(CH$_3$)$_2$C=C(CH$_3$)CH$_2$CH(C$_2$H$_5$)$_2$
5-ethyl-2,3-dimethyl-2-heptene

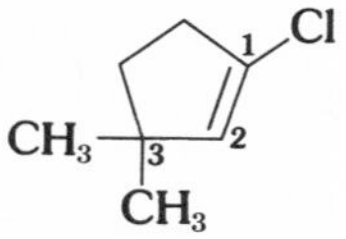

1-chloro-3,3-dimethylcyclopentene

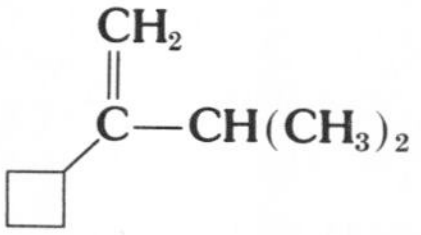

2-cyclobutyl-3-methyl-1-butene

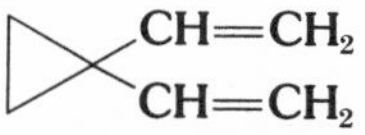

1,1-divinylcyclopropane

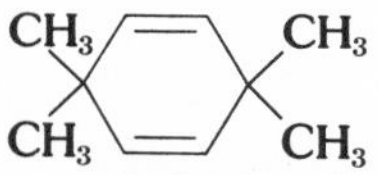

3,3,6,6-tetramethyl-1,4-cyclohexadiene

Compounds with more than one double bond are referred to in general as *polyenes*. However, the names of specific compounds include a prefix denoting the exact number of double bonds in a molecule (as in the above *diene*). It is useful to distinguish three kinds of grouping of the double

TABLE 4·1 *Alkene nomenclature*

molecular formula	structural formula	IUPAC names	common names
C$_2$H$_4$	H$_2$C=CH$_2$	ethene	ethylene
C$_3$H$_6$	CH$_3$CH=CH$_2$	propene	propylene
C$_4$H$_8$	CH$_3$CH$_2$CH=CH$_2$	1-butene	α-butylene
	CH$_3$CH=CHCH$_3$	2-butene	β-butylene
	(CH$_3$)$_2$C=CH$_2$	2-methylpropene	isobutylene
...	...	...	...
C$_n$H$_{2n}$	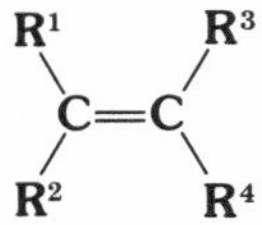 R^{1}R^{2}C=CR^{3}R^4	alkene	olefin

bonds in a diene: *cumulated double bonds* have a carbon atom in common, *conjugated double bonds* are directly joined by a single bond, and *isolated double bonds* are separated by one or more tetrahedral carbon atoms.

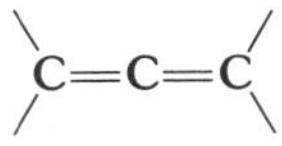

    a cumulated diene 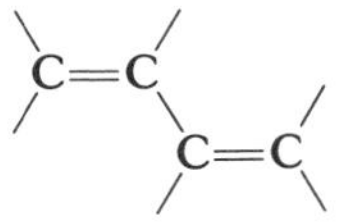   a conjugated diene 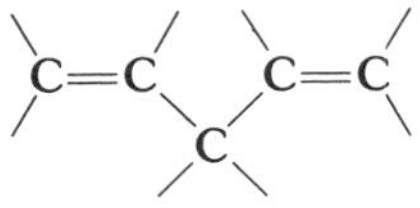   an isolated diene

According to these definitions, limonene and caryophyllene are isolated dienes, and vitamin A is a completely conjugated pentaene.

4·2 STRUCTURE AND PROPERTIES OF ALKENES

The structure of the simplest alkene, ethylene (C_2H_4), has been determined by diffraction and spectroscopic measurements. Similar examinations of a variety of double-bond compounds show a common configuration in which the six atoms directly associated with the double bond all lie essentially in the same plane. The precise dimensions of the double bond vary slightly from compound to compound, but in general the C=C bond length is about 0.2 Å shorter than the corresponding single bond, and the angle of the single bonds at the double-bonded carbon atom is usually around 118°. For example,

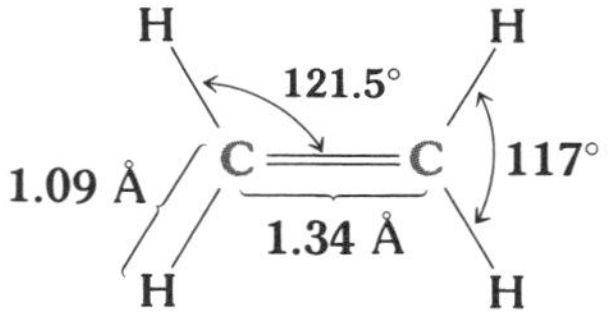

structure of ethylene

SPECTROSCOPIC PROPERTIES

Alkenes show certain distinctive infrared and pmr absorptions which appear to be characteristic of carbon-carbon double bonds. For example, the infrared spectra of 1-octene and limonene in Figure 4.1 show stretching absorptions for the C=C and C—H bonds, and the pmr spectrum of

TABLE 4·2 *Spectroscopic properties of alkenes*

absorbing group	characteristic absorption
infrared spectrum	
C=C (stretch)	1620–1680 cm^{-1}
C=C—H (stretch)	3000–3150 cm^{-1}
C=C—H (bend)	690–1420 cm^{-1}
pmr spectrum	
C=C—H	4.5–5.5 (δ)
C=C—CH$_3$	1.6–1.8 (δ)

4·2 STRUCTURE AND PROPERTIES OF ALKENES 111

limonene in Figure 4.2 exhibits typical resonance peaks for the hydrogen atoms attached to the double bond. The locations of these characteristic absorptions are listed in Table 4.2.

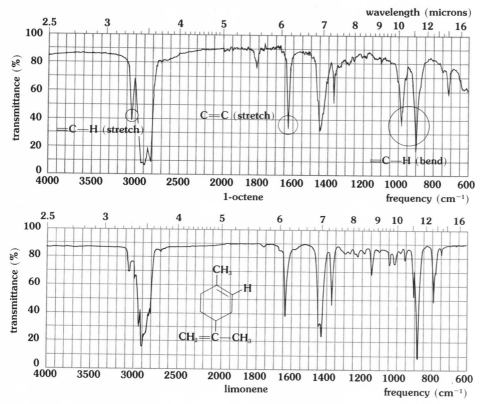

FIGURE 4·1 *Infrared spectra of 1-octene and limonene*

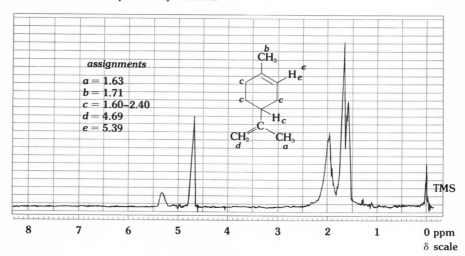

FIGURE 4·2 *Pmr spectrum of limonene*

assignments
a = 1.63
b = 1.71
c = 1.60–2.40
d = 4.69
e = 5.39

4·2 STRUCTURE AND PROPERTIES OF ALKENES

PROBLEM 4·1
The pmr spectra of α- and β-pinene are given below. Assign the proper spectrum to each isomer and indicate which hydrogen atoms give rise to the individual absorption signals.

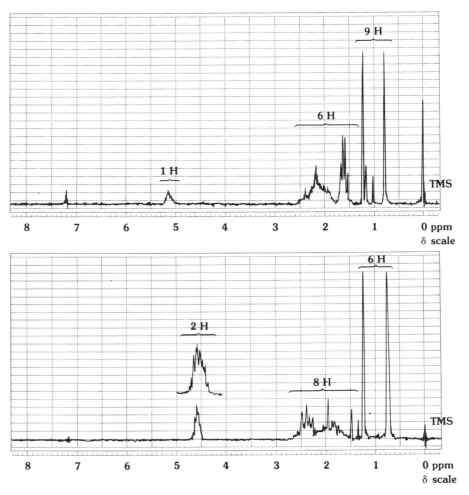

PROBLEM 4·2
a A C_6H_{12} hydrocarbon shows a single sharp peak in the pmr spectrum at $\delta = 1.7$. Write a structural formula for this compound. What isomer of this compound also exhibits a single sharp pmr signal?
b A C_7H_{12} hydrocarbon shows pmr signals at $\delta = 1.04$, 2.07, and 5.44, having relative intensities of 3:2:1. Write a structural formula for this compound.

STEREOISOMERISM OF ALKENES

Two isomers having the general bonding constitution of 2-butene have been isolated. These compounds are cis-trans stereoisomers, or geometric isomers, analogous to those observed in the cycloalkanes (see Figure 2.16). Indeed, if we view the double bond as a two-membered ring the analogy is complete. Some characteristic physical properties of these hydrocarbons are given in Table 4.3. Cis-trans isomerism also occurs in

4·2 STRUCTURE AND PROPERTIES OF ALKENES

TABLE 4·3 *Properties of stereoisomeric 2-butenes*

isomer	melting point (°C)	boiling point (°C)	infrared absorption (cm^{-1})
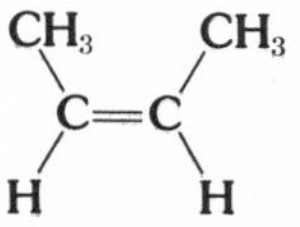 cis-2-butene (CH$_3$ and CH$_3$ on same side)	−139	3.7	1650 and 680
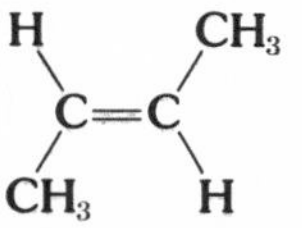 trans-2-butene	−106	0.9	1710 and 980

other alkenes. In branched hydrocarbons such as 3-methyl-2-pentene the cis-trans relationship ordinarily refers to the segment of the longest carbon chain which passes through the double bond:

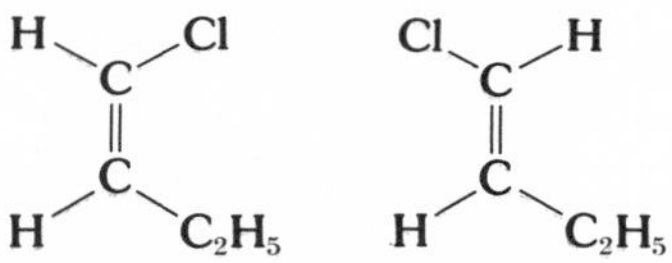

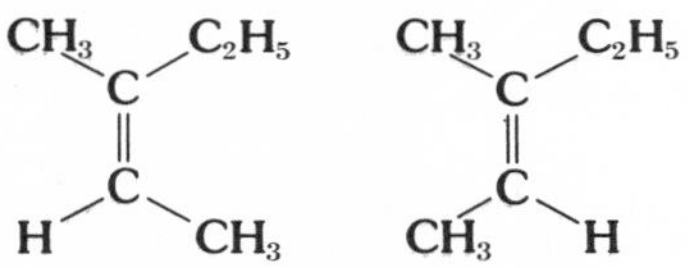

cis 1-chloro-1-butene   *trans* 1-chloro-1-butene   *cis* 3-methyl-2-pentene   *trans* 3-methyl-2-pentene

We see from these examples that only those alkenes in which each of the double-bonded carbon atoms is single-bonded to *two different substituents* can exhibit cis-trans stereoisomerism. Conversely, if one or both of the double-bonded carbon atoms bears identical single-bonded substituents, the alkene cannot show this stereoisomerism.

The properties of cis-trans double-bond isomers can exhibit striking and important differences. Natural rubber, for example, is a sticky elastic material consisting of high-molecular-weight (∼1,000,000) carbon chains characterized by a repeating five-carbon unit having a cis double bond. The corresponding trans isomer is a rather brittle, nonelastic substance called gutta-percha, which is used principally as an electrical insulator and in the manufacture of golf balls.

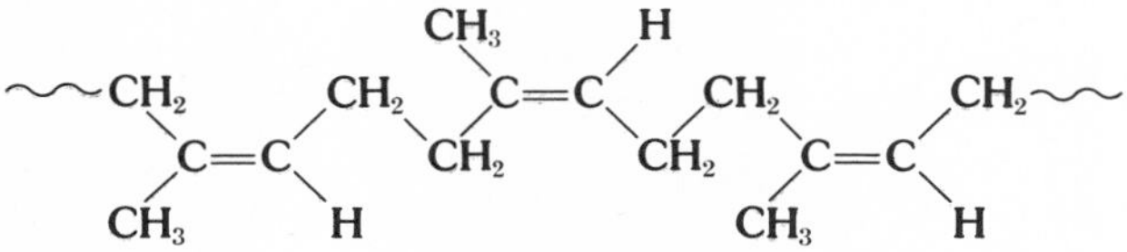

rubber

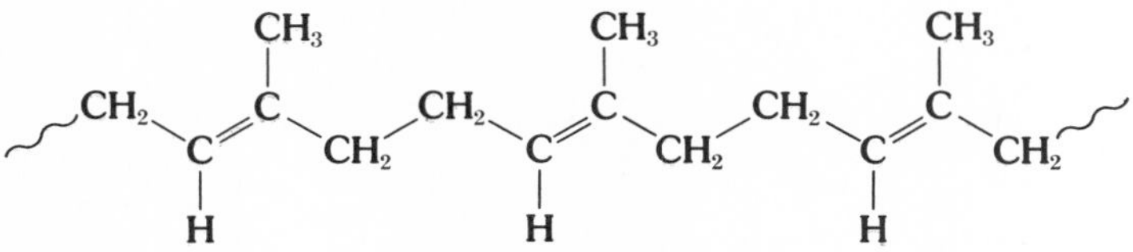

gutta-percha

4·2 STRUCTURE AND PROPERTIES OF ALKENES

Recent investigations into the nature of vision in animals have demonstrated that a light-induced cis-trans isomerization of substances related to vitamin A is fundamental to most visual systems.

PROBLEM 4·3
Which of the following alkenes are capable of stereoisomerism?
a $(CH_3)_3CCH=C(CH_3)_2$
b $CH_3CH=C(CH_3)C_2H_5$
c $CH_3CCl=CClCH_3$
d cyclohexane with =CHCH$_3$ substituent
e cyclobutane with CH$_3$ substituent and double bond

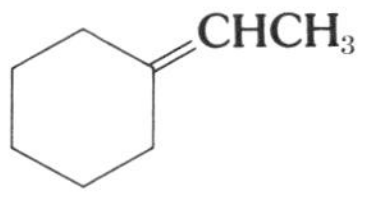

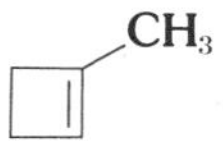

PROBLEM 4·4
How many stereoisomers are possible for each of the following structures?

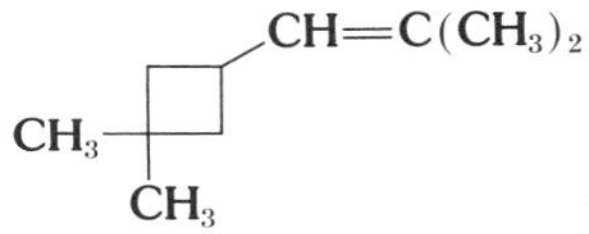

a cyclohexane with CH=CH$_2$ and Cl substituents

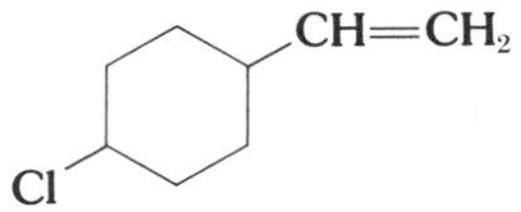

b cyclobutane with CH=C(CH$_3$)$_2$, CH$_3$, and CH$_3$ substituents

c cyclohexane with C(CH$_3$)=CHCH$_3$ and Cl substituents

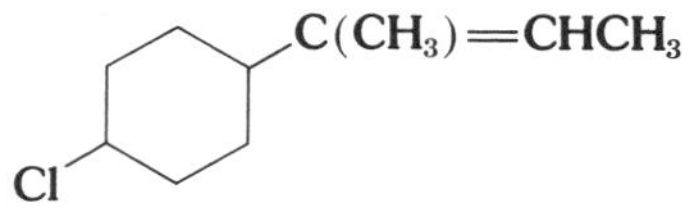

d $CH_3CH=CHCH_2CH=CHC_2H_5$

PROBLEM 4·5
Isomeric 1,2-dichloroethenes (CHCl=CHCl) have been isolated. One isomer has a dipole moment (a measure of the molecular dipole) of zero and the other has a strong dipole moment (μ = 1.85 D). Which isomer is cis and which is trans?

The very existence of stable cis-trans isomer pairs for certain alkenes forces us to conclude that twisting or rotation of the double-bonded carbon atoms with respect to each other must be severely restricted — that is, the energy barrier to rotation about a C=C bond is large. Experimental measurements confirm this interpretation and indicate that carbon-carbon double bonds have a rotational energy barrier of about 50 kcal/mole, in contrast to the barrier of approximately 3 kcal/mole observed for rotation about the single bond of ethane (see Figure 2.4). As a general rule, rotational barriers larger than 20 kcal/mole will permit the isolation of stereoisomers at normal temperatures, whereas smaller barriers allow rapid interconversion of rotational isomers at room temperature. These labile isomers are usually referred to as *conformational isomers (conformers)* or *rotamers*.

4·3 THE LCAO BONDING MODEL

The electron-pair repulsion model developed in Section 1.3 has adequately served to correlate our knowledge of molecular structure. However, when we turn our attention to the *chemical properties* of organic compounds, this model is not very helpful. For example, the alkenes undergo a variety of addition reactions which have no parallel in the chemistry of the alkanes or cycloalkanes (cyclopropane is an exception):

$$\begin{array}{c} \\ \diagdown \\ C=C \\ \diagup \end{array} \begin{array}{c} \xrightarrow{H_2SO_4} H-C-C-OSO_3H \\ \xrightarrow{Br_2} Br-C-C-Br \\ \xrightarrow{H_2} H-C-C-H \end{array} \qquad (4.3)$$

This contrast in the chemical reactivity of carbon-carbon double bonds (sharing of four electrons) in comparison with single bonds (sharing of two electrons) is not easily explained in terms of our electron-repulsion model.

If we could observe the course of a chemical reaction in slow motion, we would first notice a perturbation or shifting of the valence electrons as the reactant molecules approached each other. This polarization of the valence electrons would be followed immediately by a gradual movement of certain atomic nuclei, leading ultimately to the atomic configuration of the products. Clearly, then, a successful model for chemical reactivity must predict the distribution of bonding and nonbonding valence electrons as well as the orientation of the atomic nuclei.

One of the most widely used models for molecular structure and reactivity assumes that the bonding electrons occupy *molecular orbitals*, which can be approximated by a simple combination or mixing of atomic orbitals—that is, by a *linear combination of atomic orbitals (LCAO)*. The important atomic orbitals used in this mixing by period 1 and period 2 elements are illustrated in Figure 4.3, where the atomic nucleus is represented by the heavy dot at the origin of the arbitrary coordinate system. The s orbitals are spherically symmetrical and have the highest

FIGURE 4·3 *Electron-cloud representations of atomic orbitals*

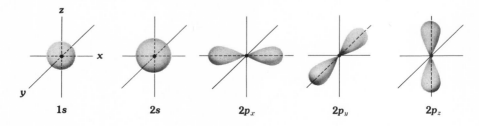

electron density at the nucleus. The three p orbitals are each cylindrically symmetrical and are directed in space at right angles to each other. Electron charge density in a p orbital is concentrated in lobes oriented on opposite sides of the nucleus, and is zero at the nucleus.

For a given atom the ground-state, or lowest-energy, electron configuration can be predicted by (1) filling the atomic orbitals in order of increasing potential energy, $1s < 2s < 2p_x = 2p_y = 2p_z$ (the aufbau principle); (2) limiting the capacity of each orbital to two electrons with opposite spins (the Pauli exclusion principle); and (3) assigning only one electron to orbitals of equal potential energy (for example, the p orbitals) until they are all half-filled (the Hund rule). The application of these rules to carbon and oxygen atoms results in the following electron configurations:

carbon: $1s$ (2), $2s$ (2), $2p_x$ (1), $2p_y$ (1)
oxygen: $1s$ (2), $2s$ (2), $2p_x$ (2), $2p_y$ (1), $2p_z$ (1)

SIGMA BONDING

The simplest illustration of the LCAO model is the hydrogen molecule H_2. As shown in Figure 4.4, the $1s$ atomic orbitals of two hydrogen atoms can interact so as to form *two* molecular orbitals—a lower-energy bonding orbital (the result of adding the atomic orbitals) and a higher-energy antibonding orbital (the result of subtracting the atomic orbitals). These two kinds of molecular orbitals are called *sigma* (σ) *orbitals* because the electron distribution in the orbitals is symmetrical about a line connecting the two nuclei, a characteristic that parallels the spherical s atomic-orbital distribution. Application of the aufbau principle indicates that the two electrons in the hydrogen molecule will occupy the lower-energy σ molecular orbital in much the same manner that two electrons occupy the $1s$ atomic orbital of a helium atom.

We can gain further insight into the LCAO model by considering the nonexistent He_2 molecule. Interaction of $1s$ orbitals of two helium atoms produces the same σ and σ^* orbitals described in Figure 4.4; however, in this case four valence electrons must be accommodated by these orbitals. Two paired electrons will occupy the bonding σ orbital, and the remaining electrons can be placed only in the antibonding σ^* orbital. Since the σ and σ^* orbitals are approximately equally spaced below and

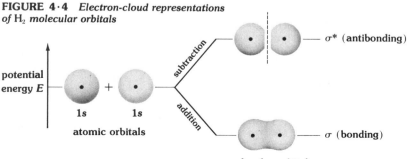

FIGURE 4·4 *Electron-cloud representations of H_2 molecular orbitals*

FIGURE 4·5 *sp³ hybrid atomic orbitals*

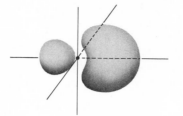

an *sp³* hybrid orbital

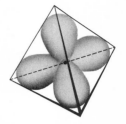

a tetrahedral grouping of four *sp³* hybrid orbitals

above the potential energy of the 1s atomic orbitals, the bonding and antibonding forces cancel each other.† The net result is that He_2 is unstable with respect to 2He.

The application of the LCAO model to molecules such as methane (CH_4) and ethylene (C_2H_4) is less straightforward than the hydrogen example. The chief problem here is that the valence atomic orbitals of carbon ($2s$, $2p_x$, $2p_y$, and $2p_z$) do not all have the same potential energy and therefore cannot individually combine with four hydrogen atoms to produce the four equivalent C—H bonds found in methane. In order to handle methane by the LCAO procedure, we must convert the $2s$, $2p_x$, $2p_y$, and $2p_z$ atomic orbitals of carbon to four equivalent *hybrid atomic orbitals,* each having 25% s and 75% p characteristics. The four orbitals formed by this hybridization process have a tetrahedral orientation and are referred to as *sp³ hybrid orbitals,* where the superscript 3 denotes the 3:1 ratio of p to s contributions. Figure 4.5 illustrates the tetrahedral grouping of *sp³* orbitals on a carbon atom.

According to an important general rule of LCAO theory, the combination of n atomic orbitals will necessarily generate n molecular orbitals. Therefore if the four *sp³* hybrid orbitals of carbon are combined with four hydrogen 1s orbitals, an LCAO treatment yields four equivalent bonding σ orbitals, as illustrated in Figure 4.6, and four antibonding σ^* orbitals (not shown). Since the eight valence electrons in methane are nicely accommodated in the four σ orbitals, the LCAO model confirms the stability and geometry of CH_4.

†Actually the σ^* molecular orbital has a slightly higher relative energy.

FIGURE 4·6 *Molecular orbitals in methane*

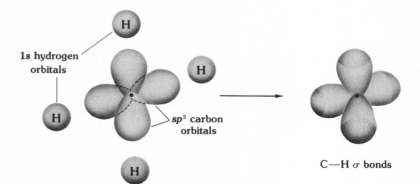

1s hydrogen orbitals

sp³ carbon orbitals

C—H σ bonds

4·3 THE LCAO BONDING MODEL

Several possible combinations of atomic orbitals can lead to σ molecular orbitals, as illustrated by Figures 4.4, 4.6, and 4.7. These molecular orbitals share a common cylindrical symmetry, and their basic role in covalent bonding is indicated by the seven σ bonds which hold together the eight atoms in ethane (C_2H_6). Indeed, the lines drawn between atoms in structural formulas most often correspond to σ bonds.

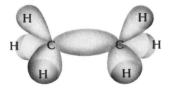

σ bonds in ethane

PI BONDING

Two LCAO models have been proposed for the carbon-carbon double bond in ethylene (C_2H_4) and other alkenes. Since each of the two carbon atoms in C_2H_4 is bonded to three other atoms (the other carbon and two hydrogen atoms), a minimum of five σ molecular orbitals are necessary to bond the six atoms together. The first LCAO model therefore requires the formation of three equivalent hybrid atomic orbitals for each carbon

FIGURE 4·7 *Formation of σ orbitals by LCAO*

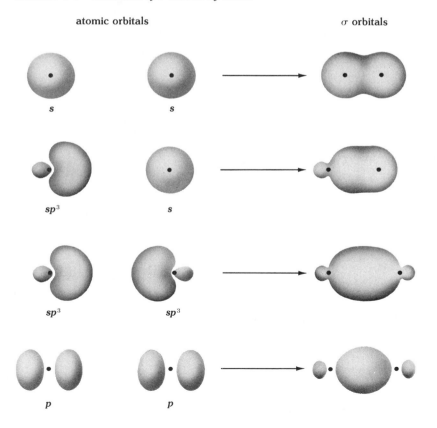

4·3 THE LCAO BONDING MODEL 119

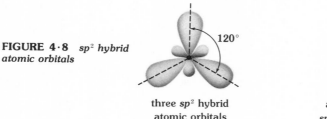

FIGURE 4·8 *sp² hybrid atomic orbitals*

three *sp²* hybrid atomic orbitals

atomic orbitals of an *sp²* hybrid carbon atom

atom: two of these will generate the C—C σ orbital and the other four will combine with the 1s orbitals of the hydrogen atoms to give four C—H σ orbitals. The required hybrid carbon orbitals are generated by mixing the 2s orbital with two of the 2p atomic orbitals. The resulting *sp² hybrid orbitals* (33.3% s and 66.7% p) have a trigonal orientation at right angles to the remaining $2p_z$ orbital, as shown in Figure 4.8.

PROBLEM 4·6
Why do we not use the three p orbitals ($2p_x$, $2p_y$, and $2p_z$) to form the three equivalent hybrid orbitals on carbon?

An LCAO treatment of two *sp²* hybridized carbon atoms and four hydrogen atoms gives the following σ-bond framework for ethylene:

σ bonds in ethylene

If we pause to add up the valence electrons and orbitals used at this stage, we find a 2p orbital and an electron remaining on each carbon atom (covalent sharing gives each carbon atom seven electrons). When these two p orbitals are parallel, they can overlap and interact as shown in Figure 4.9 to give two new molecular orbitals, oriented above and below the σ-bond framework in a manner reminiscent of p atomic

FIGURE 4·9 *Formation of π molecular orbitals*

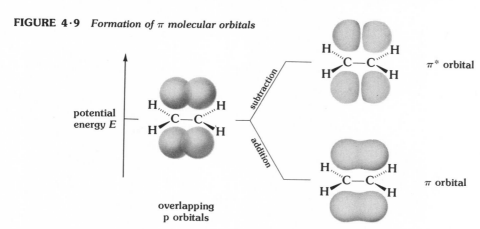

orbitals. Molecular orbitals of this kind are called *pi (π) orbitals*. The remaining two electrons in ethylene occupy the bonding π orbital and thus increase the carbon-carbon bonding energy of the double bond (the antibonding π* orbital remains empty). This additional sharing of electrons completes the valence-shell octet for each carbon atom in the double bond. To summarize, our first LCAO model for the carbon-carbon double bond consists of a *σ bond and a π bond* between the two carbon atoms.

If one of the carbon atoms of the double bond is twisted with respect to the other carbon atom, the overlap and resulting interaction of the *p* orbitals decreases, and the additional bonding energy of the π bond is lost.

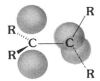

a twisted alkene

This bonding model thus provides a simple explanation of the high rotational-energy barrier for carbon-carbon double bonds. The characteristic chemical reactions of alkenes are believed to involve the polarizable high-electron-density region of the π bond, a feature we shall discuss further in Chapter 5.

TAU BONDING

The second LCAO model for ethylene combines two sp^3-hybridized carbon atoms with four hydrogen atoms to give the *bent-bond* model shown in Figure 4.10. A rehybridization of the carbon atoms improves the overlap of atomic orbitals in the bent bonds and also causes a widening of the H—C—H bond angle. The curved bonding molecular orbitals that are generated in this manner are called *tau (τ) orbitals*, but are more commonly referred to as "banana bonds."

A comparison of the important characteristics predicted by each of the two LCAO models shows many similarities (see Table 4.4). Most organic

TABLE 4·4 *Double-bond characteristics derived from LCAO models*

parameter	σ-π model	bent-bond model
molecular geometry	planar	planar
C=C bond length	shorter than single bond	shorter than single bond
C=C bond strength	stronger than single bond	not predicted
rotational barrier	large	large
high electron density above and below plane of molecule	yes	yes

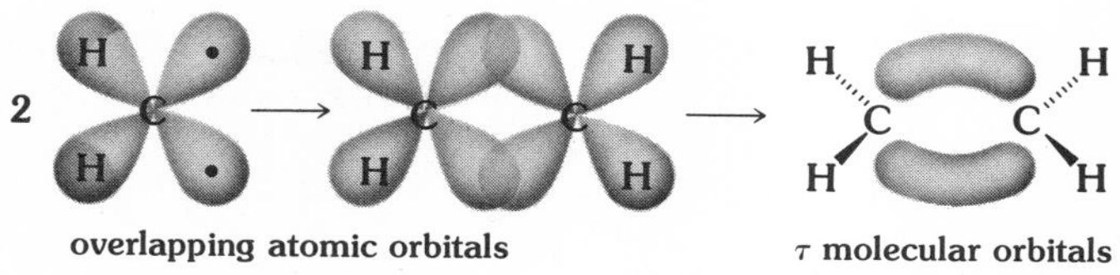

FIGURE 4·10 *Formation of τ molecular orbitals*

overlapping atomic orbitals   τ molecular orbitals

chemists prefer to use the σ-π model, which has a distinct advantage in describing systems containing several conjugated double bonds.

PROBLEM 4·7
How can the hybridization of an sp^3 carbon atom be changed to improve the overlap of atomic orbitals in the bent-bond model? *Hint:* It is convenient to consider this kind of question in terms of percentage of *s* and *p* contribution to the hybrid orbitals.

4·4 STRUCTURE AND STABILITY IN ALKENES

HEATS OF HYDROGENATION

Since alkenes (and cycloalkenes) do not contain the maximum number of hydrogen atoms commensurate with their carbon skeleton (C_nH_{2n+2} for noncyclic compounds), they are frequently called *unsaturated hydrocarbons,* in contrast to the *saturated* alkanes. In fact, bond-energy calculations indicate that the addition of molecular hydrogen to alkenes should proceed exothermically to give alkanes:

$$\text{C=C} + H_2 \longrightarrow -\overset{H}{\underset{|}{C}}-\overset{H}{\underset{|}{C}}- \qquad (4.4)$$

The bonds broken in the reactants are H—H and the π bond of C=C:†

$\Sigma_r = 104.2 + 63.2 = 167.4$ kcal/mole

The bonds formed in the products are 2C—H:

$\Sigma_p = 2 \times 98.7 = 197.4$ kcal/mole

Hence
$\Delta H = \Sigma_r - \Sigma_p = 167.4 - 197.4 = -30$ kcal/mole

Although the mixture of hydrogen and an alkene in reaction 4.4 is thermodynamically unstable in comparison to the corresponding alkane, this mixture is chemically inert in the absence of a catalyst. For the reaction to proceed, a transition-metal catalyst (**Pt, Pd, Ni,** etc.) must be added to the reaction mixture. As shown in Figure 4.11, the catalyst functions by providing an alternative, low-activation-energy reaction path and thus increases the rate of reaction without changing the heat of reaction ΔH. The experimentally determined heats of reaction for hydrogen addition to a variety of alkenes have been found to range from

†Calculated by subtracting the C—C bond energy of 82.6 from the C=C bond energy of 145.8 listed in Table 2.3.

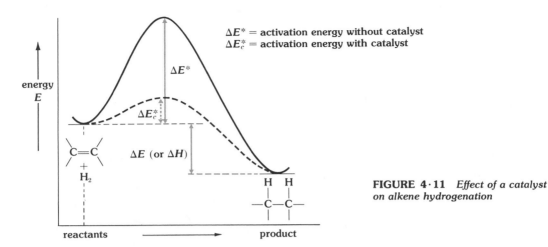

FIGURE 4·11 *Effect of a catalyst on alkene hydrogenation*

−26 to −33 kcal/mole (see Table 4.5) and are thus in good agreement with the calculated value.

THERMODYNAMIC STABILITY

From the data in Table 4.5 we see that the absolute value of the heat of hydrogenation for an alkene becomes smaller as the number of alkyl-group substituents on the double bond increases. This suggests that alkyl substituents increase the thermodynamic stability of a double bond. The validity of this conclusion can be demonstrated by comparing the heats of hydrogenation of the isomeric 1- and 2-butenes:

$$CH_2=CHCH_2CH_3 \xrightarrow{H_2,\ Pt}$$
1-butene
$$CH_3CH=CHCH_3 \xrightarrow{H_2,\ Pt} CH_3CH_2CH_2CH_3$$
2-butene

As illustrated in Figure 4.12, the additional methyl substituent in 2-butene stabilizes the double bond by 1.7 to 2.7 kcal/mole, depending on whether it is located cis or trans to the other alkyl group. The manner in which alkyl substituents effect this stabilization is not yet well understood.

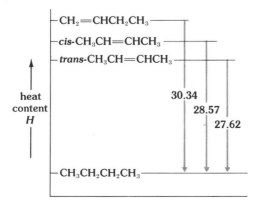

FIGURE 4·12 *Heats of hydrogenation for isomeric butenes*

4·4 STRUCTURE AND STABILITY IN ALKENES

TABLE 4·5 *Heats of hydrogenation of alkenes (as gases at 82°C)*

alkene	ΔH (kcal/mole)	alkene	ΔH (kcal/mole)
$CH_2{=}CH_2$	−32.82	$(CH_3)_2C{=}CHCH_3$	−26.92
$CH_3CH_2CH{=}CH_2$	−30.34	$(CH_3)_2C{=}C(CH_3)_2$	−26.63
cis-$CH_3CH{=}CHCH_3$	−28.57	cyclopentene	−26.92
trans-$CH_3CH{=}CHCH_3$	−27.62		
$(CH_3)_2C{=}CH_2$	−28.39	cyclohexene	−28.59

PROBLEM 4·8

Why do the isomeric 1- and 2-butenes provide better evidence for the stabilizing influence of alkyl substituents on double bonds than a comparison of ethylene and 2-methylpropene?

PROBLEM 4·9

Heat-of-combustion values for the isomeric butenes are:

	$\Delta H^{25°}$ (kcal/mole)
1-butene	−649.8
cis-2-butene	−648.1
trans-2-butene	−647.1
2-methylpropene	−646.1

Show how these values can be used to confirm the stabilizing effect of alkyl substituents and the order of cis versus trans stability.

Another important fact that is apparent from hydrogenation data is that trans disubstituted alkenes are generally more stable than the cis stereoisomers (Figure 4.12). This difference seems to be due largely to the nonbonded repulsive interactions that occur when two alkyl groups are located on the same side of the double bond. In support of this explanation, we find that replacing the methyl groups in 2-butene by the larger *t*-butyl groups increases the thermodynamic stability of the trans isomer over the cis isomer to 9.3 kcal/mole (Figure 4.13). The heats of hydrogenation also confirm that this difference is due to a decrease in the stability of the cis isomer rather than increased stabilization of the trans isomer.

FIGURE 4·13 *Crowding in stereoisomeric 2,2,5,5-tetramethyl-3-hexenes*

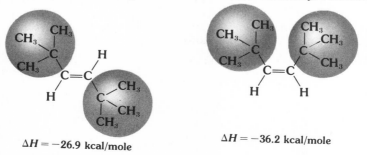

$\Delta H = -26.9$ kcal/mole $\Delta H = -36.2$ kcal/mole

The normal order of stability for alkene stereoisomers is reversed for small and medium-size cycloalkenes. In fact, the trans isomers of all cycloalkenes smaller than cyclooctene are so unstable they cannot be isolated at ordinary temperatures. The reason for this difference in stability is apparent if we consider the size of the carbon chain necessary to bridge the terminal carbon atoms of a *trans*-2-butene moiety. With the aid of molecular models such as those in Figure 2.2, we find that the distance to be bridged (about 4 Å) requires a chain of at least four carbon atoms ($n = 4$), or there would be excessive twisting of the double bond. The fact that trans stereoisomers of cyclooctene and all larger cycloalkenes can be isolated supports this simple explanation.

a trans cycloalkene

The stability order that has been observed for stereoisomers of cycloalkenes larger than cyclodecene usually favors the trans isomer. For example, heats of hydrogenation show that *trans*-cycloundecene ($C_{11}H_{20}$) is about 0.2 kcal/mole more stable than the cis isomer. However, cyclooctene, cyclononene, and cyclodecene are exceptional in that the cis isomers have smaller heats of hydrogenation and are therefore more stable than the corresponding trans isomers. We can attribute the greater stability of the cis isomer in these cases to differences in angle strain, eclipsing strain, and nonbonded repulsions (see Table 2.8) between the stereoisomers. The more flexible cis molecules can assume low-energy conformations more readily than can the relatively constrained trans molecules; this constraint is most severe for short bridging chains of four, five, or six carbon atoms.

SUMMARY

alkenes Aliphatic hydrocarbons containing one or more carbon-carbon double bonds, identified in IUPAC nomenclature by the ending *-ene* (see Table 4.1); also known as *olefins*. Such compounds are termed *unsaturated*, since they have a smaller hydrogen-to-carbon ratio than the corresponding alkanes. The carbon-carbon double bond is shorter (1.34 Å) and stronger than a corresponding single bond and has a planar configuration. Characteristic spectroscopic properties are listed in Table 4.2.

cycloalkenes Cyclic hydrocarbons containing one or more carbon-carbon double bonds.

stereoisomeric alkenes Cis-trans isomer pairs formed by unconstrained alkenes in which each carbon atom of the double bond bears different substituent atoms or groups.

LCAO model A model for chemical bonding in which molecular orbitals are described by a linear combination of atomic orbitals (LCAO). The number of molecular orbitals formed in this way is equal to the number of atomic orbitals used in the combination.

sigma (σ) molecular orbital A low-energy molecular orbital having cylindrical symmetry about the bond axis; the molecular-orbital equivalent of an s atomic orbital.

pi (π) molecular orbital A higher-energy molecular orbital (relative to σ) having a nodal plane passing through the nuclei encompassed by the orbital; the molecular-orbital equivalent of a p atomic orbital.

thermodynamic stability A qualitative term referring to the relative potential energy of a molecular system; the thermodynamic stability of alkenes generally increases with the number of alkyl substituents attached to the carbon-carbon double bond.

chemical stability A qualitative term inversely related to the relative ease of chemical transformation of a substance; compounds which are comparatively unreactive are said to be chemically stable or *inert*. Alkanes have greater chemical stability than alkenes because of the reactive π bond in alkenes.

catalyst A substance which increases the rate of a reaction by interacting with one or more of the reactants to provide an alternative low-activation-energy pathway to the products (the catalyst itself is not consumed or changed). Transition metals such as Pt, Pd, Ni, and Rh catalyze the addition of hydrogen to alkenes.

EXERCISES

4·1 Name the following alkenes:

a $CH_3CH(CH_3)CH=CH_2$

b (1-methylcyclohexene structure)

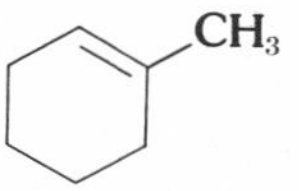

c $CH_2=CHCHClCH_3$

d (cyclopentadiene structure)

e (cyclohexyl-CH₂CH=CH₂)

f $CH_3(CH_3CH_2)C=C(CH_2CH_3)(Cl)$ — with CH_3, CH_2CH_3 above and CH_2CH_3, Cl below

g (6-chloro-6-methylcyclohex-2-ene structure)

h (1,3-cyclohexadiene structure)

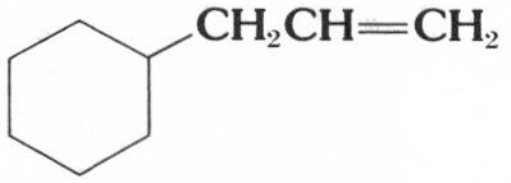

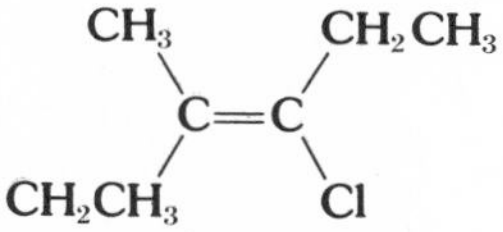

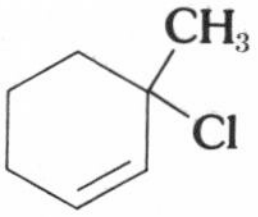

i (1,4-dimethylcyclopentene structure)

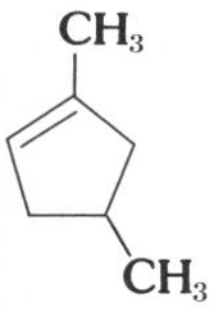

4·2 Draw and name the four possible noncyclic isomers of C_4H_8.

4·3 Draw condensed structural formulas for:
a 1-Methyl-4-ethylcyclohexene
b *trans*-3,5-Dimethyl-2-hexene

4·4 Can a single structure be unambiguously assigned to each of the following names? If not, draw all possible structures corresponding to each name given.
a 2-Methyl-2-pentene
b 3-Methylpentene
c 3,4-Diethylcyclopentene
d 3,4-Dimethyl-3-hexene

4·5 Sketch an approximate pmr spectrum showing chemical shifts (δ values) and relative areas of different signals for each of the following compounds:

a (cyclohexene structure)

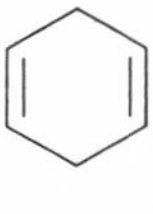

b $(CH_3)_3CC(CH_3)=C(CH_3)C(CH_3)_3$

c (1,1,2,2-tetramethylcyclopentane-like structure with four CH_3 groups)

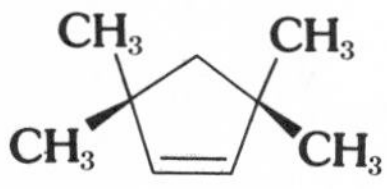

d $(CH_3)_2C=CH-CH=C(CH_3)_2$

126 EXERCISES

4·6 Using simple LCAO molecular-orbital theory, arrange the following molecules in the expected order of decreasing stability:

He_2 H_2 $He_2^{\oplus}$ $H_2^{\oplus}$

4·7 There are three isomeric pentenes (C_5H_{10}) that yield 2-methylbutane on catalytic hydrogenation. Write structural formulas for these three isomers and indicate their relative stabilities based on the ΔH values in Table 4.5.

4·8 What alkyl halide would yield each of the following pure alkenes via an elimination reaction?
a 1-Pentene
b 2-Pentene
c 2-Methyl-1-butene
d 2-Methyl-2-butene
e 3-Methyl-1-butene

4·9 How could you distinguish cyclopentene from 1-pentene by means of pmr spectroscopy?

4·10 Given the heats of hydrogenation below for isomers A and B, would you say that an exocyclic double bond (A) or an endocyclic one (B) is more stable?

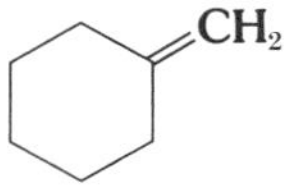

$\Delta H = -27.4$ kcal/mole
isomer A

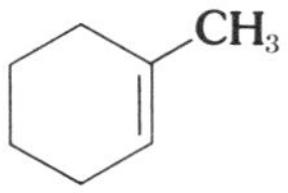

$\Delta H = -25.4$ kcal/mole
isomer B

4·11 Consider the interconversion of cis and trans isomers of 2-butene and of 1,2-dimethylcyclopentane. What must take place in each case for this interconversion? Which would require the most energy? Explain.

4·12 A compound with the molecular formula C_6H_{12} decolorizes a solution of Br_2 in CCl_4 and displays pmr signals at $\delta = 1.0, 1.7, 2.25$, and 4.7, with relative areas of 6:3:1:2. Hydrogenation of this compound gives 2,3-dimethylbutane as the only product. What is the compound?

4·13 The σ-π bonding model for carbon-carbon double bonds (Figure 4.9) accounts more satisfactorily for the structure of alkenes than the simple valence-orbital repulsion model discussed in Chapter 1. Give a specific example of the superiority of this model in explaining an aspect of alkene structure.

4·14 Calculate the heat of hydrogenation of cyclopropene from the following heats of combustion:

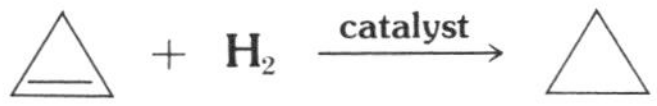

△ + H_2 $\xrightarrow{\text{catalyst}}$ △

$\Delta H^{25°}$ (kcal/mole)

cyclopropane -468.6
cyclopropene -464.6
hydrogen (H_2) -57.7

REACTIONS OF ALKENES

THE CHEMICAL reactions most characteristic of the alkenes take place at the carbon-carbon double-bond function. In general, we can regard the initial step in these reactions to be the attack of an electrophilic species on the nucleophilic π electron cloud of the double bond. Subsequent steps usually lead to addition products, but other kinds of transformations, such as bond cleavage, are also observed.

5·1 ADDITION OF BRØNSTED ACIDS TO ALKENES

Strong Brønsted acids, compounds which are proton donors, react rapidly and exothermically with the carbon-carbon double bonds of alkenes, generally giving addition products:

$$\text{C=C} + \text{H—X} \longrightarrow \text{H—C—C—X} \qquad \Delta H = -13 \text{ to } -16 \text{ kcal/mole} \qquad (5.1)$$
$$(\text{X = Cl, Br, I})$$

$$\text{C=C} + \text{H}_2\text{SO}_4 \longrightarrow \text{H—C—C—OSO}_3\text{H} \qquad \Delta H \approx -10 \text{ kcal/mole} \qquad (5.2)$$

The alkyl sulfates formed by the addition of sulfuric acid to alkenes are oily viscous liquids. It is because of this well-known property that the alkenes are sometimes referred to as *olefins* (oil-forming substances).

Weak acids such as water and alcohols (ROH) will react with alkenes provided a strong acid is present as a catalyst. Certain enzyme systems can also catalyze specific hydration reactions.

5·1 ADDITION OF BRØNSTED ACIDS TO ALKENES

$$\text{C=C} + H_3O^{\oplus} \xrightarrow{\text{dilute sulfuric acid}} H-\overset{|}{C}-\overset{|}{C}-OH \qquad \Delta H \approx -10 \text{ kcal/mole} \qquad (5.3)$$

$$\underset{\text{aconitic acid}}{\overset{HCO_2H}{\underset{HO_2CCH_2CO_2H}{\overset{C}{\underset{C}{\|}}}}} \xrightarrow[H_2O]{\text{aconitate hydratase}} \underset{\text{citric acid}}{HO-\underset{CH_2CO_2H}{\overset{CH_2CO_2H}{|}}-CO_2H} + \underset{\text{isocitric acid}}{H-\underset{CH_2CO_2H}{\overset{HOCHCO_2H}{|}}-CO_2H} \qquad (5.4)$$

PROBLEM 5·1
Suggest a method for purifying a sample of cyclohexane which contains a small amount of cyclohexene as a contaminant.

MARKOVNIKOV'S RULE

If an alkene reactant has a symmetrically located double bond, as in *cis*- or *trans*-2-butene, only one addition product is possible:

$$\begin{array}{c} \overset{H\rightarrowtail X}{\underset{\downarrow\downarrow}{CH_3CH=CHCH_3}} \\ \\ \overset{X\rightarrowtail H}{\underset{\downarrow\downarrow}{CH_3CH=CHCH_3}} \end{array} \searrow CH_3CH_2CHXCH_3 \qquad (5.5)$$

However, unsymmetrically substituted alkenes, such as propene and 2-methyl-2-butene, are open to two different modes of addition:

$$\begin{array}{c} \overset{H\rightarrowtail X}{\underset{\downarrow\downarrow}{CH_3CH=CH_2}} \longrightarrow CH_3CH_2CH_2X \\ \\ \overset{X\rightarrowtail H}{\underset{\downarrow\downarrow}{CH_3CH=CH_2}} \longrightarrow \underset{\text{major product}}{CH_3CHXCH_3} \end{array} \qquad (5.6)$$

$$\begin{array}{c} \overset{H\rightarrowtail X}{\underset{\downarrow\downarrow}{(CH_3)_2C=CHCH_3}} \longrightarrow (CH_3)_2CHCHXCH_3 \\ \\ \overset{X\rightarrowtail H}{\underset{\downarrow\downarrow}{(CH_3)_2C=CHCH_3}} \longrightarrow \underset{\text{major product}}{(CH_3)_2CXCH_2CH_3} \end{array} \qquad (5.7)$$

When these addition reactions are studied in the laboratory, a striking and important fact emerges: *The addition of Brønsted acids to unsymmetrically substituted double bonds gives almost exclusively that product in which the hydrogen atom from the acid is located at the least substituted carbon atom of the double bond.* This useful generalization is known as *Markovnikov's rule* in honor of the Russian chemist who first noted the correlation. Markovnikov's rule may be restated in several ways, one common phrasing being: *When **HX** is added to an unsymmetrical alkene, the hydrogen atom becomes bonded to that carbon atom of the double bond which already bears the most hydrogen atoms.* Or in homelier vernacular, "Them that has gits."

PROBLEM 5·2
How can the two possible products from the addition of HCl to 2-methylpropene be distinguished?

MECHANISM FOR ADDITION REACTIONS

Empirical rules play an important role in the evolution of a subject from an art into a science, but they must eventually be replaced by rational explanations of the phenomena involved. We can do this for Markovnikov's rule by considering the acid-base nature of these reactions. According to Brønsted-Lowry theory, a carbon-carbon double bond can serve as a base with respect to the acid **HX**:

$$\underset{\text{base}}{\diagdown\!\!\text{C}\!\!=\!\!\text{C}\!\!\diagup} + \underset{\text{acid}}{\text{HX}} \longrightarrow \underset{\text{acid}}{\left[-\overset{\text{H}}{\underset{|}{\text{C}}}-\overset{\oplus}{\text{C}}\diagdown \right]} + \underset{\text{base}}{:\!\ddot{\text{X}}\!:^{\ominus}} \qquad (5.8)$$

The resulting carbonium-ion intermediate is relatively unstable and rapidly undergoes further reactions, acting either as a Brønsted acid (see equation 3.21) or as a Lewis acid:

$$\left[-\overset{\text{H}}{\underset{|}{\text{C}}}-\overset{\oplus}{\text{C}}\diagdown \right] + :\!\ddot{\text{X}}\!:^{\ominus} \xrightarrow{\text{fast}} -\overset{\text{H}}{\underset{|}{\text{C}}}-\overset{\text{X}}{\underset{|}{\text{C}}}- \qquad (5.9)$$

From the energy diagram for this two-step addition mechanism (Figure 5.1) we see that the slow, or rate-determining, step is also the product-determining step; consequently, it is here that we must look for a rationale of Markovnikov's rule.

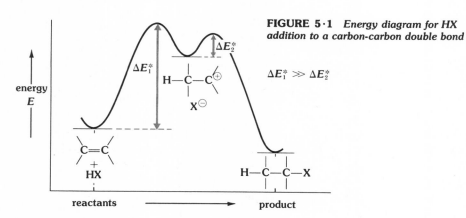

FIGURE 5·1 *Energy diagram for HX addition to a carbon-carbon double bond*

$\Delta E_1^* \gg \Delta E_2^*$

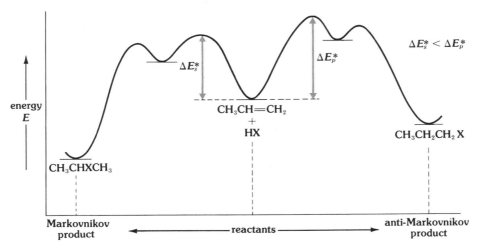

FIGURE 5·2 *Addition of* HX *to an unsymmetrical alkene*

An unsymmetrical alkene such as propene can undergo protonation at either of two sites, thereby generating two different carbonium-ion intermediates:

$$\left[\begin{array}{c} H \\ | \\ CH_3-\overset{\oplus}{\underset{|}{C}}-CH_2 \\ H \end{array} \right] X^{\ominus} \longleftarrow \begin{array}{c} CH_3CH=CH_2 \\ + \\ HX \end{array} \longrightarrow \left[\begin{array}{c} H \\ | \\ CH_3\overset{}{C}H-\overset{\oplus}{C}H_2 \end{array} \right] X^{\ominus}$$

(5.10)

From the carbonium-ion stability order in Figure 3.10 and the Hammond postulate (Section 3.7), we predict that the more highly substituted carbonium ion will be formed more rapidly ($\Delta E_s^* < E_p^*$ in Figure 5.2), and that the product derived from this intermediate will predominate. The agreement of this conclusion with Markovnikov's rule supports the mechanism suggested by equations 5.8 and 5.9.

TESTING THE MECHANISM

Chemists generally take a skeptical attitude toward a new theory and delight in devising experiments to test its assumptions and predictions. In keeping with this philosophy, let us take a critical look at our mechanism for the addition of Brønsted acids to carbon-carbon double bonds. If a carbonium ion is really formed in the rate-determining step (equation 5.8), the chemistry displayed by these intermediates, as discussed in Chapter 3, would lead us to predict the following behavior for such reactions:

1 The strongly electrophilic carbonium ion will react rapidly with any nucleophilic species present in the reaction mixture. If more than one nucleophile is present, we can expect to find products derived from each of them.

2 Carbonium-ion rearrangements may occur if a primary or secondary

carbonium ion is formed adjacent to a fully branched quaternary carbon atom (as in the hydrolysis of neopentyl bromide, described in equation 3.12).

3 Electron-donating substituents (such as alkyl groups) attached to the double bond will increase the basicity of this function and will therefore increase the speed of the slow step. Conversely, electron-withdrawing substituents will reduce the base strength of the olefin and consequently will retard the rate of carbonium-ion formation.

4 Strong acids will react faster than weak acids.

These predictions have been borne out by experimental evidence obtained from a variety of addition reactions:

$$CH_3CH=CH_2 + HCl + CH_3OH \longrightarrow CH_3\overset{Cl}{\underset{|}{C}}HCH_3 + CH_3\overset{OCH_3}{\underset{|}{C}}HCH_3 \qquad (5.11)$$

$$CH_3-\underset{\underset{CH_3}{|}}{\overset{\overset{CH_3}{|}}{C}}-CH=CH_2 + HCl \longrightarrow (CH_3)_2\overset{Cl}{\underset{|}{C}}CH(CH_3)_2 + (CH_3)_3CCHClCH_3 \qquad (5.12)$$
$$ 35\% 65\%$$

$$CH_3CH=CH_2 + H_3O^{\oplus} \xrightarrow{k_x} CH_3CH(OH)CH_3$$
$$(29.6\% \text{ HClO}_4)$$

$$(CH_3)_2C=CH_2 + H_3O^{\oplus} \xrightarrow{k_y} (CH_3)_3COH \qquad \frac{k_y}{k_x} = 8{,}000 \qquad (5.13)$$
$$(29.6\% \text{ HClO}_4)$$

PROBLEM 5·3

a What effect would an increase in chloride-ion concentration have on the product ratio in reaction 5.11?

b Explain the fact that $F_3CCH=CH_2$ reacts far more sluggishly with HCl than does propene and gives $F_3CCH_2CH_2Cl$ as the major product.

PROBLEM 5·4

a Write a detailed reaction mechanism to account for the products in reaction 5.12.

b The addition of HCl to α-pinene proceeds as shown below. Suggest an explanation for this rearrangement.

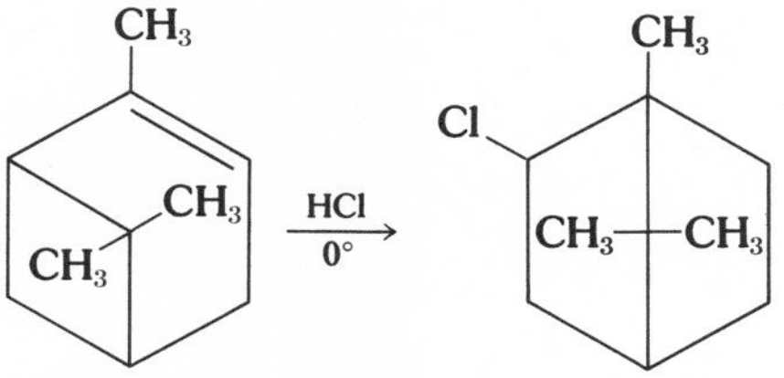

α-pinene $\xrightarrow[0°]{\text{HCl}}$ bornyl chloride

PROBLEM 5·5

The addition of HCl to ethene (C_2H_4) proceeds very slowly in the gas phase. Suggest an explanation for this fact in terms of possible ionic and free-radical mechanisms.

5·2 ADDITION OF LEWIS ACIDS TO ALKENES

From our knowledge of Brønsted acid reactions with alkenes, we might anticipate that reagents which function as Lewis acids (electrophiles) will also attack the nucleophilic carbon-carbon double bond. Such reactions in fact do occur:

$$CH_3CH_2CH=CH_2 + Br_2 \xrightarrow{CCl_4} CH_3CH_2CHBrCH_2Br \tag{5.14}$$

$$(CH_3)_2C=CH_2 + HOCl \xrightarrow{H_2O} (CH_3)_2C(OH)CH_2Cl \tag{5.15}$$

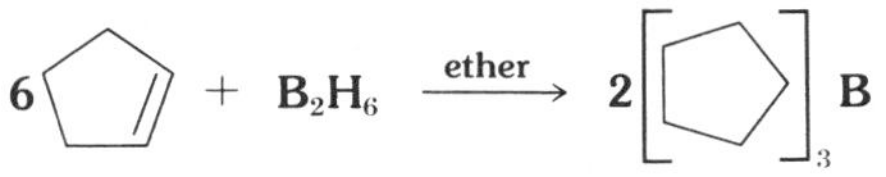

$$6\, \triangle\!\!\!\!/ \;\; + \; B_2H_6 \xrightarrow{ether} 2\,[\triangle\!\!\!\!/\,]_3 B \tag{5.16}$$

HALOGEN ADDITION

The electrophilic character of the halogens can be seen in the weakly bonded complexes they form with basic nitrogen and oxygen atoms.

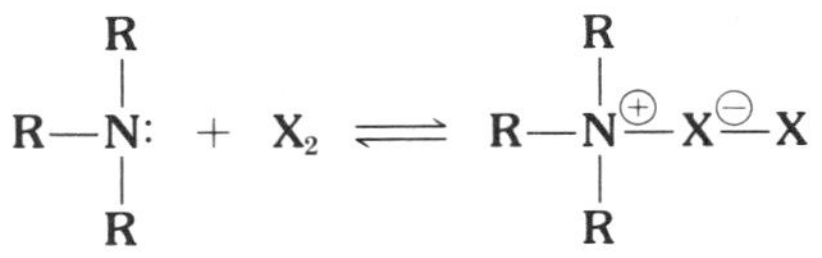

$$R_3N: + X_2 \rightleftharpoons R_3N^{\oplus}\!\!-\!\!X^{\ominus}\!\!-\!\!X \tag{5.17}$$

This interaction accounts for the fact that solutions of bromine or iodine in nonbasic solvents such as pentane and carbon tetrachloride differ in color from solutions in water, alcohols, or amines.

In contrast to the uncontrollably violent reaction of fluorine with most alkenes, chlorine and bromine add smoothly, in most cases giving excellent yields of *vicinal* dihaloalkanes (see reaction 5.14).† In fact, the addition of bromine may be used as a qualitative (and quantitative) test for unsaturation, since the addition products are colorless, whereas dilute bromine solutions have a distinct reddish-brown color. Chlorine addition is sometimes accompanied by substitution, particularly when highly branched alkenes are involved.

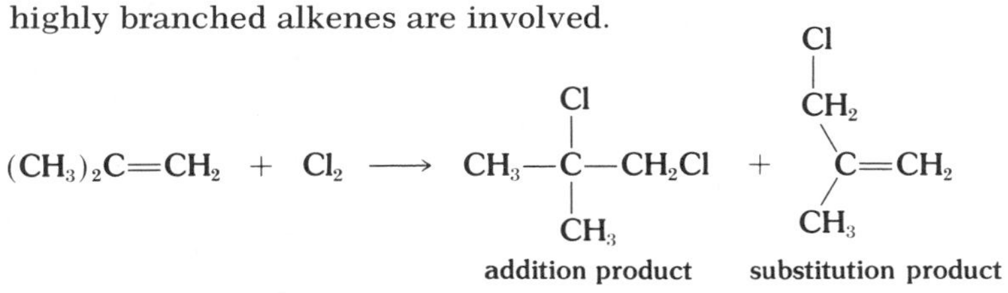

$$(CH_3)_2C=CH_2 + Cl_2 \longrightarrow \underset{\text{addition product}}{CH_3\!-\!\underset{CH_3}{\overset{Cl}{\underset{|}{C}}}\!-\!CH_2Cl} + \underset{\text{substitution product}}{\underset{CH_3}{\overset{ClCH_2}{C}}\!\!=\!\!CH_2} \tag{5.18}$$

Iodine also adds to double bonds, but the vicinal diiodide products, unlike the corresponding dichlorides and dibromides, are not ordinarily favored at equilibrium:

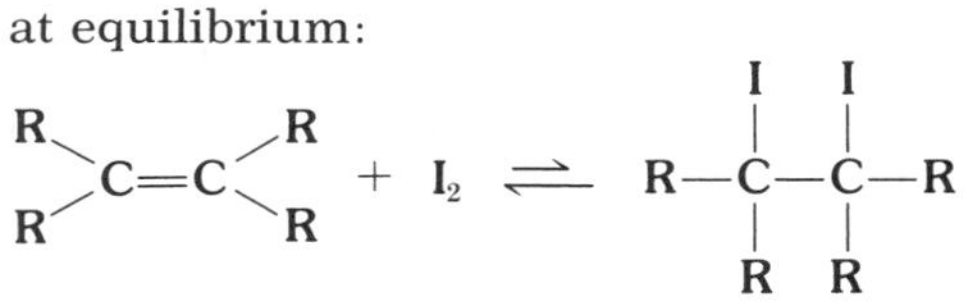

$$R_2C=CR_2 + I_2 \rightleftharpoons R_2C(I)\!-\!C(I)R_2 \tag{5.19}$$

†The term *vicinal* is from the Latin word *vicinalis*, meaning neighboring, and refers to juxtaposed functional groups. X and Y are *vicinal* in $R_2C(X)\!-\!C(Y)R_2$.

PROBLEM 5·6

Is 1,2-dibromohexane best prepared by direct bromination of hexane with two equivalents of bromine or by the addition of bromine to 1-hexene? Explain.

PROBLEM 5·7

a Calculate the heat of reaction ΔH for the addition of iodine to the double bond in ethene.

b A solution of iodine in cyclohexane has a beautiful violet color. If cyclohexene is added, the color turns a brownish red and then slowly disappears (even in the dark). Explain.

HYPOHALOUS ACID ADDITION

Hypohalous acids and their derivatives are formed when halogens dissolve in hydroxylic solvents:

$$X_2 \quad + \quad \begin{cases} H_2O \rightleftharpoons HOX + HX \\ ROH \rightleftharpoons ROX + HX \end{cases} \quad (5.20)$$

(X = Cl, Br, or I)

Although the equilibrium constant is only about 5×10^{-4} for solutions of chlorine in water, the concentration of hypochlorous acid can be increased by adding silver oxide ($Ag^{\oplus}$ reduces the chloride-ion concentration and the oxide increases the pH). Hypobromous acid may also be prepared by hydrolysis of N-bromoacetamide in wet acetone:

$$\underset{\text{N-bromoacetamide}}{CH_3CONHBr} + H_2O \longrightarrow HOBr + CH_3CONH_2 \quad (5.21)$$

Hypochlorous and hypobromous acids are weak Brønsted acids. However, when we consider the products obtained from the addition of these reagents to alkenes, it is clear that they are functioning as a source of electrophilic halogen rather than as proton donors:

$$C_2H_5CH=CH_2 + \overset{\delta\ominus}{HO}-\overset{\delta\oplus}{Br} \longrightarrow C_2H_5CH(OH)CH_2Br \quad (5.22)$$

The mechanism of reactions such as 5.15 and 5.22 will be discussed further in Section 5·3.

DIBORANE ADDITION

Diborane (B_2H_6) is a spontaneously flammable gas that is easily prepared and handled in ether solution:

$$4BF_3 + 3NaBH_4 \xrightarrow{\text{ether}} 2B_2H_6 + 3NaBF_4 \quad (5.23)$$

The structure of this compound is especially intriguing because of the unusual hydrogen bridging that bonds the boron atoms. We may think of this arrangement as a pair of three-member bonds, each consisting of

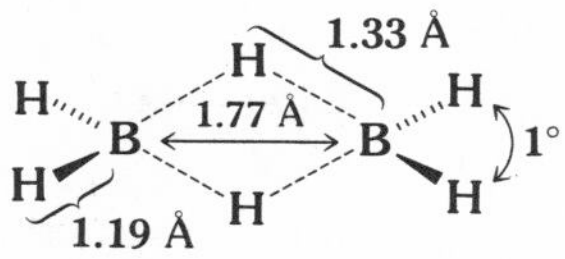

structure of diborane

5·2 ADDITION OF LEWIS ACIDS TO ALKENES

two paired electrons occupying a σ-type molecular orbital which encompasses a bridging hydrogen and both boron nuclei.

The addition reactions of diborane are simply explained by proposing an equilibrium (in ether solution) between the bridged dimer and monomeric BH_3:

$$B_2H_6 \underset{\text{dimer}}{\overset{\text{ether}}{\rightleftharpoons}} 2BH_3 \text{ (solvated)} \quad \text{monomer} \tag{5.24}$$

Since trivalent boron compounds are electron deficient (the boron atom shares only six valence electrons), the electrophilic nature of diborane can be attributed to this unstable and reactive species.

All three of the hydrogen atoms in BH_3 are potentially reactive in addition reactions with alkenes, a process called *hydroboration*:

$$\begin{array}{c} RCH=CH_2 \\ + \\ BH_3 \end{array} \longrightarrow RCH_2CH_2BH_2 \xrightarrow{RCH=CH_2} (RCH_2CH_2)_2BH \xrightarrow{RCH=CH_2} (RCH_2CH_2)_3B \tag{5.25}$$

This stoichiometry is realized, in fact, with most mono- or disubstituted alkenes. However, highly branched olefins are more crowded, and owing to steric interactions, they give only partially substituted boranes as addition products.

$$2 \begin{array}{c} CH_3 \\ \\ CH_3 \end{array}\!\!C=C\!\!\begin{array}{c} CH_3 \\ \\ H \end{array} + BH_3 \longrightarrow \left[H-\underset{\underset{CH_3}{|}}{\overset{\overset{CH_3}{|}}{C}}-\underset{\underset{CH_3}{|}}{\overset{\overset{H}{|}}{C}} \right]_2 BH \rightleftharpoons \text{dimer} \tag{5.26}$$

$$\begin{array}{c} CH_3 \\ \\ CH_3 \end{array}\!\!C=C\!\!\begin{array}{c} CH_3 \\ \\ CH_3 \end{array} + BH_3 \longrightarrow H-\underset{\underset{CH_3}{|}}{\overset{\overset{CH_3}{|}}{C}}-\underset{\underset{CH_3}{|}}{\overset{\overset{CH_3}{|}}{C}}-BH_2 \rightleftharpoons \text{dimer} \tag{5.27}$$

Alkyl boranes are important chemical intermediates, as well as interesting compounds in their own right. Two characteristic reactions of this functional group are described by the following equations:

$$(RCH_2CH_2)_3B + 3C_2H_5CO_2H \xrightarrow{\text{heat}} 3RCH_2CH_3 + (C_2H_5CO_2)_2B \tag{5.28}$$

$$(RCH_2CH_2)_3B + 3H_2O_2 + NaOH \longrightarrow 3RCH_2CH_2OH + NaB(OH)_4 \tag{5.29}$$

PROBLEM 5·8
The dialkyl borane $[(CH_3)_2CHCH(CH_3)]_2BH$ retains an active B—H bond and exists primarily as a hydrogen-bridged dimer. Write a structural formula for the product formed in the reaction of this compound with 1-hexene.

PROBLEM 5·9
Write equations showing how 2-methyl-1-butene can be transformed into the following products:
a $(CH_3)_2C(OH)CH_2CH_3$
b $CH_3CH_2CH(CH_3)CH_2OH$

5·3 ADDITION MECHANISMS AND STEREOCHEMISTRY

At this point it is tempting to explain all the addition reactions described thus far by a single two-step mechanism, with initial attack by an electrophile (Y) on the nucleophilic π electrons of an alkene, followed by addition of a nucleophile (Z) to a carbonium-ion intermediate:

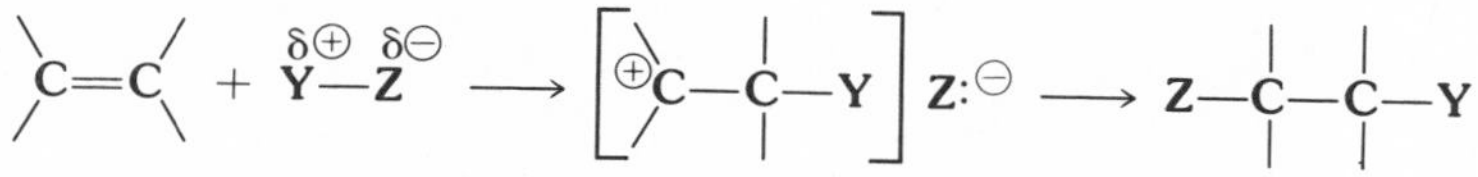

$$\text{\textbackslash}C=C\text{/} + \overset{\delta\oplus}{Y}-\overset{\delta\ominus}{Z} \longrightarrow \left[\overset{\oplus}{C}-\overset{|}{\underset{|}{C}}-Y\right] Z{:}^{\ominus} \longrightarrow Z-\overset{|}{\underset{|}{C}}-\overset{|}{\underset{|}{C}}-Y \qquad (5.30)$$

The orientation observed in the addition of an unsymmetrical reagent (Y—Z) to an unsymmetrically substituted double bond supports this interpretation. However, two further facts rule out a general application of this mechanism. First, carbonium-ion rearrangements like that in reaction 5.12 do not always take place when susceptible alkenes such as 3,3-dimethyl-1-butene are used:

$$(CH_3)_3CCH=CH_2 + X_2 \xrightarrow[(X=Cl,\,Br)]{CCl_4} (CH_3)_3CCHXCH_2X \qquad (5.31)$$

$$(CH_3)_3CCH=CH_2 + BH_3 \xrightarrow{ether} [(CH_3)_3CCH_2CH_2]_3B \qquad (5.32)$$

Second, many addition reactions are highly stereospecific, although not necessarily in the same manner:

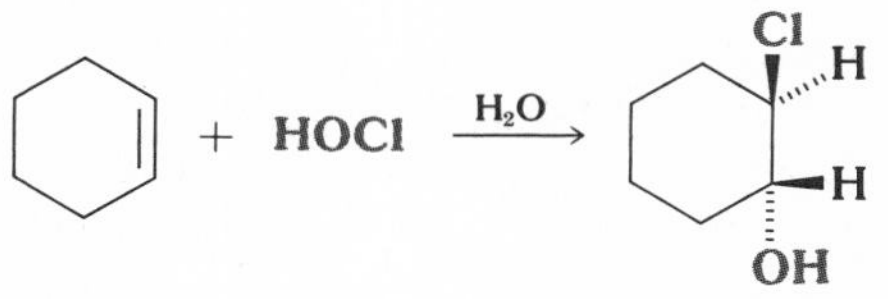

$$\text{cyclohexene} + HOCl \xrightarrow{H_2O} \text{trans addition product} \qquad (5.33)$$

trans addition

$$\text{cholesterol} + Br_2 \xrightarrow{CCl_4} \text{trans addition product} \qquad (5.34)$$

trans addition

$$\text{1,2-dimethylcyclohexene} + H_3O^{\oplus} \longrightarrow \text{trans addition} + \text{cis addition} \qquad (5.35)$$

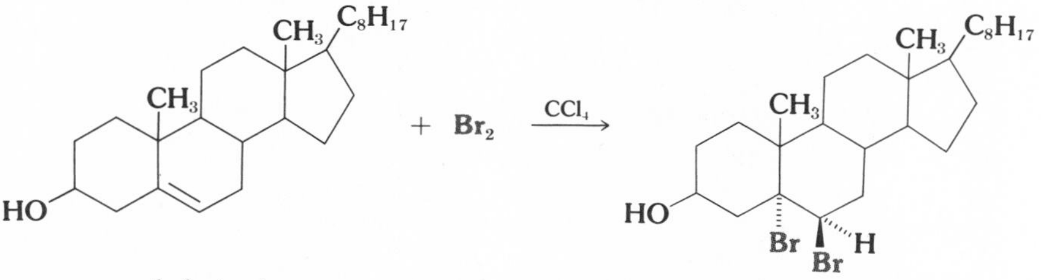

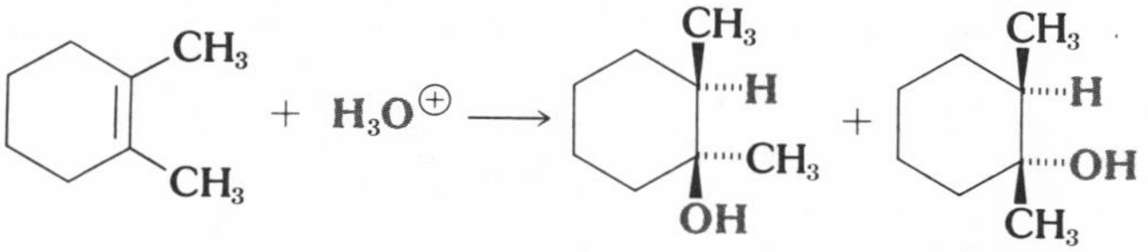

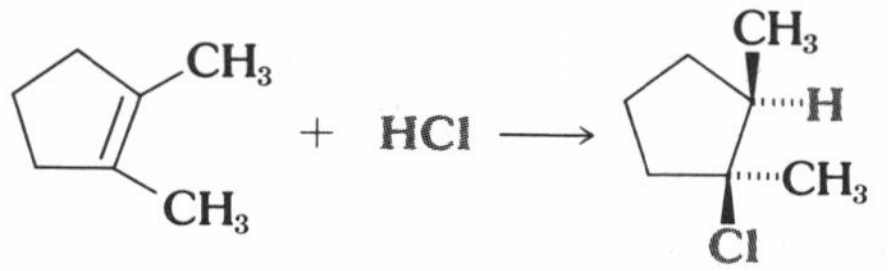

$$\text{1,2-dimethylcyclopentene} + HCl \longrightarrow \text{mainly trans addition product} \qquad (5.36)$$

mainly trans addition

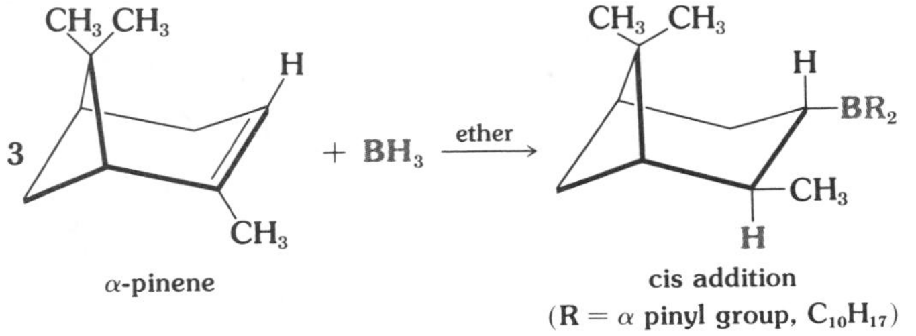

5·3 ADDITION MECHANISMS AND STEREOCHEMISTRY

$$\text{α-pinene} + BH_3 \xrightarrow{\text{ether}} \text{cis addition} \quad (R = \alpha \text{ pinyl group, } C_{10}H_{17}) \tag{5.37}$$

PROBLEM 5·10
What distribution of product stereoisomers would you expect from reaction 5.33 if the following discrete carbonium ion were an intermediate?

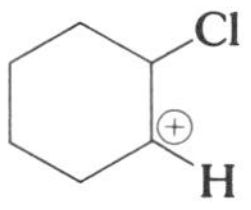

The two-step carbonium-ion mechanism proposed in equations 5.8 and 5.9 for the addition of Brønsted acids to alkenes is still reasonable, since these reactions show variable stereochemistry (as in equations 5.35 and 5.36) and are also subject to molecular rearrangement (as in equation 5.12). However, the corresponding addition reactions of other electrophilic reagents indicate that a different reaction path is followed in these cases. Furthermore, the stereochemical facts force us to propose different mechanisms for reactions initiated by electrophilic halogen species (such as trans addition of X_2 and HOX) and those involving diborane (cis addition).† These two mechanisms are outlined in Figure 5.3.

One of the most important features of the mechanism for electrophilic halogen addition is the stabilizing interaction that exists between the developing carbonium ion and the electron-rich halogen atom $:\ddot{X}$. This interaction strengthens as the neighboring halogen atom becomes larger, that is, as $X = Cl < Br < I$ (the order of increasing polarizability and decreasing ionization potential). Depending on the structure of the alkene and the strength of the halogen interaction, the ion-pair intermediates formed in these reactions may resemble an unsymmetrically bonded ion or the symmetrical *halonium ion*. The consequences are:

1 Rearrangement is less likely because the positive charge in the intermediate (either the symmetrical or unsymmetrical species) is no longer concentrated on a single carbon atom.

2 The strong bias for trans addition in these reactions (note equations 5.33 and 5.34) is understandable if we assume that the three-membered

---

†J. Traynham of Louisiana State University has drawn an apt analogy between a reaction-mechanism study and the classical murder mystery. An eye-witness account is given by the reaction kinetics, alibis are provided by stereochemical relationships, the motive is found in the energy changes during reaction, and the *corpus delicti* is the reaction product(s). All these factors are important and should be investigated; however, in some circumstances we may be forced to proceed with only limited information in one or more areas. We must always remember that a jury of our colleagues is going to hear the case and criticize our interpretation of the evidence.

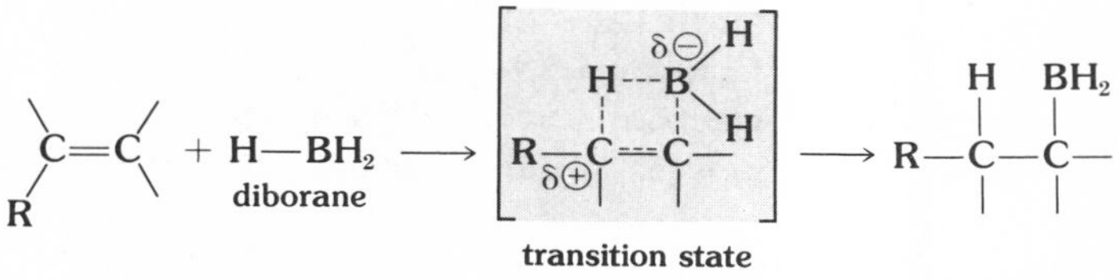

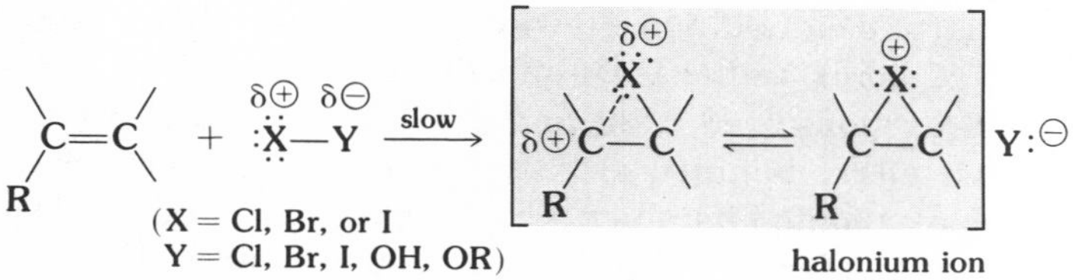

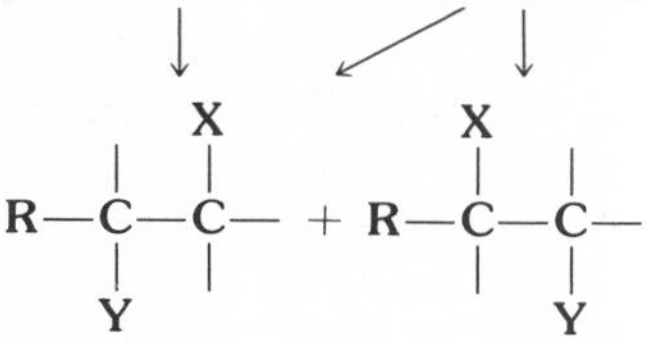

5·3 ADDITION MECHANISMS AND STEREOCHEMISTRY

FIGURE 5·3 *Two mechanisms for addition of electrophilic reagents to the carbon-carbon double bond*

cyclic intermediate is opened by attack of the nucleophile :Y$^\ominus$ from the back side (with inversion, as in $S_N 2$ displacement). For example, the following bromonium-ion intermediate has two possible modes of attack, one indicated by the solid curved arrows and the other indicated by the dashed curved arrows:

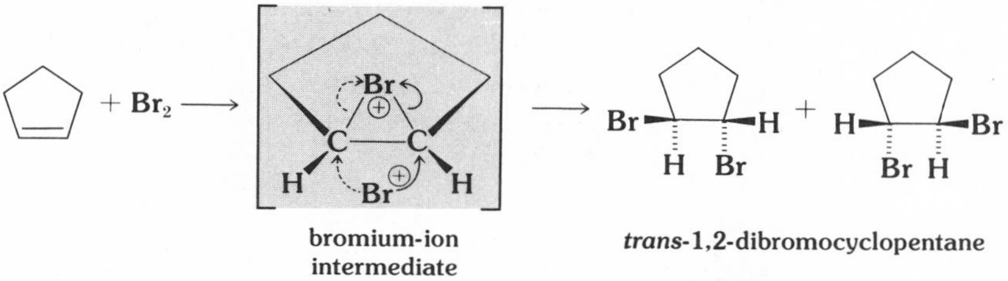

(5.38)

3 If the intermediate ion pair is unsymmetrically bonded, the product isomer having Markovnikov orientation should predominate, since a significant positive charge must reside on one of the carbon atoms of the double bond.

4 If the intermediate ion pair resembles the symmetrical halonium ion in Figure 5.3, products having anti-Markovnikov orientation may be formed:

$$(CH_3)_3CCH=CH_2 \xrightarrow[CH_3OH]{Br_2} (CH_3)_3CCHBrCH_2OCH_3 \qquad (5.39)$$

This is because nucleophilic displacement will proceed most rapidly at the least substituted carbon atom (compare Figure 3.6).

PROBLEM 5·11

Draw energy diagrams for the addition of HCl and Br_2 to 3,3-dimethyl-1-butene (reactions 5.12 and 5.31). These diagrams should indicate why rearrangement occurs in one case and not in the other.

PROBLEM 5·12
Equation 5.18 describes the liquid-phase chlorination of 2-methylpropene. This reaction is not affected by the presence of free-radical species and does not proceed in the gas phase below 150°C. Write a mechanism for this reaction.

The four-membered cyclic transition state proposed in the single-step mechanism for diborane addition to alkenes clearly restricts the geometry of these reactions to cis addition. However, to account for the highly selective bonding of boron to the least substituted carbon atom of the double bond (as in reactions 5.26, 5.32, and 5.37), we must assume that this transition state either is quite sensitive to steric-hindrance forces or has some dipolar or ionic character which favors formation of that carbon cation having the larger number of alkyl substituents. Although strong steric hindrance is undoubtedly responsible for the exclusive attack of diborane at the least hindered side of α-pinene (equation 5.37), the hydroboration of 3,3-dimethylcyclohexene is nonselective:

$$\text{3,3-dimethylcyclohexene} + BH_3 \xrightarrow{\text{ether}} \text{product (BR}_2\text{ down, 50%)} + \text{product (BR}_2\text{ up, 50%)} \tag{5.40}$$

This fact suggests that the moderate to small steric factors in most of these reactions do not significantly control the product-forming step. We can therefore conclude that the major directing factor in the diborane addition mechanism is polar in nature, as indicated in Figure 5.3. The degree of positive-charge development in the transition state is probably rather small, because no example of molecular rearrangement has yet been observed during diborane addition reactions.†

5·4 COORDINATION OF ALKENES WITH TRANSITION METALS

As we saw in Chapter 4, certain transition metals (notably **Ni**, **Pd**, **Pt**, and **Rh**) can serve as catalysts for the addition of hydrogen to the carbon-carbon double bond. This catalytic action appears to be related to the ability of these metals (and metal cations with similar electron configurations) to coordinate with (bond to) the unsaturated double-bond function. Ethene, for example, dissolves in an aqueous solution of potassium chloroplatinite (K_2PtCl_4), yielding a yellow water-soluble salt with

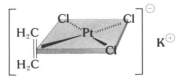

structure of $K[C_2H_4PtCl_3]$

†In contrast, the addition of bromine to α-pinene gives mostly rearrangement products.

5·4 COORDINATION OF ALKENES WITH TRANSITION METALS

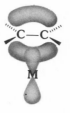

electrons move from carbon to the metal M

π-d_{z^2} overlap

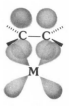

electrons move from the metal M to carbon

π^*-d_{xy} overlap

FIGURE 5·4 *Atomic-orbital overlap in transition-metal–alkene bonds*

the composition $K[C_2H_4PtCl_3]$. The structure of this compound has been established by x-ray diffraction analysis.

The two carbon atoms of the C_2H_4 grouping are equidistant from the platinum atom and are therefore considered to be equally bonded to the metal. This bonding apparently weakens the double bond, since the carbon-carbon bond length in this complex (1.47 Å) is longer than that for ethene itself.

An orbital-overlap representation of the bonding in transition metal–alkene complexes is presented in Figure 5.4 (the two major orbital interactions are shown separately for clarity). In keeping with the nucleophilic nature of the carbon-carbon double bond, we propose that the electron pair of the π bond coordinates with an empty (electrophilic) d orbital on the metal. This bonding interaction is accompanied by a unique "backbonding," wherein a filled metal d orbital coordinates with the empty π^* molecular orbital of the alkene. The cancelling effect of these electron transfers can be seen in the rather low dipole moments usually observed for such bonds.

PROBLEM 5·13
How does the theoretical bonding model described in Figure 5.4 explain the carbon-carbon bond lengthening observed in some of these alkene complexes?

Similar coordination complexes have been prepared for $Rh^{\oplus}$, Pd^{2+}, $Cu^{\oplus}$, $Ag^{\oplus}$, and Hg^{2+} cations. For example, although cyclohexene is immiscible with water, it dissolves in aqueous mercuric nitrate and will remain in solution until the mercury-alkene complex is decomposed by the addition of thiocyanate. In some cases crystalline metal-alkene complexes can be isolated:

$$\bigcirc + Hg^{2+} \cdot (H_2O)_2 \underset{}{\overset{K_{eq} \approx 5 \times 10^4}{\rightleftharpoons}} \bigcirc\!\!-\!\!\overset{Hg^{\oplus}OH}{|} + H_3O^{\oplus} \xrightarrow{2KCNS} Hg(CNS)_2 + \bigcirc \quad (5.41)$$

PROBLEM 5·14
What neutral metal atoms are isoelectronic (have the same electron configuration as) $Cu^{\oplus}$, $Ag^{\oplus}$, and Hg^{2+}?

5·5 OXIDATION REACTIONS OF ALKENES

In this section we will consider two oxidation reactions of alkenes which have one common feature — each of the double-bonded carbon atoms ends up bonded to an oxygen atom.

OXIDATION-REDUCTION TERMINOLOGY

Although the classification of chemical reactions as *oxidations* or *reductions* is more common in inorganic chemistry, this terminology is occasionally useful in organic chemistry as well. For this purpose we may define the *oxidation state* of a specific carbon atom by assuming the following oxidation numbers for the most common substituents:

H (+1)   C (0)   O— (−1)   O= (−2)   F, Cl, Br, or I (−1)
N— (−1)   N= (−2)   N≡ (−3)

The carbon oxidation numbers are assigned so that they cancel the substituent values:

$$H_3C-CH_3 \quad H_2C=CH_2 \quad HC\equiv CH \quad H_3C-Br \quad H_2C=O \quad HC\equiv N \quad O=C=O$$
$$\;-3\;\;-3 \qquad -2\;\;-2 \qquad\;\; -1\;\;-1 \qquad\;\;\; -2 \qquad\quad\; 0 \qquad\quad\; +2 \qquad\quad +4$$

A reaction is then classified according to the change in oxidation state of the reacting carbon atoms. If a carbon atom's oxidation state becomes more positive, we say it has been *oxidized*; a shift to a more negative oxidation number indicates a *reduction*:

$$CH_3CH=CH_2 + HOCl \longrightarrow CH_3CH(OH)CH_2Cl \tag{5.42}$$
$$-3\;\;-1\;\;\;-2 \qquad\qquad\qquad\qquad -3\;\;\;\;0\;\;\;\;\;\;-1$$
$$\underbrace{}_{-6} \longrightarrow \underbrace{}_{-4}$$
an oxidation

$$CH_3CH=CH_2 + H_2 \xrightarrow{Pt} CH_3CH_2CH_3 \tag{5.43}$$
$$-3\;\;-1\;\;\;-2 \qquad\qquad\quad -3\;\;-2\;\;-3$$
$$\underbrace{}_{-6} \longrightarrow \underbrace{}_{-8}$$
a reduction

PROBLEM 5·15
Calculate the change in oxidation number for each carbon atom of propene during the addition of sulfuric acid. What is the overall change in oxidation states?

HYDROXYLATION

Alkaline or neutral solutions of potassium permanganate react with alkenes to give vicinal, or neighboring, dihydroxy compounds:

$$\begin{array}{c}\\ \diagdown\diagup\\ C=C\\ \diagup\diagdown\end{array} + 2KMnO_4 + 2KOH \xrightarrow[\text{acetone}]{H_2O} \begin{array}{c} OH\;\;\;OH\\ |\quad\;\;\;|\\ -C-\!\!-C-\\ |\quad\;\;\;| \end{array} + 2K_2MnO_4 \tag{5.44}$$

In strongly alkaline solutions the manganese is reduced to a green manganate ion (MnO_4^{2-}), whereas in neutral and weakly acidic solutions

5·5 OXIDATION REACTIONS OF ALKENES

a brownish-black precipitate of manganese dioxide (MnO_2) is formed:

$$3MnO_4^{2-} + 4H^+ \longrightarrow MnO_2 + 2MnO_4^- + H_2O \tag{5.45}$$

Alkaline solutions are generally preferred for the hydroxylation reaction, since acidic conditions often cause further oxidative degradation of the organic product. Some specific examples of this reaction are:

$$CH_3(CH_2)_7CH{=}CH(CH_2)_7CO_2H \xrightarrow[H_2O,\ KOH]{KMnO_4} CH_3(CH_2)_7\overset{\overset{OH}{|}}{CH}-\overset{\overset{OH}{|}}{CH}(CH_2)_7CO_2H \tag{5.46}$$

oleic acid
(from olive oil)

(5.47)

cis addition

Further study has shown that the cis introduction of the two hydroxyl functions noted in reaction 5.47 is a general characteristic of this method.

The distinctive purple color of permanganate solutions is lost during the reaction with alkenes; this reagent is therefore widely used as a qualitative test for the carbon-carbon double bond. Since permanganate is able to oxidize other functional groups as well, this test, often called the *Baeyer test*, is best used to complement other criteria, such as bromine decolorization and spectroscopic absorption.

Alkenes may also be hydroxylated by a two-step reaction sequence using the toxic reagent osmium tetroxide (OsO_4):

(5.48)

osmium intermediate   cis addition

Despite the high reagent cost ($5,000/lb), this method is often preferred for the hydroxylation of small amounts of alkene because of the high yields and ease with which the products can be isolated.

Although cyclic manganese salts analogous to the osmium intermediate in reaction 5.48 have not been isolated from permanganate hydroxylation reactions, the cis orientation of the hydroxyl groups in the products from both reactions suggests that similar pathways are followed, with a corresponding cyclic intermediate in the permanganate reactions.

Structural investigations tell us that MnO_4^- and OsO_4 have equivalent configurations, with the metal atom occupying the center of a tetrahedral grouping of negatively charged oxygen atoms. That such a species should readily attack a carbon-carbon double bond seems at first to contradict our general hypothesis that in addition reactions to alkenes the nucleophilic double bond is most easily attacked by electrophilic reagents.

manganese intermediate

5·5 OXIDATION REACTIONS OF ALKENES

However, when we recognize that the electron configurations of manganese and osmium are, respectively, just one and two electrons below the nickel and platinum atomic configurations, a simple explanation comes to mind: an empty d orbital of the transition metal (probably d_{z^2}) overlaps with and accepts electrons from the π bond, while two of the nucleophilic oxygen atoms begin to bond to the carbon atoms of the double bond (perhaps by overlap with the empty π^* molecular orbital).

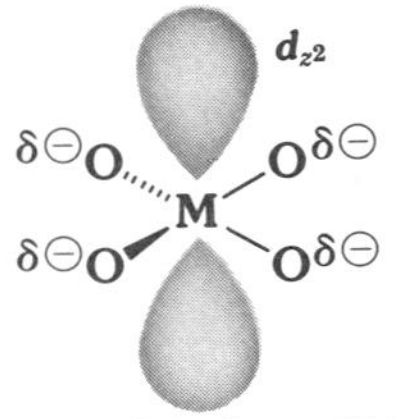

structure of OsO_4 and $MnO_4^{\ominus}$

OZONOLYSIS

OZONE REACTIONS   Ozone (O_3), an allotrope of oxygen, reacts readily with carbon-carbon double bonds, ultimately leading to cleavage products:

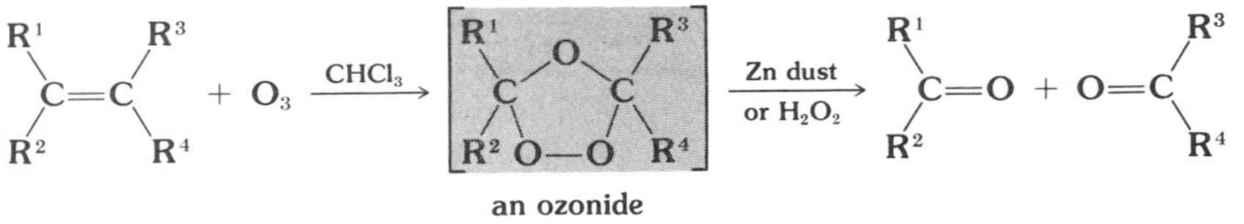

$$\underset{R^2}{\overset{R^1}{\diagdown}}C=C\underset{R^4}{\overset{R^3}{\diagup}} + O_3 \xrightarrow{CHCl_3} \left[\underset{R^2\ O-O\ R^4}{\overset{R^1\ \ \ O\ \ \ R^3}{\underset{\diagdown\diagup}{C\diagdown\ \ \diagup C}\underset{\diagup\diagdown}{O}}}\right] \xrightarrow[\text{or } H_2O_2]{\text{Zn dust}} \underset{R^2}{\overset{R^1}{\diagdown}}C=O + O=C\underset{R^4}{\overset{R^3}{\diagup}} \quad (5.49)$$

an ozonide

Although intermediate ozonides have actually been isolated from some reactions, these unstable and sometimes dangerously explosive substances are usually transformed into cleavage products without delay. From a practical standpoint, the reactivity of ozone with olefins is both a nuisance and a blessing. On one hand, natural rubber and some synthetic rubbers deteriorate (crack and harden) as a result of attack by atmospheric ozone (generated by the action of radiation or lightning discharges in the atmosphere). On the other hand, controlled ozonolysis in the laboratory provides us with a method for locating the position of carbon-carbon double bonds in molecules.

The ozonolysis reaction is normally conducted in two stages. First a mixture of ozone and oxygen, prepared by passing oxygen through an electrical discharge, is bubbled into a cooled solution of the alkene in an inert solvent such as CH_2Cl_2 or $CHCl_3$. The resulting ozonide mixture is then treated either with a mild reducing agent, such as zinc dust in water, or with an oxidizing agent, such as hydrogen peroxide, in order to decompose the ozonide intermediates. These two workup procedures will give different products if at least one of the double-bonded carbon atoms bears one or more hydrogen atoms.

The ozonolysis of 2-methyl-2-butene illustrates the value of the ozonolysis reaction for determining the location of a double bond. Although there are five isomeric pentenes (C_5H_{10}), only one compound can reasonably react with ozone to give the products shown here:

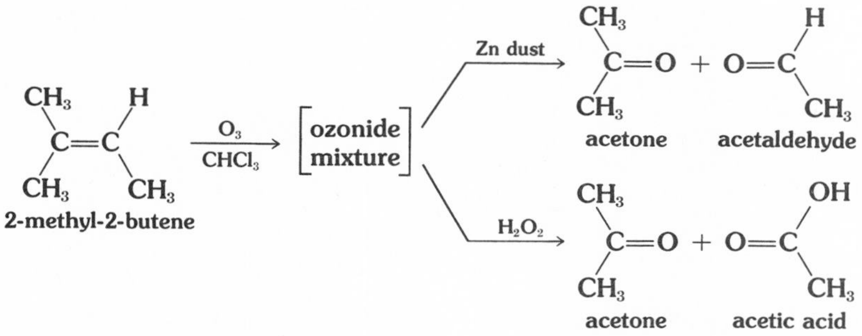

5·5 OXIDATION REACTIONS OF ALKENES

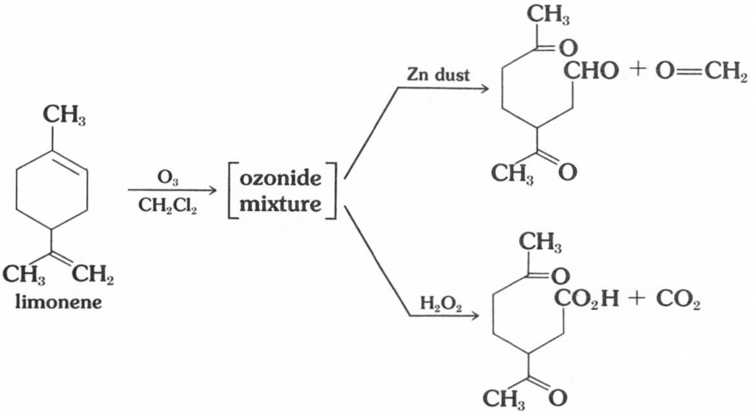

(5.50) 2-methyl-2-butene reacts with O_3/$CHCl_3$ to give an ozonide mixture, which with Zn dust gives acetone ($(CH_3)_2C=O$) + acetaldehyde (CH_3CHO), and with H_2O_2 gives acetone + acetic acid (CH_3CO_2H).

(5.51) Limonene reacts with O_3/CH_2Cl_2 to give an ozonide mixture. With Zn dust: a keto-aldehyde (4-acetyl-1-methylcyclohexane-carbaldehyde type) + $O=CH_2$. With H_2O_2: the corresponding keto-carboxylic acid + CO_2.

PROBLEM 5·16
Write structures for the expected ozonolysis products from the remaining four pentene isomers, using both workup procedures for each compound.

Ozonolysis will also distinguish cycloalkenes in which both double-bond carbon atoms are part of the same ring from isomers having the double bond outside the ring. In the first case a single ozonolysis product retaining all the carbon atoms of the reactant is formed, and in the second case cleavage at the double bond leads to two smaller fragments. The isomeric pinenes provide a good example of this application:

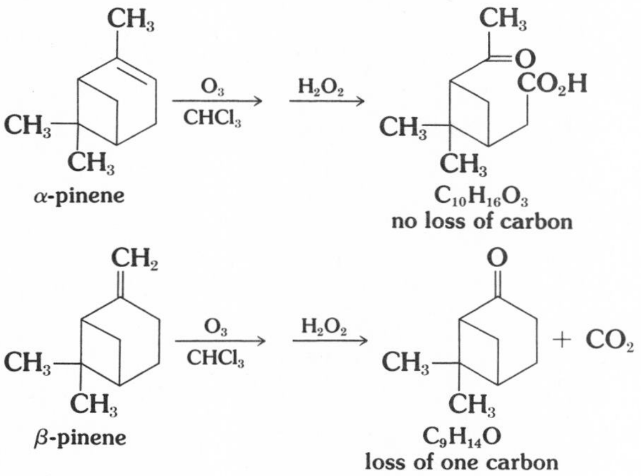

(5.52)

α-pinene $\xrightarrow[CHCl_3]{O_3, H_2O_2}$ keto-acid product $C_{10}H_{16}O_3$ (no loss of carbon)

β-pinene $\xrightarrow[CHCl_3]{O_3, H_2O_2}$ ketone + CO_2, $C_9H_{14}O$ (loss of one carbon)

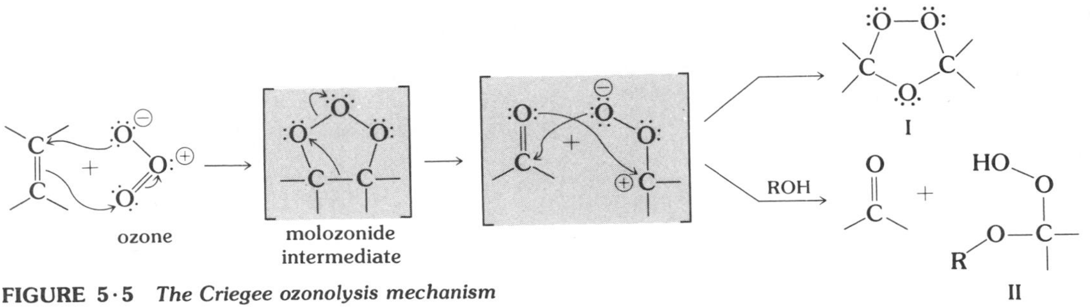

FIGURE 5·5 *The Criegee ozonolysis mechanism*

Of course, the location of the double bond in each of these isomers can also be determined by spectroscopic measurements (see, for example, Problem 4.1); however, a combination of physical and chemical evidence is generally desirable for a rigorous structure proof.

PROBLEM 5·17
There are 13 isomeric hexenes (not counting stereoisomers). Of these, only three give C_3 fragments exclusively on ozonolysis (that is, they do not give C_1, C_2, C_4, or C_5 fragments). Write equations illustrating these three reactions and suggest a spectroscopic method for distinguishing the three hexene isomers.

MECHANISM OF OZONOLYSIS Ozone is a triangular molecule, the terminal atoms of which may function as either an electrophile or a nucleophile (see Figure 5.5). These reactive sites readily bond to the carbon atoms of a double bond, giving a very unstable cyclic intermediate known as the *molozonide*. This intermediate has in fact been detected recently by low-temperature pmr measurements at −75°C. The molozonide normally decomposes rapidly to reactive fragments that may recombine to yield a more stable ozonide (I) or react with solvent hydroxyl groups to yield product II. Type I ozonides have been shown, with the proper substitution pattern, to display cis-trans stereoisomerism.

PROBLEM 5·18
a The structure for ozone in Figure 5.5 seems at first glance to indicate that the central oxygen atom is electrophilic. Explain why this is not so, and why one end oxygen is operationally electrophilic.

b Give a specific example of a substituent pattern for ozonide I that would allow cis-trans isomerism to occur. Give an example for which such stereoisomerism is not possible.

5·6 ADDITION OF RADICALS TO ALKENES

Contradictory facts concerning the orientation of hydrogen bromide additions to unsymmetrical alkenes were reported in the early chemical literature (before 1933). Most of the uncertainty was removed, however, by the investigations of Kharasch and Mayo, who demonstrated the existence of an alternative free-radical pathway for this addition reaction. Samples of alkenes often contain small amounts of peroxides (R—O—O—R) formed by exposure to the oxygen in air. The oxygen-

5·7 DIMERIZATION AND POLYMERIZATION

oxygen bond in these compounds is relatively weak, with a bond-dissociation energy of about 35 kcal/mole. Consequently, peroxides are excellent sources of free radicals and can initiate the radical-chain addition of HBr to carbon-carbon double bonds:

$$RO-OR \xrightarrow{heat} 2R-\ddot{O}\cdot \qquad \Delta H = +35 \text{ kcal/mole} \qquad (5.53)$$

$$RO\cdot + HBr \longrightarrow ROH + Br\cdot \qquad \Delta H = -23 \text{ kcal/mole} \qquad (5.54)$$

$$\left. \begin{array}{l} R'CH=CH_2 + Br\cdot \longrightarrow R'\dot{C}HCH_2Br \qquad \Delta H = -5 \text{ kcal/mole} \\ R'\dot{C}HCH_2Br + HBr \longrightarrow R'CH_2CH_2Br \qquad \Delta H = -11 \text{ kcal/mole} \end{array} \right\} \text{chain reaction} \qquad (5.55)$$

Since the addition of bromine atoms (Br·) to unsymmetrical alkenes will proceed so as to form the most stable (most highly substituted) carbon radical, an anti-Markovnikov orientation in the product will be observed whenever this radical-chain mechanism is operating. Peroxide-free samples of alkenes will, however, react with hydrogen bromide by the slower ionic or polar mechanism described in Section 5.1 to give products having a Markovnikov orientation.

PROBLEM 5·19
Show how the relative stabilities of carbon radicals (tertiary R· > secondary R· > primary R· > CH_3·) can be deduced from the corresponding C—H bond-dissociation energies.

Radical-chain addition to alkenes is not unique to hydrogen bromide, but is observed for many other reagents as well. Two examples serve to illustrate this class of reactions:

$$RCH=CH_2 + CCl_4 \xrightarrow[\Delta]{peroxide} RCHClCH_2CCl_3 \qquad (5.56)$$

$$RCH=CH_2 + HSCH_2CO_2H \xrightarrow[\Delta]{peroxide} RCH_2CH_2SCH_2CO_2H \qquad (5.57)$$

PROBLEM 5·20
a Write radical-chain mechanisms for reactions 5.56 and 5.57 and calculate ΔH for each step in these mechanisms.
b Suggest an explanation for the fact that radical-chain addition of HCl or HI to alkenes does not normally take place.

5·7 DIMERIZATION AND POLYMERIZATION

Reaction of the branched olefin isobutylene with acidic reagents can proceed in several ways. In dilute aqueous acid a reversible hydration occurs, as in reaction 5.58. When stronger acids are used, dimeric alkenes are the major products, and with anhydrous acids long chains of repeating isobutane units are formed.

$$\begin{array}{c} CH_3 \\ \diagdown \\ C=CH_2 + H_3O^{\oplus} \rightleftharpoons (CH_3)_3COH_2^{\oplus} \xrightleftharpoons[-H_2O]{H_2O} (CH_3)_3COH + H_3O^{\oplus} \\ \diagup \\ CH_3 \end{array} \qquad (5.58)$$

$$2 \underset{CH_3}{\overset{CH_3}{\diagdown}}C=CH_2 + 60\% \ H_2SO_4 \xrightarrow{80-100°} \begin{array}{c} (CH_3)_3CCH_2-\underset{CH_3}{\overset{CH_3}{\underset{|}{C}}}=CH_2 \\ + \\ (CH_3)_3CCH=C(CH_3)_2 \\ \text{dimers of isobutylene} \end{array}$$
(5.59)

$$n \underset{CH_3}{\overset{CH_3}{\diagdown}}C=CH_2 + HF \ (or \ H_2SO_4) \xrightarrow{-100°} \sim\!\!\left[\begin{array}{c} CH_3 \\ | \\ C-CH_2-\underset{CH_3}{\overset{CH_3}{\underset{|}{C}}}-CH_2 \\ | \\ CH_3 \end{array}\right]_{n/2}\!\!\sim$$

polyisobutylene
(n repeating units)
(5.60)

High-molecular-weight molecules of this kind are called *polymers* (from the Greek *polys*, meaning much, and *meros*, meaning part). The molecule which generates the repeating unit (isobutylene in this example) is termed a *monomer*.

The formation of these dimeric and polymeric hydrocarbons from isobutylene should not come as a surprise, since the tertiary butyl cation generated by protonation of the olefin is a strong electrophile and can itself attack the nucleophilic double bond of another molecule:

$$\underset{CH_3}{\overset{CH_3}{\diagdown}}C=CH_2 + H_2SO_4 \longrightarrow \left[\underset{CH_3}{\overset{CH_3}{\diagdown}}\overset{\oplus}{C}-CH_3\right]$$

t-butyl cation
(5.61)

$$\underset{CH_3}{\overset{CH_3}{\diagdown}}C=CH_2 + t\text{-butyl cation} \longrightarrow \left[\underset{CH_3}{\overset{CH_3}{\diagdown}}\overset{\oplus}{C}-CH_2-\underset{CH_3}{\overset{CH_3}{\underset{|}{C}}}-CH_3\right] \longrightarrow \begin{array}{c}\text{dimeric}\\\text{products}\end{array}$$

$$\downarrow (CH_3)_2C=CH_2$$

$$\left[\underset{CH_3}{\overset{CH_3}{\diagdown}}\overset{\oplus}{C}-CH_2-\underset{CH_3}{\overset{CH_3}{\underset{|}{C}}}-CH_2-\underset{CH_3}{\overset{CH_3}{\underset{|}{C}}}-CH_3\right] \xrightarrow{\text{repeat}} \begin{array}{c}\text{polymeric}\\\text{products}\end{array}$$
(5.62)

The resulting cation intermediate can either give dimeric products (composed of two monomer units) or react with another olefin molecule. Repetition of this latter course leads to polymers.

The usefulness and commercial value of many naturally occurring polymers such as rubber, wool, silk, and cellulose fibers (cotton, linen, etc.) has stimulated a vigorous and productive study of synthetic polymers. The simplest hydrocarbon polymer, polyethylene, can be prepared by heating ethylene containing a trace of oxygen under pressure:

$$n(CH_2=CH_2) \xrightarrow[\text{tr } O_2]{15{,}000 \text{ psi, } 100°} \ \ \!\!\text{\Large(}CH_2-CH_2\text{\Large)}_n \quad n \approx 10{,}000$$
polyethylene
(5.63)

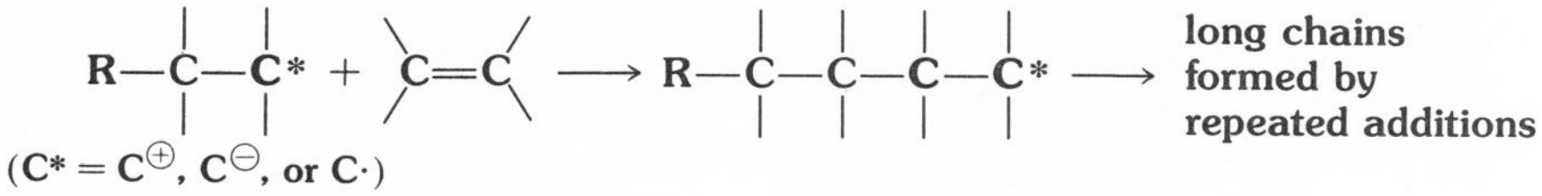

$$\underset{(C^* = C^{\oplus},\ C^{\ominus},\ or\ C\cdot)}{R-\overset{|}{\underset{|}{C}}-\overset{|}{\underset{|}{C}}*} + \overset{\diagdown}{\underset{\diagup}{C}}=\overset{\diagup}{\underset{\diagdown}{C}} \longrightarrow R-\overset{|}{\underset{|}{C}}-\overset{|}{\underset{|}{C}}-\overset{|}{\underset{|}{C}}-\overset{|}{\underset{|}{C}}* \longrightarrow \text{long chains formed by repeated additions}$$

FIGURE 5·6 *Mechanisms for polymerization*

Functional groups such as double bonds or hydroxyl groups may occupy the ends of the polymer chains, but the concentration of these active sites is so low that the waxy polyethylene molecule essentially resembles a large alkane.

Three basic polymerization mechanisms are formally possible, as summarized in Figure 5.6. The cationic mechanism ($C^* = C^{\oplus}$) operates in the formation of polyisobutylene (equation 5.60), but the polymerization of ethylene (equation 5.63) appears to proceed by a radical-chain pathway, as do the polymerizations of vinyl chloride and tetrafluoroethylene:

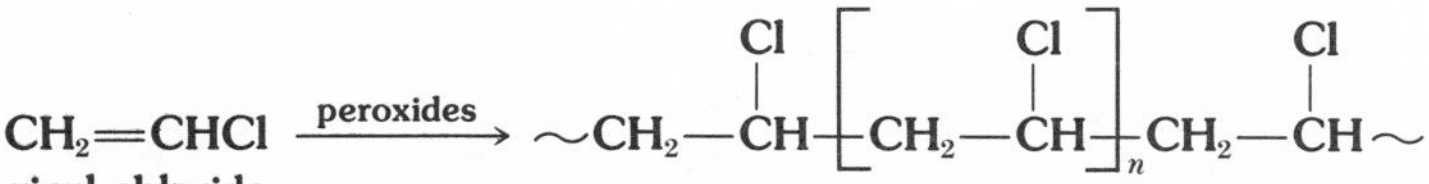

$$\underset{\text{vinyl chloride}}{CH_2=CHCl} \xrightarrow{\text{peroxides}} \sim CH_2-\overset{Cl}{\underset{|}{CH}}\left[CH_2-\overset{Cl}{\underset{|}{CH}}\right]_n CH_2-\overset{Cl}{\underset{|}{CH}}\sim \qquad (5.64)$$
$$\text{polyvinyl chloride}$$

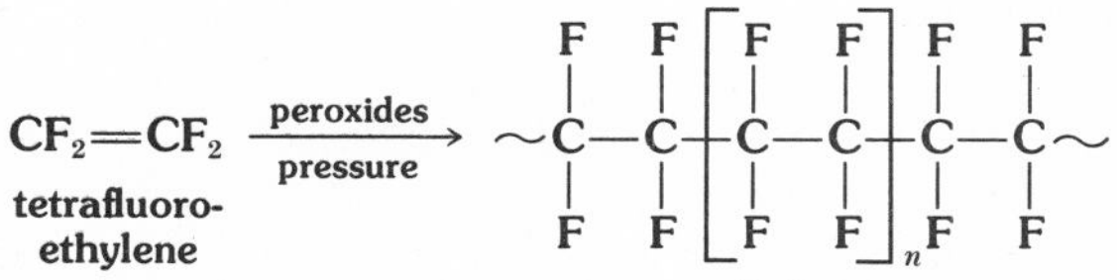

$$\underset{\substack{\text{tetrafluoro-}\\ \text{ethylene}}}{CF_2=CF_2} \xrightarrow[\text{pressure}]{\text{peroxides}} \sim \overset{F}{\underset{F}{C}}-\overset{F}{\underset{F}{C}}\left[\overset{F}{\underset{F}{C}}-\overset{F}{\underset{F}{C}}\right]_n \overset{F}{\underset{F}{C}}-\overset{F}{\underset{F}{C}}\sim \qquad (5.65)$$
$$\text{Teflon}$$

The Nobel prize–winning investigations of Karl Ziegler (Germany) and Giulio Natta (Italy) led to the recognition of a fourth polymerization mechanism, which is particularly advantageous for monomers of the type $RCH=CH_2$. In this mechanism a titanium-aluminum catalyst, called the *Ziegler catalyst*, becomes located at the active or growing end of the polymer chain. Further monomer units then attach themselves to the chain by first coordinating with the metal and then rearranging to a chain-lengthened unit.

As we come to understand better the relationship of the molecular structure of a polymer to its chemical and physical properties, we can begin to tailor-make polymeric substances to fit particular needs. A useful technique in this respect is the copolymerization of different monomers. For example, although polyvinyl chloride (equation 5.64) is a hard, brittle resin, the copolymer of vinyl chloride and acrylonitrile is easily spun into the useful synthetic fiber Dynel. The homopolymer of acrylonitrile provides us with Orlon.

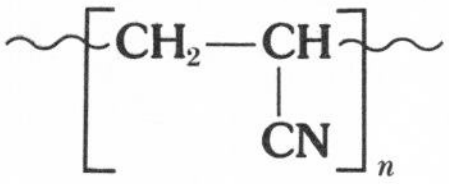

$$\sim\left[CH_2-\underset{\underset{CN}{|}}{CH}\right]_n \qquad \sim\left[CH_2-\underset{\underset{CN}{|}}{CH}-CH_2-\underset{\underset{Cl}{|}}{CH}\right]_n$$
$$\text{Orlon} \qquad\qquad\qquad \text{Dynel}$$

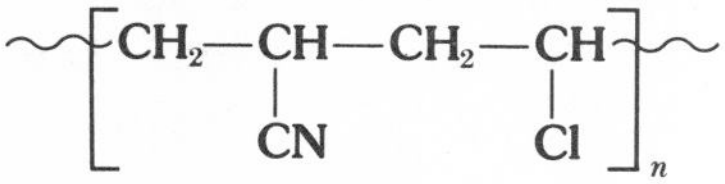

PROBLEM 5·21
Identify the monomer unit in each of the following polymers:

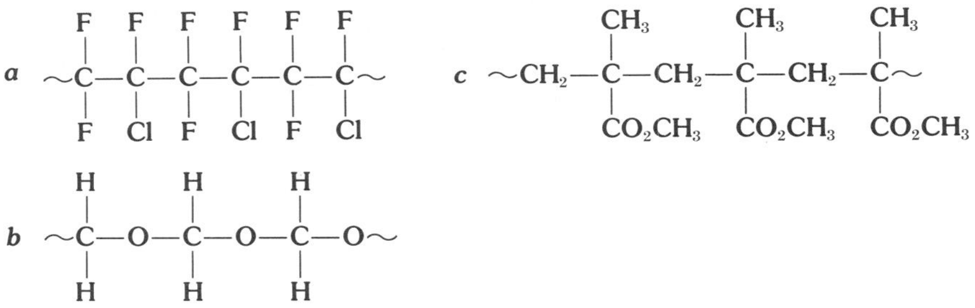

a $\quad \sim\underset{F}{\overset{F}{C}}-\underset{Cl}{\overset{F}{C}}-\underset{F}{\overset{F}{C}}-\underset{Cl}{\overset{F}{C}}-\underset{F}{\overset{F}{C}}-\underset{Cl}{\overset{F}{C}}\sim$

c $\quad \sim CH_2-\underset{CO_2CH_3}{\overset{CH_3}{C}}-CH_2-\underset{CO_2CH_3}{\overset{CH_3}{C}}-CH_2-\underset{CO_2CH_3}{\overset{CH_3}{C}}\sim$

b $\quad \sim\underset{H}{\overset{H}{C}}-O-\underset{H}{\overset{H}{C}}-O-\underset{H}{\overset{H}{C}}-O\sim$

PROBLEM 5·22
a Write a structural formula for a small segment (about eight carbon atoms long) of polypropylene, the polymer formed from propene.

b Write a structural formula for a small segment of the alternating copolymer of vinyl chloride ($CH_2{=}CHCl$) and vinylidene chloride ($CH_2{=}CCl_2$).

5·8 ADDITION OF CARBENES TO ALKENES

Recall from Chapter 1 that neutral divalent carbon intermediates, called *carbenes*, add rapidly to alkenes, thereby giving rise to cyclopropane derivatives:

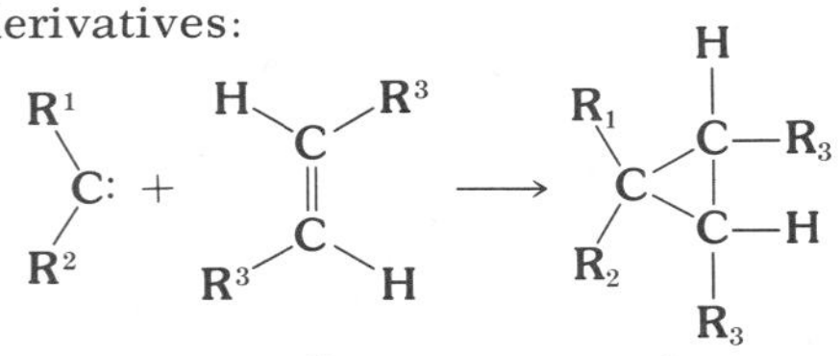

$$\underset{\text{a carbene}}{\overset{R^1}{\underset{R^2}{>}}C{:}} + \underset{\text{an alkene}}{\overset{H}{\underset{R^3}{>}}C{=}C\overset{R^3}{\underset{H}{<}}} \longrightarrow \underset{\text{a cyclopropane}}{\overset{R_1}{\underset{R_2}{>}}C\overset{H}{\underset{R_3}{<}}\underset{\underset{R_3}{|}}{C{-}R_3}\atop{C{-}H}} \quad (5.66)$$

The simplest carbene, methylene ($H_2C{:}$), can be generated by decomposition of the poisonous gas diazomethane (CH_2N_2).

$$\underset{\text{diazomethane}}{\overset{H}{\underset{H}{>}}C{=}\overset{\oplus}{N}{=}\overset{\ominus}{\underset{..}{N}}{:}} \xrightarrow{\text{heat or light}} N_2 + \underset{\text{methylene}}{H_2C{:}} \quad (5.67)$$

Dihalocarbenes are usually prepared by α elimination reactions:

$$CHBr_3 + KOR \rightleftharpoons HOR + \overset{\oplus\ \ominus}{K{:}CBr_3} \longrightarrow KBr + \underset{\text{dibromocarbene}}{:CBr_2} \quad (5.68)$$

$$(R{=}H \text{ or alkyl})$$

$$BrCCl_3 + C_4H_9Li \longrightarrow C_4H_9Br + \overset{\oplus\ \ominus}{Li{:}CCl_3} \longrightarrow LiCl + \underset{\text{dichlorocarbene}}{:CCl_2} \quad (5.69)$$

Since carbene intermediates are too reactive to be isolated and studied at our leisure, their fleeting existence is largely deduced from their characteristic addition reactions with alkenes:

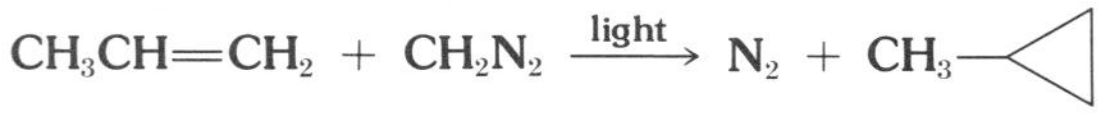

$$CH_3CH{=}CH_2 + CH_2N_2 \xrightarrow{\text{light}} N_2 + CH_3{-}\triangleleft \quad (5.70)$$

$$\text{C}_6\text{H}_{10} + \text{CBrCl}_3 + n\text{-C}_4\text{H}_9\text{Li} \longrightarrow \text{C}_6\text{H}_{10}\text{CCl}_2 + \text{C}_4\text{H}_9\text{Br} + \text{LiCl} \quad (5.71)$$

$$\text{CH}_3\text{CH}=\text{CHCH}_3 + \text{CHBr}_3 + t\text{-C}_4\text{H}_9\text{OK} \longrightarrow \underset{\text{CH}_3}{\overset{\text{CH}_3}{\triangle}}\text{CBr}_2 + t\text{-C}_4\text{H}_9\text{OH} + \text{KBr} \quad (5.72)$$

PROBLEM 5·23
Gas-phase decomposition of diazomethane by light gives, in the absence of any other organic substances, ethylene and cyclopropane. Suggest a mechanism for the formation of these products.

PROBLEM 5·24
Write equations illustrating the fact that dibromocarbene addition to 2-butene (equation 5.72) proceeds stereospecifically and with retention of configuration.

Certain organometallic reagents also undergo carbenelike addition reactions to alkenes. One such carbenoid reagent is the widely used *Simmons-Smith reagent*, ICH_2ZnI, which is readily prepared from methylene iodide by the action of a zinc-copper couple (essentially a copper-plated zinc powder) in refluxing ether:

$$\text{CH}_2\text{I}_2 + \text{Zn(Cu)} \xrightarrow{\text{ether reflux}} \underset{\text{Simmons-Smith reagent}}{\text{ICH}_2\text{ZnI}} \quad (5.73)$$

Although this reagent can be stored briefly, it is usually prepared in the presence of the alkene coreactant:

$$\text{C}_6\text{H}_{10} + \text{CH}_2\text{I}_2 + \text{Zn(Cu)} \xrightarrow[\Delta]{\text{ether}} \text{bicyclic product} + \text{ZnI}_2 \quad (5.74)$$

$$\text{isoprene-like} + \text{CH}_2\text{I}_2 + \text{Zn(Cu)} \xrightarrow[\Delta]{\text{ether}} \underset{64\%}{\text{product A}} + \underset{4\%}{\text{product B}} + \underset{32\%}{\text{product C}} + \text{ZnI}_2 \quad (5.75)$$

5·9 ADDITION REACTIONS OF CYCLOPROPANE

Cyclopropane is unique among the cycloalkanes in its ability to undergo some of the same addition reactions that characterize the alkene group. Although cyclopropane is slightly less reactive than propene toward bromine and platinum-catalyzed hydrogen addition, this three-membered cycloalkane shows far greater unsaturated behavior than any of its larger-ring homologs. Substituted cyclopropanes usually react with electrophilic reagents by ring opening at the most highly substituted carbon atom; however, catalytic hydrogen addition proceeds most rapidly at the least substituted bond:

$$\underset{\text{CH}_3}{\overset{\text{CH}_3}{\text{R}_2\text{C}}}\text{(CH}_3\text{)}_2 \xleftarrow{\text{H}_2,\ \text{Pt}} \underset{\text{R}}{\overset{\text{R}}{\triangle}} \xrightarrow{\text{HBr}} \text{R}_2\text{CBrCH}_2\text{CH}_3 \quad (5.76)$$

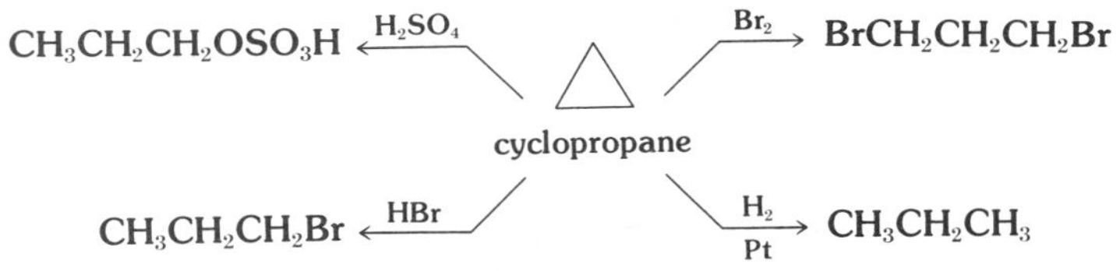

5·9 ADDITION REACTIONS OF CYCLOPROPANE

$CH_3CH_2CH_2OSO_3H \xleftarrow{H_2SO_4}$ cyclopropane $\xrightarrow{Br_2} BrCH_2CH_2CH_2Br$

$CH_3CH_2CH_2Br \xleftarrow{HBr}$ cyclopropane $\xrightarrow[Pt]{H_2} CH_3CH_2CH_3$

FIGURE 5·7 *Addition reactions of cyclopropane*

This reactivity of cyclopropane toward electrophilic reagents is understandable when we consider the valence electron distribution in the carbon-carbon bonds. The carbon atoms of a cyclopropane ring form an equilateral triangle with apparent carbon-bond angles of 60°. The angles between atomic orbitals of carbon cannot, however, be smaller than 90° (the angle between p orbitals). We see, then, that regardless of the hybridization state of carbon, the overlapping atomic orbitals that form the carbon-carbon bonds in cyclopropane must lie outside the triangle described by the three carbon nuclei.

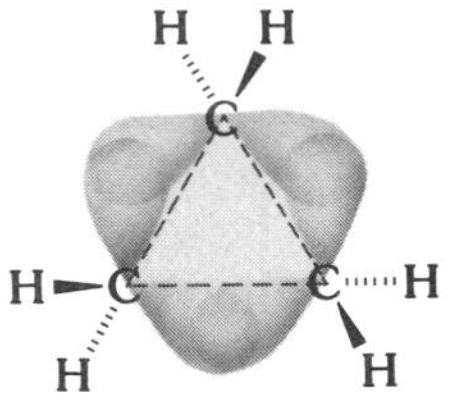

structure of cyclopropane

This concentration of bonding electron density away from the internuclear axis resembles the τ-orbital description of a double bond (Figure 4.10) and constitutes a nucleophilic region which can be attacked by electrophilic reagents:

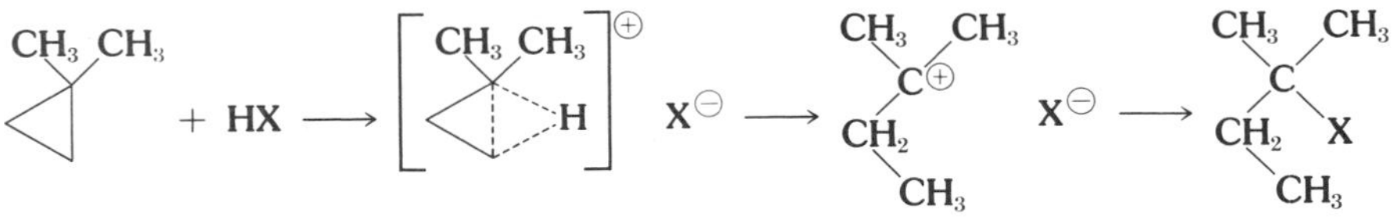

$$\triangle\!\!\!\!\!\!{}^{CH_3\ CH_3} + HX \longrightarrow \left[\triangle\!\!\!\!\!\!{}^{CH_3\ CH_3}\!\!:\!H \right]^{\oplus} X^{\ominus} \longrightarrow \underset{CH_3}{\overset{CH_3\ CH_3}{C^{\oplus}}}\!\!-\!CH_2 \quad X^{\ominus} \longrightarrow \underset{CH_3}{\overset{CH_3\ CH_3}{C}}\!\!-\!CH_2\!-\!X \quad (5.77)$$

The unsaturated character of cyclopropanes does not mean a complete mimicry of alkene chemistry. For example, the three-membered ring is normally unreactive to potassium permanganate and ozone, and this difference in behavior allows us to distinguish cyclopropanes from olefins by chemical means. Pmr measurements also provide a quick and reliable means of distinguishing cyclopropane moieties from double bonds. Protons directly bonded to cyclopropane rings generally give rise to high-field pmr signals ($\delta \approx 0.4$), whereas signals from vinyl protons are always found at much lower fields ($\delta \approx 4.5$ to 5.5). Of course, if there are no protons directly bonded to these structural units, this method of discrimination is not effective.

PROBLEM 5·25

Three C_5H_{10} isomers A, B, and C exhibit the following properties:

1 A and B decolorize a test solution of Br_2 in CCl_4, whereas C does not react.
2 A and C are inert to dilute aqueous permanganate solutions and to brief treatment with ozone solutions.

3 B reacts rapidly with ozone in CH_2Cl_2 solution, and on workup with basic hydrogen peroxide the reaction gives acetone (CH_3COCH_3) and an unidentified water-soluble carboxylic acid.

4 The pmr spectrum of A shows two sharp resonance signals, at $\delta = 0.6$ and 1.2, with a relative area of 2:3. The pmr spectrum of C shows only a single sharp signal at $\delta = 1.5$.

Write structural formulas for A, B, and C and equations for the reactions described here.

PROBLEM 5·26
Write two other ring-opening addition products that could conceivably be formed in reaction 5.77. Discuss the feasibility of these reactions in comparison with the one illustrated in equation 5.77.

SUMMARY

alkene reactions Characteristic chemical transformations involving the carbon-carbon double bond. Alkenes react with electrophilic reagents (acids, halogens, diborane, ozone, etc.) and with free radicals (Br·, Cl_3C·, etc.) and carbenes, with addition being the primary course. In some oxidation reactions initial addition may be followed by carbon-carbon bond cleavage at the site of the original double bond.

oxidation The replacement of hydrogen atoms or carbon groups by electronegative atoms or groups (F, Cl, Br, OR, SR, NR_2, etc.), or the elimination of hydrogen, as in the formation of multiple bonds.

reduction The addition of hydrogen to a multiple bond or the substitution of hydrogen for more electronegative atoms or groups.

Markovnikov's rule The principle that when a Brønsted acid (HA) is added to an unsymmetrical alkene, the hydrogen atom becomes bonded to that carbon atom of the double bond which already bears more hydrogen atoms; this rule is explained by a two-step mechanism proceeding via the more stable carbonium-ion intermediate.

carbonium-ion rearrangement A reaction in which a primary or secondary carbonium-ion intermediate having a highly branched neighboring carbon atom is transformed to a more stable carbonium ion by a 1,2 shift of one of the neighboring substituents.

electrophilic halogen addition The addition of halogens (X_2) and hypohalous acids (HOX) to alkenes via a halonium-ion intermediate. The intermediate proposed for this two-step mechanism is less prone to rearrangement than a simple carbonium ion and generally undergoes inversion of configuration in reactions with nucleophiles, resulting in a trans orientation of the entering groups.

diborane addition A concerted (single-step) reaction of diborane (B_2H_6) with alkenes which results in cis addition of hydrogen and boron; also called *hydroboration*. The alkylboranes thus formed can be oxidized to alcohols by the action of alkaline hydrogen peroxide. The result is an apparent anti-Markovnikov addition of water to the initial alkene.

hydroxylation Oxidation of an alkene to a vicinal (neighboring) dihydroxy derivative by $MnO_4^{\ominus}$ or OsO_4.

ozonolysis Oxidative cleavage of alkenes to carbonyl compounds effected by treatment with ozone (O_3), followed by oxidative (H_2O_2) or reductive (Zn dust) workup.

polymerization The stepwise combination of simple molecules (monomers) such as alkenes to form high-molecular-weight chains composed of repeating units (polymers).

carbenes Neutral divalent carbon intermediates, such as :CH$_2$, which add rapidly to alkenes to give cyclopropane products.

cyclopropane addition reactions Ring-opening attack on cyclopropane bonding orbitals by electrophilic reagents such as strong Brønsted acids and halogens, resulting in the formation of substituted propane products.

EXERCISES

5·1 Describe simple chemical tests that could be used to distinguish between:
a Pentane and 1-pentene
b 2-Pentene and ethylcyclopropane
c Ethylcyclopropane and pentane

5·2 Indicate the products of the following reactions. Specify stereochemistry when appropriate.

a decalin-type alkene + Br$_2$ / CCl$_4$

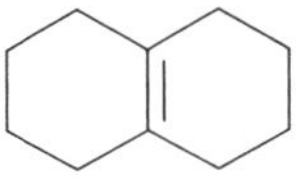

b decalin-type alkene + O$_3$, then H$_2$O, Zn

c decalin-type alkene + OsO$_4$, then NaHSO$_3$

d decalin-type alkene (with H shown) + HCl ⟶ (2 isomers)

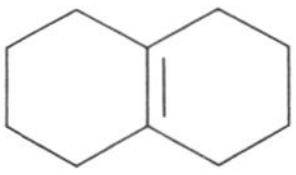

e decalin-type alkene (with H's shown) + H$_3$O$^{\oplus}$

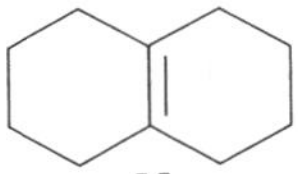

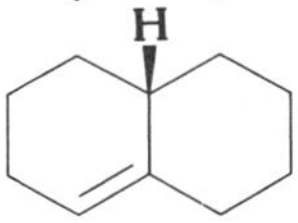

f decalin-type alkene (with H's shown) + B$_2$H$_6$, ether, then H$_2$O$_2$, NaOH

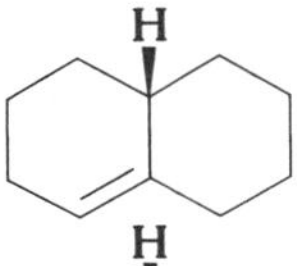

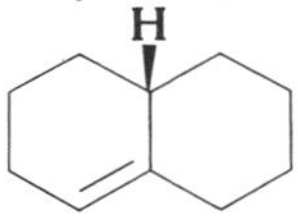

5·3 Write a balanced equation for each of the following reactions:
a CH$_3$CH=CH$_2$ + KMnO$_4$ ⟶ CH$_3$CH(OH)CH$_2$(OH) + MnO$_2$ + KOH
b CH$_3$CH$_2$CH=CH$_2$ + B$_2$H$_6$ ⟶ trialkylborane

5·4 Show the products of the following ionic and free-radical addition reactions:

a CH$_3$CH$_2$CH=CH$_2$ + HCl ⟶

b CH$_3$CH$_2$CH=CH$_2$ + HBr + ROOR ⟶

c CH$_3$CH$_2$CH=CH$_2$ + HBr (no peroxides) ⟶

d (CH$_3$)$_2$C=CH$_2$ + CCl$_4$ + ROOR ⟶

e cyclopentene + HSCH$_2$CH$_3$ + ROOR ⟶

f cyclopentene + H$_2$SO$_4$ ⟶

5·5 Identify the structures of the starting materials.

a A $\xrightarrow{\text{H}_2\text{O, Br}_2}_{\text{NaHCO}_3}$ CH$_3$CH$_2$CHCH$_2$Br
                                                                                    |
                                                                                    OH

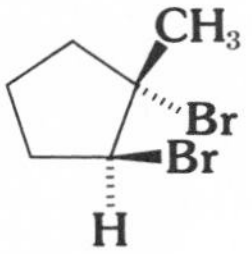

b B $\xrightarrow{\text{B}_2\text{H}_6}$ $\xrightarrow[\text{OH}^\ominus]{\text{H}_2\text{O}_2}$ (CH$_3$)$_2$CHCHCH$_3$
                                                                                         |
                                                                                         OH

c C $\xrightarrow[\text{CCl}_4]{\text{Br}_2}$ [cyclopentane with CH$_3$ (up), Br, Br (down), H]

d D $\xrightarrow{\text{OsO}_4}$ $\xrightarrow{\text{NaHSO}_3}$ [cyclopentane with H, OH, OH, H]

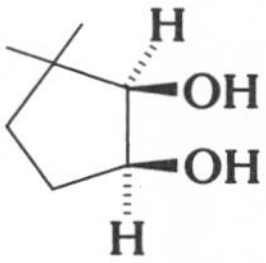

5·6 Write structural formulas for the alkenes that give the following products on treatment with ozone, followed by workup with zinc dust:

a (CH$_3$)$_2$C=O and (CH$_3$)$_2$CHCHO

b [cyclopentanone] and CH$_3$CH$_2$CHO

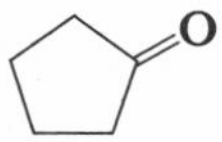

c CH$_3$ĊCH$_2$CH$_2$CH$_2$Ċ—H
      ‖   ‖
      O   O

5·7 Show the structure of the compounds identified by letters:

a A (C$_5$H$_{10}$) + CH$_2$I$_2$ + Zn(Cu) ⟶ B $\xrightarrow{\text{H}_2\text{, Pt}}$ [cyclohexane] + [methylcyclopentane]

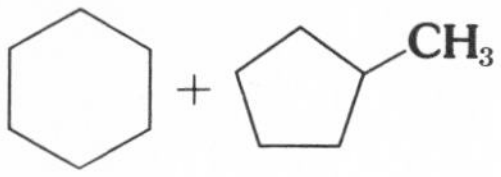

b cis-2-butene + CHBr$_3$ + t-C$_4$H$_9$OK ⟶ E
   trans-2-butene + CHBr$_3$ + t-C$_4$H$_9$OK ⟶ F

c [bicyclic compound] + Br$_2$ ⟶ C + D
                                    isomers

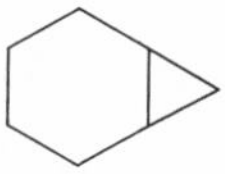

5·8 Indicate the reagents and reaction conditions necessary to transform 1-methylcyclohexene into each of the following compounds:

a [cyclohexane with CH$_3$, OH]

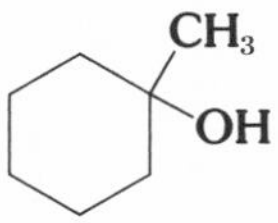

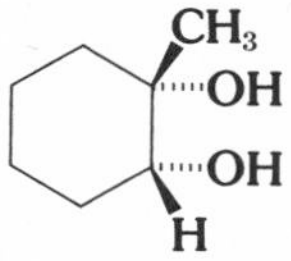

b [cyclohexane with CH$_3$, Br (up), H, Br]

c [cyclohexane with CH$_3$, OH, OH, H]

d [cyclohexane with CH$_3$, H (up), OH, H]

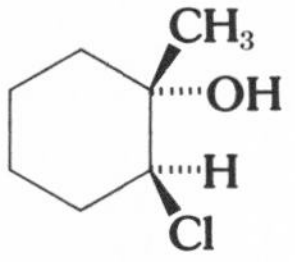

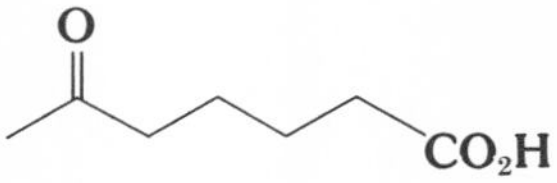

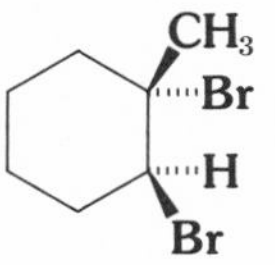

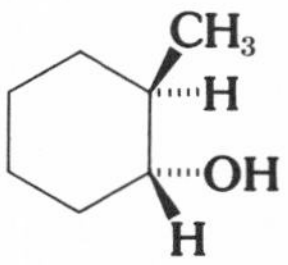

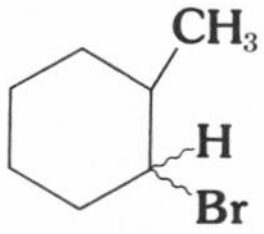

e [cyclohexane with CH$_3$, OH, H, Cl]

f [cyclohexane with CH$_3$, H, Br] mixture of isomers

g [6-oxoheptanoic acid type structure] CO$_2$H

h [cyclohexane with CH$_3$, Br]

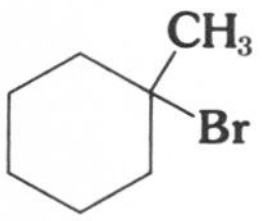

5·9 Are the ΔH values for the following reactions identical? Explain.

cis-2-butene + HCl ⟶
                           CH$_3$CH$_2$CHCH$_3$
trans-2-butene + HCl ⟶ |
                                Cl

5·10 Suggest a reason why polyisobutylene (equation 5.60) has a repeating head-to-tail monomer orientation in the polymer. Would you expect a similar head-to-tail orientation if the polymerization were free-radical initiated?

154 EXERCISES

5·11 What effect does the methanol, acting as a nucleophile, have on the rate of reaction 5.11? Will doubling the methanol concentration double the rate of disappearance of propylene?

5·12 The sex-attractant pheromone of the housefly (*Musca domestica* L.) has the molecular formula $C_{23}H_{46}$, decolorizes Br_2 in CCl_4 and $KMnO_4$ solutions, and exhibits infrared absorption at 685 cm^{-1} (cis double bond). Ozonolysis of this substance, followed by workup with H_2O_2, gives $n\text{-}C_8H_{17}CO_2H$ and $n\text{-}C_{13}H_{27}CO_2H$. What is the structure of the pheromone?

5·13 Myrcene ($C_{10}H_{16}$), a hydrocarbon isolated from oil of bayberry, decolorizes Br_2 in CCl_4 and adds three molar equivalents of hydrogen. Ozonolysis of myrcene with a Zn dust workup yields the following products:

$$CH_3\overset{O}{\overset{\|}{C}}CH_3 + H\overset{O}{\overset{\|}{C}}-CH_2CH_2-\overset{O}{\overset{\|}{C}}-\overset{O}{\overset{\|}{C}}-H + 2H-\overset{O}{\overset{\|}{C}}-H$$

What structure or structures are consistent with these facts?

5·14 Each of the following conversions can be accomplished in three or fewer steps. Indicate the intermediate compounds and the reagents necessary to effect each reaction.

a $CH_3CH_2CH_2Br \longrightarrow CH_3\overset{\overset{Cl}{|}}{C}HCH_3$

b [cyclohexane with Cl] $\longrightarrow HO_2C(CH_2)_4CO_2H$

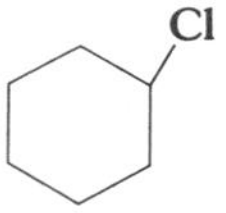

c $(CH_3)_3CCl \longrightarrow (CH_3)_2CHCH_2OH$

d [cyclopropane] $\longrightarrow CH_3CHClCH_3$

ALKYNES

THE ALKYNES are hydrocarbons containing one or more carbon-carbon triple bonds. These compounds include gases, liquids, and solids and have melting points, boiling points, and densities that increase with molecular weight in much the same way as the corresponding properties of alkanes and alkenes. In general, the alkynes have slightly higher boiling points and densities than structurally equivalent alkanes, suggesting that intermolecular forces are a bit stronger for alkynes.

The simplest alkyne is acetylene (C_2H_2), a gas which is used as a fuel in high-temperature torches. Many more complex derivatives of this functional group have been isolated from several varieties of green plants and fungi. For example, hydrocarbon I is found in some ordinary garden flowers such as dahlia and coreopsis, and compound II is one of several toxic substances that cause convulsions and death in cattle feeding on certain genera of the Umbelliferae family. An unstable antibiotic called mycomycin has been isolated from a funguslike actinomycete. Although it is highly active against tubercle bacilli, pure mycomycin does not survive storage above −40°C and is therefore of little medicinal value.

$$CH_3-CH=CH-C\equiv C-C\equiv C-C\equiv C-C\equiv C-CH=CH_2$$
I (from dahlia flowers)

$$HO-CH_2CH=CHC\equiv CC\equiv CCH=CHCH=CHCH_2CH_2CH(OH)CH_2CH_2CH_3$$
II (from water hemlock)

$$HC\equiv C-C\equiv C-CH=C=CH-CH=CH-CH=CH-CH_2-CO_2H$$
III (from the actinomycete *Nocardia acidophilus*)

6·1 NOMENCLATURE OF ALKYNES

Acetylene itself can be prepared in two steps from coke (carbon) by first effecting a high-temperature reaction with lime, followed by hydrolysis of the resulting calcium carbide:

$$3C + CaO \xrightarrow{\text{heat}} CO + \underset{\substack{\text{calcium}\\\text{carbide}}}{CaC_2} \xrightarrow{2H_2O} HC{\equiv}CH + Ca(OH)_2 \tag{6.1}$$

More complex alkynes can be synthesized by elimination reactions of dihalides:

$$\begin{array}{l}RCHBrCHBrR' \xrightarrow[-2HBr]{\text{strong base}} \searrow \\ \qquad\qquad\qquad RC{\equiv}CR' \\ RCBr_2CH_2R' \xrightarrow[-2HBr]{\text{strong base}} \nearrow \end{array} \tag{6.2}$$

Another method is the alkylation of acetylide bases, described later in the chapter.

6·1 NOMENCLATURE OF ALKYNES

The IUPAC nomenclature system treats alkynes in essentially the same way as alkenes, the characteristic ending in this case being *-yne*. Table 6.1 lists some examples, along with the common names of these compounds. Common acetylene groups have also been given names:

HC≡C—   HC≡CCH$_2$—
ethynyl   propargyl

The nomenclature rules and principles outlined in Chapter 4 are illustrated for alkynes by the following examples:

HC≡CCH$_2$Br   (CH$_3$)$_3$CC≡CCH$_3$ (CH$_2$)$_8$ with C≡C ring

3-bromopropyne   4,4-dimethyl-2-pentyne   cyclodecyne
or propargyl bromide

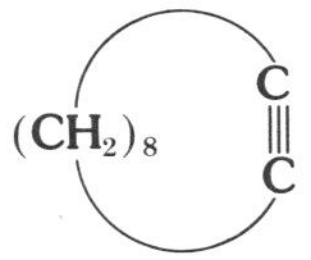

[cyclohexane with Cl and C≡CH]   HC≡CC≡CCH$_2$CH=CH$_2$

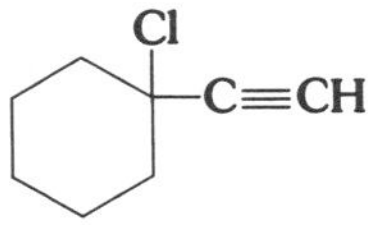

1-chloro-1-ethynyl-   6-hepten-1,3-diyne
cyclohexane

As shown in the last of these examples, the lowest possible numbers are given to the double- and triple-bonded carbon atoms.

PROBLEM 6·1

a Write a structural formula for 1,1,1-trifluoro-4-ethyl-2-hexyne.
b Write a structural formula for *cis*-1,2-diethynylcyclobutane.
c Name the dahlia extract on page 155 according to the IUPAC system.

TABLE 6·1 *Alkyne nomenclature*

molecular formula	structural formula	IUPAC name	common name
C_2H_2	H—C≡C—H	ethyne	acetylene
C_3H_4	CH_3—C≡C—H	propyne	methylacetylene
C_4H_6	C_2H_5—C≡C—H	1-butyne	ethylacetylene
C_4H_6	CH_3—C≡C—CH_3	2-butyne	dimethylacetylene
. . .	. . .	. . .	
C_nH_{2n-2}	R—C≡C—R'	alkyne	

6·2 STRUCTURE AND PROPERTIES OF ALKYNES

STRUCTURE-COMPOSITION RELATIONSHIP

The alkynes listed in Table 6.1 are the beginning of a homologous series of hydrocarbons having the general molecular formula C_nH_{2n-2}. When we compare this composition with that of the alkanes (C_nH_{2n+2}), the cycloalkanes (C_nH_{2n}), the alkenes (C_nH_{2n}), and the cycloalkenes (C_nH_{2n-2}), a useful structure-composition relationship emerges: *Hydrocarbons with no rings or double bonds (alkanes) have a maximum ratio of hydrogen to carbon, and this ratio is reduced by a factor equivalent to two hydrogen atoms per molecule for every ring or double bond that is introduced.* The effect of a triple bond is the same as that of two double bonds—that is, the number of hydrogen atoms per molecule is reduced by four. To determine the number of rings and/or double bonds that must be present in an unknown hydrocarbon, we simply note the difference between the hydrogen content of the unknown compound and that of the corresponding alkane, C_xH_{2x+2}, and divide by 2. Thus for an unknown hydrocarbon C_xH_y,

$$\frac{\text{hydrogen content of corresponding alkane} - \text{hydrogen content of unknown hydrocarbon}}{2} = \text{rings or double bonds in sample molecule}$$

$$\frac{2x + 2 - y}{2} = \text{rings} + \text{double bonds} \quad (6.3)$$

The number of double and triple bonds in a molecule can often be established by measuring the amount of hydrogen required to obtain a saturated system:

$$\begin{array}{c} \underset{H}{\overset{R}{\diagdown}}C=C\underset{R}{\overset{H}{\diagup}} \xrightarrow{H_2, Pd} \\ RC{\equiv}CR \xrightarrow{2H_2, Pd} \end{array} RCH_2CH_2R \quad (6.4)$$

Hence we can easily determine the number of rings by a simple computation: (rings + double bonds) − double bonds = rings.

6·2 STRUCTURE AND PROPERTIES OF ALKYNES

The useful relationship described here is not limited to hydrocarbons. For compounds containing divalent elements such as oxygen or sulfur there is no change in the method of calculation. However, the presence of halogen and nitrogen substituents necessitates a modified approach.

PROBLEM 6·2

a Show why the presence of the divalent element oxygen does not change the method of calculating the number of rings and double bonds in a compound.

b Calculate the number of rings and double bonds present in molecules having the following compositions:

C_6H_6 $C_5H_{10}O_3$ $C_{10}H_{16}SO$

c A compound having the molecular formula $C_7H_{10}O$ adds one equivalent of hydrogen in the presence of a palladium catalyst to give a substance with the formula $C_7H_{12}O$. What conclusions can be drawn about the structures of these compounds?

d Suggest a modified procedure for calculating the number of rings and double bonds when halogen atoms are present in a compound.

STRUCTURAL AND BONDING CHARACTERISTICS

X-ray diffraction analysis of several acetylenic compounds has shown that the triple-bond function has a linear configuration. The short carbon-carbon triple-bond length (1.20 Å) observed in acetylene indicates that this bond is quite strong, a fact confirmed by heat-of-combustion measurements.

1.06 Å
H——C≡≡≡C——H
  1.20 Å
structure of acetylene

The linear configuration of the alkynes allows no stereoisomerism about the triple bond, but it does restrict the stable cycloalkynes to rings composed of eight or more carbon atoms, since a smaller ring would severely distort the π bonds.

PROBLEM 6·3

The heat of combustion of acetylene is about 300 kcal/mole (assuming all products are gaseous). Using the bond energies for O=O, O—H, C—H, and C=O (for CO_2) in Table 2.4, calculate the triple-bond energy for acetylene.

PROBLEM 6·4

Suggest a hypothetical configuration for the triple bond that would permit stereoisomerism about this function.

The most commonly used theoretical model for the triple bond is the σ-π molecular orbital model. Since each carbon atom in this functional group is bonded to only two other atoms or groups, the σ-bonded framework can be constructed by using only *two* equivalent hybrid orbitals per carbon atom. The best orbitals for this purpose are generated from the $2s$ and one of the $2p$ atomic carbon orbitals and are therefore re-

6·2 STRUCTURE AND PROPERTIES OF ALKYNES   159

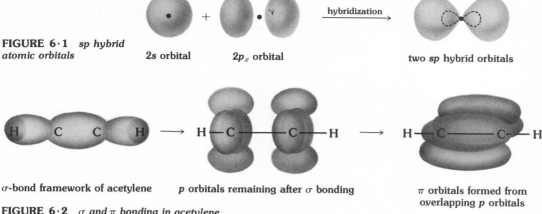

FIGURE 6·1 *sp hybrid atomic orbitals*

FIGURE 6·2 *σ and π bonding in acetylene*

ferred to as *sp hybrid orbitals;* that is, they have 50% s and 50% p characteristics. These two hybrid orbitals are oriented in opposite directions (at an angle of 180°).

Once the σ-bond framework of the triple bond has formed, the two remaining p orbitals on each carbon atom can overlap to generate four new molecular orbitals: two π orbitals and two π* orbitals. The ten valence electrons in acetylene (four electrons from each carbon atom, plus two from the hydrogen atoms) can then occupy the five lower-energy bonding molecular orbitals according to the aufbau principle (the three σ orbitals are filled first, followed by the two π orbitals), so as to give a stable electron configuration for this molecule (see Figure 6.2).

PROBLEM 6·5
Draw a representation of a bent-bond model (see Figure 4.10) for acetylene. Compare this model with the σ-π model, using the important known properties of the triple bond: linear configuration, shorter bond length than the double bond, and stronger than the double bond.

SPECTROSCOPIC PROPERTIES

Alkynes exhibit characteristic infrared and pmr absorptions which may be used to identify the triple-bond function.

TABLE 6·2 *Spectroscopic properties of alkynes*

absorbing group	characteristic absorption
infrared spectrum	
C≡C (stretch)	2100–2260 cm^{-1} (variable)
C≡C—H (stretch)	3200–3300 cm^{-1} (strong)
pmr spectrum	
C≡C—H	2.0 to 3.1 (δ)
C≡C—CH$_3$	1.7 to 2.0 (δ)

PROBLEM 6·6

a A compound having the composition C_5H_4 shows strong absorptions at 3300, 2900, and 2110 cm^{-1} in its infrared spectrum and two equally intense signals, at $\delta = 3.1$ and 2.0, in its pmr spectrum. Suggest a structure for this compound.

b How many σ and π bonds are present in the structure you have written?

c Write a structural formula for an isomer of this compound, which would give the same product as the original compound on catalytic hydrogenation. Predict the pmr spectrum of this isomer.

The spectroscopic properties listed in Table 6.2 provide us with additional information about the nature of the triple bond. For example, the carbon triple-bond stretching frequency, approximately 2200 cm^{-1}, is higher than the corresponding carbon double-bond absorption, found at about 1650 cm^{-1}, and both exceed the rather variable single-bond stretching frequency, usually observed below 1100 cm^{-1}. These facts indicate that the energy required to stretch a carbon-carbon bond increases in the order

$$C-C < C=C < C\equiv C$$

This, of course, is also the order of increasing bond strength. A similar relationship is observed for the corresponding C—H stretching vibration:

$$C-H < =C-H < \equiv C-H$$

A correspondingly simple relationship is not observed for the pmr chemical shifts of C—H, C=C—H, and C≡C—H. The relatively large downfield shift observed for olefinic hydrogen atoms ($\delta \approx 5$) is thought to be due in part to the increased electronegativity of an sp^2-hybridized carbon atom, in comparison with the sp^3-hybridized carbon atoms in alkanes. However, this line of reasoning would lead us to predict an even greater downfield chemical shift for a hydrogen bonded to an sp-hybridized carbon atom. Since these acetylene hydrogens resonate at unexpectedly high fields, it is possible that an additional shielding factor is operating near the triple bond. Specifically, it has been proposed that the applied magnetic field induces a circular motion in the cylindrical shell of π electrons surrounding the triple bond, and that this induced current generates a secondary magnetic field, as shown in Figure 6.3. This secondary field opposes the applied field in some regions (shielding) and complements it in other regions (deshielding).

ACIDITY OF ETHYNYL HYDROGEN ATOMS

Alkynes having a triple bond at the end of a carbon chain are called *terminal acetylenes*, and the hydrogen attached to a terminal triple bond is termed an *ethynyl hydrogen atom*. Ethynyl hydrogen atoms are weakly acidic, and insoluble salts of silver and copper are formed when terminal acetylenes are added to alkaline solutions of these metal ions:

$$R-C\equiv C-H + Ag(NH_3)_2^{\oplus} \xrightarrow{NH_4OH} R-C\equiv C-Ag + NH_3 + NH_4^{\oplus} \tag{6.5}$$

6·2 STRUCTURE AND PROPERTIES OF ALKYNES

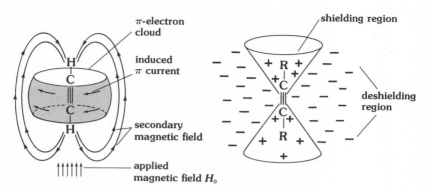

FIGURE 6·3 *Shielding and deshielding regions about a triple bond*

The acidity of ethynyl hydrogen atoms is too weak to be measured directly in aqueous solution (note problem 3.20 in Section 3.8). However, measurements in less acidic solvents, such as liquid ammonia, allow us to conclude that the ionization constant K_a in water would be smaller than 10^{-23}. Since ammonia is an even weaker acid ($K_a = 10^{-35}$), the sodium and potassium salts of terminal acetylenes can be formed by treatment with the very strong base sodium (or potassium) amide (see Figure 3.14):

$$RC\equiv CH + Na^{\oplus}NH_2^{\ominus} \xrightleftharpoons{NH_3(l)} RC\equiv C:^{\ominus}Na^{\oplus} + NH_3 \qquad (6.6)$$
stronger acid   stronger base   weaker base   weaker acid

The strong base sodium acetylide ($NaC\equiv CH$) is also a powerful nucleophile and participates in displacement reactions with primary alkyl halides. However, with the more hindered secondary and tertiary alkyl halides, acetylide ions act almost exclusively as Brønsted bases and give elimination reactions.

$$HC\equiv CNa + C_2H_5Br \xrightarrow{NH_3(l)} HC\equiv CC_2H_5 + NaBr \qquad (6.7)$$

$$C_2H_5C\equiv CNa + CH_3CH_2CH_2Br \xrightarrow{NH_3(l)} C_2H_5C\equiv CC_3H_7 + NaBr \qquad (6.8)$$

$$RC\equiv CNa + (CH_3)_3CCl \xrightarrow{NH_3(l)} \underset{CH_3}{\overset{CH_3}{\diagdown}}C=CH_2 + RC\equiv CH + NaCl \qquad (6.9)$$

Although terminal acetylenes are very weak acids in comparison with water, their acidity is nevertheless more than 10^{20} times stronger than that of the corresponding alkanes. This fact is nicely illustrated by the ease with which these alkynes undergo hydrogen-metal exchange with alkyl Grignard reagents; in fact, the preparation of acetylenic Grignard reagents is usually accomplished in this way:

$$RC\equiv CH + C_2H_5MgBr \xrightarrow{ether} RC\equiv CMgBr + C_2H_6 \qquad (6.10)$$
stronger acid   stronger base   weaker base   weaker acid

6·3 ADDITION REACTIONS OF ALKYNES

HYDROGEN ADDITION

This large difference in the acidity of carbon-bonded hydrogen atoms (carbon acids) is understandable if we consider the change in the hybridization of the carbon orbitals used in forming C—H σ bonds. An alkyne carbon atom forms such bonds by using an orbital with 50% s character (compared with 25% s character for an alkane carbon atom), and the electron pair in the σ bond is therefore closer to and more strongly held by the carbon nucleus (recall the shapes of $2s$ and $2p$ orbitals). Since an electron in an s orbital has a lower potential energy than an electron in a related p orbital, we expect the conjugate base formed by abstraction of an ethynyl proton to be more stable than the base derived from an alkane.

6·3 ADDITION REACTIONS OF ALKYNES

HYDROGEN ADDITION

The strongly exothermic addition of molecular hydrogen to a carbon-carbon triple bond, as shown in reaction 6.11, proceeds expeditiously only when a transition-metal catalyst is present. Although complete saturation of the triple-bond function normally takes place, this two-step reaction can be stopped at an intermediate cis-disubstituted alkene stage by using a deactivated, or "poisoned," catalyst, such as palladium treated with lead acetate, nitrogen bases, or sulfur-containing compounds.

$$HC \equiv CH \xrightarrow[Pd]{H_2} H_2C = CH_2 \xrightarrow[Pd]{H_2} CH_3 - CH_3 \tag{6.11}$$

overall $\Delta H = -75$ kcal/mole

$$RC \equiv CR' \xrightarrow[Pd \text{ (poisoned)}]{H_2} \begin{matrix} H & & H \\ & \diagdown \diagup \\ & C=C \\ & \diagup \diagdown \\ R & & R' \end{matrix} \tag{6.12}$$

$\Delta H \approx -40$ kcal/mole

The heat of hydrogenation for the addition of one equivalent of hydrogen to the triple bond of acetylene ($\Delta H = -42$ kcal/mole) can be measured directly, or it can be calculated by subtracting the heat of hydrogenation for ethylene ($\Delta H = -32.8$ kcal/mole) from the heat of reaction for the complete reduction of acetylene (see equation 6.11). Similar measurements with other alkynes confirm the thermodynamic instability of the carbon-carbon triple bond relative to the corresponding double and single bonds; however, we find little correlation between this thermodynamic driving force and the actual chemical reactivity of the alkynes.

The common influence of transition-metal catalysts on the rate of hydrogen addition to triple bonds and double bonds suggests that alkynes will form stable transition-metal complexes similar to those observed for some alkenes. Examples of such complexes are in fact known for an impressive array of metals, including iron, cobalt, nickel, rhodium, platinum, and manganese. In at least one case an alkyne has actually displaced the alkene from an existing olefin complex:

$$K[PtCl_3 \cdot C_2H_4] + (CH_3)_3CC \equiv CC(CH_3)_3 \xrightarrow{\text{acetone}} K[PtCl_3 \cdot C_{10}H_{18}] + H_2C = CH_2 \tag{6.13}$$

6·3 ADDITION REACTIONS OF ALKYNES

The nature of the metal-acetylene bonding in these complexes is not sufficiently well understood to warrant discussion here; however, π-d overlap, as described in Figure 5.4, is believed to be significant in many compounds. Chemists consider further study of these complexes important, since many addition reactions to alkynes employ transition-metal compounds as catalysts.

REACTIONS WITH ELECTROPHILIC REAGENTS

HALOGEN The triple bond is surprisingly sluggish in its reactions with electrophilic reagents. For example, a comparison of the rates of halogen addition to an alkyne and an alkene of comparable structure shows that the triple bond is considerably less susceptible to electrophilic halogen attack than is the double bond. This difference is sufficiently great (500 to 50,000 times) to permit selective addition of bromine to pent-4-en-1-yne:

$$CH_2=CHCH_2C\equiv CH + Br_2 \longrightarrow CH_2BrCHBrCH_2C\equiv CH \quad (6.14)$$

The reaction of chlorine or bromine with acetylene proceeds first to a 1,2-dihaloethene, followed by addition of a second equivalent of halogen to give a 1,1,2,2-tetrahaloethane. Even iodine adds slowly to acetylene, generating *trans*-1,2-diiodoethene as the major product.

$$HC\equiv CH + X_2 \longrightarrow XHC=CXH \xrightarrow{X_2} CHX_2-CHX_2 \quad (6.15)$$
$$(X = Cl, Br) \qquad\qquad \text{1,1,2,2-tetrahaloethane}$$

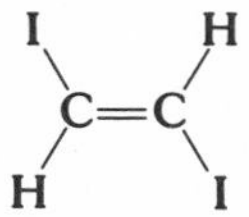

$$HC\equiv CH + I_2 \longrightarrow \underset{\text{\textit{trans}-1,2-diiodoethene}}{\overset{I\quad\quad H}{\underset{H\quad\quad I}{C=C}}} + \text{trace of cis isomer} \quad (6.16)$$

BRØNSTED ACIDS Addition reactions of Brønsted acids to alkynes do not usually proceed as readily as comparable reactions with alkenes. However, this difficulty can be circumvented by using transition-metal salts such as $Cu^{\oplus}$ and $Hg^{\oplus\oplus}$ as catalysts.

$$HC\equiv CCH_2Cl + HCl \xrightarrow{HgCl_2} CH_2=CClCH_2Cl \quad (6.17)$$

$$HC\equiv CH + CH_3CO_2H \xrightarrow{HgSO_4} \underset{\text{vinyl acetate}}{CH_2=CHOCOCH_3} \quad (6.18)$$

$$HC\equiv CH + HCN \xrightarrow{Cu_2Cl_2\ NH_4Cl} \underset{\text{acrylonitrile}}{CH_2=CHCN} \quad (6.19)$$

Noncatalyzed as well as catalyzed additions of Brønsted acids proceed in accord with the Markovnikov rule (see equation 6.17):

$$CH_3C\equiv CH + 2HF \xrightarrow{-20°} CH_3CF_2CH_3 \quad (6.20)$$

PROBLEM 6·7
Write equations showing how 2-butyne can be converted to:
a 2,3-Dibromobutane
b 2,2-Dibromobutane
c 2,2,3,3-Tetrabromobutane

6·3 ADDITION REACTIONS OF ALKYNES

HYDRATION AND TAUTOMERISM

Addition of water to a triple bond (hydration) is easily accomplished in dilute mineral acid solution if a mercuric salt is added as a catalyst:

$$HC\equiv CH + H_2O \xrightarrow[HgSO_4]{H_3O^\oplus} \underset{enol}{H_2C=C\begin{smallmatrix}OH\\H\end{smallmatrix}} \rightleftarrows \underset{acetaldehyde}{CH_3-C\begin{smallmatrix}O\\H\end{smallmatrix}} \quad (6.21)$$

$$CH_3C\equiv CH + H_2O \xrightarrow[HgSO_4]{H_3O^\oplus} \underset{enol}{CH_3-\underset{|}{\overset{OH}{C}}=CH_2} \rightleftarrows \underset{acetone}{CH_3-\overset{O}{\underset{\|}{C}}-CH_3} \quad (6.22)$$

The initially formed unsaturated alcohols, called *enols*, cannot be isolated because they rearrange rapidly and completely to the corresponding carbon-oxygen double-bonded isomers. The thermodynamic driving force for this isomerization is given by the change in free energy ΔG; this in turn is most conveniently estimated by calculating the heat of reaction ΔH from bond energies, as described in Section 2.3. The computations in Table 6.3 show that the carbonyl isomer (C=O) has a lower energy (by about 15 kcal/mole) and is therefore more stable than the enol isomer (C=C—OH).

An energy difference of this magnitude between two species in equilibrium causes the more stable one to predominate by a factor greater than 10,000:1. We must bear in mind, however, that a large thermodynamic driving force only indicates the position of equilibrium; it does not necessarily tell us how rapidly the equilibrium will be established. A good illustration of this point is a comparison of the enol-carbonyl isomerization in Table 6.3 with the following hypothetical rearrangement of the C_3H_6O isomers:

$$CH_2=CH-O-CH_3 \not\rightleftarrows CH_3-CH_2-C\begin{smallmatrix}O\\H\end{smallmatrix} \quad (6.23)$$

$\Delta H_{calc} = -24.4$ kcal/mole

TABLE 6·3 *Bond-energy changes in tautomerism*

\begin{matrix}H\\ \diagdown\\ \\ \diagup\\ H\end{matrix}C=C\begin{matrix}OH\\ \\ \\ \\ H\end{matrix}		→	H—C—C\begin{matrix}H\\ |\\ \\ |\\ H\end{matrix}\begin{matrix}O\\ \diagup\!\!\diagup\\ \\ \diagdown\\ H\end{matrix}	
3C—H	3 × 98.7		4C—H	4 × 98.7
C=C	145.8		C—C	82.6
C—O	85.5		C=O	176.0
O—H	110.6			
total bond energy 638.0			total bond energy 653.3	

$\Delta H = 638.0 - 653.3 = -15.3$ kcal/mole

6·3 ADDITION REACTIONS OF ALKYNES

Bond-energy calculations point to a large negative ΔH for this isomerization; however, the two compounds are quite stable in the absence of acids, and chemists have been unable to effect either the forward or the reverse transformation. Apparently the activation energy for the hydrogen-transfer reaction (equations 6.21 and 6.22) is much smaller than the corresponding activation energy for a methyl-group transfer (equation 6.23).

Constitutional isomers that are rapidly interconverted under normal laboratory conditions are often referred to as *tautomers*, and the isomerization reaction itself is called *tautomerization*. The interconversion of products in reactions 6.21 and 6.22 are examples of proton tautomerization. The mechanism of such tautomerizations will be discussed in Chapter 12.

PROBLEM 6·8

Three C_4H_6 isomers, A, B, and C, show the following chemical reactivity:

1. In the presence of finely divided palladium all three isomers absorb hydrogen gas. When equal samples are used, B and C absorb twice as much hydrogen as A.

2. All three isomers react with hydrogen chloride, but B and C do so most easily if a little mercuric chloride is added to the reaction mixture:

$$A + HCl \longrightarrow C_4H_7Cl$$

$$B \text{ and } C + HCl \xrightarrow{HgCl_2} C_4H_8Cl_2$$
(same product from both)

3. B and C react rapidly with dilute sulfuric acid containing mercuric sulfate and give a ketone (C_4H_8O).

4. B reacts with aqueous $Ag(NH_3)_2^{\oplus}$ to give a greyish precipitate which explodes when heated.

Write structural formulas for A, B, and C and equations for all the reactions described.

PROBLEM 6·9

The unsaturated nature of alkenes and alkynes is illustrated by the many addition reactions exhibited by the double- and triple-bond functions. When we consider the following *hypothetical* reactions of the saturated hydrocarbon ethane, we discover they are thermodynamically feasible. Why are these "addition reactions to a single bond" not observed under ordinary conditions?

$$C_2H_6 + H_2 \longrightarrow 2CH_4 \quad \Delta H_{calc} = -10.6 \text{ kcal}$$

$$C_2H_6 + Cl_2 \longrightarrow 2CH_3Cl \quad \Delta H_{calc} = -21.4 \text{ kcal}$$

PROBLEM 6·10

In the presence of peroxide initiators hydrogen bromide adds to 1-hexyne in an anti-Markovnikov fashion as shown. Write equations illustrating the mechanism of this addition reaction.

$$C_4H_9C{\equiv}CH + HBr \xrightarrow{peroxides} C_4H_9CH{=}CHBr \xrightarrow[peroxides]{HBr} C_4H_9CHBrCH_2Br$$

6·3 ADDITION REACTIONS OF ALKYNES

THE NUCLEOPHILICITY OF ALKYNES

The sluggish reactivity of alkynes (in comparison with alkenes) to common electrophilic reagents is at first glance surprising. Since addition reactions of alkynes are more exothermic than corresponding alkene additions, we might expect alkynes to be more reactive than alkenes. The error here lies again in the assumption that relative thermodynamic stabilities are the determining factor in reaction rates. However, the rate of a reaction is dependent on the activation energy of the rate-determining step, not on the overall energy released (or consumed) in going to products. Note in Figure 6.4 that the activation energy (ΔE_3^*) for addition to a triple bond (C≡C + Y—Z ⟶ Y—C=C—Z) is larger than the corresponding activation energy (ΔE_2^*) for addition to an alkene (C=C + Y—Z ⟶ Y—C—C—Z). The rate of addition by an electrophilic reagent Y—Z to an alkene will therefore be greater than the rate of addition to an alkyne, despite the fact that the latter reaction is more exothermic ($\Delta H_3 > \Delta H_2$).

This important distinction between thermodynamic and kinetic stability does not explain why alkynes are generally less susceptible than alkenes to attack by electrophilic reagents. Clearly, this fact implies that a triple bond is less nucleophilic than a double bond, even though it has twice the number of π electrons. This, in turn, must mean that the π electrons in a triple bond are more tightly held and therefore less easily perturbed by an approaching electrophile than are the π electrons in a double bond. The respective *ionization potentials* of alkenes and alkynes—that is, minimum energy required to remove an electron from a molecule of each compound—should reflect the strength with which the double- and triple-bond functions hold and restrain their π electrons. We can see from the experimental values for ethene and ethyne that π electrons are in fact more tightly held in a triple bond:

$$H_2C=CH_2 \longrightarrow [H_2C=CH_2]^{\oplus} + e^{\ominus} \quad \Delta H = +242.5 \text{ kcal}$$
ethene

$$HC\equiv CH \longrightarrow [HC\equiv CH]^{\oplus} + e^{\ominus} \quad \Delta H = +259.3 \text{ kcal}$$
ethyne

(6.24)

FIGURE 6·4 *Energy profiles for electrophilic addition to double and triple bonds*

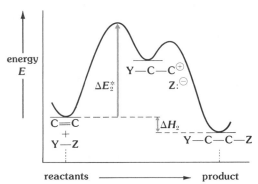
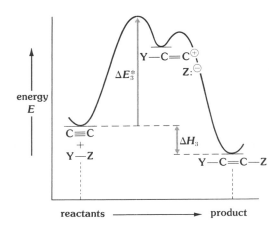

6·3 ADDITION REACTIONS OF ALKYNES

What factors, then, are responsible for this variation in binding affinity for π electrons? To answer this question we turn again to the difference in hybridization of the unsaturated carbon atoms. The greater electronegativity of sp-hybridized carbon atoms, in comparison with the sp^2- and sp^3-hybridized states, is borne out by the acidity of ethynyl hydrogen atoms. If this electron affinity extends to the π electrons, we would expect the π-electron cloud around a triple bond to be more constricted and tightly bound than that around a double bond. The lower nucleophilicity of alkynes thus becomes quite reasonable.

REACTIONS WITH NUCLEOPHILIC REAGENTS

As suggested by the preceding discussion, the competition of the two carbon atoms in a triple bond for the π electrons would tend to concentrate the π-electron density near the center of the bond, leaving the carbon atoms electron deficient. It therefore stands to reason that alkynes will be more electrophilic than alkenes, and experimental studies have confirmed that they are:

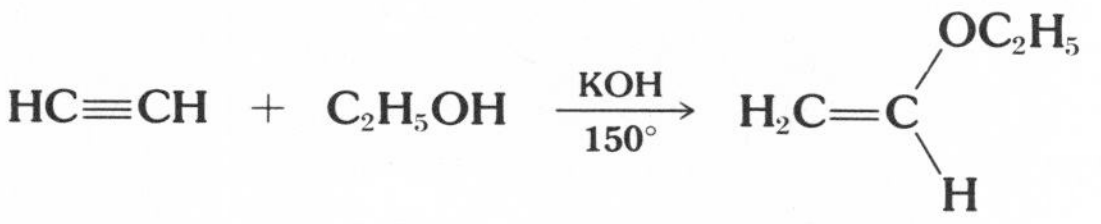

$$HC\equiv CH + C_2H_5OH \xrightarrow[150°]{KOH} H_2C=C\begin{smallmatrix}OC_2H_5\\H\end{smallmatrix} \quad (6.25)$$

$$CH_3C\equiv CCH_3 + RS^\ominus Na^\oplus \xrightarrow{alcohol} \begin{smallmatrix}H\\CH_3\end{smallmatrix}C=C\begin{smallmatrix}CH_3\\SR\end{smallmatrix} \quad (6.26)$$

$$C_4H_9C\equiv CCH_3 + 2Na \xrightarrow{NH_3(l)} \begin{smallmatrix}H\\C_4H_9\end{smallmatrix}C=C\begin{smallmatrix}CH_3\\H\end{smallmatrix} + 2NaNH_2 \quad (6.27)$$

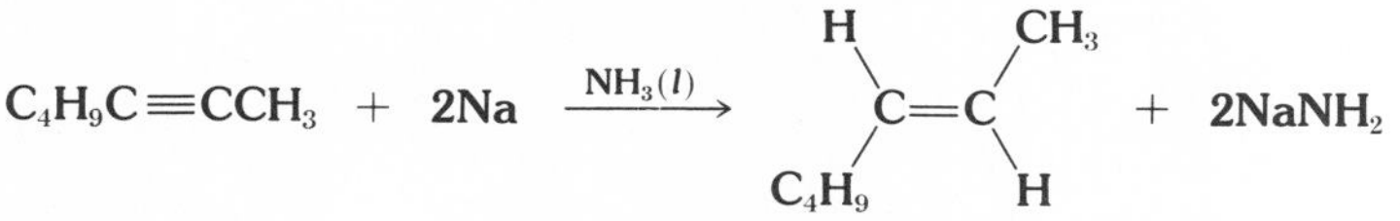

The alkali metals Li, Na, and K dissolve in liquid ammonia to give deep-blue solutions containing very mobile and loosely bound electrons. As shown in Figure 6.5, these "electron solutions" react with most electrophilic functions to give reduced products. Since isolated triple bonds are reduced to trans alkenes in this fashion (equation 6.27) and isolated double bonds are not affected, the relative electrophilicity of these func-

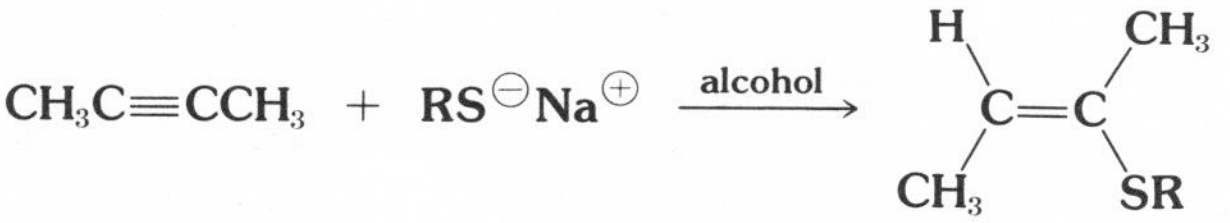

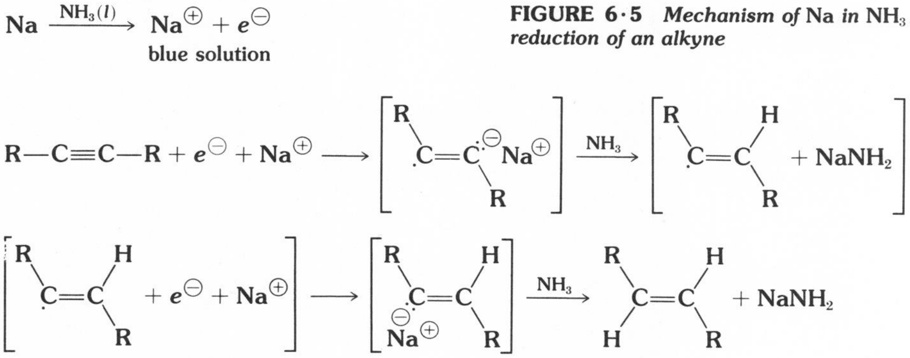

$$Na \xrightarrow{NH_3(l)} Na^\oplus + e^\ominus$$
blue solution

FIGURE 6·5 *Mechanism of Na in NH_3 reduction of an alkyne*

$$R-C\equiv C-R + e^\ominus + Na^\oplus \longrightarrow \left[\begin{smallmatrix}R\\ \\ \end{smallmatrix}C=C\begin{smallmatrix} \\ \\R\end{smallmatrix}^\ominus Na^\oplus\right] \xrightarrow{NH_3} \left[\begin{smallmatrix}R\\ \\ \end{smallmatrix}C=C\begin{smallmatrix}H\\ \\R\end{smallmatrix} + NaNH_2\right]$$

$$\left[\begin{smallmatrix}R\\ \\ R\end{smallmatrix}C=C\begin{smallmatrix}H\\ \\ \end{smallmatrix} + e^\ominus + Na^\oplus\right] \longrightarrow \left[\begin{smallmatrix}R\\ \\Na^\oplus\end{smallmatrix}C=C\begin{smallmatrix}H\\ \\R\end{smallmatrix}^\ominus\right] \xrightarrow{NH_3} \begin{smallmatrix}R\\H\end{smallmatrix}C=C\begin{smallmatrix}H\\R\end{smallmatrix} + NaNH_2$$

tions is clearly established. The trans stereochemistry of the alkene product persists even when this configuration is less stable than the cis isomer:

$$\text{cis-cyclononene} \xleftarrow{H_2 / Pd(\text{poisoned})} \text{cyclononyne} \xrightarrow{Na / NH_3(l)} \text{trans-cyclononene} \quad (6.28)$$

Consequently, this method of converting alkynes into disubstituted alkenes nicely complements the catalytic addition of hydrogen.

6·4 ADDITION REACTIONS WITH DIBORANE

Although diborane (B_2H_6) is generally considered to be an electrophilic reagent, and was in fact so classified in the previous chapter, its reactions with alkenes and alkynes are sufficiently unusual to justify a separate discussion here. For example, the exclusive cis stereochemistry of alkene hydroborations stands in sharp contrast to the predominantly trans orientation in addition reactions with other electrophilic reagents. Furthermore, the rate of diborane addition to double bonds is far less sensitive to variations in alkyl-group substitution than are electrophilic additions of halogen or Brønsted acids (see equation 5.13).

Diborane adds rapidly to the triple bond of alkynes, but the reaction often proceeds beyond the useful alkenyl borane intermediate to produce a mixture of polymeric diboronated alkanes:

$$RC{\equiv}CR + BH_3 \longrightarrow \left[\begin{array}{c} H \quad BH_2 \\ C{=}C \\ R \quad R \end{array} \right] \xrightarrow{B-H} \begin{array}{c} H \;\; H \\ B-C-C-B \\ R \;\; R \end{array} + R-CH_2-C-B \qquad (6.29)$$

an alkenyl borane

This undesirable secondary reaction can be avoided by substituting for diborane the dialkyl borane derivative prepared by hydroboration of 2-methyl-2-butene.

$$2(CH_3)_2C{=}CHCH_3 + BH_3 \longrightarrow [(CH_3)_2CHCH(CH_3)]_2BH = R_2BH \qquad (6.30)$$

The reaction of this reagent with alkynes proceeds slightly more rapidly than with similarly substituted olefins (presumably because of steric hindrance) and yields cis alkenyl boranes from disubstituted acetylenes. An anti-Markovnikov orientation is observed in the reaction with terminal acetylenes.

$$R'C{\equiv}CR' + R_2BH \longrightarrow \begin{array}{c} H \quad BR_2 \\ C{=}C \\ R' \quad R' \end{array} \begin{array}{c} \xrightarrow{NaOH, H_2O_2} R'CH{=}C(OH)R' \rightleftharpoons R'CH_2CR' \\ \\ \xrightarrow{\text{acetic acid}} \begin{array}{c} H \quad H \\ C{=}C \\ R' \quad R' \end{array} \end{array} \qquad (6.31)$$

$$R'C\equiv CH + R_2BH \longrightarrow R'CH=CHBR_2 \xrightarrow[H_2O_2]{NaOH} R'CH=C\underset{H}{\overset{OH}{\diagup}} \rightleftharpoons R'-CH_2-C\underset{H}{\overset{O}{\diagup\diagup}} \quad (6.32)$$

These unsaturated boranes are important intermediates to cis alkenes and ketones or aldehydes, as shown in equations 6.31 and 6.32. The latter reaction provides a particularly useful alternative to the mercuric-ion-catalyzed hydration of terminal acetylenes (equation 6.22) because it generates aldehydes rather than ketones from terminal acetylenes.

PROBLEM 6·11

Write equations showing how 1-hexyne or 3-hexyne (choose the isomer best suited to the desired reaction in each case) can be transformed into the following compounds:

a $C_4H_9COCH_3$
b cis-$C_2H_5CH=CHC_2H_5$
c $C_3H_7COC_2H_5$
d $C_5H_{11}C\underset{H}{\overset{O}{\diagup\diagup}}$
e $trans$-$C_2H_5CH=CHC_2H_5$

6·5 OXIDATION REACTIONS OF ALKYNES

Complete oxidation of the triple bond by ozone or potassium permanganate generally leads to bond cleavage and carboxylic acid products. Under controlled reaction conditions the permanganate oxidation can sometimes be stopped at an intermediate diketone stage.

$$R-C\equiv C-R' + O_3 \xrightarrow{Zn\ dust} RCO_2H + R'CO_2H \quad (6.33)$$

$$CH_3(CH_2)_7C\equiv C(CH_2)_7CO_2H \xrightarrow[pH\ 7.5,\ 25°]{KMnO_4} CH_3(CH_2)_7-\overset{O}{\underset{\|}{C}}-\overset{O}{\underset{\|}{C}}-(CH_2)_7CO_2H \quad (6.34)$$

6·6 DIMERIZATION, TRIMERIZATION, AND POLYMERIZATION

When acetylene is slowly added to an aqueous cuprous chloride solution, a mixture of dimers and trimers is formed. This reaction may be regarded as another example of the transition-metal-catalyzed addition of a very weak Brønsted acid ($\equiv$C—H) to a triple bond:

$$C_2H_2 \xrightarrow[H_2O]{Cu_2Cl_2,\ NH_4Cl} \underset{\text{vinylacetylene (a dimer)}}{CH_2=CHC\equiv CH} + \underset{\text{divinylacetylene (a trimer)}}{CH_2=CHC\equiv CCH=CH_2} \quad (6.35)$$

Interaction of the π electrons in a group of triple-bonded molecules can be imagined to proceed so as to give either a linear polyolefin or unsaturated carbon rings, as shown in Figure 6.6. In practice, only the cyclization mode of reaction has been accomplished with reasonable success, and this only for six- and eight-membered ring systems.†

†The cyclobutadiene ring system shown in Figure 6.6 is not stable and can be made only in the form of a transition-metal complex. This interesting compound is discussed further in Chapter 9.

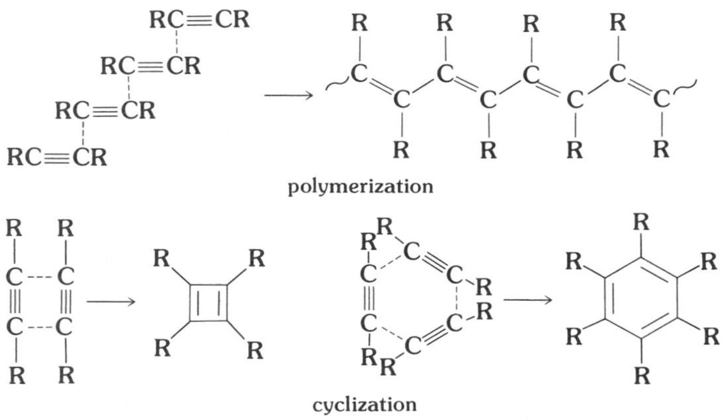

6·6 DIMERIZATION, TRIMERIZATION, AND POLYMERIZATION

FIGURE 6·6 Possible modes of π addition of alkynes

acetylene is passed through a metal tube heated to about 700°, low yields of benzene and related compounds are obtained. Fortunately, we can accomplish this and similar cyclizations in higher yields under much milder reaction conditions by using certain transition-metal catalysts. The following examples demonstrate that the nature of the catalyst used is critical to the course and success of the cyclization reaction.

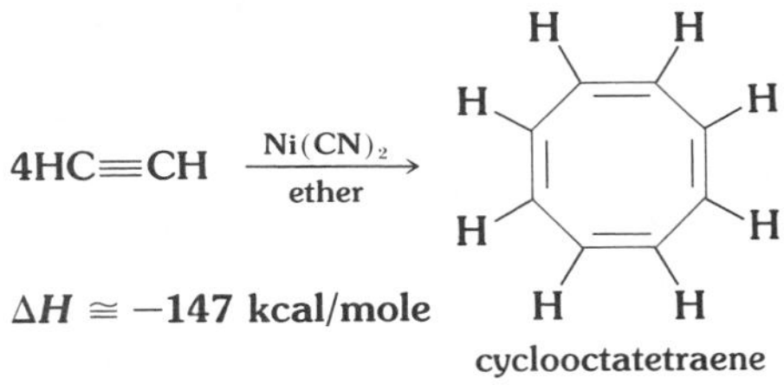

$$4HC\equiv CH \xrightarrow[\text{ether}]{Ni(CN)_2} \text{cyclooctatetraene} \quad (6.36)$$

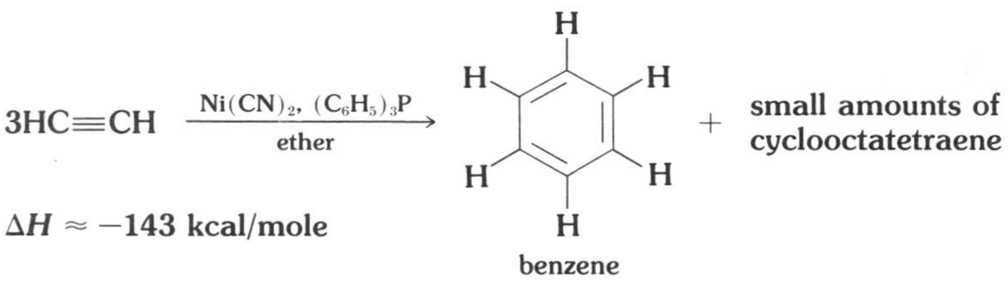

$\Delta H \cong -147 \text{ kcal/mole}$

$$3HC\equiv CH \xrightarrow[\text{ether}]{Ni(CN)_2,\ (C_6H_5)_3P} \text{benzene} + \text{small amounts of cyclooctatetraene} \quad (6.37)$$

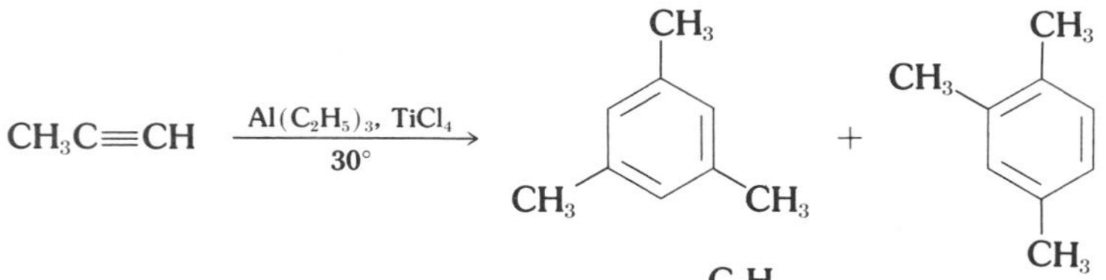

$\Delta H \approx -143 \text{ kcal/mole}$

$$CH_3C\equiv CH \xrightarrow[30°]{Al(C_2H_5)_3,\ TiCl_4} \quad (6.38)$$

$$C_2H_5C\equiv CC_2H_5 \xrightarrow[\text{ether}]{[Co(CO)_4]_2,\ Hg} \quad (6.39)$$

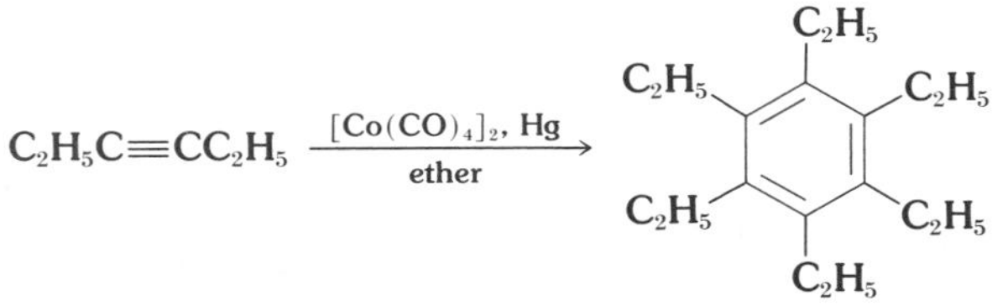

Most theories concerning these reactions assume that the triple-bond functions can replace the nucleophile groups, or ligands, that normally

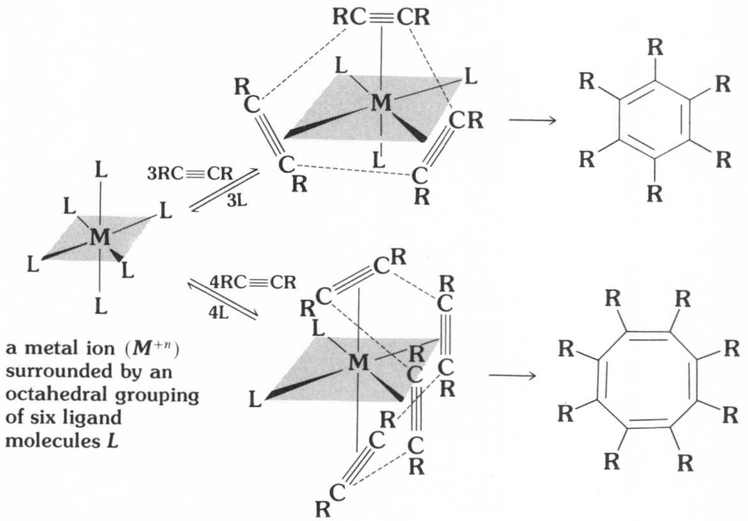

6·6 DIMERIZATION, TRIMERIZATION, AND POLYMERIZATION 171

FIGURE 6·7 *Cycloaddition of coordinated alkyne units*

surround the metal atom (π-d bonding may be involved), and that two, three, or four of these *coordinated* alkynes then cyclize by a redistribution of the π electrons (see Figure 6.7). If the ligand groups are only weakly held by the metal atom, many of the ligand molecules can be replaced by triple-bond functions; however, more tightly bound ligands will not be as easily displaced. Since the ether-solvent molecules in reaction 6.36 are rather weak ligands, we can conceive that as many as four acetylene molecules could be incorporated into an octahedral nickel complex, and the formation of cyclooctatetraene is thus reasonable (note that cyanide ligands are usually strongly held by the metal). The introduction of a strongly competing ligand into the reaction mixture (for example, a trivalent phosphorus compound) apparently reduces the number of coordinated acetylenic groups to a maximum of three, and benzene becomes the major product (equation 6.37).

PROBLEM 6·12
If the cycloaddition of substituted acetylenes proceeds in a statistical manner (that is, there are no steric or electronic orienting forces), calculate the relative proportions of 1,3,5- and 1,2,4-trisubstituted benzenes expected from RC≡CH. Would you expect any 1,2,3-trisubstituted isomer if a mechanism similar to that described in Figure 6.7 were operating?

Some intriguing and unusual isomeric cyclic trimers of acetylenes have been reported recently:

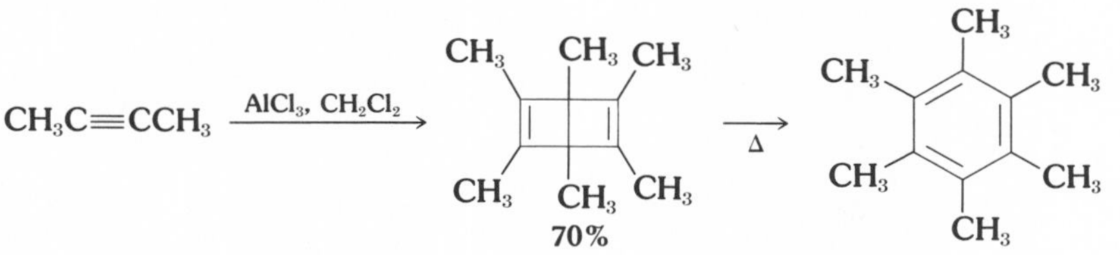

$$CH_3C{\equiv}CCH_3 \xrightarrow{AlCl_3, CH_2Cl_2} \text{[cyclobutene dimer, 70\%]} \xrightarrow{\Delta} \text{hexamethylbenzene} \quad (6.40)$$

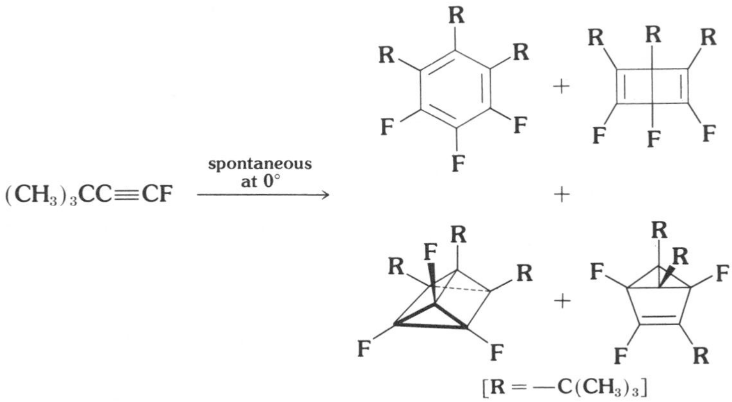

$(CH_3)_3CC\equiv CF \xrightarrow[\text{at } 0°]{\text{spontaneous}}$ [structures shown: hexasubstituted benzene, cyclobutadiene dimer, bicyclic, and cyclopentadiene-type trimers] $[R = -C(CH_3)_3]$ (6.41)

PROBLEM 6·13

The four isomeric trimers of $(CH_3)_3CC\equiv CF$ exhibit the following properties. Match these four isomers to the structures in equation 6.41.

1 Isomer A is unstable at 100°C (changes completely to B), exhibits pmr signals for hydrogen at $\delta = 1.12$ and 1.2 (ratio 1:2) and two nmr signals for fluorine (ratio 1:2), and has an infrared absorption band at 1678 cm^{-1}.
2 Isomer B is stable at 250°C and shows pmr signals for hydrogen at $\delta = 1.39$ and 1.45 (ratio 1:2) and two nmr signals for fluorine (ratio 1:2).
3 Isomer C is stable at 250°C and shows a single pmr signal for hydrogen at $\delta = 1.18$ and a single nmr signal for fluorine.
4 Isomer D is stable at 100°C, but not at 220°C and shows two hydrogen pmr signals at $\delta = 1.16$ and 1.22 (ratio 2:1) and three equal fluorine nmr signals.

SUMMARY

alkynes Hydrocarbons containing one or more carbon-carbon triple bonds, identified in IUPAC nomenclature by the ending *-yne* (Table 6.1). The carbon-carbon triple bond is shorter (1.20 Å) and stronger than corresponding double and single bonds and has a linear configuration. Characteristic spectroscopic properties are given in Table 6.2.

structure-composition relationship A correlation between the number of rings and/or double bonds in a hydrocarbon and its degree of unsaturation:

$$\frac{\text{hydrogen content of corresponding alkane} - \text{hydrogen content of unknown hydrocarbon}}{2} = \text{rings} + \text{double bonds} + 2 \times \text{triple bonds}$$

tautomer One of two or more constitutional isomers which are easily interconverted through a low-energy barrier:

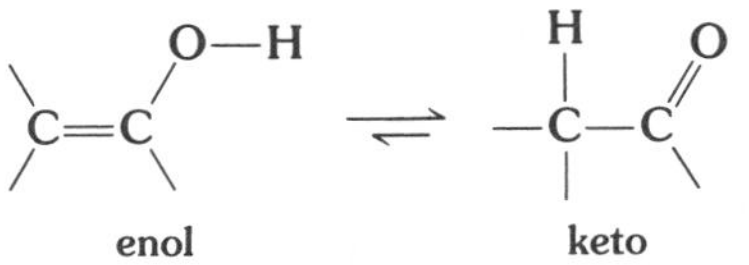

$$\underset{\text{enol}}{\overset{}{\underset{}{\text{C}=\text{C}}}}\text{O-H} \rightleftharpoons \underset{\text{keto}}{\overset{\text{H}}{\underset{}{-\text{C}-\text{C}}}}\text{O}$$

alkyne reactions Characteristic chemical transformations which usually take the form of addition reactions at the carbon-carbon triple bond.

Although alkynes are thermodynamically less stable than similarly constituted alkenes, addition reactions of electrophilic reagents are often sluggish and require transition-metal-ion catalysis. In some oxidation reactions cleavage of the triple bond takes place.

Brønsted acid addition Addition of Brønsted acids such as HCl, HBr, and $H_3O^{\oplus}$ to alkynes, with products formed according to Markovnikov's rule. Catalysis by transition-metal ions such as Hg^{2+} or Cu^{+} is usually necessary to effect a reaction.

halogen addition Chlorine and bromine addition to alkynes to give tetrahaloadducts in which the triple bond is saturated with halogen. With care dihaloalkene intermediates can be isolated.

diborane addition A rapid reaction of diborane with alkynes which gives mixtures of products. A modified borane reactant, $(C_5H_{11})_2BH$, must be used if a single anti-Markovnikov addition product is to be obtained.

catalytic hydrogenation The rapid addition of hydrogen to alkynes in the presence of Pd, Pt, or Ni catalysts, yielding alkane products. By using a poisoned Pd catalyst, the intermediate cis alkenes can be obtained from disubstituted alkynes.

dissolving-metal reduction Reduction of disubstituted alkynes to trans alkenes by a solution of sodium in liquid ammonia.

nucleophilic addition reactions The carbon-carbon triple bond is much more susceptible to attack by nucleophiles than is a corresponding double bond. Strong nucleophiles such as $RS^{\ominus}$ will add to alkynes, but not to alkenes.

oxidative cleavage Reaction of alkynes with ozone or potassium permanganate, leading to carboxylic acid cleavage products.

trimerization Transition-metal catalysts such as nickel and cobalt can induce trimerization of alkynes to benzene derivatives.

ethynyl hydrogen atom The hydrogen attached to a triple bond at the end of a carbon chain. Such terminal alkynes (R—C≡C—H) are weak Brønsted acids ($K_a \approx 10^{-25}$). The formation of insoluble silver or copper salts of such compounds is a useful test. The acidity of ethynyl hydrogen atoms is the key to a useful synthesis of alkynes:

HC≡CH + NaNH$_2$ ⟶ NH$_3$ + HC≡CNa

HC≡CNa + RX ⟶ HC≡CR + NaX

EXERCISES

6·1 Draw structural formulas for the following compounds:
a 1-Pentyne
b 2-Pentyne
c Methylacetylene
d Propyne
e Ethynylcyclohexane
f Propargyl chloride

6·2 Name the following compounds according to the IUPAC system:

a CH$_3$CH$_2$CH$_2$C≡CH
b (CH$_3$)$_3$CC≡CCH$_2$CH$_3$
c Cl(CH$_2$)$_2$C≡C(CH$_2$)$_3$CH$_3$

d ◇—C≡C—CH$_3$

e ⬠—CH$_2$C≡CH

174 EXERCISES

6·3 Correct the following names:
a 3-Butyne
b trans-2-Pentyne
c 1,1-Dimethyl-5-hexyne

6·4 Draw structural formulas for all isomeric alkynes with molecular formula C_6H_{10}.
a Name each isomer according to the IUPAC system.
b Which of these isomers will react with $Ag(NH_3)_2$?
c Write structures for the ozonolysis products expected from each isomer (H_2O_2 workup).

6·5 Arrange the two-carbon hydrocarbons ethane, ethylene, and acetylene in increasing order of:
a Carbon-hydrogen bond length
b Carbon-hydrogen bond strength
c Carbon-hydrogen infrared stretching frequency
d Carbon-carbon bond length
e Carbon-carbon bond strength
f Acidity
g Hydrogen (proton) chemical shifts in pmr spectra

6·6 Match the following values to each of the bond angles indicated below:

a H—C—C—H (with H's above and below, ? at lower left)

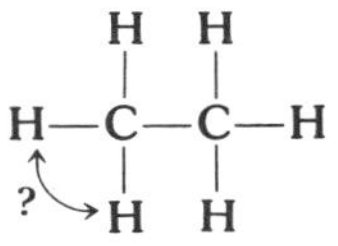

b H₂C=CH₂ (? at lower left)

c H—C≡C—H (? below)

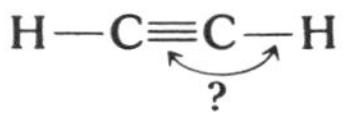

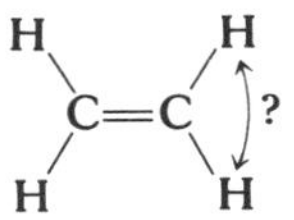

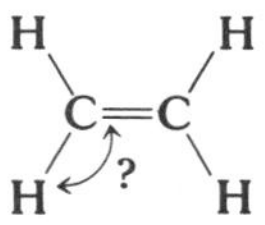

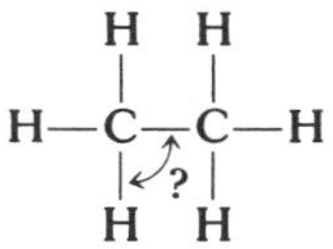

d H—C—C—H (with H's, ? in middle)

e H₂C=CH₂ (? on right)

f benzene ring (?)

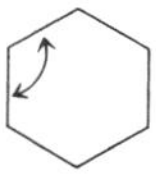

6·7 What simple chemical test could you use to distinguish between the following compounds?
a $CH_3CH_2C\equiv CH$ and $CH_3C\equiv CCH_3$
b $CH_3CH_2C\equiv CH$ and $CH_3CH_2CH=CH_2$
c —CH_3 and $CH_3C\equiv CCH_3$

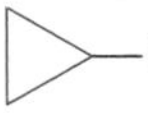

6·8 Indicate the relative acidities of acetylene, ammonia, and water from the equilibria shown below. Explain your reasoning. What are the relative base strengths of sodium acetylide, sodium amide, and sodium hydroxide?

$HC\equiv CH + Na^{\oplus}NH_2^{\ominus} \rightleftarrows NH_3 + HC\equiv C^{\ominus}Na^{\oplus}$

$H_2O + HC\equiv C^{\ominus}Na^{\oplus} \rightleftarrows HC\equiv CH + Na^{\oplus}OH^{\ominus}$

6·9 Show the products from a reaction between 2-bromo-3-methylbutane and:
a Sodium acetylide acting as a nucleophile (S_N2)
b Sodium acetylide acting as a Brønsted base ($E2$)

6·10 Starting with acetylene and any other reagents, show how to prepare the following:

a HC≡CCH$_3$
b CH$_3$CCL$_2$CH$_3$
c CH$_3$CH$_2$C≡CH
d CH$_3$C≡CCH$_3$
e cis-CH$_3$CH=CHCH$_3$
f trans-CH$_3$CH=CHCH$_3$
g CH$_2$=CHCH$_3$
h CH$_3$CHBrCH$_3$
i CH$_3$CCH$_2$CH$_3$
   $\|$
   O
j CH$_3$CH$_2$CH$_2$CHO
k CH$_3$CH$_2$C(=O)OH

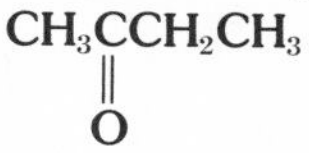

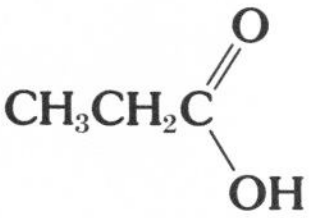

6·11 Write keto tautomers for the following compounds:

a (cyclopentylidene)=CHOH

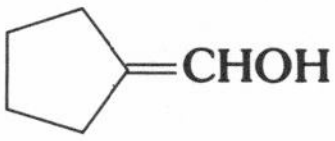

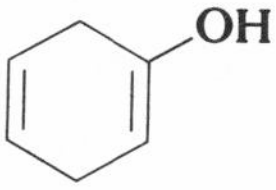

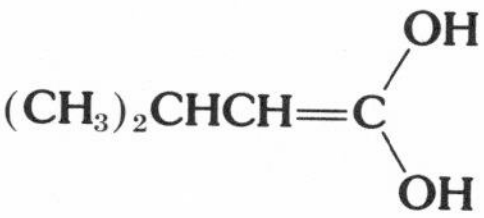

b cyclohexenol (1-hydroxycyclohexene)

c (CH$_3$)$_2$CHCH=C(OH)$_2$

d C$_2$H$_5$C(OH)=C(OH)C$_2$H$_5$

e (cyclobutylidene)C(OH)=C(OH)(cyclobutylidene)

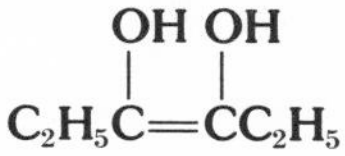

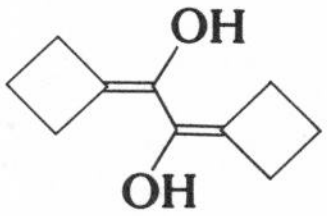

6·12 Draw an energy diagram which will account for the lack of correlation between the thermodynamic instability of alkynes and their chemical reactivity with electrophilic reagents.

6·13 An unlabeled bottle is known to contain either *n*-pentane, 1-pentene, or 1-pentyne.

a Describe a simple chemical test that would identify the contents.

b How would you distinguish the samples by means of pmr measurement? By means of infrared spectroscopy?

6·14 Some of the steps in the mechanism shown in Figure 6.5 are Brønsted acid-base reactions. Identify these steps and label the conjugate-acid–base pairs.

STEREOCHEMISTRY

STEREOISOMERS are isomers that differ only in the spatial arrangement of their atoms. Examples of stereoisomerism in disubstituted cycloalkanes and alkenes were discussed in earlier chapters. Let us now extend this concept to a more subtle form of isomerism associated with a certain kind of molecular symmetry. This new stereoisomerism, which introduces the idea of right- and left-hand molecular configurations, will play an important role in our study of organic chemistry because it is characteristic of so many naturally occurring compounds and reactions. Furthermore, such stereoisomerism is a useful tool for probing reaction mechanisms.

7·1 OPTICAL ACTIVITY

PLANE-POLARIZED LIGHT

Studies of the interaction of matter with light, or electromagnetic radiation, have provided us with much of our fundamental knowledge concerning molecular structure. In previous chapters we have focused on the useful information that can be obtained from the characteristic absorption spectra of a substance; in this chapter we shall examine a remarkable and selective interaction of matter with plane-polarized light.

A beam of ordinary monochromatic light (light consisting of only one wavelength) generates electrical and magnetic fields that oscillate rapidly in all possible directions about the axis of its path (Figure 7.1). Certain optical devices, such as a Polaroid disk or a Nicol prism (a device constructed from two pieces of Icelandic spar), will selectively transmit that portion of a light beam having electrical and magnetic field vibrations only in a specific plane. Such light is called *plane polarized*. If two polarizing devices are placed in the light path, maximum light will pass

7·1 OPTICAL ACTIVITY

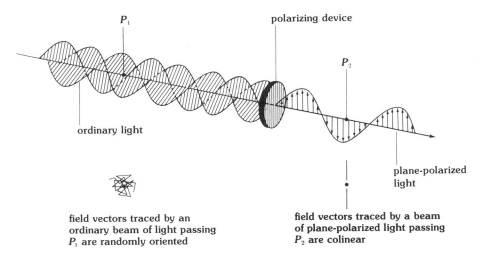

FIGURE 7·1 *Formation of plane-polarized light*

when their polarizing axes are oriented parallel to each other, but all light will be cut off if the polarizing axes are perpendicular.

Plane-polarized light passes unchanged through air, water, and a number of organic substances (chloroform, cyclohexane, ethyl alcohol, etc.). In the early nineteenth century, however, the French physicist Biot discovered that some substances, such as quartz, sugar, and turpentine, had the unique property of rotating the plane of polarization when they were inserted into a beam of polarized light. We call compounds of this kind *optically active*. An instrument called a *polarimeter* can be used to measure the direction and magnitude of rotation of the polarization plane, as shown in Figure 7.2. A substance which rotates the plane of polarization to the right, or clockwise as the observer looks toward the light source, is referred to as *dextrorotatory* (from the Latin *dexter*, meaning right). If the rotation is to the left, or counterclockwise, the optically active material is said to be *levorotatory* (from the Latin *laevus*, left). The direction and magnitude of rotation are measured in degrees (plus for dextrorotatory and minus for levorotatory) by adjusting the movable analyzer of the polarimeter to give maximum transmitted light intensity. Figure 7.2 illustrates an experiment in which the sample has rotated the plane of polarization 40° to the right. Thus the observed angle of rotation α is +40°.

The optical activity of inorganic solids such as quartz was soon shown to be unique to the crystalline state. However, the activity of most organic substances persisted in both liquid and gaseous phases, indicating that the phenomenon was molecular in nature. Of course, it is not possible to obtain the optical rotation of a single molecule; therefore all reported values of α represent measurements made with a large assembly of sample molecules (about 10^{20}), either as a pure liquid or in solution in an inert solvent.

The magnitude of α proved to be sensitive to a number of variables, including the concentration of sample molecules in the light path, the length of the sample tube, the solvent, and the wavelength of light used

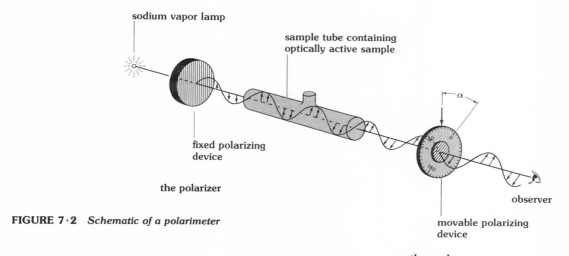

FIGURE 7·2 *Schematic of a polarimeter*

to make the observation. In order to standardize the reporting of optical activities, chemists have adopted *specific rotation* as a unit of measure:†

$$\text{specific rotation} = \frac{\text{observed rotation}}{\text{length of sample tube (dm)} \times \text{concentration (g/ml)}}$$

$$[\alpha] = \frac{\alpha}{l \times c}$$

(7.1)

The wavelength of light used and other variables can also be incorporated into the specific-rotation value. For example, a specific rotation of sucrose might be reported as follows:

sucrose $[\alpha]_D^{25°} = +66.5°, H_2O$

where the $25°$ is the temperature, D indicates 5890 Å light from a sodium lamp (D line), and H_2O is the solvent.

PROBLEM 7·1
How can an apparent observed rotation of $+40°$ be distinguished from a rotation of $+220°$, or $-140°$, or $-320°$, or any value fitting the expression $\alpha = 40 \pm n180$ (where $n = 1, 2, 3, \ldots$)? *Hint:* Remember that α is sensitive to changes in the number of sample molecules in the light path.

OPTICAL ISOMERISM

At first the phenomenon of optical activity seemed to provide chemists with little more than another physical property for characterizing compounds. However, studies by Berzelius, Pasteur, and others soon disclosed a remarkable kind of isomerism. Pairs of optically active isomers were discovered which differed only in the sign of optical rotation. For

---

†The concentration term becomes the density when a pure liquid sample is used. Most sample tubes are 1 decimeter (dm) long.

TABLE 7·1 Properties of optical isomers

compound	source	melting or boiling point (°C)	specific rotation
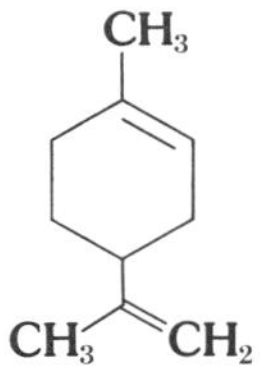 limonene	caraway seeds	bp 178	$[\alpha]_D \approx +126°$
	pine needles	bp 178	$[\alpha]_D \approx -123°$
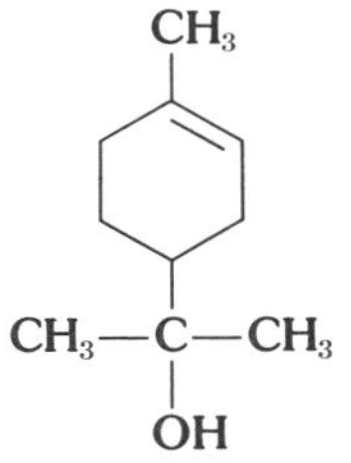 α-terpineol	neroli oil	mp 37	$[\alpha]_D = +100.5°$
	camphor oil	mp 37	$[\alpha]_D = -100.5°$

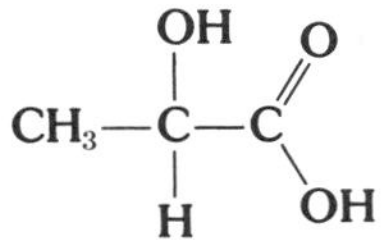

example, lactic acid isolated from muscle tissue and lactic acid from sour milk have identical properties (such as a melting point of 26°), save for the fact that the former is dextrorotatory and the latter is levorotatory. Two other examples of this phenomenon of *optical isomerism* are given in Table 7.1.

Optically inactive forms of many of these substances were also encountered. For example, an inactive limonene was isolated from some samples of lemon oil, and synthetically prepared samples of lactic acid proved to have no measurable optical rotation. The possibility that each of these substances represented a third separate and distinct isomer was excluded when later studies demonstrated that each was an equimolar mixture of the two optically active isomers. Such mixtures are now referred to as *racemic modifications* and are identified by a plus-minus (±) sign.

7·2 ENANTIOMERISM

The period 1860 to 1875 witnessed a revolutionary change in the structural theories of organic chemistry as chemists began to realize that the shapes of molecules (the orientation of the atoms in three-dimensional space) helped to determine the physical and chemical properties of compounds.† One of the most important facets of this development was the correlation of optical isomerism with molecular dissymmetry.

†This "birth" of stereochemistry was a truly international affair. Kekulé and Wislicenus (in Germany), Butlerov (a Russian), Couper (a Scot), Pasteur and LeBel (in France), and van't Hoff (from the Netherlands) all helped formulate the new ideas which were to affect organic chemistry to the same degree that Darwin's *The Origin of Species* influenced biology.

ELEMENTS OF SYMMETRY

All objects, including discrete molecular conformations, may be classified according to the number and kinds of *symmetry elements* they possess.† A symmetry element is defined by a corresponding symmetry operation, either a *rotation* or a *reflection*, the action of which returns the object to a position indistinguishable from its original orientation.

AXIS OF SYMMETRY   If rotation of an object about an axis by $360°/n$ results in that object assuming an orientation indistinguishable from the original one, it is said to have a C_n *axis of symmetry*. A regular tetrahedron, for example, has three C_2 axes, for which rotation by $180°$ ($360°/2$) about an axis bisecting opposite edges duplicates the initial orientation. It also has four C_3 axes, for which rotation by $120°$ ($360°/3$) about an axis passing through a vertex and the center of the opposite face duplicates the initial orientation. These axes of symmetry are shown in Figure 7.3. Molecules such as CH_4 and CCl_4, which have an exact tetrahedral configuration, will also have this characteristic combination of symmetry elements. If we identify the individual hydrogen atoms in methane by subscripts, as shown in Figure 7.4, the result of these symmetry operations is readily seen.

†Symmetry has esthetic as well as mathematical meaning. In the former sense it suggests a balanced or well-proportioned arrangement of parts, which some regard as particularly pleasing or beautiful.

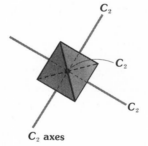

FIGURE 7·3 *Axes of symmetry in a regular tetrahedron*

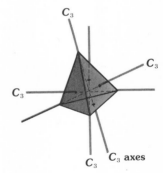

FIGURE 7·4 *Symmetry operations in methane*

7·2 ENANTIOMERISM

Every object has, of course, a C_1 axis, since a full rotation of 360° (often called an *identity operation*) restores the object to its original orientation.

> **PROBLEM 7·2**
> Identify the symmetry axes in the twist and chair conformations of cyclohexane shown in Table 2.8.

MIRROR PLANE OF SYMMETRY If a plane can be passed through an object in such a way that all the parts or features on one side of the plane are a reflection of equivalent parts on the other side, as though the plane were a mirror, we refer to such a plane as a *mirror plane of symmetry* σ. The operation of reflecting each half through this plane of symmetry generates an indistinguishable orientation of the original object; for example, for chlorobromomethane:

$$\underset{\text{initial orientation}}{\overset{\sigma}{\text{Cl–C(H}_b\text{)(H}_a\text{)–Br}}} \xrightarrow{\text{reflection in } \sigma} \underset{\text{indistinguishable orientation}}{\text{Cl–C(H}_a\text{)(H}_b\text{)–Br}}$$

The boat conformation of cyclohexane has two perpendicular mirror planes of symmetry which intersect on the C_2 symmetry axis, as shown in Figure 7.5.

CENTER OF SYMMETRY If any straight line passing through the center of an object encounters identical environments on both sides of the center point, such an object is said to have a *center of symmetry*, denoted by i. Reflection of all parts through a center of symmetry generates a different but indistinguishable orientation of the object. Any face card in a standard deck of cards has a center of symmetry, as does the chair conformation of *trans*-1,4-dichlorocyclohexane. Note in Figure 7.6 that lines passing through the center intersect equivalent points at the same distance on either side of the center.

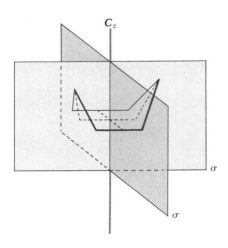

FIGURE 7·5 *Symmetry elements in boat cyclohexane*

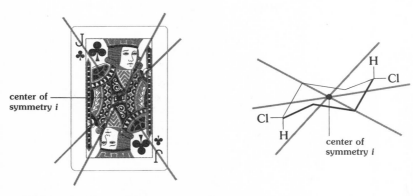

FIGURE 7·6 *Objects having a center of symmetry i*

PROBLEM 7·3
a What symmetry elements are present in the trans and cis isomers of 1,2-dichloroethene?
b What symmetry elements are present in each of the following conformations of 1,1,2,2-tetrachloroethane?

The existence of a *reflection symmetry element* (a plane or point of symmetry) in an object is sufficient to assure that the object is identical with its mirror image. Conversely, an object without a reflection symmetry element cannot be made to coincide in all respects with its mirror image. For example, our right and left hands, which have a mirror-image relationship, lack a plane or center of symmetry; hence they cannot be superimposed so that all parts—thumbs, fingers, palms, and backs—have exactly the same orientation. If an object is not identical with its mirror image it is called *dissymmetric*, or *chiral* (from the Greek *cheir*, meaning hand). Hence the simplest test for dissymmetry is to demonstrate that the object in question and its mirror image cannot be superimposed. Since this criterion makes severe demands on our ability to visualize three-dimensional structures, models such as those illustrated in Figure 2.2 are valuable study aids.

MOLECULAR CHIRALITY

We can now conceive of a new kind of stereoisomerism caused by molecular chirality. Since the mirror-image configurations of a chiral (dissymmetric) molecule are distinct species, they define a pair of stereoisomers called *enantiomers*, which have essentially identical chemical and physical properties: the same melting and boiling points; refractive indices; ultraviolet, infrared, pmr, and mass spectra; electron and x-ray diffraction patterns; and so on. One of the most important steps in the early development of stereochemistry was the recognition that optical

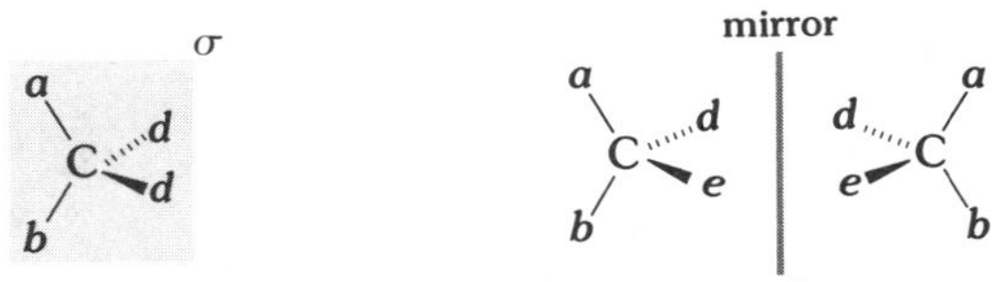

7·2 ENANTIOMERISM

FIGURE 7·7 *Examples of chiral and achiral carbon units*

isomerism is simply a manifestation of enantiomerism. Indeed, the easiest way to distinguish pure enantiomers is to measure their opposite but equal optical rotations α.

When we consider the tetrahedral configuration of saturated carbon atoms, we soon discover one type of molecular chirality. If a tetracoordinate carbon atom bears two or more identical substituents, as shown in Figure 7.7, a plane of symmetry must exist, and we say that such a structure is symmetrical—or better, it is *achiral*. If four different substituents are attached to a carbon atom, the resulting assemblage is chiral, and a molecule incorporating such a grouping will usually be chiral as well. If only one dissymmetric or asymmetric unit is present, the molecule will always be chiral.†

Chiral carbon units, sometimes referred to as asymmetric or chiral carbon atoms, are the most common source of molecular dissymmetry and are found in many (but not all) optically active compounds. The enantiomerism of α-terpineol, limonene, and lactic acid is in each case caused by a chiral carbon unit (identified by an asterisk):

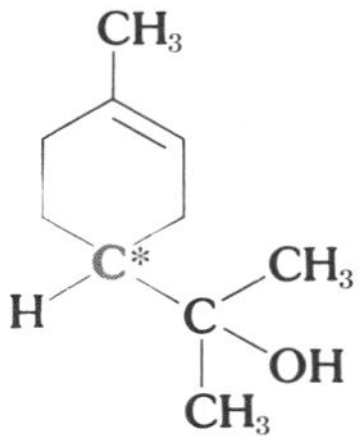

α-terpineol 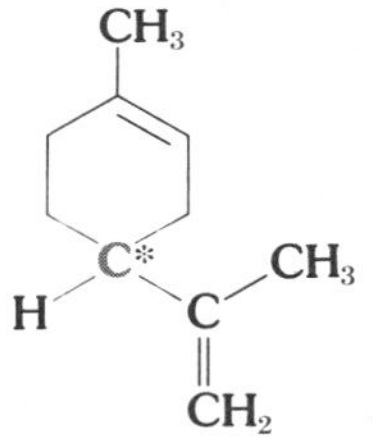   limonene 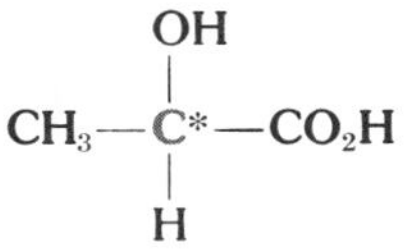   lactic acid

PROBLEM 7·4
Which of the following compounds may exist as a pair of enantiomers? Which are constitutional isomers?

a $(CH_3)_2CHCH_2OH$

b cyclohexyl—OH

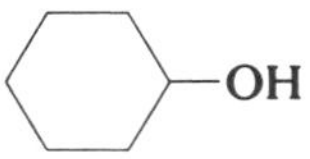

c cyclohexenyl—CH_3

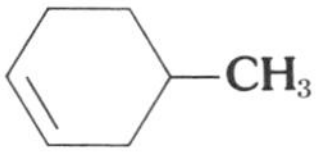

d $(CH_3)_2CHOCH_3$

e CH_3—cyclobutyl with CH_3 and OH

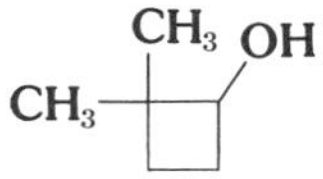

f $CH_3CH_2CH(OH)CH_3$

g cyclopentyl with CH_3 and Cl

h cyclopentyl—CH_2CH_3

i cyclopentyl—CH_2Cl

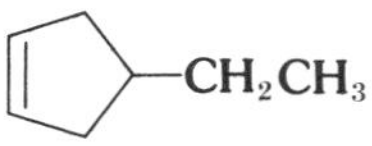

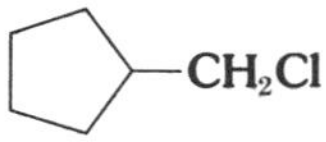

†*Asymmetry* is the complete absence of symmetry elements. *Dissymmetry* is the absence of reflection symmetry elements. All asymmetric objects are chiral, but not all chiral objects are asymmetric.

7·2 ENANTIOMERISM   185

PROBLEM 7·5
Designate all the chiral carbon units in cholesterol by asterisks.

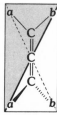

cholesterol

Chiral groupings other than a dissymmetric carbon unit can also cause molecular dissymmetry. For example, the atoms or groups attached to the ends of a 1,2-diene system—an *allene*—outline an elongated tetrahedron, as shown in Figure 7.8; this structure becomes chiral when each of the terminal carbon atoms of the diene is bonded to two different substituents. (This is the same substitution pattern required for cis-trans isomerism in simple alkenes.) The number of different substituents required to achieve chirality of an allene is less than that for chirality at a single tetrahedral atom, because the inherent symmetry of a regular tetrahedron is reduced when it is stretched or elongated. Closely related chiral configurations are found in alkylidenecycloalkanes and spiroalkanes. Some specific examples of compounds which show this kind of enantiomerism are:

PROBLEM 7·6
a What is the hybridization state of the end carbon atoms of an allene system? What is the hybridization state of the central carbon atom?
b How many π orbitals would be necessary in a σ-π model of an allene? What would be the relative orientation of these π orbitals?
c From your answer to part *b*, what orientation of the terminal substituents of the allene would you expect?

FIGURE 7·8 *Elongated tetrahedral configurations*

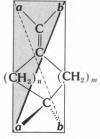

an allene   an alkylidenecycloalkane   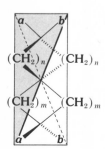 a spiroalkane

PROBLEM 7·7

Mycomycin, a naturally occurring antibiotic, is obtained in an optically active form. Identify the chiral grouping responsible for the enantiomerism of this compound:

$$HC\equiv C-C\equiv C-CH=C=CH-CH=CH-CH=CH-CH_2-CO_2H$$

PROBLEM 7·8

1,2,3-Trienes exhibit cis-trans isomerism, but not enantiomerism. Cumulated tetraenes, however, exist as a pair of enantiomers and do not show geometric (cis-trans) isomerism.

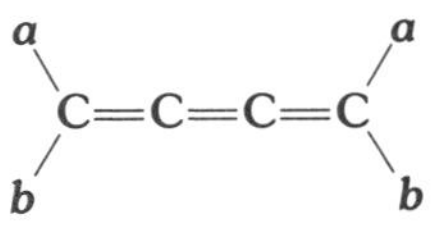

$$\underset{b}{\overset{a}{>}}C=C=C=C\underset{b}{\overset{a}{<}} \qquad \underset{b}{\overset{a}{>}}C=C=C=C=C\underset{b}{\overset{a}{<}}$$

a 1,2,3-triene 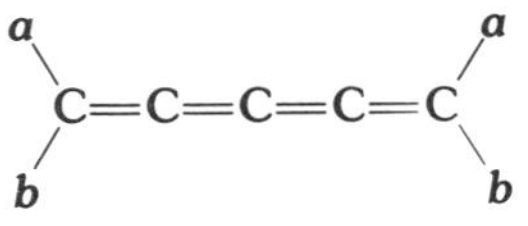   a cumulated tetraene

Discuss the reasons for these differences in isomer type, and suggest a rule for predicting the kind of isomerism one would expect from structures of the kind

$abC=(C)_n=Cab$

RACEMIC MODIFICATIONS

When chiral molecules are synthesized in the laboratory from achiral precursors, the product is inevitably an optically inactive *racemic modification*. This fact is illustrated by the hydration of 2-butene and the transformation of limonene dihydrochloride to racemic limonene:

$$CH_3CH=CHCH_3 \xrightarrow{H_3O^\oplus} CH_3CH_2CH(OH)CH_3 \qquad (7.3)$$

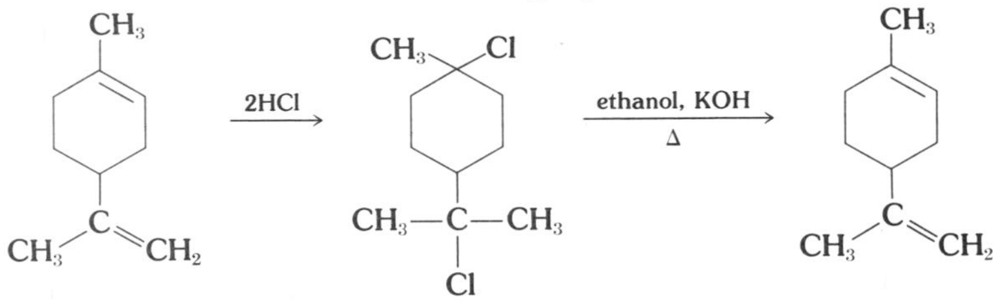

cis- or trans-2-butene   (±)-2-butanol
                                  $\alpha = 0°$

(+)-limonene $\xrightarrow{2HCl}$ limonene dihydrochloride (optically inactive) $\xrightarrow[\Delta]{ethanol, KOH}$ (±)-limonene $\alpha = 0°$ (7.4)

PROBLEM 7·9

Why is limonene dihydrochloride in reaction 7.4 optically inactive? If a substance shows no optical activity ($\alpha = 0°$), can it be composed of chiral molecules? Explain.

Reaction paths leading from symmetrical reactants to enantiomeric products must have a mirror-image relationship, as shown in Figure 7.9. It follows that the activation energies and the corresponding reaction rates are therefore identical. From this reasoning we might expect all compounds having chiral molecules to exist as 50:50 mixtures of enantiomers. Indeed, since a racemic modification is a more random or disorderly assemblage of molecules than a pure enantiomer, *racemization*, the conversion of a single enantiomer to a racemic modification, should be thermodynamically favored (the entropy increases). Racemization will, in fact, occur spontaneously, provided a low-activation-energy pathway for reversible interconversion of enantiomers is available.

It is remarkable that living systems resist the thermodynamic bias favoring racemization. Thus plants and animals selectively manufacture and use a variety of optically active compounds.† This stereochemical discrimination is due to the chiral nature of the enzyme catalysts, which play a vital role in almost all chemical reactions in living systems. In fact, enantiomers can be distinguished chemically only by the use of *chiral reagents*.

To illustrate, suppose we have the task of separating an assortment of bolts having both right- and left-hand threads. If we simply observe how well they fit into a hole drilled through a metal block, we can distinguish bolts of different sizes. However, right- and left-threaded bolts of the same size will fit into this hole equally well, so we are unable to detect the difference in threading by this method. If we make the symmetrical hole chiral by threading it in one direction or the other, the bolts with a corresponding thread can be screwed into the hole, while the mirror-image bolts cannot. This chiral device effectively distinguishes the enantiomeric forms of the bolts. It is thus conceivable that the multitude of optically active compounds we find around us owe their existence to a single primeval chiral substance.

7·3 DIASTEREOISOMERISM

Since a chiral unit can have two distinct configurations, one the mirror image of the other, we expect a molecule that incorporates two chiral units to have four stereoisomers. Indeed, if there are n different chiral units in a molecule, the number of stereoisomers should be 2^n. This is demonstrated for $n = 2$ by the stereoisomeric 3-methyl-1-chlorocyclo-

†Examples of such compounds range from glucose and lactic acid to more complex structures such as cholesterol, quinine, and penicillin.

FIGURE 7·9 *Mirror-image reaction paths for the hydration of trans-2-butene*

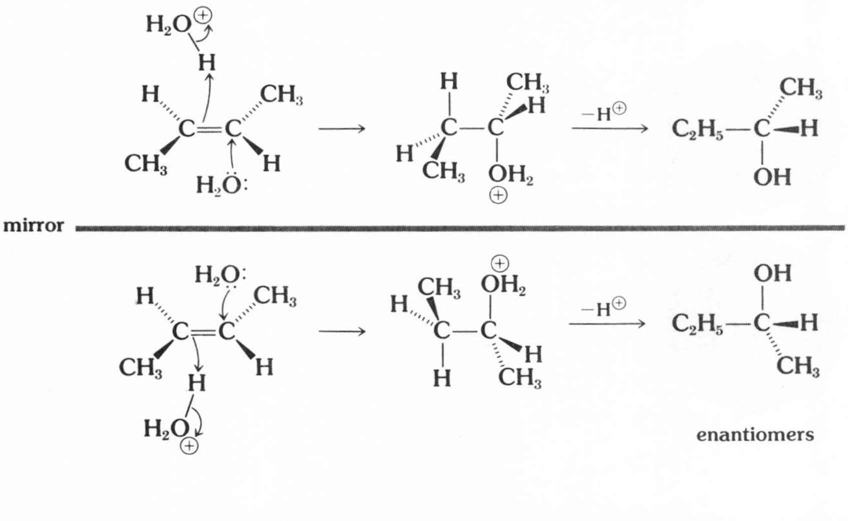

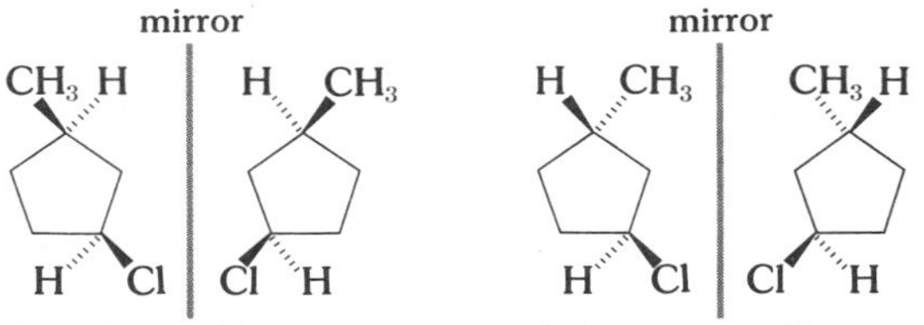

7·3 DIASTEREOISOMERISM

FIGURE 7·10 *Stereoisomeric configurations of 3-methyl-1-chlorocyclopentane*

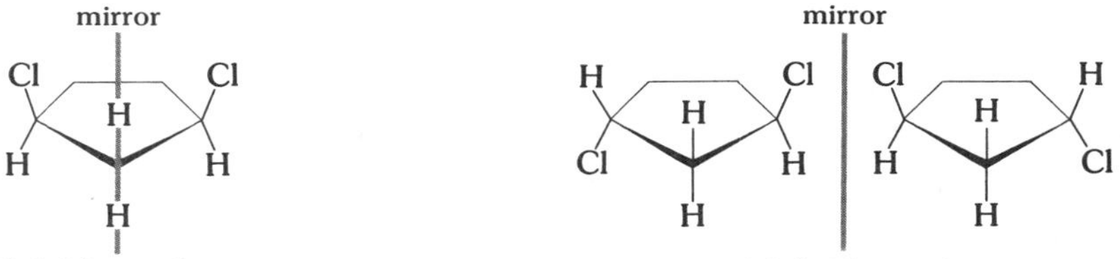

(±) cis enantiomers

(±) trans enantiomers

cis-1,3-dichlorocyclopentane
(a meso compound)

(±)-trans-1,3-dichlorocyclopentane
(a pair of enantiomers)

FIGURE 7·11 *Stereoisomers of 1,3-dichlorocyclopentane*

pentanes shown in Figure 7.10. Stereoisomers which do not bear a mirror-image relationship to each other are called *diastereoisomers* or *diastereomers*. For example, each cis enantiomer in Figure 7.10 is diastereoisomeric with both trans enantiomers. In contrast to enantiomers, the atoms and groups in diastereoisomers do not have the same relative spatial orientation; for example, the distance between the chlorine and methyl groups in the cis isomer is clearly less than the corresponding distance in the trans isomer. Consequently, diastereoisomers have different chemical and physical properties. Optically active diastereoisomers do not show a simple activity relationship corresponding to the equal but opposite specific rotations displayed by enantiomers.

The number of stereoisomers in a molecule can never exceed 2^n (for n chiral centers), but in some cases fewer isomers are observed. When two or more chiral units bearing identical sets of substituents are present in the same structure, symmetry will eliminate some of the enantiomers. Thus, although there are two chiral units in the 1,3-dichlorocyclopentanes shown in Figure 7.11, only three discrete stereoisomers are possible. The cis isomer is optically inactive because it has a plane of symmetry; such compounds are called *meso*. The chiral trans isomer exists as a pair of enantiomers, each of which is diastereoisomeric with the meso isomer.

The constraint imposed by certain ring systems can also act to reduce the number of possible stereoisomers. Despite the presence of two chiral units in camphor, this substance exists only as a pair of enantiomers

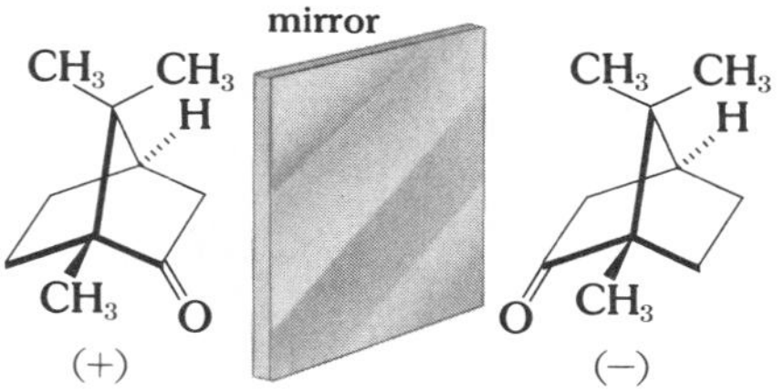

enantiomeric configurations of camphor

because the diastereoisomer having a trans-oriented $(CH_3)_2C\!\!<$ bridge is structurally impossible.

PROBLEM 7·10

a Draw stereochemical formulas (like those in Figures 7.10 and 7.11) showing all stereoisomers of menthol. Which pairs of stereoisomers are enantiomers? Which are diastereoisomers?

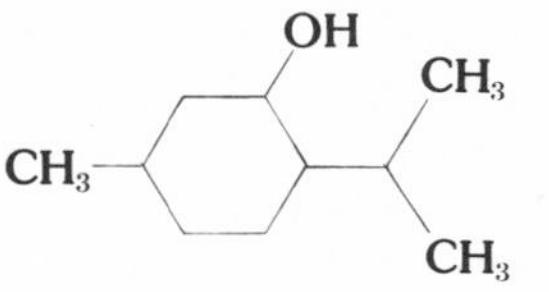

menthol

b Write stereochemical formulas for all the stereoisomers of 2-methyl-1,3-dibromocyclopentane and indicate which pairs of isomers are enantiomers. Which isomers are meso compounds (optically inactive)?

PROBLEM 7·11

Predict the total number of stereoisomers for each of the following structures and identify enantiomeric pairs:

a Cholesterol (see Problem 7.5)

b $CH_3CH{=}C{=}CHCH{=}CHCH_3$

c (cyclohexane with =CHCH$_3$ and OH substituents)

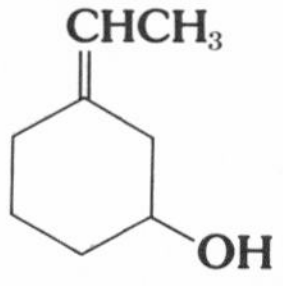

d (cyclohexane with =CHCH$_3$ and two OH substituents)

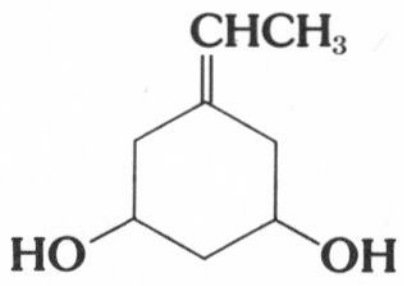

7·4 RESOLUTION OF ENANTIOMERS

The chemical and physical differences between diastereoisomers (boiling and melting points, solubility, thermodynamic stability, rates of chemical reactions, etc.) provide us with a general method for separating, or *resolving*, the enantiomers in a racemic modification. Simply stated, the underlying principle is as follows: *Reaction of a racemic modification* $(\pm)$-*Y with an optically active chiral reagent* $(+)$-*Z gives a mixture of diastereoisomers,* $(+)$-*Y*$(+)$-*Z and* $(-)$-*Y*$(+)$-*Z, which can be separated and reconverted to optically active reactants* $(+)$-*Y,* $(-)$-*Y, and* $(+)$-*Z.* This process is illustrated in Figure 7.12. Many useful chiral reagents for this purpose can be obtained from plant and animal sources.

FIGURE 7·12 *Resolution of enantiomers via diastereoisomeric derivatives*

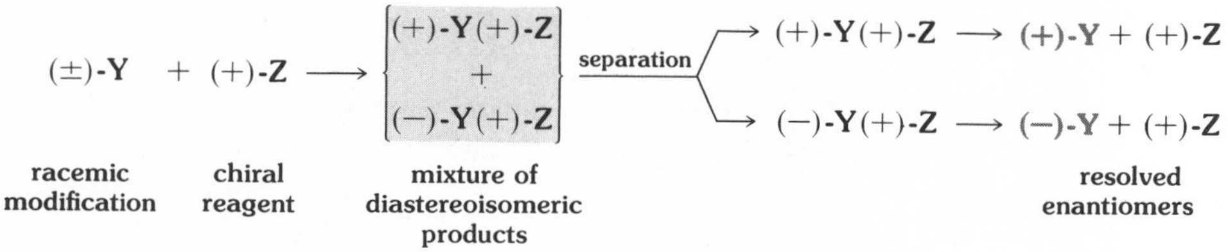

$$(\pm)\text{-Y} + (+)\text{-Z} \longrightarrow \begin{Bmatrix} (+)\text{-Y}(+)\text{-Z} \\ + \\ (-)\text{-Y}(+)\text{-Z} \end{Bmatrix} \xrightarrow{\text{separation}} \begin{matrix} (+)\text{-Y}(+)\text{-Z} \longrightarrow (+)\text{-Y} + (+)\text{-Z} \\ (-)\text{-Y}(+)\text{-Z} \longrightarrow (-)\text{-Y} + (+)\text{-Z} \end{matrix}$$

racemic modification — chiral reagent — mixture of diastereoisomeric products — resolved enantiomers

7·4 RESOLUTION OF ENANTIOMERS

PROBLEM 7·12
a Using the notation shown in Figure 7.12, write equations for all stages of the resolution of $(\pm)$-Y if $(-)$-Z is employed as the chiral reagent in this reaction.
b What is the enantiomer of $(+)$-Y$(+)$-Z? Of $(-)$-Y$(+)$-Z?
c Why is it not possible to use a racemic resolving agent such as $(\pm)$-Z to separate enantiomers?

It is important to recognize that a racemic modification can be resolved without actually isolating a pair of diastereoisomeric intermediates. For example, the $(+)$-Y$(+)$-Z and $(-)$-Y$(+)$-Z species in Figure 7.12 could represent diastereoisomeric transition states, rather than stable compounds. In this event the two reactions that take place have different activation energies and therefore proceed at different rates, resulting in the predominant formation of one of the enantiomeric products, either $(+)$-Y or $(-)$-Y. This kinetic selectivity is most pronounced in enzyme-catalyzed reactions. For example, dextrorotatory lactic acid is rapidly transformed to pyruvic acid by the action of the heart-muscle enzyme $(+)$-lactic dehydrogenase, whereas the levorotatory isomer is not affected by this enzyme,

$$\underset{(+)\text{-lactic acid}}{CH_3-\underset{\underset{H}{|}}{\overset{\overset{OH}{|}}{C}}-CO_2H} \quad \xrightarrow[-2H]{\text{enzyme system}} \quad \underset{\text{pyruvic acid}}{CH_3-\overset{\overset{O}{\|}}{C}-CO_2H} \tag{7.5}$$

The gross bilateral symmetry of the human body hides a conglomeration of chiral enzymes, hormones, vitamins, and other compounds, which in some cases can cause highly specific physiological responses to enantiomers. For example, there is a marked difference in the taste of $(+)$- and $(-)$-leucine, $(CH_3)_2CHCH_2C^*H(NH_2)CO_2H$; the former is sweet, while the latter is bitter. Other differences in the physiological properties of enantiomers will be noted in later chapters.

PROBLEM 7·13
Draw an energy diagram illustrating the fact that *cis*-1-chloro-4-methylcyclohexane (is this a chiral structure?) reacts with an optically active base $(+)$-B: to give a mixture of enantiomeric alkenes, $(+)$-A and $(-)$-A, in which $(-)$-A predominates:

$$\underset{}{\text{(cis-1-chloro-4-methylcyclohexane)}} + (+)\text{-B:} \longrightarrow \underset{(+)\text{-A}}{} + \underset{(-)\text{-A}}{} + BH^{\oplus}Cl^{\ominus}$$

Show the two reactions leading to the enantiomeric products on a single diagram (for example, as in Figure 5.2) and clearly indicate the relative activation energies ΔE^*.

7·5 CONFIGURATION

THE STRUCTURAL DILEMMA

We have established that compounds having chiral molecules exist as a pair of optically active enantiomers which can be separated by the use of chiral reagents, provided the rate of enantiomer interconversion (racemization) is not rapid. However, a question remains as to which of the mirror-image configurations about the chiral center corresponds to the dextrorotatory enantiomer (the other configuration would, of course, have to be that of the levorotatory enantiomer).

Since an unambiguous answer to this question did not seem possible from chemical studies, the first efforts in this direction concerned the establishment of *relative configurations*. The basic premise in this approach was that optically active compounds which could be interconverted by reactions that did not directly affect the center of chirality were assumed to have equivalent configurations. For example, the dimethylation of lactic acid with methyl iodide and silver oxide does not involve bond cleavage or bond formation at the chiral carbon atom:

$$\text{(+)-lactic acid} \xrightarrow{2CH_3-I,\ Ag_2O} \text{(−)-dimethyl derivative} + 2AgI + H_2O \qquad (7.6)$$

Consequently, the reaction must proceed without any change in the configuration of equivalent groups, despite the reversal of the optical rotation of the product. In fact, one of the important lessons learned from these relative-configuration studies is that in most cases the configuration and optical rotation of a compound are not related in any simple or obvious manner.

PROBLEM 7·14
From the chemical evidence given here, can the configurations shown in equation 7.6 be distinguished from the mirror-image configurations?

PROBLEM 7·15
The following three reactions have been reported. Without further information, what conclusions can you draw concerning the relative configurations of the reactants and products?

a $\ CH_3CHClCO_2H \xrightarrow{Ag_2O,\ H_2O} CH_3CH(OH)CO_2H$
    $\quad\quad (-) \quad\quad\quad\quad\quad\quad\quad\quad\quad (-)$

b (cyclohexane with CH$_2$Br and two CH$_3$ groups, (+)) + NaSH ⟶ (cyclohexane with CH$_2$SH and two CH$_3$ groups, (+)) + NaBr

7·5 CONFIGURATION

(−)-cholesterol →[oxidation] (+)-cholestenone

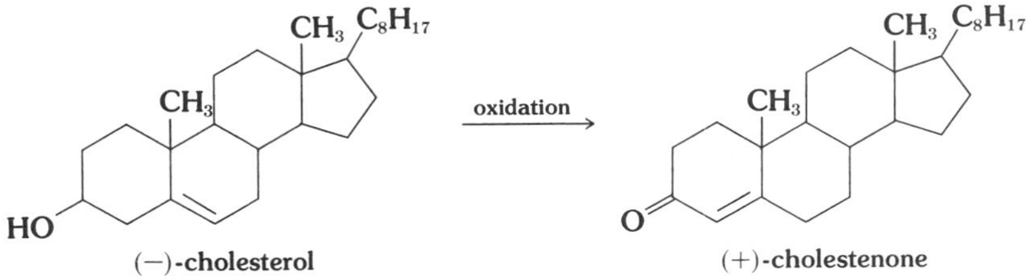

A definitive answer to the configuration question was finally obtained in 1951, when Bijvoet, Peerdeman, and van Bommel in the Netherlands were able to apply a special x-ray fluorescence technique for distinguishing mirror images to a crystalline salt of (+)-tartaric acid, $HO_2CCH(OH)CH(OH)CO_2H$. Since the configurations of many other optically active compounds had been previously correlated with tartaric acid derivatives, the absolute configurations of all these compounds were automatically established as well. Thus, the configuration of (+)-lactic acid shown in equation 7.6 is now known to be correct.

CONFIGURATIONAL NOTATION

The wedge and dashed-line formulas used in this chapter are helpful in representing three-dimensional structures, but they can be awkward and time-consuming to draw. Other more specialized configurational notations have been developed for certain classes of compounds. The *Fischer projection*, a simple representation for chains of chiral atoms, is illustrated in Figure 7.13. In the Fischer projection we imagine that the longest carbon chain containing the chiral atoms lies vertically on the plane of the paper (or blackboard) in a conformation such that all substituent groups and atoms are oriented horizontally in front of the paper. This vertical chain is then represented by a line, and the substituents are joined by crossbars.

Many polymers have different diastereoisomeric forms. Polypropylene, for example, may have the methyl substituents all oriented on the same side of the carbon chain, alternating regularly from one side to the other

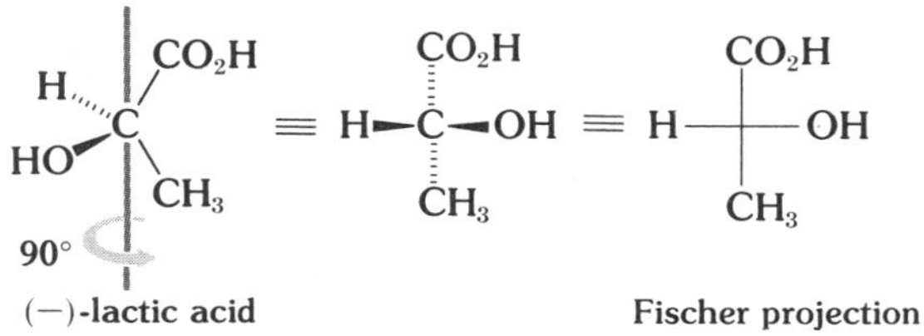

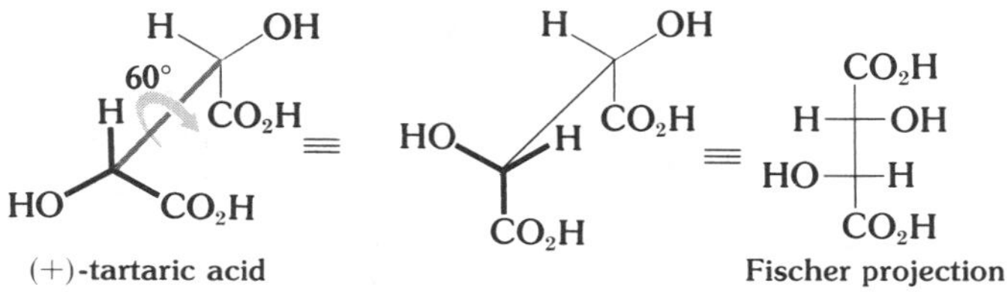

FIGURE 7·13 *Examples of Fischer representation*

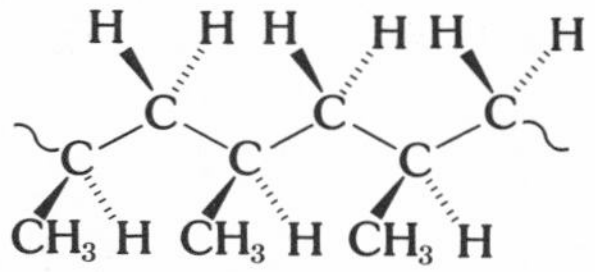

FIGURE 7·14 *Stereoisomeric polymers of propene*

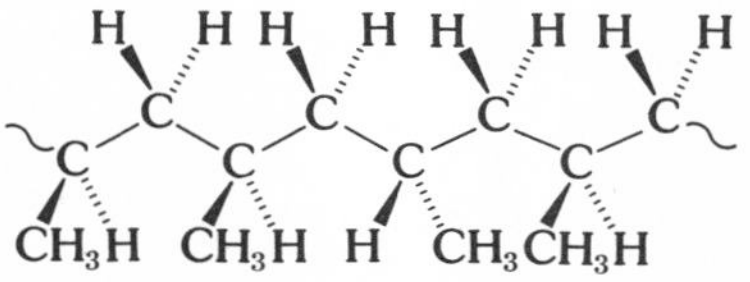

(isotactic configuration)
(syndiotactic configuration)
(atactic configuration)

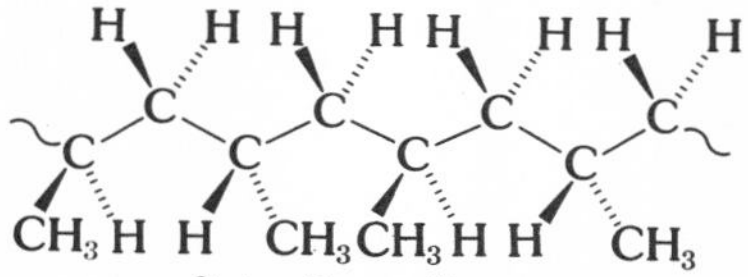

or randomly distributed. These different arrangements of substituent groups on the carbon chain are referred to, respectively, as *isotactic*, *syndiotactic*, and *atactic*. Stereoregular polymers (isotactic or syndiotactic forms) are generally more crystalline and have higher melting points than the atactic forms.

Ring systems are represented as lying in the plane of the paper, with the substituents extending either above (wedge bonds) or below (dashed bonds) this plane (Figure 7.15).

THE CAHN-INGOLD-PRELOG NOMENCLATURE RULES

In addition to a convenient two-dimensional projection of chiral molecules, we need to find a nomenclature system that will enable us to specify the configuration by a name or symbol. Several rather ambiguous schemes were proposed by early workers, but the only completely satisfactory method is the Cahn-Ingold-Prelog system, which was introduced in 1956. Under this nomenclature system we can assign the symbol R or S (from the Latin *rectus* and *sinister*, meaning right and left) to each chiral center in a molecule on the basis of a viewing procedure which defines the configuration at each site.†

†These new designations are necessary because the terms *dextro-* and *levo-* [or (+) and (−)] refer only to the sign of the observed optical rotation of 5890 Å plane-polarized light. As noted earlier, configuration and optical rotation are not related in any simple or obvious manner.

FIGURE 7·15 *Representation of ring systems in a steroid*

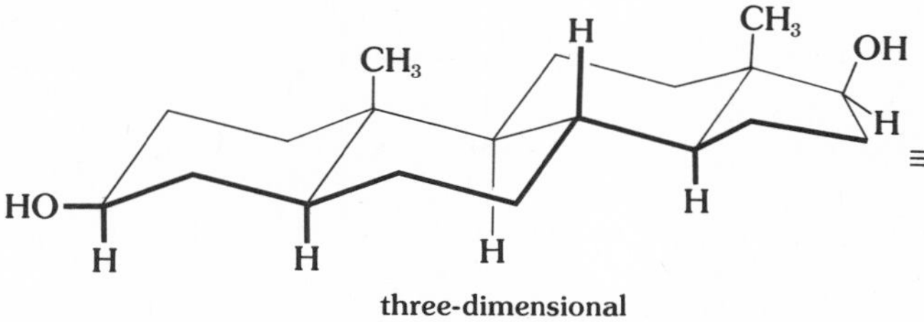

three-dimensional representation 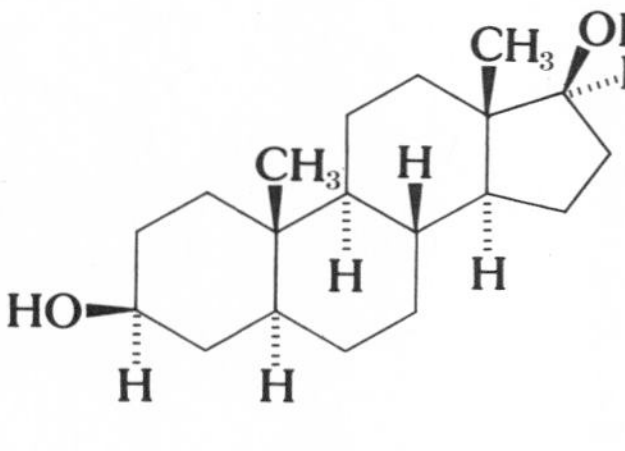 ≡ two-dimensional projection

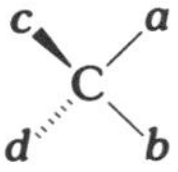

7·5 CONFIGURATION

Under the Cahn-Ingold-Prelog nomenclature system, the four atoms or groups attached to a chiral carbon atom are ordered ($a > b > c > d$) according to the following sequence rules:

1 The atoms directly attached to the chiral center are ordered according to atomic number. Thus for the compound $CH_3CHClOCH_3$ the sequence order is

$$Cl > OCH_3 > CH_3 > H$$

2 If two or more atoms attached to the chiral center are alike, then the relative priority of the groups incorporating these atoms is determined at the site of difference nearest to the chiral center. For example,

$$-CH_2CH_2OH > -CH_2CH_2CH_3 > -CH_2CH_3$$

When two substituent groups contain the same kinds of atoms, it may be necessary to count the number of nearest high-priority neighbors.

$$-C(CH_3)_3 > -CH(CH_3)_2 > -CH_2CH_2CH_3$$

3 Multiple bonds are treated as hypothetical tetracoordinate systems, with the atoms of the multiple bond replicated according to the number of extra bonds. Thus the vinyl group has a higher sequence priority than an ethyl group because the carbon atoms of the double bond are replicated.

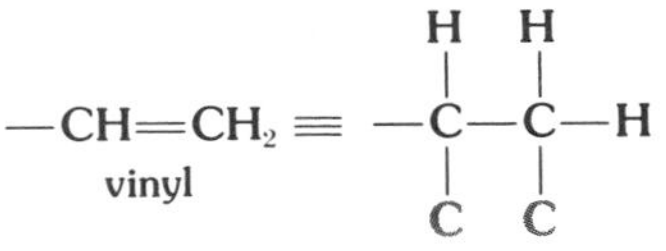

$$-CH=CH_2 \equiv -\underset{\underset{C}{|}}{\overset{\overset{H}{|}}{C}}-\underset{\underset{C}{|}}{\overset{\overset{H}{|}}{C}}-H$$
vinyl

Application of this rule would give the following sequence orders:

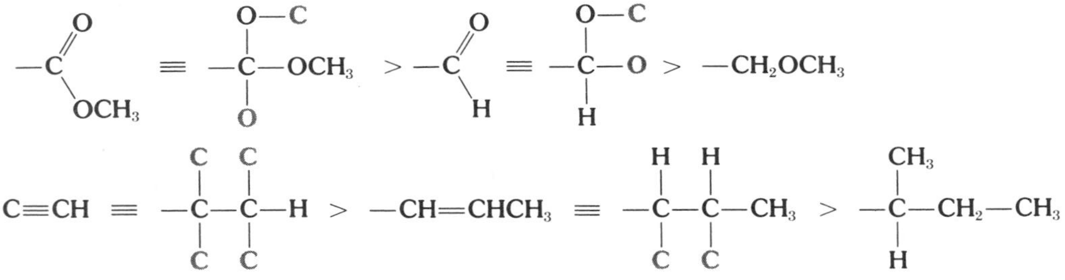

$$-\underset{OCH_3}{\overset{\overset{O}{\parallel}}{C}} \equiv -\underset{\underset{O}{|}}{\overset{\overset{O-C}{|}}{C}}-OCH_3 > -\underset{H}{\overset{\overset{O}{\parallel}}{C}} \equiv -\underset{H}{\overset{\overset{O-C}{|}}{C}}-O > -CH_2OCH_3$$

$$-C\equiv CH \equiv -\underset{\underset{C}{|}}{\overset{\overset{C}{|}}{C}}-\underset{\underset{C}{|}}{\overset{\overset{C}{|}}{C}}-H > -CH=CHCH_3 \equiv -\underset{\underset{C}{|}}{\overset{\overset{H}{|}}{C}}-\underset{\underset{C}{|}}{\overset{\overset{H}{|}}{C}}-CH_3 > -\underset{\underset{H}{|}}{\overset{\overset{CH_3}{|}}{C}}-CH_2-CH_3$$

4 When the only difference is due to isotopic substitution, the heavier atom or group takes precedence.

$$D > H \qquad -CD_3 > -CH_3$$

Once the sequence of groups around a chiral center is determined, we view this center from the side *opposite the group of lowest rank* and observe whether the remaining sequence is clockwise (an R configuration) or counterclockwise (an S configuration). The examples in Table 7.2 illustrate these rules.

The Cahn-Ingold-Prelog system can also be used for systems having axial chirality, such as allenes and spiroalkanes.†

---

†This application is described in detail by R. S. Cahn in the article An Introduction to the Sequence Rule, *J. Chem. Ed.*, 41(1964)116.

7·5 CONFIGURATION

TABLE 7·2 *The Cahn-Ingold-Prelog sequence system*

compound	sequence order	configuration
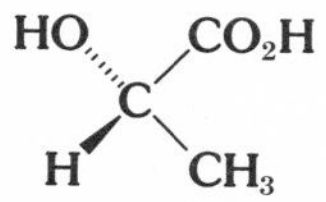 (+)-lactic acid	HO > CO₂H > CH₃ > H	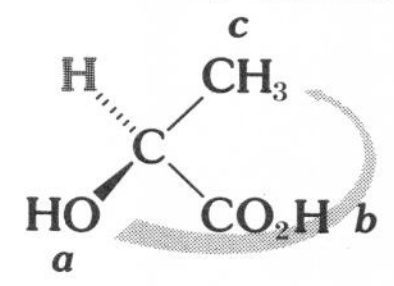 S
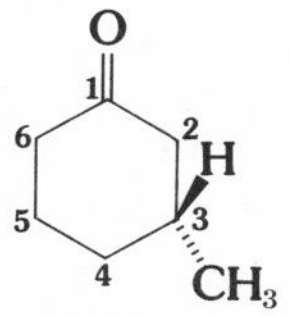 (+)-3-methylcyclohexanone	C—2 > C—4 > CH₃ > H	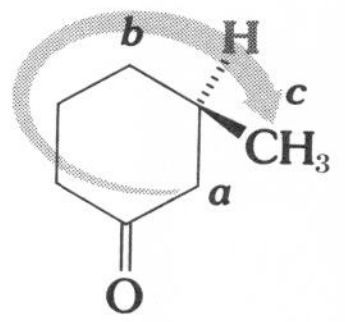 R
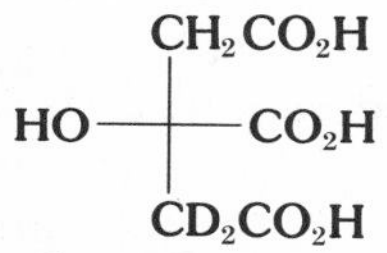 a deuterium-labeled citric acid	HO > CO₂H > CD₂ > CH₂	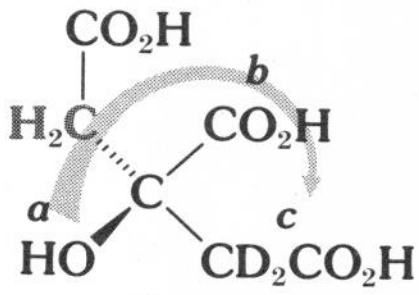 R

PROBLEM 7·16

a Draw a stereo representation (wedge and dotted-line bonds) and a Fischer projection for *R*-(−)-2-chlorobutane.

b Assign the proper configurational symbol (*R* or *S*) to each of the chiral centers in (+)-tartaric acid and (−)-menthol.

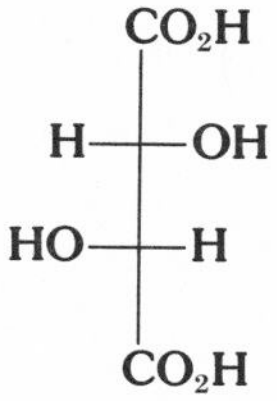

(+)-tartaric acid

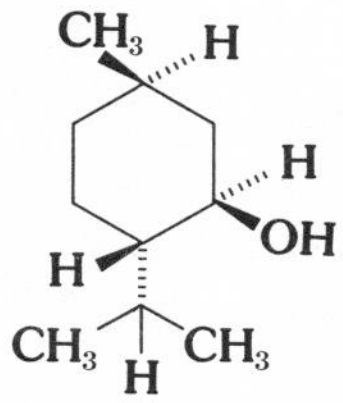

(−)-menthol

The Cahn-Ingold-Prelog sequencing rule allows us to develop an unambiguous nomenclature for stereoisomers involving double-bond configurational differences. The stereoisomers of 2,4-dichloro-3-ethyl-2-pentene, for example, cannot be clearly differentiated by a *cis*- or *trans*- prefix. However, such stereoisomerism requires that the two substituents on each of the double-bonded carbon atoms must have a different sequence order. Consequently, it is a simple matter to focus our attention on the relative configuration of the higher-order substituent at each end of the double bond. Since chlorine precedes methyl and CHClCH₃ precedes ethyl in sequence order, we may designate the two stereoisomers of 2,4-dichloro-3-ethyl-2-pentene as *seqcis* and *seqtrans* (sometimes denoted by Z and E, from the German *zusammen*, together, and *entgegen*, against or opposed).

$$\begin{array}{c}CH_3\\ \diagdown \\ C=C \\ \diagup \quad \diagdown \\ Cl \quad\quad CHClCH_3\end{array} \begin{array}{c}CH_2CH_3\\ \diagup\end{array}$$

seqcis (Z) isomer

$$\begin{array}{c}CH_3\\ \diagdown \\ C=C \\ \diagup \quad \diagdown \\ Cl \quad\quad CH_2CH_3\end{array} \begin{array}{c}CHClCH_3\\ \diagup\end{array}$$

seqtrans (E) isomer

7·6 CONFORMATIONAL ANALYSIS

With the basic concepts of enantiomerism and diastereoisomerism well in mind, we are now ready to take into consideration the additional stereochemical factor of conformational isomerism. As an illustration of this relationship consider the isomeric tartaric acids listed in Table 7.3. At the molecular level there are three tartaric acid isomers.[†] Two of these are enantiomers, and the third is an optically inactive meso isomer, resulting from the symmetry introduced by the identical sets of substituents on the two chiral atoms (as in the case of *cis*-1,3-dichlorocyclopentane, described in Figure 7.11). Although some conformationally rigid meso compounds have clearly defined elements of symmetry (note the plane of symmetry in *cis*-1,3-dichlorocyclopentane), the optical inactivity of open-chain meso compounds cannot usually be explained in this manner.

For example, *meso*-tartaric acid can assume many conformations, some achiral (I and IV) and others chiral (II and III).

The eclipsed conformation (IV) has a plane of symmetry, but it should only be present in low concentration because of the general instability of eclipsed bond conformations (see Chapter 2). The only staggered

TABLE 7·3 *Tartaric acids,* $HO_2CCH(OH)CH(OH)CO_2H$

compound†	melting point (°C)	solubility in water (g/100 ml)	specific rotation $[\alpha]_D$
(+)-tartaric acid	170	139	+12°
(−)-tartaric acid	170	139	−12°
meso-tartaric acid	140	125	0°

†A fourth, optically inactive, tartaric acid called *racemic acid* (or racemic tartaric acid) is often listed in handbooks and chemical encyclopedias. This substance is actually a special kind of racemic modification called a *racemic compound* or a *racemate*, which is a unique crystalline form composed of alternating molecules of the enantiomeric tartaric acids. The stability of this crystalline form is indicated by its high melting point (250°C) and low solubility in water.

achiral conformation is I (with a center of symmetry; we can therefore conclude that an assemblage of *meso*-tartaric acid molecules will probably have a higher concentration of chiral conformers than achiral conformers. The reason for the optical inactivity of *meso*-tartaric acid is found in the mirror-image relationship of conformers II and III. Since the energy barrier to conformational interconversions is low (about 4 kcal/mole), and since enantiomeric conformers have identical thermodynamic parameters, these two conformations exist in exactly equal amounts and their optical rotations cancel.

A conformational analysis of (+)-tartaric acid shows a different situation. All three staggered conformations are chiral, and no two of them have a mutual mirror-image relationship:

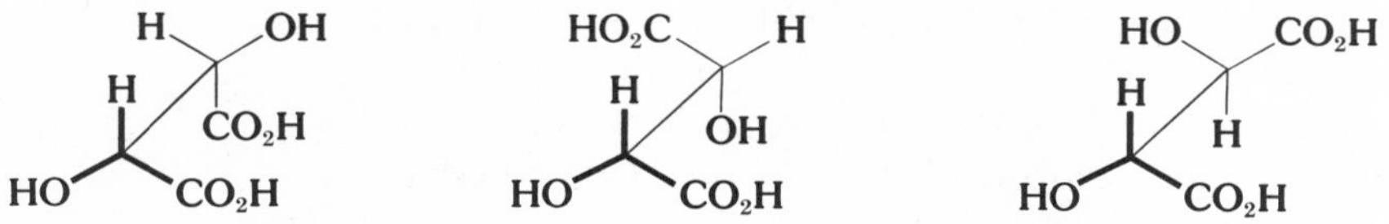

A solution of (+)-tartaric acid will therefore contain several different chiral conformations, and a corresponding solution of (−)-tartaric acid will consist of an equivalent mixture of mirror-image conformations.

PROBLEM 7·17
In view of the chirality of the following conformation, why can 1,2-dichloroethane not be resolved into a pair of enantiomers?

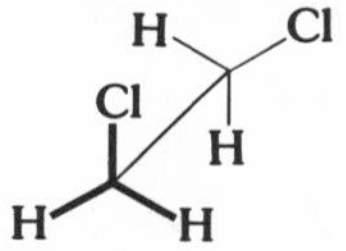

Conformational racemization also accounts for our inability to isolate enantiomers in certain cyclic meso compounds. For example, the compound *cis*-1,2-dichlorocyclohexane exists predominantly in two chair conformations:

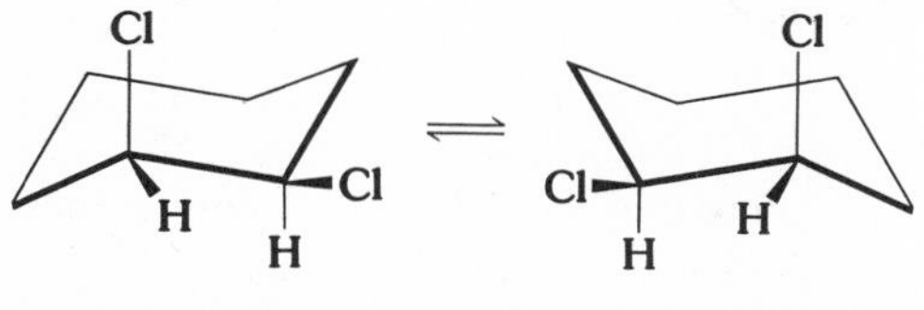

These conformers are in fact enantiomeric, and once again, a low barrier to enantiomer interconversion (about 11 kcal/mole) prevents the isolation of optically active isomers.

Compounds having enantiomeric conformations separated by an energy barrier greater than 20 kcal/mole can normally be resolved, and an interesting example of conformational enantiomerism is found in *trans*-cyclooctene. Molecular models of this compound show that it assumes a stable twisted (chiral) conformation, and that interconversion of this with its enantiomeric conformer requires a transition state having considerable angle and eclipsing strain. Resolution of *trans*-cyclooctene enantiomers has been accomplished, and the optically active form exhibits a remarkably large rotational strength, with $[\alpha]_D > 400°$.

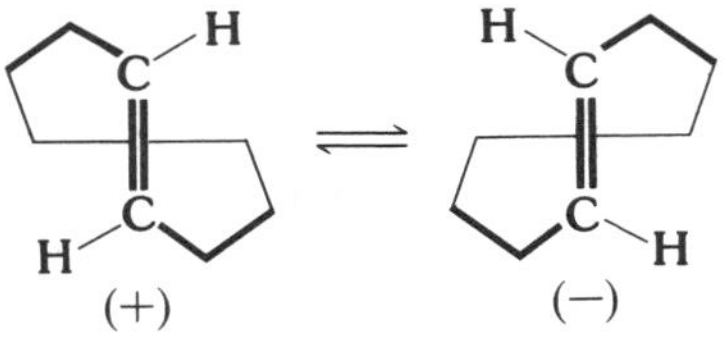

trans-cyclooctene enantiomers

It should be clear from the facts presented thus far that a chiral unit (such as an asymmetric carbon atom) is neither necessary nor sufficient to assure optical isomerism. We must therefore formulate a more precise criterion for optical activity based on molecular chirality: *A compound composed of chiral molecules which cannot be transformed into a mirror-image configuration or conformation by a low-activation-energy process (for example, rotation about a single bond) can be separated into optically active enantiomers.*

Our increased stereochemical insight now enables us to speak with greater authority about the mechanisms of many reactions. If, for example, a skeptic were to argue that the trans orientation observed in bromine addition to olefins (Equations 5.34 and 5.38) were correct only for cyclic alkenes, we could refute this claim by demonstrating the stereospecific trans addition of bromine to *cis-* and *trans-*2-butene:

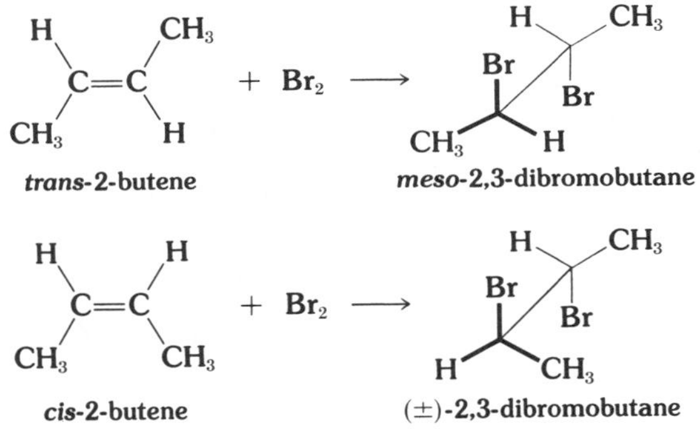

trans-2-butene + Br₂ ⟶ *meso*-2,3-dibromobutane

cis-2-butene + Br₂ ⟶ (±)-2,3-dibromobutane

(7.7)

PROBLEM 7·18
Why could we not refute the argument concerning bromine addition to alkenes by examining the addition of bromine to 1-butene?

PROBLEM 7·19
Draw an energy diagram illustrating the addition of bromine to the isomeric 2-butenes, clearly showing which product diastereoisomer results from which reactant isomer. Assuming the product diastereoisomers can be cleanly separated (say, by distillation or chromatography), how can you determine which isomer corresponds to which structure?

PROBLEM 7·20
The hydroxylation of alkenes with osmium tetroxide proceeds stereospecifically to cis products. Write equations showing the course of hydroxylation of *cis-* and *trans*-2-pentene.

SUMMARY

plane-polarized light Light having electrical and magnetic field vectors oscillating each in a single plane.

optical activity The ability of some substances to rotate the plane of polarization of plane-polarized light.

polarimeter A device for measuring optical activity. Rotation to the right (clockwise) is termed dextrorotatory (+), and rotation to the left is termed levorotatory (−).

specific rotation $[\alpha]$ The rotation α (degrees) adjusted to a standard cell length l (dm) and concentration c (g/ml): $[\alpha] = \alpha/lc$.

symmetry element A characteristic geometric feature whose presence in an object is established by a corresponding symmetry operation. Common symmetry elements are rotational axes of symmetry C_n, mirror planes of symmetry σ, and a center of symmetry i.

symmetry operation An operation which orients an object in a position indistinguishable from (superimposable on) its original position. These operations are rotation about an axis C_n by $360°/n$, reflection of each half in a bisecting plane σ, and reflection of all parts through the center point i.

asymmetry The absence of all symmetry elements.

dissymmetry The absence of reflection symmetry elements.

chiral A term referring to the handedness of a figure or object which is not identical with (cannot be superimposed on) its mirror image; such an object must be either asymmetric or dissymmetric.

achiral Lacking chirality.

asymmetric carbon unit A tetracoordinate carbon atom bearing four different substituents (ligands).

configuration The arrangement of atoms in space that characterizes a particular stereoisomer. Chiral configurations are designated as R (rectus) or S (sinister).

enantiomers A pair of stereoisomers configurationally related as a chiral object and its nonsuperimposable mirror image. Enantiomers have equal but opposite specific rotations.

diastereoisomers Stereoisomers which are not enantiomers.

meso compound An optically inactive diastereoisomer structurally related to one or more pairs of enantiomers.

racemic modification An equimolar mixture of enantiomers; such mixtures are optically inactive and are denoted by (±).

racemization The formation of a racemic modification by the reversible interconversion of enantiomers.

resolution of enantiomers The separation of a racemic modification into its component enantiomers.

REFERENCES

Stereochemical concepts and relationships play a major role in such fields as biochemistry and molecular biology. For further information about this fascinating area of chemistry, the following references are recommended.

E. L. Eliel, *Elements of Stereochemistry*, John Wiley & Sons, Inc., New York, 1969.

K. Mislow, *Introduction to Stereochemistry*, W. A. Benjamin, Inc., New York, 1965.

R. S. Cahn, An Introduction to the Sequence Rule, *J. Chem. Ed.*, 41(1964): 116 (also correction on p. 508).

L. Glasser, Teaching Symmetry, *J. Chem. Ed.*, 44(1967):502.

J. Idoux, Conformational Analysis and Chemical Reactivity, *J. Chem. Ed.*, 44(1967):495.

E. L. Eliel, *Stereochemistry of Carbon Compounds*, McGraw-Hill Book Company, New York, 1962.

E. L. Eliel, Recent Advances in Stereochemical Nomenclature, *J. Chem. Ed.*, 48(1971):163.

E. L. Eliel, N. L. Allinger, S. J. Angyal, and G. A. Morrison, *Conformational Analysis*, Interscience Division, John Wiley & Sons, Inc., New York, 1965.

M. Hanack, *Conformation Theory*, Academic Press, Inc., New York, 1965.

G. Natta and M. Farina, *Stereochemistry*, Harper and Row, Publishers, New York, 1972.

EXERCISES

7·1 Draw structural formulas for:
a The smallest alkane that is chiral
b The smallest alkene that is chiral
c The smallest alkyl chloride that is chiral

7·2 Draw the *R* configurations of the following compounds:
a 2-chloropentane
b $CH_3CH(OH)CO_2H$ (lactic acid)
c $CH_2=CHCH(OH)CH_2CH_3$
d CH_3CHDOH
e 2,3-dimethylhexane

7·3 Indicate the symmetry elements present in the following structures (assume a planar ring in all cases). Using models, verify that those structures with reflective symmetry elements are identical with their mirror images and those without reflective symmetry elements are nonsuperimposable on their mirror images.
a *cis*-1,2-dimethylcyclobutane
b *trans*-1,2-dimethylcyclobutane
c *cis*-1,3-dimethylcyclobutane
d *trans*-1,3-dimethylcyclobutane

EXERCISES

7·4 Specify the configuration of each as *R* or *S*:

a. C with Br (up), H (wedge), Cl (left), F (right)

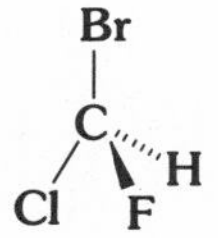

b. H—C—OH with CH₃ (up), CH₂CH₂CH₃ (down)

c. C with CH₂CH₂CH₃ (up), H (wedge), CH₃ (left), CH₂CH₃ (right)

d. Cl—C—H with CH₃ (up), OCH₃ (down)

e. CH₂=CH—C—H with CH₂CH₂CH₃ (up), CH(CH₃)₂ (down)

f. C with CH₂CH₂Br (up), CH₂CH₃ (wedge), H (left), OH (right)

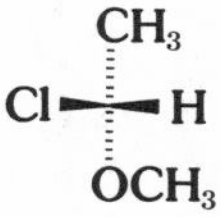

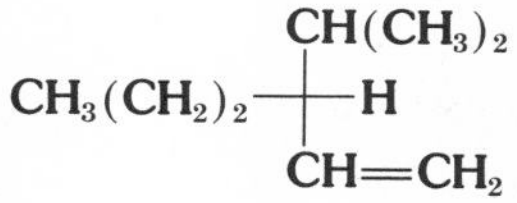

g. CH₃(CH₂)₂—C—H with CH(CH₃)₂ (up), CH=CH₂ (down)

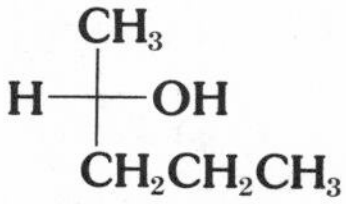

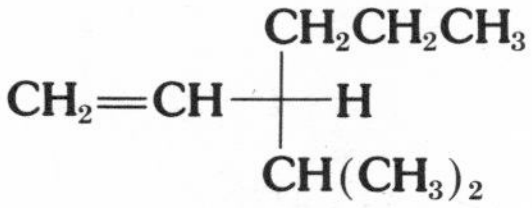

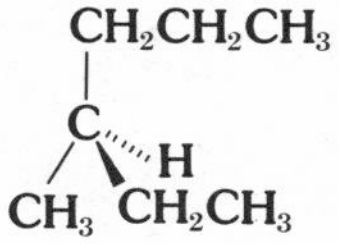

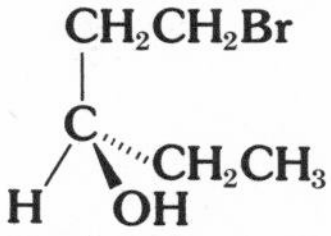

7·5 Match the items indicated by letters with identical structures indicated by numbers:

a. C with CH₂CH₃ (up), OH (wedge), H (left), CH₃ (right)

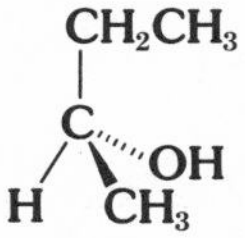

b. Newman projection: CH₃, H, H on front; CH₃, H, OH on back

c. Structure with H, OCH₃, CH₃, CH₃, Br, H

d. CH₃O—C—C—Br with CH₃, H on left C and H, CH₃ on right C

e. cyclohexane with OH and CH₃ substituents

f. cyclohexane with CH₃ and OH substituents

g. CH₃, H, Br, OCH₃ tetrahedral

1. HO—C—H with CH₃ (up), CH₂CH₃ (down)

2. cyclohexane chair with H, CH₃, OH, H

3. H—C—OH with CH₃ (up), CH₂CH₃ (down)

4. cyclohexane chair with H, OH, H, CH₃

5. H—C—OCH₃ / H—C—Br with CH₃ top and CH₃ bottom

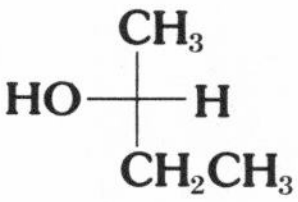

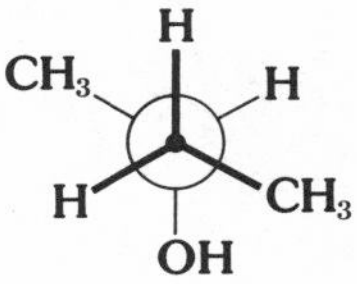

6. cyclohexane chair with H, CH₃, H, OH

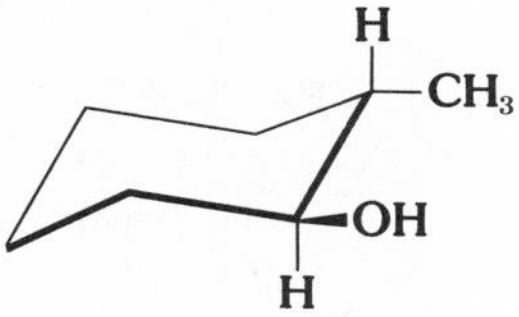

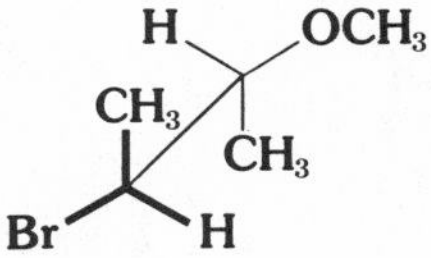

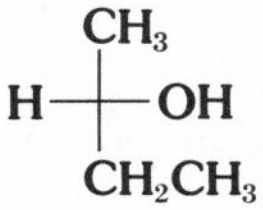

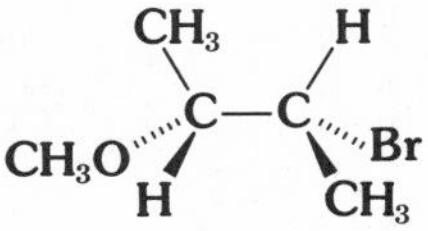

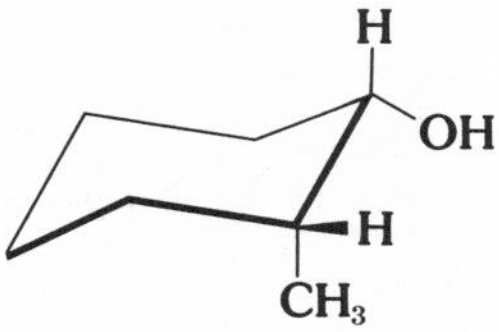

7. H—C—OCH₃ / Br—C—H with CH₃ top and CH₃ bottom

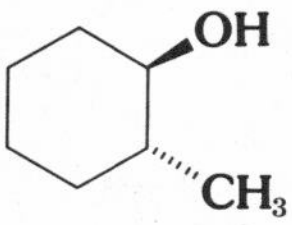

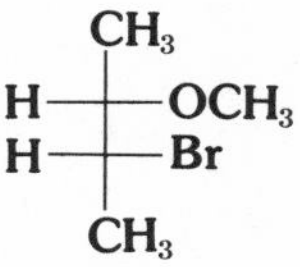

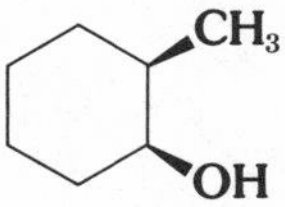

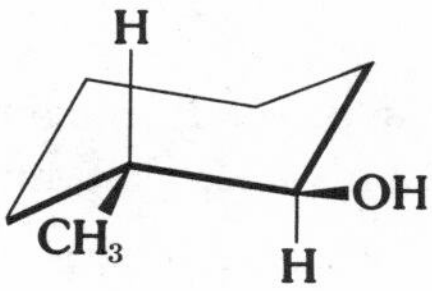

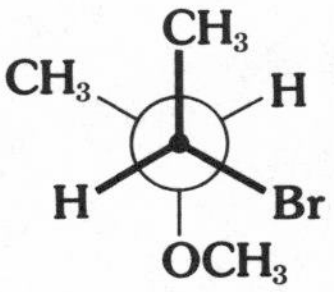

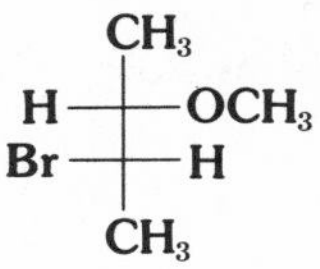

7·6 Identify the following double-bond configurational isomers as seqcis (*Z*) or seqtrans (*E*):

a (CH₃)(H)C=C(Br)(Cl)

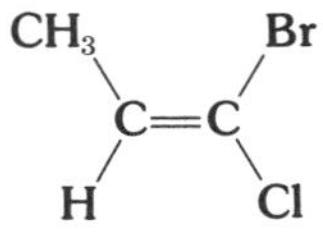

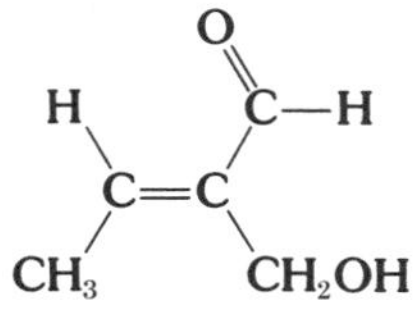

b (CH₃)(Cl)C=C(CH₂CH₂CH₃)(CH₂Br)

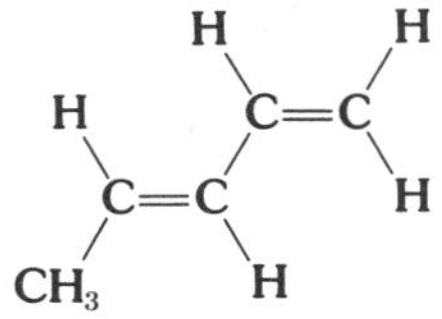

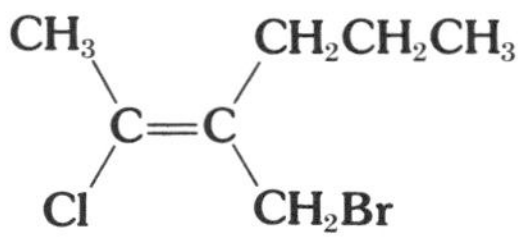

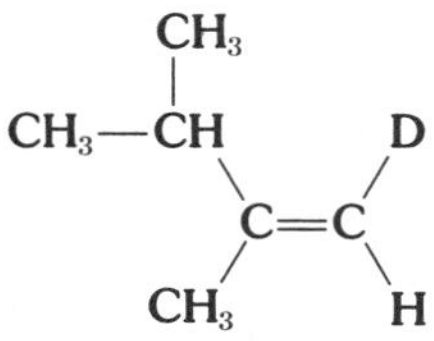

c (H)(CH₃)C=C(COH=O)(CH₂OH)

d (CH₃—CH(CH₃))(CH₃)C=C(D)(H)

e (H)(CH₃)C=C(H)(CH=CH₂)

f Br—C(CH₃)(C₂H₅)—... H—C(CH₃)—H (stereocenter structure)

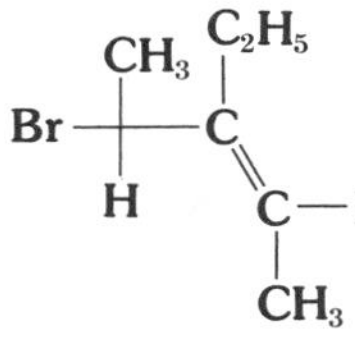

7·7 Draw structural formulas for all possible stereoisomers of each compound and indicate pairs of enantiomers and meso structures:
a 1,2-Dichlorocyclohexane
b 1,3-Dichlorocyclohexane
c 1,4-Dichlorocyclohexane

7·8 Which of the following are chiral?
a Fork
b Spoon
c Scissors
d Knife
e Shoe
f Glove
g Your ear
h Your nose
i Golf club
j Baseball bat
k Wood screw
l Screwdriver
m Light bulb

7·9 Indicate which of the following properties of (+)-2-chlorobutane would be different for (−)-2-chlorobutane:
a Boiling point
b Melting point
c Refractive index
d Specific rotation
e Solubility in water
f Pmr spectrum
g Infrared spectrum
h Rate of reaction with sodium hydroxide
i Rate of reaction with (+)-sodium lactate
j Mass spectrum
k Color

7·10 Indicate the stereochemistry of the products in each reaction and specify any racemic modifications or meso compounds that are formed:

a $CH_3C{\equiv}CCH_3 \xrightarrow[\text{Pd (poisoned)}]{H_2}$ product I $\xrightarrow{Br_2}$ product II

b $CH_3C{\equiv}CCH_3 \xrightarrow[\text{NH}_3\ (l)]{Na}$ product I $\xrightarrow{Br_2}$ product II

EXERCISES

7·11 Estimate the relative magnitude of the activation energy for each conversion:

a
$$\begin{array}{c} CH_2CH_3 \\ H{-}{|}{-}OH \\ CH_3 \end{array} \rightleftharpoons \begin{array}{c} CH_2CH_3 \\ HO{-}{|}{-}H \\ CH_3 \end{array}$$

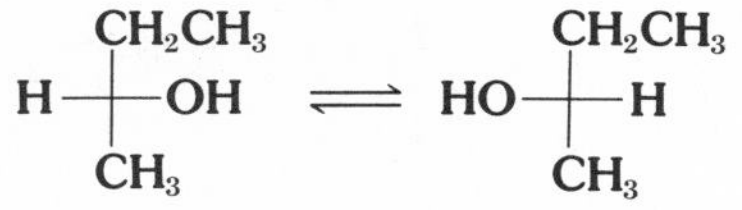

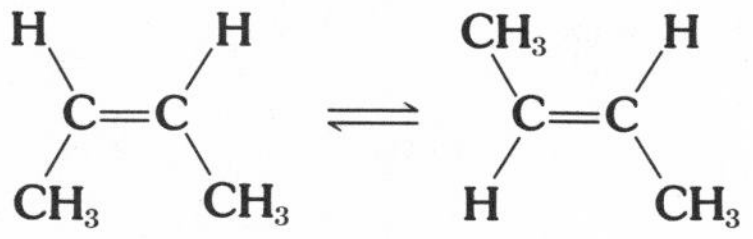

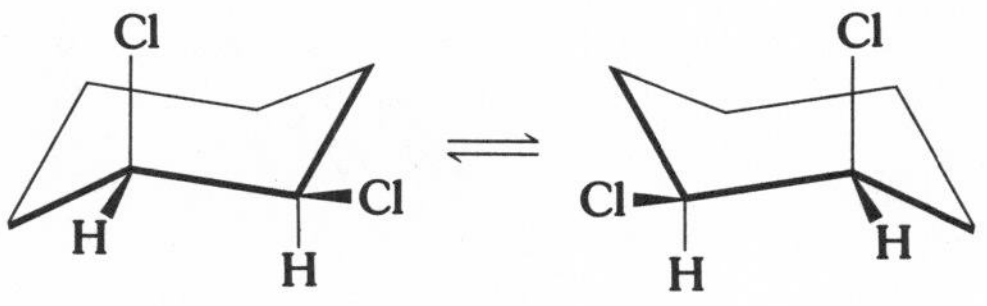

b
$$\begin{array}{c} H \\ \diagdown \\ CH_3 \end{array} C=C \begin{array}{c} H \\ \diagup \\ CH_3 \end{array} \rightleftharpoons \begin{array}{c} CH_3 \\ \diagdown \\ H \end{array} C=C \begin{array}{c} H \\ \diagup \\ CH_3 \end{array}$$

c (chair cyclohexane with Cl axial/equatorial conversions)

d (Newman projections of butane-like conformations)

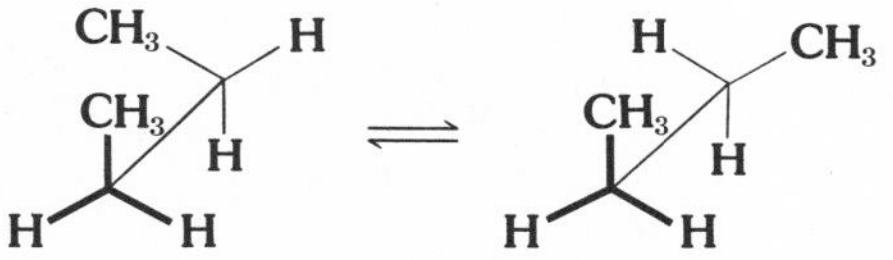

7·12 Compound A below is not optically active. It is expected to exist primarily as conformation I. Is this conformation dissymmetric? There is also an appreciable concentration of a dissymmetric conformation II in solutions of this dibromide. Why is the solution optically inactive? Both III and IV are dissymmetric conformations of compound B, which is optically active. Why are mirror-image conformations of III and IV absent from the solution?

a 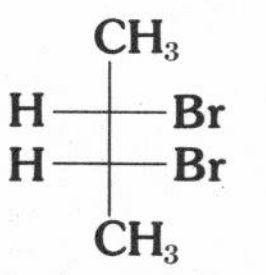
$$\begin{array}{c} CH_3 \\ H{-}{|}{-}Br \\ H{-}{|}{-}Br \\ CH_3 \end{array}$$
compound A

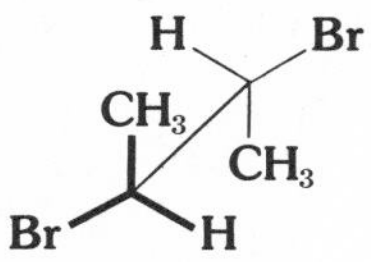

I, 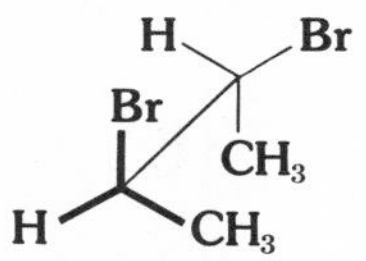 II (Newman projections)

b 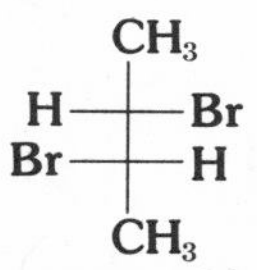
$$\begin{array}{c} CH_3 \\ H{-}{|}{-}Br \\ Br{-}{|}{-}H \\ CH_3 \end{array}$$
compound B

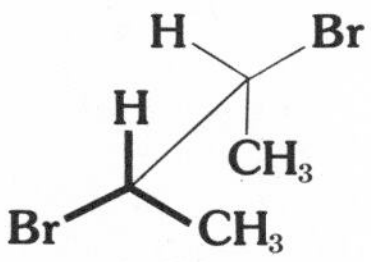

III, 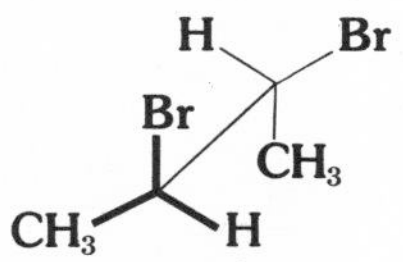 IV (Newman projections)

7·13 Determine the total number of stereoisomers possible for each compound:

a 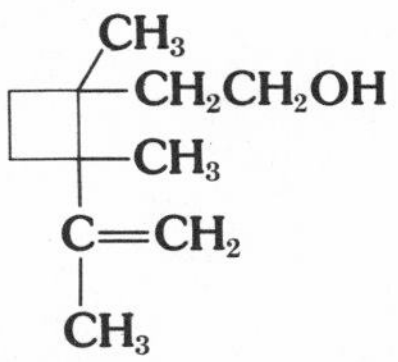
$$\begin{array}{c} CH_3 \\ |{-}CH_2CH_2OH \\ |{-}CH_3 \\ C=CH_2 \\ | \\ CH_3 \end{array}$$
sex attractant of male boll weevil

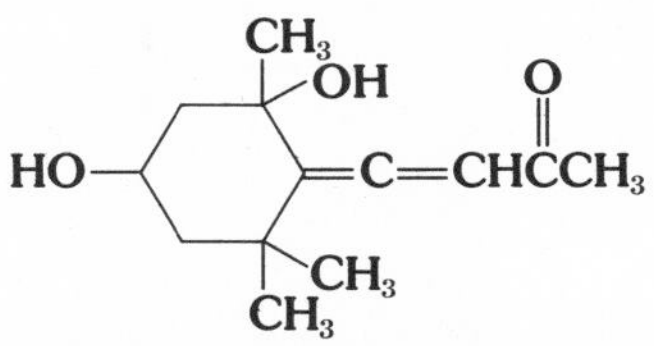

(cyclohexane derivative with OH, CH_3, =C=CHCCH_3, CH_3 substituents)
defense secretion of *Romalea* grasshopper

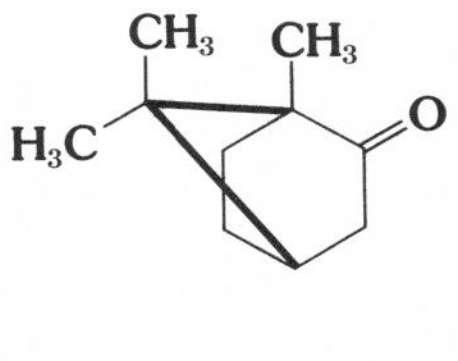

camphor

7·14 Carboxylic acids react with organic bases, such as amines, to yield crystalline salts:

$$R-CO_2H + R'-NH_2 \longrightarrow R-CO_2^{\ominus}H_3N^{\oplus}-R'$$

a carboxylic acid   an amine   a salt

Write equations showing how this reaction can be used to resolve the following racemic amine through the agency of (+)-tartaric acid as a resolving agent:

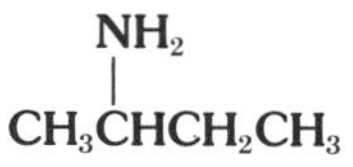

$$CH_3\overset{\overset{\displaystyle NH_2}{|}}{C}HCH_2CH_3$$

(+)-amine

$$\begin{array}{c} CO_2H \\ H\!-\!\!-\!OH \\ HO\!-\!\!-\!H \\ CO_2H \end{array}$$

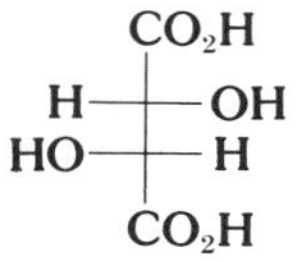

(+)-tartaric acid

ALCOHOLS AND ETHERS

TO MOST people the name *alcohol* refers to only a few substances: the ethyl alcohol (ethanol) found in intoxicating beverages, the isopropyl alcohol found in the medicine cabinet, and the wood alcohol (methanol) used as a solvent for shellac. The truth of the matter is that the alcohols are a very large class of compounds, exhibiting a rich and diverse chemical behavior. As a consequence, they are versatile intermediates in multistep syntheses. A similar misconception surrounds the name *ether*, which is commonly associated only with the anesthetic diethyl ether. Although diethyl ether and other volatile ethers are highly flammable, they tend to be much less reactive than their cousins the alcohols. In fact, ethers are sufficiently inert to chemical change that they are often used as solvents for reactions involving highly reactive species.

Alcohols and ethers are organic derivatives of water in which one or both hydrogen atoms are replaced by alkyl groups:

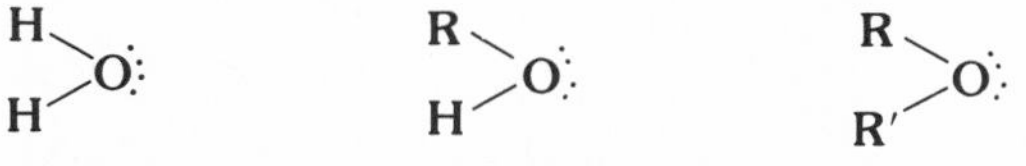

water   an alcohol   an ether

A wide variety of naturally occurring alcohols (and to a lesser degree, ethers) have been identified, from both plant and animal sources:

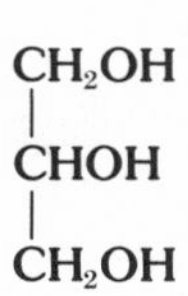

glycerol
(from fats and oils)

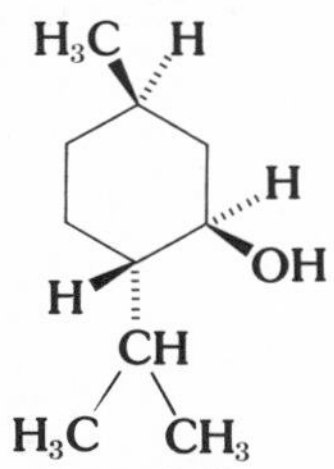

menthol (from peppermint oil)

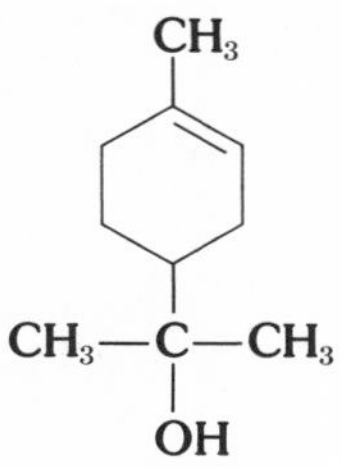

α-terpineol
(from neroli and camphor oils)

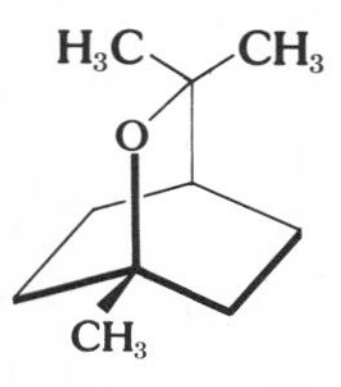

cineole (from eucalyptus oil)

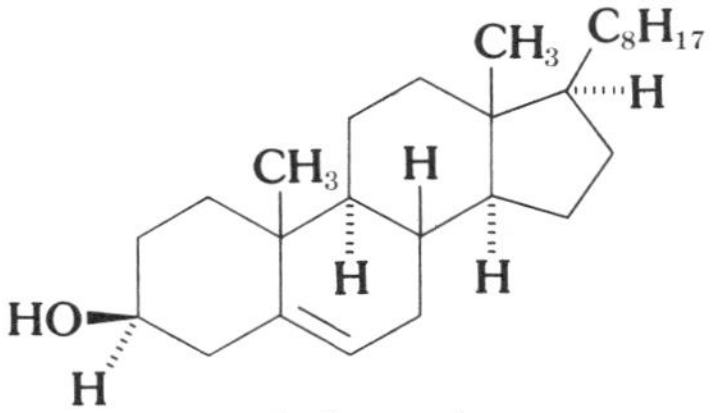

cholesterol
(from animal tissue)

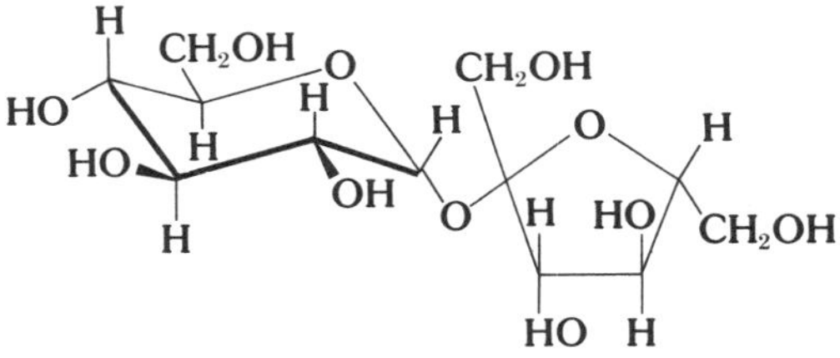

sucrose
(from sugar cane)

The fermentation of sugar and starch by yeast or bacteria is an important source of ethanol and 1-butanol. However, many alcohols that are not found in natural sources, or are present only in small amounts, can be prepared by transformations of other functional groups. Examples of such syntheses include the hydrolysis of alkyl halides (see equations 3.2 and 3.4), the hydration of alkenes (Section 5.1), and hydroboration of alkenes (Section 5.2). Other important and useful methods of making alcohols are discussed in Chapters 12 and 14. These methods involve the addition of hydrogen or alkyl moieties to the carbonyl groups of aldehydes, ketones, or carboxylic acid derivatives:

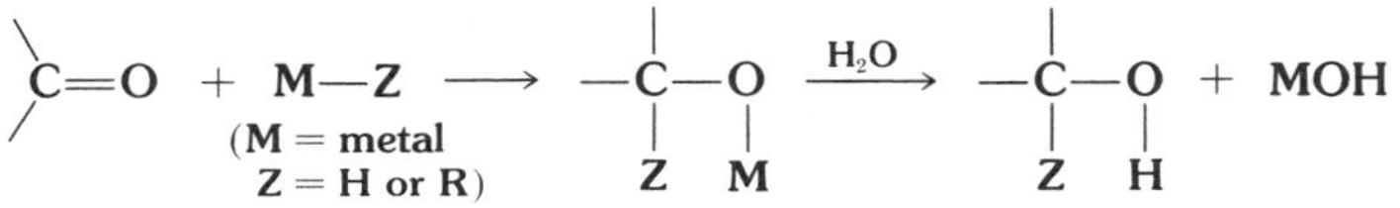

$$\begin{array}{c}\diagdown\\ C=O\\ \diagup\end{array} + M-Z \longrightarrow \begin{array}{c}|\\-C-O\\|\ \ |\\Z\ \ M\end{array} \xrightarrow{H_2O} \begin{array}{c}|\\-C-O\\|\ \ |\\Z\ \ H\end{array} + MOH \qquad (8.1)$$

(M = metal
Z = H or R)

8·1 NOMENCLATURE OF ALCOHOLS AND ETHERS

Several nomenclature systems are used for alcohols and ethers. For simple compounds the generic term *alcohol* or *ether* is preceded by the name of the alkyl group (or groups) replacing the hydrogen:

CH$_3$OH   (CH$_3$)$_3$COH   C$_2$H$_5$OC$_2$H$_5$ cyclohexyl—O—CH(CH$_3$)$_2$

methyl alcohol   *t*-butyl alcohol   diethyl ether   cyclohexyl isopropyl ether

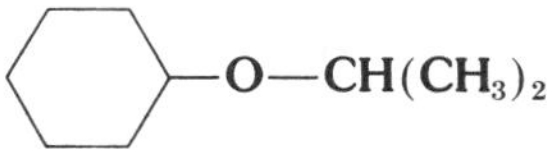

The IUPAC system for naming alcohols operates according to previously established rules, with the characteristic ending *-ol* and a number designating the location of the hydroxyl (OH) group. Some examples of this nomenclature are given in Table 8.1. We also classify alcohols as primary, secondary, or tertiary, depending on the kind of carbon atom that bears the hydroxyl group.

Oxygen-containing substituents have appropriate names: *hydroxy-* for —OH, *methoxy-* for —OCH$_3$, *ethoxy-* for —OC$_2$H$_5$, etc.:

HO(CH$_2$)$_4$OH

1,4-dihydroxybutane
or 1,4-butanediol

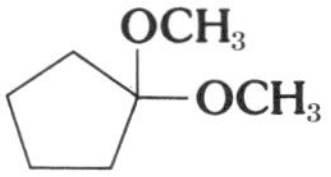

1,1-dimethoxycyclopentane

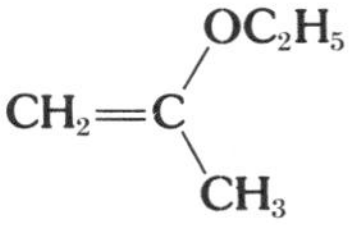

2-ethoxypropene

8·1 NOMENCLATURE OF ALCOHOLS AND ETHERS

TABLE 8·1 *IUPAC nomenclature of alcohols*

structure	IUPAC name	common name	classification
CH_3OH	methan*ol*	methyl alcohol	primary
C_2H_5OH	ethan*ol*	ethyl alcohol	primary
$CH_3(CH_2)_2OH$	1-propan*ol*	*n*-propyl alcohol	primary
$(CH_3)_2CHOH$	2-propan*ol*	isopropyl alcohol	secondary
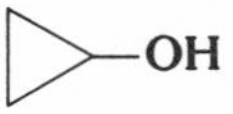 ▷—OH	cyclopropan*ol*	cyclopropyl alcohol	secondary
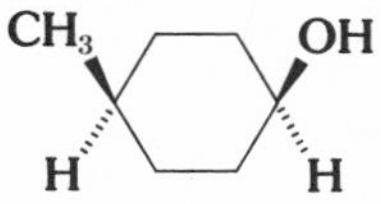 (cis-4-methylcyclohexanol structure)	*cis*-4-methylcyclohexan*ol*		secondary
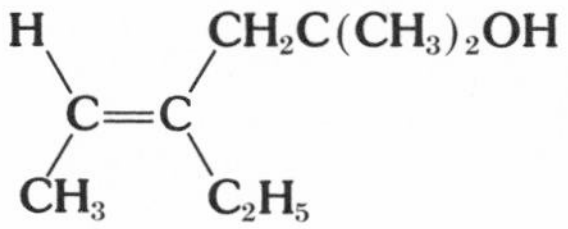 (trans-alkene structure)	*trans*-4-ethyl-2-methyl-4-hexen-2-*ol*		tertiary

Special names are given to *heterocyclic* ethers, cyclic compounds having oxygen in the ring:

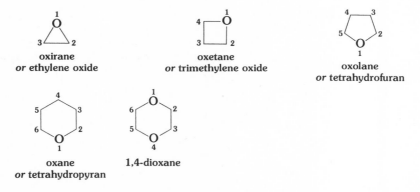

oxirane
or ethylene oxide

oxetane
or trimethylene oxide

oxolane
or tetrahydrofuran

oxane
or tetrahydropyran

1,4-dioxane

PROBLEM 8·1
a Write structural formulas for the eight constitutionally isomeric pentyl alcohols ($C_5H_{12}O$). Assign a IUPAC name to each isomer and classify it as primary, secondary, or tertiary.
b Write structures for and name the six isomeric ethers having the same molecular formula as the pentyl alcohols.

PROBLEM 8·2
a Which of the following approaches would you recommend for the synthesis of 3-pentanol: (1) alkaline hydrolysis of 3-chloropentane, (2) acid-catalyzed hydration of *trans*-2-pentene, or (3) hydroboration of *cis*-2-pentene, followed by oxidation with alkaline hydrogen peroxide?
b There are four constitutionally isomeric alcohols having the formula $C_4H_{10}O$. Write equations showing how each of these can be synthesized selectively from one (your choice) of the isomeric butenes: 1-butene, *cis*- or *trans*-2-butene, or 2-methylpropene.

8·2 PROPERTIES OF ALCOHOLS AND ETHERS

HYDROGEN-BONDING EFFECTS

The boiling points and solubilities in water of some alcohols and ethers are listed in Table 8.2. From these data we see that alcohols generally have higher boiling points than ethers (or hydrocarbons) of equivalent molecular weight, and also that the solubility in water of both alcohols and ethers increases with the relative proportion of oxygen in the molecule.

The phenomenon of hydrogen bonding, discussed in Section 2.2, provides an explanation of this behavior. The aggregation of alcohol molecules induced by hydrogen bonding makes it more difficult for individual molecules to vaporize and leave the liquid phase (see Figure 8.1). Alcohols therefore have higher boiling points than the corresponding ethers, which lack the electrophilic hydrogen atom OH necessary for hydrogen bonding. Since the nucleophilic oxygen atom of an ether or an alcohol can participate in hydrogen bonds to water molecules, these compounds have greater water solubility than hydrocarbons. The presence of more than one hydroxyl group or ether oxygen per molecule increases these effects.

PROBLEM 8·3

The boiling point of 1,2-butanediol (192°C) is 38° lower than that of the isomeric 1,4 diol (see Table 8.2), and the solubility of the 1,2 diol in water is also lower. Bearing in mind the consequences of intra- and intermolecular hydrogen bonding, suggest a reason for this difference.

TABLE 8·2 *Physical properties of alcohols and ethers*

compound	molecular weight	boiling point (°C)	water solubility
CH_3CH_2OH	46	78	soluble (∞)
CH_3OCH_3	46	−24	soluble (7 g/100 g)
$CH_3(CH_2)_4OH$	88	138	soluble (5 g/100 g)
$CH_3CH_2CH(OH)CH_2CH_3$	88	116	slightly soluble
$CH_3O(CH_2)_3CH_3$	88	71	slightly soluble
$C_2H_5O(CH_2)_2CH_3$	88	64	soluble (∞)
$HOCH_2CH_2CH_2CH_2OH$	90	230	soluble (∞)
$CH_3OCH_2CH_2OCH_3$	90	85	soluble (∞)
(dioxane)	88	101	soluble (∞)
$CH_3(CH_2)_9OH$	158	228	insoluble
$CH_3(CH_2)_4O(CH_2)_4CH_3$	158	190	insoluble
$C_2H_5OCH_2CH_2OCH_2CH_2OC_2H_5$	162	188	soluble (∞)

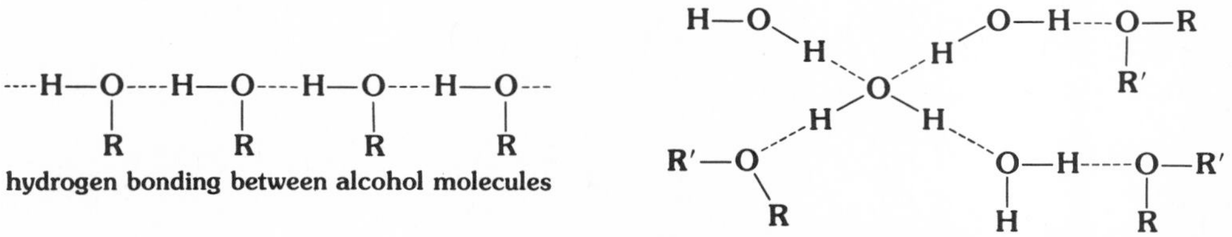

hydrogen bonding between alcohol molecules

hydrogen bonding of ether (or alcohol) in water

FIGURE 8·1 *Hydrogen bonding in alcohols and ethers*

SPECTROSCOPIC PROPERTIES

Some characteristic infrared absorptions of alcohols and ethers are listed in Table 8.3. Note that the vibrational frequency for the O—H bond depends on whether this hydroxyl group is free or hydrogen-bonded. Free, or unassociated, hydroxyl groups can be distinguished from their hydrogen-bonded counterparts by taking advantage of the fact that intermolecular hydrogen bonding is rare in very dilute solutions of alcohols in carbon tetrachloride. Thus solutions of increasing alcohol concentration give infrared and pmr spectra that reflect an increase in hydrogen bonding.

The three infrared spectra in Figure 8.2 are typical of the absorption bands exhibited by alcohols and ethers. 3-Methyl-2-buten-1-ol, for example, shows the presence of both hydroxyl and double-bond functional groups.

Hydrogen atoms bonded to or near the oxygen atom in alcohols and ethers also have characteristic pmr absorption signals (Table 8.4) which help to identify these functions. See, for example, the spectrum of methanol in Figure 8.3.

A quick and useful method of identifying hydroxyl signals in a pmr spectrum rests on the rapid exchange of active deuterium and hydrogen atoms that takes place when heavy water is added to a solution of an alcohol:

$$CH_3OH + D_2O \rightleftharpoons CH_3OD + DOH \qquad (8.2)$$

Since deuterium atoms do not display resonance signals in a spectrometer tuned to examine protons, such an isotopic exchange results in a diminution and eventual disappearance of all hydroxyl proton resonance signals.

TABLE 8·3 *Infrared absorptions of alcohols and ethers*

bond	*type of vibration*	*characteristic absorption (cm^{-1})*
O—H (free)	stretching	3600–3650 (sharp)
O—H (H-bonded)	stretching	3200–3600 (broad)
O—D	stretching	2200–2780
C—O	stretching	1050–1150

8·2 PROPERTIES OF ALCOHOLS AND ETHERS

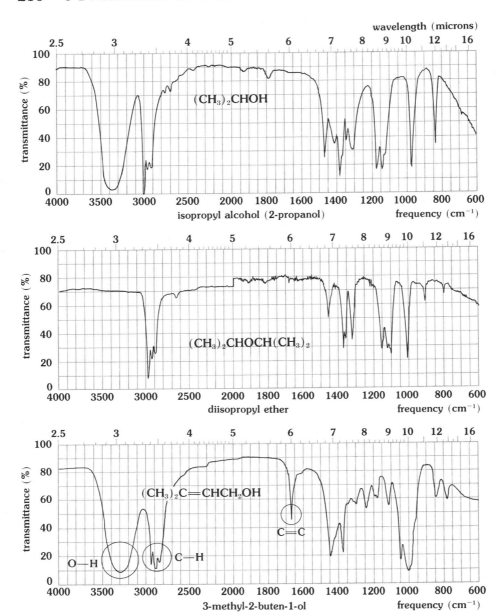

FIGURE 8·2 *Infrared spectra of alcohols and ethers*

TABLE 8·4 *PMR chemical shifts for alcohols and ethers*

hydrogen type	chemical shift (δ value)
OH (dilute solution)	0–4.0
OH (0.1–0.9 M, H-bonded)	4.0–7.0
OCH$_2$R	3.3–4.0
OCR$_2$CH$_2$R	1.2–2.0

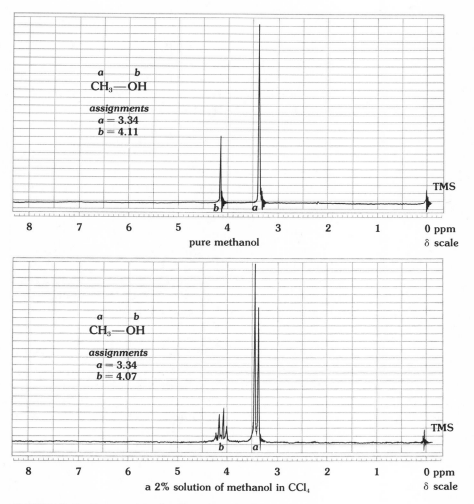

FIGURE 8·3 *Pmr spectra of methanol*

8·3 SPIN-SPIN INTERACTIONS

High-resolution pmr spectra often show a splitting of the resonance signals.† For example, the spectrum of a dilute solution of methanol in carbon tetrachloride displays a doublet and a quartet of relatively sharp peaks at $\delta = 3.3$ and 4.1, respectively (Figure 8.3). We shall go into the curious effect of methanol concentration on this pmr spectrum shortly; however, at this point let us consider the splitting phenomenon itself. Since the O-deuterated methanol analog (prepared by exchange with heavy water, as shown in equation 8.2) displays only a single sharp methyl resonance signal under similar conditions, it is clear that the signal splitting that occurs in dilute carbon tetrachloride solutions must be due to an interaction between the hydroxyl and methyl hydrogen atoms.

†A review of the discussion of nmr spectroscopy in Chapter 1 is advisable before proceeding with this topic.

PROBLEM 8·4
a If equimolar quantities of methanol and heavy water are allowed to exchange, as shown in equation 8.2, and if the hydrogen and deuterium atoms are assumed to be equally (or statistically) distributed among the reactants and products, what is the percentage of CH_3OD present at equilibrium?
b How many molar equivalents of heavy water would be necessary to give 90% CH_3OD at equilibrium?

The magnetic moment of the hydroxyl proton in methanol may be aligned with or against the applied magnetic field (H_0), as illustrated in Figure 8.4. These two nuclear-spin states exert an equal but opposite influence on the adjacent hydrogen nuclei in the methyl group, causing the resonance signal for these three equivalent protons to be split into two signals, one at slightly higher field and the other at slightly lower field than the unperturbed methyl signal. Because the two spin orientations for the hydroxyl proton occur in almost equal concentration, the two parts of the methyl doublet have nearly the same intensity.

Since spin-spin interactions between nuclei are mutual, it is clear that a similar perturbation of the hydroxyl proton by the methyl protons must accompany the splitting of the methyl resonance signal. Note in Figure 8.4 that four spin orientations are possible: all three methyl protons may be aligned with the external magnetic field, two may parallel the field while the third opposes it, one may be oriented with the field while two oppose it, or all three may be aligned against the applied field. We can therefore expect to find the hydroxyl resonance signal split into four separate peaks, as shown. Since the probability that the methyl protons will assume a *mixture* of spin orientations is greater than that for a

FIGURE 8·4 *Spin-spin interactions in methanol*

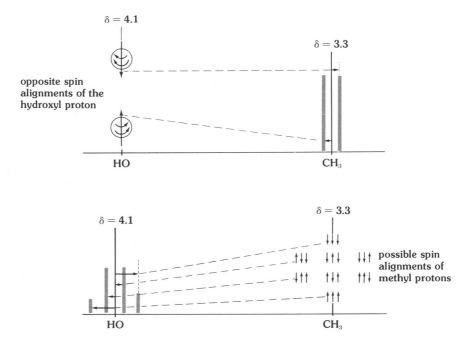

completely uniform alignment (note all the possible combinations of these spins), the central peaks of the hydroxyl quartet should be stronger than the outside peaks by a factor of 3. We see, then, that our model for spin-spin interactions of neighboring nuclei predicts the splitting pattern observed for methanol rather well.

PROBLEM 8·5

Why do we not predict that the outside peaks of the hydroxyl quartet will be stronger than the inside peaks?

PREDICTING SPIN-SPIN SPLITTING PATTERNS

Although we can also predict the splitting patterns of other compounds by a similar analysis of their spin orientations, it is easier to apply the following rule: *The observed multiplicity of a pmr signal generated by a group of protons is $n + 1$, where n is the number of neighboring protons that cause splitting.* In methanol, for example, $n = 1$ for the methyl protons (the adjacent OH proton), while $n = 3$ for the hydroxyl proton (the adjacent CH_3 protons). The center of these splitting patterns usually coincides with the chemical shift of the nuclei undergoing magnetic resonance, and the relative line intensities decrease symmetrically from the center of the multiplet, being 1:1 in a doublet, 1:2:1 in a triplet, 1:3:3:1 in a quartet, 1:4:6:4:1 in a quintet, etc.

The splitting patterns of diisopropyl ether and oxetane in Figure 8.5 are easily predicted by this rule. In the case of diisopropyl ether, for example, it is convenient and correct to regard the two isopropyl groups as completely independent (there is no spin-spin interaction between these groups). Thus we need only calculate the splitting pattern for one of these groups, and the other will coincide with it. Since the tertiary hydrogen is perturbed by six equivalent methyl protons, we expect to observe a septet near $\delta = 4.0$; however, the outside peaks of this multiplet are so weak that they can be observed only by amplification of the signal.

FIGURE 8·5 *Pmr spectra of diisopropyl ether and oxetane*

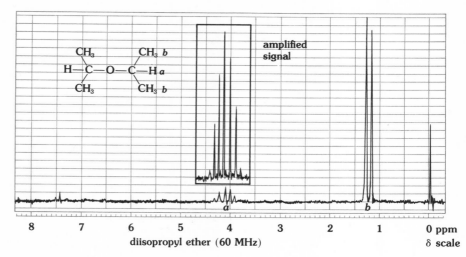

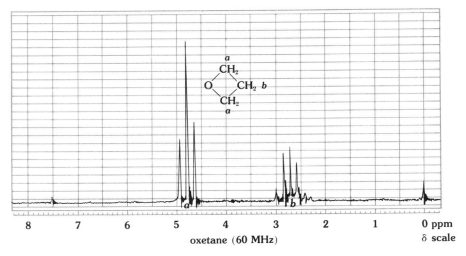

FIGURE 8·5 *(continued)*

In the case of oxetane the methylene group opposite the oxygen atom interacts with the two adjacent (and equivalent) methylene groups, causing the signal at $\delta = 4.73$ to be split into a triplet. The latter methylene groups in turn transform the higher-field signal at $\delta = 2.72$ into a quintet (there are *four* equivalent hydrogens interacting here).

THE SPIN-SPIN COUPLING CONSTANT

The magnitude of the spin-spin interactions of one group of protons with another is apparent from the spacing of the resulting multiplets. This spacing, called the *spin-spin coupling constant J,* is identical for the two sets of interacting nuclei and depends on their spatial relationship

TABLE 8·5 *Typical spin-spin coupling constants*

	saturated systems		unsaturated systems
proton group	coupling constant J (Hz)[†]	proton group	coupling constant J (Hz)
geminal H-C-H	10–15	geminal C=C (H,H)	0–3
vicinal H-C-C-H	0–9	cis C=C (H,H)	6–12
H-C-[C]$_n$-C-H, $n > 0$	0	trans C=C (H,H)	12–18

[†]Since J values do not vary with changes in the applied magnetic field or the solvents used, they are usually measured and reported in units of frequency (Hertz or cps).

8·3 SPIN-SPIN INTERACTIONS

and the number and kind of intervening bonds. Some typical coupling constants are shown in Table 8.5.

The simple symmetrical splitting pattern shown in Figure 8.5 and predicted by the $(n + 1)$ rule is called *first-order splitting*. This pattern occurs only when J is much smaller than the separation $\Delta \nu$ of the resonance signals of the interacting nuclei; for example, in the spectrum of diisopropyl ether $\Delta \nu = 160$ Hz and $J = 6$ Hz.† When $\Delta \nu$ and J are of the same order of magnitude, the pmr spectrum is more complex and becomes more difficult to interpret.

PROBLEM 8·6

a Predict the first-order splitting patterns for the two compounds described in Problem 3.4 on page 78.

b Analyze the following pmr spectrum of pure ethanol:

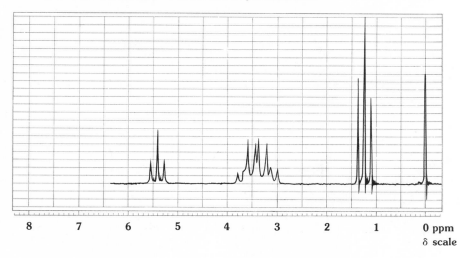

PROTON EXCHANGE AND SPIN-SPIN COUPLING

There are some circumstances in which spin-spin coupling of neighboring nuclei cannot be observed. One example is the lack of detectable splitting between groups of equivalent nuclei. Since the pmr spectrum of 1,4-dioxane consists of a single sharp signal at $\delta = 3.70$, there seems to be no spin-spin coupling between the adjacent methylene groups in this compound. We should not assume, however, that spin-spin interactions do not exist here. The lack of observable signal splitting is due to large changes in the intensities and positions of resonance lines that occur when the coupled nuclei have the same or very similar chemical shifts.

PROBLEM 8·7

Predict the pmr spectra (including splitting patterns) for 1,2-dimethoxyethane and 1,1-dimethoxyethane.

† For spectra taken at 60 MHz, to convert a chemical-shift difference $\Delta \nu$ from parts per million (δ units) to Hertz, simply multiply by 60.

8·3 SPIN-SPIN INTERACTIONS

The pmr spectra in Figure 8.3 show a puzzling situation in which spin-spin coupling seems to be absent in pure methanol but appears in a dilute solution. An explanation is suggested by the fact that the anticipated splitting pattern does emerge when the methanol sample is cooled to −40°C. If the exchange of hydroxyl protons in pure methanol is rapid, protons having both parallel and opposed (antiparallel) spin states will bond intermittently to a specific methanol molecule:

$$CH_3OH + CH_3OH \rightleftharpoons CH_3OH + CH_3OH \tag{8.3}$$

The resulting perturbation of the neighboring methyl protons is thus canceled or averaged to zero.† The rate of proton exchange in equation 8.3 can be decreased either by lowering the temperature of the sample or by employing a dilute solution of methanol in a nonacidic solvent. Under both these conditions the same spin-spin splitting of OH and CH_3 is observed.

Proton exchange in pure ethanol is slower than in methanol, and the pmr spectrum of pure ethanol in Problem 8.6 clearly exhibits spin-spin splitting involving the hydroxyl proton. However, if we add a small amount of acid or water to the ethanol sample, the rate of proton exchange increases sufficiently to prevent observation of the OH spin-coupling, and the resulting splitting pattern is the one we saw in Figure 1.10.

PROBLEM 8·8
Why do mixtures of ethanol and water show pmr spectra with only a single OH resonance signal?

In reviewing the spin-spin coupling of nuclei we can draw an analogy to a conversation being conducted between two groups of people by means of a bullhorn. An observer monitoring the communication between these groups would note the following:

1 Communication is most efficient if the groups are reasonably close to each other. Should the groups become widely separated, no communication will be possible ($J = 0$). The relative orientation of the groups is also important; communication will be poor if the bullhorns are not aimed in the right direction.

2 The efficiency of communication is the same for both groups; that is, it cannot be good for one group and poor for the other (J is identical for both sets of coupled nuclei).

3 Communication between members of the same group does not require bullhorns and is therefore not detected (equivalent protons do not *show* spin-spin splitting).

4 If the two groups are near each other for only a brief period, as when a rapidly moving bus carries one group past the other, communication will be very poor (rapidly exchanging nuclei do not show spin-spin splitting).

---

‡ In terms of the analogy used in Chapter 3, the OH proton exchange is fast compared with the "shutter speed" of our pmr "camera."

8·4 CHEMICAL REACTIVITY OF ALCOHOLS AND ETHERS

The alcohols comprise one of the most versatile groups of compounds that have been studied by organic chemists. Through a variety of relatively straightforward reactions alcohols can be transformed into (or prepared from) alkenes, alkyl halides, aldehydes, ketones, and carboxylic acids. The nucleophilic oxygen atom which is present in both alcohols and ethers is the center of chemical reactivity in these compounds, but despite the similarity of these functional groups, alcohols are generally far more reactive than ethers.

Table 8.6 summarizes some of the many different kinds of reactions observed with alcohols.

TABLE 8·6 *An outline of alcohol reactions*

---

substitution of H *in* RO—H

direct

$$ROH + Na \longrightarrow RO^{\ominus}Na^{\oplus} + \tfrac{1}{2}H_2$$

$$ROH + CH_3MgBr \longrightarrow RO^{\ominus}MgBr^{\oplus} + CH_4$$

$$ROH + CH_2{=}C{\underset{CH_3}{\overset{CH_3}{\diagup\!\!\!\diagdown}}} \xrightarrow[\text{ether}]{H_2SO_4} RO{-}C(CH_3)_3$$
an alkyl *t*-butyl ether

$$ROH + CH_3COCl \longrightarrow RO{-}\underset{\underset{\text{an ester of}}{}}{\overset{\overset{O}{\|}}{C}}{-}CH_3 + HCl$$
acetyl   an ester of
chloride   acetic acid

$$ROH + CH_3SO_2Cl \longrightarrow RO{-}\underset{\underset{O}{\|}}{\overset{\overset{O}{\|}}{S}}{-}CH_3 + HCl$$
methanesulfonyl   an ester of
chloride   methanesulfonic acid

indirect

$$RO^{\ominus}Na^{\oplus} + C_2H_5Br \longrightarrow ROC_2H_5 + NaBr$$
an alkyl ethyl ether

(continued)

TABLE 8·6 *An outline of alcohol reactions (continued)*

substitution of OH in R—OH

direct

$$ROH + HX \longrightarrow R\text{—}X + H_2O$$
$$(X = Cl, Br, \text{ or } I)$$

$$ROH + \underset{\substack{\text{thionyl} \\ \text{chloride}}}{SOCl_2} \longrightarrow R\text{—}Cl + SO_2 + HCl$$

$$3ROH + \underset{\substack{\text{phosphorus} \\ \text{tribromide}}}{PBr_3} \longrightarrow 3R\text{—}Br + H_3PO_3$$

indirect

$$R\text{—}OSO_2CH_3 + NaI \xrightarrow{\text{acetone}} R\text{—}I + CH_3SO_3Na$$

$$R\text{—}OSO_2CH_3 + NaCN \longrightarrow R\text{—}CN + CH_3SO_3Na$$

elimination

direct

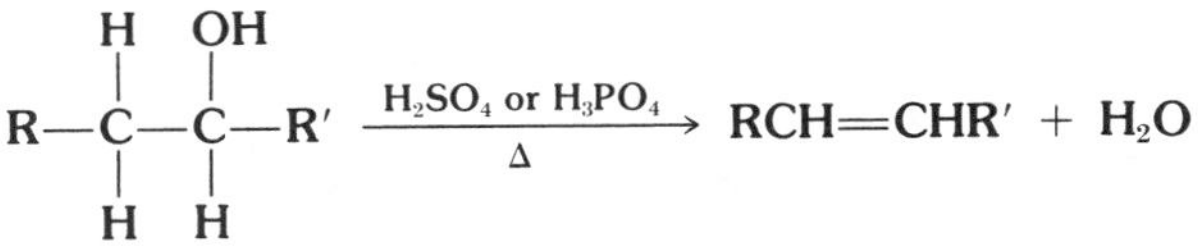

$$R\underset{\underset{H}{|}}{\overset{\overset{H}{|}}{C}}\text{—}\underset{\underset{H}{|}}{\overset{\overset{OH}{|}}{C}}\text{—}R' \xrightarrow[\Delta]{H_2SO_4 \text{ or } H_3PO_4} RCH\text{=}CHR' + H_2O$$

indirect

$$R\underset{\underset{H}{|}}{\overset{\overset{H}{|}}{C}}\text{—}\underset{\underset{H}{|}}{\overset{\overset{OSO_2CH_3}{|}}{C}}\text{—}R' + (CH_3)_3CO^{\ominus}K^{\oplus} \longrightarrow RCH\text{=}CHR' + \begin{array}{c} CH_3SO_3K \\ + \\ (CH_3)_3COH \end{array}$$

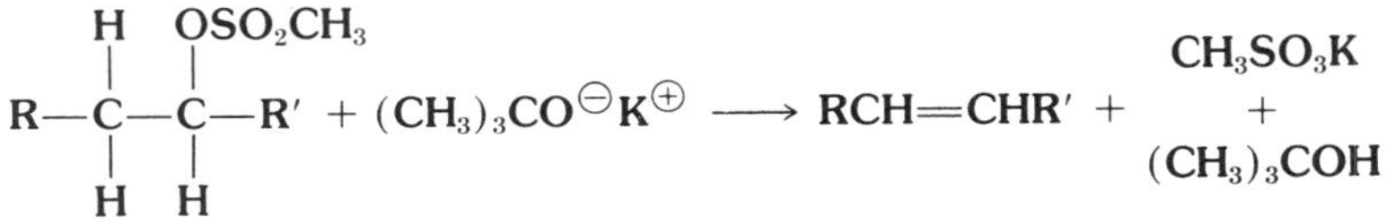

oxidation (dehydrogenation)

primary alcohols $\quad RCH_2OH \xrightarrow[H^{\oplus}]{Cr_2O_7^{2-}} R\text{—}\overset{\overset{O}{\|}}{C}\text{—}H \xrightarrow[H^{\oplus}]{Cr_2O_7^{2-}} R\text{—}\overset{\overset{O}{\|}}{C}\text{—}OH$

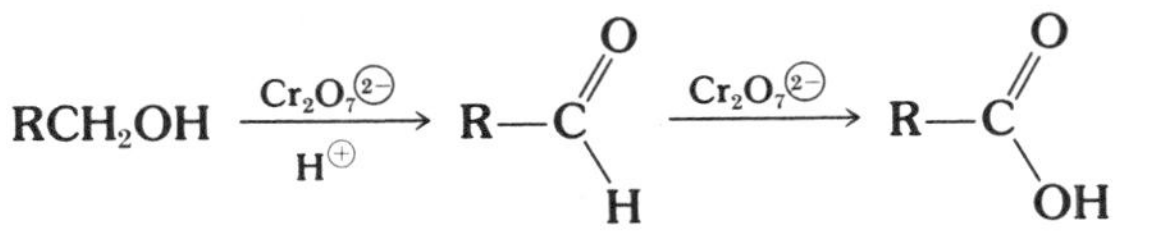

$\qquad\qquad\qquad\qquad\qquad\qquad$ an aldehyde $\qquad$ a carboxylic acid

secondary alcohols $\quad R_2CHOH \xrightarrow[H^{\oplus}]{Cr_2O_7^{2-}} \underset{R}{\overset{R}{\diagdown}}C\text{=}O$

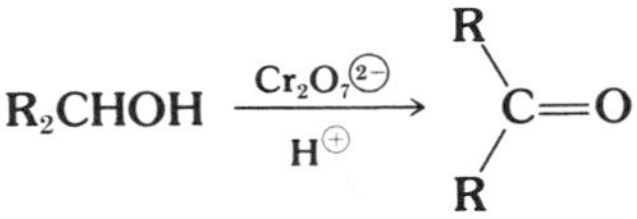

$\qquad\qquad\qquad\qquad\qquad\qquad$ a ketone

tertiary alcohols $\quad R_3COH \xrightarrow[H^{\oplus}]{Cr_2O_7^{2-}}$ no reaction

8·4 CHEMICAL REACTIVITY OF ALCOHOLS AND ETHERS

In contrast, the ethers exhibit only one generally useful reaction, the following acid-catalyzed cleavage:

$$R-O-R' + 2HX \longrightarrow R-X + R'-X + H_2O \qquad (8.4)$$
$$(X = Br \text{ or } I)$$

Indeed, simple ethers are often used as inert solvents for reactions involving species that would be entirely incompatible with alcohols. For example, diethyl ether is the solvent of choice for preparing most Grignard reagents.

The logical conclusion to be drawn from these facts is that the unique chemical reactivity of the alcohols is associated with the polar O—H bond of the hydroxyl group (in a similar sense the H bonding of this function accounts for most of the characteristic physical properties of alcohols). In the following discussions we shall see that many of the important and typical reactions of alcohols can be explained by the assumption of an initial electrophilic substitution at oxygen:

$$R-\ddot{O}-H + \overset{\delta\oplus\;\;\delta\ominus}{Y-Z} \rightleftharpoons R-\underset{\underset{H}{|}}{\overset{\oplus}{\ddot{O}}}-Y + Z:^{\ominus} \rightleftharpoons R-\ddot{O}-Y + Z-H \qquad (8.5)$$

an oxonium salt

PROBLEM 8·9
Why do reactions analogous to equation 8.5 play no important role in the chemistry of ethers?

ACIDITY AND BASICITY OF ALCOHOLS AND ETHERS

Alcohols are amphoteric and thus can function both as weak Brønsted acids and as bases (or nucleophiles), whereas ethers can only serve as bases:

$$R-\ddot{O}-H + \begin{cases} Z:^{\ominus} \rightleftharpoons R-\ddot{O}:^{\ominus} + Z-H & \text{(ROH serves as an acid)} \\ HA \rightleftharpoons R-\underset{\underset{H}{|}}{\overset{\oplus}{\ddot{O}}}-H + A:^{\ominus} & \text{(ROH serves as a base)} \end{cases} \qquad (8.6)$$

(HA = a Brønsted acid, $Z:^{\ominus}$ = a Brønsted base)

$$R-\ddot{O}-R' + HA \rightleftharpoons R-\underset{\underset{H}{|}}{\overset{\oplus}{\ddot{O}}}-R' + A:^{\ominus} \qquad (8.7)$$

The acidity of the hydroxyl group is seen in the rapid proton-deuteron exchange that takes place when alcohols are dissolved in heavy water (equation 8.2) and in the reactions of alcohols with alkali metals and organometallic reagents:

$$\text{cyclohexyl-OH} + Na \longrightarrow \text{cyclohexyl-O}^{\ominus}Na^{\oplus} + \tfrac{1}{2}H_2 \qquad (8.8)$$

8·4 CHEMICAL REACTIVITY OF ALCOHOLS AND ETHERS

$$(CH_3)_2CHCH_2OH + \begin{cases} CH_3Li \longrightarrow (CH_3)_2CHCH_2O^{\ominus}Li^{\oplus} + CH_4 \\ C_3H_7MgBr \longrightarrow (CH_3)_2CHCH_2O^{\ominus}MgBr^{\oplus} + C_3H_8 \end{cases} \quad (8.9)$$

These last reactions proceed rapidly and quantitatively; therefore the volume of gas evolved may be used to determine the number of hydroxyl groups in an unidentified alcohol, provided the molecular weight is known and no other acidic (or active) hydrogen atoms are present. The relative reactivity of different alcohols in these reactions parallels their acidity, which decreases in the order

H_2O > primary ROH > secondary ROH > tertiary ROH

PROBLEM 8·10
Write a series of equilibrium equations illustrating all steps in the exchange of cyclopentanol with heavy water (D_2O). Label the conjugate acids and bases in each separate equilibrium.

PROBLEM 8·11
Mangostin, a yellow crystalline substance obtained from the bark or latex of the mangosteen tree, has the molecular formula $C_{24}H_{26}O_6$. A 102-mg sample of mangostin dissolved in dry ether releases 16.8 ml of methane (corrected to 760 mm pressure and 0°C) when treated with an excess of methylmagnesium bromide (CH_3MgBr). How many hydroxyl groups are present in the mangostin molecule?

The nonbonding electron pairs on the oxygen atoms of alcohols and ethers make these compounds nucleophilic (and basic). Thus low-molecular-weight alcohols and ethers (1-pentanol, dibutyl ether, etc.) are soluble in strong acids such as sulfuric acid, whereas the corresponding hydrocarbons (hexane and nonane) are not. Trialkyl oxygen salts, called *oxonium salts*, have in fact been isolated in several cases.

$$R-\ddot{O}-R' + H_2SO_4 \rightleftharpoons R-\overset{\oplus}{\underset{H}{\ddot{O}}}-R' + HSO_4^{\ominus} \quad (8.10)$$

base   acid   acid   base

$$CH_3-\ddot{O}-CH_3 + CH_3-F \xrightarrow{BF_3} CH_3-\overset{\oplus}{\underset{CH_3}{\ddot{O}}}-CH_3 \; BF_4^{\ominus} \quad (8.11)$$

a trimethyloxonium salt

Divalent oxygen atoms can also coordinate with Lewis acids such as metal cations, and this feature, together with the absence of acidic hydrogen atoms, makes ethers excellent solvents for organometallic reagents.

8·5 ELECTROPHILIC SUBSTITUTION AT OXYGEN

Since alcohols are not strong nucleophiles, direct electrophilic substitution reactions at oxygen require strongly electrophilic reagents such as halogens or carbonium ions:

$$(CH_3)_3COH + Cl_2 \xrightarrow{NaOH, H_2O} (CH_3)_3C-O-Cl + NaCl \qquad (8.12)$$

$$\text{cyclopentyl-OH} + (CH_3)_3C^{\oplus} \rightleftharpoons \text{[protonated intermediate]} \underset{H^{\oplus}}{\overset{-H^{\oplus}}{\rightleftharpoons}} \text{cyclopentyl-O-C(CH_3)_3} \qquad (8.13)$$

Reaction 8.13 plays an important role in the general synthesis of *t*-butyl ethers from alcohols and isobutylene:

$$(CH_3)_2C=CH_2 + ROH \underset{}{\overset{H_2SO_4 \text{ catalyst}}{\rightleftharpoons}} (CH_3)_3COR \qquad (8.14)$$

The energy profile for this transformation is outlined in Figure 8.6.

ESTER FORMATION

An important class of electrophilic substitutions at oxygen is exemplified by the reaction of alcohols with electrophilic derivatives of carboxylic acids and sulfonic acids:

$$R-C(=O)O-H \qquad R-C(=O)Cl \qquad R-C(=O)-O-C(=O)-R \qquad R-S(=O)(=O)-OH \qquad R-S(=O)(=O)-Cl$$

a carboxylic acid   an acid chloride   an acid anhydride   a sulfonic acid   a sulfonyl chloride

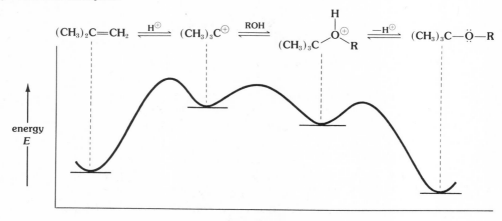

FIGURE 8·6 *Energy profile for the acid-catalyzed addition of alcohols to isobutylene*

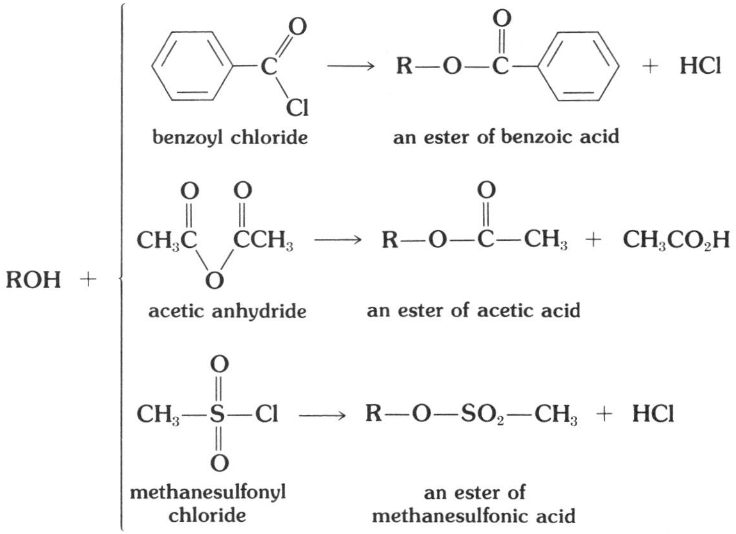

8·5 ELECTROPHILIC SUBSTITUTION AT OXYGEN

ROH +
- benzoyl chloride (C₆H₅COCl) → R—O—CO—C₆H₅ + HCl
  an ester of benzoic acid
- acetic anhydride (CH₃CO)₂O → R—O—C(O)—CH₃ + CH₃CO₂H
  an ester of acetic acid
- methanesulfonyl chloride CH₃—SO₂—Cl → R—O—SO₂—CH₃ + HCl
  an ester of methanesulfonic acid

FIGURE 8·7 *Esterification reactions*

In the ester-forming, or *esterification*, reactions in Figure 8.7 we see that the hydroxyl proton of an alcohol is replaced by an acyl group (**RCO**) or a sulfonyl group (**RSO₂**) characteristic of the electrophilic reagent being used. The manner in which these important reactions take place will be discussed further in Chapter 14.

PROBLEM 8·12
Write structural formulas for the esters formed from the following reactants:

a cyclohexanol (with H and OH shown) + C_4H_9COCl

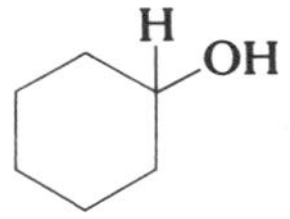

b $(CH_3)_2CHCH_2OH$ + CH_3SO_2Cl

c cholesterol + $(CH_3CO)_2O$

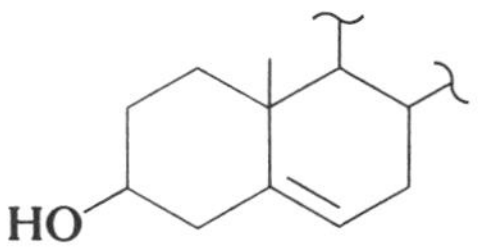

d cyclopentyl-CH₂OH (with H shown) + CH_3—C₆H₄—SO_2Cl

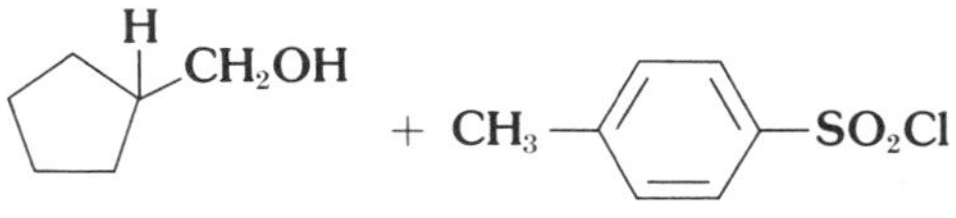

WILLIAMSON ETHER SYNTHESIS

Although alcohols do not normally react with weak electrophiles such as alkyl halides (see Figure 3.3), it is often possible to achieve substitution and elimination reactions by first converting the alcohols to their more reactive conjugate bases. The increased nucleophilicity and basicity

8·5 ELECTROPHILIC SUBSTITUTION AT OXYGEN

of alkoxide ions is illustrated by the following substitution and elimination reactions:

$$RO^\ominus Na^\oplus + R'CH_2-X \xrightarrow{S_N 2} RO-CH_2R' + NaX \quad (8.15)$$
$$(X = Cl, Br, or I)$$

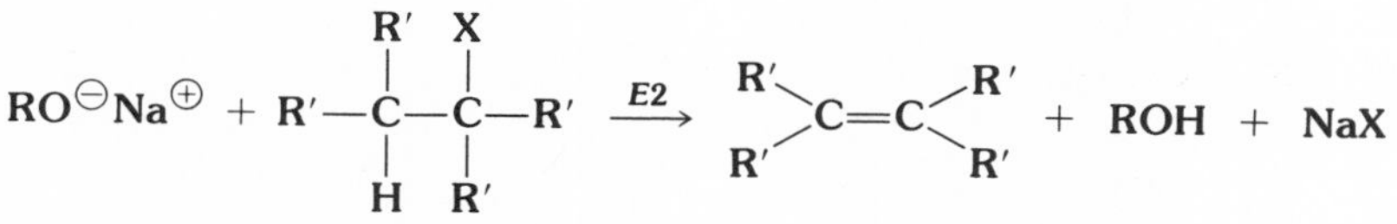

$$RO^\ominus Na^\oplus + R'-\underset{\underset{H}{|}}{\overset{\overset{R'}{|}}{C}}-\underset{\underset{R'}{|}}{\overset{\overset{X}{|}}{C}}-R' \xrightarrow{E2} \underset{R'}{\overset{R'}{>}}C=C\underset{R'}{\overset{R'}{<}} + ROH + NaX \quad (8.16)$$

The nucleophilic displacement ($S_N 2$) of halide by alkoxide (equation 8.15) provides a useful general method for the preparation of ethers which is called the *Williamson ether synthesis*:

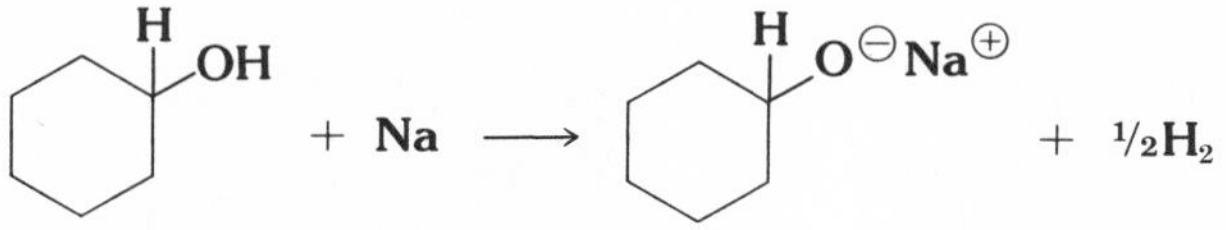

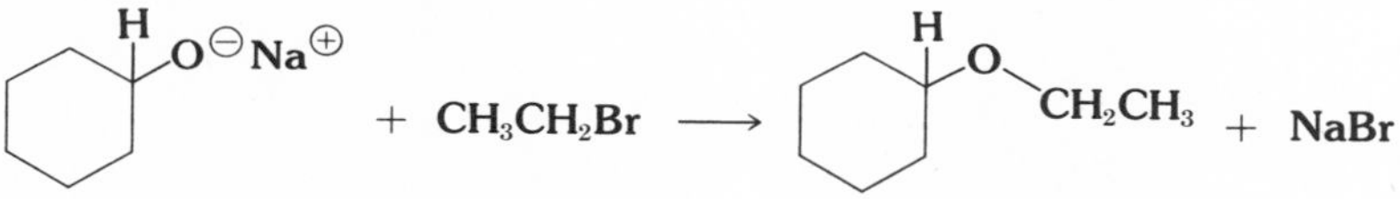

[cyclohexanol] + Na ⟶ [cyclohexyl O⁻Na⁺] + ½H₂

[cyclohexyl O⁻Na⁺] + CH₃CH₂Br ⟶ [ethyl cyclohexyl ether] + NaBr (8.17)

ethyl cyclohexyl ether

Since alkoxide ions can also function as strong bases, elimination products often accompany the ethers formed by the synthesis. When unsymmetrical ethers are being prepared by this method, it is therefore best to choose that combination of reagents which will give the most favorable $S_N 2$ reaction (see Table 3.4). For example, the preparation of ethyl cyclohexyl ether by the procedure in equation 8.17 is much superior to the alternative reaction of sodium ethoxide with cyclohexyl bromide:

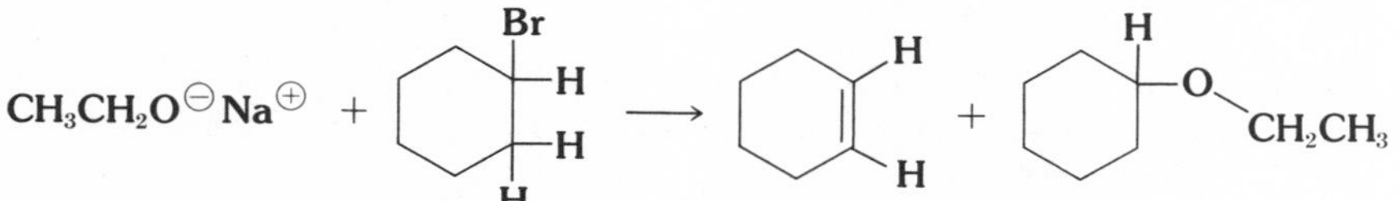

$$CH_3CH_2O^\ominus Na^\oplus + [\text{cyclohexyl bromide}] \longrightarrow [\text{cyclohexene}] + [\text{ethyl cyclohexyl ether}] \quad (8.18)$$

main product

Since the ratio of substitution to elimination products is usually higher for primary halides than for secondary and tertiary halides, the Williamson synthesis should employ primary halides whenever possible.

PROBLEM 8·13

a Write equations showing the best combination of reagents for preparing $CH_3OCH(CH_3)_2$ and $(CH_3)_3CCH_2OCH_2CH_3$ by the Williamson synthesis.

b Predict the major product (or products) in the following reactions:

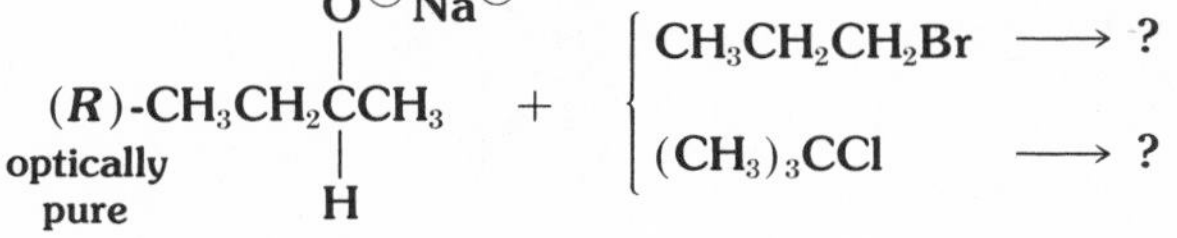

$$(R)\text{-}CH_3CH_2\underset{\underset{H}{|}}{\overset{\overset{O^\ominus Na^\oplus}{|}}{C}}CH_3 + \begin{cases} CH_3CH_2CH_2Br \longrightarrow ? \\ (CH_3)_3CCl \longrightarrow ? \end{cases}$$

optically pure

8·5 ELECTROPHILIC SUBSTITUTION AT OXYGEN

OXIRANE FORMATION

The oxirane or epoxide ring system is readily formed from 1,2-halohydrins by an *intra*molecular displacement reaction:

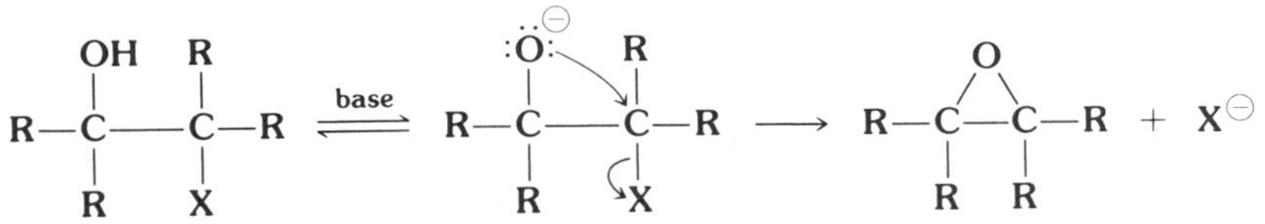

$$\text{R-C(OH)(R)-C(R)(X)-R} \xrightleftharpoons{\text{base}} \text{R-C(O}^{\ominus}\text{)(R)-C(R)(X)-R} \longrightarrow \text{R-C(R)-O-C(R)-R} + X^{\ominus} \quad (8.19)$$

An anti orientation of oxygen and halogen is required during the displacement step (this is an S_N2 process). This fact is well illustrated by the rapid reaction of *trans*-2-chlorocyclohexanol with aqueous base to give cyclohexene oxide, in contrast to the slower and completely different transformation of the cis isomer:

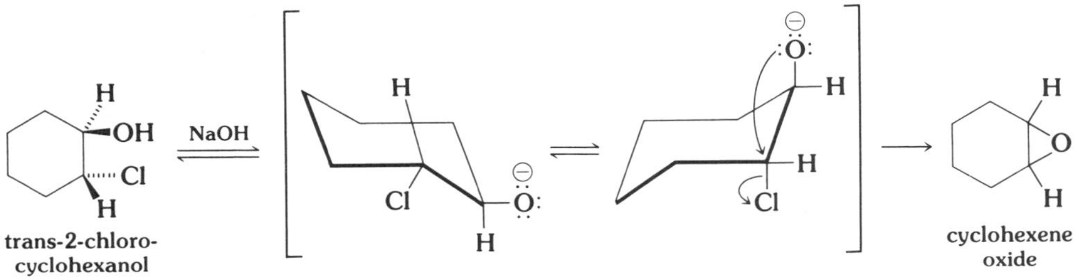

trans-2-chloro-cyclohexanol $\xrightleftharpoons{\text{NaOH}}$ [chair conformations] $\longrightarrow$ cyclohexene oxide (8.20)

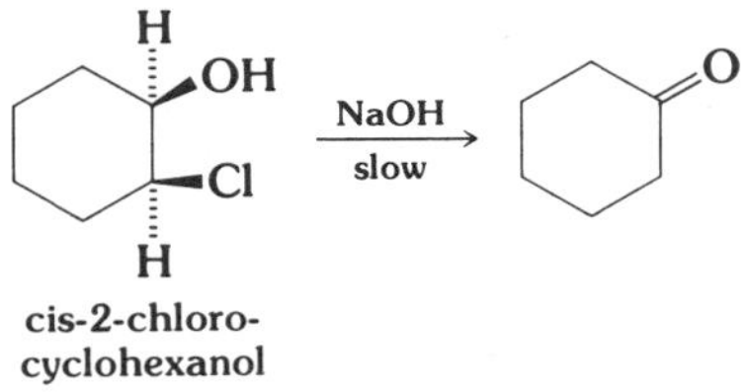

cis-2-chloro-cyclohexanol $\xrightarrow[\text{slow}]{\text{NaOH}}$ cyclohexanone (8.21)

PROBLEM 8·14
How can *trans*-2-chlorocyclohexanol be prepared from cyclohexane as a starting material?

PROBLEM 8·15
Draw the chair conformations expected for the *cis*-2-chlorocyclohexanol conjugate base. Discuss the probability of epoxide formation from these conformations (as in equation 8.20).

PROBLEM 8·16
Starting with *cis*- and/or *trans*-2-butene, write equations for the stereospecific preparation of *cis*- and *trans*-2,3-dimethyloxirane:

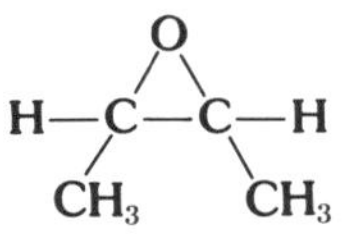

$$\text{H-C(CH}_3\text{)-O-C(CH}_3\text{)-H}$$

8·6 NUCLEOPHILIC SUBSTITUTION REACTIONS

Alcohols and ethers do not normally undergo nucleophilic displacement of $OH^\ominus$ or $OR^\ominus$ even when powerful nucleophiles such as $I^\ominus$ and $SCN^\ominus$ are used. The contrast between this behavior and the facile reactions of alkyl halides with these nucleophiles clearly indicates that the strongly basic hydroxide and alkoxide ions are much poorer leaving groups than halide ions.† Fortunately, we can easily modify the OH function to overcome its inherent reluctance to serve as a leaving group.

SULFONATE ESTERS AS LEAVING GROUPS

Sulfonate ester derivatives of *primary* and *secondary* alcohols (see Figure 8.7) are observed to undergo nucleophilic displacement with almost the same facility as comparable reactions of alkyl halides, indicating that sulfonate anions ($RSO_3^\ominus$) are relatively good leaving groups (in comparison with $OH^\ominus$):

$$n\text{-}C_4H_9OH + CH_3SO_2Cl \longrightarrow CH_3CH_2CH_2CH_2OSO_2CH_3 \xrightarrow[\text{acetone}]{NaI} n\text{-}C_4H_9I + CH_3SO_3^\ominus Na^\oplus \quad (8.22)$$

Kinetic measurements of many of these displacement reactions show second-order behavior, suggesting that an S_N2 mechanism is operating. The inversion of configuration that should accompany an S_N2 reaction (see Section 3.4) has been clearly demonstrated by using simple optically active alcohol derivatives designed so that the site of asymmetry is the carbon atom involved in the displacement reaction:

$$\begin{array}{c}
\text{CH}_3 \\
\text{H} \blacktriangleright \text{C} - \text{OH} \\
\text{C}_6\text{H}_5\text{CH}_2 \\
[\alpha] = +33°
\end{array}
+ C_7H_7SO_2Cl \xrightarrow{\text{esterification}}
\begin{array}{c}
\text{CH}_3 \\
\text{H} \blacktriangleright \text{C} - \text{OSO}_2\text{C}_7\text{H}_7 \\
\text{C}_6\text{H}_5\text{CH}_2 \\
[\alpha] = +31°
\end{array}$$

esterification ↓ CH_3COCl S_N2 ↓ $\begin{array}{l}CH_3CO_2^\ominus Na^\ominus\\ CH_3CO_2H\end{array}$

$$\begin{array}{c}
\text{CH}_3 \quad \text{O} \\
\text{H} \blacktriangleright \text{C} - \text{O} - \overset{\|}{\text{C}} - \text{CH}_3 \\
\text{C}_6\text{H}_5\text{CH}_2 \\
[\alpha] = +7°
\end{array}
\qquad
\begin{array}{c}
\text{O} \quad \text{CH}_3 \\
\text{CH}_3 - \overset{\|}{\text{C}} - \text{O} - \text{C} \blacktriangleleft \text{H} \\
\text{CH}_2\text{C}_6\text{H}_5 \\
[\alpha] = -7°
\end{array}
\quad (8.23)$$

Direct esterification of the dextrorotatory alcohol would give the corresponding sulfonate and carboxylate esters, without a change in the

---

†We find that differences in the reactivity of various functions as anionic leaving groups are matched by corresponding differences in the ionization constants of their conjugate acids (water and alcohol are about 10^{20} times weaker acids than HCl, HBr, or HI). This is not surprising, inasmuch as both factors reflect the relative stabilities of the anions. Since the ionization constants of sulfonic acids are similar to those of the halogen acids, we predict that sulfonate esters should function as good leaving groups in nucleophilic displacement reactions.

8·6 NUCLEOPHILIC SUBSTITUTION REACTIONS

configuration at the asymmetric carbon unit; hence inversion of configuration to the levorotatory acetate must occur in the nucleophilic-displacement step.

PROBLEM 8·17
a Write equations showing how 1-propanol can be converted to $CH_3(CH_2)_2CN$ and $CH_3(CH_2)_2SH$.
b Suggest the order of reactivity anticipated in the S_N2 reaction of the following sulfonate esters with sodium iodide in acetone:

$(CH_3)_2CHOSO_2CH_3$ $CH_3OSO_2CH_3$ $(CH_3)_3CCH_2OSO_2CH_3$

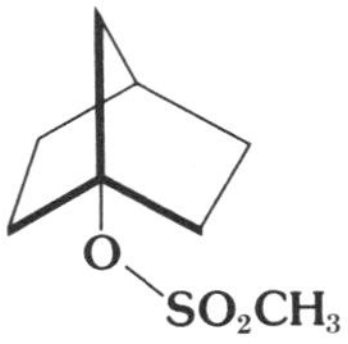

[norbornyl-O-SO₂CH₃ structure]

PROBLEM 8·18
Optically active 2-methyl-1-butanol is converted first to a methanesulfonate derivative, and then to a bromide by reaction with sodium bromide in ethanol. What changes in configuration will take place during this sequence of reactions? If this alkyl bromide is converted to a Grignard reagent which is then treated with ethanol, what can you predict about the optical activity of the major organic product?

If the sulfonate ester derivatives of most alcohols are dissolved in ionizing solvents and heated in the absence of strong nucleophiles, the corresponding sulfonic acids are formed along with substitution and elimination products:†

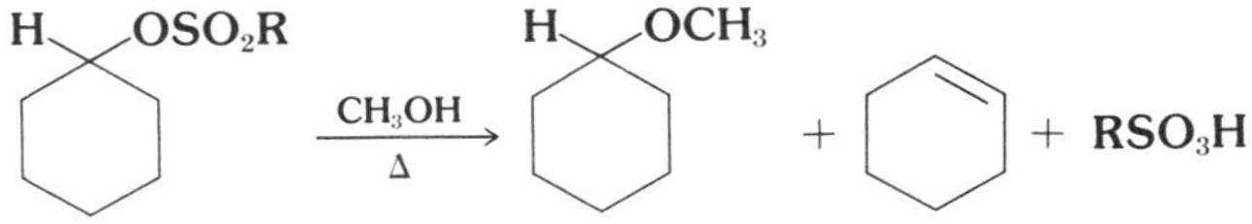

$$\text{cyclohexyl-OSO}_2R \xrightarrow[\Delta]{CH_3OH} \text{cyclohexyl-OCH}_3 + \text{cyclohexene} + RSO_3H \qquad (8.24)$$

The appearance of the sulfonic acid can be followed by the titration of aliquot samples (small samples withdrawn from the reaction mixture at regular intervals). In this manner the rate of reaction 8.24 was shown to be first order in sulfonate ester concentration. These *solvolysis reactions*, so called because the solvent becomes incorporated in some of the products, sometimes lead to molecular rearrangements:

$$(CH_3)_3CCH_2OSO_2CH_3 \xrightarrow[\Delta]{C_2H_5OH} \begin{array}{c} (CH_3)_3CCH_2OC_2H_5 + (CH_3)_2\overset{|}{C}CH_2CH_3 \\ 9\% \qquad\qquad OC_2H_5 \\ 31\% \\ + \\ (CH_3)_2C=CHCH_3 + CH_2=C(CH_3)CH_2CH_3 \\ 30\% \qquad\qquad 30\% \end{array} \qquad (8.25)$$

†Tertiary sulfonate esters are relatively unstable and have not been extensively studied.

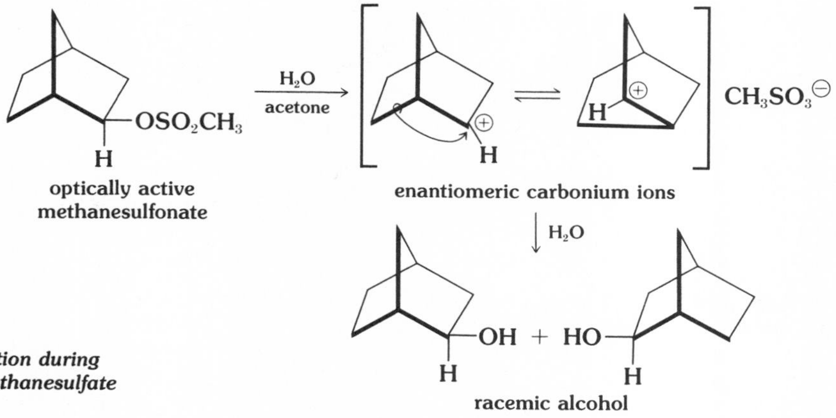

8·6 NUCLEOPHILIC SUBSTITUTION REACTIONS

FIGURE 8·8 *Racemization during solvolysis of norbornyl methanesulfate*

They are therefore believed to proceed via carbonium-ion intermediates — that is, by an S_N1 mechanism — as illustrated in Figure 8.8.†

PROBLEM 8·19
Write equations showing how a carbonium-ion intermediate (or intermediates) can explain the formation of all the products given in equation 8.25.

SULFITE AND PHOSPHITE ESTER INTERMEDIATES AS LEAVING GROUPS

Alcohols can be converted to alkyl halides in good yield by treatment with phosphorus trihalides and thionyl halides:

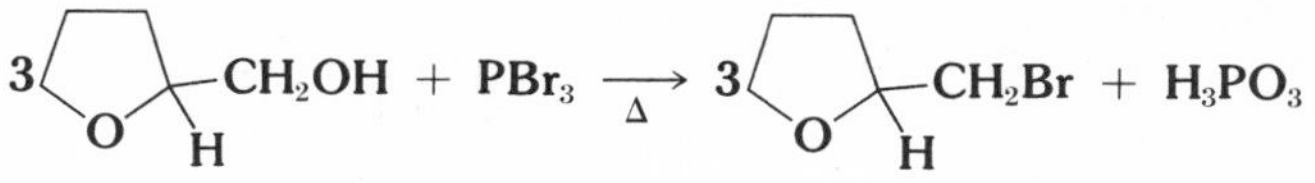

$$3 \text{ (tetrahydrofurfuryl)-CH}_2\text{OH} + \text{PBr}_3 \xrightarrow{\Delta} 3 \text{ (tetrahydrofurfuryl)-CH}_2\text{Br} + \text{H}_3\text{PO}_3 \tag{8.26}$$

$$\text{CH}_3(\text{CH}_2)_5\text{CH(OH)CH}_3 + \text{SOCl}_2 \xrightarrow[\Delta]{\text{dioxane}} \text{CH}_3(\text{CH}_2)_5\text{CHClCH}_3 + \text{SO}_2 + \text{HCl} \tag{8.27}$$

Although detailed mechanisms for these reactions have not been clearly established, there is convincing evidence that they proceed by way of phosphite or sulfite ester intermediates. Indeed, reactions of phosphorus trichloride with primary alcohols give trialkyl phosphites as one of the major products:

$$3 \ n\text{-C}_4\text{H}_9\text{OH} + \text{PCl}_3 \longrightarrow (n\text{-C}_4\text{H}_9\text{O})_3\text{P} + 3\text{HCl} \tag{8.28}$$

For this reason thionyl chloride is generally the preferred reagent for transforming alcohols to alkyl chlorides.

†The distinction between S_N1 and S_N2 processes is a very useful generalization. However, many reactions are borderline, and there is even some question of whether any real reaction systems fit exactly into one category or the other. We do know that *ion pairs* play an important role in many of these nucleophilic displacement reactions. If, for example, the bicycloheptyl methanesulfonate hydrolysis described in Figure 8.8 is stopped after 50% conversion, the remaining sulfonate ester is found to be partially racemized. We can explain this fact by assuming that ionization proceeds first to a close (or intimate) ion pair which can either return to starting material (before or after rearrangement) or be intercepted by the nucleophilic solvent.

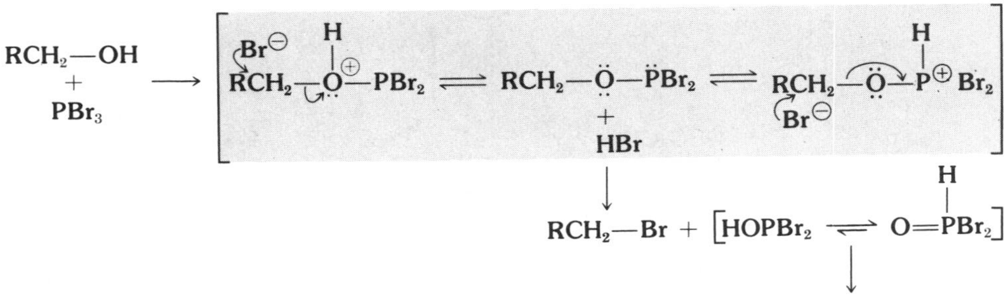

$$\begin{array}{c}
RCH_2-OH \\
+ \\
PBr_3
\end{array} \longrightarrow \left[RCH_2 \overset{Br^\ominus}{\underset{\cdot\cdot}{-O^\oplus-}}\overset{H}{\underset{|}{P}}Br_2 \rightleftarrows RCH_2-\overset{\cdot\cdot}{O}-\ddot{P}Br_2 \rightleftarrows RCH_2-\overset{H}{\underset{Br^\ominus}{\overset{|}{\ddot{O}}}}-P^\oplus Br_2 \right]$$

$$\begin{array}{c} + \\ HBr \end{array}$$

$$\downarrow$$

$$RCH_2-Br + [HOPBr_2 \rightleftarrows O=\overset{H}{\underset{|}{P}}Br_2]$$

FIGURE 8·9 *A mechanism for alcohol-phosphorus tribromide reactions*

similar displacement of remaining bromine atoms

In contrast, phosphorus tribromide and phosphorus triiodide (prepared by mixing iodine and red phosphorus) give high yields of alkyl bromides and iodides, presumably by an S_N2 displacement of primary and secondary phosphite esters or by an S_N1 reaction of the corresponding tertiary intermediates. Since trivalent phosphorus is a Lewis base (note the position of phosphorus in the periodic table), it is likely that conjugate acids of the phosphite intermediates play an important role in these reactions (see Figure 8.9). The superiority of PBr_3 and PI_3 over PCl_3 is explained by the relative nucleophilicities of $I^\ominus$, $Br^\ominus$, and $Cl^\ominus$.

Thionyl chloride and thionyl bromide react with most primary and secondary alcohols at room temperature to give alkyl halosulfites or dialkyl sulfites, depending on the molal ratio of the reactants:

$$ROH + SOX_2 \xrightarrow[-HX]{} RO-\underset{\underset{O}{\|}}{S}-X \xrightarrow[-HX]{ROH} RO-\underset{\underset{O}{\|}}{S}-OR \qquad (8.29)$$

(X = Cl or Br) an alkyl halosulfite a dialkyl sulfite

Alkyl chlorides and bromides are easily and cleanly prepared from alcohols by heating with one or more equivalents of the appropriate thionyl halide. Under these conditions the intermediate alkyl halosulfites smoothly decompose to sulfur dioxide and the corresponding alkyl halide. Ionization of the alkyl halosulfites to an intimate ion pair has been proposed to account for the molecular rearrangements that are sometimes observed, as well as for the variations in reaction stereochemistry.†

$$(CH_3)_2\underset{\underset{OSOCl}{|}}{CH}CHCH_3 \longrightarrow \left[(CH_3)_2\underset{\underset{OSOCl^\ominus}{}}{CH\overset{\oplus}{C}HCH_3} \xrightarrow{\text{rearrangement}} (CH_3)_2\underset{\underset{OSOCl^\ominus}{}}{\overset{\oplus}{C}CH_2CH_3} \right]$$

$$\downarrow \qquad\qquad\qquad\qquad \downarrow$$

$$\begin{array}{c}(CH_3)_2CHCHClCH_3 \\ + \\ SO_2\end{array} \quad + \quad \begin{array}{c}(CH_3)_2CClCH_2CH_3 \\ + \\ SO_2\end{array} \qquad (8.30)$$

†If an equivalent of an organic base is added, the reaction of thionyl chloride with optically active 2-butanol gives 2-chlorobutane with *inversion* of configuration. In the absence of added base, the reaction proceeds with predominant retention of configuration. For an explanation of these facts see more advanced works such as J. March, *Advanced Organic Chemistry*, 2d ed., McGraw-Hill Book Company, 1977.

8·6 NUCLEOPHILIC SUBSTITUTION REACTIONS

CONJUGATE ACIDS AS LEAVING GROUPS

Nucleophilic displacement reactions of alcohols and ethers are facilitated by strong acids, which serve to convert the oxygen-containing functions to their conjugate acids:

$$RCH_2OR' + HBr \rightleftharpoons RCH_2-\overset{H}{\underset{..}{O}}{}^{\oplus}-R'\ Br^{\ominus} \quad (8.31)$$
(R' = H or alkyl)

$$RCH_2-\overset{H}{\underset{..}{O}}{}^{\oplus}-R'\ Br^{\ominus} \xrightarrow{S_N2} RCH_2Br + R'OH$$

$$(CH_3)_3COR' + HCl \rightleftharpoons (CH_3)_3C-\overset{H}{\underset{..}{O}}{}^{\oplus}-R'\ Cl^{\ominus} \quad (8.32)$$
(R' = H or alkyl)

$$(CH_3)_3C-\overset{H}{\underset{..}{O}}{}^{\oplus}-R' \xrightarrow{S_N1} R'OH + \left[CH_3-\overset{CH_3}{\underset{CH_3}{C^{\oplus}}}\right] \xrightarrow{Cl^{\ominus}} (CH_3)_3CCl$$

The greater reactivity of these conjugate acids undoubtedly reflects the fact that ROH and H_2O are better leaving groups than $RO^{\ominus}$ and $OH^{\ominus}$. Or to put it more precisely, conjugate-acid displacement requires a smaller activation energy for S_N2 and S_N1 pathways, as illustrated for the S_N2 case in Figure 8.10.

Primary and secondary alcohols usually react with HX by an S_N2 mechanism, as evidenced both by the order of reactivity (HI > HBr > HCl), which parallels the nucleophilicity of the corresponding halide ions,

FIGURE 8·10 *Effect of conjugate-acid intermediates in substitution reactions*

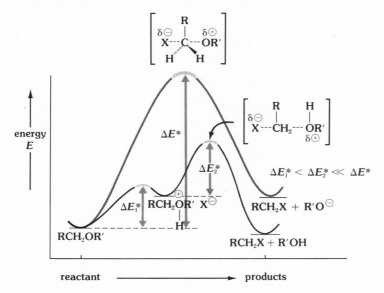

8·6 NUCLEOPHILIC SUBSTITUTION REACTIONS

and by the predominant inversion of configuration that often occurs. An example is the conversion of optically active 2-butanol to 2-bromobutane on treatment with **HBr**:

$$\underset{R\,(+)}{\underset{C_2H_5}{\overset{CH_3\quad H}{>}}C-OH} + HBr \xrightarrow[48\%]{\Delta} \underset{S\,(-)}{Br-\underset{C_2H_5}{\overset{H\quad CH_3}{<}}C} \tag{8.33}$$

The use of strong mineral acids to effect displacement of OH does not require formation of a reactive ester intermediate (as in the case of PX_3 and SOX_2). Consequently, the ease with which ethers can be cleaved by hot **HBr** or **HI** comes as no surprise (see Figure 8.10).

$$(CH_3)_2CHOCH_3 \xrightarrow[\Delta]{HI} CH_3I + (CH_3)_2CHOH \xrightarrow[\Delta]{HI} (CH_3)_2CHI + H_2O \tag{8.34}$$

Tertiary alcohols and tertiary alkyl ethers are particularly sensitive to acid treatment:

$$(CH_3)_3COH + HCl \underset{36\%\ in\ H_2O}{\longrightarrow} (CH_3)_3CCl \tag{8.35}$$

$$\underset{}{\bigcirc\!\!-O\!-\!C(CH_3)_3} \xrightarrow[\Delta]{H_2SO_4\ (1\ drop)} \bigcirc\!\!-OH + (CH_3)_2C{=}CH_2 \tag{8.36}$$

An S_N1 mechanism (equation 8.32) is assumed for these reactions, which may also give elimination products by an $E1$ pathway.

PROBLEM 8·20
If one equivalent of HI is used in reaction 8.34, the major products are 2-propanol and methyl iodide, rather than methyl alcohol and 2-iodopropane. Explain.

PROBLEM 8·21
Reaction of 2,2-dimethyl-1-propanol with hot HBr gives 2-bromo-2-methylbutane as the major product. Explain.

PROBLEM 8·22
Write equations showing all the important steps in reaction 8.36.

RING-OPENING REACTIONS OF OXIRANES

The three-membered oxirane ring system is more reactive than other cyclic (and acyclic) ethers and is observed to undergo a variety of acid- and base-catalyzed addition reactions (Figure 8.11). The facility of these reactions is apparently due to a relief of angle strain on opening the small ring (a similar effect was noted for cyclopropane in Section 5.9).

The simplest oxirane, ethylene oxide, is easily prepared by catalytic oxidation of ethylene:

$$H_2C{=}CH_2 \xrightarrow[\substack{Ag\ catalyst,\\ 250°}]{O_2\ (air)} \underset{ethylene\ oxide}{\overset{O}{\overset{/\ \ \backslash}{CH_2-CH_2}}} \tag{8.37}$$

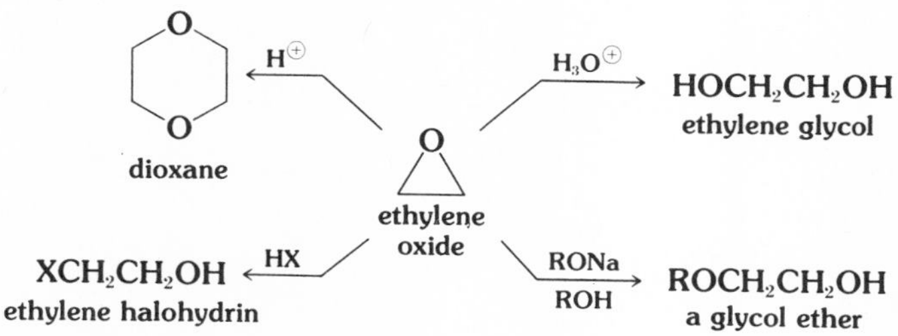

8·7 ELIMINATION REACTIONS   231

FIGURE 8·11 *Addition reactions of ethylene oxide*

More complex oxiranes can be synthesized from the corresponding alkenes via halohydrin intermediates (equation 8.19) or by more direct methods which will be discussed in Chapter 14.

Since the opening of an oxirane ring normally proceeds with inversion of configuration (an S_N2 mechanism), it is possible to obtain glycols that are stereoisomeric with those produced by direct hydroxylation of alkenes (reaction with $MnO_4^{\ominus}$ or OsO_4):

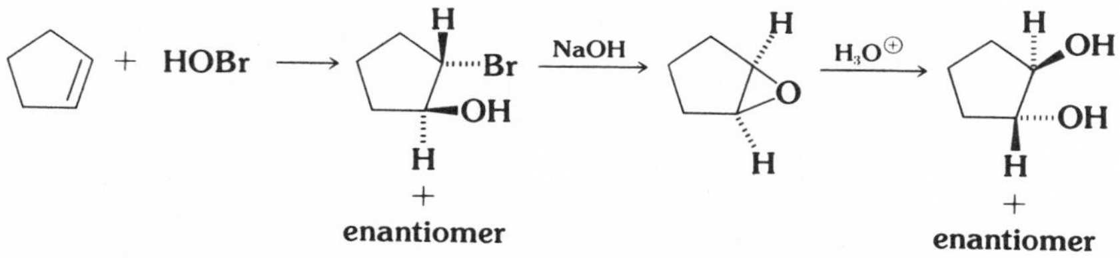

(8.38)

PROBLEM 8·23
Suggest a method for distinguishing *cis*- and *trans*-cyclopentane-1,2-diol.

8·7   ELIMINATION REACTIONS

DEHYDRATION

When alcohols are heated with strong acids, such as H_2SO_4 and H_3PO_4, alkenes are frequently formed by the elimination of water:

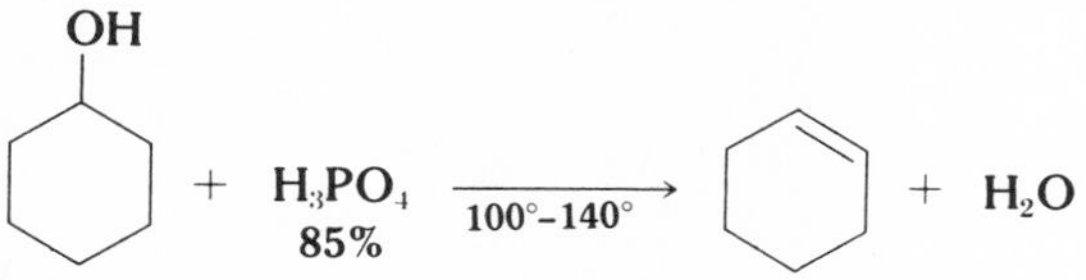

(8.39)

Tertiary alcohols undergo this dehydration reaction more readily than secondary alcohols, which are in turn more reactive than their primary counterparts. From this order of reactivity we are led to suspect that carbonium-ion intermediates are involved in these reactions. The mechanism for the acid-catalyzed dehydration of alcohols is, in fact, essentially

8·7 ELIMINATION REACTIONS

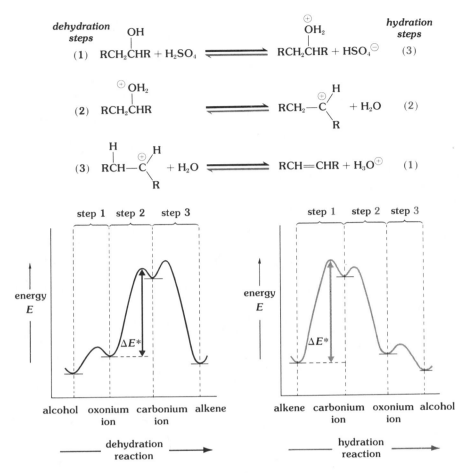

FIGURE 8·12 *Acid-catalyzed dehydration of alcohols and hydration of alkenes*

the reverse of that for the acid-catalyzed hydration of alkenes, as illustrated in Figure 8.12.

In dehydration reactions that can lead to more than one alkene product, that isomer having the more highly substituted double bond normally predominates:

$$(CH_3)_2CHCH(OH)CH_2CH_3 \xrightarrow[-H_2O]{H^\oplus} \left[CH_3-\underset{CH_3}{\overset{H}{C}}-\overset{\oplus}{C}\underset{H}{\overset{H}{\underset{CHCH_3}{}}} \right] \begin{matrix} \xrightarrow{-H^\oplus} & \underset{CH_3}{\overset{CH_3}{}}C=CHCH_2CH_3 \\ & \text{major product} \\ \xrightarrow{-H^\oplus} & (CH_3)_2CHCH=CHCH_3 \\ & \text{minor product} \end{matrix}$$

carbonium-ion intermediate

(8.41)

A significant factor in establishing this Saytzeff orientation is probably the greater thermodynamic stability of the more highly substituted alkene (see Section 4.6). In this respect note in Figure 8.12 that the rate-determining step in alcohol dehydration *precedes* the product-determining step, whereas the rate- and product-determining steps are the same in the hydration of alkenes.

PROBLEM 8·24
Indicate the conjugate acids and bases for each of the three steps shown in Figure 8.12. Which of these steps is the product-determining step in alcohol dehydration?

PROBLEM 8·25
The acid-catalyzed dehydration of 3,3-dimethyl-2-butanol gives a single alkene, which on ozonolysis yields acetone (CH_3COCH_3) as the only organic product. Write a structural formula for this alkene, predict its pmr spectrum, and write a mechanism showing how it is formed.

Most organic reactions do not proceed quantitatively to a single product, but give instead a more or less complex mixture of compounds as a result of competition between concurrent reaction paths. Since it is often possible to improve the yield of a desired product by adjusting the reaction conditions, the importance of understanding and controlling competitive reaction systems poses a challenging, stimulating, and occasionally frustrating problem to organic chemists. As an illustration of the manner in which products can be controlled, consider the reactions of ethanol with sulfuric acid. Ethanol dissolves readily in cold concentrated sulfuric acid to give an oxonium bisulfate salt, which converts slowly to ethyl hydrogen sulfate:

$$C_2H_5OH + H_2SO_4 \rightleftharpoons C_2H_5OH_2^{\oplus} + HSO_4^{\ominus} \xrightarrow{S_N2} C_2H_5OSO_3H + H_2O \quad (8.42)$$

If the ethanol-sulfuric acid mixture is heated above 150°C, elimination reactions give ethene; however, at slightly lower temperatures diethyl ether is the major product:

$$C_2H_5OH_2^{\oplus} + HSO_4^{\ominus} \xrightarrow[E2]{150°-180°} CH_2=CH_2 + H_2SO_4 + H_2O \quad (8.43)$$

$$C_2H_5OSO_3H \xrightarrow{150°-180°} CH_2=CH_2 + H_2SO_4 \quad (8.44)$$

$$\underset{\text{(or } C_2H_5OSO_3H)}{C_2H_5OH_2^{\oplus}} + C_2H_5OH \underset{S_N2}{\overset{130°}{\rightleftharpoons}} C_2H_5-\overset{\overset{H}{|}}{\underset{}{O}}{}^{\oplus}-C_2H_5 + \underset{\text{(or } HSO_4^{\ominus})}{H_2O} \underset{H^{\oplus}}{\overset{-H^{\oplus}}{\rightleftharpoons}} \underset{\text{diethyl ether}}{C_2H_5OC_2H_5} \quad (8.45)$$

OXIDATION AND DEHYDROGENATION

The simple endothermic dehydrogenation of a primary or secondary alcohol to an aldehyde or ketone can be accomplished directly by distilling the alcohol over a hot copper catalyst or by the action of certain enzyme systems:

$$CH_3-\overset{\overset{H}{|}}{\underset{\underset{H}{|}}{C}}-O-H \xrightarrow[300°]{Cu} CH_3-C\overset{\nearrow O}{\underset{\searrow H}{}} + H_2 \quad \Delta H = +15 \text{ kcal/mole} \quad (8.46)$$

8·7 ELIMINATION REACTIONS

$$R-CH_2OH \xrightarrow{\text{alcohol-dehydrogenase enzyme system}} R-C\overset{\displaystyle O}{\underset{\displaystyle H}{\diagdown}} \qquad (8.47)$$

If a chemical oxidizing agent, denoted in a general way as [O], is used to effect this transformation, the hydrogen is eliminated as water and the reaction becomes substantially exothermic:

$$\underset{R'}{\overset{R}{\diagdown}}C\underset{O}{\overset{H}{\diagup}}H \xrightarrow{[O]} \underset{R'}{\overset{R}{\diagdown}}C=O + H_2O \qquad \Delta H \approx -100 \text{ kcal/mole} \qquad (8.48)$$

Chromium VI compounds are widely used for this purpose, the color change from the red-orange dichromate reagent to the green chromic salts providing a useful qualitative test for primary and secondary alcohols:

$$3RCH_2OH + Na_2Cr_2O_7 + 4H_2SO_4 \longrightarrow 3RCHO + Na_2SO_4 + Cr_2(SO_4)_3 + 7H_2O \qquad (8.49)$$

Chromate esters are intermediates in these alcohol oxidations, and there is convincing evidence that the decomposition of these esters is rate determining (Figure 8.13). Since ethers and tertiary alcohols lack one or the other of the two hydrogen atoms essential to this reaction, they are relatively inert to these oxidizing reagents.

PROBLEM 8·26
Write equations for the reactions required to effect the following two-step conversions:

a   cyclopentene → → cyclopentanone

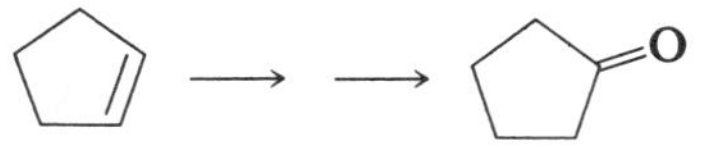

b $(CH_3)_2CHCH_2OCH_3 \longrightarrow \longrightarrow (CH_3)_2CHC\overset{\displaystyle O}{\underset{\displaystyle H}{\diagdown}}$

c $(CH_3)_3COH \longrightarrow \longrightarrow (CH_3)_2\underset{\underset{\displaystyle OH}{|}}{C}-CH_2Br$

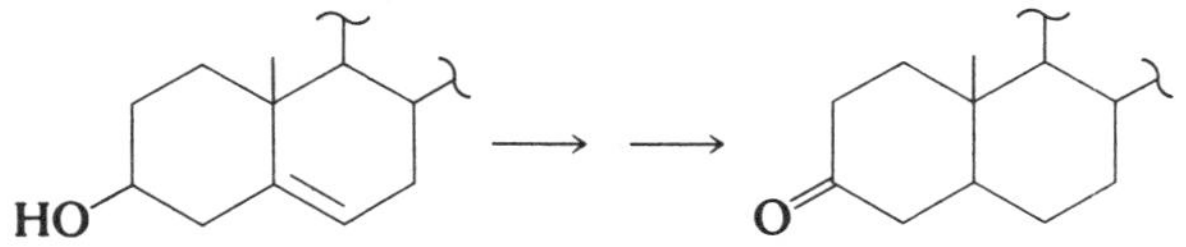

d   cholesterol (HO-) → → cholestanone (O=)

PROBLEM 8·27
Three $C_4H_{10}O$ isomers (A, B, and C) exhibit the following properties:

1   A and B react with CH_3MgBr to liberate a flammable gas.
2   B reacts with acidic dichromate reagent; A and C do not.
3   A and B give the same product when heated with phosphoric acid.
4   The pmr spectrum of C shows a doublet at $\delta = 1.1$ ($J = 7$ Hz), a singlet at $\delta = 3.45$, and a multiplet at $\delta = 3.6$ ($J = 7$ Hz), with relative intensities of 6:3:1.

Write structural formulas for compounds A, B, and C and equations for all the reactions described above.

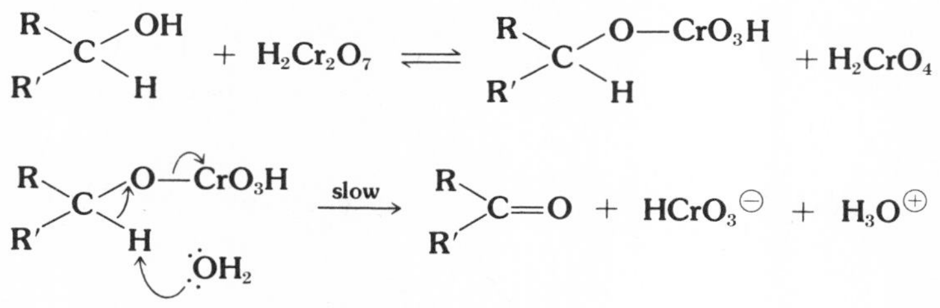

$$\underset{R'}{\overset{R}{>}}C\underset{H}{\overset{OH}{<}} + H_2Cr_2O_7 \rightleftharpoons \underset{R'}{\overset{R}{>}}C\underset{H}{\overset{O-CrO_3H}{<}} + H_2CrO_4$$

$$\underset{R'}{\overset{R}{>}}C\underset{\underset{:OH_2}{H}}{\overset{O-CrO_3H}{<}} \xrightarrow{slow} \underset{R'}{\overset{R}{>}}C=O + HCrO_3^{\ominus} + H_3O^{\oplus}$$

FIGURE 8·13 A mechanism for chromium VI oxidation of alcohols

$$3H_2CrO_3 + 3H_2SO_4 \longrightarrow CrO_3 + Cr_2(SO_4)_3 + 6H_2O$$

8·8 THIOLS, SULFIDES, AND PEROXIDES

It is instructive to compare the properties of alcohols and ethers with their sulfur analogs, *thiols* and *sulfides*. The thiols and sulfides can be considered derivatives of hydrogen sulfide (just as alcohols and ethers are derivatives of water), and it is perhaps not surprising that many of these compounds have extremely disagreeable odors.

$CH_3CH_2CH_2CH_2SH$
1-butanethiol
or n-butyl mercaptan

$CH_3SCH_2CH_3$
methyl ethyl sulfide

$HSCH_2CH(NH_2)CO_2H$
cysteine,
a thiol

$CH_3SCH_2CH_2CH(NH_2)CO_2H$
methionine,
a sulfide

Thiols and sulfides are found in animals (skunks), vegetables (onions), and petroleum; in fact, the small amounts of sulfur compounds in crude petroleum are pollutants and must be removed in the refining process. The thiol group of the amino acid cysteine plays an important role in determining the structure and chemistry of proteins. Another sulfur-containing amino acid, methionine, is essential to good nutrition.

PROBLEM 8·28
Natural sources of thiols and sulfides are limited. How would you suggest preparing these compounds in the laboratory?

Although the thiols have some properties in common with the alcohols (for example, both form stable esters with organic-acid derivatives), the fact that sulfur atoms are larger, have lower electronegativity, and are more polarizable than oxygen gives rise to important physical and chemical differences. Thus the much weaker hydrogen bonding in thiols is reflected by their relatively lower boiling points (ethanethiol boils at 35°C) and the absence of large shifts in the S—H stretching frequency or proton resonance with changes in concentration (compare Table 8.7 with the corresponding data for alcohols). The rate of exchange of SH protons is generally slow with respect to the pmr time scale, as may be seen from the spin-spin splitting patterns in 2-propanethiol, shown in Figure 8.14.

TABLE 8·7 *Spectroscopic properties of thiols and sulfides*

absorbing group	characteristic absorption
infrared spectrum	
RS—H (stretch)	2500–2600 cm^{-1}
pmr spectrum	
RS—H	1.2–1.6 (δ)
CH$_3$SR'	2.0–2.2 (δ)
RCH$_2$SR'	2.4–2.7 (δ)
R$_2$CHSR'	3.0–3.2 (δ)

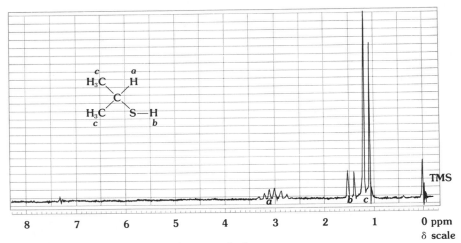

FIGURE 8·14 *Pmr spectrum of 2-propanethiol*

Thiols are stronger Brønsted acids than alcohols (the ionization constant of ethanethiol is about 10^{-11}, compared with 10^{-17} for ethanol), and the strong nucleophilicity of the thiol conjugate bases provides a useful approach to the synthesis of dialkyl sulfides. Indeed, dialkyl sulfides are themselves readily alkylated to give relatively stable sulfonium salts.

$$C_2H_5S^\ominus Na^\oplus + \text{(cyclopentyl)}\text{-Br} \xrightarrow{\text{alcohol}} C_2H_5\text{-S-(cyclopentyl)} + NaBr \tag{8.50}$$

$$CH_3\ddot{S}CH_3 + CH_3I \rightleftharpoons (CH_3)_3S^\oplus I^\ominus \tag{8.51}$$

The carbon-sulfur bonds in thiols and sulfides are more susceptible to catalytic cleavage than the corresponding carbon-oxygen bonds in alcohols and ethers. Two examples of this useful *hydrogenolysis reaction* are the desulfurization of 1-butanethiol and of tetramethylene sulfide:

$$C_4H_9SH + H_2 \xrightarrow{Ni} C_4H_{10} + NiS \tag{8.52}$$
$$\text{(or } H_2S\text{)}$$

$$\underset{S}{\bigcirc} + H_2 \xrightarrow{Ni} C_4H_{10} + NiS \qquad (8.53)$$

Finally, it is important to recognize that oxidation reactions of thiols take a completely different course from oxidations of alcohols. Whereas the oxidation of alcohols to aldehydes and ketones increases the oxidation state of the adjacent carbon atom (a carbon-oxygen double bond is formed), the oxidation of thiols is essentially restricted to the sulfur atom. Thus mild oxidizing agents such as the halogens transform thiols into disulfides, while more vigorous reagents produce a variety of oxy-sulfur acids:

$$2RSH + I_2 \longrightarrow \underset{\text{a disulfide}}{R\text{—}S\text{—}S\text{—}R} + 2HI \qquad (8.54)$$

$$RSH \xrightarrow{[O]} \underset{\text{a sulfenic acid}}{R\text{—}S\text{—}OH} \xrightarrow{[O]} \underset{\text{a sulfinic acid}}{R\overset{O}{\underset{\|}{\text{—}S\text{—}}}OH} \xrightarrow{[O]} \underset{\text{a sulfonic acid}}{R\overset{O}{\underset{\underset{O}{\|}}{\overset{\|}{\text{—}S\text{—}}}}OH} \qquad (8.55)$$

$([O] = MnO_4^{\ominus}, HNO_3, H_2O_2,$ etc.$)$

Sulfides are similarly oxidized to sulfoxides and/or sulfones:

$$R\text{—}S\text{—}R' \xrightarrow{[O]} \underset{\text{a sulfoxide}}{R\overset{O}{\underset{\|}{\text{—}S\text{—}}}R'} \xrightarrow{[O]} \underset{\text{a sulfone}}{R\overset{O}{\underset{\underset{O}{\|}}{\overset{\|}{\text{—}S\text{—}}}}R'} \qquad (8.56)$$

The oxygen analogs of disulfides are called *peroxides* and are customarily prepared by displacement reactions involving hydrogen peroxide. Peroxides are excellent sources of free radicals, since the oxygen-oxygen bond is rather weak (35 kcal/mole) in comparison with other covalent bonds (note that the S—S bond energy is about 54 kcal/mole).

$$R\text{—}\ddot{O}\text{—}\ddot{O}\text{—}R \longrightarrow 2R\text{—}\ddot{O}\cdot \qquad (8.57)$$

SUMMARY

alcohol An organic derivative of water in which one hydrogen atom is replaced by an alkyl group (R—OH). Alcohols are identified in IUPAC nomenclature by the ending *-ol* (see Table 8.1).

ether An organic derivative of water in which both hydrogen atoms are replaced by alkyl (or aryl) groups (R—O—R'). Ethers have no characteristic IUPAC nomenclature suffix. Protons near the oxygen atom of an ether or alcohol have large (about 3.0 to 4.0 ppm) pmr chemical shifts (see Table 8.4).

hydrogen bonding A bonding force between a polar Z—H function and a nucleophilic or basic group Y:, usually represented by a dashed line, as in Z—H---:Y (Z = N, O, F, Cl, Br, or I and Y = N, O, F, Cl, Br, I, C=C, etc.). As a consequence of hydrogen bonding, alcohols have higher boiling points than ethers or hydrocarbons of similar molecular weight, and the hydroxyl proton undergoes absorption shifts in both the infrared and pmr spectra (see Tables 8.3 and 8.4).

spin-spin splitting The splitting of a pmr signal from a set of equivalent protons into a multiplet of $n + 1$ lines, as a result of interaction between these protons and a neighboring group of n protons different from the first group. The magnitude of the spin-spin interaction is given by the coupling constant J, which is reflected by the spacing between the lines of a multiplet. No coupling is observed between (1) remote protons, (2) protons of equal chemical shifts, or (3) protons undergoing rapid exchange.

reactions of alcohols and ethers Chemical transformations involving the oxygen atom of alcohols and ethers. Alcohols are much more reactive than ethers, because the hydrogen atom of a hydroxyl function is more readily removed than is the alkyl group of an ether.

electrophilic substitution at oxygen A reaction of alcohols in which an attacking electrophile becomes bonded to the nucleophilic oxygen of the alcohol (see Table 8.6). Examples are isotope exchange (alcohols are amphoteric), salt formation, ether formation, and ester formation:

$$ROH + D_2O \rightleftharpoons ROD + HOD$$

$$ROH + C_2H_5MgBr \text{ (or } C_2H_5Li) \longrightarrow RO^{\ominus}MgBr^{\oplus} \text{ (or } RO^{\ominus}Li^{\oplus}) + C_2H_6$$
an ionic salt

$$ROH + CH_2=C(CH_3)_2 \xrightarrow{H^{\oplus}} R-O-C(CH_3)_3$$

$$ROH \xrightarrow{Na} RO^{\ominus}Na^{\oplus} + R'-X \xrightarrow{S_N2} R-O-R' + NaX$$
Williamson ether synthesis

$$ROH + CH_3COCl \longrightarrow R-O-\overset{O}{\overset{\|}{C}}-CH_3 + HCL$$
an acetate ester

$$ROH + CH_3SO_2Cl \longrightarrow R-O-SO_2CH_3 + HCl$$
a methanesolfonate ester

nucleophilic substitution at carbon A reaction of alcohols and ethers in which the hydroxyl group of an alcohol or the alkoxyl group of an ether is displaced by a nucleophile. Since $OH^{\ominus}$ and $OR^{\ominus}$ are very poor leaving groups (compared, for example, with halide ions), these functions must be modified before S_N2 or S_N1 reactions can take place. With *alcohols*, esters of strong acids may serve as leaving groups:

$$ROH + CH_3SO_2Cl \longrightarrow ROSO_2CH_3 \xrightarrow{Nu:^{\ominus}} R-Nu + CH_3SO_3^{\ominus}$$

$$ROH + PX_3 \longrightarrow [R-O-PX_2] + HX \longrightarrow R-X + H_3PO_3$$
$$(X = Br \text{ or } I)$$

$$ROH + SOCl_2 \xrightarrow{-HCl} [ROSOCl] \longrightarrow R-Cl + SO_2$$

With *alcohols and ethers,* conjugate acids may serve as leaving groups

$$ROH + HBr \longrightarrow ROH_2^{\oplus}Br^{\ominus} \xrightarrow{\Delta} R-Br + H_2O$$

$$ROR' + HI \longrightarrow R-\underset{\oplus}{\overset{H}{O}}-R'\ I^{\ominus} \xrightarrow{\Delta} R-I + R'-OH \xrightarrow{HI}{\Delta} R'I + H_2O$$

elimination reactions The formation of alkenes from alcohols (or their ester derivatives) by loss of water (or the corresponding acid):

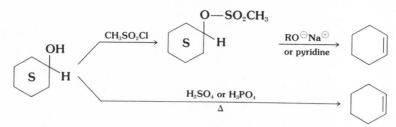

oxidation Conversion of primary or secondary alcohols to aldehydes or ketones, respectively. The most common oxidizing reagents for this purpose are Cr^{+6} compounds such as CrO_3 and $Na_2Cr_2O_7$ (usually used in acidic solutions).

oxiranes (epoxides) Three-membered cyclic ethers formed by intramolecular S_N2 reactions of vicinal halohydrins. The angle strain in the three-membered ring makes these cyclic ethers much more reactive than other ethers.

thiols and sulfides Sulfur analogs of alcohols and ethers:

$$\begin{array}{ccc} R-S-H & R-S-R' & R-S-S-R' \\ \text{a thiol or mercaptan} & \text{a sulfide} & \text{a disulfide} \end{array}$$

Sulfur is nucleophilic in these compounds and is easily oxidized under mild conditions.

peroxides Alkyl derivatives of hydrogen peroxide (R—O—O—H and R—O—O—R'). The oxygen–oxygen bond in these compounds is relatively weak.

EXERCISES

8·1 Name the following compounds by any accepted nomenclature system:

a $CH_3CH_2CH_2OH$

b $CH_3CH(OH)CH_3$

c $CH_3CH(CH_3)CH_2CH_2OH$

d cyclohexane with CH_3 and OH on the same carbon

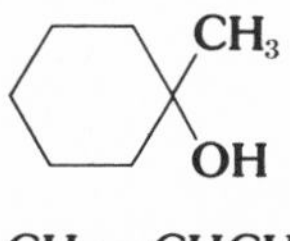

e $CH_2{=}CHCH_2CH(OH)CH_3$

f epoxide with CH_3 and H

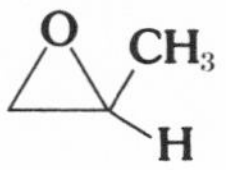

g $HOCH_2CH_2CH_2OH$

h cyclopentyl ether (tetrahydropyran)

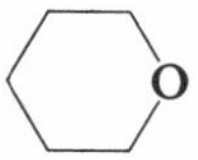

i CH_3OCH_3

j $CH_3CH_2OCH_2CH_2CH_2CH_3$

8·2 Classify each of the alcohols listed in Exercise 8.1 as primary, secondary, or tertiary.

8·3 Match the names for identical structures:

a n-Butyl alcohol   1 Glycerol
b 1,2-Propanediol   2 Tetrahydrofuran
c Isobutyl alcohol   3 2-Methyl-1-propanol
d Oxolane   4 1-Butanol
e 1,2,3-Propanetriol   5 Ethanol
f Ethyl alcohol   6 1,2-Dihydroxypropane
g t-Butyl alcohol   7 Ethyl ether
h Diethyl ether   8 2-Methyl-2-propanol

8·4 Draw structural formulas for all isomers having the composition $C_4H_{10}O$ and label each isomer as an alcohol or an ether. For each alcohol, provide the IUPAC name and classify it as primary, secondary, or tertiary.

8·5 Show the products of the reaction (if any) of n-propyl alcohol with:

a Na   f $SOCl_2$
b CH_3CH_2MgBr g PBr_3
c CH_3COCl h H_2SO_4 (concentrated, 0°)
d CH_3SO_2Cl i $Na_2Cr_2O_7$ ($H^{\oplus}$)
e HBr   j H_3PO_4 (150°)

8·6 Repeat Exercise 8.5 for isopropyl alcohol as the starting material.

8·7 Repeat Exercise 8.5 (except part j) for diethyl ether as the starting material.

8·8 Show the products of the reaction of t-butyl alcohol with:

a Na   e $Na_2Cr_2O_7$ ($H^{\oplus}$)
b CH_3COCl f CH_3MgI
c HBr   g PBr_3
d H_2SO_4 (concentrated)

8·9 Show the major organic products of the reaction of sodium n-propoxide ($CH_3CH_2CH_2ONa$) with:

a H_2O d $CH_3CH_2CH_2CH_2Br$
b HCl   e $(CH_3)_3CBr$
c CH_3SO_2Cl f $(CH_3)_2CHCHCH_3$
                                |
                                OSO_2CH_3

8·10 Show the products of the reaction (if any) of 1-butanethiol with:

a I_2 d $MnO_4^{\ominus}$ (one equivalent)
b Na   e $MnO_4^{\ominus}$ (two equivalents)
c H_2 in Ni   f $MnO_4^{\ominus}$ (three equivalents)

8·11 Show the products of the reaction (if any) of diethyl sulfide with:

a H_2, Ni
b $MnO_4^{\ominus}$ (one equivalent)
c $MnO_4^{\ominus}$ (two equivalents)
d CH_3I

8·12 In the nomenclature of organic sulfur compounds the prefix *thia-* is occasionally used to indicate replacement of carbon by sulfur. For example, $CH_3CH_2SCH_2CH_3$ could be named 3-thiapentane. Give alternative names for thiacyclobutane and 2-thiahexane.

8·13 Given the following boiling points and solubilities of compounds of similar molecular weight, explain:
a The relative boiling points of *n*-butyl alcohol and diethyl ether
b The relative boiling points of pentane and diethyl ether
c The relative boiling points of *n*-butyl alcohol and *n*-butyl thiol
d The almost identical water solubilities of diethyl ether and *n*-butyl alcohol, despite their widely differing boiling points

	boiling point (°C)	water solubility (g/100 g)
pentane	35	0.04
diethyl ether	35	7.8
n-butyl alcohol	118	7.9
1-butanethiol	98	slightly soluble

8·14 Regarding Brønsted acids and bases:
a Explain the relative acidity of primary, secondary, and tertiary alcohols in aqueous solution.
b Arrange the following compounds in order of acidity:

CH_3CH_2OH CH_3CH_2SH HOH ($K_a = 10^{-16}$)

c Arrange the following compounds in order of basicity:

$CH_3CH_2O^{\ominus}$ $CH_3CH_2S^{\ominus}$ $OH^{\ominus}$

8·15 Describe simple chemical tests that could be used to distinguish:
a Ethyl alcohol from diethyl ether
b 2-Methyl-2-propanol from 2-methyl-1-propanol
c Octane from 1-octanol
d Cyclohexene from cyclohexanol

8·16 What features of their infrared spectra could be used to make the distinctions in Exercise 8.15?

8·17 What is the maximum number of equivalent protons on adjacent carbon atoms that can interact with a given proton? What splitting pattern would the pmr signal for this proton show?

8·18 Expand the binomial $(a+b)^n$ for $n = 1, 2, 3$. How do the coefficients of the terms compare with the relative intensities of the pmr signals in a doublet, triplet, and a quartet. What is the physical significance of *n* in pmr spectroscopy?

8·19 An unknown substance has the molecular formula $C_5H_{10}O$ and does not react with sodium metal or bromine. What structures are possible for this compound?

8·20 Using the information available in reactions 1 and 2 below, determine whether a base was present in the reaction of thionyl chloride with the alcohol, as shown in the first step of reaction 1. Assume that all other displacement reactions proceed by S_N2 mechanisms.

(+)-2-butanol $\xrightarrow{SOCl_2}$ (−)-2-chlorobutane $\xrightarrow[CH_3OH]{NaOCH_3}$ (+)-2-methoxybutane   (1)

(−)-2-butanol $\xrightarrow[\text{ether}]{Na}$ [alkoxide salt] $\xrightarrow{CH_3Br}$ (−)-2-methoxybutane   (2)

242 EXERCISES

8·21 Explain the difference in the course of the following two reactions:

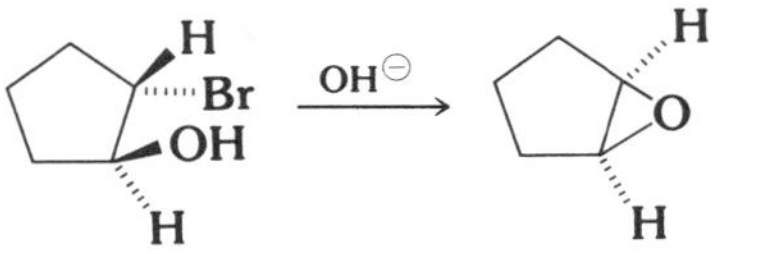

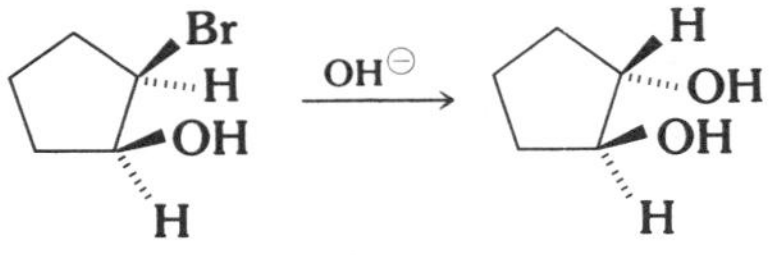

[structures: trans-2-bromocyclopentanol + OH⁻ → cyclopentene oxide; cis-2-bromocyclopentanol + OH⁻ → trans-cyclopentane-1,2-diol]

8·22 What prominent feature in the indicated spectra could be used to distinguish the following pairs of compounds?
a $CH_3CH_2CH_2OH$ and $CH_3CH_2CH_2CH_3$; pmr spectrum
b CH_3CH_2OH and CH_3CH_2Cl; infrared spectrum
c $CH_3CH_2CH_2CH_2OH$ and $CH_3CH_2OCH_2CH_3$; infrared spectrum
d $CH_3CH_2CH_2CH_2OH$ and $CH_3CH_2OCH_2CH_3$; pmr spectrum

8·23 Predict the relative ease with which acid-catalyzed elimination (dehydration) of the following alcohols takes place and write a structural formula for the major product:
a $CH_3CH_2CH(CH_3)CH(OH)CH_3$
b [cyclopentane with OH and H on same carbon]
c $(CH_3)_3CCH(CH_3)CH_2OH$

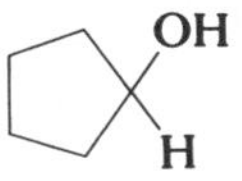

8·24 In view of the nonbonded electron pairs of the oxygen atom in ethers, propose a structure for solvated BH_3 in the following reaction, assuming that the ether is tetrahydrofuran:

$$B_2H_6 \underset{\text{ether}}{\rightleftharpoons} 2BH_3 \text{ (solvated)}$$

8·25 Equation 8.50 employs ethyl alcohol as the solvent. Explain why this reaction does *not* follow the course shown below:

$C_2H_5S^\ominus Na^\oplus + C_2H_5OH \longrightarrow C_2H_5O^\ominus Na^\oplus + C_2H_5SH$

$C_2H_5O^\ominus Na^\oplus +$ [cyclopentyl bromide] $\longrightarrow$ [cyclopentyl ethyl ether] $+ NaBr$

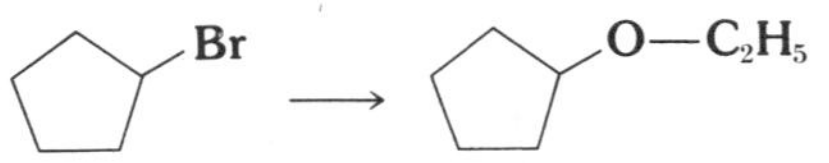

8·26 Each of the following conversions can be accomplished in three steps or fewer. Indicate the intermediate compounds and the reagents necessary for the reactions.

a [trans-4-methylcyclohexanol → cis-4-methylcyclohexyl cyanide]

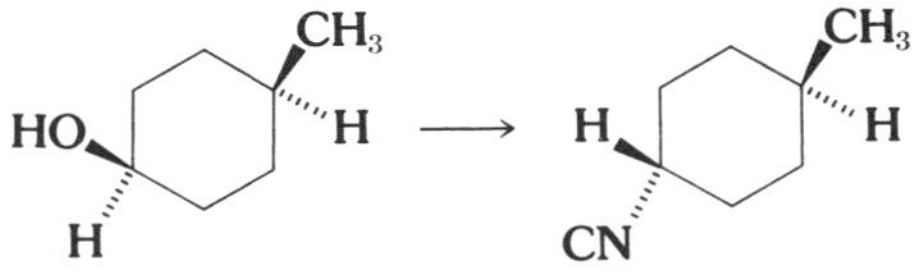

b $CH_3CH=C(CH_3)CH_2CH_3 \longrightarrow CH_3\overset{O}{\overset{\|}{C}}CH(CH_3)CH_2CH_3$

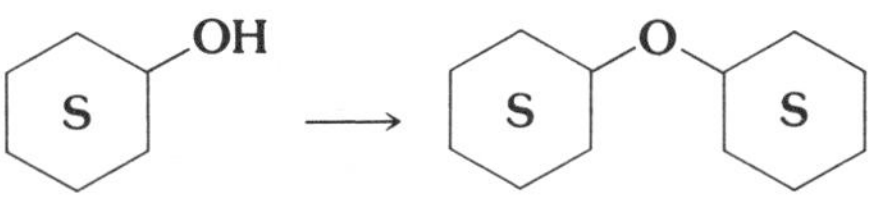

c [2-hydroxy cyclohexyl sulfide → bis-cyclohexyl ether sulfide] d ![cyclopentene] → ![cyclopentyl-SCH2CH3] SCH₂CH₃

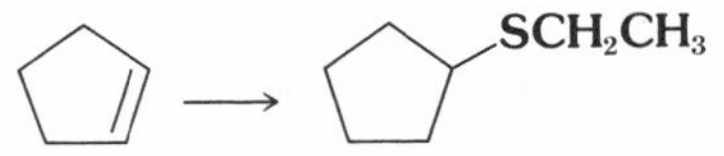

e HOCH₂CH₂CH=CHCH₃ → ![2-methyltetrahydrofuran with CH₃]

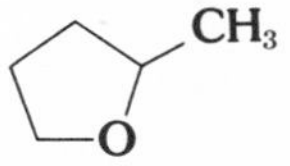

8·27 Write a plausible mechanism for each of the following reactions:

a CH₃CH₂CH₂OH + ![dihydropyran] $\xrightarrow{H^{\oplus}}$ CH₃CH₂CH₂—O—![tetrahydropyranyl]

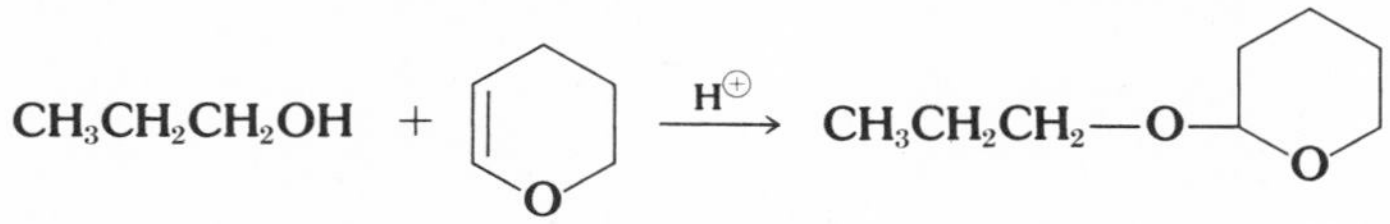

b ![cyclopentylmethanol with CH₂OH] $\xrightarrow[\Delta]{H^{\oplus}}$![cyclohexene]

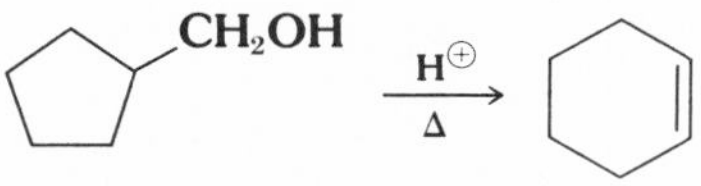

8·28 The sulfonium salt shown below reacts with potassium *t*-butoxide to yield isobutylene. What kind of reaction is it? Show the mechanism.

CH₃SCH₃ + (CH₃)₂CHCH₂I ⟶ [(CH₃)₂S$^{\oplus}$—CH₂CH(CH₃)₂]I$^{\ominus}$

$\xrightarrow{(CH_3)_3COK}$ (CH₃)₂S + (CH₃)₂C=CH₂

8·29 Bombykol, the sex attractant of the female silk moth (*Bombyx mori*), has been isolated and identified. From 2 tons of pupae, 12 mg of biologically active material was obtained and characterized as follows:

1 Molecular formula C₁₆H₃₀O, with strong infrared absorption at 3580 cm⁻¹.
2 Hydrogenation gives 1-hexadecanol.
3 Reacts with acetic anhydride to yield an ester, and ozonolysis (O₃, H₂O₂) of this ester gives the products

CH₃CH₂CH₂CO₂H   CH₃CO—(CH₂)₉—COH   HOC—COH
                        ‖   ‖   ‖   ‖
                        O   O   O   O

What is the structure of Bombykol? What structural features are still in doubt? Write equations for all chemical reactions.

8·30 How does the order of reactivity of HX (HI > HBr > HCl) with primary and secondary alcohols substantiate an S_N2 mechanism and rule out an S_N1 mechanism? Explain your reasoning.

8·31 Show a possible mechanism for each of the transformations illustrated in Figure 8.11. What is responsible for the fact that oxiranes undergo fairly facile reactions with strong base, whereas ordinary ethers are inert to these conditions?

8·32 Sketch the predicted pmr spectra of the following compounds:
a BrCH₂CH₂OH
b the three isomers of C₃H₈O
c CH₃SH and CH₃SCH₂CH₃
d *cis*- and *trans*-1-Bromo-3,3-dimethyl-1-butene
e *cis*- and *trans*-1,2-Dichloroethene
f 1,1-Dichloroethene

8·33 Three isomeric compounds have the molecular formula $C_4H_{10}O$. From the pmr data given below, indicate their structures.

a A triplet at $\delta = 1.1$ (6H) and a quartet at $\delta = 3.4$ (4H)
b A singlet at $\delta = 1.3$ (9H) and a singlet at $\delta = 4.0$ (1H)
c A doublet at $\delta = 1.1$ (6H), a singlet at $\delta = 3.2$ (3H), and a septet at $\delta = 3.7$ (1H)
d A doublet at $\delta = 0.9$ (6H), a multiplet at $\delta = 1.7$ (1H), a doublet at $\delta = 3.3$ (2H), and a singlet at $\delta = 4.0$ (1H)

8·34 Compound A has the molecular formula C_5H_{10} and yields product B when treated with hot aqueous H_2SO_4. The infrared spectrum of B shows absorption at 3600 cm^{-1}, and the pmr spectrum shows a triplet at $\delta = 0.9$ ($J = 7$ Hz), a singlet at $\delta = 1.2$, a quartet at $\delta = 1.4$ ($J = 7$ Hz), and a broad singlet at $\delta = 3.0$, with relative areas of 3:6:2:1. Dehydration of compound B with concentrated H_2SO_4 gives C, an isomer of compound A. What are likely structures for all three compounds?

9 AROMATIC COMPOUNDS

THE IMPORTANCE of perfumes and spices in early trade and commerce led to an intensive study of the fragrant essential oils that could be isolated from these sources. Since these investigations preceded the development of spectroscopic methods of structure analysis, it was often necessary to degrade the complex natural substances into smaller identifiable fragments by chemical means. The oxidative cleavage of geraniol is an example of this technique:

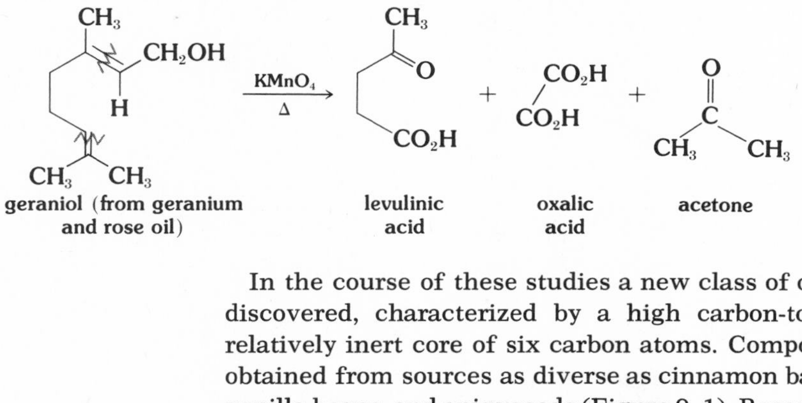

CH_3
CH_2OH
H
CH_3 CH_3
geraniol (from geranium and rose oil)
$KMnO_4$
Δ
CH_3
O
CO_2H
levulinic acid
+
CO_2H
CO_2H
oxalic acid
+
O
C
CH_3 CH_3
acetone

In the course of these studies a new class of organic compounds was discovered, characterized by a high carbon-to-hydrogen ratio and a relatively inert core of six carbon atoms. Compounds of this kind were obtained from sources as diverse as cinnamon bark, wintergreen leaves, vanilla beans, and anise seeds (Figure 9.1). Because of the pleasant odors of these naturally occurring substances, early chemists referred to them as *aromatic*. Not all the aromatic compounds have such pleasant associations. For example, benzene, the simplest, is rather toxic, and several polycyclic compounds (including some amines) have proved to be strongly carcinogenic and must be handled with care.

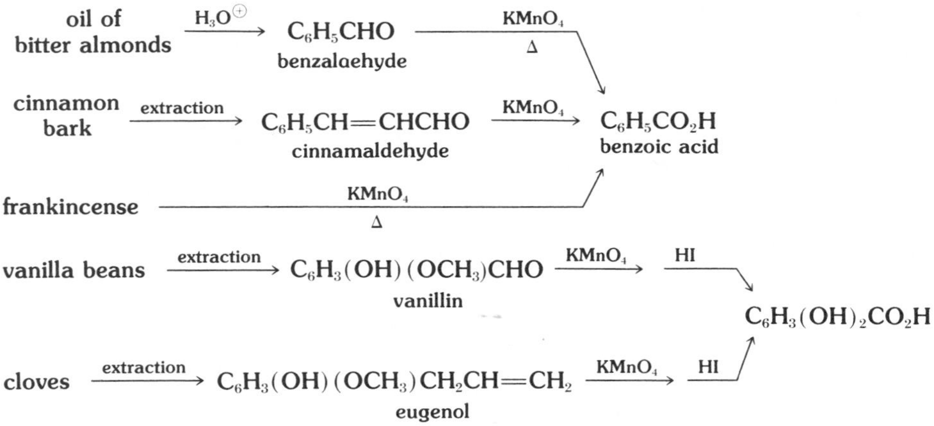

9·1 THE BENZENE PUZZLE

oil of bitter almonds $\xrightarrow{H_3O^\oplus}$ C_6H_5CHO (benzaldehyde) $\xrightarrow[\Delta]{KMnO_4}$

cinnamon bark $\xrightarrow{extraction}$ $C_6H_5CH{=}CHCHO$ (cinnamaldehyde) $\xrightarrow[\Delta]{KMnO_4}$ $C_6H_5CO_2H$ (benzoic acid)

frankincense $\xrightarrow[\Delta]{KMnO_4}$

vanilla beans $\xrightarrow{extraction}$ $C_6H_3(OH)(OCH_3)CHO$ (vanillin) $\xrightarrow{KMnO_4}$ $\xrightarrow{HI}$ $C_6H_3(OH)_2CO_2H$

cloves $\xrightarrow{extraction}$ $C_6H_3(OH)(OCH_3)CH_2CH{=}CH_2$ (eugenol) $\xrightarrow{KMnO_4}$ $\xrightarrow{HI}$

FIGURE 9·1 *Oxidative degradation of some aromatic essential oils*

9·1 THE BENZENE PUZZLE

The simplest compound containing the fundamental six-carbon aromatic unit is the hydrocarbon benzene (C_6H_6), which was first obtained over a century ago by heating benzoic acid with lime or by the destructive distillation of coal. Despite the high ratio of carbon to hydrogen in benzene, this hydrocarbon proved to be remarkably inert to the common chemical tests for unsaturation:

$$C_6H_6 \text{ benzene} \begin{cases} \xrightarrow[CCl_4]{Br_2} \text{ no reaction} \\ \xrightarrow[H_2O]{KMnO_4} \text{ no reaction} \end{cases} \tag{9.1}$$

In fact, when forced to undergo chemical transformation, benzene generally suffers substitution reactions rather than addition reactions:

$$C_6H_6 + Br_2 \xrightarrow[\Delta]{FeBr_3} C_6H_5Br + HBr \tag{9.2}$$
benzene, bromobenzene

$$C_6H_5Br + Br_2 \xrightarrow[\Delta]{FeBr_3} C_6H_4Br_2 + HBr \tag{9.3}$$
bromobenzene, dibromobenzene

$$C_6H_6 + HNO_3 \xrightarrow[\Delta]{H_2SO_4} C_6H_5NO_2 + H_2O \tag{9.4}$$
nitrobenzene

$$C_6H_6 + H_2SO_4 \xrightarrow{\Delta} C_6H_5SO_3H + H_2O \tag{9.5}$$
benzene-sulfonic acid

Since the monosubstituted benzenes did not show isomerism, whereas the disubstituted benzenes (as in equation 9.3) existed in three isomeric forms, early workers concluded that the six carbon atoms were symmetrically arranged.

9·1 THE BENZENE PUZZLE

PROBLEM 9·1

a From the molecular formula of benzene, calculate the number of rings and/or double bonds that must be present.

b Write structural formulas for three possible C_6H_6 isomers.

PROBLEM 9·2

How many of the C_6H_6 isomers proposed in Problem 9.1 would have a single monosubstituted derivative? How many of these would have three isomeric disubstituted derivatives?

Despite the reluctance of benzene to give addition reactions, its slow destruction on exposure to ozone and the sluggish catalytic addition of hydrogen established the unsaturated nature of this hydrocarbon.

$$C_6H_6 \;+\; 3H_2 \xrightarrow[150°, \,100 \text{ atm } H_2]{\text{Pt catalyst}} \text{cyclohexane} \tag{9.6}$$

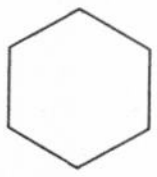

benzene

In 1865, from an evaluation of the fragmentary and incomplete data on hand at that time, F. A. Kekulé, a professor at the University of Bonn, proposed that the six carbon atoms of benzene were arranged in a six-membered ring and that one hydrogen atom was bonded to each carbon.† This suggestion nicely accounted for the numbers of isomeric substituted benzenes (Figure 9.2). However, it violated the established tetravalence of carbon. To dispose of the valency problem Kekulé subsequently postulated that benzene consisted of a rapidly equilibrating 1,3,5-cyclohexatriene mixture. This proposal sparked a storm of controversy, criticism, and counterproposals that lasted well into the twentieth century.

†According to Kekulé, the idea came to him during a dream: "I was sitting, writing at my textbook, but the work did not progress; my thoughts were elsewhere. I turned my chair to the fire and dozed. Again the atoms were gambolling before my eyes. This time the smaller groups kept modestly in the background. My mental eye, rendered more acute by repeated visions of the kind, could now distinguish larger structures of manifold conformation: long rows, sometimes more closely fitted together; all twining and twisting in snake-like motion. But look! What was that? One of the snakes had seized hold of its own tail, and the form whirled mockingly before my eyes. As if by a flash of lightning I awoke; and this time also I spent the night in working out the consequences of the hypothesis."

FIGURE 9·2 *Di- and trisubstituted isomers expected for a regular hexagonal structure*

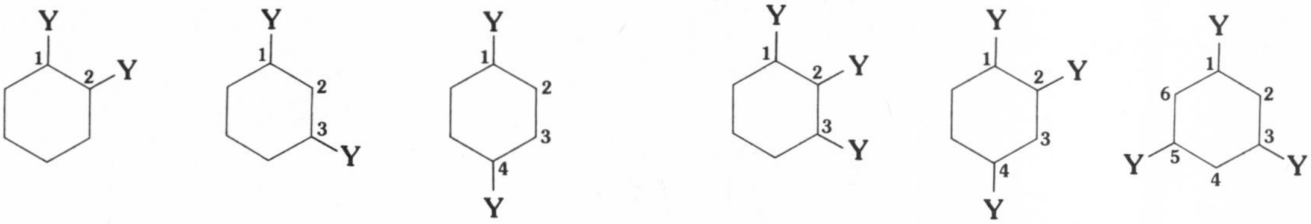

| 1,2 | 1,3 | 1,4 | 1,2,3 | 1,2,4 | 1,3,5 |

disubstituted derivatives   trisubstituted derivatives

PROBLEM 9·3
How many di- and trisubstituted isomers would you expect from a single 1,3,5-cyclohexatriene? How many would you expect from the equilibrium mixture proposed by Kekulé?

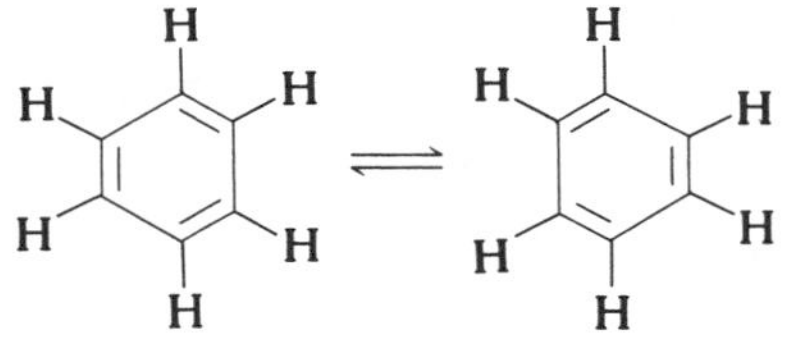

A major deficiency of the Kekulé description of benzene is that it fails to account for the unusual chemical behavior of this aromatic hydrocarbon. For example, the three double bonds in 1,3,5-cyclohexatriene would ordinarily undergo most of the alkene addition reactions described in Chapter 5. As we have seen, however, benzene is relatively inert, and there is no convincing evidence to support the idea that this lack of reactivity is due to the rapid equilibrium shown above.

The only remaining feature of the Kekulé triene that could conceivably account for the unique properties of benzene is the *cyclic conjugation* of the three double bonds. To evaluate the importance of this circumstance, we must first examine the consequences of conjugation in simple acyclic systems.

9·2 CONJUGATED DIENES AND POLYENES

THERMODYNAMIC STABILITY

The observed heats of hydrogenation of the pentadienes in Table 9.1 clearly demonstrate that the isomer with conjugated double bonds is thermodynamically more stable (by about 7 kcal/mole) than the nonconjugated dienes. Since part of this increased stability is due to the greater alkyl substitution of the 1,3-diene double bonds (compared with the 1,4-diene), we ought to compare the experimentally determined heats of hydrogenation with those expected from the ΔH values of equivalently substituted monoenes (compounds with one double bond). From the data in Table 4.5 we see that the isolated diene behaves essentially as expected; however, the conjugated diene is *more* stable (by about 4 kcal/mole) and the cumulated diene is *less* stable (by 10 kcal/mole) than anticipated. Experimental measurements also show that the planar conformations of conjugated dienes are generally more stable than twisted conformations, as evidenced by the unexpectedly large barrier to rotation about the C-2—C-3 bond in 1,3-butadiene (5 kcal/mole compared with 2 kcal/mole for rotation about C-2—C-3 in propene).

The enhanced thermodynamic stability of conjugated π-electron systems has been chiefly attributed to two factors, the increased delocalization of π electrons brought about by the overlap of adjacent π orbitals and the increase in the bond strength of carbon-carbon σ bonds as the s contribution of the component hybrid atomic orbitals increases:

$$^{sp}C-^{sp}C > {}^{sp}C-^{sp^2}C > {}^{sp^2}C-^{sp^2}C > {}^{sp^2}C-^{sp^3}C > {}^{sp^3}C-^{sp^3}C$$

9·2 CONJUGATED DIENES AND POLYENES

TABLE 9·1 *Heats of hydrogenation of isomeric pentadienes*

compound	ΔH observed (kcal/mole)	ΔH calculated (kcal/mole)†
1,2-pentadiene	−68	−58
1,4-pentadiene	−61	−60
1,3-pentadiene	−54	−58

†From the heats of hydrogenation of correspondingly substituted simple alkenes, given in Table 4.5.

For example, the two carbon-carbon single bonds in 1,4-pentadiene join an sp^3 carbon atom to two sp^2 carbon atoms, while the isomeric 1,3 diene has one sp^2-sp^2 single bond and one sp^2-sp^3 bond. Despite these stabilizing factors, the chemical reactivity of conjugated dienes is not noticeably decreased. Indeed, both 1,3-pentadiene and 1,4-pentadiene react rapidly with bromine or potassium permanganate test solutions.

These facts are nicely accounted for by the molecular-orbital description of a coplanar conjugated diene shown in Figure 9.3. Simple LCAO theory indicates that the four overlapping $2p$ atomic orbitals remaining after the σ-bonded framework of the diene has been formed will combine to generate four π-molecular orbitals (two bonding and two antibonding). Each of these can hold a maximum of two paired electrons. In the case of 1,3-butadiene (CH_2=CH—CH=CH_2), 18 of the 22 valence electrons

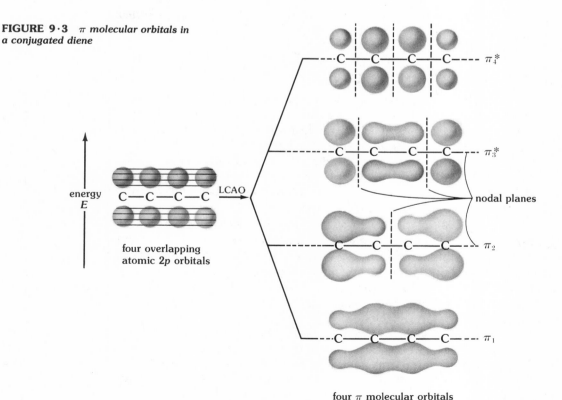

FIGURE 9·3 π *molecular orbitals in a conjugated diene*

four π molecular orbitals

will occupy the nine bonding σ orbitals, and the remaining four electrons are accommodated by the two bonding π orbitals (π_1 and π_2).

Note the shape of the molecular orbitals in Figure 9.3. With the exception of σ orbitals, all molecular orbitals are divided by *nodal regions* (usually planes) in which the electron density of the occupied orbital is essentially zero. As a rule, the number of nodal planes increases with the energy of the orbitals. For example, the lowest-energy π-molecular orbital, π_1, has a single nodal plane (the plane passing through all four carbon atoms of the conjugated diene), π_2 has two perpendicular nodal planes, π_3^* has three nodal planes, and π_4^* has four. The wave equations which describe the molecular orbitals undergo a change in sign at the node, a characteristic which is sometimes described as follows:

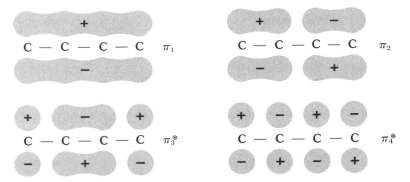

These signs do *not* indicate electrical charge, but merely a mathematical phase change.

ULTRAVIOLET-VISIBLE SPECTROSCOPY

Conjugated dienes and polyenes display characteristic absorption spectra in the visible and/or ultraviolet part of the electromagnetic spectrum. These absorption bands result from the promotion (or excitation) of molecular valence electrons into higher-energy (empty) molecular orbitals (see Section 1.3). The energies required to effect some typical electronic excitations are shown in Figure 9.4.

The substantial energy required to excite the tightly bound σ electrons can be supplied only by high-energy ultraviolet light (wavelengths shorter than 100 nm). Since the spectrophotometers commonly available to organic chemists do not function well at wavelengths shorter than 200 nm, no $\sigma \longrightarrow \sigma^*$ excitations are ever observed, and saturated hydrocarbons such as heptane and cyclohexane are transparent throughout the accessible ultraviolet ($\lambda = 200$–400 nm) and visible ($\lambda = 400$–800 nm) spectrum. As a result, these compounds make excellent solvents for spectroscopic studies.

It is clear from Figure 9.4 that $\pi \longrightarrow \pi^*$ excitations require less energy and will therefore be observed at longer wavelengths than $\sigma \longrightarrow \sigma^*$ transitions. Thus simple alkenes and alkynes absorb strongly at $\lambda = 180$–200 nm ($\Delta E \approx 160$ kcal/mole), and the position of *maximum absorption* λ_{max} is shifted to still longer wavelengths by conjugation with

9·2 CONJUGATED DIENES AND POLYENES 251

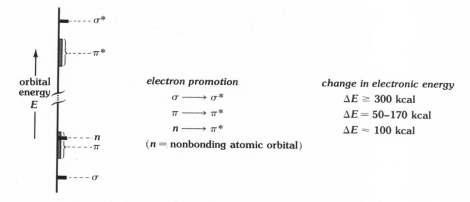

FIGURE 9·4 *Approximate energy changes for electron excitations*

additional unsaturated functions. For example, 1,3-butadiene shows maximum absorption at $\lambda_{max} = 217$ nm. This decrease in the electronic excitation energy of conjugated dienes is not surprising once we realize that an electron is being promoted from the *highest-energy occupied π orbital* (π_2 in Figure 9.3) to the *lowest-energy unoccupied π^* orbital* (π_3). A typical ultraviolet absorption spectrum of a conjugated diene is shown in Figure 9.5.

Since the absorption shift to longer wavelengths conveniently bridges the 200-nm low-wavelength limit of most commercial spectrophotometers, we can conclude that strong absorption above 200 nm in the ultraviolet-visible region usually indicates the presence of a conjugated π-electron system. As the number of conjugated double bonds increases, the absorption maximum and the intensity of absorption also increase.

FIGURE 9·5 *Ultraviolet absorption spectrum of 2,5-dimethyl-2,4-hexadiene in cyclohexane*

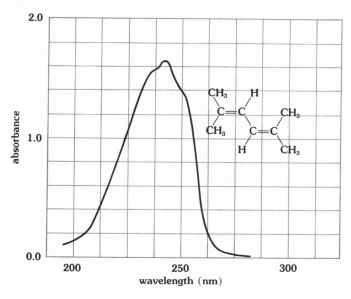

9·2 CONJUGATED DIENES AND POLYENES

Eventually the absorption band moves into the visible spectrum, and a substance incorporating such a conjugated system, called a *chromophore*, becomes colored. For example, vitamin A is yellow, and the plant pigment β-carotene is deep red.

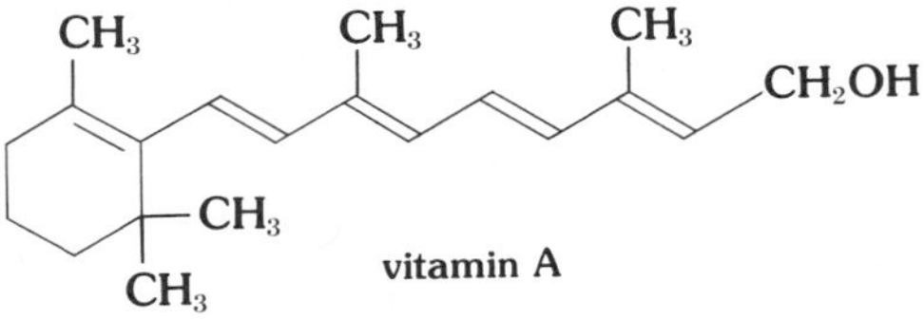

[Structure of vitamin A]

[Structure of β-carotene]

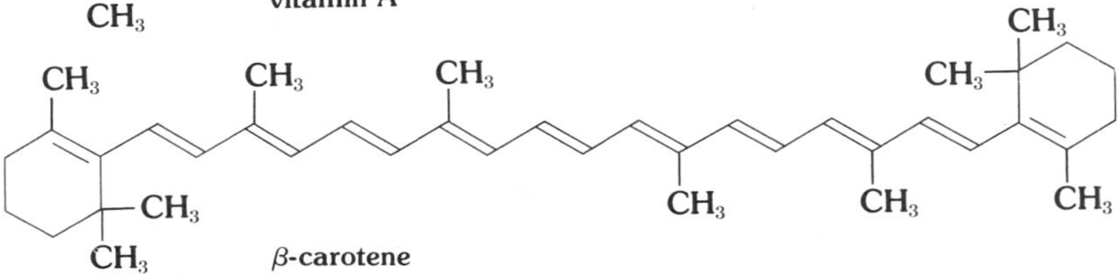

The mapping of an absorption spectrum requires a determination of the absorption intensity, as well as the wavelengths at which light absorption occurs. For this purpose we use the *molar absorptivity* or *extinction coefficient* ϵ, which is defined as

$$\epsilon = \frac{A}{lM} \tag{9.7}$$

where A is the measured absorbance of the sample solution, l is the path length of light through the sample (in cm), and M is the molar concentration of the sample solution. Some characteristic molar absorptivities are listed in Table 9.2.

PROBLEM 9·4
Two isomeric C_6H_8 hydrocarbons, A and B, both give substance C on catalytic hydrogenation. Compound C shows a single sharp pmr signal at $\delta = 1.4$, while A and B have similar pmr spectra consisting of two equally strong signals, one at $\delta = 1.5$–2.0 and the other at $\delta = 5.0$–5.7. Compound C is transparent above 200 nm; compound B shows increasing absorption near 200 nm, although no maximum occurs above that wavelength; and compound A has a strong well-defined absorption maximum between 250 and 260 nm. Write structures for compounds A, B, and C.

TABLE 9·2 *Typical molar absorptivities of conjugated double-bond systems*

compound	maximum absorption λ_{max} (nm)	molar absorptivity ϵ
$CH_2{=}CH{-}CH{=}CH_2$	217	21,000
$CH_2{=}C(CH_3){-}C(CH_3){=}CH_2$	226	21,400
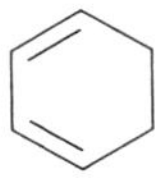 (cyclohexadiene)	256	8,000
$CH_2{=}CH{-}CH{=}CH{-}CH{=}CH_2$	~260	52,000

9·2 CONJUGATED DIENES AND POLYENES

TABLE 9·3 *Heats of hydrogenation of unsaturated hydrocarbons related to cyclohexane*

compound	ΔH observed (kcal/mole)	ΔH calculated (kcal/mole)†
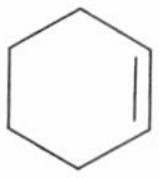 cyclohexene	−28.6	—
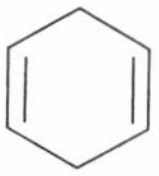 1,4-cyclohexadiene	−59.3	−57.2
 1,3-cyclohexadiene	−55.4	−57.2
benzene	−49.8	−85.8

†Assuming that each double bond makes an equal contribution of −28.6 kcal/mole to the heat of hydrogenation.

CONJUGATION IN BENZENE

Thus far we have noted three general properties of conjugated systems of double bonds:

1 The heat content of a conjugated system is lower than that of the corresponding unconjugated isomer (it is thermodynamically more stable).

2 Two or more unsaturated functions show a strong tendency to be coplanar when they are conjugated (although small deviations from coplanarity are readily tolerated).

3 Conjugated double bonds show strong absorption in the ultraviolet spectrum at wavelengths greater than 200 nm.

Similar characteristics would be expected for the conjugated 1,3,5-cyclohexatriene structure proposed by Kekulé, and we must therefore determine how well these properties correspond with those of benzene. The heats of hydrogenation of cyclohexene, the two isomeric cyclohexadienes, and benzene are given in Table 9.3. A moderate stabilizing effect due to conjugation is apparent from the 3.9 kcal/mole difference between 1,3-cyclohexadiene and 1,4-cyclohexadiene. Benzene, however, shows a remarkably small heat of hydrogenation, which corresponds to a stabilization energy of roughly 36 kcal/mole (85.8 − 49.8). From the magnitude of this effect we see that the addition of the first molecule of hydrogen to benzene must be *endothermic*. Clearly, the small conjugative stabilization we have noted for dienes is quite different from this extraordinary decrease in the heat content of benzene, which is reflected in its chemical stability.

X-ray crystallographic analysis of benzene discloses a *planar regular hexagonal structure*, as illustrated in Figure 9.6. The carbon-carbon bonds are all of the same length, which is slightly greater than that of an isolated double bond. We may argue whether this evidence supports Kekulé's suggestion of equilibrating cyclohexatrienes. Of course, a

9·3 THEORETICAL MODELS FOR BENZENE

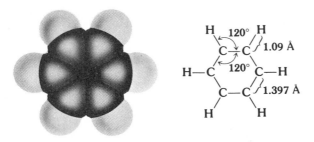

FIGURE 9·6 *The configuration of benzene*

static 1,3,5-cyclohexatriene molecule would be expected to have alternating long and short (that is, single and double) bonds.

Benzene is colorless, but its ultraviolet spectrum shows moderately strong absorption extending to 280 nm, with maximum absorption at 257 nm. Although this is also in agreement with the conjugated cyclohexatriene formulation, it does not clarify or explain the exceptional thermodynamic and chemical stability of benzene.

PROBLEM 9·5
Draw an energy diagram illustrating the changes in ΔH during the hydrogenation reactions listed in Table 9.3.

PROBLEM 9·6
Calculate ΔH for the following reaction: *Hint:* Use the data in Table 9.3.

$$\text{cyclohexene} \xrightarrow[\Delta]{\text{Pd}} \text{benzene} + H_2$$

9·3 THEORETICAL MODELS FOR BENZENE

THE MOLECULAR-ORBITAL MODEL

The shortcomings of Kekulé's description of benzene become apparent when we examine a molecular-orbital model of this hydrocarbon. Of the 30 valence electrons in benzene, 24 are accommodated in a hexagonal σ-bonded framework of six sp^2-hybridized carbon atoms and six hydrogen atoms. The six remaining unoccupied $2p$ atomic orbitals (one on each carbon atom) overlap when the ring is planar, giving six π molecular orbitals by an LCAO treatment, as shown in Figure 9.7. Three of these six molecular orbitals are bonding (π_1, π_2, and π_3) and three are antibonding (π_4^*, π_5^* and π_6^*). By placing the six remaining valence electrons in the three *bonding π orbitals* (two in each orbital), we create a plausible model of the benzene molecule. This model, in addition to possessing the structural features of the Kekulé proposal, lends itself to simple molecular-orbital calculations which account moderately well for the remarkable thermodynamic stability of benzene. However, the most

9·3 THEORETICAL MODELS FOR BENZENE 255

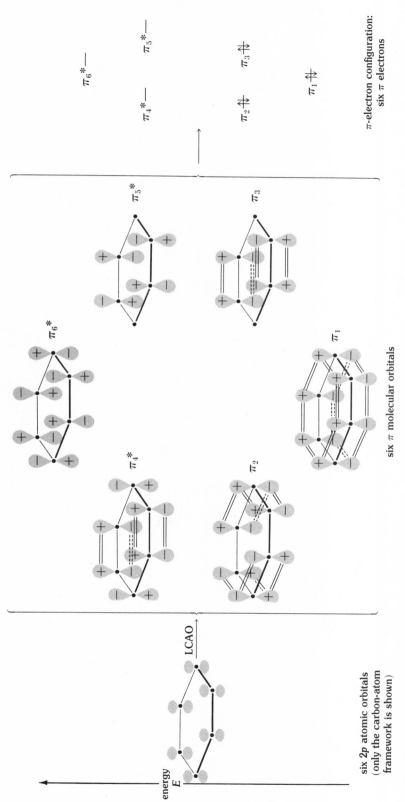

FIGURE 9·7 *A molecular-orbital description of benzene*

delocalized-bond representation

cyclohexatriene representation

FIGURE 9·8 *Two representations of a benzene ring*

important and significant change introduced in this approach is that it abandons Kekulé's equilibrium concept and treats benzene as a single discrete molecular species.

Two structural notations are commonly used to represent benzene (see Figure 9.8). The form containing a circle emphasizes the molecular-orbital viewpoint, whereas the classical cyclohexatriene representation is probably more widely used.† In the next section we shall consider the merits of these symbolic formulas in greater detail, but regardless of which is used, it is important to remember that they both represent the same compound.

THE RESONANCE MODEL

The localized bond notation (a line between two atoms) employed in writing structural formulas proved to be so useful during the early development of organic chemistry that it is now firmly ingrained in our thinking. Difficulties arise, however, when we must write formulas for certain kinds of unsaturated molecules and ions, such as $NO_3^{\ominus}$, CO, and benzene. In the case of benzene, for example, the cyclohexatriene notation is not strictly correct and could be misleading, while the delocalized-bond notation requires the adoption of a new symbol.

The theory of *resonance* has provided a useful method of describing molecules such as benzene by regarding them as hybrids of two or more classical structures (structures with localized bonds). This refinement of the traditional bonding description came originally from a mathematical treatment or orbital wave equations which was related to the mathematics of resonating coupled harmonic oscillators such as tuning forks and pendulums. As used by organic chemists, the concept of resonance has become a completely qualitative descriptive device. Thus we describe benzene as a *resonance hybrid* of two cyclohexatrienes:

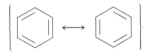

In using resonance to describe chemical compounds, we must observe the following rules:

1 Two or more contributing structures (not necessarily of real or possible compounds) comprise a hybrid structure which approximates the actual molecule. These contributors are connected by a characteristic

†To avoid possible confusion, cyclohexane is sometimes written with an *S* (saturated) as ⬡ S

double-headed arrow ⟷. This double-headed arrow is not to be confused with the double arrows ⇌ used to show equilibrium.

$$NO_3^{\ominus} \equiv \left\{ \begin{array}{c} \overset{\overset{\cdot\cdot}{O}}{\underset{\overset{\cdot\cdot}{O}^{\ominus}}{\overset{\|}{N}^{\oplus}}}-\overset{\cdot\cdot}{O}^{\ominus} \longleftrightarrow \overset{\overset{\cdot\cdot}{O}^{\ominus}}{\underset{\overset{\cdot\cdot}{O}^{\ominus}}{\overset{|}{N}^{\oplus}}}=\overset{\cdot\cdot}{O} \longleftrightarrow \overset{\overset{\cdot\cdot}{O}^{\ominus}}{\underset{\overset{\cdot\cdot}{O}^{\ominus}}{\overset{|}{N}^{\oplus}}}=\overset{\cdot\cdot}{O} \end{array} \right\}$$

2 The relative positions of atomic nuclei must not change from one contributing structure to another. In other words, resonance is not an equilibrium involving movement or shifts of atoms; the resonance hybrid is a single discrete molecular species, and only the electrons change location in the contributors that make up the hybrid. Thus the contributors used to describe benzene are not classical 1,3,5-cyclohexatrienes with alternating single and double bonds; they are *hypothetical* trienes having the symmetrical hexagonal configuration shown in Figure 9.6.

3 Resonance is a stabilizing factor. Molecules which are best described as resonance hybrids generally have high thermodynamic stability as a consequence of the delocalization of bonding electrons. The term *resonance energy* is used synonymously with *stabilization energy*, and is defined as the difference between the potential energy calculated for the most stable of the hypothetical contributing structures and the potential energy of the molecule itself.

4 The greatest stability occurs in those resonance hybrids having energetically equivalent contributors. For example, benzene is more stable than 1,3-pentadiene:

{ ⬡ ⟷ ⬡ } E_{res} = 36 kcal/mole
benzene

Contributors are structurally and energetically equivalent.
Resonance energy is large.
Benzene is *not* adequately described by any single contributor.

$$\left\{ \begin{array}{c} CH_3-CH=CH-CH=CH_2 \\ \updownarrow \\ CH_3-\overset{\oplus}{C}H-CH=CH-\overset{\cdot\cdot}{C}H_2^{\ominus} \\ \updownarrow \\ CH_3-\overset{\ominus}{C}H-CH=CH-\overset{\oplus}{C}H_2 \end{array} \right\} E_{res} \approx 4 \text{ kcal/mole}$$

1,3-pentadiene

Contributors are energetically nonequivalent.
Resonance energy is small.
1,3-Pentadiene is adequately described by the lowest-energy contributor ($CH_3CH=CH-CH=CH_2$).

There are some useful corollaries to this rule:
a Contributors having greatly distorted bond angles or lengths will not usually significantly influence the hybrid structure. For example, the

cyclohexatriene contributors to benzene are more important than any of the following contributors:

b Contributors having fewer covalent bonds (lower total bond energies) than conventional localized bond structures will be less important in a resonance description. For example, the ionic contributors to 1,3-pentadiene (shown above) have one less carbon-carbon π bond than the conventional structure and are therefore relatively unimportant.

c Contributors exhibiting charge separation will exert their maximum influence if the charge distribution agrees with that predicted from electronegativity differences:

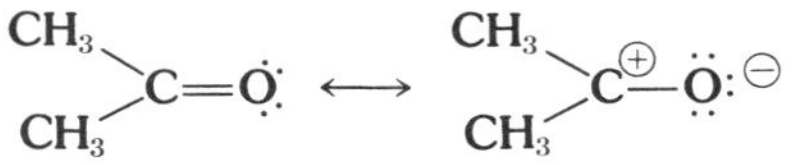

$$\mathrm{CH_3}\!\!>\!\!\mathrm{C}\!=\!\ddot{\mathrm{O}}\!: \quad \longleftrightarrow \quad \mathrm{CH_3}\!\!>\!\!\overset{\oplus}{\mathrm{C}}\!-\!\ddot{\ddot{\mathrm{O}}}\!:^{\ominus}$$

is preferred over

$$\mathrm{CH_3}\!\!>\!\!\mathrm{C}\!=\!\ddot{\mathrm{O}}\!: \quad \longleftrightarrow \quad \mathrm{CH_3}\!\!>\!\!\overset{\ominus}{\mathrm{C}}\!-\!\dot{\ddot{\mathrm{O}}}\!:^{\oplus}$$

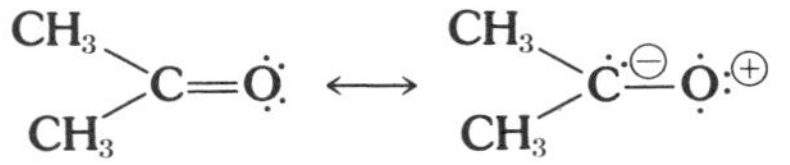

5 The number of *unpaired* electrons must remain the same in all contributors:

$$\underset{\text{one unpaired electron}}{\dot{\mathrm{C}}\mathrm{H}_2\!-\!\mathrm{CH}\!=\!\mathrm{CH}_2} \quad \longleftrightarrow \quad \underset{\text{one unpaired electron}}{\mathrm{CH}_2\!=\!\mathrm{CH}\!-\!\dot{\mathrm{C}}\mathrm{H}_2}$$

The broad descriptive powers of the resonance approach will become increasingly apparent as we proceed with our study of organic chemistry. However, the following analogy should help to emphasize the underlying concept, and also to call attention to the most common pitfall in the use of resonance terminology. Suppose you have a mongrel dog you wish to describe to a friend. You can, of course, tell your friend the dog's height, weight, length, color, and other measurements. If your friend is familiar with dogs, however, you can communicate more quickly and effectively by telling him that your dog is part beagle and part cocker spaniel. You would naturally assume that your friend understands you are talking about a perfectly normal dog, and that he would not make the mistake of thinking that the animal is changing back and forth from a beagle to a cocker spaniel.† Using our chemical notation, you might write

dog = {beagle $\longleftrightarrow$ cocker spaniel}

but you would not write

dog = beagle $\rightleftharpoons$ cocker spaniel

---

†This is not a completely faithful analogy, since the contributing structures for a molecular description are not real molecules, whereas the contributing species beagles and cocker spaniels are capable of independent existence.

9·4 NOMENCLATURE AND PROPERTIES OF AROMATIC COMPOUNDS

PROBLEM 9·7

Which of the following pairs of structures satisfy the requirements governing contributors to a resonance hybrid? Introduce the appropriate arrow symbol ⟷ or ⇌ between each pair of structures (note that catalysts may be necessary for equilibrium).

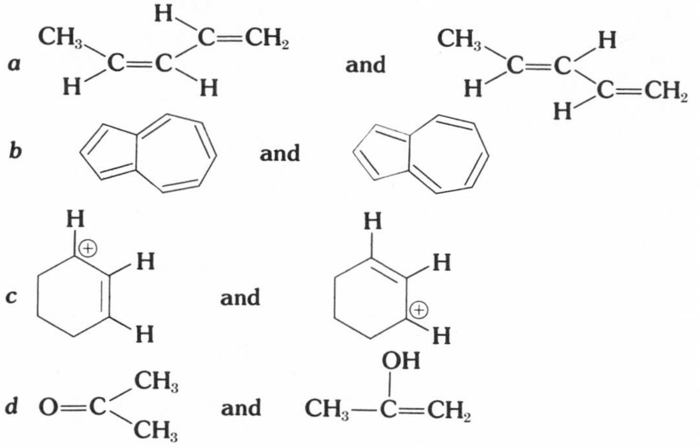

a) $\text{CH}_3\text{-C=C-H}$ with H-C=CH_2 (H on top) and $\text{CH}_3\text{-C=C-H}$ with H-C=CH_2

b) azulene and isomeric bicyclic structure

c) cyclohexenyl cation (H⁺ on one carbon) and cyclohexenyl cation (H⁺ on adjacent carbon)

d) $\text{O=C(CH}_3\text{)}_2$ and $\text{CH}_3\text{-C(OH)=CH}_2$

PROBLEM 9·8

Write contributing structures for a resonance description of (*a*) carbonate anion CO_3^{2-} and (*b*) carbon monoxide CO.

PROBLEM 9·9

Propylene reacts with chlorine at 300°C to give allyl chloride. What other isomeric substitution products are theoretically possible from this reaction? Write radical-chain mechanisms accounting for each of these products and suggest a reason for the exclusive formation of allyl chloride in this reaction.

9·4 NOMENCLATURE AND PROPERTIES OF AROMATIC COMPOUNDS

COMPOUNDS RELATED TO BENZENE

An enormous variety of organic substances include benzenelike six-membered rings. Many of these compounds can be obtained by the destructive distillation of coal, and the others are readily prepared by substitution reactions that will be discussed later in this chapter.

Aromatic hydrocarbons are often given common names or are named according to their alkyl substituents:

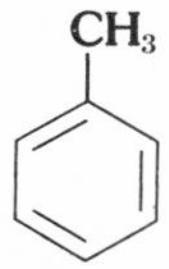

methylbenzene or toluene (CH_3 substituent)

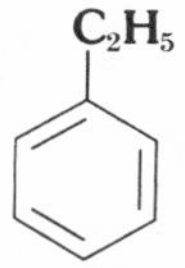

ethylbenzene (C_2H_5 substituent)

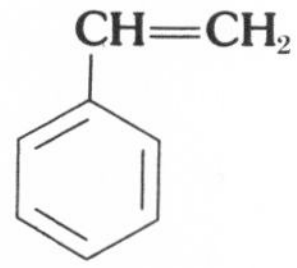

styrene ($CH=CH_2$ substituent)

The prefixes *ortho-*, *meta-*, and *para-* (abbreviated *o-*, *m-*, and *p-*) are used to specify disubstituted benzene derivatives. *Ortho* substituents have a neighboring (1,2) relationship, *meta* refers to 1,3 substituents, and 1,4 substituents are termed *para*.

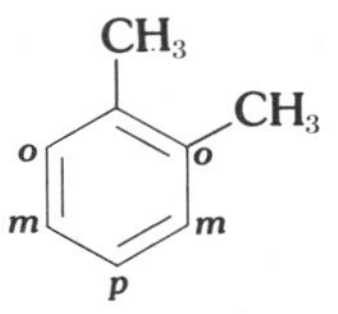

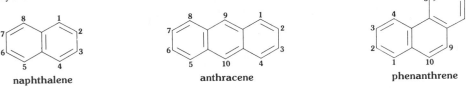

ortho-xylene 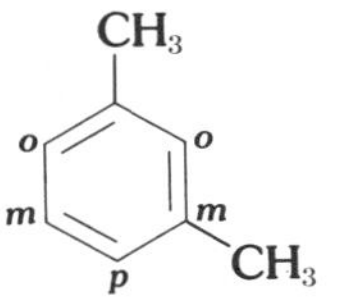   meta-xylene 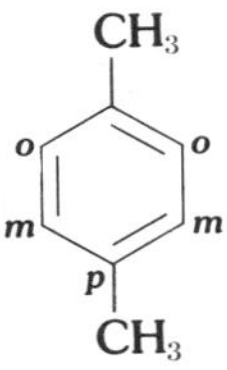   para-xylene

Two common aromatic groups are given special names, and compounds that incorporate these groups are named accordingly:

$C_6H_5-\ \equiv\ \phi-\ \equiv$ ⟨phenyl⟩— $C_6H_5CH_2-\ \equiv\ \phi CH_2-\ \equiv$ ⟨phenyl⟩—CH_2—

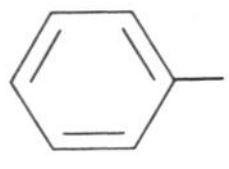

phenyl group   benzyl group

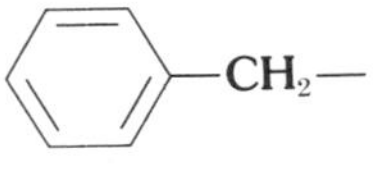

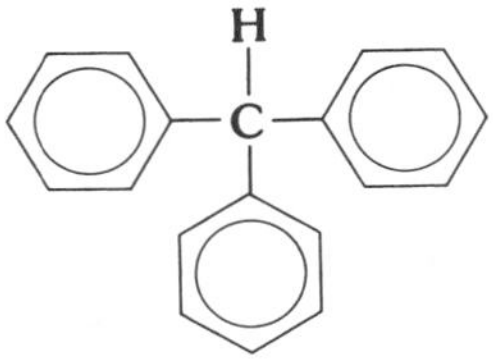

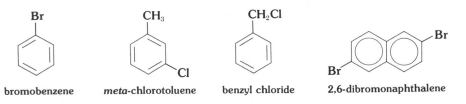

triphenylmethane 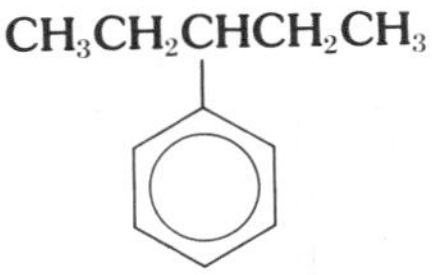   3-phenylpentane 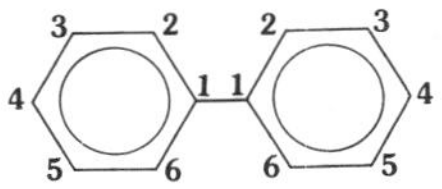   biphenyl

The fusion of two or more benzene rings creates other aromatic ring systems:

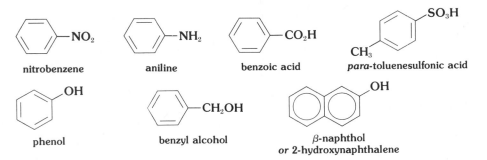

naphthalene   anthracene   phenanthrene

Halogen derivatives of benzene are named on the same basis:

bromobenzene   meta-chlorotoluene 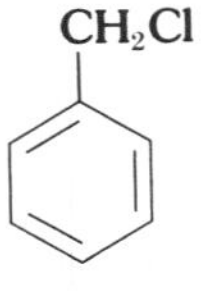   benzyl chloride   2,6-dibromonaphthalene

Examples of other functional derivatives are:

nitrobenzene   aniline   benzoic acid   para-toluenesulfonic acid

phenol 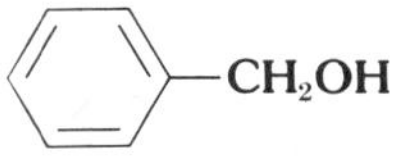   benzyl alcohol   β-naphthol *or* 2-hydroxynaphthalene

9·4 NOMENCLATURE AND PROPERTIES OF AROMATIC COMPOUNDS   261

PROBLEM 9·10
How many dichloro isomers are there for each of the following?
a para-Xylene
b Naphthalene
c Biphenyl

PROPERTIES OF AROMATIC HYDROCARBONS

Aromatic hydrocarbons are water-insoluble liquids or solids, having characteristic, and in some cases pleasant, odors. As noted earlier, benzene itself has toxic qualities, and some of the fused-ring systems are highly carcinogenic:

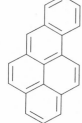
benz[a]pyrene

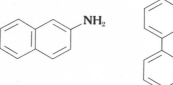

7,12-dimethylbenz[a]anthracene   β-naphthylamine   4-aminobiphenyl

In general, aromatic compounds are detected and identified by their characteristic spectroscopic properties.

ULTRAVIOLET-VISIBLE SPECTROSCOPY   Benzene and the alkyl benzenes are characterized by two electronic absorption bands in the ultraviolet-visible range, one near 200 nm and the other near 260 nm. The moderately strong long-wavelength band is shifted to even longer wavelengths (a bathochromic shift) as the number of alkyl substituents attached to the ring increases, or as additional unsaturated functions are conjugated with the ring (Figure 9.9). Fused-ring aromatic hydrocarbons show similar bathochromic shifts, even to the extent of becoming colored when the conjugation is extensive. The very strong short-wavelength band also experiences a bathochromic shift in conjugated systems. For example, styrene has a band at $\lambda_{max} = 244$ nm which is six times more intense than the 282-nm absorption.

FIGURE 9·9 *Long-wavelength electronic absorption maxima for some aromatic hydrocarbons*

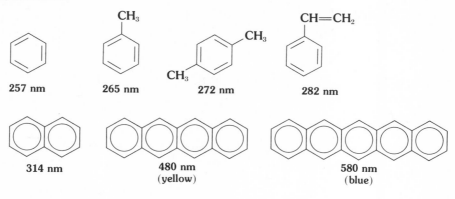

262   9·4 NOMENCLATURE AND PROPERTIES OF AROMATIC COMPOUNDS

INFRARED SPECTROSCOPY   In infrared spectra aromatic C—H stretching vibrations appear as sharp bands in the 3000 to 3100 cm^{-1} region. Other absorption bands between 1650 and 2000 cm^{-1}, between 1400 and 1600 cm^{-1}, and below 1000 cm^{-1} are associated with the number and positions of ring substituents and are widely used for structural analysis.† The infrared spectra of benzene and p-cymene, illustrated in Figure 9.10, are typical of aromatic compounds.

PMR SPECTROSCOPY   Hydrogen atoms bonded directly to a benzene ring give rise to characteristic resonance signals, with chemical shifts about 2 ppm lower field than the hydrogen atoms attached to the double bond of an alkene. In toluene, for example, the aromatic protons produce a sharp signal at $\delta = 7.2$ and the methyl protons appear as an equally sharp signal at $\delta = 2.3$ (about 0.5 ppm lower field than most allylic methyl protons). Other examples are shown in Figure 9.11. Note that structurally nonequivalent aromatic protons may display either separate signals, as in $para$-methylanisole, or a single signal, as in toluene and $para$-cymene.

†These assignments are catalogued in most reference texts dealing with molecular spectroscopy.

FIGURE 9·10   *Infrared spectra of benzene and p-cymene*

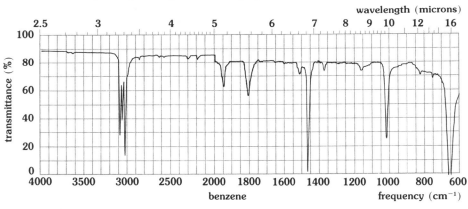

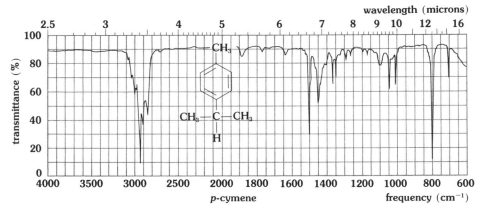

9·4 NOMENCLATURE AND PROPERTIES OF AROMATIC COMPOUNDS

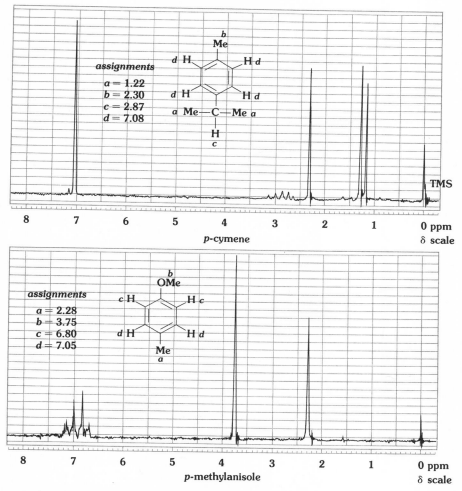

FIGURE 9·11 *Pmr spectra of p-cymene and p-methylanisole*

PROBLEM 9·11

a Write structural formulas for and show the structurally nonequivalent protons in each of the isomeric xylenes.

b Sketch the pmr spectrum you would expect from each of the xylenes if the aromatic protons all had the same chemical shift.

The unusually low-field resonance signals of aromatic protons can be explained by the deshielding effect of a ring current generated by the applied magnetic field H_0. As illustrated in Figure 9.12, this ring current induces a secondary magnetic field in each molecule such that the induced field *opposes* the large applied field in the space immediately above or below the ring and *reinforces* the applied field at the periphery of the ring. Induced ring currents are strongest in cyclic π-electron systems and are generally regarded as relatively unimportant in saturated rings such as cyclopentane and cyclohexane. Cyclopropane, because of its unsaturated character, is a notable exception to this rule. Instead of a

9·4 NOMENCLATURE AND PROPERTIES OF AROMATIC COMPOUNDS

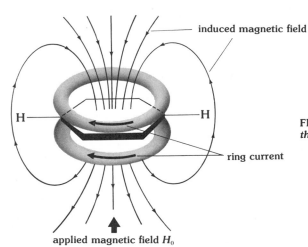

FIGURE 9·12 *Deshielding of aromatic protons by the magnetic field induced by a ring current*

deshielding effect, cyclopropyl protons experience a shielding effect of roughly 1 ppm (the chemical shift of the protons in cyclopropane is found to be $\delta = 0.22$). Cyclopropyl hydrogen atoms lie above and below the plane containing the three carbon atoms; consequently, the ring-current effect acts to shield the protons from the applied field. The wide variation in chemical shifts that has been observed in hydrocarbons is summarized in Table 9.4.

TABLE 9·4 *Chemical shifts for hydrocarbons*

type of hydrocarbon	type of hydrogen	δ value (ppm)
alkanes and cycloalkanes	R—CH$_3$	0.8–1.0
	R—CH$_2$—R	1.2–1.4
	R$_3$CH	1.4–1.6
cyclopropanes	cyclopropane (H, H$_2$, R, H$_2$)	0.2–0.9
alkenes	R$_2$C=CRH	4.5–5.5
	R$_2$C=C(R)—CH$_3$	1.6–1.8
alkynes	R—C≡C—H	2.0–2.8
	R—C≡C—CH$_3$	1.7–2.0
benzene and derivatives	C$_6$H$_5$—H	7.3
	C$_6$H$_5$—CH$_3$	2.3

9·5 CONFORMATIONAL ENANTIOMERISM IN BIPHENYLS

PROBLEM 9·12

Two hydrocarbons, both with molecular formula C_9H_{12}, show the following pmr signals. Write a structural formula for each compound.

a $\delta = 1.25$ (doublet, $J = 7$ Hz), $\delta = 2.95$ (septet, $J = 7$ Hz), and $\delta = 7.25$ (singlet), with relative areas of 6:1:5

b $\delta = 2.25$ (singlet) and $\delta = 6.78$ (singlet), with relative areas of 3:1

PROBLEM 9·13

Two hydrocarbons with the molecular formula C_9H_{10} show the following pmr signals. Write structural formulas for these compounds.

a $\delta = 2.04$ (quintet, $J = 8$ Hz), $\delta = 2.91$ (triplet, $J = 8$ Hz), and $\delta = 7.17$ (singlet), with relative areas of 1:2:2

b $\delta = 0.5$–0.9 (multiplet), $\delta = 1.6$–2.0 (multiplet), and $\delta = 7.08$ (multiplet), with relative areas of 4:1:5

9·5 CONFORMATIONAL ENANTIOMERISM IN BIPHENYLS

Several unique and interesting examples of conformational isomerism have been discovered in the course of studies dealing with aromatic compounds.

Biphenyl derivatives having bulky substituents in the ortho positions are more stable in twisted conformations than in the planar conformation, which suffers serious nonbonded compressions from the juxtaposed substituents.† The loss of conjugation in the twist conformations

†Although biphenyl itself is slightly twisted, the angle of twist is so small that conjugation between the rings is not affected ($\lambda_{max} = 250$ nm, $\epsilon = 18{,}000$).

FIGURE 9·13 *Potential-energy change for rotation about the central bond in substituted biphenyls*

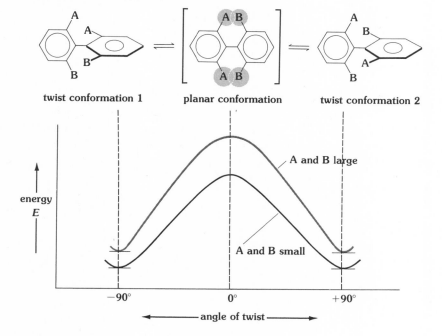

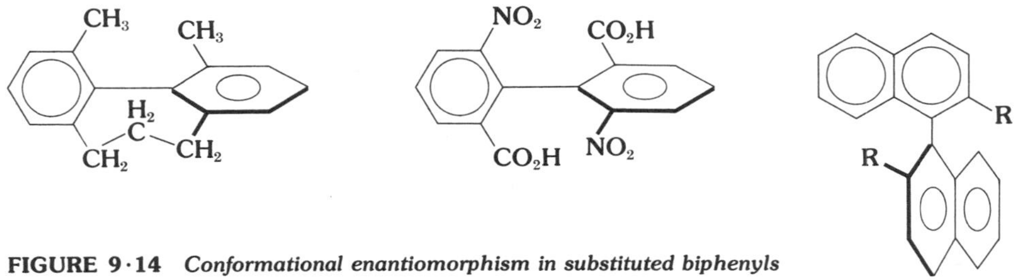

9·5 CONFORMATIONAL ENANTIOMERISM IN BIPHENYLS

FIGURE 9·14 *Conformational enantiomorphism in substituted biphenyls*

($R = H$ or CH_3)

provides one way of detecting this effect. Thus 3,3'-dimethylbiphenyl shows a very intense absorption band at 255 nm in the ultraviolet spectrum which fades to only 5% of this intensity in the isomeric 2,2'-dimethylbiphenyl ($A = CH_3$ and $B = H$) in Figure 9.13.

Even more convincing evidence of stable twist conformations in these molecules comes from the resolution of *unsymmetrically* substituted biphenyls. If the twist conformations have two different ortho substituents (**A** and **B** in Figure 9.13), they are enantiomeric—that is, they are nonsuperimposable mirror images. Furthermore, if these substituents are sufficiently bulky, the barrier to interconversion becomes large enough to prevent the two enantiomers from rapidly establishing equilibrium at normal temperatures. In this event they can be *resolved*, or separated, by the general method described in Section 7.4. For example, the substituted biphenyl compounds in Figure 9.14 have been resolved. As expected, the optical stability of these enantiomers is proportional to the bulk of the ortho substituents (biphenyls with smaller substituents racemize more rapidly):

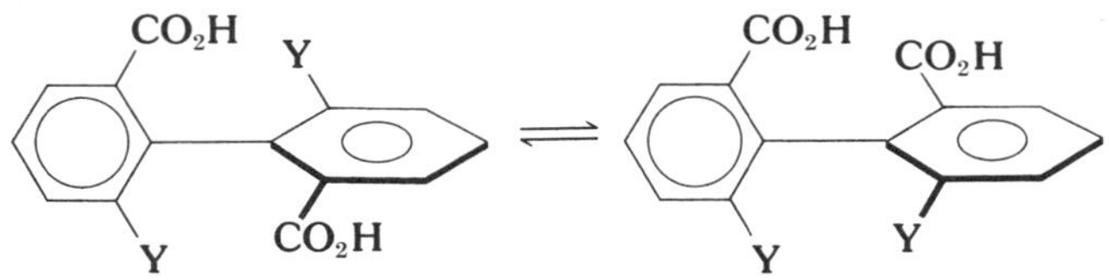

(9.8)

rate of racemization: $Y = F > OCH_3 > NO_2$

It is important to remember that the optical activity observed in such biphenyls is due to *molecular chirality* and is not an intrinsic property of these twisted molecules. For example, if the substituent pattern on either ring were symmetrical, the resulting twisted biphenyl would have a molecular plane of symmetry and would not exhibit enantiomorphism:

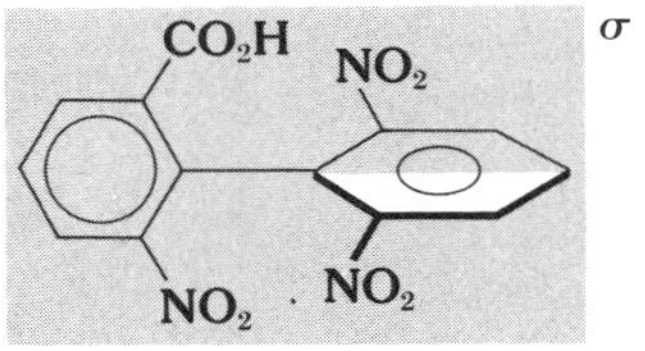

twisted biphenyl with
a plane of symmetry

9·5 CONFORMATIONAL ENANTIOMERISM IN BIPHENYLS 267

PROBLEM 9·14

Which of the 1,1'-binaphthyls in Figure 9.14 would be expected to racemize more rapidly, R = H or R = CH$_3$?

PROBLEM 9·15

X is an optically active biphenyl derivative isolated from tannin. On standing, it is slowly transformed into optically inactive Y.

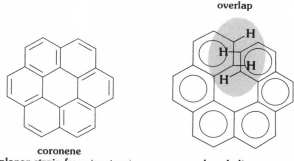

(9.9)

Why is Y optically inactive? If X is reformed by hydrolysis of Y, will it be optically active? Explain.

One of the most remarkable examples of conformational enantiomorphism yet discovered is the aromatic hydrocarbon hexahelicene. If we compare the structure of hexahelicene with that of the strain-free planar hydrocarbon coronene, we see that portions of the end rings in hexahelicene are forced to overlap in space:

To accommodate the overlapping atoms, the hexahelicene molecules adopt a helical conformation, as shown in Figure 9.15, and this mixture of chiral molecules can be resolved by appropriate laboratory methods. A sample of optically pure hexahelicene has an optical rotation of $[\alpha]_D^{25°} = 3700°$.

FIGURE 9·15 *Helical conformations of hexahelicene*

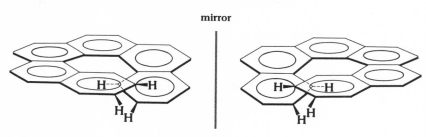

9·6 ELECTROPHILIC SUBSTITUTION REACTIONS OF BENZENE

The chemical reactions of benzene are predominantly of the substitution type, as illustrated by equations 9.2 to 9.5. In view of the large resonance stabilization of the benzene ring (about 36 kcal/mole), it comes as no surprise that products retaining this feature — substitution products — are favored. Nevertheless, large as this stabilization energy is, it is smaller than the C—H σ-bonding energy (about 103 kcal/mole for benzene). We must therefore consider the possibility that aromatic conjugation may be temporarily interrupted during the course of these reactions. In this section we shall examine a general mechanism for aromatic substitution reactions which will prove helpful in later discussions of more complex reaction systems.

ELECTROPHILIC ATTACK ON THE BENZENE RING

Four of the most common benzene substitution reactions are outlined in Figure 9.16, and a convincing body of evidence has accumulated that these reactions proceed by an initial electrophilic attack on the aromatic ring.

HALOGENATION Benzene halogenation reactions show the same order of reactivity ($F_2 \gg Cl_2 > Br_2 \gg I_2$) as the electrophilic-halogen addition reactions to alkenes, discussed in Chapter 5. Since benzene is less reactive than most alkenes, it is necessary to catalyze the chlorination and bromination reactions with Lewis acids such as FeX_3 and AlX_3 (where X = Cl or Br).† This requirement makes sense only if a strongly electrophilic species is needed for an attack on the aromatic ring:

$$X_2 + FeX_3 \rightleftharpoons X-X^{\oplus}-FeX_3^{\ominus} \rightleftharpoons X^{\oplus} \; FeX_4^{\ominus} \quad (9.9)$$
$$\text{weak} \qquad\qquad \text{strong} \qquad\qquad \text{stronger}$$
$$\text{electrophile} \qquad \text{electrophile} \qquad \text{electrophile}$$

NITRATION Aromatic hydrocarbons are efficiently nitrated by treatment with a 1:2 mixture of concentrated nitric and sulfuric acids. The active electrophilic species in these mixed acids has been identified as the nitronium ion $NO_2^{\oplus}$, formed by the following reaction:

$$HNO_3 + 2H_2SO_4 \rightleftharpoons NO_2^{\oplus} + 2HSO_4^{\ominus} + H_3O^{\oplus} \quad (9.10)$$

Salts of the nitronium cation can be isolated and have been shown to be powerful nitrating agents:

$$NO_2^{\oplus} \; BF_4^{\ominus}$$
nitronium tetrafluoroborate

† Fluorine reacts vigorously with benzene in the absence of catalysts, while iodine is unreactive under conditions that give complete reaction with chlorine and bromine.

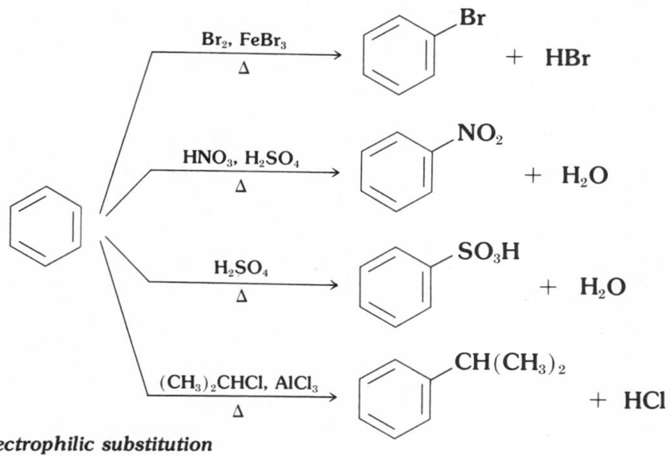

9·6 ELECTROPHILIC SUBSTITUTION REACTION OF BENZENE

[Figure 9·16: Benzene reacting via four pathways:
- Br$_2$, FeBr$_3$, Δ → C$_6$H$_5$Br + HBr
- HNO$_3$, H$_2$SO$_4$, Δ → C$_6$H$_5$NO$_2$ + H$_2$O
- H$_2$SO$_4$, Δ → C$_6$H$_5$SO$_3$H + H$_2$O
- (CH$_3$)$_2$CHCl, AlCl$_3$, Δ → C$_6$H$_5$CH(CH$_3$)$_2$ + HCl]

FIGURE 9·16 *Electrophilic substitution reactions of benzene*

SULFONATION Sulfur trioxide is believed to be the active electrophilic agent in sulfonation reactions, since fuming sulfuric acid (H$_2$SO$_4$·SO$_3$) accomplishes the sulfonation of benzene more rapidly and at lower temperatures than does concentrated sulfuric acid.

PROBLEM 9·16
Explain the electrophilic nature of SO$_3$.

FRIEDEL-CRAFTS ALKYLATION Although alkyl halides are only weakly electrophilic, Lewis acid catalysts act to generate strongly electrophilic species related to carbonium ions:

$$\text{R—X} + \text{AlX}_3 \rightleftharpoons \overset{\delta\oplus}{\text{R}}\text{---X---}\overset{\delta\ominus}{\text{AlX}_3} \rightleftharpoons \text{R}^\oplus \text{ AlX}_4^\ominus \quad (9.11)$$

weak electrophile — strong electrophile — very strong electrophile

The use of a mixture of alkyl halides and aluminum trihalide for the alkylation of benzene and related aromatic hydrocarbons is called *Friedel-Crafts alkylation*, in honor of Charles Friedel (French) and James Crafts (American), who collaborated in the initial investigations.

If carbonium ions are generated by other methods, a similar alkylation of benzene can be effected:

$$(\text{CH}_3)_2\text{C}=\text{CH}_2 \xrightarrow{\text{H}_3\text{PO}_4} [(\text{CH}_3)_3\text{C}^\oplus \text{H}_2\text{PO}_4^\ominus] \xrightarrow{\text{C}_6\text{H}_6} \text{C}_6\text{H}_5\text{C}(\text{CH}_3)_3 \quad (9.12)$$

Because of the carbonium-ion-like character of the alkylating species in the Friedel-Crafts reaction, molecular rearrangements are sometimes observed:

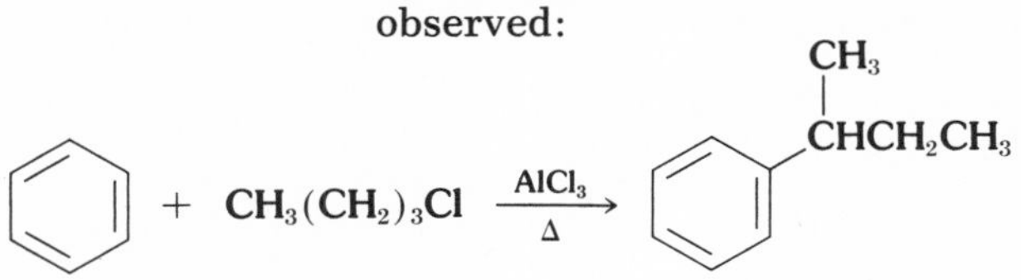

$$\text{C}_6\text{H}_6 + \text{CH}_3(\text{CH}_2)_3\text{Cl} \xrightarrow[\Delta]{\text{AlCl}_3} \text{C}_6\text{H}_5\text{CH}(\text{CH}_3)\text{CH}_2\text{CH}_3 \quad (9.13)$$

9·6 ELECTROPHILIC SUBSTITUTION REACTION OF BENZENE

A MECHANISM FOR AROMATIC SUBSTITUTION

As we saw in Chapter 3, there are three fundamentally different mechanisms by which substitution reactions can proceed:

1 A two-step mechanism in which bond breaking of the leaving group precedes bond making to the entering group

2 A two-step mechanism in which bond making precedes bond breaking

3 A single-step mechanism in which bond breaking and bond making are approximately concerted.

Now, it is unlikely that benzene substitution reactions proceed by the first mechanism, since a strong and relatively nonpolar bond (C—H) must be broken in the initial step. Indeed, the resistance of the strong C—H bonds in benzene to direct hydrogen abstraction may be seen in the fact that, in contrast to the chlorination reactions of alkanes, photochemical chlorination of benzene gives *addition* rather than substitution products:

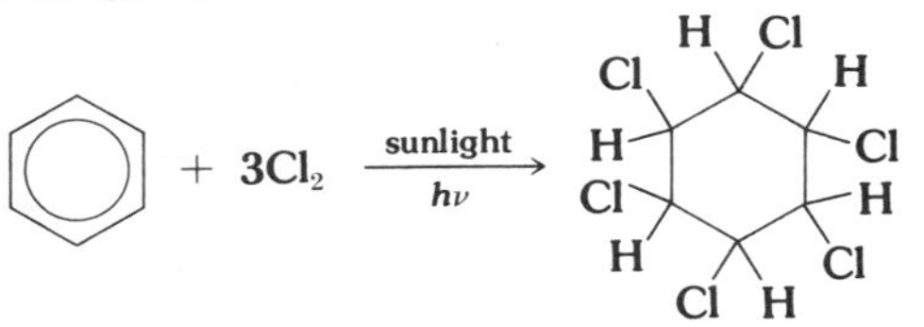

$$\text{C}_6\text{H}_6 + 3\text{Cl}_2 \xrightarrow[h\nu]{\text{sunlight}} \text{C}_6\text{H}_6\text{Cl}_6 \quad (9.14)$$

Furthermore, such a mechanism would not clearly account for the role of the electrophilic reagent in these reactions.

Of the remaining two possibilities, mechanism 2 seems to be the more plausible, because the π-electron clouds above and below the plane of the benzene ring should be susceptible to electrophilic attack in much the same fashion as the π bond in alkenes (see Chapter 5). We can therefore propose a general mechanism for electrophilic aromatic substitution reactions which involves a temporary disruption of the aromatic π-electron system (Figure 9.17).

Since the first step in this mechanism destroys the aromatic conjugation of the benzene ring, we expect this to be the slowest or rate-determining step, as shown in Figure 9.18.† To conserve space, the three contributing formulas in the resonance description of the reactive inter-

†The full 36 kcal/mole of resonance energy is not lost, since significant conjugative charge delocalization still exists in the intermediate ion.

FIGURE 9·17 *A mechanism for electrophilic substitution of benzene*

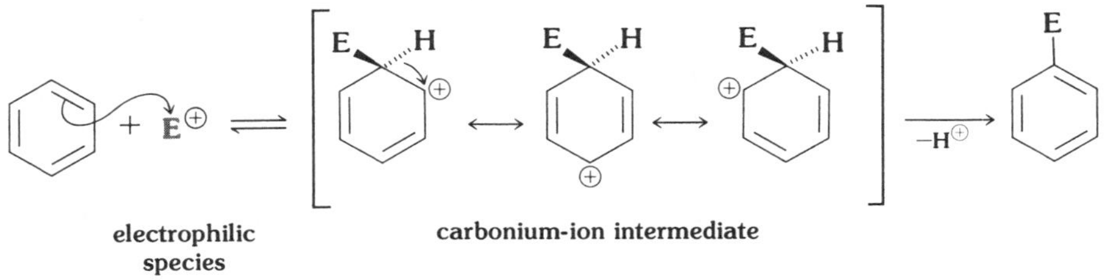

electrophilic species   carbonium-ion intermediate

9·6 ELECTROPHILIC SUBSTITUTION REACTION OF BENZENE 271

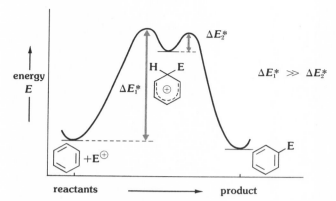

FIGURE 9·18 *Energy diagram for electrophilic aromatic substitution*

mediate are represented by a single partial-bond formula which emphasizes the charge delocalization in this ion:

It is helpful at this point to compare our mechanism for electrophilic aromatic substitution in Figure 9.17 with the mechanism proposed earlier to account for Markovnikov's rule in alkene addition reactions (Figures 5.1 and 5.2). In both cases a π-electron system is attacked by an electrophile to give a carbonium-ion intermediate. This intermediate usually reacts by combining with a nucleophile (an overall addition reaction), but in the case of an aromatic π-electron system, a proton loss from the intermediate regenerates the aromatic ring and thus offers an alternative low-activation-energy reaction path (an overall substitution reaction). The exceptional stability of the benzene ring therefore diverts the course of the reaction at some point after the rate-determining step.

PROBLEM 9·17
Write mechanisms showing all the steps for (*a*) the nitration of benzene and (*b*) the sulfonation of *p*-xylene.

If the mechanism proposed here is essentially correct, then substituents which increase or decrease the nucleophilicity of the aromatic ring should change the activation energy of the first step (ΔE_1^*) in a predictable way (the Hammond postulate in Section 3.6). Alkyl substituents, for example, will act to stabilize the cationic intermediate, thus decreasing the activation energy for electrophilic attack, as evidenced by the fact that 1,3,5-trimethylbenzene (mesitylene) reacts rapidly with bromine in the *absence* of a Lewis acid catalyst:

$$\text{mesitylene} + Br_2 \xrightarrow{10°} \text{bromomesitylene} + HBr \tag{9.15}$$

9·6 ELECTROPHILIC SUBSTITUTION REACTION OF BENZENE

In contrast to the effect of alkyl substituents, nitro groups act to decrease the nucleophilicity of the aromatic ring, making it much less reactive to electrophilic substitution than benzene (ΔE_i^* is higher for nitrobenzene):

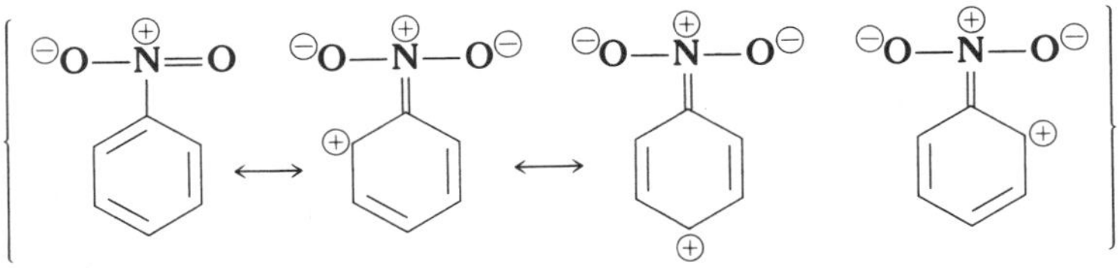

[resonance structures of nitrobenzene σ-complex]

In fact, nitrobenzene is often used as an inert solvent for Friedel-Crafts alkylation reactions.

PROBLEM 9·18
a Draw the resonance contributors to the reactive intermediate formed in the bromination of mesitylene (equation 9.15). Is this intermediate more or less stable than the corresponding intermediate formed in the bromination of benzene? Why?
b Draw an energy diagram showing the relative activation energies (ΔE_i^*) for the bromination of mesitylene and benzene.
c Would you expect mesitylene to be nitrated more or less rapidly than benzene?

PROBLEM 9·19
When benzene is treated with deuterosulfuric acid, deuterated benzene derivatives are formed much more rapidly than sulfonation occurs. Write a mechanism for this hydrogen-exchange reaction:

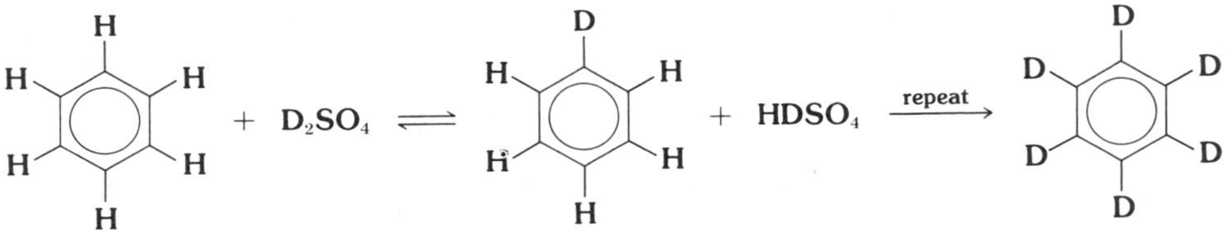

[benzene + D_2SO_4 ⇌ monodeuteriobenzene + $HDSO_4$ → repeat → hexadeuteriobenzene]

PROBLEM 9·20
Sulfonation is reversible at high temperatures. Write a mechanism for the following reaction.

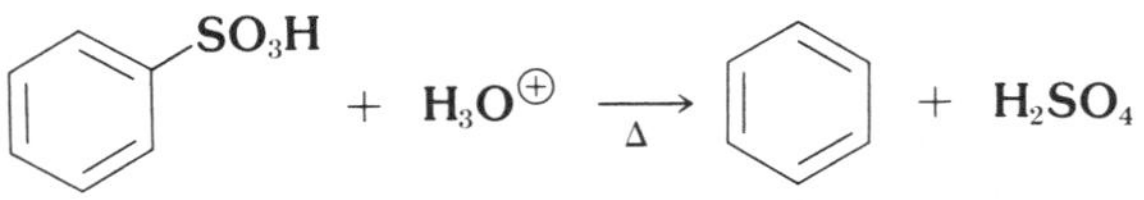

[PhSO$_3$H + H_3O^+ $\xrightarrow{\Delta}$ benzene + H_2SO_4]

OXIDATION OF ALKYL SIDE CHAINS

Although the benzene ring is relatively resistant to oxidative degradation, alkyl side chains attached to the ring are attacked by hot permanganate solutions or other oxidizing agents:

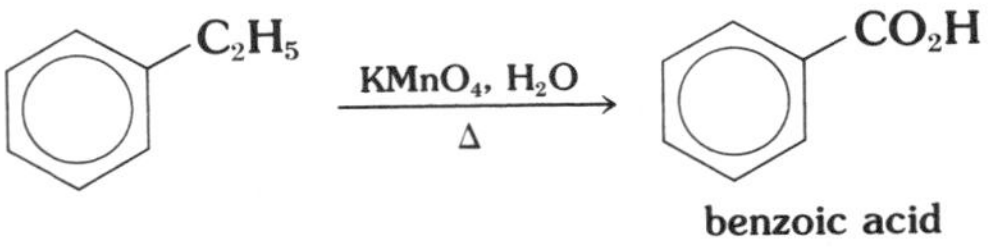

[PhC$_2$H$_5$ $\xrightarrow{KMnO_4, H_2O, \Delta}$ PhCO$_2$H (benzoic acid)] (9.16)

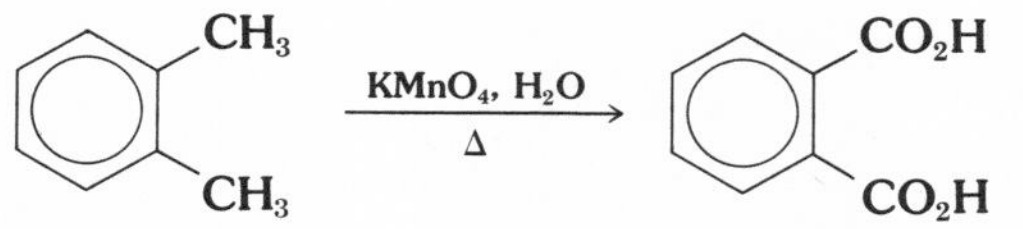
phthalic acid

(9.17)

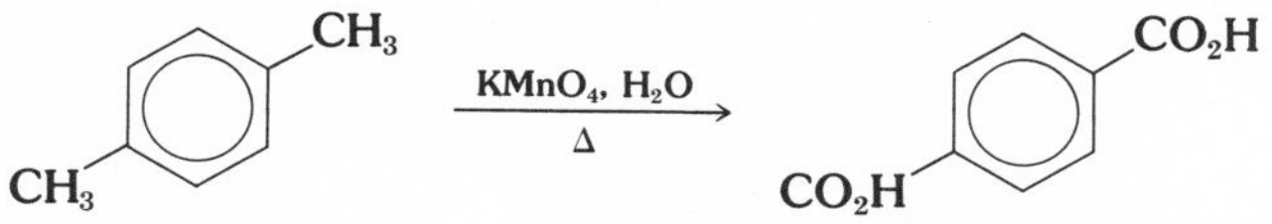
terephthalic acid

(9.18)

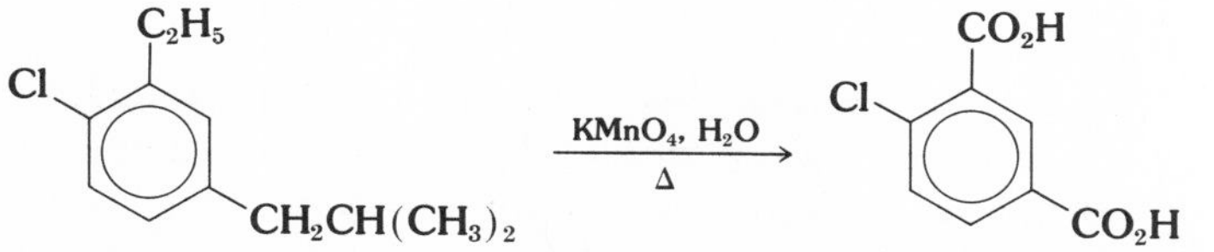
4-chloroisophthalic acid

(9.19)

9·7 CRITERIA OF AROMATICITY

The thermodynamic stability and unique chemical behavior of benzene and related hydrocarbons define what is commonly called the *aromatic character* or *aromaticity* of these compounds. At this point we are naturally led to ask if benzene is really unique, or whether other cyclic conjugated molecules also have aromatic properties.

MINOR CHANGES IN THE BENZENE RING

A relatively small change in the benzene ring, such as the introduction of a nitrogen atom in place of a CH group, does not destroy the aromaticity of the molecule.

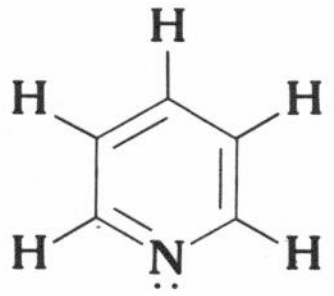
pyridine
or azine

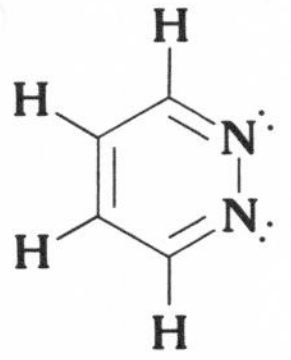
pyridazine
or 1,2-diazine

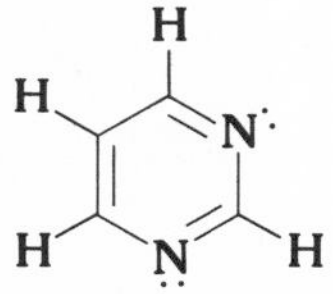
pyrimidine
or 1,3-diazine

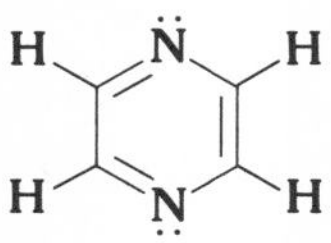
pyrazine
or 1,4-diazine

The hybridization of each nitrogen incorporated into the aromatic ring is sp^2 (the same as the carbon atoms), and the remaining nonbonding electron pair on each nitrogen occupies an sp^2 orbital oriented in the nodal plane of the π-orbital system. This electron pair is thus available for further bonding:

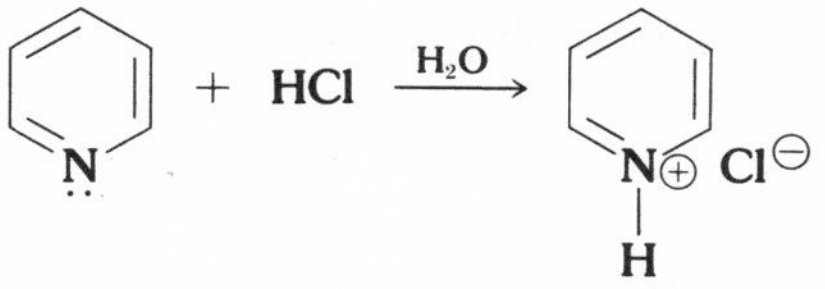
base acid a salt

(9.20)

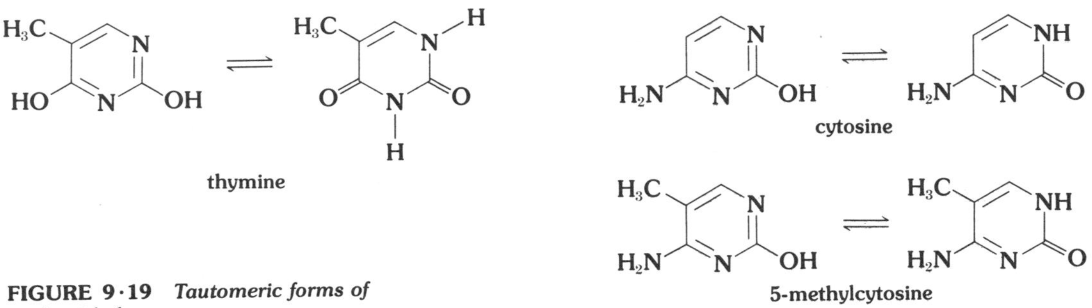

9·7 CRITERIA OF AROMATICITY

FIGURE 9·19 *Tautomeric forms of nucleoside bases*

(thymine tautomers; cytosine tautomers; 5-methylcytosine tautomers)

Although the resonance stabilization energy of pyridine (21 kcal/mole) is slightly less than that of benzene, pyridine is actually less reactive than benzene in electrophilic substitution reactions such as nitration, sulfonation, and bromination. This is due partly to salt formation under the strongly acidic reaction conditions and partly to the strong molecular dipole of pyridine, described by the following resonance contributers:

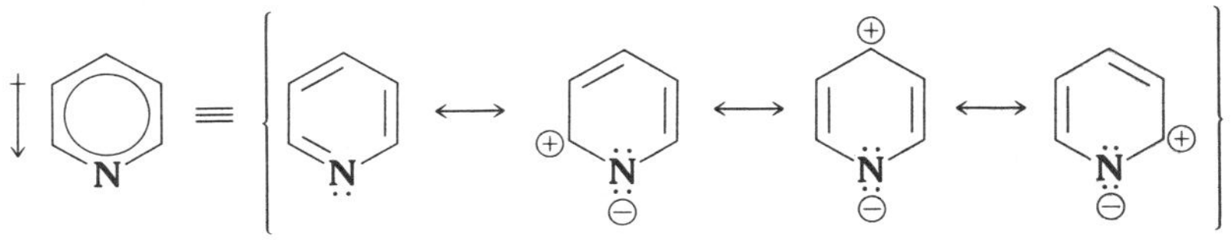

PROBLEM 9·21
Why should salt formation retard the rate of electrophilic substitution of pyridine?

Aromatic nitrogen heterocycles are widely distributed in nature. In fact, pyrimidine derivatives are present in all living cells as an integral part of nucleic acids (see Figure 9.19).

OTHER CYCLIC CONJUGATED POLYENES

The name *annulene* is given to monocyclic conjugated polyenes having the general formula

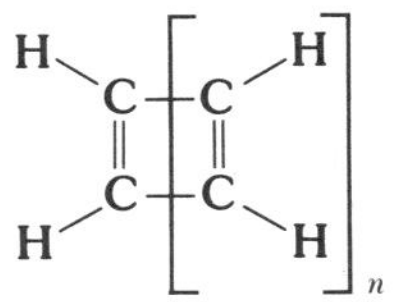

The most common annulene, the benzene ring ($n=2$), has been found in many natural products other than coal and petroleum. No other annulenes occur naturally, but many have been synthesized in the laboratory.

Cyclooctatetraene, a conjugated eight-carbon ring, shows none of the aromatic characteristics of the benzene hydrocarbons.

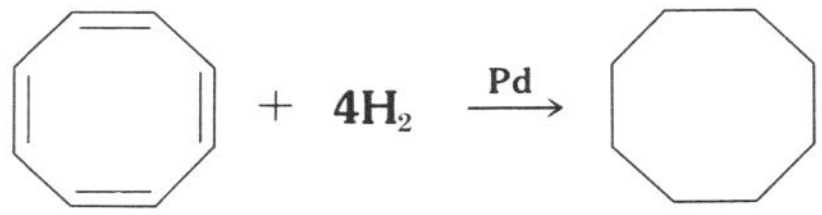

$$\text{cyclooctatetraene} + 4H_2 \xrightarrow{Pd} \text{cyclooctane}$$

$\Delta H = -120$ kcal/mole
($\Delta H_{calc} = 4 \times -28 = -112$ kcal/mole)

(9.21)

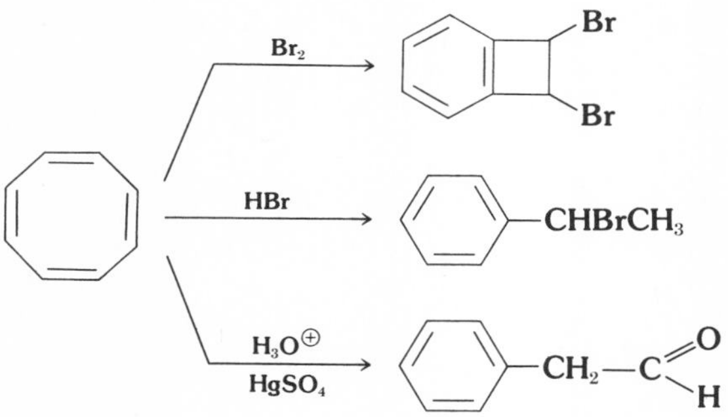

9·7 CRITERIA OF AROMATICITY

FIGURE 9·20 *Reactions of cyclooctatetraene*

The heats of hydrogenation and combustion indicate no exceptional thermodynamic stabilization due to the cyclic conjugation. Moreover, treatment of cyclooctatetraene with common electrophilic reagents leads to addition products, many of which have undergone extensive molecular rearrangement (see Figure 9.20).

One of the factors which may be responsible for the lack of aromatic stabilization in cyclooctatetraene is its nonplanarity. A planar octagon of sp^2-hybridized carbon atoms would necessarily have bond angles of 135°, resulting in an angle strain of 15° per carbon atom. Molecular models indicate, and spectroscopic measurements confirm, that molecules of cyclooctatetraene avoid this angle strain by adopting a tublike conformation:

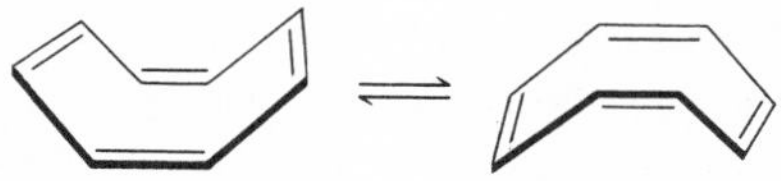

This bending of the molecule will necessarily decrease the amount of overlap of adjacent double bonds, making aromatic conjugation impossible.

PROBLEM 9·22
Would you expect the following conformation to be present in a sample of cyclooctatetraene?

Explain your reasoning.

We can avoid the complication of nonplanarity by shifting our attention to the smaller annulene 1,3-cyclobutadiene. Here again, though, we find no evidence of aromatic stabilization despite the favorable conjugation of the double bonds. In fact, cyclobutadiene is so unstable it cannot be isolated and studied by the usual techniques. Reactions in which cyclobutadiene is generated from its stable transition-metal complexes (Fe or Ni) lead invariably to dimeric and polymeric products:

$$\text{a stable complex} \xrightarrow{Ce^{+4}} \text{cyclobutadiene} \longrightarrow \square\square + \text{polymer} \tag{9.22}$$

9·7 CRITERIA OF AROMATICITY

Our first response might be to attribute the instability of cyclobutadiene to the angle strain in this small ring. However, this explanation is not consistent with the existence and stability of such highly strained molecules as

Thus it appears not only that benzene is unique among the annulenes in its extraordinary aromaticity, but that cyclobutadiene may in fact suffer a conjugative *destabilization*, termed *antiaromaticity*.

THE HÜCKEL RULE

Some forty years ago the German physical chemist Erich Hückel showed that simple molecular-orbital calculations for conjugated monocyclic systems predicted stable (closed-shell) electron configurations when the π electrons occupying the bonding molecular orbitals numbered $4n + 2$ (where $n = 0, 1, 2, 3, \ldots$). Although the importance of this hypothesis was not immediately recognized, within the last two decades it has furnished the chief stimulus for exciting new studies of aromaticity.

According to the *Hückel rule*, benzene with its six π electrons is a stable or closed-shell annulene ($n = 1$), while cyclobutadiene with four π electrons and cyclooctatetraene with eight π electrons are not. In the following discussion we shall examine some other $4n + 2$ annulene systems to test the credibility of Hückel's proposal. Because we have already established the aromaticity of benzene, let us begin by considering some other six-π-electron systems.

SIX-π-ELECTRON SYSTEMS ($n = 1$) The heterocyclic compounds pyrrole, furan, and thiophene are considered aromatic because they resist addition reactions, undergo electrophilic substitution, and have significant resonance stabilization energies.

pyrrole
$E_{res} = 21$ kcal/mole

furan
$E_{res} = 16$ kcal/mole

thiophene
$E_{res} = 28$ kcal/mole

To form a six-π-electron conjugated system, the heteroatom (N, O, or S) must assume sp^2 hybridization and a nonbonding electron pair must be added to the four π electrons in the two double bonds, as shown in Figure 9.21. From the following resonance descriptions of pyrrole and furan (thiophene is equivalent to furan) it should be clear that only one of the nonbonding electron pairs in furan (or thiophene) is incorporated into the π-orbital system:

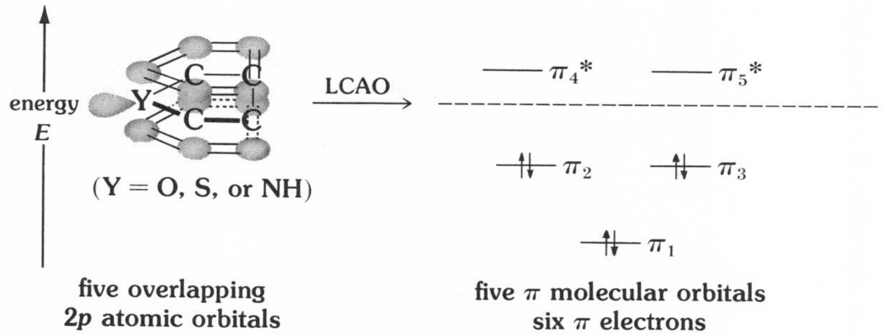

9·7 CRITERIA OF AROMATICITY   277

FIGURE 9·21 *A molecular-orbital description of a six-π-electron heterocycle*

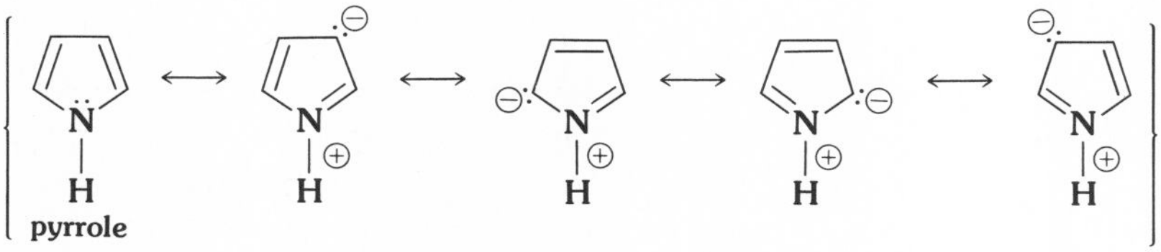

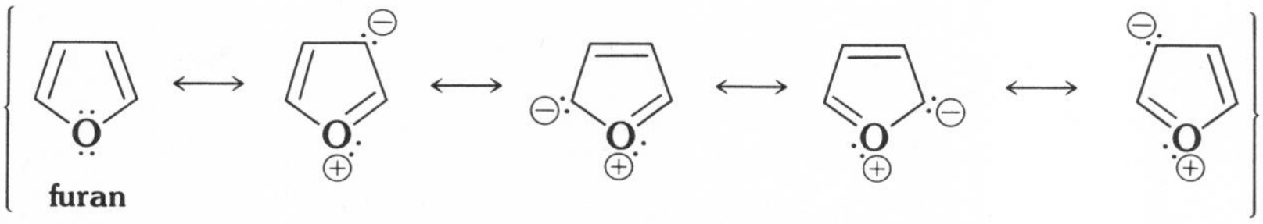

[Diagram: five overlapping 2p atomic orbitals (Y = O, S, or NH) → LCAO → five π molecular orbitals, six π electrons, with π_1, π_2, π_3 filled and π_4^*, π_5^* empty]

[Resonance structures of pyrrole and furan]

This description of pyrrole leads us to expect that it will be a weaker Brønsted base and a stronger Brønsted acid than pyrrolidine, the equivalent saturated amine. A difference in acidity of about 10^{17} between these two compounds is in fact observed:

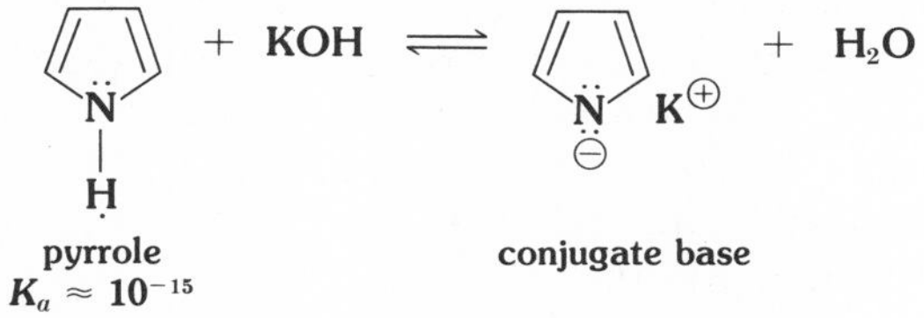

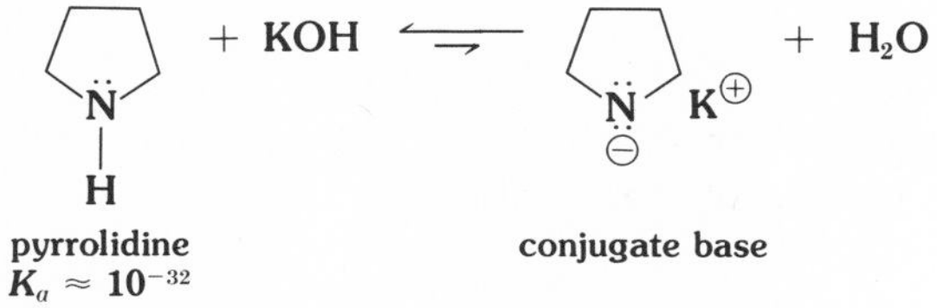

pyrrole + KOH ⇌ conjugate base + H_2O (9.23)
$K_a \approx 10^{-15}$

pyrrolidine + KOH ⇌ conjugate base + H_2O (9.24)
$K_a \approx 10^{-32}$

PROBLEM 9·23

a Which of the following heterocyclic systems may be considered aromatic? For each compound you select, circle the nonbonding electron pairs that will become part of the aromatic π-electron system:

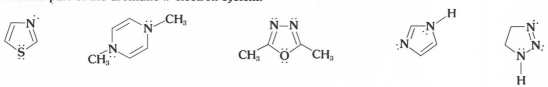

9·7 CRITERIA OF AROMATICITY

b Write two equations showing the proton-transfer equilibria involved when the amines pyrrole and pyrrolidine function as Brønsted bases in aqueous solution (water serves as the acid). Predict the relative magnitude of the equilibrium constants for the two equilibria.

The conjugate base of cyclopentadiene can be regarded as a negatively charged six-π-electron annulene. However, the chemical reactivity of this base greatly exceeds that of benzene; for example, cyclopentadienyl anion is neutralized by dilute aqueous acid:

$$\text{cyclopentadienyl}^- \text{Na}^{\oplus} + \text{H}_3\text{O}^{\oplus}\text{Cl}^{\ominus} \longrightarrow \text{cyclopentadiene} + \text{NaCl} + \text{H}_2\text{O} \qquad (9.25)$$

In coming to grips with the question of whether or not this anion exhibits aromatic stabilization, we must recognize the importance of making meaningful comparisons. An anion will clearly have rather different chemical and physical properties from a neutral molecule. Consequently, the cyclopentadienyl anion should be compared not with benzene, but with other charge-delocalized cyclic and acyclic carbanions, such as the conjugate bases in Figure 9.22.

The relative stabilities of a series of carbon anions are proportional to the corresponding ionization constants of the conjugate acids, measured with respect to a standard reference base (usually water):

$$\text{HA} + \text{H}_2\text{O} \rightleftharpoons \text{H}_3\text{O}^{\oplus} + \text{A}^{\ominus} \qquad K_a = \frac{[\text{H}_3\text{O}^{\oplus}][\text{A}^{\ominus}]}{[\text{HA}]} \qquad (9.26)$$

Thus, if one acid (HA_1) has a larger ionization constant than another acid (HA_2), the relative stability of the conjugate bases is $A_1^{\ominus} > A_2^{\ominus}$. Although the ionization constants of 1,3-pentadiene and cycloheptatriene are not known precisely, they are clearly very much smaller ($K_a \approx 10^{-31}$) than the ionization constant for cyclopentadiene ($K_a \approx 10^{-16}$). The extraordinary stabilization of the cyclopentadienyl anion (by a factor of 10^{15} over similar delocalized carbon anions) is attributed to its aromatic character.

PROBLEM 9·24
If the relative stability of each of the three anions in Figure 9.22 were proportional to the degree of charge delocalization, what order of stability would you expect?

By determining the relative stability of *cationic* five- and seven-membered annulenes, we can put our previous arguments to a critical test. Whereas the six-π-electron cyclopentadienyl anion proved to be much more stable than the eight-π-electron cycloheptatrienyl anion, the corresponding cations will represent cyclic four-π-electron and six-π-electron systems. If the aromatic stabilization of six-π-electron annulenes is a predominant factor, we would expect the cycloheptatrienyl cation, often referred to as *tropylium ion*, to show exceptional stability in comparison with other conjugated cations.

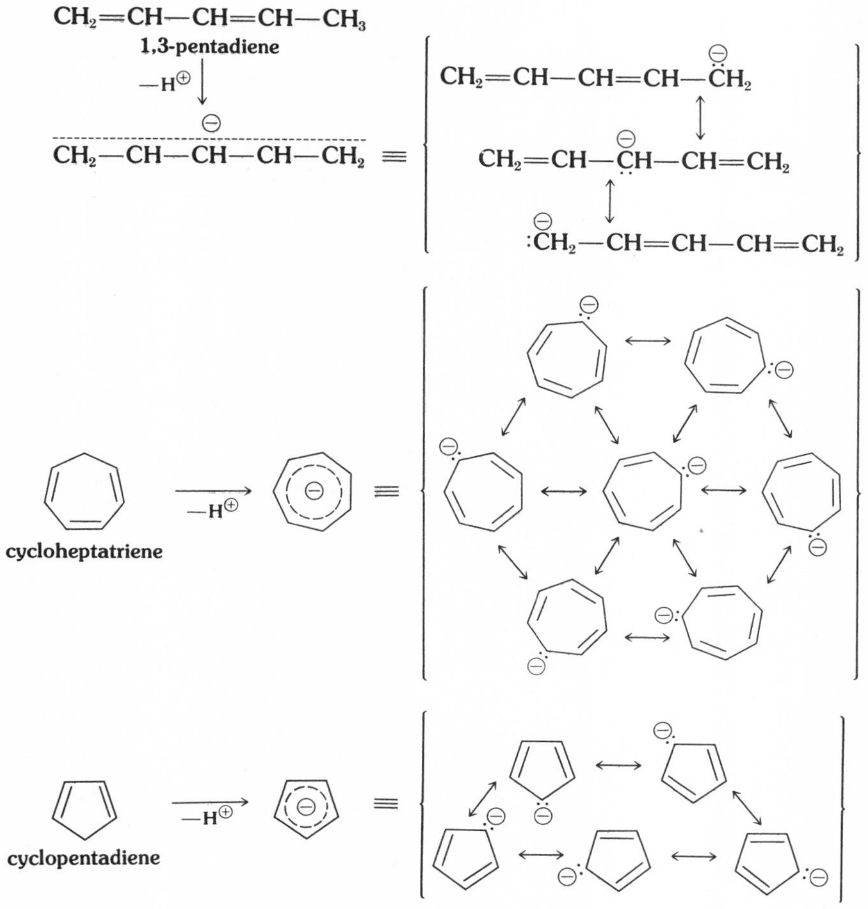

9·7 CRITERIA OF AROMATICITY 279

FIGURE 9·22 *Charge delocalization in unsaturated conjugate bases*

In hydroxylic solvents cyclopentadienyl and cycloheptatrienyl cations will exist in a state of equilibrium with the corresponding alcohols or ethers:

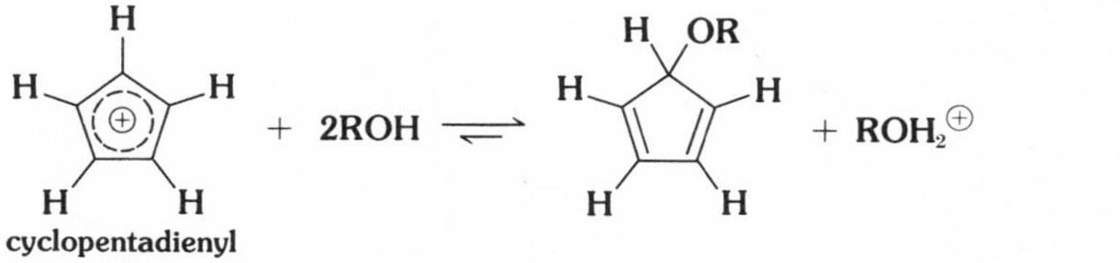

cyclopentadienyl cation + 2ROH ⇌ [cyclopentadiene–OR] + $ROH_2^\oplus$ (9.27)

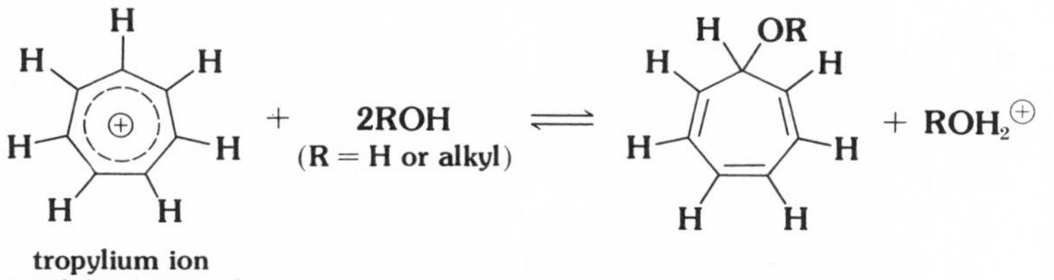

tropylium ion (cycloheptatrienyl cation) + 2ROH ⇌ [cycloheptatriene–OR] + $ROH_2^\oplus$ (9.28)

(R = H or alkyl)

280 9·7 CRITERIA OF AROMATICITY

Once again, the relative stability of these ions can be roughly gauged from the constants for these equilibria (because the cations appear on the left side of the equations, a large value of K_a reflects low cation stability). The experimental observations involving these systems are in complete agreement with the proposed aromaticity of six-π-electron annulenes. The tropylium ion is stable in dilute aqueous mineral acid solutions (pH < 4) and can be isolated as perchlorate ($ClO_4^{\ominus}$) or tetrafluoroborate ($BF_4^{\ominus}$) salts. In contrast, cyclopentadienyl cations are extremely unstable and can be formed only in nonnucleophilic solvents containing very strong acids.

PROBLEM 9·25
Write a resonance description of the tropylium ion:

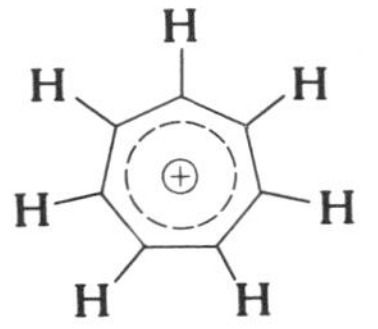

[structure of tropylium ion with 7 H's around a 7-membered ring with ⊕]

TWO-π-ELECTRON SYSTEMS ($n = 0$) Despite the angle strain in small rings, a stable two-π-electron cyclic cation has been prepared. Crystalline salts of the trimethylcyclopropenium cation can be prepared and stored without deterioration, and since this cation survives in neutral aqueous solution, it is estimated to be over a hundred times more stable than the tropylium ion.

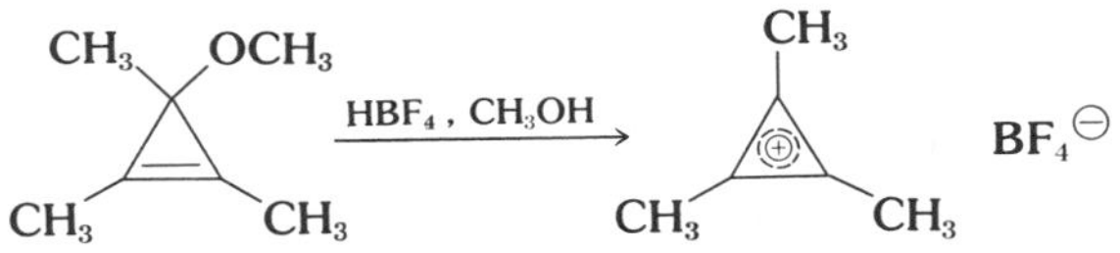

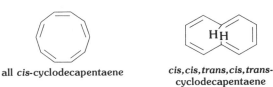

(9.29)

The remarkable stability of cyclopropenium cations (first demonstrated in 1957) caused a rebirth of interest in the Hückel concept of aromaticity and stimulated investigations of other annulenes.

TEN-π-ELECTRON SYSTEMS ($n = 2$) Several stereoisomeric cyclodecapentaenes are possible, but since angle strain and/or steric repulsions force them to assume severely bent nonplanar conformations, conjugation is poor and there is no evidence of aromatic character.

all cis-cyclodecapentaene   cis,cis,trans,cis,trans-cyclodecapentaene

By bonding or bridging across the ten-membered ring, it is possible to form planar or nearly planar cyclic conjugated pentaenes that show clear-cut aromatic properties, such as resonance stabilization and electrophilic substitution:

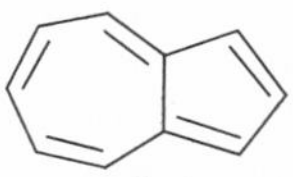

9·7 CRITERIA OF AROMATICITY 281

azulene
E_{res} = 42 kcal/mole

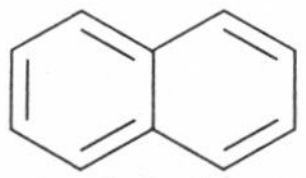

naphthalene
E_{res} = 71 kcal/mole

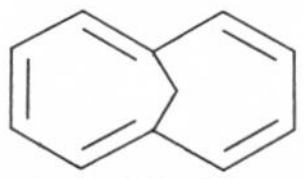

1,6-methanocyclodecapentaene

Two novel anionic ten-π-electron annulenes have been prepared and are considered aromatic:

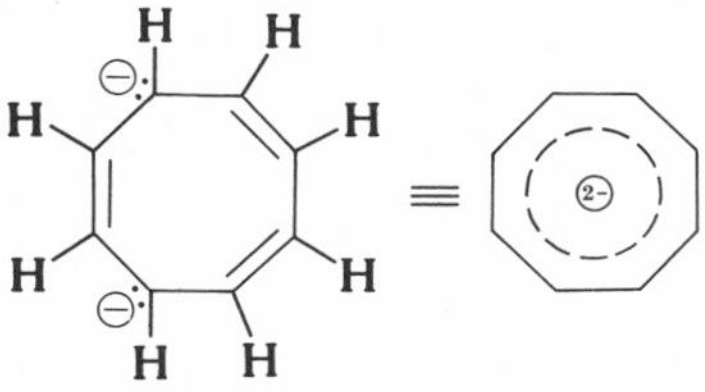

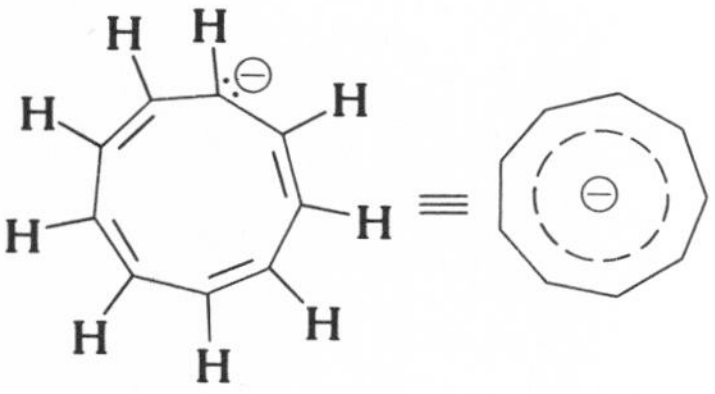

FOURTEEN-π-ELECTRON SYSTEMS ($n = 3$) A 14-π-electron cyclotetradecaheptaene has been prepared, but it is rather unstable and shows no aromatic characteristics, possibly because the crowding of the four internal hydrogen atoms prevents the ring from being planar. Coplanarity of the conjugated double bonds in the related tetracyclic heptaene is fixed, and this stable green crystalline substance undergoes typical aromatic electrophilic substitution reactions.

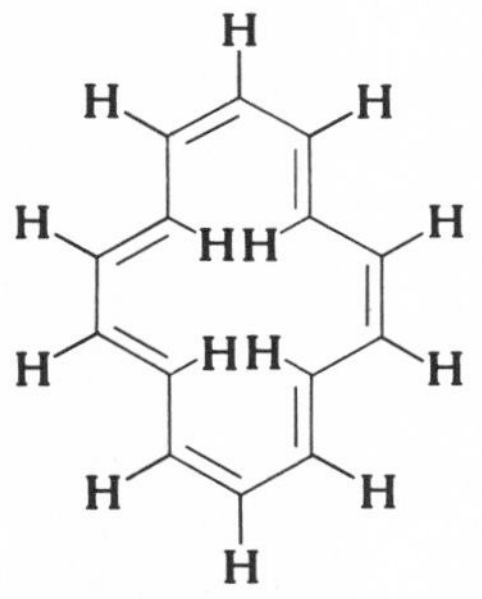

unstable

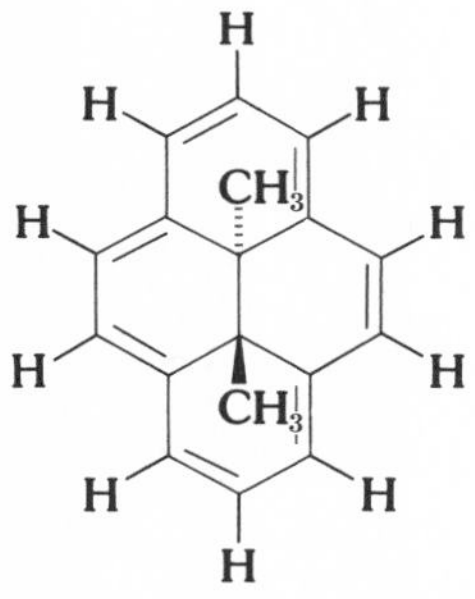

stable

If the ring-current model of benzene in Figure 9.12 is generalized to all aromatic annulenes, we can further test the aromaticity of the methyl-substituted system by measuring the pmr chemical shifts of the two methyl groups, as well as those of the hydrogen atoms on the periphery of the ring. Since the methyl protons are located just above and below the center of the π-electron ring, these nuclei should be strongly shielded by the induced magnetic field generated by the π ring current. Of course, the olefinic protons should suffer a corresponding deshielding effect. The experimental measurements support such an interpretation: this annulene exhibits two sharp signals in the pmr spectrum, with an area ratio of 5:3. The stronger signal appears at $\delta = 8.1$, while the weaker one is found at $\delta = -3.9$, or at 3.9 ppm *higher field* than tetramethylsilane.

PROBLEM 9·26
Predict the pmr spectrum of the ten-π-electron annulene 1,6-methanocyclodecapentaene.

PROBLEM 9·27
If the cyclic heptaenyne shown below is planar, would you expect it to be aromatic? Why? How would you test the aromatic character of this compound?

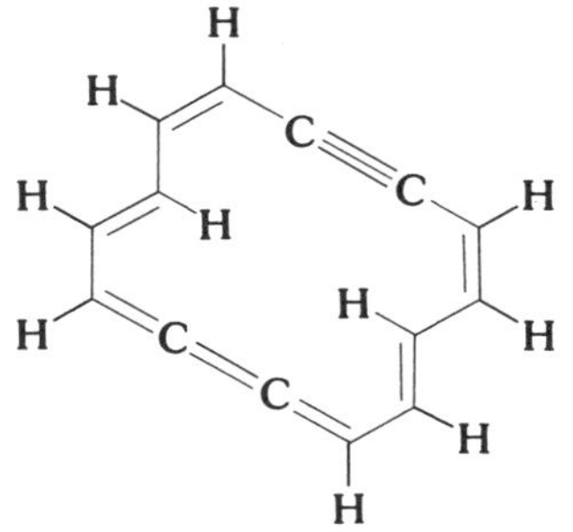

EIGHTEEN-π-ELECTRON SYSTEMS ($n = 4$) Although the 18-π-electron annulene below is more stable than the corresponding 14-π-electron compound, it readily adds bromine and from a chemical point of view could hardly be considered aromatic.

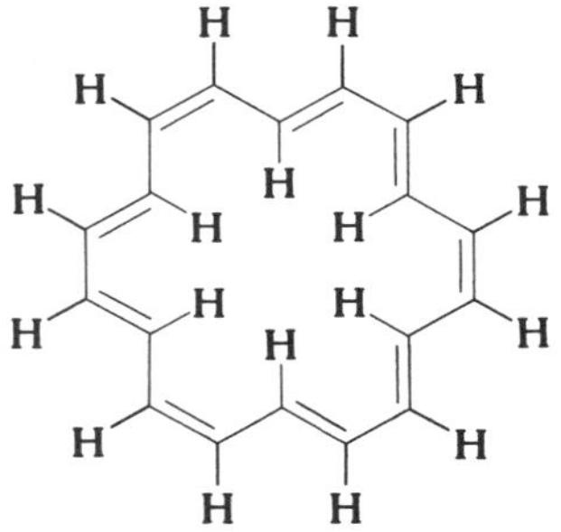

However, the pmr criterion tells a different story. Two sharp signals having an area ratio of 2:1 are observed, and these are assigned respectively to the "outside" and "inside" hydrogen atoms on the ring. The stronger outside signal appears at $\delta = 9.0$, while the smaller inside signal is upfield at $\delta = -2.0$. Such a marked difference in chemical shifts is expected if this annulene has the same kind of aromatic π-electron ring current proposed for benzene.

SUMMARY

aromatic compounds A large and important group of organic compounds, characterized by a relatively high carbon-to-hydrogen ratio and consisting of three general types of cyclic hydrocarbons: (*a*) benzene derivatives, such as toluene and the xylenes; (*b*) fused benzene-ring systems, such as naphthalene and anthracene; and (*c*) nonbenzenoid systems, such as azulene and some annulenes.

benzene The simplest aromatic hydrocarbon (C_6H_6). Benzene has exceptional thermodynamic stability and shows characteristic electronic absorption at 257 nm and a pmr signal at $\delta = 7.4$ ppm. Isomeric disubstituted benzene derivatives are designated as *ortho* (1,2), *meta* (1,3), or *para* (1,4).

electronic-absorption spectroscopy A method of analysis in which electron transitions of conjugated π-electron systems are induced by absorption of light having a wavelength of 200 to 800 nm (ultraviolet-visible). The electronic spectrum of a compound is characterized by one or more absorption maxima (λ_{max}) and a corresponding intensity measure, given by the molar absorptivity ϵ.

conjugation A relationship of neighboring π-electron functions in which the p orbitals on all the atoms of such systems overlap when oriented parallel to each other. The LCAO treatment of n overlapping p orbitals generates n π-molecular orbitals, half of which are bonding and half antibonding.

resonance A means of describing the bonding in certain unsaturated molecules for which localized covalent-bond notations are inadequate. In this approach the real molecule is considered a hybrid of two or more contributing structures which differ only in the arrangement of electrons. Contributing structures are linked by double-headed arrows ($\longleftrightarrow$).

resonance energy A stabilization of a resonance hybrid, given by the difference between the actual energy of a hybrid structure and the calculated energy of the most stable contributing structure.

electrophilic substitution reactions A class of reactions, common to aromatic compounds, in which strong electrophilic reagents effect a substitution on the aromatic ring. Examples of such reactions are *halogenation* ($X_2 + FeX_3$), *nitration* ($HNO_3 + H_2SO_4$), *sulfonation* (H_2SO_4 or SO_3), and *Friedel Crafts alkylation* ($R—X + AlX_3$).

aromaticity A concept that is useful in correlating the properties of conjugated unsaturated ring systems. It is necessary (but not sufficient) that an aromatic system have *cyclic conjugation;* that is, the overlapping p orbitals must be oriented in a ring without beginning or end. The cyclic conjugation must be continuous; that is, every atom considered to be a member of the aromatic ring must contribute a properly aligned p orbital to the π system. If the conjugation is interrupted at any point, the special characteristics expected for an aromatic system will not result. The cyclic π-orbital system will contain all the π electrons in those conjugated double bonds that are part of the aromatic ring. It may also include nonbonding electron pairs, provided such pairs occupy one of the overlapping p orbitals in the ring. Not all nonbonding (unshared) electron pairs held by ring atoms will necessarily be part of the aromatic π-electron system.

annulene A monocyclic conjugated polyene having the general formula

$$\begin{bmatrix} H & & H \\ \diagdown & & \diagup \\ C & \!\!\!\!-\!\!\!\! & C \\ \| & & \| \\ C & \!\!\!\!-\!\!\!\! & C \\ \diagup & & \diagdown \\ H & & H \end{bmatrix}_m$$

the Hückel rule A rule for predicting the aromaticity of annulenes. Aromatic (closed-shell) electron configurations are predicted for annulenes having $4n + 2$ π electrons ($n = 0, 1, 2, 3, \ldots$).

conformational enantiomers Enantiomers whose chirality lies in restricted rotation or twisting about bonds. Examples include certain substituted biphenyls and hexahelicene.

EXERCISES

9·1 Draw structural formulas for the following compounds:
a Chlorobenzene
b p-Xylene
c 1,1-Diphenylethane
d m-Nitrophenol
e o-Nitrobenzoic acid
f Butylbenzene
g 4,4'-Dinitrobiphenyl
h 1,5-Dichloronaphthalene
i 9,10-Dibromoanthracene
j α-Naphthol
k 3,3-Diphenyl-1-hexanol
l 9,10-Dibromophenanthrene
m 2,4,6-Trinitrotoluene (TNT)

9·2 Name each compound:

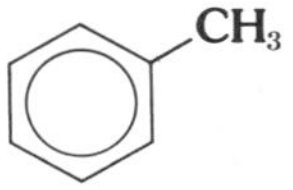

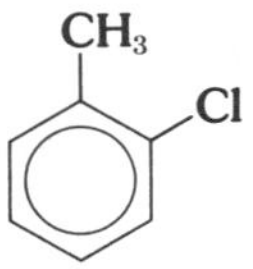

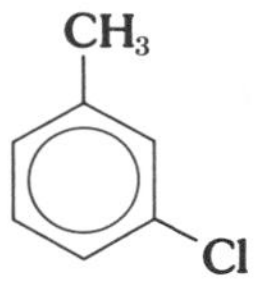

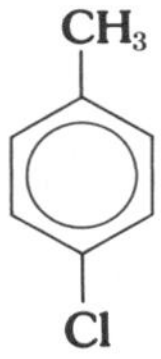

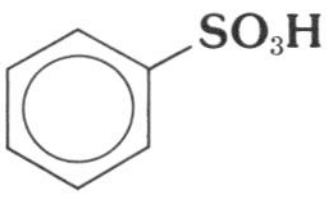

9·3 Match identical structures in each column:
a Ethylbenzene
b α-Naphthol
c m-Chlorobromobenzene
d m-Bromotoluene
e o-Bromotoluene
f β-Naphthol
g p-Bromotoluene

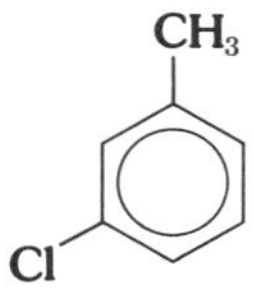

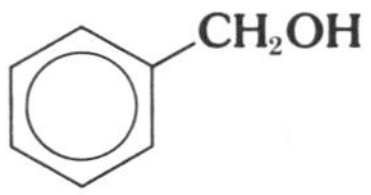

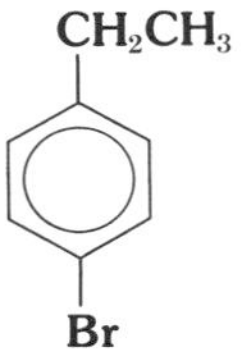

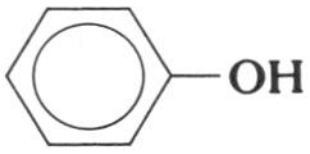

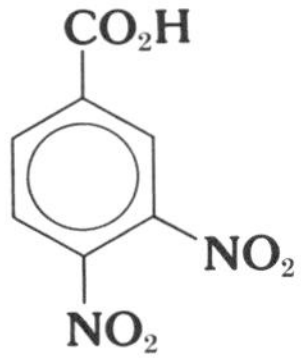

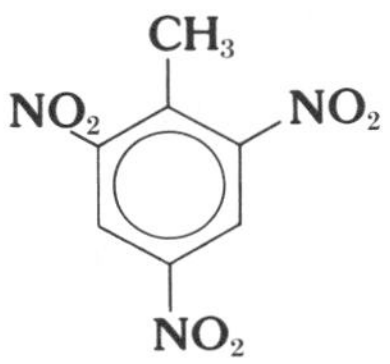

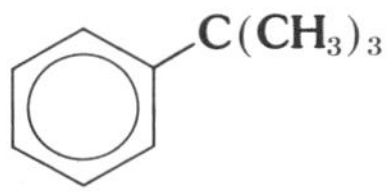

1 4-Bromotoluene
2 2-Bromotoluene
3 2-Naphthol
4 Phenylethane
5 3-Bromotoluene
6 1-Naphthol
7 m-Bromochlorobenzene

9·4 Compounds A, B, and C are the three isomeric dibromobenzenes. Identify which is ortho, para, and meta from the number of mononitration products:

a Compound A $\xrightarrow{HNO_3, H_2SO_4}$ two mononitration products

b Compound B $\xrightarrow{HNO_3, H_2SO_4}$ three mononitration products

c Compound C $\xrightarrow{HNO_3, H_2SO_4}$ one mononitration product

9·5 Draw the three Kekulé-type resonance contributors for naphthalene and the five Kekulé-type resonance contributors for phenanthrene.

9·6 What is the hybridization state of the oxygen and nitrogen atoms in furan and pyrrole? *Hint:* See the resonance descriptions of these compounds.

9·7 Draw structural formulas for the starting materials in the following reactions:

a C_8H_{10} $\xrightarrow{KMnO_4, H_2O, \Delta}$ benzoic acid (mono-CO_2H)

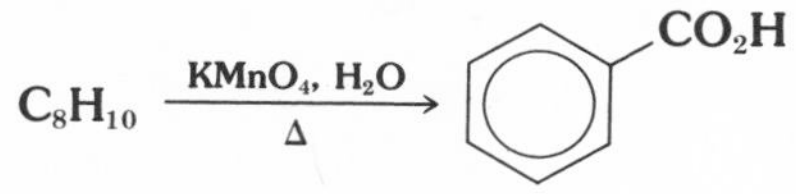

b C_8H_{10} $\xrightarrow{KMnO_4, H_2O, \Delta}$ ortho-di-CO_2H benzene

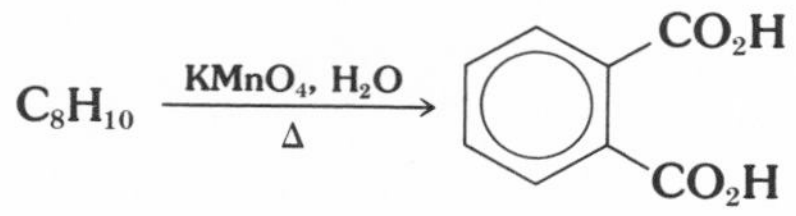

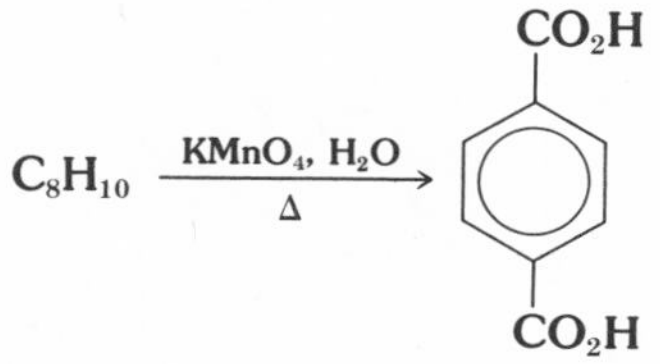

c C_8H_{10} $\xrightarrow{KMnO_4, H_2O, \Delta}$ para-di-CO_2H benzene

d C_9H_{12} $\xrightarrow{KMnO_4, H_2O, \Delta}$ mono-CO_2H benzene

e C_9H_{12} $\xrightarrow{KMnO_4, H_2O, \Delta}$ meta-di-CO_2H benzene

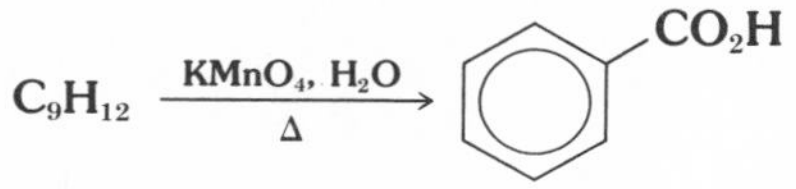

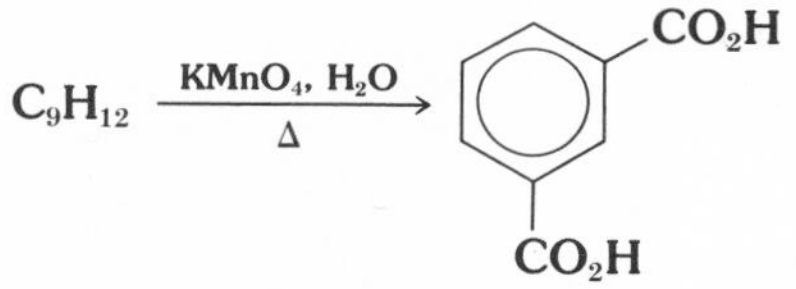

9·8 Indicate the reagents and conditions necessary to convert benzene into the following products:
a Bromobenzene
b Benzenesulfonic acid
c Nitrobenzene
d Cyclohexane
e Ethylbenzene
f Isopropylcyclohexane (two steps)
g Benzoic acid (two or three steps)
h 1-Phenyl-1-propanol (three steps)

9·9 Arrange the compounds nitrobenzene, toluene, and benzene in order of their reactivity:
a To nitration
b To bromination

9·10 In an acid-base equilibrium involving nitric and sulfuric acids, write structural formulas for two possible conjugate acids of nitric acid (formed when nitric acid functions as a Brønsted base). Which of these conjugate acids is a likely precursor of the nitronium ion $NO_2^{\oplus}$?

9·11 Do naphthalene, anthracene, and vitamin A fit the Hückel rule? Explain. Which of the following compounds would you predict to have aromatic character according to the Hückel rule?

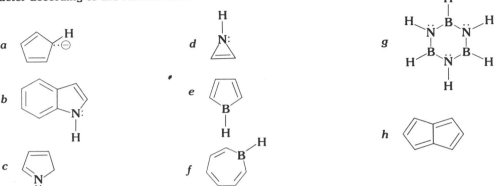

9·12 Predict the preferred position of monosubstitution in the nitration of naphthalene by HNO_3 in H_2SO_4. Explain your reasoning.

9·13 Calculate the conjugative stabilization of the double bond in styrene from the following heats of hydrogenation:

	ΔH (kcal/mole)
styrene	−76.9
benzene	−49.8
propene	−28.6

9·14 Explain the formation of the rearranged product in equation 9.13.

9·15 A theory involving delocalization of σ electrons (called hyperconjugation) is exemplified by resonance structures for the ethyl cation as shown:

$$\underset{\text{I}}{\text{H--}\underset{\underset{H}{|}}{\overset{\overset{H}{|}}{\underset{\oplus}{C}}}\text{--}\underset{\underset{H}{|}}{\overset{\overset{H}{|}}{C}}\text{--H}} \longleftrightarrow \underset{\text{II}}{\text{H--}\overset{\overset{H}{|}}{C}\text{=}\overset{\overset{H}{|}}{C}\text{--H}\ \ H^{\oplus}} \longleftrightarrow \underset{\text{III}}{H^{\oplus}\ \overset{\overset{H}{|}}{C}\text{=}\underset{\underset{H}{|}}{C}\text{--H}} \longleftrightarrow \underset{\text{IV}}{\overset{\overset{H^{\oplus}}{\ }}{\text{H--}\underset{\underset{H}{|}}{C}\text{=}\underset{\underset{H}{|}}{C}\text{--H}}}$$

a Do these structures follow the rules of resonance?
b Evaluate the relative contribution to the resonance hybrid of structures II, III, and IV.
c Would you expect a large resonance energy? Explain.

9·16 Two compounds of molecular formula C_9H_{12} are each allowed to react with hot potassium permanganate solution. Compound A yields terephthalic acid and B gives 1,3,5-benzenetricarboxylic acid. What are the structures of A and B?

9·17 Using only the simple rules of resonance in Section 9.3, predict the relative stability of these ions. Taking into account the Hückel rule, which would you predict to be more stable?

9·18 A compound of molecular formula C₇H₈O displays a strong absorption at 3300 cm⁻¹ in the infrared range and three singlets in the pmr spectrum at δ = 3.65, 4.4, and 7.2, with relative areas of 1:2:5. Write a reasonable structural formula for this compound.

9·19 Write structural formulas for two isomers of C₁₀H₁₄ which display the following pmr spectra:
a Singlets at δ = 1.3 and 7.3, with relative areas of 9:5
b A doublet at δ = 1.2, a singlet at δ = 2.3, a septet at δ = 2.8, and a singlet at δ = 7.2, with relative areas of 6:3:1:4

9·20 What prominent feature of the indicated spectrum could be used to distinguish the following pairs of compounds?
a Benzene and cyclohexane; ultraviolet spectrum
b Benzene and cyclohexane; pmr spectrum
c Benzene and cyclohexane; infrared spectrum
d CH₂=CH—CH=CHCH₃ and CH₂=CH—CH₂CH=CH₂; ultraviolet spectrum
e △ and CH₂=CHCH₃; pmr spectrum

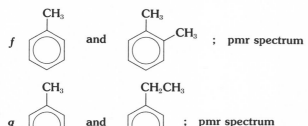

9·21 In the course of an independent study project a student attempted to prepare (±)-1,2-dimethyl-1-phenylpropane by means of a Friedel-Crafts alkylation of benzene with 3-methyl-1-butene and phosphoric acid. The major product from this reaction had a molecular weight of 148 (mass spectrometry) and exhibited four sets of resonance signals in the pmr spectrum: a triplet at δ = 0.8 (3H, *J* = 7 Hz), a singlet at δ = 1.2 (6H), a quartet at δ = 1.6 (2H, *J* = 7 Hz), and a singlet at δ = 7.2 (5H). What compound did the student make?

9·22 Three isomers of C₈H₁₀O exhibit the following spectroscopic properties. Write a structural formula for each compound.
a Infrared absorptions at 3600 and 3080 cm⁻¹; a pmr spectrum showing triplets at δ = 2.8 (2H, *J* = 6 Hz) and δ = 3.7 (2H, *J* = 6 Hz), a singlet at δ = 4.8 (1H) which shifts with dilution, and a singlet at δ = 7.2 (5H)
b Infrared absorptions at 3610 and 3100 cm⁻¹; a pmr spectrum showing a doublet at δ = 1.3 (3H, *J* = 7 Hz), a singlet at δ = 4.2 (1H) which shifts with dilution, a quartet at δ = 4.6 (1H, *J* = 7 Hz), and a singlet at δ = 7.2 (5H)
c No infrared absorptions from 3200 to 3700 cm⁻¹ and a strong absorption band at 3100 cm⁻¹; a pmr spectrum showing a triplet at δ = 1.3 (3H, *J* = 7 Hz), a quartet at δ = 3.9 (2H, *J* = 7 Hz), and a multiplet at δ ≈ 7.0 (5H)

9·23 A hydrocarbon X with a molecular weight of 118 gives benzoic acid on vigorous oxidation with $KMnO_4$. The pmr spectrum of X shows signals at $\delta = 2.1$, 5.5, 5.4, and 7.3, with an area ratio of 3:1:1:5. Reaction of X with HCl gives compound Y, which exhibits only two pmr signals, at $\delta = 1.7$ and 7.3, area ratio 6:5. Write structural formulas for X and Y.

ARYL HALIDES AND PHENOLS

THE UNIQUE properties of benzene and related hydrocarbons were described in the last chapter. Let us now compare the chemical behavior of aromatic ring systems that include halogen or hydroxyl groups with the reactions we have observed for alkyl halides (Chapter 3) and alcohols (Chapter 8). As we shall see, interaction between the aromatic nucleus and these functional groups creates a new and distinctive chemistry.

10·1 ARYL HALIDES

Compounds in which a halogen is bonded directly to a benzene or other aromatic ring are called *aryl halides*. The aryl group in such compounds is denoted by the general symbol Ar— to distinguish it from an alkyl group (R—). The nomenclature of aryl halides is straightforward and was described in the last chapter. Although these compounds do not as a rule occur naturally, they are manufactured in large quantities by the chemical industries. Thanks to the recent development of special preparation techniques, a wide variety of fluorine-containing aromatic compounds have joined the readily synthesized chloride and bromide derivatives as useful intermediates. Aryl iodides are usually made by indirect routes.

Aryl halides are characterized by a shorter and stronger carbon-halogen bond and a smaller bond dipole than their alkyl halide counterparts.

10·1 ARYL HALIDES

We attribute these differences to the sp^2 hybridization of the aryl carbon atoms and to delocalization of nonbonding electron pairs (p-π conjugation):

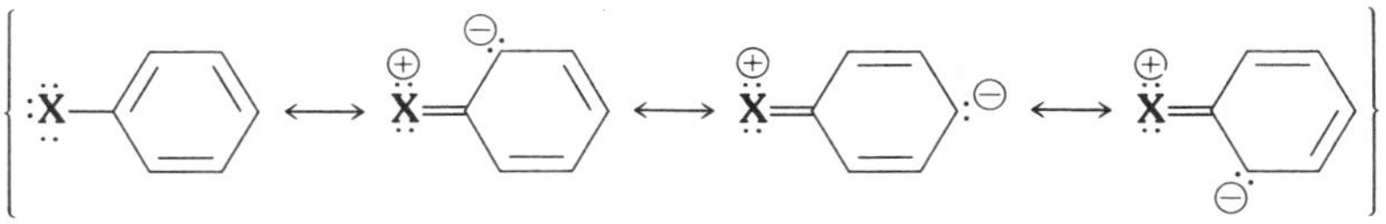

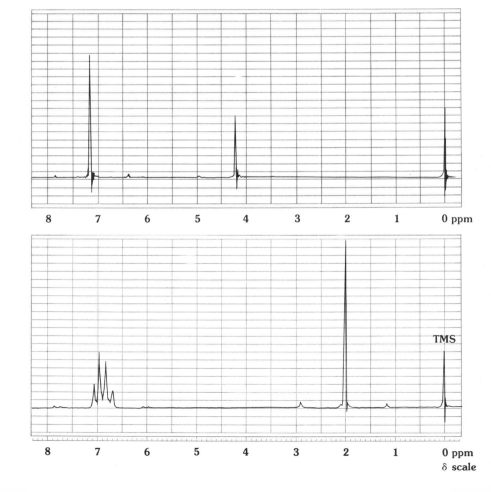

Similar trends are observed for vinyl halides, which also have an sp^2 carbon-halogen bond and exhibit p-π conjugation. The proposed hybridization effect appears to be reasonable, since the electrons in a σ bond joining a carbon atom to some other atom or group will be more strongly attracted by the carbon if it is sp^2 hybridized than if it has sp^3 hybridization (a σ orbital of an sp^2 hybrid has a higher percentage of carbon $2s$ character).

Infrared, ultraviolet, and pmr spectroscopic measurements may be used to obtain information about the molecular structure of specific halide compounds.

PROBLEM 10·1
The pmr spectra of two C_7H_7Cl isomers are shown below. Suggest structural formulas for these compounds.

10·2 REACTIONS OF ARYL HALIDES: THE HALIDE GROUP

TABLE 10·1 *Relative reactivity of alkyl and aryl halides*

reaction type	alkyl halides (R—X)	aryl halides (Ar—X)
nucleophilic substitution S_n1	reaction (R = 3° > 2° > 1°)	no reaction†
S_N2	reaction (R = 1° > 2° > 3°)	no reaction†
elimination $E1$	reaction (R = 3° > 2° > 1°)	no reaction†
$E2$	reaction (R = 3° > 2° > 1°)	no reaction†
active-metal substitution	reaction (X = I > Br > Cl)	reaction (X = I > Br > Cl)

†For most compounds under moderate reaction conditions.

The general pattern of aryl halide reactions is, with one exception, quite different from the reactions of alkyl halides (see Table 10.1). We shall explore some of these reactions more fully in the following sections, however, the difficulty in effecting nucleophilic substitution and elimination reactions of aryl halides by the mechanisms described in Chapter 3 should be apparent.

10·2 REACTIONS OF ARYL HALIDES: THE HALIDE GROUP

NUCLEOPHILIC DISPLACEMENT OF HALIDE

In discussing the nucleophilic substitution reactions of alkyl halides we considered two mechanisms: a two-step process beginning with a rate-determining ionization (S_N1) and a single-step process in which bond making and bond breaking were concerted (S_N2). Efforts to effect similar reactions with simple aryl halides, using typical S_N1 or S_N2 reaction conditions, have generally been unsuccessful:

$$\text{CH}_3\text{-C}_6\text{H}_4\text{-Br} \xrightarrow[\text{AgNO}_3, \Delta]{\text{H}_2\text{O, C}_2\text{H}_5\text{OH}} \text{no reaction} \quad (10.1)$$

$$\text{C}_6\text{H}_5\text{-Br} \xrightarrow[\Delta]{\text{NaI, acetone}} \text{no reaction} \quad (10.2)$$

Since most vinyl halides (such as 2-chloropropene) show a similar resistance to nucleophilic displacement, we can again attribute the characteristic behavior of these compounds to the sp^2 hybridization of the carbon atom bearing the halogen. Carbon atoms having an sp^2 hybridization are more electronegative than their sp^3-hybridized counterparts. Consequently, the carbonium-ion intermediates formed by the ionization of aryl and vinyl halides are thermodynamically less stable than alkyl carbonium ions, and the resulting increase in the activation energy for this ionization prevents the S_N1 mechanism from operating.

The lack of aryl halogen displacement by an S_N2 mechanism is also understandable, since the necessary inversion transition state is clearly impossible for these compounds (the center of the aromatic ring cannot accommodate the nucleophile, and the ring carbon atoms cannot change their relative positions). A front-side displacement involving a transition state like the following apparently has a prohibitively high activation energy:

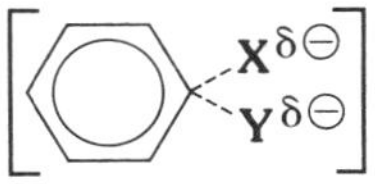

$$\left[\begin{array}{c} \cdots X^{\delta\ominus} \\ C_6H_5 \\ \cdots Y^{\delta\ominus} \end{array} \right]$$

PROBLEM 10·2
a Suggest a chemical test to distinguish cyclohexyl chloride from chlorobenzene.
b Draw a possible structure for an inversion transition state if an S_N2 reaction between vinyl chloride and iodide ion were to take place. Indicate the hybridization change at the carbon atom involved in the displacement.

The presence of certain electron-withdrawing groups (NO_2, CN, CF_3, etc.) at the ortho or para position of an aryl halide greatly enhances the rate of nucleophilic displacement of the halogen:

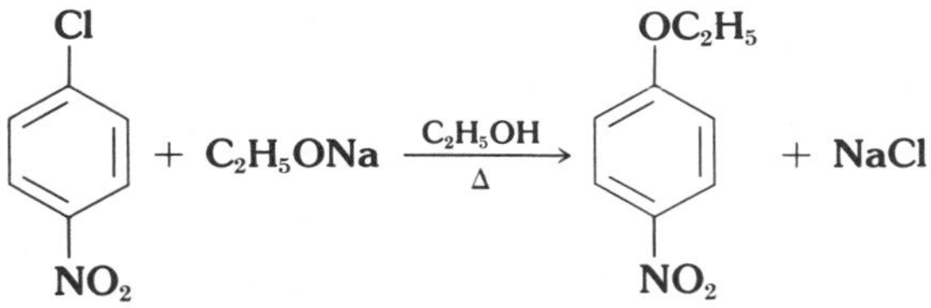

$$\text{p-}O_2N\text{-}C_6H_4\text{-}Cl + C_2H_5ONa \xrightarrow[\Delta]{C_2H_5OH} \text{p-}O_2N\text{-}C_6H_4\text{-}OC_2H_5 + NaCl \tag{10.3}$$

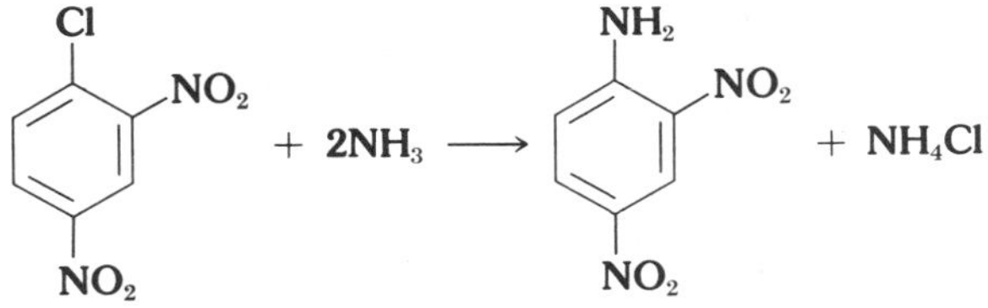

$$\text{2,4-}(NO_2)_2C_6H_3Cl + 2NH_3 \longrightarrow \text{2,4-}(NO_2)_2C_6H_3NH_2 + NH_4Cl \tag{10.4}$$

$$\text{2,4,6-}(NO_2)_3C_6H_2Cl + NaHCO_3 \xrightarrow{H_2O} \text{2,4,6-}(NO_2)_3C_6H_2OH + NaCl \tag{10.5}$$

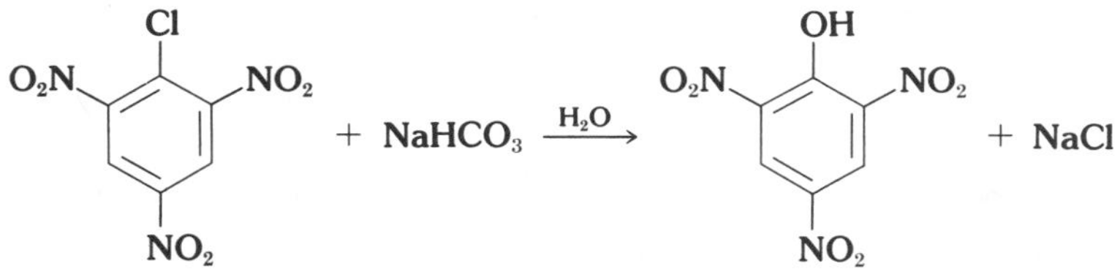

These results present us with an intriguing puzzle. We have just seen that simple aryl halides do not normally undergo nucleophilic substitution and are unlikely in any event to react by either an S_N1 or an S_N2 mechanism. Now we are confronted with the fact that nucleophilic displacement can occur in the presence of certain substituents adjacent to, or even remote from, the reaction site. Clearly, we must consider a new mechanism for nucleophilic displacement if we hope to account for this behavior.

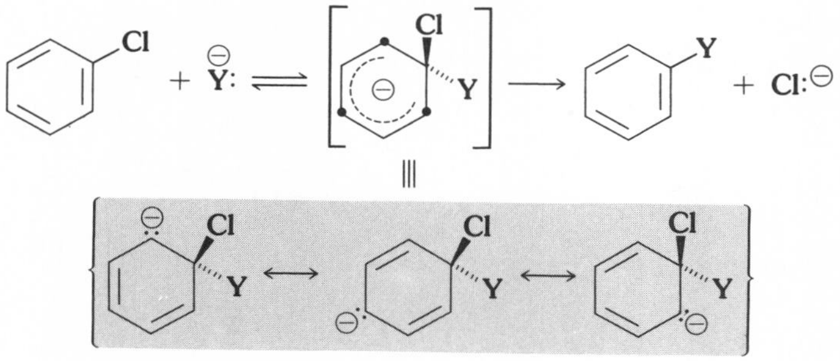

10·2 REACTIONS OF ARYL HALIDES: THE HALIDE GROUP

FIGURE 10·1 *A mechanism for nucleophilic substitution of aryl halides*

As we have noted in previous chapters, substitution reactions can proceed in three fundamentally different ways:

1. A two-step process in which bond breaking precedes bond making (the S_N1 mechanism belongs to this class)
2. A two-step process in which bond making precedes bond breaking
3. A one-step concerted process in which bond breaking and bond making occur simultaneously (the S_N2 mechanism belongs to this class).

We did not consider class 2 mechanisms for alkyl halide substitutions because an intermediate having a five-bonded carbon atom would have been formed in the first step. In the present situation, however, the unsaturated nature of the benzene ring would permit an initial bond-making step, as illustrated in Figure 10.1. Of course, the activation energy for the first (rate-determining) step in this reaction must be prohibitively large, since chlorobenzene and other simple aryl halides usually do not undergo nucleophilic displacement. However, the real merit of this mechanism is that it accounts for the unique rate-enhancing effect of ortho and para electron-withdrawing substituents. Because of the alternating fashion in which the negative charge is delocalized in the reactive intermediate (see Figure 10.2), these substituents can

FIGURE 10·2 *Charge delocalization in the intermediate formed during nucleophilic displacement of 2,4-dinitrochlorobenzene*

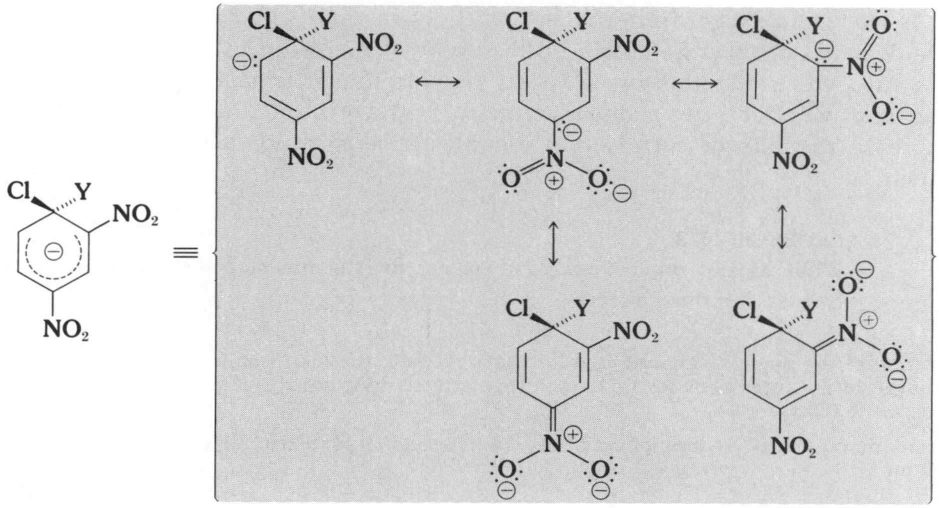

10·2 REACTIONS OF ARYL HALIDES: THE HALIDE GROUP

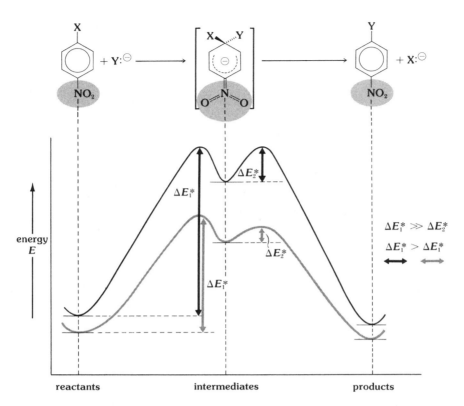

FIGURE 10·3 *Rate-enhancing effect of a para nitro substituent on nucleophilic substitution of aryl halides*

provide further stabilization through charge delocalization *only* when they are located at ortho or para positions relative to the halogen being displaced.†

The energy diagram in Figure 10.3 shows how these stabilizing substituents reduce the activation energy of the rate-determining step and, as a result, cause an increase in the overall rate of the substitution reaction.†† As we would expect from the reasoning above, *meta*-chloronitrobenzene does not display the enhanced reactivity of the ortho and para isomers. Also, electron-withdrawing substituents in the latter positions have an additive effect; two such substituents in the ortho or para positions increase the rate of substitution roughly twice as much as one substituent.

PROBLEM 10·3
What kinetic order would you expect for the two-step displacement of aryl halides described above?

---

†This distribution of the negative charge is exactly analogous to the distribution of positive charge in the intermediates formed during electrophilic substitution reactions of benzene (Chapter 9).

††This rate-enhancing effect is an application of the Hammond postulate, discussed in Section 3.6.

10·2 REACTIONS OF ARYL HALIDES: THE HALIDE GROUP   295

PROBLEM 10·4
Write resonance contributors showing the complete delocalization of charge possible in the displacement intermediate formed when 1-bromo-4-cyano-2,5-dinitrobenzene reacts with sodium amide ($NaNH_2$).

ELIMINATION REACTIONS

We should not be surprised to find that base-catalyzed elimination reactions of aryl halides are much more difficult to achieve than the corresponding transformation of alkyl halides to alkenes. The acetylenic or cumulated triene grouping that would necessarily be formed in such an elimination reaction is structurally incompatible with the geometry of a six-membered carbon ring, since four of the carbon atoms would require a linear geometry (see Section 6.2).

In practice, vigorous treatment of chlorobenzene with a strong aqueous solution of sodium hydroxide causes a reaction, but the product is characteristic of substitution rather than elimination:

$$\text{C}_6\text{H}_5\text{Cl} \xrightarrow{\text{NaOH, H}_2\text{O}}_{340°, \text{ pressure}} \text{C}_6\text{H}_5\text{O}^\ominus \text{Na}^\oplus \xrightarrow{\text{H}_3\text{O}^\oplus} \text{C}_6\text{H}_5\text{OH} \quad (10.6)$$

phenol

For many years this transformation was the main industrial source of phenol, but its mechanism remained obscure until the late 1950s. Efforts to prepare substituted phenols by this method gave curious results which implied that these substitution reactions did not belong to any of the three general classes described earlier. For example, *p*-chlorotoluene yielded an equimolar mixture of the *para*- and *meta*-cresol isomers:

$$p\text{-ClC}_6\text{H}_4\text{CH}_3 \xrightarrow{\text{NaOH, H}_2\text{O}}_{\Delta, \text{ pressure}} [\text{salts}] \xrightarrow{\text{H}_3\text{O}^\oplus} p\text{-cresol} + m\text{-cresol} \quad (10.7)$$

Results of this kind provided a clue to the mechanism of these interesting reactions, and an important experiment involving the reaction of isotopically labeled bromobenzene (^{14}C) with potassium amide settled the question of an elimination intermediate once and for all:

$$\text{[}^{14}C\text{]-bromobenzene} \xrightarrow{\text{KNH}_2}_{\text{NH}_3(l)} \text{benzyne} \xrightarrow{\text{NH}_3}_{\text{fast}} \text{50\%} + \text{50\%} \quad (10.8)$$

The complete scrambling of the radioactive carbon atom between the position of amino substitution and the ortho positions can be convincingly explained only by the assumption that a highly reactive *aryne* inter-

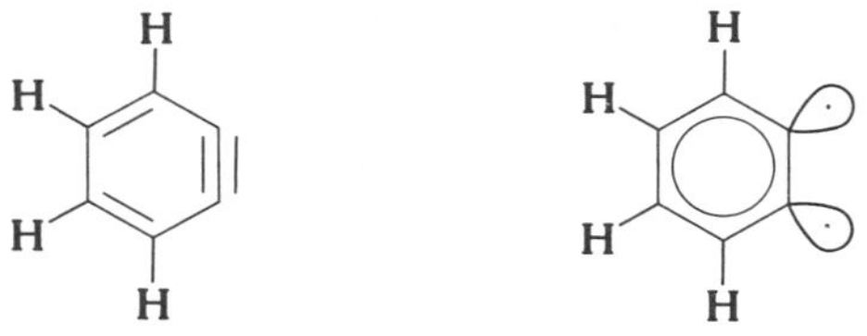

FIGURE 10·4 *Two representations of benzyne*

mediate—an aromatic ring with a triple bond—is formed in an initial elimination reaction. We call this intermediate *benzyne*, and it is most commonly represented by the triple-bonded structure in equation 10.8; another representation is shown in Figure 10.4.

Although benzyne has been trapped and studied at temperatures of about −200°C, it is exceedingly reactive and cannot be isolated under normal laboratory working conditions. We must be careful, therefore, to distinguish the triple bond in benzyne from that in an ordinary unstrained alkyne. Addition reactions to benzyne apparently proceed with very low activation energies; consequently, unsymmetrically substituted benzynes show very little selectivity in the fast product-forming step:

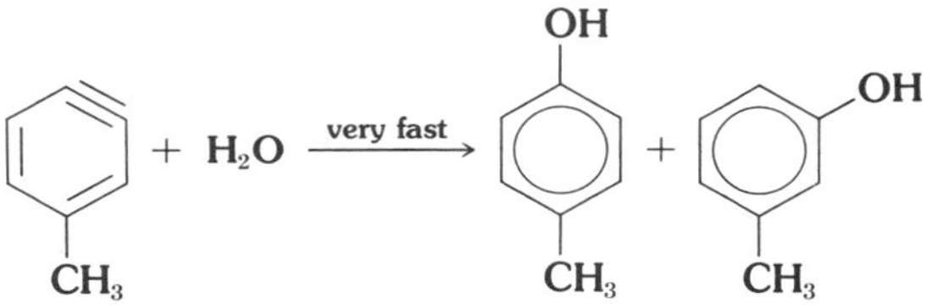

$$\text{(benzyne-CH}_3\text{)} + H_2O \xrightarrow{\text{very fast}} \text{(4-methylphenol)} + \text{(3-methylphenol)} \tag{10.9}$$

FORMATION OF ORGANOMETALLIC COMPOUNDS

Grignard reagents and aryl lithium compounds can be prepared directly from aryl halides, and both react rapidly and exothermically with active hydrogen functions such as OH and NH:

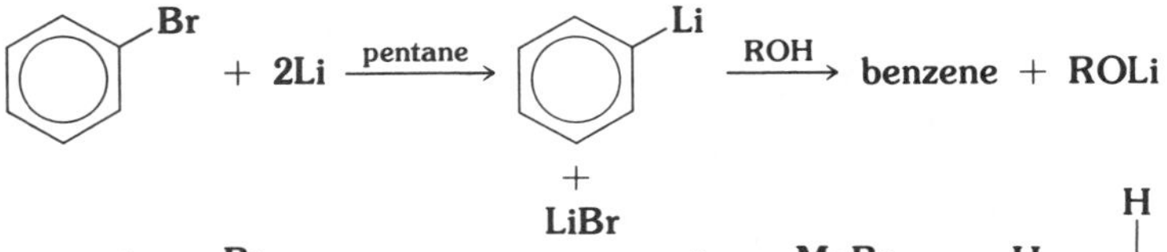

$$\text{PhBr} + 2Li \xrightarrow{\text{pentane}} \text{PhLi} + LiBr \xrightarrow{\text{ROH}} \text{benzene} + ROLi \tag{10.10}$$

$$\text{1,4-Br}_2C_6H_4 + 2Mg \xrightarrow{\text{ether}} \text{1,4-(BrMg)}_2C_6H_4 \xrightarrow{D_2O} \text{1,4-D}_2C_6H_4 + 2MgBrOD \tag{10.11}$$

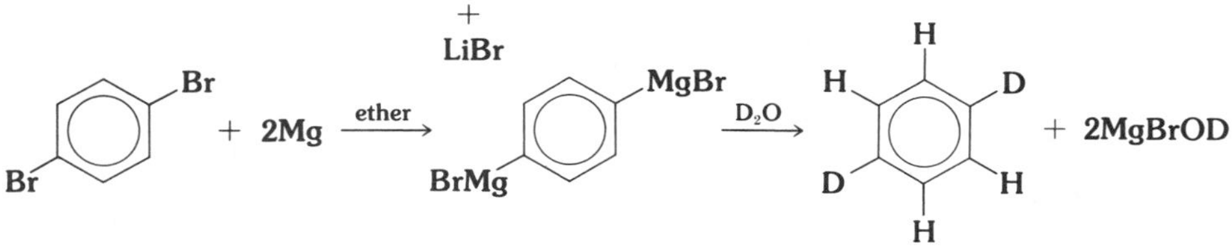

In ether solution aryl chlorides form Grignard reagents much less readily than do aryl bromides, although both halides react readily in tetrahydrofuran solution.

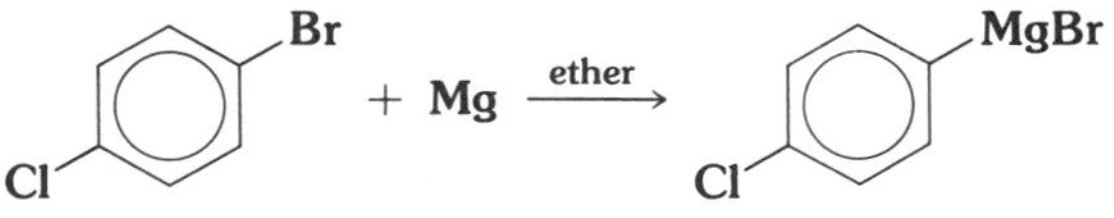

$$\text{4-Cl-C}_6H_4\text{-Br} + Mg \xrightarrow{\text{ether}} \text{4-Cl-C}_6H_4\text{-MgBr} \tag{10.12}$$

10·3 REACTIONS OF ARYL HALIDES AT THE AROMATIC RING

The formation of *bis*-Grignard reagents from dihalides (as in equation 10.11) usually proceeds well if the halogens are remote. However, halogen atoms ortho to each other often undergo elimination reactions which produce benzyne intermediates:

$$\text{o-BrC}_6\text{H}_4\text{F} \xrightarrow[\text{Li, pentane}]{\text{Mg, THF}} [\text{benzyne}] \longrightarrow \text{biphenylene, a stable dimer of benzyne} + \text{other products} \quad (10.13)$$

A similar elimination reaction was noted for alkyl halides in equation 3.26.

10·3 REACTIONS OF ARYL HALIDES AT THE AROMATIC RING

The aromatic ring of aryl halides is subject to the same electrophilic substitution reactions described for benzene itself in Chapter 9. In the absence of any electronic interaction with the halogen atom, we would expect a statistical distribution of the product isomers (40% ortho, 40% meta, and 20% para). Experimental studies clearly show this is not the case, indicating that the halogen substituent has a dramatic influence on the course of these substitution reactions. For example, chlorobenzene and bromobenzene are nitrated about one-thirtieth as rapidly as benzene and give predominantly the ortho and para nitro derivatives. A similar discrimination in favor of ortho and para product isomers is noted in sulfonation and halogenation reactions of aryl halides.

$$\text{C}_6\text{H}_5\text{X} \xrightarrow[\Delta]{\text{HNO}_3,\ \text{H}_2\text{SO}_4} \text{para-O}_2\text{N-C}_6\text{H}_4\text{X} + \text{ortho-O}_2\text{N-C}_6\text{H}_4\text{X} + \text{meta-O}_2\text{N-C}_6\text{H}_4\text{X} \quad (10.14)$$

(X = Cl or Br)   63–69%   30–36%   ~1%

$$\text{C}_6\text{H}_5\text{Br} \xrightarrow[\Delta]{\text{Cl}_2,\ \text{FeCl}_3} \text{para-Cl-C}_6\text{H}_4\text{Br} + \text{ortho-Cl-C}_6\text{H}_4\text{Br} + \text{meta-Cl-C}_6\text{H}_4\text{Br} \quad (10.15)$$

51%   42%   7%

We can explain these results by considering the general mechanism for electrophilic aromatic substitution described in Figures 9.16 and 9.17. Since the rate-determining step in these reactions is assumed to be the attack of an electrophilic species $E^{\oplus}$ on the electron-rich aromatic ring, any factor which decreases the nucleophilicity of the ring should decrease the rate of electrophilic substitution. Now, the experimentally measured dipole moment of chloro- or bromobenzene indicates that the

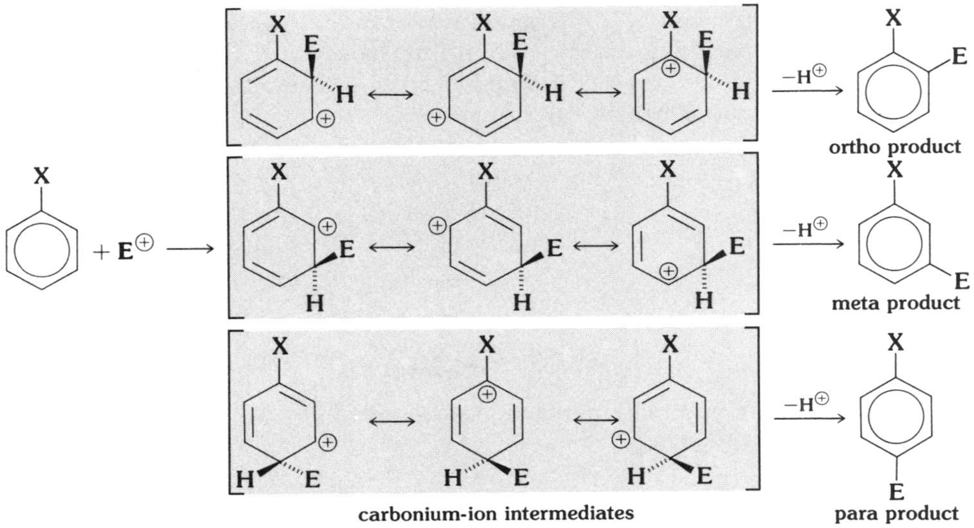

10·3 REACTIONS OF ARYL HALIDES AT THE AROMATIC RING

FIGURE 10·5 *Electrophilic substitution of a halobenzene*

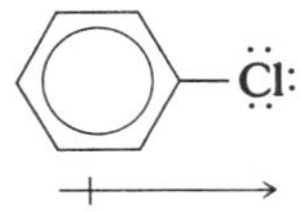

electronegative halogen atom has caused a shift of electrons away from the aromatic ring, leaving it less nucleophilic than benzene. Consequently, aryl halides should be less reactive than benzene toward electrophilic reagents, and this is generally found to be true.

The next step in our analysis of these reactions is to consider the relative activation energies required for electrophilic attack at the ortho, meta, and para positions of an aryl halide. Once again we assume that the more stable reaction intermediate is generated more rapidly (the Hammond postulate). The three intermediates formed during electrophilic substitution at the ortho, meta, and para positions of a halobenzene are shown in Figure 10.5.

At first glance, these carbonium-ion intermediates seem to be more or less identical; however, note that the intermediates leading to ortho and para products have the halogen atom bonded directly to one of the three electron-deficient carbon atoms, while the precursor to the meta product does not. In the former cases, the halogen atom is able to effect a further delocalization of charge by sharing one of its nonbonding electron pairs with the neighboring positively charged carbon atom (Figure 10.6). The result of this additional stabilization is that electrophilic substitution occurring ortho and para to the halogen atom in an aryl halide proceeds more rapidly than substitution at the meta position; thus the ortho and para products predominate. The fact that ortho substitution does not normally exceed para substitution can be explained in part by steric hindrance to ortho attack.

PROBLEM 10·5

a Write resonance contributors for all possible intermediates in the bromination of chlorobenzene.

b Draw energy diagrams comparing substitution at the para position with substitution meta to the chlorine.

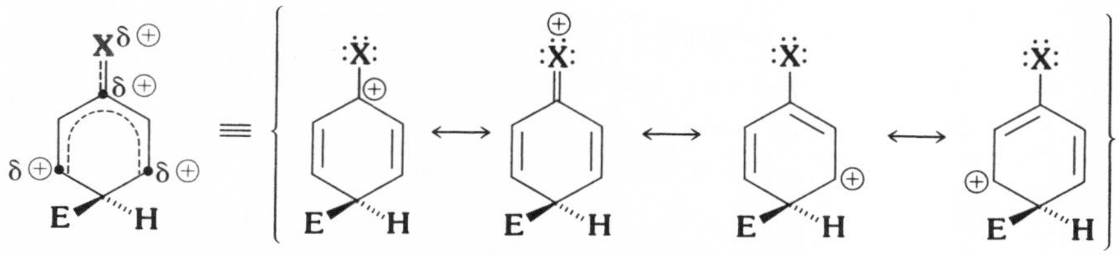

FIGURE 10·6 *Charge delocalization in an intermediate cation leading to para substitution of an aryl halide*

PROBLEM 10·6

The following reactions proceed in good yield. Discuss their mechanism in light of the preceding discussion.

a $CH_2=CHBr + HBr \xrightarrow{\text{no peroxides}} CH_3CHBr_2$

b $(CH_3)_3CCl + CH_2=CHCl \xrightarrow{FeCl_3} (CH_3)_3CCH_2CHCl_2$

We have noted in the previous paragraphs what might be termed a "schizophrenic" influence of the halogen atom on electrophilic substitution in aryl halides. Since halogen is more electronegative than carbon, it attracts electrons from the ring through its σ bond—the *inductive effect*. However, the halogen then helps to stabilize an adjacent electron-deficient center by sharing a pair of its valence shell electrons in a π bond—the *resonance effect*. In the competition between electron withdrawal and electron donation the former wins, and we find the aryl halides to be less reactive than benzene. The resonance effect, however, softens the inductive deactivation at the ortho and para positions; consequently, electrophilic substitution takes place preferentially at these sites. We shall encounter this competition again later in the chapter, but the outcome will not be the same.

10·4 PHENOLS AND THEIR DERIVATIVES

Phenols are compounds having a hydroxyl group bonded directly to an aromatic ring. In contrast to the aryl halides, phenols are widely distributed in nature. Phenol itself and the simple methyl-substituted phenols (cresols and xylenols) occur in coal tar, wood tar, and petroleum distillates. Thymol is found in thyme, eugenol and vanillin are important constituents of the essential oils from cloves and vanilla beans, and the urushiols are the active vesicants in poison ivy and related toxic plants of the genus *Anacardiacae*.

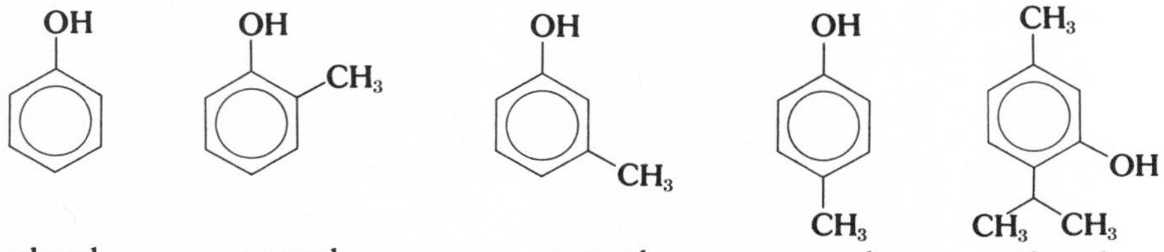

phenol   o-cresol   m-cresol   p-cresol   thymol

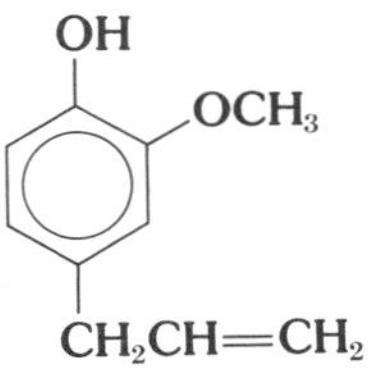

10·4 PHENOLS AND THEIR DERIVATIVES

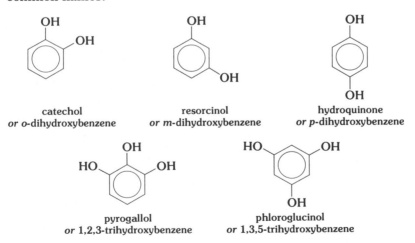

eugenol

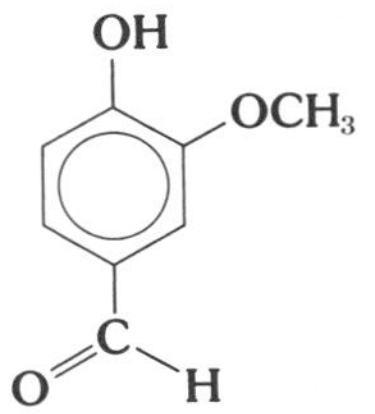

vanillin

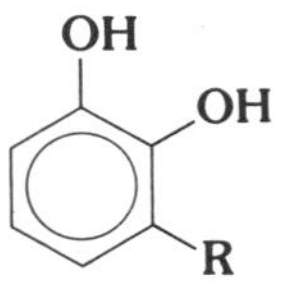

the urushiols
[R = (CH$_2$)$_7$CH=CH(CH$_2$)$_5$CH$_3$
or (CH$_2$)$_7$CH=CHCH$_2$CH=CHC$_2$H$_5$]

Many more complex phenolic compounds have also been isolated and identified. For example, the female hormone estradiol and most antibiotics belonging to the tetracycline family (aureomycin, terramycin, etc.) have a phenolic ring as part of their molecular structure.

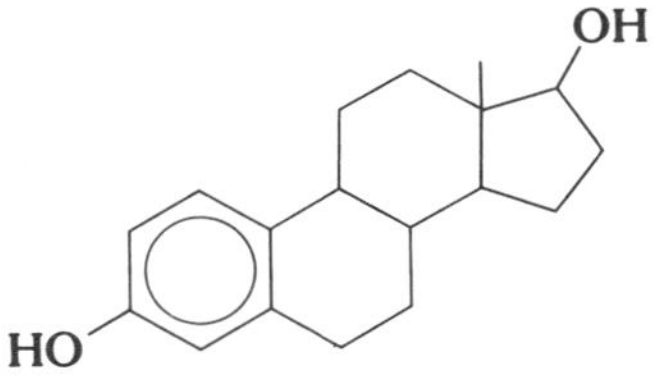

estradiol

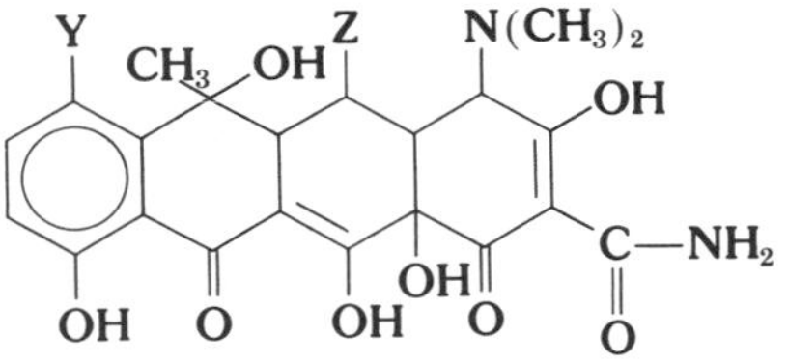

aureomycin (Y = Cl, Z = H)
terramycin (Y = H, Z = OH)

Many derivatives of di- and trihydroxybenzenes occur naturally (for example, eugenol and urushiols), and the parent phenols can be obtained by hydrolysis and/or pyrolysis of the gums and resins of certain trees. The most frequently encountered isomers are usually referred to by their common names:

catechol
or o-dihydroxybenzene

resorcinol
or m-dihydroxybenzene

hydroquinone
or p-dihydroxybenzene

pyrogallol
or 1,2,3-trihydroxybenzene

phloroglucinol
or 1,3,5-trihydroxybenzene

It is interesting to compare the keto-enol tautomerism of phenol with that of cyclohexanone. In Chapter 6 (Table 6.3) we saw that the thermodynamically more stable keto tautomers of simple aldehydes and ketones predominate overwhelmingly at equilibrium. However, the resonance stabilization of the aromatic ring in phenol is large enough to reverse the normal stability order of the tautomers, thereby tipping the equilibrium in favor of the enol isomer (the phenol).

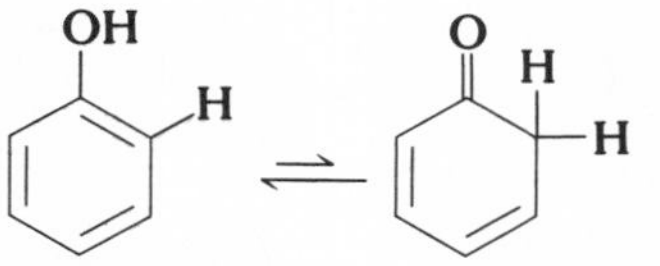

$$\text{phenol} \rightleftharpoons \text{keto tautomer} \qquad (10.16)$$

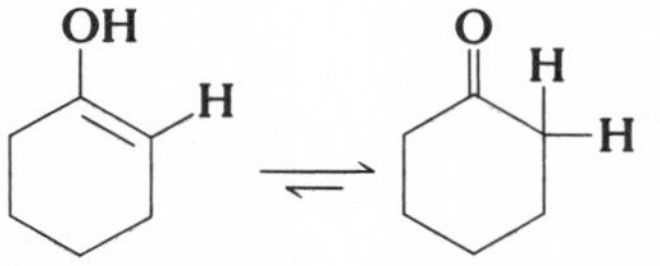

$$\text{enol tautomer} \rightleftharpoons \text{cyclohexanone} \qquad (10.17)$$

PROBLEM 10·7

a Write structural formulas for all possible keto tautomers of phloroglucinol.
b Catalytic hydrogenation of resorcinol results in the consumption of one molar equivalent of hydrogen. Write a plausible formula for the hydrogenation product and explain the stoichiometry of this reaction.

Phenols and other stable enol tautomers often give characteristic colors ranging from green to purple when mixed with an alcohol solution of ferric chloride. The nature of these reactions is not well understood, but they constitute a useful diagnostic test for the enol function.

Phenol is highly toxic to all kinds of cells and can cause severe burns if allowed to remain in contact with the skin.† In 1867 Lister used a dilute solution of phenol as an antiseptic. Subsequent studies showed that alkyl-substituted phenols had greater antiseptic power, as well as a slightly reduced cell toxicity. Consequently, synthetic phenol derivatives such as 4-*n*-hexylresorcinol are now widely used as mild antiseptics and disinfectants.

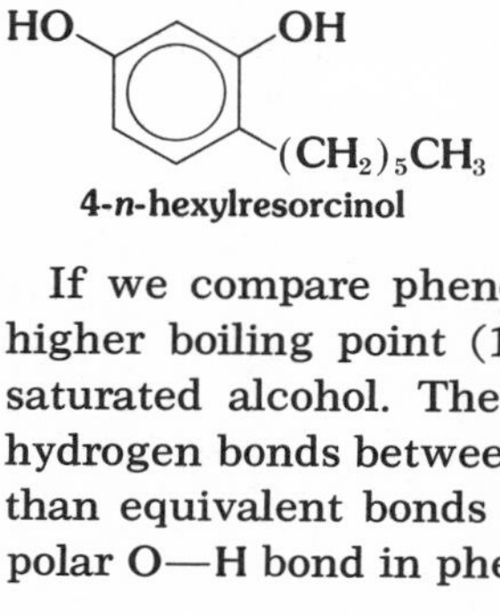

4-*n*-hexylresorcinol

If we compare phenol and cyclohexanol, we find that phenol has a higher boiling point (182°C) and greater solubility in water than the saturated alcohol. These differences suggest that the intermolecular hydrogen bonds between phenol hydroxyl functions are usually stronger than equivalent bonds involving alcohols. This in turn implies a more polar O—H bond in phenols than in alcohols:

polarity: $\overset{\delta\ominus}{R}O\text{—}\overset{\delta\oplus}{H}$ < $\overset{\delta\ominus}{Ar}O\text{—}\overset{\delta\oplus}{H}$

hydrogen-bond strength: $RO\text{—}H\text{---}O\overset{H}{\underset{H}{\diagup\diagdown}}$ < $ArO\text{—}H\text{---}O\overset{H}{\underset{H}{\diagup\diagdown}}$

†Recent investigations have disclosed that phenol, despite its toxicity, is produced and used as a sex attractant by the female grass grub beetle, *Costelytra Zealandica* (White).

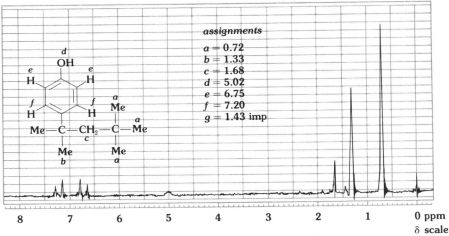

FIGURE 10·7 *Pmr spectrum of 2,4,4-trimethyl-2-(p-hydroxyphenyl)pentane*

In agreement with these ideas, we also find that phenols are stronger Brønsted acids than alcohols and exhibit lower-field hydroxyl-proton resonance signals ($\delta = 5.0$ to 9.0), as shown in Figure 10.7. The O—H stretching frequency in the infrared spectra of phenols, however, is very similar to that observed with alcohols (3200 to 3650 cm^{-1}, depending on the concentration of the sample solution).

10·5 REACTIONS OF PHENOLS

ACIDITY OF PHENOLS

Phenol is a much stronger Brønsted acid than cyclohexanol, as may be seen from the ionization constants determined (or extrapolated) in aqueous solution:

$$\text{phenol} + H_2O \rightleftharpoons \text{phenoxide} + H_3O^{\oplus} \qquad (10.18)$$

phenol
$K_a \approx 10^{-10}$

$$\text{cyclohexanol} + H_2O \rightleftharpoons \text{cyclohexoxide} + H_3O^{\oplus} \qquad (10.19)$$

cyclohexanol
$K_a \approx 10^{-18}$

Furthermore, this acidity is greatly enhanced by electron-withdrawing substituents such as NO$_2$ or CN located at ortho or para positions on the aromatic ring (but not at a meta position).

We can explain these increases in acidity by focusing our attention on

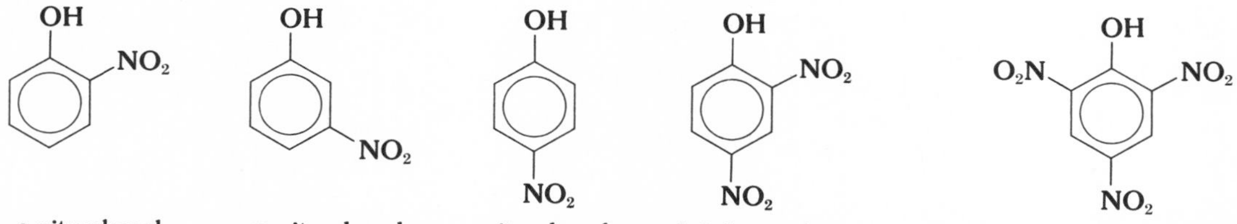

10·5 REACTIONS OF PHENOLS

o-nitrophenol
$K_a = 6.8 \times 10^{-8}$

m-nitrophenol
$K_a = 5.3 \times 10^{-9}$

p-nitrophenol
$K_a = 6.5 \times 10^{-8}$

2,4-dinitrophenol
$K_a = 8.3 \times 10^{-5}$

2,4,6-trinitrophenol or picric acid
$K_a = 0.4$

the relative stabilities of the corresponding conjugate bases. As a rule, strong acids give relatively stable conjugate bases on ionization, while weak acids have unstable (and reactive) conjugate bases. Since phenoxide anions are stabilized by the charge delocalization described in Figure 10.8, we would expect to find that these anions are weaker bases than alkoxide ions and that phenol compounds are stronger acids than alcohols.

As a consequence of the aromatic π-electron perturbation resulting from this charge delocalization, we should observe a significant change in the $\pi \longrightarrow \pi^*$ electronic absorption spectra (see Figure 9.4) when phenols are converted to their conjugate bases. The data in Table 10.2

FIGURE 10·8 *Charge delocalization in phenoxide anions*

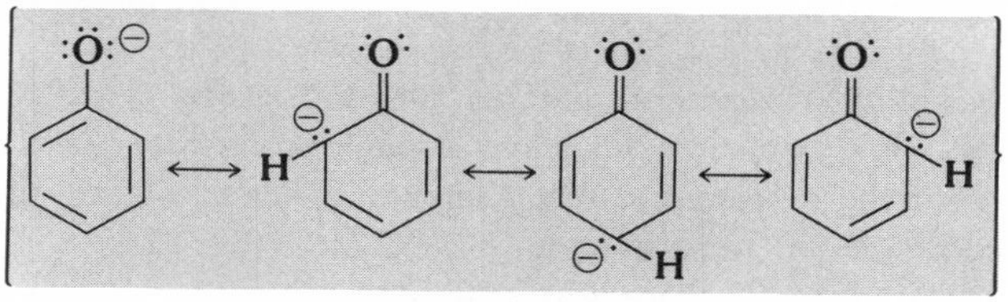

phenoxide anion

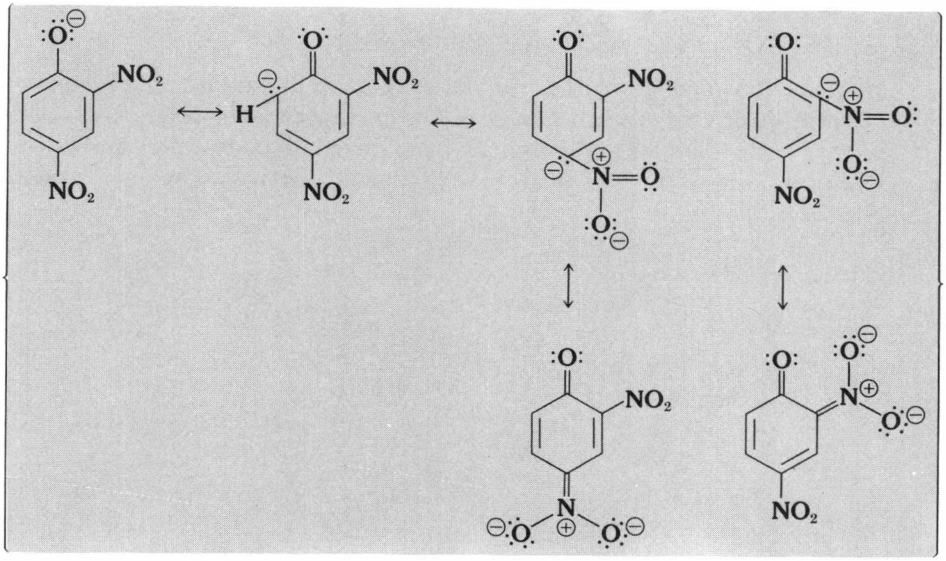

2,4-dinitrophenoxide anion

10·5 REACTIONS OF PHENOLS

TABLE 10·2 *Electronic spectra of phenol derivatives*

compound	solvent	λ_{max} (nm)	molar absorptivity ϵ
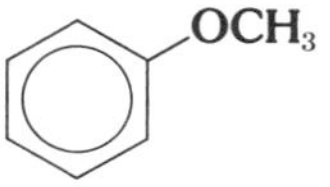 C₆H₅—OCH₃	CH_3OH	269	1,500
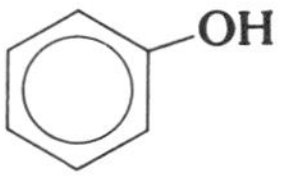 C₆H₅—OH	H_2O	270	1,450
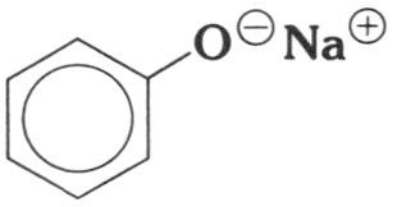 C₆H₅—O⁻Na⁺	H_2O	287	2,600

clearly show this to be the case. Phenol itself also exhibits p-π conjugation, as we see by comparing its absorption with that of toluene (λ_{max} = 261 nm, ϵ = 225); however, the electron-pair delocalization is less extensive than in the anion.

PROBLEM 10·8

The two ionization constants for carbonic acid are

$$H_2CO_3 + H_2O \rightleftharpoons HCO_3^- + H_3O^+ \qquad K_{a1} = 4.3 \times 10^{-7}$$
$$HCO_3^- + H_2O \rightleftharpoons CO_3^{2-} + H_3O^+ \qquad K_{a2} = 5.6 \times 10^{-11}$$

a Keeping in mind the general rule that strong acids have weak conjugate bases, while weaker acids have stronger conjugate bases, describe the acid-base equilibria resulting when phenol is added (1) to an aqueous $NaHCO_3$ solution and (2) to an aqueous Na_2CO_3 solution.
b Do the same for 2,4-dinitrophenol.

SUBSTITUTION REACTIONS AT OXYGEN

Substitution reactions at the phenol oxygen atom parallel the reactions described for alcohols in Chapter 8. For example, nucleophilic displacement of alkyl halides and sulfonate esters by phenoxide anions provides a useful route for the synthesis of aryl alkyl ethers (the Williamson ether synthesis):

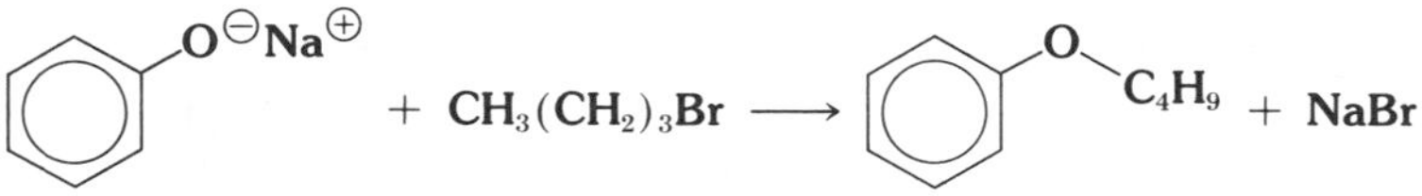

$$C_6H_5\text{—}O^-Na^+ + CH_3(CH_2)_3Br \longrightarrow C_6H_5\text{—}O\text{—}C_4H_9 + NaBr \qquad (10.20)$$

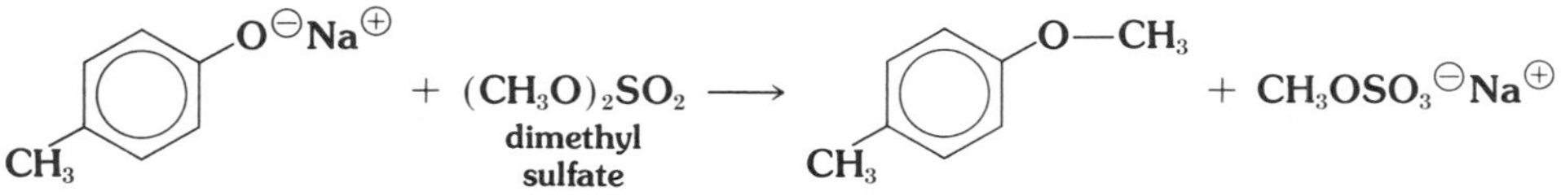

$$\text{4-}CH_3\text{-}C_6H_4\text{—}O^-Na^+ + (CH_3O)_2SO_2 \longrightarrow \text{4-}CH_3\text{-}C_6H_4\text{—}O\text{—}CH_3 + CH_3OSO_3^-Na^+ \qquad (10.21)$$

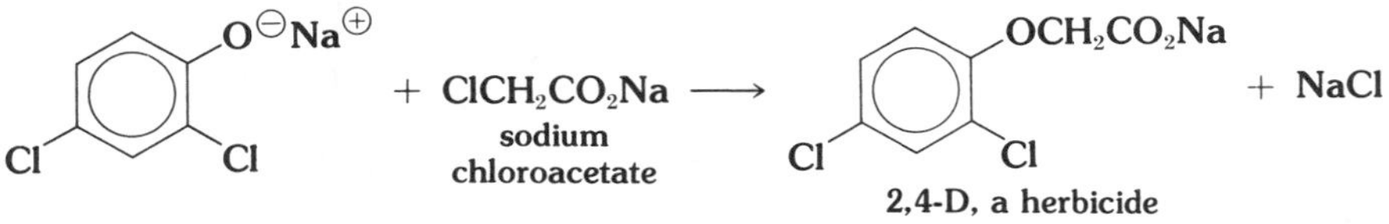

dimethyl sulfate

$$\text{2,4-}Cl_2\text{-}C_6H_3\text{—}O^-Na^+ + ClCH_2CO_2Na \longrightarrow \text{2,4-}Cl_2\text{-}C_6H_3\text{—}OCH_2CO_2Na + NaCl \qquad (10.22)$$

sodium chloroacetate

2,4-D, a herbicide

10·5 REACTIONS OF PHENOLS

As noted in Section 10.2, ether formation involving displacement of aryl halide by alkoxide (or phenoxide) anions is generally not possible unless suitably located electron-withdrawing substituents activate the aryl halide:

$$\text{PhO}^- K^+ + \text{(4-NO}_2\text{-2-Cl-C}_6\text{H}_3\text{)} \longrightarrow \text{O}_2\text{N-C}_6\text{H}_3\text{-O-C}_6\text{H}_5 + \text{KCl} \qquad (10.23)$$

Esterification of phenols can generally be accomplished by the action of acid chlorides or acid anhydrides. The preparation of tricresyl phosphate, a useful plasticizer and fuel additive, can also be considered an esterification of this type if we regard phosphorus oxychloride as the *tris*-acid chloride of phosphoric acid.

$$\text{PhOH} + \text{CH}_3\text{COCl} \longrightarrow \text{Ph-O-CO-CH}_3 + \text{HCl} \qquad (10.24)$$
phenyl acetate

$$3\, \text{(4-CH}_3\text{-C}_6\text{H}_4\text{OH)} + \text{POCl}_3 \longrightarrow [\text{(4-CH}_3\text{-C}_6\text{H}_4\text{-O)}]_3\text{PO} + 3\text{HCl} \qquad (10.25)$$
phosphorus oxychloride — tricresyl phosphate (TCP)

SUBSTITUTION AT CARBON

Phenols and their ethers seldom undergo substitution or elimination reactions involving cleavage of the aryl carbon-oxygen bond. Thus acid-catalyzed cleavage of aryl alkyl ethers usually proceeds with formation of phenols and alkyl halides, rather than aryl halides and alcohols. In this respect the behavior of the phenols parallels the relatively inert character of the aryl halides.

$$\text{(2-OCH}_3\text{-C}_6\text{H}_4\text{-OH)} + \text{HBr} \xrightarrow{\Delta} \text{(2-OH-C}_6\text{H}_4\text{-OH)} + \text{CH}_3\text{Br} \qquad (10.26)$$

REACTIONS AT THE AROMATIC RING

We saw earlier that halogen substituents on a benzene ring strikingly influence the course of electrophilic substitution reactions at the ring. It comes as no surprise, therefore, to find that hydroxyl and alkoxyl substituents also exert a strong perturbing effect on these reactions. However, in contrast to the aryl halides, phenols and their ether derivatives strongly *activate* the aromatic ring to electrophilic substitution. Indeed, phenol is so reactive (bromination proceeds about 10^{11} times faster than

10·5 REACTIONS OF PHENOLS

with benzene) that many reactions are difficult to control and often lead to complex mixtures of products. As a general rule, electrophilic attack on phenols and their derivatives occurs most rapidly at the ortho and para positions:

$$\text{PhOH} + \text{Br}_2 \xrightarrow{\text{CCl}_4, 0°} \text{p-BrC}_6\text{H}_4\text{OH (67\%)} + \text{o-BrC}_6\text{H}_4\text{OH (33\%)}$$

$$\text{PhOH} + \text{Br}_2 \xrightarrow{\text{H}_2\text{O}, 25°} \text{2,4,6-tribromophenol}$$

(10.27)

$$\text{PhOH} + \text{H}_2\text{SO}_4 \text{ (concentrated)} \longrightarrow \text{o-HOC}_6\text{H}_4\text{SO}_3\text{H (main product at 25°C)} + \text{p-HO}_3\text{SC}_6\text{H}_4\text{OH (main product at 100°C)}$$

(10.28)

$$\text{PhOH} + \text{HNO}_3 \text{ (dilute)} \longrightarrow \text{p-O}_2\text{NC}_6\text{H}_4\text{OH} + \text{o-O}_2\text{NC}_6\text{H}_4\text{OH} + \text{mixture of oxidation products}$$

(10.29)

PROBLEM 10·9

ortho-Nitrophenol has a lower boiling point (214°C) and is less soluble in water than the para isomer (bp > 250°C with decomposition). These differences do not appear in the *ortho*- and *para*-nitroanisoles (methyl ether derivatives of the phenols), which both boil at about 260°C. Explain this behavior of the isomeric nitrophenols. *Hint:* Consider the hydrogen-bonding possibilities for both phenols.

The fact that hydroxyl and alkoxyl substituents on a benzene ring favor electrophilic attack at the ortho and para positions can again be explained in terms of stabilization of the reactive intermediate through charge delocalization (see Figure 10.9). However, the extraordinary enhancement of the reactivity of phenols and their ethers toward common electrophilic reagents is less easily explained. Since oxygen is more electronegative than either chlorine or bromine, hydroxyl and alkoxyl substitutents should cause a greater inductive withdrawal of electrons from the aromatic ring. This, of course, would be expected to decrease the reactivity of phenols and aryl ethers. Only by assuming an even greater electron release to the π-electron system from the oxygen (the resonance effect) can we bring our theory into agreement with the facts. This is not as arbitrary a suggestion as it might seem, since the π bonding necessary for p-π delocalization of an adjacent nonbonding electron pair

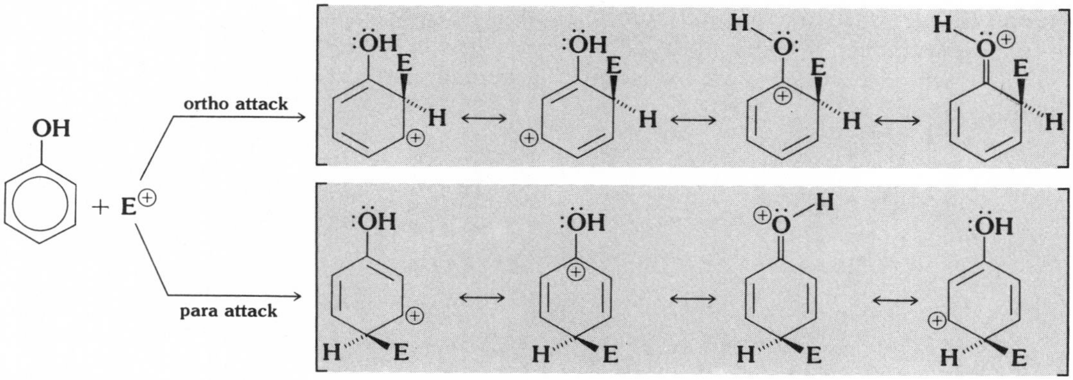

FIGURE 10·9 *Charge delocalization in intermediate cations leading to ortho and para substitution of phenol*

(as in Figure 10.9) is more effective between atoms of similar size, such as carbon and oxygen, than between a small atom such as carbon and a larger one such as bromine or chlorine.

The predominance of electron release over electron withdrawal by oxygen substituents on an aromatic ring is confirmed by the experimentally determined dipole moments of phenol and anisole, which are roughly equivalent to, but opposite in direction from, the dipole moments of chloro- and bromobenzene.

10·6 QUINONES

Quinones are cyclic conjugated diketones which, although not aromatic themselves, are easily transformed into dihydroxy derivatives of aromatic systems. The complex mixtures that are formed when phenols are oxidized often contain quinones; however, mild oxidation of *ortho-* or *para*-dihydroxy benzenes provides a more efficient route to these compounds:

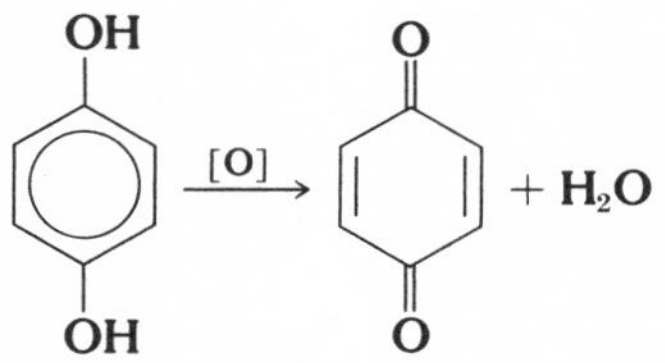

hydroquinone → p-benzoquinone + H_2O (10.30)

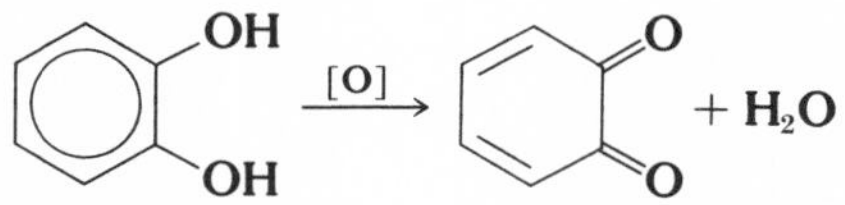

catechol → o-benzoquinone + H_2O (10.31)

10·6 QUINONES

Vitamin K, the antihemorrhagic factor in blood, is a naturally occurring 1,4-naphthoquinone. Walnut hulls contain a trihydroxy naphthalene derivative which undergoes air oxidation to a similar naphthoquinone called juglone).

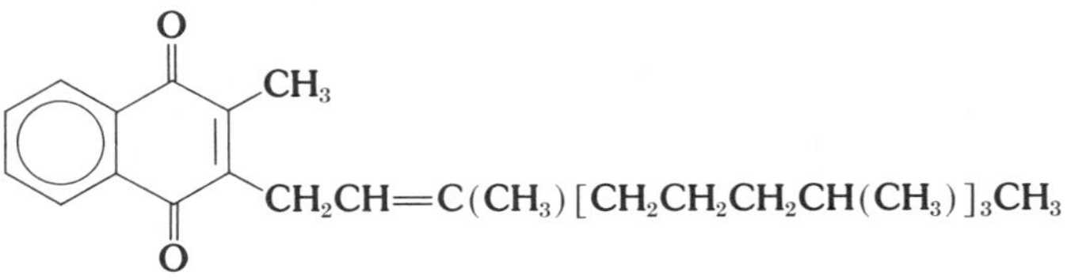

vitamin K

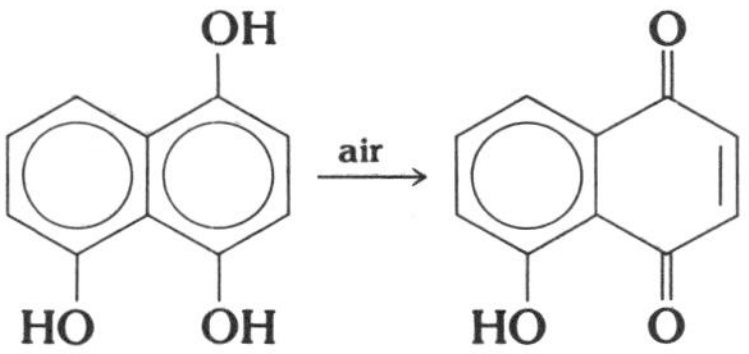

from walnuts → air → juglone

The reduction of quinone and the oxidation of hydroquinone are sufficiently rapid and reversible to give reproducible electrode potentials, which are essentially a measure of the free energy liberated in the quinone reduction:

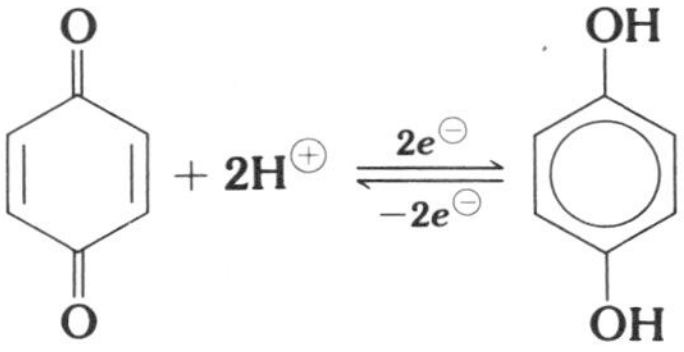

$$\text{quinone} + 2H^{\oplus} \underset{-2e^{\ominus}}{\overset{2e^{\ominus}}{\rightleftarrows}} \text{hydroquinone} \tag{10.32}$$

Different quinones have different reduction potentials, and from these potentials we can obtain information about the relative stabilities of variously substituted ring systems. Since the position of the quinone-hydroquinone equilibrium is proportional to the square of the hydrogen-ion concentration, this electrode potential may also be used to measure pH value.

PROBLEM 10·10
Equations 10.30 and 10.31 show the formation of *ortho-* and *para-*quinones. Can you write a reasonable structure for a *meta-*quinone?

PROBLEM 10·11
The energy released in the reduction of quinone I is roughly 14 kcal/mole greater than that generated in the reduction of quinone II. What bearing does this have on the question of cyclobutadiene stability and aromaticity (see Figure 9.19)?

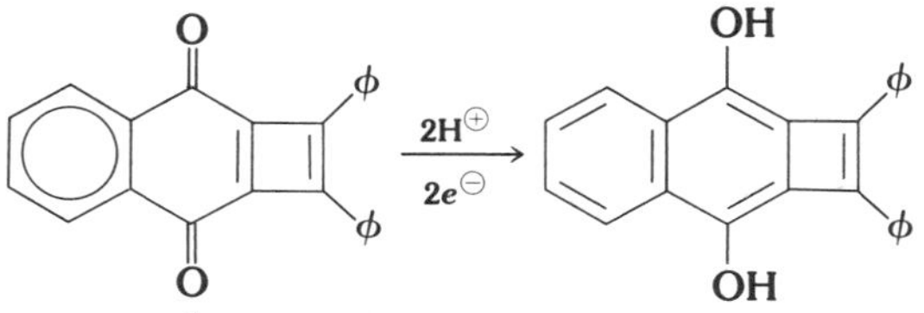

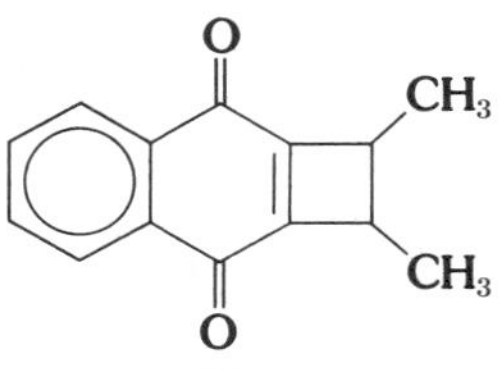

Certain large beetles of the genus *Brachinus* employ a unique defense mechanism in which they spray attackers with a boiling aqueous solution of quinone from a pair of glands at the tip of their abdomens. An inner compartment of each gland contains an aqueous solution of hydroquinone and hydrogen peroxide. When this solution is forced into an outer chamber of the gland, enzymes catalyze the decomposition of the hydrogen peroxide and the oxidation of the hydroquinone:

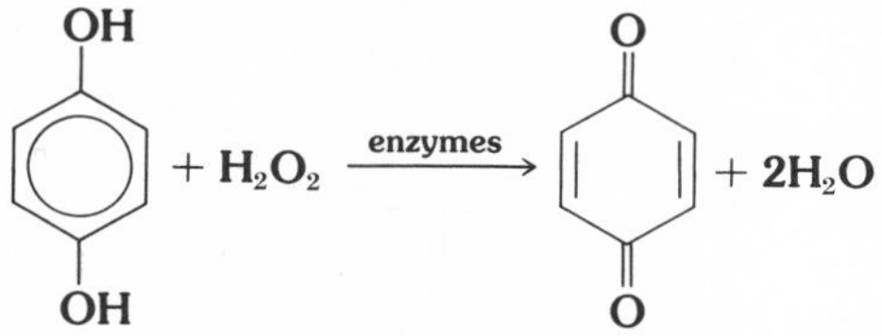

$$\text{hydroquinone} + H_2O_2 \xrightarrow{\text{enzymes}} \text{quinone} + 2H_2O \qquad \Delta H \approx -48 \text{ kcal/mole} \tag{10.33}$$

The heat evolved in these reactions is sufficient to bring the solution to its boiling point, causing it to discharge from an opening in the outer chamber with an audible report—hence the popular name "bombardier beetle" for this insect.

10·7 NONBENZENOID COMPOUNDS RELATED TO PHENOLS

A number of unsaturated cyclic alcohols have been identified which are not properly classified as phenols, but have high acidity ($K_a > 10^{-8}$) and significant resonance stabilization. These compounds fall roughly into two classes.

The *tropolones* are analogs of phenols in which the aromatic ring has been expanded to seven members by the inclusion of a carbonyl group:

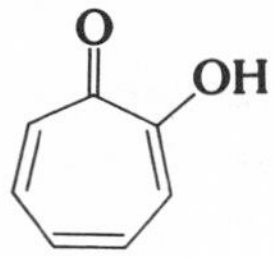

tropolone
$K_a \approx 10^{-7}$

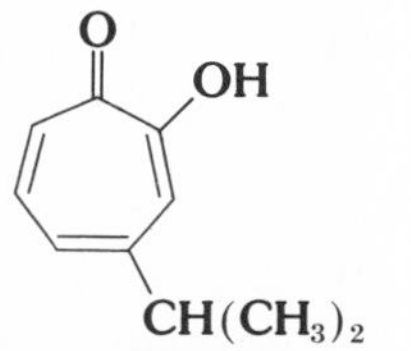

β-thujaplicin
(from the red cedar)

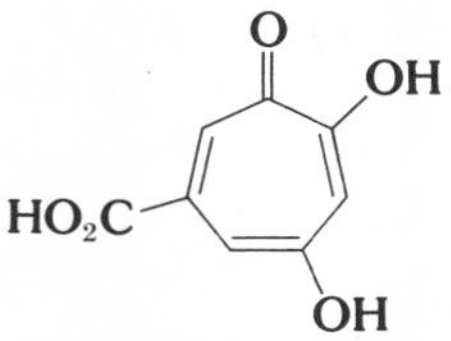

stipitatic acid
(from a mold)

The alkaloid colchicine (from the autumn crocus) is a naturally occurring tropolone methyl ether. It is used in the treatment of gout and has intrigued biologists because of its ability to arrest cell division in plants and animals.

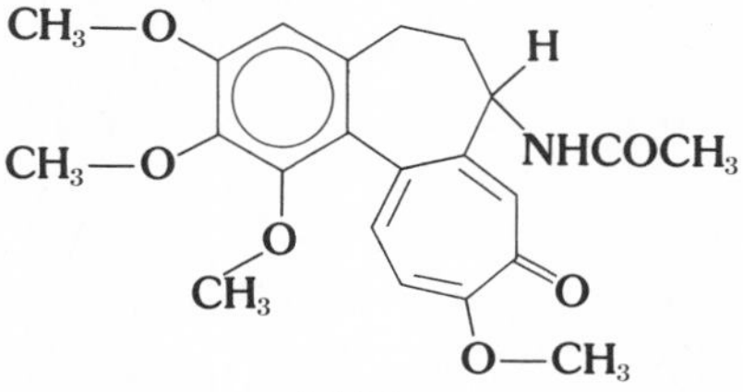

colchicine

The *oxocarbons* are a group of highly oxidized cyclic compounds having the general formula $C_nH_2O_n$, where n is an integer equal to or larger than 3:

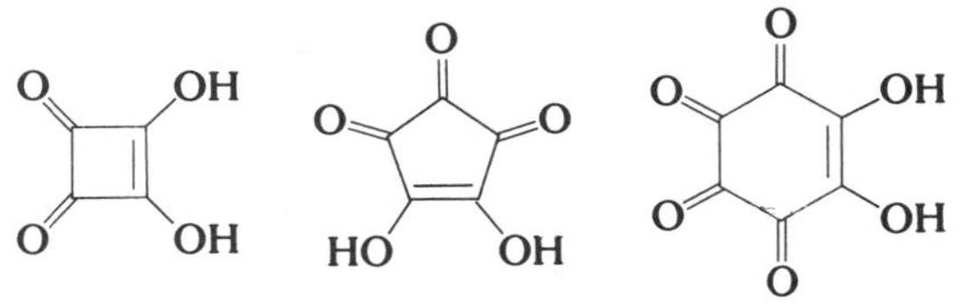

squaric acid
$K_{a^1} = 6 \times 10^{-2}$
$K_{a^2} = 3 \times 10^{-4}$

croconic acid
$K_{a^1} = 0.2$
$K_{a^2} = 1 \times 10^{-2}$

rhodizonic acid
$K_{a^1} = 10^{-3}$
$K_{a^2} = 8 \times 10^{-4}$

The high acidity of these compounds (croconic acid is about as strong as sulfuric acid) results from the extensive charge delocalization that is possible in their symmetrical planar anions. Indeed, we may regard the oxocarbon dianions having the formula $C_nO_n^{-2}$ as an aromatic series. Two examples are

10·8 BENZYL AND ALLYL HALIDES AND ALCOHOLS

Halogen atoms and hydroxyl groups directly bonded to an aromatic ring show marked resistance to nucleophilic substitution and elimination reactions. As we shall see, however, compounds having a halogen or hydroxyl group bonded to a *benzylic* carbon atom, a carbon atom adjacent to an aromatic ring, generally exhibit greater reactivity than simple alkyl halides and alcohols. If the functional group is in a more remote position, the aromatic ring does not normally affect its chemical behavior. This general pattern is summarized in Figure 10.10.

S_N2 REACTIONS OF BENZYL HALIDES

The enhanced reactivity of benzyl halides under S_N2 reaction conditions is apparent from the relative-rate measurements shown in Table 10.3. Here the combination of a poor ionizing solvent (acetone) and a powerful nucleophile ($I^{\ominus}$) provides an ideal system for examining structure-reactivity relationships in S_N2 reactions (see Section 3.4). Kinetic and stereochemical evidence supporting an S_N2 mechanism for many other reactions of benzyl halides has been obtained. For example, the reaction of optically active 1-chloro-1-phenylethane with sodium ethoxide (a poorer nucleophile than iodide ion) is kinetically second order and pro-

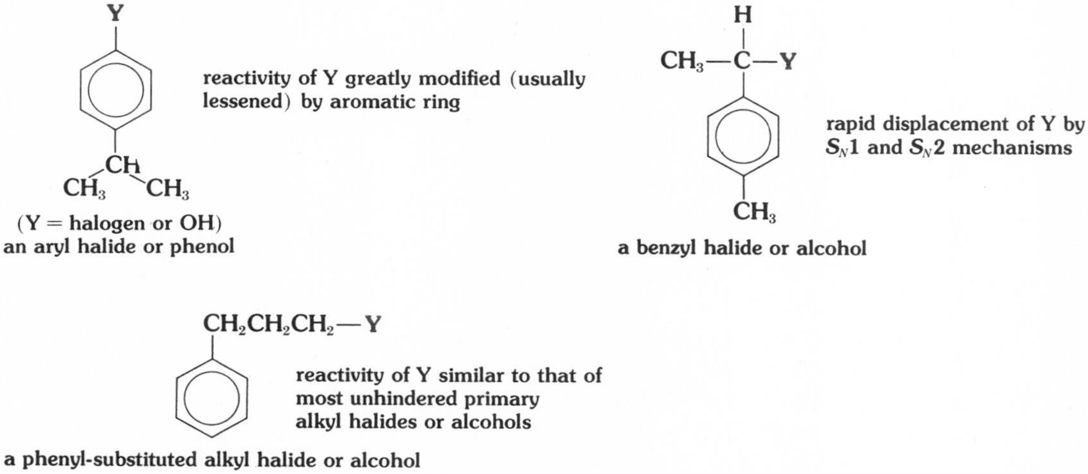

10·8 BENZYL AND ALLYL HALIDES AND ALCOHOLS

reactivity of Y greatly modified (usually lessened) by aromatic ring

(Y = halogen or OH)
an aryl halide or phenol

rapid displacement of Y by $S_N 1$ and $S_N 2$ mechanisms

a benzyl halide or alcohol

reactivity of Y similar to that of most unhindered primary alkyl halides or alcohols

a phenyl-substituted alkyl halide or alcohol

FIGURE 10·10 *Functional-group interaction with aromatic rings*

ceeds with complete inversion of configuration at the carbon atom bearing the substituents:

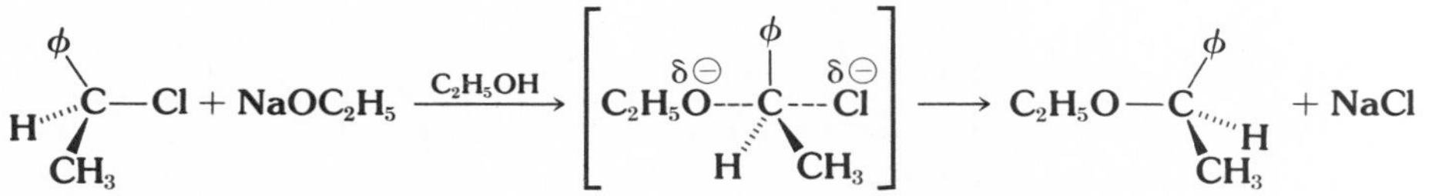

$$\phi\text{C}(H)(CH_3)\text{-Cl} + \text{NaOC}_2H_5 \xrightarrow{C_2H_5OH} \left[C_2H_5O\overset{\delta\ominus}{\cdots}\underset{H\ CH_3}{\overset{\phi}{C}}\cdots\overset{\delta\ominus}{Cl} \right] \longrightarrow C_2H_5O\text{-}C(\phi)(H)(CH_3) + \text{NaCl} \quad (10.34)$$

$(\phi = C_6H_5)$

This corresponds, of course, to the inversion transition state described in Figure 3.5.

It is interesting to note that para-oriented electron-withdrawing or electron-donating substituents on the benzylic aromatic ring do not exert a strong influence on the rates of $S_N 2$ reactions. For example, the rate of iodide-ion displacement of *p*-nitrobenzyl chloride is about six times greater than that of the unsubstituted benzyl chloride, and a *p*-methoxy substituent shows an even smaller effect (about a fourfold increase in rate for $Z = OCH_3$ over $Z = H$):

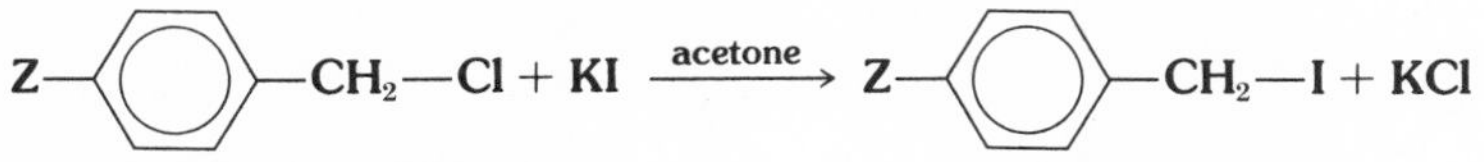

$$Z\text{-}\phi\text{-}CH_2\text{-}Cl + KI \xrightarrow{\text{acetone}} Z\text{-}\phi\text{-}CH_2\text{-}I + KCl \quad (10.35)$$

$(Z + H, NO_2, CH_3, OCH_3)$

TABLE 10·3 *Relative nucleophilic substitution rates*

$$R\text{-}Br + NaI \xrightarrow{\text{acetone}} R\text{-}I + NaBr$$

compound	relative rate
C_2H_5Br	1.00 (standard)
$(CH_3)_2CHBr$	0.03
CH_3Br	30.0
ϕCH_2Br	120.0

ϕ—CH$_2$—Y + Z:$^\ominus$ ⟶ [structure with Z$^{\delta\ominus}$ and Y$^{\delta\ominus}$] ⟶ ϕ—CH$_2$—Z + Y:$^\ominus$

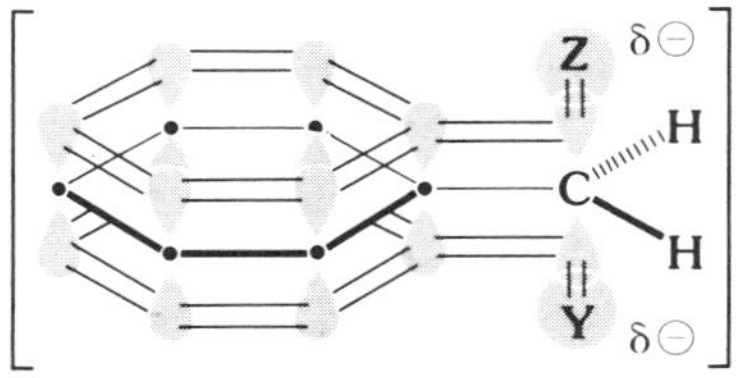

FIGURE 10·11 *Overlap of p atomic orbitals in S_N2 reactions of benzyl halides*

We shall observe much larger substituent effects, when we examine the S_N1 reactions of benzyl halides.

The factors responsible for the enhanced S_N2 reactivity of benzyl halides are not yet clearly understood. However, overlap of the developing and breaking bonds with the π orbitals of the aromatic ring is believed to lower the energy of the transition state, thus increasing the rate of the substitution reaction. The atomic orbitals involved in this extended molecular-orbital system are shown in Figure 10.11.

S_N2 REACTIONS OF ALLYL AND PROPARGYL HALIDES

If S_N2 transition states are in fact stabilized by overlap with adjacent π orbitals, then rate enhancements similar to those observed with benzyl halides should also occur with allyl and propargyl halides. In this case we would expect a transition state like that illustrated in Figure 10.12. The experimentally determined relative rates listed in Table 10.4 provide strong support for this hypothesis.

The fact that allyl halides react so much more rapidly than vinyl halides allows a striking selectivity in the displacement reactions of certain dihalides:

$$HC\equiv C^\ominus Na^\oplus + BrCH_2-\underset{\underset{Br}{|}}{C}=CH_2 \longrightarrow HC\equiv CCH_2-\underset{\underset{Br}{|}}{C}=CH_2 + NaBr \qquad (10.36)$$

TABLE 10·4 *Relative nucleophilic substitution rates*

$$R-Cl + NaOC_2H_5 \longrightarrow R-OC_2H_5 + NaCl$$

compound	relative rate
n-C$_3$H$_7$Cl	1.00
H$_2$C=CHCH$_2$Cl	40.0
CH$_3$CH=CHCH$_2$Cl	1.04×10^2
ϕCH=CHCH$_2$Cl	2.64×10^2
(CH$_3$)$_2$C=CHCH$_2$Cl	8.00×10^2
HC≡CCH$_2$Cl	4.12×10^3

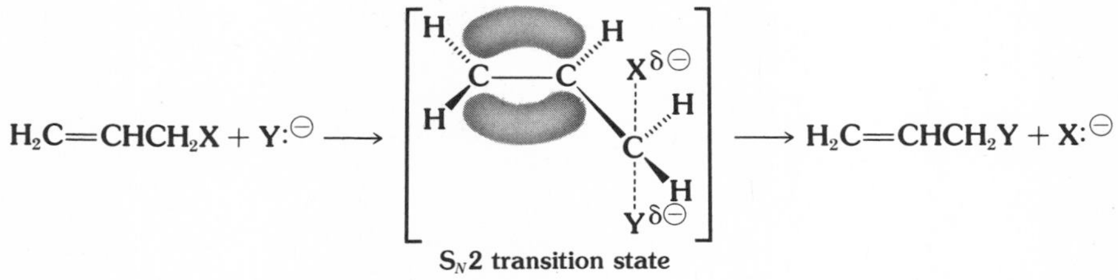

10·8 BENZYL AND ALLYL HALIDES AND ALCOHOLS   313

$$H_2C=CHCH_2X + Y:^{\ominus} \longrightarrow \left[\begin{array}{c} \text{transition state} \end{array} \right] \longrightarrow H_2C=CHCH_2Y + X:^{\ominus}$$

S_N2 transition state

FIGURE 10·12 *S_N2 reaction of an allyl halide*

PROBLEM 10·12
Predict the major product from each of the following reactions (in two cases no reaction is expected):

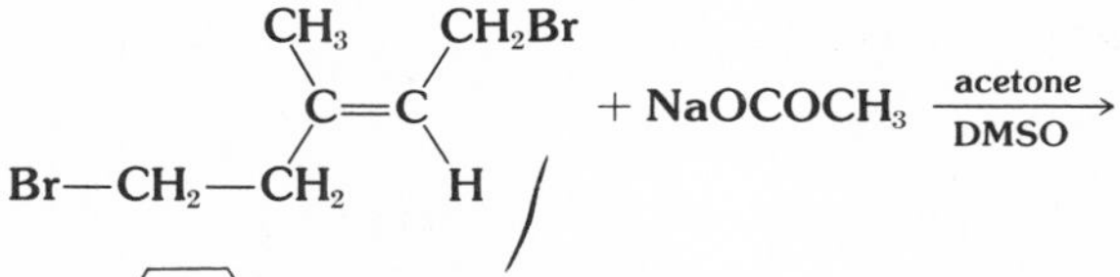

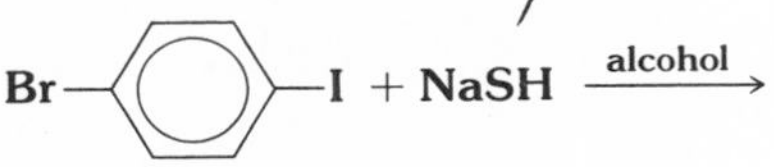

a. 4-Br-C$_6$H$_4$-CH$_2$Cl + KCN $\xrightarrow{\text{alcohol}}$

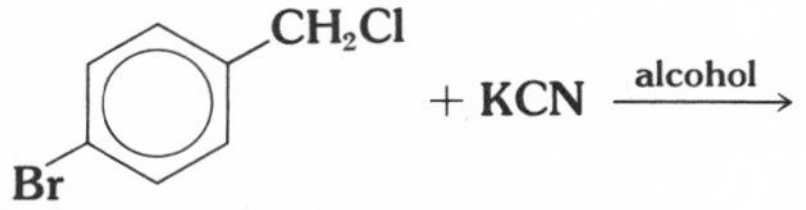

b. (BrCH$_2$-CH$_2$)(CH$_3$)C=C(CH$_2$Br)(H) + NaOCOCH$_3$ $\xrightarrow[\text{DMSO}]{\text{acetone}}$

c. Br-C$_6$H$_4$-I + NaSH $\xrightarrow{\text{alcohol}}$

d. (4-Br-2,6-(CH$_3$)$_2$-C$_6$H$_2$)-CH(Cl)-C(CH$_3$)$_3$ + KI $\xrightarrow{\text{acetone}}$

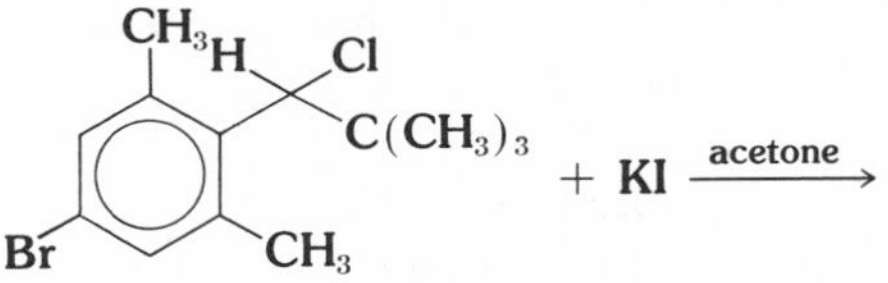

S_N1 REACTIONS OF BENZYL DERIVATIVES

A solution of optically active 1-chloro-1-phenylethane in an 80:20 water–acetone mixture undergoes hydrolysis to racemic 1-phenyl-1-ethanol at a rate roughly equal to the corresponding hydrolysis of *t*-butyl chloride:

$$\phi-\underset{H}{\overset{Cl}{\underset{|}{C}}}-CH_3 + 2H_2O \xrightarrow{\text{aqueous acetone}} \phi-\underset{H}{\overset{OH}{\underset{|}{C}}}-CH_3 + H_3O^{\oplus} + Cl^{\ominus} \quad (10.37)$$
$$(+) \qquad\qquad (\pm)$$

$$(CH_3)_3CCl + 2H_2O \xrightarrow{\text{aqueous acetone}} (CH_3)_3COH + H_3O^{\oplus} + Cl^{\ominus} \quad (10.38)$$

The rate of this *solvolysis reaction* is proportional to the alkyl chloride concentration but is not affected by moderate changes in hydroxide-ion concentration (note that OH$^{\ominus}$ is a stronger nucleophile than H$_2$O); hence an S_N1 mechanism is probably operating.

The fact that the secondary alkyl chloride in reaction 10.37 ionizes at about the same rate as *t*-butyl chloride can be attributed to resonance

10·8 BENZYL AND ALLYL HALIDES AND ALCOHOLS

stabilization of the benzyl cation which is formed as an intermediate in this reaction:

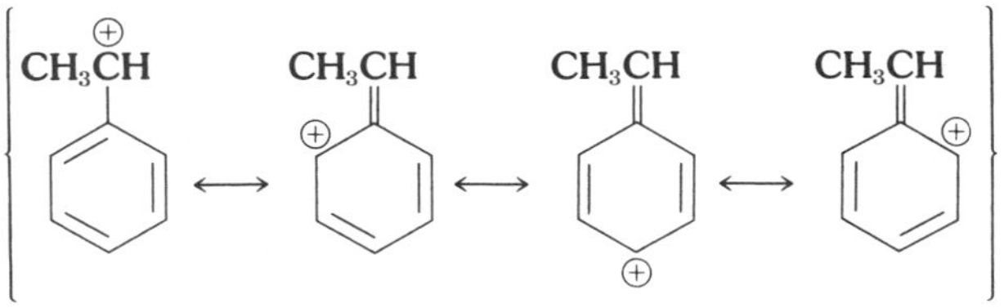

[resonance structures of benzyl cation]

Extending this explanation, we would expect a second phenyl substituent on the benzylic carbon atom to provide additional stabilization (as should a third phenyl substituent), and it is gratifying to find that solvolysis experiments with phenyl-substituted methyl halides show exactly this order of reactivity. The relative reaction rates of these compounds are listed in Table 10.5.

PROBLEM 10·13
Write a step-by-step mechanism and draw an energy diagram illustrating the hydrolysis of benzyl bromide in 80% aqueous acetone.

PROBLEM 10·14
Write the important contributing structures for a resonance description of the triphenylmethyl cation ($\phi_3 C^\oplus$).

PROBLEM 10·15
Is the stereochemical result of reaction 10.37 in accordance with an $S_N 1$ mechanism?

Since the delocalization of positive charge in a benzyl cation is most pronounced at the ortho and para positions of the aromatic ring, substituents at these sites should strongly influence the rate of isomerization of benzyl derivatives. The relative hydrolysis rates of para-substituted benzyl tosylates, shown in Table 10.6, illustrate this effect.

PROBLEM 10·16
a Explain the relative effects of the Z substituents in Table 10.6.
b The solvolysis of *p*-methoxybenzyl chloride in a 40:60 ethanol-ether solvent proceeds roughly 10,000 times more rapidly than solvolysis of the meta isomer. Write equations for both these reactions and explain the difference in reaction rates.

TABLE 10·5 *Relative reaction rates for solvolysis of phenyl-substituted methyl halides*

$$R-Cl + 2C_2H_5OH \longrightarrow R-OC_2H_5 + C_2H_5OH_2^\oplus + Cl^\ominus$$

compound	relative rate
$\phi CH_2 Cl$	1.0
$\phi_2 CHCl$	2.0×10^3
$\phi_3 CCl$	3.0×10^7

TABLE 10·6 *Relative reaction rates for hydrolysis of benzyl tosylates*

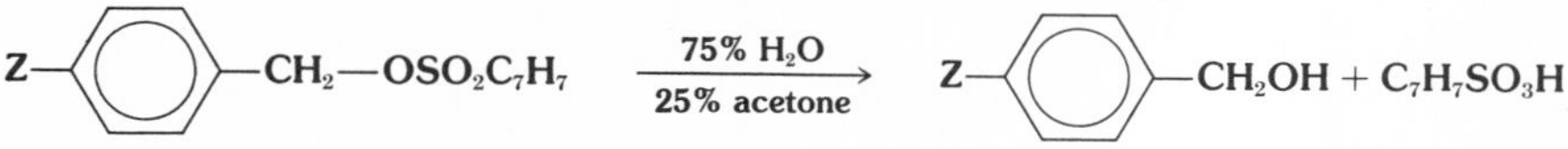

$$Z\text{-}C_6H_4\text{-}CH_2\text{-}OSO_2C_7H_7 \xrightarrow[25\% \text{ acetone}]{75\% \text{ H}_2\text{O}} Z\text{-}C_6H_4\text{-}CH_2OH + C_7H_7SO_3H$$

Z substituent	relative rate
H	1.00
CH_3	30.0
Br	0.41
NO_2	0.02
OCH_3	2.5×10^4

Reactions in which benzyl cations are formed as intermediates can also give rise to elimination products:

$$\phi_2\overset{OH}{\underset{|}{C}}CH_3 \xrightarrow{H_2SO_4} [\phi_2\overset{\oplus}{C}\text{-}CH_3] \xrightarrow{HSO_4^\ominus} \phi_2C\text{=}CH_2 \qquad (10.39)$$

S_N1 REACTIONS OF ALLYL DERIVATIVES

The observed solvolysis rates of allyl halides (but not propargyl halides) indicate a degree of reactivity that parallels that of the benzyl halides (see Table 10.7). We can account for the enhanced ionization rates of allylic groups by assuming that charge delocalization stabilizes allyl-cation intermediates:

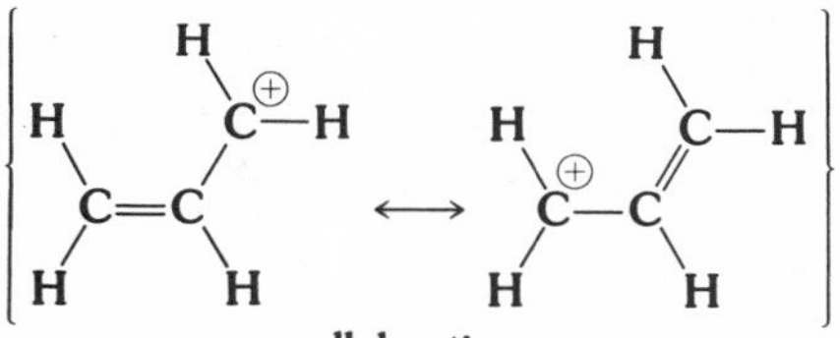

[allyl cation resonance structures]

allyl cation

This explanation suggests that allyl cations will have *two* reactive electrophilic sites. If we examine reactions in which the allylic intermediate is not symmetrical, we do in fact find products corresponding to

TABLE 10·7 *Relative reaction rates for hydrolysis of alkyl chlorides*

$$R\text{-}Cl + 2H_2O \xrightarrow[45°]{H_2O, \text{ dioxane}} R\text{-}OH + H_2O^\oplus + Cl^\ominus$$

compound	relative rate
$n\text{-}C_3H_7Cl$	1.00
$CH_2\text{=}CHCH_2Cl$	1.4×10
$CH_3CH\text{=}CHCH_2Cl$	1.3×10^3
$\phi CH\text{=}CHCH_2Cl$	1.1×10^5
$(CH_3)_2C\text{=}CHCH_2Cl$	1.9×10^6
$HC\text{≡}CCH_2Cl$	7.1×10^{-1}

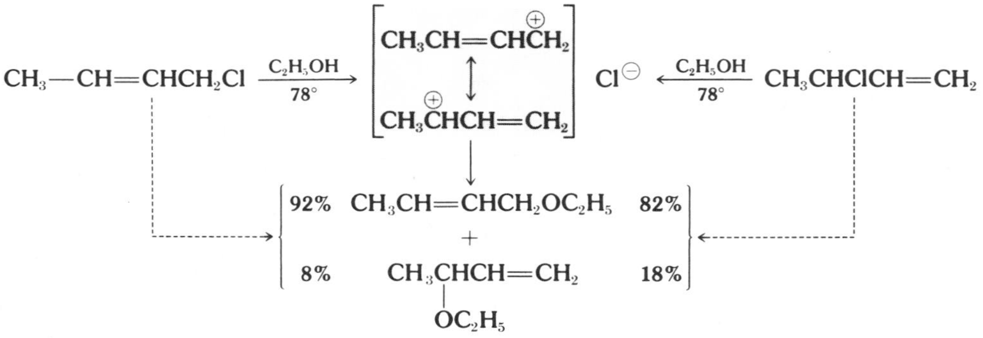

10·8 BENZYL AND ALLYL HALIDES AND ALCOHOLS

$$CH_3-CH=CHCH_2Cl \xrightarrow{C_2H_5OH}_{78°} \begin{bmatrix} CH_3CH=\overset{\oplus}{C}HCH_2 \\ \updownarrow \\ CH_3\overset{\oplus}{C}HCH=CH_2 \end{bmatrix} Cl^{\ominus} \xleftarrow{C_2H_5OH}_{78°} CH_3CHClCH=CH_2$$

$$\begin{Bmatrix} 92\% & CH_3CH=CHCH_2OC_2H_5 & 82\% \\ & + & \\ 8\% & CH_3CHCH=CH_2 & 18\% \\ & \underset{OC_2H_5}{|} & \end{Bmatrix}$$

FIGURE 10·13 *Solvolysis of isomeric allyl chlorides*

nucleophilic attack at each site. The slightly different product mixtures resulting from the solvolysis of the isomeric chlorobutenes in Figure 10.13 may be due to a competing S_N2 reaction path, or perhaps to uneven solvation of the carbonium ion as it is being formed.

PROBLEM 10·17
a Suggest a reason for the absence of ring substitution products in S_N1 reactions of benzyl halides (that is, products from the attack of nucleophiles at the ortho or para carbon atoms, which are involved in the charge delocalization).
b Suggest a reason for the slow ionization of propargyl halides.

PROBLEM 10·18
Reaction of 1-phenyl-2-propen-1-ol with concentrated hydrochloric acid gives a mixture of two C_9H_9Cl isomers. Write structures for these compounds. Which isomer would be thermodynamically more stable?

STABILITY OF BENZYL AND ALLYL RADICALS

The exceptional stability of the benzyl and allyl cations is also found in intermediates other than carbonium ions. The bond-dissociation energies of the primary C—H bonds in ethane, propene, and toluene clearly show, for example, that allyl and benzyl radicals are more stable than their alkyl counterparts (see Table 10.8). The reactions outlined in Figures 10.14 and 10.15 will serve to illustrate the enhanced stability and delocalization of the odd electron in allyl and benzyl radicals.

TABLE 10·8 *Approximate bond-dissociation energies*

type of bond	ΔH (kcal/mole at 25°C)
CH_3CH_2—H	97
$H_2C=CHCH_2$—H	78
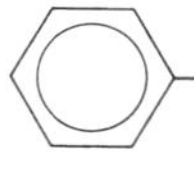 C_6H_5—CH_2—H	79

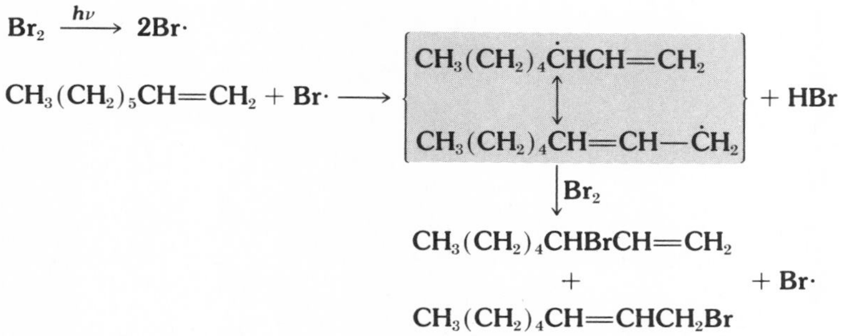
Br2 hν 2Br·
CH3(CH2)5CH=CH2 + Br·
CH3(CH2)4ĊHCH=CH2
CH3(CH2)4CH=CH—ĊH2
+ HBr
Br2
CH3(CH2)4CHBrCH=CH2
+
+ Br·
CH3(CH2)4CH=CHCH2Br

FIGURE 10·14 *Allylic bromination of 1-octene*

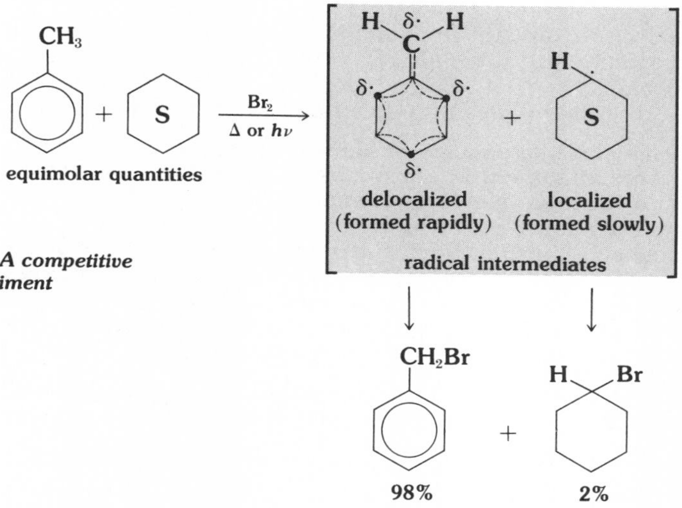
CH3
S
equimolar quantities
Br2
Δ or hν
H δ· H
C
δ· δ·
δ·
H
S
delocalized
(formed rapidly)
localized
(formed slowly)
radical intermediates
CH2Br
H Br
98%
2%

FIGURE 10·15 *A competitive bromination experiment*

SUMMARY

aryl halides Aromatic compounds having a halogen atom bonded to an aromatic ring. In their physical properties the aryl halides are similar to aromatic hydrocarbons of equivalent molecular weight. Aryl halides are much less reactive than alkyl halides, the only generally applicable reaction being the formation of aryl organometallic derivatives.

nucleophilic substitution of aryl halides A reaction in which the halogen substituent of an aryl halide is replaced by a nucleophile. Such reactions are not normally observed, because the S_N1 and S_N2 reaction paths, so common for alkyl halides, are unfavorable for aryl halides. Ortho and/or para electron-withdrawing substituents such as NO_2 and CN enhance the rate of nucleophilic substitution by a two-step bimolecular mechanism.

elimination reactions of aryl halides Loss of adjacent substituents on an aromatic ring, resulting in an aryne intermediate such as benzyne. Dehydrohalogenation (loss of HX) requires forcing reaction conditions because the aryne intermediates formed in such reactions are thermodynamically unstable.

benzyne The simplest aryne intermediate, usually represented as

Such intermediates are very unstable and react rapidly to give addition products (for example, with the solvent).

aryl organometallic derivatives Aryl lithium and Grignard reagents, formed by reaction of aryl halides with **Li**, **Mg**, and other active metals. The reactions of these reagents parallel reactions of corresponding alkyl organometallic reagents.

electrophilic substitution of aryl halides Nitration, sulfonation, halogenation, and Friedel-Crafts alkylation reactions, proceeding via electrophilic intermediates as described for benzene (Chapter 9). Aryl halides are less reactive than benzene and give predominant ortho and para substitution.

phenols Aromatic compounds having a hydroxyl function bonded to the aromatic ring. Phenols ($K_a \approx 10^{-10}$) are stronger Brønsted acids than alcohols ($K_a \approx 10^{-17}$), and electron-withdrawing substituents ortho and para to the hydroxyl group enhance this acidity.

electrophilic substitution of phenols Reactions in which electrophilic reagents attack either the nucleophilic oxygen atom or the aromatic ring of a phenol, yielding characteristic substitution products. Phenols undergo most of the ether and ester-forming reactions reported for alcohols. Furthermore, the hydroxyl function strongly activates the aromatic ring to electrophilic attack, thereby enhancing the rate of ring substitution reactions to a nearly uncontrollable degree. The major products from halogenation and sulfonation reactions are the ortho and para isomers; nitration destroys most phenols.

oxidation of phenols The destruction of phenols with oxidizing reagents such as Cr^{+6} and Mn^{+7}, giving numerous products, including quinones.

quinones Conjugated oxidation products of 1,2- and 1,4-dihydroxybenzenes:

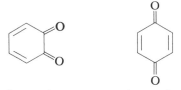

o-benzoquinone   p-benzoquinone

reactions of allyl and benzyl derivatives Rapid nucleophilic substitution of allyl and benzyl halides and alcohols, facilitated by stabilization of carbonium-ion intermediates by charge delocalization:

$$\left\{ H_2C=CH-\overset{\oplus}{C}H_2 \longleftrightarrow H_2\overset{\oplus}{C}-CH=CH_2 \right\}$$
$$\text{allyl cation}$$

$$\left\{ \text{Ph}-\overset{\oplus}{C}H_2 \longleftrightarrow \overset{\oplus}{\text{Ph}}=CH_2 \longleftrightarrow \oplus\text{Ph}=CH_2 \longleftrightarrow \text{Ph}=CH_2 \right\}$$
$$\text{benzyl cation}$$

Allyl and benzyl radicals are similarly stabilized.

EXERCISES

10·1 Arrange each group of compounds in order of increasing acidity:

a. C₆H₅—OH, (thiophenol) S—OH, O₂N—C₆H₄—OH

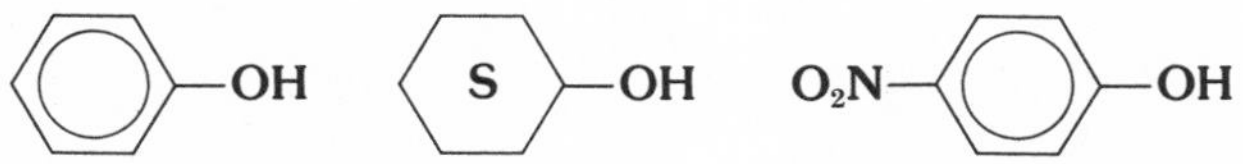

b. 4-nitrophenol, 3-nitrophenol, 2,4-dinitrophenol

c. Cl—C₆H₄—OH, tetrahydrofuran, H₂O

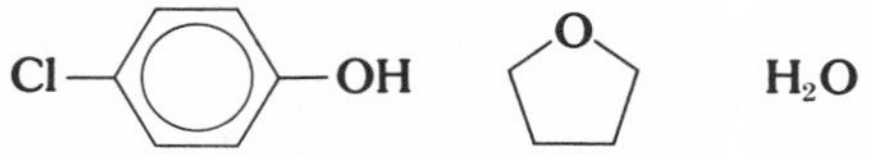

10·2 Arrange each group of compounds in order of increasing reactivity toward nucleophilic substitution by hydroxide ion:

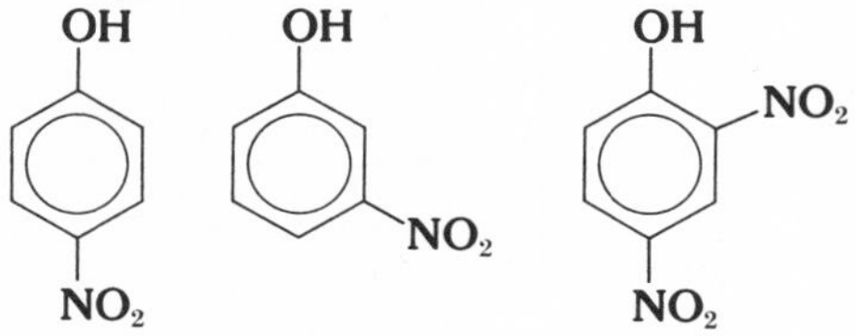

a. O₂N—C₆H₄—Br,   O₂N—C₆H₃(NO₂)—Br,   C₆H₅—Br

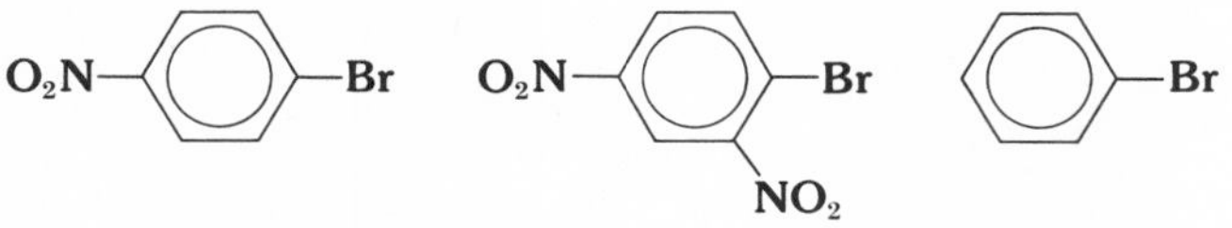

b. CH₃—C₆H₄—Cl,   C₆H₅—Cl,   NC—C₆H₄—Cl

c. 3-bromonitrobenzene, 2-bromonitrobenzene

d. 4-bromoethylbenzene, 3-(bromomethyl)toluene, (2-bromoethyl)benzene

e. CH_2—CH=CH—CH_2CH_3 CH_3—CH=CH—CH_2CH_3 $CH_3CH=CH—CH_2CH_2$
   | | |
  Cl   Cl   Cl

                  Cl
                  |
$CH_3CH=CH—CHCH_3$

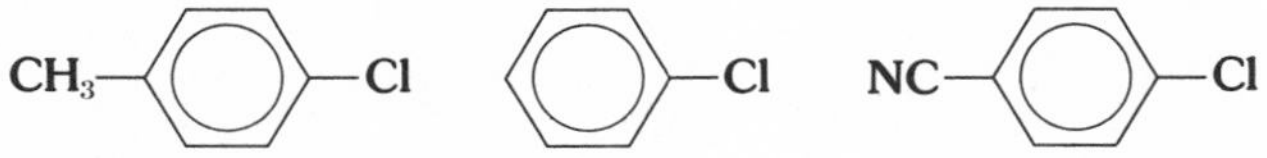

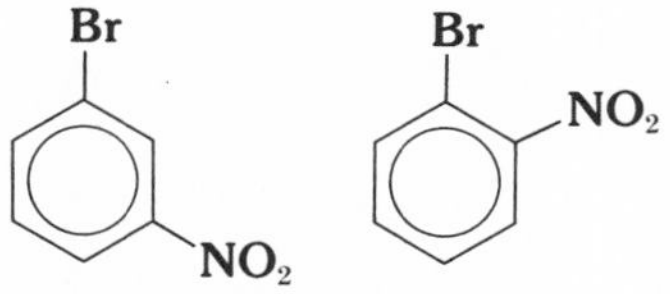

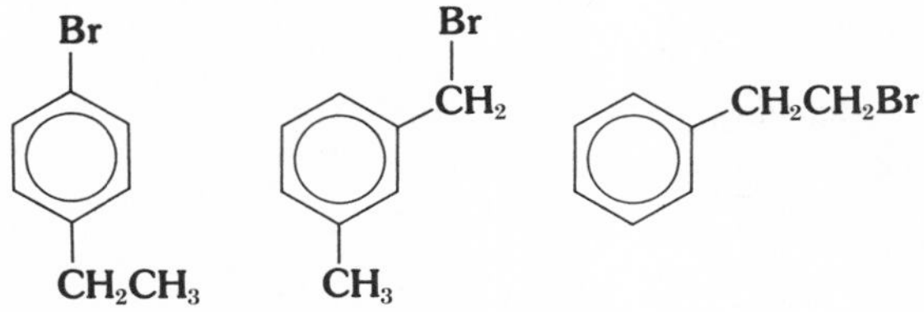

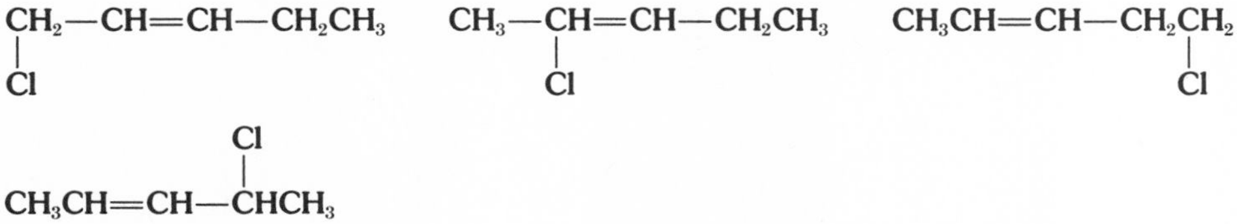

320 EXERCISES

10·3 Indicate the reagents and conditions you would use to prepare the following compounds from phenol:

a φ—O—CH$_2$CH$_3$

b 2,4-dinitrophenyl phenyl ether (O—φ attached to benzene ring with NO$_2$ groups)

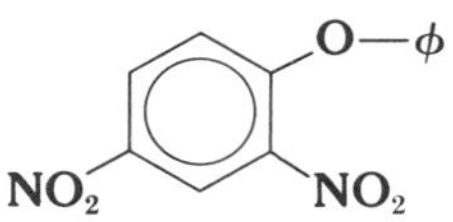

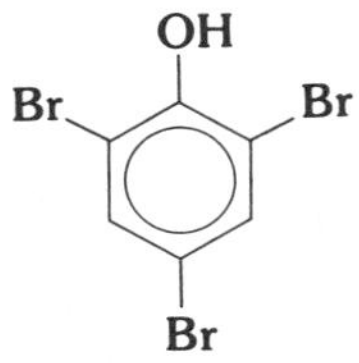

c phenyl acetate (φ—O—CCH$_3$ with C=O)

d 2,4,6-tribromophenol

e 2-hydroxybenzenesulfonic acid (OH and SO$_3$H on benzene)

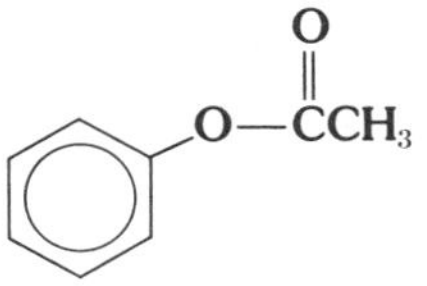

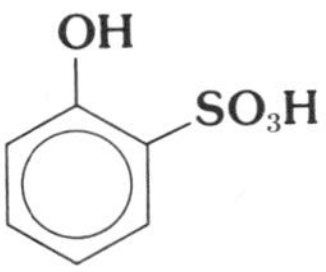

10·4 Draw the structures for all products anticipated from a benzyne reaction mechanism:

a 4-chlorotoluene $\xrightarrow[\Delta,\text{ pressure}]{\text{NaOH}}$ $\xrightarrow{\text{H}_3\text{O}^{\oplus}}$

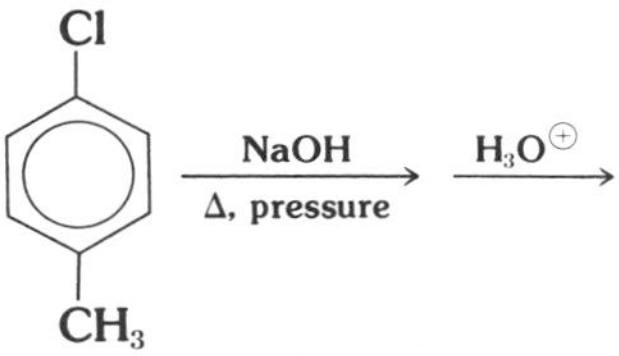

b 3-bromo-4,5-dimethyl(?) — bromobenzene with two CH$_3$ groups $\xrightarrow{\text{KNH}_2,\text{ NH}_3(l)}$

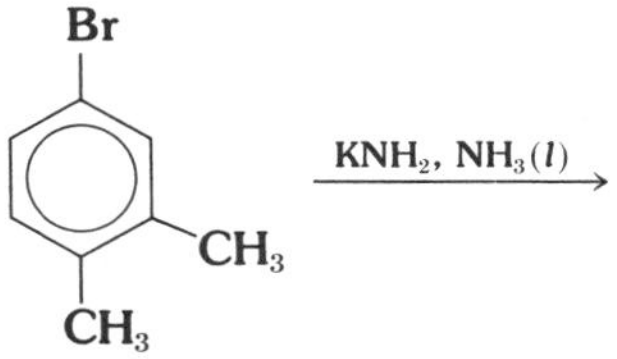

10·5 Explain the formation of the products in the following reaction:

CH$_2$=CH—CH=CH$_2$ + HCl (1 mole) ⟶ CH$_3$—CH—CH=CH$_2$ + CH$_3$—CH=CH—CH$_2$Cl
                                                                                |
                                                                                Cl

10·6 Explain the high acidity of tropolone:

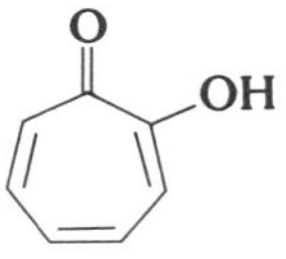

(tropolone structure: seven-membered ring with C=O and OH on adjacent carbons)

10·7 Summarize the order of events (for example, nucleophile attack and leaving-group departure) and their effect on the overall rate of reaction for nucleophilic substitution of the S_N1, S_N2, and aryl halide types.

10·8 A compound with the molecular formula C$_7$H$_8$O is insoluble in water and dilute aqueous NaHCO$_3$, but dissolves in dilute aqueous NaOH. Treatment with Br$_2$ in H$_2$O rapidly produces a compound of formula C$_7$H$_5$OBr$_3$. What is the structure of the starting compound?

10·9 The second reaction in the following sequence, called the Fries rearrangement, is often used to prepare phenolic ketones. Suggest a reasonable mechanism for this rearrangement.

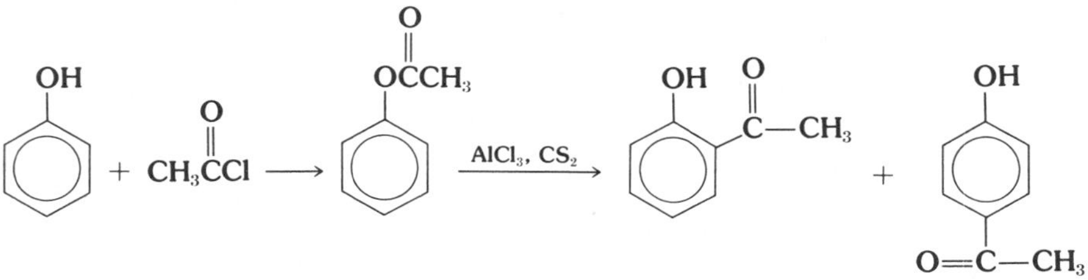

phenol + CH$_3$CCl (with C=O) ⟶ phenyl acetate (OCCH$_3$) $\xrightarrow{\text{AlCl}_3,\text{ CS}_2}$ ortho-hydroxyacetophenone (OH and C(=O)—CH$_3$) + para-hydroxyacetophenone (OH and O=C—CH$_3$)

EXERCISES

10·10 Write the molecular formulas for hydroquinone and *p*-benzoquinone. What is the difference between the two formulas? Does the conversion of *p*-benzoquinone to hydroquinone fit our definition of a reduction process?

10·11 Draw a potential-energy diagram showing the keto-enol tautomerism of phenol and cyclohexanone (see equations 10.16 and 10.17).

10·12 Suggest a mechanism for the following reaction. *Hint:* What is the polarity of the bonds in CO_2?

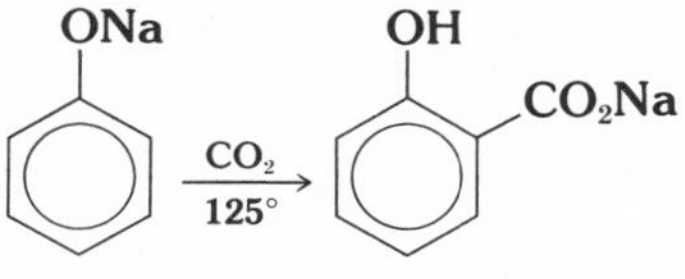

C$_6$H$_5$ONa $\xrightarrow{CO_2, 125°}$ *o*-HOC$_6$H$_4$CO$_2$Na

10·13 Write a mechanism for the following reaction:

m-ClC$_6$H$_4$CH$_2$CH$_2$NHCH$_3$ $\xrightarrow[\text{ether}]{C_6H_5Li}$ *N*-methylindoline + benzene + LiCl

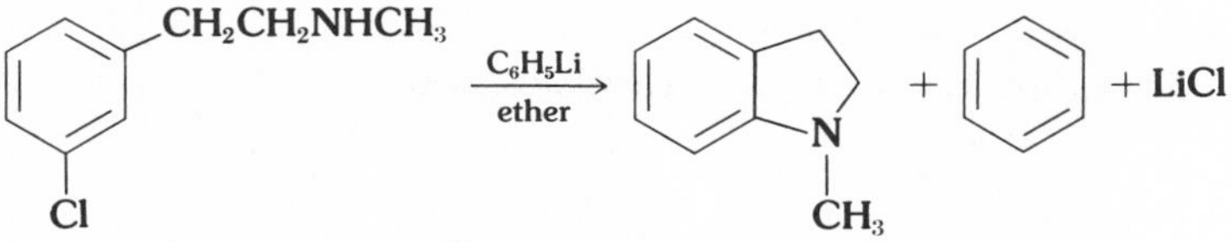

What function is the phenyl lithium serving?

10·14 The reaction of sodium phenoxide and benzyl chloride in the solvent tetrahydrofuran yields only benzyl phenyl ether. With water as the solvent the products are, in addition to the ether, *ortho*- and *para*-benzylphenol. Explain the course of this reaction in these two solvent systems. To what general class of reactions does each mechanism belong?

10·15 The following conversions can be accomplished in three steps or fewer. Write equations showing all necessary reagents and conditions for each step:

a. C$_6$H$_5$CH$_2$Br → C$_6$H$_5$CH$_2$OCH$_2$CH$_3$

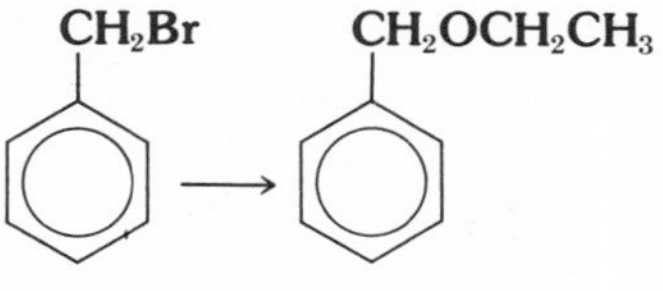

b. chlorobenzene → *p*-chlorobenzoic acid

c. phenol → *o*-hydroxybenzenesulfonic acid

d. (R)-1-phenyl-1-chloroethane → (R)-1-phenyl-1-methoxyethane (with H—Cl and H—OCH$_3$ configurations shown)

e. bromobenzene → *p*-nitroanisole

f. phenol → cyclohexanone

g. bromobenzene → aniline

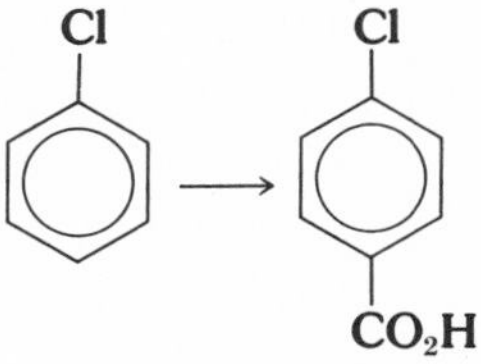

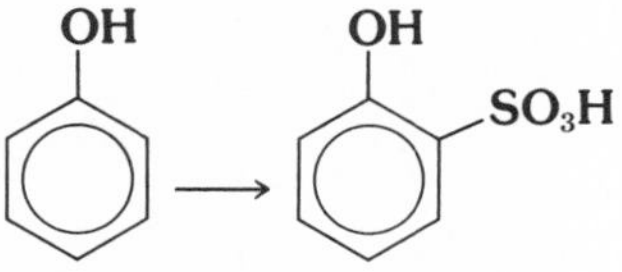

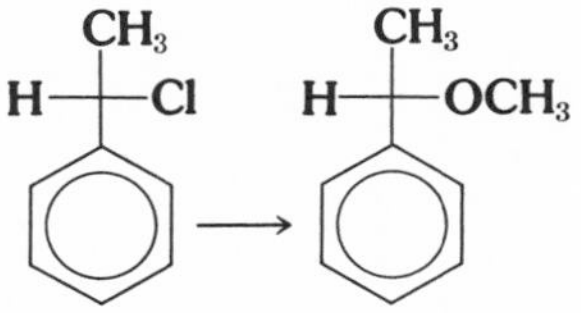

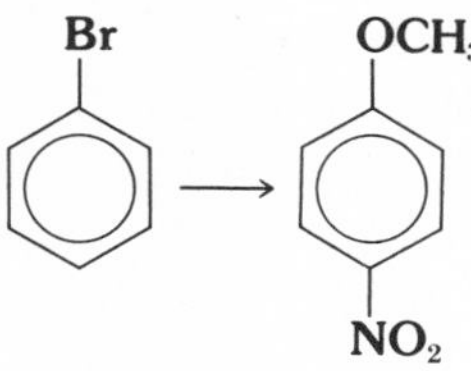

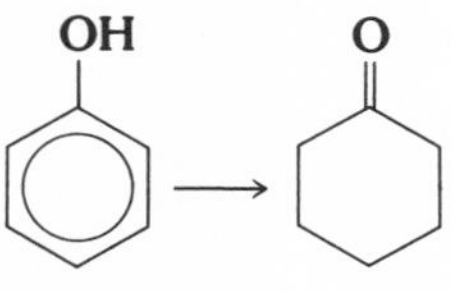

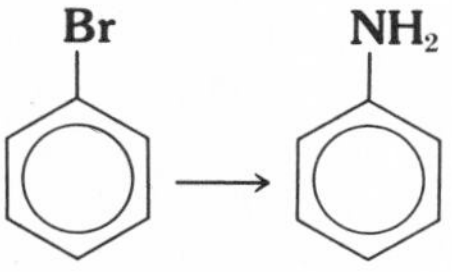

10·16 Of the carbonium ions shown below:
a Which are isomers?
b Which do not have significant charge delocalization?
c Which are contributors to the same resonance hybrid?

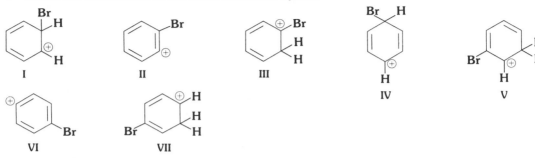

10·17. We find that 2,4-dinitrofluorobenzene is much more reactive in nucleophilic substitution reactions than the corresponding chloride compound. For example,

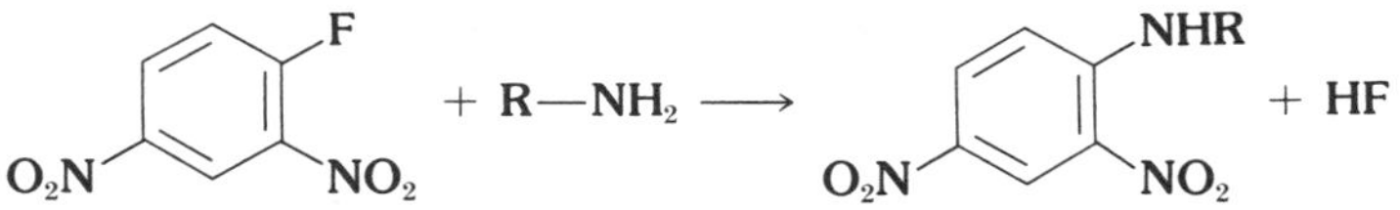

$$\text{2,4-(O_2N)_2C_6H_3F} + \text{R—NH}_2 \longrightarrow \text{2,4-(O_2N)_2C_6H_3NHR} + \text{HF}$$

This conflicts with the order of reactivity usually observed for alkyl halides (I > Br > Cl ≫ F). Suggest a reason for this difference.

10·18 Suggest structures for three C_7H_8O isomers, A, B, and C, from the pmr below. Both B and C show strong infrared absorption in the 3600 cm^{-1} region; isomer C is the only compound soluble in aqueous NaOH.
a Compound A shows a singlet at $\delta = 3.8$ (3H) and a multiplet at $\delta \approx 7.1$ (5H).
b Compound B shows singlets at $\delta = 3.0$ (1H), 4.6 (2H), and 7.3 (5H).
c Compound C shows singlets at $\delta = 2.2$ (3H) and 6.0 (1H) and doublets at $\delta = 6.7$ (2H, $J = 8.5$) and 6.9 (2H, $J = 8.5$).
Note: The coupling constants for aromatic protons vary with their relative locations on the ring:

	J (Hz)
ortho	6–9
meta	1–3
para	0–1

AMINES

IN THIS chapter we turn our attention to organic nitrogen compounds, and specifically to that important group of substances called *amines*. Amines are alkyl or aryl derivatives of ammonia, and as such they may be regarded as the nitrogen analogs of alcohols and ethers:

$$\begin{array}{cccc}
\text{H}\diagdown & \text{R}\diagdown & \text{R}^1\diagdown & \text{R}^1\diagdown \\
\text{H—N:} & \text{H—N:} & \text{R}^2\text{—N:} & \text{R}^2\text{—N:} \\
\diagup & \diagup & \diagup & \diagup \\
\text{H} & \text{H} & \text{H} & \text{R}^3 \\
\text{ammonia} & & \text{amines} &
\end{array}$$

Gaseous and liquid amines generally have noxious odors and are therefore unpleasant to work with.

11·1 NOMENCLATURE OF AMINES

Several unusual features appear in the nomenclature of amines. To begin with, amines are classified as primary, secondary, or tertiary according to the number of alkyl substituents bonded to the *nitrogen atom*. Since this usage conflicts with the established designations for alkyl groups, it is important to remember that these classifications are unrelated to the names of specific compounds.

$$\begin{array}{ccc}
\text{CH}_3 & & \\
| & & \\
\text{CH}_3\text{—C—NH}_2 & \text{CH}_3\text{CH}_2\text{—N—CH}_3 & (\text{CH}_3)_3\text{N:} \\
| & | & \\
\text{CH}_3 & \text{H} & \\
\textit{t-butylamine,} & \text{methylethylamine,} & \text{trimethylamine,} \\
\text{a primary amine} & \text{a secondary amine} & \text{a tertiary amine}
\end{array}$$

11·1 NOMENCLATURE OF AMINES

As long as the alkyl substituents on the amino nitrogen atom remain relatively simple, amines can be named by listing the substituent groups and adding the suffix *-amine*.

$(C_2H_5)_2NCH(CH_3)C_2H_5$
 diethyl-*s*-butylamine

Since there is no characteristic IUPAC ending for amines, the common nitrogen functions are given distinctive names. The *amino group* NH_2 is present in all primary amines. Alkyl or aryl substituents on the amino group are designated by the prefix *N*-.

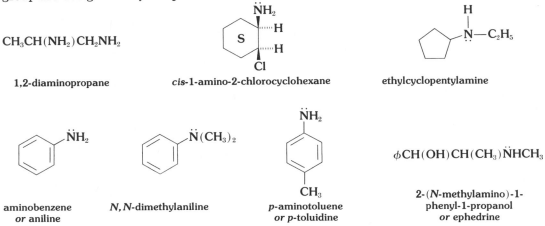

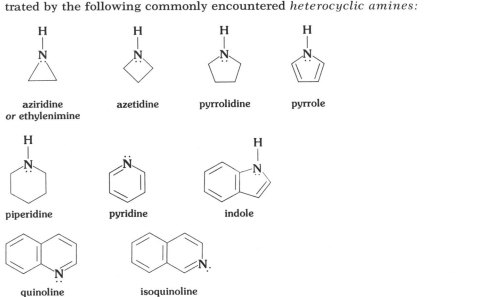

Many of the names associated with these heterocycles have been inherited from earlier workers, and the reasons for them have in some cases been forgotten or have no modern significance. Although it is not universally used, a special nomenclature system for heterocyclic compounds has been devised. According to this system we begin each name with a prefix which indicates the heteroatom (or atoms) present in the ring: *oxa-* (O), *thia-* (S), or *aza-* (N). Following this prefix is a stem that represents the ring size: *-ir-* for three-membered rings, *-et-* for four-membered rings, *-ol-* for five-membered rings, *-in-* for six-membered rings, etc. Finally, an ending indicates whether the ring is saturated (*-ane*) or unsaturated (*-e*).

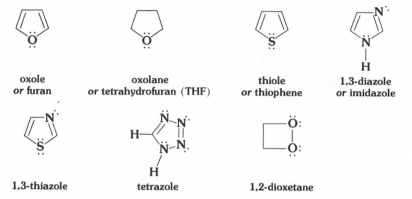

oxole *or* furan

oxolane *or* tetrahydrofuran (THF)

thiole *or* thiophene

1,3-diazole *or* imidazole

1,3-thiazole

tetrazole

1,2-dioxetane

The nomenclature of alkyl and aryl derivatives of ammonium salts is similar to that of the amines, with an *-ium* suffix, followed by the name of the anion:

anilinium sulfate

pyridinium chloride

trimethylbenzylammonium hydroxide *or* Triton-B

$(CH_3)_3 \overset{\oplus}{N} CH_2CH_2OH \ OH^\ominus$
choline

Tetraalkylammonium hydroxides (also called quaternary ammonium bases) find use as strong organic bases; Triton-B is an example. Choline is of particular interest because of the role it and its acetyl derivative play in the transmission of nerve impulses. We shall discuss the nature of these compounds and the mechanism of synaptic transmission in Chapter 14.

11·2 NATURALLY OCCURRING AMINES

ALKALOIDS

Amines and their derivatives are more widely distributed in nature than any other functional group. Even if we do not count the ubiquitous proteins and nucleic acids, the variety of naturally occurring amines is truly impressive. For example, many higher plants contain basic nitrogenous substances called *alkaloids*. The function of these compounds in the biochemical processes of the living plant is not yet well understood; however, a number of plant alkaloids exert a marked physiological action when they are administered to animals. Some of these substances are

11·2 NATURALLY OCCURRING AMINES

toxic to the human organism and others, such as quinine, have important medicinal value.

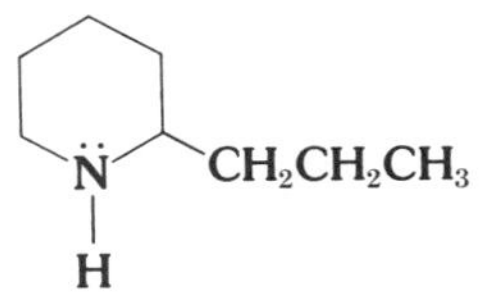

coniine (from poison hemlock, *Conium masculatum*)

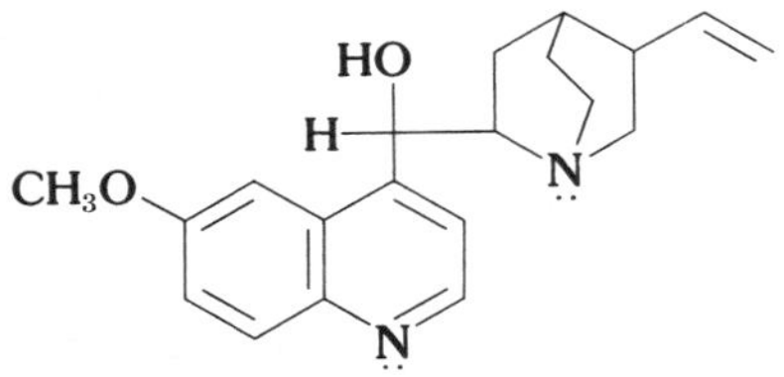

quinine (from the bark of *Cinchona officinalis*)

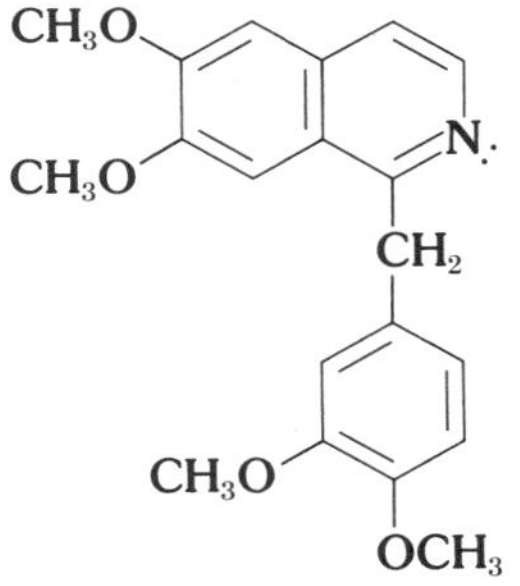

papaverine

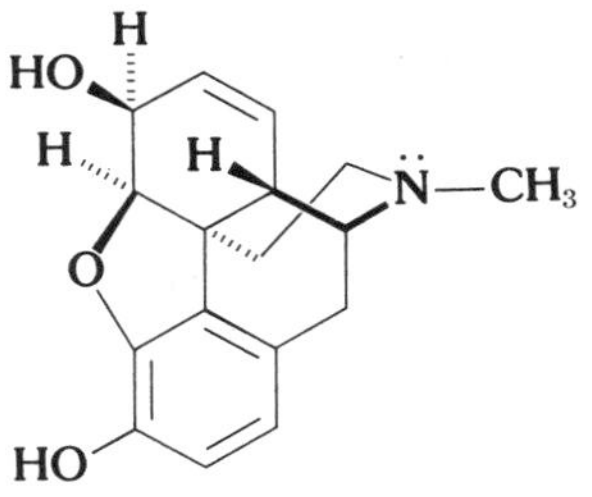

morphine

(two analgesics obtained from the opium poppy)

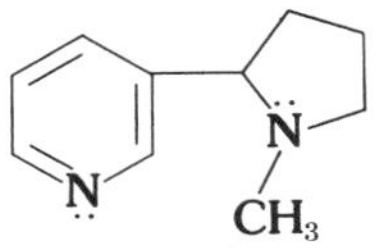

nicotine (from tobacco, *Nicotiana tabacum*)

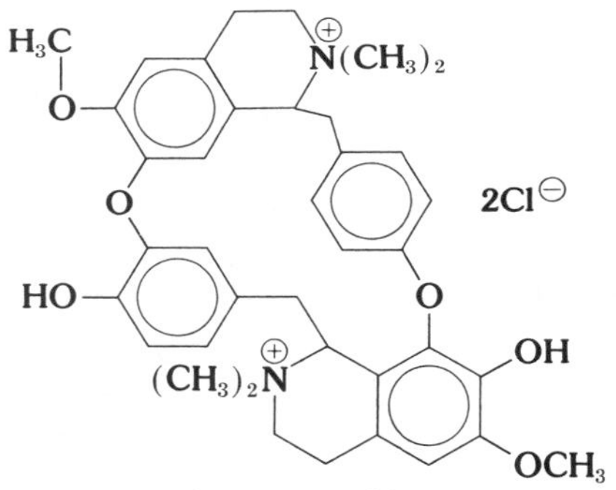

tubocurarine chloride (from curare)

The group of substances known as the vitamin B complex contains many nitrogen bases. A deficiency of thiamine in the diet of people in Japan, India, China, and other parts of the Orient led at one time to an endemic form of polyneuritis called beriberi.

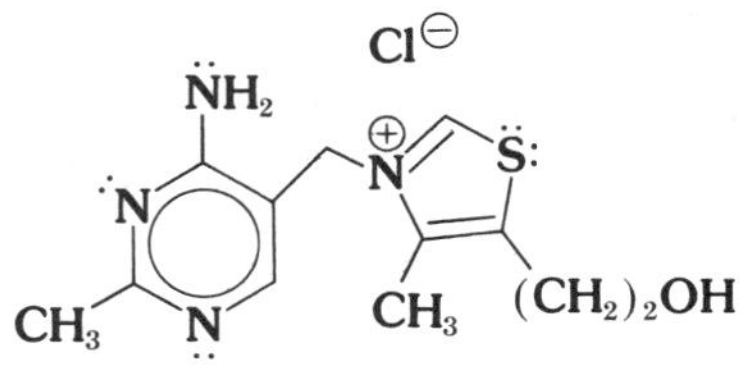

thiamine
or vitamin B_1

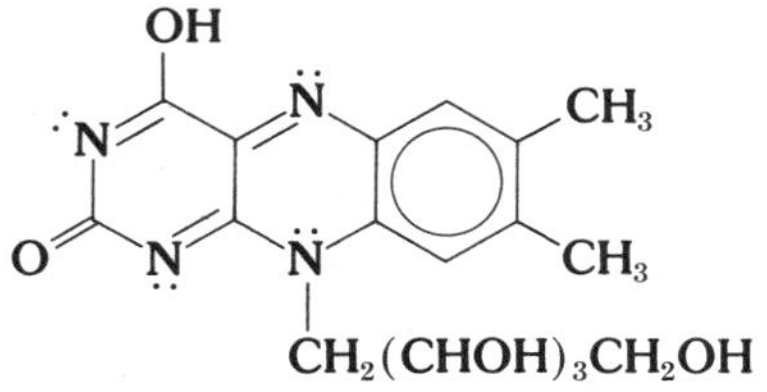

riboflavin
or vitamin B_2

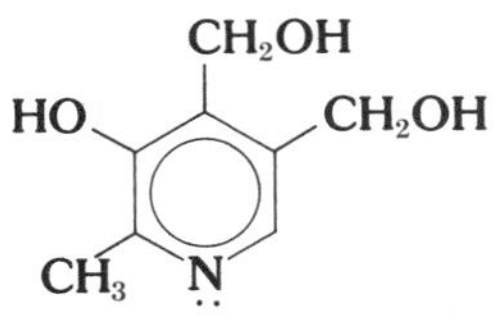

pyridoxine
or vitamin B_6

HORMONES AND DRUGS

The hormones epinephrine (adrenaline) and norepinephrine are derivatives of 2-phenylethylamine which are secreted by the adrenal medulla (the inner part of the adrenal gland). They serve as transmitting agents for part of the sympathetic nervous system and function as a chemical alarm when the body must be aroused to face a stressful situation.† In response to stimulation from the sympathetic nervous system, the adrenal medulla releases massive amounts of these hormones, which put every organ on ready-for-action footing.

HO—C$_6$H$_3$(OH)—CH(OH)CH$_2$NHCH$_3$
epinephrine

HO—C$_6$H$_3$(OH)—CH(OH)CH$_2$NH$_2$
norepinephrine

The naturally occurring levorotatory enantiomers are over 20 times more active than the dextrorotatory isomers.

Powerful physiological properties have been found for many other 2-arylethylamine derivatives, both naturally occurring and synthetic. The Chinese herb ma huang yields the drug ephedrine, used in the treatment of bronchial asthma and nasal congestion. Mescaline, from the mescal buttons on the cactus *Lophophora williamsii*, is a hallucinogen used in religious ceremonies by some Indian tribes of the American Southwest. The indole derivative serotonin occurs in both plants and animals and is of special interest because of its role in establishing stable mental processes. The amphetamines are a group of synthetic 2-amino-1-arylpropanes which affect the central nervous system. Benzedrine is the simplest example (dexedrine is the dextrorotatory enantiomer). The temporary alertness resulting from a dose of benzedrine is usually accompanied by increased irritability, followed by mental depression and eventually extreme fatigue.

ϕCH(OH)CH(CH$_3$)NHCH$_3$
ephedrine

(CH$_3$O)$_3$C$_6$H$_2$—CH$_2$CH$_2$NH$_2$
mescaline

5-HO-indole—CH$_2$CH$_2$NH$_2$
serotonin

ϕCH$_2$CH(CH$_3$)NH$_2$
benzedrine

Histamine is a toxic amine which is present, combined in some manner with proteins, in all the tissues of the body. The release of free histamine is believed to cause the symptoms of most allergies and the common cold.

†The relationship of the sympathetic and parasympathetic nervous systems and the nature of neural transmission in the latter are discussed in Chapter 14.

11·3 STRUCTURE AND PROPERTIES OF NITROGEN COMPOUNDS

Drugs capable of moderating and relieving the effects of histamine have been synthesized; two such antihistamines are benadryl and pyribenzamine.

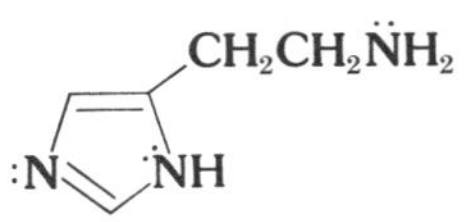

histamine

benadryl: $\phi_2\text{CHOCH}_2\text{CH}_2\ddot{\text{N}}(\text{CH}_3)_2$

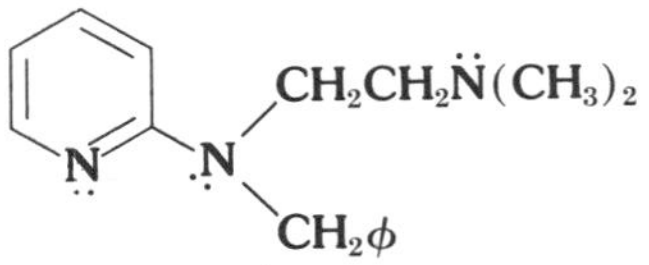

pyribenzamine

PROBLEM 11·1
Identify and name the heterocyclic unit or units present in coniine, nicotine, papaverine, serotonin, and histamine (see Section 11.1).

11·3 STRUCTURE AND PROPERTIES OF NITROGEN COMPOUNDS

BONDING AND STEREOCHEMISTRY

The different hybridization states of covalently bonded nitrogen atoms parallel the sp^3, sp^2, and sp hybridization states of carbon discussed in earlier chapters. As shown in Figure 11.1, saturated amines and ammonium salts have roughly tetrahedral bond angles at nitrogen (counting the nonbonding electron pair as one of the substituents). Nitrogen atoms involved in double bonds have a trigonal bonding geometry and sp^2 hybridization, and triple-bonded nitrogen is sp hybridized.

PROBLEM 11·2
Indicate the hybridization state of each nitrogen atom in nicotine, quinine, thiamine, histamine, and phenyl isocyanide ($\phi\overset{\oplus}{-}\text{N}\equiv\text{C}:^{\ominus}$).

Since an sp^3-hybridized nitrogen atom has a tetrahedral bonding configuration, we might expect to observe enantiomorphism in chiral secondary and tertiary amines:

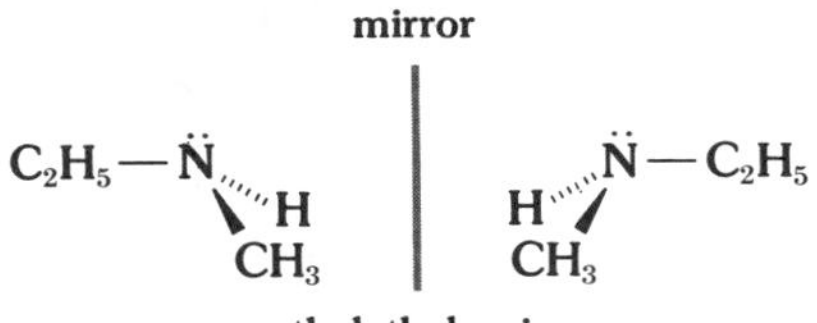

methylethylamine,
(a chiral secondary amine)

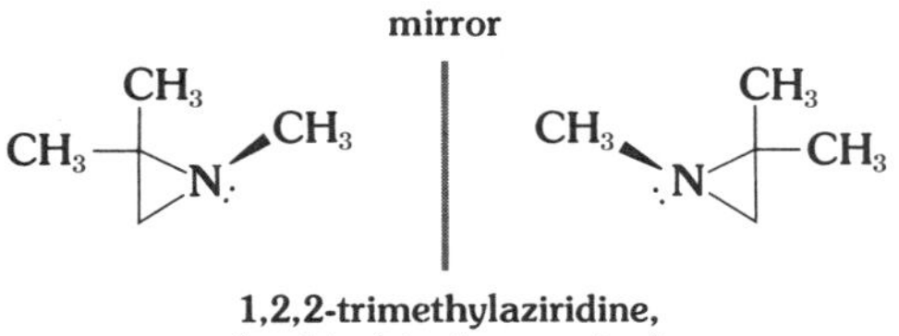

1,2,2-trimethylaziridine,
(a chiral tertiary amine)

Attempts to resolve such compounds have been fruitless, however, owing to the rapid interconversion of the enantiomers:

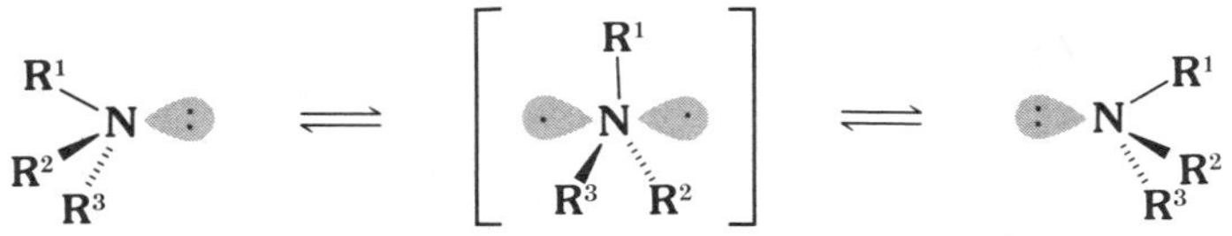

planar transition state

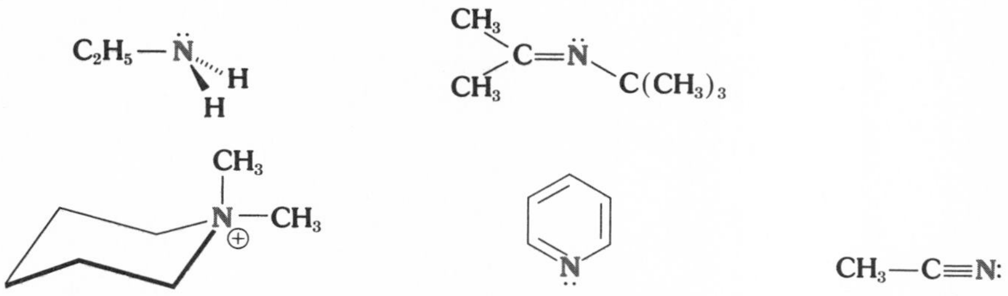

*sp*³ hybrid *sp*² hybrid *sp* hybrid

FIGURE 11·1 *Hybridization states of nitrogen*

At room temperature most amines undergo inversion of configuration 10^3 to 10^8 times per second (ammonia inverts 4×10^{10} times per second), making it impossible to obtain a pure sample of either enantiomer.

PROBLEM 11·3

a Aziridines undergo nitrogen inversion less rapidly than acyclic amines. Suggest a reason for the difference.

b The pmr spectrum of 1,2,2-trimethylaziridine (above) at 25°C consists of five discrete signals: three 3H singlets at $\delta = 2.2$, 1.2, and 1.05 and two 1H singlets at $\delta = 1.5$ and 0.85. When the sample is heated to 110°C, the spectrum changes so as to display three sharp signals at $\delta = 2.2$ (3H), 1.2 (2H), and 1.1 (6H). Explain these observations.

In contrast to the configurational mobility of amines, quaternary ammonium salts have rigid tetrahedral structures. This is demonstrated by the resolution of salts having four different nitrogen substituents, as shown in Figure 11.2.

A MODIFIED STRUCTURE-COMPOSITION RELATIONSHIP

The general structure-composition relationship outlined in Section 6.2 must be modified before we can use it to determine the number of rings and/or double bonds in nitrogen-containing compounds. The modification is a simple one. Each nitrogen atom in the molecular formula is

FIGURE 11·2 *Configurational integrity of ammonium salts*

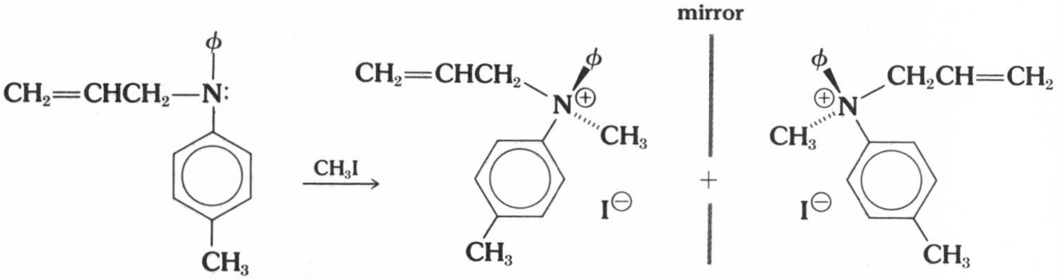

configurationally mobile (cannot be resolved)

enantiomeric salts (resolvable)

11·3 STRUCTURE AND PROPERTIES OF NITROGEN COMPOUNDS

replaced by a CH unit. The new molecular formula arrived at in this way is then treated by the rule as previously described.

To illustrate, suppose we have a compound with the molecular formula $C_4H_{11}N$. Replacing N by CH, we arrive at the formula of the alkane pentane:

$$C_4H_{11}N \xrightarrow{\text{replace N by CH}} C_5H_{12}$$

According to our previous rule, this nitrogen-containing compound cannot have any rings or double bonds; in other words, it is saturated. Examples are $CH_3CH_2N(CH_3)_2$, $CH_3(CH_2)_2NH_2$, and three other isomers.

Suppose we have a compound with the molecular formula $C_6H_8N_2$:

$$C_6H_8N_2 \xrightarrow{\text{replace N by CH}} C_8H_{10}$$

Octane has the formula C_8H_{18}; hence this nitrogen-containing compound must have a combination of rings and double bonds (or triple-bond equivalents) totaling four:†

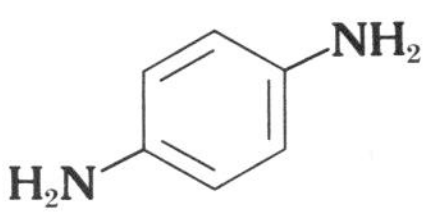

[structures: para-diaminobenzene, $N\equiv C(CH_2)_4C\equiv N$, 2,3-dimethylpyrazine, and a bicyclic H—N...N—H compound] and many more

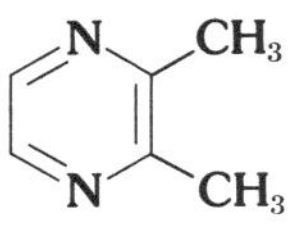

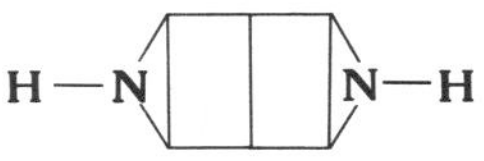

HYDROGEN-BONDING EFFECTS IN AMINES

Although hydrogen bonds of the type N—H----N are weaker than the corresponding oxygen bonds (see Section 8.1), they are strong enough to cause primary and secondary amines to have higher boiling points and greater water solubilities than nonpolar compounds of similar molecular weight. The data in Table 11.1 illustrate this fact.

SPECTROSCOPIC PROPERTIES

Primary and secondary amines exhibit a characteristic N—H stretching absorption at 3200 to 3500 cm^{-1} in the infrared spectrum (see Figure 11.3). Since this absorption band overlaps the corresponding O—H band in alcohols, there is a possibility of confusing these two functional groups. Primary amines can usually be distinguished by the presence of two N—H stretching bands, which result from coupled vibrations of the two equivalent bonds:

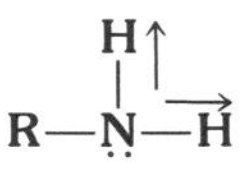

R—N—H (symmetric stretching)   R—N—H (asymmetric stretching)

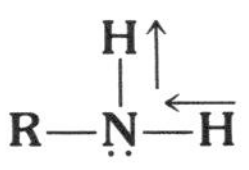

---

†Note that this rule cannot be used with *-onium* salts (compounds incorporating positively charged nitrogen atoms), since such compounds have more hydrogen than our rule predicts.

11·3 STRUCTURE AND PROPERTIES OF NITROGEN COMPOUNDS

TABLE 11·1 *Boiling points and water solubilities of amines, alcohols, and hydrocarbons*

compound	name	molecular weight	boiling point (°C)	water solubility
$CH_3CH_2CH_2NH_2$	n-propylamine	59	49	very soluble
$C_2H_5NHCH_3$	methylethylamine	59	35	very soluble
$(CH_3)_3N$	trimethylamine	59	3	very soluble
$H_2NCH_2CH_2NH_2$	1,2-diaminoethane	60	117	very soluble
$CH_3CH_2CH_2OH$	n-propyl alcohol	60	97	very soluble
$CH_3CH_2CH_2CH_3$	butane	58	0	insoluble
⬡—NH_2	cyclohexylamine	99	134	very soluble
⬡N—CH_3	N-methylpiperidine	99	107	soluble
⌬—NH_2	aniline	93	184	moderately soluble
⬡—OH	cyclohexanol	100	160	moderately soluble
⌬—OH	phenol	94	182	moderately soluble
CH_3—⌬N	4-methylpyridine	93	143	very soluble
⬡—CH_3	methylcyclohexane	98	100	insoluble

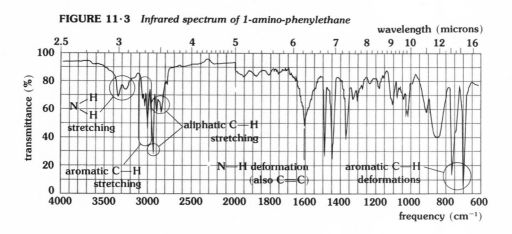

FIGURE 11·3 *Infrared spectrum of 1-amino-phenylethane*

11·3 STRUCTURE AND PROPERTIES OF NITROGEN COMPOUNDS

Hydrogen atoms bonded to or near the nitrogen atom of an amine have characteristic pmr absorptions, as shown in Figure 11.4. As we would expect from the greater electronegativity of oxygen in comparison with nitrogen, the magnitude of the downfield shifts in amines (Table 11.2) is less than that for alcohols and ethers (Table 8.4). The NH-proton chemical shifts show a concentration dependency similar to that of the hydroxyl group (a consequence of hydrogen bonding), and these protons

TABLE 11·2 *Typical chemical shifts of amines*

type of hydrogen	δ values (ppm)
R—NH_2	0–5.0 (concentration dependent)
R'—CH_2—NR_2	2.0–3.0

FIGURE 11·4 *Pmr spectra of 1-amino-phenylethane and morpholine*

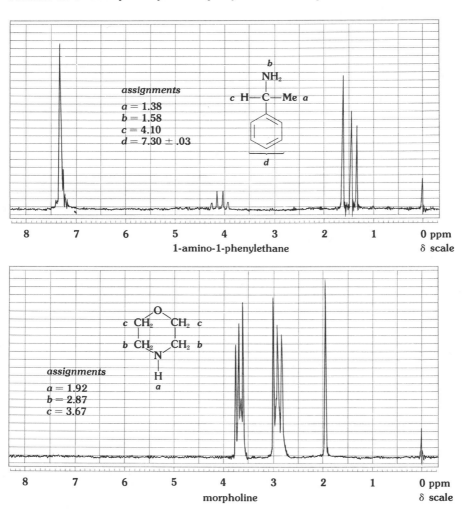

normally exchange so rapidly that spin-spin coupling to N—H is often not observed. We can eliminate NH signals from a pmr spectrum by deuterium exchange with heavy water, as described in equation 8.2 for elimination of the OH signals in alcohols.

BASICITY OF AMINES

Since amines can be regarded as derivatives of ammonia, we should not be surprised to find that they are also moderate to weak Brønsted bases. The base strengths of simple alkyl amines are, in fact, rather close to that of ammonia; they turn litmus paper blue and form stable water-soluble salts with strong mineral acids. However, adjacent unsaturated functions may reduce the basicity of more complex amines by factors of more than 10^{10}. We can measure the relative basicities of all but the weakest amine bases in terms of the equilibrium constant K_b for the reversible proton transfer from the weak acid water:

$$R-\ddot{N}H_2 + H_2O \rightleftharpoons R-NH_3^{\oplus} + OH^{\ominus} \qquad K_b = \frac{[R-NH_3^{\oplus}][OH^{\ominus}]}{[RNH_2]} \qquad (11.1)$$

Some typical basicity constants, determined in or extrapolated to aqueous solutions, are given in Table 11.3.

Relative basicities may also be deduced from the acidity constants K_a of the corresponding conjugate acids:†

$$RNH_3^{\oplus} + H_2O \rightleftharpoons R\ddot{N}H_2 + H_3O^{\oplus} \qquad K_a = \frac{[RNH_2][H_3O^{\oplus}]}{[RNH_3^{\oplus}]} \qquad (11.2)$$

†These must not be confused with the very weak acidities of free primary or secondary amines. Thus ammonium ion ($NH_4^{\oplus}$) has an acidity constant of $K_a = 5.6 \times 10^{-10}$, but K_a for ammonia itself is about 10^{-34} (extrapolated to water solution).

TABLE 11.3 *Basicity constants of some nitrogen compounds*

compound	K_b	compound	K_b
$:NH_3$	1.8×10^{-5}	$O_2N-\langle\bigcirc\rangle-\dot{N}H_2$	$\sim 10^{-13}$
$C_2H_5\dot{N}H_2$	5.0×10^{-4}	$O_2N-\langle\bigcirc\rangle-\dot{N}H_2$ (with NO_2)	$\sim 10^{-18}$
$(CH_3)_3N:$	6.0×10^{-5}		
$H_2\ddot{N}-\ddot{N}H_2$	3.0×10^{-6}	$\langle\bigcirc\rangle N:$	$\sim 10^{-9}$
$\langle\bigcirc\rangle-\ddot{N}H_2$	4.0×10^{-10}	$\langle\bigcirc\rangle:NH$	$\sim 10^{-14}$
$CH_3O-\langle\bigcirc\rangle-\dot{N}H_2$	1.5×10^{-9}	$CH_3C\equiv N:$	$\sim 10^{-24}$

11·3 STRUCTURE AND PROPERTIES OF NITROGEN COMPOUNDS

Since both K_a and K_b are used in the literature, it is important to differentiate between them. K_b values are directly proportional to the strength of the base, whereas the corresponding K_a values are *inversely* proportional. Fortunately, it is a simple matter to convert a K_a value to K_b by means of the relationship

$$K_b = \frac{1}{K_a} \times 10^{-14} \tag{11.3}$$

PROBLEM 11·4
Suggest a method other than fractional distillation for separating a mixture of the following compounds into the pure components:

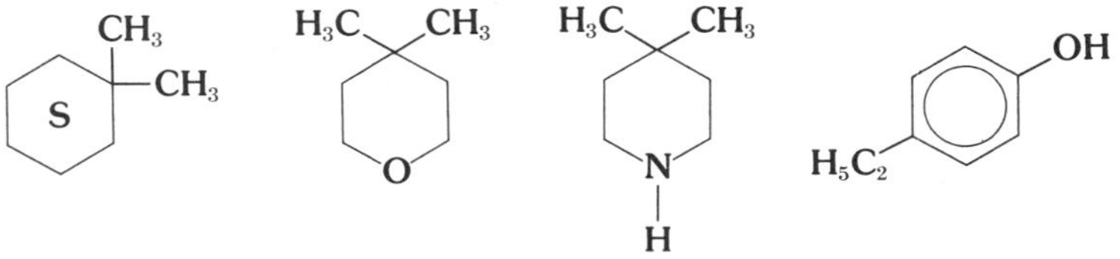

PROBLEM 11·5
From the data in Table 11.3 calculate K_a for the following ions:

$C_2H_5NH_3^{\oplus}$ $\phi NH_3^{\oplus}$ (pyrrolidinium $NH_2^{\oplus}$)

The deviation of many amines from ammonialike basicities can be attributed to two factors. As the hybridization of nitrogen changes from sp^3 to sp^2 to sp, the proportion of s character assumed by the nonbonding electron pair increases. Since s orbitals are spherically symmetrical and lie closer to the nucleus than p orbitals, this increase in s character results in a decrease in the nucleophilicity or basicity of the nonbonding electron pair. The electron pair is more tightly held by the nitrogen nucleus and therefore cannot coordinate strongly with an approaching electrophile, such as a proton. This effect is dramatically demonstrated by the relative basicities of trimethylamine, pyridine, and acetonitrile (CH_3CH) given in Table 11.3.

The other factor is delocalization of the nitrogen nonbonding electron pair into adjacent unsaturated functions, which results in resonance stabilization of amines such as pyrrole and aniline (Figure 11.5).† This stabilization ranges from a few kilocalories per mole for aniline (in addi-

---
†A similar stabilization of the phenolate anion was proposed in Figure 10.8 to explain the enhanced acidity of phenol.

FIGURE 11·5 *Delocalization of the nitrogen electron pair in aniline*

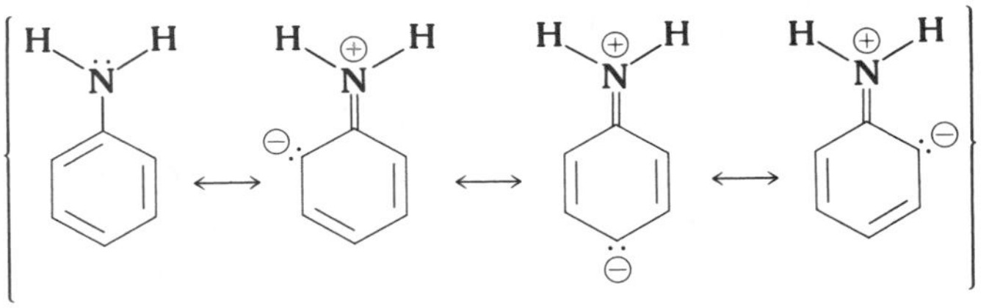

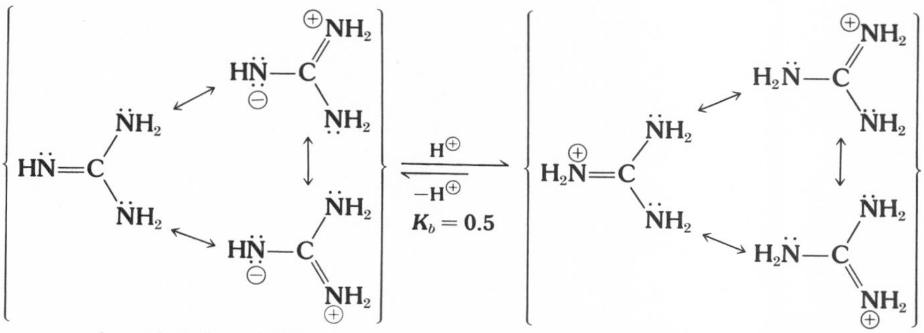

[Figure 11.6: Resonance structures showing guanidine (left, slightly stabilized by resonance delocalization of nitrogen electrons) in equilibrium with guanidinium cation (right, strongly stabilized by resonance delocalization of positive charge), with $K_b = 0.5$.]

FIGURE 11·6 *Resonance stabilization of guanidine and its conjugate acid*

tion to the aromatic-ring resonance energy) to 21 kcal/mole for pyrrole (see the discussion of six-π-electron systems in Section 9.7).

Since the conjugate acids of these amines no longer have a nonbonding electron pair available for delocalization, the additional resonance stabilization is lost in the protonation step, and the base strength of the amines is reduced correspondingly. Indeed, the additional electron-pair delocalization that occurs in triphenylamine ($\phi_3\text{N}\colon$) decreases its basicity to approximately the level of an alcohol or ether ($K_b \approx 10^{-17}$).

PROBLEM 11·6
a Draw a rough energy diagram illustrating the effect of nitrogen electron-pair delocalization on the energy change occurring when aniline and cyclohexylamine are protonated by the weak acid water.
b Explain the effect of an ortho or para nitro substituent on the base strength of aniline. What relative basicity would you predict for *m*-nitroaniline?

In some cases resonance stabilization of a conjugate acid exceeds that of the parent amine, making it a stronger base than ammonia. A good illustration of this point is the compound guanidine (Figure 11.6), which is nearly as strong a base as sodium hydroxide. The energy equivalence of the contributors to the conjugate-acid resonance hybrid indicates an exceptionally large resonance stabilization of the guanidinium cation (see rule 4 of the rules of resonance in Section 9.3).

11·4 ELECTROPHILIC SUBSTITUTION AT NITROGEN

N-ALKYLATION

Nitrogen atoms are generally more nucleophilic than equivalently substituted oxygen atoms. Consequently, we find that weak electrophilic reagents (such as alkyl halides) attack amines more readily than they do alcohols and ethers.

The reaction of amines (or ammonia) with alkyl halides can in principle be repeated until complete conversion to a quaternary ammonium salt has taken place (Figure 11.7). By varying the proportion of alkyl

11·4 ELECTROPHILIC SUBSTITUTION AT NITROGEN

$$:NH_3 + RX \xrightarrow{S_N2} RNH_3^{\oplus}X^{\ominus} \xrightarrow{-HX} RNH_2$$

$$R\ddot{N}H_2 + RX \xrightarrow{S_N2} R_2NH_2^{\oplus}X^{\ominus} \xrightarrow{-HX} R_2NH$$

$$R_2\ddot{N}H + RX \xrightarrow{S_N2} R_3NH^{\oplus}X^{\ominus} \xrightarrow{-HX} R_3N$$

$$R_3N: + RX \xrightarrow{S_N2} R_4N^{\oplus}X^{\ominus}$$

FIGURE 11·7 *Alkylation reactions of ammonia and amines*

halide to amine we can favor a high or low degree of N-alkylation. However, the accompanying side reactions detract from the usefulness of these reactions for synthesizing specific amines. The following example shows an exceptionally favorable reaction:

$$\phi NH_2 + \phi CH_2Cl \xrightarrow{NaHCO_3, H_2O} \phi CH_2NH\phi + NaCl \tag{11.4}$$

Quaternary ammonium salts can be prepared in good yield from tertiary amines, or from primary and secondary amines if an excess of the alkyl halide is used:

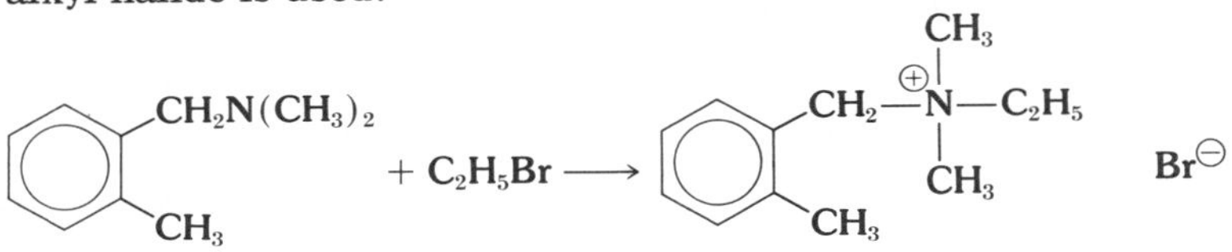

(11.5)

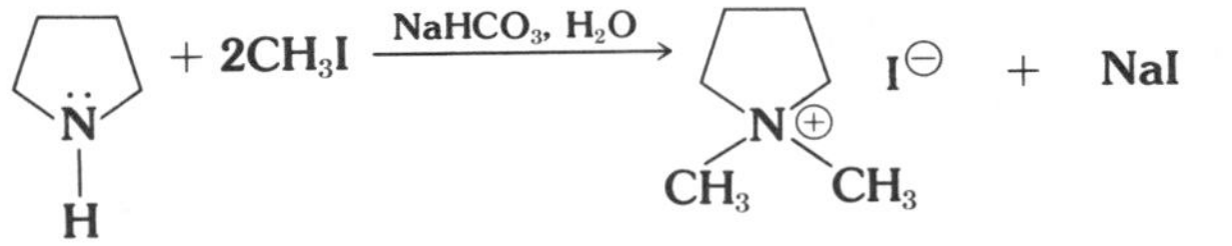

(11.6)

Onium salts have also been obtained from sp^2-hybridized nitrogen compounds:

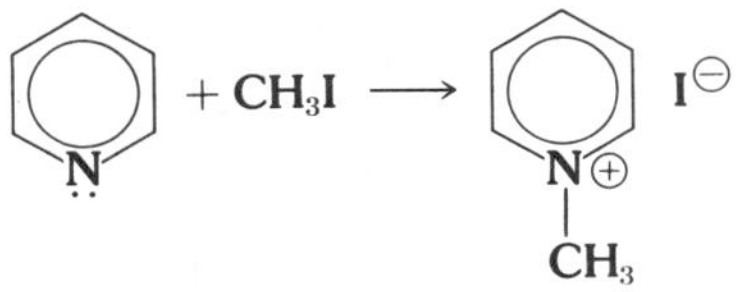

(11.7)

Amines do not ordinarily react with aryl halides unless the halogen atom is activated by ortho and/or para electron-withdrawing substituents. For example, the fluorine atom in 2,4-dinitrofluorobenzene is readily displaced by primary amines, a reaction which is useful for marking the free amino groups in peptides and proteins.

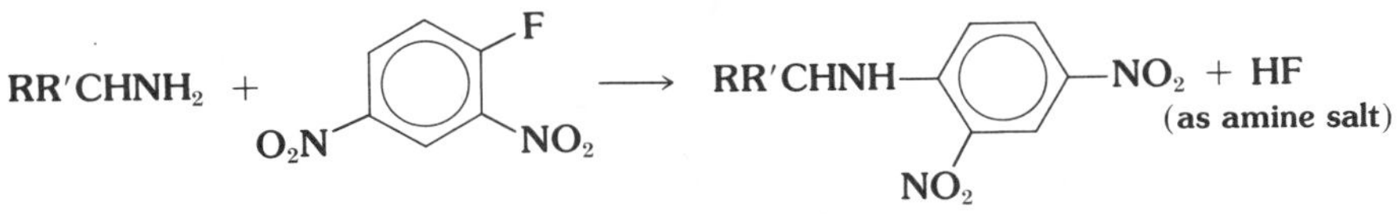

(11.8)

11·4 ELECTROPHILIC SUBSTITUTION AT NITROGEN 337

PROBLEM 11·7
Write a structural formula for the compound you would expect to find contaminating the major product in equation 11.4, assuming equimolar quantities of the reactants.

N-ACYLATION AND SULFONATION

Carboxylic and sulfonic acyl halides and anhydrides react with primary and secondary amines even more readily than they acylate (or sulfonate) alcohols (see Figure 8.7). Some examples of these reactions are given in the following equations. Experience indicates that tertiary amines do not normally react in this manner to form stable reaction products.

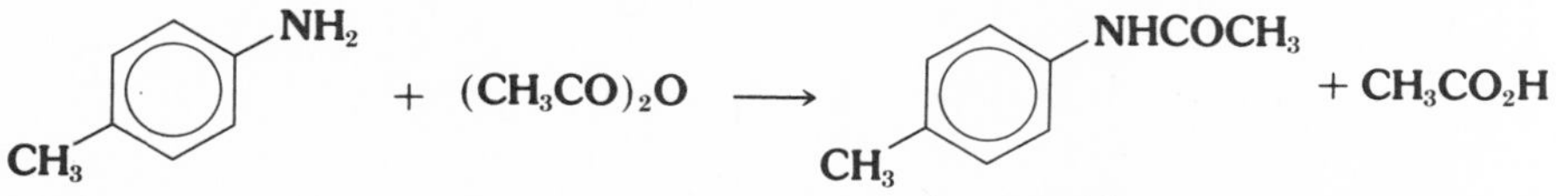

$$\text{4-CH}_3\text{-C}_6\text{H}_4\text{-NH}_2 + (\text{CH}_3\text{CO})_2\text{O} \longrightarrow \text{4-CH}_3\text{-C}_6\text{H}_4\text{-NHCOCH}_3 + \text{CH}_3\text{CO}_2\text{H} \quad (11.9)$$
an amide

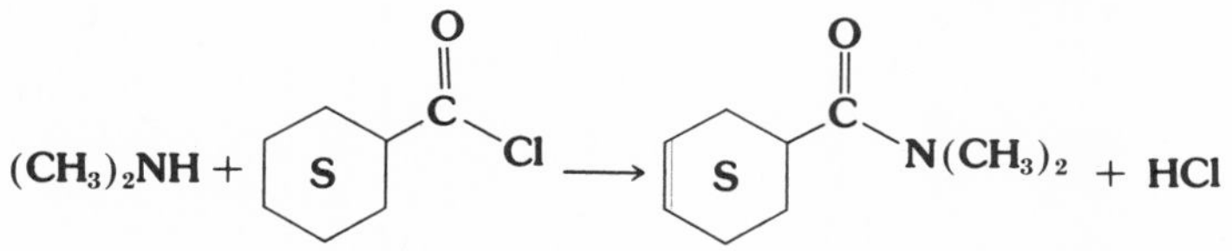

$$(\text{CH}_3)_2\text{NH} + \text{C}_6\text{H}_{10}\text{S-COCl} \longrightarrow \text{C}_6\text{H}_{10}\text{S-CON}(\text{CH}_3)_2 + \text{HCl} \quad (11.10)$$
an amide (as amine salt)

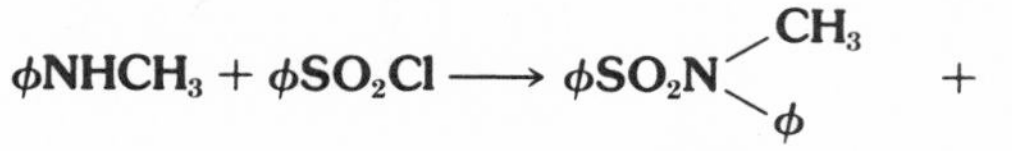

$$\phi\text{NHCH}_3 + \phi\text{SO}_2\text{Cl} \longrightarrow \phi\text{SO}_2\text{N}(\text{CH}_3)(\phi) + \text{HCl} \quad (11.11)$$
a sulfonamide (as amine salt)

Disubstitution at nitrogen is not a problem in these acylation reactions because the nitrogen electron pair is delocalized into the adjacent acyl or sulfonyl group of the amide or sulfonamide products, rendering them relatively unreactive to further electrophilic attack:

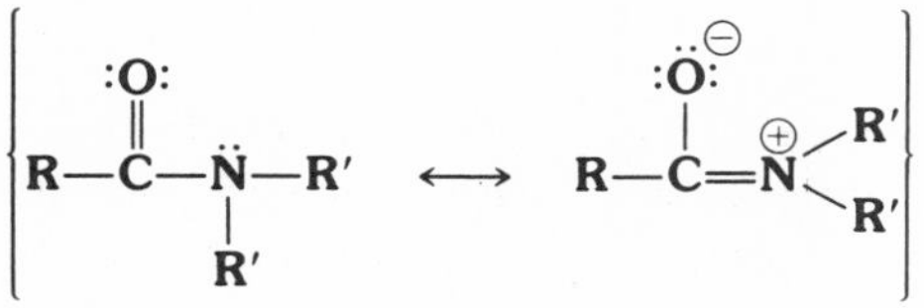

$$\left\{ \begin{array}{c} :\!\ddot{\text{O}}\!: \\ \| \\ \text{R}-\text{C}-\ddot{\text{N}}-\text{R}' \\ | \\ \text{R}' \end{array} \longleftrightarrow \begin{array}{c} :\!\ddot{\text{O}}\!:^{\ominus} \\ | \\ \text{R}-\text{C}=\overset{\oplus}{\text{N}}\!\!\diagup^{\text{R}'}_{\text{R}'} \end{array} \right\}$$

FIGURE 11·8 *The Hinsberg amine test*

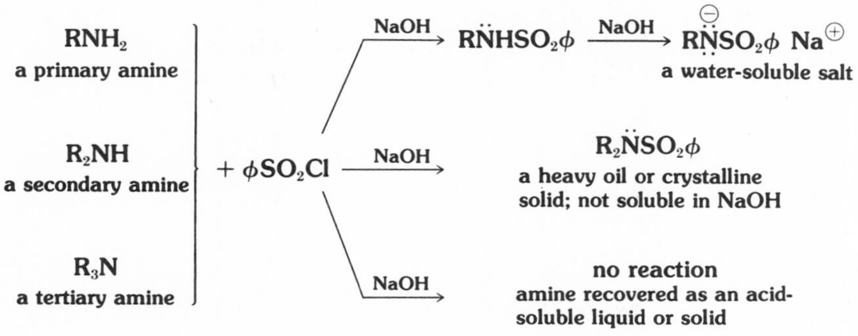

$$\text{RNH}_2 \text{ (a primary amine)} \xrightarrow{\phi\text{SO}_2\text{Cl}} \text{R}\ddot{\text{N}}\text{HSO}_2\phi \xrightarrow{\text{NaOH}} \text{R}\overset{\ominus}{\text{N}}\text{SO}_2\phi \text{ Na}^{\oplus} \text{ (a water-soluble salt)}$$

$$\text{R}_2\text{NH} \text{ (a secondary amine)} \xrightarrow{\phi\text{SO}_2\text{Cl}} \text{R}_2\ddot{\text{N}}\text{SO}_2\phi \xrightarrow{\text{NaOH}} \text{a heavy oil or crystalline solid; not soluble in NaOH}$$

$$\text{R}_3\text{N} \text{ (a tertiary amine)} \xrightarrow{\phi\text{SO}_2\text{Cl}} \xrightarrow{\text{NaOH}} \text{no reaction; amine recovered as an acid-soluble liquid or solid}$$

The sulfonyl group is, in fact, sufficiently electrophilic to make sulfonamides derived from primary amines weakly *acidic*. It is thus possible to distinguish primary, secondary, and tertiary amines by their different behavior toward benzenesulfonyl chloride in dilute sodium hydroxide. This diagnostic method is known as the *Hinsberg test* (Figure 11.8).

N-NITROSATION

Nitrous acid (HNO_2) is a moderately strong Brønsted acid ($K_a = 5 \times 10^{-4}$) which, because of its instability, must be freshly prepared by treatment of sodium nitrite with a cold mineral acid solution:

$$NaNO_2 + H_2SO_4 \xrightarrow[0°]{H_2O} HONO + NaHSO_4 \tag{11.12}$$

The reactions of nitrous acid with amines are of two distinct types. The first is simply the rapid and reversible proton transfer characteristic of Brønsted acid-base interactions; the second involves an initial nitrosation at nitrogen:

$$(C_2H_5)_2NH + NaNO_2 \xrightarrow{H_2SO_4,\ H_2O} \underset{\text{N-nitrosodiethylamine}}{(C_2H_5)_2N-N=O} + H_2O \tag{11.13}$$

> **PROBLEM 11·8**
> Write an equation showing the acid-base equilibrium between diethylamine and nitrous acid.

Although the nature of the products from these nitrosation reactions depends on the number and kind of amino nitrogen substituents, the first stage of reaction in most cases appears to be electrophilic attack by the nitroso cation or an equivalent species (such as N_2O_3 or $NOCl$):

$$\underset{\substack{(X = HSO_4 \\ \text{or Cl})}}{HONO + HX} \longrightarrow \underset{\substack{\text{nitroso} \\ \text{cation}}}{NO^{\oplus}} + X^{\ominus} + H_2O \tag{11.14}$$

In practice, only secondary amines give stable N-nitroso derivatives on reaction with nitrous acid. Primary aliphatic amines undergo subsequent transformations (see Figure 11.9) which rapidly produce alcohols, alkenes, and molecular nitrogen:

$$CH_3CH_2CH_2NH_2 \xrightarrow{NaNO_2,\ H_2SO_4} N_2 + CH_3CH_2CH_2OH + (CH_3)_2CHOH + CH_3CH=CH_2 \tag{11.15}$$

Primary aryl amines give moderately stable diazonium ions, and tertiary amines, in addition to forming water-soluble salts, may suffer decomposition of an N-nitroso ammonium intermediate to alcohols, carbonyl compounds, and secondary amine derivatives.

$$\phi-NH_2 \xrightarrow{NaNO_2,\ HCl} \underset{\textbf{benzenediazonium chloride (stable at 0°)}}{\langle\bigcirc\rangle-\overset{\oplus}{N}\equiv N: Cl^{\ominus}} \tag{11.16}$$

$$R_3N: + NO^{\oplus} \rightleftharpoons [R_3\overset{\oplus}{N}-N=O] \longrightarrow \text{complex product mixtures} \tag{11.17}$$

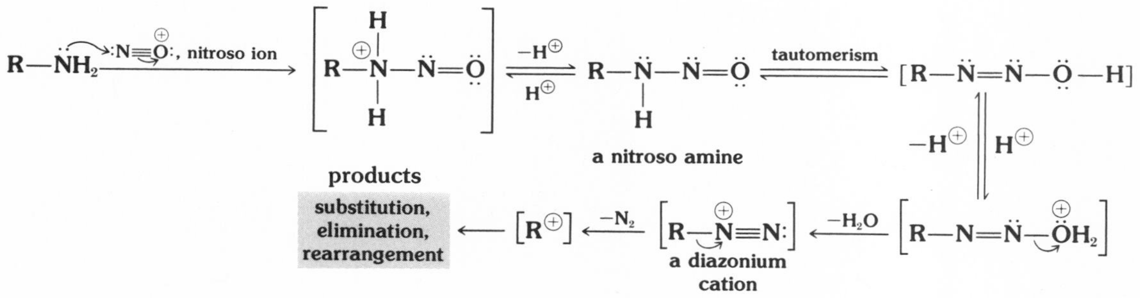

11·4 ELECTROPHILIC SUBSTITUTION AT NITROGEN

$$R-\ddot{N}H_2 \xrightarrow{:N\equiv\overset{\oplus}{O}:,\text{ nitroso ion}} \left[R-\overset{\oplus}{\underset{H}{\overset{H}{N}}}-\ddot{N}=\ddot{O}: \right] \underset{H^\oplus}{\overset{-H^\oplus}{\rightleftharpoons}} R-\underset{H}{\overset{}{\ddot{N}}}-\ddot{N}=\ddot{O} \overset{\text{tautomerism}}{\rightleftharpoons} [R-\ddot{N}=\ddot{N}-\ddot{O}-H]$$

a nitroso amine

$$\text{products} \xleftarrow[\text{rearrangement}]{\text{substitution, elimination,}} [R^\oplus] \xleftarrow{-N_2} [R-\overset{\oplus}{N}\equiv N:] \xleftarrow{-H_2O} [R-N=\overset{\oplus}{\ddot{N}}-\overset{}{\ddot{O}}H_2]$$

a diazonium cation

FIGURE 11·9 *A mechanism for primary amine nitrosation*

Tertiary aromatic amines usually undergo ring nitrosation, discussed in the next section.

The nitrogen evolution from primary aliphatic amines is easily discerned and the *N*-nitroso derivatives of secondary amines are acid-insoluble yellow liquids or solids; hence these reactions have on occasion been used to distinguish primary, secondary, and tertiary amines.

PROBLEM 11·9
What formal relationship do N_2O_3 and NOCl have to nitrous acid? *Hint:* Examine the derivatives of carboxylic and sulfonic acids shown in Section 8.5.

PROBLEM 11·10
Suggest an explanation for the low basicity of *N*-nitrosoamines.

The greater stability of aryl diazonium ions over that of their aliphatic counterparts is due largely to the high energy (thermodynamic instability) of the aryl cations necessarily formed by the loss of molecular nitrogen (see chapter 10). As we shall see, however, these intermediates can be induced to undergo a variety of useful substitution reactions in which the diazonium function is replaced by other atoms or groups.

N-OXIDATION

Hydrogen peroxide and its simple alkyl and acyl derivatives react with tertiary amines, yielding a class of compounds known as the *amine oxides*:

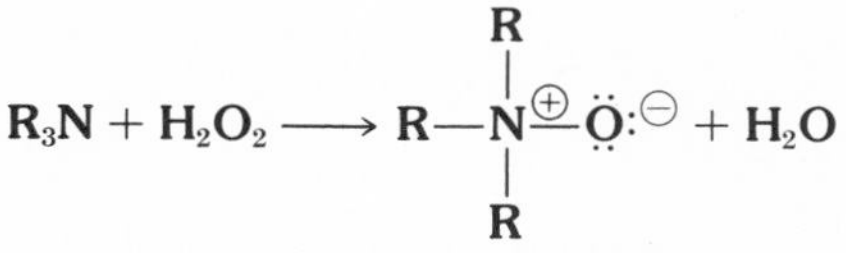

$$R_3N + H_2O_2 \longrightarrow R-\underset{R}{\overset{R}{\underset{|}{\overset{|}{N}}}}\overset{\oplus}{-}\ddot{O}:^\ominus + H_2O$$

an amine oxide

The highly polar nature of the nitrogen-oxygen bond in these compounds is evidenced by the fact that trimethylamine oxide (molecular weight 75) fails to boil at temperatures up to 180°C, at which point it begins to decompose (compare this with the boiling point of trimethylamine and similar compounds in Table 11.2).

Since amine oxides having three different alkyl substituents have been resolved, these compounds undoubtedly have a rigid tetrahedral configuration:

$$\underset{\substack{CH_3 \\ C_2H_5}}{\overset{CH_2=CHCH_2}{\diagdown}}\!\!\!\!\overset{\oplus}{N}\!\!-\!\!\overset{..}{\underset{..}{O}}:^{\ominus} \quad \Big| \quad ^{\ominus}:\overset{..}{\underset{..}{O}}-\overset{\oplus}{N}\underset{\substack{CH_3 \\ C_2H_5}}{\diagup}\!\!\!\!\!\!\overset{CH_2CH=CH_2}{}$$

mirror

a resolvable chiral amine oxide

Primary and secondary amines undergo initial oxidation to hydroxylamines; on more vigorous treatment primary amines may undergo subsequent transformation to nitroso or nitro compounds:

$$R-NH_2 \xrightarrow{H_2O_2} R-\underset{\underset{H}{|}}{N}-OH \xrightarrow{[O]} R-N=O \xrightarrow{[O]} R-NO_2 \qquad (11.19)$$

PROBLEM 11·11
a Why does aniline not form an amine oxide derivative?
b Amine oxides are weakly basic ($K_b \approx 10^{-10}$). Write a structural formula for the conjugate acid of triethylamine oxide.

11·5 DISPLACEMENT AND ELIMINATION OF NITROGEN FROM ALIPHATIC AMINES

Experience has shown that amine functions seldom serve as leaving groups in nucleophilic substitution or base-catalyzed elimination reactions. Indeed, they are even less effective in this role than the hydroxyl or alkoxyl group (see Section 8.6). In order to displace or eliminate an amino group we must transform it into an intermediate form that can be more easily lost. The reaction of primary amines with nitrous acid, described in Figure 11.9, is an excellent example of this stratagem, whereby NH_2 (a very poor leaving group) is converted to $\overset{\oplus}{-}N\equiv N:$ (an excellent leaving group), followed by S_N1 and $E1$ reactions.

QUATERNARY SALTS AS LEAVING GROUPS

Another method of enhancing the ability of $-\ddot{N}R_2$ to serve as a nucleophilic leaving group is to convert it to a quaternary salt ($-NR_3^{\oplus}X^{\ominus}$). This approach is illustrated by the following substitution reactions:

$$\underset{Br^{\ominus}}{\underset{CH_2\overset{\oplus}{N}(CH_3)_2C_2H_5}{\underset{CH_3}{\text{[Ar]}}}} \xrightarrow{CH_3CO_2Na, \ CH_3CO_2H}{\Delta} \underset{CH_3}{\underset{\text{[Ar]}}{}}\!\!\!\overset{OCOCH_3}{\underset{CH_2}{|}} + C_2H_5N(CH_3)_2 + NaBr \qquad (11.20)$$

$$\phi N(CH_3)_3^{\oplus}Br^{\ominus} + \phi S^{\ominus}Na^{\oplus} \xrightarrow[\Delta]{acetone} \phi-S-CH_3 + \phi N(CH_3)_2 + NaBr \qquad (11.21)$$

$$(CH_3)_4N^{\oplus}OH^{\ominus} \xrightarrow[\Delta]{H_2O} CH_3OH + (CH_3)_3N: \qquad (11.22)$$

11·5 DISPLACEMENT AND ELIMINATION OF NITROGEN

To avoid undesired elimination reactions, it is usually best to use only weakly basic nucleophilic reagents.

Elimination reactions of quaternary ammonium salts may also be effected. In particular, pyrolytic elimination of quaternary ammonium hydroxide salts constitutes the final step of a useful alkene synthesis known as the *Hofmann elimination*. The example outlined in Figure 11.10 also illustrates the preparation of the quaternary hydroxide precursor via *exhaustive methylation* followed by anion exchange.

When the alkyl substituents on the quaternary nitrogen are such that two or more competing eliminations can take place, we find that tertiary alkyl groups are usually eliminated more rapidly than secondary or primary substituents:

$$\underset{\underset{CH_3}{|}}{\overset{\overset{CH_3}{|}}{CH_3-C}}-\underset{\underset{CH_3}{|}}{\overset{\overset{CH_3}{|}}{\overset{\oplus}{N}}}-CH_2-CH_3 \quad \underset{OH^{\ominus}}{\xrightarrow{150°}} \quad \begin{array}{c} (CH_3)_2C=CH_2 + C_2H_5N(CH_3)_2 \\ 93\% \\ + \\ CH_2=CH_2 + (CH_3)_3CN(CH_3)_2 \\ 7\% \end{array} \quad (11.23)$$

However, competitive eliminations within a given alkyl group normally favor that alkene isomer having the smallest number of double-bond alkyl substituents:

$$\underset{\underset{\overset{\oplus}{N}(CH_3)_3}{|}}{\overset{}{CH_3CH_2CHCH_3}} \xrightarrow{180°} CH_3CH_2CH=CH_2 + CH_3CH=CHCH_3 + (CH_3)_3N \quad (11.24)$$
$$OH^{\ominus} \qquad\qquad\qquad\qquad 95\% \qquad\qquad 5\%$$

This tendency of Hofmann eliminations to yield the less highly substituted alkene as the major product is referred to as the *Hofmann rule*.

The direction and the specificity of this reaction contrasts strikingly with the Saytzeff mode of elimination described for alkyl halides in Section 3.8:

$$CH_3CH_2CHBrCH_3 \xrightarrow[\Delta]{C_2H_5OH, KOH} \underset{81\%}{CH_3CH=CHCH_3} + \underset{19\%}{C_2H_5CH=CH_2} \quad (11.25)$$

The reason for a shift in E2 elimination products from the Saytzeff rule to the Hofmann rule as the leaving group is changed from halide ion to amine is not entirely clear. One speculation is that the E2 transition

FIGURE 11·10 *The Hofmann elimination*

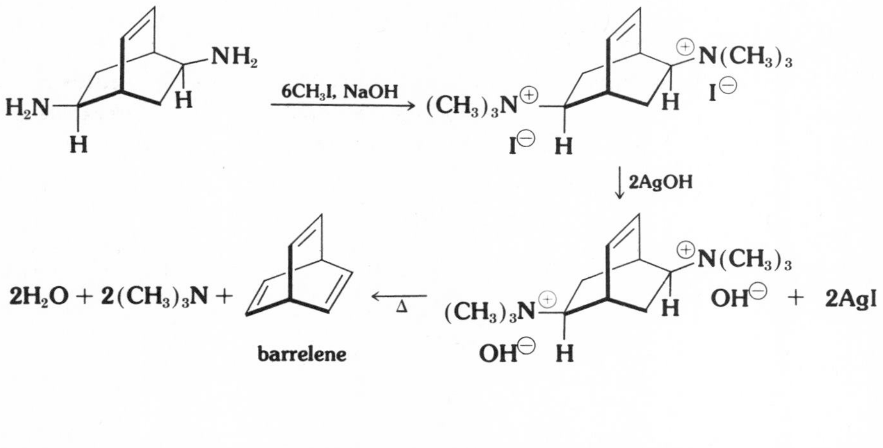

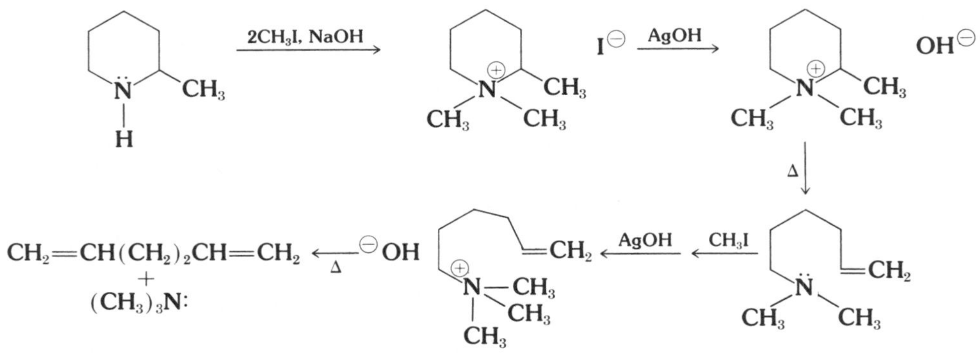

11·5 DISPLACEMENT AND ELIMINATION OF NITROGEN

FIGURE 11·11 *Sequential Hofmann eliminations of a cyclic amine*

state in a Hofmann elimination probably has less double-bond development and greater carbanion character at the β carbon atom than the equivalent transition state for a Saytzeff elimination:

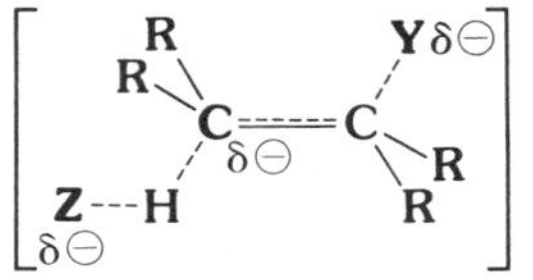

(Y = halide or tertiary amine
Z = OH or OR)
   E2 transition state

If the amine nitrogen atom is part of a ring system, more than one Hofmann elimination will be required to remove it from the molecule, as illustrated in Figure 11.11.

PROBLEM 11·12
a Why do we speculate that the transition state for a Hofmann elimination has less double-bond development than the corresponding Saytzeff transition state?
b How many successive Hofmann eliminations would be required to remove the nitrogen atom from each of the following amines?

AMINE OXIDES AS LEAVING GROUPS

Another useful method of preparing alkenes from primary amines involves pyrolysis of a dimethylamine oxide derivative at temperatures of 150° to 200°C. This transformation is often referred to as the *Cope elimination reaction*:

   (11.26)

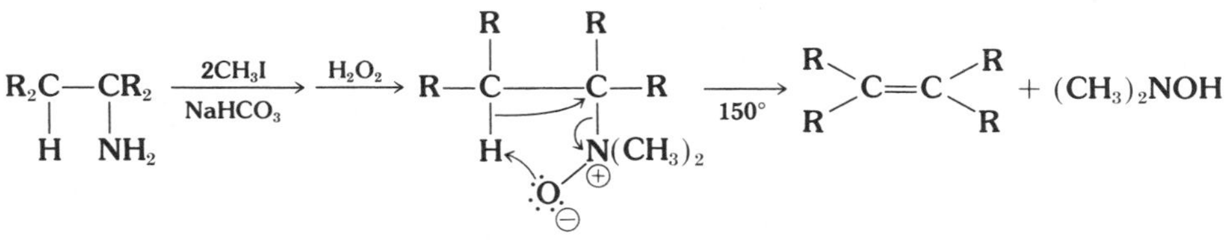

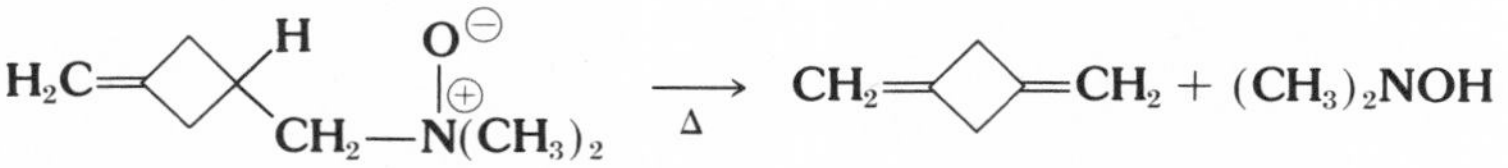

$$H_2C=\underset{CH_2-\overset{\oplus}{N}(CH_3)_2}{\overset{H\quad \overset{\ominus}{O}}{\diamond}} \xrightarrow{\Delta} CH_2=\diamond=CH_2 + (CH_3)_2NOH \quad (11.27)$$

Experimental evidence suggests that these reactions proceed by a cyclic reorganization of atoms and electrons, rather than by means of an $E1$ or $E2$ mechanism.

11·6 RING-SUBSTITUTION REACTIONS OF AROMATIC AMINES

We accounted for the reduced basicity of aromatic amines by assuming that these compounds are stabilized by delocalization of the nitrogen electron pair into the aromatic ring (Figure 11.5). If this assumption is correct, the resulting increase in ring nucleophilicity would be expected to enhance electrophilic substitution rates at the ortho and para carbon atoms. In other words, an amino substituent should activate the aromatic ring and direct electrophilic attack to the ortho and para positions.

Experimental evidence supports this viewpoint. Indeed, amino groups share with the hydroxyl group the distinction of being the most powerful activating substituents commonly encountered in aromatic compounds. The nitration and bromination of aniline in fact, takes a course very similar to the equivalent reactions of phenol (equations 10.27 and 10.29):

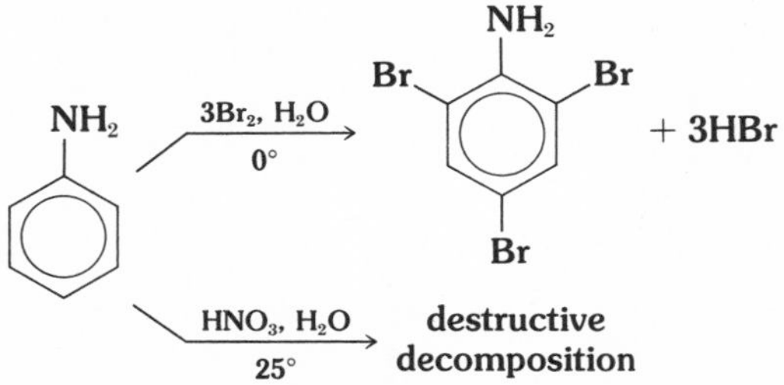

$$\text{PhNH}_2 \xrightarrow[0°]{3Br_2, H_2O} \text{2,4,6-tribromoaniline} + 3HBr \quad (11.28)$$

$$\text{PhNH}_2 \xrightarrow[25°]{HNO_3, H_2O} \text{destructive decomposition}$$

Weak electrophilic reagents are more selective in their reactions. The following examples are unique for arylamines (and phenols):

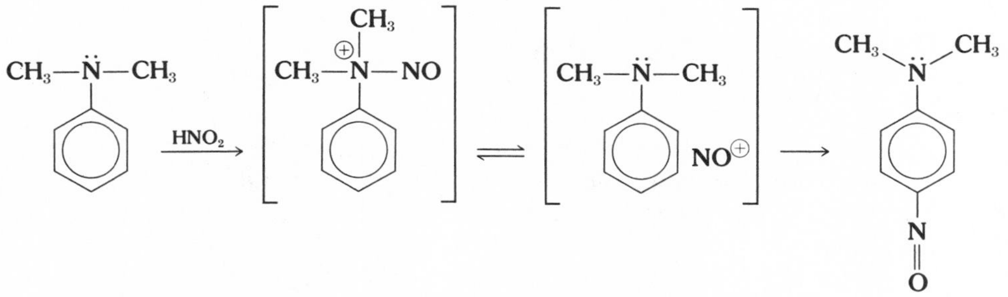

$$\text{PhN(CH}_3)_2 \xrightarrow{HNO_2} \left[\text{CH}_3\text{-}\overset{\oplus}{\underset{|}{N}}\text{-NO on ring} \right] \rightleftharpoons \left[\text{CH}_3\text{-}\ddot{N}\text{-CH}_3 \text{ with } NO^{\oplus} \right] \longrightarrow p\text{-nitroso-}N,N\text{-dimethylaniline} \quad (11.29)$$

$$\text{1,2,3,4-tetrahydroquinoline} + Br(CH_2)_3Cl \longrightarrow [\text{intermediate}] \longrightarrow \text{julolidine} + HCl \quad (11.30)$$

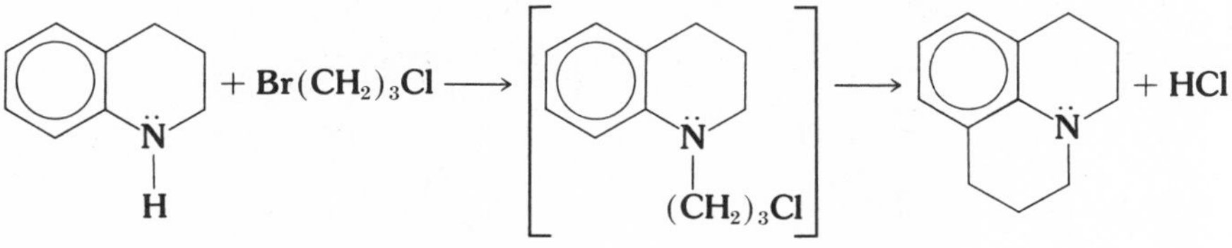

$$2\,\phi\text{-NH}_2 + I_2 \longrightarrow I\text{-}\phi\text{-NH}_2 + \phi\text{NH}_3^{\oplus}I^{\ominus} \tag{11.31}$$

PROBLEM 11·13
a Write an equation showing the shift of electrons occurring during the aromatic substitution step in reaction 11.30.
b What will be the major product from the reaction of 2,6-dimethylphenol with nitrous acid?

The excessive reactivity of aniline and other aromatic amines in common electrophilic substitution reactions often makes such reactions difficult to control. Fortunately, we can diminish the activating influence of a nitrogen atom by introducing substituents that interfere with electron-pair delocalization into the aromatic ring. An extreme example of this effect is seen in the reactions of quaternary salts, where the absence of an unshared electron pair on nitrogen, combined with its formal positive charge, changes the substituent effect to deactivating and meta directing:

$$\phi\text{-}\overset{\oplus}{N}(CH_3)_3\ NO_3^{\ominus} \xrightarrow[\Delta]{HNO_3,\ H_2SO_4} O_2N\text{-}\phi\text{-}\overset{\oplus}{N}(CH_3)_3\ NO_3^{\ominus} \tag{11.32}$$

By acylating the amine nitrogen atom (that is, by forming an amide derivative) we create an intermediate situation in which the activating influence of the amine is moderated but not lost.† Since the acyl group can be removed by hydrolysis, it represents a useful protective group for the amine function:

$$\phi\text{-NH}_2 \xrightarrow{(CH_3CO)_2O} \phi\text{-NHCOCH}_3 \xrightarrow[\Delta]{Br_2,\ Fe} \underset{Br}{\phi\text{-NHCOCH}_3} \xrightarrow[\Delta]{H_3O^{\oplus}} \underset{Br}{\phi\text{-NH}_2} \tag{11.33}$$

+ ortho isomer

11·7 REACTIONS OF ARYL DIAZONIUM SALTS

DISPLACEMENT OF NITROGEN

Reactions in which a covalently bonded N_2 function is lost as molecular nitrogen are strongly exothermic because of the very large $N{\equiv}N$ bond energy (225.8 kcal/mole). Thus substitution and elimination of N_2 from aliphatic diazonium salts was found to be fast and irreversible. In this section we shall consider an important group of reactions which involve the substitution of aryl diazonium functions by a wide variety of nucleophilic species. The scope and utility of these reactions is outlined in Figure 11.12.

---
†This decreased reactivity of the amide nitrogen electron pair was noted earlier in connection with reactions 11.9 to 11.11.

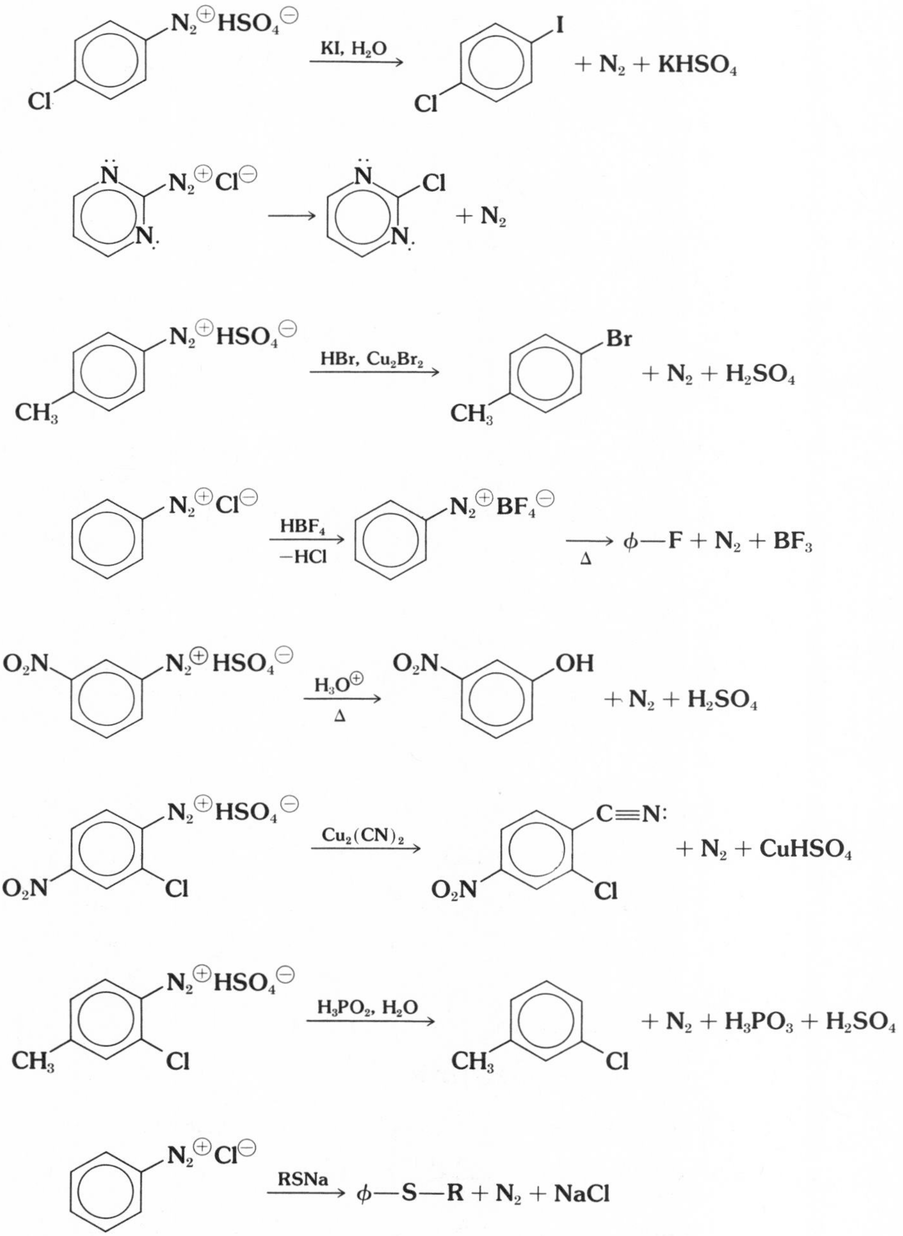

11·7 REACTIONS OF ARYL DIAZONIUM SALTS

FIGURE 11·12 *Substitution reactions of aryl diazonium salts*

Although these reactions all appear to take roughly the same course, we must not be lured into assuming that they proceed by a common reaction mechanism. In fact, there is substantial evidence that several different mechanisms can operate.

THE S_N1 MECHANISM The difficulties inherent in generating aryl cations are largely overcome by the extraordinary stability of molecular nitrogen as a leaving group. Consequently, the activation energy for

11·7 REACTIONS OF ARYL DIAZONIUM SALTS

decomposition of the aryl diazonium cation is lowered sufficiently to allow an S_N1 mechanism to operate:

$$ArN_2^{\oplus} X^{\ominus} \xrightarrow{2H_2O} [Ar^{\oplus}] \begin{array}{c} \xrightarrow{X^{\ominus}} Ar-X \\ \xrightarrow{2H_2O} ArOH + H_3O^{\oplus} \\ \xrightarrow{Z^{\ominus}} Ar-Z \end{array} \quad (11.34)$$

As expected, diazonium-salt decompositions proceeding by an S_N1 mechanism generally show a first-order kinetic relationship and are essentially unaffected by changes in either the concentration or the nature of the nucleophile:

$$\text{rate of aryl diazonium-ion decomposition} = k_1[ArN_2^{\oplus}] \quad (11.35)$$

Since the rate at which *p*-nitrobenzenediazonium bisulfate decomposes in aqueous solution increases slightly with added bromide or iodide salts, it is likely that in these cases there is also a slow bimolecular reaction.

RADICAL DECOMPOSITION The use of copper salts as catalysts for diazonium-ion substitutions characterizes a common group of reactions known as *Sandmeyer reactions*. Figure 11.12 includes several examples of this type, in which the entering substituent is chloride, bromide, or cyanide. A radical-chain mechanism has been proposed as an an explanation of these reactions:

$$ArN_2^{\oplus} + Cu^I Cl_2^{\ominus} \longrightarrow Ar\cdot + N_2 + Cu^{II}Cl_2$$
$$Ar\cdot + Cu^{II}Cl_2 \longrightarrow ArCl + Cu^I Cl \xrightarrow{Cl^{\ominus}} Cu^I Cl_2^{\ominus} \quad (11.36)$$

Hypophosphorous acid (H_3PO_2) participates in a radical-chain process which results in replacement of the diazonium cation by a hydrogen atom:

$$ArN_2^{\oplus} X^{\ominus} + H_3PO_2 + H_2O \longrightarrow Ar-H + N_2 + H_3PO_3 + HX \quad (11.37)$$

$$Ar\cdot + H_3PO_2 \longrightarrow Ar-H + H_2PO_2\cdot$$
$$H_2PO_2\cdot + ArN_2^{\oplus} \longrightarrow Ar\cdot + N_2 + H_2PO_2^{\oplus} \quad \text{radical-chain reaction}$$
$$H_2PO_2^{\oplus} + 2H_2O \longrightarrow H_3PO_3 + H_3O^{\oplus}$$

The reactions outlined in Figure 11.12 enable us to synthesize a host of polysubstituted aromatic compounds which would be difficult or impossible to prepare by direct electrophilic substitution. For example,

$$\phi-NH_2 \xrightarrow{3Br_2} \underset{Br}{\underset{|}{\text{2,4,6-Br}_3C_6H_2NH_2}} \xrightarrow{NaNO_2, H_2SO_4} \underset{Br}{\underset{|}{\text{2,4,6-Br}_3C_6H_2N_2^{\oplus}HSO_4^{\ominus}}} \xrightarrow{H_3PO_2} \text{1,3,5-tribromobenzene} \quad (11.38)$$

11·7 REACTIONS OF ARYL DIAZONIUM SALTS

$$\text{p-NO}_2\text{-C}_6\text{H}_4\text{-NH}_2 \xrightarrow{2I_2} \text{(2,6-diiodo-4-nitroaniline)} \xrightarrow{\text{NaNO}_2, \text{HCl}} \text{(diazonium salt)} \xrightarrow{\text{KI, H}_2\text{O}} \text{3,4,5-triiodonitrobenzene} \quad (11.39)$$

NUCLEOPHILIC ATTACK AT NITROGEN

Although the positive charge carried by an aryl diazonium ion is delocalized over both nitrogen atoms (and, to a lesser extent, over the aromatic ring), direct bonding reactions with nucleophiles are observed only at the terminal nitrogen:

$$\{\phi-\ddot{N}=\ddot{N}^{\oplus} \longleftrightarrow \phi-N\overset{\oplus}{\equiv}N\!:\} + CN^{\ominus} \longrightarrow \phi-\ddot{N}=\ddot{N}-CN \quad (11.40)$$
$$\text{phenyldiazocyanide}$$

It is noteworthy, in fact, that diazonium ions can function as dibasic acids in aqueous solution:

$$Ar-N_2^{\oplus} + 2H_2O \underset{}{\overset{K_1}{\rightleftarrows}} Ar-N=N-OH + H_3O^{\oplus} \quad (11.41)$$
$$\text{an aryl diazoic acid}$$

$$Ar-N=N-OH + H_2O \underset{}{\overset{K_2}{\rightleftarrows}} Ar-N=N-\ddot{O}\!:^{\ominus} + H_3O^{\oplus} \quad (11.42)$$
$$\text{an aryl diazoate anion}$$

Since K_2 is over 1000 times larger than K_1, only the salts of diazoic acids can be isolated, and these can be obtained in stereoisomeric forms analogous to the cis-trans configurations of certain alkenes.

PROBLEM 11·14
a Write a resonance description of the diazoate anion and suggest an explanation for the fact that the diazoic acid derived from benzenediazonium chloride is a weaker acid than that derived from its *p*-nitro derivative.
b Write structural formulas showing the configurational difference between the stereoisomeric potassium benzenediazoates.

Primary alkyl amines react with aryl diazonium salts in acidic media to give deamination products similar to those obtained with nitrous acid treatment. A mechanism involving initial nucleophilic attack at nitrogen has been proposed for this transformation:

$$R-NH_2 + \phi-N_2^{\oplus}Cl^{\ominus} \longrightarrow \phi-NH_2 + [R-N_2^{\oplus}]Cl^{\ominus} \longrightarrow R-Cl + \text{alkenes} + N_2$$
$$\Updownarrow \qquad\qquad\qquad \Updownarrow$$
$$\left[RNH_2^{\oplus}-N=N\phi \rightleftarrows R\overset{\delta\oplus}{N}H-N-\overset{\delta\oplus}{N}H\phi \rightleftarrows RN=N-NH_2^{\oplus}\phi\right]Cl^{\ominus} \quad (11.43)$$

The following deceptively simple reaction appears at first glance to be a simple case of nitrogen displacement:

11·7 REACTIONS OF ARYL DIAZONIUM SALTS

$$\phi N_2^{\oplus} Cl^{\ominus} + NaN_3 \longrightarrow \phi-N_3 + N_2 + NaCl$$

sodium azide · phenyl azide

However, when we use a diazonium cation isotopically labeled at the terminal nitrogen atom, we find that at least half the labeled nitrogen atoms remain in the phenyl azide. This surprising result can be explained if we assume a mechanism in which nucleophilic attack at the terminal nitrogen atom takes place more rapidly than the S_N1 or S_N2 displacement reaction:

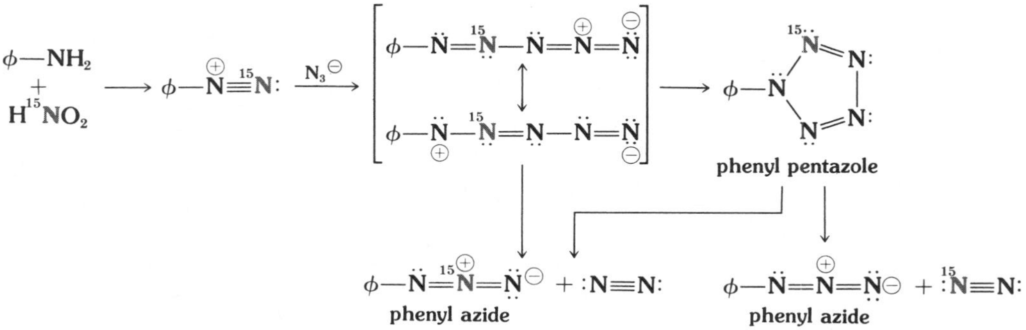

The fact that phenyl pentazole can be isolated from a reaction run at $-30°C$ increases the plausibility of this proposal. This reaction is unique and is discussed here only as an interesting variation of the previous mechanisms.

DIAZO COUPLING REACTIONS

The diazonium ion is a relatively weak electrophile; it therefore tends to be highly selective in its reactions with weak nucleophiles. Electrophilic aromatic substitution, for example, occurs only if the aromatic ring is strongly activated by a hydroxyl or an amino group, as shown in Figure 11.13.

Such reactions are termed *diazo coupling reactions* and give rise to highly colored products known as *azo dyes*. Coupling normally occurs at a position para to the activating group, but it will proceed at an ortho position if the para is blocked by another substituent. If both an amino group and a hydroxyl group are present, the amino group will usually direct the coupling in weakly acidic solutions, while the hydroxyl group directs in weakly alkaline media.

FIGURE 11·13 *Diazo coupling reactions*

$$Ar-N_2^{\oplus} \; X^{\ominus} \quad + \quad \langle\!\bigcirc\!\rangle-Y \quad \longrightarrow \quad Ar-N=N-\langle\!\bigcirc\!\rangle-Y + HX$$

electrophilicity increases as electron-withdrawing groups are introduced on Ar

activation increases:
$Y = O^{\ominus} > NR_2 > NRH > OR$

an azo compound (brightly colored)

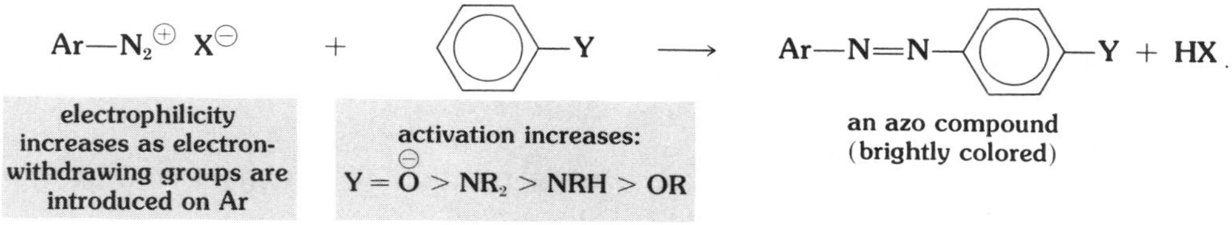

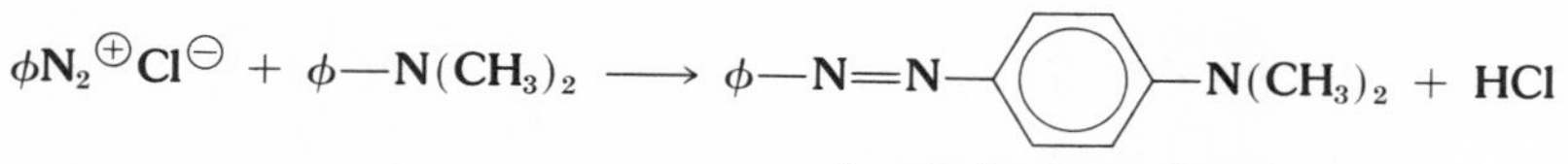

11·7 REACTIONS OF ARYL DIAZONIUM SALTS

$$\phi N_2^{\oplus} Cl^{\ominus} + \phi-N(CH_3)_2 \longrightarrow \phi-N=N-\!\!\left\langle\!\!\bigcirc\!\!\right\rangle\!\!-N(CH_3)_2 + HCl \qquad (11.44)$$

p-dimethylaminoazobenzene
(a light orange)

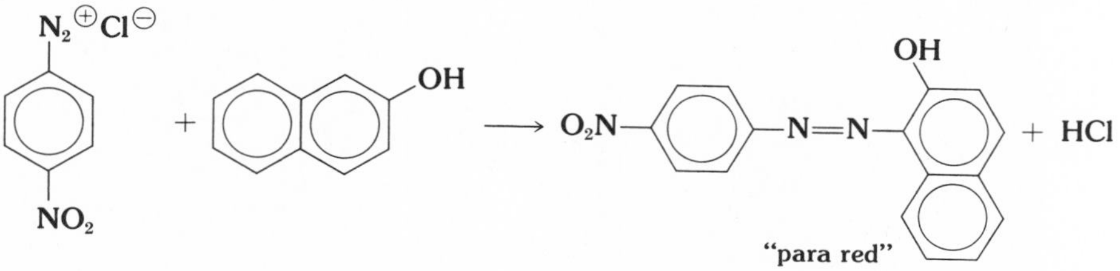

(11.45)

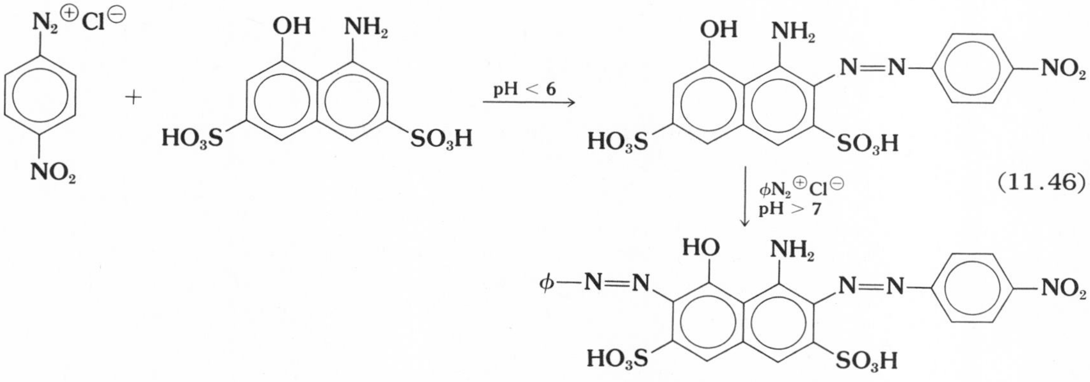

(11.46)

"naphthol blue-black B"

PROBLEM 11·15

Write structural formulas for the coupling components that would lead to the following azo compounds:

a

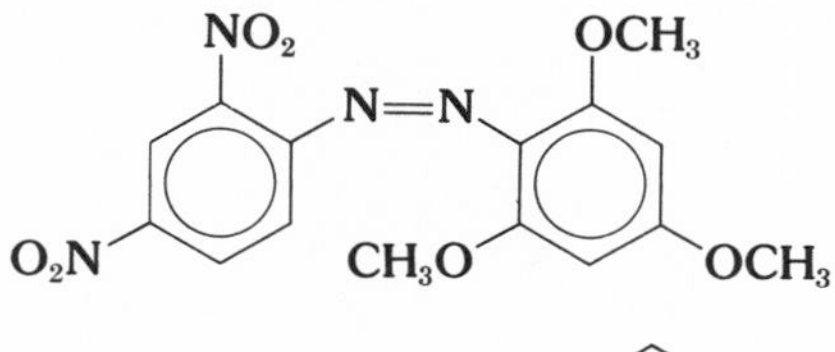

b

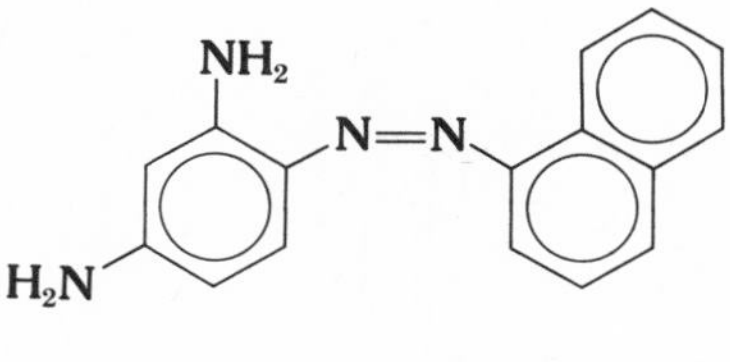

c 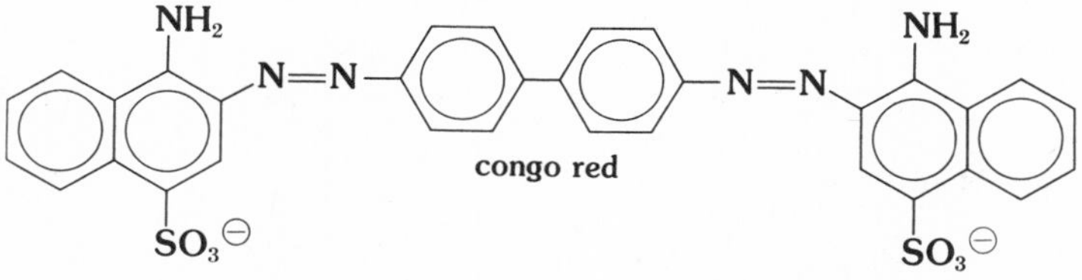 congo red

REDUCTION OF THE DIAZONIUM CATION

Mild reducing agents such as sodium bisulfite, stannous chloride, and zinc dust in acetic acid reduce aryl diazonium salts to the corresponding hydrazine derivatives:

$$ArN_2^{\oplus}X^{\ominus} \xrightarrow{NaHSO_3} [Ar-N=N-SO_3H] \xrightarrow{NaHSO_3} [ArNHNHSO_3H] \xrightarrow{H_3O^{\oplus}} ArNHNH_2 \quad (11.47)$$
an arylhydrazine

The simplest of these arylhydrazines is phenylhydrazine. Its discovery by Emil Fischer in 1875 (he was twenty-three at the time) opened the door for his monumental work concerning the structure of sugars, for which he received a Nobel Prize.

11·8 METHODS OF PREPARING AMINES

The widespread occurrence of amine functions in alkaloids and other drugs has made the controlled synthesis of such groups a much-sought goal. Furthermore, the fact that aryl amines can be transformed into a variety of other functional groups makes them exceptionally versatile as intermediates in the synthesis of many other classes of compounds. It is therefore important to consider some of the methods we can use to prepare amines in the laboratory (a complete outline of these methods appears at the end of Chapter 19).

A straightforward but often marginally successful approach to making amines is the direct nucleophilic substitution of alkyl halides by ammonia or other amines, as described in Figure 11.7.

$$\underset{\text{electrophile}}{R-X} + \underset{\text{nucleophile}}{R'-\ddot{N}H_2} \longrightarrow R-\underset{|}{\overset{H}{\underset{..}{N}}}-R' + HX \quad (11.48)$$

Among the limitations and difficulties of this method are:

1. Double (or higher) alkylation by **RX**
2. Poor reactivity of hindered **RX** reactants
3. Elimination induced by the base **R′ṄH₂**
4. Limited application to aryl halides (see equation 11.8)

The first three of these problems can be overcome by using a more nucleophilic and less basic nitrogen-containing nucleophile which can be converted in a subsequent step to an amine function. Examples of this strategy are shown in the following equations:

$$R-\underset{H}{\overset{CH_3}{\underset{|}{\overset{|}{C}}}}-Br + \underset{\text{sodium azide}}{Na^{\oplus}N_3^{\ominus}} \xrightarrow{S_N2} R-\underset{H}{\overset{CH_3}{\underset{|}{\overset{|}{C}}}}-N_3 \xrightarrow{H_2, Pt} R-\underset{H}{\overset{CH_3}{\underset{|}{\overset{|}{C}}}}-NH_2 \quad (11.49)$$

$$R-\underset{H}{\overset{CH_3}{\underset{|}{\overset{|}{C}}}}-I + \underset{\text{silver nitrite}}{AgNO_2} \longrightarrow R-\underset{H}{\overset{CH_3}{\underset{|}{\overset{|}{C}}}}-NO_2 \xrightarrow[H_3O^{\oplus}]{Fe, FeSO_4} R-\underset{H}{\overset{CH_3}{\underset{|}{\overset{|}{C}}}}-NH_3^{\oplus}HSO_4^{\ominus} \quad (11.50)$$

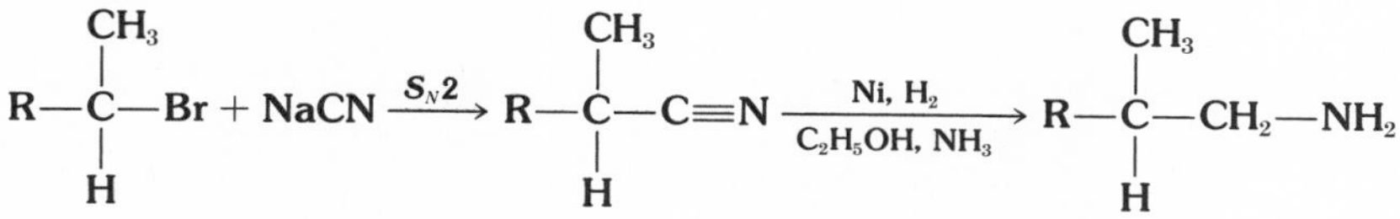

11·8 METHODS OF PREPARING AMINES

$$R-\underset{\underset{H}{|}}{\overset{\overset{CH_3}{|}}{C}}-Br + NaCN \xrightarrow{S_N2} R-\underset{\underset{H}{|}}{\overset{\overset{CH_3}{|}}{C}}-C\equiv N \xrightarrow[C_2H_5OH, NH_3]{Ni, H_2} R-\underset{\underset{H}{|}}{\overset{\overset{CH_3}{|}}{C}}-CH_2-NH_2 \quad (11.51)$$

The catalytic reduction of the nitrile function in reaction 11.51 may also give some secondary amine, $[RCH(CH_3)CH_2]_2NH$, unless the reaction mixture is saturated with ammonia.† A better way of avoiding this undesirable byproduct is to use the versatile reducing agent lithium aluminum hydride ($LiAlH_4$):

$$R-C\equiv N + LiAlH_4 \xrightarrow{ether} R-CH_2-NH_2 \quad (11.52)$$

This reagent is also useful in reducing amides to amines, a procedure we shall examine in Chapter 14:

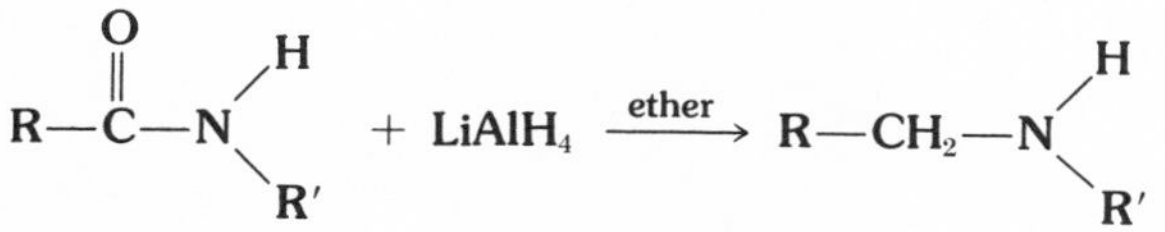

$$R-\overset{\overset{O}{\|}}{C}-N\overset{H}{\underset{R'}{\diagdown}} + LiAlH_4 \xrightarrow{ether} R-CH_2-N\overset{H}{\underset{R'}{\diagdown}} \quad (11.53)$$

The ready availability of aromatic nitro compounds by direct nitration (Sections 9.6 and 10.3) makes their reduction to aryl amines an important synthesis procedure. This can be accomplished in high yield by catalytic hydrogenation or by chemical reduction in aqueous acid:

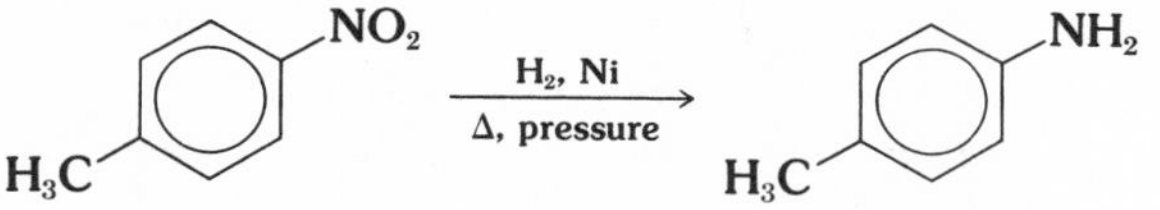

$$H_3C-C_6H_4-NO_2 \xrightarrow[\Delta, \text{ pressure}]{H_2, Ni} H_3C-C_6H_4-NH_2 \quad (11.54)$$

$$Br-C_6H_4-NO_2 \xrightarrow{H_2, Pt, \text{ alcohol}} Br-C_6H_4-NH_2 \quad (11.55)$$

$$m\text{-}(NO_2)_2C_6H_4 \xrightarrow[\Delta]{Fe, HCl, \text{ alcohol}} m\text{-}(NH_3^{\oplus}Cl^{\ominus})_2C_6H_4 \quad (11.56)$$

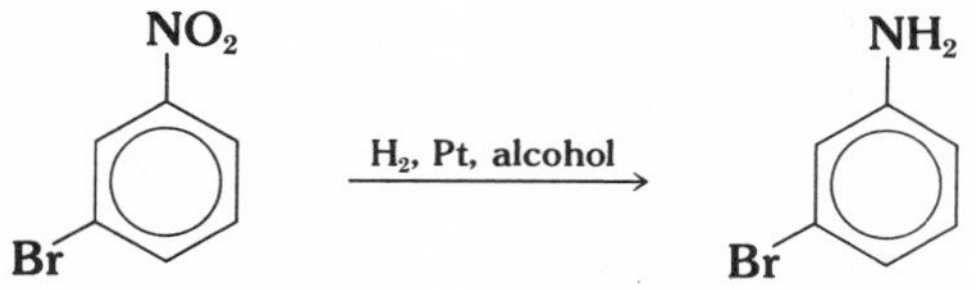

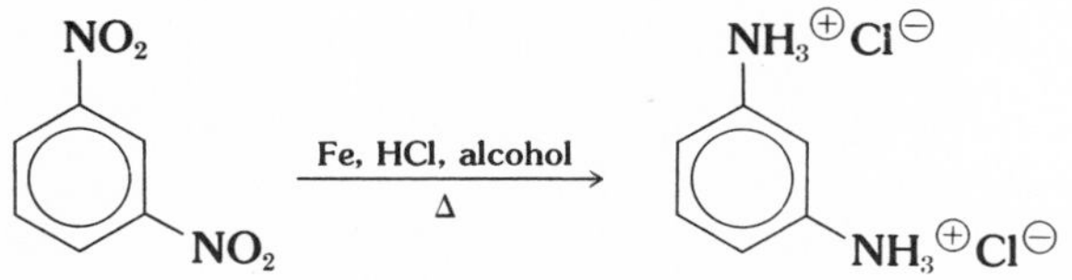

$$HO_2C-C_6H_4-NO_2 \xrightarrow{SnCl_2, HCl} HO_2C-C_6H_4-NH_3^{\oplus}Cl^{\ominus} \quad (11.57)$$

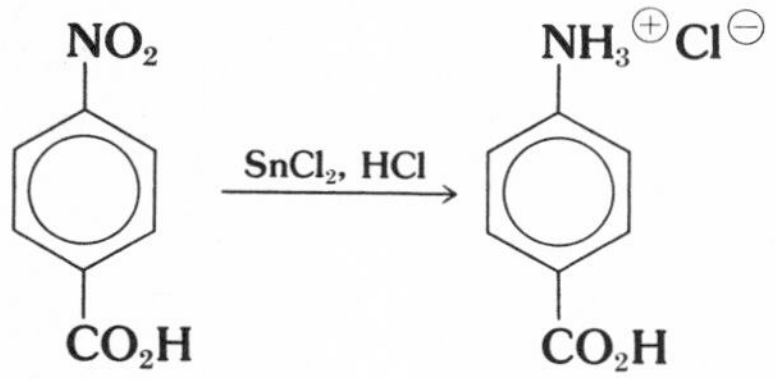

†The formation of this byproduct can be explained by a mechanism similar to that described in Chapter 12 for the reductive amination of aldehydes and ketones.

11·9 PHOSPHINES

PROBLEM 11·16
The following transformations require two or more steps. Write equations for these steps, assuming that isomeric products from electrophilic aromatic substitution can be separated. *Hint:* See Sections 10.3, 10.5, and 11.7; also note that a nitro group is a meta-directing substituent on a benzene ring.

a. C₆H₆ ⟶ C₆H₅I

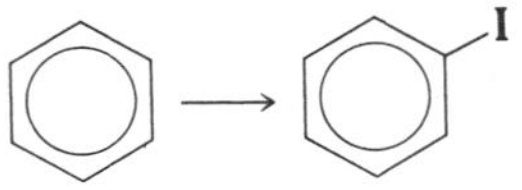

b. $(CH_3)_2CHBr \longrightarrow (CH_3)_2CHCH_2NH_2$

c. C₆H₆ ⟶ 4-Br-C₆H₄-CN (Br and CN para)

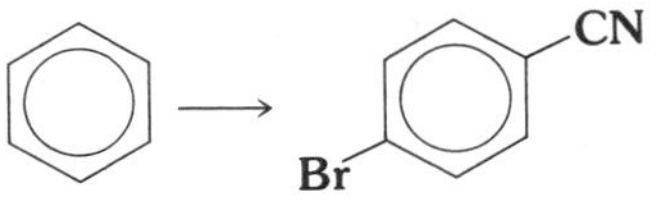

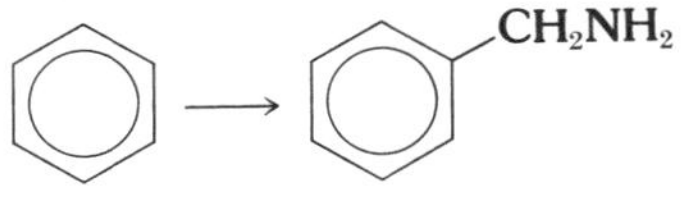

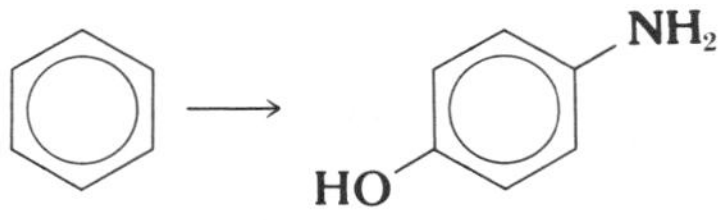

d. C₆H₆ ⟶ C₆H₅CH₂NH₂

e. C₆H₆ ⟶ 4-HO-C₆H₄-NH₂

f. C₆H₆ ⟶ 3,5-dibromoaniline (Br, Br, NH₂)

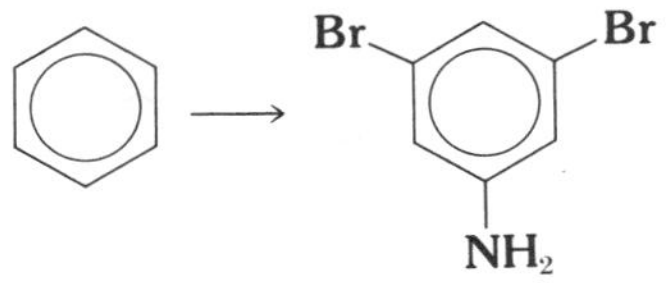

11·9 PHOSPHINES

The Group V element immediately following nitrogen in the periodic table is phosphorus, and in view of their similar valence-shell configurations, we might expect the chemical behavior of organic phosphorus derivatives to parallel that of the amines. There are indeed some similarities in behavior, but we also find that important differences exist between the chemistry of nitrogen and phosphorus compounds—differences which generally parallel those between oxygen and sulfur derivatives. For example, the low-energy d orbitals in the valence shell of phosphorus and sulfur allow both of these elements to expand their covalent coordination beyond the limit of the octet rule:

$PCl_3 + Cl_2 \longrightarrow PCl_5$

$SF_4 + F_2 \longrightarrow SF_6$

Phosphorus analogs of amines are called *phosphines* and can be prepared by the action of Grignard reagents on phosphorus trichloride:

$$3RMgCl + PCl_3 \longrightarrow R_3P + 3MgCl_2 \qquad (11.58)$$
a trialkylphosphine

The phosphines have a pyramidal configuration, with a C—P—C angle of about 109°. Since their inversion rate is much slower than that of the amines, phosphines having three different alkyl or aryl substituents can be resolved.

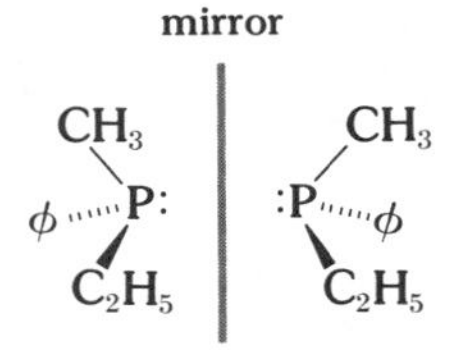

mirror

ϕ—P(CH₃)(C₂H₅) : | : P(CH₃)(C₂H₅)—ϕ

enantiomeric phosphines

Although phosphine itself (PH_3) is only weakly basic ($K_b \approx 10^{-27}$), its alkyl derivatives have basicities closer to those of corresponding amines; $(C_2H_5)_3P$, for example, has a basicity of $K_b \approx 10^{-7}$. However, because phosphorus is larger and more polarizable than nitrogen, phosphines generally have greater nucleophilicities than similarly substituted amines:

$$\phi_3P: + CH_3I \longrightarrow \phi_3\overset{\oplus}{P}-CH_3 \; I^{\ominus} \qquad (11.59)$$
$$\text{a phosphonium salt}$$

(No comparable reaction takes place with $\phi_3N:$.)

Since phosphorus forms unusually strong covalent bonds with oxygen and weak bonds with hydrogen, phosphine and its mono- or dialkyl derivatives are very easily oxidized, often burning spontaneously on exposure to oxygen. Trisubstituted phosphines form oxides, and the exceptional stability of these compounds provides a driving force for many interesting and useful reactions:

$$(C_2H_5)_3\overset{\oplus}{N}-\overset{..}{\underset{..}{O}}:^{\ominus} + \phi_3P: \xrightarrow{\Delta} (C_2H_5)_3N: + \left\{ \phi_3\overset{\oplus}{P}-\overset{..}{\underset{..}{O}}:^{\ominus} \longleftrightarrow \phi_3P=\overset{..}{\underset{..}{O}} \right\} \qquad (11.60)$$
$$\text{triphenylphosphine oxide}$$

$$\underset{\overset{O-O}{RCH\underset{O}{\diagdown}CHR}}{} + \phi_3P: \longrightarrow 2 \; RC\overset{O}{\underset{H}{\diagup}} + \phi_3PO \qquad (11.61)$$

$$ArOH + \phi_3PBr_2 \xrightarrow{-HBr} \left[\phi_3\overset{\oplus}{P}-O-Ar \right] Br^{\ominus} \longrightarrow ArBr + \phi_3PO \qquad (11.62)$$

The methyl phosphonium salt shown in equation 11.59 reacts with strong bases (such as alkyl lithium reagents) to give an internal salt known as an *ylide*. The adjacent charges are stabilized by a *p-d* orbital overlap of the sort operating in triphenylphosphine oxide:

$$\phi_3\overset{\oplus}{P}-CH_3 \; Br^{\ominus} + C_4H_9Li \longrightarrow \{\phi_3\overset{\oplus}{P}-CH_2^{\ominus} \longleftrightarrow \phi_3P=CH_2\} + C_4H_{10} + LiBr \qquad (11.63)$$
$$\text{triphenylphosphonium methylide}$$

We shall examine some useful reactions of phosphorus ylides with carbonyl compounds in Chapter 12.

SUMMARY

amines Alkyl or aryl derivatives of ammonia, classified as primary, secondary, or tertiary according to the number of substituents on the nitrogen atom. Hydrogen bonding between primary and secondary amines manifests itself in high boiling points (see Table 11.1) and characteristic N—H absorption in the infrared and pmr spectra (see Table 11.2).

structure-composition relationship A correlation between the number of rings and/or double bonds in a nitrogen-containing compound and its degree of unsaturation. The relationship given in equation 6.3 can be applied if the molecular formula is first adjusted by substituting a CH increment for each nitrogen atom.

stereochemistry of amines The bonding configurations of nitrogen compounds. Nitrogen atoms bonded to three other atoms or groups have rapidly inverting pyramidal configurations; consequently, chiral secondary and tertiary amines cannot be resolved. Double- and triple-bonded nitrogen atoms have trigonal and linear configurations, respectively. A nitrogen atom bearing four substituents must be positively charged, and chiral salts of this kind can be resolved.

basicity of amines The ability of the nitrogen atom of an amine to accept a proton from a Brønsted acid. Most amines are moderate to weak bases ($K_b = 10^{-3}$ to 10^{-15}). Increased s character of the nonbonding electron pair or the presence of adjacent unsaturated functions usually causes a decrease in basicity and nucleophilicity.

electrophilic substitution at nitrogen Replacement of a proton on the nitrogen atom of a primary or secondary amine by an electrophilic moiety. *Alkylation* by alkyl halides, *acylation* by acyl halides and anhydrides, *sulfonation* by sulfonic acid derivatives (the Hinsberg test, for example), *nitrosation* by nitrous acid (usually followed by subsequent reactions), and *oxidation* of tertiary amines.

elimination reactions of amines The conversion of amines or their derivatives into alkenes by loss of the nitrogen function and a neighboring hydrogen atom. Since amino functions are very poor leaving groups, they are methylated and transformed to quaternary salt or amine oxides before elimination is effected. *Hofmann elimination* is the E2 elimination of quaternary salts and usually gives the less highly substituted alkene as the major product. *Pyrolysis* of amine oxides yields alkenes by a cyclic shifting of atoms and electrons.

electrophilic ring substitution of aryl amines Halogenation, nitration, sulfonation, or alkylation of an aromatic ring bearing an amino group. These reactions are difficult to control because the amine function strongly activates the ring to electrophilic attack. If the influence of the amine function is diminished by formation of an amide derivative, ring substitution proceeds smoothly at the ortho and para positions.

aryl diazonium salts Stable ionic intermediates formed by the reaction of primary aryl amines with cold nitrous acid:

$$[Ar\!-\!\overset{\oplus}{N}\!\equiv\!N\!:\:\longleftrightarrow\:Ar\!-\!\overset{\oplus}{\ddot{N}}\!=\!N\!:]X^{\ominus}$$

Displacement of molecular nitrogen from aryl diazonium salts by various nucleophiles and radicals occurs readily (for example, substitution of N_2 by I, Br, Cl, F, OH, CN and H). Nucleophilic attack may also take place at the terminal nitrogen atom of a diazonium ion, as in diazo coupling. Reduction of diazonium salts with sodium bisulfite gives arylhydrazines.

phosphines Phosphorus analogs of amines. Phosphines are generally weaker bases and stronger nucleophiles than the corresponding amines.

EXERCISES

11·1 Name the following amines and classify each as primary, secondary, tertiary, or quaternary:

a $(CH_3)_3N$

b $CH_3CH_2CH_2CH_2NH_2$

c $CH_3NHCH_2CH_3$

d (thiophene)—NH_2

e (cyclohexane)N—H

f (phenyl)—$N(CH_2CH_3)_2$

g $(CH_3)_3NH^{\oplus}Cl^{\ominus}$

h $(CH_3)_4N^{\oplus}I^{\ominus}$

11·2 Draw structures for and name the following compounds:
a The eight isomeric amines of molecular formula $C_4H_{11}N$
b Five isomeric amines of formula C_7H_9N that contain a benzene ring

11·3 Write electron-dot formulas showing all valence-shell electrons for the significant resonance contributors to the following:
a Azide ion ($N_3^{\ominus}$)
b Nitroso cation ($NO^{\oplus}$)
c Methyl isonitrile (CH_3-NC)

11·4 Determine whether the compounds represented by the molecular formulas below are saturated. If not, how many rings and/or multiple bonds are present in each case?
a C_3H_9N
b $C_5H_{11}NO_2$
c $C_7H_{11}NS$
d $C_4H_{10}N_2$
e $C_8H_{12}N_3BrO$

11·5 Hydrogen bonding has been used to explain the moderately high boiling points of amines and their relatively high solubilities in water. How, then, can the great difference in boiling point between *n*-propyl amine and trimethyl amine, both of molecular weight 59, be explained when each is equally very soluble in water (see Table 11.1)?

11·6 Arrange each group of compounds in order of decreasing base strength:
a $CH_3CH_2\ddot{N}H_2$ $\phi\ddot{N}H_2$ $:NH_3$
b $(CH_3)_3N:$ $\phi\ddot{N}(CH_3)_2$ $\phi C\equiv N:$

c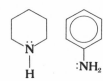

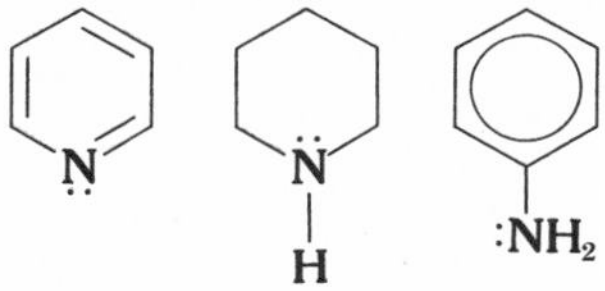

11·7 Compare the behavior of aniline, *N*-methylaniline, and *N,N*-dimethylaniline toward:
a $(CH_3CO)_2O$
b $NaNO_2$ in aqueous HCl
c CH_3I
d ϕSO_2Cl in aqueous NaOH

11·8 Compare the behavior of ethylamine, diethylamine, and triethylamine toward the reactants listed in Exercise 11.7.

11·9 Write equations for the following reactions:
a *n*-Butylamine + HCl $\longrightarrow$
b Methylethylamine + H_2SO_4 $\longrightarrow$
c Pyridinium chloride + NaOH $\longrightarrow$
d Piperidine + $2CH_3I$ $\xrightarrow{NaHCO_3}$
e Dimethylamine + ϕSO_2Cl $\xrightarrow{NaOH}$
f Trimethylamine + CH_3COCl $\longrightarrow$
g *n*-Propylamine + $(CH_3CO)_2O$ $\longrightarrow$
h Aniline + $NaNO_2$ $\xrightarrow[0°]{HCl, H_2O}$
i *N,N*-Diethylaniline + $NaNO_2$ $\xrightarrow{HCl, H_2O}$
j Trimethylamine + H_2O_2 $\longrightarrow$
k ϕNH_2 + HNO_3 (dilute) $\longrightarrow$
l ϕNH_2 + $3Br_2$ $\xrightarrow{H_2O}$
m ϕ_3P + CH_3I $\longrightarrow$

356 EXERCISES

11·10 In the alkylation of ammonia in Figure 11.7 the use of a large excess of alkyl halide yields the quaternary ammonium salt. What major organic product would you expect if a large excess of ammonia were used? Explain your reasoning.

11·11 Classify the following unknown amines as primary, secondary, or tertiary from the results of the reactions shown:

a Amine A $\xrightarrow[\text{NaOH, H}_2\text{O}]{\phi\text{SO}_2\text{Cl}}$ a clear solution $\xrightarrow{\text{H}_3\text{O}^\oplus}$ an insoluble solid

b Amine B $\xrightarrow[\text{NaOH, H}_2\text{O}]{\phi\text{SO}_2\text{Cl}}$ an insoluble oil $\xrightarrow{\text{H}_3\text{O}^\oplus}$ a clear solution

c Amine C $\xrightarrow[\text{NaOH, H}_2\text{O}]{\phi\text{SO}_2\text{Cl}}$ an insoluble solid $\xrightarrow{\text{H}_3\text{O}^\oplus}$ no change

11·12 Predict the major product from the following E2 elimination reactions:

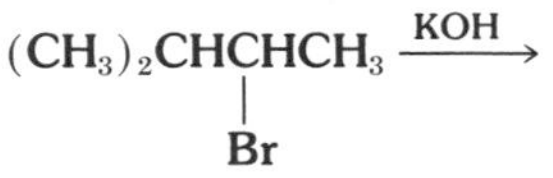

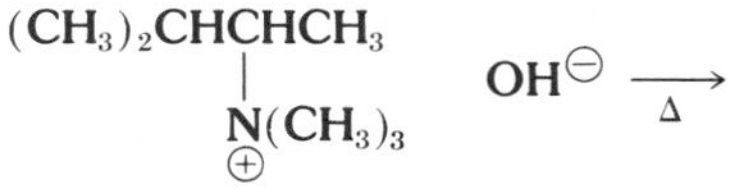

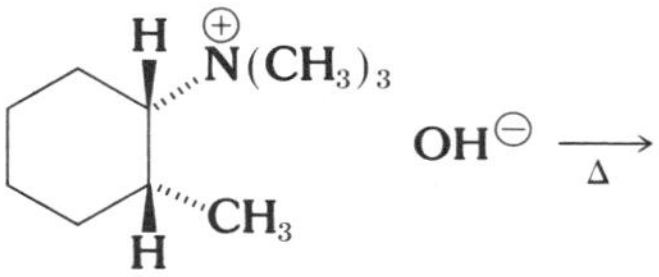

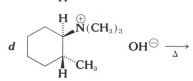

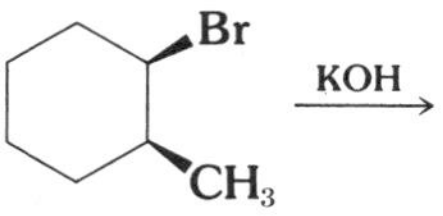

11·13 Write equations for reactions that will transform aniline into the following compounds.

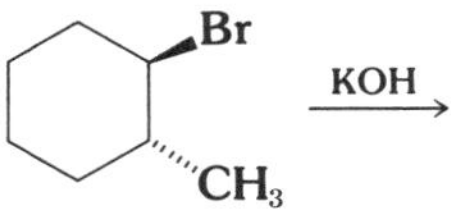

a Chlorobenzene
b Phenol
c Cyanobenzene
d Fluorobenzene
e Benzene
f Benzoic acid
g p-Bromoaniline
h m-Dichlorobenzene
i p-Iodophenol
j 1,2,4-Tribromobenzene
k Phenyl methyl sulfide

11·14 Can the direction of a Hofmann elimination be explained on the basis of steric factors? Note that the quaternary amine group is much larger than the halide leaving group. As an example use the compound

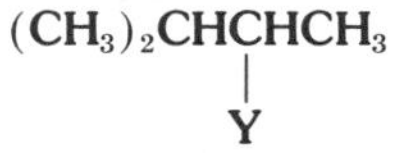

$(\text{CH}_3)_2\text{CHCHCH}_3$
           |
           Y

where Y = Cl or $\text{N(CH}_3)_3^\oplus$ with the base $\text{OH}^\ominus$ and explain your reasoning. *Hint:* What are the geometry requirements for the eliminations?

11·15 Identify intermediates F, G, and H in the two reaction sequences shown below. What does the formation of only the trans product in the Hofmann elimination tell us about the stereochemistry of this reaction? What is the stereochemistry of the Cope elimination? What stereoisomer of the starting material should be used to obtain the cis alkene via Hofmann elimination?

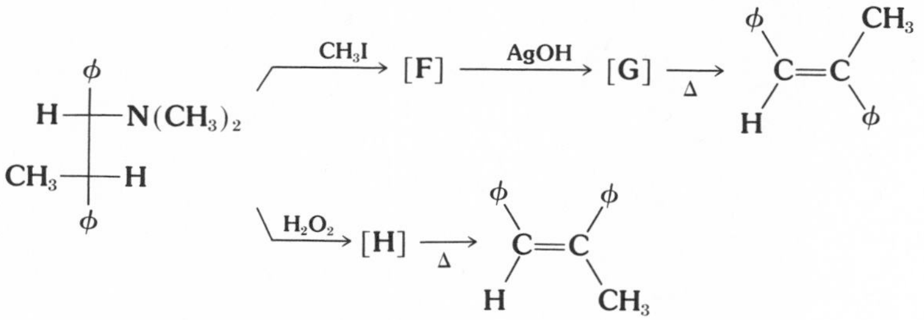

$$\begin{array}{c} \phi \\ H-\!\!\!\!\!-\!\!\!\!\!-N(CH_3)_2 \\ CH_3-\!\!\!\!\!-\!\!\!\!\!-H \\ \phi \end{array} \quad \xrightarrow{CH_3I} [F] \xrightarrow{AgOH} [G] \xrightarrow{\Delta} \begin{array}{c} \phi \\ \diagdown \\ H \end{array} C\!=\!C \begin{array}{c} CH_3 \\ \diagup \\ \phi \end{array}$$

$$\xrightarrow{H_2O_2} [H] \xrightarrow{\Delta} \begin{array}{c} \phi \\ \diagdown \\ H \end{array} C\!=\!C \begin{array}{c} \phi \\ \diagup \\ CH_3 \end{array}$$

11·16 Draw structural formulas for pyridine, pyrrole, and pyrrolidine. Indicate the geometry and hybridization of the nitrogen in each, and the location of the nonbonding electrons. Explain the relative base strengths of pyrrolidine ($K_b \approx 10^{-3}$) and pyrrole ($K_b \approx 10^{-14}$). Why is quinoline a stronger base than indole?

11·17 Phenacetin, an antipyretic (fever-reducing) agent, is used in APC tablets in combination with aspirin and caffeine. Draw structural formulas to complete the following synthesis for phenacetin:

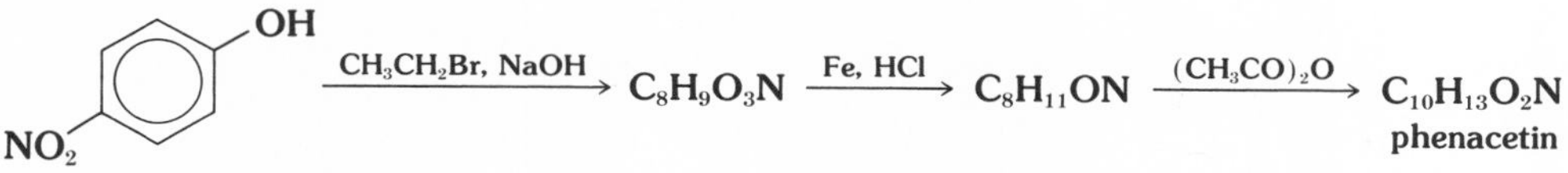

$$NO_2\text{-}C_6H_4\text{-}OH \xrightarrow{CH_3CH_2Br,\ NaOH} C_8H_9O_3N \xrightarrow{Fe,\ HCl} C_8H_{11}ON \xrightarrow{(CH_3CO)_2O} C_{10}H_{13}O_2N \text{ phenacetin}$$

11·18 Alkaloid X has the molecular formula $C_8H_{17}N$ and reacts with benzenesulfonyl chloride to yield an insoluble material, which remains insoluble on acidification. X consumes 2 moles of methyl iodide under exhaustive-methylation conditions. On heating the methylated product with $Ag_2O + H_2O$, the following olefinic amine Y is obtained:

$$CH_2\!=\!CHCH_2CH_2CHCH_2CH_2CH_3$$
$$\quad\quad\quad\quad\quad\quad\quad\ \ |$$
$$\quad\quad\quad\quad\quad\quad\quad\ \ N(CH_3)_2$$

a What are two possible structures for X?
b X can be prepared by the reaction of ammonia with 1,5-dibromooctane under carefully controlled conditions. What is the structure of alkaloid X?

11·19 Indicate prominent features in the specified spectra that could be used to distinguish the following pairs of compounds:
a CH_3CH_2OH and $CH_3CH_2NH_2$; pmr spectrum
b $CH_3(CH_2)_4CH_3$ and $CH_3(CH_2)_4NH_2$; infrared spectrum

c $\phi\text{-}NH_2$ and (thiophene)$\text{-}NH_2$ pmr spectrum

d $CH_3CH_2CH_2NH_2$ and $(CH_3)_3N$; infrared spectrum

358 EXERCISES

11·20 A compound of unknown structure (X) is readily soluble in dilute hydrochloric acid and has the molecular formula $C_{10}H_{15}N$. Although it does not react with benzenesulfonyl chloride, it does react with nitrous acid, yielding a compound of molecular formula $C_{10}H_{14}ON_2$. The pmr spectrum of X displays a triplet at $\delta = 1.1$, a quartet at $\delta = 3.3$, and a multiplet at $\delta \approx 6.8$, with relative areas of 6:4:5. What is the structure of compound X?

11·21 The poisonous component of the venom of the red fire ant has the molecular formula $C_{17}H_{35}N$. It reacts with 2 moles of methyl iodide under exhaustive-methylation conditions. When the methylated product is heated with AgOH, the three products shown below are obtained. What is the structure of this poison?

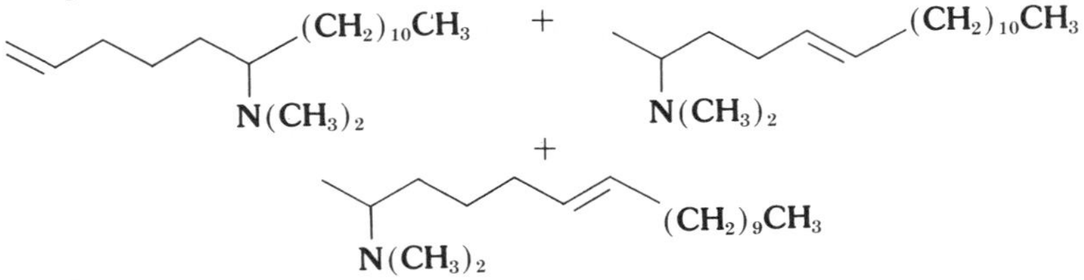

11·22 The following isomeric salts have different colors. Why would their visible absorption spectra be different?

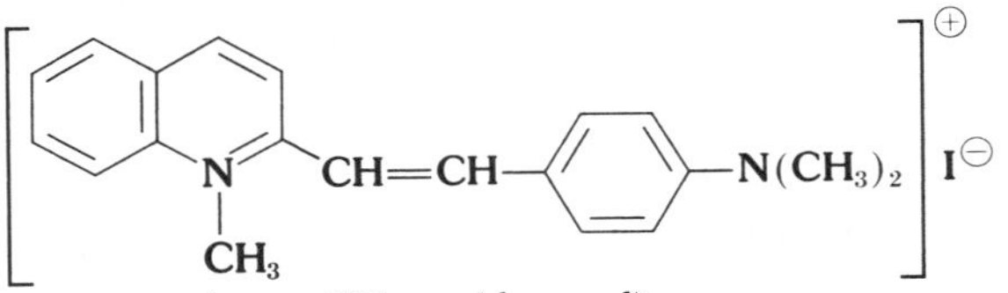

$\lambda_{max} = 526$ nm (deep red)

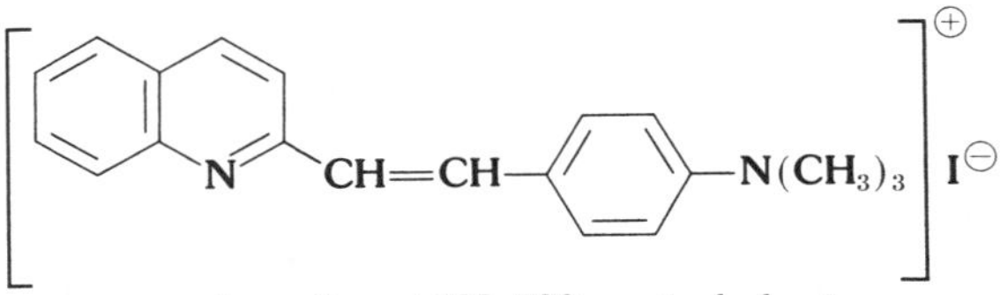

no absorption at 400–800 nm (colorless)

11·23 The dye "crystal violet" is prepared in the following manner. Write a mechanism for both steps in this synthesis.

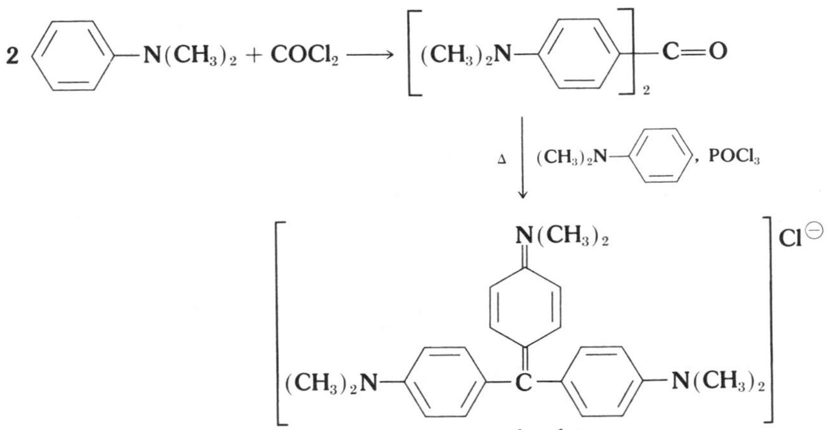

crystal violet

11·24 Suggest mechanisms for the following reactions:

a $(C_6H_5)_3P + I_2 \longrightarrow (C_6H_5)_3PI_2$

   $(C_6H_5)_3PI_2 + (CH_3)_2CHCH_2OH \longrightarrow (C_6H_5)_3PO + (CH_3)_2CHCH_2I + HI$

b $(C_6H_5)_3P + C_6H_5CH_2Br \longrightarrow$

   $[(C_6H_5)_3PCH_2C_6H_5]^{\oplus} Br^{\ominus} \xrightarrow{\text{NaOH, H}_2\text{O}} (C_6H_5)_3PO + C_6H_5CH_3$

11·25 Four isomeric amines with the molecular formula $C_4H_{11}N$ have the following pmr signals:

S: $\delta = 0.8$ (1H, singlet), $\delta = 1.1$ (6H, triplet), $\delta = 2.6$ (4H, quartet)
T: $\delta = 1.1$ (3H, triplet), $\delta = 2.2$ (6H, singlet), $\delta = 2.3$ (2H, quartet)
U: $\delta = 1.1$ (9H, singlet), $\delta = 1.3$ (2H, singlet)
V: $\delta = 0.9$ (6H, doublet), $\delta = 1.6$ (1H, multiplet), $\delta = 1.8$ (2H, singlet), $\delta = 2.5$ (2H, doublet)

From your answer to Exercise 11.2 above select four structures that best agree with these facts.

11·26 Compound X ($C_{11}H_{17}N$) is soluble in dilute aqueous hydrochloric acid and reacts with nitrous acid to give compound Y ($C_{11}H_{16}N_2O$). The infrared spectrum of X shows prominent absorptions at 3020 and 2990 cm^{-1}, but nothing higher than 3080 cm^{-1}. The pmr spectrum of X consists of five distinct groups of signals: $\delta = 1.0$ (doublet, $J = 6.8$ Hz), $\delta = 2.6$ (septet, $J = 6.8$ Hz), $\delta = 3.0$ (singlet), $\delta = 6.5$ (doublet, $J = 9$ Hz), and $\delta = 7.1$ (doublet, $J = 9$ Hz); the area ratio of these signals is 6:1:6:2:2. Write structures for X and Y.

12

ALDEHYDES AND KETONES

THE REACTIVE unit consisting of a carbon double-bonded to oxygen is known as the *carbonyl group,* and this functional group is responsible for the characteristic reactivity of aldehydes and ketones. The carbonyl group is also found, modified by other functions, in carboxylic acids and their derivatives.

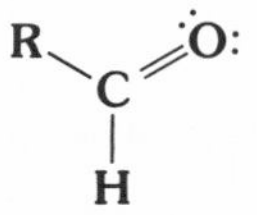

an aldehyde

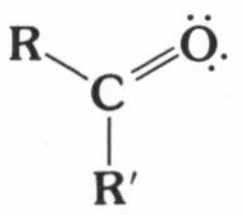

a ketone

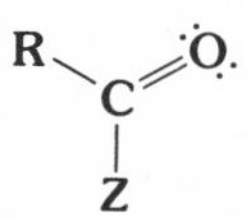

a carboxylic acid: Z = OH
an ester: Z = OR'
an acid chloride: Z = Cl
an amide: Z = NH_2

A wide variety of aldehydes and ketones have been found in plants and animals, sometimes in combination with sugars, but more commonly as the free carbonyl compound. Many of these natural substances are important as flavorings and perfume ingredients. Steroidal hormones often have carbonyl groups, as in the case of the male sex hormone testosterone, the female hormone progesterone, and cortisone, a hormone produced by the adrenal glands.

$C_6H_5CH=CHCHO$
cinnamaldehyde
(from cinnamon)

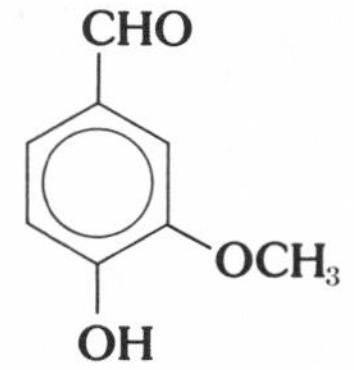

vanillin
(from the vanilla bean)

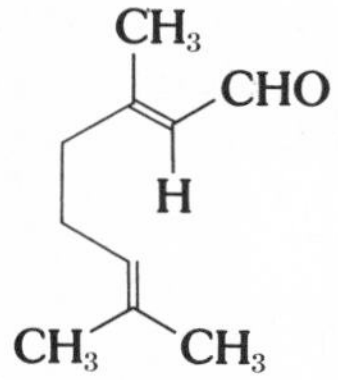

citral
(from lemon grass)

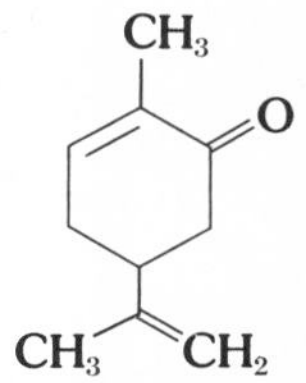

carvone
(from spearmint)

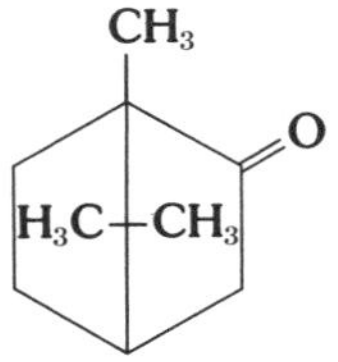

362 ALDEHYDES AND KETONES

camphor (from the camphor tree)

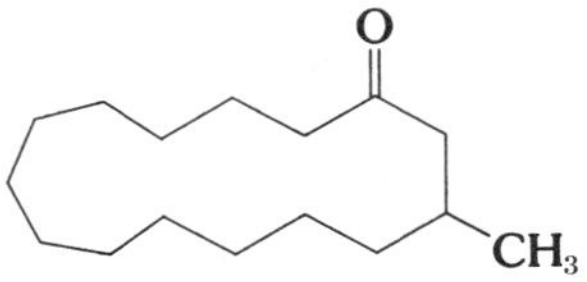

muscone (from the musk deer)

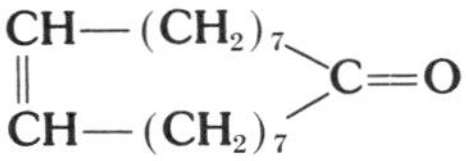

civetone (from the civet cat)

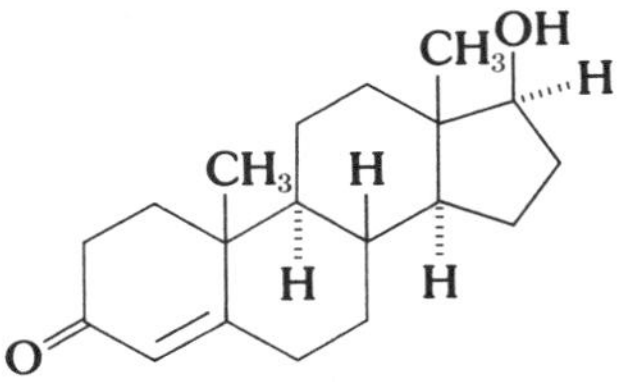

testosterone

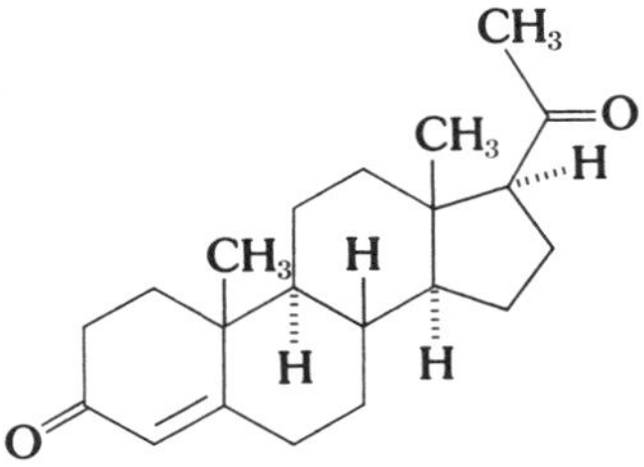

progesterone

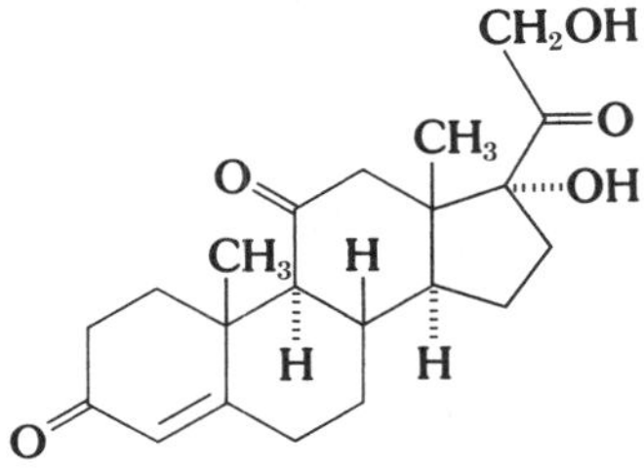

cortisone

Since large quantities of the simpler aldehydes and ketones are used as solvents and synthetic intermediates, efficient methods of manufacturing these compounds have been developed. Formaldehyde and acetaldehyde are made by catalytic air oxidation of methanol and ethanol, respectively:

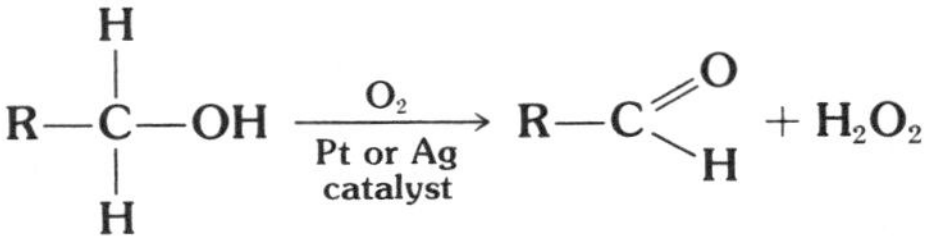

$$R-\underset{\underset{H}{|}}{\overset{\overset{H}{|}}{C}}-OH \xrightarrow[\text{Pt or Ag catalyst}]{O_2} R-C\overset{O}{\underset{H}{\diagdown}} + H_2O_2 \tag{12.1}$$

Acetaldehyde is also prepared by the hydration of acetylene (see equation 6.20). Acetone can be manufactured in a similar manner by the oxidation of isopropyl alcohol, but a fermentation of molasses (equation 12.2) and rearrangement of cumene hydroperoxide (described in Chapter 15) are more important industrial sources of this useful ketone:

$$\text{molasses} \xrightarrow{\substack{\text{clostridium} \\ \text{acetobutylicum}}} \underset{\sim 70\%}{\text{1-butanol}} + \underset{\sim 25\%}{\text{acetone}} + \underset{\sim 5\%}{\text{ethanol}} + CO_2 + H_2 \tag{12.2}$$

A summary of generally useful methods for preparing aldehydes and ketones in the laboratory is given in Table I on page 670. Some of these procedures, such as the oxidation of alcohols (Section 8.7) and the hydration or hydroboration of alkynes (Section 6.4), have already been described. Another group of important procedures involves substitution reactions of carboxylic acid derivatives, discussed in Chapter 14. Among these are Friedel-Crafts reactions with acyl halides and nucleophilic substitution by metal-hydride or metal-alkyl reagents:

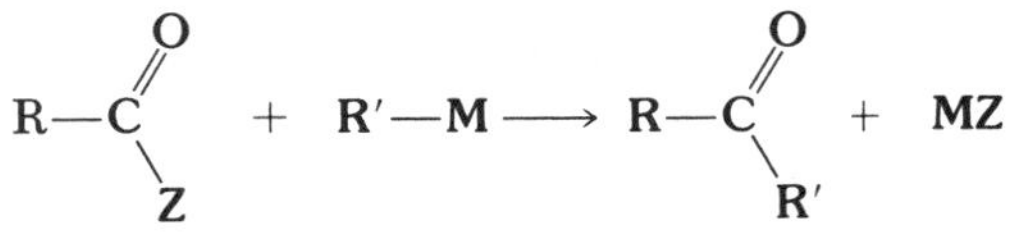

$$\underset{\substack{\text{a carboxyl} \\ \text{derivative}}}{R-C\overset{O}{\underset{Z}{\diagdown}}} + \underset{\substack{\text{a metal-alkyl or} \\ \text{-hydride reagent}}}{R'-M} \longrightarrow R-C\overset{O}{\underset{R'}{\diagdown}} + MZ \tag{12.3}$$

12·1 NOMENCLATURE OF ALDEHYDES AND KETONES

Many aldehydes and ketones are normally referred to by common names. However, the IUPAC nomenclature system can also be applied to these compounds. The IUPAC suffix for *aldehydes* is *-al*, and the numbering starts from the carbonyl end of the chain:

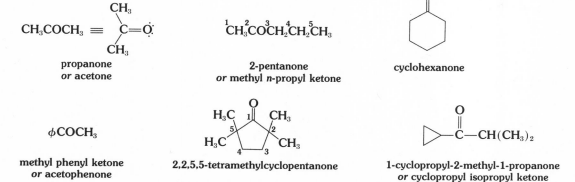

The IUPAC suffix for naming *ketones* is *-one*. Numbering starts from the end nearest the carbonyl group, and in cyclic ketones it starts from the carbonyl group:

In the naming of more complex structures the following *carbonyl groups* are given characteristic names:

acetyl group   formyl group   benzoyl group

Compounds containing these groups are then named accordingly:

2-formyl-2,4,4-trimethylcyclohexanone

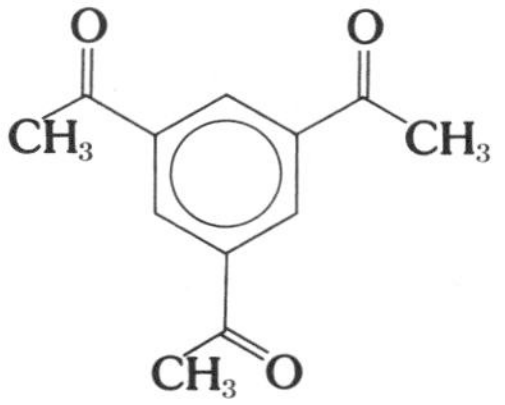

1,3,5-triacetylbenzene

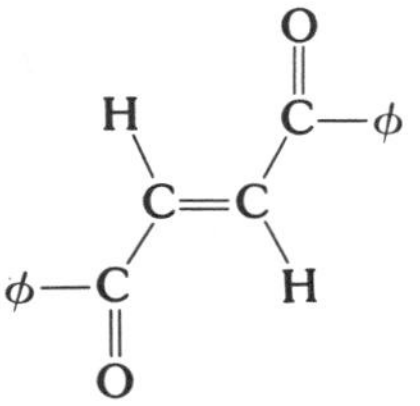

1,4-diphenyl-*trans*-2-butene-1,4-dione
or *trans*-1,2-dibenzoylethylene

The carbon atoms near a carbonyl group are sometimes identified by Greek letter prefixes, with the atom immediately adjacent to the group designated as alpha (α):

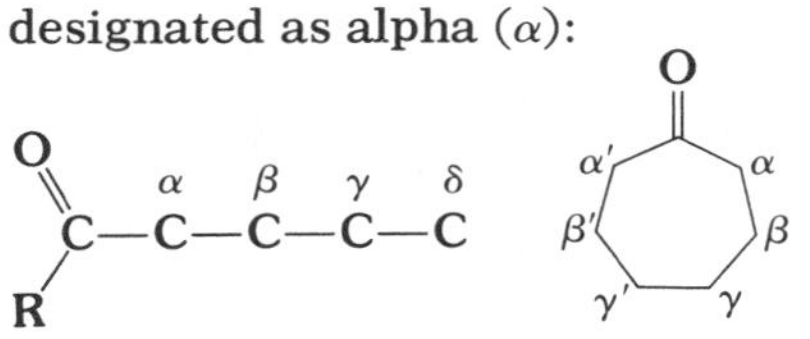

For example,

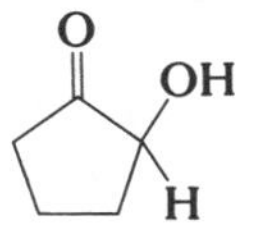

an α-hydroxyketone
or an α-ketol

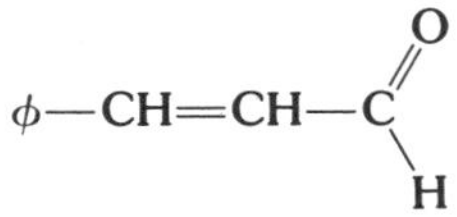

an α,β-unsaturated aldehyde

12·2 PROPERTIES OF ALDEHYDES AND KETONES

STRUCTURE AND BONDING OF THE CARBONYL GROUP

The carbonyl group and the two atoms directly bonded to it are coplanar, with dimensions and bond angles similar to those in Figure 12.1. A σ-π molecular-orbital model like that developed for the carbon-carbon double bond provides a good description of the electron distribution in this functional group. Since oxygen is more electronegative than carbon, we would expect a greater π-electron density at the oxygen end of the carbonyl group; this is confirmed by a large bond dipole (2.3) directed as shown for this structure.

Carbonyl double bonds are over 30 kcal/mole stronger than carbon-carbon double bonds, perhaps because of greater resonance stabilization involving an ionic contributor:

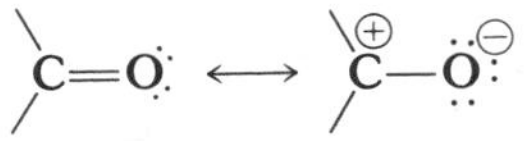

$$\verb|\|C=\ddot{O}: \longleftrightarrow \verb|\|\overset{\oplus}{C}-\ddot{\ddot{O}}:^{\ominus}$$

This large bond energy is even more striking when compared with the corresponding single-bond energy. The bond energy of a carbonyl double bond, 176 to 179 kcal/mole in ketones and aldehydes other than formaldehyde, is more than twice that of a carbon-oxygen single bond (about 86 kcal/mole). In contrast, the strength of a carbon-carbon double bond (about 146 kcal/mole) is considerably less than twice that of the corresponding single bond (about 83 kcal/mole).

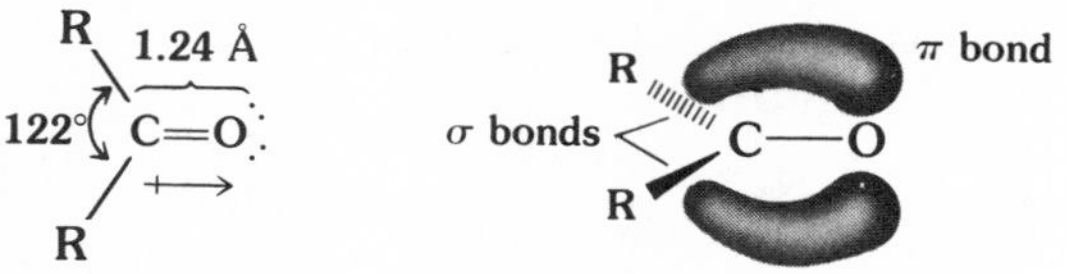

12·2 PROPERTIES OF ALDEHYDES AND KETONES

FIGURE 12·1 *Bonding in a typical carbonyl group*

The chemical consequences of this bond-energy relationship are interesting and important. Because of its large dipole, the carbonyl double bond adds polar reagents such as water rapidly. However, the reverse elimination is also very fast, and in most cases the equilibrium constant favors the aldehyde or ketone over its hydrated form; that is, the strong carbonyl double bond makes the aldehyde or ketone the thermodynamically favored reactant:

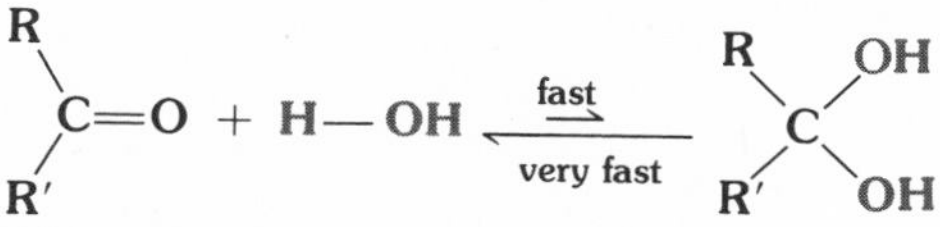

$$\underset{R'}{\overset{R}{\diagdown}}C=O + H-OH \underset{\text{very fast}}{\overset{\text{fast}}{\rightleftarrows}} \underset{R'}{\overset{R}{\diagdown}}C\underset{OH}{\overset{OH}{\diagup}} \tag{12.4}$$

$\Delta H \approx +6$ kcal/mole

The hydration of alkenes stands in striking contrast to this situation:

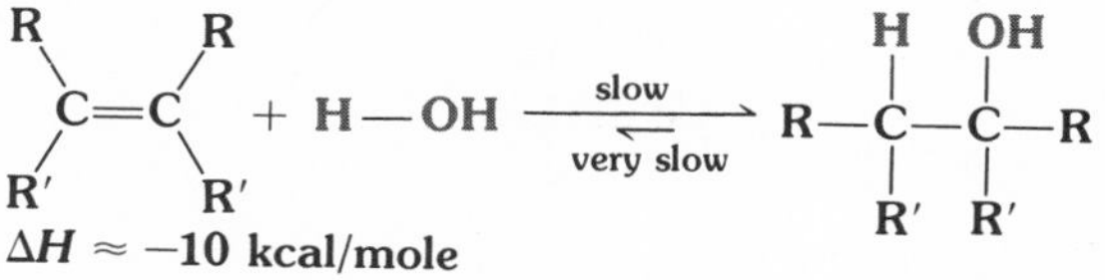

$$\underset{R'}{\overset{R}{\diagdown}}C=C\underset{R'}{\overset{R}{\diagup}} + H-OH \underset{\text{very slow}}{\overset{\text{slow}}{\rightleftarrows}} R-\underset{R'}{\overset{H}{\underset{|}{C}}}-\underset{R'}{\overset{OH}{\underset{|}{C}}}-R \tag{12.5}$$

$\Delta H \approx -10$ kcal/mole

Here the reaction is normally quite slow and must be catalyzed by strong Brønsted acids, as discussed in Chapter 5. Once water has been added to the carbon-carbon double bond, however, the product (an alcohol) is stable, and even more vigorous reaction conditions must be used to reverse the addition (see Figure 8.12). In other words, the addition product to the alkene is thermodynamically favored even though the rate of addition is slow because of a large activation energy.

PROBLEM 12·1
How could we detect the rapid addition-elimination of water to a carbonyl function, as shown in equation 12.4?

PROBLEM 12·2
Draw energy diagrams illustrating the two hydration-dehydration equilibria described in equations 12.4 and 12.5. Clearly label these diagrams.

PHYSICAL PROPERTIES

Because of the large bond dipole of the carbonyl group, increased intermolecular attraction significantly raises the boiling points of aldehydes and ketones over those of similarly constituted hydrocarbons (see Table 12.1). The water solubility of carbonyl compounds is also greater than that of the corresponding hydrocarbons, since the carbonyl oxygen can serve as a hydrogen-bond acceptor (but not as a donor).

12·2 PROPERTIES OF ALDEHYDES AND KETONES

TABLE 12·1 *Physical constants of aldehydes, ketones, and hydrocarbons*

compound	molecular weight	boiling point (°C)	solubility in water (g/100 g)
CH_3COCH_3	58	56.5	completely miscible
$(CH_3)_3CH$	58	−11.7	0.03
$CH_3CH_2CH_2CHO$	72	76.0	7
$CH_3(CH_2)_3CH_3$	72	36.1	0.03
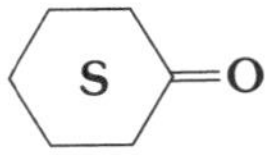 cyclohexanone (⬡=O)	98	155.6	5
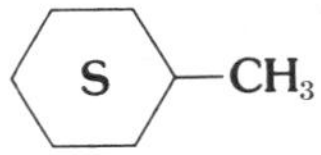 methylcyclohexane (⬡–CH₃)	98	101.0	insoluble

SPECTROSCOPIC PROPERTIES

ULTRAVIOLET-VISIBLE SPECTROSCOPY We can distinguish two kinds of electronic excitations involving the carbonyl group. One is a $\pi \longrightarrow \pi^*$ electron promotion analogous to that described in Section 9.4 for carbon-carbon double bonds. The other is an $n \longrightarrow \pi^*$ excitation in which one of the four nonbonding valence electrons on the carbonyl oxygen atom is promoted to the π^* molecular orbital (see Figure 9.4). The first of these produces an intense absorption band with maximum absorption at about 180 nm in the ultraviolet spectrum, which is below the range of commercial spectrophotometers. The $n \longrightarrow \pi^*$ excitation, however, gives a very weak absorption band at λ_{max} = 280 to 320 nm which can easily be observed in concentrated solutions of aldehydes and ketones. As in the case of alkenes, the strong $\pi \longrightarrow \pi^*$ absorption band is shifted to longer wavelengths when the carbonyl group is conjugated with a double bond or an aromatic ring (Table 12.2).

TABLE 12·2 *Electron-absorption bands of aldehydes and ketones*

compound	$\pi \longrightarrow \pi^*$ λ_{max} (nm)	molar absorptivity ϵ	$n \longrightarrow \pi^*$ λ_{max} (nm)	molar absorptivity ϵ
CH_3COCH_3	~180	10^4	271	16
$(CH_3)_2C=CHCOCH_3$	237	10^4	310	55
$CH_3CH=CHCHO$	220	10^4	320	31
ϕCHO	243	10^4	320	48
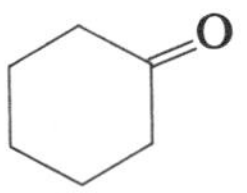 cyclohexanone	~180	10^4	283	16
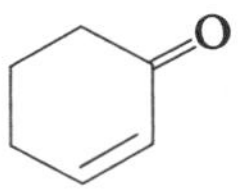 2-cyclohexenone	225	10^4	320	36

12·2 PROPERTIES OF ALDEHYDES AND KETONES

PROBLEM 12·3

From the data in Table 12.2 and the following experimentally observed absorption bands ($\pi \longrightarrow \pi^*$), suggest a general rule for predicting λ_{max} for unsaturated ketones bearing different numbers of double-bond substituents:

	λ_{max} (nm)
3-buten-2-one	215
3-penten-2-one	224
2-methylcyclohex-2-en-1-one	235
3-methylcyclohex-2-en-1-one	235

INFRARED SPECTROSCOPY Carbon-oxygen double bonds exhibit a characteristic strong absorption at 1680 to 1750 cm^{-1} in the infrared spectrum which is very useful as a diagnostic tool for identifying this functional group. Of particular interest and importance is the sensitivity of this absorption band to conjugation with alkenes or aromatic rings and to variations in the ring size of cyclic ketones. Table 12.3 shows some examples of this sensitivity.

Aldehydes may also be identified by two characteristic stretching vibrations of the aldehyde carbon-hydrogen bond, usually found in the 2650 to 2850 cm^{-1} region of the infrared spectrum (Figure 12.2).

FIGURE 12·2 Infrared spectra of benzaldehyde and progesterone

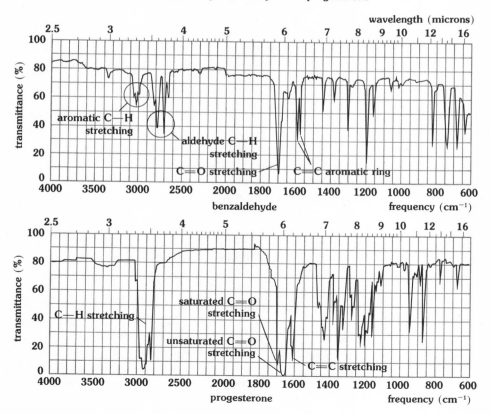

12·2 PROPERTIES OF ALDEHYDES AND KETONES

TABLE 12·3 *Carbonyl stretching frequencies*

compound	$\tilde{\nu}_{max}$ (cm^{-1})†	compound	$\tilde{\nu}_{max}$ (cm^{-1})†
RCHO	1720–1740		
RCOCH$_3$	1710–1725	cyclopentanone	1740–1750
ϕCOR	1680–1700		
cyclohexanone	1710–1725	cyclopentenone	1710–1725
cyclohexenone	1665–1685	cyclobutanone	~1775
		cyclopropanone	~1825

†$\tilde{\nu}_{max}$ = wavenumber, ν/c

PROBLEM 12·4
From the carbonyl absorption frequencies in Table 12.3 suggest general rules for predicting the effect of introducing α,β-unsaturation into an aldehyde or ketone or of changing the ring size of a cyclic ketone.

PMR SPECTROSCOPY Hydrogen atoms bonded to a carbonyl group or to its α carbon atom experience downfield shifts in the pmr spectrum similar to those noted earlier for vinyl and allyl protons (see Table 9.4). In fact, this deshielding effect of the carbonyl group is actually greater than that of an isolated carbon-carbon double bond because of the strongly electronegative oxygen atom:

R—CHO, $\delta \approx 9.5$ R—COCH$_3$, $\delta \approx 2.3$

Aldehydes are particularly noteworthy in this respect, since the aldehyde proton is shifted so far from the TMS reference signal that the routine 8.0-ppm downfield sweep of most pmr spectrometers fails to bring it into resonance. For this reason wider field sweeps must always be used if aldehydes are suspected.

PROBLEM 12·5
Three isomers of C$_4$H$_8$O have the following infrared and pmr absorptions. Write structural formulas for each compound and indicate the nature of the information that each spectrum provides.

a Infrared absorptions at 2900–2940, 1710, 1460, 1410, 1170, 940, and 758 cm^{-1}; pmr signals at $\delta = 1.05$ (triplet, $J = 7$ Hz), $\delta = 2.13$ (singlet), and $\delta = 2.47$ (quartet, $J = 7$ Hz), with relative intensities of 3:3:2

b Infrared absorptions at 2850–2950, 1465, 1065, and 910 cm^{-1}; pmr signals at $\delta = 1.85$ and $\delta = 3.75$, with equally intense broad multiplets

c Infrared absorptions at 2800–2940, 2700, 1725, 1470, 1380, 1125, and 913 cm^{-1}; pmr signals at $\delta = 1.2$ (doublet, $J = 6.5$ Hz), $\delta = 3.3$ (broad septet, $J \approx 7$ Hz), and $\delta = 9.7$ (doublet, $J \approx 1$ Hz), with relative intensities of 6:1:1

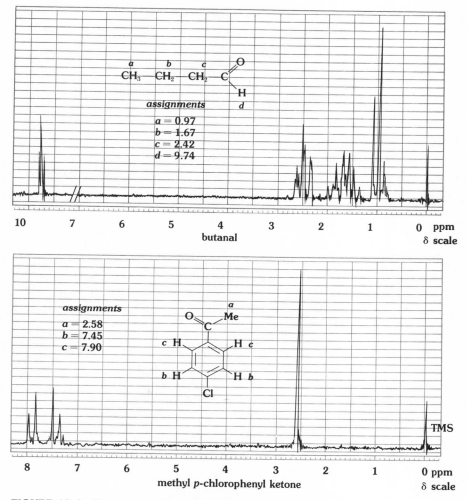

FIGURE 12·3 *Pmr spectra of butanal and methyl p-chlorophenyl ketone*

12·3 REVERSIBLE ADDITION REACTIONS

Although the chemistry of the carbonyl group is more diverse than that of any functional group we have discussed thus far, almost all the important reactions of aldehydes and ketones fall into one of two general classes: (1) carbonyl addition or addition-elimination reactions or (2) reactions of enol intermediates.

Reactions of the first type begin with an addition of polar reagents to the carbon-oxygen double bond:

$$\underset{R'}{\overset{R}{\diagdown}}\overset{\delta\oplus}{C}=\overset{\delta\ominus}{O} + \overset{\delta\oplus}{Y}-\overset{\delta\ominus}{Z} \rightleftharpoons \underset{R'}{\overset{R}{\diagdown}}\underset{Z}{\overset{O-Y}{\diagup}} \quad (12.6)$$

As we saw in Section 12.2, these additions are usually fast and in many cases are also reversible (as when the nucleophilic atom in $Y^{\ominus}$ is O, N, F, Cl, Br, I, or a stable carbanion such as CN). Both the rate of addition and the equilibrium constant for reaction 12.6 are adversely affected by an increase in steric hindrance at the carbonyl function. Thus the rate of addition decreases in the order

$R = H, R' = CH_3 > R, R' = (CH_2)_5 > R = R' = CH_3 \gg R = R' = C(CH_3)_3$

FORMATION OF HYDRATES, HEMIACETALS, AND HEMIKETALS

With a few exceptions, the addition of water to aldehydes and ketones, although rapid, is not favored thermodynamically. This reversible addition reaction can nevertheless be confirmed by using isotopically labeled water:

$$\underset{R'}{\overset{R}{>}}C=O + H_2^{18}O \rightleftharpoons \underset{R'}{\overset{R}{>}}C\underset{OH}{\overset{^{18}OH}{<}} \rightleftharpoons \underset{R'}{\overset{R}{>}}C=^{18}O + H_2O \quad (12.7)$$

Careful study of these oxygen-exchange reactions has shown that the hydration step is catalyzed by traces of acids or bases. This is reasonable, since hydroxide ion is a more reactive nucleophile than water:

$$\underset{R'}{\overset{R}{>}}C=\ddot{O}: + :\ddot{O}H^{\ominus} \rightleftharpoons \left[\underset{R'}{\overset{R}{>}}C\underset{OH}{\overset{\ddot{O}:^{\ominus}}{<}} \right] + H_2O \rightleftharpoons \underset{R'}{\overset{R}{>}}C\underset{\ddot{O}H}{\overset{\ddot{O}H}{<}} + :\ddot{O}H^{\ominus} \quad (12.8)$$

Furthermore, the conversion of a carbonyl group to its conjugate acid generates a more reactive electrophile:

$$\underset{R'}{\overset{R}{>}}\overset{\delta\oplus \ \delta\ominus}{C=\ddot{O}:} + H_3\ddot{O}:^{\oplus} \rightleftharpoons \left[\underset{R'}{\overset{R}{>}}C=\overset{H}{\underset{\oplus}{\ddot{O}}} \leftrightarrow \underset{R'}{\overset{R}{>}}\overset{\oplus}{C}-\overset{H}{\ddot{O}:} \right] + H_2\ddot{O} \rightleftharpoons \underset{R'}{\overset{R}{>}}C\underset{\ddot{O}H_2^{\oplus}}{\overset{\ddot{O}-H}{<}} \quad (12.9)$$

weak
electrophile

stronger
electrophile

If alcohols are substituted for water in the preceding reactions we find that analogous addition products called *hemiacetals* (from aldehydes) and *hemiketals* (from ketones) are formed. As in the case of hydration, the equilibrium usually favors the carbonyl reactant.

$\underset{H}{\overset{R^1}{>}}C\underset{OH}{\overset{O-R^2}{<}}$
a hemiacetal

$\underset{R^2}{\overset{R^1}{>}}C\underset{OH}{\overset{O-R^3}{<}}$
a hemiketal

PROBLEM 12·6
Write equations illustrating mechanisms for both acid- and base-catalyzed hemiketal formation from cyclohexanone and methanol.

When strong electron-withdrawing groups are bonded to the carbonyl carbon atom (as in chloral) or the carbonyl bond angle is under extreme compression (as in cyclopropanone), the reactant bearing the carbonyl

12·3 REVERSIBLE ADDITION REACTIONS

group becomes less stable, and the hydrate or hemiketal form may predominate at equilibrium:

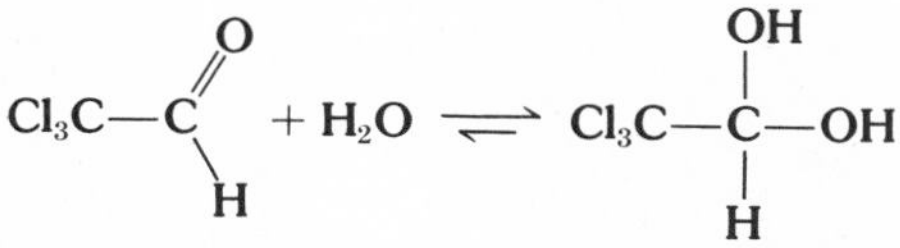

$$Cl_3C-CHO + H_2O \rightleftharpoons Cl_3C-CH(OH)_2 \quad (12.10)$$

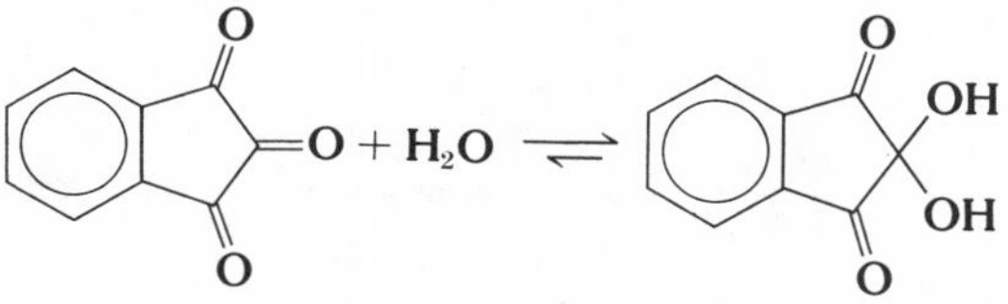

chloral → chloral hydrate (mickey finn)

$$\text{(indane-1,2,3-trione)} + H_2O \rightleftharpoons \text{ninhydrin} \quad (12.11)$$

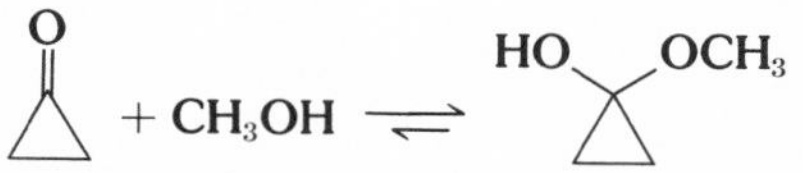

$$\text{(cyclopropanone)} + CH_3OH \rightleftharpoons \text{HO-C(OCH}_3\text{)-cyclopropane} \quad (12.12)$$

Most sugars tend to adopt a cyclic hemiacetal structure:

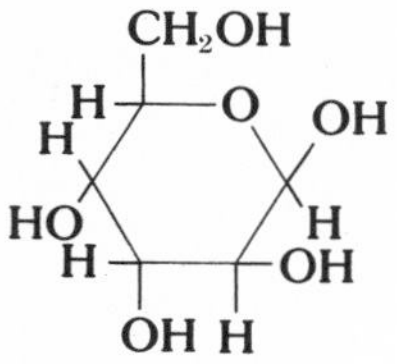

(cyclic glucose hemiacetal structure)

PROBLEM 12·7
Explain the destabilizing effect of neighboring electron-withdrawing substituents on a carbonyl group.

FORMATION OF ACETALS AND KETALS

The acid-catalyzed elimination of a divalent oxygen group from a hemiacetal or hemiketal can proceed in one of two ways, as illustrated in Figure 12.4 for the methanol hemiketal of cyclohexanone. The first pathway simply generates the conjugate acid of the initial aldehyde or ketone, but the second pathway produces an O-alkyl carbonyl cation, which in the presence of excess alcohol is converted to a geminal diether, better known as an *acetal* or a *ketal*. Since all the steps in the ketalization of a ketone are reversible, it is possible to control the direction of the reaction by adding an excess of one of the reactants or by removing one of the products as it is formed. Thus a ketone (or aldehyde) may be transformed into a ketal (or acetal) by selectively removing the water generated during this reaction, and the original carbonyl compound can be re-formed by treating the ketal with excess aqueous acid.

PROBLEM 12·8
Copy the equations for pathway 2 in Figure 12.4 and indicate the conjugate-acid–base relationship of the reactants and products in each step.

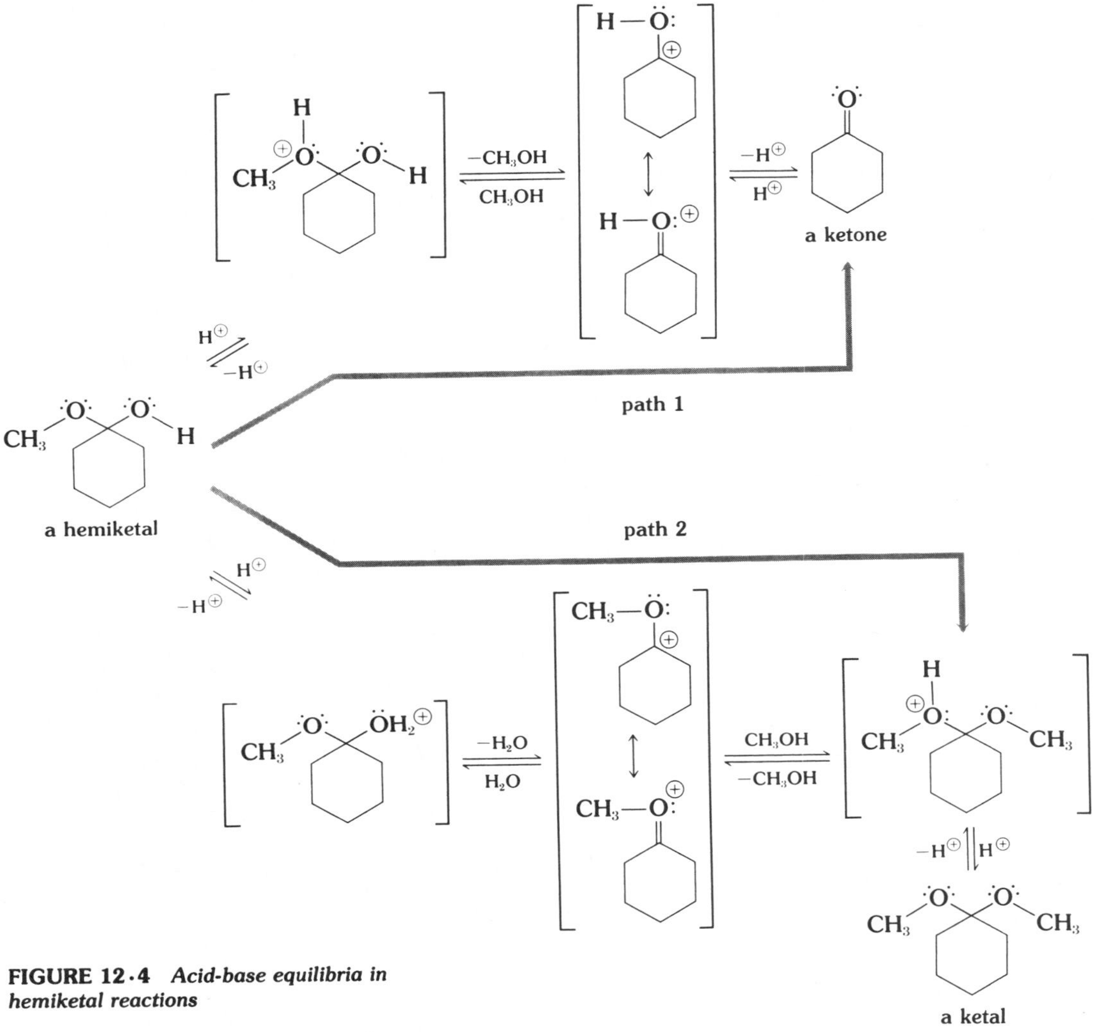

FIGURE 12·4 *Acid-base equilibria in hemiketal reactions*

PROBLEM 12·9

An industrially important synthesis of phenol and acetone involves the following acid-catalyzed rearrangement of cumene hydroperoxide:

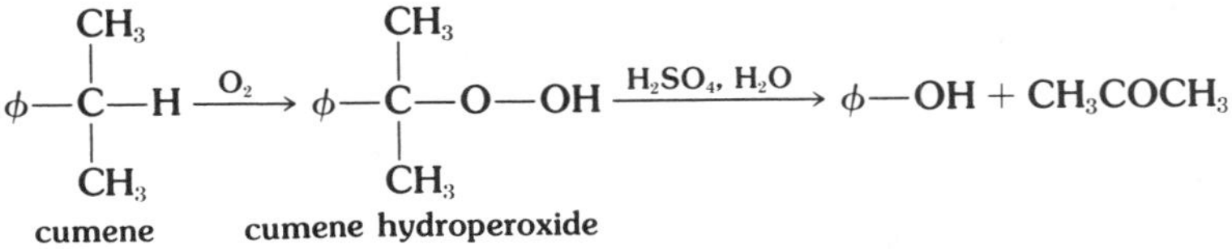

$$\phi-\underset{\underset{CH_3}{|}}{\overset{\overset{CH_3}{|}}{C}}-H \xrightarrow{O_2} \phi-\underset{\underset{CH_3}{|}}{\overset{\overset{CH_3}{|}}{C}}-O-OH \xrightarrow{H_2SO_4,\ H_2O} \phi-OH + CH_3COCH_3$$

cumene   cumene hydroperoxide

Write a mechanism for the second reaction. *Hint:* Aryl groups migrate readily to adjacent electron-deficient atoms, giving rearrangements similar to those discussed in Section 3.7

Acetals and ketals differ chemically from hemiacetals and hemiketals in a manner which reflects the fact that the former are ethers and the latter are alcohols. All four classes of carbonyl derivatives react readily in the presence of acid catalysts, but only hemiacetals and hemiketals

12·3 REVERSIBLE ADDITION REACTIONS

are attacked by bases. As a result, acetals and ketals are much more easily isolated and purified than the corresponding hemi derivatives (the reaction mixture can be worked up under mild alkaline conditions). Furthermore, the chemical stability of acetals and ketals to bases of varying strength, coupled with the ease of acid-catalyzed hydrolysis of these derivatives, makes them ideal protective functions for the carbonyl group. Cyclic ketals are generally preferred for this purpose because they are easily prepared in high yield:

$$\underset{R'}{\overset{R}{>}}C=O + \underset{|}{\overset{OH}{CH_2}}-\underset{|}{\overset{OH}{CH_2}} \xrightarrow[\Delta]{H^\oplus} \underset{R'}{\overset{R}{>}}C\underset{O-}{\overset{O-}{<}} + H_2O \tag{12.13}$$

Sulfur analogs of ketals and acetals are called *thioketals* and *thioacetals* (the prefix *thio-* indicates that sulfur has replaced oxygen in a compound). These compounds are easily synthesized, although strong acid catalysts such as BF_3 or $HClO_4$ are usually necessary, even for the preparation of cyclic thioketals:

$$\text{O}\diagup\diagdown\text{O} + 2\begin{bmatrix}-SH\\-SH\end{bmatrix} \xrightarrow{BF_3,\ ether} \begin{bmatrix}-S\\-S\end{bmatrix}\diagup\diagdown\begin{bmatrix}S-\\S-\end{bmatrix} + 2H_2O \tag{12.14}$$

Thioketals and thioacetals show greater chemical stability than their oxygen counterparts and are difficult to hydrolyze back to the original carbonyl compounds. Vigorous refluxing of an aqueous alcohol solution containing mercuric sulfate and sulfuric acid is usually required to reverse reaction 12.14.

PROBLEM 12·10
a Write a detailed mechanism for the base-catalyzed decomposition of the hemiacetal $CH_3CH(OH)OC_2H_5$ and show why such a mechanism cannot operate with the acetal $CH_3CH(OC_2H_5)_2$.
b In general, intramolecular reactions are faster than intermolecular reactions. How would this affect the ease of formation of a cyclic ketal in comparison with a noncyclic ketal?
c Is the formation of a thioketal an exothermic or an endothermic process?

PROBLEM 12·11
Optically active 2,3-butanedithiol has been used as a resolving agent for chiral ketones. Write out the steps for such a resolution of 3-phenylcyclohexanone.

CYANOHYDRIN FORMATION

The highly poisonous weak acid hydrogen cyanide ($K_a \approx 10^{-9}$) adds to many carbonyl compounds, giving products called *cyanohydrins*:

$$\text{cyclohexanone} + HCN \underset{}{\overset{H_2O}{\rightleftarrows}} \text{cyclohexanone cyanohydrin} \tag{12.15}$$

12·3 REVERSIBLE ADDITION REACTIONS

Bond-energy calculations indicate that this reaction should be slightly exothermic. Consequently, the equilibrium usually favors the addition product, although it is very sensitive to structural changes in the carbonyl reactant.†

PROBLEM 12·12
The equilibrium constant for the addition of HCN to cyclohexanone (equation 12.15) is $K_{eq} = 11,000$. This contrasts sharply with the value $K_{eq} = 15$ observed for the ketone menthone. Explain this difference.

menthone

Since cyanohydrin formation is catalyzed by traces of base, cyanide ion ($CN^\ominus$) is clearly the reactive nucleophile in these reactions:

$$\begin{array}{c} R \\ \diagdown \\ C=\ddot{O} \\ \diagup \\ R \end{array} + :C\equiv N:^\ominus \rightleftharpoons \begin{array}{c} R \quad O^\ominus \\ \diagdown \diagup \\ C \\ \diagup \diagdown \\ R \quad CN: \end{array} \underset{-H_2O}{\overset{H_2O}{\rightleftharpoons}} \begin{array}{c} R \quad OH \\ \diagdown \diagup \\ C \\ \diagup \diagdown \\ R \quad CN \end{array} + OH^\ominus \quad (12.16)$$

It is surprising, therefore, to find that an aqueous solution of potassium cyanide is a relatively ineffective reagent for cyanohydrin formation, despite the high concentration of cyanide ion. We can explain this peculiarity by considering the change in pH that may occur during reaction 12.16 if it is not buffered. As the addition of cyanide ion proceeds, a weak base ($CN^\ominus$) is replaced by a strong base ($OH^\ominus$) and the pH of the reaction mixture increases. However, if excess hydrogen cyanide is present the hydroxide ion will be neutralized, resulting in the overall conversion of a moderately weak acid (HCN) into a weaker acid (the cyanohydrin). Since the weaker acid (or base) is generally favored at equilibrium (see Figure 3.13), cyanohydrin formation is best accomplished in a weakly acidic reaction medium.

FORMATION OF IMINES AND RELATED COMPOUNDS

The reaction of ammonia with most aldehydes and ketones establishes an equilibrium analogous to that for hydration and oxygen exchange and generates a new chemical species called an *imine*:

$$\begin{array}{c} R \\ \diagdown \\ C=\ddot{O} \\ \diagup \\ R' \end{array} + :NH_3 \rightleftharpoons \begin{array}{c} R \quad \ddot{O}-H \\ \diagdown \diagup \\ C \\ \diagup \diagdown \\ R' \quad \ddot{N}H_2 \end{array} \rightleftharpoons \begin{array}{c} R \quad H \\ \diagdown \diagup \\ C=\ddot{N} \\ \diagup \\ R' \end{array} + H_2O \quad (12.17)$$

an imine

---

†The reversibility of cyanohydrin formation is used by the millipede *Apheloria corrugata* in a remarkable defense mechanism. This arthropod releases mandelonitrile, the cyanohydrin of benzaldehyde, from an inner storage gland into an outer chamber, where it is broken down by enzyme action to benzaldehyde and hydrogen cyanide before being ejected at an enemy.

$$\phi CH(OH)CN \xrightarrow{enzymes} \phi CHO + HCN$$

12·3 REVERSIBLE ADDITION REACTIONS

TABLE 12·4 *Common ammonia reagents and their carbonyl derivatives*

ammonia derivative (H$_2$N—Y)	carbonyl derivative
H$_2$N—OH hydroxylamine	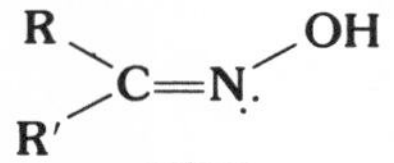 $\underset{\text{oxime}}{\text{R}\text{R}'\text{C}=\text{N—OH}}$
H$_2$N—NH$_2$ hydrazine	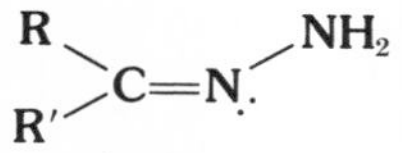 $\underset{\text{hydrazone}}{\text{R}\text{R}'\text{C}=\text{N—NH}_2}$
H$_2$N—NHϕ phenylhydrazine	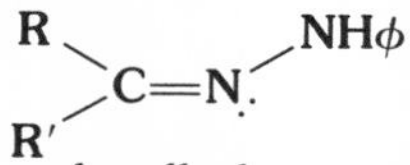 $\underset{\text{phenylhydrazone}}{\text{R}\text{R}'\text{C}=\text{N—NH}\phi}$
H$_2$N—NHCONH$_2$ semicarbazide	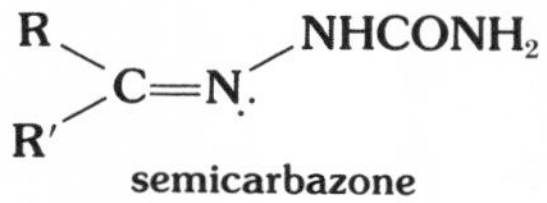 $\underset{\text{semicarbazone}}{\text{R}\text{R}'\text{C}=\text{N—NHCONH}_2}$

Primary amines also give imine products, but the reaction of formaldehyde with ammonia takes an unexpectedly complex course.†

OXIMES, HYDRAZONES, AND SEMICARBAZONES Since imines are often difficult to isolate and purify, chemists commonly employ certain derivatives of ammonia (here denoted by Y—NH$_2$) when they wish to prepare stable iminelike compounds. Some of these ammonia derivatives and their carbonyl reaction products are listed in Table 12.4.

With the exception of the hydrazones, carbonyl derivatives of this kind are easily prepared and are often crystalline solids—even in cases where the parent aldehyde or ketone is a liquid. Since melting points can be determined more precisely than boiling points, unknown carbonyl substances can sometimes be identified by comparing the melting points of their oxime or semicarbazone derivatives with recorded values. 2,4-Dinitrophenylhydrazones (DNP) are generally more useful for this purpose than unsubstituted phenylhydrazones because they show a stronger tendency to crystallize and also have characteristic colors.

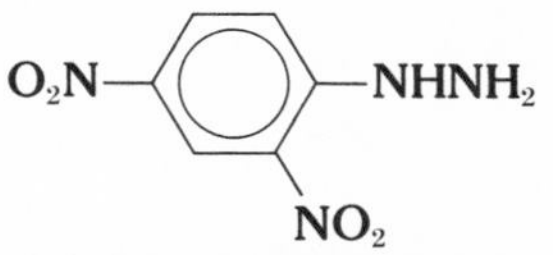

2,4-dinitrophenylhydrazine
red-orange, mp = 198°

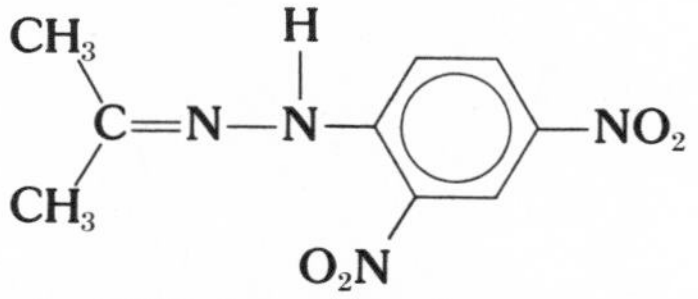

the DNP derivative of acetone
bright yellow, mp = 126°

†Formaldehyde and ammonia combine to give the tricyclic compound hexamethylenetetramine:

$$6CH_2O + 4NH_3 \longrightarrow \text{(hexamethylenetetramine)} + 6H_2O$$

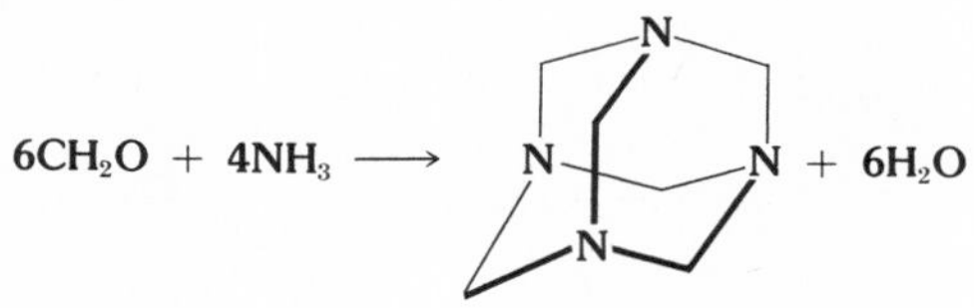

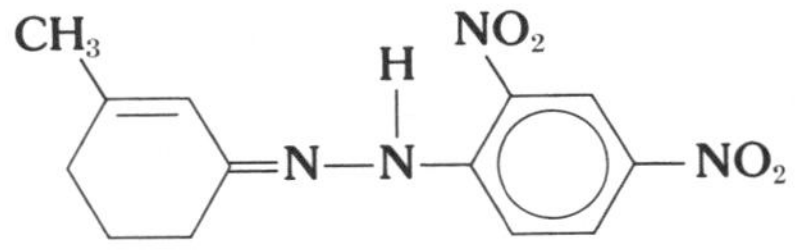

the DNP derivative of 3-methylcyclohex-2-en-1-one
red, mp = 177°–178°

PROBLEM 12·13
a Why would hydrazones be less desirable than the other carbonyl derivatives listed in Table 12.4 for characterizing aldehydes and ketones?
b There are two primary amino groups in semicarbazide. Why does the reaction with carbonyl compounds produce only the product shown in Table 12.4?

The conversion of aldehydes and ketones into iminelike derivatives is mildly exothermic ($\Delta H \approx -4$ kcal/mole) and proceeds at a rate which reaches a maximum when the pH of the reaction mixture is about 5 or 6. We interpret this pH dependence to mean that the rate-determining step of these reactions involves a bonding of the nucleophilic nitrogen reagent to the conjugate acid of the carbonyl reactant:

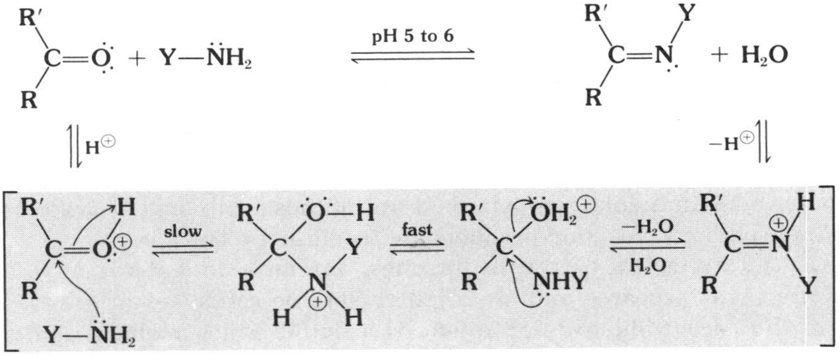

$$\begin{array}{c} R' \\ R \end{array}\!\!C=\ddot{O} + Y-\ddot{N}H_2 \xrightleftharpoons[]{\text{pH 5 to 6}} \begin{array}{c} R' \\ R \end{array}\!\!C=\ddot{N}\!\!-\!\!Y + H_2O \qquad (12.18)$$

If the pH is too low the concentration of the nitrogen nucleophile is reduced by salt formation ($Y-NH_3^{\oplus}$), but at higher pH the concentration of the electrophilic carbonyl conjugate acid drops sharply.

The carbon-nitrogen double bond in these derivatives is a potential locus of stereoisomerism. The oxime of benzaldehyde, for example, has been obtained in two stereoisomeric forms, *syn* and *anti*. In the case of aldehydes these prefixes refer to the relative positions of the aldehyde hydrogen atom and the hydroxyl group and correspond to our previous *cis-trans* notation.

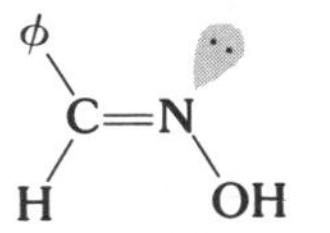

syn-benzaldoxime
mp = 129°

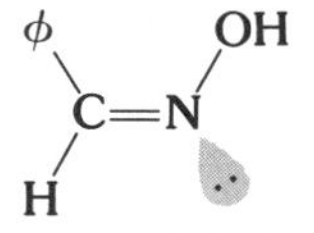

anti-benzaldoxime
mp = 35°

The other derivatives, such as phenylhydrazones and semicarbazones, seem to have less configurational stability than the oximes, and although their stereoisomerism is easily observed by pmr spectroscopy, both pure isomers are not as readily isolated.

12·3 REVERSIBLE ADDITION REACTIONS

Although the reactions leading to these carbonyl derivatives are reversible, as in equations 12.17 and 12.18, equilibrium is not established instantaneously. A striking example of this is the competitive reaction illustrated in Figure 12.5. If an aqueous solution containing one molar equivalent of cyclohexanone and one molar equivalent of 2-furfuraldehyde is shaken for a few minutes with exactly one molar equivalent of semicarbazide at pH = 6, the semicarbazone of cyclohexanone precipitates and can be obtained in good yield by filtration. However, if the reaction mixture is heated, the initially formed cyclohexanone semicarbazone slowly dissolves and is converted (in about an hour) to the semicarbazone of 2-furfuraldehyde, which precipitates when the solution is cooled.

In order to explain these results, we must consider the relative rates of the competitive reactions. The unusually high reactivity of the carbonyl group in cyclohexanone is not surprising, since the distortion introduced into the normally strain-free chair conformation of cyclohexane by the sp^2-hybridized carbon atom is relieved in the addition product (the carbon hybridization becomes sp^3, as shown in equation 12.18). In contrast, the carbonyl double bond in 2-furfuraldehyde is stabilized by conjugation with the aromatic ring (about 5 kcal/mole), and addition reactions to this function are relatively slow. Consequently, the initial product in this experiment will be that formed in the fastest reaction (*kinetic control*). However, if the reaction conditions allow equilibrium to be established, then the predominant product will be that having the

FIGURE 12·5 *Kinetic and thermodynamic control in the formation of semicarbazones*

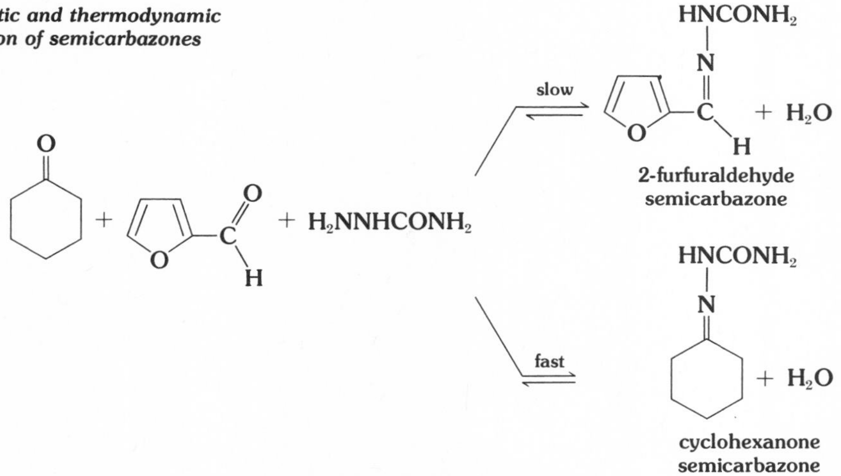

competitive semicarbazone formation

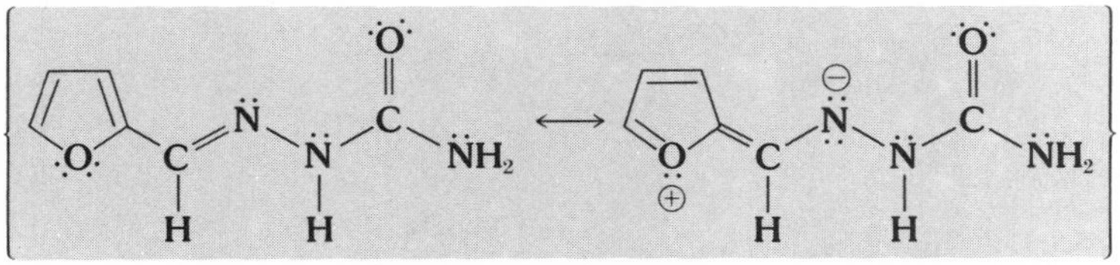

conjugative stabilization of furfuraldehyde semicarbazone

12·3 REVERSIBLE ADDITION REACTIONS

$$C=Y + H_2 \xrightarrow{\text{catalyst}} \overset{H}{\underset{|}{C}}-\overset{H}{\underset{|}{Y}}$$

Y = C Hydrogen adds readily with Pt, Pd, Rh, and Ni catalysts
Y = N Hydrogen adds readily with Pt, Pd, Rh, and Ni catalysts
Y = O Hydrogen addition generally sluggish; best catalysts are Pt, Ni, and Cu

FIGURE 12·6 *Catalytic addition of hydrogen to double bonds*

greatest thermodynamic stability (*equilibrium control*). Since the carbon-nitrogen double bond in the semicarbazone of 2-furfuraldehyde is also stabilized by conjugation with the furan ring, the formation of this product under conditions of equilibrium control is reasonable. Other examples of kinetic and thermodynamic control of reactions will be considered later in the text. Competitions of this kind are encountered frequently in biochemical transformations.

REDUCTION OF IMINES TO AMINES Although catalytic hydrogenation of carbonyl functions is sluggish in comparison with alkenes and alkynes, the carbon-nitrogen double bond in imines and related compounds adds hydrogen rapidly (and exothermically) in the presence of transition-metal catalysts (see Figure 12.6). We can use this relatively large difference in reactivity between imines and carbonyl compounds to advantage in developing a practical synthesis of amines, as shown in the following equations:

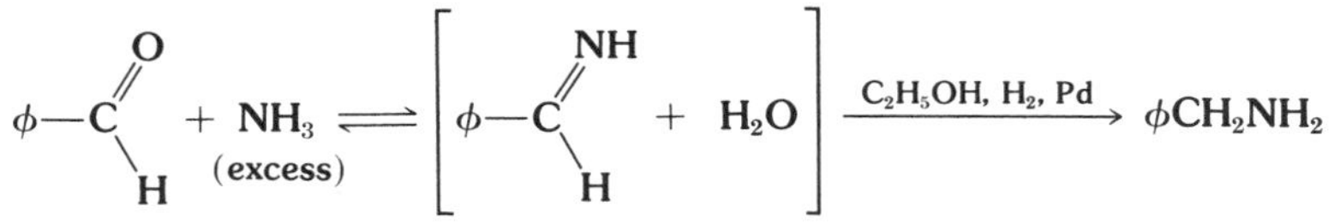

$$\phi-\overset{O}{\underset{H}{C}} + NH_3 \text{ (excess)} \rightleftharpoons \left[\phi-\overset{NH}{\underset{H}{C}} + H_2O\right] \xrightarrow{C_2H_5OH, H_2, Pd} \phi CH_2NH_2 \quad (12.19)$$

$$\underset{CH_3}{\overset{CH_3}{\diagdown}}C=O + H_2NCH_2CH_2OH \text{ (excess)} \rightleftharpoons \left[\underset{CH_3}{\overset{CH_3}{\diagdown}}C=N\underset{(CH_2)_2OH}{} + H_2O\right] \quad (12.20)$$

$$\downarrow {}^{C_2H_5OH,}_{H_2, Pd}$$

$$(CH_3)_2CHNH(CH_2)_2OH$$

The intermediate imines in these *reductive alkylation reactions* are not normally isolated. Consequently, the ratio of ammonia (or amine) to the carbonyl reactant is important in determining whether primary, secondary, or tertiary amines are obtained:

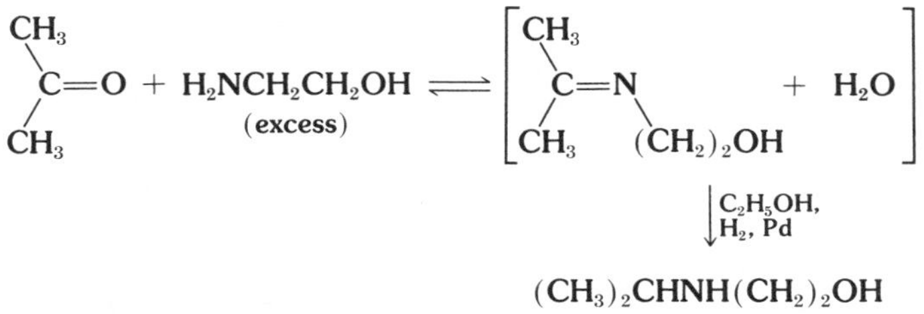

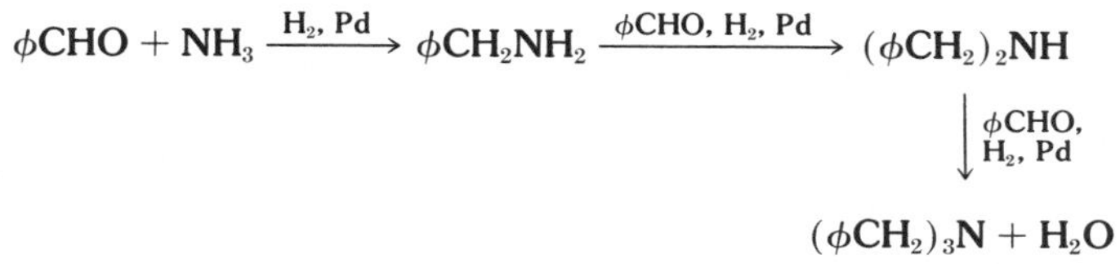

$$\phi CHO + NH_3 \xrightarrow{H_2, Pd} \phi CH_2NH_2 \xrightarrow{\phi CHO, H_2, Pd} (\phi CH_2)_2NH$$

$$\downarrow {}^{\phi CHO,}_{H_2, Pd}$$

$$(\phi CH_2)_3N + H_2O \quad (12.21)$$

12·4 IRREVERSIBLE ADDITION TO THE CARBONYL GROUP

We expect strong, highly reactive nucleophilic reagents to add rapidly to the carbonyl function of aldehydes and ketones. If the alkoxide product from this reaction is much more stable than the nucleophilic reagent, the addition reaction will be essentially *irreversible*, in contrast to the example in equation 12.6. The products from hydrolytic workup of such irreversible addition reactions will be alcohols:

$$\underset{\substack{\text{nucleophilic reagent}\\ \text{M = active metal}}}{\overset{\delta\ominus\;\;\delta\oplus}{\text{Nu—M}}} + \underset{R}{\overset{R}{\diagdown}}\!\!C\!\!=\!\!\ddot{O}: \longrightarrow \text{Nu}-\underset{R}{\overset{R}{\underset{|}{C}}}-\ddot{O}:^{\ominus} M^{\oplus} \xrightarrow{H_2O} \text{Nu}-\underset{R}{\overset{R}{\underset{|}{C}}}-\ddot{O}-H \qquad (12.22)$$

REDUCTION BY COMPLEX METAL HYDRIDES

During World War II chemists at the University of Chicago discovered a group of complex (mixed) metal hydrides which have since been shown to be versatile reducing agents for many functional groups. Two of the most widely used reagents of this kind are lithium aluminum hydride (**LiAlH$_4$**) and sodium borohydride (**NaBH$_4$**). Lithium aluminum hydride is a light grey ether-soluble solid which reacts vigorously with the active hydrogens in alcohols, carboxylic acids, and amines, evolving hydrogen and giving the corresponding salts:

$$4C_2H_5OH + LiAlH_4 \xrightarrow{\text{ether}} 4H_2 + (C_2H_5O)_3Al + \underset{\text{ethoxide salts}}{C_2H_5OLi} \qquad (12.23)$$

In effect, the reagent behaves as though it were a source of hydride ion H:$^\ominus$, although it is doubtful that such a highly reactive species is ever actually present during these reactions. Sodium borohydride, a white water- and alcohol-soluble solid, reacts slowly with active hydrogens and is used in the laboratory preparation of diborane (see equation 5.23). Both LiAlH$_4$ and NaBH$_4$ reduce ketones and aldehydes to secondary and primary alcohols, respectively:

$$4\;\underset{}{\bigcirc}\!\!=\!\!O + LiAlH_4 \xrightarrow{\text{ether}} 4\;\underset{(M = Al\ or\ Li)}{\bigcirc\!\!\overset{H}{\underset{}{\diagup}}\!\!\overset{O^\ominus M^\oplus}{\diagdown}} \xrightarrow{H_2O} 4\;\underset{(n = 1\ or\ 3)}{\bigcirc\!\!\overset{H}{\underset{}{\diagup}}\!\!\overset{OH}{\diagdown}} \quad M(OH)_n \qquad (12.24)$$

$$4\phi CHO + LiAlH_4 \xrightarrow{\text{ether}} \underset{(M = Al\ or\ Li)}{4\phi CH_2\ddot{O}:^\ominus M^\oplus} \xrightarrow{H_2O} 4\phi CH_2\ddot{O}H + M(OH)_n \qquad (12.25)$$

$$4\;\triangleright\!\!\overset{\ddot{O}}{\underset{}{\overset{\|}{C}}}\!\!-\!CH_3 + NaBH_4 \xrightarrow{H_2O,\ CH_3OH} 4\;\triangleright\!\!\overset{:\ddot{O}:^\ominus M^\oplus}{\underset{H}{\overset{|}{C}}}\!\!-\!CH_3 \xrightarrow{H_2O} 4\;\triangleright\!\!\overset{:\ddot{O}H}{\underset{H}{\overset{|}{C}}}\!\!-\!CH_3 \qquad (12.26)$$
$$(M = Na\ or\ B)$$

12·4 IRREVERSIBLE ADDITION TO THE CARBONYL GROUP

When we consider the polar nature of metal-hydrogen bonds (metals have lower electronegativities than hydrogen), it is clear that these reactions proceed by attack of a nucleophilic hydrogen atom at the electrophilic carbon atom of the carbonyl group. Furthermore, since the alkoxide salts thus formed cannot decompose by ejecting a stable anionic moiety, this addition reaction is irreversible:

$$\text{—M}^{\delta+}\text{—H}^{\delta-} + \underset{R'}{\overset{\overset{\ddot{O}^{\delta-}}{\|}}{C^{\delta+}}}\text{—R} \longrightarrow \text{—M}^{\oplus}\ :\ddot{O}:^{\ominus} \quad \underset{R'}{\overset{|}{\underset{|}{C}}}\text{—R} \atop H \tag{12.27}$$

REDUCTION BY DIBORANE

Diborane (B_2H_6) reduces aldehydes and ketones to alcohols, and such reactions may be competitive with the hydroboration of alkenes (see Section 5.2). In contrast to the nucleophilic character of the metal hydrides $AlH_4^\ominus$ and $BH_4^\ominus$, diborane is electrophilic, as may be seen from its selectivity (relative to $NaBH_4$) in the reduction of a trimethylacetaldehyde-chloral mixture:

$$3RCHO + BH_3 \longrightarrow (RCH_2\text{—O})_3B \xrightarrow{H_2O} 3RCH_2OH \tag{12.28}$$

$$(CH_3)_3CC\overset{\overset{O}{\|}}{\underset{H}{}} \qquad\qquad Cl_3CC\overset{\overset{O}{\|}}{\underset{H}{}}$$

preferentially reduced by B_2H_6 preferentially reduced by $NaBH_4$

REACTIONS WITH ORGANOMETALLIC REAGENTS

Grignard reagents, alkyl-lithium compounds, and alkali metal salts of terminal acetylenes all have polar carbon-metal bonds in which the organic group is strongly nucleophilic. It comes as no surprise, therefore, to find that these organometallic reagents combine rapidly with the polar carbonyl group in aldehyde and ketone compounds to give the salts of alcohols:

$$C_2H_5^{\delta-}MgBr^{\delta+} + \phi\text{—}\overset{\overset{\ddot{O}^{\delta-}}{\|}}{\underset{\delta+}{C}}\text{—}CH_3 \xrightarrow{\text{ether}} \phi\text{—}\underset{C_2H_5}{\overset{:\ddot{O}:^\ominus MgBr^\oplus}{\underset{|}{C}}}\text{—}CH_3 \xrightarrow{H_2O} \phi\text{—}\underset{C_2H_5}{\overset{OH}{\underset{|}{C}}}\text{—}CH_3 + MgBrOH \tag{12.29}$$

$$\phi\text{—}Li^{\delta+} + CH_3\text{—}\underset{\text{(aryl)}}{\bigcirc}\text{—}\overset{\delta+}{C}\overset{\delta-}{=}\ddot{O}: \xrightarrow{\text{pentane}} \underset{H}{\bigcirc}\text{—}\overset{:\ddot{O}:^\ominus Li^\oplus}{\underset{|}{C}}\text{—}\phi \xrightarrow{H_2O} CH_3\text{—}\underset{H}{\bigcirc}\text{—}\overset{OH}{\underset{|}{C}}\text{—}\phi + LiOH \tag{12.30}$$

$$\underset{S}{\bigcirc}\text{—}MgCl^{\delta+} + \underset{H}{\overset{H}{\underset{}{C}}}\overset{\delta+}{=}\overset{\delta-}{\ddot{O}:} \xrightarrow{\text{ether}} \underset{S}{\bigcirc}\text{—}CH_2\text{—}\ddot{O}:^\ominus Mg^\oplus Cl \xrightarrow{H_2O} \underset{S}{\bigcirc}\text{—}CH_2OH + MgClOH \tag{12.31}$$

12·4 IRREVERSIBLE ADDITION TO THE CARBONYL GROUP

$$\phi C\equiv CNa + \text{cyclohexanone} \longrightarrow \text{cyclohexyl-}O^-Na^+\text{-}C\equiv C\phi \xrightarrow{H_2O} \text{cyclohexyl-}OH\text{-}C\equiv C\phi + NaOH \quad (12.32)$$

By choosing the appropriate carbonyl reactant we can prepare primary alcohols (from formaldehyde), secondary alcohols (from other aldehydes), or tertiary alcohols (from ketones). Indeed, these reactions constitute one of our most versatile methods for synthesizing alcohols.

PROBLEM 12·14

Suggest one or more combinations of reactants (a carbonyl compound and an organometallic reagent or complex hydride) which would serve to synthesize each of the following alcohols:

a 5-Methyl-5-nonanol
b 1-Cyclopropyl-2-methyl-1-propanol
c 2-Phenylethanol
d 1-Phenylcyclopentanol
e 2,5-Dimethyl-2,5-dihydroxy-3-hexyne

The ability of Grignard reagents and related organometallic compounds to add to polarized double bonds has been used to further advantage in reactions with carbon dioxide and oxirane:

$$\text{(2,4,6-trimethylphenyl)MgBr} + CO_2 \longrightarrow \text{ArC(=O)O}^-MgBr^+ \xrightarrow{H_3O^+} \text{ArCOOH} \quad (12.33)$$

$$\text{2-pyridyl-CH}_2\text{Li} + CO_2 \longrightarrow \text{2-pyridyl-CH}_2\text{-C(=O)O}^-Li^+ \xrightarrow{H_3O^+} \text{2-pyridyl-CH}_2\text{-COOH} \quad (12.34)$$

$$n\text{-}C_4H_9MgBr + \text{oxirane} \longrightarrow n\text{-}C_4H_9\text{-}CH_2\text{-}CH_2\text{-}O^-MgBr^+ \xrightarrow{H_2O} n\text{-}C_6H_{13}OH \quad (12.35)$$

Although oxirane (ethylene oxide) is usually classified as an ether, the large angle strain in the small ring causes it to exhibit many carbonyl traits, just as cyclopropane has some alkenelike properties (Section 5.9). Only one of the two carbon-oxygen double bonds in carbon dioxide suffers addition under the customary reaction conditions (addition of the organometallic reagent to excess CO_2 at low temperature). The salts formed in these reactions are relatively stable and require treatment by mineral acids in order to generate a carboxylic acid.

THE WITTIG REACTION

Recall from Section 11.8 that alkyl triphenylphosphonium salts react with strong bases to give vicinal ionic intermediates called *ylides*:

$$\phi_3\overset{+}{P}\text{-}CH_2R \; X^- \xrightarrow[-H^+]{\text{strong base}} \{\phi_3\overset{+}{P}\text{-}\overset{-}{C}HR \longleftrightarrow \phi_3P=CHR\} \quad (12.36)$$

a phosphonium salt   an ylide

12·4 IRREVERSIBLE ADDITION TO THE CARBONYL GROUP

Most ylides are themselves strong nucleophiles (and bases) and can be alkylated or condensed with aldehydes and/or ketones:

$$\overset{\delta\oplus}{\phi_3 P}=\overset{\delta\ominus}{CHR} + R'X \xrightarrow{S_N 2} \phi_3\overset{\oplus}{P}-CHRR' \; X^{\ominus} \tag{12.37}$$

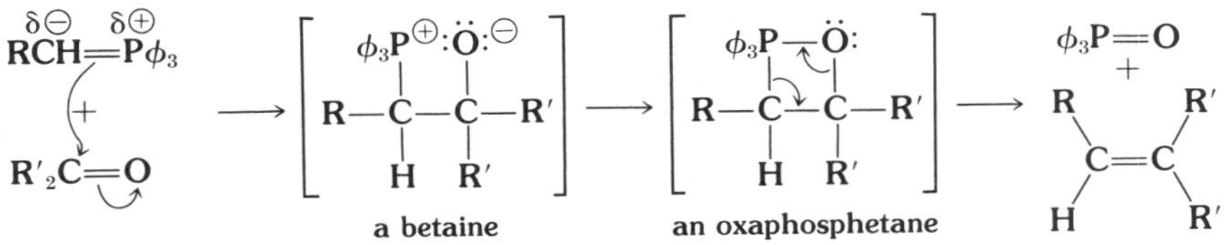

$$\begin{array}{c} \overset{\delta\ominus}{RCH}=\overset{\delta\oplus}{P\phi_3} \\ (+) \\ R'_2 C=O \end{array} \longrightarrow \left[\begin{array}{c} \phi_3\overset{\oplus}{P}:\overset{\ominus}{\ddot{O}}: \\ | \quad | \\ R-C-C-R' \\ | \quad | \\ H \quad R' \end{array} \right] \longrightarrow \left[\begin{array}{c} \phi_3 P-\ddot{O}: \\ | \quad | \\ R-C-C-R' \\ | \quad | \\ H \quad R' \end{array} \right] \longrightarrow \begin{array}{c} \phi_3 P=O \\ + \\ R \quad \quad R' \\ \diagdown\diagup \\ C=C \\ \diagup\diagdown \\ H \quad \quad R' \end{array} \tag{12.38}$$

a betaine   an oxaphosphetane

The transformation in equation 12.38, which provides an exceptionally versatile means of synthesizing alkenes, is commonly referred to as the *Wittig reaction*. Extensive study of this reaction has produced evidence of one or more intermediates, such as the betaine and oxaphosphetane shown.

The usefulness of the Wittig reaction is illustrated by the following examples.

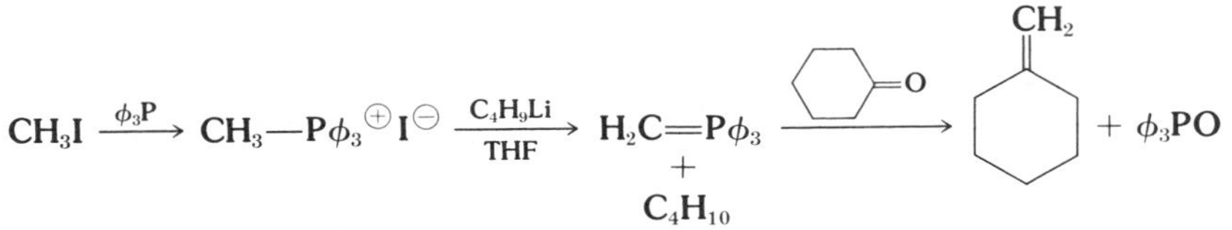

$$CH_3I \xrightarrow{\phi_3 P} CH_3-\overset{\oplus}{P}\phi_3 \; I^{\ominus} \xrightarrow[THF]{C_4H_9Li} \begin{array}{c} H_2C=P\phi_3 \\ + \\ C_4H_{10} \end{array} \xrightarrow{\text{cyclohexanone}} \text{methylenecyclohexane} + \phi_3 PO \tag{12.39}$$

$$\phi CH=CHCH_2Cl \xrightarrow{\phi_3 P} \phi CH=CHCH_2 \overset{\oplus}{P}\phi_3 \; Cl^{\ominus} \xrightarrow[ether]{C_2H_5OLi} \phi CH=CH-CH=P\phi_3 + C_2H_5OH$$

$$\phi CH=CHCH=P\phi_3 + \phi CHO \longrightarrow \phi CH=CH-CH=CH\phi + \phi_3 PO \tag{12.40}$$

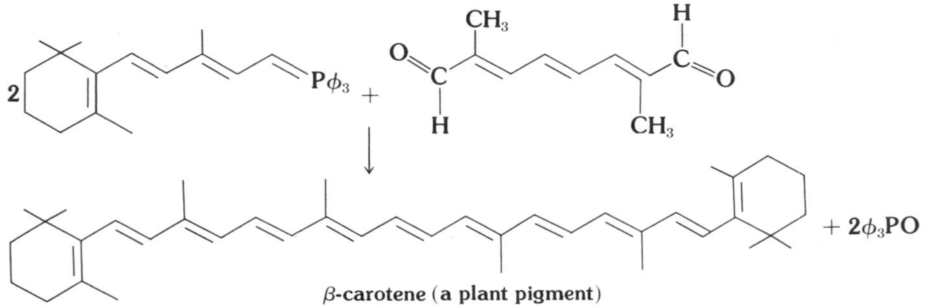

$$2 \; [\text{retinyl phosphorane}] \; + \; [\text{dialdehyde}] \longrightarrow \beta\text{-carotene (a plant pigment)} + 2\phi_3 PO \tag{12.41}$$

PROBLEM 12·15

a Triphenylphosphine dissolves in but does not react significantly with chloroform at 0°C. If a strong base is added to this solution, a reactive ylide appears to be generated, as suggested by the following reaction:

$$\phi_3 P + CHCl_3 \xrightarrow[0°]{C_4H_9OK, \; ether} [X] \xrightarrow{C_6H_5CHO} C_6H_5CH=CCl_2 + \phi_3 PO$$

Suggest a structure for the intermediate X and a mechanism by which it could be formed.

b Suggest a structure for the intermediate Y in the following reaction and a mechanism for its formation:

$$\overset{\oplus}{\phi_3 P}CH_2CH_2CH_2CH_2Br \xrightarrow{2C_6H_5Li, \; ether} [Y] \xrightarrow{\phi_2 C=O} \phi_2 C= \text{cyclobutylidene}$$

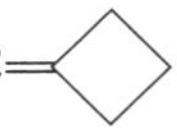

12·5 REDUCTIVE DEOXYGENATION OF CARBONYL COMPOUNDS

Any of three different procedures may be used to transform a carbonyl function into a methylene group:

$$\begin{array}{c} R \\ \diagdown \\ R' \end{array} C=O \longrightarrow \begin{array}{c} R \\ \diagdown \\ R' \end{array} CH_2 \tag{12.42}$$

The first method, known as the *Wolff-Kishner reduction,* involves heating the carbonyl compound or its hydrazone with excess hydrazine and a strong base (such as KOH) in a high-boiling solvent. A tautomer of the initially formed hydrazone is believed to play an important role in the mechanism of this useful reaction.

$$\begin{array}{c} R \\ \diagdown \\ R' \end{array}\!\!C=O \xrightarrow[\Delta]{N_2H_4} \left[\begin{array}{c} R \\ \diagdown \\ R' \end{array}\!\!C=\ddot{N} \\ \ddot{N}H_2 \\ \Updownarrow OH^{\ominus} \\ \begin{array}{c} R \\ \diagdown \\ R' \end{array}\!\!\begin{array}{c} H \\ \diagup \\ C \\ N{=}NH \end{array} \right] \xrightarrow{-N_2} \begin{array}{c} R \\ \diagdown \\ R' \end{array}\!\!\begin{array}{c} H \\ \diagup \\ C \\ \diagdown \\ H \end{array} \tag{12.43}$$

The second procedure, usually referred to as the *Clemmensen reduction,* involves heating an aqueous mineral-acid solution of the carbonyl compound in the presence of zinc metal which has been activated by amalgamation. This interesting reaction is thought to proceed by means of an organozinc moiety located on the surface of the metal.

$$\begin{array}{c} R \\ \diagdown \\ R' \end{array}\!\!C=O \xrightleftharpoons{H_3O^{\oplus}} \left[\begin{array}{c} R \\ \diagdown \\ R' \end{array}\!\!\begin{array}{c} H \\ \diagup \\ C=\overset{\oplus}{O} \end{array} \right] \xrightarrow{Zn(Hg)} \left[\begin{array}{c} OH \\ | \\ R{-}C{-}R' \\ | \\ Zn \end{array} \right] \xrightarrow[-H_2O]{H_3O^{\oplus}} R{-}CH_2{-}R' + Zn^{2+} \tag{12.44}$$

The third method requires two steps: formation of a thioketal derivative followed by catalytic hydrogenolysis of this intermediate, as decribed for alkyl sulfides in Section 8.8.

$$\begin{array}{c} R \\ \diagdown \\ R' \end{array}\!\!C=O + \begin{array}{c} {-}SH \\ {-}SH \end{array} \xrightarrow{BF_3} \begin{array}{c} R \\ \diagdown \\ R' \end{array}\!\!\begin{array}{c} S{-} \\ C \\ S{-} \end{array} \xrightarrow[\Delta]{H_2, Ni \atop acetone} \begin{array}{c} R \\ \diagdown \\ R' \end{array}\!\!CH_2 + 2H_2S + C_2H_6 \tag{12.45}$$

PROBLEM 12·16
a Write equations illustrating the Wolff-Kishner reduction of cyclooctanone, the Clemmensen reduction of ethyl phenyl ketone, and desulfurization of the ethylene thioketal of 2,2-dimethylcyclohexanone.
b An optically active ketone $C_7H_{12}O$ is stable to acid or base treatment but yields an optically inactive hydrocarbon on Wolff-Kishner reduction. Write two possible structures for such a ketone.

12·5 REDUCTIVE DEOXYGENATION OF CARBONYL COMPOUNDS

OXIDATION REACTIONS

Aldehydes and ketones are easily distinguished by chemical tests, since aldehydes oxidize readily to carboxylic acids under mild conditions, whereas ketones are relatively resistant to most oxidizing reagents. Two common oxidizing agents [O] are potassium permanganate ($KMnO_4$) and chromic acid (H_2CrO_4).

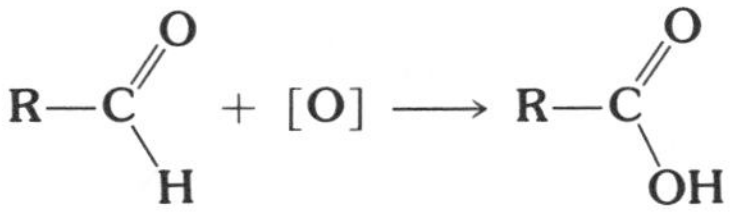

$$R-\underset{H}{\overset{O}{C}} + [O] \longrightarrow R-\underset{OH}{\overset{O}{C}} \qquad (12.46)$$

$$\text{piperonal-CHO} + KMnO_4 \xrightarrow{H_3O^{\oplus}} \text{piperonal-CO}_2H \qquad (12.47)$$

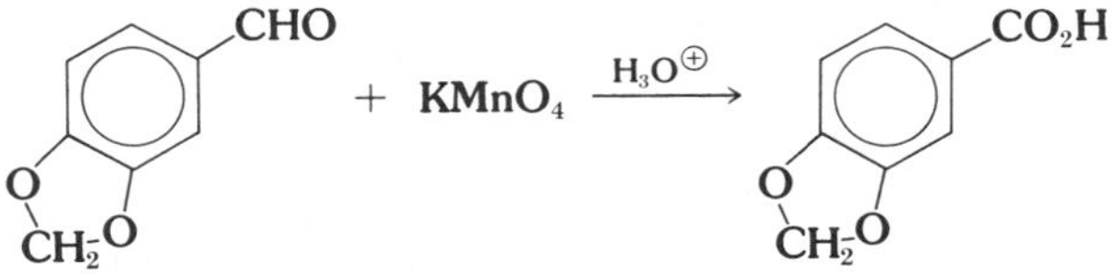

In the case of aldehydes very mild oxidizing agents can be used, thus providing us with a highly selective test for this functional group. Two such oxidizing systems are *Tollens reagent*, an alkaline solution of silver cation (complexed with ammonia), and *Fehling* or *Benedict's reagent*, an alkaline solution of cupric ion (complexed with tartrate or citrate anions, respectively). During the course of oxidations with Tollens reagent, metallic silver is deposited on the walls of the reaction flask and a characteristic mirror develops, as in the following balanced reaction for butyraldehyde:

$$\begin{array}{c} C_3H_7CHO \\ + \\ 2\underbrace{\left[Ag(NH_3)_2^{\oplus}OH^{\ominus}\right]}_{\text{Tollens reagent}} \end{array} \longrightarrow C_3H_7CO_2^{\ominus}NH_4^{\oplus} + \underset{\text{mirror}}{2Ag} + NH_4OH + 2NH_3 \qquad (12.48)$$

Oxidations with Fehling or Benedict's solution also show a characteristic color change from the blue-green cupric-ion complex to a reddish-brown cuprous precipitate.

The mechanisms of aldehyde oxidations probably involve hydrated intermediates analogous to those shown earlier for the oxidation of secondary alcohols. For example, an intermediate such as that in the following reaction would be formed in a permanganate oxidation of an aldehyde:

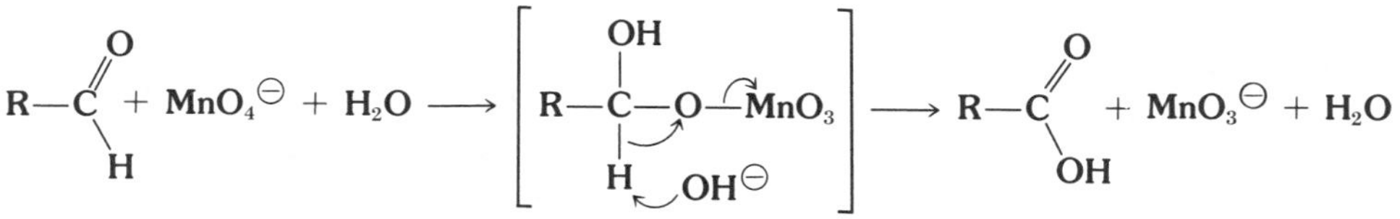

$$R-\underset{H}{\overset{O}{C}} + MnO_4^{\ominus} + H_2O \longrightarrow \left[R-\underset{H\ \curvearrowleft OH^{\ominus}}{\overset{OH}{\underset{|}{C}}-O-MnO_3} \right] \longrightarrow R-\underset{OH}{\overset{O}{C}} + MnO_3^{\ominus} + H_2O \qquad (12.49)$$

PROBLEM 12·17
Consider the permanganate oxidation mechanism illustrated in equation 12.49 and suggest an explanation for the low reactivity of ketones with Cr^{+6} and Mn^{+7} oxidizing agents.

12·6 REACTIONS VIA ENOL INTERMEDIATES

Many aldehydes and ketones undergo surprisingly facile substitution reactions at the α carbon atoms. We find, for example, that cyclohexanone and acetophenone are readily halogenated at moderate temperatures in the absence of peroxides or other free-radical initiators:

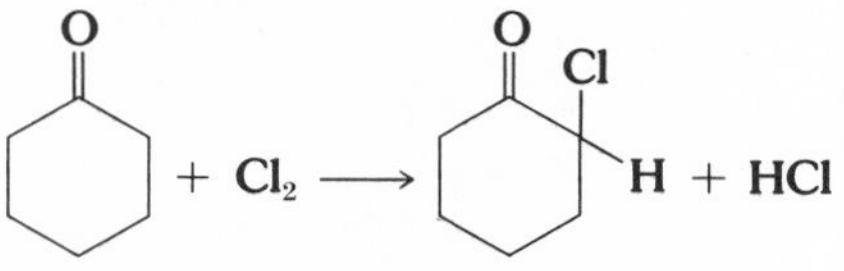

$$\text{cyclohexanone} + Cl_2 \longrightarrow \text{2-chlorocyclohexanone} + HCl \qquad (12.50)$$

$$\phi\text{—CO—}CH_3 + Br_2 \longrightarrow \phi\text{—CO—}CH_2Br + HBr \qquad (12.51)$$

Acid or base catalysis is usually necessary, and under equivalent reaction conditions the rates of chlorination, bromination, and iodination are approximately the same. In fact, although these reactions exhibit a first-order rate relationship with respect to the ketone concentration, they are relatively unaffected by changes in the halogen concentration.

An insight into the nature of these halogen substitution reactions and similar transformations is provided by some investigations with optically active s-butyl phenyl ketone, outlined in Figure 12.7. Observe that acids and bases catalyze not only bromination at the α carbon (racemic bromide is formed), but also racemization and deuterium exchange with heavy water. Since the rates of all these reactions are the same, we are immediately led to suspect the existence of a common intermediate which must be formed in a slow (rate-determining) step, before the final substitution.

Our previous experience suggests that the common intermediate in these reactions may be the *enol tautomer* of the carbonyl reactant (see equations 6.21 and 6.22). A convincing set of arguments and facts can be marshalled in support of this contention.

To begin with, carbonyl compounds which lack an α hydrogen atom, and therefore cannot enolize, do not undergo any of these reactions:

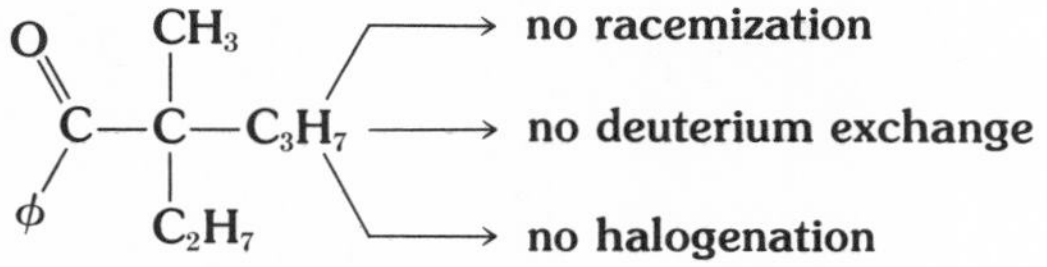

$$\phi\text{—CO—}C(CH_3)(C_2H_7)(C_3H_7) \longrightarrow \begin{cases} \text{no racemization} \\ \text{no deuterium exchange} \\ \text{no halogenation} \end{cases} \qquad (12.52)$$

FIGURE 12·7 *Reactions of optically active s-butyl phenyl ketone*

$$(+)\text{-}s\text{-butyl phenyl ketone} \xrightarrow[\text{slow}]{\text{acid or base catalysis}} [\text{intermediate}] \begin{cases} \xrightarrow{H_2O} (\pm)\text{-}C_2H_5CH(CH_3)CO\phi \\ \xrightarrow{D_2O} (\pm)\text{-}C_2H_5CD(CH_3)CO\phi \\ \xrightarrow{Br_2} (\pm)\text{-}C_2H_5CBr(CH_3)CO\phi \end{cases}$$

Moreover, if the enolization step is rate determining, then for a given catalyst concentration, the overall rate of reaction (halogenation, racemization, or deuterium exchange) will depend only on the concentration of the carbonyl reactant:†

rate of reaction = k[carbonyl reactant] (12.53)

This is in fact the rate equation that is observed for reactions of this kind.

As further evidence, we find that the ease with which different carbonyl compounds can be halogenated is proportional to the concentration of the enol tautomer present. Although simple aldehydes and ketones usually contain only very small amounts of enol (less than 0.1%), certain β-dicarbonyl compounds which are found to be largely enolic undergo rapid bromination. In fact, the keto and enol tautomers of ethyl acetoacetate ($CH_3COCH_2CO_2C_2H_5$) have actually been separated. As expected, the enol tautomer reacted instantaneously with bromine, whereas the keto form proved to be relatively unreactive on initial mixing of the reactants.

The carbonyl compounds listed in Table 12.5 illustrate the wide variation in enol composition. It is important to note, however, that these values may change dramatically when the compounds are dissolved in certain solvents; for example, 2,4-pentanedione is 20% enolic in aqueous solution, but over 90% enolic in hexane. Pmr spectroscopy has proved to be useful in determining keto-enol compositions, as can be seen from the chemical-shift assignments for 2,4-pentanedione described in Figure 12.8.

Enol-keto tautomerism easily accounts for the racemization of simple optically active ketones having an enolizable hydrogen atom at the asymmetric center. Since the enol tautomer has a plane of symmetry passing through the carbon atoms that constitute the double bond, it is achiral and is subject to protonation equally well from either side of this sym-

†If the halogen is present in very low concentration it can also appear in the rate equation.

TABLE 12·5 *Approximate enol content of carbonyl compounds*

compound	enol content (%)†
CH_3COCH_3	10^{-4}
cyclohexanone	10^{-2}
$CH_3COCH_2CO_2C_2H_5$	8
$CH_3COCH_2COCH_3$	80
HCOCH$_2$CHO	100

†Equilibrium composition of pure liquids.

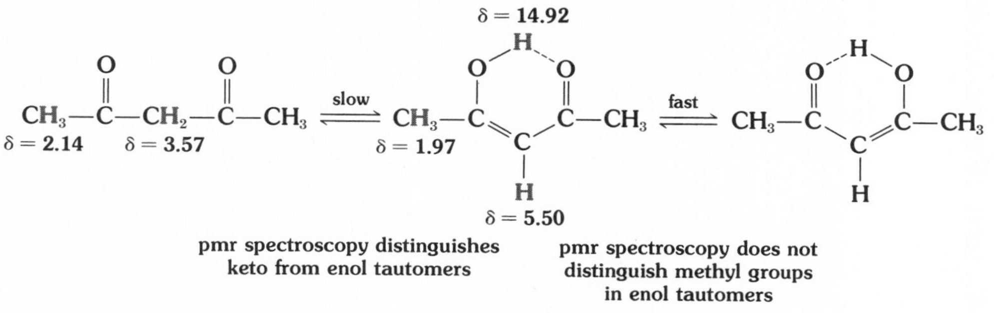

12·6 REACTIONS VIA ENOL INTERMEDIATES

$$CH_3-\underset{\delta=2.14}{\overset{O}{\underset{\|}{C}}}-\underset{\delta=3.57}{CH_2}-\overset{O}{\underset{\|}{C}}-CH_3 \underset{\text{slow}}{\rightleftharpoons} CH_3-C\underset{\underset{\delta=5.50}{H}}{\overset{\overset{\delta=14.92}{\overset{H}{\underset{O\cdots\cdots O}{\diagdown\diagup}}}}{\diagdown\diagup}}C-CH_3 \underset{\text{fast}}{\rightleftharpoons} CH_3-C\underset{\underset{H}{\diagup}}{\overset{\overset{H}{\underset{O\cdots\cdots O}{\diagdown\diagup}}}{\diagdown\diagup}}C-CH_3$$

$$\delta=1.97$$

pmr spectroscopy distinguishes keto from enol tautomers

pmr spectroscopy does not distinguish methyl groups in enol tautomers

FIGURE 12·8 *Pmr signals for 2,4-pentanedione*

metry plane. The resulting product is a 50:50 mixture of enantiomeric keto tautomers:

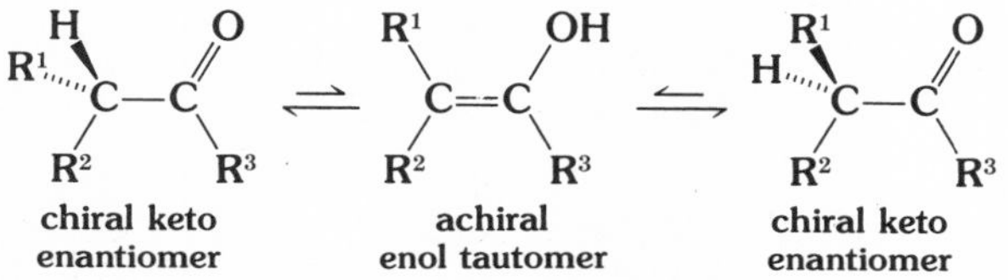

$$\underset{\substack{\text{chiral keto}\\\text{enantiomer}}}{R^1_{\diagup}\overset{H}{\underset{R^2}{\diagdown}}C-\overset{O}{\underset{R^3}{\diagdown}}C} \rightleftharpoons \underset{\substack{\text{achiral}\\\text{enol tautomer}}}{\overset{R^1}{\underset{R^2}{\diagdown}}C=\overset{OH}{\underset{R^3}{\diagdown}}C} \rightleftharpoons \underset{\substack{\text{chiral keto}\\\text{enantiomer}}}{\overset{R^1}{H_{\diagup}\underset{R^2}{\diagdown}}C-\overset{O}{\underset{R^3}{\diagdown}}C} \quad (12.54)$$

Compounds having more than one chiral center, only one of which is enolizable, are converted to diastereoisomers that differ in configuration only at this one asymmetric site. Such diastereoisomers are called *epimers*, and their interconversion is referred to as *epimerization*. The isomerization of menthone and isomenthone is an example of epimerization. Optically active (−)-menthone is converted to a 70:30 mixture of itself and (+)-isomenthone; this mixture remains optically active, although the specific rotation changes. (±)-Menthone is transformed to an equivalent equilibrium mixture with (±)-isomenthone; this mixture is, of course, optically inactive.

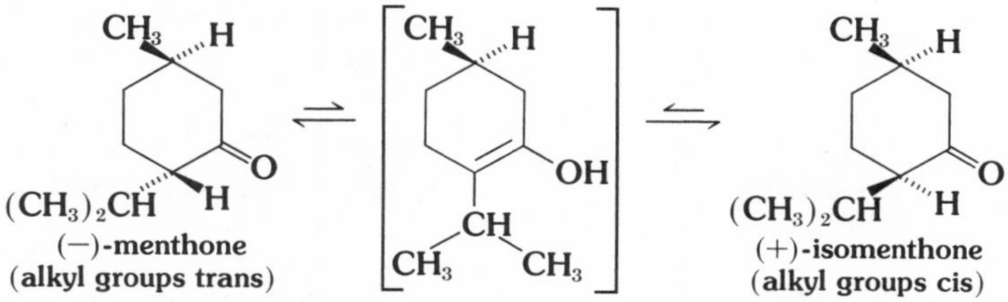

(12.55)

(−)-menthone (alkyl groups trans)   (+)-isomenthone (alkyl groups cis)

PROBLEM 12·18
Would a 50:50 mixture of (−)-menthone and (+)-isomenthone be a racemic modification? Would it be optically active?

CONJUGATE ACIDS AND BASES OF CARBONYL COMPOUNDS

Keto-enol tautomerism is catalyzed by acids or bases, and in Figure 12.9 we see that the conjugate-acid and conjugate-base intermediates in these interconversions are charge-delocalized species which require a resonance description.

Since ketones and aldehydes are very weak bases (about a million times weaker than water) and very weak acids (about 10,000 times

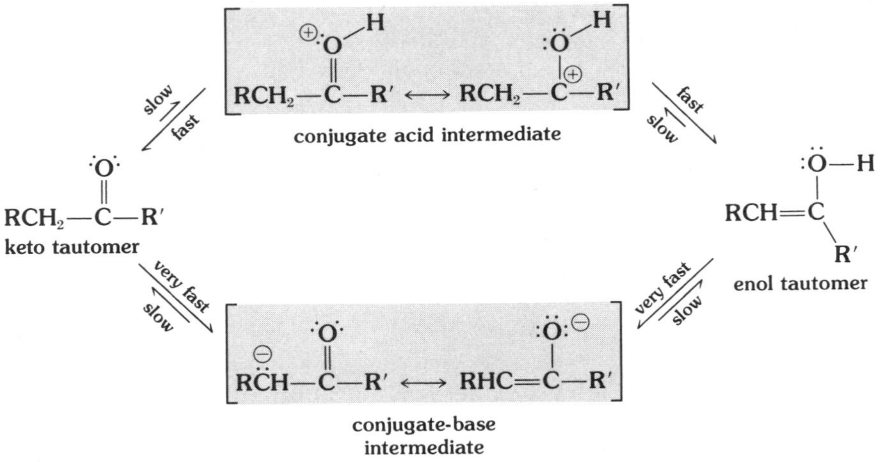

12·6 REACTIONS VIA ENOL INTERMEDIATES

FIGURE 12·9 *Establishment of a keto-enol equilibrium via conjugate-acid or conjugate-base intermediates*

weaker than water), the concentration of their conjugate acids or bases in water or alcohol solutions is practically zero. However, solutions of the conjugate acids can be formed in a strongly acidic solvent such as fluorosulfonic acid (FSO_3H) and studied by methods such as pmr spectroscopy. In a very weakly acidic solvent such as ammonia it is possible to prepare stable solutions of the conjugate bases, usually referred to as *enolate anions*, by reaction with a strong base:

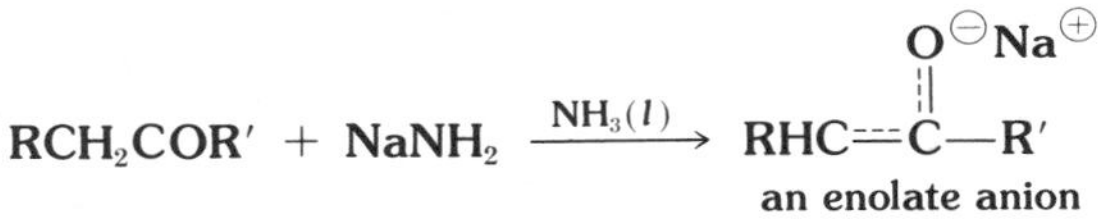

$$RCH_2COR' + NaNH_2 \xrightarrow{NH_3(l)} RHC=\!\!\!=\!\!\underset{\text{an enolate anion}}{\underset{|}{\overset{O^\ominus Na^\oplus}{C}}}-R' \qquad (12.56)$$

PROBLEM 12·19
a Draw an energy diagram illustrating the base-catalyzed enolization of acetone. Clearly label each reactant, intermediate, and transition state in your diagram.
b Base-catalyzed halogenation of ketones is usually about 1000 times faster than the corresponding acid-catalyzed reactions. Suggest a reason for this.

PROBLEM 12·20
An ether solution of acetophenone ($\phi COCH_3$) reacts with bromine in the presence of a trace of $AlCl_3$ to give $\phi COCH_2Br$ in good yield. If 2.5 molar equivalents of $AlCl_3$ are mixed with the acetophenone immediately before addition of the bromine, this mixture slowly reacts (heating is required) to give *m*-bromoacetophenone. Suggest a reason for this change in reaction.

From the approximate ionization constants of some weak to very weak acids, listed in Table 12.6, we see that the strengths of carbon (C—H) acids vary by almost *40 powers of 10*. This, of course, reflects a corresponding variation in the stability of the conjugate bases, with the strongest acids generating the most stable conjugate bases. The factor primarily responsible for increased carbanion stability appears to be charge delocalization, particularly delocalization involving atoms of

12·6 REACTIONS VIA ENOL INTERMEDIATES

TABLE 12·6 *Approximate dissociation constants for some weak acids*

compound	K_a†	compound	K_a†
$CH_3-\overset{O}{\underset{\parallel}{C}}-CH_2-\overset{O}{\underset{\parallel}{C}}-CH_3$	10^{-9}	$CH_3SO_2CH_3$	10^{-23}
CH_3-NO_2	10^{-10}	$HC\equiv CH$	10^{-25}
$CH_3COCH_2CO_2C_2H_5$	10^{-11}	ϕ_3CH	10^{-32}
H_2O	10^{-15}	CH_3SOCH_3 (DMSO)	10^{-33}
cyclopentadiene	10^{-16}	NH_3	10^{-34}
C_2H_5OH	10^{-17}	H_2	$<10^{-37}$
$(CH_3)_3COH$	10^{-19}	C_2H_4	$<10^{-40}$
CH_3COCH_3	10^{-20}	C_2H_6	$<10^{-44}$

† Extrapolated to water solution.

higher electronegativity than carbon. For example, the conjugate bases of nitromethane and acetone are described as:

$$CH_3NO_2 \xrightarrow{-H^{\oplus}} \left\{ H_2\overset{\ominus}{C}-\overset{\oplus}{N}\overset{\overset{\cdot\cdot}{O}\cdot}{\underset{\overset{\cdot\cdot}{O}:\ominus}{}} \longleftrightarrow H_2\overset{\ominus}{C}-\overset{\oplus}{N}\overset{\overset{\cdot\cdot}{O}:\ominus}{\underset{\overset{\cdot\cdot}{O}\cdot}{}} \longleftrightarrow H_2C=\overset{\oplus}{N}\overset{\overset{\cdot\cdot}{O}\cdot}{\underset{\overset{\cdot\cdot}{O}:\ominus}{}} \right\} \quad (12.57)$$

$$CH_3-\overset{\overset{\cdot\cdot}{O}\cdot}{\underset{\parallel}{C}}-CH_3 \xrightarrow{-H^{\oplus}} \left\{ H_2\overset{\ominus}{C}-\overset{\overset{\cdot\cdot}{O}\cdot}{\underset{\parallel}{C}}-CH_3 \longleftrightarrow H_2C=\overset{:\overset{\cdot\cdot}{O}:\ominus}{\underset{|}{C}}-CH_3 \right\} \quad (12.58)$$

Among atoms of similar size, a negative charge is best accommodated on that atom having the highest electronegativity. Thus acid strength decreases markedly in the order $H_2O > NH_3 > CH_4$. The ability of an atom to support a negative charge also changes with the hybridization state, as evidenced by the enhanced acidity of acetylene (see Section 6.6).

PROBLEM 12·21

a Which of the following contributors makes the larger contribution to the resonance hybrid of an enolate anion?

$\overset{\ominus}{C}-\overset{\overset{\cdot\cdot}{O}\cdot}{\underset{\parallel}{C}}$; $C=\overset{:\overset{\cdot\cdot}{O}:\ominus}{\underset{|}{C}}$

b Suggest an explanation for the remarkably enhanced acidity of 2,4-pentanedione.

It should be clear from our previous discussions of acid-base equilibria that the conjugate base of an acid can be efficiently generated by a proton transfer to the conjugate base of a weaker acid as described by the following equation:

12·6 REACTIONS VIA ENOL INTERMEDIATES

$$H—A + Na^{\oplus} :B^{\ominus} \rightleftharpoons B—H + Na^{\oplus} :A^{\ominus} \qquad (12.59)$$

stronger acid and base — weaker acid and base

Some useful bases for preparing enolate anions from carbonyl compounds are NaH, $NaNH_2$, $NaCH_2SOCH_3$, and $NaC\phi_3$. Sodium hydride and sodium amide generate gaseous products in the proton exchange, so that even unfavorable equilibria can be forced to completion:

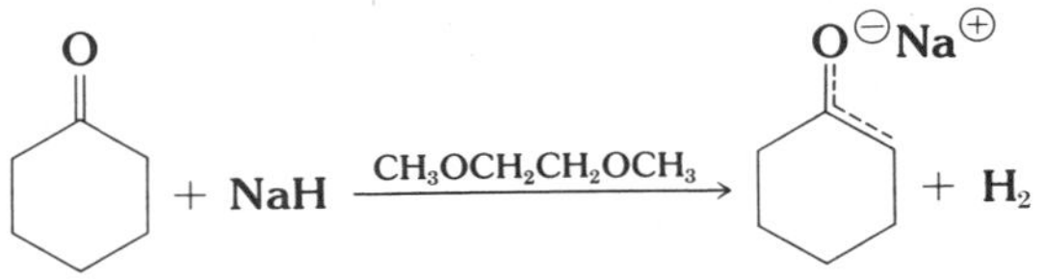

$$\text{cyclohexanone} + NaH \xrightarrow{CH_3OCH_2CH_2OCH_3} \text{sodium enolate} + H_2 \qquad (12.60)$$

The increased acidity of β-dicarbonyl compounds such as 2,4-pentanedione reflects the extended charge delocalization in the corresponding enolate anion (both carbonyl oxygen atoms help to support the negative charge):

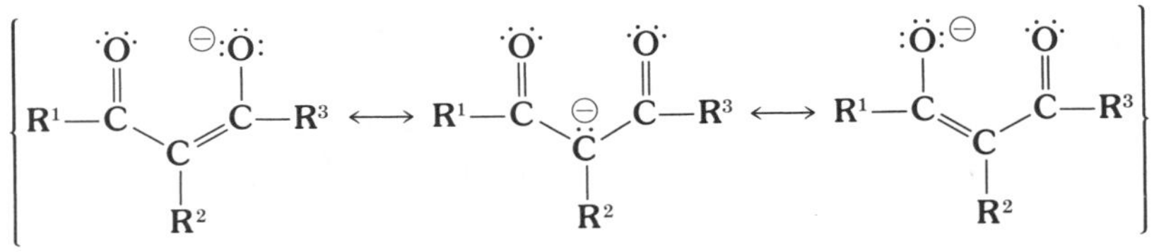

$$\left\{ R^1-\underset{\underset{R^2}{|}}{\overset{\overset{\ominus\ddot{O}:}{||}}{C}}-\underset{C}{\overset{:\ddot{O}:}{|}}-C-R^3 \longleftrightarrow R^1-\underset{\underset{R^2}{|}}{\overset{\overset{\cdot\ddot{O}\cdot}{||}}{C}}-\underset{C}{\overset{\cdot\ddot{O}\cdot}{\ominus}}-C-R^3 \longleftrightarrow R^1-\underset{\underset{R^2}{|}}{\overset{\overset{:\ddot{O}:^{\ominus}}{|}}{C}}=\underset{C}{\overset{\cdot\ddot{O}\cdot}{||}}-C-R^3 \right\}$$

Stable metal salts of these anions have been prepared with a variety of cations ($Na^{\oplus}$, $Tl^{\oplus}$, $Ni^{2\oplus}$, $Cu^{2\oplus}$, $Fe^{3\oplus}$, $Sc^{3\oplus}$, $Eu^{3\oplus}$, $Pr^{3\oplus}$). The metal cations in these salts are held between the two nucleophilic oxygen atoms

FIGURE 12·10 *Metal complexes of 2,4-pentanedione*

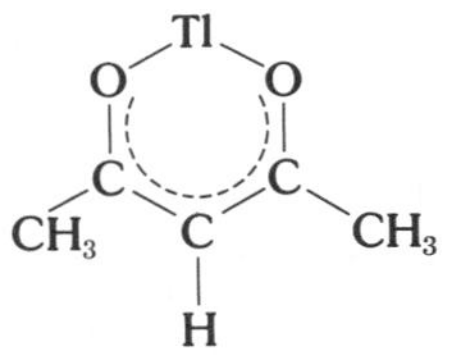

thalium complex

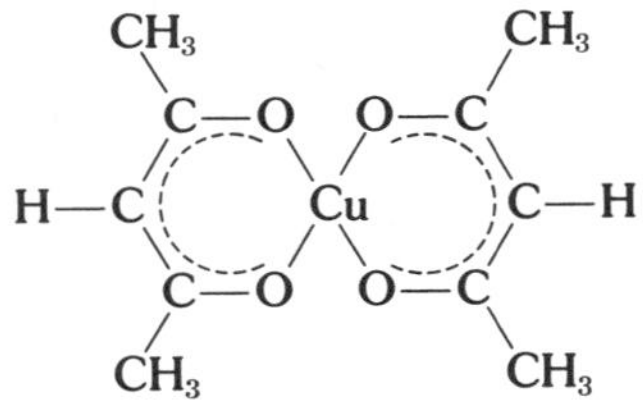

cupric complex

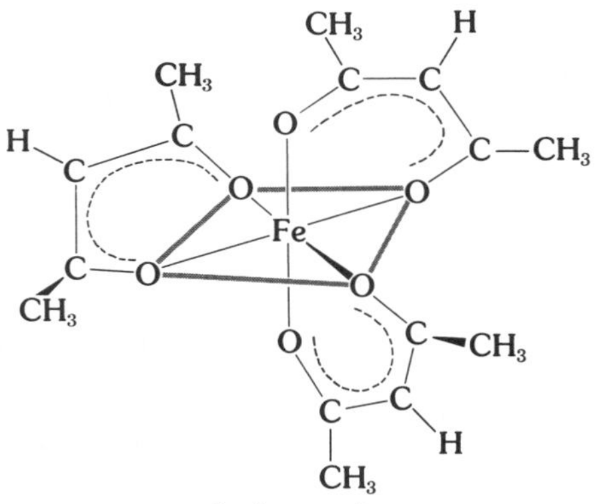

ferric complex

of the organic anion by a bonding process called *chelation* (from the Greek *chele*, meaning claw). Depending on the valence state and the size of the cation, the chelated complex may include one (*mono-*), two (*bis-*), three (*tris-*), or more of the β-dicarbonyl ligands. These possibilities are illustrated for the 2,4-pentanedione ligand by the thallium salt, the cupric complex (a beautiful sky-blue solid having a *bis*-square-planar configuration), and the octahedral ferric complex.

12·7 REACTIONS OF ENOLATE ANIONS

Enolate anions are highly reactive nucleophilic (and basic) intermediates which combine rapidly with a variety of electrophilic functional groups, including such weak electrophiles as alkyl halides and carbonyl compounds. Since the nucleophilic character of these intermediates is delocalized over two atoms (oxygen and carbon), it is not surprising to find reactions with electrophilic reagents at both sites:

$$\text{cyclohexanone enolate Na}^+ + (CH_3)_3SiCl \xrightarrow{THF} \text{1-(trimethylsilyloxy)cyclohexene} + NaCl \quad (12.61)$$

$$\text{cyclohexanone enolate Li}^+ + CH_3I \xrightarrow{THF, DMSO} \text{2-methylcyclohexanone} + LiI \quad (12.62)$$

The relative proportion of electrophilic attack at oxygen versus carbon in an enolate anion is influenced by many factors—the solvent, the metal cation, the nature of the electrophilic reagent, and so on. However, we shall simplify our present study of this important reaction class to a consideration of two general principles.

First of all, the negative charge of an enolate anion rests predominantly on oxygen, the atom of greater electronegativity. Therefore, in the absence of other factors, we would expect reactions to be faster at oxygen than at carbon. This is apparently true for very reactive electrophilic reagents such as $(CH_3)_3SiCl$ and $(C_2H_5)_3O^+BF_4^-$, which undergo highly exothermic and irreversible reactions via transition states resembling the enolate anion (the Hammond postulate). Proton transfer to the enolate anion is also about 10^{10} times faster at oxygen than at carbon. However, the reversibility of enol formation (see Figure 12.5) results in a final preponderance of the more stable keto tautomer—another example of competing kinetic and thermodynamic effects.

The other main consideration is the relative stability of the products. The combination of less reactive electrophilic reagents with enolate anions generally proceeds via an S_N2 transition state having sufficient productlike character that the relative product stabilities influence the activation energies. Since ketones and aldehydes are generally more stable than their enol derivatives (see Section 6.7), this factor almost

12·7 REACTIONS OF ENOLATE ANIONS

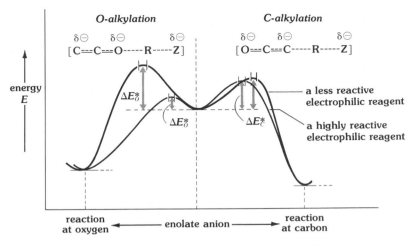

FIGURE 12·11 *Displacement reactions of a carbon electrophile with an enolate anion*

always favors reaction at carbon. Both these principles are illustrated by the energy diagram in Figure 12.11.

As a further illustration, recall from Chapter 10 that the enol-like structure of phenol is thermodynamically favored over its keto tautomer as a result of the resonance energy associated with the benzene ring. From our two principles above, we would predict that alkylation of phenolate anions will occur predominantly at oxygen, and experimental observations support this expectation:

$$\text{PhOH} \xrightarrow{\text{NaOH, H}_2\text{O, acetone}} \text{PhO}^{\ominus}\text{Na}^{\oplus} \xrightarrow{\text{CH}_2=\text{CHCH}_2\text{Br}} \phi-\text{O}-\text{CH}_2\text{CH}=\text{CH}_2 + \text{NaBr} \quad (12.63)$$

ALKYLATION REACTIONS

Since the alkylation of enolate anions proceeds almost exclusively and irreversibly at carbon, this reaction provides a useful general method for transforming simple carbonyl compounds into more complex structures which retain the versatile carbonyl function. From the following examples we observe that aryl and vinyl halides are generally unreactive as alkylating agents (equations 12.64 and 12.65), intramolecular alkylation reactions can take place with certain difunctional compounds (equations 12.68 and 12.69), and sulfonate esters can serve as well as or better than halogen in the role of a leaving group (equation 12.69). These features all support the S_N2 mechanism outlined in Figure 12.11 for reactions of this kind.

$$\text{(α-tetralone)} \xrightarrow{\text{NaH, THF}} \text{(enolate)} \xrightarrow{\text{CH}_2=\overset{\text{Cl}}{\text{C}}-\text{CH}_2\text{Cl}} \text{(α-alkylated product)} + \text{NaCl}$$

(12.64)

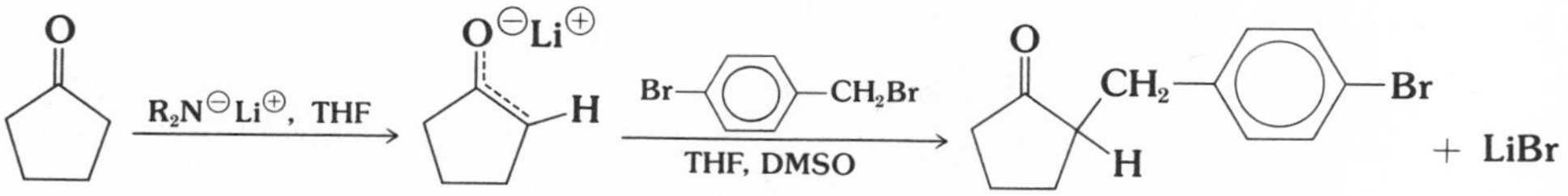

12·7 REACTIONS OF ENOLATE ANIONS

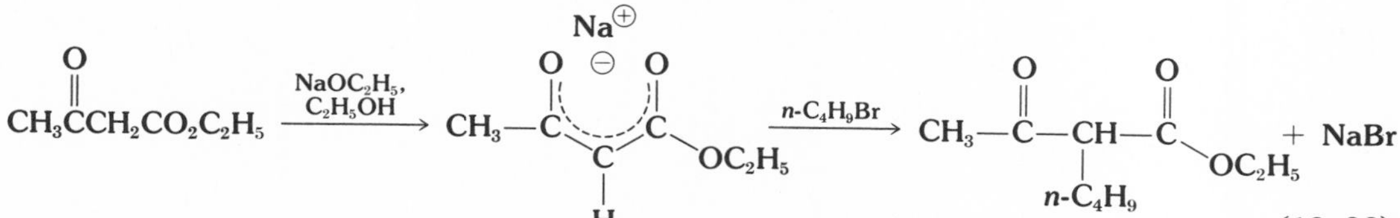

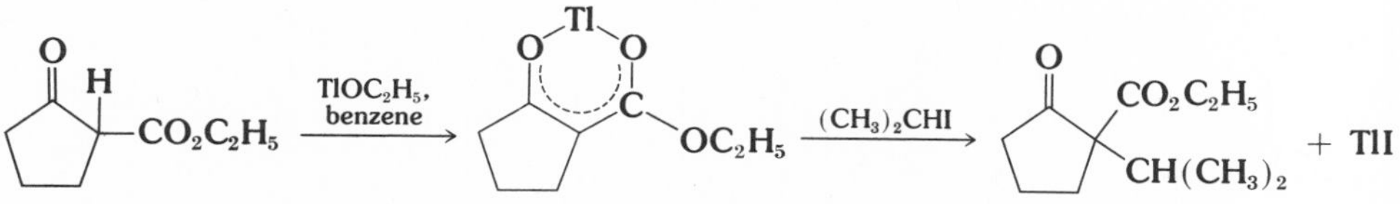

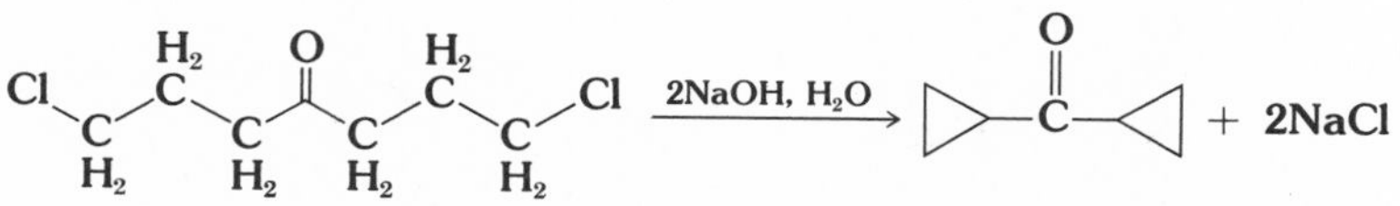

[Reaction 12.65: Cyclopentanone + R$_2$N$^-$Li$^+$, THF → lithium enolate + Br-C$_6$H$_4$-CH$_2$Br (THF, DMSO) → 2-(4-bromobenzyl)cyclopentanone + LiBr]

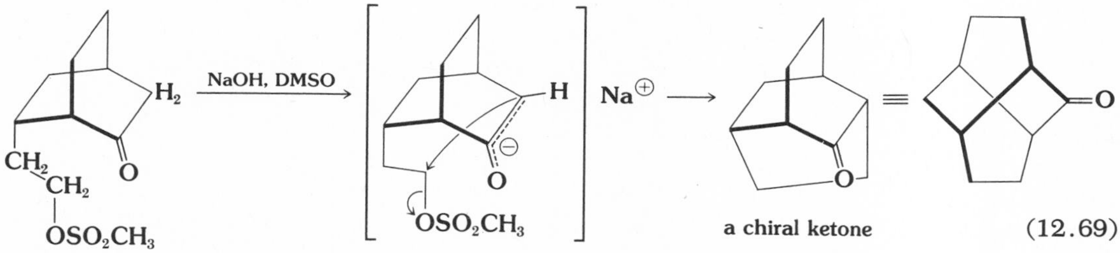

[Reaction 12.66: CH$_3$COCH$_2$CO$_2$C$_2$H$_5$ + NaOC$_2$H$_5$, C$_2$H$_5$OH → sodium enolate + n-C$_4$H$_9$Br → CH$_3$-CO-CH(n-C$_4$H$_9$)-CO-OC$_2$H$_5$ + NaBr]

[Reaction 12.67: 2-carbethoxycyclopentanone + TlOC$_2$H$_5$, benzene → thallium enolate + (CH$_3$)$_2$CHI → 2-carbethoxy-2-isopropylcyclopentanone + TlI]

[Reaction 12.68: Cl-CH$_2$CH$_2$CH$_2$-CO-CH$_2$CH$_2$CH$_2$-Cl + 2NaOH, H$_2$O → dicyclopropyl ketone + 2NaCl]

[Reaction 12.69: bicyclic mesylate ketone + NaOH, DMSO → intramolecular alkylation → a chiral ketone]

PROBLEM 12·22

a Write equations for the reactions expected if the product from equation 12.64 is treated with sodium hydride in THF, followed by methyl iodide.

b Reaction of the enolate anion derived from 2,4-pentanedione with *t*-butyl bromide does not give the anticipated C-alkylation product. The alkyl bromide is consumed, but most of the diketone is recovered unchanged. Suggest an explanation.

PROBLEM 12·23

The conjugate base of 2-hydroxynaphthalene reacts with methyl iodide as shown below. Write a mechanism showing all steps in the formation of the 1-methyl derivative and suggest a reason for alkylation at C-1 rather than C-3.

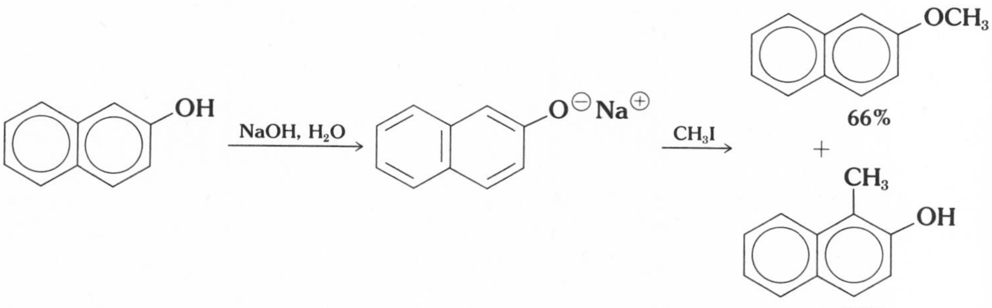

[Scheme: 2-naphthol + NaOH, H$_2$O → 2-naphthoxide Na$^+$ + CH$_3$I → 2-methoxynaphthalene (66%) + 1-methyl-2-naphthol (30%)]

12·7 REACTIONS OF ENOLATE ANIONS

PROBLEM 12·24
Dialkylation of ketones is sometimes a troublesome side reaction in the alkylation procedures described above. Suggest a mechanism for the formation of the dialkylated product in the following reaction:

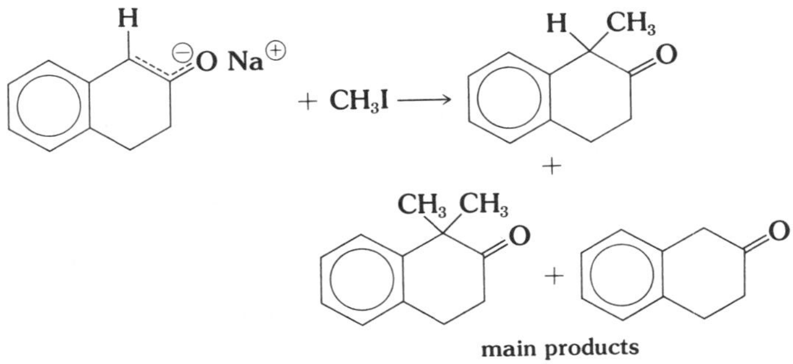

[structure: sodium enolate of α-tetralone] + CH₃I ⟶ [monomethylated α-tetralone]

+

[dimethylated α-tetralone (CH₃, CH₃)] + [α-tetralone]

main products

THE ALDOL CONDENSATION

Since the carbonyl group of an aldehyde or a ketone is polarized so that the carbon atom becomes electrophilic, we might expect to observe a bonding interaction when this function encounters a strongly nucleophilic enolate anion. The dimerization of acetaldehyde described in Figure 12.12 is an example of this important reaction class, commonly referred to as the *aldol condensation* (the 3-hydroxybutanal product was originally called aldol).

In contrast to the alkylation reactions of enolate anions, aldol condensations are reversible. Although this equilibrium normally favors the dimeric product from simple aldehydes, most ketones are not appreciably dimerized under similar conditions:

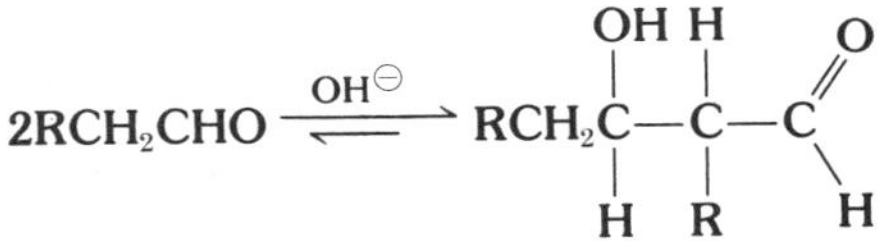

$$2RCH_2CHO \xrightleftharpoons{OH^\ominus} RCH_2\underset{\underset{H}{|}}{\overset{\overset{OH}{|}}{C}}-\underset{\underset{R}{|}}{\overset{\overset{H}{|}}{C}}-\overset{O}{\underset{H}{C}} \qquad (12.70)$$

$$2CH_3COCH_3 \xrightleftharpoons{OH^\ominus} CH_3-\underset{\underset{CH_3}{|}}{\overset{\overset{OH}{|}}{C}}-CH_2\overset{O}{\overset{\|}{C}}CH_3 \qquad (12.71)$$

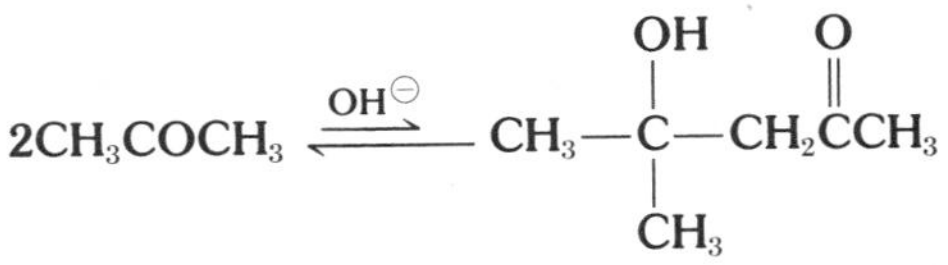

FIGURE 12·12 *The aldol condensation*

$$CH_3-\overset{O}{\overset{\|}{C}}\diagdown_H + CH_3-\overset{O}{\overset{\|}{C}}\diagdown_H \xrightleftharpoons{NaOH} CH_3-\underset{\underset{H}{|}}{\overset{\overset{OH}{|}}{C}}-CH_2-\overset{O}{\overset{\|}{C}}\diagdown_H$$

$$-H^\oplus \Updownarrow \qquad\qquad\qquad H_2O \parallel OH^\ominus$$

$$\left[CH_3-\overset{O:}{\overset{\|}{C}}\diagdown_H + CH_2=\overset{O^\ominus}{\overset{\|}{C}}\diagdown_H \rightleftharpoons CH_3-\underset{\underset{H}{|}}{\overset{:\ddot{O}:^\ominus}{C}}-CH_2-\overset{O}{\overset{\|}{C}}\diagdown_H \right]$$

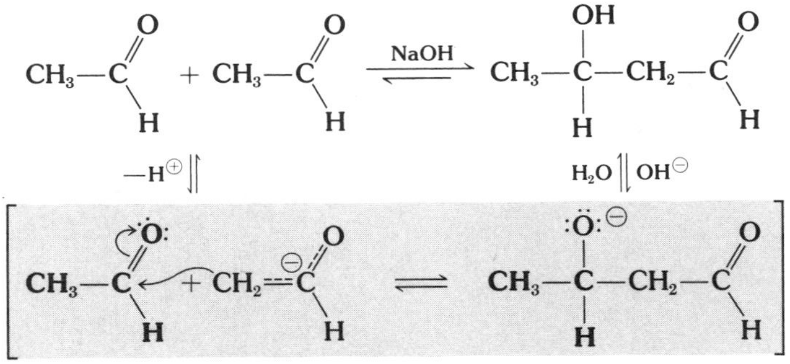

12·7 REACTIONS OF ENOLATE ANIONS

A combination of steric hindrance in the aldol dimer and the slightly increased carbonyl bond energy of ketones over that of aldehydes may account for the unfavorable equilibrium constants that are observed with ketones.

PROBLEM 12·25
a Use the average bond energies given in Table 2.4 to calculate the approximate heats of reaction (ΔH) for reactions 12.70 and 12.71.
b Since an enolate anion has two nucleophilic reactive sites, the reaction with a carbonyl compound could conceivably take the following course:

$$CH_3-\overset{O}{\underset{H}{C}}-H \;+\; \overset{\overset{\ominus}{O}}{\underset{H}{C}}=CH_2 \;\rightleftharpoons\; CH_3-\overset{OH}{\underset{H}{\underset{|}{C}}}-O-CH=CH_2$$

Calculate ΔH for this reaction and suggest an explanation for the absence of this product in reaction 12.70.

The β-hydroxyaldehydes formed by the aldehyde condensation in equation 12.70 are versatile intermediates that can be transformed into α,β-unsaturated aldehydes by acid- or base-catalyzed dehydration, into 1,3-diols by carbonyl reduction, and into an assortment of saturated and unsaturated aldehydes, acids, alcohols, and hydrocarbons, all derived from the same carbon skeleton. Indeed, we shall see in Chapter 14 that similar condensation reactions are widely used by living organisms to construct the complex molecules necessary for their survival.

$$RCH_2\underset{R}{\underset{|}{\overset{OH}{\overset{|}{C}}}}HCHCHO \xrightarrow[\Delta]{H^{\oplus} \text{ or } OH^{\ominus}} RCH_2CH=\underset{R}{\underset{|}{C}}-\overset{O}{\underset{H}{C}} \qquad (12.72)$$

mixture of stereoisomers

$$RCH_2\underset{R}{\underset{|}{\overset{OH}{\overset{|}{C}}}}HCHCHO \xrightarrow{NaBH_4} RCH_2\underset{R}{\underset{|}{\overset{OH}{\overset{|}{C}}}}HCHCH_2OH \qquad (12.73)$$

PROBLEM 12·26
Acid-catalyzed dehydration of alcohols is a well-known reaction, but simple alcohols are generally stable under treatment with bases. Suggest a reason for the facile dehydration of β-hydroxycarbonyl compounds by base catalysts.

PROBLEM 12·27
The synthetic insect repellent "6-12" is 2-ethyl-1,3-hexanediol. Suggest a means of preparing this compound from butanal.

If an aldol product undergoes a subsequent transformation, such as dehydration, during the course of the condensation reaction, an unfavorable equilibrium may be displaced in favor of bimolecular products. The condensation of acetophenone is an example of the dehydration route,

12·7 REACTIONS OF ENOLATE ANIONS

which becomes the rule when aryl ketones or aldehydes are used as reagents:

$$2\phi-\overset{O}{\underset{\|}{C}}-CH_3 \xrightarrow{NaOCH_3} \left[\phi-\overset{HO}{\underset{CH_3}{\overset{|}{C}}}-\overset{}{\underset{H}{CH}}-\overset{O}{\underset{\|}{C}}-\phi \atop \underset{\ominus}{\overset{\uparrow}{:\!O\!-\!CH_3}} \right] \longrightarrow \underset{CH_3}{\overset{\phi}{\underset{}{C}}}=CH\overset{O}{\underset{\|}{C}}\phi + OH^\ominus + CH_3OH \quad (12.74)$$

mixture of isomers

Apparently the resulting conjugation of the unsaturated functions provides a strong driving force for the elimination.

Crossed aldol condensations, in which the carbonyl precursor to the enolate anion (the donor reactant) is different from the electrophilic carbonyl reactant (the acceptor reactant), are useful only when at least one of the reactants is clearly disposed to function in only one capacity, as either donor or acceptor. For example, the condensation of benzaldehyde with cyclohexanone proceeds smoothly to the desired product because the aldehyde can serve only as an enolate acceptor (it has no α hydrogen atoms), while the cyclic ketone functions poorly in this role:

$$\underset{\text{donor}}{\text{cyclohexanone}} + \underset{\text{acceptor}}{\phi-\overset{O}{\underset{H}{\overset{\|}{C}}}} \xrightarrow{KOH, H_2O} \left[\text{intermediate} \right] \xrightarrow{-H_2O} \text{product} \quad (12.75)$$

A crossed condensation between propanal and butanal would not be effective, since each reactant can serve equally well in each role, and the result would be four different condensation products.

> **PROBLEM 12·28**
> Write structural formulas for and name the four possible condensation products from propanal and butanal. In each case indicate which aldehyde served as the donor and which as the acceptor.

There are several other facets of the versatile aldol condensation that deserve mention. For example, reactant molecules having more than one active site may undergo multiple condensations or cyclization:

$$CH_3COCH_3 + 2\phi CHO \xrightarrow{NaOH} \phi\text{-CH=CH-CO-CH=CH-}\phi + 2H_2O \quad (12.76)$$

$$\text{(cyclohexanone with CH}_2\text{-CH}_2\text{-C(=O)-CH}_2\text{ side chain)} \underset{}{\overset{R_2NH}{\rightleftharpoons}} \text{(bicyclic alcohol-ketone)} \xrightarrow{\Delta} \text{(enone)} + H_2O \quad (12.77)$$

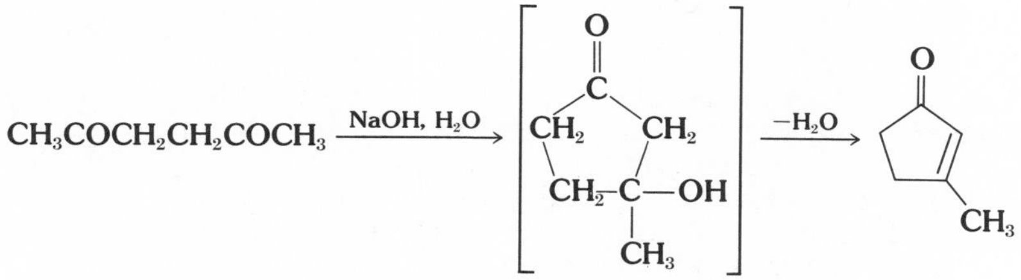

12·7 REACTIONS OF ENOLATE ANIONS

$$CH_3COCH_2CH_2COCH_3 \xrightarrow{NaOH, H_2O} \left[\begin{array}{c} \text{intermediate} \end{array} \right] \xrightarrow{-H_2O} \text{cyclopentenone with CH}_3 \tag{12.78}$$

Only five- or six-membered carbon rings appear to be formed in this manner. The angle strain in smaller rings causes an unfavorable aldol equilibrium, and intermolecular condensations predominate when the reactive sites in the reactant molecule are far apart.

Donor molecules are not restricted to aldehydes and ketones, the chief requisite being that a highly nucleophilic enol-like conjugate base be formed as an intermediate:

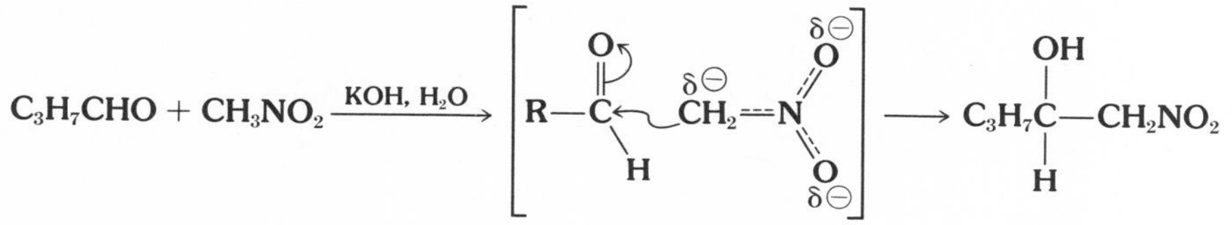

$$C_3H_7CHO + CH_3NO_2 \xrightarrow{KOH, H_2O} \left[R-\underset{H}{\overset{O^{\delta-}}{C}}\cdots CH_2=N\underset{O^{\delta-}}{\overset{O}{}} \right] \longrightarrow C_3H_7\underset{H}{\overset{OH}{C}}-CH_2NO_2 \tag{12.79}$$

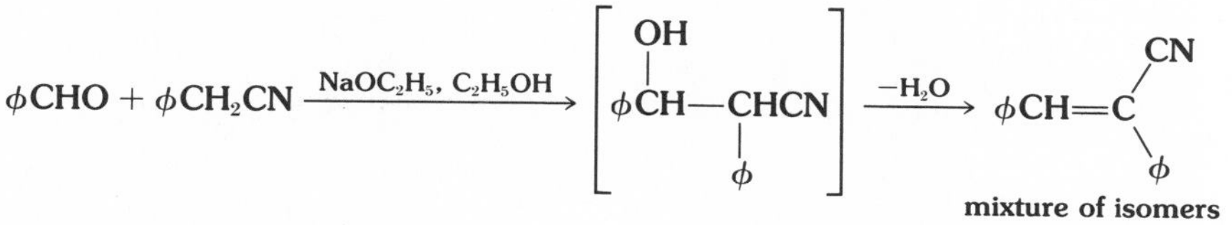

$$\phi CHO + \phi CH_2CN \xrightarrow{NaOC_2H_5,\ C_2H_5OH} \left[\phi\underset{}{\overset{OH}{CH}}-\underset{\phi}{CHCN} \right] \xrightarrow{-H_2O} \phi CH=C\underset{\phi}{\overset{CN}{\diagdown}} \tag{12.80}$$

mixture of isomers

Many aldol condensations may also be catalyzed by acids such as HF, HCl, H_2SO_4, and $AlCl_3$. The acid-catalyzed reaction cannot, of course, proceed via an enolate anion. Instead, the weakly nucleophilic enol tautomer of the donor reactant is attacked by the strongly electrophilic conjugate acid of the acceptor. An example of this mechanism is the crossed condensation of benzaldehyde with 2-octanone in Figure 12.14.

FIGURE 12·3 *Crossed aldol condensation of benzaldehyde with 2-octanone*

$$\phi CHO + CH_3(CH_2)_5COCH_3 \xrightarrow[\Delta]{HCl} \phi CH=C\underset{(CH_2)_4CH_3}{\overset{\overset{\cdot\cdot}{O}\cdot}{\underset{|}{C}-CH_3}}$$

the more stable enol tautomer

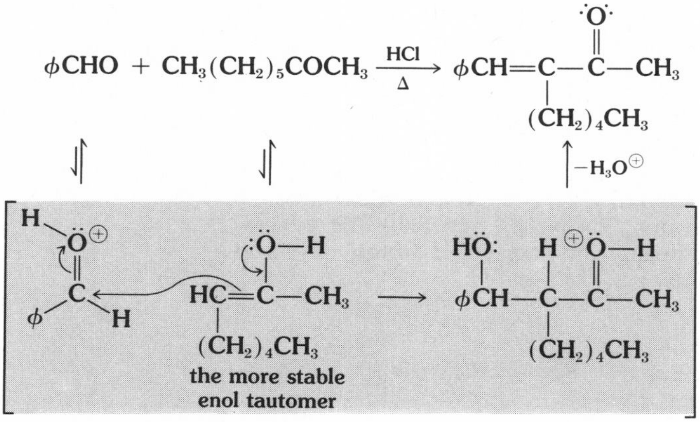

SUMMARY

the carbonyl group A functional group consisting of a carbon-oxygen double bond, found in aldehydes and ketones. Aldehydes (RCHO) are identified in IUPAC nomenclature by the ending *-al* and show a characteristic low-field pmr signal at δ-9.5. Ketones (RCOR′) are identified in IUPAC nomenclature by the ending *-one*. Characteristic infrared and ultraviolet absorptions due to the carbonyl group are given in Tables 12.2 and 12.3.

reactions of aldehydes and ketones Characteristic chemical transformations of the carbonyl group or its enol tautomer. The carbon-oxygen double bond is polarized, so that the carbon is electrophilic and the oxygen nucleophilic:

$$\{R_2C=O \longleftrightarrow R_2\overset{\oplus}{C}-\overset{..}{\underset{..}{O}}:^{\ominus}\}$$

reversible addition reactions The addition of polar reagents (H—Z) to the carbon-oxygen double bond to produce hemiacetals and hemiketals (Z=OR′), hydrates (Z=OH), cyanohydrins (Z=CN), and unstable imines (Z = HN$_2$ or HN$_3$):

$$R_2C=O + H-Z \rightleftharpoons R_2C\diagdown_{Z}^{OH}$$

irreversible addition reactions The addition of very strong nucleophilic reagents to the carbon-oxygen double bond, converting aldehydes and ketones to alkoxide salts:

$$R_2C=O + Z-M \longrightarrow R_2C\diagdown_{Z}^{O^{\ominus}M^{\oplus}}$$

Common reagents of this kind are metal hydrides (**LiAlH$_4$** or **NaBH$_4$**), diborane (**B$_2$H$_6$**), organolithium compounds (**R′—Li**), and Grignard reagents (**R′—MgX**). Hydrolysis of the salts formed in these reactions produces alcohols.

addition-elimination reactions A two-step process in which addition to the carbonyl group is followed by a subsequent elimination reaction. *Reversible addition* with elimination converts hemiacetals and hemiketals to acetals and ketals:

$$R_2C\diagdown_{OR'}^{OH} \xrightleftharpoons{H^{\oplus}, -H_2O} [R_2C=\overset{\oplus}{O}R'+] \xrightleftharpoons{R'OH, -H^{\oplus}} R_2C\diagdown_{OR'}^{OR'}$$

Reversible addition-elimination of ammonia derivatives (Y—ṄH$_2$) also produces stable imine forms such as oximes (Y = OH), semicarbazones (Y = NHCONH$_2$), and phenylhydrazones (Y = NHφ); see Table 12.4. *Irreversible addition* followed by elimination converts aldehydes and ketones to alkenes (the *Wittig reaction*):

$$R_2C=O + H_2\overset{..}{\overset{\ominus}{C}}\overset{\oplus}{-}P\phi_3 \longrightarrow R_2C\diagdown_{CH_2}^{O^{\ominus}}\overset{\oplus}{P\phi_3} \longrightarrow R_2C=CH_2 = \phi_3P$$

reductive deoxygenation Conversion of aldehydes and ketones to corresponding hydrocarbons. Methods include the *Clemmensen reduction* (**Zn-Hg** amalgam, $H_3O^{\oplus}$) *Wolff-Kishner reduction* (N_2H_4, Δ, strong base), thioketal desulfurization ($HSCH_2CH_2SH$, BF_3; Ni catalyst, H_2).

aldehyde oxidation Conversion of an aldehyde to a carboxylic acid by the action of oxidizing agents such as $KMnO_4$, H_2CrO_4, and $Ag(NH_3)_2^{\oplus}$ (Tollens reagent).

epimers Diastereoisomers differing in configuration at only one chiral site. The interconversion of epimers is called *epimerization*.

enolate anions Conjugate bases of aldehydes or ketones (or their enol tautomers). These reactive intermediates are stable in ether solution if protected from moisture and oxygen.

reactions of enolate anions *Alkylation* of enolate anions (at both oxygen and carbon) is irreversible and requires one equivalent of the base. The *aldol condensation* is a reversible reaction requiring catalytic amounts of the base:

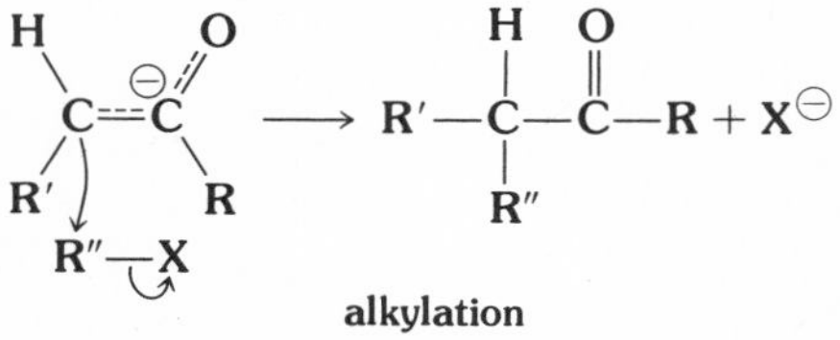

EXERCISES

12·1 Name the following compounds according to the IUPAC system of nomenclature:

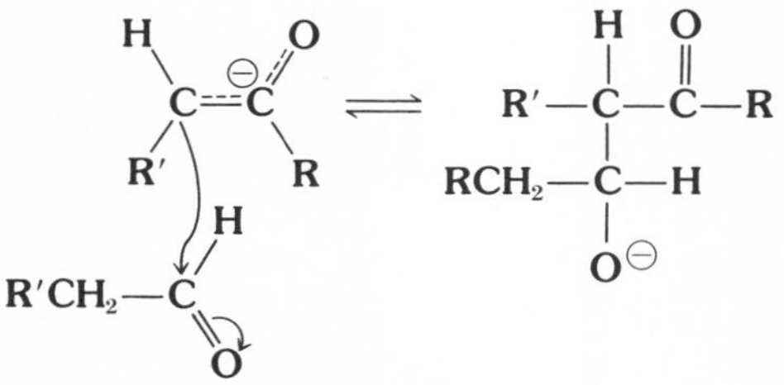

a CH_3CHO
b $CH_3CH(CH_3)CH_2CH_2CHO$
c CH_3COCH_3
d $CH_3COCH_2CH_3$
e   (4-chlorophenyl)-CHO
f   2,2-dimethylcyclohexanone (structure shown)
g $CH_3CH(CH_3)CH_2C(CH_3)_2COCH_3$
h $CH_3COCHCH_2CH_3$ with CHO substituent
i $CH_3CHClCHO$

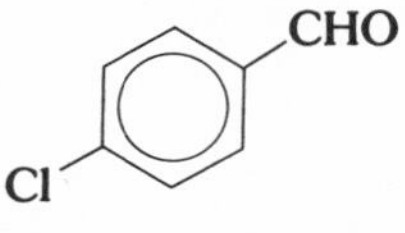

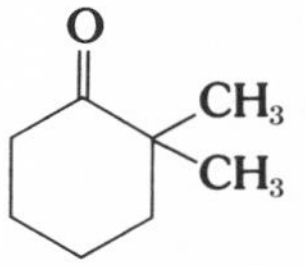

12·2 Give structural formulas, common names, and IUPAC names for the seven carbonyl compounds of molecular formula $C_5H_{10}O$ and the seven isomers of C_3H_6O.

12·3 Draw structural formulas for the following compounds:
a   Methyl isobutyl ketone
b   3-Methyl-2-pentanone
c   Acetophenone
d   Cinnamaldehyde
e   γ-Chlorocycloheptanone
f   p-Methoxybenzaldehyde
g   3-Methyl-2-cyclohexenone
h   2,2,4,4-Tetramethylcyclobutan-1,3-dione

400   EXERCISES

12·4 Write equations for the reactions (if any) of butanone with:

a   HCN
b   HOCH$_2$CH$_2$OH, H$^\oplus$
c   LiAlH$_4$, ether
d   Product of part c + H$_2$O
e   NaBH$_4$, H$_2$O
f   N$_2$H$_4$, KOH, Δ
g   KMnO$_4$
h   NaH, THF
i   Product of part h + CH$_3$I
j   NH$_2$NHφ
k   CH$_3$MgBr, ether
l   Product of part k + H$_2$O
m   2CH$_3$SH, BF$_3$, ether
n   Product of part m + H$_2$, Ni
o   Ag(NH$_3$)$_2$$^\oplus$OH$^\ominus$
p   NaOH, Δ
q   CH$_3$CH$_2$NH$_2$
r   Product of part q + H$_2$, Pd
s   CH$_3$C≡C:$^\ominus$Na$^\oplus$
t   φ$_3$P=CH$_2$
u   C$_6$H$_5$Li
v   Product of part u + H$_2$O
w   (CH$_3$)$_3$N:

12·5 Repeat Exercise 12.4 for butanal as the starting material.

12·6 From the following list of organic reactants and any additional solvents or inorganic reagents you may require, write equations to illustrate each of the reactions below:

reactants: 3-methylcyclohexanone, 4-methylcyclohexanone, p-bromobenzaldehyde, 3-methyl-2-butanone, semicarbazide, 2,4-dinitrophenylhydrazine

a   Reduction of a chiral ketone to an achiral hydrocarbon
b   Formation of a semicarbazone derivative of a cyclic ketone
c   Reduction of an achiral ketone to a chiral alcohol
d   A crossed aldol condensation
e   Conversion of an aldehyde into a 2,4-dinitrophenylhydrazone derivative

12·7 Arrange the following compounds in order of reactivity toward the addition of hydrogen cyanide:

acetaldehyde   di-*t*-butyl ketone   methyl-*t*-butyl ketone   acetone

12·8 Write equations for simple chemical tests that would easily distinguish the following pairs of compounds:

a   CH$_3$CH$_2$CHO + CH$_3$ĊCH$_3$ (O)

b   CH$_3$CH$_2$CH$_2$OH + CH$_3$ĊCH$_3$ (O)

c   CH$_3$CH$_2$ĊCH$_2$Br + CH$_3$ĊCH$_2$CH$_2$Br (O, O)

d   CH$_3$ĊCH$_3$ + CH$_3$CHCH$_3$ (O, OH)

12·9 Suggest a reason that the boiling points of butanal and 1-butanol are very different, whereas their water solubilities are quite similar.

	molecular weight	boiling point (°C)	water solubility (g/100 ml)
butanal	72	76	7.0
1-butanol	74	118	7.9

12·10 Arrange the following compounds in order of decreasing acidity:

H$_2$O   CH$_3$CH$_3$ CH$_3$CCH$_2$CCH$_3$ (O O)   NH$_3$ CH$_3$CH$_2$OH   CH$_3$COCH$_3$

12·11 Write structural formulas for compounds U to Z:

a $CH_3COCH_3 \xrightarrow{KOH} U(C_6H_{12}O_2)$

b $U + H^\oplus \xrightarrow{\Delta} V(C_6H_{10}O)$

c $U + LiAlH_4 \xrightarrow{ether} \xrightarrow{H_2O} W(C_6H_{14}O_2)$

d $V + NaBH_4 \xrightarrow{H_2O} X(C_6H_{12}O)$

e $V + H_2, Pt \longrightarrow Y(C_6H_{12}O)$

f $Y + NH_2NH_2 \xrightarrow{KOH} \xrightarrow{\Delta} Z(C_6H_{14})$

12·12 The following conversions can be accomplished in three or fewer steps. Indicate the intermediate products, reagents, and conditions necessary for each:

a $CH_3COCH_3 \longrightarrow CH_3CH=CH_2$

b $n\text{-}C_3H_7CH(OH)CH_3 \longrightarrow n\text{-}C_3H_7\overset{\overset{\displaystyle NOH}{\|}}{C}CH_3$

c $CH_3CH_2CHO \longrightarrow CH_3CH_2CH(OH)CH_3$

d $C_6H_5Br \longrightarrow C_6H_5CO_2H$

e cyclopentanol with H → cyclopentanol with C≡CH

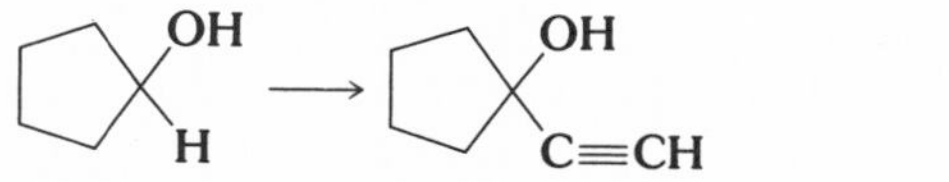

f $CH_3CH_2Br \longrightarrow CH_3CH_2CH_2CH_2OH$

g $CH_3CO(CH_2)_4COCH_3 \longrightarrow$ cyclopentene with COCH_3 and CH_3 substituents

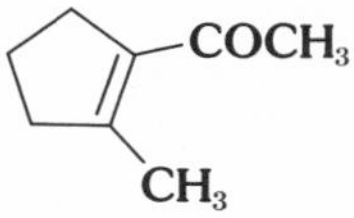

h $\phi CHO \longrightarrow \phi CH(OH)CO_2H$

i $CH_3CH_2CHO \longrightarrow CH_3CH_2CH=\overset{\overset{\displaystyle CH_3}{|}}{C}-CHO$

j $\phi COC_2H_5 \longrightarrow \phi-\overset{\overset{\displaystyle C_2H_5}{|}}{C}=CH_2$... wait

j $\phi COC_2H_5 \longrightarrow \phi-C(C_2H_5)=CH_2$

k $\phi-\overset{\overset{\displaystyle CH_3}{|}}{C}=CH_2 \longrightarrow \phi COCH_3$

l α-tetralone → 2,2-dimethyl-α-tetralone

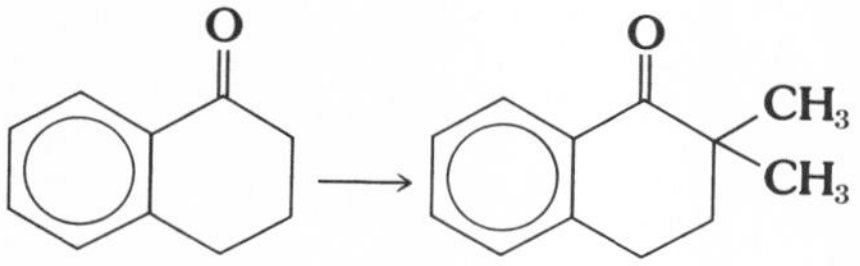

m $\phi CHO \longrightarrow \phi CH=CH-\overset{\overset{\displaystyle O}{\|}}{C}CH_3$

12·13 Using any reagents you wish, show how each of the following compounds could be prepared from cyclohexanone:

a cyclohexyl–OH

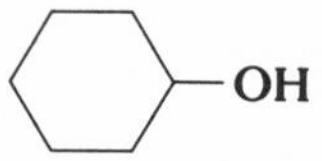

b cyclohexene

c cyclohexane

d 2-bromocyclohexanone

e 2-methylcyclohexanone

f methylenecyclohexane

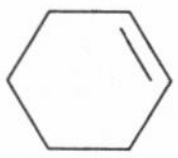

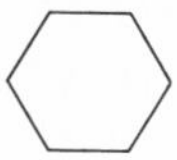

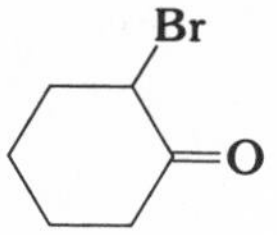

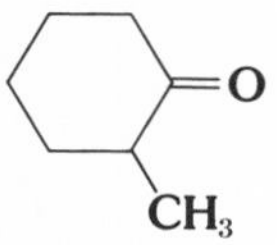

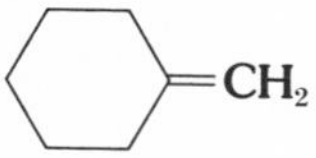

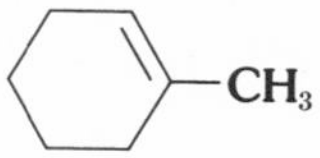

g 1-methyl-1-hydroxycyclohexane

h methylcyclohexane

i 2-methylenecyclohexanone

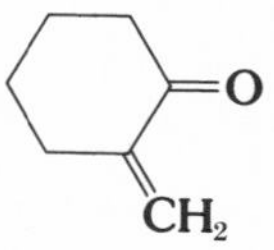

12·14 Show all steps in a plausible mechanism for the following reaction:

$HOCH_2CH_2CH_2CH_2CHO \xrightarrow{acid}$ 2-hydroxytetrahydropyran ($\pm$)

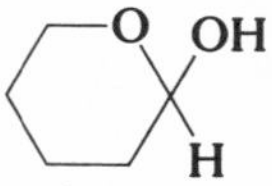

12·15 Chloral (Cl_3CCHO) adds methanol rapidly to give a hemiacetal, but it reacts very slowly in the presence of acid catalysts and excess methanol to give the corresponding acetal. Suggest a reason for this.

402   EXERCISES

12·16 Indicate reagents and conditions that would be suitable for the conversion of propiophenone ($C_6H_5COCH_2CH_3$) to the compounds shown. Assume the availability of any deuterated reagent you need.

a $\phi COCD_2CH_3$

b $\phi CDCH_2CH_3$
      |
      OH

c $\phi CH_2CD_2CH_3$

d $\phi CHCH_2CH_3$
      |
      OD

e $\phi CD_2CH_2CH_3$

f $\phi-\underset{\underset{CD_2CD_3}{|}}{\overset{\overset{OH}{|}}{C}}-CH_2CH_3$

12·17 Write structural formulas for the major products expected from the following aldol condensations:

a $CH_3CH_2CH_2CHO \xrightarrow{NaOH}$

b $\phi COCH_3 \xrightarrow{HCl}$

c [furan]—CHO + $CH_3COCH_3 \xrightarrow{NaOH}$

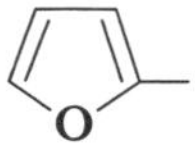

d $CH_3NO_2 + 3CH_2O \xrightarrow{Ca(OH)_2}$

e [cyclohexane with H (wedge) CH_2-CO- and H (dash) CH_2CHO substituents] $\xrightarrow{(C_2H_5)_2NH}$

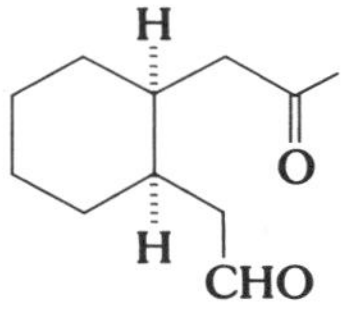

12·18 From the following list of organic reagents and any additional solvents or inorganic reagents you may require, write equations showing the preparation of each product below:

reactants: acetone, benzaldehyde, 2-methylpropanal, methyl iodide, ethyl bromide, bromobenzene

a 2-Phenyl-2-propanol
b 3-Methyl-2-butanol
c 1-Phenyl-1-propanol
d 2-Methyl-1-phenyl-1-propanol
e 2-Methylpropanoic acid, $(CH_3)_2CHCO_2H$

12·19 What prominent features of the spectroscopic method indicated could be used to readily distinguish the following pairs of compounds?

a CH_3CH_2CHO and CH_3COCH_3; infrared spectrum
b CH_3CH_2CHO and CH_3COCH_3; pmr spectrum
c $CH_3CH_2CH_2OH$ and CH_3CH_2CHO; pmr spectrum
d [cyclohexane=CH_2] and [cyclohexanone] ; pmr spectrum

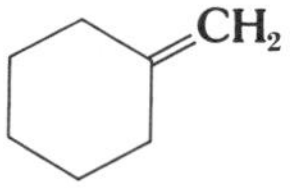

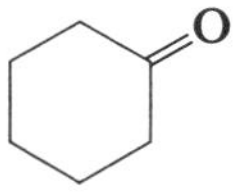

e ![cyclohexanone] and ![cyclohexanol] ; infrared spectrum

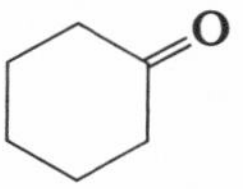

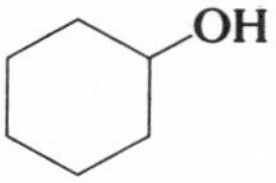

f $CH_3CH=CHCOCH_3$ and $CH_3CH_2CH_2COCH_3$; ultraviolet spectrum

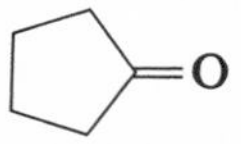

g ![cyclohexanone]=O and ![cyclopentanone]=O ; infrared spectrum

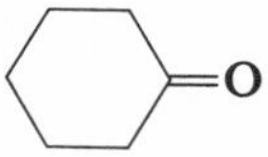

12·21 Give a reasonable explanation for the difference in the course of the aldol condensations shown here. In both cases a mixture of cis-trans isomers is formed.

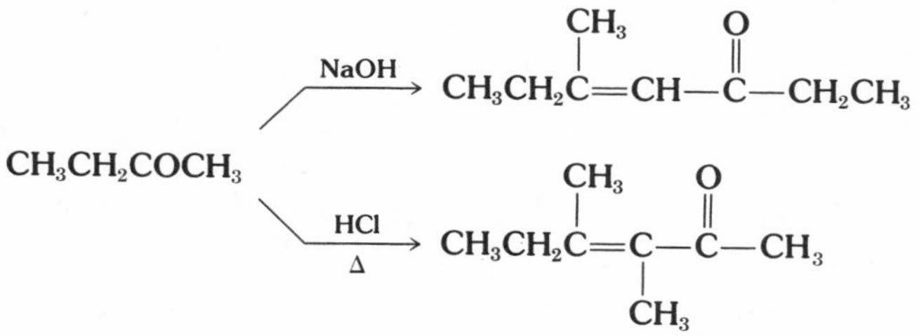

$$CH_3CH_2COCH_3 \xrightarrow{NaOH} CH_3CH_2\underset{CH_3}{\overset{|}{C}}=CH-\overset{O}{\underset{\|}{C}}-CH_2CH_3$$

$$CH_3CH_2COCH_3 \xrightarrow[\Delta]{HCl} CH_3CH_2\underset{CH_3}{\overset{CH_3}{\overset{|}{C}}}=\underset{CH_3}{\overset{|}{C}}-\overset{O}{\underset{\|}{C}}-CH_3$$

12·21 The oral contraceptive Enovid can be prepared by the following sequence of reactions. Indicate the reagents that could be used in each step. Why was the ketone converted to a ketal during this sequence?

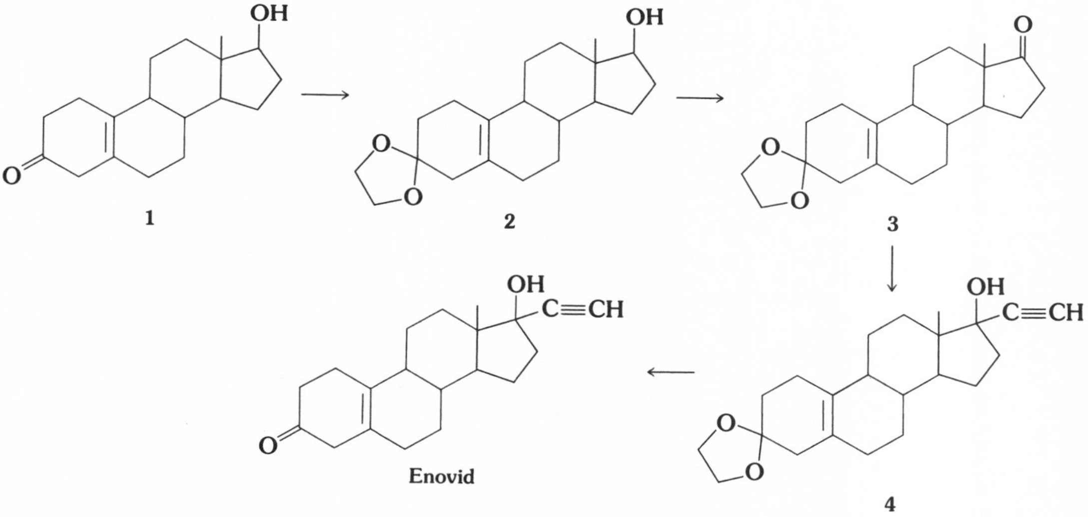

[steroid structures 1, 2, 3, 4, and Enovid]

12·22 An organic chemist attempted to add one molar equivalent of methyl magnesium iodide (anhydrous ether solution) to 1,3-pentanedione. His plan was to dehydrate the resulting $C_6H_{12}O_2$ ketol to an unsaturated ketone. During the Grignard reaction, our chemist noticed an evolution of gas, and when he worked up the reaction he recovered all his starting material (the diketone).
a Write a structural formula for the $C_6H_{12}O_2$ ketol that was the target of the Grignard reaction.
b Write an equation for the reaction that actually took place, explaining the gas evolution and the recovery of starting material.
c How could the intended $C_6H_{12}O_2$ ketol be prepared in one simple step?

12·23 The reaction (if any) of amines with aldehydes and ketones usually gives imines or enamines as the major product. Explain the course of the following reactions of amines with cyclohexanone:

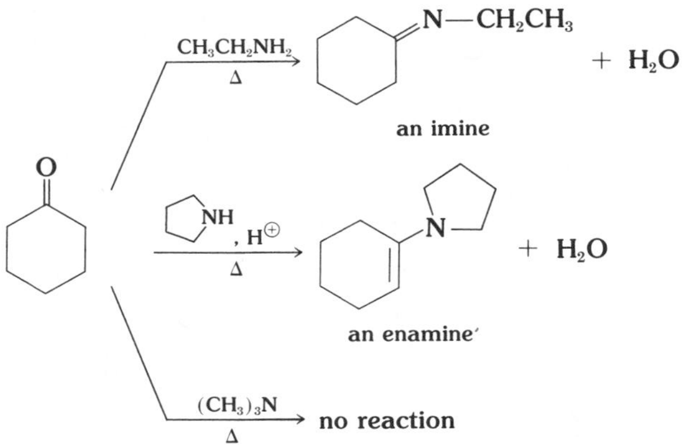

cyclohexanone + $CH_3CH_2NH_2$, Δ → N-cyclohexylidene ethylamine (an imine) + H_2O

cyclohexanone + pyrrolidine, $H^{\oplus}$, Δ → 1-(1-pyrrolidinyl)cyclohexene (an enamine) + H_2O

cyclohexanone + $(CH_3)_3N$, Δ → no reaction

12·24 Which of the conditions suggested for each of the following transformations would be most effective in bringing about the desired change?

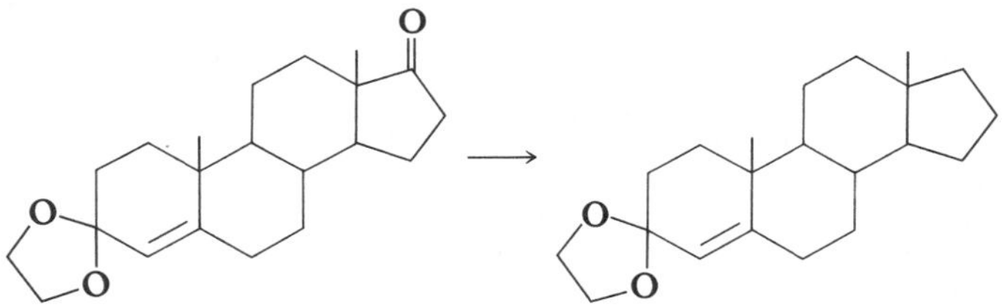

a cyclopentanone → sodium cyclopentenolate   (1) NaH, THF, Δ or (2) NaOH, H_2O, Δ?

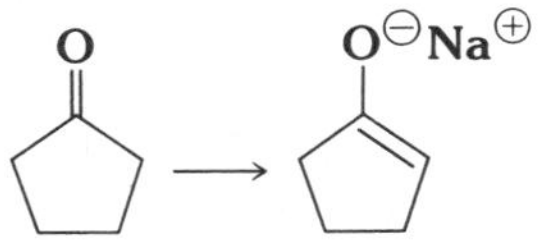

b (steroid with ketone and dioxolane) → (steroid with CH₂ and dioxolane)   (1) N_2H_4, KOH, $\begin{array}{c}\text{—OH}\\\text{—OH}\end{array}$, Δ or (2) Zn(Hg), HCl, toluene, Δ?

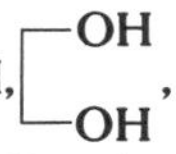

c $CH_3CH_2CH_2CHO \longrightarrow CH_3CH_2CH_2CH(OH)CN$ (1) KCN, H_2O or (2) HCN, H_2O?

d (decalone) → (decalone with =CHC₆H₅)   (1) $C_6H_5CH=P(C_6H_5)_3$ or (2) C_6H_5CHO, NaOH?

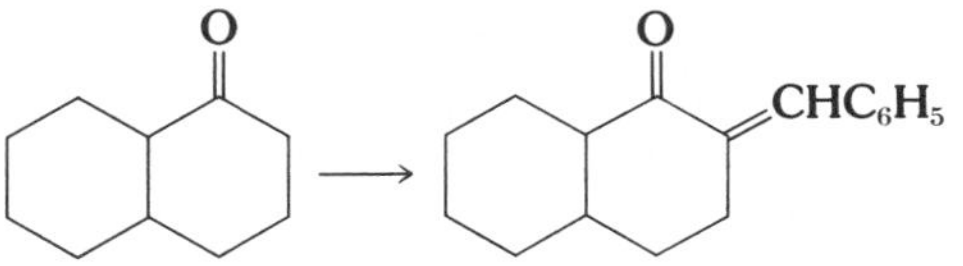

12·25 Structures I and II are contributors to the resonance-hybrid description of the conjugate acid from cyclohexanone (see Section 12.3).

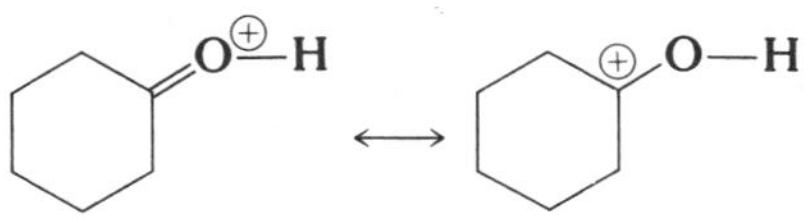

  I: cyclohexanone with $O\overset{\oplus}{=}H$ II: cyclohexanone with $\overset{\oplus}{C}$—O—H

a According to the octet rule, which structure would you expect to make the greater contribution to the hybrid?
b On the basis of electronegativity, which structure would you expect to make the greater contribution to the hybrid?
c Assuming equal contributions of I and II, at which positively charged site would attack by a nucleophile be favored? Why?

12·26 Explain the large difference in the acidity of these ketones:

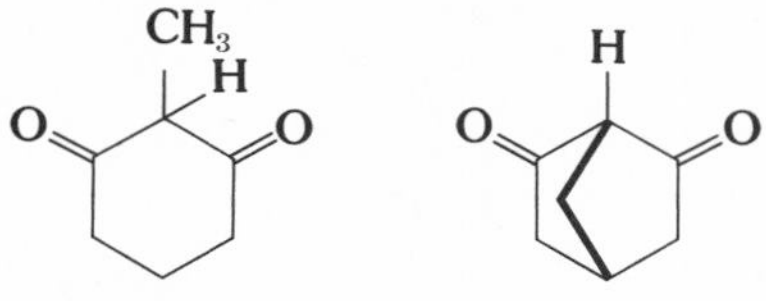

2-methyl-2H-1,3-cyclohexanedione ($K_a \approx 10^{-10}$) vs. bicyclic 1,3-diketone ($K_a \approx 10^{-20}$)

12·27 Write equations showing all the steps by means of which ammonia and an excess of benzaldehyde react in the presence of hydrogen and a palladium catalyst to give tribenzylamine, $(C_6H_5CH_2)_3N$.

12·28 Cyanohydrin formation from aldehydes and ketones is reversible, whereas Grignard addition reactions to carbonyl functions are irreversible. What are some of the relevant factors in this difference?

12·29 The following reactions represent proposed synthetic transformations. Consider each one and indicate whether you believe the reaction will proceed as written. Give reasons. If you believe side reactions will occur, write equations for each.

a $CH_3CHO + C_6H_5CH_2CHO \xrightarrow{NaOH} C_6H_5CH_2CH(OH)CH_2CHO$

b $CH_3CH(OCH_3)_2 + 2NaOC_2H_5 \xrightarrow[\Delta]{C_2H_5OH} CH_3CH(OC_2H_5)_2 + 2NaOCH_3$

c $CH_3CO(CH_2)_2COCH_3 + 2C_6H_5CHO \xrightarrow{NaOH} (C_6H_5CH=CHCOCH_2)_2$

d trans-1,2-cyclohexanediol $+ CH_3COCH_3 \xrightarrow[\Delta]{H^{\oplus}}$ cyclohexane-fused acetonide $+ H_2O$

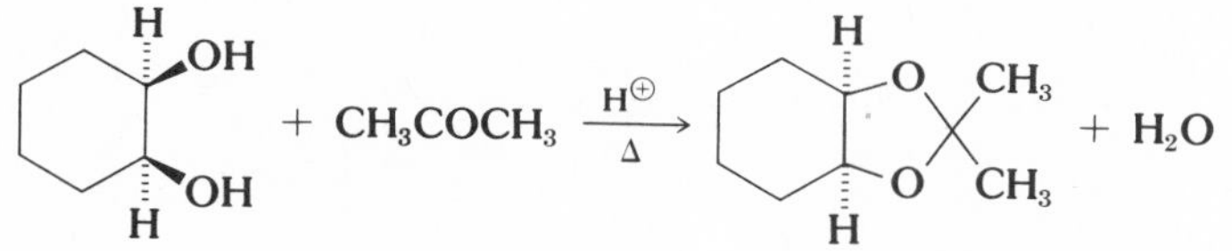

e 3-cyclohexen-1-ol $+ CrO_3 \xrightarrow{CH_3CO_2H, H_3O^{\oplus}}$ 2-cyclohexen-1-one

f $(CH_3)_2C\overset{O}{\underset{}{-}}CH_2 + CH_3Li \xrightarrow{pentane} (CH_3)_3CCH_2OH$
   (after hydrolysis)

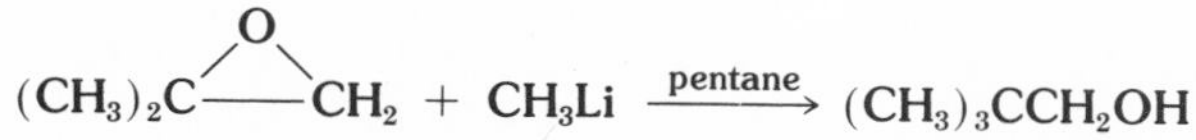

12·30 A pair of optically active isomeric alcohols A and B with the molecular formula $C_5H_{10}O$ give the same optically active ketone E on oxidation with a chromic trioxide–pyridine complex. E is rapidly racemized on treatment with base and gives a d_3-labeled analog when warmed with heavy water (D_2O) containing NaOD. Two other $C_5H_{10}O$ alcohols, C and D, which prove to be achiral, both give the achiral ketone F on oxidation. Base-catalyzed exchange of F in heavy water yields a d_4-labeled analog. Both E and F show strong absorption at 1775 cm^{-1} in the infrared spectrum and give the same C_5H_{10} hydrocarbon G on Wolff-Kishner reduction.

Write structures for A, B, C, D, E, F, and G (note that configuration assignments for A, B, C, and D will be arbitrary) and write equations for all the reactions mentioned above.

12·31 Five isomeric $C_9H_{10}O$ compounds all give crystalline 2,4-dinitrophenylhydrazone derivatives. With the aid of the following spectroscopic data, propose a reasonable structure for each compound:

a Ultraviolet absorption at 256 nm (strong, $\epsilon > 10^4$); infrared absorption at 1700 cm^{-1}; pmr signals at $\delta = 2.60$ (3H, singlet), 2.72 (3H, singlet), 7.70 (2H, doublet, $J = 8.5$ Hz) and 8.1 (2H, doublet, $J = 8.5$ Hz)

b Ultraviolet absorptions at 254 nm (moderate) and 280 nm (very weak); infrared absorption at 1720 cm^{-1}; pmr signals at $\delta = 2.0$ (3H, singlet), 3.5 (2H, singlet), and 7.2 (5H, singlet)

c Ultraviolet absorptions at 274 nm (moderate) and 290 nm (very weak); infrared absorptions at 1730, 2700, and 2810 cm^{-1}; pmr signals at $\delta = 2.7$ (3H, singlet), 3.5 (2H, doublet, $J = 1.5$ Hz), 7.1 (4H, singlet), and 9.8 (1H, doublet, $J = 1.5$ Hz)

d Ultraviolet absorption at 243 (strong, $\epsilon > 10^4$); infrared absorption at 1700 cm^{-1}; pmr signals at $\delta = 1.3$ (3H, triplet, $J = 7$ Hz), 2.8 (2H, quartet, $J = 7$ Hz), 7.35 (3H, multiplet), and 7.85 (2H, multiplet)

e Ultraviolet absorption at 255 (strong, $\epsilon > 10^4$); infrared absorptions at 1705, 2730, 2870 cm^{-1}; pmr signals at $\delta = 1.3$ (3H, triplet, $J = 7$ Hz), 3.05 (2H, quartet, $J = 7$ Hz), 7.70 (2H, doublet, $J = 8.5$ Hz), 8.1 (2H, doublet, $J = 8.5$ Hz), and 10.3 (1H, singlet)

12·32 An unknown substance X has a molecular ion at 116 mass/charge units in its mass spectrum, is transparent from 200 to 800 nm in the ultraviolet-visible spectrum, and shows no hydroxyl or carbonyl absorptions in the infrared spectrum. The pmr spectrum of X is shown below. Hydrolysis of X in dilute aqueous sulfuric acid yields two organic compounds, Y and Z, having the following properties:

1 Compound Y has a molecular ion at 72 units in the mass spectrum; ultraviolet absorptions at 2850–3010 cm^{-1} (strong), 1320–1460 cm^{-1}, and 1175 cm^{-1}; and pmr signals at $\delta = 1.0$ (3H, triplet, $J = 7$ Hz), 2.1 (3H, singlet), and 2.45 (2H, quartet, $J = 7$ Hz).

2 Compound Z has a molecular ion at 62 units in the mass spectrum; ultraviolet absorptions at 3100–3600 cm^{-1} (strong, broad), 2850–2950 cm^{-1}, and 1000–1100 cm^{-1} (strong); and pmr signals for pure liquid at $\delta = 3.7$ (singlet) and $\delta = 4.7$ (singlet), with an area ratio of 2:1.

Write structures for compounds X, Y, and Z.

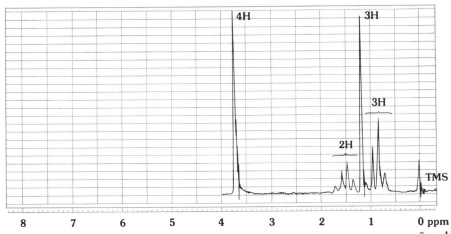

CARBOXYLIC ACIDS AND THEIR DERIVATIVES

The carboxylic acids are a widely distributed and important class of compounds distinguished by the *carboxyl functional group* CO_2H:

$$\underset{1.36\ \text{Å}}{\overset{1.23\ \text{Å}}{}} \begin{array}{c} \ddot{\text{O}}: \\ \diagup \\ -\text{C} \quad 120° \\ \diagdown \\ :\ddot{\text{O}}-\text{H} \end{array}$$

Although this function appears to be a simple combination of a carbonyl group (C=O) and a hydroxyl group (OH), the interaction between them generates some unique properties, such as an acidity more than 10^{11} times that of the hydroxyl group in an alcohol.

13·1 NOMENCLATURE OF CARBOXYLIC ACIDS

Many of the most common carboxylic acids were first obtained from natural sources, particularly from fats and oils. Consequently, the normal (unbranched-chain) acids are called *fatty acids*. The common names of some of these acids are given in Table 13.1. Those having an odd number of carbon atoms greater than 9 are not included because they are not naturally abundant and therefore never acquired common names.

The IUPAC nomenclature system can be applied to branched-chain (and normal) acids by substituting *-oic acid* for the final *-e* in the name of the alkane corresponding to the longest chain incorporating the carboxyl group.

$CH_3CH_2CH_2CO_2H$
 butanoic acid

13·1 NOMENCLATURE OF CARBOXYLIC ACIDS

TABLE 13·1 *Common names of some normal carboxylic acids,* $H(CH_2)_{n-1}CO_2H$

number of carbon atoms (n)	name	derivation of name	boiling point (°C)	melting point (°C)
1	formic acid	Latin *formica*, ant	100.7	8.4
2	acetic acid	Latin *acetum*, vinegar	118.2	16.6
3	propionic acid	Greek *proto*, first + *pion*, fat	141.4	−20.8
4	butyric acid	Latin *butyrum*, butter	164.1	−5.5
5	valeric acid	Latin *valere*, valerian root	186.4	−34.5
6	caproic acid	Latin *caper*, goat	205.4	−3.9
7	enanthic acid	Greek *oenanthe*, vine blossom	223.0	−7.5
8	caprylic acid	Latin *caper*, goat	239.3	16.3
9	pelargonic acid	pelargonium	253.0	12.0
10	capric acid	Latin *caper*, goat	218.7	31.3
12	lauric acid	laurel	−	43.4
14	myristic acid	myristica (nutmeg)	−	54.4
16	palmitic acid	palm oil	−	62.8
18	stearic acid	Greek *stear*, tallow or suet	−	69.6
20	arachidic acid	arachis (peanut)	−	75.4

The carboxyl function must always lie at the end of a carbon chain, and the numbering of the longest such chain begins at that point:

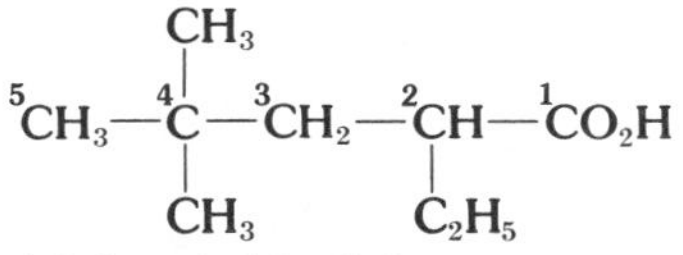

$$^5CH_3-\overset{CH_3}{\underset{CH_3}{\overset{|}{\underset{|}{C}}}}-^3CH_2-\overset{2}{\underset{C_2H_5}{\overset{|}{CH}}}-^1CO_2H$$

4,4-dimethyl-2-ethylpentanoic acid

Alternatively, the atoms of a fatty-acid chain may be designated by the Greek-letter notation used with aldehydes and ketones (see Section 12.1):

$\overset{\gamma}{CH_3}\overset{\beta}{CH(OH)}\overset{\alpha}{CH(NH_2)}CO_2H$
α-amino-β-hydroxybutyric acid
or threonine

In naming cyclic compounds, it is sometimes advantageous to consider the carboxyl function as a substituent:

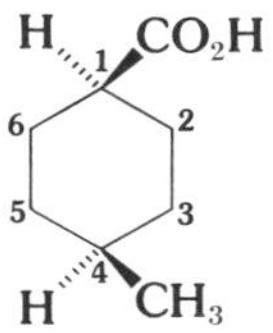

cis-4-methylcyclohexanecarboxylic acid 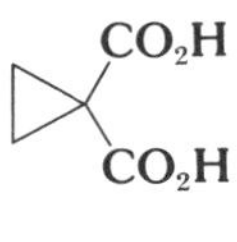   1,1-cyclopropanedicarboxylic acid

Aromatic carboxylic acids having a variety of ring substituents are known. Some of these are customarily referred to by common names:

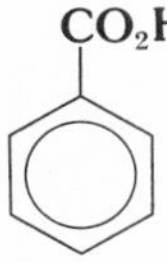

benzoic acid    p-fluorobenzoic acid 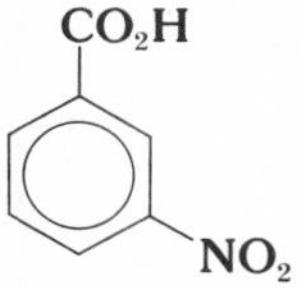   m-nitrobenzoic acid

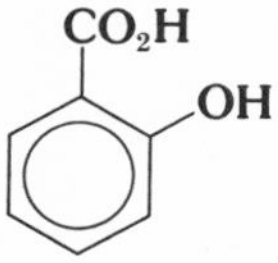

o-hydroxybenzoic acid   o-aminobenzoic acid   3,4,5-trihydroxybenzoic acid
or salicylic acid   or anthranilic acid   or gallic acid

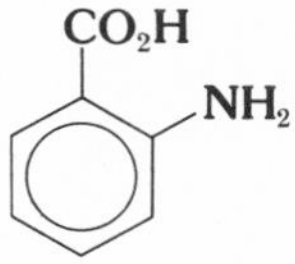

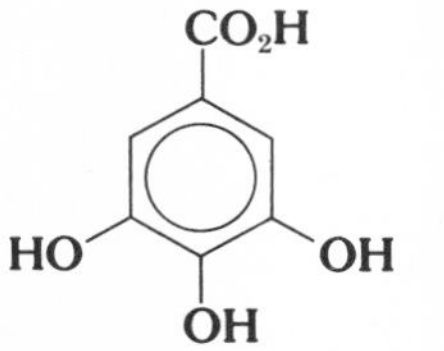

Carboxylic acids incorporating other functional groups, such as double bonds, hydroxyl, amino, and carbonyl groups, are often referred to by common (nonsystematic) names:

$H_2C=CHCO_2H$
propenoic acid
or acrylic acid

$\overset{4}{C}H_3\overset{3}{C}H=\overset{2}{C}H\overset{1}{C}O_2H$
2-butenoic acid
or crotonic acid
(may be cis or trans)

$CH_3(CH_2)_7CH=CH(CH_2)_7CO_2H$
oleic acid
(the cis isomer is derived from olive oil)

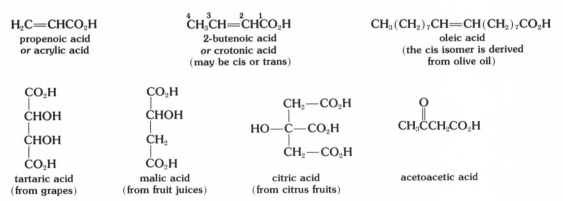

tartaric acid
(from grapes)

malic acid
(from fruit juices)

citric acid
(from citrus fruits)

acetoacetic acid

FCH_2CO_2H
fluoroacetic acid
(from *Dichapetalum cymorum*, one of the most poisonous plants in South Africa)

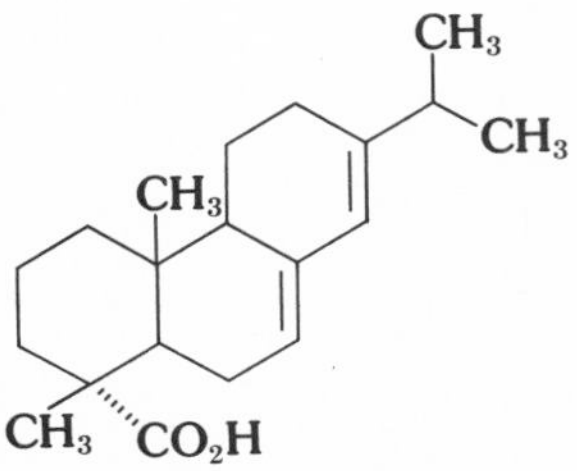

abietic acid
(from pine resin)

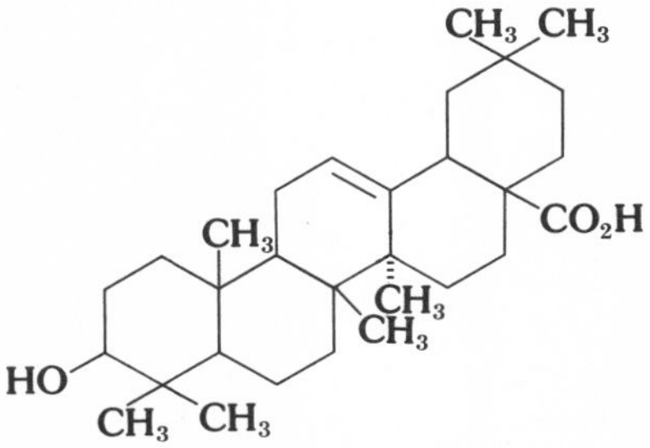

oleanolic acid
(from olive leaves, mistletoe, cloves, and grape skins)

$CH_3CH(OH)CO_2H$
lactic acid
(from sour milk)

$CH_3NHCH_2CO_2H$
sarcosine
(from muscle tissue)

$HC\equiv CCO_2H$
propynoic acid
or propiolic acid

13·1 NOMENCLATURE OF CARBOXYLIC ACIDS

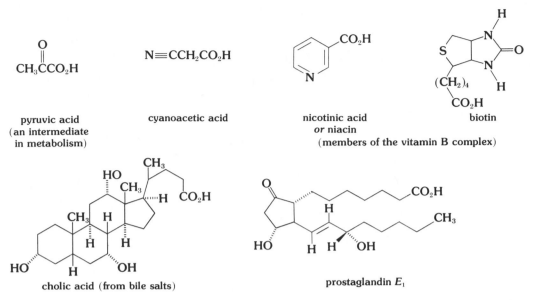

pyruvic acid (an intermediate in metabolism)

cyanoacetic acid

nicotinic acid *or* niacin (members of the vitamin B complex)

biotin

cholic acid (from bile salts)

prostaglandin E_1

The simple dicarboxylic acids are usually referred to by common names:

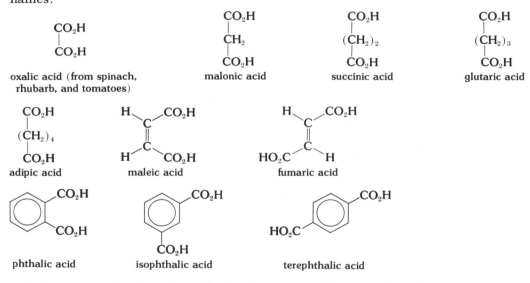

oxalic acid (from spinach, rhubarb, and tomatoes)

malonic acid

succinic acid

glutaric acid

adipic acid

maleic acid

fumaric acid

phthalic acid

isophthalic acid

terephthalic acid

In the names of carboxylic acid salts, the name of the cationic moiety is followed by a name designating the carboxylate anion (the -*ic* suffix is replaced by -*ate*):

ϕ—$CO_2^{\ominus}$ $Na^{\oplus}$
sodium benzoate

$CH_3CO_2^{\ominus}$ $NH_4^{\oplus}$
ammonium acetate

$Ca^{\oplus}$ $C_2O_4^{\ominus}$
calcium oxalate

Compounds having anionic and cationic sites in the same molecule are called *internal salts* or *zwitterions*. Beets, for example, contain an internal quaternary ammonium salt known as betaine:

$(CH_3)_3N^{\oplus}$—$CH_2CO_2^{\ominus}$
betaine
or N,N,N-trimethylglycine

13·2 SOURCES OF CARBOXYLIC ACIDS AND THEIR DERIVATIVES

The diazonium benzoate derived from anthranilic acid is of interest because it smoothly decomposes to benzyne with the evolution of carbon dioxide and nitrogen:

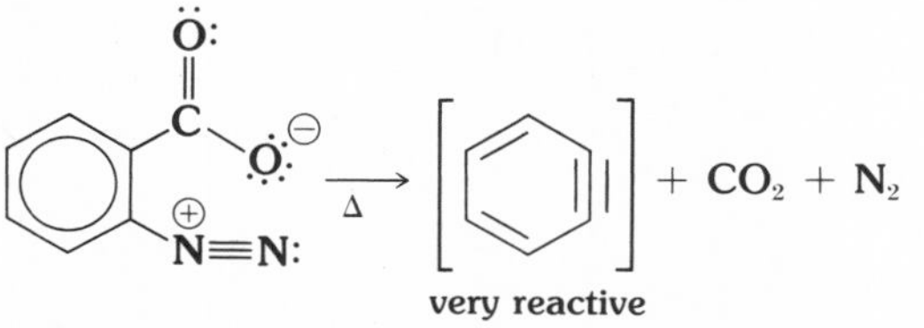

$$\text{(anthranilic acid diazonium benzoate)} \xrightarrow{\Delta} [\text{benzyne}] + CO_2 + N_2 \quad (13.1)$$

very reactive

METHODS OF PREPARING CARBOXYLIC ACIDS

Many methods of preparing carboxylic acids in the laboratory have been discussed in previous chapters. One procedure, as we saw in Section 8.7, is the oxidation of primary alcohols. Oxidizing reagents such as dichromate (Cr^{+6}) and permanganate (Mn^{+7}) are particularly effective in converting primary alcohols to carboxylic acids:

$$R-CH_2-OH + 2[O] \longrightarrow R-C(=O)O-H + H_2O \quad (13.2)$$

$$3RCH_2OH + 2Na_2Cr_2O_7 + 8H_2SO_4 \longrightarrow 3RCO_2H + 2Cr_2(SO_4)_3 + 2Na_2SO_4 + 11H_2O \quad (13.3)$$

Ordinarily we do not balance reduction-oxidation reactions of this kind, but simple use the symbol [O] to represent chemical oxidizing agents.

Aldehydes are readily oxidized to carboxylic acids (Section 12.5) and are probably intermediates in the oxidations of primary alcohols discussed above:

$$R-C(=O)H + [O] \longrightarrow R-C(=O)O-H \quad (13.4)$$

Although Cr^{+6} and Mn^{+7} reagents serve to effect this oxidation, it can also be carried out with milder oxidizing agents such as Ag^{+1}:

$$R-C(=O)H + 2Ag(NH_3)_2OH \longrightarrow R-C(=O)O^{\ominus}NH_4^{\oplus} + 2Ag + H_2O + 3NH_3 \quad (13.5)$$

Mild oxidation procedures such as this have the advantage of not disturbing other sensitive functional groups, such as double and triple bonds, that might be present in the aldehyde molecule.

A third procedure, discussed in Section 9.6, is the oxidation of aryl side chains:

$$ArCH_2CH_3 + 6[O] \longrightarrow Ar-C(=O)O-H + CO_2 + 2H_2O$$

13·2 SOURCES OF CARBOXYLIC ACIDS AND THEIR DERIVATIVES

Alkyl chains attached to an aromatic ring are oxidized at the benzylic position by Cr^{+6} and Mn^{+7}, and vigorous reaction conditions ultimately give aryl carboxylic acids as the major product:

$$ArCH_3 + 2KMnO_4 \xrightarrow{\Delta} ArCO_2H + 2MnO_2 + 2KOH \tag{13.7}$$

Oxidative cleavage of alkenes and alkynes was discussed in Sections 5.5 and 6.5:

$$\underset{\text{(or RC}\equiv\text{CR)}}{RCH=CHR} + 4[O] \longrightarrow 2RCO_2H \tag{13.8}$$

In most cases cleavage is effected by ozone or permanganate:

$$RCH=CHR + O_3 \longrightarrow RCH\underset{O}{\overset{O-O}{\diagup\diagdown}}CHR \xrightarrow{H_2O_2} 2RCO_2H \tag{13.9}$$

$$RCH=CHR + KMnO_4 \longrightarrow \underset{OH\ \ OH}{RCH-CHR} \xrightarrow{KMnO_4} 2RCO_2H \tag{13.10}$$

This is not a widely used procedure for synthesizing carboxylic acids, but in certain situations it is useful for quick preparation of specific acids or diacids (see Chapter 19).

A more common procedure is the hydrolysis of nitriles, discussed in the next chapter:

$$R-C\equiv N + 2H_2O \xrightarrow[\Delta]{\text{acid or base}} R-\overset{O}{\underset{O-H}{C}} + NH_3 \tag{13.11}$$

Many nitriles are easily made by S_N2 reactions of alkyl halides with cyanide ion. Their hydrolysis to carboxylic acids is therefore a general and useful synthesis method.

As we saw in Section 12.4, organometallic reagents may be used to add a carboxyl function to alkyl and aryl halides:

$$\underset{\text{(or R--Li)}}{R-MgX} + CO_2 \xrightarrow{\text{ether}} R-\overset{O}{\underset{O-MgX}{C}} \xrightarrow{H_2O} R-\overset{O}{\underset{O-H}{C}} + MgXOH \tag{13.12}$$

The reaction is not particularly sensitive to steric-hindrance effects, making it more generally useful than the nitrile hydrolysis synthesis.

OXIDATION STATES OF CARBON

In most of the reactions of other functional groups which lead to carboxylic acid products an inorganic oxidizing agent such as Cr^{+6}, Mn^{+7}, or O_3 is reduced, while the organic substance undergoes oxidation (see Section 5.5):

$$\overset{-1}{R}CH=\overset{-1}{C}HR + O_3 \xrightarrow{H_2O,\ H_2O_2} 2\ \overset{+3}{R}CO_2H \tag{13.13}$$

$$\underset{\text{(or Na}_2\text{Cr}_2\text{O}_7)}{\underset{\overset{-3}{C}H_3}{\bigcirc}} + KMnO_4 \xrightarrow[\Delta]{H_2O} \underset{\text{(or Cr}^{+3})}{\underset{\overset{+3}{C}O_2H}{\bigcirc}} + Mn^{+2},\ Mn^{+4} \tag{13.14}$$

13·2 SOURCES OF CARBOXYLIC ACIDS AND THEIR DERIVATIVES

$$\overset{-1}{R CH_2}OH + CrO_3 \xrightarrow{H_3O^{\oplus}} R\overset{+3}{-}CO_2H + Cr^{+3} \quad (13.15)$$

$$R-\overset{+1}{\underset{H}{C}}\!\!\overset{O}{\diagup}\!\! + Ag^{\oplus} \xrightarrow[OH^{\ominus}]{H_2O} R\overset{+3}{-}CO_2H + Ag \quad (13.16)$$

From this we can conclude that the carbon atom of a carboxyl group has a high oxidation state; as indicated by the oxidation numbers in the products above. For example, the oxidation state of the alkene carbon atoms in equation 13.13 changes from -1 to $+3$, and that of the carbonyl carbon atom in equation 13.16 changes from $+1$ to $+3$. The $+3$ oxidation state of a carboxyl carbon atom is, in fact, the highest common oxidation state of carbon, with the exception of carbon dioxide, carbon tetrachloride, and related compounds, which have an oxidation number of $+4$.

PROBLEM 13·1
a Calculate the change in oxidation number for the methyl carbon atom in equation 13.14.
b Indicate the oxidation numbers for all the carbon atoms in biotin.
c Carboxylic acids can be prepared by the reaction of Grignard reagents with carbon dioxide (equation 12.33). Indicate the change in oxidation numbers occurring during this reaction.

PROBLEM 13·2
Write equations showing all steps in each of the following synthesis transformations:
a 1-Butene to butanoic acid
b 1-Butene to propanoic acid
c 1-Butene to pentanoic acid
d 1-Butene to ($\pm$)-2-methylbutanoic acid
e Benzene to benzoic acid
f Benzene to *p*-nitrobenzoic acid
g Benzene to *p*-hydroxybenzoic acid
h *p*-Bromotoluene to terephthalic acid

ORGANIC ACID DERIVATIVES

As we study the chemistry of carboxylic acids we will encounter a number of compounds with a terminal functional group, whose oxidation state is identical to that of the corresponding carboxylic acid. In most cases these compounds can be prepared from or transformed into the related acid by straightforward reactions, and we refer to them as *derivatives of carboxylic acids*. Table 13.2 lists the general formulas for some of these derivatives. In all but the last two cases the *acyl group* (RCO) remains intact.

Each of the carboxylic acid derivatives has a distinct but simple nomenclature. *Esters* are named like salts (with the R' alkyl group replacing the cation). *Anhydrides* use the same names as the acids. The acyl group of an *acyl halide* is named by replacing *-ic* with *-yl* in the corresponding acid name; the names of amides are derived from the related acids by replacing the *-ic* (or *-oic*) ending with *-amide*. Cyclic esters and amides are called *lactones* and *lactams*, respectively.

Organic derivatives of sulfur and phosphorus acids are classified and named in an analogous fashion, as outlined in Table 13.3.

TABLE 13·2 *Functional derivatives of carboxylic acids*

type of derivative	structure	example†
acyl (or acid) halides	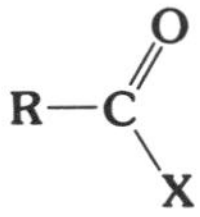 $R-\underset{X}{\overset{O}{\overset{\|\|}{C}}}-$ (X = F, Cl, Br, or I)	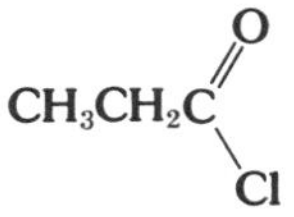 $CH_3CH_2-\underset{Cl}{\overset{O}{\overset{\|\|}{C}}}-$ propionyl chloride *or* propanoyl chloride
anhydrides	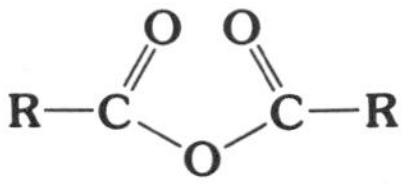 $R-\overset{O}{\overset{\|\|}{C}}-O-\overset{O}{\overset{\|\|}{C}}-R$	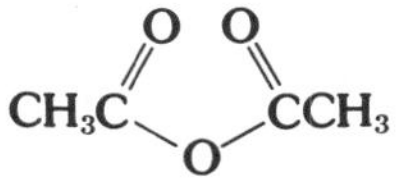 $CH_3-\overset{O}{\overset{\|\|}{C}}-O-\overset{O}{\overset{\|\|}{C}}-CH_3$ acetic anhydride *or* ethanoic anhydride
esters	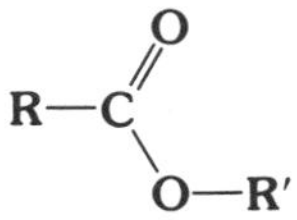 $R-\underset{O-R'}{\overset{O}{\overset{\|\|}{C}}}-$	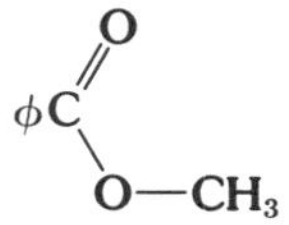 $\phi-\underset{O-CH_3}{\overset{O}{\overset{\|\|}{C}}}-$ methyl benzoate
amides primary	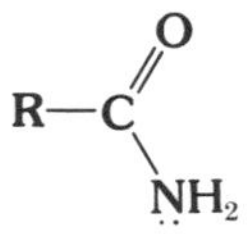 $R-\underset{\ddot{N}H_2}{\overset{O}{\overset{\|\|}{C}}}-$	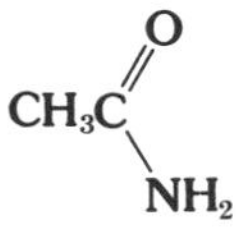 $CH_3-\underset{NH_2}{\overset{O}{\overset{\|\|}{C}}}-$ acetamide *or* ethanamide
secondary	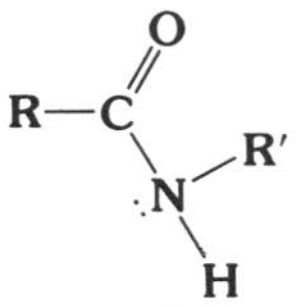 $R-\underset{\underset{H}{\overset{..}{N}}-R'}{\overset{O}{\overset{\|\|}{C}}}-$	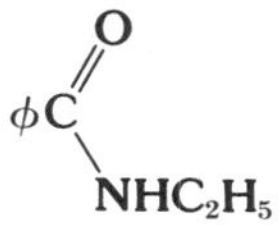 $\phi-\underset{NHC_2H_5}{\overset{O}{\overset{\|\|}{C}}}-$ N-ethylbenzamide
tertiary	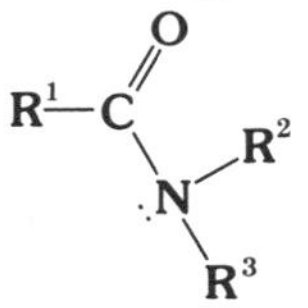 $R^1-\underset{\underset{R^3}{\overset{..}{N}}-R^2}{\overset{O}{\overset{\|\|}{C}}}-$	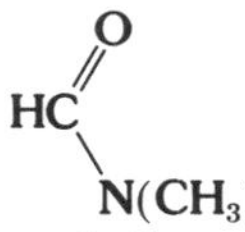 $H-\underset{N(CH_3)_2}{\overset{O}{\overset{\|\|}{C}}}-$ N,N-dimethylformamide
imides	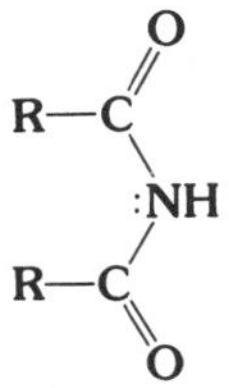 $\begin{array}{c}R-\overset{O}{\overset{\|\|}{C}}\\ \ddot{N}H\\ R-\underset{O}{\underset{\|\|}{C}}\end{array}$	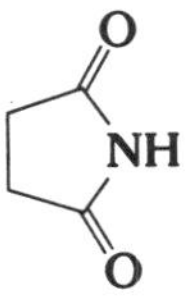 succinimide (name derived from succinic acid)
nitriles	$R-C\equiv N:$	$CH_3C\equiv N$ acetonitrile (name derived from acetic acid)
ortho esters	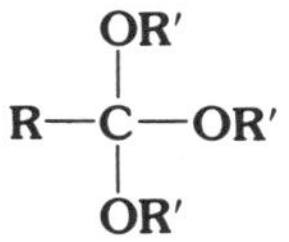 $R-\underset{OR'}{\overset{OR'}{\overset{\|}{\underset{\|}{C}}}}-OR'$	$HC(OC_2H_5)_3$ ethyl orthoformate

†The less widely used IUPAC names are given as alternatives.

13·2 SOURCES OF CARBOXYLIC ACIDS AND THEIR DERIVATIVES

TABLE 13·3 *Functional derivatives of sulfur and phosphorus acids*

inorganic acid	organic derivative	type of derivative
H_2SO_4	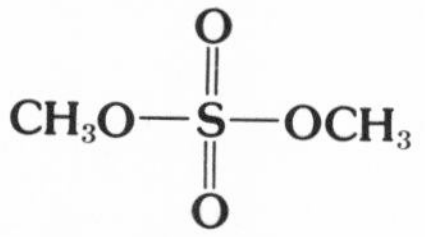 $CH_3O-\underset{\underset{O}{\|\|}}{\overset{\overset{O}{\|\|}}{S}}-OCH_3$ dimethyl sulfate	sulfate ester
	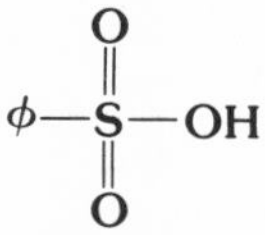 $\phi-\underset{\underset{O}{\|\|}}{\overset{\overset{O}{\|\|}}{S}}-OH$ benzenesulfonic acid	sulfonic acid
	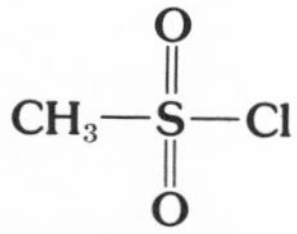 $CH_3-\underset{\underset{O}{\|\|}}{\overset{\overset{O}{\|\|}}{S}}-Cl$ methanesulfonyl chloride	sulfonyl halide
H_2SO_3	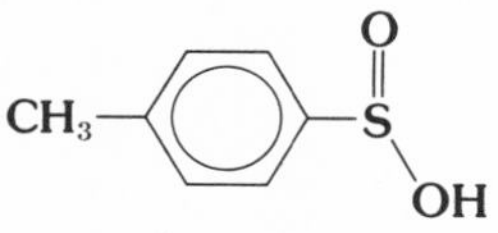 $CH_3-C_6H_4-S(=O)-OH$ p-toluenesulfinic acid	sulfinic acid
	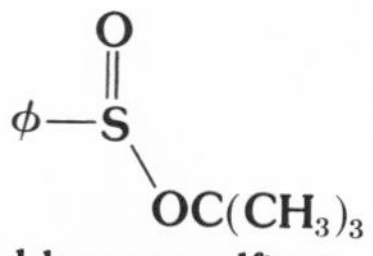 $\phi-S(=O)-OC(CH_3)_3$ t-butyl benzenesulfinate	sulfinate ester
H_3PO_4	$(C_2H_5O)_3P=O$ triethyl phosphate	phosphate ester
	$[(CH_3)_2N]_3PO$ hexamethyl phosphoric triamide	phosphoric amide
	$(CH_3)_2P(O)OH$ dimethylphosphinic acid	phosphinic acid
	$\phi HP(O)OC_2H_5$ ethyl phenylphosphinate	phosphinate ester
	$\phi PO(OH)_2$ phenylphosphonic acid	phosphonic acid
H_3PO_3	$(C_4H_9O)_3P$ tri-n-butylphosphite	phosphite ester

13·3 PROPERTIES OF CARBOXYLIC ACIDS AND THEIR DERIVATIVES

HYDROGEN-BONDING EFFECTS

Carboxylic acids and their amide derivatives have unusually high boiling points and water solubilities in comparison with related compounds of similar molecular weight (Table 13.4). Since esters and aldehydes do not have especially high boiling points, this property in the case of the acids has been ascribed to a hydrogen-bonding association involving the acidic carboxyl proton. Indeed, the boiling points of the acids are even higher than those of equivalent alcohols, suggesting that dimeric association is particularly strong:

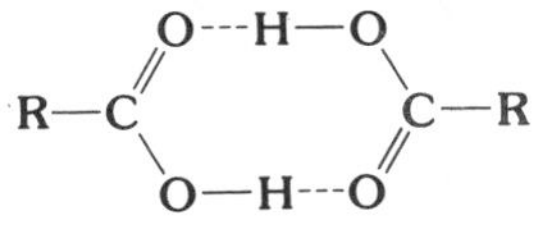

$$R-C\begin{matrix}O\cdots H-O\\ \diagdown \quad\quad\diagup\\ \diagup\quad\quad\diagdown\\ O-H\cdots O\end{matrix}C-R$$

The high boiling points of the amides are thought to result from the highly polar nature of this functional group. Dipolar attractions are consequently strong and may be buttressed by hydrogen bonding in those compounds having N—H bonds.

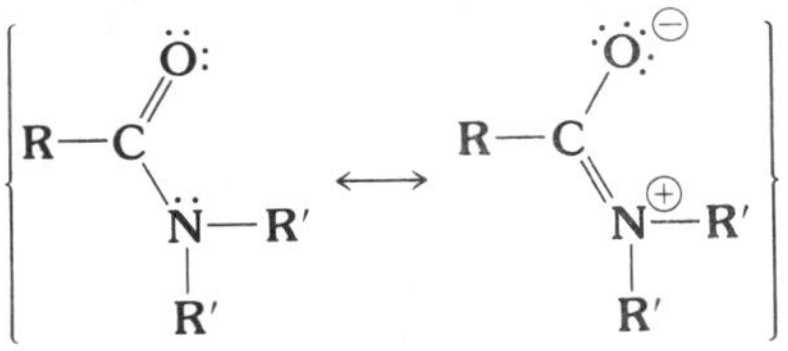

$$\left[R-C\begin{matrix}\ddot{O}:\\ \diagdown\\ N-R'\\ |\\ R'\end{matrix} \longleftrightarrow R-C\begin{matrix}\ddot{O}:^{\ominus}\\ \diagdown\\ \overset{\oplus}{N}-R'\\ |\\ R'\end{matrix} \right]$$

SPECTROSCOPIC PROPERTIES

INFRARED SPECTROSCOPY It is apparent from the data in Table 13.5 that carbonyl stretching frequencies of carboxylic acid derivatives (RCOZ) are sensitive to changes in the functional substituent (Z). If we choose the carbonyl absorption of simple saturated aldehydes as our

TABLE 13·4 *Physical constants of some organic compounds*

compound	molecular weight	boiling point (°C)	water solubility
$CH_3CH_2CH_2CO_2H$	88	164	very soluble
$CH_3(CH_2)_4OH$	88	138	slightly soluble
$CH_3(CH_2)_3CHO$	86	103	slightly soluble
$CH_3CO_2C_2H_5$	88	77	moderately soluble
$C_2H_5CO_2CH_3$	88	80	slightly soluble
$CH_3CH_2CH_2CONH_2$	87	216	soluble
$CH_3CON(CH_3)_2$	87	165	very soluble
$CH_3(CH_2)_4NH_2$	87	103	very soluble

13·3 PROPERTIES OF CARBOXYLIC ACIDS AND THEIR DERIVATIVES

TABLE 13·5 *Carbonyl stretching absorptions of carboxylic acid derivatives*

functional group	stretching frequencies (cm^{-1})
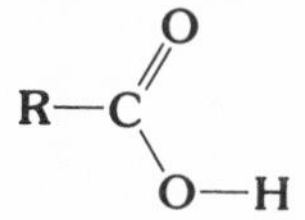 R—C(=O)—O—H	1705–1725 (strong)
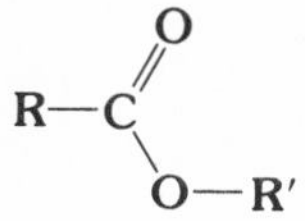 R—C(=O)—O—R'	1735–1750 (strong)
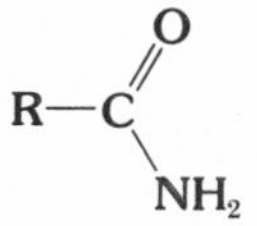 R—C(=O)—NH$_2$	1630–1680 (strong)
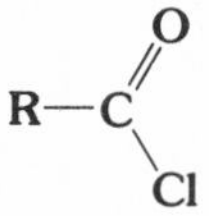 R—C(=O)—Cl	~1795 (strong)
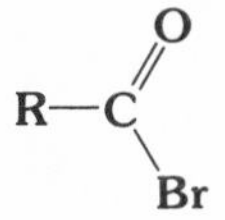 R—C(=O)—Br	~1810 (strong)
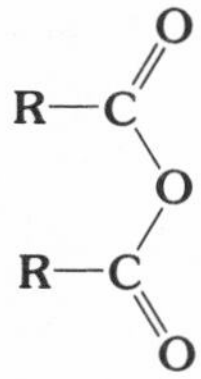 R—C(=O)—O—C(=O)—R	1800–1850 and 1740–1760 (both strong)

reference (1720 to 1740 cm^{-1}), we find that acyl halides and anhydrides have higher stretching frequencies, while amides are shifted to lower frequencies. Acids and esters are relatively unchanged.

Anhydrides exhibit two carbonyl stretching absorptions due to a coupling of the two carbonyl vibrations, which may be "in phase" or "out of phase." The effect of conjugated double bonds or ring strain (in lactones, lactams, or cyclic anhydrides) on the general values in Table 13.5 is similar to that noted earlier for aldehydes and ketones (see Section 12.4). Thus α,β-unsaturated esters have stretching frequencies in the range 1715 to 1730 cm^{-1} (a shift to lower frequencies) and γ-lactones display strong absorption at 1760 to 1780 cm^{-1} (a shift to higher frequencies). As seen in the spectrum of acetic acid in Figure 13.1, carboxylic acids also exhibit a strong, broad hydroxyl absorption at 2500 to 3300 cm^{-1}, which overlaps the normally sharp C—H stretching band.

PROBLEM 13·3

The variations of carbonyl stretching frequencies with Z (in RCOZ) can be explained by the influence of inductive and resonance effects on the bonding of the carbonyl group. Recall the influence of OH, NH$_2$, OR, Cl, and Br substit-

13·3 PROPERTIES OF CARBOXYLIC ACIDS AND THEIR DERIVATIVES

uents on the reactivity of aromatic rings toward electrophilic reagents. How would these groups be expected to affect the resonance hybrid shown below? Would a change in the hybrid composition affect the stretching frequency of the carbonyl group?

$$\left[\begin{array}{c} \ddot{\text{O}}: \\ \| \\ \text{R}-\text{C} \\ | \\ \text{Z} \end{array} \longleftrightarrow \begin{array}{c} \ddot{\text{O}}:^{\ominus} \\ | \\ \text{R}-\overset{\oplus}{\text{C}} \\ | \\ \text{Z} \end{array} \right]$$

FIGURE 13·1 *Infrared spectra of acetic acid, ethyl acetate, and acetic anhydride*

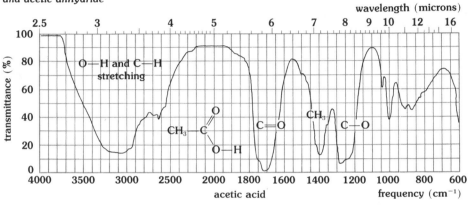

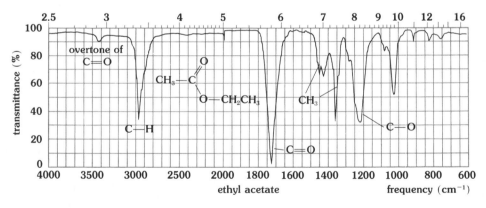

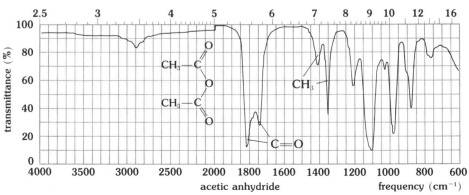

13·3 PROPERTIES OF CARBOXYLIC ACIDS AND THEIR DERIVATIVES

PMR SPECTROSCOPY The acidic protons of carboxylic acids give rise to sharp low-field absorption signals at $\delta = 10.0$ to 13.0. A typical example is the pmr spectrum of 2-bromobutyric acid in Figure 13.2. This downfield shift of hydroxyl protons is usually the result of hydrogen bonding (see Table 8.3) and would be expected for the dimeric cluster of acid molecules shown on page 416.

In our considerations of amide properties we have attributed the low basicities, high boiling points, and low carbonyl stretching frequencies of amides to an exceptionally large dipolar character resulting from nitrogen electron-pair delocalization:

$$\left[\begin{array}{c} \ddot{\text{O}}\text{:} \\ \| \\ \text{R}-\text{C} \\ | \\ \ddot{\text{N}}\text{H}_2 \end{array} \longleftrightarrow \begin{array}{c} \ddot{\text{O}}\text{:}^{\ominus} \\ | \\ \text{R}-\text{C} \\ \| \\ \overset{\oplus}{\text{N}}\text{H}_2 \end{array} \right]$$

We can test this hypothesis by examining the pmr spectra of amide compounds. If the dipolar species is an important contributor to the hybrid amide structure, then the hydrogen atoms attached to the nitrogen atom should be shifted downfield from the TMS reference signal, as in the spectrum of p-ethoxyacetanilide in Figure 13.3. Furthermore, the added double-bond character between carbon and nitrogen will increase the barrier to rotation about that bond, perhaps sufficiently to permit observation of individual rotamers. The spectrum of N,N-dimethylformamide does in fact show this effect, although it disappears when the temperature is raised from 30° to 170°C.

The pmr spectra of esters and anhydrides do not in general show any unexpected or surprising features. The spectrum of methyl methacrylate in Figure 3.3 is typical.

PROBLEM 13·4
Why do the two methyl signals in the pmr spectrum of N,N-dimethylformamide coalesce into a single signal as the temperature is raised?

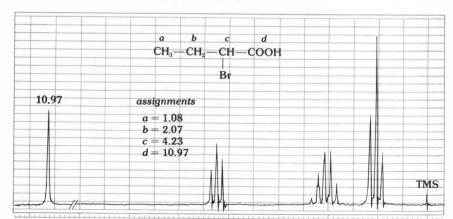

FIGURE 13·2 Pmr spectrum of 2-bromobutyric acid

$\text{CH}_3\!-\!\text{CH}_2\!-\!\text{CH}\!-\!\text{COOH}$
           $|$
           Br

assignments
$a = 1.08$
$b = 2.07$
$c = 4.23$
$d = 10.97$

420   13·3 PROPERTIES OF CARBOXYLIC ACIDS AND THEIR DERIVATIVES

FIGURE 13·3 *Pmr spectra of p-ethoxyacetanilide, N,N-dimethylformamide, and methyl methacrylate*

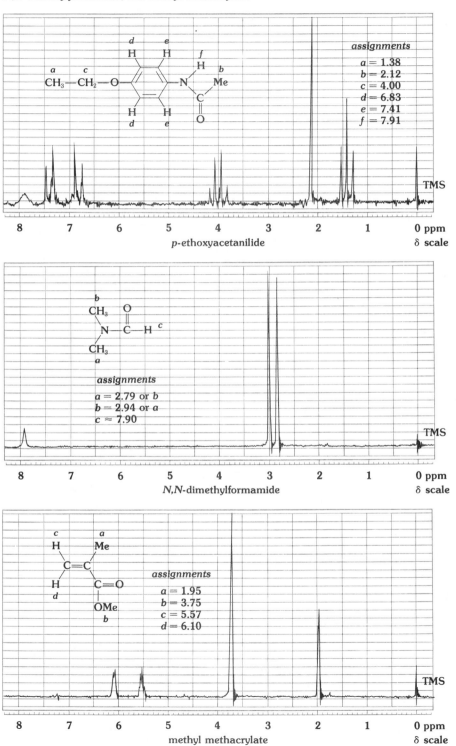

13·4 NATURALLY OCCURRING CARBOXYL DERIVATIVES

PROBLEM 13·5

Three $C_4H_8O_2$ isomers have the following spectroscopic properties. Write structural formulas for these compounds and indicate the nature of the information provided by each spectroscopic technique.

a Infrared absorptions at 2900–3000, 1740 (very strong), 1370, 1240 (strong), and 1050 cm^{-1}; pmr signals at $\delta = 1.25$ (triplet, $J = 7$ Hz), $\delta = 2.03$ (singlet), and $\delta = 4.12$ (quartet, $J = 7$ Hz), with relative intensities of 3:3:2

b Infrared absorptions at 2850–3000, 2725, 1725 (strong), 1160–1220 (strong), and 1100 cm^{-1}; pmr signals at $\delta = 1.29$ (doublet, $J = 6$ Hz), $\delta = 5.13$ (septet, $J = 6$ Hz), and $\delta = 8.0$ (singlet), with relative intensities of 6:1:1

c Infrared absorptions at 2500–3200 (broad and strong), 1715 (strong), and 1230 cm^{-1} (strong); pmr signals at $\delta = 1.2$ (doublet, $J = 6$ Hz), $\delta = 2.7$ (septet, $J = 6$ Hz), and $\delta = 11.0$ (singlet, vanishes on treatment with D_2O), with relative intensities of 6:1:1

13·4 NATURALLY OCCURRING CARBOXYL DERIVATIVES

Since acyl halides and anhydrides are highly reactive substances, they do not as a rule occur naturally (cantharidin, produced by the blister beetle, is an exception). However, the more stable ester and amide derivatives are abundantly distributed in nature.

WAXES, FATS, OILS, AND PHOSPHOLIPIDS

Waxes are esters of high-molecular-weight monohydric alcohols (one hydroxyl group) and the higher fatty acids. Natural waxes are usually mixtures of esters and may also contain hydrocarbons. Spermaceti, from the head of the sperm whale, is chiefly cetyl palmitate (molecular formula $nC_{15}H_{31}CO_2nC_{16}H_{33}$).

The most valuable natural wax is Carnauba wax, obtained from the leaves of a Brazilian palm. The toughness and water resistance of this substance, actually a mixture of high-molecular-weight esters, make it an important component of automobile and floor waxes.

The solid (or semisolid) fats and liquid oils obtained from both animal and vegetable sources are mixtures of esters derived from fatty acids and the trihydric alcohol glycerol; such esters are called *glycerides*. Fats generally have fewer unsaturated functions and higher molecular weights than the lower-melting oils. Catalytic hydrogenation of the double bonds in vegetable oils produces solid glycerides, used chiefly in the manufacture of margarine and cooking fat. Although people living in temperate and cooler climates prefer solid fats to oils in their diets, recent medical research suggests that these saturated glycerides may be one of the factors involved in a high incidence of atherosclerosis in populations such as that of the United States.

Certain polyunsaturated fatty acids are termed "essential" because their absence in the human diet appears to cause scaley skin, stunted growth, and increased water loss through the skin. These acids are also known to be precursors of the prostaglandins, a family of physiologically potent lipid acids found in most body tissues in minute amounts. The

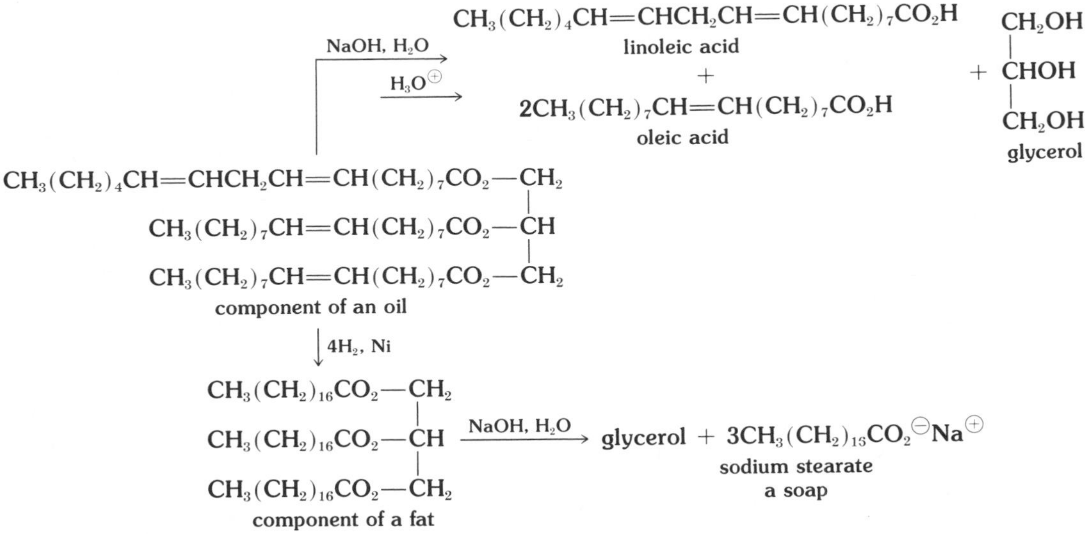

13·4 NATURALLY OCCURRING CARBOXYL DERIVATIVES

$$\text{CH}_3(\text{CH}_2)_4\text{CH}=\text{CHCH}_2\text{CH}=\text{CH}(\text{CH}_2)_7\text{CO}_2\text{H}$$
linoleic acid
+
$$2\text{CH}_3(\text{CH}_2)_7\text{CH}=\text{CH}(\text{CH}_2)_7\text{CO}_2\text{H}$$
oleic acid

$$\begin{array}{l}\text{CH}_2\text{OH}\\|\\\text{CHOH}\\|\\\text{CH}_2\text{OH}\end{array}$$
glycerol

via NaOH, H$_2$O / H$_3$O$^\oplus$ from:

$$\begin{array}{l}\text{CH}_3(\text{CH}_2)_4\text{CH}=\text{CHCH}_2\text{CH}=\text{CH}(\text{CH}_2)_7\text{CO}_2-\text{CH}_2\\\text{CH}_3(\text{CH}_2)_7\text{CH}=\text{CH}(\text{CH}_2)_7\text{CO}_2-\text{CH}\\\text{CH}_3(\text{CH}_2)_7\text{CH}=\text{CH}(\text{CH}_2)_7\text{CO}_2-\text{CH}_2\end{array}$$
component of an oil

↓ 4H$_2$, Ni

$$\begin{array}{l}\text{CH}_3(\text{CH}_2)_{16}\text{CO}_2-\text{CH}_2\\\text{CH}_3(\text{CH}_2)_{16}\text{CO}_2-\text{CH}\\\text{CH}_3(\text{CH}_2)_{16}\text{CO}_2-\text{CH}_2\end{array}\xrightarrow{\text{NaOH, H}_2\text{O}}\text{glycerol}+3\text{CH}_3(\text{CH}_2)_{15}\text{CO}_2^\ominus\text{Na}^\oplus$$
component of a fat, sodium stearate a soap

FIGURE 13·4 *Some reactions of glycerides*

formation of the prostaglandins from unsaturated C$_{20}$ fatty acids such as arachidonic acid proceeds by enzymatic oxidation to a very active *endoperoxide* precursor:

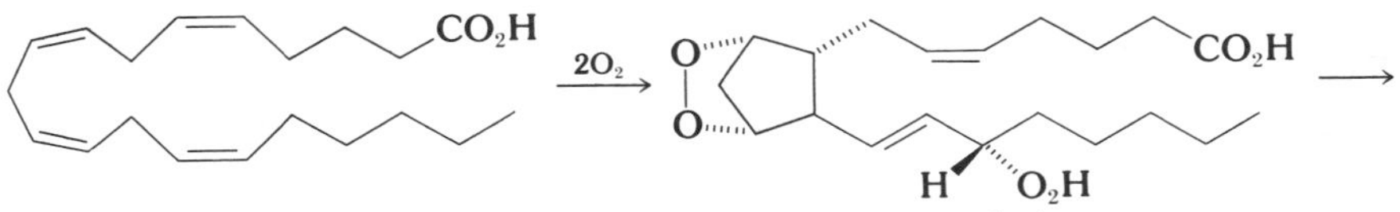

arachidonic acid $\xrightarrow{2\text{O}_2}$ an endoperoxide derivative $\longrightarrow$ prostaglandins (13.17)

Prostaglandins can act to regulate menstruation, prevent conception, induce childbirth or abortion, lower blood pressure, prevent blood clotting, and possibly even as long-lasting decongestants.† The common drug aspirin appears to alter prostaglandin metabolism.

Phospholipids, such as the lecithins and cephalins, are found in all animal and vegetable cells and are particularly abundant in the brain, spinal cord, eggs, and soybeans. The lecithins are mixed esters of fatty acids and phosphoric acid with glycerol and choline.

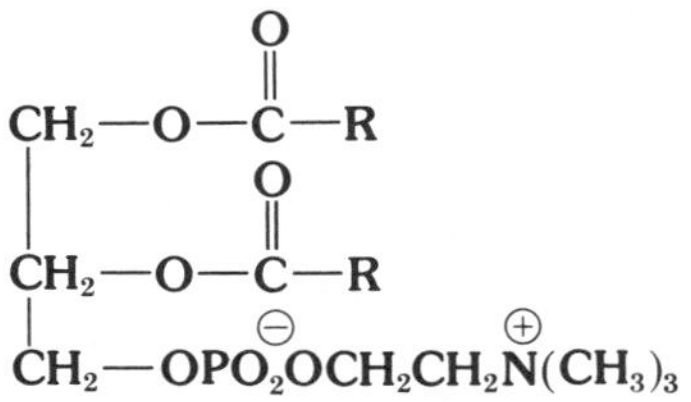

$$\begin{array}{l}\text{CH}_2-\text{O}-\overset{\text{O}}{\underset{\|}{\text{C}}}-\text{R}\\|\\\text{CH}_2-\text{O}-\overset{\text{O}}{\underset{\|}{\text{C}}}-\text{R}\\|\\\text{CH}_2-\text{OPO}_2^\ominus\text{OCH}_2\text{CH}_2\overset{\oplus}{\text{N}}(\text{CH}_3)_3\end{array}$$
a lecithin

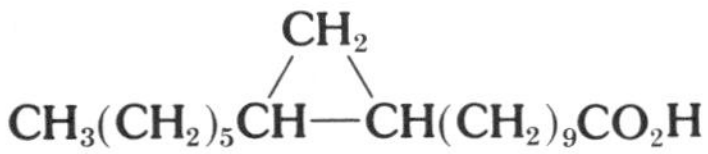

$$\text{CH}_3(\text{CH}_2)_5\overset{\text{CH}_2}{\overset{/\,\,\backslash}{\text{CH}-\text{CH}}}(\text{CH}_2)_9\text{CO}_2\text{H}$$
lactobacillic acid

Most acids isolated from naturally occurring fats and oils have an unbranched even-numbered chain of carbon atoms (see Table 13.1). However, there are some interesting exceptions. Lactobacillic acid, from

†For a further discussion of these remarkable lipic acids see J. E. Pike, *Scientific American*, 225:84 (1971).

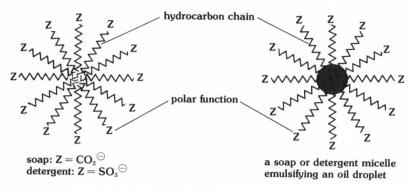

soap: $Z = CO_2^{\ominus}$
detergent: $Z = SO_3^{\ominus}$

a soap or detergent micelle emulsifying an oil droplet

FIGURE 13·5 *Soap and detergent structures*

Lactobacillus arabinosus, has 19 carbon atoms and a cyclopropane ring, and tuberculostearic acid, from the fatty capsule of the tuberculosis bacillus, is 10-methylstearic acid.

SOAPS AND DETERGENTS

Alkali-metal salts of long-chain fatty acids (C_{10} to C_{18}) are known as *soaps*. The combination of a long, oil-soluble hydrocarbon chain attached to a water-soluble function, such as a carboxylate anion, results in emulsifying and wetting properties that make these substances powerful cleaning agents.

Soap solutions differ from solutions of other ionic salts in that the organic ions are clustered in spherical groups called *micelles* (see Figure 13.5). The polar functions have an affinity for water and are said to be *hydrophilic;* the hydrocarbon chains avoid water (remember that alkanes are generally immiscible with water) and are termed *hydrophobic*. By grouping together, these hydrophobic chains can avoid an aqueous environment, and the resulting spherical array of soap molecules will have all the hydrophilic groups on the outside, where solvation can take place. Electrostatic repulsion of like charges will, of course, keep these micelles dispersed in the aqueous medium.

The cleaning and emulsifying action of soap and detergents is due in large part to the ability of the hydrophobic center of the micelle to accept nonpolar particles such as dirt or oil.

FLAVOR AND PERFUME COMPONENTS

In contrast with the acrid and decidedly unpleasant odors of the lower-molecular-weight carboxylic acids, most volatile esters have pleasant fruity fragrances, and many are used as synthetic flavoring agents and in perfumes. For example, the odors of isopentyl acetate, butyl butyrate, isopentyl valerate, methyl salicylate, and methyl anthranilate resemble the aromas of banana, pineapple, apple, wintergreen, and grape juice, respectively. Natural aromas and flavors, however, are usually derived from a complex mixture of compounds. Over 200 compounds have been identified in the characteristic aroma of freshly ground coffee, and the bouquet of fine wines is due to the formation of esters and other compounds during the aging process.

13·4 NATURALLY OCCURRING CARBOXYL DERIVATIVES

PROBLEM 13·6
Write structural formulas for the five esters named in the preceding paragraph.

The lactones coumarin (from the tonka bean) and ambrettolide (from ambrette seeds) have odors resembling new-mown hay and musk, respectively, and are widely used in perfumery. Catnip's remarkable effect on members of the cat family is caused by a substance called nepetalactone.

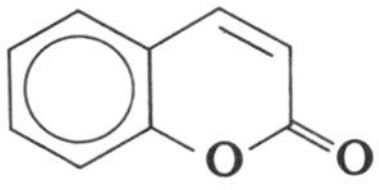

coumarin 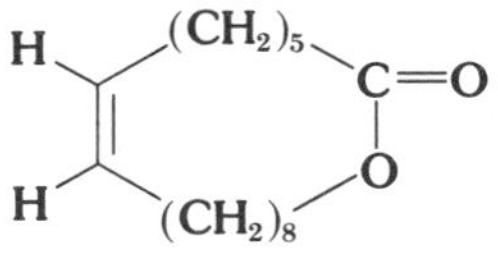   ambrettolide 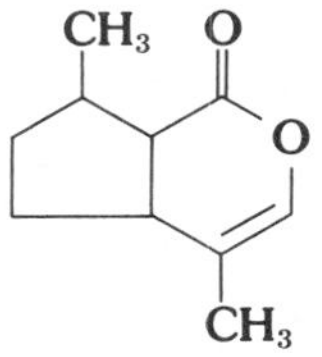   nepetalactone

The amides capsaicin and piperine are found in red and black pepper, the former being responsible for the pungent quality of tabasco and cayenne.

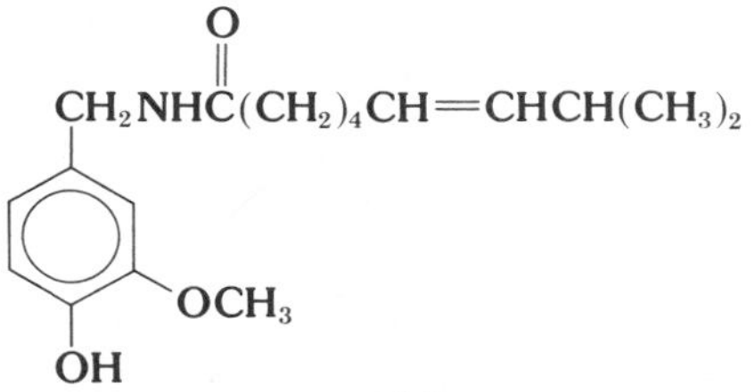

capsaicin 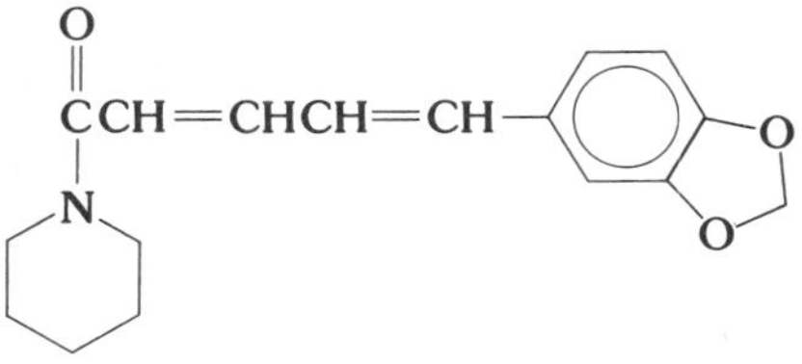   piperine

PROTEINS

The amide function is probably best known for the prominent role it plays in protein constitution. Proteins are polyamides of α-amino carboxylic acids and are found in all living cells. Indeed, protein is the principle substance not only of skin, muscles, nerves, and blood, but also of hormones and enzymes.

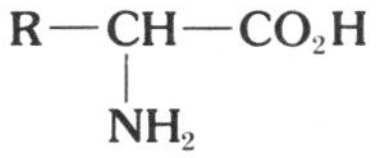

an α-amino acid 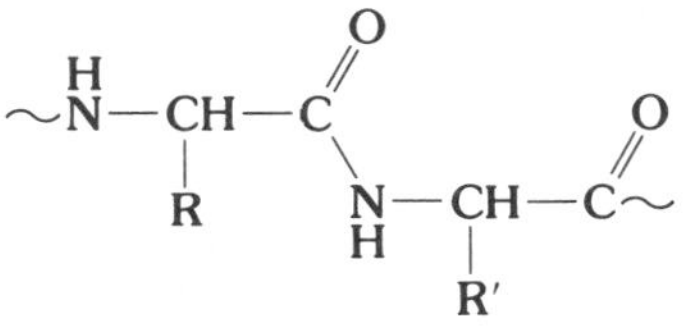   part of a protein

The structure and properties of proteins are determined largely by the nature and distribution of the amide groups. A single protein molecule may contain hundreds or thousands of amino acid units, and in turn many thousands of different proteins go to make up a living organism. Although the number of naturally occurring amino acids is relatively small (25 to 30), the possible combinations of these biochemical building blocks in polymeric molecules is very large. We shall discuss protein structure and chemistry more extensively in Chapter 17.

13·4 NATURALLY OCCURRING CARBOXYL DERIVATIVES

ALKALOIDS AND OTHER PHYSIOLOGICALLY ACTIVE COMPOUNDS

As we saw in Chapter 11, many of the nitrogenous plant substances called alkaloids exert powerful physiological effects on animal organisms. In addition to the examples mentioned in Section 11.2, there are numerous important alkaloids which incorporate carboxylic acid (or derivative) functions.

Cocaine, the chief alkaloid in the leaves of the coca bush (not to be confused with the cocoa bean), is a central-nervous-system stimulant which bears an obvious structural relationship to atropine, an alkaloid found in belladonna, henbane, and deadly nightshade.

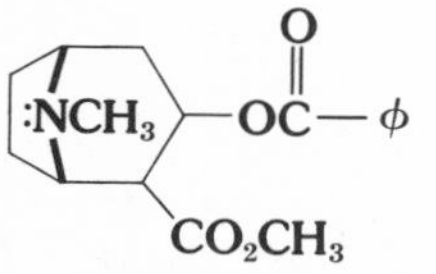

(structures of cocaine and atropine)

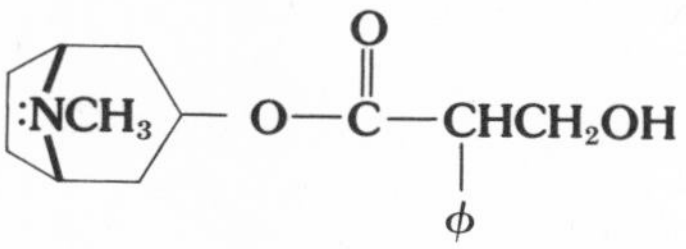

The mild stimulant effect of coffee and tea is due to caffeine, and that of cocoa is due to the closely related substance theobromine. Anthropologists have found that wherever plants with a high caffeine content are indigenous to a region, extracts of these plants are used as a native beverage.

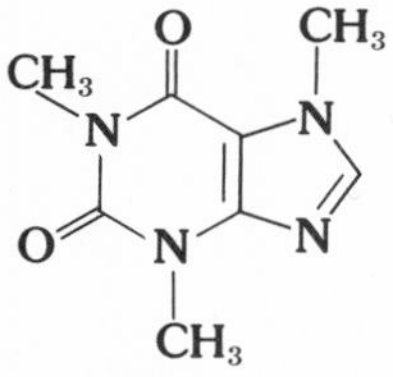

(structures of caffeine and theobromine)

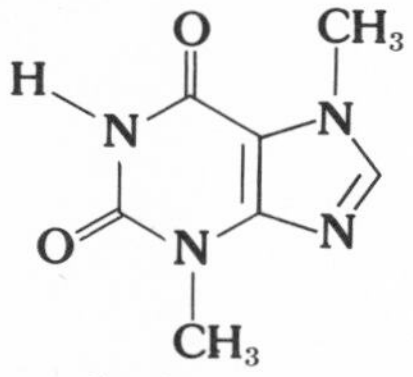

The human emotional state can be profoundly influenced by certain substances which appear to affect the concentration of serotonin in the brain. Reserpine, from the Indian snake root (*Rauwolfia serpentina*), is used extensively as a tranquilizer, especially in the control of psychotic states, and also seems to have some effect in treating the withdrawal symptoms of extreme schizophrenia. Ironically, the diethylamide of lysergic acid, LSD, products a psychosis resembling schizophrenia. LSD

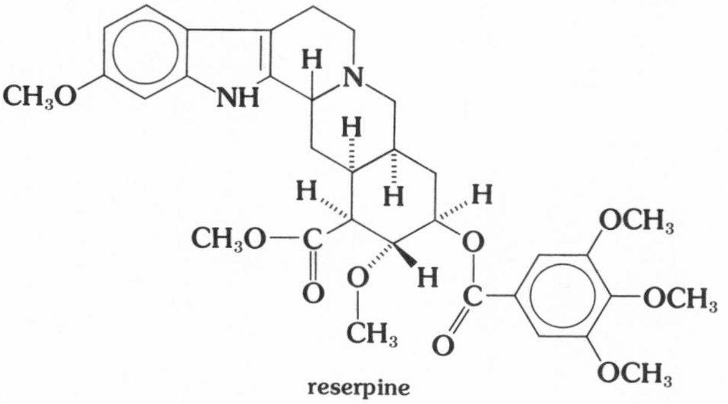

(structures of reserpine and LSD)

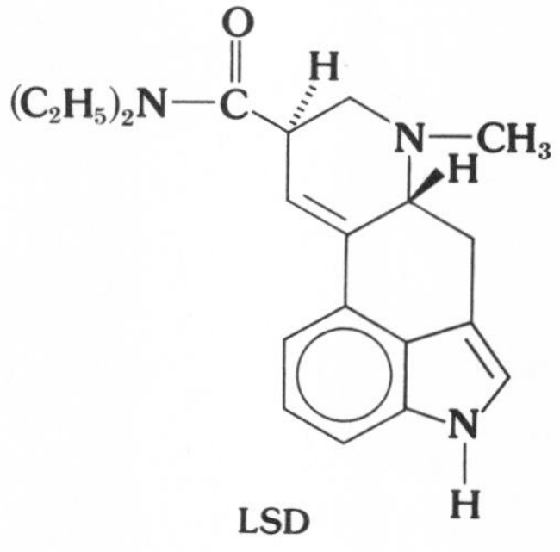

13·4 NATURALLY OCCURRING CARBOXYL DERIVATIVES

does not occur naturally and must be manufactured from alkaloids derived from ergot, a poisonous fungus that infects rye and other grains. Ergot alkaloids have also been used to induce childbirth and in the treatment of migraine headache.

The seeds of the strychnos plant yield two poisonous alkaloids, strychnine and brucine, the former being a favorite device of mystery writers. In the realm of poisons, however, the plant kingdom must yield honors to the animals. Even if we ignore the protein-based snake venoms and bacterial toxins, the poisonous substances found in the viscera of the puffer fish and the skin of the frog *Phyllobates surotaenia*, among others, far exceed the toxicity of strychnine.

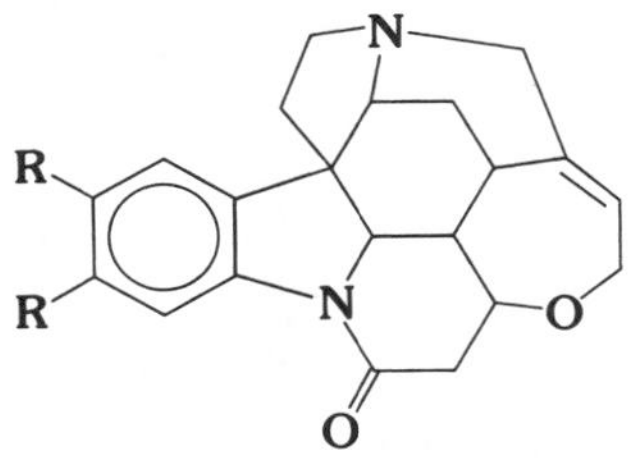

strychnine: R = H
brucine: R = CH$_3$O

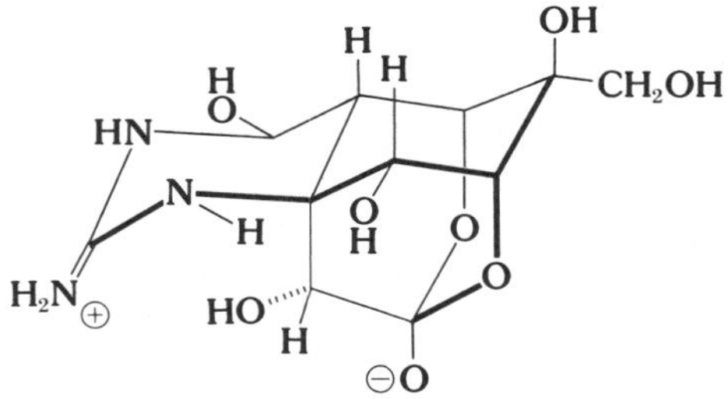

tetrodotoxin (from the puffer fish and the California newt)

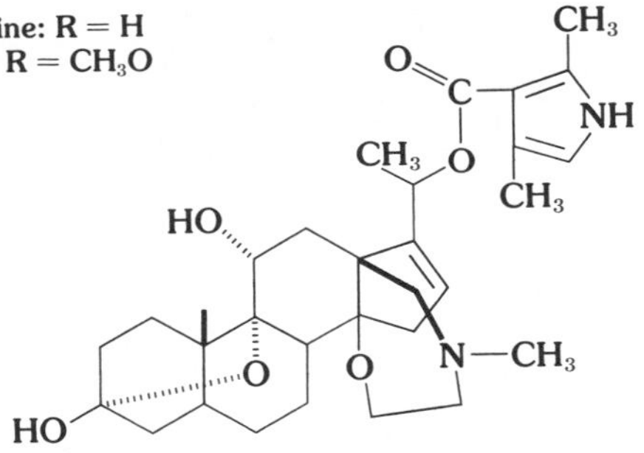

batrachotoxin (from a Columbian arrow-poison frog)

Comparisons of poisons require a quantitative and reproducible means of estimating a compound's toxicity. In practice, this is accomplished by giving different doses known to be in the toxic range to large groups of test animals. A graph relating the mortality rate for each dose is then

TABLE 13·6 *LD$_{50}$ for toxic substances administered subcutaneously to mice*

substance	LD$_{50}$ (μg/kg)
sodium cyanide	10,000
strychnine	500
curare	500
fluoroacetic acid	300
tetrodotoxin	8
batrachotoxin	2

13·4 NATURALLY OCCURRING CARBOXYL DERIVATIVES

prepared, and by interpolation the dose expected to be lethal to 50% of the animals (LD_{50}) is determined. These doses are cited in milligrams (1 mg = 10^{-3} g) or micrograms (1 μg = 10^{-6} g) per kilogram of body weight of the animal (Table 13.6). The type of animal used is always specified, since the toxicity of a substance and even its mode of action can vary enormously from one species to another. For example, morphine causes strychninelike convulsions in frogs, excitement in cats, vomiting in dogs, and depression in man.

Some of the poisonous manmade substances match the most powerful natural toxins in their lethal character. The nerve gas sarin, for example, was first prepared in Germany during World War II, and related compounds have been studied and stockpiled by the military establishments of several different countries. Sarin, the isopropyl ester of methylfluorophosphonic acid, has an LD_{50} rating of 10 μg per kg in man.

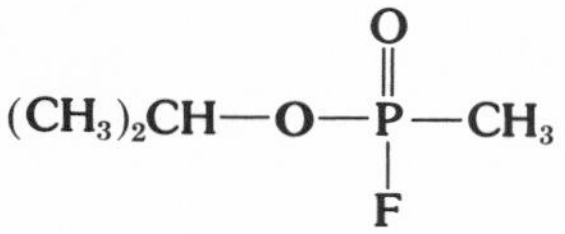

$$(CH_3)_2CH-O-\underset{F}{\overset{\overset{O}{\|}}{P}}-CH_3$$

sarin

Related phosphorus compounds of lower toxicity have been used as insecticides (they are much less persistent than DDT), but they have nevertheless caused several fatalities among farm workers.

Synthetic toxins have sometimes been inadvertently introduced into our environment. In the manufacture of 2,4,5-trichlorophenol, for example, failure to control reaction temperature can lead to formation of the highly toxic and teratogenic substance 2,3,7,8-tetrachlorodibenzo-p-dioxin (TCDD). This thermally stable and persistent compound has an LD_{50} rating of 1 μg per kg in the guinea pig. Great care must be taken,

FIGURE 13·6 *Derivatives of 2,4,5-trichlorophenol*

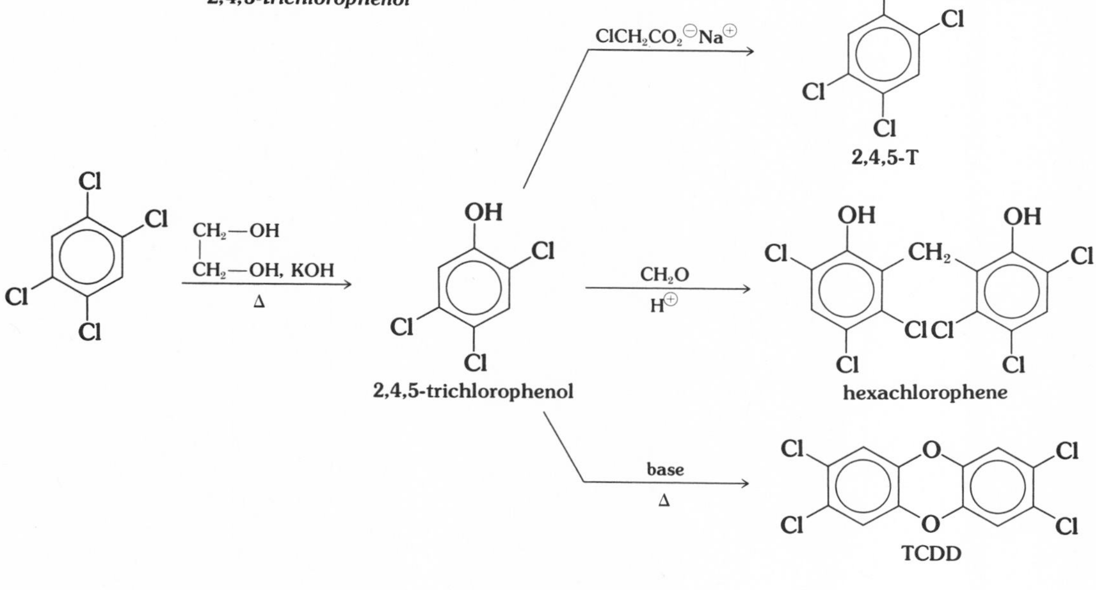

13·4 NATURALLY OCCURRING CARBOXYL DERIVATIVES

therefore, to avoid TCDD contamination of products derived from this trichlorophenol, such as the herbicide 2,4,5-T and the mild disinfectant hexachlorophene. In 1976 a malfunctioning reactor at the Icmesa chemical plant in northern Italy exploded, scattering about 2 kilograms of TCDD over an area which included the small town of Seveso. As a consequence of this accident, all persons were evacuated from a 130-acre contaminated zone which will probably not be habitable for five to ten years.

PROBLEM 13·7
List the different functional groups in reserpine, tetrodotoxin, and bactrachotoxin and indicate which can be considered derivatives of carboxylic acids.

Fortunately for us, nature's toxins and venoms are matched by beneficial drugs such as reserpine and antibiotics such as the penicillins, cephalosporins, macrolides, and tetracyclines. The flower heads of *Chrysanthemum cinerariaefolium* even yield relatively nontoxic (to mammals) insecticides called the pyrethrums.

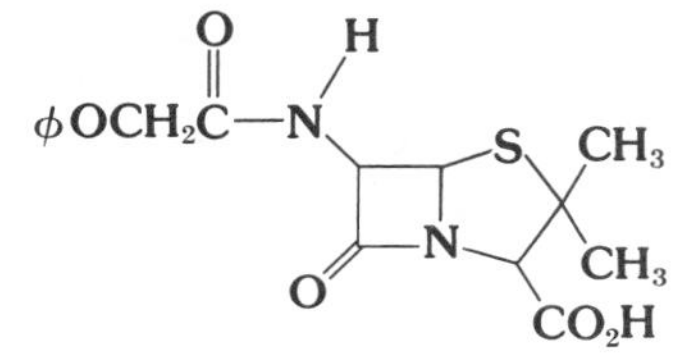

penicillin V

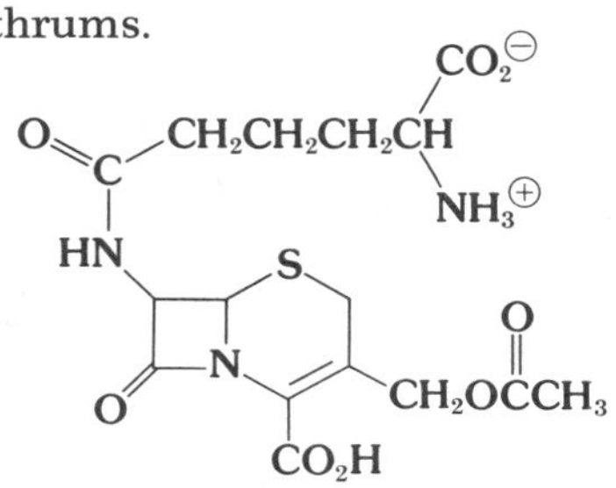

cephalosporin C

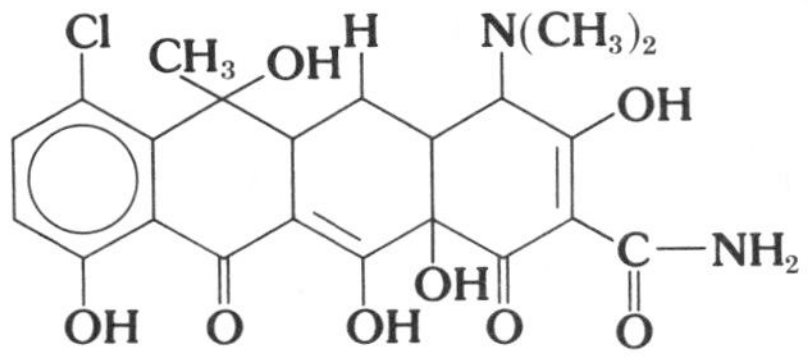

aureomycin, a tetracycline

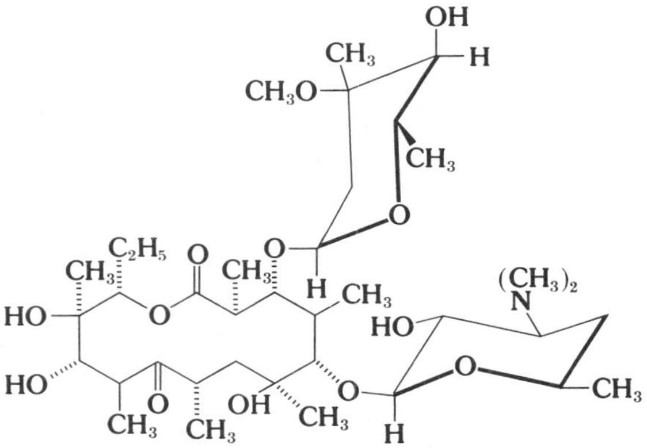

erythromycin, a macrolide

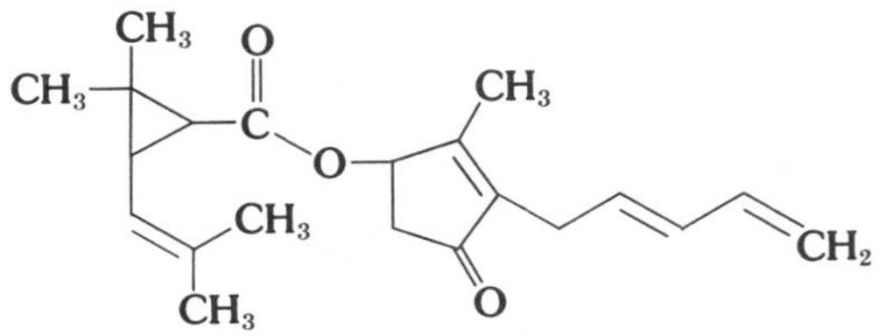

pyrethrin I, a pyrethrum

13·4 NATURALLY OCCURRING CARBOXYL DERIVATIVES

The importance of the *porphin* (or porphyrin) *ring system* in the vital processes of most living organisms can be seen in the structural formulas of hemin (formed by acid hydrolysis of hemoglobin, in which it exists as the ferrous analog) and chlorophyll A in Figure 13.7. Vitamin B_{12} (cyanocobalamin), the antianemia factor in liver, is one of the most complex organic compounds of known structure (other than proteins and nucleic acids). Its structural formula, elucidated by X-ray analysis, incorporates seven amide functions.

FIGURE 13·7 *The porphin ring system and some of its derivatives*

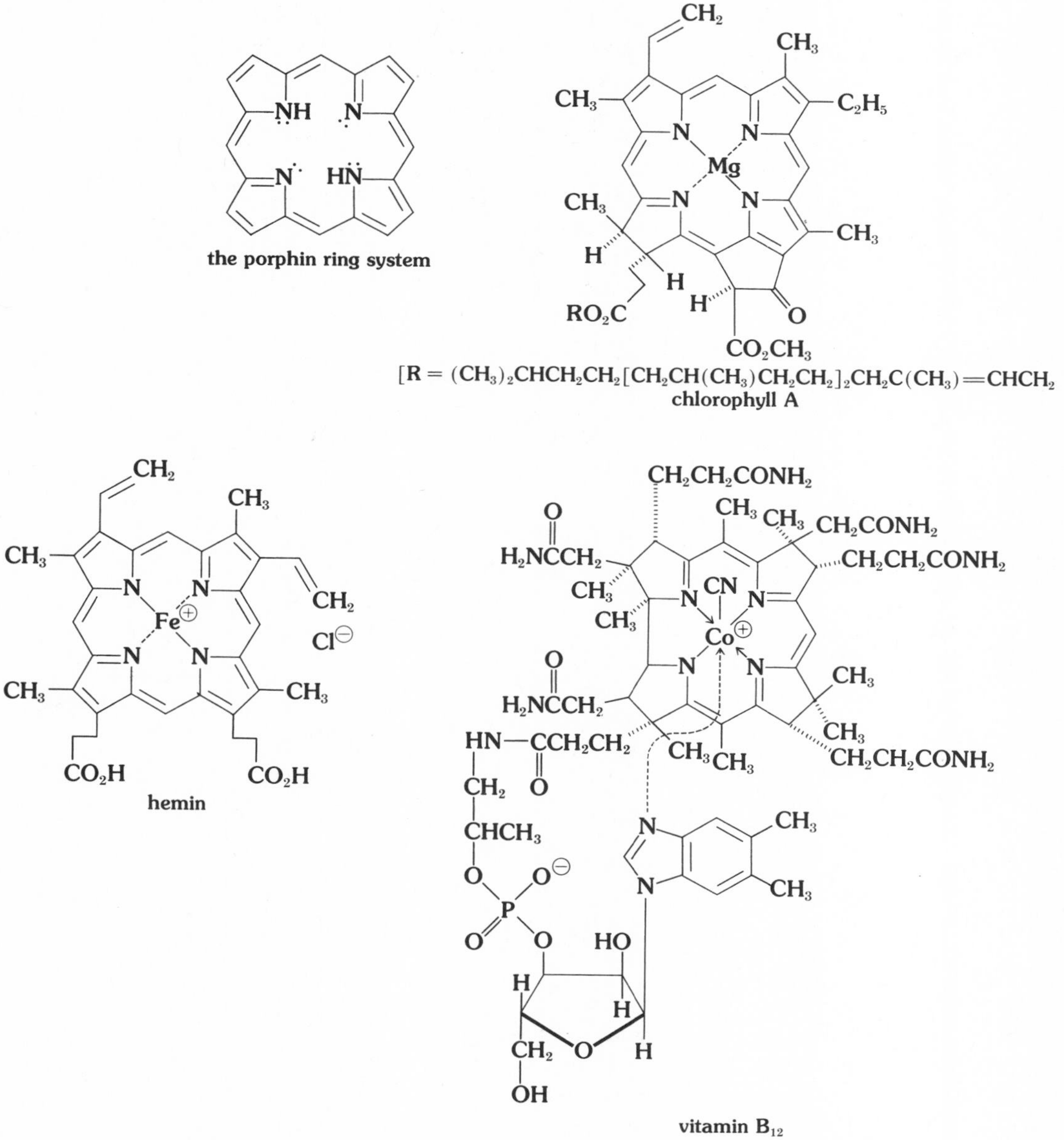

the porphin ring system

$[R = (CH_3)_2CHCH_2CH_2[CH_2CH(CH_3)CH_2CH_2]_2CH_2C(CH_3)=CHCH_2$
chlorophyll A

hemin

vitamin B_{12}

INSECT HORMONES AND PHEROMONES

The normal developmental cycle of insects from larva to pupa to adult is controlled by several hormones. The "juvenile hormone," secreted by the corpora allata, maintains the larval character of the growing insect until it is ready for metamorphosis. The "molting hormone," from the prothoracic gland, initiates transformation to the pupa state, provided juvenile-hormone levels are low enough, and also controls imaginal (adult) development. A poorly characterized substance called the "brain hormone" apparently stimulates the release of the molting hormone, among other functions. Thanks to the efforts of several research groups in this country and abroad, we not only know the molecular structures of the juvenile and molting hormones, but can also synthesize them in the laboratory. This latter accomplishment was important for further studies of the action of these hormones, because their isolation from insects is extremely arduous.

Larvae treated with dilute solutions of juvenile hormone die without completing their development. Since insects can scarcely develop a resistance to their own hormones (as they do to DDT), the insecticidal potential of this substance and related compounds is being explored. The discovery of the juvenile-hormone activity of juvabione is a fascinating story. A colony of *Pyrrhocoris apterus*, which had been brought to Harvard University by a young Czech entomologist, failed to undergo normal metamorphosis and died. The difficulty was subsequently traced to the newspaper used to line the rearing jars (the "paper factor"). After an investigation of various paper samples, it was found that the *New York Times*, the *Wall Street Journal*, and *Science* had this hormonelike effect, while *The Times* (London) and *Nature* (British) were inactive. American paper usually contains pulp from the balsam fir, *Abies balsamea*, and the active substance juvabione was eventually isolated and identified from this source.

juvenile hormone (C_{18})

juvabione

molting hormone
ecdysone: Y = H
crustecdysone: Y = OH

The molting hormone crustecdysone, first isolated from the crayfish, has also been found in the Australian tree *Podocarpus elatus*. The function and origin of these hormonal substances in plants is currently being debated.

13·4 NATURALLY OCCURRING CARBOXYL DERIVATIVES 431

Many insects employ simple chemicals called *pheromones* as signals for reproductive activity, social recognition, defense, and trail marking. For example, the boll weevil, *Anthonomus grandis* (Boheman), uses a compound christened "grandisol" as one of its sex attractants, and the equally unpopular pink boll worm moth, *Pectinophora gossypiella* (Saunders), employs "gossyplure," a mixture of stereoisomeric 7,11-hexadecadienyl acetates, for the same purpose (the trans isomer is more than 10^6 times as active as the cis isomer). Since up to 70% of the DDT once used for agricultural purposes in the United States was applied to cotton, an ecologically more acceptable manner of controlling these pests might be to use these compounds to lure them to strategically located chemical traps. Such a plan has been suggested for control of the gypsy moth, *Porthetria dispar* (L.).

grandisol, sex attractant of the boll weevil

$n\text{-}C_4H_9CH=CHCH_2CH_2CH=CH(CH_2)_6-O-\overset{\overset{\displaystyle O}{\|}}{C}-CH_3$
gossyplure, sex attractant of the pink boll worm moth

$(CH_3)_2CH(CH_2)_4\overset{\overset{\displaystyle O}{\diagdown\diagup}}{CH\!-\!CH}(CH_2)_9CH_3$
disparlure, sex attractant of the gypsy moth

The sex attractants of the moth *Bryotopha similis* (Gelechüdae) and a closely related sibling species, only recently recognized, show a remarkable stereospecificity. The pheromone of the former species is *cis*-9-tetradecenyl acetate, whereas the trans isomer is used by the latter. Furthermore, each pheromone actually repels moths of the other species, thus reinforcing reproductive isolation.

When provoked, the whip scorpion *Mastigoproctua giganteus*, an arachnid 2 to 5 cm in length, sprays a mixture of acetic acid (84%), caprylic acid (5%), and water (11%) at its enemies. The caprylic acid apparently enhances the effectiveness of the defensive agent, CH_3CO_2H, by acting as a wetting agent. Similarly, the blister beetle (family *Meloidae*) secretes a powerful irritant or blistering agent called cantharidin which renders it unappetizing to most insectivores. Cantharidin is a rare example of a naturally occurring anhydride.

cantharidin

Slavemaker ants, *Formica sanguinea*, direct their raids on the nests of other species of the same or related genus by means of odor trails. During the raids they discharge decyl, dodecyl, and tetradecyl acetates at the defending workers. These acetates produce very efficient alarm responses that attract the attackers and at the same time disperse the defenders.

432 EXERCISES

Attine ants of the Acromyrmex and Atta genera cultivate fungus gardens in their nests as a primary source of food. The fungus, which is slow growing and delicate, is cultivated on a compost prepared by the ants from fragments of leaves and flowers. At least one South American species, *Atta sexdens*, protects its fungus gardens from invasion by faster-growing and more viable microorganisms by secreting a selective herbicide. This substance has been identified as β-hydroxydecanoic acid, and preliminary experiments suggest that it might serve as a useful food preservative.

SUMMARY

carboxylic acid A compound incorporating the carboxyl functional group (CO_2H), one of the highest commonly encountered oxidation states of carbon. Carboxylic acids are identified by the IUPAC nomenclature ending *-oic acid*.

fatty acids Normal or unbranched acyclic carboxylic acids. Ester derivatives of fatty acids make up fats, oils, and waxes.

properties of carboxylic acids Substantially higher boiling points than other organic compounds of equivalent molecular weight as a result of hydrogen-bonded association of acid molecules. This hydrogen bonding also shifts the carboxyl proton resonance to very low fields ($\delta = 10$ to 13). Characteristic carbonyl stretching frequencies in the infrared range are useful for distinguishing different derivatives of carboxylic acids (see Table 13.6).

carboxylic acid derivatives Esters (RCO_2R'), acyl halides (RCOX), amides ($RCONR'_2$), and anhydrides (RCO_2COR') (see Table 13.2).

synthesis of carboxylic acids The laboratory preparation of carboxylic acids from other functional groups. Preparation methods include

1 Oxidation of primary alcohols (Cr^{+6} or Mn^{+7} reagents)
2 Oxidation of aldehydes (Ag^{+1}, Cr^{+6}, and Mn^{+7} reagents)
3 Oxidation of aryl side chains (Mn^{+7} or Cr^{+6} reagents)
4 Oxidative cleavage of disubstituted alkenes and alkynes
5 Reaction of Grignard reagents or alkyl lithium reagents with carbon dioxide (carboxylation)
6 Hydrolysis of nitriles ($RCN + 2H_2O \xrightarrow{\text{acid or base}} RCO_2H + NH_3$)

EXERCISES

13·1 Write a structural formula and give an alternative name for each of the following:

a Butyric acid
b Ethanoic acid
c Isovaleric acid
d Salicylic acid
e Fumaric acid
f Sodium caproate
g Acetic anhydride
h Propanamide
i Butanoyl chloride
j Isoamyl acetate
k α-Chlorocaprylic acid
l N-ethylphenylacetamide

13·2 Name the following compounds:

a $CH_3CH_2\overset{\overset{O}{\|}}{C}OCH_2CH_3$

b $H-\overset{\overset{O}{\|}}{C}-OH$

c $CH_3CH_2CH_2CH_2\overset{\overset{H_2N}{|}}{CH}-\overset{\overset{O}{\|}}{C}-NH_2$

d $CH_3\overset{\overset{O}{\|}}{C}NHCH_3$

e $CH_3CH_2\overset{\overset{O}{\|}}{C}-N\overset{CH_2CH_3}{\underset{CH_2CH_3}{\diagdown}}$

f $\phi\overset{\overset{O}{\|}}{C}-Cl$

g (benzene ring with CO_2H and $N(CH_3)H$ ortho substituents)

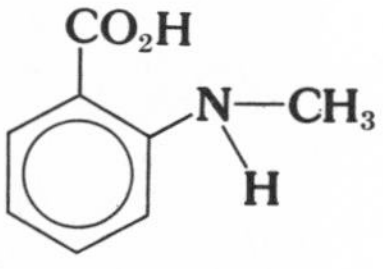

h (benzene ring with $C(=O)OCH_3$ and NH_2 ortho substituents)

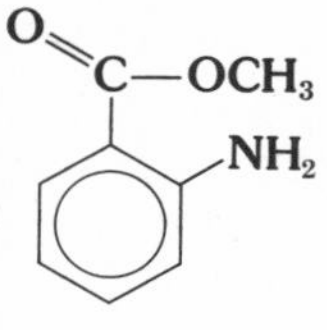

i $HO_2CCH_2CO_2H$

j (cyclohexane with CO_2H and CH_3 substituents, stereochemistry shown)

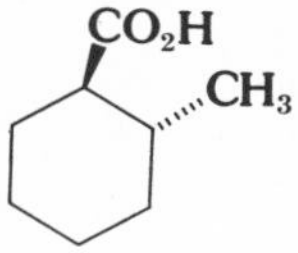

13·3 Write structural formulas for the following:
a Methanesulfonic acid
b p-Toluenesulfonyl chloride
c Diethyl sulfate
d Benzanilide
e Benzenesulfinic acid
f Trimethyl phosphite
g Benzenesulfonamide
h Phenylphosphonic acid
i Phenyl p-toluenesulfinate
j Trimethylphosphine

13·4 Classify each of the following as either a wax, a fat or oil, or a phospholipid:

a Cetyl myristate (cetyl = $n\text{-}C_{16}H_{33}$)

b $CH_3(CH_2)_{24}CO_2(CH_2)_{29}CH_3$

c $CH_2-OCO(CH_2)_{17}CH_3$
   $|$
   $CH-OCO(CH_2)_{11}CH_3$
   $|$
   $CH_2-OCO(CH_2)_{13}CH_3$

d Glyceryl tripalmitate

e $CH_2-O-\overset{\overset{O}{\|}}{C}-(CH_2)_{17}CH_3$
   $|$
   $CH-O-CO(CH_2)_{17}CH_3$
   $|$
   $CH_2-O-\overset{\overset{O}{\|}}{\underset{\overset{|}{O^{\ominus}}}{P}}-O-CH_2CH_2\overset{\oplus}{N}(CH_3)_3$

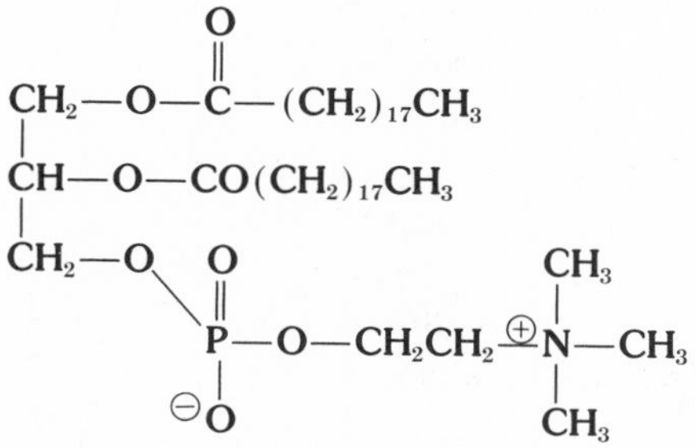

434 EXERCISES

13·5 Arrange each set of compounds in order of increasing boiling point:

a $CH_3CH_2CH_2CH_2OH$ $CH_3CH_2OCH_2CH_3$ $CH_3CH_2CH_2CHO$ $CH_3CH_2CO_2H$

b $CH_3CH_2CO_2H$ $CH_3\overset{O}{\overset{\|}{C}}OCH_3$ $CH_3\overset{O}{\overset{\|}{C}}-NHCH_3$

13·6 For the following physiologically active compounds indicate the oxidation state for each carbon atom bonded to one or more heteroatoms (atoms other than carbon or hydrogen). Which of these carbon atoms have the same oxidation state as a carboxyl carbon atom?

a   tetrahydrozoline, a nasal decongestant

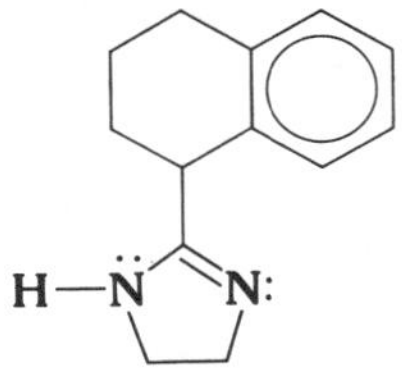

b   thalidomide, a sedative and hypnotic useful in treating leprosy; causes fetal abnormality

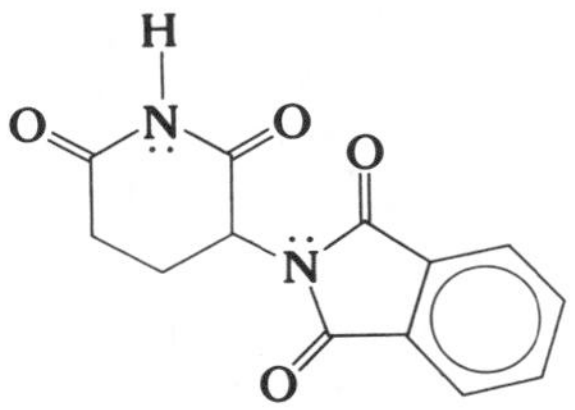

c   mecloqualone, a sedative

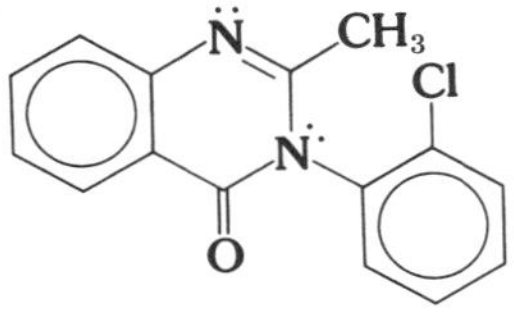

d   saxitoxin hydrochloride, seasonal toxin from clams

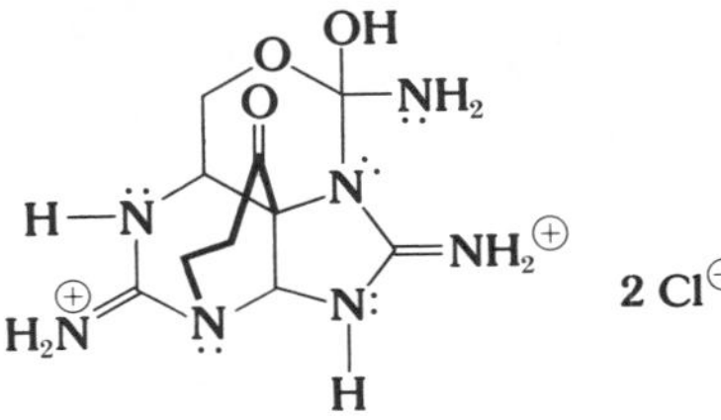

13·7 Suggest a mechanism for the formation of tetrachlorodioxin in the manufacture of 2,4,5-T by a Williamson synthesis.

13·8 Explain the difference in water solubility of butyric acid and methyl propanoate as shown in Table 13.4.

13·9 Aspirin is acetylsalicylic acid, and oil of wintergreen (used in liniments) is methylsalicylate. Write equations showing how each could be prepared from salicylic acid.

13·10 Treatment of anthranilic acid with aqueous HCl and $NaNO_2$ yields salicyclic acid and phenol. Show the pathway by which these products are formed.

13·11 Compound A (C_8H_9Cl), when treated with hot concentrated aqueous $KMnO_4$, gives compound B ($C_7H_5ClO_2$). Reaction of an ether solution of A with magnesium metal, followed by the addition of dry ice, yields a salt, which then gives compound C ($C_9H_{10}O_2$) on acidification. Oxidation of C with $KMnO_4$ gives isophthalic acid. Name compounds A, B, and C and write equations for the reactions described.

13·12 Which prominent features of the spectra indicated could be used to distinguish the following pairs of compounds?

a CH$_3$CH$_2$CH$_2$COCH$_3$ and CH$_3$CH$_2$CH$_2$COH;   pmr spectrum

b CH$_3$CO$_2$H and CH$_3$CH$_2$OH;   infrared spectrum

c CH$_3$COCOCH$_3$ and CH$_3$COCH$_3$;   infrared spectrum

d ϕ—CO$_2$H and ϕ—C(=O)—H;   infrared spectrum

e (γ-butyrolactone) and CH$_3$CH$_2$COCH$_3$;   infrared spectrum

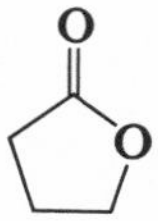

f CH$_3$CH$_2$NH$_2$ and CH$_3$CH$_2$CO$_2$H;   pmr spectrum

g C$_6$H$_5$—COCH$_3$ and (2-thienyl)—COCH$_3$;   infrared spectrum

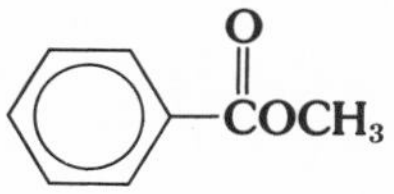

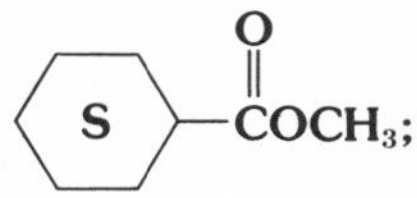

h CH$_3$CH$_2$NH$_2$ and CH$_3$CH$_2$C(=O)—NH$_2$;   infrared spectrum

13·13 Sketch the pmr spectra you would predict for the following compounds. What feature or features of these spectra could be used to distinguish between each pair?

a Ethyl acetate and methyl propionate
b Propionic acid and propionamide
c Acetic acid and acetaldehyde
d Benzoic acid and benzoic anhydride
e Isobutyric acid and isopropyl alcohol
f Maleic acid and succinic acid
g Maleic acid and fumaric acid
h Acetamide and methyl amine

13·14 The "queen bee substance," a C$_{10}$H$_{16}$O$_3$ compound from the mandibular glands of the queen bee, inhibits the development of ovaries in worker bees. This compound (Q) is soluble in sodium bicarbonate solution and gives a crystalline 2,4-dinitrophenylhydrazone derivative. Catalytic hydrogenation of Q, followed by Wolff-Kishner reduction, yields decanoic acid as the only organic product. Ozonolysis of Q (hydrogen peroxide workup) gives a C$_8$H$_{14}$O$_3$ compound P which is converted to octanoic acid by Wolff-Kishner reduction. Compound P can be prepared from cycloheptanone in three steps as follows: (1) reaction with methyl magnesium iodide, (2) dehydration with phosphoric acid (assume Saytzeff specificity), and (3) ozonolysis followed by hydrogen peroxide workup. Write structures for Q and P and equations for all the reactions described.

13·15 Suggest a structure for each compound consistent with the data given:

a $C_5H_{10}O_2$; infrared absorption at 1735 cm^{-1}; pmr signals at $\delta = 1.2$ (6H, doublet), $\delta = 2.0$ (3H, singlet), and $\delta = 5.0$ (1H, septet)

b $C_3H_6O_2$; infrared absorptions at 1740 and 1200 cm^{-1}; pmr signals at $\delta = 1.3$ (3H, triplet), $\delta = 4.2$ (2H, quartet), and $\delta = 8.0$ (1H, singlet)

c $C_8H_{14}O_4$; infrared absorption at 1740 cm^{-1}; pmr signals at $\delta = 1.2$ (3H, triplet), $\delta = 2.6$ (2H, singlet), and $\delta = 4.1$ (2H, quartet)

d $C_6H_{12}O_2$; infrared absorption at 1735 cm^{-1}; pmr signals at $\delta = 1.45$ (3H, singlet) and $\delta = 1.96$ (1H, singlet)

13·16 Cinnamic acid ($C_6H_5CH=CH-CO_2H$) and its esters are found in oil of cinnamon. The pmr spectrum of cinnamic acid is shown below.

a What is the configuration of substituents about the double bond?

b How can cinnamic acid be prepared from benzaldehyde?

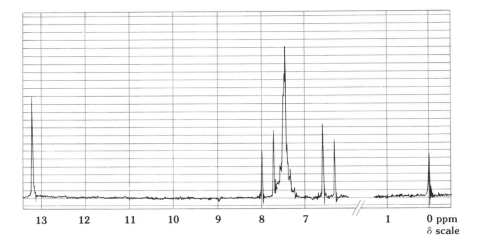

REACTIONS OF CARBOXYLIC ACIDS AND THEIR DERIVATIVES

A CARBOXYLIC acid derivative is characterized by a carbonyl group bonded directly to another functional group, such as NH_2, OR, Cl, or other halogens. This combined functional group exhibits a chemical behavior quite different from that predicted from the separate properties of the component groups. We shall first examine some of these differences for the carboxyl group itself (CO_2H) and then extend our study to other combinations. The chemistry of carboxylic acid derivatives is a storehouse of useful transformations that enable the chemist to build complex molecules from simple starting materials. In this respect we are only mimicking nature, for reactions of this kind are employed by living organisms to synthesize the macromolecules that are vital to all life on this planet.

14·1 SUBSTITUTION OF HYDROGEN IN THE CARBOXYL GROUP

ACIDITY AND SALT FORMATION

As we have observed in previous chapters, carboxylic acids are roughly a trillion (10^{12}) times more acidic than similarly constituted alcohols, and alcohols in turn are far more acidic than ketones, amines, and hydrocarbons. This wide range of acidities is most conveniently expressed on a logarithmic scale. Accordingly, we define an exponential function pK_a such that

$$pK_a = -(\log K_a) = \log \frac{1}{K_a} \tag{14.1}$$

where K_a is the dissociation constant.

14·1 SUBSTITUTION OF HYDROGEN IN THE CARBOXYL GROUP

TABLE 14·1 *Acidity constants of some common organic and inorganic acids*

compound	K_a†	pK_a	compound	K_a†	pK_a
HCO_2H	1.7×10^{-4}	3.75	H_2S	1.0×10^{-7}	7.00
CH_3CO_2H	1.7×10^{-5}	4.75	ϕOH	1.3×10^{-10}	9.89
FCH_2CO_2H	2.2×10^{-3}	2.67	CH_3NO_2	6.2×10^{-11}	10.21
$ClCH_2CO_2H$	1.6×10^{-3}	2.87	H_2O_2	2.5×10^{-12}	11.60
$BrCH_2CO_2H$	1.4×10^{-3}	2.90	H_2O	$\sim 10^{-15}$	15
Cl_2CCO_2H	1.7×10^{-1}	0.77	C_2H_5OH	$\sim 10^{-17}$	17
ϕCO_2H	6.3×10^{-5}	4.21	$HC\equiv CH$	$\sim 10^{-25}$	25
H_2CO_3	K_{a1} 4.3×10^{-7}	6.36	CH_3COCH_3	$\sim 10^{-20}$	20
	K_{a2} 5.6×10^{-11}	10.25	NH_3	$\sim 10^{-34}$	34

†Extrapolated to aqueous solution.

Table 14.1 illustrates this relationship and together with Table 12.6 provides a comparison of the acidities of some common acids as measured by their dissociation constants.† The pK_a of an acid is, in effect, the pH of an aqueous solution in which the acid is half-ionized.

PROBLEM 14·1
Write an expression for the acidity constant (K_a) of benzoic acid and show that pK_a is equal to the pH of a half-ionized aqueous solution.

The extraordinary acidity of the hydroxyl group in the carboxylic acids can be explained in much the same way that we accounted for the acidity of the phenols (Section 10.5). Since X-ray diffraction measurements of carboxylate salts show that the two carbon-oxygen bonds are of equal length, it is reasonable to describe carboxylate ions as a resonance hybrid of two structurally (and energetically) equivalent contributors (see Figure 14.1). The resulting stabilization of this species should be very great, and as we found in earlier chapters, that acid having the most stable conjugate base will generally be the strongest.

In comparing the relative acidities of different acids we must consider the enthalpy and entropy changes that occur during ionization. The ionization reactions of ethanol and acetic acid will serve to illustrate this point:

$$CH_3CH_2OH + H_2O \rightleftharpoons CH_3CH_2O^\ominus + H_3O^\oplus \qquad (14.2)$$
$$K_a \approx 10^{-17}$$

†The extraordinary range of acidities represented by the compounds listed in Tables 12.6 and 14.1 is often overlooked. From trichloroacetic acid ($pK_a = 0.77$) to ethane ($pK_a \approx 45$) the acidity constants span almost 50 powers of 10. In terms of a more familiar variable, such as distance (length), the span of distances ranging from the diameter of an atomic nucleus (about 10^{-12} cm) to that of the farthest galaxy (about 10^{28} cm) covers approximately 40 powers of 10, less than one-millionth of the acidity spread.

14·1 SUBSTITUTION OF HYDROGEN IN THE CARBOXYL GROUP

$$CH_3CO_2H + H_2O \rightleftharpoons CH_3-C\begin{matrix}O^{1/2\ominus}\\O^{1/2\ominus}\end{matrix} + H_3O^{\oplus} \quad (14.3)$$

$$K_a \approx 10^{-5}$$

Since charge dispersal (or delocalization) in an ion reduces the charge density at a given atom, the potential energy (enthalpy) of the acetate ion is correspondingly lowered with respect to ethoxide anion, which supports a full negative charge on the single oxygen atom. Furthermore, ionization is greatly enhanced by ion solvation (see Figure 3.7), and the degree of solvation, or solvent-molecule ordering, about a single charged atom such as ethoxide oxygen will necessarily be greater than for a delocalized ion such as acetate. This orientation of solvent molecules in the space around an ion is achieved only at the cost of a decrease in entropy (equation 2.21); consequently, the delocalized ion is again favored, as shown in Figure 14.1.

Electron-withdrawing groups near a carboxyl function usually increase its acidity. The pK_a values of the monochlorobutyric acids in Table 14.2 clearly show how this effect gradually disappears as the distance from the carboxyl group increases.

Since the ionization constants of most carboxylic acids are larger than K_{a1} for carbonic acid (H_2CO_3), they readily dissolve in aqueous sodium

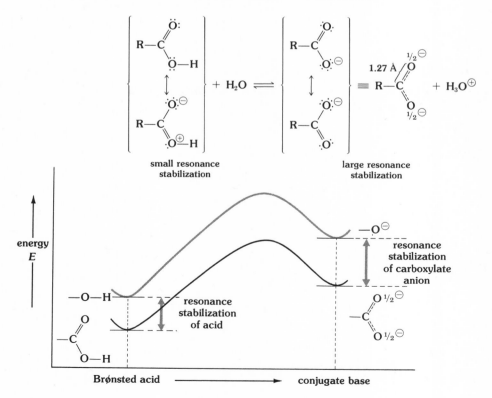

FIGURE 14·1 *Resonance stabilization of the carboxylate anion*

14·1 SUBSTITUTION OF HYDROGEN IN THE CARBOXYL GROUP

TABLE 14·2 *Acidity of chlorobutyric acids*

compound	pK_a†
$CH_3CH_2CHClCO_2H$	2.85
$CH_3CHClCH_2CO_2H$	4.05
$ClCH_2CH_2CH_2CO_2H$	4.53
$CH_3CH_2CH_2CO_2H$	4.82

† In the case of α-chlorobutyric acid the change in entropy is found to have a greater effect on acidity than changes in enthalpy.

bicarbonate solutions, accompanied by a characteristic effervescence caused by the evolution of carbon dioxide. We are thus able to distinguish and separate carboxylic acids from alcohols and phenols.

$$RCO_2H + NaHCO_3 \longrightarrow RCO_2^{\ominus}Na^{\oplus} + H_2O + CO_2 \tag{14.4}$$

PROBLEM 14·2
Show how a mixture of benzoic acid, 3,5-dimethyl phenol, and 1-hexanol can be separated by chemical means (that is, by methods other than distillation, chromatography, etc.)

PROBLEM 14·3
A 156-mg sample of a pure carboxylic acid requires 12 ml of a 0.1 N sodium hydroxide solution for titration to a phenolphthalein endpoint. What is the equivalent weight (proportional molecular weight per carboxyl function) of this acid?

CARBOXYLATE OXYGEN AS A NUCLEOPHILE

Esters can be prepared from alkyl halides and carboxylate salts by both S_N2 and S_N1 mechanisms:

$$\phi CO_2Na + CH_2=CHCH_2Br \xrightarrow[S_N2]{H_2O} \phi-C\begin{smallmatrix}\nearrow O \\ \searrow O-CH_2CH=CH_2\end{smallmatrix} \tag{14.5}$$

$$\square\!\!-\!CO_2Na + \phi_3CCl \xrightarrow[S_N1]{H_2O} \square\!\!-\!C\begin{smallmatrix}\nearrow O \\ \searrow O-C\phi_3\end{smallmatrix} \tag{14.6}$$

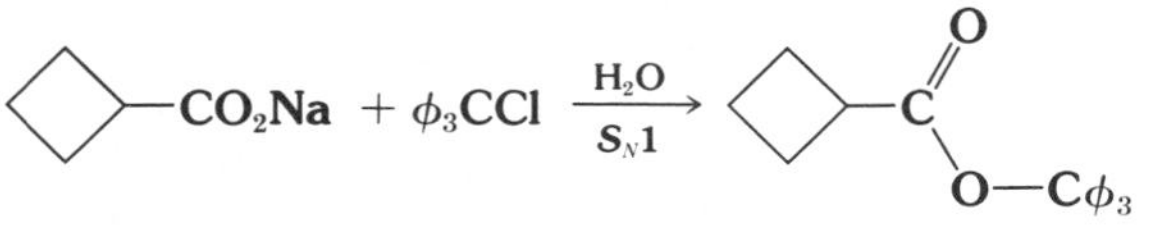

Simple alkyl halides react more readily with the corresponding silver salts owing to the silver cation's ability to assist halide ionization (see page 87).

$$R-CO_2Ag + R'-X \longrightarrow R-C\begin{smallmatrix}\nearrow O \\ \searrow O-R'\end{smallmatrix} + AgX \tag{14.7}$$

(R' = primary or secondary alkyl group
X = Cl, Br, or I)

Acid-catalyzed addition of carboxylic acids to double and triple bonds also generates esters in good yield:

$$\text{C}_5\text{H}_9\text{-CO}_2\text{H} + \text{CH}_2\text{=C(CH}_3)_2 \xrightarrow[\text{catalyst}]{\text{H}_2\text{SO}_4} \text{C}_5\text{H}_9\text{-C(=O)-O-C(CH}_3)_3 \tag{14.8}$$

$$2\text{CH}_3\text{CO}_2\text{H} + \text{HC} \equiv \text{CH} \xrightarrow{\text{HgSO}_4} (\text{CH}_3\text{CO}_2)_2\text{CHCH}_3 \tag{14.9}$$

$$\text{CH}_2\text{=CHCH}_2\text{CH}_2\text{CO}_2\text{H} \xrightarrow{\text{ArSO}_3\text{H}} \text{(methyl-γ-butyrolactone)} \tag{14.10}$$

One of the quickest and most efficient methods of preparing methyl esters from carboxylic acids involves treating a solution or suspension of the acid with an ether solution of the poisonous and explosive substance diazomethane (CH_2N_2):

$$\text{R-CO}_2\text{H} + \left\{ \underset{\text{diazomethane}}{\text{CH}_2\text{=N}^{\oplus}\text{=}\ddot{\text{N}}^{\ominus}} \longleftrightarrow {}^{\ominus}\text{:CH}_2\text{-N}^{\oplus}\text{≡N:} \right\} \xrightarrow{\text{ether}} \text{R-C(=O)-O-CH}_3 + \text{N}_2 \tag{14.11}$$

This bright-yellow compound is usually prepared immediately prior to use and is always handled in solution (the pure compound is a gas at room temperature).

PROBLEM 14·4
Write equations illustrating reasonable mechanisms for reactions 14.10 and 14.11. *Hint:* In the latter case consider an initial acid-base proton transfer.

14·2 SUBSTITUTION OF HYDROXYL IN THE CARBOXYL GROUP

The hydroxyl function is, generally speaking, a very poor anionic leaving group. However, it can easily be modified or transformed into derivatives which are good leaving groups.

ESTERIFICATION

Two fundamentally different approaches may be used in transforming carboxylic acids into their ester derivatives. Either the acidic proton of the acid can be replaced by an alkyl group, as described above, or the hydroxyl moiety of the acid can be replaced by an alkoxyl group.

If we allow a carboxylic acid to react with an alcohol in the presence of a mineral-acid catalyst, we find that in most cases reversible ester formation occurs:

$$\text{C}_3\text{H}_7\text{CO}_2\text{H} + \text{C}_2\text{H}_5\text{OH} \underset{K_{eq} \approx 2}{\overset{\text{H}_2\text{SO}_4}{\rightleftarrows}} \text{C}_3\text{H}_7\text{-C(=O)-O-C}_2\text{H}_5 + \text{H}_2\text{O} \tag{14.12}$$

This reaction can be conducted so as to favor either esters or acids by increasing or decreasing the concentration of appropriate reactants or products (such as H_2O).

14·2 SUBSTITUTION OF HYDROXYL IN THE CARBOXYL GROUP

Steric hindrance in either the alcohol or acid reactants *decreases* the rate of esterification. Thus trimethylacetic acid (2,2-dimethylpropanoic acid) reacts with methanol roughly one-fortieth as fast as acetic acid does under similar conditions:

$$CH_3-C(=O)OH \quad\quad CH_3-C(CH_3)(CH_3)-C(=O)OH$$

unhindered   hindered

In all cases a strong acid catalyst is required, and bases such as sodium hydroxide are effective only for the hydrolysis of esters, not for their formation. Bifunctional molecules having hydroxyl and carboxyl functions oriented so that five- and six-membered lactones can form usually undergo spontaneous lactonization, illustrating again that intramolecular reactions are generally faster than their intermolecular counterparts:

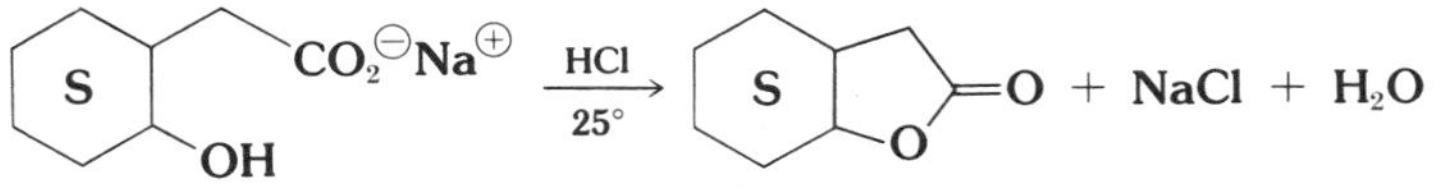

$$\text{(S)-CO}_2^\ominus\text{Na}^\oplus\text{-OH} \xrightarrow[25°]{HCl} \text{(S)-lactone}=O + NaCl + H_2O \quad\quad (14.13)$$

PROBLEM 14·5
a Write an expression for the equilibrium constant of reaction 14.12. What effect would doubling the concentration of ethanol have on the conversion to ester?
b How would you achieve complete (100%) conversion of butyric acid to ethyl butyrate?

Although esterification superficially resembles the neutralization of a base by an acid (assuming that alcohols are sources of hydroxide ion, which, of course, they are not), nothing could be further from the actual state of affairs. If we probe the mechanism of esterification by using an isotopically labeled alcohol as a reactant, we find that the ^{18}O label is retained in the ester, rather than being lost to the water. The mechanism depicted in Figure 14.2 not only accounts for the fate of the labeled oxygen atom, but also explains the steric hindrance and the acid catalysis.

FIGURE 14·2 *Acid-catalyzed esterification*

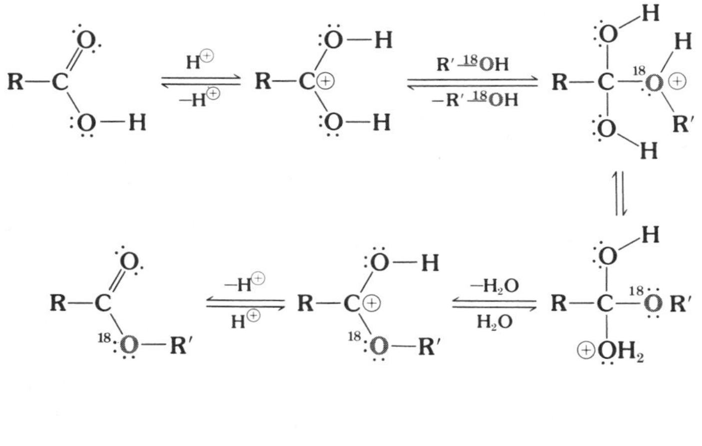

$$R-C(=\ddot{O})\ddot{O}-H \underset{-H^\oplus}{\overset{H^\oplus}{\rightleftharpoons}} R-C^\oplus(\ddot{O}-H)(\ddot{O}-H) \underset{-R'^{18}OH}{\overset{R'^{18}OH}{\rightleftharpoons}} R-C(^{18}\overset{H}{\underset{R'}{\ddot{O}}}{}^\oplus)(\ddot{O}-H)(\ddot{O}\overset{H}{_{H}})$$

$$\updownarrow$$

$$R-C(=\ddot{O})\,^{18}\ddot{O}-R' \underset{H^\oplus}{\overset{-H^\oplus}{\rightleftharpoons}} R-C^\oplus(\ddot{O}-H)(^{18}\ddot{O}-R') \underset{H_2O}{\overset{-H_2O}{\rightleftharpoons}} R-C(\overset{H}{\ddot{O}})(^{18}\ddot{O}R')(\overset{\oplus}{\ddot{O}H_2})$$

14·2 SUBSTITUTION OF HYDROXYL IN THE CARBOXYL GROUP

It is worth noting that these esterification reactions proceed under much milder conditions than acid-catalyzed conversion of primary and secondary alcohols into symmetrical ethers (see Figure 8.10). Consequently, this mode of reaction—a preliminary addition to the carbon-oxygen double bond, followed by proton tautomerism and loss of a small stable fragment such as water—seems to represent a particularly advantageous pathway for the transformations of carboxylic acids and their derivatives. Indeed, we shall find that the following general mechanism operates in many of the reactions to be discussed in the rest of this chapter:

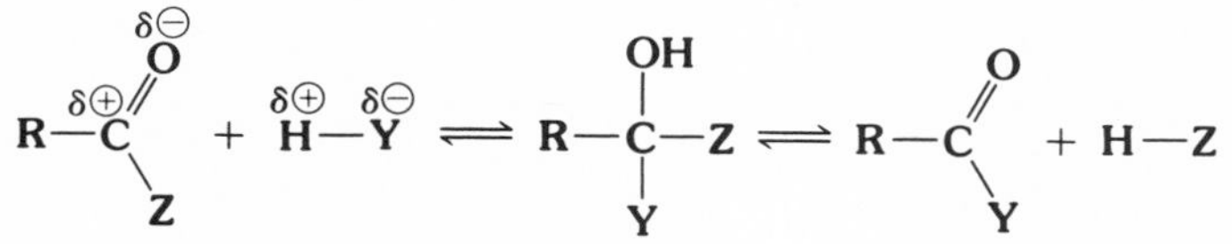

$$\overset{\delta\ominus}{\underset{Z}{R-\overset{\delta\oplus}{\overset{\|}{C}}}}\overset{O}{} + \overset{\delta\oplus}{H}-\overset{\delta\ominus}{Y} \rightleftharpoons R-\underset{Y}{\overset{OH}{\underset{|}{\overset{|}{C}}}}-Z \rightleftharpoons R-\underset{Y}{\overset{O}{\overset{\|}{C}}} + H-Z \quad (14.14)$$

PROBLEM 14·6
Mesitoic acid (2,4,6-trimethylbenzoic acid) is inert to ordinary esterification reaction conditions. However, if mesitoic acid is dissolved in cold concentrated sulfuric acid and then poured into cold methanol, a good yield of the methyl ester results. Explain the reason for this. *Hint:* A completely different mechanism is operating here.

ACYL HALIDE FORMATION

Acyl chlorides (the bromides and iodides are seldom used) are easily prepared from carboxylic acids by reaction with thionyl chloride, phosphorus trichloride, or phosphorus pentachloride. Thionyl chloride ($SOCl_2$) is the preferred reagent for making acyl chlorides because it gives only gaseous byproducts.

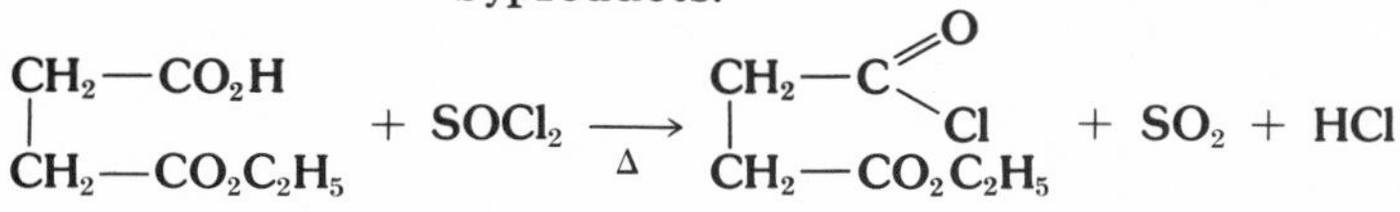

$$\begin{array}{c} CH_2-CO_2H \\ | \\ CH_2-CO_2C_2H_5 \end{array} + SOCl_2 \xrightarrow{\Delta} \begin{array}{c} CH_2-C\overset{O}{\underset{Cl}{\diagup}} \\ | \\ CH_2-CO_2C_2H_5 \end{array} + SO_2 + HCl \quad (14.15)$$

$$O_2N-\underset{}{\bigcirc}-CO_2H + PCl_5 \xrightarrow{\Delta} O_2N-\underset{}{\bigcirc}-C\overset{O}{\underset{Cl}{\diagdown}} + POCl_3 + HCl \quad (14.16)$$

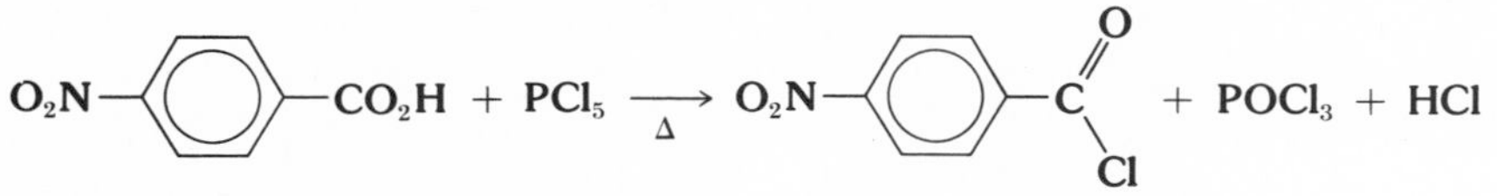

ANHYDRIDE FORMATION

Symmetrical anhydrides are conveniently prepared by dehydration of carboxylic acids with phosphorus pentoxide. This approach is not well suited for the preparation of mixed anhydrides, because it yields a mixture that includes the two symmetrical anhydrides.

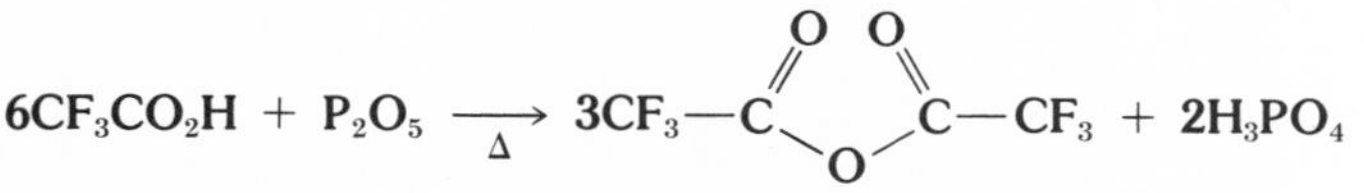

$$6CF_3CO_2H + P_2O_5 \xrightarrow{\Delta} 3CF_3-\overset{O}{\overset{\|}{C}}\underset{O}{\diagdown}\overset{O}{\overset{\|}{C}}-CF_3 + 2H_3PO_4 \quad (14.17)$$

Substituted succinic and glutaric acids yield five- and six-membered cyclic anhydrides on being heated with acetic anhydride, as illustrated by the following reaction:

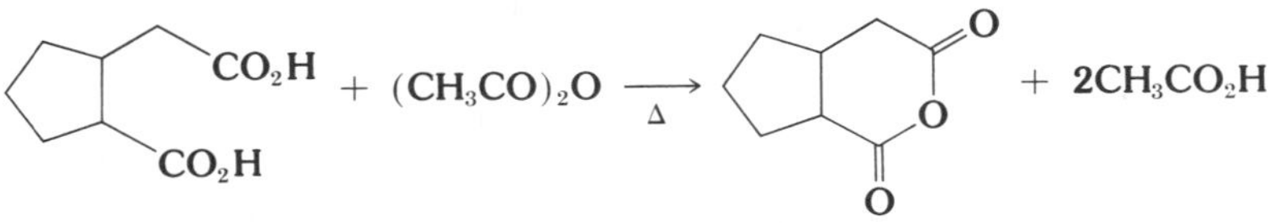

14·3 DECARBOXYLATION REACTIONS

$$\text{(cyclopentane-CH}_2\text{-CO}_2\text{H, -CO}_2\text{H)} + (CH_3CO)_2O \xrightarrow{\Delta} \text{(bicyclic anhydride)} + 2CH_3CO_2H \quad (14.18)$$

In order to prepare mixed anhydrides, we can employ a strategy similar to that used in the Williamson ether synthesis, wherein an acyl halide is allowed to react with a carboxylic acid salt:

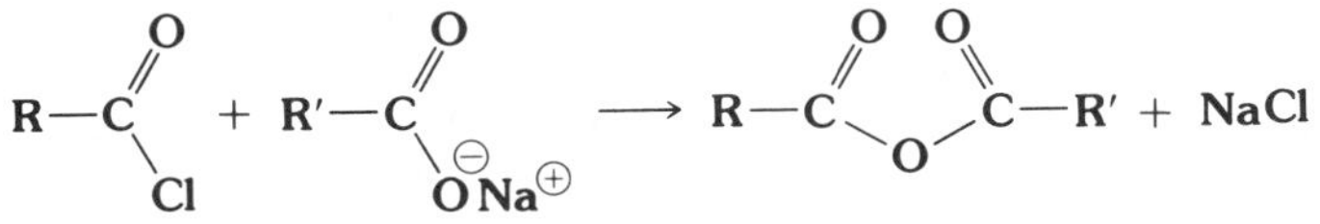

$$R-\underset{Cl}{\overset{O}{\underset{\|}{C}}} + R'-\underset{\overset{\ominus}{O}\text{Na}^{\oplus}}{\overset{O}{\underset{\|}{C}}} \longrightarrow R-\overset{O}{\underset{\|}{C}}-O-\overset{O}{\underset{\|}{C}}-R' + NaCl \quad (14.19)$$

14·3 DECARBOXYLATION REACTIONS

Carboxylic acids having a β carbonyl function or a β,γ double bond often suffer loss of carbon dioxide, called *decarboxylation*, on being heated. A cyclic transition state involving a concerted hydrogen transfer and double-bond shift is proposed for these reactions:

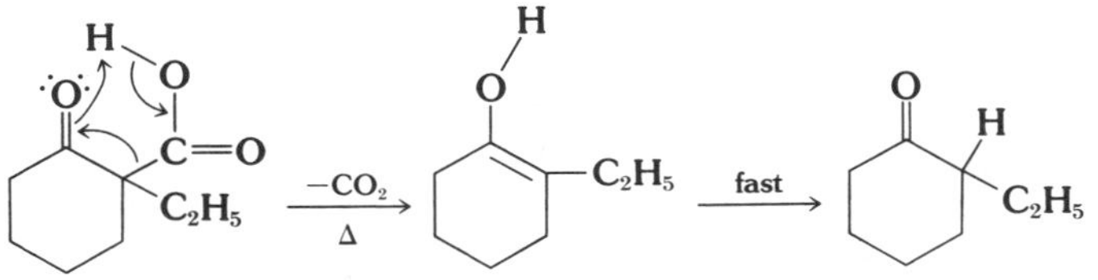

$$\text{(cyclohexanone with C}_2\text{H}_5\text{ and CO}_2\text{H)} \xrightarrow[\Delta]{-CO_2} \text{(enol, C}_2\text{H}_5\text{)} \xrightarrow{fast} \text{(cyclohexanone, C}_2\text{H}_5\text{)} \quad (14.20)$$

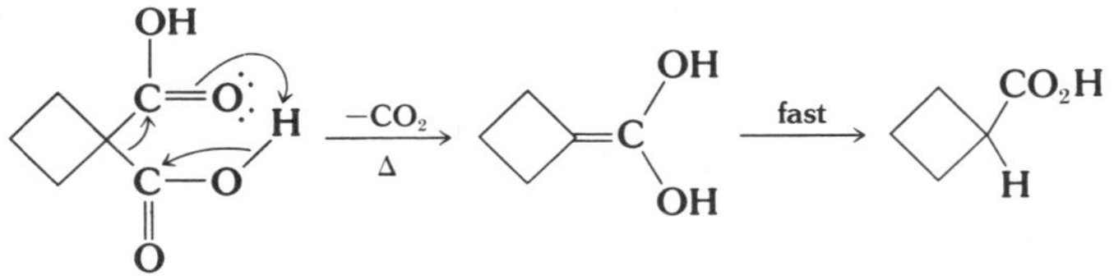

$$\text{(cyclobutane spiro malonic acid)} \xrightarrow[\Delta]{-CO_2} \text{(cyclobutane=C(OH)}_2\text{)} \xrightarrow{fast} \text{(cyclobutane-CO}_2\text{H)} \quad (14.21)$$

$$H_2C=CH-CH_3, \text{ with } O-C(CH_3)=O \xrightarrow[\Delta]{-CO_2} CH_3CH=C(CH_3)_2 \quad (14.22)$$

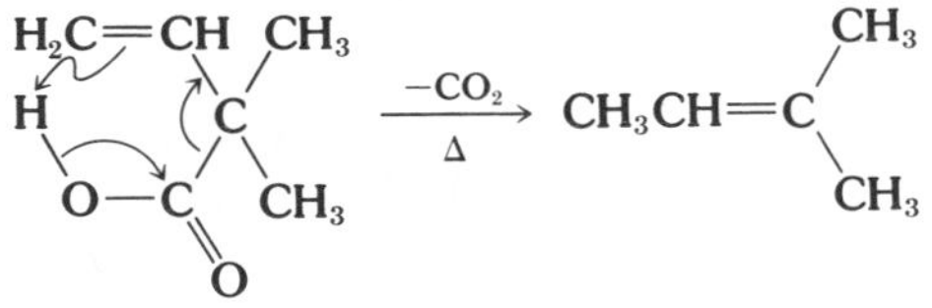

PROBLEM 14·7

a Predict the major product if reaction 14.20 is conducted in heavy water (D_2O).

b The bicyclic β-keto acid shown below does not suffer decarboxylation under conditions which bring about reaction 14.20. Suggest an explanation.

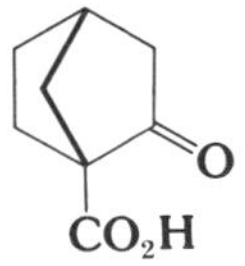

(bicyclic β-keto acid structure with CO₂H)

Oxidative decarboxylation of carboxylic acids to alkyl halides may be accomplished by metal ions such as Pb^{+4} and Hg^{+2} in the presence of halogen (or in some cases, halide ions). Silver and thallous salts of the acids may also be used. A free-radical mechanism has been suggested for these transformations:

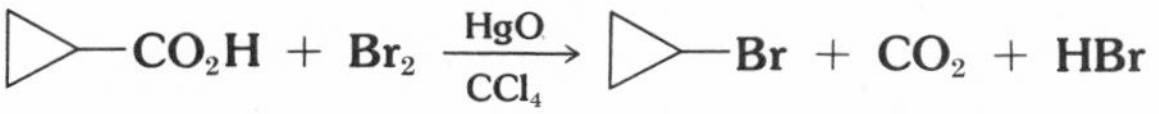

$$\triangleright\!\!-\!CO_2H + Br_2 \xrightarrow[CCl_4]{HgO} \triangleright\!\!-\!Br + CO_2 + HBr \quad (14.23)$$

$$n\text{-}C_{17}H_{35}CO_2Tl + Br_2 \longrightarrow n\text{-}C_{17}H_{35}Br + CO_2 + TlBr \quad (14.24)$$

Substituted succinic acids suffer oxidative bis-decarboxylation to alkenes on treatment with Pb^{+4} oxidizing agents. This useful reaction was instrumental in the synthesis of an interesting isomer of benzene called *Dewar benzene*:

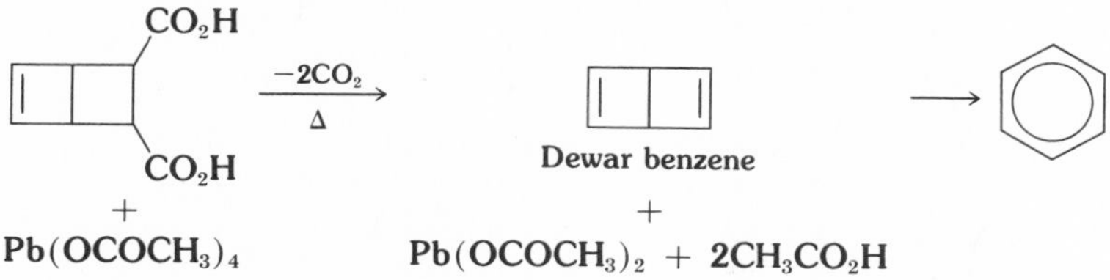

(14.25)

14·4 REDUCTION OF THE CARBOXYL GROUP

LITHIUM ALUMINUM HYDRIDE

Carboxylic acids are reduced to primary alcohols under mild conditions by the action of lithium aluminum hydride solutions. One equivalent of the hydride is consumed by the active hydrogen of the acid:

$$4RCO_2H + 3LiAlH_4 \xrightarrow{\text{ether or THF}} 4H_2 + 4RCH_2OM + \text{metal oxides} \xrightarrow{H_2O} 4RCH_2OH \quad (14.26)$$
$$(M = Li \text{ or } Al)$$

Sodium borohydride reacts with the acidic hydrogen atom but does not cause any reduction of carboxylic acids.

DIBORANE

Diborane rapidly reduces carboxylic acids to primary alcohols. Consequently, the presence of this function would complicate the hydroboration of an alkene elsewhere in the molecule. Fortunately, this reaction differs from the lithium aluminum hydride reductions in one important respect: the salts of carboxylic acids are not reduced. This fact can be used to protect a carboxyl function from the action of diborane:

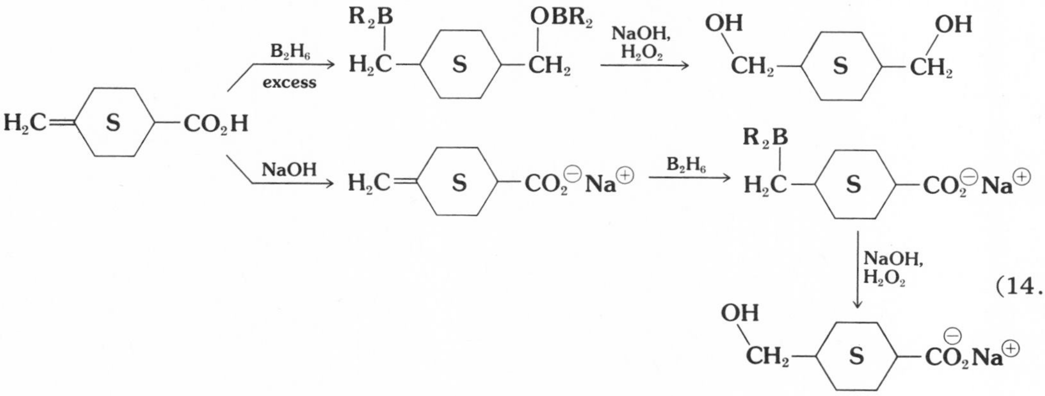

(14.27)

14·5 ACYLATION REACTIONS OF CARBOXYLIC ACID DERIVATIVES

REACTIVITY TO NUCLEOPHILIC SUBSTITUTION

The most common and important reaction of carboxylic acids and their derivatives is nucleophilic substitution:

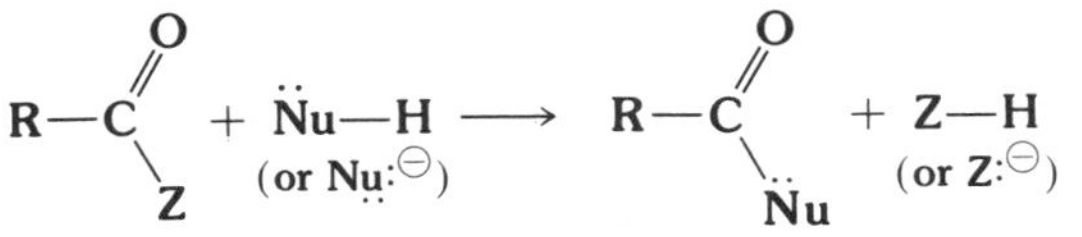

$$R-\underset{Z}{\overset{O}{\underset{\|}{C}}} + Nu-H \text{ (or Nu}^{\ominus}) \longrightarrow R-\underset{Nu}{\overset{O}{\underset{\|}{C}}} + Z-H \text{ (or Z}^{\ominus}) \tag{14.28}$$

Transformations of this kind are often referred to as *acylation reactions*, a term which refers to the bonding of an acyl group (RCO—) to a nucleophilic moiety. Acylation reactions involving common nucleophiles such as water, alcohols, and ammonia are usually termed *hydrolysis*, *alcoholysis*, and *ammonolysis*, respectively. In most cases these substitution reactions do not proceed by the familiar $S_N 1$ and $S_N 2$ mechanisms favored by saturated systems, but by a two-step *addition-elimination* pathway.

We shall examine many examples of acylation reactions. In all of them, however, two general relationships are worth remembering:

1 An acyl group greatly increases the reactivity to substitution of a function Z over that of the corresponding alkyl or aryl derivative; that is, RCO—Z reacts with nucleophiles faster than R—Z or Ar—Z. Thus acyl halides are more reactive than alkyl halides, anhydrides are more reactive than acids, acids are more reactive than alcohols, esters are more reactive than ethers, and amides are more reactive than amines.

2 The relative reactivity of various acyl derivatives, or *acylating agents*, decreases in the order RCOCl (or RCOBr) > (RCO)$_2$O ≫ RCO$_2$H ≈ RCO$_2$R' ≫ RCONH$_2$. Interestingly, this order parallels the carbonyl stretching frequencies of the different acyl derivatives (Table 13.6) and the ionization (acidity) constants of the Brønsted acids H—Z (K_a for HCl and HBr > RCO$_2$H > ROH > RNH$_2$).

The acylation reaction of carboxylic acid derivatives in equation 14.28 represents a different kind of transformation from the many aldehyde and ketone reactions we studied in Chapter 12. Since all these compounds have carbonyl groups, this difference in chemical behavior suggests that the nature of substituent Z must determine the eventual course of their reactions with nucleophilic substances. We can, in fact, accommodate most of these features with the following simple mechanism:

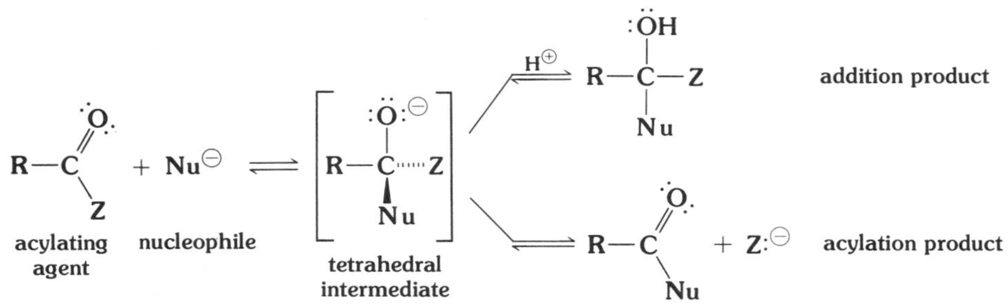

$$R-\underset{Z}{\overset{O}{\underset{\|}{C}}} + Nu^{\ominus} \rightleftharpoons \left[R-\underset{Nu}{\overset{\overset{:\ddot{O}:^{\ominus}}{|}}{\underset{|}{C}}}\cdots Z\right] \begin{array}{c} \overset{H^{\oplus}}{\rightleftharpoons} R-\underset{Nu}{\overset{\overset{:\ddot{O}H}{|}}{\underset{|}{C}}}-Z \quad \text{addition product} \\ \\ \rightleftharpoons R-\underset{Nu}{\overset{O}{\underset{\|}{C}}} + Z:^{\ominus} \quad \text{acylation product} \end{array} \tag{14.29}$$

acylating agent · nucleophile · tetrahedral intermediate

14·5 ACYLATION REACTIONS OF CARBOXYLIC ACID DERIVATIVES

The initial step in most carbonyl reactions appears to be a rapid and sometimes reversible addition of nucleophilic agents to the carbon-oxygen double bond. For aldehydes and ketones (Z = H or R) subsequent loss of the Z substituent is seldom observed. However, in the case of carboxylic acid derivatives, $Z^{\ominus}$ is a much better nucleophilic leaving group than $H^{\ominus}$ or $R^{\ominus}$, and acylation products predominate. This dependence on the ease of $Z^{\ominus}$ formation is consistent with the previously noted correspondence of HZ acidity with RCOZ reactivity.

PROBLEM 14·8

Predict the major products from the reaction of the following compounds with (1) 2,4-dinitrophenylhydrazine and (2) $H_2^{18}O$ (water isotopically labeled with ^{18}O):

a Benzoyl chloride
b Acetic anhydride
c Methyl butanoate
d Benzaldehyde
e Acetone

REACTIONS OF ACYL HALIDES

AMINES, ALCOHOLS, WATER, AND HYDROGEN PEROXIDE The acyl halides are the most powerful acylating agents commonly used by organic chemists. As such, they react rapidly and exothermically with strong to moderate nucleophiles such as carboxylate salts (equation 14.19); other common reactions involve primary and secondary amines, alcohols, phenols, water, and hydrogen peroxide:

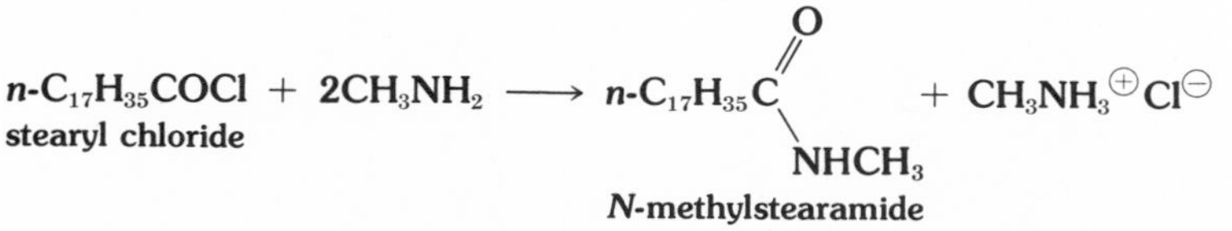

$$n\text{-}C_{17}H_{35}COCl + 2CH_3NH_2 \longrightarrow n\text{-}C_{17}H_{35}C(=O)NHCH_3 + CH_3NH_3^{\oplus}Cl^{\ominus} \quad (14.30)$$

stearyl chloride → N-methylstearamide

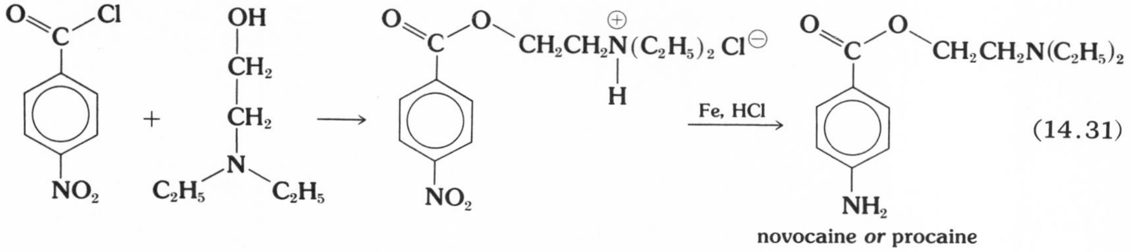

$$(14.31)$$

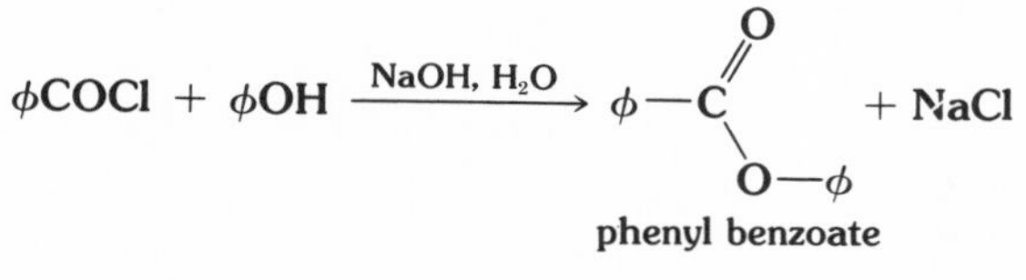

→ novocaine or procaine

$$\phi COCl + \phi OH \xrightarrow{NaOH, H_2O} \phi\text{-}C(=O)\text{-}O\text{-}\phi + NaCl \quad (14.32)$$

phenyl benzoate

$$\phi COCl + H_2O_2 \xrightarrow{NaOH, H_2O} \phi\text{-}C(=O)\text{-}O\text{-}O\text{-}H + \phi\text{-}C(=O)\text{-}O\text{-}O\text{-}C(=O)\text{-}\phi \quad (14.33)$$

perbenzoic acid   benzoyl peroxide

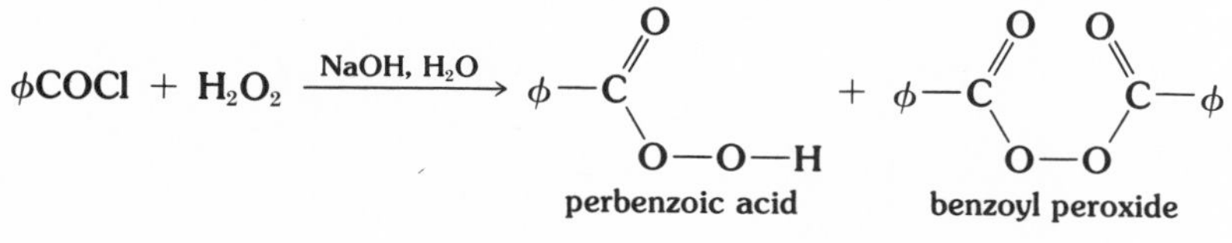

14·5 ACYLATION REACTIONS OF CARBOXYLIC ACID DERIVATIVES

PROBLEM 14·9
a Write general equations (using R and R' groups) which illustrate the reactions of acyl chlorides with water, alcohols, and secondary amines.
b Why does the acylation in equation 14.31 proceed to an ester rather than an amide?
c Suggest a reason why phenols and hydrogen peroxide can be acylated in aqueous base without losing a major portion of the acyl halide to hydrolysis. *Hint:* Consider the acidity constants in Table 14.1.

PROBLEM 14·10
The reaction of acyl chlorides with diazomethane takes one of two courses, depending on the relative proportion of the reactants. If excess diazomethane (or some other base such as triethylamine) is present, the major product is an α-diazoketone. In the absence of additional bases, the reactants are transformed into an α-chloroketone:

$$R-\underset{Cl}{\overset{O}{\overset{\|}{C}}} + CH_2N_2 \begin{cases} \xrightarrow{\text{excess } CH_2N_2} R-\underset{CHN_2}{\overset{O}{\overset{\|}{C}}} + CH_3Cl + N_2 \\ \text{an } \alpha\text{-diazo ketone} \\ \xrightarrow{1:1} R-\underset{CH_2-Cl}{\overset{O}{\overset{\|}{C}}} + N_2 \\ \text{an } \alpha\text{-chloro ketone} \end{cases}$$

Write equations for a mechanism that will explain this result.

FRIEDEL-CRAFTS ACYLATION OF AROMATIC RINGS AND ALKENES

Although weak nucleophiles such as aromatic rings and alkenes are not normally attacked by acyl halides, such reactions can be catalyzed by Lewis acids such as $AlCl_3$ and $SnCl_4$, which serve to activate the acylating reagent. Catalyzed reactions of this kind are called *Friedel-Crafts acylations*:

$$\text{C}_6\text{H}_5\text{CH}_3 + n\text{-}C_3H_7COCl \xrightarrow[\Delta]{AlCl_3} CH_3\text{-}C_6H_4\text{-}COCH_2CH_2CH_3 + \text{(ortho isomer)} + HCl \quad (14.34)$$

$$\text{thiophene} + \phi COCl \xrightarrow{AlCl_3} \text{2-benzoylthiophene} + HCl \quad (14.35)$$

$$\text{anthracene} + CH_3COCl \xrightarrow{AlCl_3} \text{9-acetylanthracene} + HCl \quad (14.36)$$

$$\begin{array}{c}(CH_3)_2C=CHCH_3 \\ + \\ CH_3COCl\end{array} \xrightarrow{SnCl_4} (CH_3)_2C=\underset{\underset{O}{\overset{\|}{C}}-CH_3}{\overset{CH_3}{\overset{|}{C}}} + \underset{CH_3}{\overset{CH_2}{\overset{\|}{C}}}-CH(CH_3)\underset{CH_3}{\overset{O}{\overset{\|}{C}}} + HCl \quad (14.37)$$

14·5 ACYLATION REACTIONS OF CARBOXYLIC ACID DERIVATIVES

PROBLEM 14·11
a Suggest the best combination of reagents for preparing *m*-nitrobenzophenone by a Friedel-Crafts acylation.
b Write a mechanism showing all important steps in the acylation of 2-methyl-2-butene by acetyl chloride and stannic chloride. Would the acetylation of 1,1-diphenylethylene be faster or slower? Explain.
c Why does anthracene undergo Friedel-Crafts acylation preferentially at C-9 (see equation 14.36)?

Friedel-Crafts acylations of aromatic systems have two important advantages over the corresponding alkylation reactions, discussed in Section 9.6. First, the carbonyl function in the product deactivates the aromatic ring, effectively preventing the formation of di- and triacylated products through subsequent reaction. Second, molecular rearrangements of the acylating groups do not occur, probably because the acylium-ion intermediates are more stable than any of the alkyl cations that could be formed by 1,2 shifts:

$$R-\underset{\underset{Cl}{|}}{\overset{\overset{\delta\ominus}{O}}{C}} + AlCl_3 \rightleftharpoons \left\{ R-\overset{\oplus}{C}=\ddot{O}: \longleftrightarrow R-C\equiv\overset{\oplus}{O}: \right\} AlCl_4^{\ominus} \qquad (14.38)$$

a weak electrophile ... an acylium cation, a strong electrophile

FORMYL CHLORIDE Although formyl chloride (HCOCl) is unstable with respect to its decomposition products (CO and HCl), formylation of aromatic rings can be accomplished by a mixture of carbon monoxide and hydrogen chloride in the presence of aluminum chloride and a trace of cuprous halide:

$$\phi-\phi + CO + HCl \xrightarrow[\Delta]{AlCl_3(Cu_2Cl_2)} \phi-C_6H_4-CHO + HCl \qquad (14.39)$$

PHOSGENE Phosgene, the *bis*-acyl chloride of carbonic acid, reacts with a variety of nucleophiles to give characteristic products. When phosgene reacts with primary amines, *alkyl isocyanates* are formed as intermediates; they in turn act as acylating agents for other nucleophiles, as shown in Figure 14.3.

Reaction of phenyl isothiocyanate (a sulfur analog of the isocyanates) with ammonia gives phenylthiourea, a compound which serves as an interesting genetic probe.

$$\phi-N=C=S + NH_3 \longrightarrow \underset{H}{\overset{\phi}{\underset{|}{N}}}-\overset{\overset{S}{\|}}{C}-NH_2 \qquad (14.40)$$

phenyl isothiocyanate ... phenylthiourea

An inherited ability to taste this substance—it is either intensely bitter or tasteless—is unevenly distributed among human populations. In some Indian tribes in the American Southwest, for example, almost the entire population are tasters, while in a group of Welsh subjects more than half

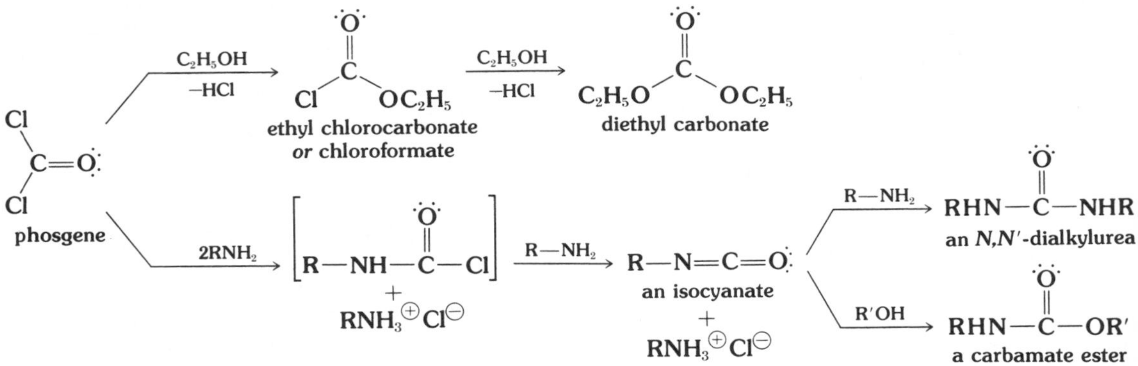

14·5 ACYLATION REACTIONS OF CARBOXYLIC ACID DERIVATIVES

FIGURE 14·3 *Reactions of phosgene*

the males (but only one-fourth of the females) proved to be nontasters. A related substance, α-naphthylthiourea (ANTU), owes its effectiveness as a rat poison to the fact that most rodents cannot taste it.

KETENES The ketenes, a group of very reactive acylating agents structurally related to the isocyanates, can be prepared from carboxylic acid derivatives by elimination reactions:

$$\underset{\underset{H}{|}}{CH_2}-\underset{OH}{\overset{\overset{O}{\|}}{C}} \xrightarrow[\Delta]{AlPO_4} H_2C=C=O + H_2O \qquad (14.41)$$
$$\text{ketene}$$

$$\underset{\underset{H}{|}}{R_2C}-\underset{Cl}{\overset{\overset{O}{\|}}{C}} + (C_2H_5)_3N \longrightarrow \underset{R}{\overset{R}{|}}C=C=O + (C_2H_5)_3NH^{\oplus}Cl^{\ominus} \qquad (14.42)$$
$$\text{a substituted ketene}$$

The carbonyl stretching frequencies of these intermediates are exceptionally high, about 2100 cm^{-1}. Most ketenes react very rapidly with nucleophiles to give a variety of carboxylic acid derivatives (Figure 14.4).

In the absence of nucleophiles, ketene and its substituted derivatives dimerize to unsaturated β-lactones or 1,3-cyclobutanediones (the latter being favored by disubstituted ketenes). Because the bulk of its substituents shield it from attack, dibutylketene is stable in the monomeric form, reacting only slowly with water and, in the absence of catalysts, not at all with aniline.

FIGURE 14·4 *Reactions of ketene*

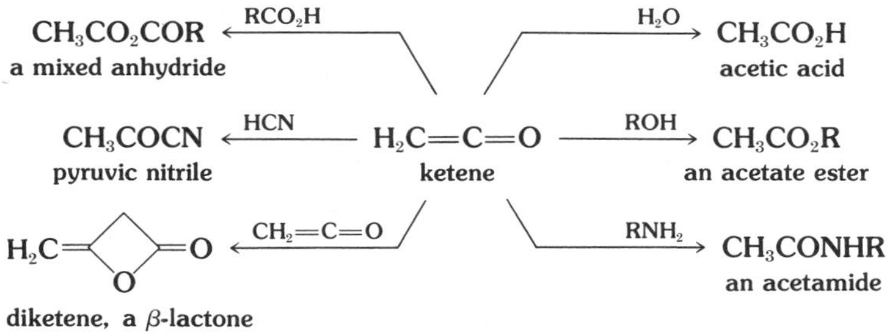

14·5 ACYLATION REACTIONS OF CARBOXYLIC ACID DERIVATIVES

REACTIONS OF ANHYDRIDES

As a general rule, anhydrides may be used instead of acyl halides in most of the acylation reactions described above. In fact, they are often preferred as acylating agents because they are easier to handle and are less expensive than acyl chlorides. Examples of acylation reactions using anhydrides are given in Figure 14.5.

FIGURE 14·5 *Acylation reactions of anhydrides*

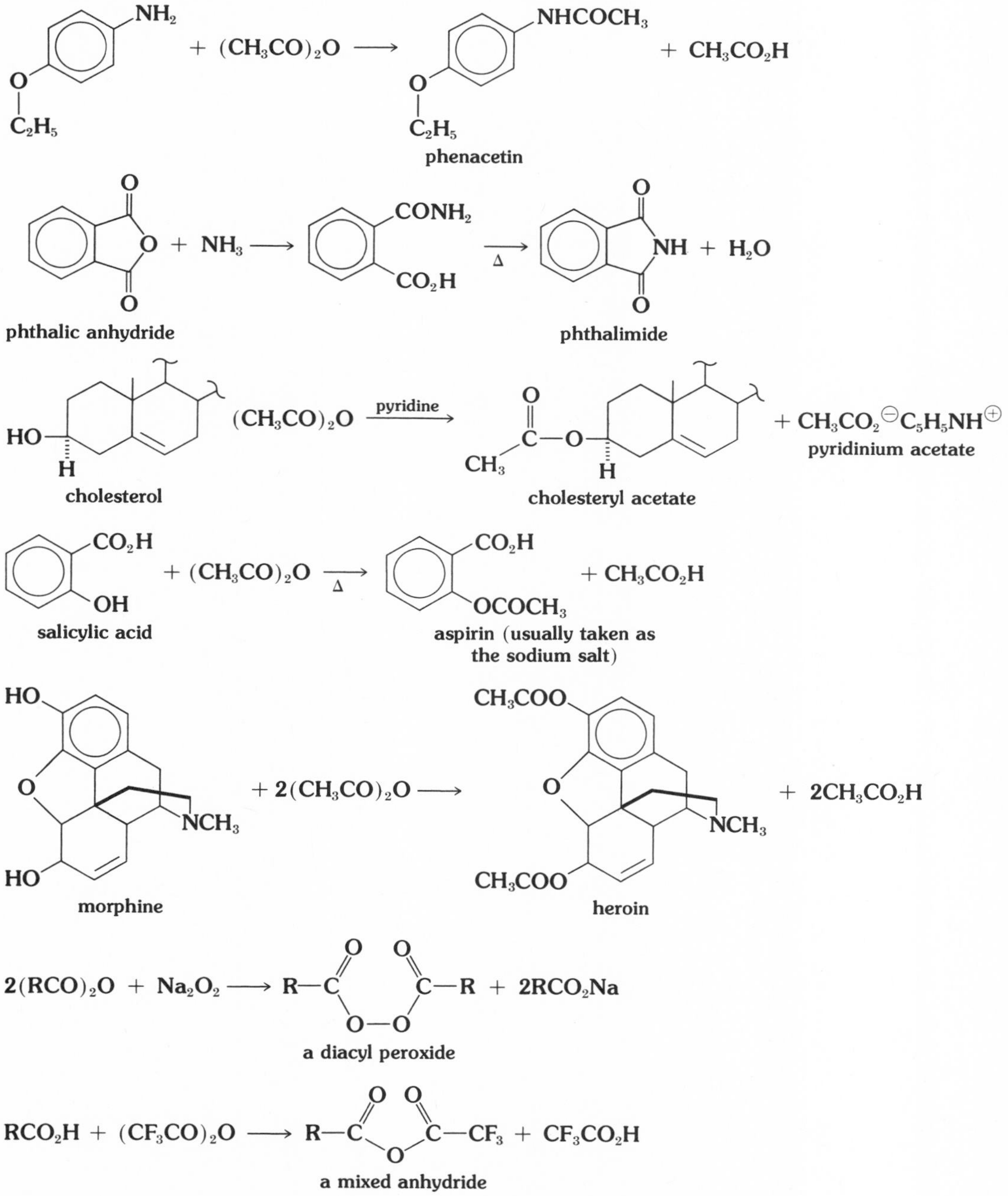

14·5 ACYLATION REACTIONS OF CARBOXYLIC ACID DERIVATIVES

Friedel-Crafts acylation of aromatic rings and alkenes by anhydrides can be effected in good yield, provided generous amounts of Lewis acid catalysts are used to activate the acylating reagent:

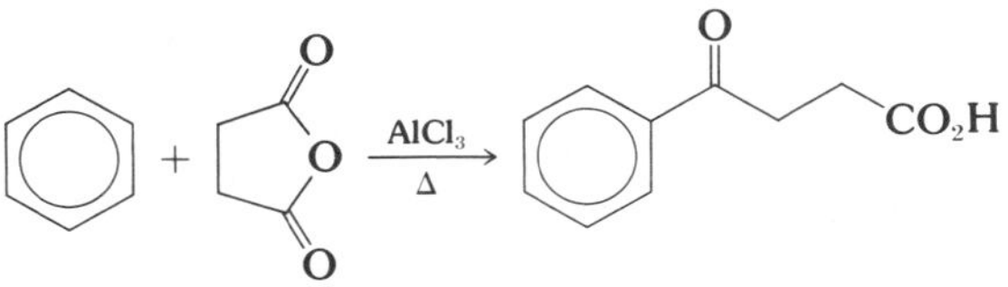

$$\text{C}_6\text{H}_6 + \text{succinic anhydride} \xrightarrow[\Delta]{\text{AlCl}_3} \text{C}_6\text{H}_5\text{COCH}_2\text{CH}_2\text{CO}_2\text{H} \tag{14.43}$$

$$\text{furan} + (\text{CH}_3\text{CO})_2\text{O} \xrightarrow[\Delta]{\text{BF}_3} \text{2-acetylfuran} + \text{CH}_3\text{CO}_2\text{H} \tag{14.44}$$

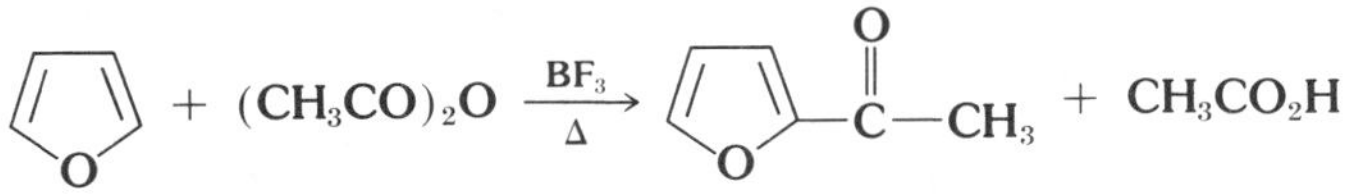

$$\text{cyclohexene} + (\text{CH}_3\text{CO})_2\text{O} \xrightarrow{\text{SnCl}_4} \text{1-acetylcyclohexene} + \text{CH}_3\text{CO}_2\text{H} \tag{14.45}$$

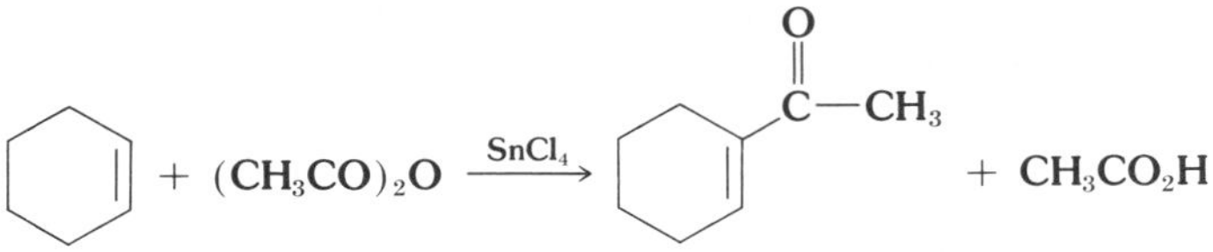

PROBLEM 14·12
The reaction of acetic anhydride with alcohols can be catalyzed by dry hydrogen chloride. Suggest a mechanism for these acetylation reactions.

REACTIONS OF ESTERS

Esters are, for the most part, much weaker acylating reagents than anhydrides or acyl halides. Consequently, although strong nucleophiles such as amines can be directly acylated, weaker nucleophiles react with esters only in the presence of acid or base catalysts:

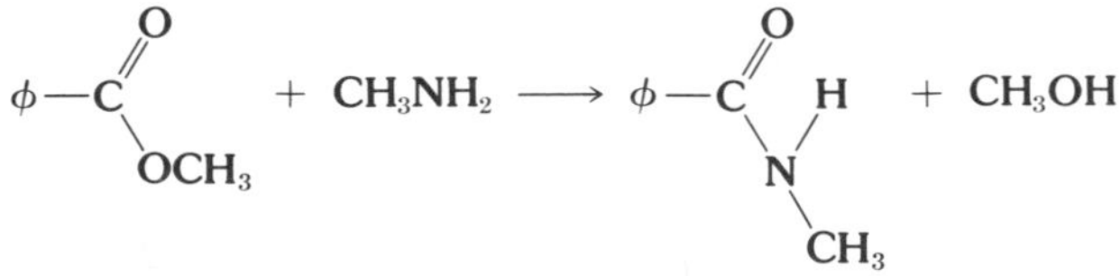

$$\phi\text{-CO-OCH}_3 + \text{CH}_3\text{NH}_2 \longrightarrow \phi\text{-CO-NH(CH}_3) + \text{CH}_3\text{OH} \tag{14.46}$$

$$(\text{CH}_3)_2\text{CH-CO-OC}_2\text{H}_5 + \text{H}_2\text{NOH} \longrightarrow (\text{CH}_3)_2\text{CHCO-NHOH} + \text{C}_2\text{H}_5\text{OH} \tag{14.47}$$
a hydroxamic acid

$$\text{CH}_2(\text{CO}_2\text{C}_2\text{H}_5)_2 + \text{H}_2\text{NCONH}_2 \xrightarrow[\text{C}_2\text{H}_5\text{OH}]{\text{NaOC}_2\text{H}_5,} \text{barbituric acid} \rightleftharpoons \text{tautomer} \tag{14.48}$$

$$\begin{array}{l}\text{CH}_2\text{-O-CO-R}^1\\\text{CH-O-CO-R}^2\\\text{CH}_2\text{-O-CO-R}^3\end{array} + 3\text{CH}_3\text{OH} \xrightarrow[\Delta]{\text{NaOCH}_3} \begin{array}{l}\text{CH}_2\text{-OH}\\\text{CH-OH}\\\text{CH}_2\text{-OH}\end{array} + \begin{array}{l}\text{R}^1\text{CO}_2\text{CH}_3\\+\\\text{R}^2\text{CO}_2\text{CH}_3\\+\\\text{R}^3\text{CO}_2\text{CH}_3\end{array} \tag{14.49}$$
a fat or oil \hspace{2cm} glycerol \hspace{1cm} fatty esters

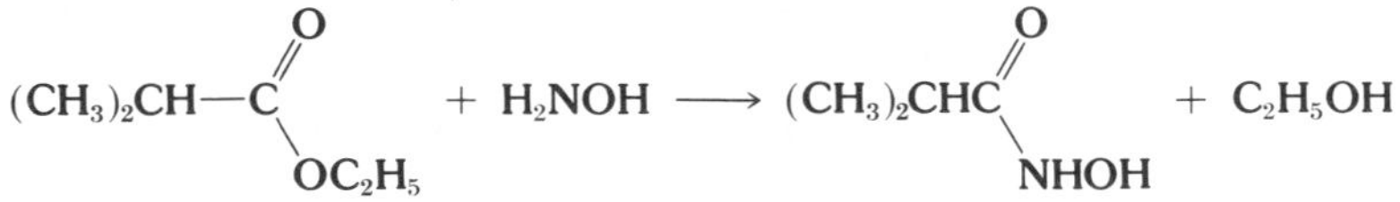

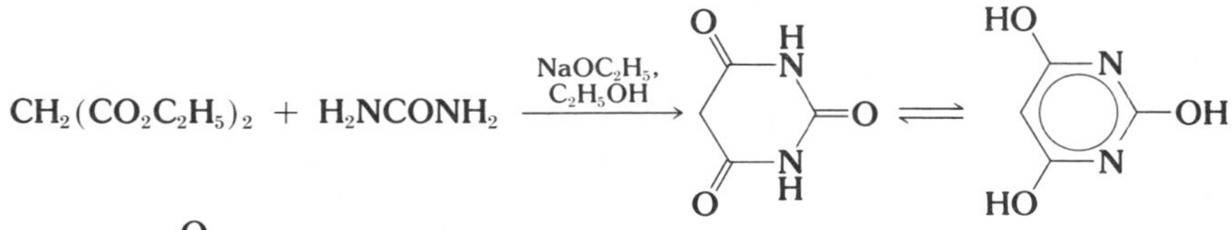

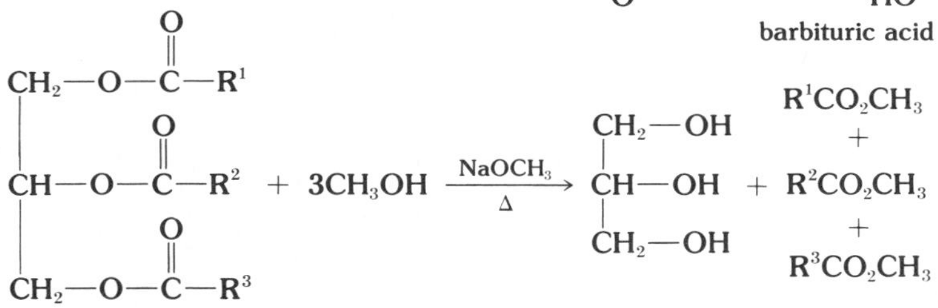

14·5 ACYLATION REACTIONS OF CARBOXYLIC ACID DERIVATIVES

The direct conversion of one ester to another, as in equation 14.49, is called *transesterification* and can be catalyzed by either acids or bases. Useful polyesters such as Dacron are often prepared by transesterification reactions:

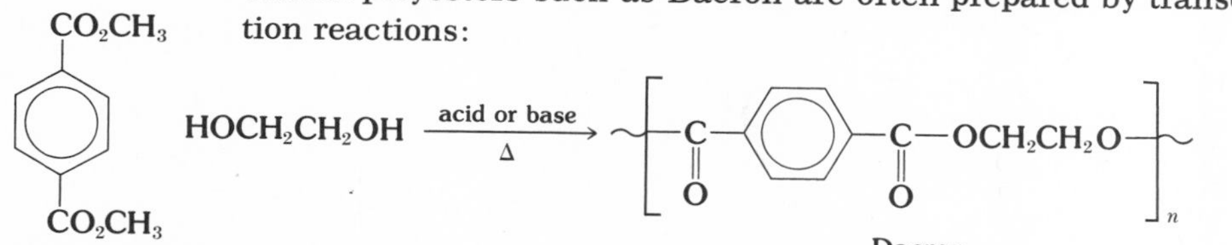

$$\text{dimethyl terephthalate} + HOCH_2CH_2OH \xrightarrow[\Delta]{\text{acid or base}} \text{Dacron} \quad (14.50)$$

Ester hydrolysis, which may be regarded as an acylation of water, is catalyzed by acids and bases. We can conceive of several possible reaction pathways for these transformations. The major points of distinction are whether the acyl (AC) carbon-oxygen bond or the alkyl (AL) carbon-oxygen bond is cleaved, and whether the reaction is bimolecular or unimolecular.

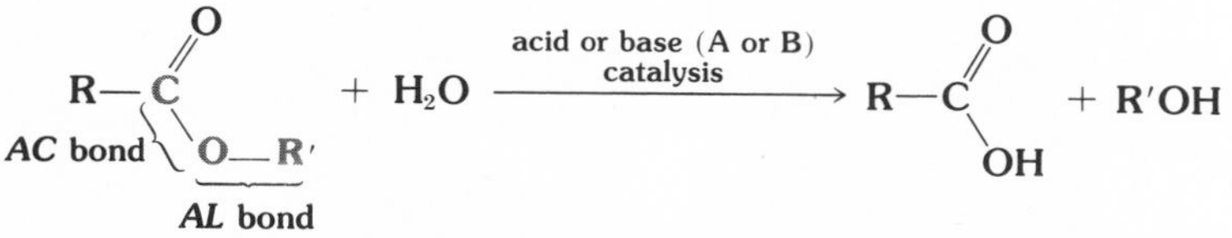

$$R-C(=O)-O-R' + H_2O \xrightarrow{\text{acid or base (A or B) catalysis}} R-C(=O)-OH + R'OH \quad (14.51)$$

If we conduct a base-catalyzed hydrolysis of ethyl benzoate in isotopically labeled (^{18}O) water, the ethanol produced is completely devoid of the tracer isotope, making it clear that this reaction is proceeding by cleavage of the acyl carbon-oxygen bond. Furthermore, if we stop the hydrolysis short of completion, the recovered ester contains substantial amounts of the oxygen isotope. This suggests that a slow reversible addition of water to the carbonyl double bond is followed by decomposition of the resulting tetrahedral intermediate by one of three closely related steps, as outlined in Figure 14.6. Most base-catalyzed ester hydrolyses are thought to proceed by this $B_{AC}2$ *mechanism* (B for base catalysis, AC for acyl carbon-oxygen bond cleavage, and 2 for bimolecular). Base-catalyzed ester hydrolysis is called *saponification* because of its long-established use in the manufacture of soap (from the Latin *sapon*) from fats and oils. Acid-catalyzed ester hydrolyses appear to be slower than the equivalent base-catalyzed reactions and usually follow the $A_{AC}2$ *mechanism* described in Figure 14.2.

FIGURE 14·6 *Base-catalyzed ester hydrolysis*

$$\phi-C(=O)OC_2H_5 + H_2^{18}O \xrightleftharpoons{^{18}OH^{\ominus}} \left[\phi-C(OH)(^{18}OH)OC_2H_5\right] \xrightleftharpoons{OH^{\ominus}} \phi-C(=^{18}O)OC_2H_5 + H_2O$$

$$\downarrow {^{18}OH^{\ominus}}$$

$$\phi-C(=O)(^{18}O^{\ominus}) + C_2H_5OH + H_2^{18}O$$

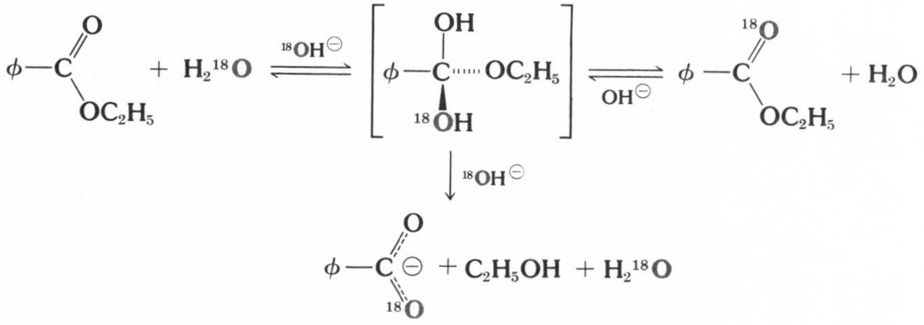

14·5 ACYLATION REACTIONS OF CARBOXYLIC ACID DERIVATIVES

The reactivity of esters can be substantially altered by steric hindrance and ring strain. For example, methyl mesitoate (methyl 2,4,6-trimethylbenzoate) is recovered unchanged after several days of exposure to refluxing aqueous base:

steric crowding → greater steric crowding

In contrast, ketene dimer, a β-lactone, reacts rapidly with water, alcohols, or amines to yield derivatives of acetoacetic acid:

$$\text{H}_2\text{C} \underset{\text{O}}{\overset{\text{O}}{\square}} + \text{ROH} \longrightarrow \text{CH}_3-\overset{\text{O}}{\underset{\|}{\text{C}}}-\text{CH}_2-\overset{\text{O}}{\underset{\|}{\text{C}}}\text{OR} \tag{14.53}$$

This reaction is facilitated by relief of angle strain.

PROBLEM 14·13
a Why does reaction 14.48 require base catalysis while reaction 14.47 does not?
b Is it possible to obtain double-labeled benzoate ion in the base-catalyzed hydrolysis of ethyl benzoate in $\text{H}_2{}^{18}\text{O}$? Explain your reasoning.
c Write equations showing the $A_{AC}2$ mechanism for acid-catalyzed transesterification in the reaction of $\text{R}^1\text{CO}_2\text{R}^2$ and R^3OH.

PROBLEM 14·14
a Methyl mesitoate is easily hydrolyzed by dissolving it in cold concentrated sulfuric acid, followed by slow addition of this solution to ice water. Methyl benzoate, however, is recovered unchanged from this treatment. Explain the reason for this.
b Acid-catalyzed hydrolysis of *t*-butyl benzoate proceeds by an $A_{AL}1$ mechanism. Write equations for this reaction.

REACTIONS OF AMIDES

Amides, the least reactive class of carboxyl-derived acylating agents, owe their chemical stability to the previously noted interaction between the nitrogen electron pair and the carbonyl group. As expected from this effect, amides are weaker bases (by a factor of about 10^{-12}) and stronger acids (about 10^{10} times) than similarly constituted amines. Experimental evidence indicates that the conjugate acids formed by protonation at the amide oxygen atom are more stable than their N-protonated isomers.

$$\left\{ \text{R}-\overset{\overset{..}{\text{O}}:}{\underset{\text{NH}_2}{\text{C}}} \longleftrightarrow \text{R}-\overset{\overset{..}{\text{O}}:^{\ominus}}{\underset{\overset{+}{\text{NH}_2}}{\text{C}}} \right\} + \text{H}_3\text{O}^{\oplus} \rightleftharpoons \left\{ \text{R}-\overset{\overset{\oplus}{\text{O}}-\text{H}}{\underset{\text{NH}_2}{\text{C}}} \longleftrightarrow \text{R}-\overset{\overset{..}{\text{O}}-\text{H}}{\underset{\overset{+}{\text{NH}_2}}{\text{C}}} \right\} + \text{H}_2\text{O} \tag{14.54}$$

14·5 ACYLATION REACTIONS OF CARBOXYLIC ACID DERIVATIVES

Hydrolysis of amides usually requires vigorous and prolonged heating with aqueous acid or base. Unsubstituted amides, however, react readily with cold nitrous acid, yielding the corresponding carboxylic acids. This exceptional reactivity is undoubtedly due to the transformation of a poor leaving group (NH_2) into an excellent leaving group ($^\oplus N \equiv N$).

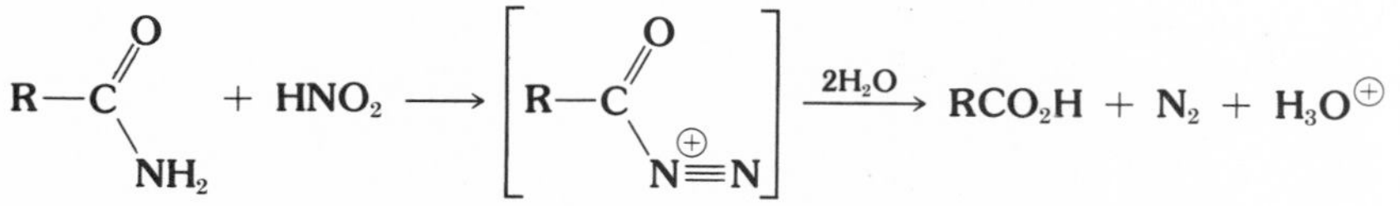

$$R-C(=O)-NH_2 + HNO_2 \longrightarrow [R-C(=O)-N\overset{\oplus}{=}N] \xrightarrow{2H_2O} RCO_2H + N_2 + H_3O^\oplus \tag{14.55}$$

The amide bonds in proteins are rapidly and selectively hydrolyzed under mild conditions (room temperature and pH $\approx$ 7) by enzymes such as pepsin, trypsin, and chymotrypsin, which are found in the digestive systems of mammals.

Studies of the penicillin and cephalosporin antibiotics provide an interesting example of how the acylating power of an amide function is decreased by nitrogen electron-pair delocalization and increased by angle strain:

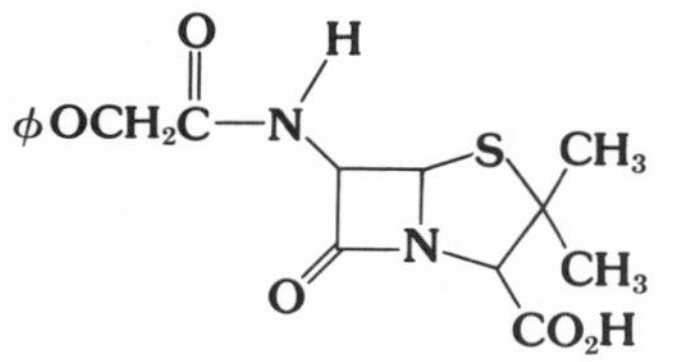

penicillin V

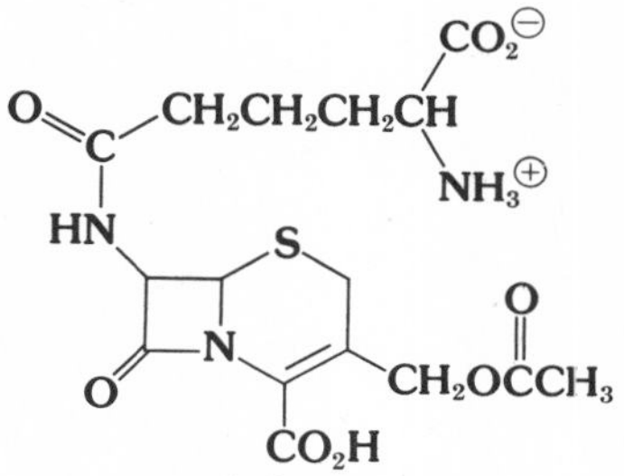

cephalosporin C

The antibiotic activity of these compounds is believed to result primarily from the β-lactam moiety, which serves to acylate irreversibly—and thereby inactivate—an enzyme necessary for bacterial cell-wall formation. With the growth of the microorganisms thus hindered, the body's natural defenses can operate more effectively. This seems to be a reasonable explanation, since even simple β-lactams show greater hydrolysis rates and have higher carbonyl stretching frequencies (1730 to 1760 cm^{-1}) than unstrained amides.

On further study, however, we find that not all β-lactams of the penicillin or cephalosporin type are effective antibiotics. For example, cephaloridine is a powerful antibiotic, but the cephalosporin derivative shown below is essentially inactive:

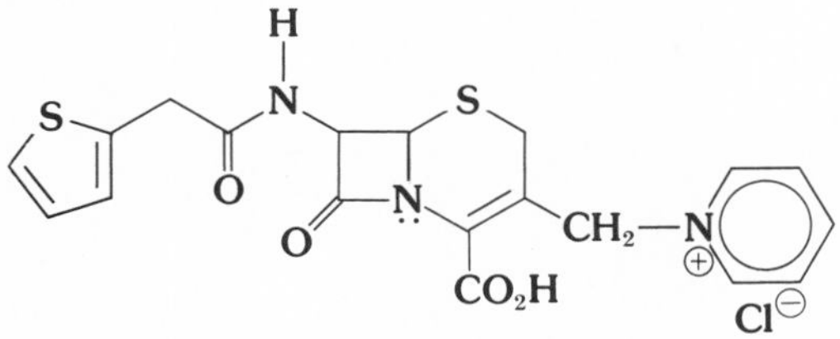

cephaloridine
$\tilde{\nu}_{max} = 1770\text{--}1775$ cm^{-1}

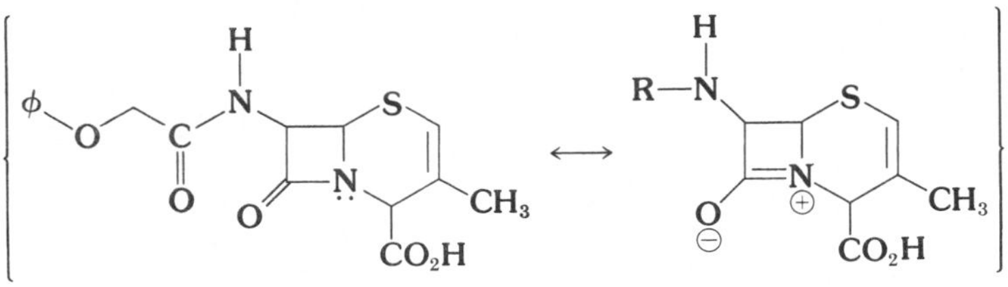

14·5 ACYLATION REACTIONS OF CARBOXYLIC ACID DERIVATIVES

a cephalosporin derivative
$\tilde{\nu}_{max} = 1755\text{–}1760 \text{ cm}^{-1}$

If we analyze the structures of these compounds by x-ray diffraction, we find that the lactam nitrogen atom in cephaloridine is pyramidal, while the inactive compound has a planar lactam grouping with a shorter acyl carbon-nitrogen bond. This difference suggests a greater delocalization of the nitrogen electron pair in the latter compound, as further reflected by its lower carbonyl stretching frequency. Antibiotic activity, in other words, correlates roughly with the acylating reactivity of the β-lactam function.

PROBLEM 14·15

a Write a structural formula for an N-protonated amide and suggest an explanation for the greater stability of the O-protonated form.

b Write an equation for the base-catalyzed hydrolysis of nylon 66:

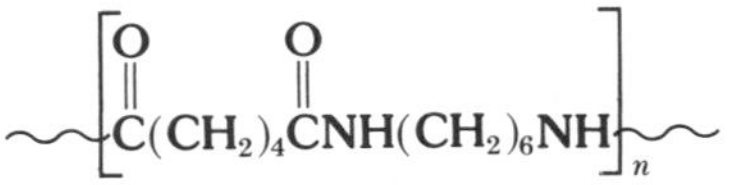

$$\sim\!\!\left[\overset{O}{\overset{\|}{C}}(CH_2)_4\overset{O}{\overset{\|}{C}}NH(CH_2)_6NH\right]_n\!\!\sim$$

c The bicyclic lactam is much more rapidly hydrolyzed than the monocyclic lactam. Explain why.

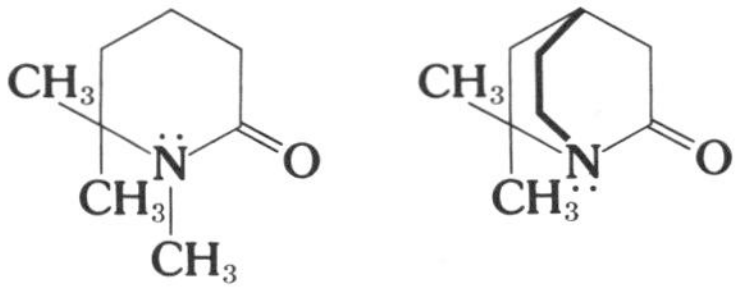

REACTIONS OF NITRILES

Nitriles are alkyl or aryl cyanides (R—C≡N: or Ar—C≡N:) and are most commonly prepared by nucleophilic displacement reactions of alkyl halides or sulfonate esters with cyanide salts, or by dehydration of unsubstituted amides with phosphorus pentoxide or thionyl chloride:

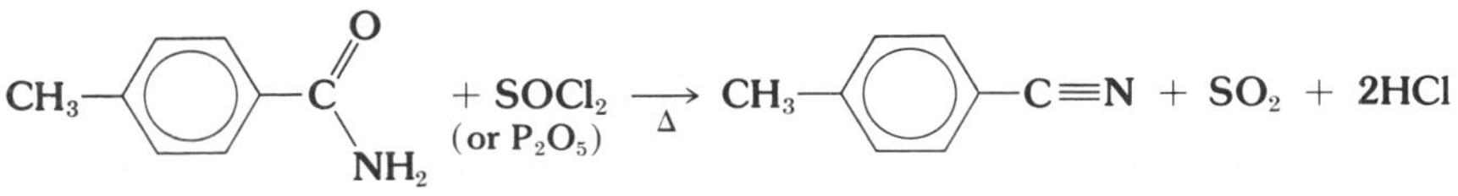

$$CH_3\text{–}\phi\text{–}C(=O)NH_2 + SOCl_2 \text{ (or } P_2O_5\text{)} \xrightarrow{\Delta} CH_3\text{–}\phi\text{–}C\equiv N + SO_2 + 2HCl \qquad (14.56)$$

As illustrated in Figure 14.7, acid-catalyzed addition of alcohols to the triple bond of nitriles leads to *imido esters, esters,* or *ortho esters,* depending on the workup conditions.

$$\phi CH_2CN + H_2O + NaOH \xrightarrow{\Delta} \phi CH_2CO_2^{\ominus}Na^{\oplus} + NH_3 \qquad (14.57)$$

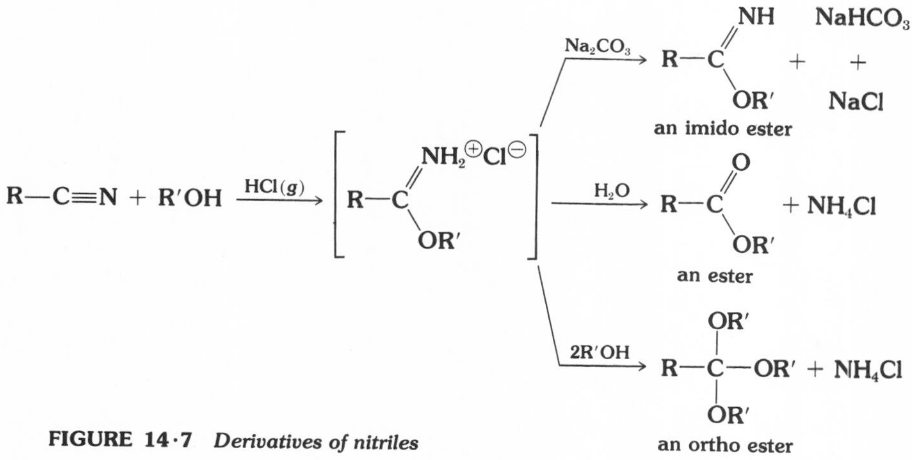

14·6 BIOLOGICAL ACYLATION REACTIONS

$$R-C \equiv N + R'OH \xrightarrow{HCl(g)} \left[R-C \begin{matrix} NH_2^{\oplus}Cl^{\ominus} \\ OR' \end{matrix} \right]$$

with outcomes:

- $\xrightarrow{Na_2CO_3}$ $R-C(=NH)(OR')$ + NaHCO$_3$ + NaCl   (an imido ester)
- $\xrightarrow{H_2O}$ $R-C(=O)(OR')$ + NH$_4$Cl   (an ester)
- $\xrightarrow{2R'OH}$ $R-C(OR')_3$ + NH$_4$Cl   (an ortho ester)

FIGURE 14·7 *Derivatives of nitriles*

Acylation reactions conducted in the laboratory usually require highly reactive acylating reagents, such as acyl halides or anhydrides, or forcing reaction conditions, such as strong acid or base catalysts, elevated temperature, and/or selective removal of a reaction product. Although the aqueous nature and narrow pH and temperature tolerance of biological systems might seem incompatible with such reactions, biological acylations—particularly those in which amide bonds are formed or broken—play a vital role in the chemical processes of living organisms. As a result, the manner in which these acyl transfers are accomplished has been a subject of extensive study by chemists and biochemists.

ENZYME CATALYSIS

The remarkable efficiency and selectivity of biological acylation reactions is due largely to highly specific catalysts called *enzymes*. These complex proteins operate on a restricted range of reactions involving reactant compounds commonly referred to as *substrates*. The most specific enzymes catalyze only a single reaction of a particular substrate (or two reactions, if we take into account the reverse transformation). Other enzymes, however, may act on a variety of substrates. One of the chief proteolytic enzymes secreted by the pancreas in mammals, α-chymotrypsin, normally catalyzes protein hydrolysis but is also active as a hydrolysis catalyst for simple esters and amides. Most of the evidence regarding the mechanism of enzyme action suggests that enzymes initially form complexes with the substrate molecules, and these complexes then undergo a rate-determining decomposition to products.

Some enzymes require the presence of a cofactor for activity. A substance called *coenzyme A*, for example, is the acyl transfer agent for many acylation reactions (see Figure 14.8). Experiments with simple thiol esters (RCOSR') confirm the superiority of this function over normal esters as acylating reagents.

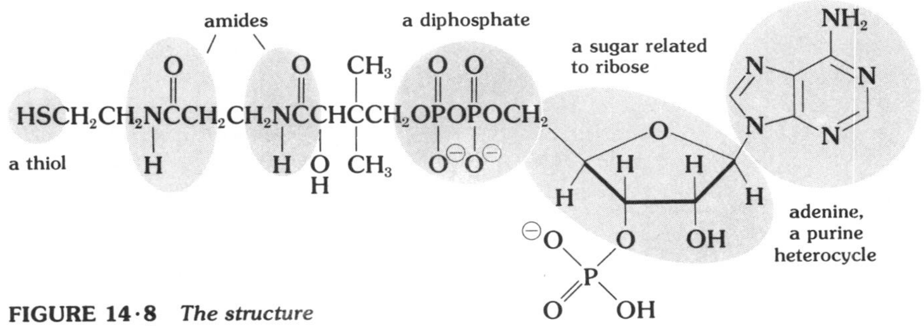

14·6 BIOLOGICAL ACYLATION REACTIONS

FIGURE 14·8 The structure of coenzyme A

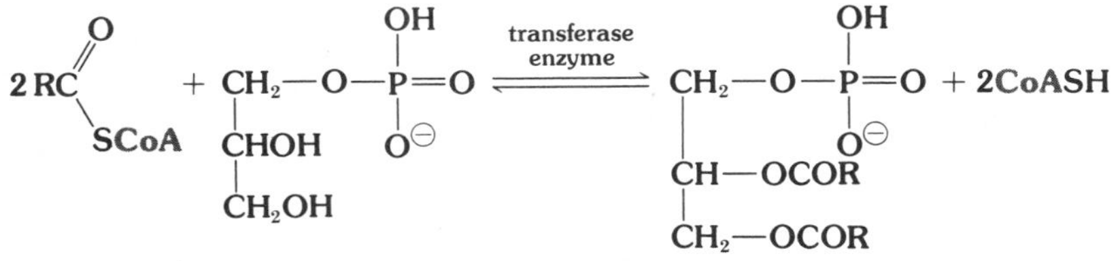

(Structure shows: a thiol HSCH$_2$CH$_2$—amides—diphosphate—a sugar related to ribose—adenine, a purine heterocycle)

$$2\,\text{RC(O)SCoA} + \begin{array}{c}\text{CH}_2-\text{O}-\text{P(=O)(O}^\ominus\text{)OH} \\ \text{CHOH} \\ \text{CH}_2\text{OH}\end{array} \xrightleftharpoons[\text{enzyme}]{\text{transferase}} \begin{array}{c}\text{CH}_2-\text{O}-\text{P(=O)(O}^\ominus\text{)OH} \\ \text{CH}-\text{OCOR} \\ \text{CH}_2-\text{OCOR}\end{array} + 2\,\text{CoASH} \quad (14.58)$$

NERVE-IMPULSE TRANSMISSION

A specific enzymatic ester hydrolysis plays a prominent role in the transmission of signals by the autonomic nervous system.† Nerve cells, or *neurons*, have filamentous extensions called *axons* and *dendrites* which may reach over a meter in length. Nerve impulses travel from the central nervous system (the brain and spinal cord) through collections of neurons called *ganglia* to various organs and muscles at speeds ranging from a fraction of a meter per second to over 60 meters per second. Where the axon of one cell ends, its electrical message must be transferred to the dendrite of a neighboring nerve cell or to an operator cell, such as smooth or striated muscle. The gap to be bridged (about 200Å) is called a *synapse*, or *synaptic cleft,* and the transmission across this gap is effected by chemical agents.

As noted in Chapter 11, part of the sympathetic nervous system employs the hormonal substances epinephrine and norepinephrine as synaptic transfer agents. In the parasympathetic system and the remainder of the sympathetic system the substance *acetylcholine* functions as the transmitting agent.

Upon stimulation by a nerve impulse, the terminus of an axon releases thousands of molecules of acetylcholine into the synapse. This substance

---

†The autonomic nervous system exerts nonvoluntary control over visceral organs and has two divisions, *sympathetic* and *parasympathetic,* classified on the basis of function and anatomy. One division is antagonistic toward the other; thus the sympathetic system accelerates heart beat, dilates the eye pupils, and relaxes the stomach and bladder, while the parasympathetic system returns the body to a more restful state. These two systems usually act together to provide normal (balanced) conditions in the body. For further information about the chemistry of the nervous systems see E. Brand and T. Westfall, Neuropharmacology, in A. Burger (ed.), *Medicinal Chemistry,* vol. 2, Wiley-Interscience, 1970, and A. Burger (ed.), *Drugs Affecting the Peripheral Nervous System,* vol. 1, Marcel Dekker, Inc., 1967.

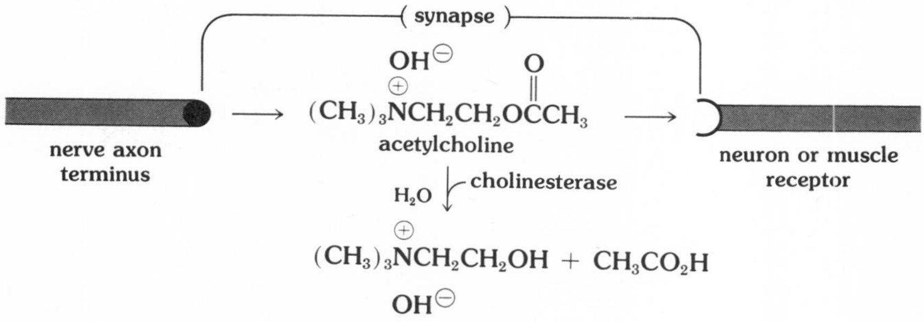

FIGURE 14·9 *Signal transmission over a neural synapse*

$$(CH_3)_3\overset{\oplus}{N}CH_2CH_2O\overset{O}{\overset{\|}{C}}CH_3 \quad OH^{\ominus}$$
acetylcholine

$$\xrightarrow{H_2O,\ \text{cholinesterase}} (CH_3)_3\overset{\oplus}{N}CH_2CH_2OH + CH_3CO_2H$$
$$OH^{\ominus}$$
choline

interacts with specific receptor sites, causing a new impulse to be generated in a neighboring nerve cell or initiating a change in muscle tension. The remaining acetylcholine molecules are hydrolyzed in about a millionth of a second by the action of a specific enzyme, *cholinesterase*, and the inactive choline formed by this reaction is reassimilated, thus preparing the system to receive another impulse (as many as several hundred signals a second can be transmitted over short periods). This process is illustrated in Figure 14.9.

The chemistry of the synapse can be perturbed by foreign substances, often with disastrous results. For example, the drugs curare, atropine, and procaine block synaptic transmission by competing for the receptor sites and frustrating an interaction with acetylcholine. Nerve gases such as sarin operate by inhibiting the action of cholinesterase, thus preventing the enzymatic destruction of acetylcholine. As a result, the receptors are left in an activated state, and progressively intense convulsions and death normally follow. In contrast, botulin, the toxin of the bacillus *Clostridium botulinum,* blocks the release of acetylcholine from the axon terminals. In this case no signals reach the receptors, resulting in general muscle weakness, disturbance of vision and speech, and eventually paralysis and death.

14·7 SULFONYLATION AND PHOSPHORYLATION

The acylation reactions of carboxylic acid derivatives are matched by equivalent transformations of sulfuric and phosphoric acid derivatives. Alkyl and arylsulfonyl halides, for example, react with oxygen and nitrogen nucleophiles to give sulfonate esters and sulfonamides, respectively:

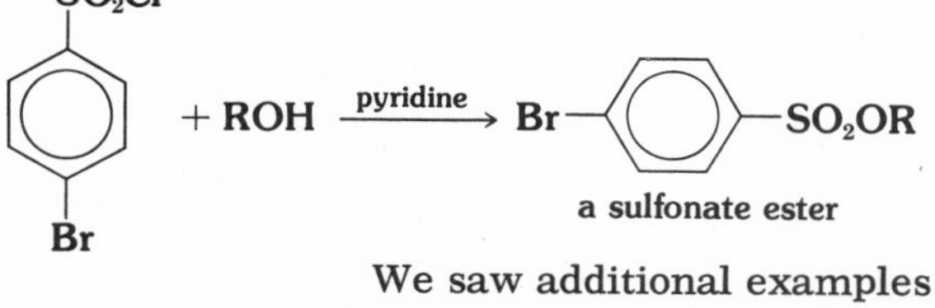

$$Br\text{–}C_6H_4\text{–}SO_2Cl + ROH \xrightarrow{\text{pyridine}} Br\text{–}C_6H_4\text{–}SO_2OR \qquad (14.59)$$
a sulfonate ester

We saw additional examples of this reaction in Figure 8.7 and equation 11.11. Indeed, such *sulfonylation reactions* are the foundation of the Hinsberg amine test (Figure 11.8).

Sulfonation of aromatic rings can be effected by sulfonyl halides in the presence of Lewis acid catalysts:

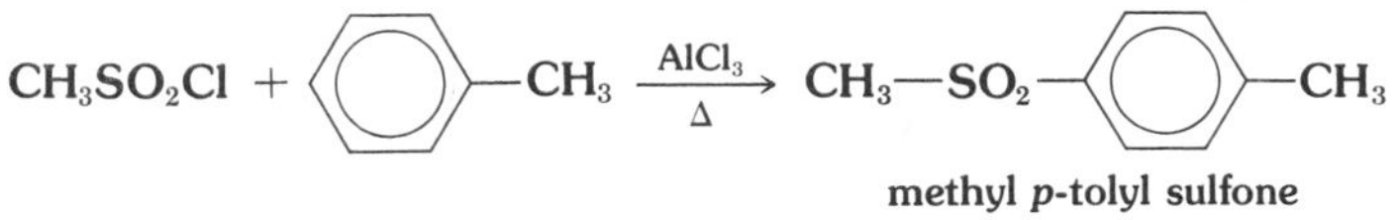

14·7 SULFONYLATION AND PHOSPHORYLATION

$$CH_3SO_2Cl + \phi\text{-}CH_3 \xrightarrow[\Delta]{AlCl_3} CH_3\text{-}SO_2\text{-}\phi\text{-}CH_3 \quad \text{methyl } p\text{-tolyl sulfone} \tag{14.60}$$

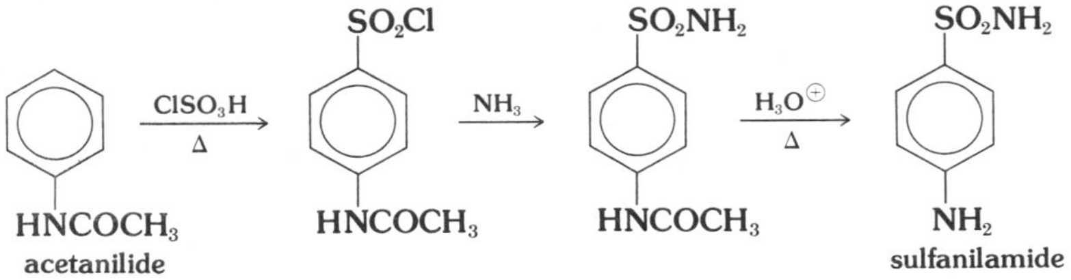

$$\underset{\text{acetanilide}}{\phi\text{-HNCOCH}_3} \xrightarrow[\Delta]{ClSO_3H} \underset{\text{HNCOCH}_3}{\phi(SO_2Cl)} \xrightarrow{NH_3} \underset{\text{HNCOCH}_3}{\phi(SO_2NH_2)} \xrightarrow[\Delta]{H_3O^{\oplus}} \underset{\substack{NH_2 \\ \text{sulfanilamide}}}{\phi(SO_2NH_2)} \tag{14.61}$$

Derivatives of p-aminobenzenesulfonamide (sulfanilamide) have pronounced bactericidal properties, which appear to be due to an inhibition of folic acid synthesis in the bacterial organisms (folic acid is an essential B vitamin). These sulfa drugs are particularly effective against streptococcal infections such as pneumonia and gonorrhea. Extensive structural modifications of the basic sulfanilamide molecule have been made in an effort to minimize, hazardous side effects, such as deposits in the kidney as a consequence of low water solubility. Some of the beneficial drugs produced by these studies are the antibiotics sulfisoxazole and sulfathiazole and orinase, an antidiabetic compound that lowers blood sugar.

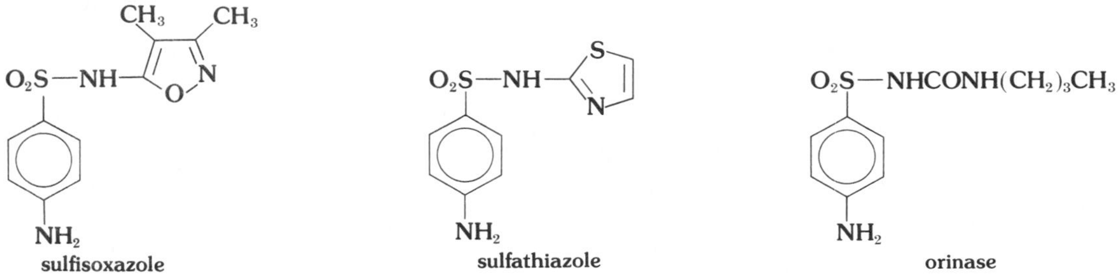

sulfisoxazole   sulfathiazole   orinase

Mixed anhydrides of sulfonic and carboxylic acids can be prepared and are found in many cases to be more powerful acylating agents than acyl halides:

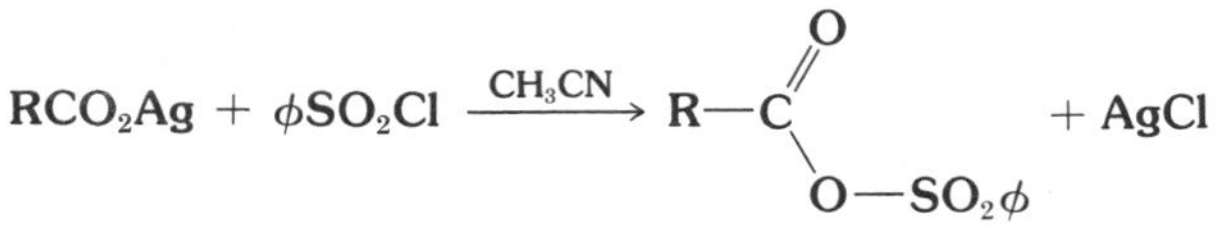

$$RCO_2Ag + \phi SO_2Cl \xrightarrow{CH_3CN} R\text{-}C(=O)\text{-}O\text{-}SO_2\phi + AgCl \tag{14.62}$$

$$R\text{-}C(=O)\text{-}O\text{-}SO_2CH_3 \begin{cases} \xrightarrow{2\phi NH_2} R\text{-}C(=O)\text{-}NH\phi + \phi NH_3^{\oplus} CH_3SO_3^{\ominus} \\ \xrightarrow{\phi OH} R\text{-}C(=O)\text{-}O\text{-}\phi + CH_3SO_3H \end{cases} \tag{14.63}$$

$$CH_3\text{-}C(=O)\text{-}OSO_2\phi + \text{(THF)} \xrightarrow{3 \text{ hr}, 25°} CH_3C(=O)\text{-}O(CH_2)_4OSO_2\phi \tag{14.64}$$

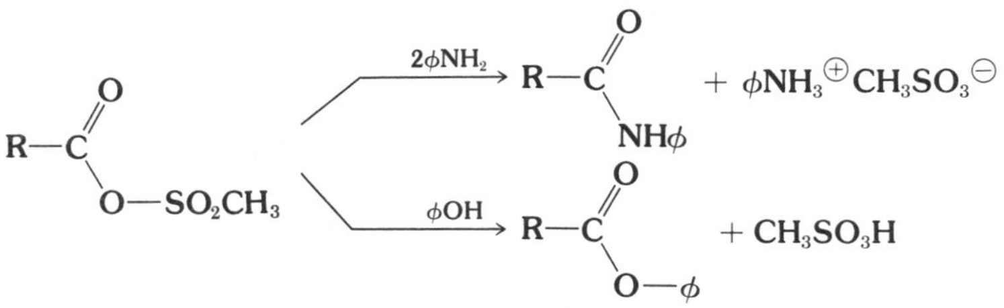

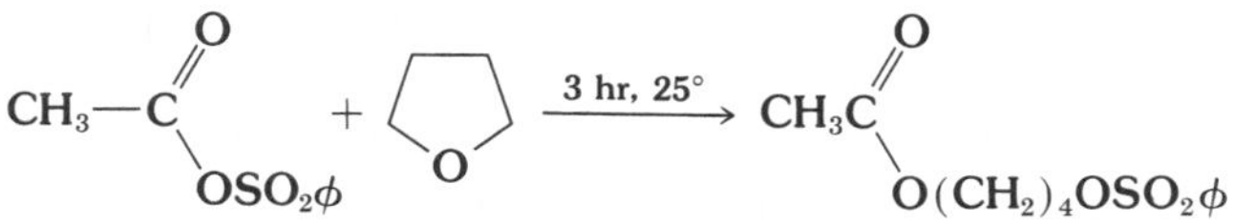

14·7 SULFONYLATION AND PHOSPHORYLATION

The cyclic mixed anhydride of o-carboxybenzenesulfonic acid is converted to the cyclic imide saccharin when it is heated with ammonia. Saccharin is a strong acid ($pK_a = 1.6$), and its sodium salt is widely used as an artificial sweetening agent (it is 500 to 700 times sweeter than cane sugar).

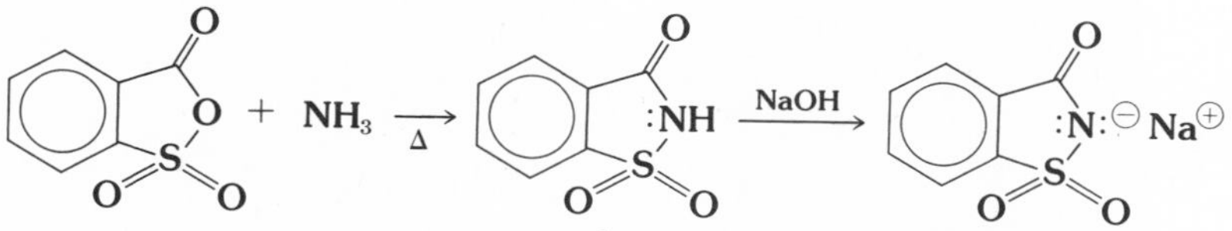

$$\text{[o-carboxybenzenesulfonic anhydride]} + NH_3 \xrightarrow{\Delta} \text{saccharin} \xrightarrow{NaOH} \text{sodium saccharin} \quad (14.65)$$

The parallel nature of *phosphorylation reactions* to acylation and sulfonylation is apparent from the following reactions of dialkylphosphinic chlorides (equation 14.66), alkylphosphonic dichlorides (equation 14.67), and dialkylchlorophosphonates (equation 14.68):

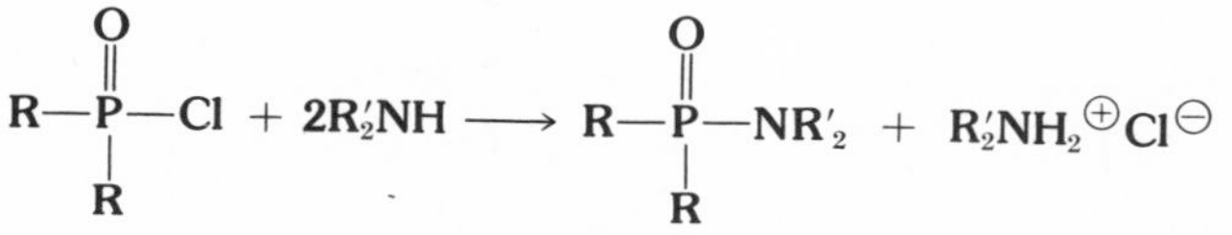

$$\underset{R}{\overset{O}{\underset{\|}{R-P-Cl}}} + 2R'_2NH \longrightarrow \underset{R}{\overset{O}{\underset{\|}{R-P-NR'_2}}} + R'_2NH_2^{\oplus}Cl^{\ominus} \quad (14.66)$$

a dialkylphosphinamide

$$\underset{Cl}{\overset{O}{\underset{\|}{R-P-Cl}}} + 2R'OH \xrightarrow{\text{pyridine}} \underset{OR'}{\overset{O}{\underset{\|}{R-P-OR'}}} + \text{pyridinium chloride} \quad (14.67)$$

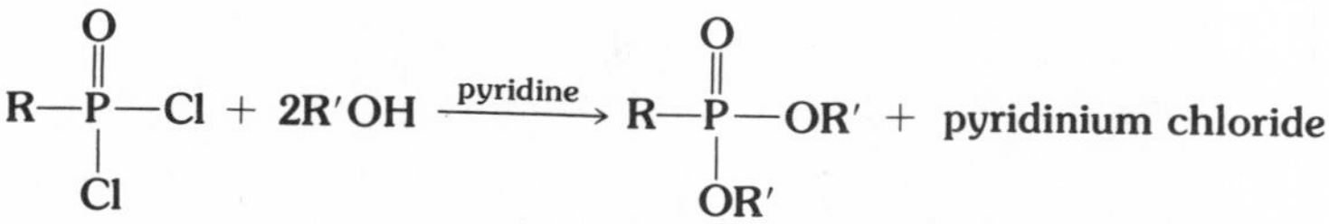

an alkylphosphonate ester

$$(RO)_2\overset{O}{\underset{\|}{P}}{-}Cl \xrightarrow{R'OH} (RO)_2\overset{O}{\underset{\|}{P}}{-}OR' + HCl \quad (14.68)$$

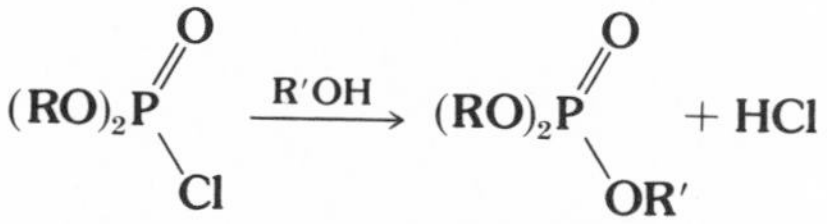

a trialkylphosphate ester

$$(RO)_2\overset{O}{\underset{\|}{P}}{-}Cl + \overset{O}{\underset{OR'}{\underset{\|}{R'O-P-OAg}}} \longrightarrow R-O-\overset{O}{\underset{O-R}{\underset{\|}{P}}}-O-\overset{O}{\underset{O-R'}{\underset{\|}{P}}}-O-R' \quad (14.69)$$

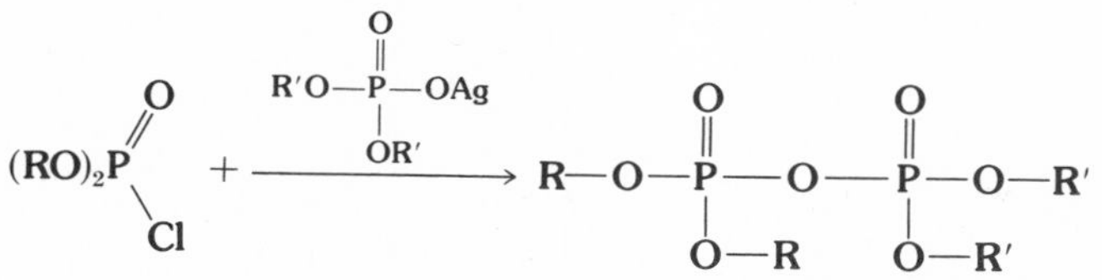

a tetraalkyldiphosphate ester

Phosphate derivatives and phosphorylation reactions are involved in many important processes taking place in living organisms. For example, the nucleotide units of DNA and RNA are joined by phosphate ester groupings, and polyphosphates such as adenosine triphosphate (ATP) serve as key transfer agents in the metabolic process (see Figures 14.10 and 14.11). Acyl phosphates, mixed anhydrides of carboxylic and phosphoric acids, are powerful acylating and phosphorylating reagents under the characteristically mild conditions found in biological systems.

The efficiency of phosphorylation and esterification reactions depends on the direction and magnitude of the energy changes taking place. The thermodynamic driving force for a reaction is best represented by the accompanying change in standard free energy $\Delta G°$ (see Section 2.4).

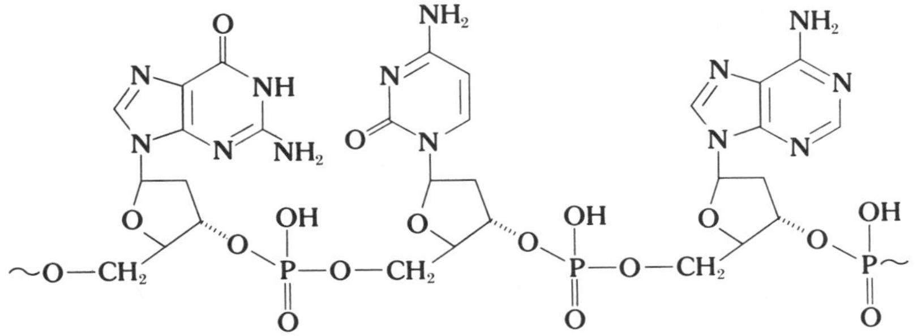

14·7 SULFONYLATION AND PHOSPHORYLATION

FIGURE 14·10 *A DNA nucleotide triplet sequence*

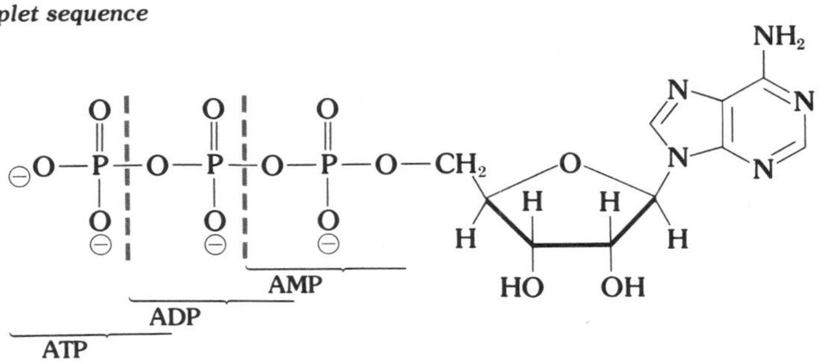

FIGURE 14·11 *Adenosine triphosphate (ATP), diphosphate (ADP), and monophosphate (AMP) at pH = 7.0*

A comparison of the free energies of hydrolysis for some phosphate compounds, listed in Table 14.3, should therefore help us to evaluate their phosphorylation capabilities.

Equations 14.70 and 14.71 illustrate the manner in which ATP serves to transfer phosphate groups from high-energy donors to low-energy acceptors:

$$ATP + \begin{array}{c} CH_2OH \\ | \\ CHOH \\ | \\ CH_2OH \end{array} \xrightleftharpoons{\text{enzymes}} \begin{array}{c} CH_2-OPO_3^{2-} \\ | \\ CHOH \\ | \\ CH_2OH \end{array} + ADP \quad (14.70)$$

glycerol 1-phosphate

$$\underset{\substack{\text{phosphocreatine,} \\ \text{a phosphoramide}}}{{}^{\ominus}O-\overset{O}{\underset{O_\ominus}{\overset{\|}{P}}}-\overset{H}{\underset{}{\overset{|}{N}}}-\overset{NH}{\underset{CH_3}{\overset{\|}{C}}}-N-CH_2CO_2^{\ominus}} + ADP \xrightleftharpoons{\text{enzymes}} H_2N-\overset{NH}{\overset{\|}{C}}-\underset{\text{creatine}}{N(CH_3)CH_2CO_2^{\ominus}} + ATP \quad (14.71)$$

Mixed carboxylate-phosphate anhydrides may also serve as phosphorylation reagents:

$$R-\overset{O}{\underset{S-\text{enzyme}}{\overset{\|}{C}}} + HPO_4^{2-} \rightleftharpoons R-\overset{O}{\underset{OPO_3^{2-}}{\overset{\|}{C}}} + \text{enzyme}-SH \quad (14.72)$$

$$R-\overset{O}{\underset{OPO_3^{2-}}{\overset{\|}{C}}} + ADP \rightleftharpoons RCO_2^{\ominus} + ATP \quad (14.73)$$

TABLE 14·3 *Free-energy changes during hydrolysis of phosphate derivatives*

$$Z-PO_3^{2\ominus} + H_2O \xrightleftharpoons[]{pH\ 7,\ 25°} H-Z + HPO_4^{2\ominus}$$

compound	Z substituent	ΔG^0 (kcal/mole)†
phosphocreatine	$-NHC(=NH)N(CH_3)CH_2CO_2^{\ominus}$	-10.3
acetyl phosphate	$-O_2CCH_3$	-10.1
ATP	$-OPO_2OPO_2OR$	-7.3
monoalkyl phosphates	$-OR$	≈ -3.0

†As ΔG^0 becomes more negative, the further these reactions proceed to completion. Common esters and thiol esters have ΔG^0 (hydrolysis) values of about -4 and -7 kcal/mole, respectively.

14·8 REDUCTION OF CARBOXYLIC ACID DERIVATIVES

REACTIONS WITH METAL HYDRIDES

Lithium aluminum hydride in ether or THF solution rapidly reduces most carboxylic acid derivatives. Esters, anhydrides, and acyl halides are usually converted to primary alcohols, and amides and nitriles are generally reduced to amines:

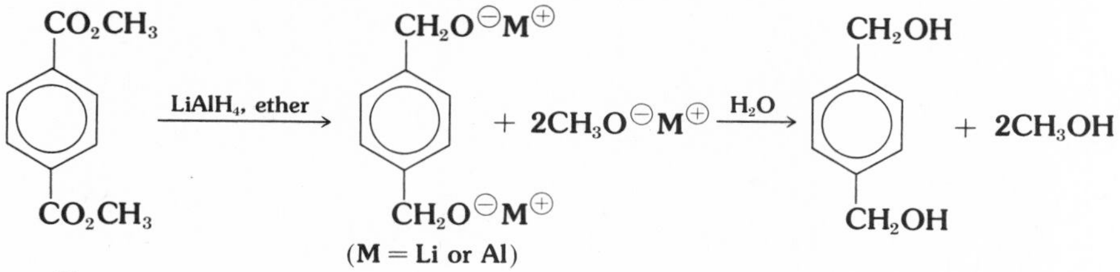

$$p\text{-}C_6H_4(CO_2CH_3)_2 \xrightarrow{LiAlH_4,\ ether} p\text{-}C_6H_4(CH_2O^{\ominus}M^{\oplus})_2 + 2CH_3O^{\ominus}M^{\oplus} \xrightarrow{H_2O} p\text{-}C_6H_4(CH_2OH)_2 + 2CH_3OH \quad (14.74)$$

$$(M = Li\ or\ Al)$$

$$\text{succinic anhydride} \xrightarrow{LiAlH_4,\ ether} M^{\oplus}O^{\ominus}-(CH_2)_4-O^{\ominus}M^{\oplus} \xrightarrow{H_2O} HO(CH_2)_4OH \quad (14.75)$$

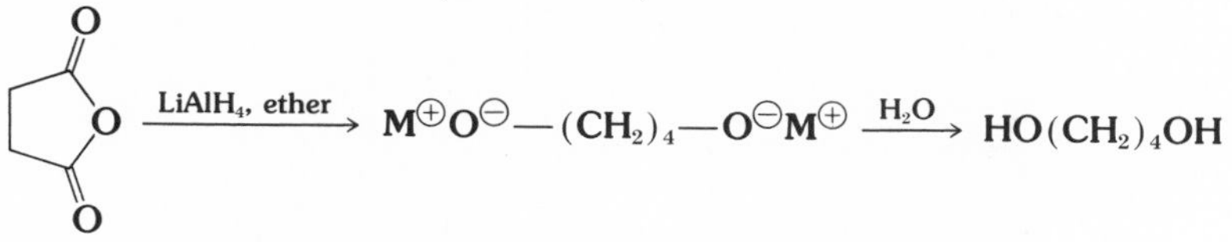

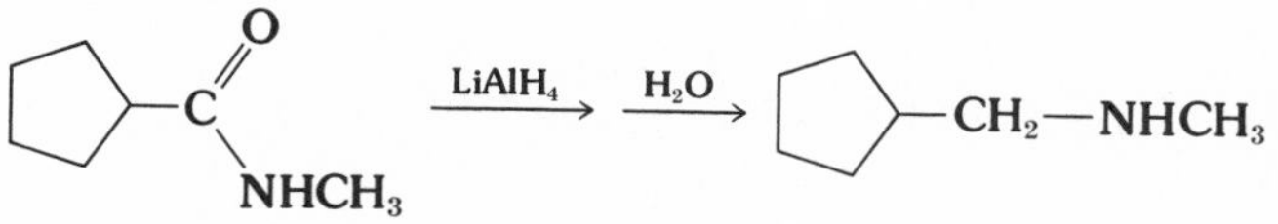

$$\text{cyclopentyl-C(=O)-NHCH}_3 \xrightarrow{LiAlH_4} \xrightarrow{H_2O} \text{cyclopentyl-CH}_2-NHCH_3 \quad (14.76)$$

$$\phi-CN \xrightarrow{LiAlH_4} \xrightarrow{H_2O} \phi-CH_2NH_2 \quad (14.77)$$

PROBLEM 14·16
Suggest a plausible mechanism for the reduction of an ester (RCO_2R') by $AlH_4^{\ominus}$ in ether solution.

Experiments with less powerful reducing agents demonstrate that esters and amides are significantly less reactive than aldehydes and ketones. Sodium borohydride, for example, does not reduce esters and amides. Diborane does not normally react with many of the common

14·8 REDUCTION OF CARBOXYLIC ACID DERIVATIVES

carboxylic acid derivatives, although it rapidly reduces acids themselves.

When lithium aluminum hydride is allowed to react with t-butyl alcohol, three of the hydride hydrogen atoms are consumed by the acidic hydroxyl protons:

$$\text{LiAlH}_4 + 3(\text{CH}_3)_3\text{COH} \xrightarrow[25°]{\text{ether}} 3\text{H}_2 + \text{LiAl}[\text{OC}(\text{CH}_3)_3]_3\text{H} \tag{14.78}$$

The resulting lithium t-butoxyaluminum hydride still retains one active hydride unit and can be used as a selective reducing agent. At low temperatures it smoothly converts acyl halides to aldehydes, a reaction which can be viewed as an acylation of hydrogen:

$$\underset{\underset{\text{Cl}}{|}}{\text{R}-\overset{\overset{\text{O}}{\|}}{\text{C}}} + \text{LiAl}[\text{OC}(\text{CH}_3)_3]_3\text{H} \xrightarrow[-78°]{\text{ether}} \underset{\underset{\text{H}}{|}}{\text{R}-\overset{\overset{\text{O}}{\|}}{\text{C}}} + \text{LiCl} + \text{Al}[\text{OC}(\text{CH}_3)_3]_3 \tag{14.79}$$

REACTIONS WITH ORGANOMETALLIC REAGENTS

In the light of our previous findings, the ease with which acyl halides and anhydrides attack and acylate Grignard reagents and alkyllithium compounds comes as no surprise. Since the ketones formed in this manner are further attacked by these highly nucleophilic organometallic reagents, the major products from reactions of this kind are usually tertiary alcohols:

$$\begin{array}{c} \text{R}-\overset{\overset{\overset{\delta\ominus}{\text{O}}}{\|}}{\underset{\underset{\text{Z}}{\delta\oplus}}{\text{C}}} \\ (\text{Z} = \text{Cl or OCOR}) \\ + \\ \underset{\delta\ominus \ \ \delta\oplus}{\text{R}'-\text{M}} \\ (\text{M} = \text{MgX or Li}) \end{array} \longrightarrow \left[\begin{array}{c} \overset{\ominus}{:}\overset{\oplus}{\text{O}}\cdots\text{M} \\ \text{R}-\overset{|}{\underset{\underset{\text{Z}}{\downarrow}}{\text{C}}}-\text{R}' \end{array} \right] \longrightarrow \underset{\text{R}'}{\overset{\text{R}}{\underset{\delta\oplus \ \delta\ominus}{\text{C}}=\text{O}}} + \text{MZ} \xrightarrow{\text{R}'\text{M}} \xrightarrow{\text{H}_2\text{O}} \underset{\underset{\text{R}'}{|}}{\text{R}-\overset{\overset{\text{OH}}{|}}{\underset{}{\text{C}}}-\text{R}'} \tag{14.80}$$

Esters, because they are less expensive and easier to handle, are generally preferred for synthesis applications of this nature:

$$\begin{array}{c} \text{CH}_3-\overset{\overset{\text{O}}{\|}}{\underset{\underset{\text{OC}_2\text{H}_5}{}}{\text{C}}} \\ + \\ 2\phi\text{CH}_2\text{MgBr} \end{array} \longrightarrow \begin{array}{c} \text{OMgBr} \\ | \\ \text{CH}_3-\overset{|}{\underset{\underset{\text{CH}_2\phi}{|}}{\text{C}}}-\text{CH}_2\phi \\ + \\ \text{C}_2\text{H}_5\text{OMgBr} \end{array} \xrightarrow{\text{H}_2\text{O}} \begin{array}{c} \text{OH} \\ | \\ \text{CH}_3-\overset{|}{\underset{\underset{\text{CH}_2\phi}{|}}{\text{C}}}-\text{CH}_2\phi \\ + \\ \text{C}_2\text{H}_5\text{OH} \end{array} \tag{14.81}$$

The less reactive organocadmium reagents, prepared as shown in equation 14.82, are sufficiently nucleophilic to react with acyl halides but are relatively unreactive toward ketones and esters; thus high-yield conversions of acyl halides to ketones are possible. Nitriles react with Grignard reagents to give imine derivatives, which yield ketones on hydrolysis:

$$2\text{RMgBr} + \text{CdCl}_2 \longrightarrow \text{R}_2\text{Cd} + \text{MgBr}_2 + \text{MgCl}_2 \tag{14.82}$$

14·9 REACTIONS AT THE α CARBON ATOM

$$[(CH_3)_2CHCH_2CH_2]_2Cd + 2Cl-\overset{O}{\underset{\|}{C}}CH_2CH_2CO_2CH_3 \xrightarrow{\phi H} 2(CH_3)_2CHCH_2CH_2\overset{O}{\underset{\|}{C}}CH_2CH_2CO_2CH_3$$
$$+ CdCl_2 \qquad (14.83)$$

$$\overset{\delta\ominus\;\;\delta\oplus}{\phi-MgBr} + \overset{\delta\oplus\;\;\delta\ominus}{CH_3OCH_2C\equiv N:} \longrightarrow \phi-\overset{:N-MgBr}{\underset{\|}{C}}CH_2OCH_3 \xrightarrow{H_2O} \phi\overset{O}{\underset{\|}{C}}CH_2OCH_3$$
$$+ NH_3 + MgBrOH \qquad (14.84)$$

PROBLEM 14·17
Suggest reagents and conditions for accomplishing each of the following transformations. (In parts c and d note the strategy used to avoid subsequent reaction of the carbonyl function in the products.)

a $\phi-\overset{O}{\underset{\|}{C}}-OC_2H_5 \longrightarrow \phi-\overset{OH}{\underset{|}{C}}(CH_2CH_3)_2$

b (piperidine)$N-\overset{O}{\underset{\|}{C}}-CH_3 \longrightarrow$ (piperidine)$N-CH_2CH_3$

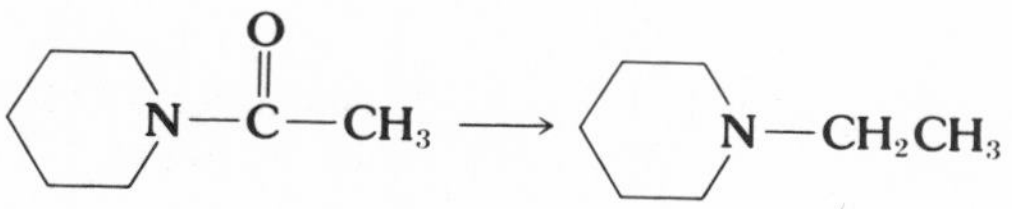

c $CH_3(CH_2)_{16}CO_2H \xrightarrow{\text{2 steps}} CH_3(CH_2)_{16}\overset{O}{\underset{H}{C}}$

d (norbornyl)$\overset{H}{\underset{}{C}}\overset{O}{\underset{OH}{\|C}} \xrightarrow{\text{2 or more steps}}$ (norbornyl)$\overset{H}{\underset{}{C}}\overset{O}{\underset{}{\|C}}-CH_3$

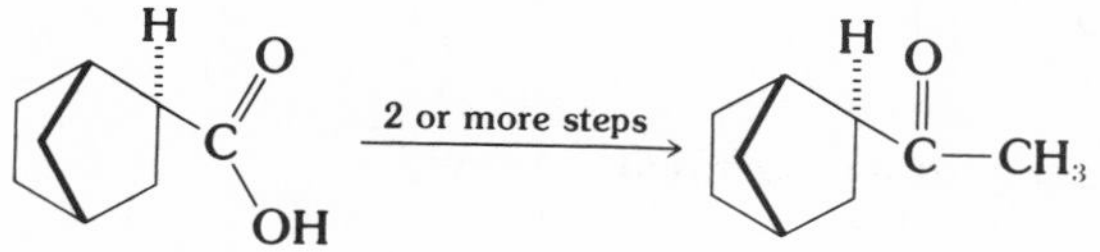

14·9 REACTIONS AT THE α CARBON ATOM

Carboxylic acid derivatives having α hydrogen atoms undergo α-substitution reactions similar to those described in Chapter 12 for aldehydes and ketones.

HALOGENATION

Just as with ketones, halogenation at the α carbon atom undoubtedly proceeds via an enol-like intermediate. Moreover, since acyl halides are more easily halogenated than acids or esters, the activating influence of carboxyl derivatives on the α carbon atom seems to be roughly proportional to their reactivity as acylating agents (see Section 14.5). The inconvenience and expense of forming acyl halides prior to halogenation can be avoided by using a catalytic amount of phosphorus tribromide (or trichloride) in the reaction, thus taking advantage of the acyl exchange shown in equation 14.86.

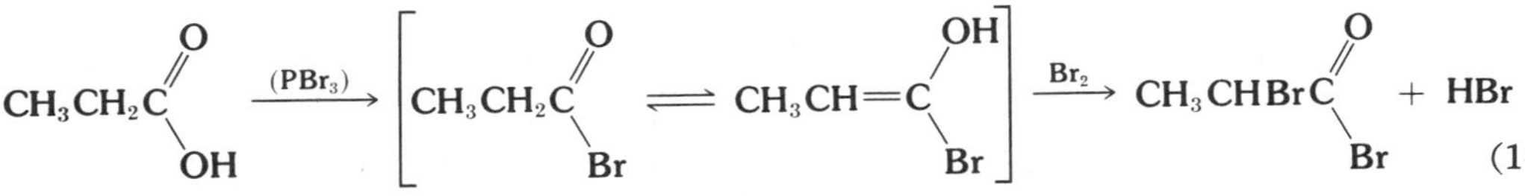

14·9 REACTIONS AT THE α CARBON ATOM

$$CH_3CH_2\overset{O}{\underset{OH}{C}} \xrightarrow{(PBr_3)} \left[CH_3CH_2\overset{O}{\underset{Br}{C}} \rightleftharpoons CH_3CH=\overset{OH}{\underset{Br}{C}} \right] \xrightarrow{Br_2} CH_3CHBr\overset{O}{\underset{Br}{C}} + HBr \quad (14.85)$$

$$CH_3CHBr\overset{O}{\underset{Br}{C}} + CH_3CH_2\overset{O}{\underset{OH}{C}} \rightleftharpoons CH_3CHBr\overset{O}{\underset{OH}{C}} + CH_3CH_2\overset{O}{\underset{Br}{C}} \quad (14.86)$$

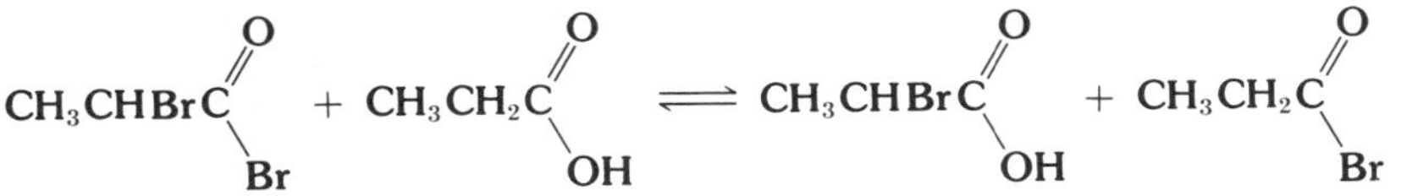

PROBLEM 14·18
Write equations illustrating a mechanism for reaction 14.86.

Subsequent displacement reactions of α-halo acids and esters can lead to a variety of useful and interesting compounds. The following reactions illustrate the utility of these intermediates in synthesis:

$$\phi CH_2COCl \xrightarrow{Br_2, CCl_4} \phi CHBrCOCl \xrightarrow{RNH_2} \phi CHBr\overset{O}{\underset{NHR}{C}} \xrightarrow{C_4H_9O^\ominus K^\oplus} \underset{\text{an α-lactam}}{\overset{\phi}{\underset{H}{\diagup}}\overset{O}{\underset{N}{\diagdown}}\overset{}{\underset{R}{}}} + \begin{matrix} C_4H_9OH \\ + \\ KBr \end{matrix} \quad (14.87)$$

$$BrCH_2CO_2C_2H_5 \xrightarrow{NaCN} N{\equiv}C{-}CH_2CO_2C_2H_5 \xrightarrow{C_2H_5OH, H_2SO_4} \underset{\text{diethyl malonate}}{CH_2(CO_2C_2H_5)_2} + (NH_4)_2SO_4 \quad (14.88)$$

THE CLAISEN ESTER CONDENSATION

Base-catalyzed deuterium exchange and/or racemization of esters is presumed, by analogy with the aldehydes and ketones, to proceed by way of enolate conjugate bases:

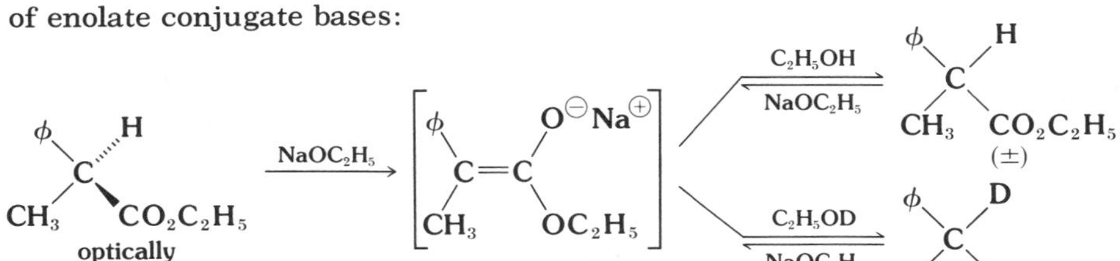

$$\underset{\substack{\text{optically}\\ \text{active}}}{\overset{\phi}{\underset{CH_3}{\diagup}}\overset{H}{\underset{CO_2C_2H_5}{\diagdown}}C} \xrightarrow{NaOC_2H_5} \underset{\text{an ester enolate}}{\left[\overset{\phi}{\underset{CH_3}{\diagup}}C{=}\overset{O^\ominus Na^\oplus}{\underset{OC_2H_5}{\diagdown}}C\right]} \overset{\underset{C_2H_5OH}{\overset{}{\underset{NaOC_2H_5}{\rightarrow}}}}{\underset{\underset{NaOC_2H_5}{\overset{C_2H_5OD}{\rightarrow}}}{}} \begin{matrix} \overset{\phi}{\underset{CH_3}{\diagup}}\overset{H}{\underset{CO_2C_2H_5}{\diagdown}}C \\ (\pm) \\ \overset{\phi}{\underset{CH_3}{\diagup}}\overset{D}{\underset{CO_2C_2H_5}{\diagdown}}C \\ (\pm) \end{matrix} \quad (14.89)$$

In practice, however, simple ester enolate salts cannot be prepared and used with the same ease and efficiency as ketone enolates, the major difficulty being self-condensation to a stable β-ketoester enolate base (Figure 14.12). This reaction, which is essentially an acylation of an enolate anion (at carbon), is known as the *Claisen ester condensation*. The beginning stage of the Claisen condensation is similar to the aldol condensation (Figure 12.13), but the initially formed β-hydroxy carbonyl compound is actually a hemiketal, and it subsequently decomposes to a β-keto ester. The rather high acidity of β-ketoesters ($pK_a \approx 11$) in comparison with alcohols ($pK_a \approx 16$) and esters ($pK_a \approx 26$) provides a driving force for this reaction, since the stable conjugate base will be favored over the other (less stable) bases at equilibrium.

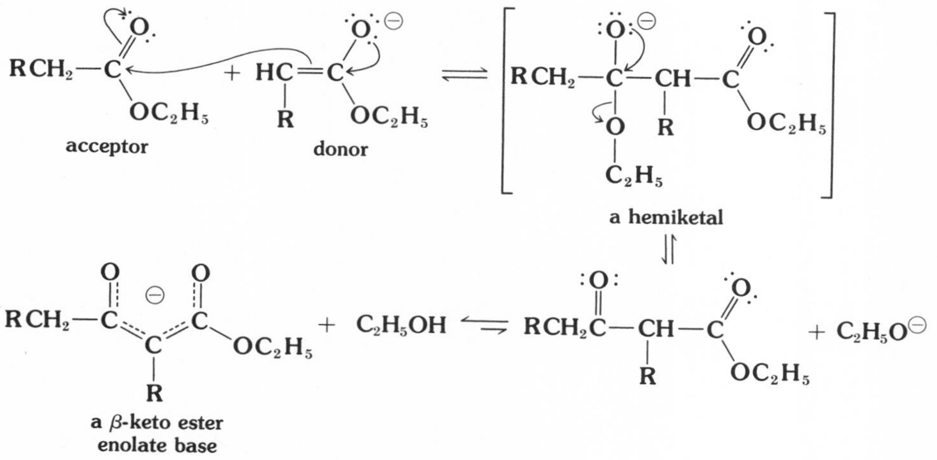

14·9 REACTIONS AT THE α CARBON ATOM

FIGURE 14·12 *The Claisen ester condensation*

The most straightforward example of the Claisen condensation is that in which both the enolate donor and the acceptor reactants are the common ester ethyl acetate (R = H in Figure 14.12). In this case the product is ethyl acetoacetate. As we shall see in the next section, many plants and animals use an enzyme-catalyzed reaction equivalent to the Claisen condensation for building carbon chains and rings.

By modifying and extending the basic concept of the Claisen condensation we can greatly enhance its versatility and usefulness. There is, for example, no reason to restrict the enolate donor role to esters or to limit the catalysts to bases:

$$CH_3CH_2CO_2C_2H_5 + \begin{array}{c} CO_2C_2H_5 \\ | \\ CO_2C_2H_5 \end{array} \xrightarrow[\Delta]{NaOC_2H_5} \begin{array}{c} CH_3CHCO_2C_2H_5 \\ | \\ COCO_2C_2H_5 \end{array} + C_2H_5OH \qquad (14.90)$$

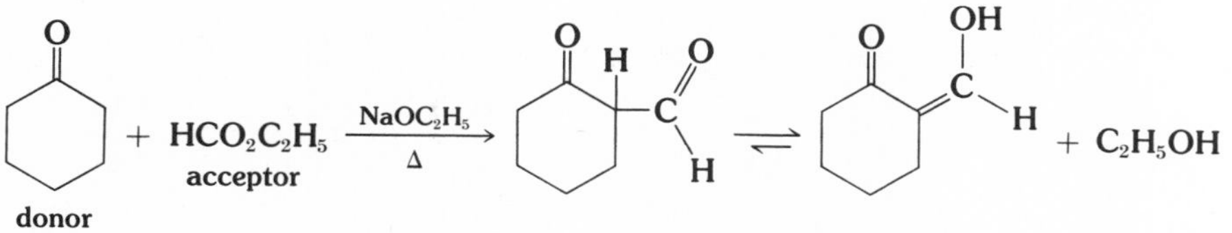

(14.91)

$$\phi CH_2CN + O{=}C(OC_2H_5)_2 \xrightarrow[\Delta]{NaOC_2H_5} \begin{array}{c} \phi CHCN \\ | \\ CO_2C_2H_5 \end{array} + C_2H_5OH \qquad (14.92)$$

$$\phi COCH_3 + (CH_3CO)_2O \xrightarrow{BF_3} \phi COCH_2COCH_3 + CH_3CO_2H \qquad (14.93)$$

Mixed condensations are particularly effective when the donor reactant is significantly more acidic than the acceptor; for example, ketones are roughly a million times more acidic than esters and therefore make better enolate donors. They are also more effective when the acceptor has no α hydrogen atoms and cannot also function as a donor; benzoate, formate, carbonate, and oxalate esters, for example, make good acceptors.

14·9 REACTIONS AT THE α CARBON ATOM

Diester cyclizations to five- and six-membered rings proceed efficiently and are known as *Dieckmann condensations* (a special case of the Claisen condensation):

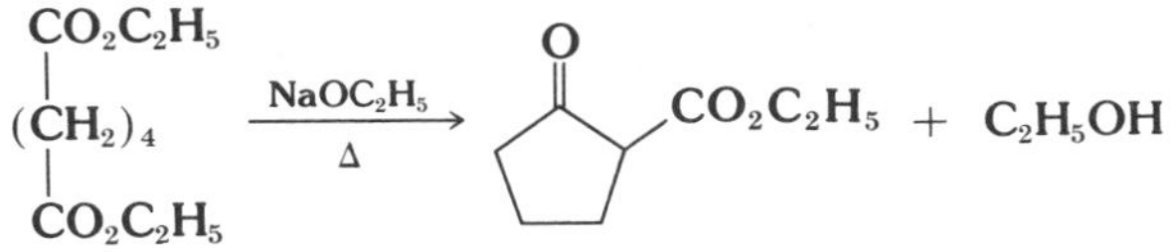

$$\underset{\underset{CO_2C_2H_5}{|}}{\overset{\overset{CO_2C_2H_5}{|}}{(CH_2)_4}} \xrightarrow[\Delta]{NaOC_2H_5} \text{cyclopentanone-}CO_2C_2H_5 + C_2H_5OH \qquad (14.94)$$

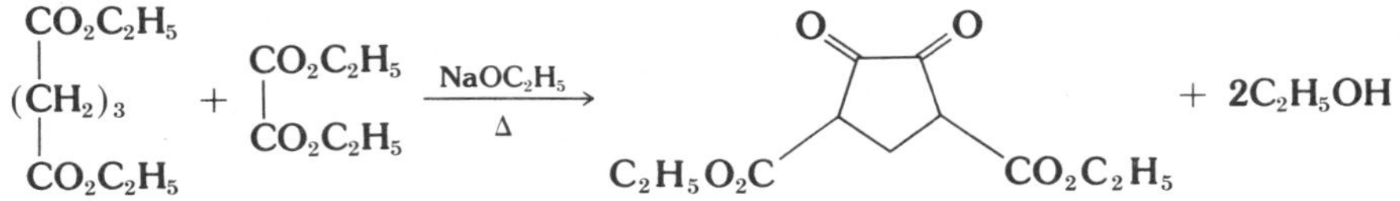

$$\underset{\underset{CO_2C_2H_5}{|}}{\overset{\overset{CO_2C_2H_5}{|}}{(CH_2)_3}} + \underset{CO_2C_2H_5}{\overset{CO_2C_2H_5}{|}} \xrightarrow[\Delta]{NaOC_2H_5} \text{diketone} + 2C_2H_5OH \qquad (14.95)$$

$$2\underset{\underset{CO_2C_2H_5}{|}}{\overset{\overset{CO_2C_2H_5}{|}}{(CH_2)_2}} \xrightarrow[\Delta]{NaOC_2H_5} [\text{intermediate}] \xrightarrow{NaOC_2H_5} \text{cyclohexanedione} + 2C_2H_5OH \qquad (14.96)$$

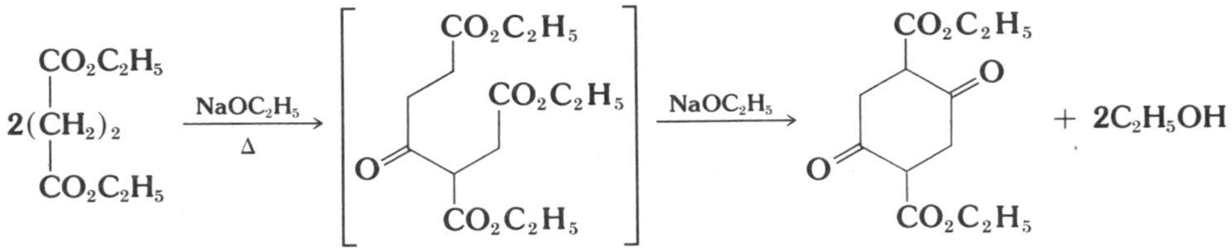

Small rings are usually not formed by Dieckmann condensations because products having low thermodynamic stability (due to angle strain or other factors) are not favored under equilibrating conditions.

PROBLEM 14·19

a Write mechanisms for reactions 14.91, 14.93, and 14.95.

b Claisen condensation of ethyl isobutyrate, $(CH_3)_2CHCO_2C_2H_5$, with sodium ethoxide as a base is not productive. Write an equation for this reaction and suggest a reason why it fails. What steps might circumvent the difficulties?

c Suggest appropriate donor and acceptor components which would be expected to give the following condensation products:

$C_2H_5COCH(CH_3)CO_2C_2H_5$ cyclohexanone-COCO$_2$C$_2$H$_5$ $\phi COCH_2NO_2$

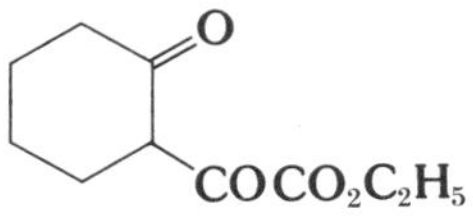

β-DICARBONYL COMPOUNDS IN SYNTHESIS

The β-keto esters formed by Claisen condensations represent an important class of organic intermediates which can be further elaborated by alkylation, decarboxylation, or reduction. Ethyl acetoacetate, for example, serves as the starting material for a useful general synthesis of branched alkyl methyl ketones, as illustrated in Figure 14.13.

Several other doubly activated methylene derivatives also have acidities greater than water and can be easily alkylated:

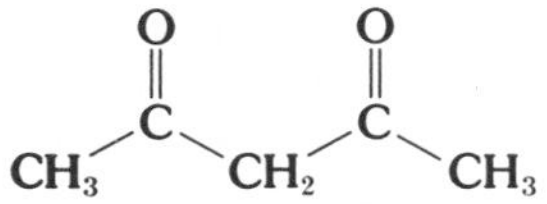

$CH_3-CO-CH_2-CO-CH_3$ $C_2H_5O-CO-CH_2-CO-OC_2H_5$ $N{\equiv}C-CH_2-C{\equiv}N$
acetylacetone   diethyl malonate   malononitrile
$pK_a = 8.8$ $pK_a = 13.3$ $pK_a = 12$

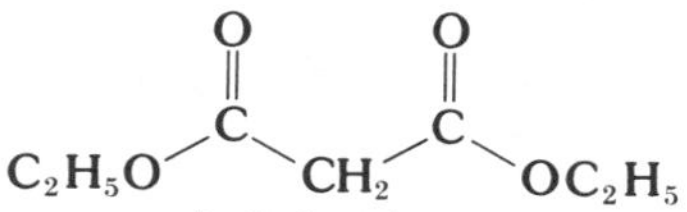

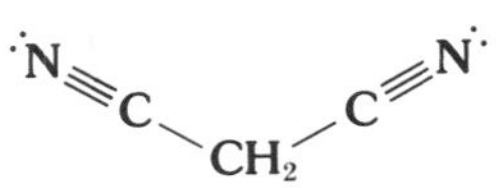

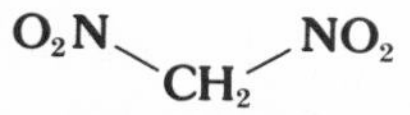

14·9 REACTIONS AT THE α CARBON ATOM

$$O_2N\text{-}CH_2\text{-}NO_2$$
dinitromethane
$pK_a = 3.6$

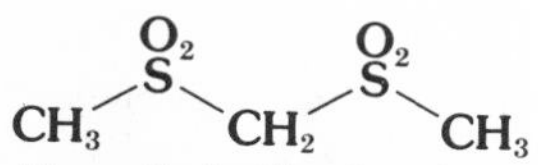

$$CH_3\text{-}\underset{O_2}{S}\text{-}CH_2\text{-}\underset{O_2}{S}\text{-}CH_3$$
bis-methylsulfonylmethane
$pK_a = 14$

These compounds and others like them offer a wide range of applications in the synthesis of complex organic molecules. Diethyl malonate, for example, can be transformed into a variety of substituted acids:

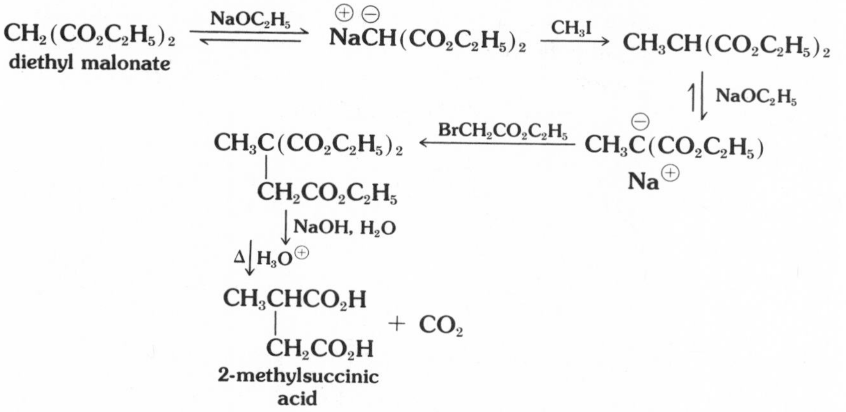

$$CH_2(CO_2C_2H_5)_2 \underset{\text{diethyl malonate}}{\xrightleftharpoons{NaOC_2H_5}} \overset{\oplus\,\ominus}{Na}CH(CO_2C_2H_5)_2 \xrightarrow{CH_3I} CH_3CH(CO_2C_2H_5)_2$$

$$\Updownarrow NaOC_2H_5$$

$$CH_3C(CO_2C_2H_5)_2 \xleftarrow{BrCH_2CO_2C_2H_5} CH_3\overset{\ominus}{C}(CO_2C_2H_5)\;Na^\oplus$$
$$|$$
$$CH_2CO_2C_2H_5$$

$$\downarrow NaOH, H_2O$$
$$\Delta \downarrow H_3O^\oplus$$

$$\underset{\text{2-methylsuccinic acid}}{CH_3CHCO_2H \atop |\;\;CH_2CO_2H} + CO_2 \qquad (14.97)$$

Small (strained) ring systems can be prepared from diethyl malonate by irreversible carbon-carbon bond-forming reactions such as alkylations, but not by reversible reactions such as condensations.

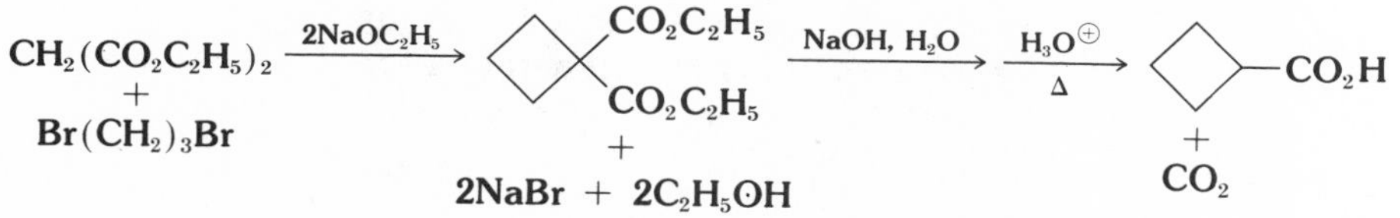

$$\underset{+\;Br(CH_2)_3Br}{CH_2(CO_2C_2H_5)_2} \xrightarrow{2NaOC_2H_5} \underset{+\;2NaBr + 2C_2H_5OH}{\square\!\!\!<\!\!{CO_2C_2H_5 \atop CO_2C_2H_5}} \xrightarrow{NaOH, H_2O} \xrightarrow[\Delta]{H_3O^\oplus} \underset{+\;CO_2}{\square\!\!\!-\!CO_2H} \qquad (14.98)$$

FIGURE 14·13 *Synthesis of ketones from ethyl acetoacetate*

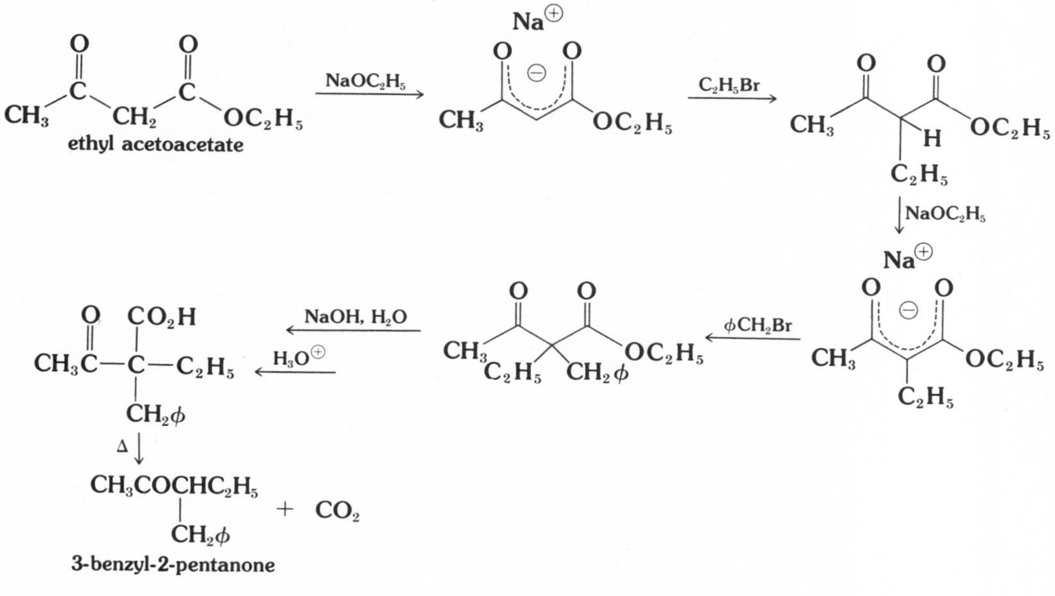

$$\underset{\text{ethyl acetoacetate}}{CH_3\overset{O}{\overset{\|}{C}}CH_2\overset{O}{\overset{\|}{C}}OC_2H_5} \xrightarrow{NaOC_2H_5} CH_3\overset{O\;\;\;Na^\oplus\;\;O}{\overset{\|\;\;\;\;\;\;\;\;\;\|}{C\text{-}\overset{\ominus}{C}\text{-}C}}OC_2H_5 \xrightarrow{C_2H_5Br} CH_3\overset{O\;\;\;\;\;\;O}{\overset{\|\;\;\;\;\;\;\;\;\;\|}{C\text{-}\underset{C_2H_5}{\overset{H}{C}}\text{-}C}}OC_2H_5$$

$$\downarrow NaOC_2H_5$$

$$\underset{\text{3-benzyl-2-pentanone}}{CH_3COCHC_2H_5 \atop |\;\;CH_2\phi} + CO_2 \xleftarrow{\Delta} \underset{\Delta}{CH_3\overset{O}{\overset{\|}{C}}\text{-}\underset{CH_2\phi}{\overset{CO_2H}{\overset{|}{C}}}\text{-}C_2H_5} \xleftarrow[H_3O^\oplus]{NaOH, H_2O} CH_3\overset{O\;\;\;\;\;\;\;\;\;\;O}{\overset{\|\;\;\;\;\;\;\;\;\;\;\;\|}{C\text{-}\underset{C_2H_5\;\;CH_2\phi}{C}\text{-}C}}OC_2H_5 \xleftarrow{\phi CH_2Br} CH_3\overset{O\;\;\;Na^\oplus\;O}{\overset{\|\;\;\;\;\;\;\;\;\;\;\|}{C\text{-}\underset{C_2H_5}{\overset{\ominus}{C}}\text{-}C}}OC_2H_5$$

PROBLEM 14·20
Write equations showing how the following synthetic preparations might be accomplished:
a 3-Phenyl-2-benzylpropanoic acid from diethyl malonate and benzyl bromide
b 2-Ethylcyclopentanone from diethyl adipate and ethyl iodide
c Methyl cyclopentyl ketone from ethyl acetoacetate and 1,4-dibromobutane

PROBLEM 14·21
a What are two important factors that interfere with attempts to alkylate simple monoesters in the same fashion as malonic ester?
b The following equation gives conditions recently used to circumvent these difficulties. Indicate some specific advantages of this procedure.

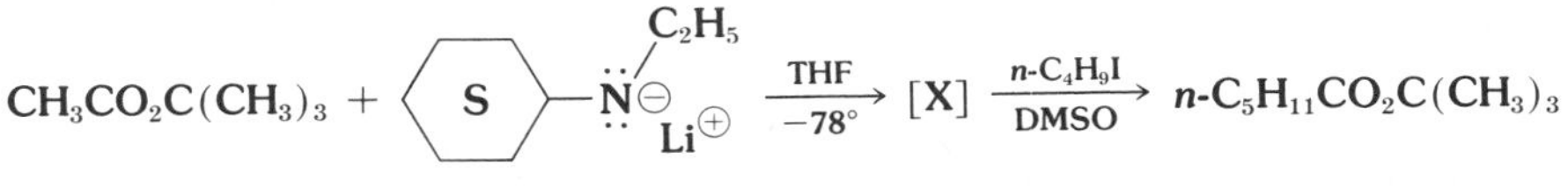

$$CH_3CO_2C(CH_3)_3 + \langle S \rangle\text{-}\ddot{N}^{\ominus}\underset{Li^{\oplus}}{\overset{C_2H_5}{|}} \xrightarrow[-78°]{THF} [X] \xrightarrow[DMSO]{n\text{-}C_4H_9I} n\text{-}C_5H_{11}CO_2C(CH_3)_3$$

14·10 BIOSYNTHESIS OF COMPLEX MOLECULES

The impressive variety and complexity of molecular structures found in naturally occurring substances indicate that living organisms possess one or more synthetic mechanisms for constructing complex arrays of carbon atoms. Condensations of the Claisen and aldol type appear to be widely used for this purpose. However, it should be clear that the strong bases ($NaOR$, NaH, or $LiNR_2$) or strong acids (BF_3 or HX) generally used to accomplish these condensations in the laboratory are not compatible with biological systems.

ACETATE CONDENSATION REACTIONS

The fundamental building block for the carbon skeletons of most natural products is the acetate unit. Condensation of these units, in the form of thiol esters, is accomplished by enzyme systems called *synthetases*. The enhanced activation of the α methylene groups in such compounds is reflected in the acidity constants of ethyl acetoacetate and its thio analog:

$CH_3COCH_2CO_2C_2H_5$
$pK_a = 10.7$; 8% enol tautomer

$CH_3COCH_2COSC_2H_5$
$pK_a = 8.5$; 30% enol tautomer

Carbon chains are constructed, two carbon atoms at a time, via the reactive malonic acid intermediate shown in the following reaction scheme:

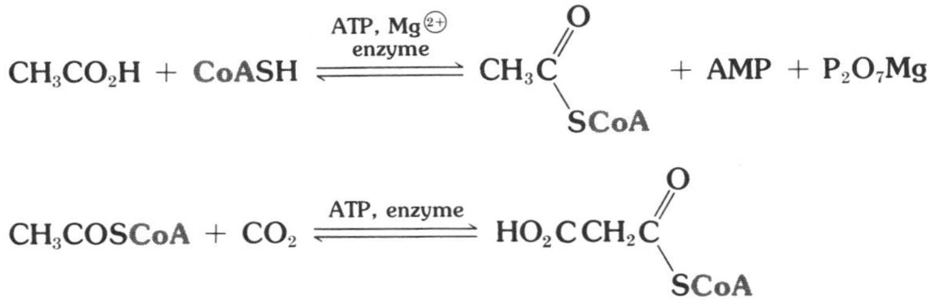

$$CH_3CO_2H + CoASH \xrightleftharpoons[\text{enzyme}]{ATP,\ Mg^{2+}} CH_3C\overset{O}{\underset{SCoA}{\diagdown}} + AMP + P_2O_7Mg \tag{14.99}$$

$$CH_3COSCoA + CO_2 \xrightleftharpoons{ATP,\ enzyme} HO_2CCH_2C\overset{O}{\underset{SCoA}{\diagdown}}$$

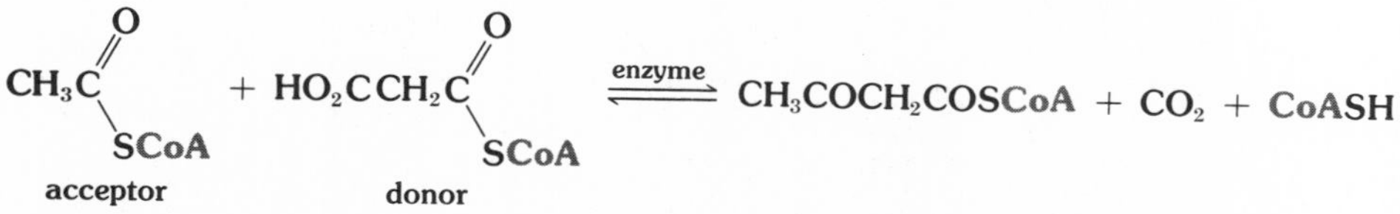

14·10 BIOSYNTHESIS OF COMPLEX MOLECULES

$$\underset{\text{acceptor}}{CH_3C(=O)SCoA} + \underset{\text{donor}}{HO_2CCH_2C(=O)SCoA} \xrightleftharpoons{\text{enzyme}} CH_3COCH_2COSCoA + CO_2 + CoASH \quad (14.100)$$

The resulting acyl coenzyme A intermediate then serves as the acceptor moiety in another condensation. In the thiol esters a carrier protein may replace coenzyme A.

Enzymatic reduction of the β-carbonyl function is accomplished by a biological hydride transfer agent called reduced *nicotinamide adenine dinucleotide* (NADH):

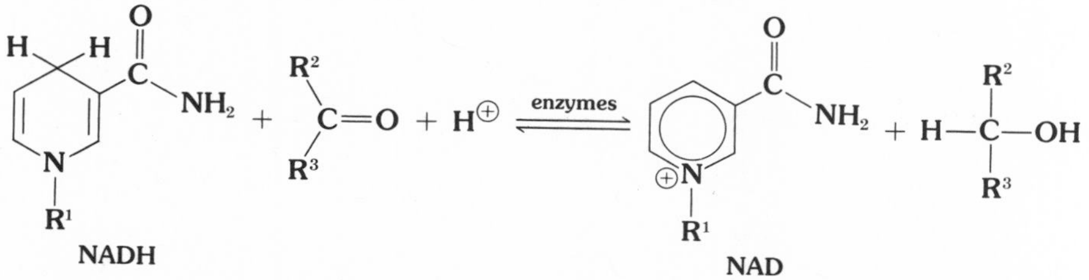

$$\text{NADH} + \underset{R^3}{\overset{R^2}{C}}=O + H^{\oplus} \xrightleftharpoons{\text{enzymes}} \text{NAD} + H-\underset{R^3}{\overset{R^2}{C}}-OH \quad (14.101)$$

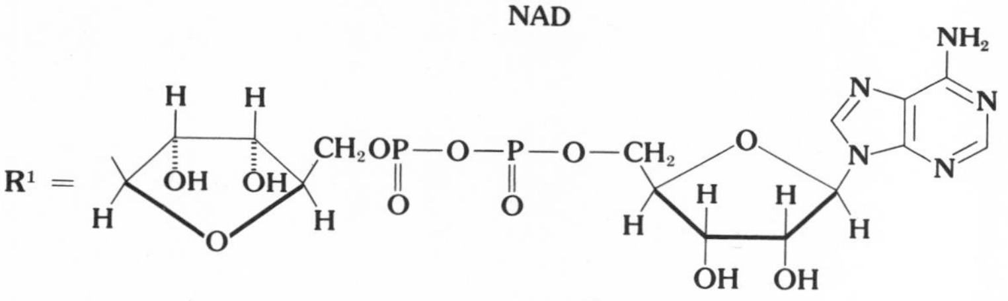

The role of NADH as a reducing agent is illustrated in Figure 14.14.

PROBLEM 14·22
Classify the first three reactions in Figure 14.14 in terms of simple laboratory transformations discussed in earlier chapters.

FIGURE 14·14 *The role of NADH as a reducing agent in the biosynthesis of fatty acids*

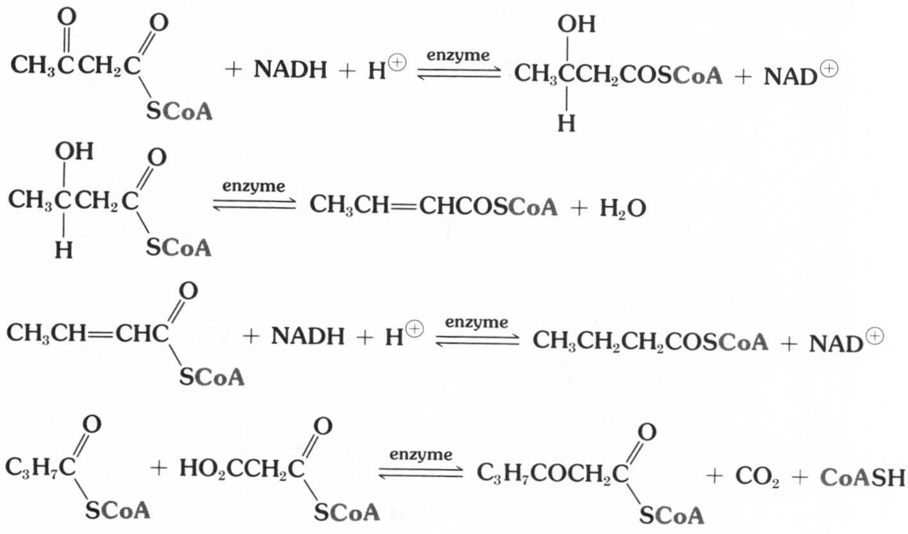

$$CH_3COCH_2C(=O)SCoA + NADH + H^{\oplus} \xrightleftharpoons{\text{enzyme}} CH_3CH(OH)CH_2COSCoA + NAD^{\oplus}$$

$$CH_3CH(OH)CH_2C(=O)SCoA \xrightleftharpoons{\text{enzyme}} CH_3CH=CHCOSCoA + H_2O$$

$$CH_3CH=CHC(=O)SCoA + NADH + H^{\oplus} \xrightleftharpoons{\text{enzyme}} CH_3CH_2CH_2COSCoA + NAD^{\oplus}$$

$$C_3H_7C(=O)SCoA + HO_2CCH_2C(=O)SCoA \xrightleftharpoons{\text{enzyme}} C_3H_7COCH_2C(=O)SCoA + CO_2 + CoASH$$

472   14·10 BIOSYNTHESIS OF COMPLEX MOLECULES

Repetition of these reactions generates fatty acids having an even number of carbon atoms; indeed, most naturally occurring acids are of this kind. Furthermore, a reverse set of similar reactions (carbon dioxide is not a necessary reactant) operates in the metabolic breakdown of the fatty acids. Since chain branching interferes with these reactions, we find that biodegradability of synthetic detergents improves as branching decreases.

Many natural fatty-acid derivatives having alkene, acetylene, and hydroxyl groups are known. The existence of such compounds is an indication of the many variations of the general synthesis scheme that are possible. In fact, it is useful—and in many cases correct—to assume that many aromatic oxygen-containing natural products are formed by cyclization of long-chain polyketo acids:

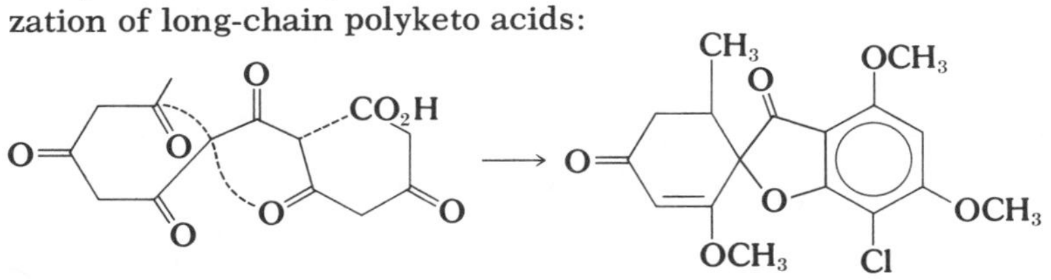

(14.102)

griseofulvin, an antibiotic from
Penicillium griseofulvum

PROBLEM 14·23

a If acetic acid labeled with ^{14}C isotope only in the carboxyl group is fed to an organism, and the linoleic acid produced by this organism is isolated after a suitable period, what distribution of radioactive carbon would you expect to find in this molecule $[CH_3(CH_2)_4CH=CHCH_2CH=CH(CH_2)_7CO_2H]$?

b Why doesn't the Claisen condensation of ethyl acetate give rise to long-chain polyketones similar to those proposed for the biosynthesis of complex molecules?

THE ISOPRENE RULE

The preponderance in nature of fatty acids having an even number of carbon atoms clearly reflects their formation from a basic two-carbon unit. However, other classes of natural products do not appear to have such a simple genesis. A very large group of plant and animal substances called *terpenes* are apparently derived from a branched five-carbon unit related to isoprene $(CH_2=C(CH_3)—CH=CH_2)$. The following structural formulas illustrate this *isoprene rule*:

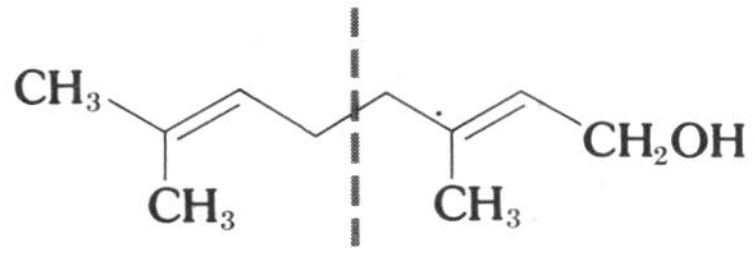

geraniol
$(C_{10}H_{18}O)$

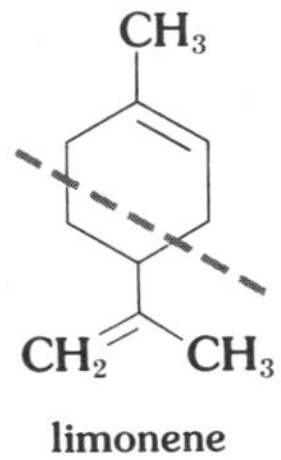

limonene
$(C_{10}H_{16})$

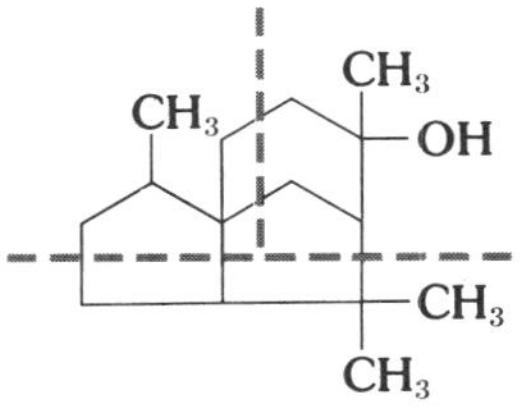

cedrol
$(C_{15}H_{26}O)$

As illustrated in Figure 14.16, the initial five-carbon unit is generated by a modified acetate condensation path. However, subsequent combinations of these isopentenyl intermediates take place by an olefin alkylation process. In the laboratory this process usually requires powerful

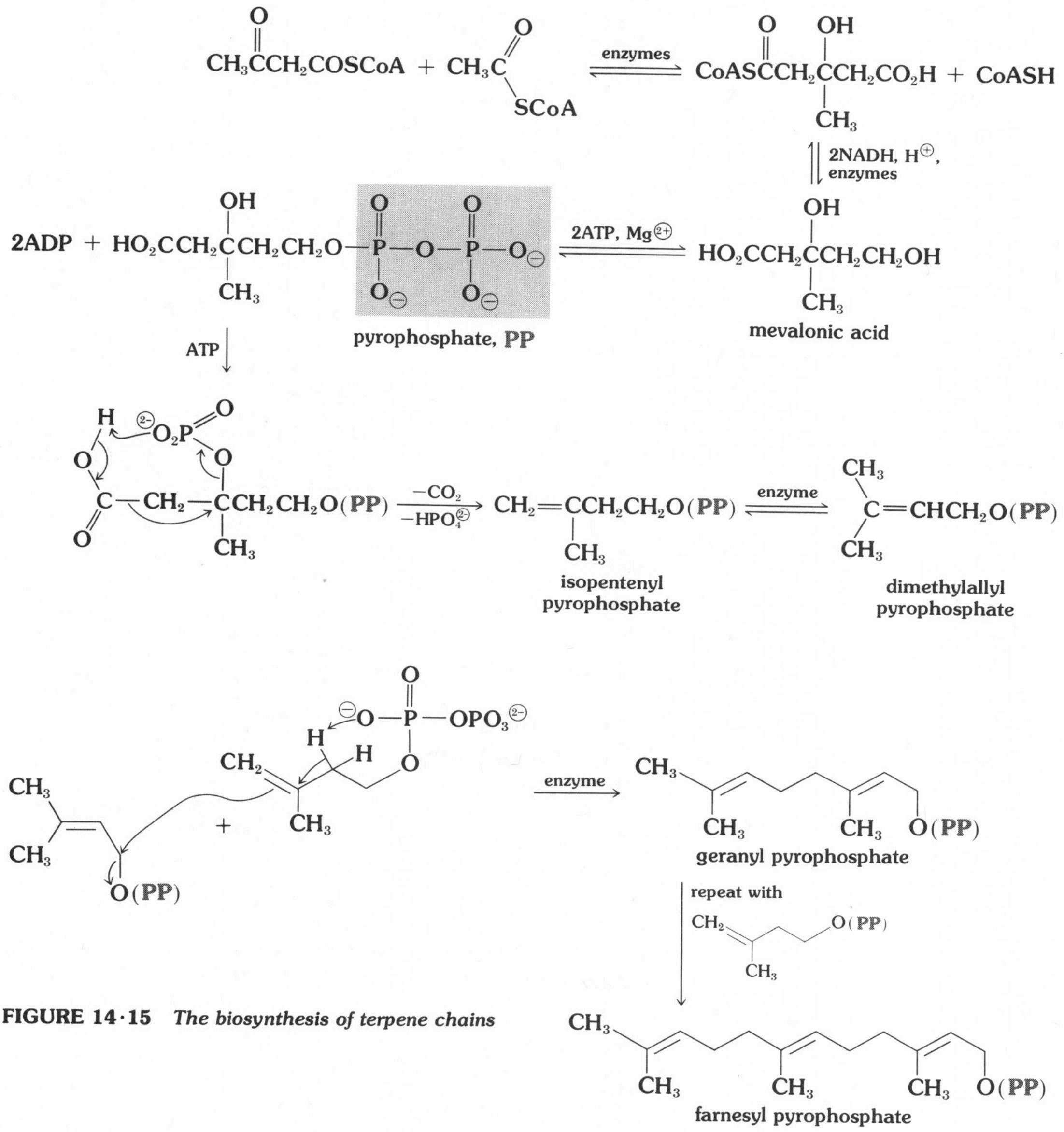

14·10 BIOSYNTHESIS OF COMPLEX MOLECULES 473

$$CH_3\overset{O}{\underset{\|}{C}}CH_2COSCoA + CH_3\overset{O}{\underset{\|}{C}}{-}SCoA \xrightleftharpoons{\text{enzymes}} CoAS\overset{O}{\underset{\|}{C}}CH_2\overset{OH}{\underset{CH_3}{\overset{|}{C}}}CH_2CO_2H + CoASH$$

$$\Updownarrow \text{2NADH, H}^{\oplus}, \text{enzymes}$$

$$2ADP + HO_2CCH_2\underset{CH_3}{\overset{OH}{\underset{|}{C}}}CH_2CH_2O{-}\underset{O_{\ominus}}{\overset{O}{\underset{\|}{P}}}{-}O{-}\underset{O_{\ominus}}{\overset{O}{\underset{\|}{P}}}{-}O_{\ominus} \xrightleftharpoons{\text{2ATP, Mg}^{2+}} HO_2CCH_2\underset{CH_3}{\overset{OH}{\underset{|}{C}}}CH_2CH_2OH$$

pyrophosphate, PP   mevalonic acid

ATP ↓

<!-- cyclic intermediate -->
$$\underset{O}{\overset{O}{\underset{\|}{C}}}{-}CH_2{-}\underset{CH_3}{\overset{|}{C}}CH_2CH_2O(PP) \xrightarrow[-HPO_4^{2-}]{-CO_2} CH_2{=}\underset{CH_3}{\overset{|}{C}}CH_2CH_2O(PP) \xrightleftharpoons{\text{enzyme}} \underset{CH_3}{\overset{CH_3}{\underset{|}{C}}}{=}CHCH_2O(PP)$$

isopentenyl pyrophosphate   dimethylallyl pyrophosphate

$$\xrightarrow{\text{enzyme}}$$ geranyl pyrophosphate

repeat with $CH_2{=}\underset{CH_3}{\overset{|}{C}}CH_2CH_2O(PP)$

↓

farnesyl pyrophosphate

FIGURE 14·15 *The biosynthesis of terpene chains*

electrophilic intermediates (such as carbonium ions) and weakly basic solvent systems which will not compete with the modestly nucleophilic double bond.† Hence the reaction in Figure 14.15 provides a striking example of rate enhancement as a result of enzyme catalysts. Instead of the halogen or sulfonate ester leaving groups favored as laboratory alkylating agents, we find that biological systems make use of pyrophosphate derivatives. These serve as good leaving groups for substitution and elimination reactions with a variety of nucleophiles and bases, and in the case of dimethylallyl pyrophosphate they are further activated by the allylic double bond.

†Alkylation of enolate derivatives is curiously rare in biosynthesis, in contrast to the acylation processes described above.

The structures of cyclic terpenes can be similarly constructed by intramolecular alkylations and rearrangements:

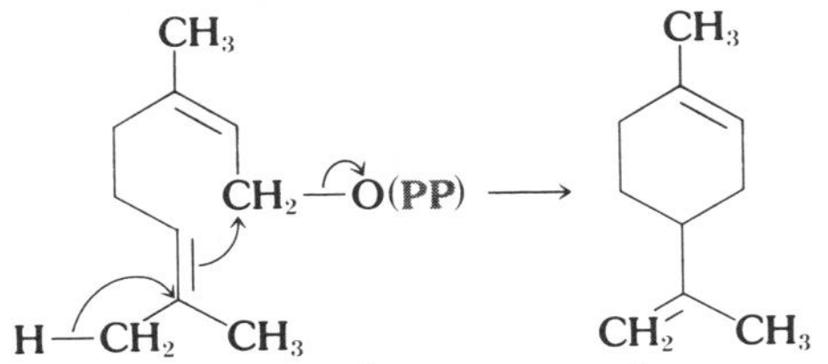

(14.103)

geranyl pyrophosphate → limonene

(14.104)

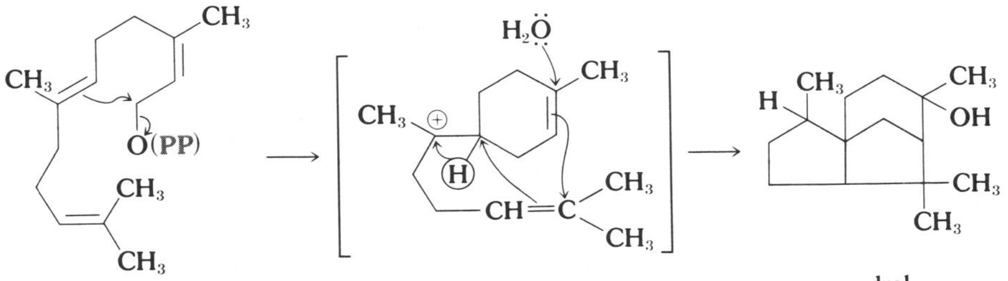

farnesyl pyrophosphate → cedrol

14·11 PEROXY ACIDS AND THEIR DERIVATIVES

Peroxy acids, often called *peracids*, are formed by an acid-catalyzed exchange reaction of hydrogen peroxide with carboxylic acids:

$$RCO_2H + H_2O_2 \xrightleftharpoons{H^\oplus} R-C(=O)O-OH + H_2O \qquad (14.105)$$

a peroxy acid

High-yield conversions to pure peroxy acids are best accomplished, however, by reacting the hydrogen peroxide with acyl halides (equation 14.33) or with anhydrides:

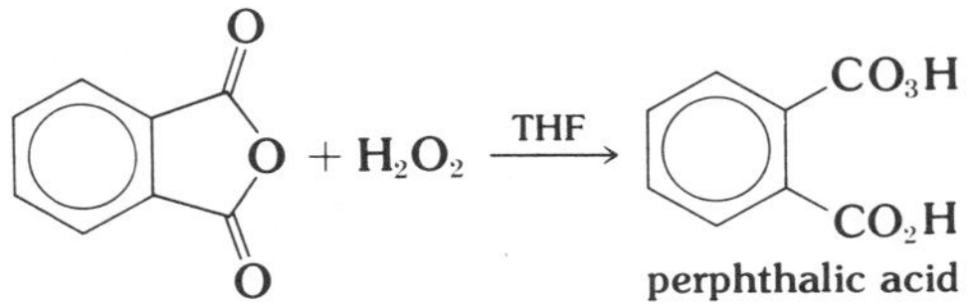

(14.106)

phthalic anhydride + H_2O_2 $\xrightarrow{THF}$ perphthalic acid (o-CO_3H, CO_2H)

Peroxy analogs of esters and anhydrides are readily prepared in a similar fashion:

$$RCOCl + R'OONa \longrightarrow R-C(=O)O-OR' + NaCl \qquad (14.107)$$

a perester

$$2\phi COCl + Na_2O_2 \longrightarrow \phi-C(=O)O-O C(=O)-\phi + 2NaCl \qquad (14.108)$$

benzoyl peroxide

14·11 PEROXY ACIDS AND THEIR DERIVATIVES

One of the most useful and general reactions of peroxy acids is their ability to oxidize alkenes to epoxides, as illustrated by the following reactions:

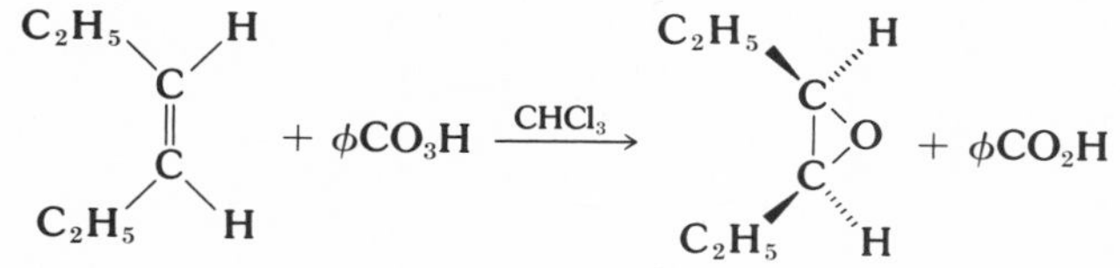

$$\underset{C_2H_5\ \ \ H}{\overset{C_2H_5\ \ \ H}{C=C}} + \phi CO_3H \xrightarrow{CHCl_3} \underset{C_2H_5\ \ \ H}{\overset{C_2H_5\ \ \ H}{C-C}}\!\!\diagdown\!O + \phi CO_2H \quad (14.109)$$

$$\underset{H\ \ \ C_2H_5}{\overset{C_2H_5\ \ \ H}{C=C}} + \phi CO_3H \xrightarrow{CHCl_3} \underset{H\ \ \ C_2H_5}{\overset{C_2H_5\ \ \ H}{C-C}}\!\!\diagdown\!O + \phi CO_2H \quad (14.110)$$

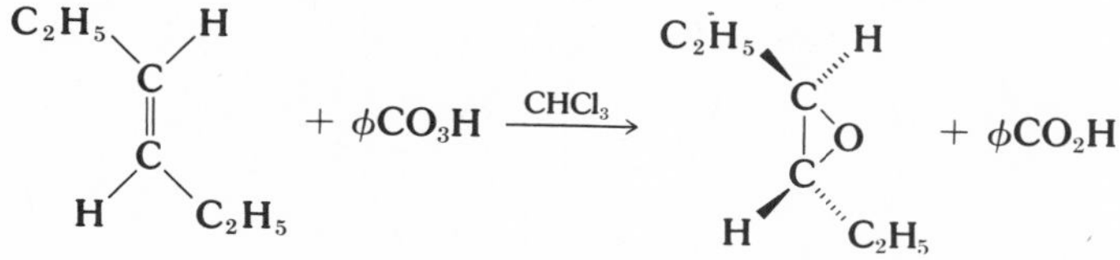

The epoxidation proceeds stereospecifically (with retention of configuration), is relatively insensitive to changes in solvent polarity, and generally follows a second-order kinetic relationship:

$$\text{rate of epoxidation} = k[\text{alkene}][\text{peroxy acid}] \quad (14.111)$$

If strong acids are present in the epoxidation reaction mixture, the initially formed epoxide may suffer acid-catalyzed ring opening to hydroxy esters or diols:

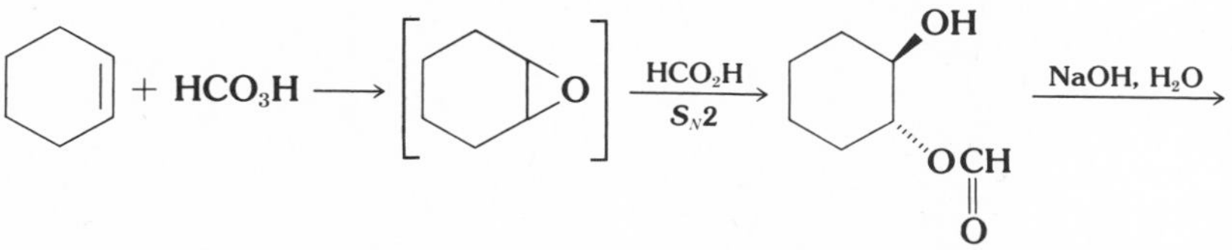

<chemical equation 14.112: cyclohexene + HCO₃H → [epoxide] → (HCO₂H, S_N2) trans-hydroxy formate ester → (NaOH, H₂O) trans-diol>

(14.112)

By conducting such reactions in the presence of sodium bicarbonate or sodium carbonate buffers, the ring opening can be avoided:

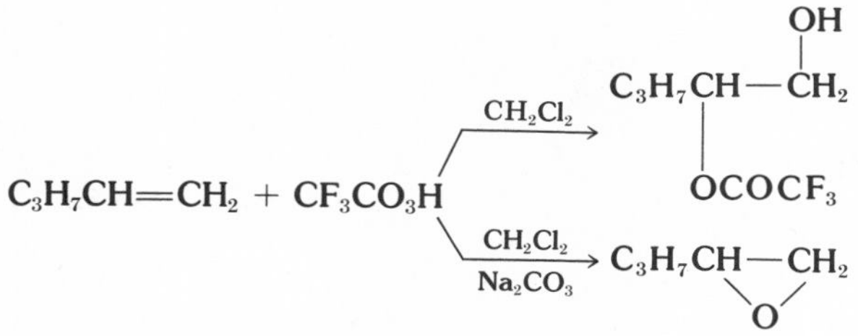

$$C_3H_7CH{=}CH_2 + CF_3CO_3H \quad \begin{array}{l}\xrightarrow{CH_2Cl_2}\ C_3H_7\underset{\underset{OCOCF_3}{|}}{CH}{-}\underset{\underset{OH}{|}}{CH_2} \\ \xrightarrow[Na_2CO_3]{CH_2Cl_2}\ C_3H_7CH{-}CH_2\diagdown\!O \end{array} \quad (14.113)$$

Peresters and diacyl peroxides are used extensively as free-radical initiators because of their facile peroxide-bond cleavage. The free-radical-initiated polymerization of methyl methacrylate is a typical example.

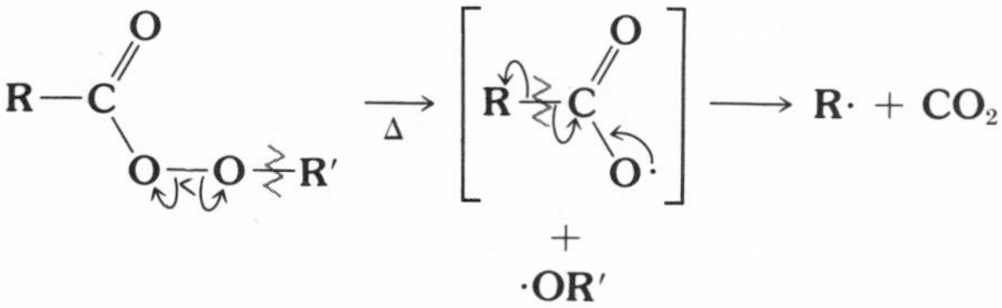

$$R{-}\underset{O{-}O{-}R'}{\overset{O}{\underset{\|}{C}}}\ \xrightarrow{\Delta}\ \left[R{-}\underset{O\cdot}{\overset{O}{\underset{\|}{C}}}\right] \longrightarrow R\cdot + CO_2 \quad (14.114)$$

$$+$$
$$\cdot OR'$$

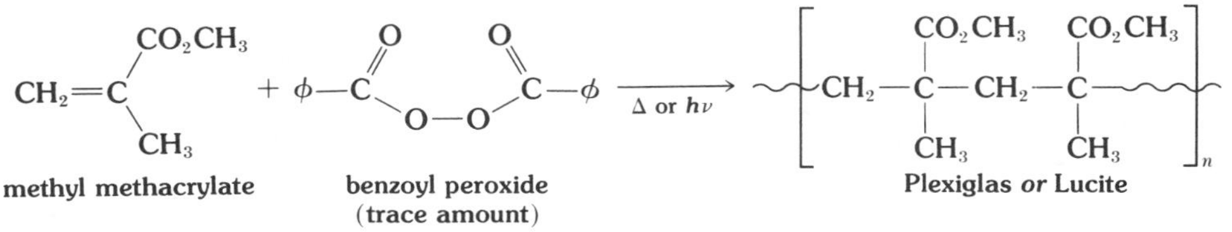

$$CH_2=C\begin{matrix}CO_2CH_3\\ \\CH_3\end{matrix} + \phi-C\underset{O-O}{\overset{O\quad\quad O}{\diagdown\quad\diagup}}C-\phi \xrightarrow{\Delta \text{ or } h\nu} \left[\begin{matrix}CO_2CH_3 & CO_2CH_3\\ | & |\\CH_2-C-CH_2-C\\ | & |\\CH_3 & CH_3\end{matrix}\right]_n \quad (14.115)$$

methyl methacrylate   benzoyl peroxide   Plexiglas or Lucite
                           (trace amount)

PROBLEM 14·24
a Peroxy acids are considerably weaker than the corresponding carboxylic acids ($pK_a = 8.2$) for CH_3CO_3H. Explain the reason for this.
b Why does one stereoisomeric epoxide predominate in the following reaction?

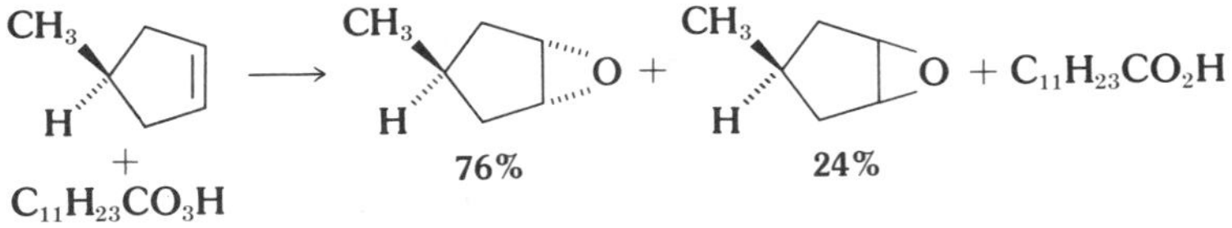

[cyclopentene with CH₃ and H substituents + $C_{11}H_{23}CO_3H$ → epoxide (76%) + epoxide (24%) + $C_{11}H_{23}CO_2H$]

SUMMARY

acidity of carboxylic acids A measure of the ability of the carboxyl group to transfer its proton to a base. Most carboxylic acids are moderately acidic ($K_a \approx 10^{-5}$) but are strengthened by electron-withdrawing substituents close to the carboxyl function.

pK_a A logarithmic scale of acidities: $pK_a = -\log K_a$.

acylation A reaction in which an acyl group (RCO—) is transferred to a nucleophile. The electrophilic source of the acyl group is called an *acylating agent* (RCOZ). The following table shows the products of some common acylation reactions:

	\multicolumn{5}{c}{acylating agent}				
nucleophile	RCOCl acyl halide	(RCO)$_2$O anhydride	RCO$_2$R'' ester	RCO$_2$H acid	RCONHR'' amide
R'NH$_2$ amine	RCONHR' (fast)	RCONHR' (fast)	RCONHR' (moderate)	RCO$_2^\ominus$R'NH$_3^\oplus$ (salt formation)	RCONHR' (slow)
R'OH alcohol	RCO$_2$R' (fast)	RCO$_2$R' (moderate)	RCO$_2$R' (needs acid or base catalysis)	RCO$_2$R' (needs acid catalysis)	RCO$_2$R' (slow, needs acid catalysis)
H$_2$O	RCO$_2$H (fast)	RCO$_2$H (moderate)	RCO$_2$H (needs acid or base catalysis)	RCO$_2$H (exchange of H and O)	RCO$_2$H (needs acid or base catalysis)
R'$_2$C=CHR' alkene	RCOCR'=CR'$_2$ (AlCl$_3$ catalysis)	RCOCR'=CR'$_2$ (AlCl$_3$ catalysis)	no reaction	no reaction	no reaction
 aromatic ring	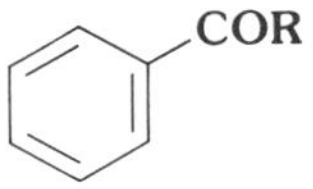 Ar-COR (AlCl$_3$ catalysis)	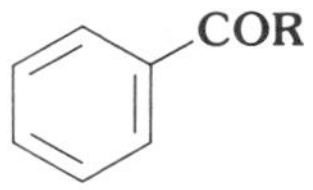 Ar-COR (AlCl$_3$ catalysis)	no reaction	no reaction	no reaction

sulfonation and phosphorylation The transfer of a sulfonyl group [RSO_2—] or a phosphoryl group [$(RO)_2P(O)$—] to a nucleophile in a fashion analogous to acyl-group transfer in acylation.

decarboxylation A reaction of a carboxylic acid in which carbon dioxide is lost:

$$\underset{R}{\overset{H}{\underset{Z}{\diagdown}}}\!\!\!\underset{O}{\overset{O}{\diagdown}}\!\!\!\overset{\Delta}{\longrightarrow}\;\;\underset{R}{\overset{Z-H}{\diagdown}}\!\!\!\overset{}{\underset{CH_2}{\diagdown}} + CO_2$$

($Z=O$ or CH_2)

thermal decarboxylation

$RCO_2M + Br_2 \longrightarrow R-Br + CO_2 + MBr$

($M = Ag^{+1}$, Tl^{+1}, or Hg^{+2})

oxidative decarboxylation

reduction Reactions of carboxylic acids and their derivatives in which the oxidation state of the carboxyl carbon atom is lowered. Complex metal hydrides are the usual reducing agents. *Lithium aluminum hydride* ($LiAlH_4$) reduces acids, esters, and anhydrides to primary alcohols and reduces amides and nitriles to amines. *Sodium borohydride* ($NaBH_4$) does not normally reduce acids or esters. *Diborane* (B_2H_6) reduces acids to primary alcohols but does not reduce carboxylate salts or esters. *Lithium tri-t-butoxyaluminum hydride* [$LiAl(OC_4H_9)_3H$] reduces acyl chlorides to aldehydes at low temperature ($<0°C$).

reactions with organometallic reagents Reactions of carboxylic acids and their derivatives with Grignard reagents and related organometallic derivatives. The major products are usually tertiary alcohols or ketones:

$RCO_2C_2H_5 + 2R'MgX \longrightarrow R'_2RCOH$
    (or $2R'Li$)

$2RCOCl + R'_2Cd \longrightarrow 2RCOR' + CdCl_2$

$RCN + R'MgX \longrightarrow [RR'C=NMgX] \xrightarrow{H_2O} RCOR'$

reactions at the α carbon atom Reactions of enol tautomers of carboxylic acid derivatives or their conjugate bases:

$RCH_2CO_2H + Br_2 \xrightarrow{(PBr_3)} RCHBrCO_2H + HBr$
   halogenation

$2RCH_2CO_2C_2H_5 \xrightarrow[\Delta]{NaOC_2H_5} RCH_2COCHRCO_2C_2H + C_2H_5OH$
Claisen condensation

$CH_2(CO_2C_2H_5)_2 \xrightarrow{NaOC_2H_5} \xrightarrow{R-X} RCH(CO_2C_2H_5)_2$
   alkylation

(Equivalent alkylation of acetoacetic ester may also be effected.)

peracids Derivatives of carboxylic acids formed by the acylation of hydrogen peroxide. Peracids convert alkenes to epoxides (oxiranes) with retention of configuration.

EXERCISES

14·1 Describe simple chemical tests that would serve to distinguish between the following pairs of compounds:

a $CH_3\overset{O}{\overset{\|}{C}}OH$ and $CH_3CH_2CH_2OH$

b φ—$\overset{O}{\overset{\|}{C}}OH$ and φ—OH

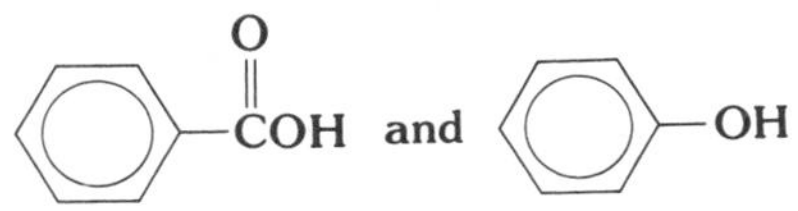

c $CH_3CH_2CO_2H$ and CH_3CH_2COCl

d $CH_3\overset{O}{\overset{\|}{C}}O\overset{O}{\overset{\|}{C}}CH_3$ and $CH_3CH_2OCH_2CH_3$

e $CH_3\overset{O}{\overset{\|}{C}}OCH_3$ and $CH_3CH_2\overset{O}{\overset{\|}{C}}OH$

14·2 Given the acidity constants K_a for the following acids, calculate the pK_a value for each:
a ϕCH_2CO_2H, $K_a = 4.9 \times 10^{-5}$
b Cl_2CHCO_2H, $K_a = 5.5 \times 10^{-2}$
c $CH_2{=}CHCO_2H$, $K_a = 5.5 \times 10^{-5}$

14·3 Acid-catalyzed hydrolysis of esters is a reversible reaction (see equation 14.12). Explain why base-catalyzed hydrolysis is irreversible.

14·4 Arrange the following groups of compounds in order of reactivity toward the reagent indicated:
a Substitution by NH_3:

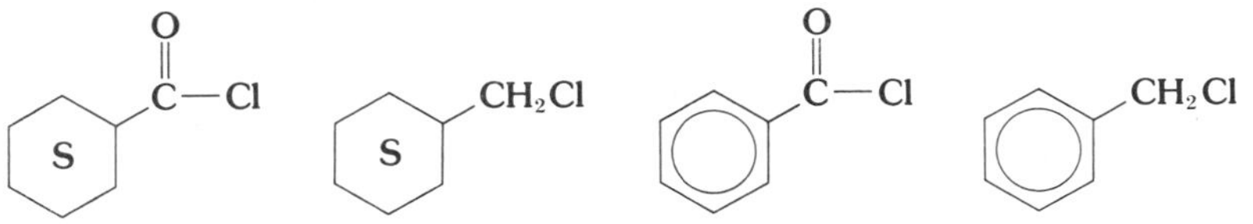

[cyclohexane-S with C(=O)—Cl]   [cyclohexane-S with CH$_2$Cl]   [benzene with C(=O)—Cl]   [benzene with CH$_2$Cl]

b Hydrolysis by $H_3O^{\oplus}$:

$CH_3\overset{O}{\overset{\|}{C}}{-}NH_2$ $CH_3\overset{O}{\overset{\|}{C}}{-}Cl$ $CH_3\overset{O}{\overset{\|}{C}}{-}OCH_2CH_3$ $CH_3\overset{O}{\overset{\|}{C}}{-}O{-}\overset{O}{\overset{\|}{C}}CH_3$

14·5 Indicate the product or products of the reaction (if any) of acetic acid with the following:
a $LiAlH_4$ (excess) in ether, then $H_3O^{\oplus}$
b $NaBH_4$ (excess) in THF, then $H_3O^{\oplus}$
c $NaOH$ in H_2O
d $NaHCO_3$ in H_2O
e H_2SO_4, H_2O, heat
f CH_3CH_2OH (excess), heat, acid catalysis
g CH_3MgBr (excess) in ether, then $H_3O^{\oplus}$
h $CH_3CH_2NH_2$, mild heat

14·6 Repeat Exercise 14.5 for acetyl chloride.

14·7 Repeat Exercise 14.5 for acetic anhydride.

14·8 Repeat Exercise 14.5 for methyl acetate.

14·9 Repeat Exercise 14.5 for N-methylacetamide.

14·10 Repeat Exercise 14.5 for acetonitrile.

14·11 Write equations for the reaction (if any) of each compound with excess quantities of the following reagents:

a NaOH in H_2O
b $LiAlH_4$, then $H_3O^{\oplus}$
c CH_3MgBr, then $H_3O^{\oplus}$
d CH_3OH

14·12 Prepare a table showing the products of the reaction (if any) of the reducing agents $LiAlH_4$, $NaBH_4$, and B_2H_6 with the following types of compounds:
a RCO_2H
b $RCO_2^{\ominus}$
c $RCOCl$
d $RCOOR'$
e $RCONHR'$
f $R-C\equiv N$
g $R-CHO$
h $R-COR'$
i ROH
j $RCH=CH_2$

14·13 Complete the following reactions:

a S—⟨⟩—CO_2H + CH_2N_2 $\xrightarrow{\text{ether}}$

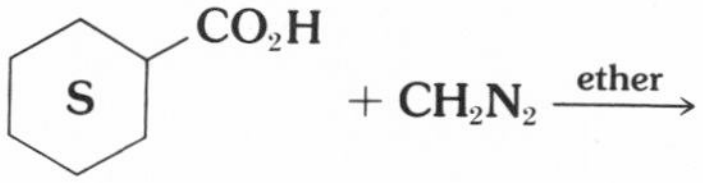

b ⟨⟩(CO_2H)($CO_2CH_2CH_3$) + $SOCl_2$ $\xrightarrow{\Delta}$

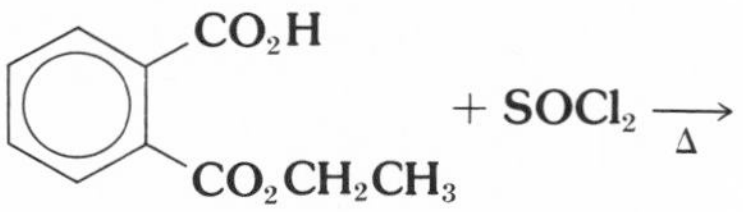

c $CH_3CH_2COCl + N(CH_3)_3 \longrightarrow$

d ⟨⟩—CO_2H $\xrightarrow{\Delta}$

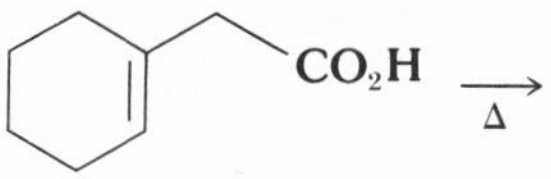

e $CH_3CH_2CH_2CH_2CO_2H + Br_2 \xrightarrow{HgO}$

f $BrCH_2CH_2CH_2CH_2CO_2H \xrightarrow[\text{warm}]{NaHCO_3}$

g $\phi CONH_2 \xrightarrow[H_2O]{HNO_2}$

h $\phi COCl + Li[(CH_3)_3CO]AlH \longrightarrow$

i $CH_3CH_2COCl + (CH_3)_2Cd \longrightarrow$

j $CH_3CH_2COCl + $ ⟨⟩ $\xrightarrow{AlCl_3}$

k $CH_3C\equiv N \xrightarrow[\Delta]{H_3O^{\oplus}}$

14·14 Comment on the use of different kinds of arrows in the following equations:

$$R-\overset{O}{\underset{\|}{C}}-OH \rightleftharpoons R-\overset{O-H}{\underset{|}{C}}=O \qquad R-\overset{O}{\underset{\|}{C}}-O^{\ominus} \longleftrightarrow R-\overset{O^{\ominus}}{\underset{|}{C}}=O$$

Write a structure that could be associated by a double-headed arrow with the carboxylic acid formula on the left.

14·15 Arrange the compounds in each group in order of increasing acidity:

a $CH_3\overset{O}{\underset{\|}{C}}OH \quad CH_3CH_2OH \quad CH_3\overset{O}{\underset{\|}{C}}NH_2 \quad CH_3CH_2NH_2$

b [cyclohexane with S and C(=O)OH]   [benzene with C(=O)OH]   [cyclohexane with S, H, and OH]   [phenol]

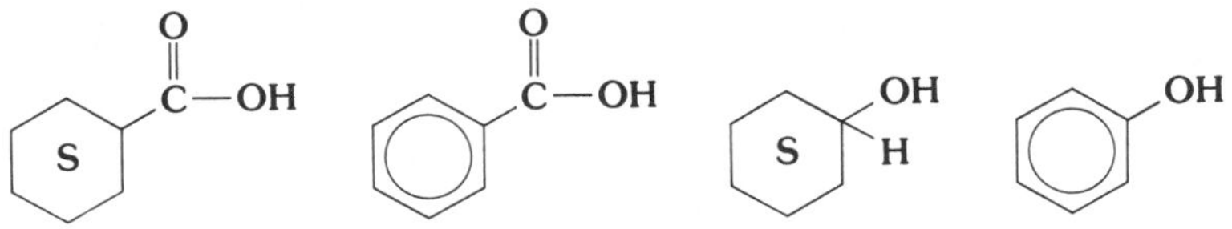

c [C₆H₅-C(=O)-OH]   [NO₂-C₆H₄-C(=O)-OH]   [CH₃O-C₆H₄-C(=O)-OH]

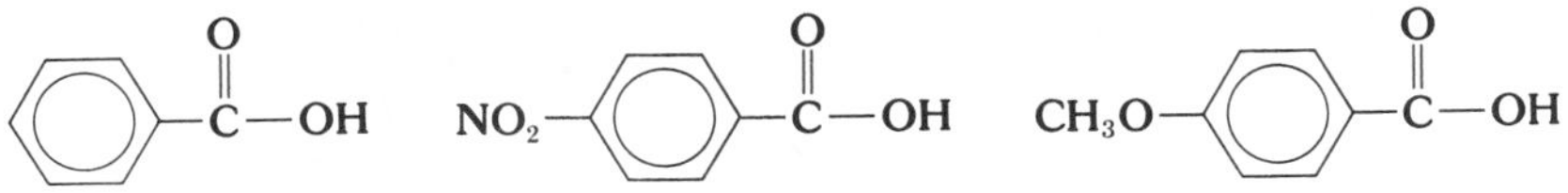

14·16 Explain the following relative acidities:
a Benzoic acid ($pK_a = 4.21$) > p-methoxy benzoic acid ($pK_a = 4.48$)
b Methoxyacetic acid ($pK_a = 3.48$) > acetic acid ($pK_a = 4.75$)

14·17 Account for the relative acidities of the hydrogen atoms specified in each set of compounds:

a $CH_3CCH_2COCH_2CH_3$ $CH_3COCH_2CH_3$
    $K_a \approx 10^{-11}$ $K_a \approx 10^{-26}$

b CH_3COH CH_3C-NH_2 CH_3-C-CH_3
   $K_a \approx 10^{-5}$ $K_a \approx 10^{-15}$ $K_a \approx 10^{-20}$

c $CH_3CH_2NH_2$ CH_3C-NH_2 $CH_3C-NH-CCH_3$
   $K_a \approx 10^{-33}$ $K_a \approx 10^{-15}$ $K_a \approx 10^{-10}$

d $CH_3CCH_2COC_2H_5$ $CH_3CCH_2CCH_3$
   $K_a \approx 10^{-11}$ $K_a \approx 10^{-9}$

14·18 Write structures for all the anhydrides expected from the reaction of an equimolar mixture of benzoic and propionic acids with P_2O_5.

14·19 Write equations illustrating the hydrolysis, alcoholysis (C_2H_5OH), and ammonolysis of acetic anhydride and succinic anhydride.

14·20 Why does the following Friedel-Crafts reaction proceed by acylation rather than alkylation?

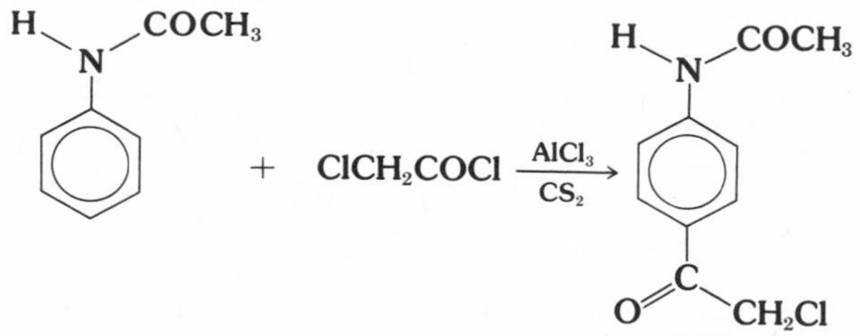

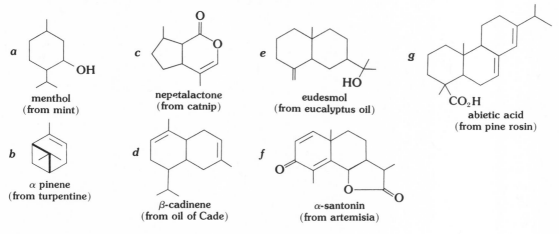

14·21 In the Claisen condensation of ethyl acetate illustrated in Figure 14.12 (R = H), the enolate donor was generated by using sodium ethoxide as a base.

$$CH_3CO_2CH_2CH_3 + {}^{\ominus}OCH_2CH_3 \rightleftharpoons CH_2=\overset{\overset{\ominus O}{|}}{C}-OCH_2CH_3 + HOCH_2CH_3$$

a In light of the conjugate base strengths of ethanol ($pK_a \approx 17$), water ($pK_a \approx 15$), and ethylacetate ($pK_a \approx 26$), is sodium hydroxide a sufficiently strong base to abstract the α proton of ethyl acetate?
b Explain why Claisen condensation of ethyl acetate does not take place when sodium hydroxide is used as the base.
c Why is sodium ethoxide an effective base in this condensation reaction?
d Are there any other suitable base catalysts for this reaction? What properties must a suitable catalyst have?

14·22 *Perkin condensations* are modifications of the Claisen condensation which utilize anhydrides as the enolate donor:

$$\phi CHO + {}^{\ominus}CH_2-\overset{\overset{O}{\|}}{C}-O-\overset{\overset{O}{\|}}{C}-CH_3 \xrightarrow{H_3O^{\oplus}} \phi-CH=CHCO_2H + CH_3CO_2H$$

a How could you generate the enolate-anion donor from acetic anhydride (for example, what base would be suitable)?
b Show a plausible mechanism to account for the products in the above reaction.

14·23 Identify isoprene units in the following terpenes:

a menthol (from mint)
b α pinene (from turpentine)
c nepetalactone (from catnip)
d β-cadinene (from oil of Cade)
e eudesmol (from eucalyptus oil)
f α-santonin (from artemisia)
g abietic acid (from pine rosin)

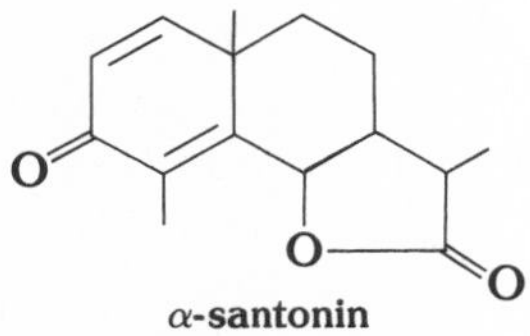

482 EXERCISES

14·24 Indicate the total number of stereoisomers possible for each compound:
a Menthol
b β-Cadinene
c Abietic acid

14·25 Suggest a reason why the imide barbituric acid in Equation 14.48 exists primarily in the enol form.

14·26 Write mechanisms for the following two substitution reactions:

$$CH_3CO_2H + CH_3CH_2OH \xrightarrow{H^\oplus} CH_3CO_2CH_2CH_3 + H_2O$$

$$2CH_3CH_2OH \xrightarrow{H^\oplus} CH_3CH_2OCH_2CH_3 + H_2O$$

a How do the reaction pathways differ with respect to the nucleophilic species? With respect to the electrophilic species?
b Suggest a reason why the esterification reaction proceeds under much milder conditions.
c Which of the two could be classed as a substitution resulting from an addition-elimination sequence?

14·27 Sulfonamides are the sulfonic acid analogs of amides. On the basis of the hydrolysis step shown in equation 14.61, what is the relative rate of hydrolysis of sulfonamides and carboxylic acid amides? Suggest a reason for this rate difference.

14·28 Although peracids are electrophilic reagents, as evidenced by their ready attack on double bonds to form epoxides, they are sluggish in their reactions with the double bonds of α,β-unsaturated carbonyl compounds. However, epoxides of these unsaturated compounds are easily prepared by reaction with sodium peroxide (H_2O_2 + NaOH):

$$R_2C=CH-\overset{O}{\overset{\|}{C}}-CH_3 + NaOOH \longrightarrow R_2\overset{O}{\overset{\diagdown\diagup}{C}}-CH-\overset{O}{\overset{\|}{C}}-CH_3 + NaOH$$

a Suggest a reason why peracids react slowly with α,β-unsaturated carbonyl compounds.
b Write a reasonable mechanism for the epoxidation of double bonds by peracids (use equation 14.108 as the example).
c Suggest a mechanism for the epoxidation of α,β-unsaturated carbonyl compounds by sodium peroxide. *Hint:* Consider nucleophilic attack on the β carbon.

14·29 Ketones having α hydrogen atoms undergo rapid base-catalyzed halogenation. If excess halogen is used, these reagents transform methyl ketones into trihalomethane (haloform) and the corresponding carboxylic acid:

$$R-\overset{O}{\overset{\|}{C}}-CH_3 + 3X_2 \xrightarrow{NaOH, H_2O} R-\overset{O}{\overset{\diagup\!\!\diagdown}{C}}_{O^\ominus Na^\oplus} + HCX_3 + 3NaX$$

This reaction is known as the *haloform reaction,* and when X = I the characteristic yellow color and odor of the crystalline iodoform product (HCI_3) can be used as a diagnostic test for methyl ketones.
a Suggest a mechanism for the haloform reaction.

b Account for the fact that secondary alcohols which can be oxidized to methyl ketones also give the haloform reaction:

$$R-CH(OH)CH_3 + 4X_2 \xrightarrow{NaOH, H_2O} R-C(=O)O^{\ominus}Na^{\oplus} + HCX_3 + 5NaX$$

14·30 The following transformations may be accomplished in three steps or fewer. Show the reagents and conditions which could be used to accomplish each:

a benzene → 3-bromobenzoic acid (m-Br-C₆H₄-CO₂H)

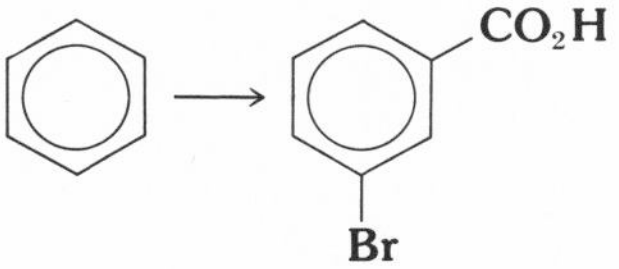

b (2-thienyl-like ring with S)—CH₂C(=O)—OH → (same ring)—CH₂Br

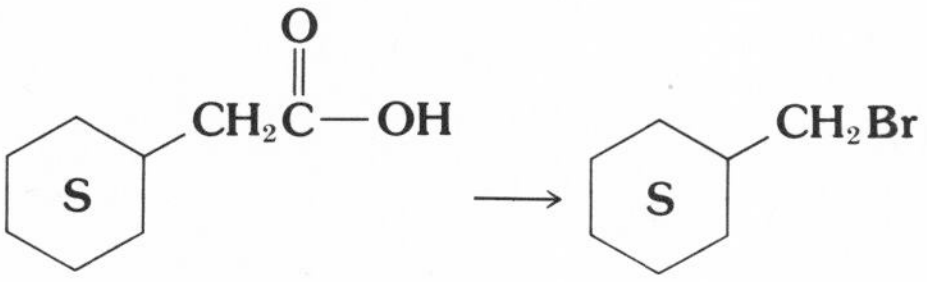

c $(CH_3)_2C=CH_2 \longrightarrow (CH_3)_2CHCO_2H$

d $CH_3CH_2CO_2CH_2CH_3 \longrightarrow CH_3CH_2CH(OH)C(CH_3)-C(=O)OCH_2CH_3$
  (product: CH₃CH₂CH(OH)–C(CH₃)H–CO–CH₂CH₃, with OH on one C and CH₃ on adjacent C)

e $\phi CH_3 \longrightarrow \phi CH_2OC(=O)CH_3$

f benzene → 2-(4-hydroxybutyl)phenol (o-HO-C₆H₄-CH₂CH₂CH₂CH₂-OH)

g (3-methylcyclohexenyl)-CH₂-C(=O)OH → 3-methylcyclohexanone (with C=O)

h 4-oxocyclohexanecarboxylic acid → methyl 4-hydroxycyclohexanecarboxylate

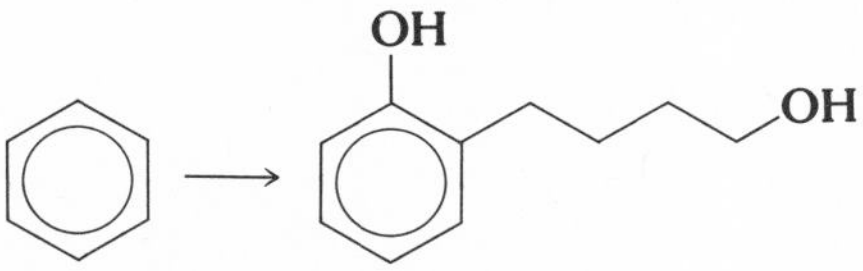

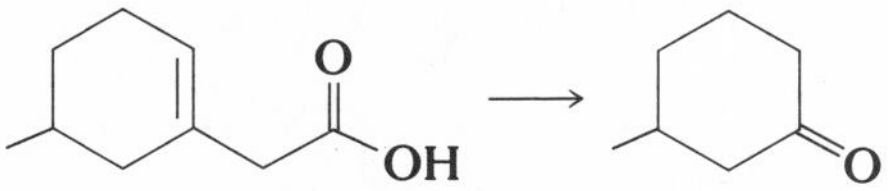

i bromocyclopentane → trans-1,2-cyclopentanediol + enantiomer

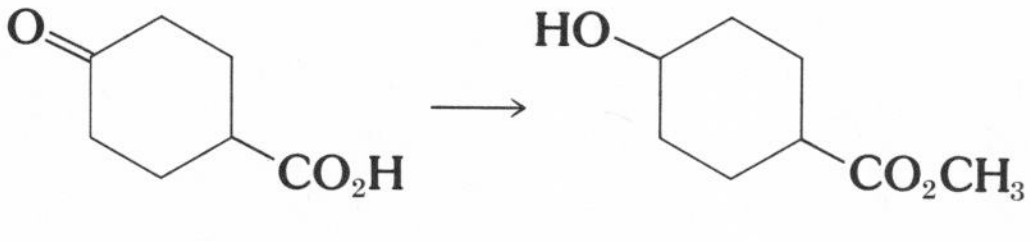

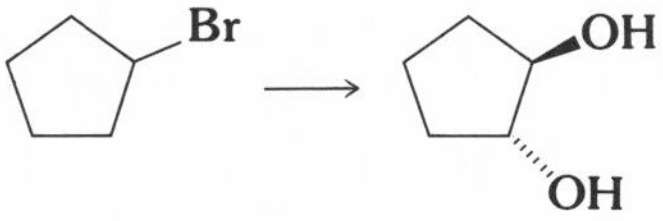

j $(CH_3)_2CHCH_2CO_2H \longrightarrow (CH_3)_2CHCH_2CH_2NHCH_3$

k γ-butyrolactone → $(CH_3)_2C=CH-CH_2CH_2OH$

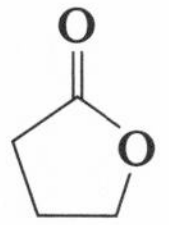

l 2-(ethoxycarbonyl)cyclohexanone ⟶ 2-ethylcyclohexanone

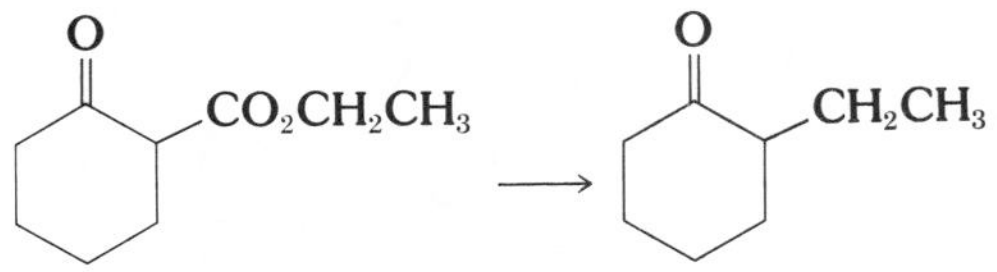

m $CH_3NHCH_2CH_2CH_2CO_2CH_3 \longrightarrow$ N-methylpyrrolidine (1-methylpyrrolidin-2-one)

INTERACTING FUNCTIONAL GROUPS

IN THE preceding chapters we have examined some distinctive properties and reactions of many commonly encountered functional groups. The power and usefulness of this organization of organic chemistry lies in the fact that most compounds incorporating a given functional group show a similar and characteristic reactivity. For example, the trifunctional steroid described in Figure 15.1 undergoes esterification with benzoyl chloride and oxidation with chromium trioxide like other secondary alcohols (Chapter 8), addition of hydrogen and bromine to the double bond like other alkenes (Chapter 5), and oxime formation or carbonyl reduction by sodium borohydride just like other ketones (Chapter 12).

The functional groups in this case behave more or less independently of each other, although occasionally one group will interfere with a reaction at another site; for example, CH_3MgI reacts with the active hydrogen of the hydroxyl group faster than it adds to the carbonyl function. If two functional groups are close together, however, they may lose some of their independence. In fact, one of the most interesting and challenging facets of organic chemistry lies in discovering and elucidating the interactions between neighboring functional groups.

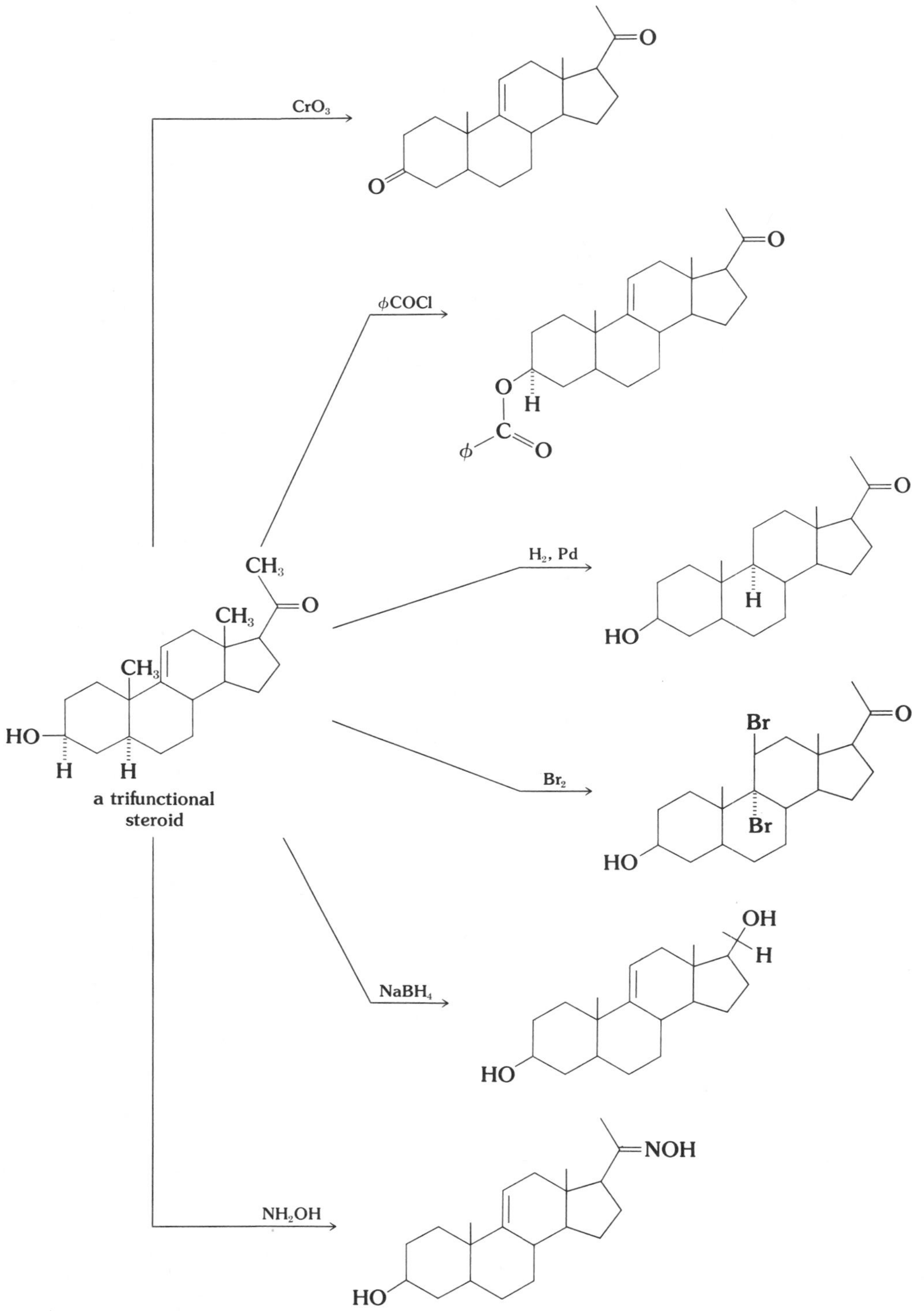

FIGURE 15·1 *Selective reactions of a polyfunctional compound*

15·1 INTERACTIONS INVOLVING THE HYDROXYL GROUP

The modifying influence of one functional group on another located nearby in the same molecule has been noted several times in our previous discussions. As an illustration, let us briefly review the effect of various neighboring functions on the hydroxyl group.

TWO HYDROXYL FUNCTIONS

Although most diols behave much as we might expect from the properties of simple alcohols, *geminal* diols undergo a rapid dehydration to aldehydes or ketones (see Section 12.5):

$$\underset{R}{\overset{R'}{>}}C\underset{OH}{\overset{OH}{<}} \underset{\text{fast}}{\rightleftharpoons} \underset{R}{\overset{R'}{>}}C=O + H_2O \qquad (15.1)$$

A unique oxidation reaction of *vicinal* diols can be effected by lead tetraacetate, periodic acid, or acidic permanganate. The action of these reagents on isolated hydroxyl functions is much slower, and when a reaction takes place it generally does not involve carbon-carbon bond cleavage.

$$R-\underset{R^2}{\overset{OH}{\underset{|}{C}}}-\underset{R^3}{\overset{OH}{\underset{|}{C}}}-R^4 + Pb(OCOCH_3)_4 \longrightarrow \underset{R^2}{\overset{R^1}{>}}C=O + \underset{R^3}{\overset{R^4}{>}}C=O \qquad (15.2)$$
(or HIO_4 or $HMnO_4$)

HYDROXYL AND CARBONYL FUNCTIONS

The most striking example of interaction between hydroxyl and carbonyl groups is undoubtedly the carboxylic acid function. Indeed, the unique properties of this combination of groups are sufficiently different from those of either component group that the carboxylic acids are generally regarded as a discrete functional group in their own right (compare the reactions described in Chapter 14 with those described in Chapters 8 and 12). A neighboring carbonyl function also modifies the chemical behavior of chloride, bromide, alkoxyl, and amino groups in an analogous fashion.

If a carbonyl function is located one carbon atom away from a hydroxyl group, the resulting α-ketol may be subject to rapid isomerization via an enediol intermediate, facile oxidation to an α-diketone (as by the action of Tollens reagent), and reductive removal of the hydroxyl group (or a derived ester) by zinc dust in acidic media:

$$R-\overset{O}{\overset{\|}{C}}-\underset{H}{\overset{OH}{\underset{|}{C}}}-R' \rightleftharpoons \left[R-\overset{OH}{\underset{|}{C}}=\overset{OH}{\underset{|}{C}}-R'\right] \rightleftharpoons R-\underset{H}{\overset{OH}{\underset{|}{C}}}-\overset{O}{\overset{\|}{C}}-R' \qquad (15.3)$$
enediol intermediate

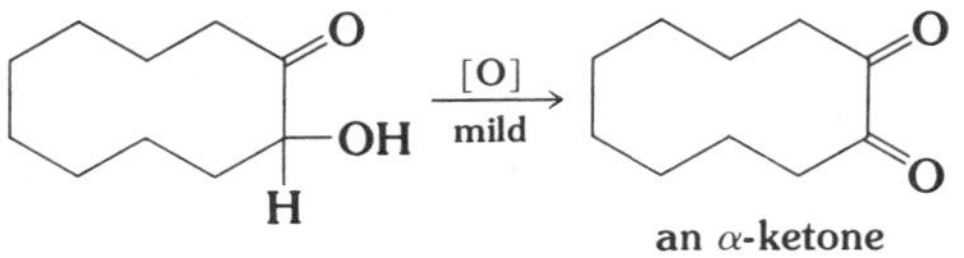

15·1 INTERACTIONS INVOLVING THE HYDROXYL GROUP

[Structure: cyclodecanone with –OH and H substituent] $\xrightarrow[\text{mild}]{[O]}$ [Structure: cyclodecane-1,2-dione] *an α-ketone* (15.4)

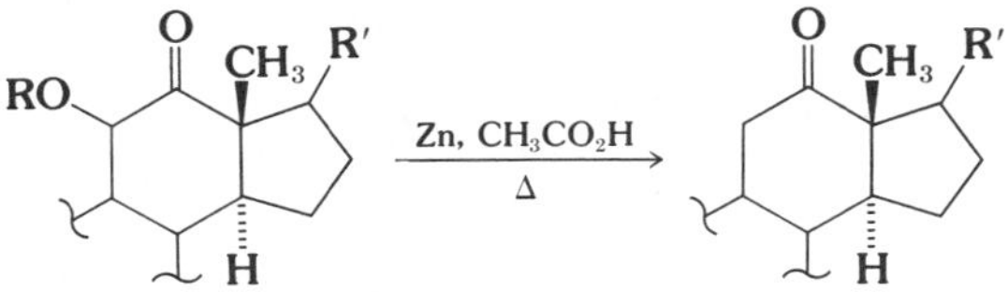

[Structure: bicyclic ketone with RO, CH₃, R', H substituents] $\xrightarrow[\Delta]{Zn,\ CH_3CO_2H}$ [Structure: bicyclic ketone with CH₃, R', H] (15.5)

(R = H or COCH₃
R' = H or alkyl)

These reactions do not proceed with simple alcohols.

Even β-ketols, despite the additional separation of functional groups, exhibit some reactions which clearly involve functional-group interaction. For example, the *retroaldol cleavage* (the reverse of the aldol condensation) not only requires the presence of both functional groups, but is also dependent on their relative position:

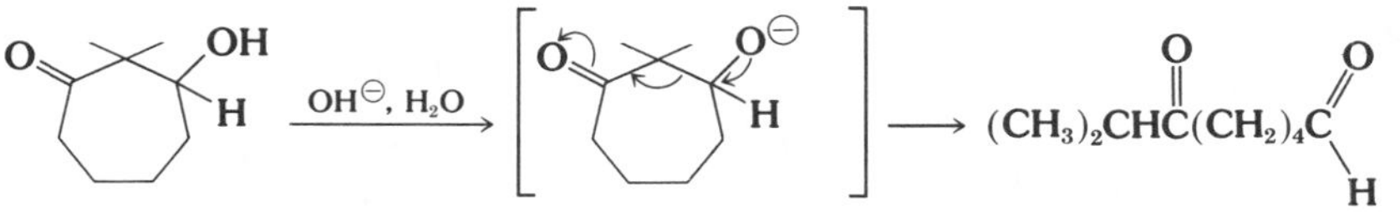

[Structure: cycloheptanone with gem-dimethyl and OH] $\xrightarrow{OH^{\ominus},\ H_2O}$ [bracketed intermediate with arrow showing ring opening] ⟶ $(CH_3)_2CHC(CH_2)_4C{\overset{O}{\underset{H}{\|}}}$ with O above C (15.6)

Dehydration of β-ketols to α,β-unsaturated ketones generally proceeds more rapidly and under milder conditions than for similarly constituted simple alcohols, suggesting that these elimination reactions are facilitated by the carbonyl function (actually by its enol or enolate anion derivative). The corresponding α-ketols (previous section) are dehydrated only under more forcing reaction conditions.

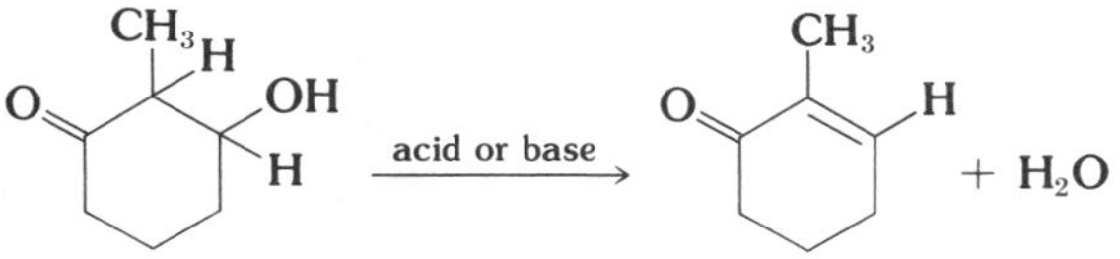

[Structure: 2-methyl-3-hydroxycyclohexanone] $\xrightarrow{\text{acid or base}}$ [Structure: 2-methylcyclohex-2-enone] + H_2O (15.7)

PROBLEM 15·1

a Compare the expected properties of 4-aminocyclohexanone with those of cyclopentane carboxamide,

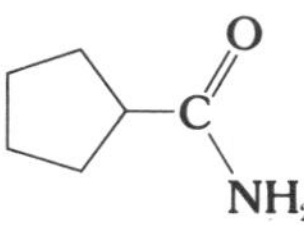

[Structure: cyclopentyl–C(=O)–NH₂]

b Write formulas showing how the carbonyl function of a β-ketol could facilitate acid- or base-catalyzed dehydration to a conjugated unsaturated ketone.

c Explain the following reaction:

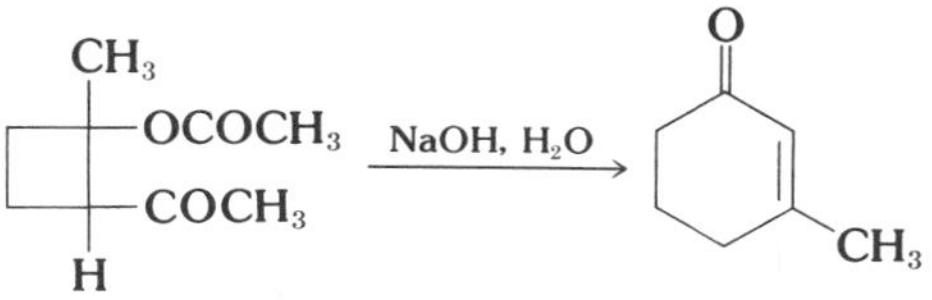

[Structure: cyclobutane with CH₃, OCOCH₃, COCH₃, H substituents] $\xrightarrow{NaOH,\ H_2O}$ [Structure: 3-methylcyclohex-2-enone]

15·1 INTERACTIONS INVOLVING THE HYDROXYL GROUP

HYDROXYL AND DOUBLE-BOND FUNCTIONS

The directly bonded combination of a double bond with a hydroxyl group, the *enol function*, shows unique properties indicative of substantial interaction between these groups. Thus most enols are rapidly and reversibly transformed to a keto tautomer which does not exhibit the characteristic properties of either alkenes or alcohols:

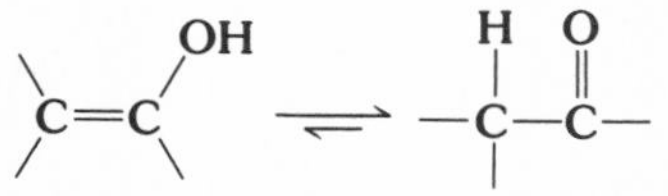

$$\overset{\diagdown}{\underset{\diagup}{C}}=C\overset{OH}{\diagdown} \rightleftharpoons -\overset{H}{\underset{|}{C}}-\overset{O}{\overset{\|}{C}}- \qquad (15.8)$$

Stabilization of the enol tautomer can be achieved by inclusion in an aromatic ring, as in phenol, or by conjugation with a carbonyl group, as observed for many β-diketones:

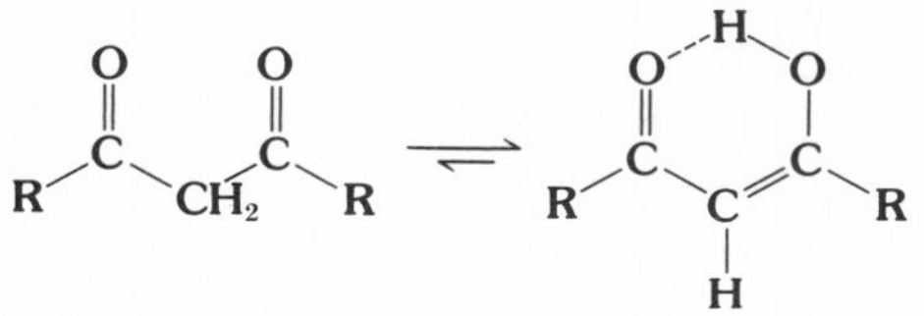

$$\underset{R}{}\overset{O}{\overset{\|}{C}}\diagdown_{CH_2}\overset{O}{\overset{\|}{C}}\diagdown_{R} \rightleftharpoons \underset{R}{}\overset{O}{\overset{\|}{C}}\diagdown_{\underset{H}{\overset{|}{C}}}\overset{H\cdots O}{\overset{\diagdown}{C}}\diagdown_{R} \qquad (15.9)$$

Although the hydroxyl and double-bond functions of an allyl alcohol are separated by a saturated carbon atom, the rates of their substitution reactions (and those of the derived halides, ethers, esters, and amines) are clearly enhanced over those of their saturated-carbon analogs. This allylic interaction under S_N1 and S_N2 reaction conditions was described in Section 10.8.

AROMATIC RINGS AND THE HYDROXYL FUNCTION

The characteristic properties of phenols, such as enhanced acidity and electrophilic ring substitution, point to an interaction between the hydroxyl group and the aromatic ring (Section 10.5). This interaction can be extended to substituent groups in the ortho and para positions, as evidenced by the increased acidity of *p*-nitrophenol and its ortho isomer:

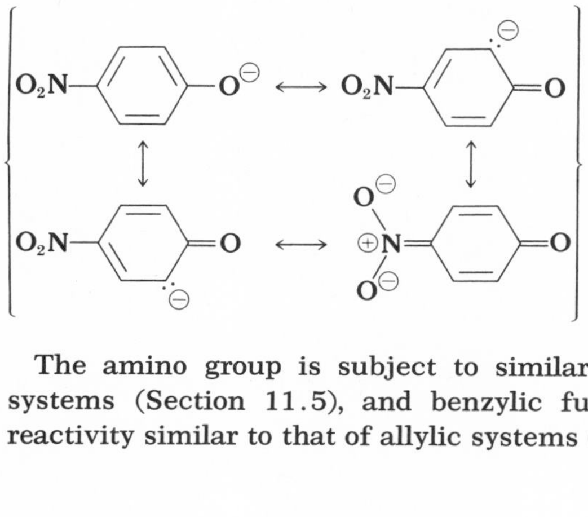

The amino group is subject to similar interactions with aromatic systems (Section 11.5), and benzylic functions show an enhanced reactivity similar to that of allylic systems (Section 10.8).

15·2 SUBSTITUENT EFFECTS ON ELECTROPHILIC AROMATIC SUBSTITUTION

CHANGES IN REACTIVITY

As we saw in Chapter 10, atoms or groups directly bonded to a benzene ring may greatly alter the reactivity of the aromatic ring in electrophilic substitution reactions. An amino substituent, for example, increases the rate of nitration about a million times; NH_2 is therefore said to be an *activating* substituent. A nitro substituent decreases the nitration rate to less than one-millionth that of benzene itself; hence NO_2 is classified as a *deactivating* substituent. Even larger effects have been noted for the more selective electrophilic reactants used in bromination reactions (see Table 15.1).

There are in general two methods we can use to obtain information about relative reactivities. One method is to treat different benzene derivatives with a given reagent mixture under identical reaction conditions for varying lengths of time. The most reactive ring systems will, of course, react quickly, while those which are deactivated will be sluggish. With this procedure we find, for example, that toluene is sulfonated about 20 times faster than benzene, but that chlorobenzene has only one-thirtieth the reactivity of benzene under nitration conditions. The methyl group is thus an activating substituent, and chlorine is considered to be deactivating.

A second method is to treat an equimolar mixture of two aromatic compounds with half an equivalent amount of an electrophilic reagent. The resulting competitive substitution reactions will favor the more reactive compound. This approach for a mixture of benzene and toluene shows that toluene has 25 times the reactivity of benzene:

$$\text{C}_6\text{H}_6 + \text{C}_6\text{H}_5\text{CH}_3 + \text{HNO}_3 \xrightarrow[\Delta]{\text{H}_2\text{SO}_4} \text{C}_6\text{H}_5\text{NO}_2 + \text{CH}_3\text{C}_6\text{H}_4\text{NO}_2 \text{ (para)} + \text{CH}_3\text{C}_6\text{H}_4\text{NO}_2 \text{ (ortho)}$$

(15.10)

    1 : 1   1 : 25

Studies of this kind have led to the general classification of common substituents outlined in Table 15.2. Dipole-moment measurements

TABLE 15·1 *Relative rates of ring bromination of C_6H_5—Z*

Z substituent	relative rate	Z substituent	relative rate
H	1.0	CH_2Cl	0.8
CH_3	3×10^2	Cl	0.1
OCH_3	1×10^9	Br	8×10^{-2}
OH	1×10^{11}	$CO_2C_2H_5$	5×10^{-4}
$N(CH_3)_2$	5×10^{18}	NO_2	2×10^{-6}

15·2 SUBSTITUENT EFFECTS ON ELECTROPHILIC AROMATIC SUBSTITUTION

TABLE 15·2 *Activating and deactivating substituents*

activating			deactivating	
strongly activating	moderately activating	weakly activating	weakly deactivating	strongly deactivating
NR_2	OR	$CH_3(R)$	Cl	$\overset{\oplus}{N}(CH_3)_3$, NO_2
NH_2	$NHCOCH_3$	ϕ	Br	CN, CO_2R
OH			I	SO_3R, COR

indicate that activating substitutents, as a rule, donate electrons to an aromatic ring, whereas deactivating substituents withdraw electrons from the ring:

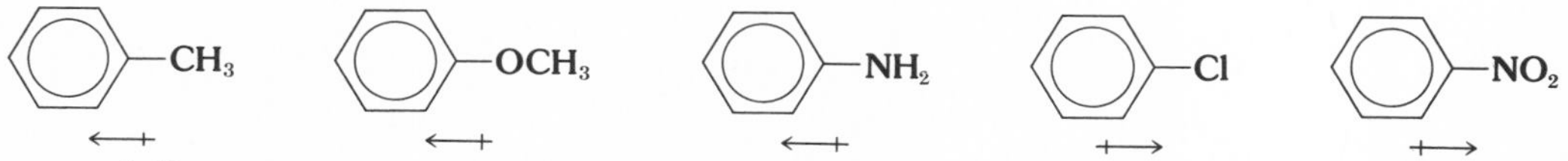

$\phi\text{—}CH_3$ $\phi\text{—}OCH_3$ $\phi\text{—}NH_2$ $\phi\text{—}Cl$ $\phi\text{—}NO_2$
$\mu = 0.40$ $\mu = 1.1$ $\mu = 1.53$ $\mu = 1.55$ $\mu = 3.95$

The direction of the dipoles can be deduced from the simple additive relationship that must operate in para-disubstituted benzenes:

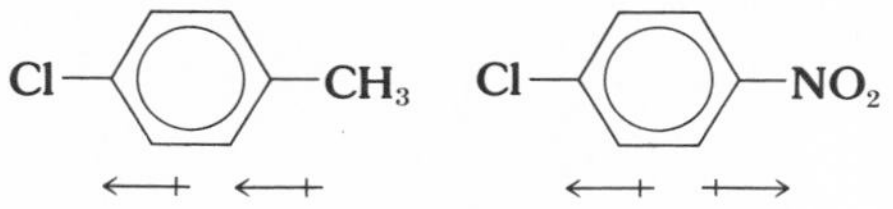

$Cl\text{—}\phi\text{—}CH_3$ $Cl\text{—}\phi\text{—}NO_2$
$\mu = 1.90$ $\mu = 2.50$
$(0.40 + 1.55 = 1.95)$ $(3.95 - 1.55 = 2.40)$

ORIENTATION OF INCOMING GROUPS

In the absence of a significant substituent effect, electrophilic substitution of a monosubstituted benzene derivative should produce a statistical distribution of product isomers—40% ortho, 40% meta, and 20% para. However, this product ratio is seldom observed, and it is clear from the data in Table 15.3 that ring substituents can influence product development in a striking manner.

If we compare these product distributions with the reactivity classification in Table 15.2, we find that all activating substituents are ortho- and para-directing, and that deactivating substituents, with the exception of the halogens, are meta-directing. We can account for this important relationship by considering the inductive and resonance effects described in Chapter 10 to explain the course of substitution reactions of aryl halides and phenols.

The *inductive effect* operates for all substituents and is probably the primary influence in the substitution reactions of trimethylanilinium salts. Recall from Figures 9.17 and 9.18 that positively charged intermediates are formed in the rate-determining step of electrophilic ring substitution. Since these intermediates are destabilized by a trimethylammonium substituent (like charges repel), the rate of electrophilic substitution is greatly reduced when this powerfully electronegative

TABLE 15·3 *Product composition from the nitration of monosubstituted benzenes*

Z substituent†	% meta	% ortho	% para
OH	—	55	45
OCH$_3$	—	100	
NHCOCH$_3$	2	19	79
CH$_3$	4	58	38
φ	—		100
F	—	12	88
Cl	—	30	70
Br	—	40	60
I	—	41	59
COCH$_3$	55	45	—
CHO	72	19	9
CO$_2$H	82	17	1
CN	81	17	2
SO$_3$H	72	19	9
NO$_2$	93	6	1
$^{\oplus}$N(CH$_3$)$_3$	100	—	—

†The amino group is not included here because aniline suffers oxidative degradation on treatment with nitric acid. Other substitution reactions of aniline give exclusively ortho and para products, except with very acidic reaction systems. The anilinium salt formed in such cases gives substantial meta substitution.

group is present on the aromatic ring. This destabilizing effect is particularly strong for attack in positions ortho or para to the positively charged nitrogen, as Figure 15.2 clearly indicates. We therefore find predominant or exclusive meta orientation in the products obtained from the vigorous reaction conditions required to effect electrophilic substitution of trimethylanilinium salts.

PROBLEM 15·2

a Explain the following product distribution for the aromatic-ring nitration of quaternary ammonium salts:

compound	% meta	% ortho + para
φCH$_2\overset{\oplus}{\text{N}}$(CH$_3$)$_3$	88	12
φCH$_2$CH$_2\overset{\oplus}{\text{N}}$(CH$_3$)$_3$	19	81
φCH$_2$CH$_2$CH$_2\overset{\oplus}{\text{N}}$(CH$_3$)$_3$	5	95

15·2 SUBSTITUENT EFFECTS ON ELECTROPHILIC AROMATIC SUBSTITUTION

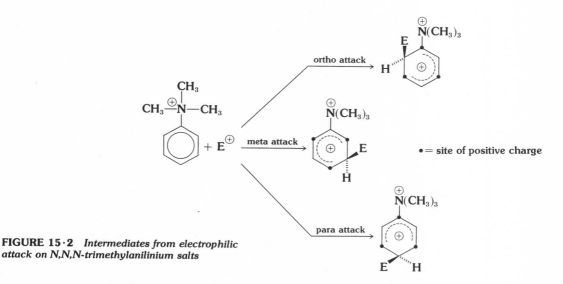

FIGURE 15·2 *Intermediates from electrophilic attack on N,N,N-trimethylanilinium salts*

b Explain the product distribution for nitration of halotoluenes:

compound	% meta	% ortho + para
φCH$_2$Cl	14	86
φCHCl$_2$	34	66
φCCl$_3$	64	36
φCF$_3$	100	0

The *resonance effect* operates in two ways, as illustrated in Figure 15.2. Meta-directing substituents such as those found in nitrobenzene and methyl benzoate exert a *conjugative* (sometimes called *mesomeric*) electron withdrawal which complements the inductive effect and favors meta substitution. In contrast, the nitrogen atom in acetanilide conjugatively releases electrons to the aromatic ring, thus stabilizing the delocalized cationic intermediates formed in ortho or para attack. The former interaction is usually associated with functional groups having polarized double or triple bonds such as

$$\underset{+\longrightarrow}{C=O} \quad \underset{+\longrightarrow}{C\equiv N} \quad \underset{+\longrightarrow}{N=O}$$

conjugated with the aromatic ring. The latter is most commonly observed when an atom directly bonded to the aromatic ring (usually O, N, Cl, Br, S, etc.) possesses a nonbonding pair of valence electrons which can interact conjugatively with the aromatic π-electron system.

Inductive electron withdrawal often competes with electron-pair donation by the resonance effect. In such cases preferential ortho-para substitution is always observed, but the degree of reactivity may vary from strong activation (as for OH and NR$_2$) to weak deactivation (as for Cl and Br).

Substituents on di- and trisubstituted aromatic systems may exert mutually supportive or antagonistic directive effects. The first situation

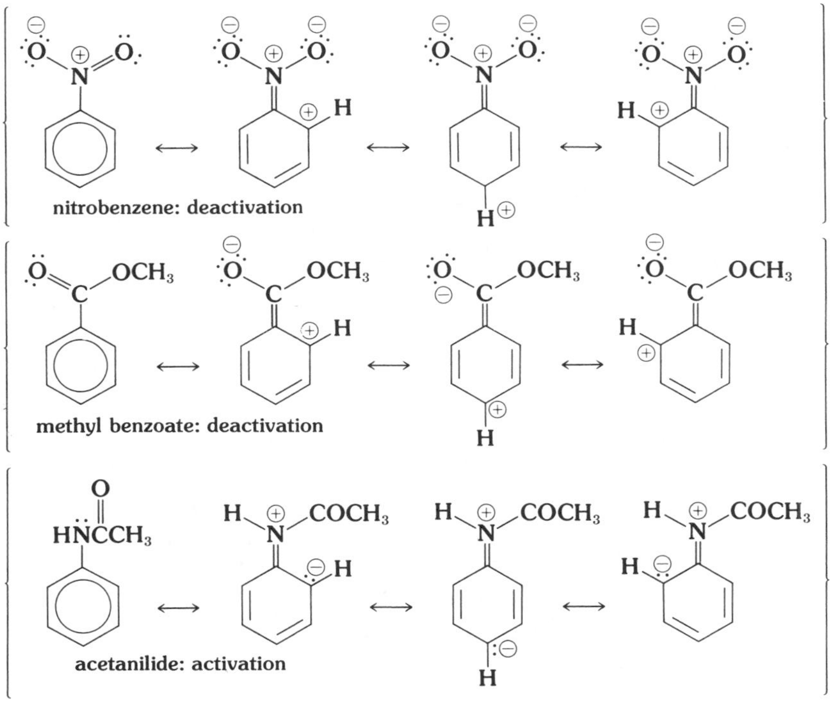

15·2 SUBSTITUENT EFFECTS ON ELECTROPHILIC AROMATIC SUBSTITUTION

FIGURE 15·3 *Resonance effects in substituted benzenes*

poses no difficulty in interpretation and is illustrated by the first three of the following examples:

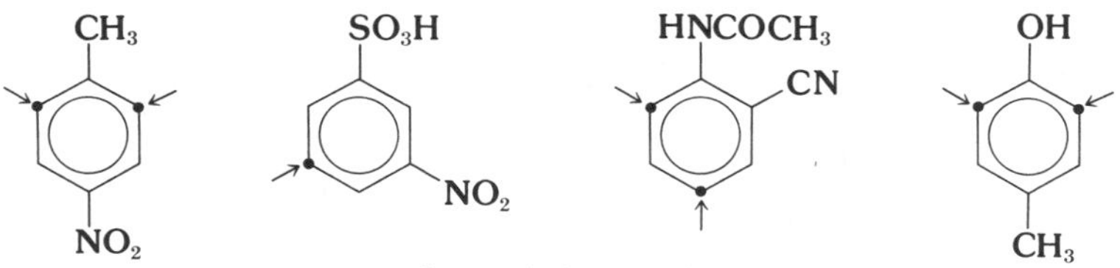

preferential electrophilic attack

In the case of antagonistic effects, a strongly activating group competing with a deactivating or weakly activating substituent will usually dominate the direction of substitution. Two substituents of roughly equal activating influence will give mixtures of products:

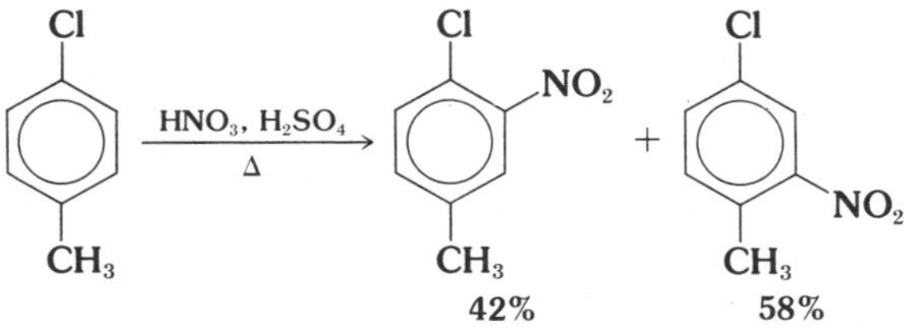

(15.11)

The presence of more than one deactivating substituent on a benzene ring generally lowers its reactivity to such a degree that few electrophilic substitution reactions take place at all.

STERIC EFFECTS

Very little substitution is found between meta-oriented substituents, regardless of their directive influence, primarily because of steric hindrance due to the size or bulk of the substituents. For example, toluene and butylbenzene have roughly equivalent reactivities under nitration reaction conditions, but the distribution of their product isomers is quite different. The bulky alkyl substituent in t-butylbenzene has clearly increased the steric hindrance to ortho substitution.

$$\text{PhR} \xrightarrow{\text{HNO}_3, \text{H}_2\text{SO}_4, \Delta} \text{o-NO}_2 + \text{m-NO}_2 + \text{p-NO}_2 \quad (15.12)$$

R	ortho	meta	para
$R = CH_3$	56.5%	3.5%	40.0%
$R = C(CH_3)_3$	12.0%	8.5%	79.5%

PROBLEM 15·3

a Nitration reactions of aromatic systems usually give only mononitro substitution products. Suggest a reason for the absence of di- and trinitro derivatives among the nitration products, even when an excess of the nitrating reagent is used.

b Durene (1,2,4,5-tetramethylbenzene) is exceptional and undergoes rapid nitration to the dinitro derivative. Very little mononitro product can be obtained even with careful control. Suggest a reason for this anomalous behavior. *Hint:* Consider the structural requirements for resonance electron withdrawal by the first nitro substituent.

15·3 ADDITION OF ELECTROPHILIC REAGENTS TO CONJUGATED DIENES

The markedly lower absorption frequencies (smaller excitation energies) of the $\pi \longrightarrow \pi^*$ electron transition in conjugated dienes, as opposed to isolated dienes, was attributed in Chapter 9 to significant π-electron interaction between the two double bonds. Some chemical manifestations of these interactions are apparent in the following addition reactions:

$$\overset{1}{C}H_2=\overset{2}{C}H-\overset{3}{C}H=\overset{4}{C}H_2 \begin{cases} \xrightarrow{Cl_2} ClCH_2CHClCH=CH_2 + ClCH_2CH=CHCH_2Cl \\ \xrightarrow{Br_2} BrCH_2CHBrCH=CH_2 + BrCH_2CH=CHCH_2Br \\ \xrightarrow{HCl} CH_3CHClCH=CH_2 + CH_3CH=CHCH_2Cl \end{cases} \quad (15.13)$$

$$\quad \text{1,2 addition} \qquad \text{1,4 addition}$$

$$CH_3CH=CHCH=CHCH_3 \xrightarrow{HBr} CH_3CH_2CHBrCH=CHCH_3 + CH_3CH_2CH=CHCHBrCH_3$$

$$\text{1,2 product} \qquad \text{1,4 product}$$

$$(15.14)$$

15·3 ADDITION OF ELECTROPHILIC REAGENTS TO CONJUGATED DIENES

The formation of both 1,2 and 1,4 addition products is understandable when we consider these reactions in the context of the two-step mechanism described in Sections 5.1 and 5.3. Electrophilic attack at one of the diene's terminal carbon atoms is favored, since it leads to a relatively stable allyl-cation intermediate. As we saw in Section 10.8, charge delocalization in the allyl cation results in two electrophilic reactive sites, and subsequent reactions with nucleophilic species lead to the observed products:

$$
\text{C=C—C=C} + \text{X}^{\oplus} \longrightarrow \left[\begin{array}{c} \text{X} \\ | \\ \text{C—C—C=C} \end{array} \overset{\oplus}{} \longleftrightarrow \begin{array}{c} \text{X} \\ | \\ \text{C—C=C—C} \end{array} \overset{\oplus}{} \right]
$$

$$
\left[\begin{array}{c} \text{X} \\ | \\ \text{C—C}\overset{\delta\oplus}{\text{---}}\text{C}\overset{\delta\oplus}{\text{---}}\text{C} + \text{Y:}^{\ominus} \end{array} \right] \longrightarrow \begin{array}{cc} \text{X} & \text{Y} \\ | & | \\ \text{C—C—C=C} \end{array} + \begin{array}{cc} \text{X} & \text{Y} \\ | & | \\ \text{C—C=C—C} \end{array}
$$

1,2 addition   1,4 addition

(15.15)

The exact composition of the product mixture from these diene addition reactions often depends on the experimental conditions. In the addition of hydrogen bromide to 1,3-butadiene, for example, a lower reaction temperature favors the 1,2 addition product:

$$
\text{CH}_2\text{=CHCH=CH}_2 + \text{HBr} \begin{array}{c} \xrightarrow{40°} \\ \\ \xrightarrow{-80°} \end{array} \begin{array}{c} \text{CH}_3\text{CHBrCH=CH}_2 + \text{CH}_3\text{CH=CHCH}_2\text{Br} \\ 20\% \qquad\qquad\qquad 80\% \\ \text{CH}_3\text{CHBrCH=CH}_2 + \text{CH}_3\text{CH=CHCH}_2\text{Br} \\ 80\% \qquad\qquad\qquad 20\% \end{array}
$$

(15.16)

On heating, both isomers give the same 4:1 mixture of 1,4 product and 1,2 product obtained at 40°C; consequently, this apparently represents the *equilibrium mixture* of the bromoalkenes. At lower temperatures, however, equilibrium is not established, and the product ratio reflects the *relative rates* of the two final steps, as shown in Figure 15.4. This rate difference corresponds to that expected if bonding of the nucleophilic bromide ion with the allyl-cation intermediate (equation 15.15)

FIGURE 15·4 *Product-determining steps in the reaction of 1,3-butadiene with hydrogen bromide*

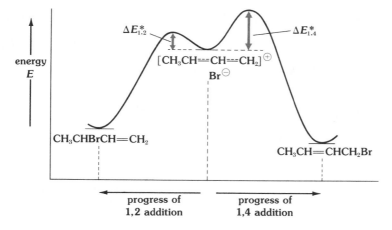

proceeds most rapidly at that carbon atom bearing the greatest positive charge (the secondary carbon atom). The equilibrium product reflects the greater thermodynamic stability of a disubstituted double bond over that of a monosubstituted double bond (see Section 4.4).

PROBLEM 15·4

Explain the preponderance of the product isomers shown in the following addition reactions:

a $CH_2=C(CH_3)CH=CH_2 \xrightarrow{HCl} (CH_3)_2C=CHCH_2Cl + (CH_3)_2CClCH=CH_2$

b $CH_2=CHCH=CHCH=CH_2 \xrightarrow{Br_2} \begin{array}{c} CH_2=CHCH=CHCHBrCH_2Br \\ + \\ BrCH_2CH=CHCH=CHCH_2Br \end{array}$

c $\phi CH=CHCH=CH_2 \xrightarrow{Br_2} \phi CH=CHCHBrCH_2Br$

15·4 THE DIELS-ALDER REACTION

Many conjugated dienes undergo addition reactions with unsaturated compounds, called *dienophiles*, to give six-member cyclic adducts as shown below. The German chemists Otto Diels and Kurt Alder received the Nobel Prize in 1950 for their discovery and study of this versatile and important reaction, which is commonly referred to as the *Diels-Alder diene synthesis*.

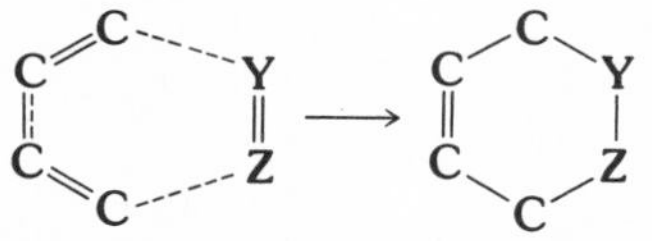

diene dienophile   adduct

Since two π bonds are converted to σ bonds in the course of these transformations, the resulting increase in thermodynamic stability causes most Diels-Alder reactions to be exothermic. At high temperatures many Diels-Alder reactions are reversible.

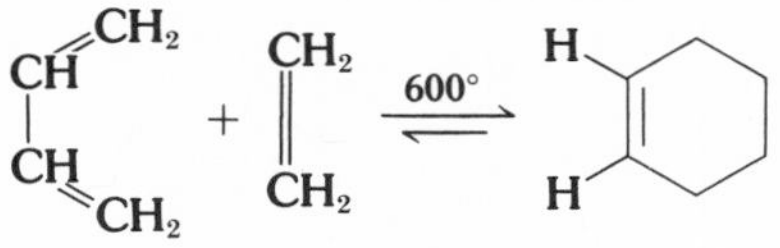

$\Delta H = -30$ kcal/mole (15.17)

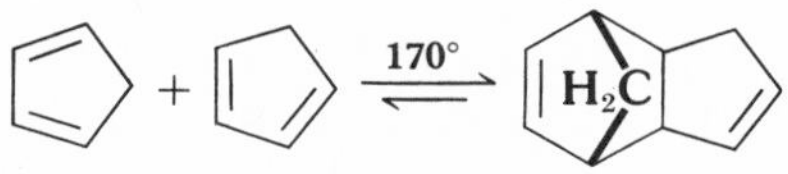

$\Delta H = -17$ kcal/mole (15.18)

DIENE CONFORMATIONS

Acyclic dienes can assume *cisoid* and *transoid* planar conformations, the latter generally predominating at equilibrium. Experimental evidence indicates that only the cisoid conformers undergo the Diels-Alder reaction, suggesting that simultaneous 1,4 bonding of the diene to the dienophile takes place (note that the distance from C-3 to C-6 in cyclohexene is about 3.0 Å).

15·4 THE DIELS-ALDER REACTION

cisoid conformer ⇌ transoid conformer (2.8 Å ⇌ 3.8 Å)

The essential role of cisoid diene conformers in Diels-Alder reactions is manifest in the unreactive nature of fixed transoid dienes such as

as well as the low reactivity of *cis*-1,3-pentadiene in comparison with its trans isomer:

cis isomer + maleic anhydride → 5% product (15.19)

trans isomer + maleic anhydride → 100% product (15.20)

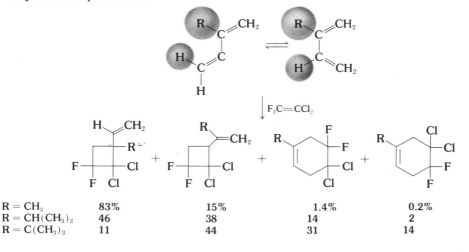

FIGURE 15·5 *Diene conformation effects on cycloaddition product ratios*

R = CH_3	83%	15%	1.4%	0.2%
R = $CH(CH_3)_2$	46	38	14	2
R = $C(CH_3)_3$	11	44	31	14

Since nonbonded compressions are more severe in the cisoid conformer of the cis isomer than in the corresponding conformer of the trans isomer, we expect the latter diene to react with dienophiles more rapidly.

A few alkenes, such as $Cl_2C=CF_2$ and $H_2C=C=O$ are observed to give 1,2 addition products with dienes, and these competitive addition reactions (1,2 versus 1,4) provide us with an additional means of probing the nature of the Diels-Alder reaction. For example, the effect of a 2-alkyl substituent on the proportion of 1,2 adducts to 1,4 adducts obtained from substituted acyclic dienes and 1,1-dichloro-2,2-difluoroethene further illustrates the necessity of a cisoid diene conformation in these reactions. As the diene substituent R increases in size, nonbonded compressions shift the conformational equilibrium in favor of the cisoid conformer — just the reverse of the influence of R in equation 15.19.

SUBSTITUENT EFFECTS

As a rule, Diels-Alder reactions proceed most easily when the dienophile is substituted by electron-withdrawing groups and the diene bears electron-donating groups—provided these groups do not interfere with the adoption of a cisoid conformation. Some highly reactive dienophiles are maleic anhydride, quinone, tetracyanoethylene, and acetylene-dicarboxylic esters:

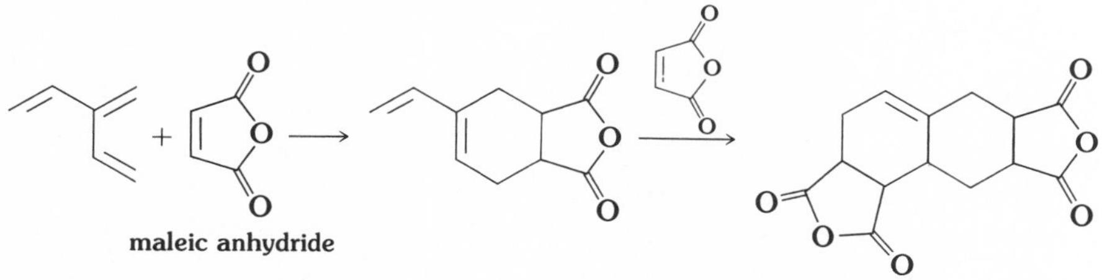

(15.21)

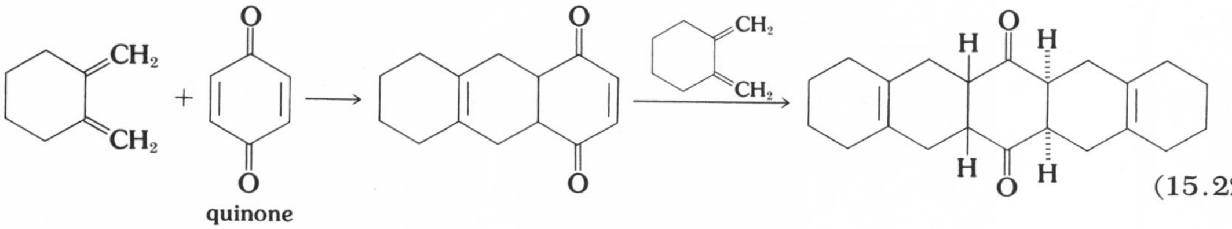

(15.22)

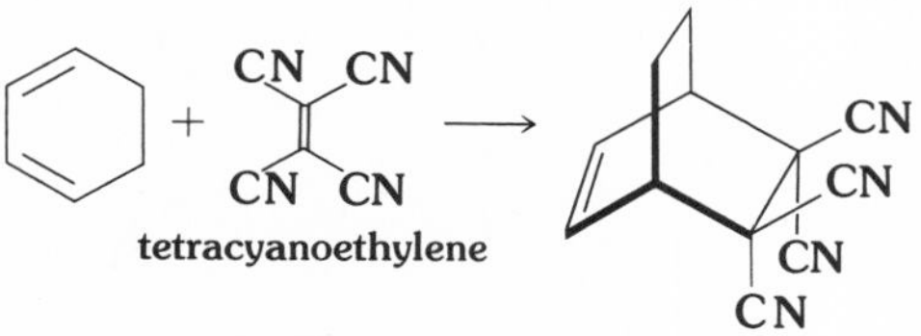

(15.23)

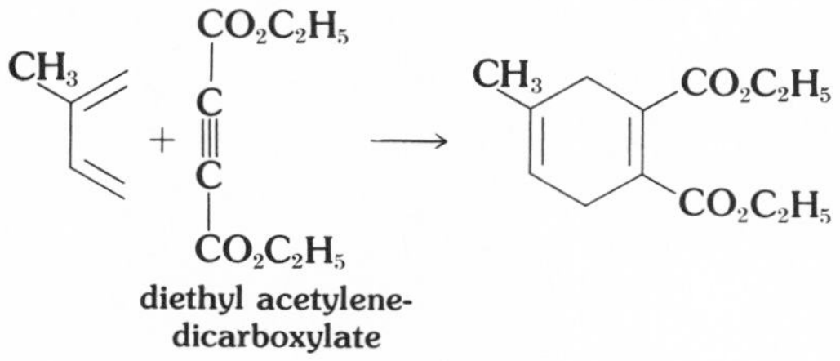

(15.24)

15·4 THE DIELS-ALDER REACTION

PROBLEM 15·5

a Write structures for the Diels-Alder adducts expected from the following combinations of dienes and dienophiles:

[structures: 2,3-dimethyl-1,3-butadiene + 1,1-dicyanoethene; furan + maleic anhydride; 1,3-cyclohexadiene + diethyl azodicarboxylate]

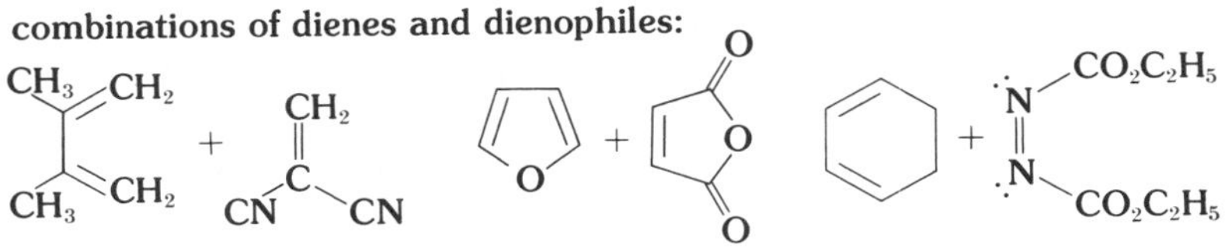

b Supply the two missing structures in the following synthesis of the insecticidal compound aldrin:

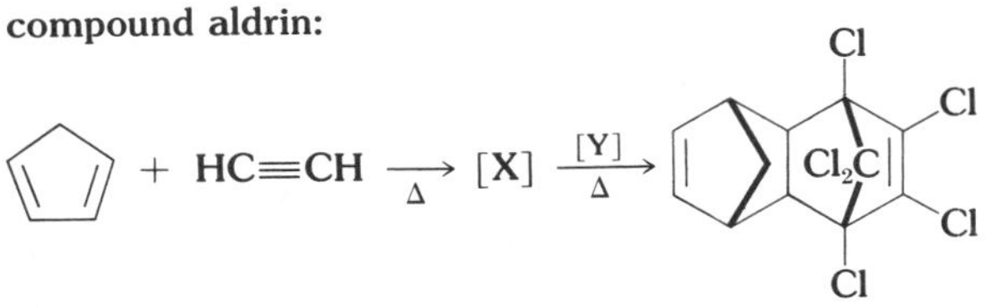

cyclopentadiene + HC≡CH $\xrightarrow{\Delta}$ [X] $\xrightarrow{[Y]}{\Delta}$ aldrin

c Write structural formulas for the dienes and dienophiles that would be expected to give the following adducts:

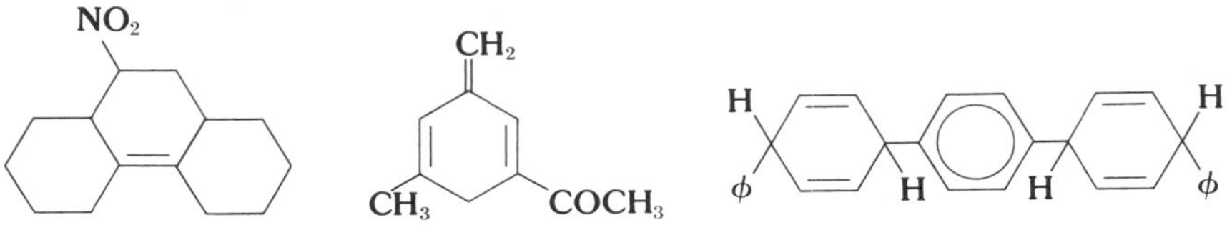

[three adduct structures shown]

STEREOCHEMISTRY

Diels-Alder reactions are generally highly stereospecific—that is, the configurational relationship of substituents on the diene and the dienophile are preserved in the adduct. This feature is well illustrated by reactions of stereoisomeric dienophiles such as maleic and fumaric esters and dienes such as *trans,trans*-2,4-hexadiene:

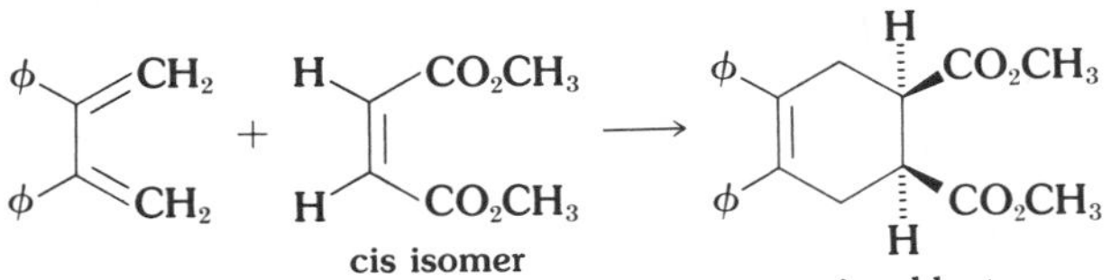

[reaction 15.25: diene + cis isomer → cis adduct]   (15.25)

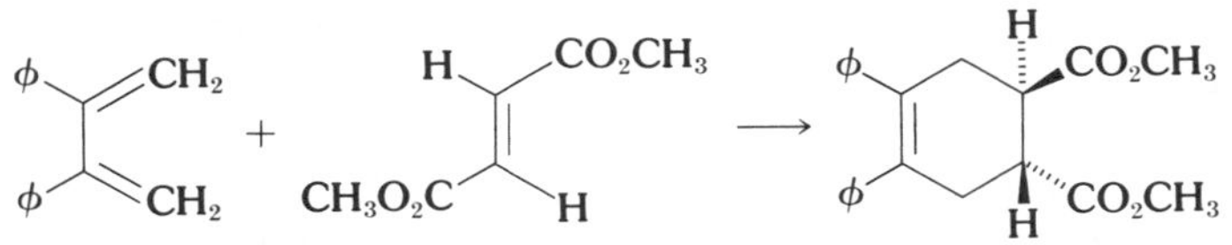

[reaction 15.26: diene + trans isomer → trans adduct]   (15.26)

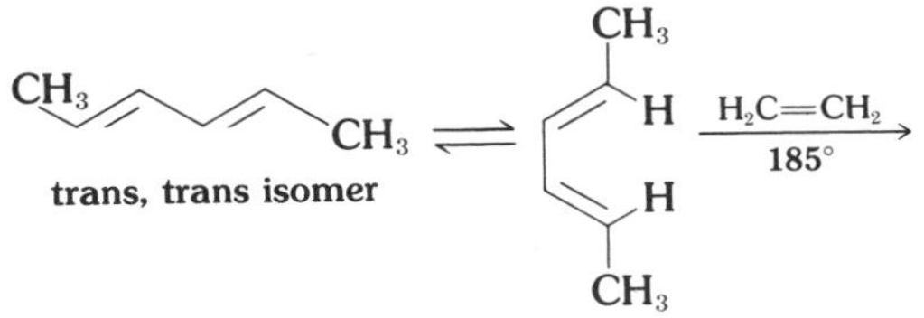

[reaction 15.27: trans,trans isomer + H₂C=CH₂ at 185°]   (15.27)

The *cis,trans*-hexadiene isomer does not give a Diels-Alder adduct with ethylene.

Monosubstituted and cis-disubstituted dienophiles may combine with cyclic dienes in either of two stereoisomeric orientations:

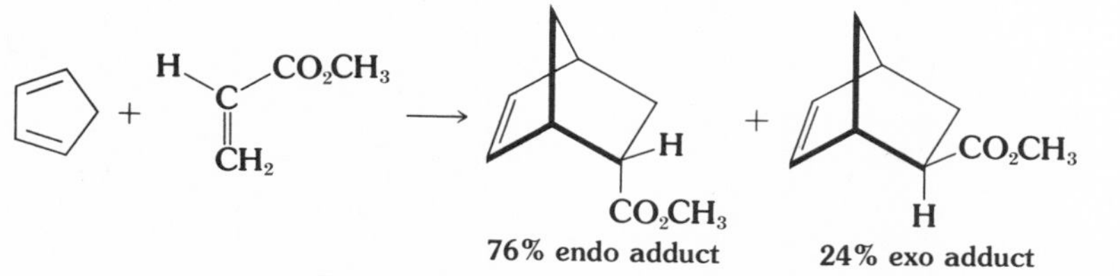

$$\text{(15.28)}$$

76% endo adduct   24% exo adduct

The designations *endo* and *exo* indicate whether substituents are located inside (concave side) or outside (convex side) of the rigidly puckered six-membered ring formed in the reaction.

Most Diels-Alder reactions of this kind give predominately *endo* adducts (for example, equations 15.18, 15.28, and 15.29); however, this generalization is limited by the ease with which some adducts undergo thermal isomerization (for example, equation 15.30).

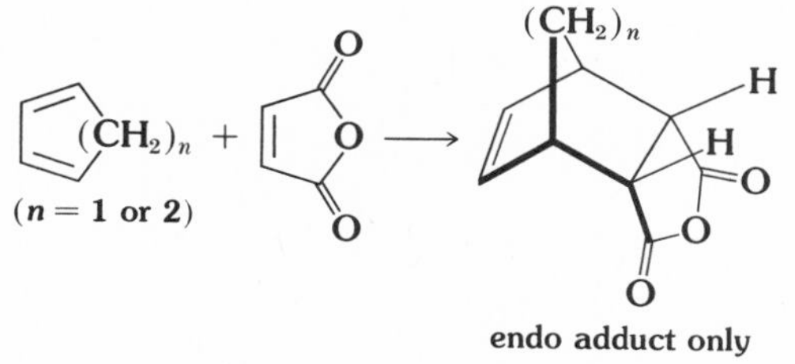

$$\text{(15.29)}$$

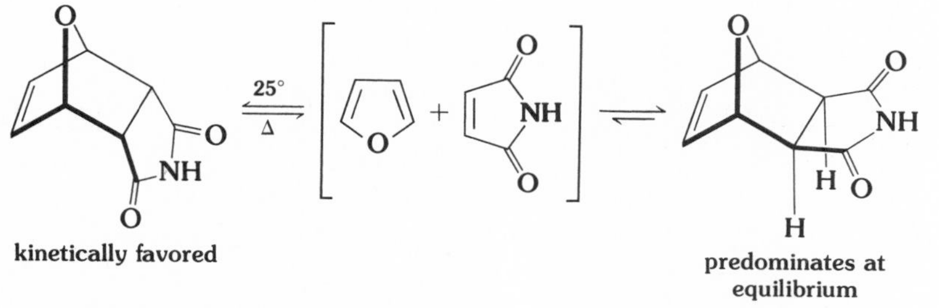

($n = 1$ or 2)

endo adduct only

$$\text{(15.30)}$$

kinetically favored   predominates at equilibrium

PROBLEM 15·6

a Cyclopentadienone dimerizes spontaneously.

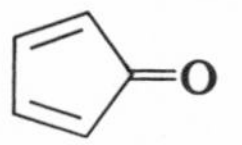

It also behaves both as a dienophile and a diene in its reactions with cyclopentadiene. Write projection formulas for the dimer and the two mixed adducts that are formed.

b Explain the following reaction:

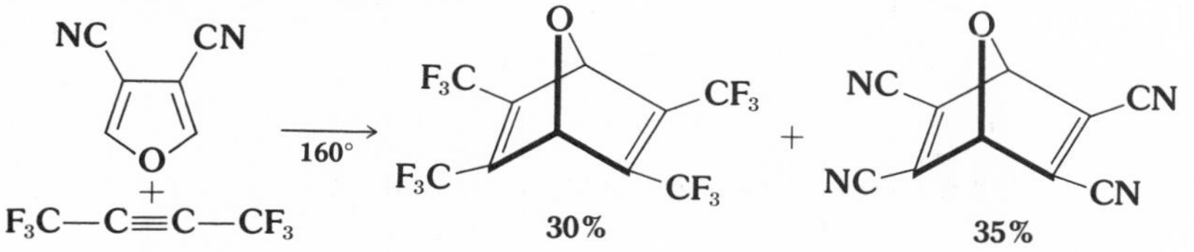

30%   35%

DIENELIKE SYSTEMS

Conjugated trienes generally do not give 1,6 addition to dienophiles, tending to react instead as dienes:

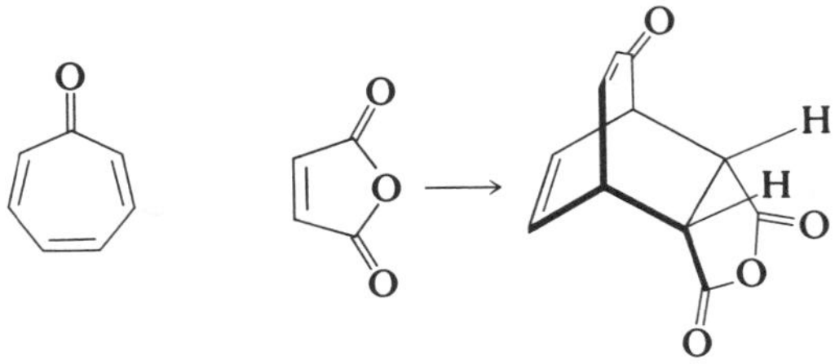

(15.31)

Simple benzene derivatives also do not as a rule undergo Diels-Alder cycloadditions with dienophiles. Styrene, for example, copolymerizes with maleic anhydride instead of giving a 1:1 adduct:

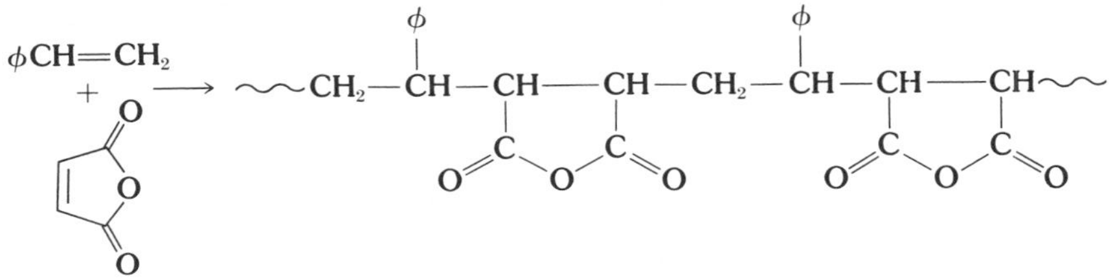

(15.32)

However, fused-ring aromatic systems and highly substituted aromatic rings do exhibit 1,4 addition reactions with reactive dienophiles:

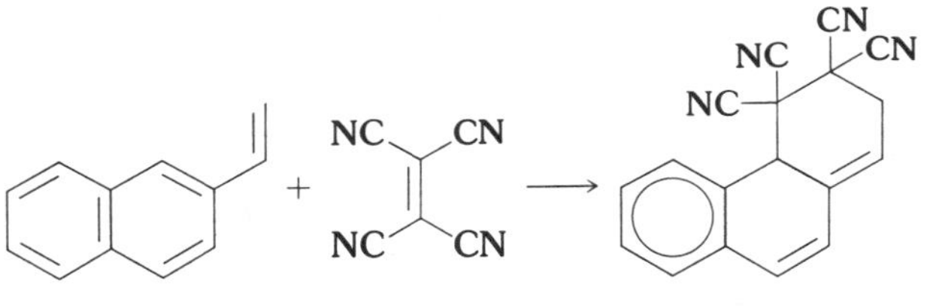

(15.33)

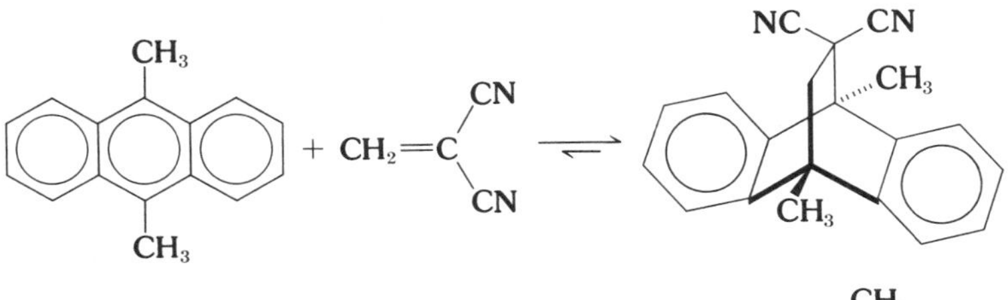

(15.34)

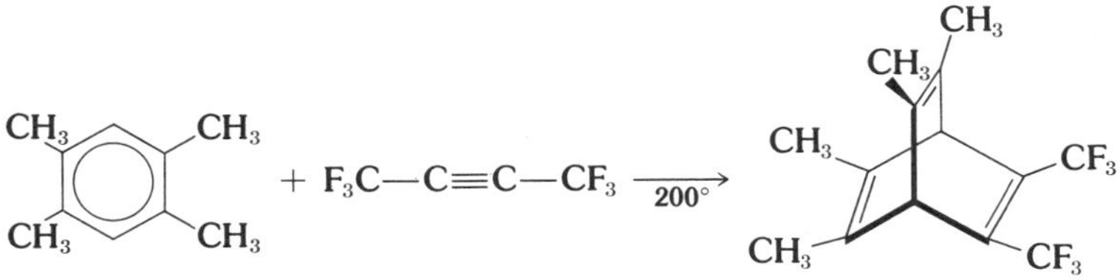

(15.35)

PROBLEM 15·7

trans,trans,cis-2,4,6-Octatriene reacts with maleic anhydride to give a 1:1 adduct. Suggest a reasonable structure for this product.

15·5 THE MICHAEL CONDENSATION

When a double or triple bond is conjugated with a polar unsaturated function such as a carbonyl group, a cyano group, or a nitro group, it becomes polarized in the following fashion:

$$\{C=C-Y=Z \longleftrightarrow \overset{\oplus}{C}-C=Y-\ddot{Z}^{\ominus}\}$$

(Y = C, Z = O; Y = C, Z = N; Y = N, Z = O; or Y = S, Z = O)

As a consequence of this polarization, addition reactions to such double bonds usually proceed by nucleophilic attack at the β carbon atom:

$$\underset{\beta}{CH_3}\overset{\alpha}{CH}=CHCO_2C_2H_5 + C_2H_5\overset{\delta\ominus}{O}\overset{\delta\oplus}{\rightleftharpoons}H \xrightarrow{NaOC_2H_5} \underset{\underset{OC_2H_5}{|}}{CH_3CHCH_2CO_2C_2H_5} \quad (15.36)$$

The addition of strong acids to conjugated alkenes of this kind follows an equivalent course, even though the product from β addition of the nucleophilic moiety may not be that predicted by the Markovnikov rule (Section 5.1):

$$CH_2=CHCO_2CH_3 + HBr \xrightarrow[0°]{\text{ether}} BrCH_2CH_2CO_2CH_3 \quad (15.37)$$

$$-H^{\oplus} \updownarrow H^{\oplus} \qquad\qquad \uparrow Br^{\ominus}$$

$$\left[CH_2=CH-\overset{OH}{\underset{OCH_3}{\overset{\oplus}{C}}} \longleftrightarrow {}^{\oplus}CH_2-CH=\overset{OH}{\underset{OCH_3}{C}} \right]$$

The scope and versatility of these conjugate addition reactions are well illustrated by the following examples:

$$\begin{array}{c} CH_2=CHCO_2C_2H_5 \\ + \\ CH_3\ddot{N}H_2 \end{array} \longrightarrow CH_3\ddot{N}HCH_2CH_2CO_2C_2H_5 \xrightarrow{CH_2=CHCO_2C_2H_5} CH_3\ddot{N}(CH_2CH_2CO_2C_2H_5) \quad (15.38)$$

$$\begin{array}{c} (CH_3)_2C=CH\overset{O}{\overset{\|}{C}}CH=C(CH_3)_2 \\ + \\ :NH_3 \end{array} \longrightarrow \left[\begin{array}{c} O \\ \| \\ \text{6-membered ring with } CH_3, CH_3, NH_2, CH_3, CH_3 \end{array} \right] \longrightarrow \begin{array}{c} O \\ \text{6-membered ring with N-H} \end{array} \quad (15.39)$$

$$\begin{array}{c} \phi CH=CHCHO \\ + \\ N_2H_4 \end{array} \longrightarrow [H_2\ddot{N}NHCH\phi CH_2CHO] \longrightarrow \underset{\underset{H}{N}}{HN}\overset{\phi}{\diagdown} + H_2O \quad (15.40)$$

$$\begin{array}{c} 2CH_2=CHCN \\ + \\ H_2S \end{array} \xrightarrow{NaOH} N\equiv CCH_2CH_2SCH_2CH_2C\equiv N \quad (15.41)$$

15·5 THE MICHAEL CONDENSATION

$$CH_3CH=CHCO_2C_2H_5 + NaCN \xrightarrow[\Delta]{H_2O} \underset{+\ C_2H_5OH}{CH_3CH(CN)CH_2CO_2Na} \xrightarrow[H_3O^{\oplus}]{Ba(OH)_2, \Delta} \underset{CH_2-CO_2H}{CH_3-CH-CO_2H} \quad (15.42)$$

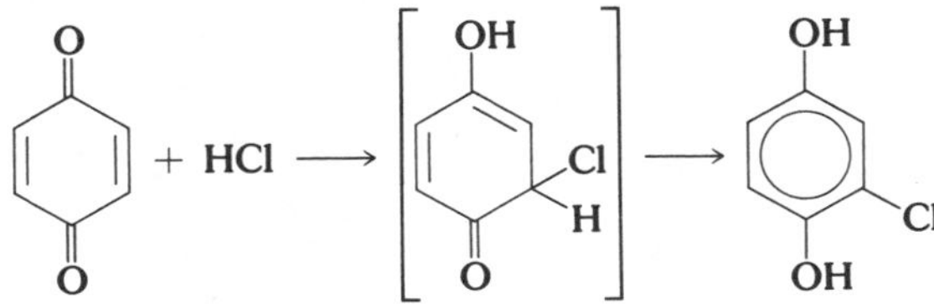

(15.43)

Carbonyl enolate anions and related intermediates may also serve as nucleophilic donors in a versatile and useful modification of conjugate addition known as the *Michael reaction* (see Figure 15.6). Possible Michael reaction donors include esters, ketones, nitriles, nitro compounds, and cyclic polyene precursors of aromatic anions such as cyclopentadienylide. The base catalysts used to generate the nucleophilic conjugate bases of the donors are usually alkoxide salts (NaOR) or amines:

$$\underset{\text{donor}}{CH_3COCH_3} + \underset{\text{acceptor}}{3CH_2=CHCN} \xrightarrow{NaOC_2H_5} \underset{\text{adduct}}{CH_3COC(CH_2CH_2CN)_3} \quad (15.44)$$

$$\underset{\text{donor}}{(CH_3)_2CHNO_2} + \underset{\text{acceptor}}{CH_2=CHCO_2CH_3} \xrightarrow{KOH} \underset{\text{adduct}}{(CH_3)_2C(NO_2)CH_2CH_2CO_2CH_3} \quad (15.45)$$

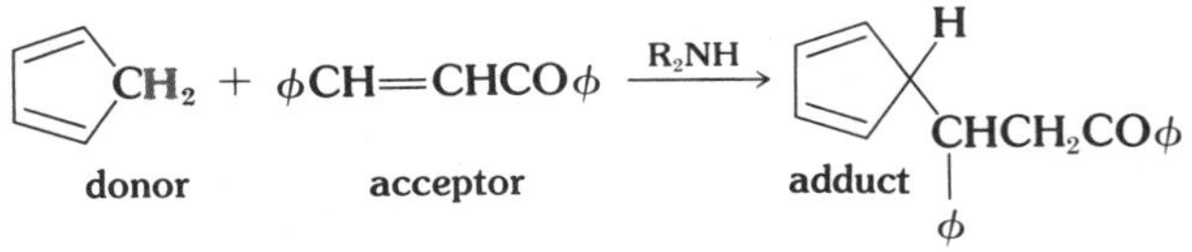

(15.46)

It is important to note, however, that lithium and sodium acetylide salts do not normally give conjugate addition to unsaturated ketones, but instead directly attack the carbonyl group:

$$CH_3OC\equiv CLi + CH_3CH=CHCHO \xrightarrow[-15°]{\text{ether}} CH_3OC\equiv C-\underset{\underset{OH}{|}}{C}HCH=CHCH_3 \quad (15.47)$$

Since Michael condensations often generate dicarbonyl compounds, subsequent intramolecular aldol or Claisen condensations may take place, as shown by the following reactions:

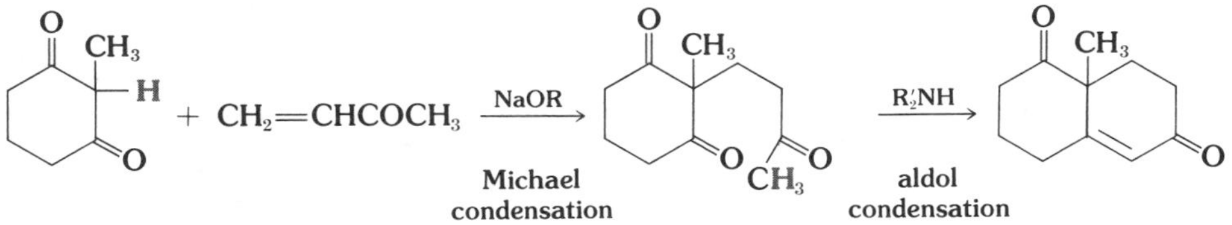

(15.48)

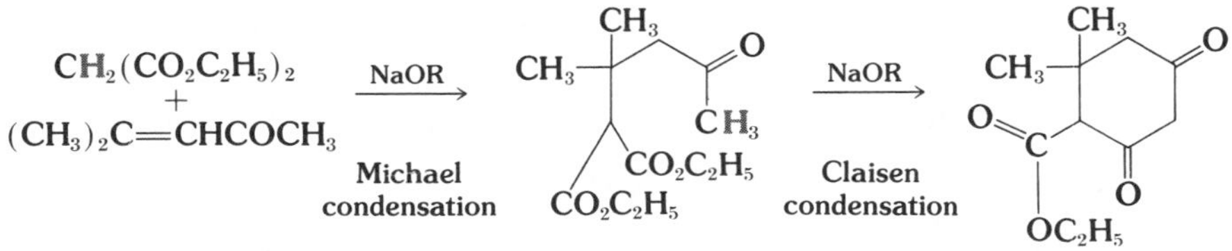

(15.49)

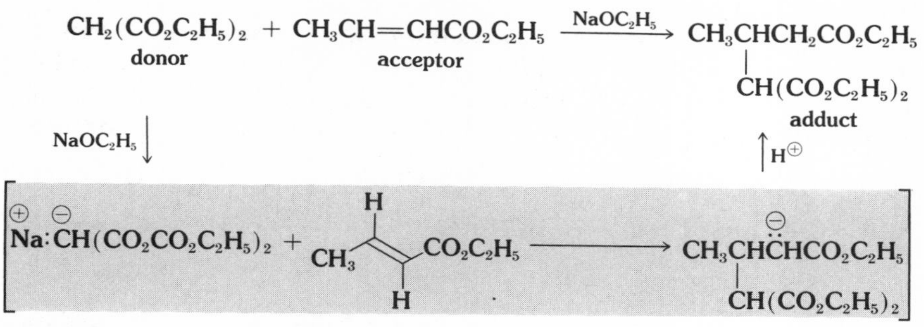

15·5 THE MICHAEL CONDENSATION

$$CH_2(CO_2C_2H_5)_2 + CH_3CH=CHCO_2C_2H_5 \xrightarrow{NaOC_2H_5} \underset{\underset{\text{adduct}}{CH(CO_2C_2H_5)_2}}{CH_3CHCH_2CO_2C_2H_5}$$

donor, acceptor

$NaOC_2H_5 \downarrow \qquad\qquad\qquad\qquad\qquad \uparrow H^{\oplus}$

$$\left[\overset{\oplus}{Na}\overset{\ominus}{:CH}(CO_2CO_2C_2H_5)_2 + \underset{H}{\overset{H}{\underset{CH_3}{>}C=C\overset{CO_2C_2H_5}{<}}} \longrightarrow \underset{\underset{}{CH(CO_2C_2H_5)_2}}{CH_3CH\overset{\ominus}{C}HCO_2C_2H_5} \right]$$

FIGURE 15·6 *The Michael reaction*

The 5,5-dimethyl-1,3-cyclohexanedione formed by saponification and decarboxylation of the product from reaction 15.49 reacts with simple aldehydes to give crystalline 2:1 condensation products which are useful for identifying and analyzing these volatile substances:

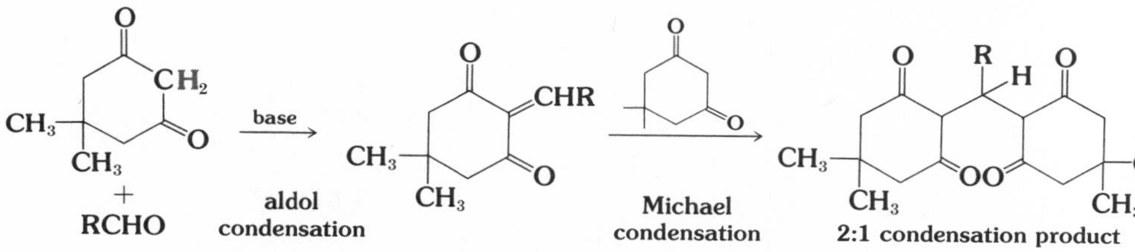

(15.50)

Consecutive Michael reactions were recently used to effect a total synthesis of the antibiotic griseofulvin:

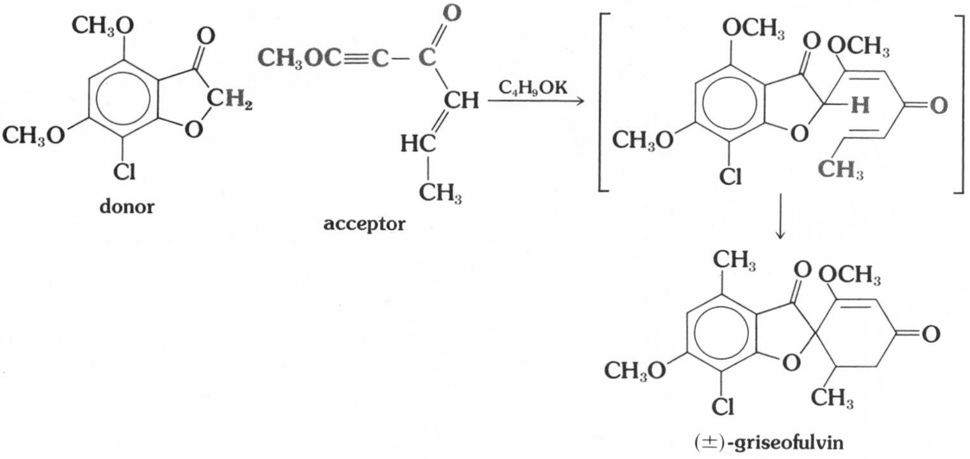

(15.51)

(±)-griseofulvin

PROBLEM 15·8

Suggest a suitable combination of reagents for preparing the following compounds by conjugate addition reactions:

a $\phi SCH_2CH_2CO_2C_2H_5$ *b* (pyrazoline structure) *c* $\phi CH(CH_2CO_2H)_2$

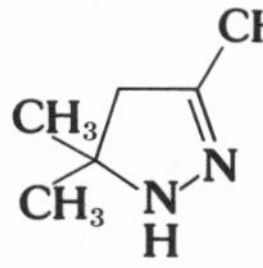

15·5 THE MICHAEL CONDENSATION

The manner in which organometallic reagents react with α,β-unsaturated ketones depends to a large extent on the metal involved. Thus organolithium compounds invariably react at the carbonyl carbon atom, while Grignard reagents may give significant amounts of 1,4 addition:

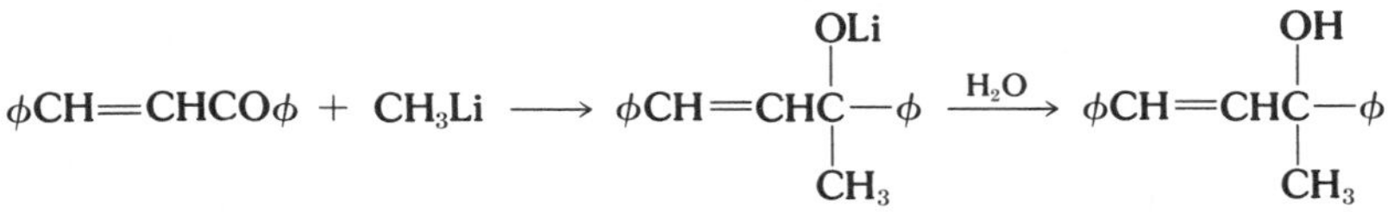

$$\phi CH=CHCO\phi + CH_3Li \longrightarrow \phi CH=CH\underset{CH_3}{\overset{OLi}{\underset{|}{C}}}-\phi \xrightarrow{H_2O} \phi CH=CH\underset{CH_3}{\overset{OH}{\underset{|}{C}}}-\phi \quad (15.52)$$

$$\phi CH=CHCOC_2H_5 + \phi MgBr \longrightarrow \phi CH=CH\underset{\phi}{\overset{OMgBr}{\underset{|}{C}}}-C_2H_5 + \phi_2CHCH=\underset{C_2H_5}{\overset{OMgBr}{C}}$$

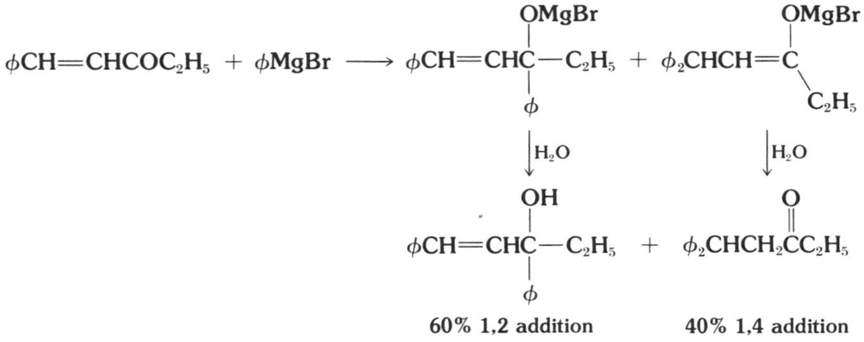

$$\downarrow H_2O \qquad\qquad\qquad \downarrow H_2O$$

$$\phi CH=CH\underset{\phi}{\overset{OH}{\underset{|}{C}}}-C_2H_5 \quad + \quad \phi_2CHCH_2\overset{O}{\overset{\|}{C}}C_2H_5 \quad (15.53)$$

60% 1,2 addition 40% 1,4 addition

Treatment of these reagents with cuprous salts converts them to complex organocuprous reagents, which shift the previous reactions to the conjugate addition mode. These 1,4 additions are, however, sensitive to steric hindrance at the β carbon atom:

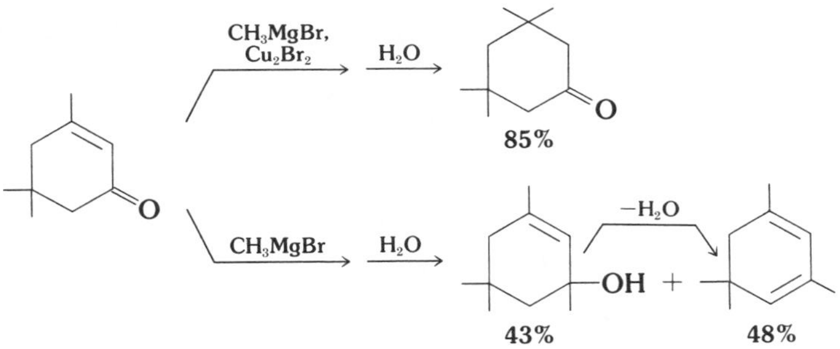

(15.54)

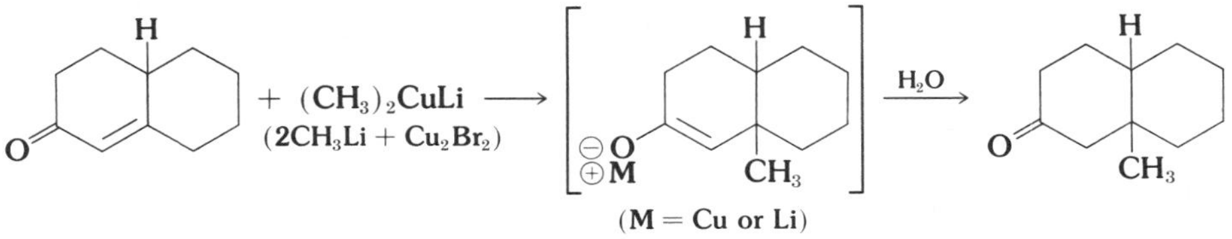

(15.55) (M = Cu or Li)

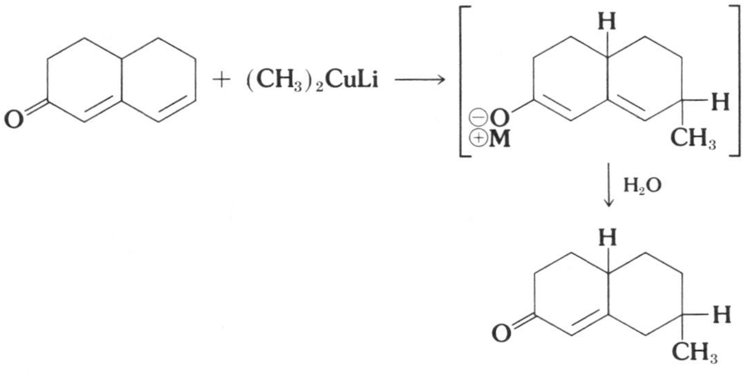

(15.56)

15·6 NEIGHBORING-GROUP PARTICIPATION IN DISPLACEMENT REACTIONS   507

PROBLEM 15·9

Explain the following reaction and account for the product shown:

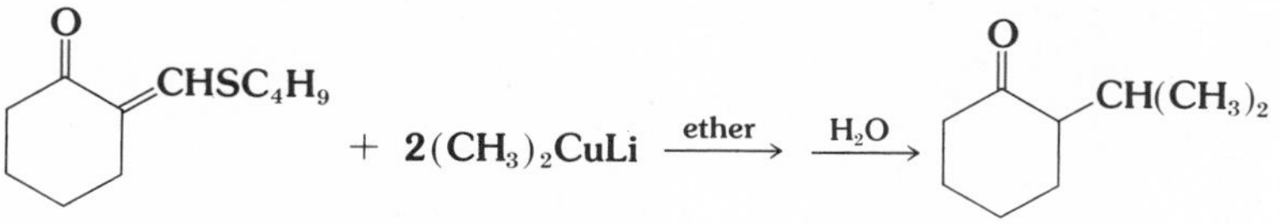

[cyclohexanone with =CHSC$_4$H$_9$ substituent] + 2(CH$_3$)$_2$CuLi $\xrightarrow{\text{ether}}$ $\xrightarrow{\text{H}_2\text{O}}$ [cyclohexanone with —CH(CH$_3$)$_2$ substituent]

15·6 NEIGHBORING-GROUP PARTICIPATION IN DISPLACEMENT REACTIONS

Nucleophilic substitution reactions at sp^3-hybridized carbon atoms, discussed in Chapter 3, are among the most widely studied classes of reactions in organic chemistry. When other functional groups are present, there is in many cases substantial evidence for internal participation of these groups in the displacement process. Neighboring-group participation such as that outlined in Figure 15.7 usually gives rise to one or more observable effects. The most common of these are isomerization, configurational changes, and rate enhancement.

Since bridged intermediates can be opened by nucleophilic attack at two different sites (a and b in Figure 15.7), an isomeric substitution product (II) may be formed in addition to the product of direct substitution (I). Isolation and identification of such product isomers from displacement reactions provides firm evidence that a bridged intermediate is formed in the product-determining stages of these reactions.

Even if the only product formed is one corresponding to direct displacement, evidence of a bridged intermediate can sometimes be obtained from stereochemical studies. Note that the direct-substitution product in this reaction can be formed either by a direct displacement of $X^\ominus$ from the starting material or by ring opening of the bridged ion through attack by $Z^\ominus$ at carbon a. The configuration at a will undergo inversion or scrambling in the case of direct substitution (assuming S_N2 or S_N1 mechanisms, respectively) and will be unchanged in the case of a bridged intermediate (the initial inversion during formation of the intermediate is followed by a second inversion on ring opening). Thus we can distinguish these alternative reaction paths by using reactants having an asymmetric center at a.

FIGURE 15·7 *Neighboring-group participation in substitution*

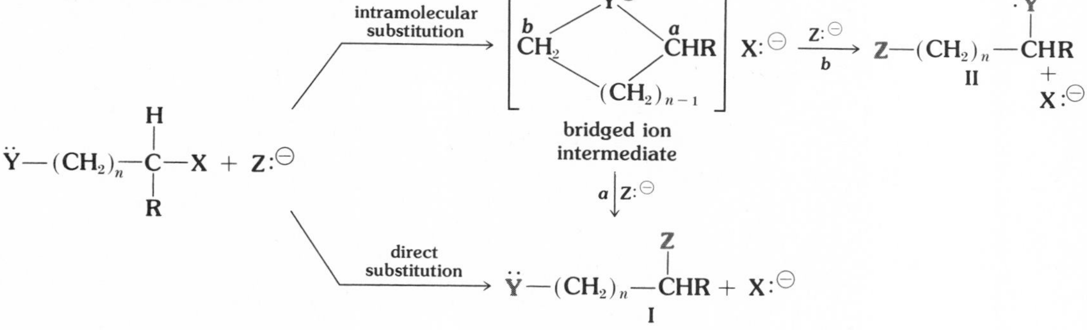

$\ddot{Y}-(CH_2)_n-\underset{\underset{R}{|}}{\overset{\overset{H}{|}}{C}}-X + Z^\ominus$

intramolecular substitution → $\left[\begin{array}{c} \overset{Y^\oplus}{\underset{(CH_2)_{n-1}}{\overset{b}{CH_2}\diagup\diagdown\overset{a}{CHR}}} \end{array}\right]$ $X^\ominus$ $\xrightarrow[b]{Z:^\ominus}$ $Z-(CH_2)_n-\underset{\underset{\text{II}}{|}}{\overset{\overset{:Y}{|}}{CHR}}$ + $X:^\ominus$

bridged ion intermediate

$a \downarrow Z:^\ominus$

$\ddot{Y}-(CH_2)_n-\underset{\underset{\text{I}}{|}}{\overset{\overset{Z}{|}}{CHR}} + X:^\ominus$

direct substitution →

15·6 NEIGHBORING-GROUP PARTICIPATION IN DISPLACEMENT REACTIONS

Substitution reactions that involve neighboring-group participation are often substantially faster than equivalent monofunctional reactions.† Rate enhancements of this sort, called *anchimeric assistance*, will be observed only when there is significant neighboring-group interaction in the transition state of the rate-determining step. According to the Hammond postulate (Chapter 3), the initial product from such an anchimerically assisted displacement reaction should be the bridged ion shown in Figure 15.7. Hence we would also expect to find constitutional and stereochemical evidence of neighboring-group participation in conjunction with the enhanced rate. Formation of a bridged intermediate may, of course, occur after the rate-determining step. In this event the neighboring substituent will have little or no effect on the reaction rate (except for an inductive retardation), although it can still strongly influence product composition and stereochemistry.

As we examine some examples of common neighboring nucleophilic groups, bear in mind that the mere existence of a nearby function does not mean that it must necessarily participate in any reactions. Neighboring-group interactions occur only when they are energetically more favorable than alternative intermolecular paths. Two of the most important variables affecting these intramolecular interactions are the nucleophilicity of the neighboring group and its orientation with respect to the reaction site.

NEIGHBORING HALOGEN

An example of a neighboring bromine effect was described in Section 5.3, where we observed that alkene addition reactions initiated by electrophilic halogen species usually proceed exclusively to trans addition products. We interpreted this to mean that a bridged halonium ion, or its unsymmetrically coordinated analog, undergoes nucleophilic ring opening in the product-determining step:

$$\overset{\delta\oplus\,\cdot\cdot}{\underset{\delta\oplus}{\text{X}:}}\quad\overset{\oplus}{\underset{}{\text{X}}}$$

$$-\overset{|}{\underset{|}{\text{C}}}-\overset{|}{\underset{|}{\text{C}}}- \;\rightleftharpoons\; -\overset{|}{\underset{}{\text{C}}}\diagup\diagdown\overset{|}{\underset{}{\text{C}}}-$$

halonium ion

Stereochemical evidence of the formation of such intermediates in substitution reactions is found in the stereospecific conversion of the diastereoisomeric 3-bromo-2-butanols to isomeric 2,3-dibromobutanes,

$$CH_3CHBrCH(OH)CH_3 + HBr \longrightarrow CH_3CHBrCHBrCH_3 \tag{15.56}$$

as illustrated in Figure 15.8. Since a racemic mixture of the R,R' and S,S' enantiomers, commonly referred to as the *threo diastereomer*, gives an equimolar (racemic) mixture of R,R and S,S product enantiomers, it is clear that substitution of the hydroxyl group has proceeded with retention of configuration. The R,S' and S,R' racemic mixture,

---

†This is another illustration of the general rule that intramolecular reactions are usually faster than similar intermolecular reactions.

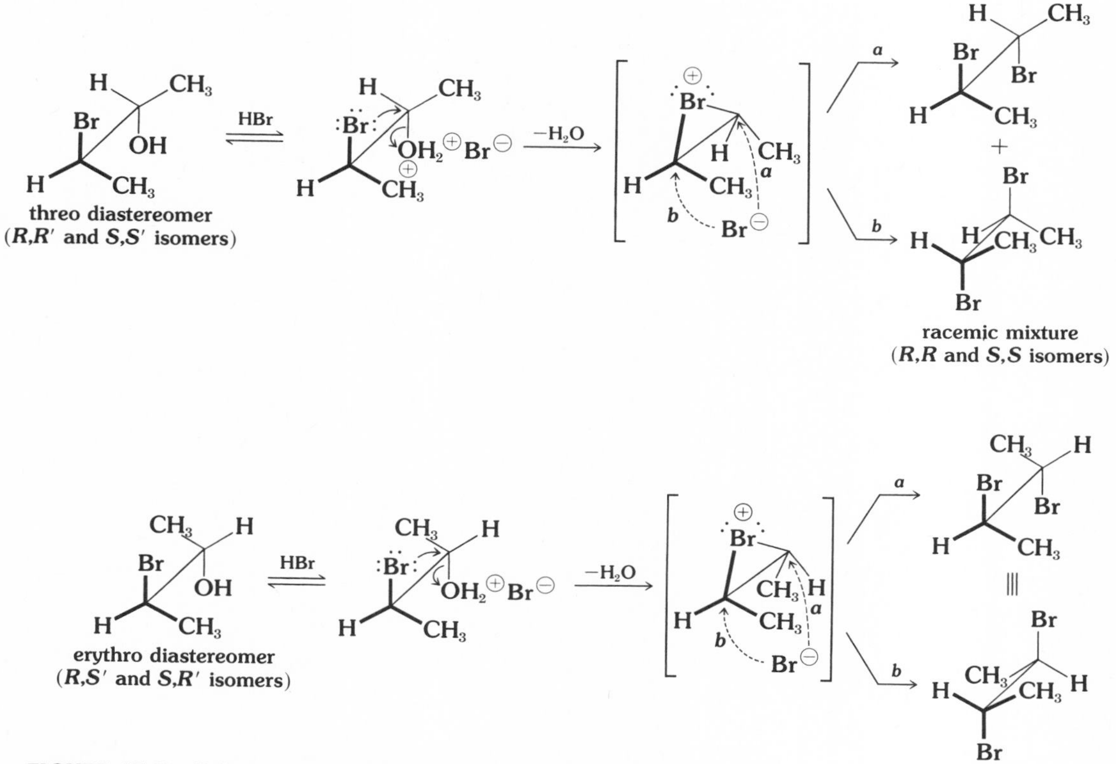

15·6 NEIGHBORING-GROUP PARTICIPATION IN DISPLACEMENT REACTIONS

FIGURE 15·8 *Substitution reactions of isomeric 3-bromo-2-butanols*

known as the *erythro diastereomer*, also reacts with retention of configuration, giving an optically inactive *meso*-dibromide, as shown. These results are nicely explained by a cyclic bromonium-ion intermediate.

PROBLEM 15·10
If optically active R,R'-3-bromo-2-butanol is treated with hydrogen bromide, will the 2,3-dibromobutane formed be optically active? Why?

The product-isomer criterion implicates a similar chloronium-ion intermediate in the solvolysis (S_N1 substitution by solvent) of 2-chloropropyl p-nitrobenzenesulfonate in trifluoroacetic acid solution:

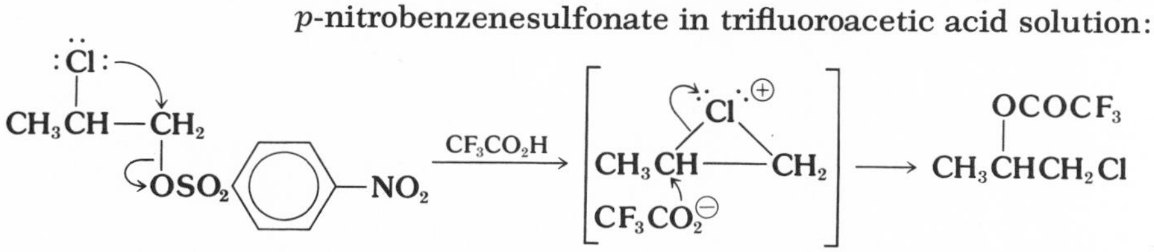

$$\text{CH}_3\text{CH}-\text{CH}_2\text{OSO}_2\text{C}_6\text{H}_4\text{NO}_2 \xrightarrow{\text{CF}_3\text{CO}_2\text{H}} \left[\begin{array}{c} \text{CH}_3\text{CH}-\text{CH}_2 \\ \text{CF}_3\text{CO}_2^\ominus \end{array} \right] \longrightarrow \text{CH}_3\text{CHCH}_2\text{Cl} \quad (15.57)$$

$$\text{OCOCF}_3$$

Comparative rate measurements for such reactions and their nonhalogenated analogs usually do not show any acceleration due to anchimeric assistance, because in most cases the inductive effect of a nearby halogen atom will act to retard the ionization. Since the inductive effect should, however, be much less sensitive to configurational changes than the neighboring-group participation described here, we can use the

15·6 NEIGHBORING-GROUP PARTICIPATION IN DISPLACEMENT REACTIONS

relative reaction rates of stereoisomers to determine whether anchimeric assistance is involved. Figure 15.9 illustrates this approach for the acetolysis of *cis*- and *trans*-2-halocyclohexyl *p*-bromobenzenesulfonates. The latter clearly shows anchimeric assistance by bromine and iodine.

When the neighboring halogen atom is shifted progressively farther away from the reaction site, we see an interesting variation in its influence:

$$CH_3CHClCH_2CH_2OSO_2Ar \xrightarrow{CF_3CO_2H, NaO_2CCF_3} CH_3CHClCH_2CH_2OCOCF_3 + ArSO_3H \quad (15.58)$$

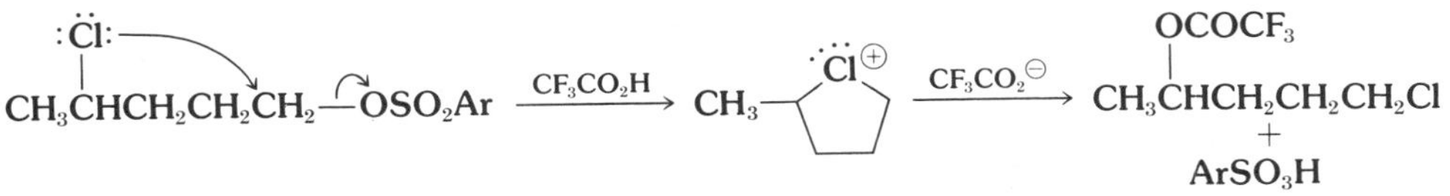

$$CH_3CHClCH_2CH_2CH_2-OSO_2Ar \xrightarrow{CF_3CO_2H} CH_3-\overset{\oplus}{\underset{}{Cl}}\text{(ring)} \xrightarrow{CF_3CO_2^{\ominus}} CH_3CHCH_2CH_2CH_2Cl + ArSO_3H \quad (15.59)$$
(with OCOCF$_3$ substituent)

Apparently, four-membered cyclic halonium ions form only with difficulty, whereas their less strained five- and six-membered analogs play an important role in the solvolysis reactions shown here. Halogen atoms located more than five carbon atoms away from another function seldom exert any significant influence on its reactions, since the probability that cyclic intermediates or transition states will form from acyclic precursors is very low. This probability is at its highest for three-membered rings, despite their angle strain, as the large number of such neighboring-group effects demonstrates (see equation 15.57).

FIGURE 15·9 *Anchimeric assistance in acetolysis reactions of 2-halocyclohexyl sulfonate esters*

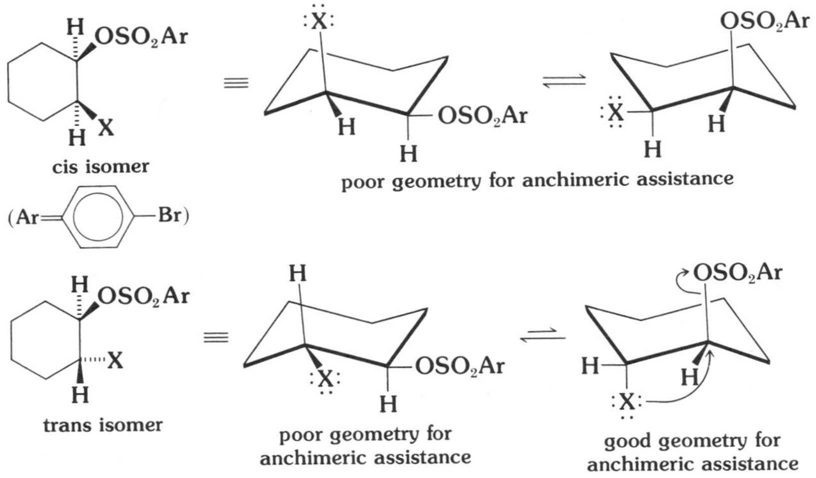

(Ar = C$_6$H$_4$—Br)

cis isomer — poor geometry for anchimeric assistance

trans isomer — poor geometry for anchimeric assistance ⇌ good geometry for anchimeric assistance

halogen	trans-cis rate ratio for solvolysis in acetic acid
Cl	4.0
Br	800.0
I	2.7×10^6

15·6 NEIGHBORING-GROUP PARTICIPATION IN DISPLACEMENT REACTIONS

NEIGHBORING OXYGEN AND SULFUR

The ease with which neighboring hydroxyl groups participate in intramolecular displacement reactions leading to oxiranes has been well documented (see, for example, equations 8.19 to 8.21). Since base catalysts are important, it is likely that the active neighboring function is actually the alkoxide ion:

$$\underset{\underset{R^2}{|}}{\overset{\underset{R^1}{|}}{C}}\text{OH} \; \underset{\underset{R^4}{|}}{\overset{\underset{R^3}{|}}{C}}\text{Cl} \; \xrightleftharpoons{\text{base}} \; \ldots \longrightarrow \; \text{oxirane} + \text{Cl}^{\ominus} \tag{15.60}$$

Alkoxyl groups do not easily form oxy anions, and we find little evidence for participation or anchimeric assistance by oxygen in solvolysis reactions of β-haloethers. In contrast to the negligible participation by alkoxyl groups, the corresponding sulfides show strong anchimeric assistance, as evidenced by the results in Table 15.4. The cases in which substantial rate increases are observed are undoubtedly proceeding by way of three- and five-membered cyclic sulfonium ions, which have the general formula:

$$(\text{CH}_2)_{n-2} \underset{\text{CH}_2}{\overset{\text{CH}_2}{<}} \overset{\oplus}{\text{S}} - \phi$$

Bridged sulfonium-ion intermediates are also implicated in reactions exhibiting product isomerization:

$$\text{C}_2\text{H}_5\ddot{\text{S}}\text{CH}(\text{CH}_3)\text{CH}_2\text{OH} \; \xrightarrow{\text{HCl}} \; \left[\text{C}_2\text{H}_5 - \overset{\oplus}{\text{S}} \overset{\text{CH}_3}{<}\right] \text{Cl}^{\ominus} \; \longrightarrow \; \text{C}_2\text{H}_5\ddot{\text{S}}\text{CH}_2\text{CHClCH}_3 \tag{15.61}$$

Mustard gas, a powerful vesicant (blistering agent) used as an offensive weapon in World War I, owes its destructive effect on tissue to its ability to alkylate nucleophiles—such as the amide function in proteins—under extremely mild conditions:

$$\underset{\text{mustard gas}}{\ddot{\text{S}}(\text{CH}_2\text{CH}_2\text{Cl})_2} \; \longrightarrow \; \left[\text{ClCH}_2\text{CH}_2\overset{\oplus}{\ddot{\text{S}}}<\right] \text{Cl}^{\ominus} \; \xrightarrow{\text{Nu:}} \; \text{Cl}(\text{CH}_2)_2\ddot{\text{S}}\text{CH}_2\text{CH}_2\overset{\oplus}{\text{Nu}}\text{Cl}^{\ominus} \tag{15.62}$$

TABLE 15·4 *Hydrolysis rates of oxygen- and sulfur-substituted alkyl chlorides*

$$\phi - \ddot{\text{Y}} - (\text{CH}_2)_n\text{Cl} \; \xrightarrow[100°]{\text{H}_2\text{O, dioxane}} \; \phi - \ddot{\text{Y}} - (\text{CH}_2)_n\text{OH} + \text{HCl}$$

	relative reaction rate	
ring size n	Y = O	Y = S
2	0.28	2.6×10^2
3	1.07	1.37
4	2.40	53.0

PROBLEM 15·11

The tritium-labeled *trans*-4-methoxycyclohexyl sulfonate ester gives a 3:1 mixture of cis and trans acetate esters on acetolysis. The trans product consists of 57.5% of 1-tritium-labeled material and 42.5% of the 4-labeled compound. Suggest an explanation for the retention of configuration in the trans acetate and the apparent migration of the isotopic hydrogen label.

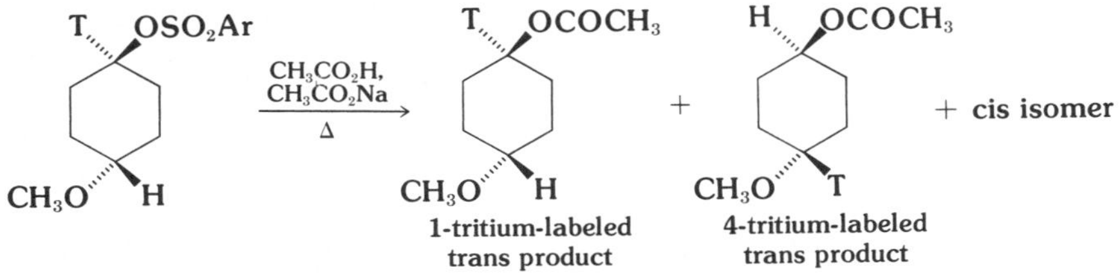

NEIGHBORING NITROGEN

Carbon chains having a primary or secondary amino group at one end and a bromine atom at the other undergo intramolecular alkylation reactions at rates which vary markedly with the size of the ring being formed (see Table 15.5). Rate enhancement is particularly evident in those reactions leading to three-, five-, and six-membered rings.

Tertiary amines may also provide anchimeric assistance for nucleophilic substitution reactions. The nitrogen analogs of mustard gas (nitrogen mustards) are, for example, a particularly reactive class of primary chlorides. These compounds have physiological effects similar to those of high-energy radiation; they cause mutations, destroy rapidly dividing cells such as leucocytes, and are carcinogenic. Consequently, they are sometimes referred to as *radiomimetic compounds*. An intermediate *aziridinium ion* apparently serves to alkylate protein and nucleic acid chains, and the difunctional nature of the nitrogen mustards results in a cross-linking of the macromolecules which make up genes and chromosomes. This, of course, prevents the genetic material from becoming properly distributed during cell division.

TABLE 15·5 *Internal-alkylation rates of amino bromides*

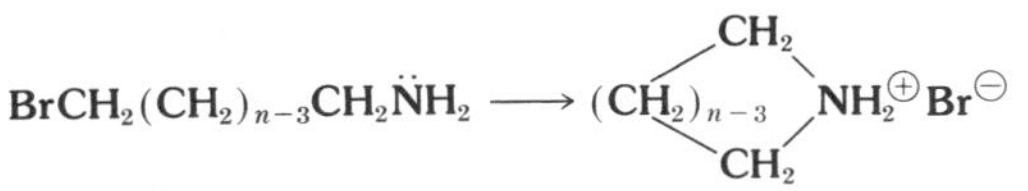

$$BrCH_2(CH_2)_{n-3}CH_2\ddot{N}H_2 \longrightarrow (CH_2)_{n-3}\begin{matrix}CH_2\\ \diagdown \\ \diagup \\ CH_2\end{matrix}NH_2^{\oplus}Br^{\ominus}$$

ring size n	relative reaction rate
3	0.12
4	2.0×10^{-3}
5	100
6	1.7
7	0.03
10	10^{-8}
12	10^{-5}

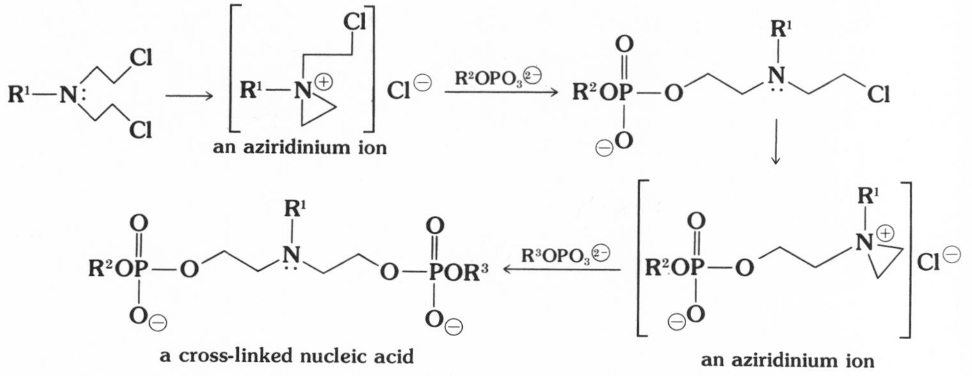

15·6 NEIGHBORING-GROUP PARTICIPATION IN DISPLACEMENT REACTIONS 513

(15.63)

PROBLEM 15·12
The thiol ester shown below undergoes hydrolysis (to the acid) 1 million times faster than similar esters lacking the imidazole substituent. Suggest a mechanism to account for this remarkable rate enhancement.

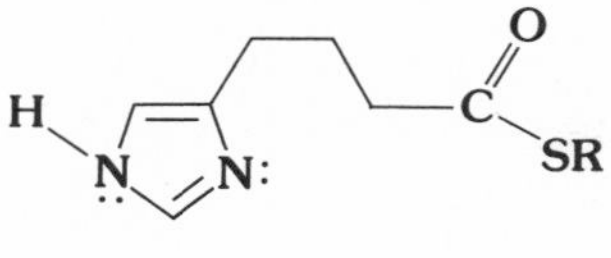

NEIGHBORING CARBOXYLATE DERIVATIVES

In contrast to the weak nucleophilic participation of neighboring alkoxyl groups in displacement reactions (Section 14.1), the corresponding acetate and benzoate esters often show substantial participation. For example, the enhanced rate and stereospecificity of *trans*-2-acetoxycyclohexyl benzenesulfonate, which undergoes solvolysis 10^3 times faster than the cis isomer, is convincingly explained by the anchimeric assistance shown in Figure 15.10. Participation by the ester function

FIGURE 15·10 *Anchimeric assistance by a neighboring acetoxy group*

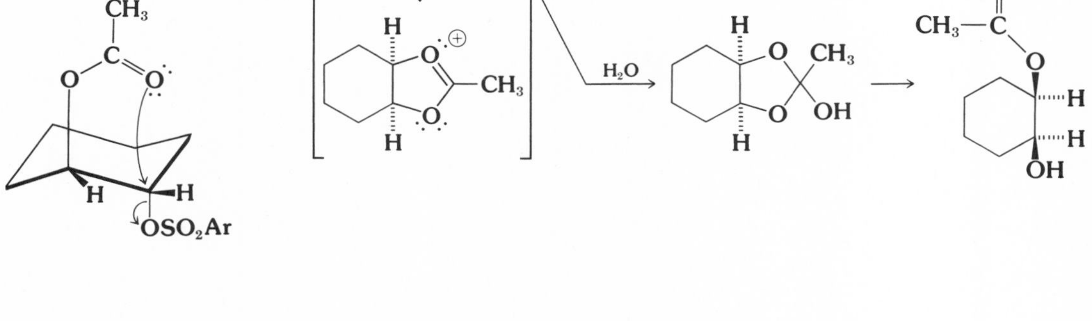

15·6 NEIGHBORING-GROUP PARTICIPATION IN DISPLACEMENT REACTIONS

leads to a heterocyclic cation intermediate which is subject to S_N2 ring opening by rearside attack of acetate ion in anhydrous solution or formation of an unstable hemiorthoester when water is present.

Substitution reactions of α-halocarboxylic acids often proceed with retention of configuration, which suggests an α-lactone intermediate:

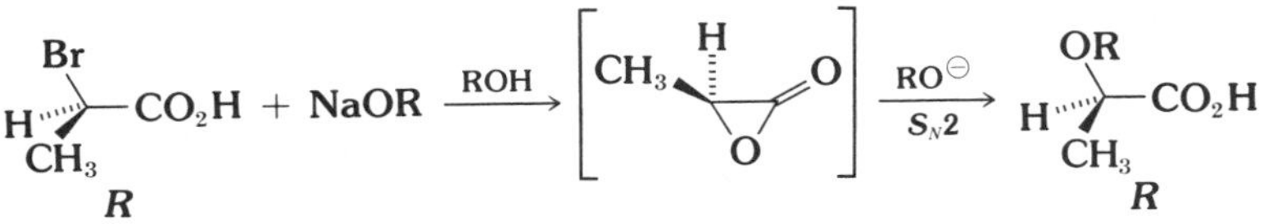

$$\underset{R}{\underset{CH_3}{H\cdots}}\!\!\!\!\!\!\!\!\!\!\diagdown\!\!-CO_2H + NaOR \xrightarrow{ROH} \left[CH_3 \overset{H}{\underset{O}{\diagup\!\!\diagdown}} O \right] \xrightarrow[S_N2]{RO^\ominus} \underset{R}{\underset{CH_3}{H\cdots}}\!\!\!\!\!\!\!\!\!\!\overset{OR}{\diagdown}\!\!-CO_2H \quad (15.64)$$

NEIGHBORING CARBON-CARBON DOUBLE BONDS

In Chapter 10 we examined the special reactivity of allylic functional groups, a feature that is further illustrated in Figure 15.11 by the S_N1 acetolysis reaction of 3-chloro-3-methyl-1-butene in acetic acid. The allyl-cation intermediate not only combines with solvent to give mixtures of isomeric allyl acetates, but also undergoes an ion-pair recombination resulting in isomerization of the starting material. Allyl alcohols undergo a structurally specific (regiospecific) allylic rearrangement on reaction with thionyl chloride:

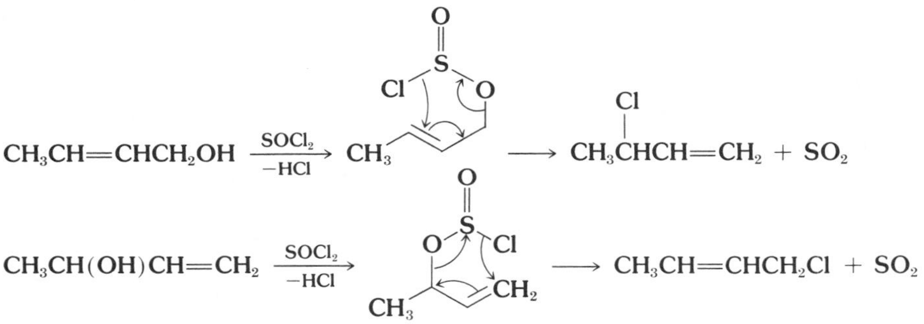

$$CH_3CH{=}CHCH_2OH \xrightarrow[-HCl]{SOCl_2} \cdots \longrightarrow CH_3\overset{Cl}{\underset{|}{C}}HCH{=}CH_2 + SO_2$$

$$CH_3CH(OH)CH{=}CH_2 \xrightarrow[-HCl]{SOCl_2} \cdots \longrightarrow CH_3CH{=}CHCH_2Cl + SO_2 \quad (15.65)$$

Double bonds which do not have an allylic orientation with respect to the functional group undergoing displacement, but which are suitably

FIGURE 15·11 *Acetolysis of isomeric allyl chlorides*

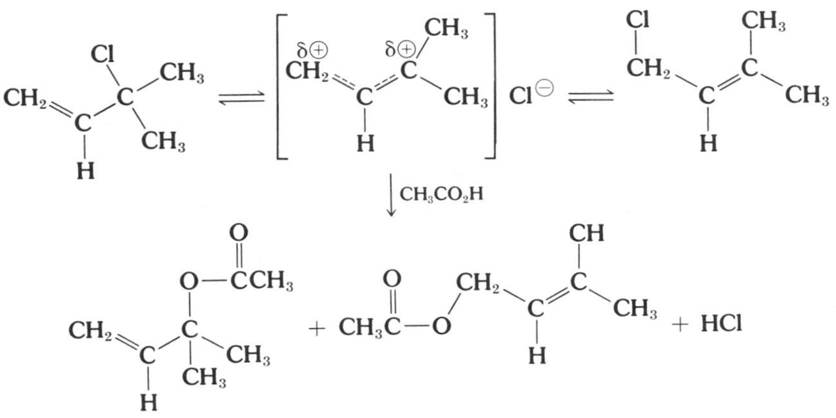

15·6 NEIGHBORING-GROUP PARTICIPATION IN DISPLACEMENT REACTIONS 515

positioned nearby, may also participate in the ionization process. For example, the bridged sulfonate ester in equation 15.66 shows an extraordinary rate enhancement and stereospecificity in acetolysis; the related reaction described in equation 15.67 shows a similar rate enhancement, as well as a characteristic isomerization:

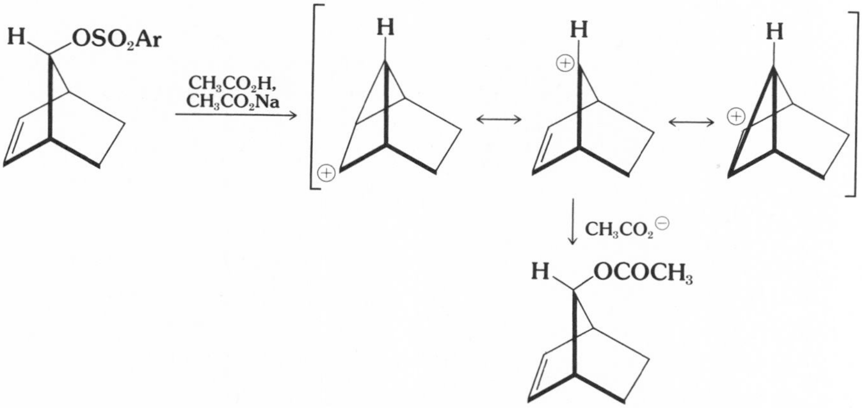

(15.66)

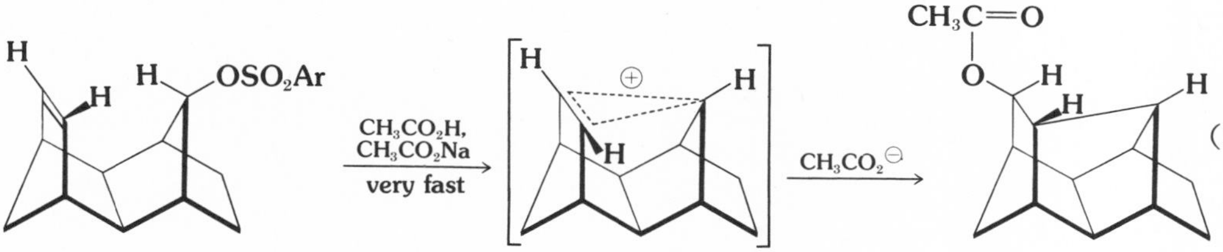

(15.67)

In cases where the functional groups are more widely separated the optimum orientation of a developing p orbital in relation to the neighboring double bond changes from the parallel orbital configuration found in allylic systems to an orthogonal arrangement, similar to that of an internal S_N2 displacement:

parallel orbitals
(allylic orientation)

orthoganol orbitals
(homoallylic orientation)

PROBLEM 15·13
Write mechanisms showing how double- or triple-bond participation can account for the products of the following two reactions:

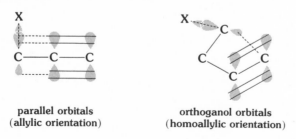

a $(CH_3)_2C=CHCH_2CH_2Cl \xrightarrow{NaHCO_3, H_2O}$ [cyclopropyl]–$\underset{CH_3}{\overset{CH_3}{C}}$–OH

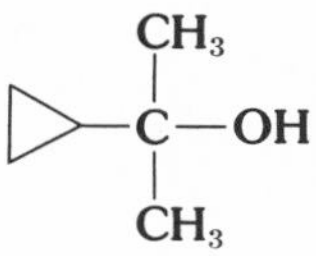

b

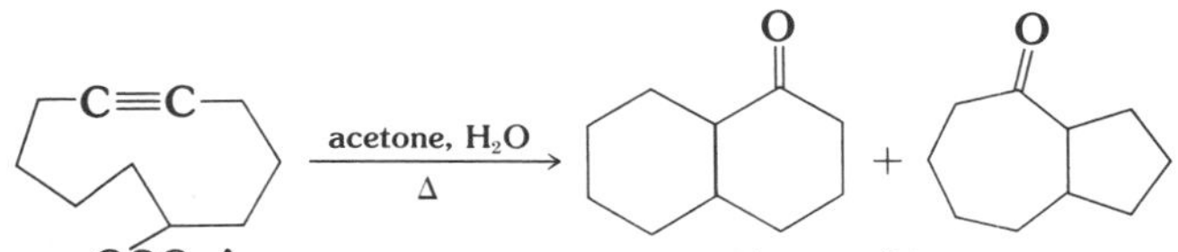

[reaction scheme: cyclodecyne with OSO₂Ar group, acetone, H₂O, Δ → two bicyclic ketones, mixture of isomers]

PROBLEM 15·14

Reaction of *trans,cis*-1,5-cyclodecadiene with one equivalent of bromine gives a mixture of isomeric ($C_{10}H_{16}Br_2$) *saturated* dibromides. Suggest structures for these products and a mechanism for their formation.

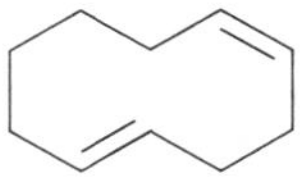

trans,cis-cyclodecadiene

PROBLEM 15·15

a The rate of acetolysis of a group of sulfonate esters varies as shown. Explain these variations:

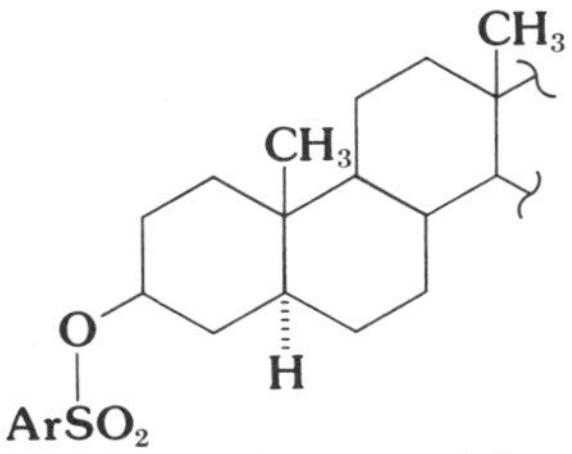

acetolysis rate: 1.0 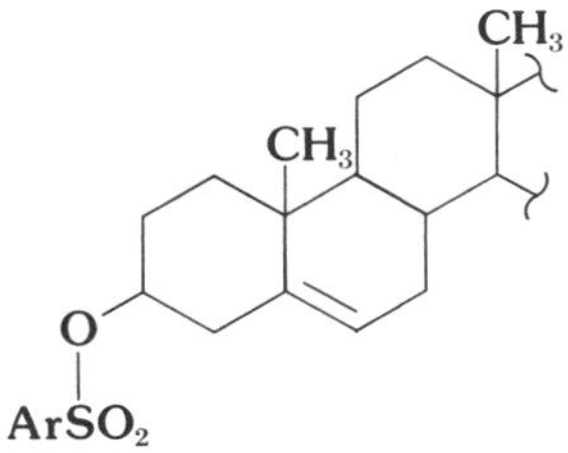   100

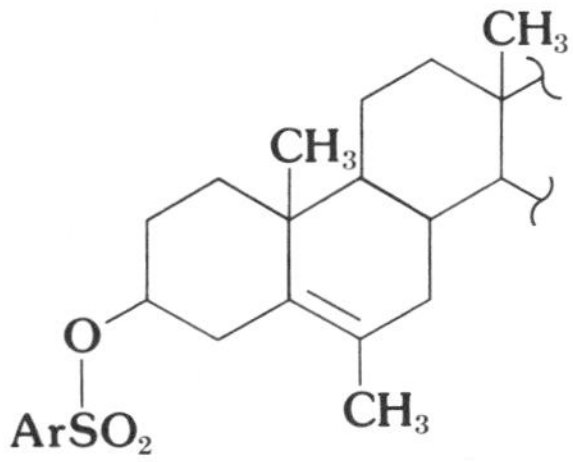

7,500 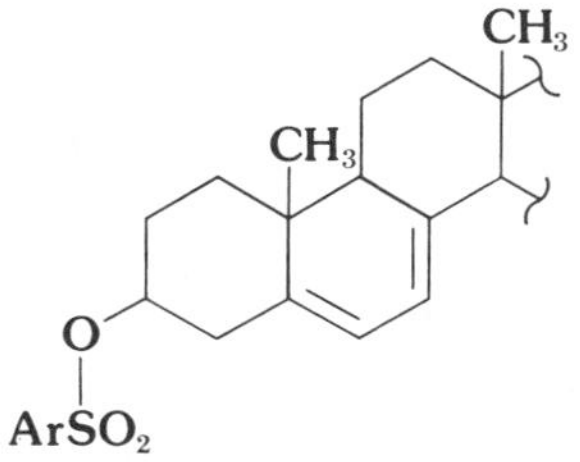   3,000

b The dibromobicycloheptene below is poisonous and has proved very dangerous to handle. Keeping in mind the toxic properties of mustard gas and related compounds, suggest a reason for this behavior.

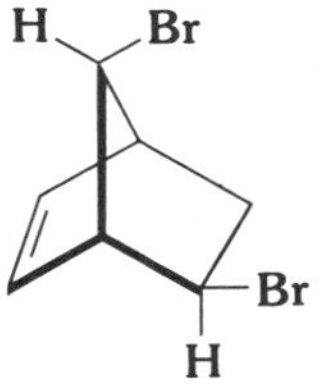

15·7 CARBONIUM-ION-INDUCED REARRANGEMENTS

In Sections 3.7 and 8.6 we saw that carbonium-ion intermediates in which the positively charged center is adjacent to a highly branched (substituted) carbon atom may rearrange by a shift of one of the neighboring substituents:

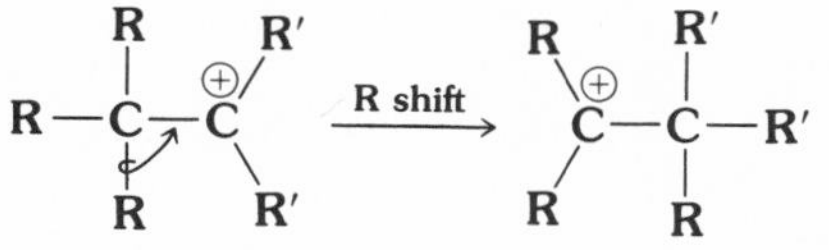

$$R-\underset{\underset{R}{|}}{\overset{\overset{R}{|}}{C}}\overset{R'}{\underset{R'}{-\overset{\oplus}{C}}} \xrightarrow{R \text{ shift}} \overset{R}{\underset{R}{\overset{\oplus}{C}}}-\underset{R}{\overset{R'}{\underset{|}{C}}}-R'$$

Such carbonium-ion rearrangements are most often observed when they result in an improvement in ion stability or in a relief of ring strain or nonbonded repulsions:

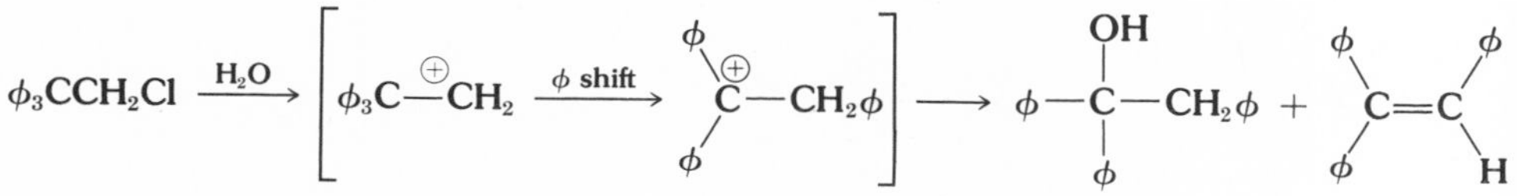

$$\phi_3CCH_2Cl \xrightarrow{H_2O} \left[\phi_3\overset{\oplus}{C}-CH_2 \xrightarrow{\phi \text{ shift}} \overset{\phi}{\underset{\phi}{\overset{\oplus}{C}}}-CH_2\phi\right] \longrightarrow \phi-\underset{\phi}{\overset{OH}{\underset{|}{C}}}-CH_2\phi + \overset{\phi}{\underset{H}{C}}=\overset{\phi}{\underset{}{C}} \quad (15.68)$$

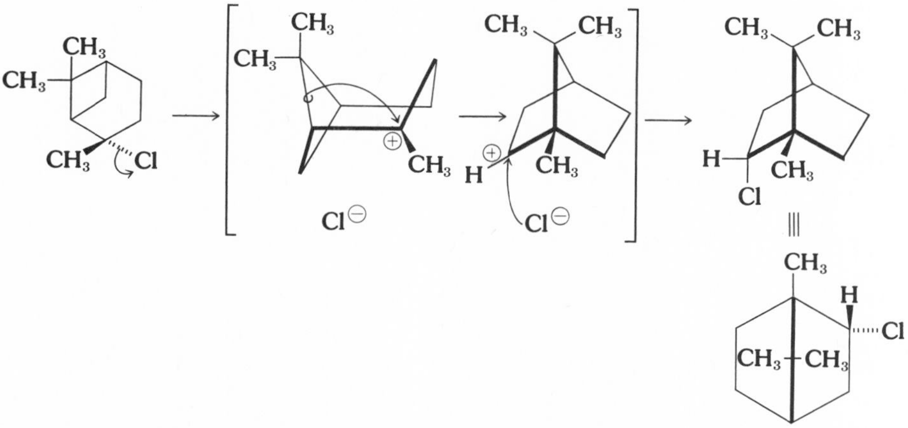

(15.69)

These rearrangements do not usually involve more than one functional group in the commonly used sense of that term. However, these reactions are related to the cases discussed above and are of special interest because of the controversy that surrounds carbon-carbon σ-bond participation in the ionization process.

PINACOL REARRANGEMENT

Vicinal diols, commonly called *pinacols*, undergo a facile dehydrative rearrangement when treated with acids. This *pinacol rearrangement* is illustrated by the following reactions:

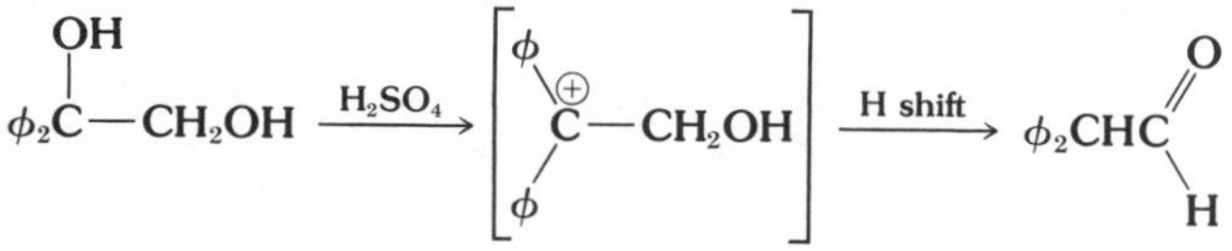

$$\phi_2\underset{|}{\overset{OH}{\underset{}{C}}}-CH_2OH \xrightarrow{H_2SO_4} \left[\overset{\phi}{\underset{\phi}{\overset{\oplus}{C}}}-CH_2OH\right] \xrightarrow{H \text{ shift}} \phi_2CH\overset{O}{\underset{H}{\overset{\|}{C}}} \quad (15.70)$$

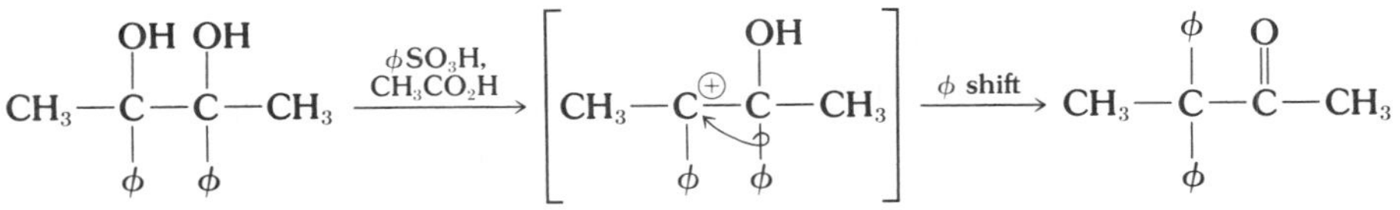

15·7 CARBONIUM-ION-INDUCED REARRANGEMENTS

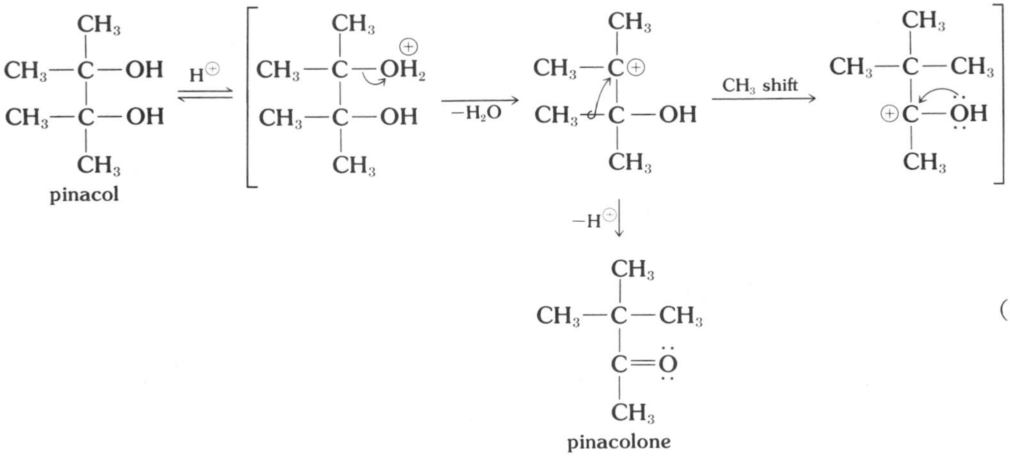

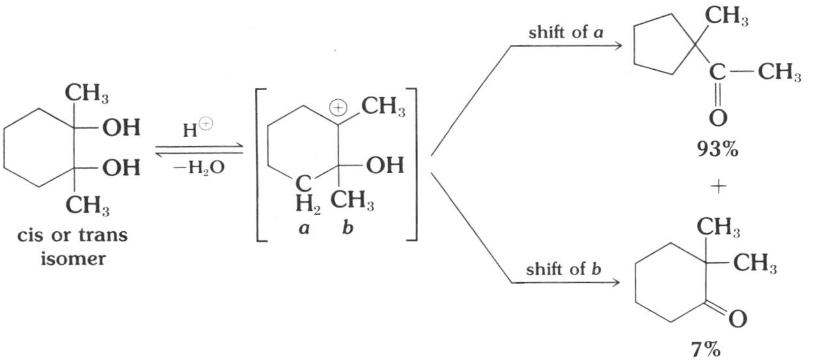

$$\text{CH}_3-\underset{\phi}{\underset{|}{\overset{\text{OH}}{\overset{|}{C}}}}-\underset{\phi}{\underset{|}{\overset{\text{OH}}{\overset{|}{C}}}}-\text{CH}_3 \xrightarrow[\text{CH}_3\text{CO}_2\text{H}]{\phi\text{SO}_3\text{H},} \left[\text{CH}_3-\underset{\phi}{\underset{|}{\overset{\text{OH}}{\overset{|}{C^{\oplus}}}}}-\underset{\phi}{\underset{|}{\overset{}{\overset{}{C}}}}-\text{CH}_3\right] \xrightarrow{\phi\text{ shift}} \text{CH}_3-\underset{\phi}{\underset{|}{\overset{\phi}{\overset{|}{C}}}}-\underset{}{\overset{\text{O}}{\overset{\|}{C}}}-\text{CH}_3 \quad (15.71)$$

(Equation 15.72: pinacol rearrangement to pinacolone)

(Equation 15.73: 1,2-dimethylcyclohexane-1,2-diol rearrangement, shift of *a* gives 93%, shift of *b* gives 7%)

In cases where the initial dehydration can generate two different carbonium-ion intermediates (as in equation 15.70) or where rearrangements of two different substituents is possible (as in equation 15.71), we usually find that the more stable ion serves as the precursor to rearrangement and that aryl groups such as phenyl migrate preferentially in competition with alkyl groups. Since the cis and trans isomers of 1,2-dimethylcyclohexane-1,2-diol are rapidly interconverted under the conditions of rearrangement, it is not surprising that both isomers give the same product mixture (equation 15.73).

PROBLEM 15·16

a What rearrangement product would you expect from reaction 15.70 if the other possible carbonium-ion intermediate were predominant in the initial stage?

b The pinacol $\text{CF}_3\text{C}\phi(\text{OH})\text{C}\phi(\text{OH})\text{CF}_3$ is relatively unreactive under conditions which normally give pinacol rearrangements. Suggest an explanation for this fact.

15·7 CARBONIUM-ION-INDUCED REARRANGEMENTS

PROBLEM 15·17
Write a reasonable mechanism for the following reaction. *Hint:* Begin with the reverse of a pinacol rearrangement.

[structure: spiro[5.3] cyclohexanone-cyclobutane] $\xrightarrow{H^\oplus}$ [structure: spiro[4.4] cyclopentanone-cyclopentane]

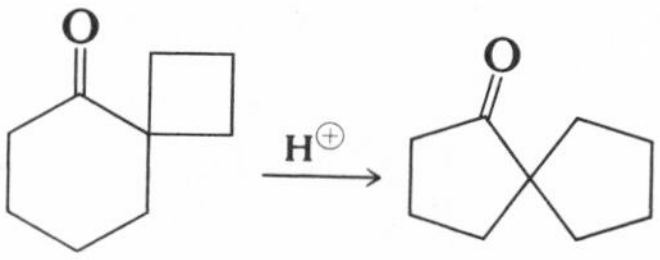

It is possible to force formation of a carbonium ion corresponding to the less stable pinacol intermediate by nitrous acid deamination of the corresponding amine. This reaction is called the *amino-pinacol rearrangement*:

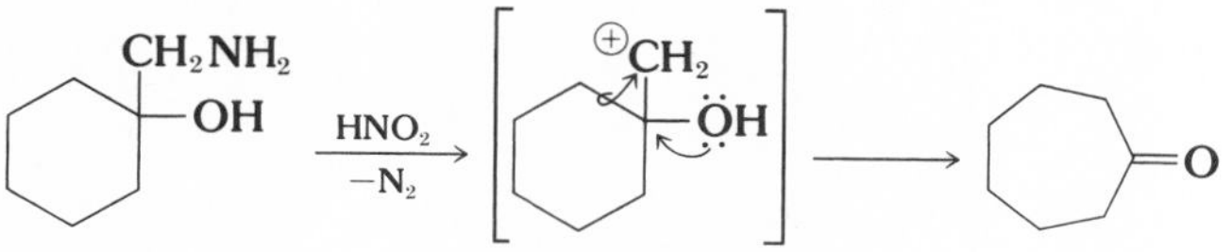

[structure: 1-(aminomethyl)cyclohexanol] $\xrightarrow[-N_2]{HNO_2}$ [bridged cation intermediate] $\longrightarrow$ [cycloheptanone] (15.74)

BRIDGED IONS INVOLVING π-ELECTRON SYSTEMS

If we compare the acetolysis reactions of 2,2,2-triphenylethyl toluenesulfonate and neopentyl toluenesulfonate, we find that while rearrangement takes place in both cases, the former reaction proceeds about 8,000 times faster than the latter:

$$\phi_3CCH_2OSO_2C_7H_7 + CH_3CO_2H \xrightarrow[\text{buffer}]{NaO_2CCH_3} \phi-\underset{\phi}{\overset{OCOCH_3}{\underset{|}{C}}}-CH_2\phi + \underset{\phi}{\overset{\phi}{C}}=CH\phi \quad (15.75)$$

$$(CH_3)_3CCH_2OSO_2C_7H_7 + CH_3CO_2H \xrightarrow[\text{buffer}]{NaO_2CCH_3} CH_3-\underset{CH_3}{\overset{OCOCH_3}{\underset{|}{C}}}-CH_2CH_3 + \underset{CH_3}{\overset{CH_3}{C}}=CHCH_3 \quad (15.76)$$

Up to this point we have been writing these molecular rearrangements in a stepwise fashion, with an initial rate-determining ionization preceding the migration step, as in equation 15.68. However, the large difference in reaction rate noted above requires a modification of this view. Since phenyl groups are more inductively electron withdrawing than methyl substituents, the resulting destabilization of the primary carbonium ion initially formed in such a stepwise rearrangement mechanism would clearly cause a *decrease* in the rate of this reaction, not an increase. If, however, we accept the point of view that phenyl participation can *assist* ionization, the dilemma is resolved. This idea also fits in nicely with the neighboring-group interactions discussed above, particularly with respect to double-bond participation in ionization. Thus aromatic rings and double bonds both have π electrons which can bond to neighboring cations without seriously disrupting the σ-bonded framework.

As shown in Figure 15.12, such an interaction could generate an intermediate bridged ion, the *phenonium ion,* similar to the inter-

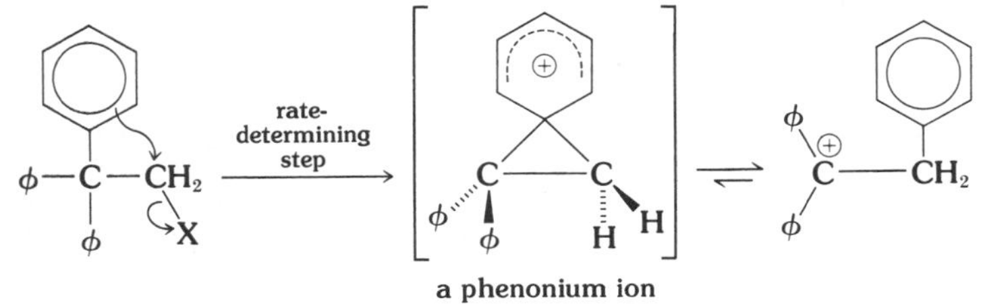

15·7 CARBONIUM-ION-INDUCED REARRANGEMENTS

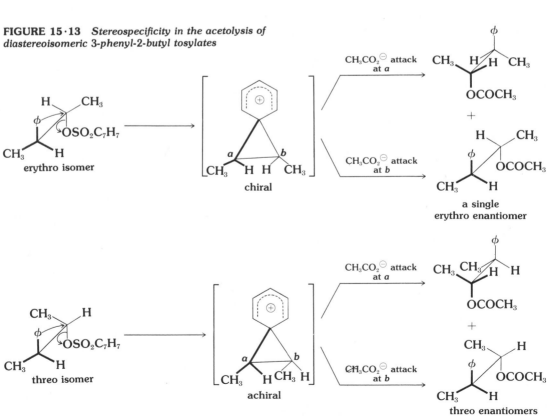

FIGURE 15·12 *Phenyl participation in ionization*

mediate ions formed during electrophilic substitution of aromatic rings. Evidence supporting the role of phenonium-ion intermediates in reactions involving aryl migrations is found in the stereochemical outcome of acetolysis reactions of *threo-* and *erythro-*3-phenyl-2-butylsulfonate esters:

$$\phi CH(CH_3)CH(CH_3)OSO_2Ar \xrightarrow[NaOCOCH_3]{CH_3CO_2H,} \phi CH(CH_3)CH(CH_3)OCOCH_3 + \underset{\phi}{\overset{CH_3}{C}}=CHCH_3$$

(+) threo isomer ⟶ 59% (±)-*threo*-acetate
+
3% *erythro*-acetate + 38%

(+) erythro isomer ⟶ 65% (+)-*erythro*-acetate
+
3% *threo*-acetate + ≈20%

(15.77)

FIGURE 15·13 *Stereospecificity in the acetolysis of diastereoisomeric 3-phenyl-2-butyl tosylates*

15·7 CARBONIUM-ION-INDUCED REARRANGEMENTS

The predominant retention of configuration in the substitution products obtained from both diastereoisomeric arylsulfonates, as well as the complete racemization of the *threo*-acetate and retained activity of the *erythro*-acetate, argues strongly for a bridged intermediate (phenonium ion), as illustrated in Figure 15.13. A compound having the phenonium-ion skeleton has actually been isolated in reactions with 2-*p*-phenoxyethyl bromide:

$$\text{HO-C}_6\text{H}_4\text{-CH}_2\text{-CH}_2\text{-Br} \xrightarrow{\text{NaOCH}_3, \text{CH}_3\text{OH}} [\text{phenonium intermediate}] \xrightarrow{\text{CH}_3\text{OH}, \text{H}^{\oplus}} \text{HO-C}_6\text{H}_4\text{-CH}_2\text{-CH}_2\text{-OCH}_3$$

(15.78)

PROBLEM 15·18

Electron-donating or -withdrawing substituents in the para position of the phenyl ring exert a striking influence on the rate and products of reaction 15.77. The para nitro compound below undergoes slow acetolysis to 68% alkene and 13% configurationally inverted *erythro*-acetate. The para methoxy analog reacts about 80 times faster than the unsubstituted phenyl compound in equation 15.77, giving the configurationally unchanged *threo*-acetate as the sole product (99.7%). Suggest an explanation for these facts.

$$\text{O}_2\text{N-C}_6\text{H}_4\text{-CH(CH}_3)\text{CH(CH}_3)\text{OSO}_2\text{Ar}$$
threo isomer

$$\text{CH}_3\text{O-C}_6\text{H}_4\text{-CH(CH}_3)\text{CH(CH}_3)\text{OSO}_2\text{Ar}$$
threo isomer

BRIDGED IONS INVOLVING σ-ELECTRON SYSTEMS

One of the most controversial and widely studied areas of modern organic chemistry has grown out of an extrapolation of the neighboring-group participation concept to include σ-bond participation leading to alkyl group shifts.

The *exo*-toluenesulfonate ester in Figure 15.14 undergoes acetolysis 350 times more rapidly than the corresponding endo-tosylate, and the exclusive product from this optically active compound is a racemic *exo*-acetate mixture. A number of chemists contend that this rate enhancement and stereospecificity indicate a rate-determining formation of a symmetrical bridged cation intermediate characterized by delocalization of σ-bonding electrons. The bonding represented by the dotted lines in this intermediate is unusual in that sufficient electrons are not available for these to be ordinary σ bonds. We refer to such cations as "nonclassical," and their existence has been the subject of hot debate.

Other chemists argue that the exo stereospecificity observed in this

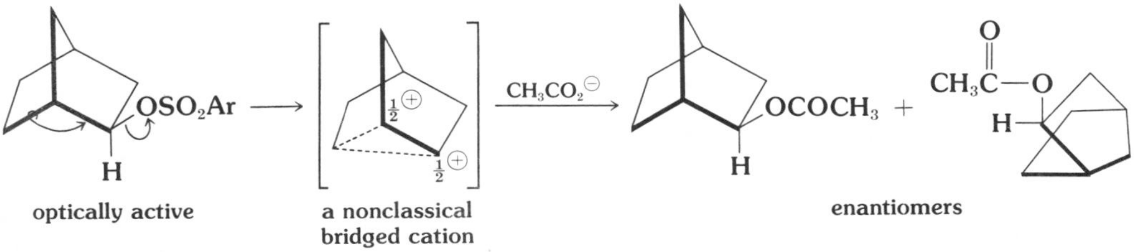

15·7 CARBONIUM-ION-INDUCED REARRANGEMENTS

FIGURE 15·14 *Acetolysis of 2-exo-norbornyl tosylate*

case is due to the lower steric hindrance of the exo face of the molecule, and in any event could be explained by a rapidly interconverting pair of enantiomeric classical carbonium ions (the "windshield-wiper" effect):

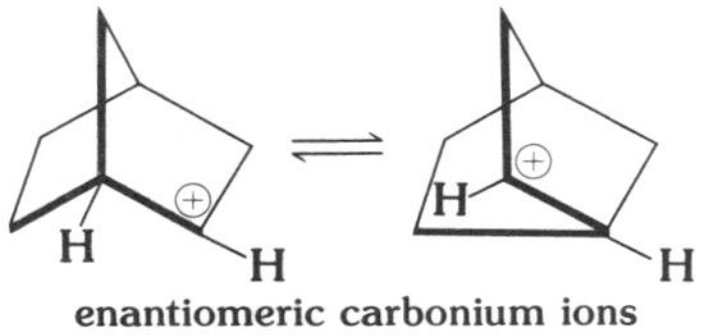

enantiomeric carbonium ions

The intermediate structure in Figure 15.14 would then be the transition state for this ion equilibrium, and not a discrete intermediate in its own right. Those who favor this viewpoint attribute the rate difference between the *exo-* and *endo-*tosylates to steric retardation of endo ionization, rather than anchimeric assistance of exo ionization.

Low-temperature pmr measurements have been made of the 2-norbornyl cation generated from the corresponding fluoride:

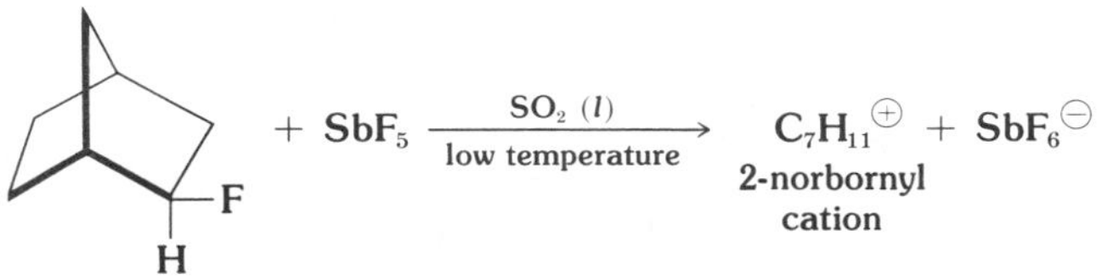

$$\text{norbornyl-F} + SbF_5 \xrightarrow[\text{low temperature}]{SO_2 \ (l)} C_7H_{11}^{\oplus} + SbF_6^{\ominus} \tag{15.79}$$

At $-143°C$ the spectrum of this cation consists of three broad signals at $\delta = 5.1$ (4H), 3.2 (1H), and 2.2 (6H). As the solution is warmed these signals first broaden and then coalesce, so that above $-60°C$ the spectrum consists of a single sharp signal at $\delta = 4.0$. In order to explain this remarkable change, we must find a mechanism by which all 11 hydrogen atoms in this system experience the same average environment. Such an averaging is in fact accomplished by a combination of the proposed nonclassical structure of the norbornyl cation (or by the alternative rapid 1,2 alkyl shift shown above) with rapid 3,2 hydrogen shifts and 6,2 hydrogen shifts:

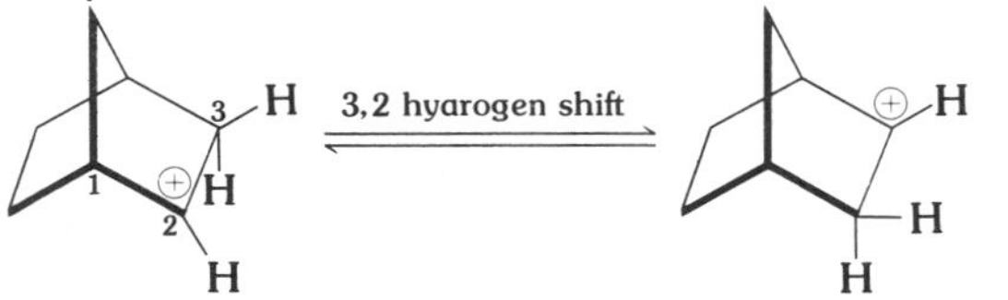

$$\tag{15.80}$$

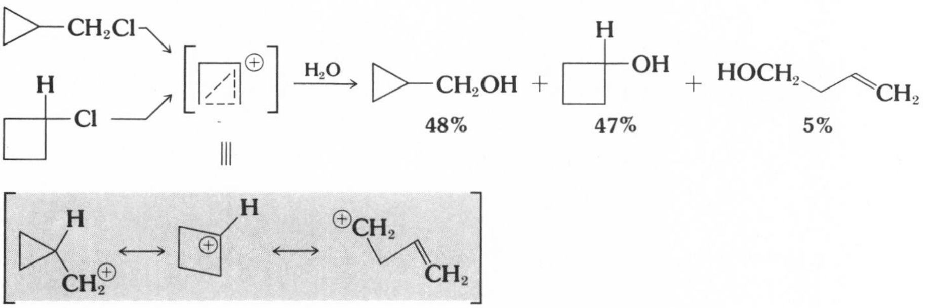

FIGURE 15·15 *Hydrolysis of cyclobutyl- and cyclopropylcarbinyl chlorides*

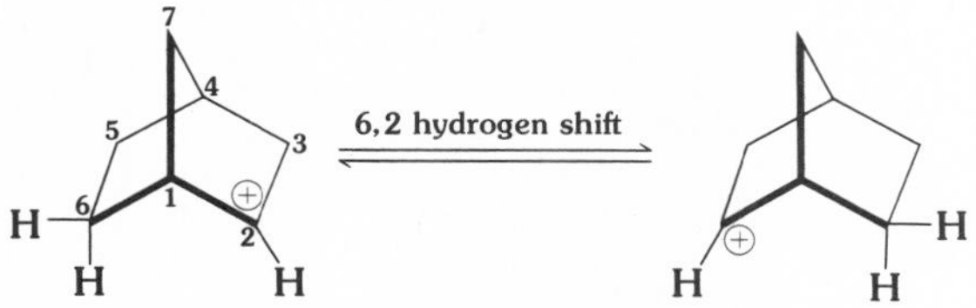

(15.81)

As the temperature of the sample is lowered below −60°C, the 3,2 hydrogen shift slows down sufficiently to permit the front and back parts of the norbornyl cation to be distinguished by pmr spectroscopy. The 6,2 hydrogen shift and the controversial 1,2 alkyl shift remain rapid on the pmr time scale even at −140°. Consequently, we observe a time average of the protons on C-1, C-2, and C-6 as the lowest-field signal at $\delta = 5.1$, with the single hydrogen atom on C-4 at $\delta = 3.2$ and an average of the protons on C-3, C-5, and C-7 at $\delta = 2.2$.†

Other alkyl rearrangements, such as that shown in Figure 15.15, have also been interpreted as proceeding by means of nonclassical ion intermediates.

15·8 REARRANGEMENTS TO ELECTRON-DEFICIENT NITROGEN ATOMS

BECKMANN REARRANGEMENT

In 1886 E. Beckmann (Germany) observed that oximes were transformed into amides on treatment with strong acids:

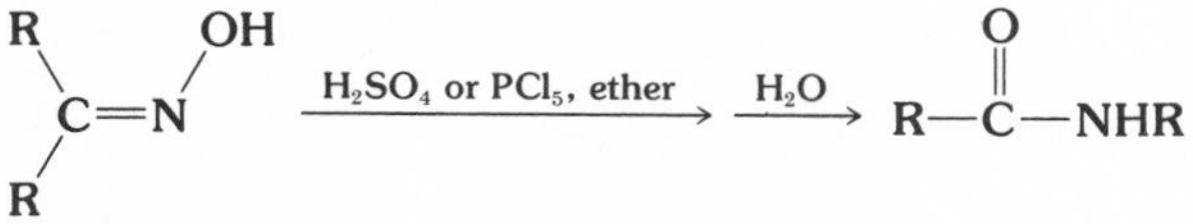

$$\begin{array}{c} R \\ \diagdown \\ C=N \\ \diagup \\ R \end{array} \begin{array}{c} OH \\ \diagdown \\ \end{array} \xrightarrow{H_2SO_4 \text{ or } PCl_5, \text{ ether}} \xrightarrow{H_2O} R-\underset{\underset{O}{\parallel}}{C}-NHR \qquad (15.82)$$

This reaction clearly requires the shift of an alkyl (or aryl) group from carbon to nitrogen, and since chiral groups migrate without loss of their configurational integrity, it seems likely that the migrating group is

†Since carbon atoms C-1, C-2, and C-6 bear the positive charge, protons bonded to them will be deshielded in relation to other hydrogen atoms in the molecule.

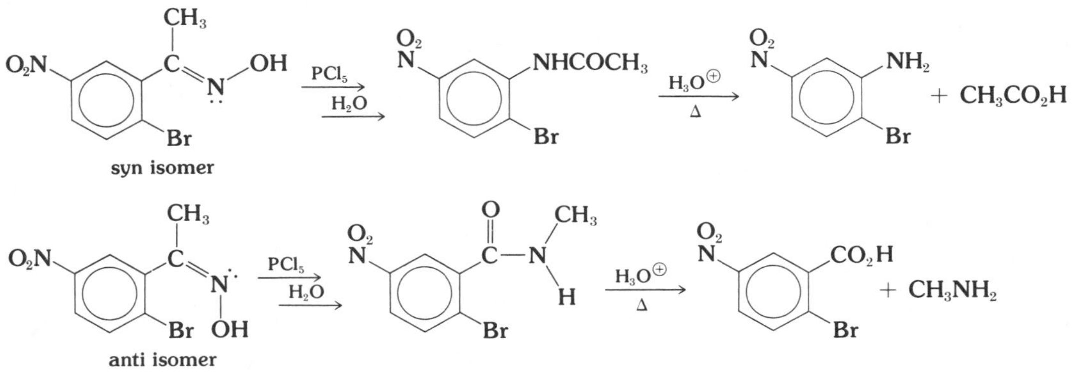

524   15·8 REARRANGEMENTS TO ELECTRON-DEFICIENT NITROGEN ATOMS

FIGURE 15·16 *Beckmann rearrangement of isomeric oximes*

never completely separated from the remainder of the molecule:

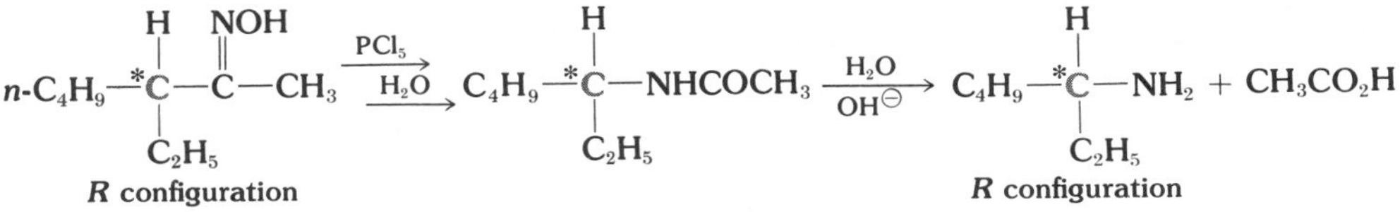

$$n\text{-}C_4H_9\overset{*}{-}\underset{C_2H_5}{\overset{H}{C}}-\underset{\parallel}{\overset{NOH}{C}}-CH_3 \xrightarrow[H_2O]{PCl_5} C_4H_9\overset{*}{-}\underset{C_2H_5}{\overset{H}{C}}-NHCOCH_3 \xrightarrow[OH^\ominus]{H_2O} C_4H_9\overset{*}{-}\underset{C_2H_5}{\overset{H}{C}}-NH_2 + CH_3CO_2H \quad (15.83)$$

R configuration   *R* configuration

Studies with stereoisomeric oximes of known configuration demonstrate that the alkyl or aryl group anti (trans) to the oxime hydroxyl group shifts preferentially in the rearrangement, as shown in Figure 15.16.† Since an initial loss of the oxime hydroxyl group would render the oxime substituents structurally equivalent, the stereospecificity noted above requires that the migration step be concerted with the hydroxyl loss:

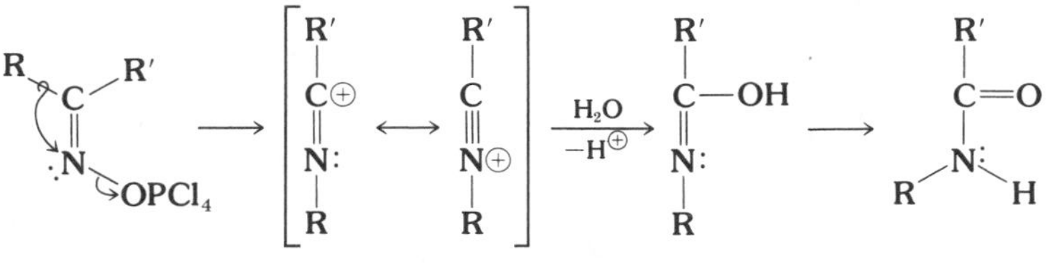

$$(15.84)$$

THE SCHMIDT REACTION

The acid-catalyzed reaction of ketones with hydrazoic acid takes a course similar to the Beckmann rearrangement and is sometimes referred to as the *Schmidt reaction*. A plausible mechanism for this transformation is outlined in Figure 15.17. Note that the last two steps in this mechanism are analogous to those proposed for the Beckmann rearrangement.

†The configurational assignments shown here rest on the selective cyclization of the stereoisomer having syn (or cis) hydroxyl and aryl groups:

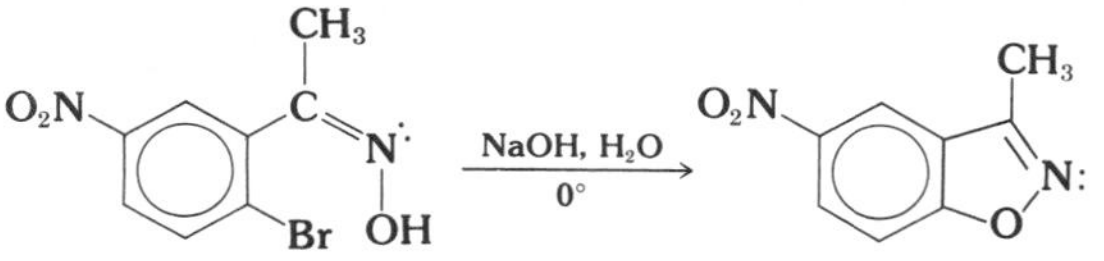

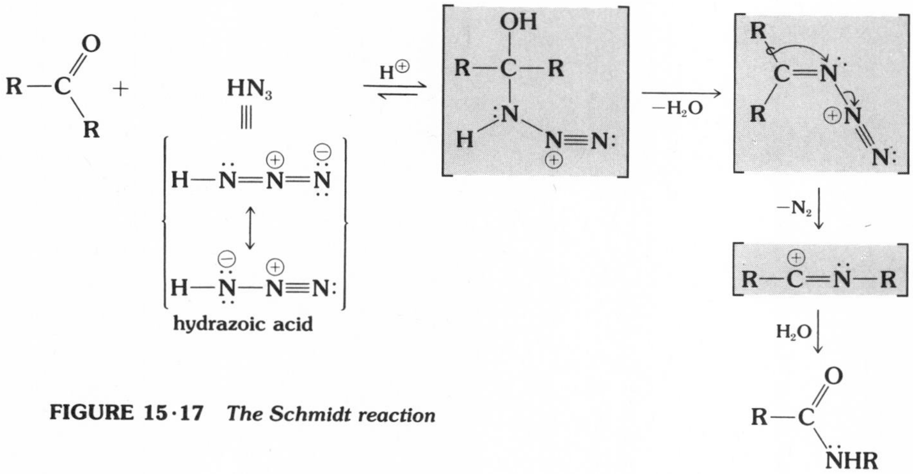

15·8 REARRANGEMENTS TO ELECTRON-DEFICIENT NITROGEN ATOMS

FIGURE 15·17 *The Schmidt reaction*

In the presence of excess hydrazoic acid the Schmidt reaction sometimes leads to substantial amounts of substituted tetrazoles:

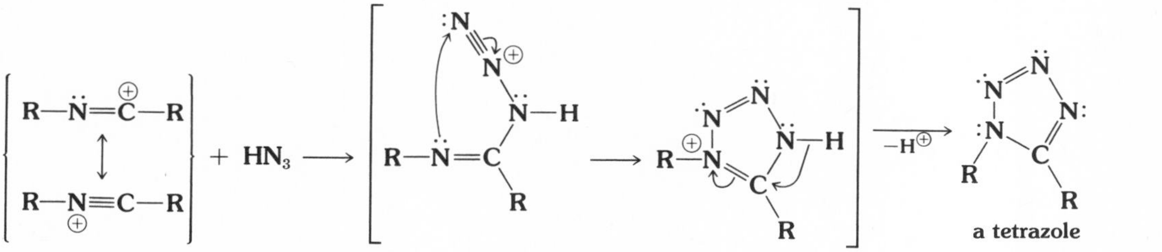

(15.85)

REACTIONS OF AMIDE DERIVATIVES

Like the Schmidt reaction of carboxylic acids with hydrazoic acid, the *Curtius rearrangement* of acyl azides (prepared by the reaction of sodium azide with acyl halides or anhydrides) and the *Hofmann reaction* of amides proceed as shown in Figure 15.18 to transform a carboxylic acid derivative into the corresponding primary amine having one less carbon atom.† Isocyanates have been identified as intermediates in these reactions, and in aqueous solutions these reactive substances are rapidly hydrated to unstable carbamic acids, which spontaneously decarboxylate.

The usefulness of these reactions is enhanced by their stereospcificity. Of course, chiral centers remote from the reaction site are not affected. However, even if the alkyl moiety (designated **R** in Figure 15.18) has a chiral unit at the bonding site, the rearrangements still proceed to the corresponding amines with *retention of configuration.*†

†For example,

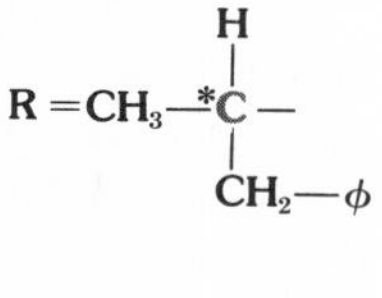

$$R = CH_3 - \overset{H}{\underset{CH_2-\phi}{\overset{*}{C}}} -$$

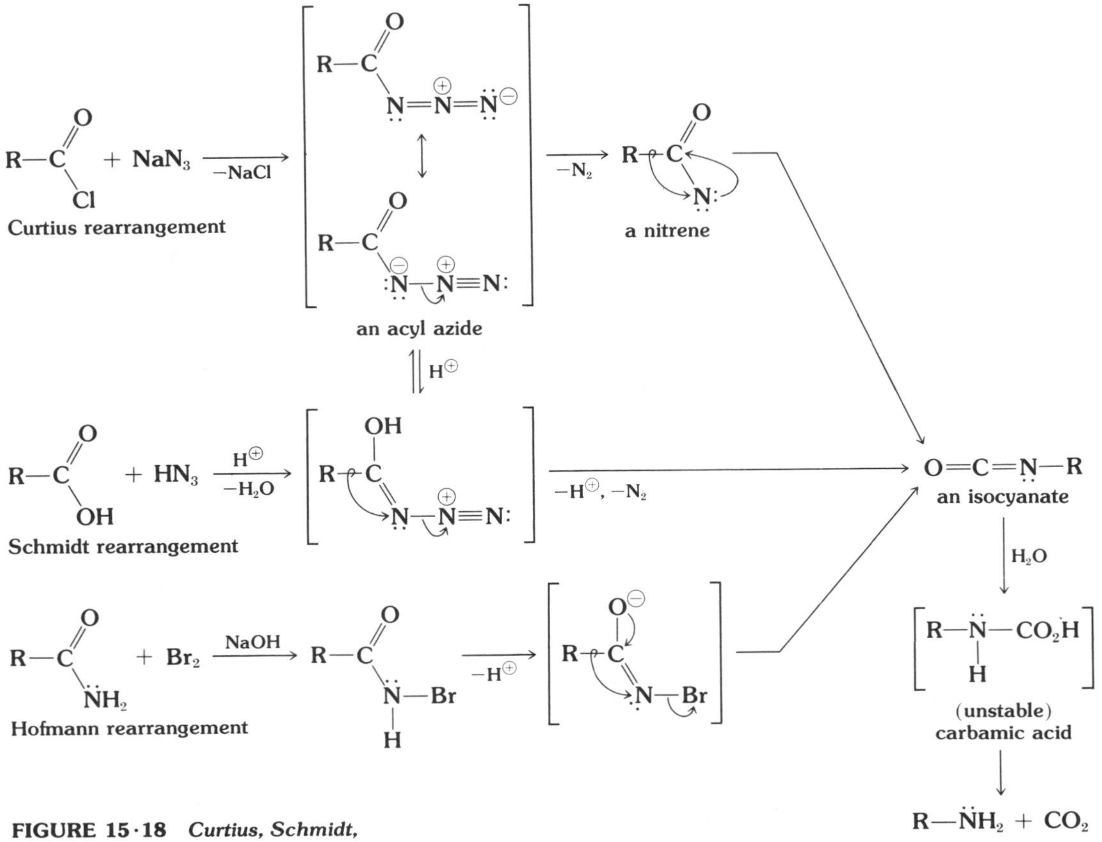

15·9 REARRANGEMENTS TO ELECTRON-DEFICIENT OXYGEN ATOMS

FIGURE 15·18 *Curtius, Schmidt, and Hofmann rearrangements*

15·9 REARRANGEMENTS TO ELECTRON-DEFICIENT OXYGEN ATOMS

The oxidative rearrangement of ketones to esters or lactones on treatment with peracids is known as the *Baeyer-Villiger reaction*. Strong acids catalyze this reaction, which is normally slower than the peracid epoxidation of alkenes (see Section 14.12):

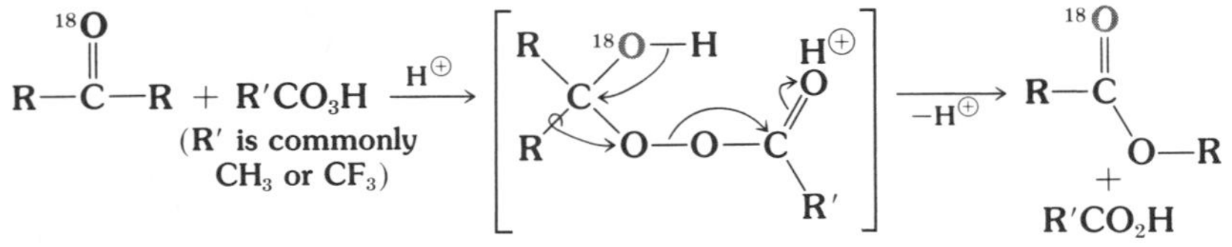

$$R-\overset{^{18}O}{\underset{\|}{C}}-R + R'CO_3H \xrightarrow{H^{\oplus}} \left[\begin{array}{c}R\\C\\R\end{array}\overset{^{18}O-H}{\underset{O-O-C}{\diagdown}}\overset{H^{\oplus}}{\underset{R'}{\diagup O}}\right] \xrightarrow{-H^{\oplus}} R-\overset{^{18}O}{\underset{\|}{C}}-O-R + R'CO_2H \qquad (15.86)$$

(R' is commonly CH$_3$ or CF$_3$)

The acid-catalyzed decomposition of cumene hydroperoxide into acetone and phenol involves a similar rearrangement to an electron-deficient oxygen atom, as shown in Figure 15.19. This example again illustrates the superior migratory aptitude of aryl groups over alkyl groups.

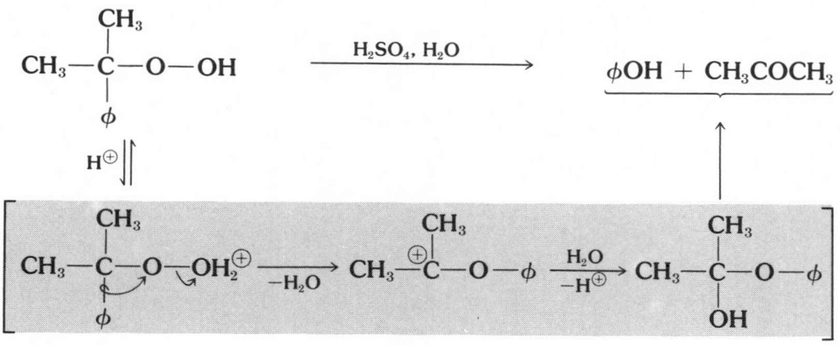

$$CH_3-\underset{\underset{\phi}{|}}{\overset{\overset{CH_3}{|}}{C}}-O-OH \xrightarrow{H_2SO_4,\ H_2O} \phi OH + CH_3COCH_3$$

$$H^{\oplus} \updownarrow$$

$$\left[CH_3-\underset{\underset{\phi}{|}}{\overset{\overset{CH_3}{|}}{C}}-O-OH_2^{\oplus} \xrightarrow{-H_2O} CH_3-\overset{\overset{CH_3}{|}}{\overset{\oplus}{C}}-O-\phi \xrightarrow[-H^{\oplus}]{H_2O} CH_3-\underset{\underset{OH}{|}}{\overset{\overset{CH_3}{|}}{C}}-O-\phi \right]$$

FIGURE 15·19 *Rearrangement of cumene hydroperoxide*

PROBLEM 15·19
a Write the formula of the product expected from the reaction of cyclopentanone with trifluoroperoxyacetic acid.
b Write the formula of the product expected from the Schmidt reaction of cyclohexanone with hydrazoic acid.
c The oxime of a ketone gives an amide on treatment with PCl_5. Hydrolysis of this amide gives cyclohexylamine and phenylacetic acid (ϕCH_2CO_2H) as the only organic products. Write structures for the ketone, the oxime, and the amide.

REFERENCES

P. Sykes, *A Guidebook to Mechanism in Organic Chemistry*, Longmans, Green, London, 1970.

P. Sykes, *The Search for Organic Reaction Pathways*, Longmans, Green, London, 1972.

R. A. Jackson, *Mechanism*, Clarendon Press, Oxford, 1972.

†R Stewart, *The Investigation of Organic Reactions*, Prentice-Hall, Inc., Englewood Cliffs, N.J., 1966.

†R. Breslow, *Organic Reaction Mechanisms*, W. A. Benjamin, Inc., New York, 1969.

†R. Alder, R. Baker, and J. Brown, *Mechanism in Organic Chemistry*, Interscience, John Wiley & Sons, Inc., New York, 1971.

J. Shorter, *Correlation Analysis in Organic Chemistry*, Clarendon Press, Oxford, 1973.

L. P. Hammett, *Physical Organic Chemistry*, McGraw-Hill Book Company, New York, 1970.

†J. Hine, *Physical Organic Chemistry*, McGraw-Hill Book Company, New York, 1962.

E. S. Gould, *Mechanism and Structure in Organic Chemistry*, Holt, Rinehart and Winston, Inc., New York, 1959.

†C. K. Ingold, *Structure and Mechanism in Organic Chemistry*, Cornell University Press, Ithaca, N.Y., 1969.

P. D. Bartlett, *Nonclassical Ions*, W. A. Benjamin, Inc., New York, 1965.

†These references also contain material pertinent to the discussion of nonionic organic reactions in Chapter 16.

EXERCISES

15·1 Give structures for the principle products expected from monobromination (Br$_2$, Fe) of the following compounds:

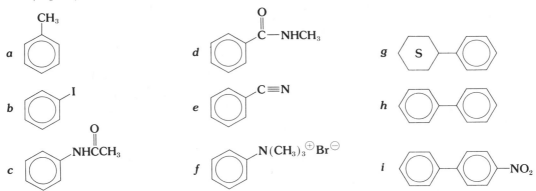

15·2 Write structures for the principal products expected from mononitration (HNO$_3$, H$_2$SO$_4$) of the following:

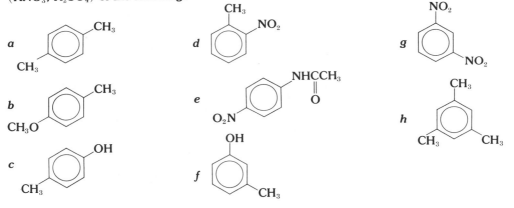

15·3 Arrange the following groups of compounds in order of increasing reactivity toward the reagents indicated:

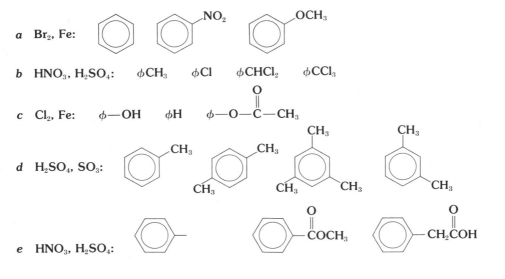

15·4 Show all steps in a reasonable synthesis of each of the following compounds from benzene, toluene, or phenol and any needed reagents. Assume that ortho and para isomers are readily separable.

a 4-bromotoluene (Br para to CH₃)
b 4-nitroanisole (O₂N para to OCH₃)
c 4-formylbenzoic acid (O₂HC—C₆H₄—CO₂H, para)
d 3-nitrobenzoic acid
e 4-nitrobenzoic acid
f 3-bromoacetophenone (COCH₃ and Br meta)
g 3-bromo(dichloromethyl)benzene (Br and CHCl₂ meta)
h 3-bromoaniline (NH₂ and Br meta)
i 3-bromophenol (OH and Br meta)
j 3-chlorobenzenesulfonic acid (SO₃H and Cl meta)
k 4-chlorobenzenesulfonic acid (SO₃H and Cl para)
l 3-dicyanobenzene (1,3-dicyanobenzene)

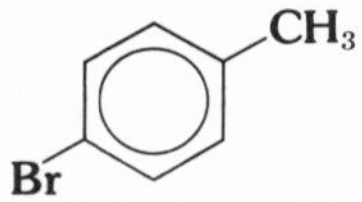

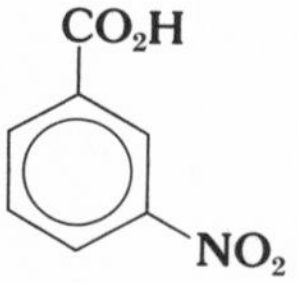

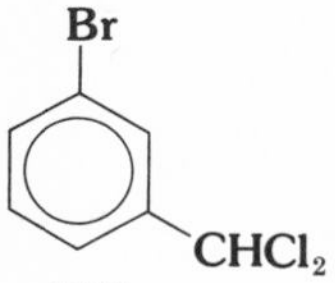

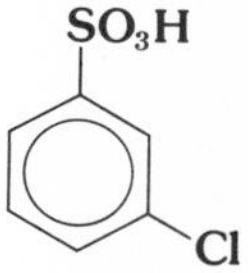

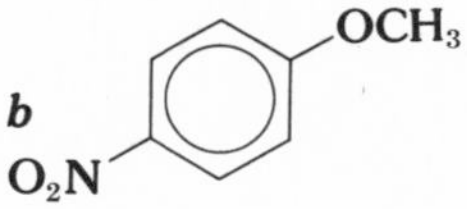

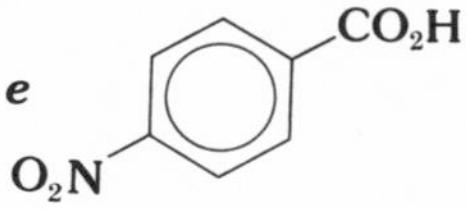

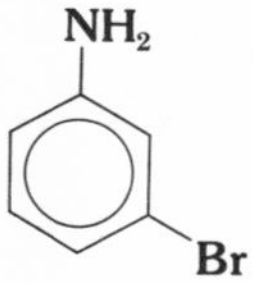

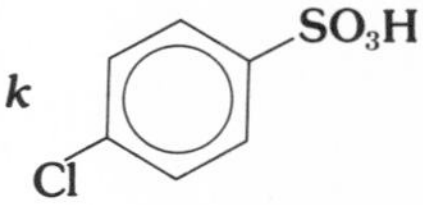

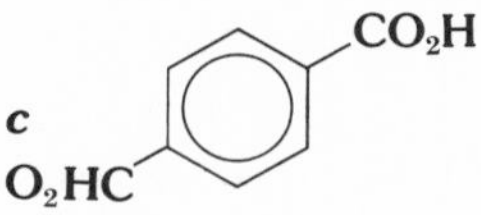

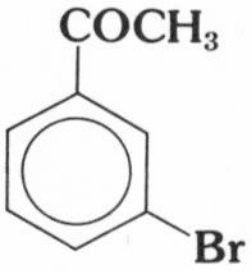

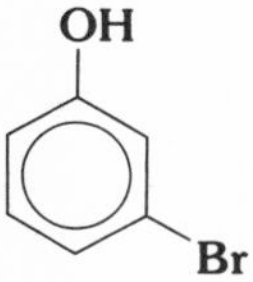

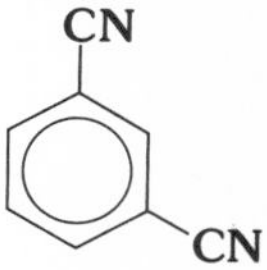

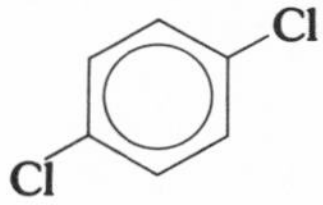

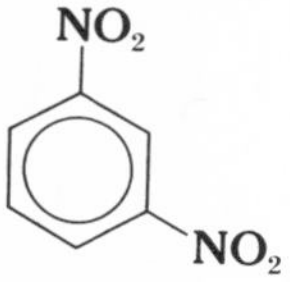

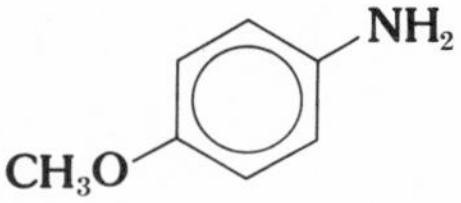

15·5 Calculate the molecular dipole moments of the compounds shown below (see Section 15.2):

a 4-nitroaniline (H₂N—C₆H₄—NO₂, para)
b 4-methoxyaniline (CH₃O—C₆H₄—NH₂, para)
c 1,4-dichlorobenzene
d 1,3,5-trimethylbenzene
e 1,3-dinitrobenzene

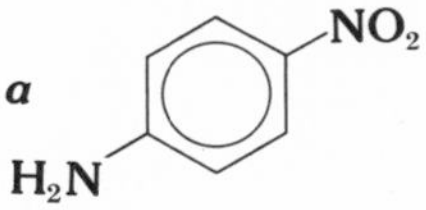

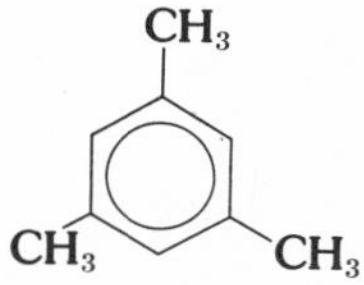

15·6 Electrophilic substitution in naphthalene occurs preferentially at the α position. Account for this directive influence in the bromination reaction:

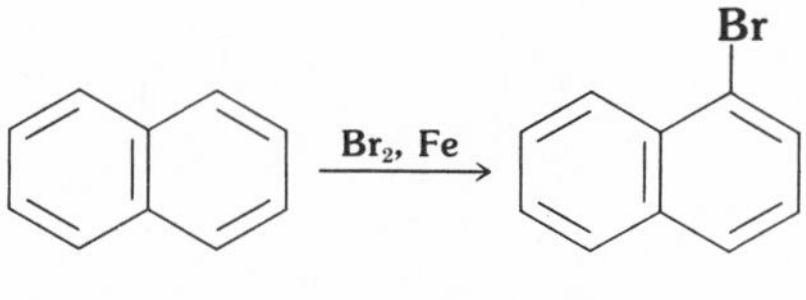

naphthalene + Br₂, Fe → 1-bromonaphthalene

15·7 Give structures for the products expected from the reaction of ethyl vinyl ketone (CH₂=CH—COCH₂CH₃) with each of the following reagents:

a H₂, Pd
b NaBH₄, then H₂O
c CH₃MgBr, then H₂O
d CH₃Li, then H₂O
e (CH₃)₂CuLi, then H₂O
f CH₃C≡C$^{\ominus}$:Na$^{\oplus}$, then H₂O
g CH₃NH₂
h cyclopentadienide anion Na$^{\oplus}$, then H₂O
i HCl
j 1,3-butadiene
k cyclopentadiene
l (CH₃CH₂OC)₂CH$^{\ominus}$Na$^{\oplus}$ (diethyl malonate anion), then H₂O

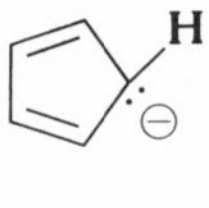

15·8 Indicate the reagent or reagents and the conditions necessary to accomplish the following *one-step* reactions:

a ![cyclohexene] → ![3-chlorocyclohexene]

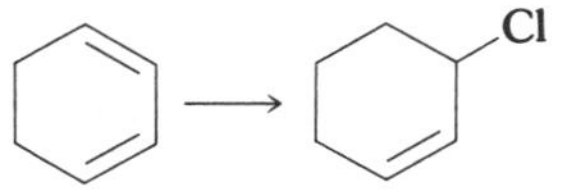

b ![butadiene] → ![dimethyl phthalate with CO$_2$CH$_3$ groups]

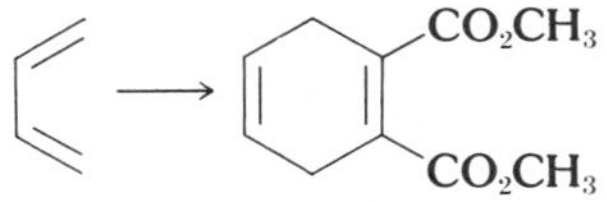

c CH$_3$NH$_2$ ⟶ CH$_3$NHCH$_2$CH$_2$COCH$_3$

d ![maleic anhydride] ⟶ ![norbornene-fused anhydride]

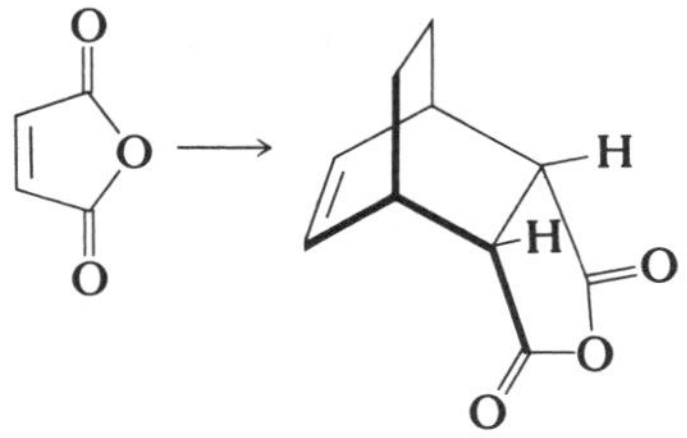

e CH$_3$CH=CHCCH$_3$ ⟶ CH$_3$CH=CH—C(CH$_3$)$_2$
      ‖ |
      O   OH

f CH$_3$CH=CHCCH$_3$ ⟶ (CH$_3$)$_2$CHCH$_2$CCH$_3$
      ‖   ‖
      O   O

g CH$_2$=CH—COCH$_3$ ⟶ ![norbornene with H and CO$_2$CH$_3$]
           ‖
            O

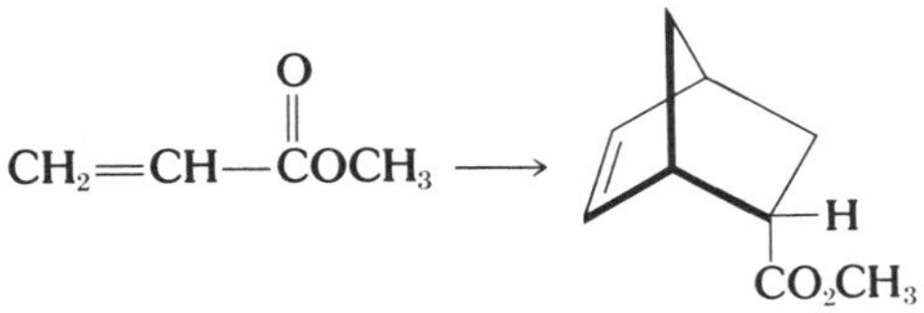

15·9 Show reagents and conditions which could be used to effect the transformations shown. Each may be accomplished in three steps or fewer.

a ![cyclohexane-1,2-diol] → OHC~~~~~CHO

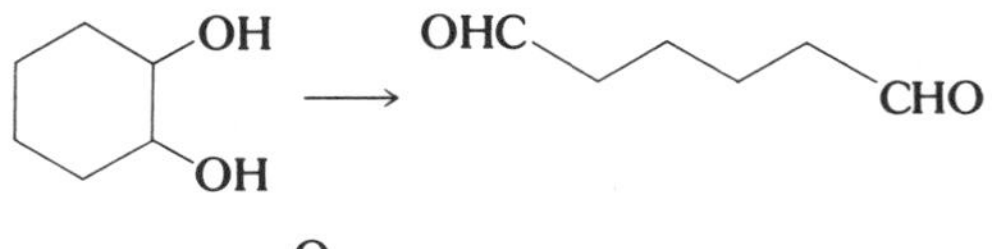

b ![cyclohexanone 2-acetoxy] → ![cyclohexanone]

c ![isopropyl-substituted aldehyde with OH] → ![isopropyl aldehyde]

d ![α,β-unsaturated ester] ⟶ CH$_3$OC(CH$_3$)$_2$CH$_2$COCH$_3$

e ![cyclohexanecarboxylic acid] ⟶ ![cyclohexylamine]

f φ—C—CH$_3$ ⟶ φOH
   ‖
   O

g φ—C—φ ⟶ φ—C—NH—φ
   ‖   ‖
   O   O

h ![cyclohexanone] → ![caprolactam]

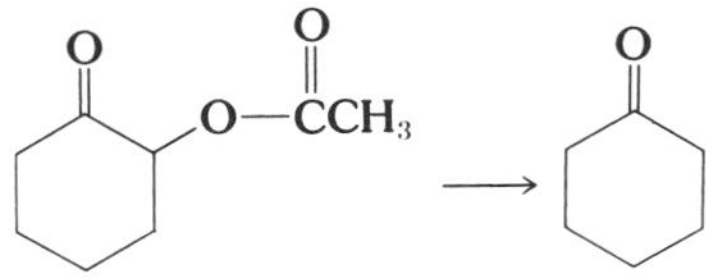

i CH$_2$=CHCO$_2$CH$_2$CH$_3$ ⟶ ![1-(2-hydroxypropan-2-yl)cyclohexene]

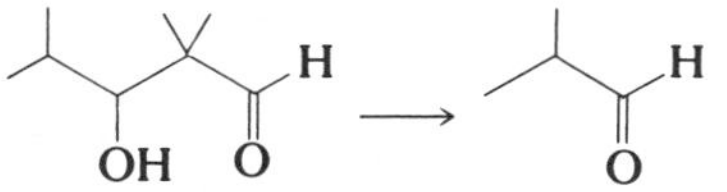

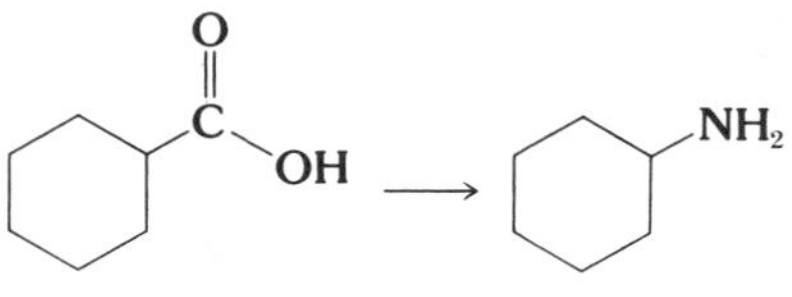

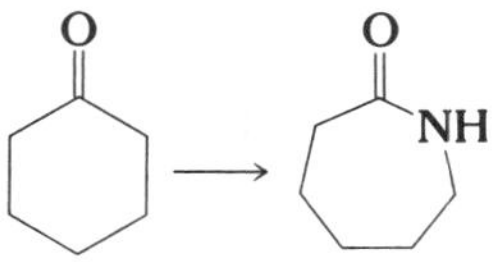

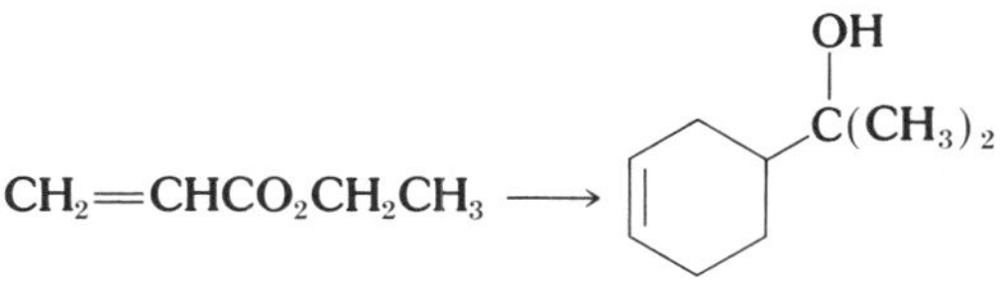

15·10 The isomeric hydroxybenzoic acids have two acidity constants, one for the carboxyl group and one for the phenolic hydroxyl group. Both constants differ for the ortho and meta isomers, as the following pK_a values indicate:

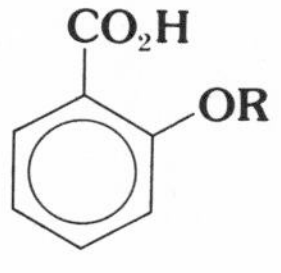

ortho isomer (2-OR-benzoic acid): R = H, $pK_{a1} = 3.0$, $pK_{a2} = 12.4$; R = CH$_3$, $pK_a = 4.1$

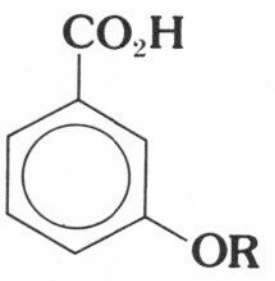

meta isomer (3-OR-benzoic acid): R = H, $pK_{a1} = 4.1$, $pK_{a2} = 9.9$; R = CH$_3$, $pK_a = 4.1$

a Why is the first ionization constant larger for the ortho isomer, whereas the meta isomer has the larger second ionization constant? The pK_a values of the corresponding methoxy isomers are provided for reference.

b Explain the large difference in acidity between 2,4-pentanedione ($pK_a \approx 10$) and 1,3-cyclohexanedione ($pK_a \approx 6$).

15·11 Write mechanisms for the following conversions:

a 1-(1-hydroxycyclopentyl)-1-hydroxy compound $\xrightarrow{H^\oplus}$ 2,2-dimethylcyclohexanone + 1-acetylcyclopentane (acyl migration / pinacol)

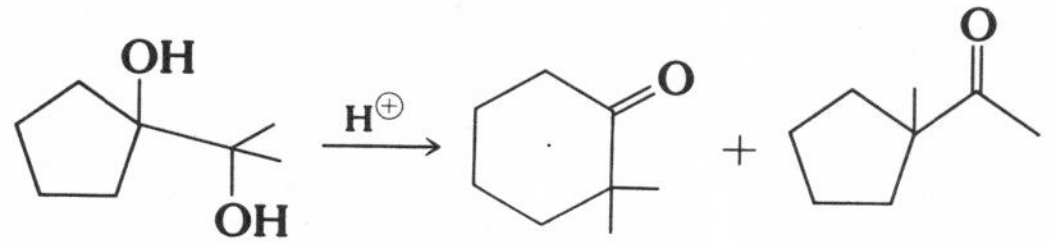

b $CH_3-\underset{\underset{CH_3}{|}}{\overset{\overset{CH_3}{|}}{C}}-CH=CH_2 \xrightarrow{HCl} CH_3-\underset{\underset{CH_3\ Cl}{|\ \ |}}{\overset{\overset{CH_3}{|}}{C}}-CH-CH_3 + CH_3-\underset{\underset{Cl\ CH_3}{|\ \ |}}{\overset{\overset{CH_3}{|}}{C}}-CH-CH_3$

c (decalin derivative with CH$_3$, OH) $\xrightarrow{SOCl_2}$ (decalin derivative with CH$_3$, Cl)

d HO_2C-(cyclopentane with gem-dimethyl)$-CNH_2 \xrightarrow[\Delta]{NaOBr, H_2O}$ bicyclic lactam (N–H)

e $\phi-\underset{\underset{OH}{|}}{CH}-\underset{\underset{OH}{|}}{CHCH_3} \xrightarrow{H^\oplus} \phi-CH_2\overset{\overset{O}{\|}}{C}-CH_3$

f CH_3O-(C$_6$H$_4$)$-\underset{\underset{\phi}{|}}{\overset{\overset{OH}{|}}{C}}-CH_2NH_2 \xrightarrow{HONO}$ (C$_6$H$_5$)$-\overset{\overset{O}{\|}}{C}-CH_2-$(C$_6H_4$)$-OCH_3$

g (decalin with CH$_3$, OSO$_2$CH$_3$, OH) $\xrightarrow{(CH_3)_3COK}$ (cyclodecenone with CH$_3$)

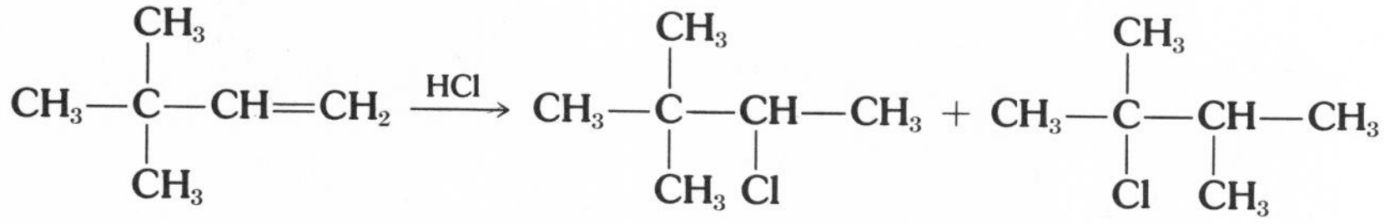

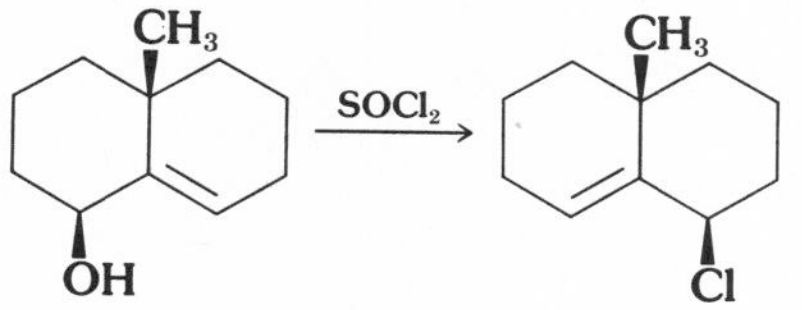

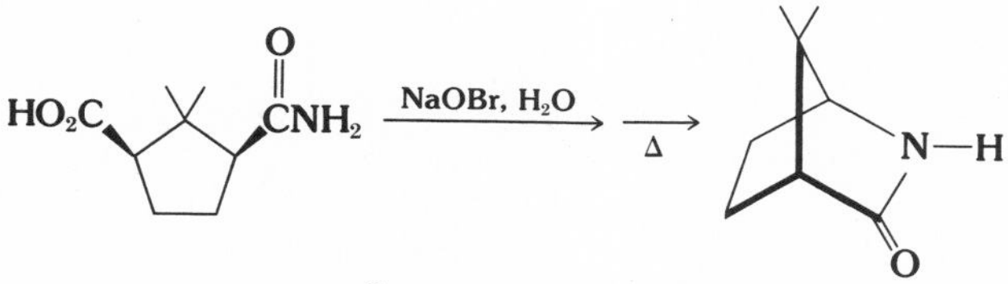

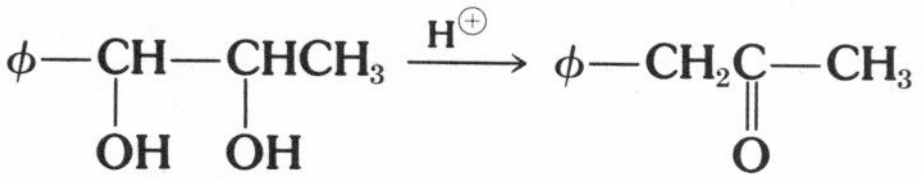

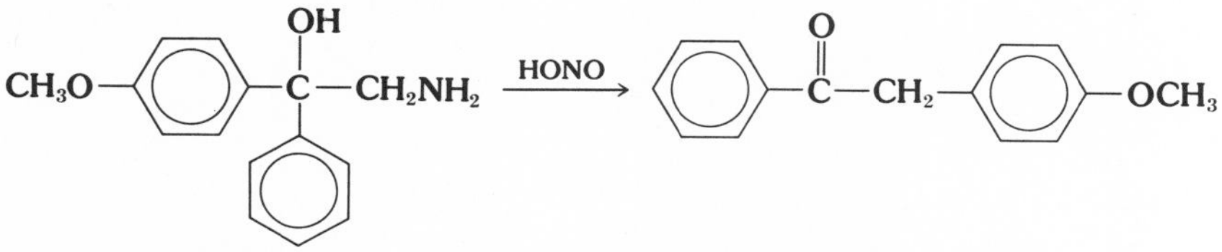

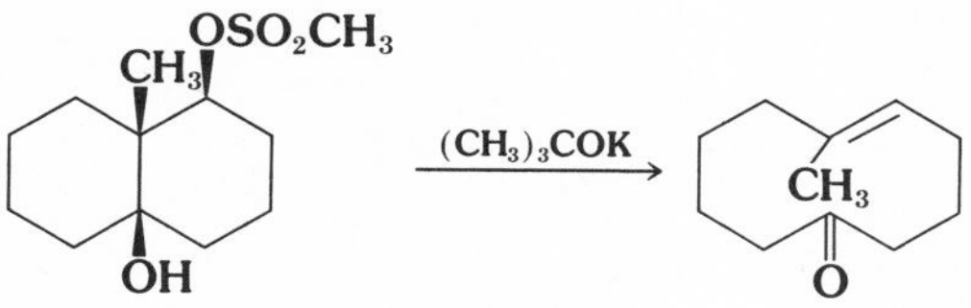

15·12 Write mechanisms for the reactions shown below:

a. [2,2-dimethylcyclohexanol] + HBr ⟶ [1-bromo-2-methyl-... cyclohexane with methyl migration]

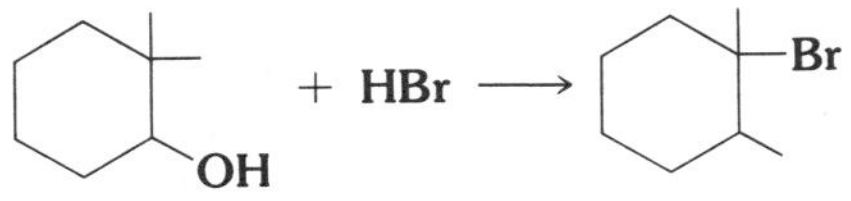

b. [N-methyl piperidine derivative with axial Cl] $\xrightarrow{OH^\ominus,\ H_2O}$ [N-methyl piperidine derivative with equatorial OH]

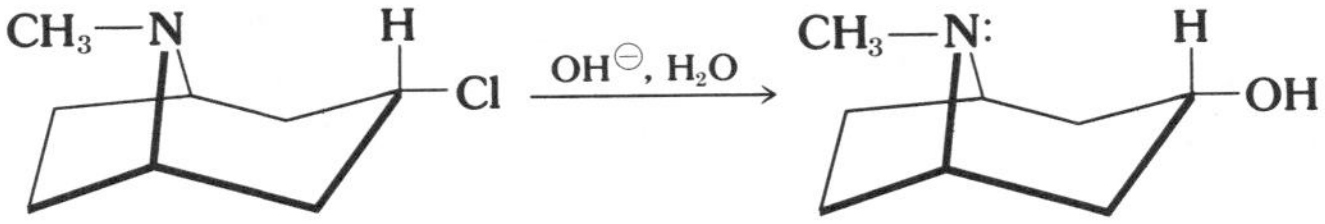

c. φ—COCO—φ $\xrightarrow{NaOH}$ φ—C(OH)(φ)—CO$_2$H

   benzylic acid rearrangement

d. [cyclohexanone with exocyclic isopropylidene and axial H] $\xrightarrow{NaOH,\ H_2O}$ [cyclohexanone with axial H] + CH$_3$COCH$_3$

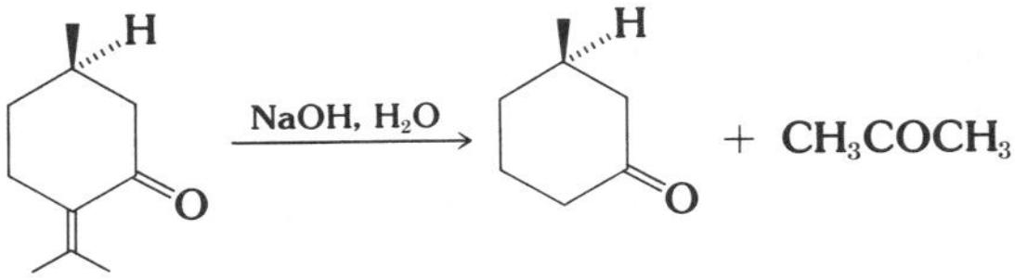

e. [2-bromocyclohexanone] $\xrightarrow{OH^\ominus}$ [cyclopentanecarboxylic acid, CO$_2$H]

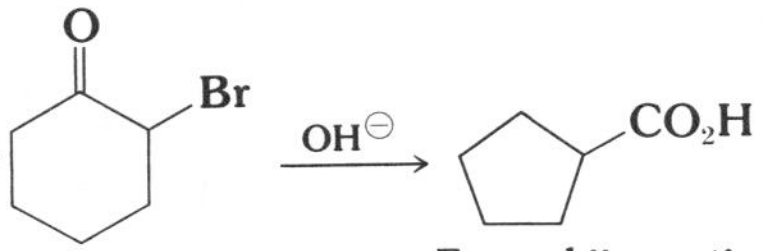

   Favorskii reaction

f. 2 CH$_2$=CHCH=O $\xrightarrow{\Delta}$ [dihydropyran-2-carboxaldehyde, CHO]

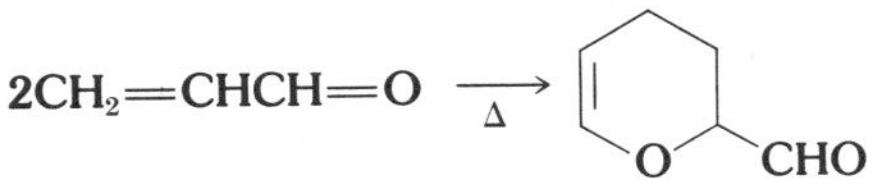

g. [decalone with CH$_3$, H, and CH(CH$_3$)CO$_2$CH$_2$CH$_3$ substituents] $\xrightarrow{CH_3CH_2ONa}$ $\xrightarrow{H_3O^\oplus}$ [tricyclic diketone with CO$_2$H]

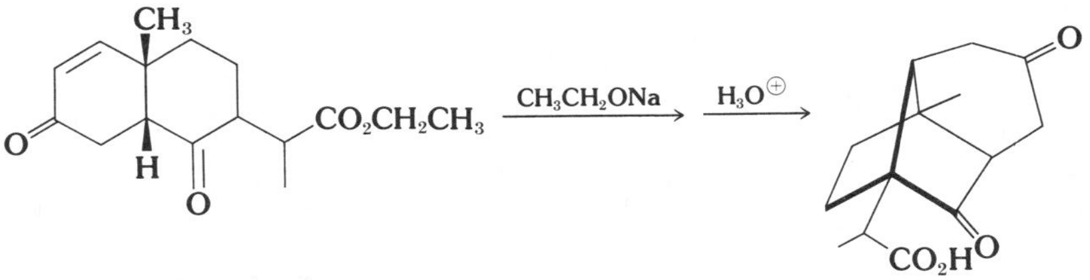

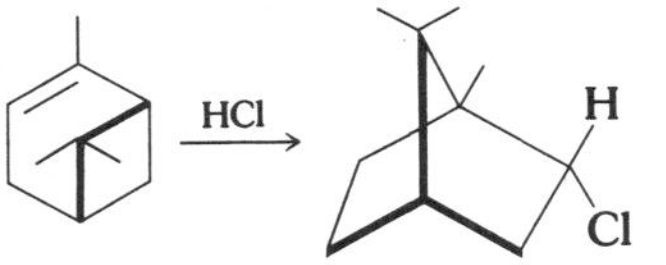

h. [α-pinene] $\xrightarrow{HCl}$ [bornyl chloride, H, Cl]

i. [3-methyl-2-cyclohexenone] $\xrightarrow{NaNH_2}$ [bicyclic diketone with CH$_3$]

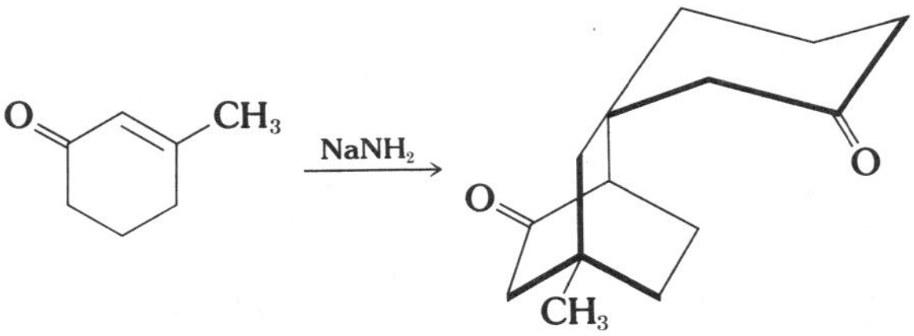

j. HO$_2$CCH$_2$—CO—CO—CH$_2$CO$_2$H $\xrightarrow[\Delta]{NaOH,\ H_2O}$ $\xrightarrow{H^\oplus}$ HO$_2$CCH$_2$C(OH)(CO$_2$H)CH$_2$CO$_2$H

15·13 Treatment of an ice-cold solution of sodium 2,2,4-trimethyl-3-pentenoate [(CH$_3$)$_2$C=CHC(CH$_3$)$_2$CO$_2$Na] with iodine gives a C$_8$H$_{13}$O$_2$I product having a strong infrared absorption at 1863 cm^{-1}. When a solution of this substance is warmed, it rearranges to an isomer with a strong absorption band at 1770 cm^{-1}. Discuss these transformations.

15·14 The compounds shown below were all prepared by a Diels-Alder reaction, followed by hydrogenation of double bonds. Suggest the best combination of diene and dieneophile which would lead to each product. Justify your choice of starting materials.

a, b, c, d, e, f

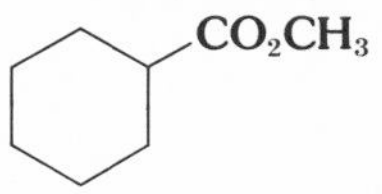

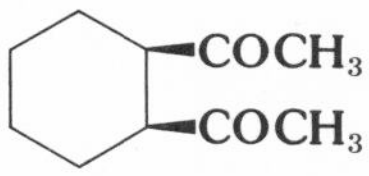

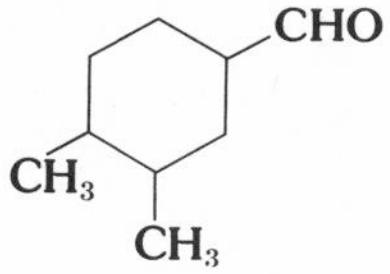

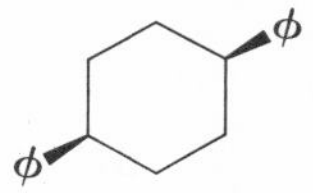

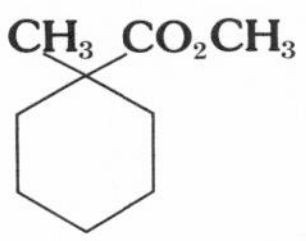

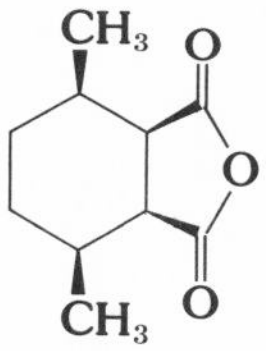

15·15 In the following Diels-Alder reactions the product isomer shown predominates. Consider alternative constitutional and stereoisomers that might have been formed and suggest reasons for the specificity of the reactions.

a, b, 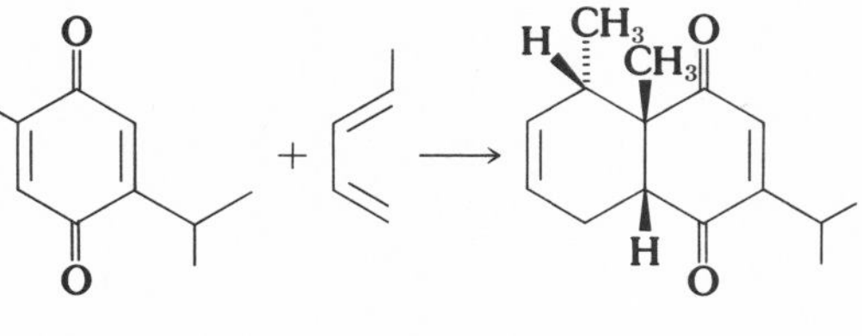 c, 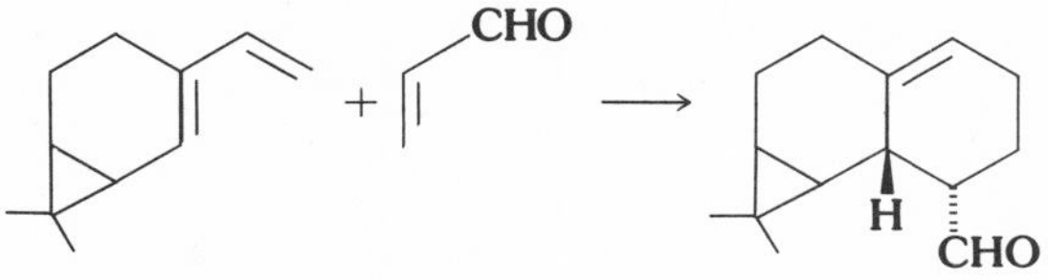 d

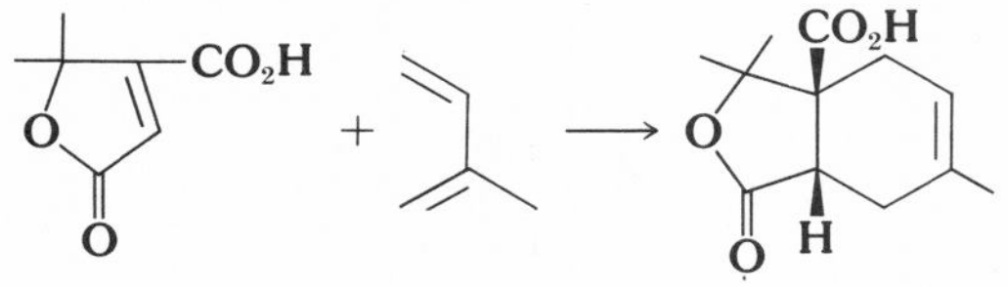

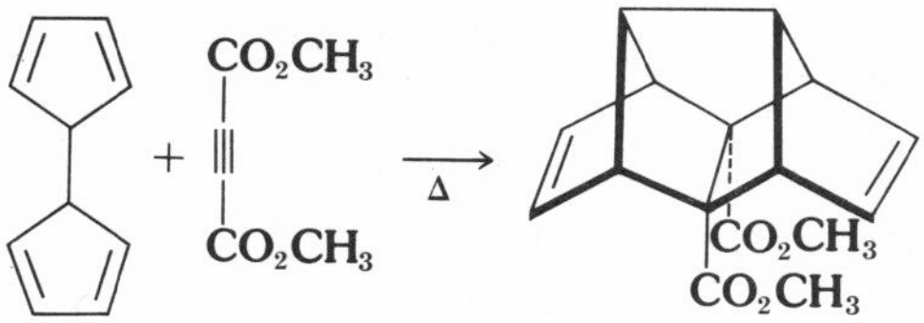

NONIONIC ORGANIC REACTIONS

WITH A FEW exceptions, the many organic chemical reactions we have studied in the preceding chapters have shared certain fundamental characteristics. Nucleophilic and electrophilic regions in the reactants normally bond to each other. Ionic intermediates such as carbonium ions, carbanions, conjugate acids, and conjugate bases are often involved, and in such cases the overall transformation requires two or more discrete steps. Changes in the solvents used often dramatically affect the rates, and sometimes even the course, of the reactions. Reactions of this kind are commonly referred to as *ionic organic reactions*.

In this chapter we shall consider two other classes of organic reactions. The first of these incorporates those transformations proceeding by way of radical intermediates, such as the alkane halogenation described in Section 2.3. These *free-radical reactions* have received increasing attention since the 1930s, thanks chiefly to their tremendous importance in industrial chemistry. The second class of reactions we shall examine consists of a large and varied group of *concerted* (single-step) thermal and photochemical transformations.

16·1 THE DISCOVERY OF STABLE ORGANIC RADICALS

A *radical* is an atomic or molecular species having an unpaired, or odd, electron. Some radicals, such as nitrogen dioxide (NO_2) and nitric oxide (NO), are relatively stable, but most are so reactive that their isolation and long-term study are not possible under normal laboratory conditions. Molecular oxygen is a rare example of a stable biradical.

Early organic chemists used the term *radical* for nomenclature purposes (much as we now speak of *groups*), and many doubted that such species could actually exist. This skepticism was repudiated in 1900 by Moses Gomberg, a young instructor at the University of Michigan. Gomberg had set about the preparation of hexaphenylethane by

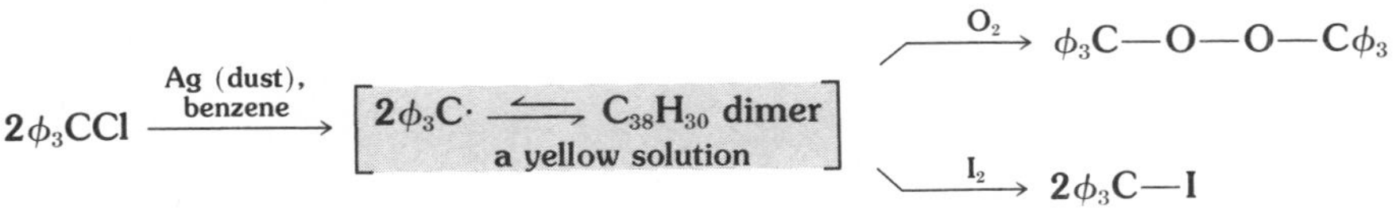

16·1 THE DISCOVERY OF STABLE ORGANIC RADICALS

$$2\phi_3CCl \xrightarrow{\text{Ag (dust), benzene}} \left[2\phi_3C \cdot \rightleftarrows C_{38}H_{30} \text{ dimer} \atop \text{a yellow solution} \right] \begin{array}{c} \xrightarrow{O_2} \phi_3C-O-O-C\phi_3 \\ \xrightarrow{I_2} 2\phi_3C-I \end{array}$$

FIGURE 16·1 *The Gomberg experiment*

attempting to couple two triphenylmethyl groups through the action of finely divided zinc or silver on triphenylchloromethane:

$$2(C_6H_5)_3CCl + Zn \xrightarrow{?} (C_6H_5)_3C-C(C_6H_5)_3 + ZnCl_2 \quad (16.1)$$
$$\text{(or Ag)} \qquad\qquad\qquad\qquad\qquad\qquad \text{(or AgCl)}$$

This reaction gave, after filtration of the metal chloride salt and evaporation of the benzene solvent, a white crystalline solid which Gomberg at first thought was hexaphenylethane. Analysis of the new compound showed it to have the composition $C_{38}H_{30}O_2$, and it was eventually identified as *bis*-triphenylmethyl peroxide (Figure 16.1)! Where did the oxygen come from?

Gomberg answered this important question by carefully excluding oxygen from the reaction system. From the yellow solution thus obtained, he was able to isolate a colorless $C_{38}H_{30}$ compound (all this work was carried out under a blanket of CO_2 gas). The yellow solution that resulted when this new dimeric hydrocarbon was dissolved in benzene reacted rapidly with the oxygen in air or with iodine, as shown in Figure 16.1. These remarkable facts forced Gomberg to conclude that the colored solutions contained triphenylmethyl radicals formed by dissociation of the dimeric hydrocarbon (K_{eq} for dissociation = 2×10^{-4} at 25°C). He wrote: "On this assumption alone do the results described above become intelligible and receive an adequate explanation."

Gomberg assumed the crystalline dimer to be hexaphenylethane, as have most chemists over the ensuing years. However, recent work in the Netherlands has shown that the dimer actually results from coupling at the para position of one of the benzene rings:

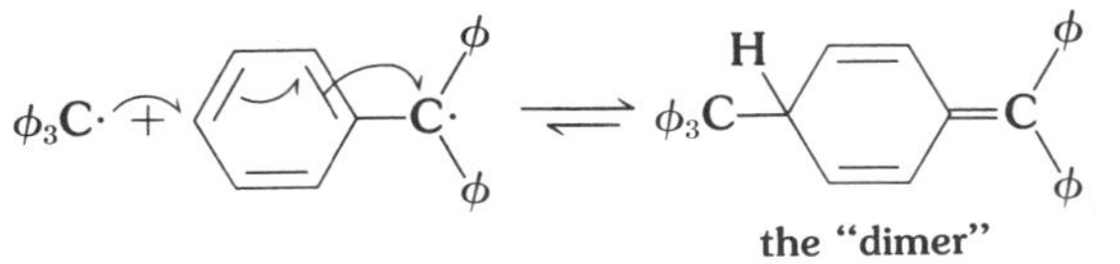

$$\phi_3C \cdot + \underset{\phi}{\overset{\phi}{\underset{|}{\overset{|}{C}}}}\!\!\!\!\bigcirc\!\!\!\!\underset{}{} \rightleftarrows \phi_3C-\underset{}{\overset{H}{\bigcirc}}=\underset{\phi}{\overset{\phi}{\underset{|}{\overset{|}{C}}}} \quad (16.2)$$
the "dimer"

If the para positions of all the phenyl groups are blocked by large substituents, such as a *t*-butyl group, this dimerization process is blocked, and higher concentrations of radicals are observed.

PROBLEM 16·1
a Write equations showing the reaction of triphenylmethyl radicals with oxygen (O_2) and iodine (I_2).
b What measurements would you recommend be made on the dimeric hydrocarbon to distinguish the para-coupled structure in equation 16.2 from hexaphenylethane?

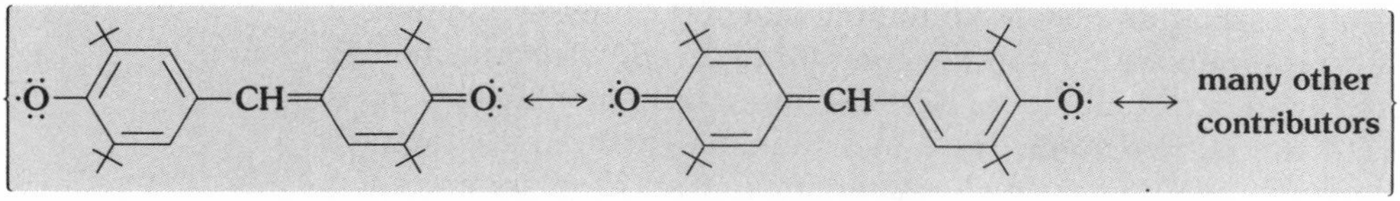

16·1 THE DISCOVERY OF STABLE ORGANIC RADICALS

[galvinoxyl structure] ⟷ [resonance structure] ⟷ many other contributors

galvinoxyl (deep blue solution)

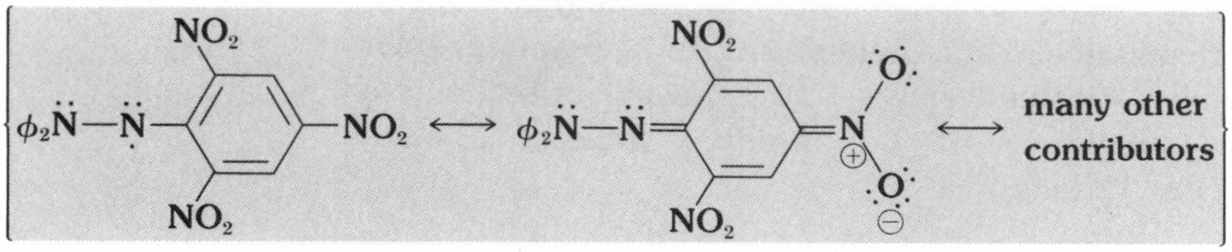

[diphenyl-β-picrylhydrazyl structure] ⟷ [resonance structure] ⟷ many other contributors

diphenyl-β-picrylhydrazyl (deep violet solution)

FIGURE 16·2 *Two stable radicals*

Although organic chemists originally referred to relatively stable radicals, such as triphenylmethyl, galvinoxyl, and diphenyl-β-picrylhydrazyl, as *free radicals*, this term is now rather loosely applied to all radical intermediates. The large and striking variation in the stabilities of free radicals can be explained in many cases by resonance delocalization of the unpaired electron, as illustrated in Figure 16.2.

DETECTION AND OBSERVATION OF FREE RADICALS

Only a few radicals, such as triphenylmethyl, are sufficiently stable to be generated and handled in concentrations suitable for examination by traditional methods. Fortunately, the same unpaired (odd) electron that causes most radical intermediates to be highly reactive and unstable, can often be induced to leave its characteristic "calling card." As we saw in Section 1.2, a spinning charged particle will occupy one of two quantum states (assuming a spin of $\pm 1/2$) when exposed to an external magnetic field. The energy difference ΔE between these states will be proportional to the magnitude of the magnetic field H and dependent on the magnetic characteristics of the particle in question. This is illustrated for an electron in Figure 16.3.

FIGURE 16·3 *Energy states of a spinning electron in a magnetic field*

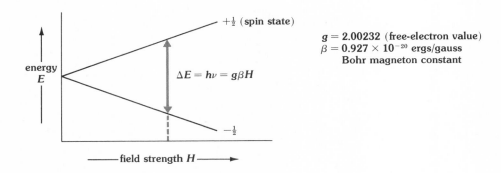

$g = 2.00232$ (free-electron value)
$\beta = 0.927 \times 10^{-20}$ ergs/gauss
Bohr magneton constant

16·1 THE DISCOVERY OF STABLE ORGANIC RADICALS

Here we find a situation closely paralleling that of the nuclear-magnetic-resonance phenomenon. If a sample containing an odd-electron species is subjected to an external magnetic field, we should be able to induce excitation of the electron spin state by irradiation with an appropriate electromagnetic radiation. The characteristics of an electron are such that for a magnetic field strength of 3000 gauss, excitation (or resonance) occurs at a frequency of 9×10^9 Hz—about 500 times higher than the frequency for proton resonance at this field strength. In practice, a concentration of odd-electron species less than 10^{-8} molar can be detected by this *electron-spin-resonance (esr)* technique, sometimes called *paramagnetic resonance.*†

PROBLEM 16·2
Why do paired electrons not give esr signals?

Just as not all protons resonate at the same frequency (the chemical shift in pmr spectroscopy), not all odd-electron species give identical spin-resonance signals. The *g factor*, which is a measure of the magnetic moment of the molecule bearing the odd electron, differs from the free-electron value given in Figure 16.3 because of spin-orbit interactions. The *g* factors of organic free radicals are normally found in the range 2.002 to 2.016 and are characteristic of the individual molecular structures of these species. Since many radicals have very short lifetimes in solution, they are often best studied in the solid glassy matrix that results when their solutions in certain mixed-solvent systems, such as ether-pentane, are cooled to $-196°C$ (the boiling point of liquid nitrogen) or lower. The reduced mobility imposed by such a rigid matrix extends the lifetime of incorporated radicals and permits a more leisurely examination of their characteristics.

†Esr spectrometers are available commercially at a cost of $20,000 to $100,000.

FIGURE 16·4 *Esr spectrum of triphenylmethyl radical*

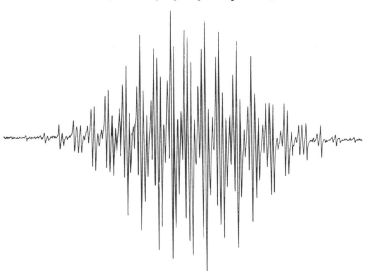

Esr spectra often appear to be composed of many lines, rather than a single signal. This *hyperfine splitting* is due to interactions of the odd electron with neighboring spinning nuclei, in a manner analogous to the spin-spin splitting phenomenon in nmr spectroscopy. The methyl radical, for example, exhibits a hyperfine structure consisting of four equally spaced lines, with relative intensities of 1:3:3:1. The triphenylmethyl radical gives a much more complex esr spectrum (see Figure 16.4) because it contains so many protons. At least 21 groups, containing four lines each, can be discerned.

16·2 METHODS OF GENERATING FREE RADICALS

COVALENT-BOND HOMOLYSIS

Covalent bonds can be ruptured by homolytic cleavage (see Table 1.5) if sufficient vibrational energy is introduced. This usually occurs either through collision with "hot" (high-kinetic-energy) molecules, which converts kinetic energy to vibrational energy, or through the absorption of light, which results in inter- or intramolecular conversion of the added electronic energy into vibrational energy.

CRACKING At temperatures higher than 500°C, and in the absence of oxygen, alkanes break into smaller alkane and alkene fragments. This *thermal-cracking process* has achieved immense importance in the petroleum industry as a means of converting higher-boiling hydrocarbon fractions into lower-boiling gasoline fractions. Free-radical intermediates, formed by homolytic cleavage of covalent bonds, are known to be involved in these reactions. However, the large number of potentially reactive bonds at these high temperatures and the complexity of the resulting product mixtures make a systematic study of thermal cracking very difficult.

PEROXIDES Organic peroxides are an exceptionally good source of free radicals because the weak O—O bond (with a dissociation energy of about 35 kcal/mole at 25°C) undergoes a unimolecular homolysis at relatively low temperatures (50° to 150°C):

$$R-\ddot{O}-\ddot{O}-R \xrightarrow{\Delta} 2R-\ddot{O}\cdot \qquad (16.3)$$

The organic substituents on the peroxide moiety may be alkyl groups (dialkyl peroxides), acyl groups (diacyl peroxides), or a combination of an alkyl and an acyl group (a perester). It is possible, therefore, to study the reactions of a wide variety of alkoxyl (or acyloxyl) radicals, generated under conditions mild enough to preclude any competing homolytic bond cleavages.

AZO COMPOUNDS The thermal decomposition of azo compounds into molecular nitrogen and free radicals requires temperature ranging from 50° to 400°C, depending on the structure of the organic substituents:

$$R-\ddot{N}=\ddot{N}-R \xrightarrow{\Delta} 2R\cdot + :N\equiv N: \qquad (16.4)$$

16·2 METHODS OF GENERATING FREE RADICALS

TABLE 16·1 *Decomposition energies for organic peroxides*

$$R-\ddot{O}-\ddot{O}-R \longrightarrow 2R-\ddot{O}\cdot$$

R substituent	activation energy (kcal/mole)
$(CH_3)_3C$	+37
$\phi\overset{O}{\underset{\|}{C}}$	+30
$CH_3\overset{O}{\underset{\|}{C}}$	+30

The activation energies listed in Tables 16.1 and 16.2 raise two questions:

1 Why does the substituent R have a more dramatic influence on the decomposition of azo compounds than on the decomposition of peroxides?

2 Since the C—N covalent bond has a bond-dissociation energy of 72.8 kcal/mole, how can the activation energy for azo-compound decomposition be less than this?

The *concerted mechanism* described by the following transition state provides an attractive answer to both questions:

$$\left[\overset{\delta\cdot}{R} \text{------} N \!=\!\!=\!\! N \text{------} \overset{\delta\cdot}{R} \right]$$

A critical feature of this mechanism is that the simultaneous rupture of both carbon-nitrogen bonds is partly compensated by concurrent formation of the very stable nitrogen molecule. The thermodynamic advantage here lies chiefly in a bond-energy gain of about 126 kcal/mole for the conversion of N=N (ΔH = 100 kcal/mole) to N≡N (ΔH = 225.8 kcal/mole). Thus the overall energy change during reaction 16.4 is

$$\Delta H = 2(72.8) + 100 - 225.8 = +19.8 \text{ kcal/mole}$$

Since the concerted transition state also requires each organic substituent

TABLE 16·2 *Decomposition energies for organic azo compounds*

$$R-\ddot{N}=\ddot{N}-R \longrightarrow 2R\cdot + :N\equiv N:$$

R substituent	activation energy (kcal/mole)
CH_3	+51
$(CH_3)_3C$	+43
$(CH_3)_2\overset{CN}{\underset{\|}{C}}$	+31
ϕ_3C	+27

to accept a partial radical character, any structural features in the substituent which act to stabilize the corresponding free radical will *lower* the activation energy of this process, thereby increasing the rate of decomposition.

PHOTOLYTIC BOND HOMOLYSIS   Compounds having absorption bands in the visible or near-ultraviolet region of the spectrum may absorb sufficient energy to break covalent bonds. (This is a necessary, but not a sufficient, condition for photochemical homolysis.) Some important examples of these radical sources are halogens (X = Cl, Br, or I), hypohalites, nitrite esters, and ketones:

$$X_2 \xrightarrow{\text{sunlight}} 2X\cdot \quad \text{halogens} \tag{16.5}$$

$$(CH_3)_3C-\ddot{O}-\ddot{C}l: \xrightarrow{\text{sunlight}} (CH_3)_3C-\ddot{O}\cdot + :\ddot{C}l\cdot \quad \text{hypohalites} \tag{16.6}$$

$$R-\ddot{O}-\dot{N}=\ddot{O}: \xrightarrow{300-380 \text{ nm}} R-\ddot{O}\cdot + \cdot\dot{N}=\ddot{O}: \quad \text{nitrite esters} \tag{16.7}$$

$$CH_3-\overset{\overset{\displaystyle :O:}{\|}}{C}-CH_3 \xrightarrow{300 \text{ nm}} CH_3-\overset{\overset{\displaystyle :O:}{\|}}{C}\cdot + H_3C\cdot \tag{16.8}$$

ELECTRON TRANSFER

The action of inorganic oxidizing and reducing agents on organic compounds often involves electron transfers that give rise to radical or radical-ion intermediates. Ferrous ion, for example, catalyzes the decomposition of hydrogen peroxide and organic hydroperoxides:

$$R-\ddot{O}-\ddot{O}-H + Fe^{2\oplus} \longrightarrow R-\ddot{O}\cdot + :\overset{\ominus}{\ddot{O}}H + Fe^{3\oplus} \tag{16.9}$$

Some of the intermediates thus formed are sufficiently stable to permit examination and can sometimes be isolated in the absence of oxygen:

$$\text{X-C}_6\text{H}_2\text{X}_2\text{-}\ddot{O}\text{-H} + Fe(CN)_6^{3\ominus} \longrightarrow \text{X-C}_6\text{H}_2\text{X}_2\text{-}\ddot{O}\cdot + Fe(CN)_6^{4\ominus} + H^{\oplus} \tag{16.10}$$

ferricyanide → ferrocyanide

a phenoxy radical

If the substrates or solvents cannot function as Brønsted acids, radical anions or cations may result:

$$\phi_2C=\ddot{O}: + Na \xrightarrow{\text{THF}} \left\{ \underset{\phi}{\overset{\phi}{\diagdown}}\dot{C}-\ddot{O}:^{\ominus} \longleftrightarrow \underset{\phi}{\overset{\phi}{\diagdown}}\overset{\ominus}{C}-\ddot{O}\cdot \right\} Na^{\oplus} \tag{16.11}$$

a ketyl (deep blue)

16·2 METHODS OF GENERATING FREE RADICALS

$$2 \; (\text{(CH}_3)_2\text{N-C}_6\text{H}_4\text{-N(CH}_3)_2) + \text{Br}_2 \xrightarrow{\text{pH} \approx 4} 2 \; [\text{Würster salt cation}] \; \text{Br}^{\ominus} \quad (16.12)$$

a Würster salt

PROBLEM 16·3
a Classify reactions 16.10 to 16.12 as oxidation or reduction reactions of the organic reactant. What is the reducing or oxidizing agent in each case?
b Draw two other resonance contributors to the Würster salt in equation 16.12.

HYDROGEN ATOM ABSTRACTION

An indirect but commonly used method of generating organic free radicals involves attack by an initiating radical at one of the hydrogen atoms in the reactants. Peroxides make very good initiators of this kind, because the hydrogen-abstraction reaction is moderately exothermic and, as noted above, the required alkoxy radicals can be formed under relatively mild conditions:

$$\text{R-H} + \cdot\ddot{\text{O}}\text{-C(CH}_3)_3 \longrightarrow \text{R}\cdot + \text{H-O-C(CH}_3)_3 \quad \Delta H \approx -12 \text{ kcal/mole} \quad (16.13)$$

The rate of hydrogen abstraction by radical initiators depends partly on the bond-dissociation energy of the C—H bond being attacked (see Table 2.5) and partly on the nature of the attacking radical. The differing selectivities of radical intermediates toward primary, secondary, and tertiary-benzylic hydrogen atoms is shown in Table 16.3.

It is also possible to use the hydrogen-abstraction reaction in an opposite role. Thus exceptionally good hydrogen donors, such as thiols (RSH) and alkyl stannanes (R_3SnH or R_2SnH$_2$), can be used to intercept, or "scavenge," free radicals before they have a chance to undergo other kinds of reactions:

$$\text{C}_6\text{H}_5\text{-S-H} + \cdot\text{R} \longrightarrow \text{R-H} + \text{C}_6\text{H}_5\text{S}\cdot \quad (16.14)$$

TABLE 16·3 *Relative reactivities of radicals toward benzylic hydrogen atoms*

	reactivity per hydrogen atom (toluene reference)		
radical	ϕ—CH$_3$	ϕ—CH$_2$CH$_3$	ϕ—CH(CH$_3$)$_2$
CH$_3\cdot$	1.0	4.1	12.9
(CH$_3$)$_3$CO·	1.0	3.2	6.9
Cl·	1.0	2.5	5.5
Br·	1.0	17.2	37.0

16·3 REACTIONS OF FREE RADICALS

ABSTRACTION AND FRAGMENTATION

Most organic free radicals are very reactive species that undergo a variety of different reactions. One important class of radical reactions is the abstraction process discussed in the previous section. For example, the chain reaction operating in alkane halogenations (equations 2.13) consists of two abstraction reactions:

$$R-H + X\cdot \longrightarrow R\cdot + H-X$$
hydrogen abstraction

(16.15)

$$R\cdot + X-X \longrightarrow R-X + X\cdot$$
halogen abstraction

Highly branched free radicals or those capable of breaking down into stable small molecules (such as nitrogen and carbon dioxide) often undergo unimolecular decomposition, or *fragmentation*, before slower bimolecular reactions can take place:

$$\left[R-C\overset{O}{\underset{O}{\diagdown}}\right]_2 \xrightarrow{\Delta} R\overset{O}{\diagdown}C\overset{O}{\underset{O\cdot}{\diagdown}} \xrightarrow{\text{fragmentation}} 2R\cdot + 2CO_2 \qquad (16.16)$$

$$R-C\overset{O}{\underset{H}{\diagdown}} + R_1\cdot \longrightarrow R_1-H + R\overset{O}{\diagdown}C\cdot \xrightarrow{\text{fragmentation}} R\cdot + CO \qquad (16.17)$$
  initiating radical

$$(CH_3)_3C-O-O-C(CH_3)_3 \xrightarrow{\Delta} 2CH_3-\underset{\underset{CH_3}{|}}{\overset{\overset{CH_3}{|}}{C}}-O\cdot \xrightarrow{\text{fragmentation}} CH_3\cdot + \underset{CH_3}{\overset{CH_3}{\diagdown}}C=O \qquad (16.18)$$

$$\downarrow \text{solvent H} \qquad \qquad \downarrow \text{solvent H}$$
$$(CH_3)_3C-OH \qquad \qquad CH_4$$

If the di-*t*-butyl peroxide in equation 16.18 is decomposed in different solvents, the proportion of acetone and *t*-butyl alcohol obtained in each case reflects the hydrogen-donating ability of the solvent. In a medium such as chlorobenzene or acetonitrile, which is a poor source of hydrogen, more acetone is formed than *t*-butyl alcohol. In a good hydrogen-donating solvent such as dioxane or tri-*n*-butylamine, the yield of alcohol is much greater than that of acetone.

ADDITION TO MULTIPLE BONDS

The anti-Markovnikov addition of HBr to alkenes, discussed in Section 5.6, is a well-known example of free-radical addition to multiple bonds. Other reagents are also observed to give free-radical addition reactions, and in every case a chain mechanism is believed to operate, as illustrated by the following general equation:

16·3 REACTIONS OF FREE RADICALS

$$R-CH=CH_2 + Y-Z \longrightarrow R-CHY-CH_2Z \qquad \text{overall reaction}$$

$$\left. \begin{array}{l} Z\cdot + R-CH=CH_2 \longrightarrow R-\dot{C}H-CH_2-Z \\ Y \\ | \\ R-\dot{C}H-CH_2Z + Y-Z \longrightarrow R-CH-CH_2-Z + Z\cdot \end{array} \right\} \text{chain reaction} \qquad (16.19)$$

Peroxides often serve as initiators for these reactions:

$$\text{peroxide} \longrightarrow RO\cdot \xrightarrow{\text{fragmentation}} R'\cdot \qquad (16.20)$$
$$ \downarrow Y-Z \downarrow Y-Z$$
$$ RO-Y + Z\cdot R'-Y + Z\cdot$$

PROBLEM 16·4

a The peroxide-initiated addition of chloroform to 1-octene yields 1,1,1-trichlorononane as the major product. Write equations illustrating this reaction and explain the absence of the alternative addition product, 2-methyl-1,1,1,-trichlorooctane.

b Peroxide-initiated addition of bromotrichloromethane to 1,3-butadiene yields a mixture of 3-bromo-5,5,5-trichloro-1-pentene and 1-bromo-5,5,5-trichloro-2-pentene (cis and trans). Write equations for all the steps in this reaction and explain the formation of these two products.

The most widely applied and economically important radical addition reaction is *polymerization*. In our day-to-day lives we constantly use fabrics, utensils, structural materials, and finishes made from synthetic polymers. The impact of these substances on our culture is widely discussed, but contrary to popular view, they offer far more than the advantage of convenience. It has been estimated, for example, that the output of a modern synthetic-fiber plant could be matched in natural fibers only at the expense of 10 million acres committed to raising sheep or about 350,000 acres devoted to cotton. The implications of this land requirement for a starving world are obvious. Similar estimates indicate that the manufacture of plastic containers requires 20 to 30 times less energy consumption than equivalent glass or aluminum containers. The primary disadvantage of these synthetic materials seems to lie in the pollution caused by their low rate of environmental decomposition. This problem can be solved, however, by using modified polymeric substances that are sensitized to chemical and/or biological decomposition through the action of environmental forces such as sunlight. Modified plastics of this kind have been developed.

The mechanism of free-radical polymerization reactions is essentially equivalent to the first step in equation 16.19, provided we define $Z\cdot$ as a growing polymer chain:

$$\sim\!\!CH_2-\underset{R}{\overset{H}{\underset{|}{\overset{|}{C}}}}\!\cdot\; +\; H_2C=\underset{R}{\overset{H}{\overset{|}{C}}} \longrightarrow \sim\!\!CH_2-\underset{R}{\overset{H}{\underset{|}{\overset{|}{C}}}}-CH_2-\underset{R}{\overset{H}{\underset{|}{\overset{|}{C}}}}\!\cdot \qquad (16.21)$$

a growing chain monomer a longer growing chain

16·3 REACTIONS OF FREE RADICALS   545

Repetition of this reaction propagates the polymer chain at rates as high as thousands of additions per second. Ideally, a polymer chain would continue to grow until all the monomer was consumed. In practice, however, chain lengths do not usually exceed 10^4 or 10^5 monomer units because of certain termination processes:

$$\sim\sim CH_2-\underset{R}{\underset{|}{\overset{H}{\overset{|}{C}}}}\cdot \;+\; \cdot\underset{R}{\underset{|}{\overset{H}{\overset{|}{C}}}}-CH_2\sim\sim \xrightarrow{\text{very fast}} \sim\sim CH_2-\underset{R}{\underset{|}{\overset{H}{\overset{|}{C}}}}-\underset{R}{\underset{|}{\overset{H}{\overset{|}{C}}}}-CH_2\sim\sim \quad (16.22)$$
$$\text{a coupling product}$$

$$\sim\sim CH_2-\underset{R}{\underset{|}{\overset{H}{\overset{|}{C}}}}\cdot \;+\; \sim\sim\underset{H}{\underset{|}{\overset{H}{\overset{|}{C}}}}-\underset{R}{\underset{|}{\overset{H}{\overset{|}{C}}}}\cdot \xrightarrow{\text{very fast}} \sim\sim CH_2CH_2R \;+\; \sim\sim CH=CHR \quad (16.23)$$
$$\text{disproportionation products}$$

Some useful polymers produced by free-radical polymerization are described in Table 16.4. Decades of effort devoted to studying polymerization processes and developing new polymeric substances have produced an extensive and complex technology which is beyond the scope of this text. However, the references at the end of the chapter provide further information about this important area of organic chemistry.

RADICAL COUPLING OR RECOMBINATION

When two radicals collide, they generally combine very rapidly to form a covalent bond (the activation energy for such reactions is zero or very close to it). Since these *radical-coupling reactions* consume free-radical intermediates, they are often responsible for the termination of chain reactions and polymerization chains. Indeed, the only reason radical-coupling reactions do not dominate free-radical chemistry is that radicals usually have very short lifetimes and are present in very low concentra-

TABLE 16·4 *Useful free-radical polymerization products*

monomer	polymer	properties
$CH_2=CH_2$	polyethylene	waxy, inert, nontoxic
$CH_3CH=CH_2$	polypropylene	more rigid than polyethylene
$\phi CH=CH_2$	polystyrene (Styron)	brittle, easily foamed
$CF_2=CF_2$	Teflon	inert, soft, low friction
$CF_2=CFCl$	Kel-F	more rigid than teflon
$CH_2=CCl_2$	polyvinylidene chloride	tough, rigid
$CH_2=CHCl$	polyvinyl chloride	hard, rigid, easily plasticized
$CH_2=\underset{CH_3}{\underset{|}{C}}CO_2CH_3$	Lucite or Plexiglas	clear, rigid
$CH_2=CHCN$	polyacrylonitrile (Orlon)	tough, low solubility

16·3 REACTIONS OF FREE RADICALS

tions. The probability of a coupling reaction in a randomly generated assemblage of radicals is therefore low. Of course, relatively stable radicals such as benzyl ($\phi\dot{C}H_2$) often have lifetimes sufficiently long to permit coupling reactions to predominate.

QUENCHING Radical-coupling reactions should be retarded or completely quenched by introducing highly reactive substances capable of intercepting and reacting with radical intermediates as soon as they are formed. These *radical traps*, or *scavengers*, may themselves be radicals, as in the case of oxygen and galvinoxyl (Figure 16.2), or they may be ordinary compounds which undergo very facile abstraction or addition reactions with radicals (iodine, quinones, and styrene are examples of this group). A methyl radical, for example, can be diverted from common reactions such as hydrogen abstraction by the introduction of iodine or styrene into the reaction system, as illustrated in Figure 16.5.

PROBLEM 16·5
Write an equation showing the reaction of an ethyl radical with galvinoxyl.

SOLVENT-CAGE EFFECTS Thermal decomposition of dilute diacetyl peroxide solutions in octane at 65°C yields methane, ethane, carbon dioxide, and methyl acetate:

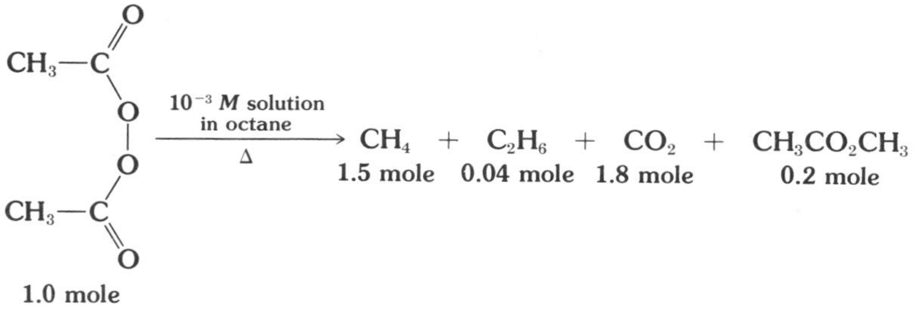

$$\begin{array}{c} CH_3-C(=O)-O-O-C(=O)-CH_3 \end{array} \xrightarrow[\Delta]{\substack{10^{-3}\,M\text{ solution} \\ \text{in octane}}} CH_4 + C_2H_6 + CO_2 + CH_3CO_2CH_3 \qquad (16.24)$$

1.0 mole   →   1.5 mole   0.04 mole   1.8 mole   0.2 mole

These products appear to be formed by initial homolytic cleavage of the peroxide, followed by fragmentation of the resulting acetoxy radicals (R = CH_3 in equation 16.16) and coupling of various intermediates. However, the effect of scavengers on the product ratio in reaction 16.24 is unexpected. The addition of styrene strongly quenches the formation

FIGURE 16·5 *Quenching reactions of a methyl radical*

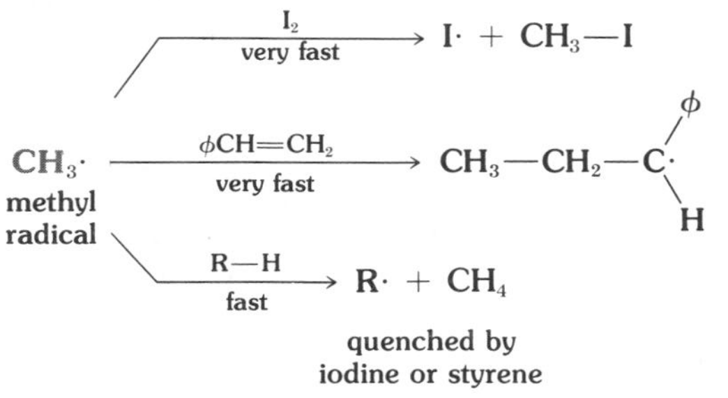

$$CH_3\cdot \text{ (methyl radical)} \begin{cases} \xrightarrow[\text{very fast}]{I_2} I\cdot + CH_3-I \\ \xrightarrow[\text{very fast}]{\phi CH=CH_2} CH_3-CH_2-\dot{C}(\phi)(H) \\ \xrightarrow[\text{fast}]{R-H} R\cdot + CH_4 \end{cases}$$

quenched by iodine or styrene

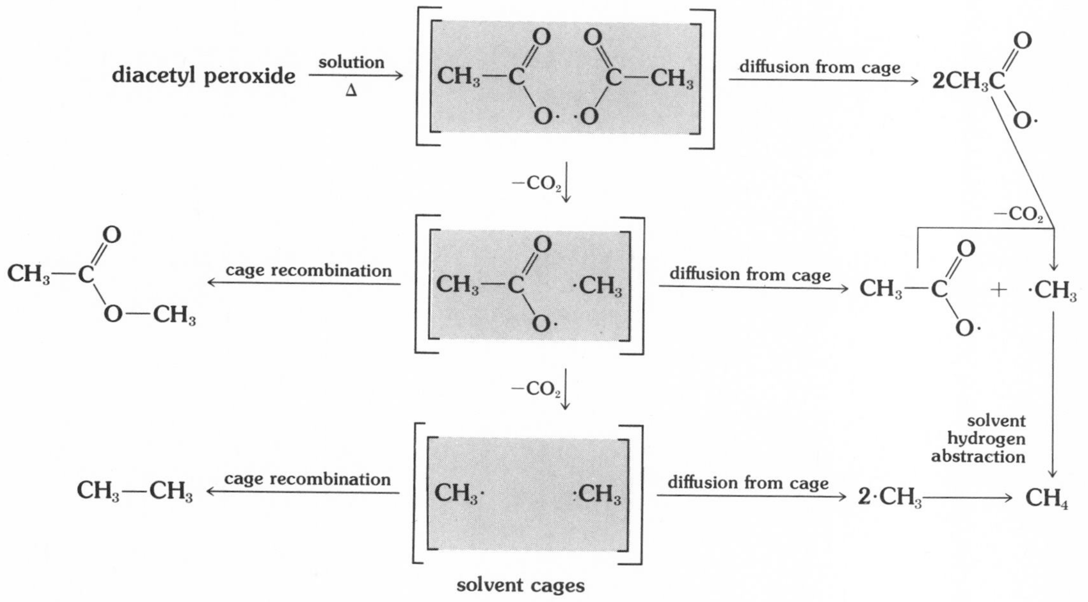

16·3 REACTIONS OF FREE RADICALS

FIGURE 16·6 *Cage effects in diacetyl peroxide decomposition*

of methane, but it does not significantly change the amounts of the other products produced.

This curious result makes sense when we recognize that the peroxide-homolysis step produces pairs of adjacent acetoxy radicals, rather than a random distribution of these intermediates. In solution, these neighboring radical pairs are held near each other by a wall of solvent molecules. This *solvent cage*, illustrated in Figure 16.6, confines the radical pair for about 10^{-10} second, which is long enough to introduce a bias favoring coupling of derived fragment radicals before they can react with solvent or scavenger molecules. Once the radicals diffuse from the solvent cage into the bulk solution, their chance to recombine becomes very small. In summary, methyl acetate and ethane are formed only in cage recombination reactions, whereas methane is formed from hydrogen-abstraction reactions of free methyl radicals with the solvents.

SOME USEFUL RADICAL-RECOMBINATION REACTIONS

Radical-recombination reactions are found in some important synthesis procedures. One such reaction is *Kolbe electrolysis*. The electrolysis of carboxylic acid salts generates carboxy radicals at the anode (an electron is transferred from the carboxylate anion to the electrode) and these intermediates lose carbon dioxide and then couple:

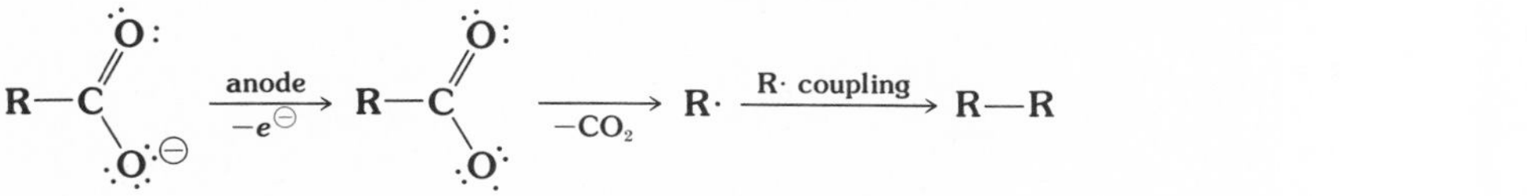

$$R-C(\ddot{O}:)(\ddot{O}:^{\ominus}) \xrightarrow[-e^{\ominus}]{\text{anode}} R-C(\ddot{O}:)(\ddot{O}\cdot) \xrightarrow{-CO_2} R\cdot \xrightarrow{R\cdot \text{ coupling}} R-R \qquad (16.25)$$

Since the radicals are all formed at the anode surface, their local concentration can be relatively high, thus favoring coupling reactions. This

548   16·3 REACTIONS OF FREE RADICALS

reaction is useful for synthesizing long-chain dicarboxylic acids:

$$\underset{\text{ethyl sodium adipate}}{\begin{array}{c}CO_2^{\ominus}Na^{\oplus}\\|\\(CH_2)_4\\|\\CO_2C_2H_5\end{array}} \xrightarrow[-2e^{\ominus}]{\text{anode}} \begin{array}{c}CO_2C_2H_5\\|\\(CH_2)_8\\|\\CO_2C_2H_5\end{array} + 2CO_2 \tag{16.26}$$

Another useful reaction is *phenol coupling*. The oxidation of phenols with mild oxidizing agents such as $Fe(CN)_6^{-3}$, MnO_2, $VOCl_3$, and certain enzymes often leads to ortho- and para-coupled products:

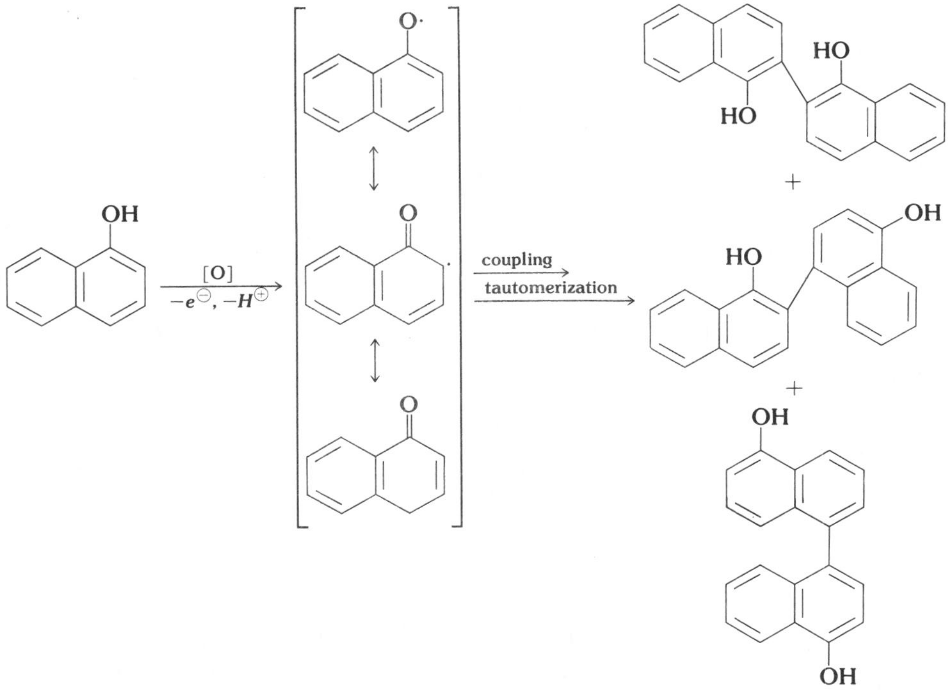

(16.27)

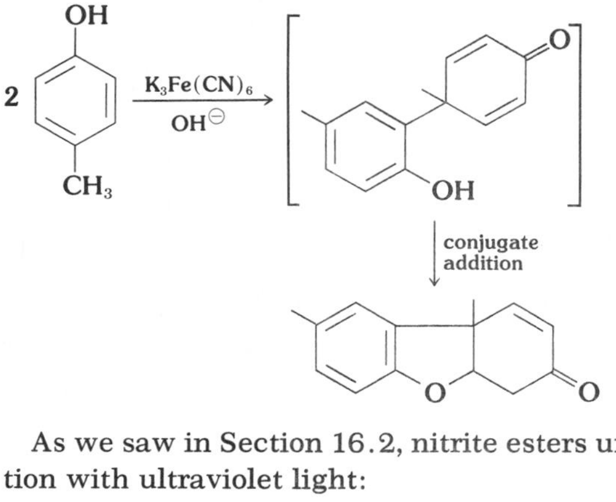

(16.28)

As we saw in Section 16.2, nitrite esters undergo homolysis on irradiation with ultraviolet light:

$$R-\ddot{O}-\dot{N}=\ddot{O}: \xrightarrow{h\nu} R-\ddot{O}\cdot + \cdot\dot{N}=\ddot{O}:$$

16·3 REACTIONS OF FREE RADICALS

If the organic substrate is so constituted that the oxy radical produced in this homolysis is located close to an alkyl group, an intramolecular hydrogen atom abstraction may take place:

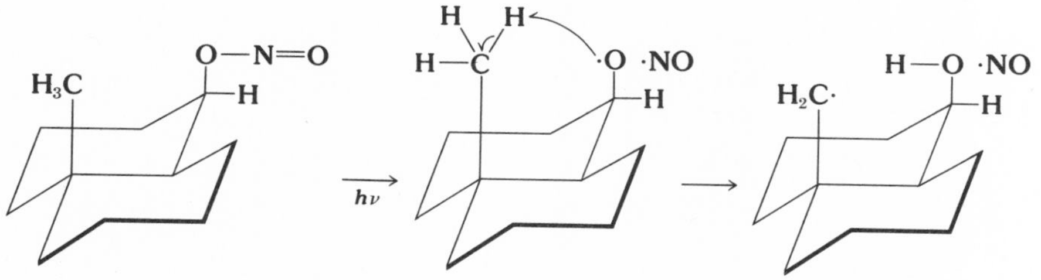

(16.29)

Since the resulting radical pair is in a solvent cage, recombination to an alkyl nitroso compound or the more stable oxime tautomer may occur:

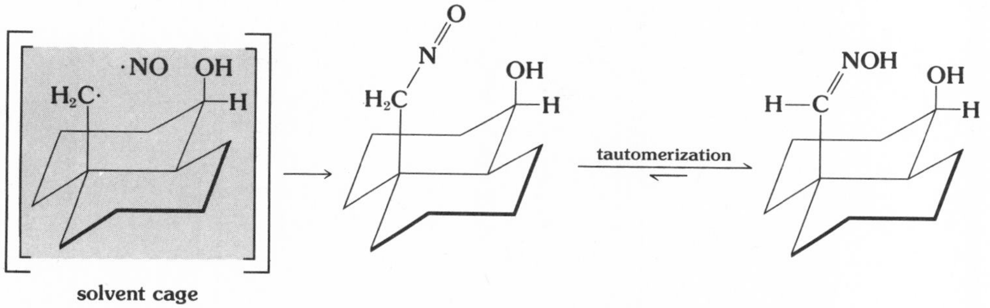

(16.30)

solvent cage

PROBLEM 16·6

The important steroid hormone aldosterone has been prepared in two steps from the nitrite ester of corticosterone acetate. The aldosterone oxime resulting from the first step is accompanied by an isomer having the structure shown. Write equations showing how these products are formed.

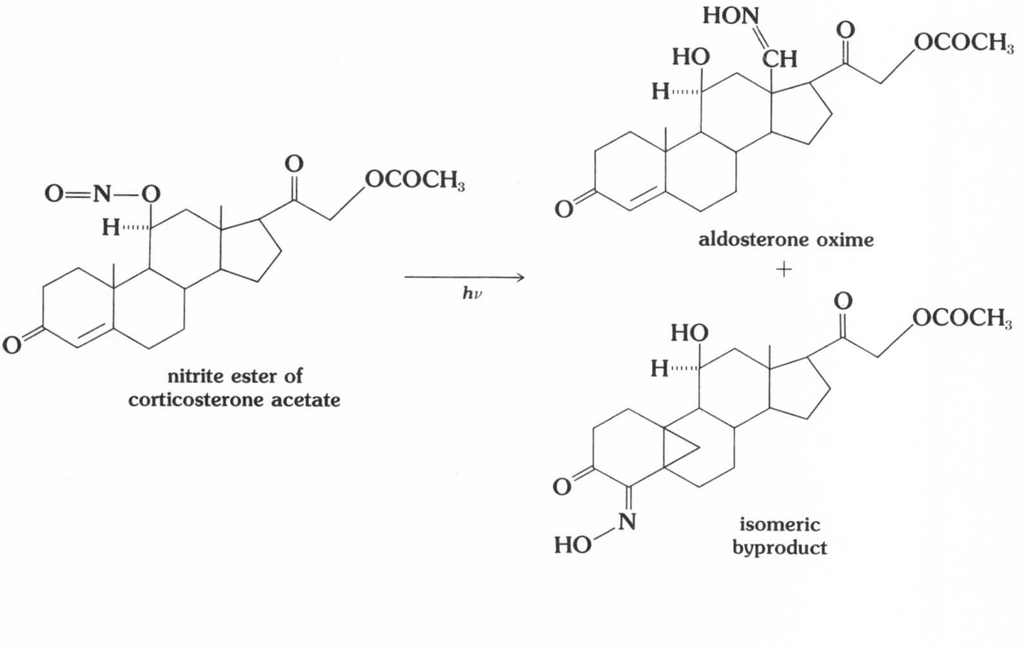

nitrite ester of corticosterone acetate

aldosterone oxime

+

isomeric byproduct

16·4 THE STEREOCHEMISTRY OF RADICAL REACTIONS

The preferred geometry of unstrained carbon radicals has not been rigorously defined. The methyl radical appears to be almost planar, but an experimental distinction between a planar configuration and rapidly interconverting pyramidal configurations cannot be made in most other cases:

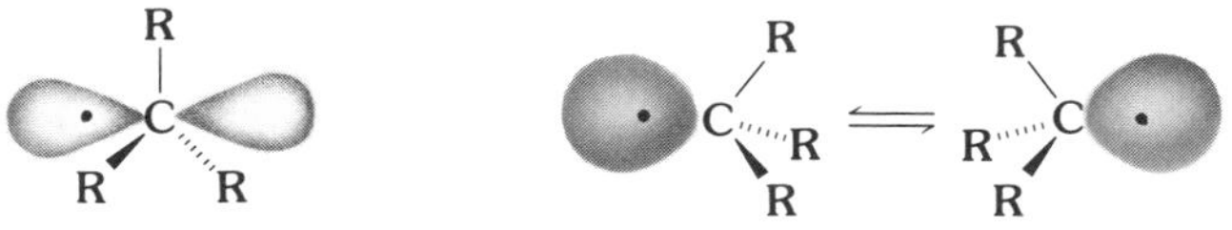

The difference in energy between a planar radical and a nonplanar one seems to be very small. Thus apocamphoyl peroxide, which can give rise only to pyramidal radicals, undergoes a fairly normal homolytic decomposition. In contrast, the corresponding chloride is essentially unreactive under S_N1 conditions because of the high potential energy of nonplanar carbonium-ion intermediates.

Radical reactions taking place at chiral centers generally give racemic products, indicating that the radical intermediates are either achiral or consist of rapidly interconverting enantiomeric species:

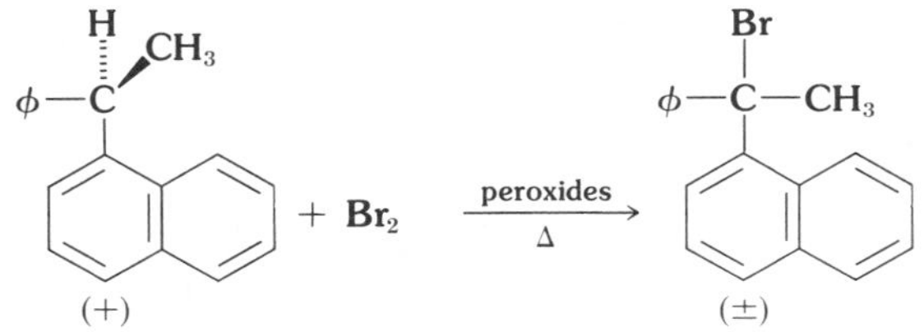

$$\phi-\underset{\underset{(+)}{\text{naphthyl}}}{\overset{H}{\underset{CH_3}{C}}} + Br_2 \xrightarrow[\Delta]{\text{peroxides}} \phi-\underset{\underset{(\pm)}{\text{naphthyl}}}{\overset{Br}{\underset{CH_3}{C}}} \quad (16.31)$$

FIGURE 16·7 *Contrasting radical and ionic reactions of apocamphoyl derivatives*

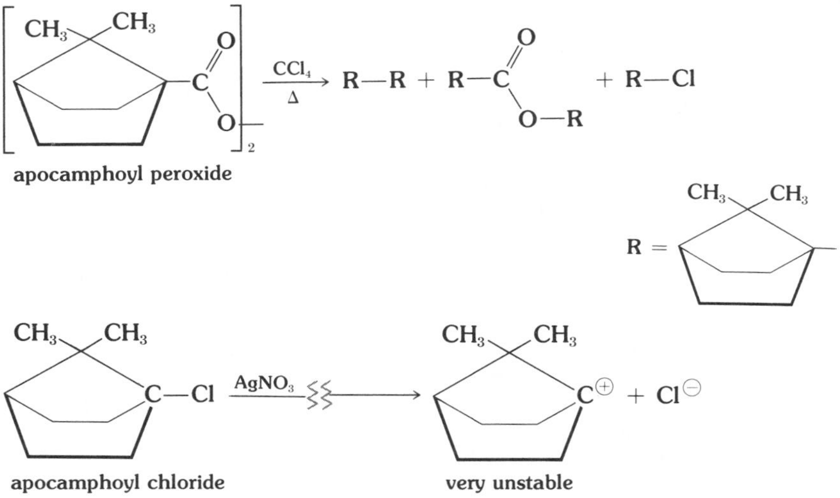

apocamphoyl peroxide $\xrightarrow[\Delta]{CCl_4}$ R—R + R—C(=O)O—R + R—Cl

apocamphoyl chloride $\xrightarrow{AgNO_3}$ ≁ ⟶ C$^{\oplus}$ + Cl$^{\ominus}$ (very unstable)

However, coupling reactions of radicals in a solvent cage may proceed so quickly that retention of configuration results, as illustrated by the ester coupling product from diacyl peroxide decomposition:

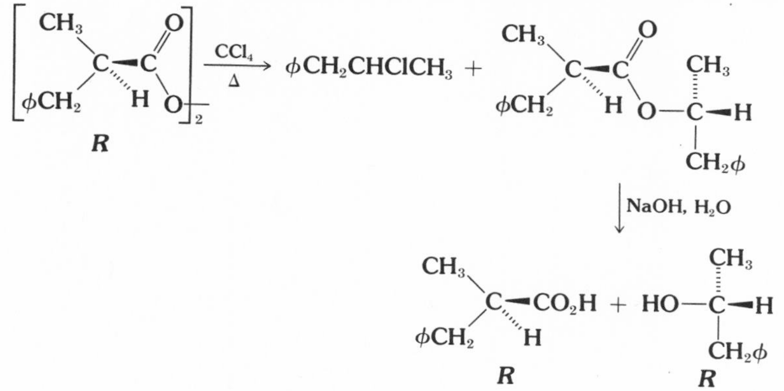

$$\left[\begin{array}{c}CH_3\\ \phi CH_2 \overset{|}{\underset{H}{C}}\text{---}\overset{O}{\underset{O}{C}}\\ R\end{array}\right]_2 \xrightarrow[\Delta]{CCl_4} \phi CH_2CHClCH_3 + \underset{\phi CH_2 \overset{|}{\underset{H}{C}}\text{---}\overset{O}{\underset{O\text{---}\overset{|}{\underset{CH_2\phi}{C}}\text{---}H}{C}}\,CH_3}{}\quad (16.32)$$

↓ NaOH, H₂O

$$\underset{\phi CH_2\quad H}{\overset{CH_3}{C}\text{---}CO_2H} + HO\text{---}\overset{CH_3}{\underset{CH_2\phi}{C}}\text{---}H$$
$\qquad R\qquad\qquad\qquad R$

16·5 PERICYCLIC REACTIONS

An important body of chemical reactions, differing from ionic or free-radical reactions in a number of respects, has been identified and studied extensively during the past decade. Among the characteristics shared by these reactions, there are three in particular that set them apart from the chemical transformations we have examined thus far:

1 They are relatively unaffected by changes in solvent, the presence of radical initiators or scavengers, or by electrophilic or nucleophilic catalysts.

2 No discernible ionic or free-radical intermediates appear to be involved in their mechanisms.†

3 They appear to proceed by a simultaneous or concerted making and breaking of two or more covalent bonds, often to stereospecific products.

The last feature is especially significant, since many of these reactions take place under milder conditions (for example, at lower temperatures) than might be expected from the energy required to break the necessary covalent bonds. In a concerted mechanism, however, this energy requirement can be partially balanced by the energy released in covalent-bond formation, as in the case of S_N2 reactions. Reactions of this kind usually have transition states involving a cyclic interchange of bonds around a ring of atoms, and we term them *pericyclic reactions*.

TYPES OF PERICYCLIC REACTIONS

CYCLOADDITION REACTIONS The concerted combination of two unsaturated (π-electron) moieties to form a ring of atoms having two new σ bonds and two fewer π bonds is called a *cycloaddition reaction;* the reverse reaction is termed *cycloreversion*. The number of π electrons

†Because these reactions do not proceed via observable intermediates, early mechanism studies met with frustrations which led some workers to christen them "no-mechanism reactions."

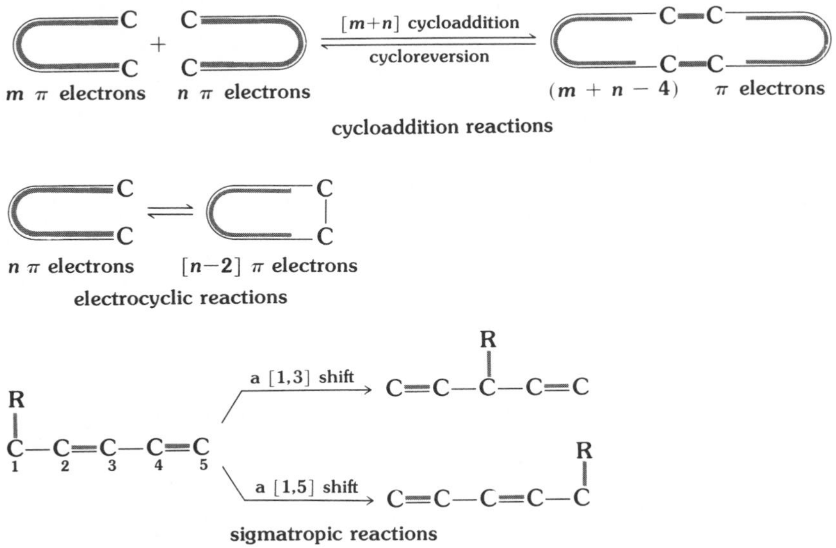

16·5 PERICYCLIC REACTIONS

FIGURE 16·8 *Types of pericyclic reactions*

in each component is customarily given in brackets preceding this designation, as described in Figure 16.8. We saw one example of a cycloaddition reaction in the Diels-Alder diene synthesis (Figure 15.5 and equation 16.33). The [6+4] cycloaddition shown in equation 16.34 is another example.

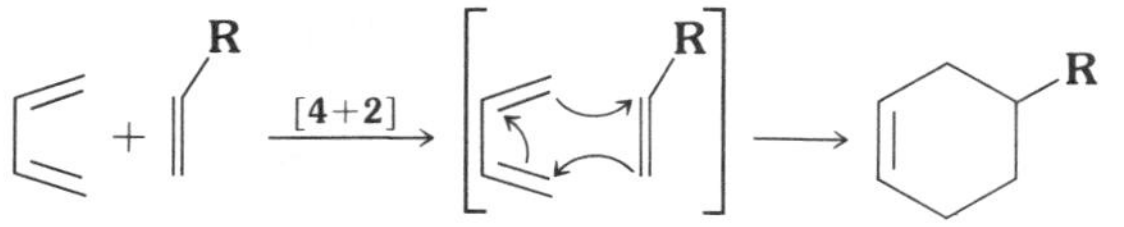

(16.33)

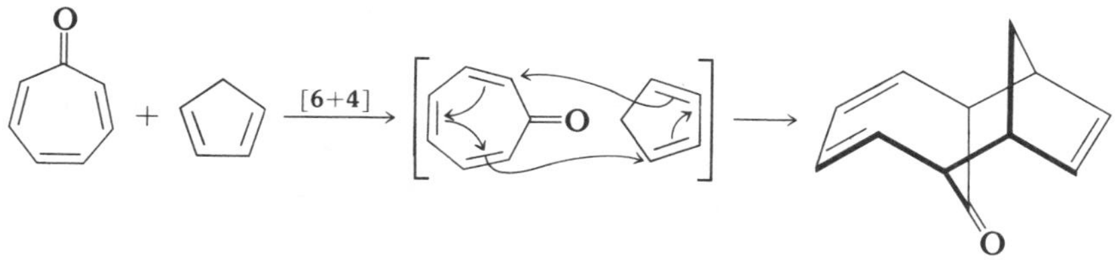

(16.34)

PROBLEM 16·7
What has happened to the four missing π electrons in the general equation for cycloaddition reactions in Figure 16.8? Classify the following cycloaddition reactions:

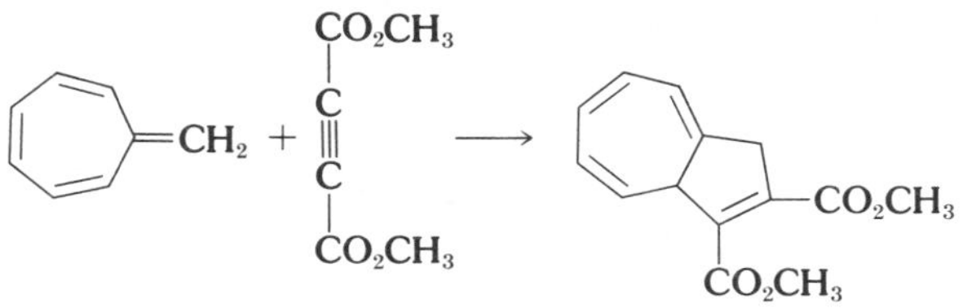

16·5 PERICYCLIC REACTIONS   553

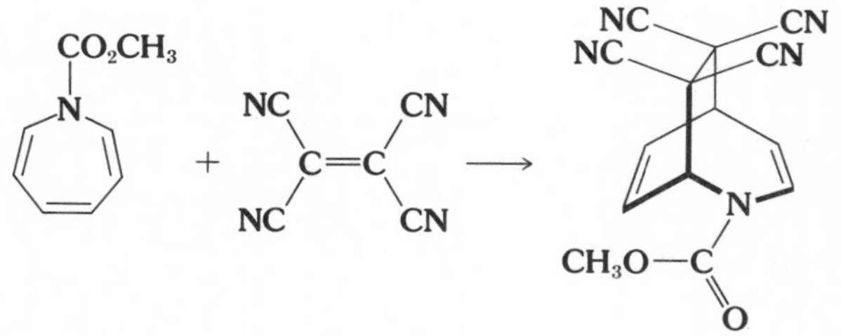

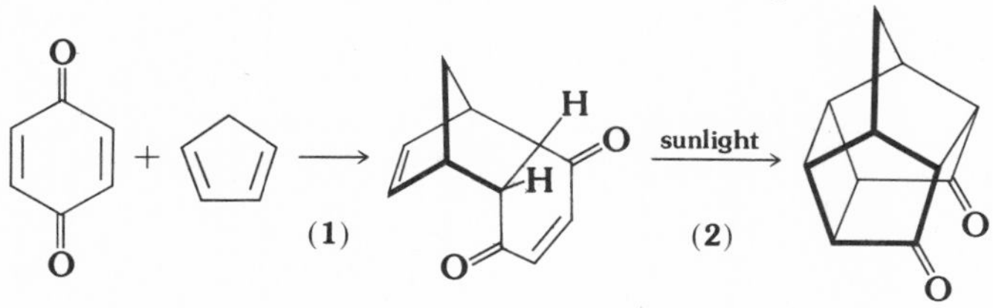

ELECTROCYCLIC REACTIONS Another form of pericyclic reaction is the concerted cyclization of a conjugated π-electron system to a ring having a new σ bond between the terminal atoms and two fewer π electrons. This process, also illustrated in Figure 16.8, is termed an *electrocyclic reaction*. The thermal cyclization of conjugated trienes and the cleavage of cyclobutenes are examples of electrocyclic reactions:

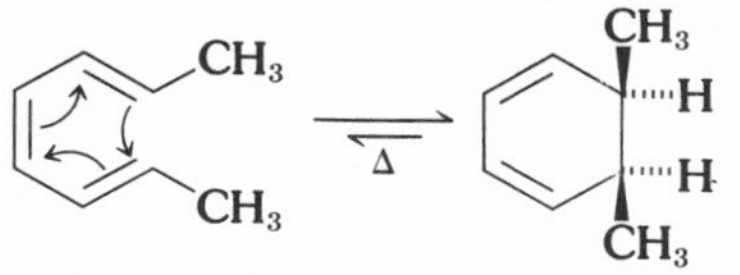

$$\text{trans,cis,trans-2,4,6-octatriene} \quad \rightleftharpoons_\Delta \quad \textit{cis}\text{-5,6-dimethyl-1,3-cyclohexadiene} \tag{16.35}$$

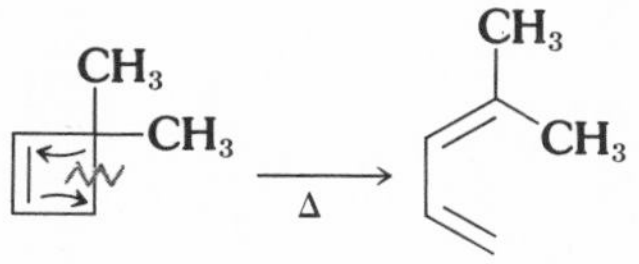

$$\tag{16.36}$$

SIGMATROPIC REACTIONS Molecular rearrangements in which a σ bond, flanked by one or more π-electron systems, shifts to a new position are called *sigmatropic reactions*. These rearrangements are classified by two numbers set in brackets, which refer to the relative distance (in atoms) each end of the migrating σ bond has moved. The *Cope rearrangement* is an important example of a [3,3] sigmatropic reaction:

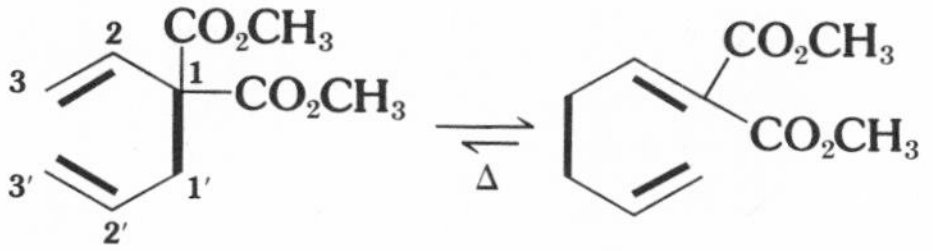

$$\tag{16.37}$$

PROBLEM 16·8
Compare reactions 16.35 and 16.37 and state a formal distinction between sigmatropic and electrocyclic reactions.

ENERGY REQUIREMENTS OF PERICYCLIC REACTIONS

Pericyclic reactions do not normally proceed spontaneously, even though many are exothermic. The isomerization of bicyclobutane to butadiene, for example, has an activation energy of 41 kcal/mole despite the fact that about 69 kcal/mole of strain energy is released in the course of the transformation:

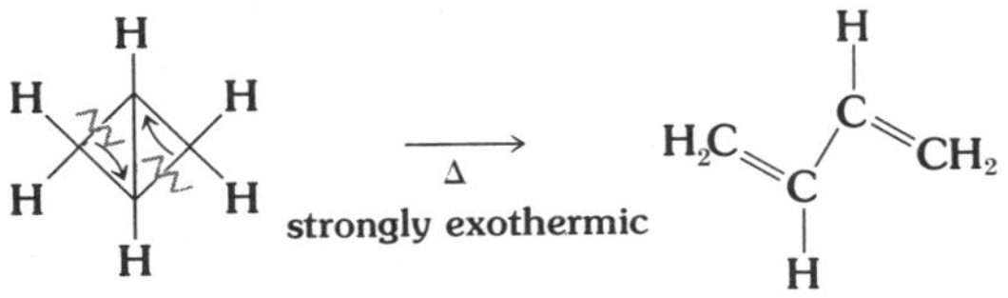

$$\text{bicyclobutane} \xrightarrow[\text{strongly exothermic}]{\Delta} \text{butadiene} \quad (16.38)$$

The activation-energy requirements of pericyclic reactions can be met in one of two ways. Either kinetic (thermal) energy can be introduced by heating the reaction system, as in most of the previously described reactions, or ultraviolet or visible light may be used to excite to a higher electron state those molecules capable of absorbing the radiation provided. Examples of such *photochemical reactions* are the [2+2] cycloaddition in equation 16.39 and the electrocyclic isomerization described in equation 16.40.

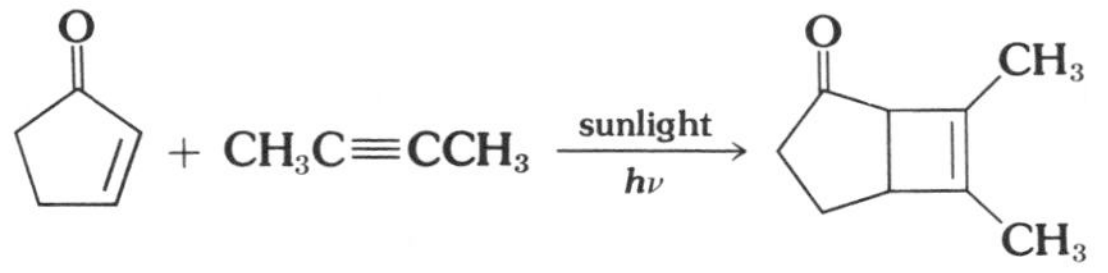

$$\text{cyclopentenone} + CH_3C \equiv CCH_3 \xrightarrow[h\nu]{\text{sunlight}} \text{bicyclic product} \quad (16.39)$$

$$\text{(2E,4E)-hexadiene} \xrightarrow{h\nu} \text{cyclobutene product} \quad (16.40)$$

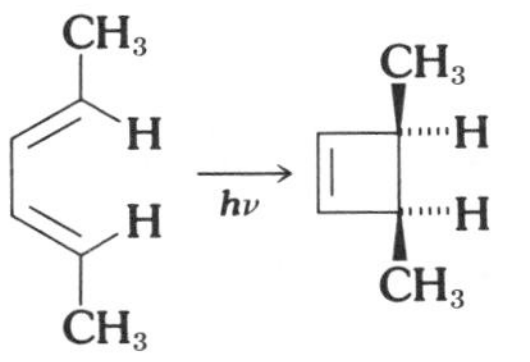

CONCERTED VERSUS NONCONCERTED REACTIONS

Before we proceed to some intriguing aspects of pericyclic reactions, it is necessary to emphasize that we are restricting our attention to *concerted reactions*. The reason for doing so is that some stepwise reactions may give products equivalent to those expected from a pericyclic reaction. Certain apparent cycloaddition reactions have in fact been found to take place via discrete reactive intermediates. Results such as the following, therefore, should not be allowed to mislead us as we probe the nature of concerted reactions.

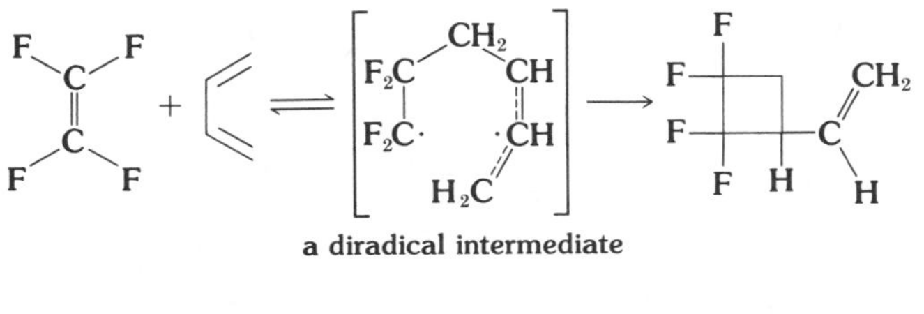

$$F_2C=CF_2 + CH_2=CH-CH=CH_2 \rightleftharpoons \left[\begin{array}{c} F_2C-CH_2 \\ | \quad \cdot CH \\ F_2C\cdot \quad \| \\ \quad H_2C \end{array}\right] \longrightarrow \text{cyclobutane product} \quad (16.41)$$

a diradical intermediate

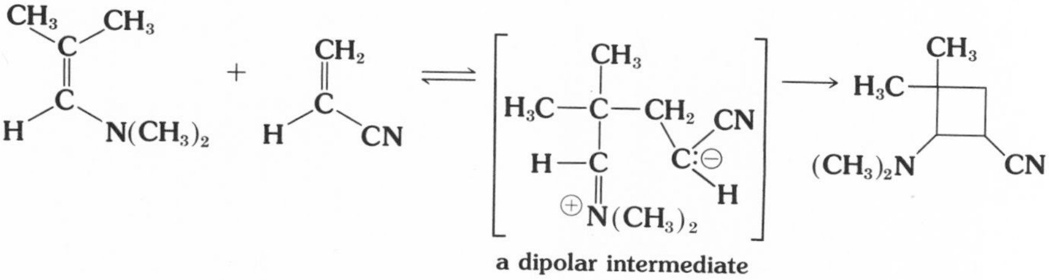

16·5 PERICYCLIC REACTIONS

$$\underset{H}{\overset{CH_3}{\underset{|}{C}}}\!\!=\!\!\underset{N(CH_3)_2}{\overset{CH_3}{\underset{|}{C}}} + \underset{H}{\overset{CH_2}{\underset{\|}{C}}}\!\!=\!\!\underset{CN}{\overset{}{C}} \rightleftharpoons \left[\begin{array}{c} CH_3 \\ H_3C\!-\!\underset{|}{\overset{|}{C}}\!-\!CH_2\ \ CN \\ H\!-\!\underset{\underset{\oplus}{N(CH_3)_2}}{\overset{\|}{C}}\ \ \ \ \underset{H}{\overset{\ominus}{C}} \end{array} \right] \longrightarrow \underset{(CH_3)_2N}{\overset{H_3C}{\square}}\underset{CN}{\overset{CH_3}{\ \ }} \quad (16.42)$$

a dipolar intermediate

Unfortunately, there is no simple way of determining whether or not a given reaction is concerted. In general, concerted reactions are stereospecific, are insensitive to solvent changes, and are not subject to interception at intermediate stages by extraneous reagents. However, these properties by themselves, although strongly suggestive, still do not unequivocally prove a reaction to be concerted. The most commonly cited evidence in this respect is often of a negative kind—the inability to trap or observe any reaction intermediate.

PROBLEM 16·9
Benzaldehyde reacts with simple alkenes under photochemical conditions to give four-membered ring ethers called oxetanes. The reaction with either *cis*- or *trans*-2-butene yields all four possible diastereoisomeric oxetanes:

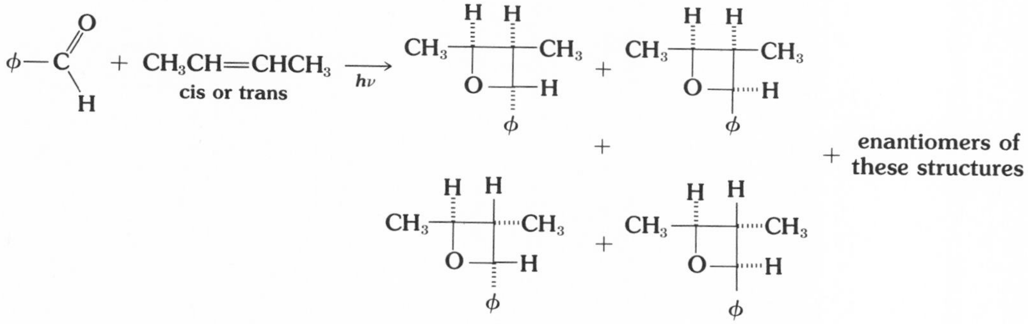

$$\phi\!-\!\overset{O}{\underset{H}{\overset{\|}{C}}} + CH_3CH\!\!=\!\!CHCH_3 \xrightarrow{h\nu}$$ [four oxetane products] + enantiomers of these structures

cis or trans

What conclusions would you draw regarding the mechanism of this reaction?

SOME PERPLEXING ASPECTS OF PERICYCLIC REACTIONS

As we study pericyclic reactions in greater detail, several curious and perplexing facts emerge. First of all, these reactions appear to be highly specific and often do not tolerate moderate structural changes. Thus the relative rates of similar reactions can vary markedly; indeed, some anticipated reactions do not occur at all. For example, the following thermal [8+2] cycloaddition proceeds smoothly, whereas the equivalent [6+2] cycloaddition does not take place:

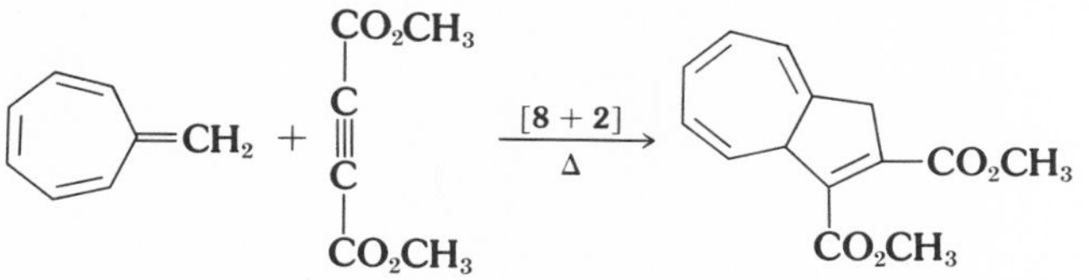

$$\text{(cycloheptatriene)}\!=\!CH_2 + \underset{CO_2CH_3}{\overset{CO_2CH_3}{\underset{|}{\overset{|}{C}}\!\!\equiv\!\!\underset{|}{\overset{|}{C}}}} \xrightarrow[\Delta]{[8+2]} \text{(azulene-type)}\!-\!CO_2CH_3 \quad (16.43)$$

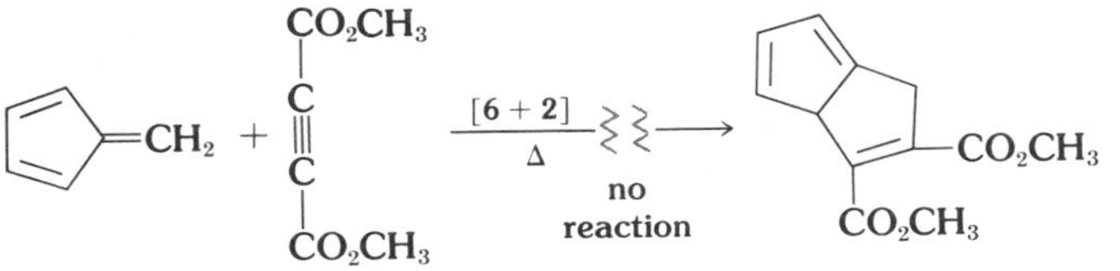

16·5 PERICYCLIC REACTIONS

$$\text{(cyclopentadienylidene)=CH}_2 + \underset{\text{CO}_2\text{CH}_3}{\overset{\text{CO}_2\text{CH}_3}{\underset{|}{\overset{|}{\underset{C}{\overset{C}{|||}}}}}} \xrightarrow[\Delta]{[6+2]} \text{no reaction} \quad \xrightarrow{} \quad \text{indene-CO}_2\text{CH}_3, \text{CO}_2\text{CH}_3 \qquad (16.44)$$

The electrocyclic isomerization of the following trans-fused cyclobutene to the more stable conjugated diene takes place readily below 100°C, but the equivalent transformation of the cis isomer requires temperatures in excess of 200°C, despite the fact it is an exothermic reaction:

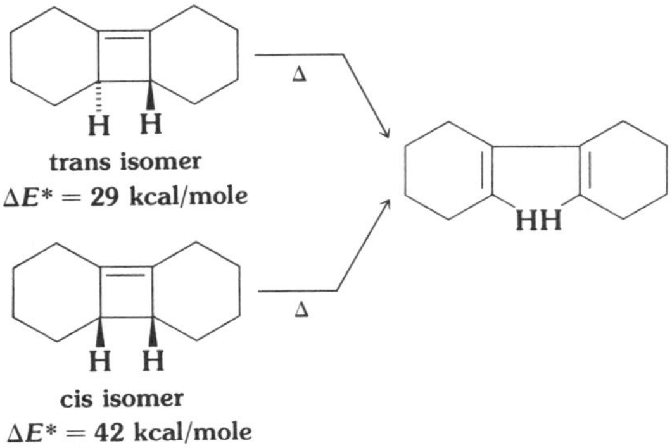

trans isomer
$\Delta E^* = 29$ kcal/mole

cis isomer
$\Delta E^* = 42$ kcal/mole

(16.45)

Deuterium-labeled cycloheptatriene undergoes a specific [1,5] sigmatropic shift of deuterium, but alternative [1,3] and [1,7] shifts do not occur:

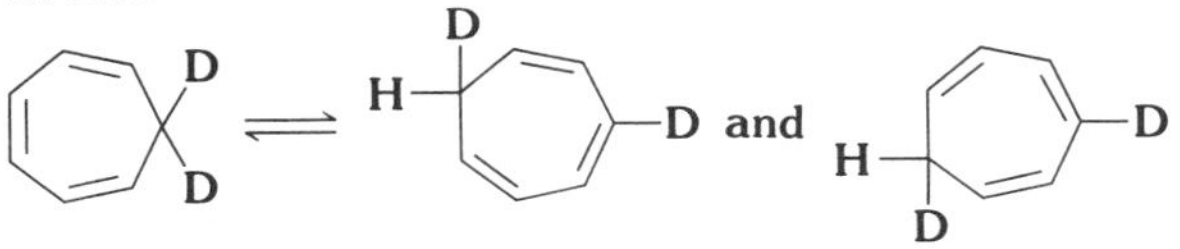

(16.46)

Another puzzling fact is that pericyclic reactions are normally stereospecific, but not always in the same sense. For example, in contrast to the high degree of retention of configuration found in the [4+2] thermal cycloaddition (see equations 15.25 to 15.27), the following [14+2] cycloaddition takes place with specific trans addition:

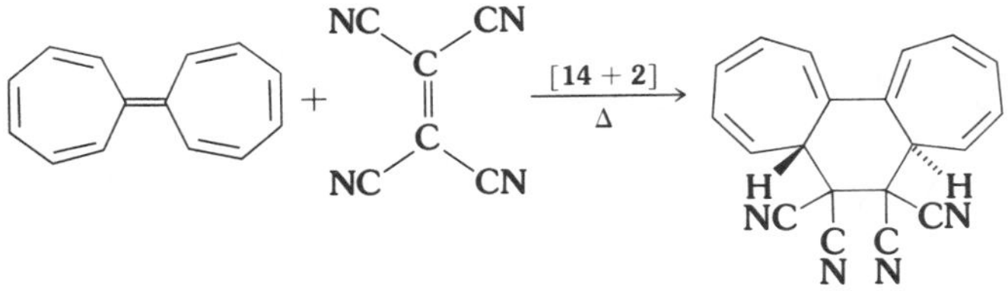

(16.47)

The electrocyclic isomerization of 1,3,5-hexatrienes to 1,3-cyclohexadienes also shows high stereospecificity:

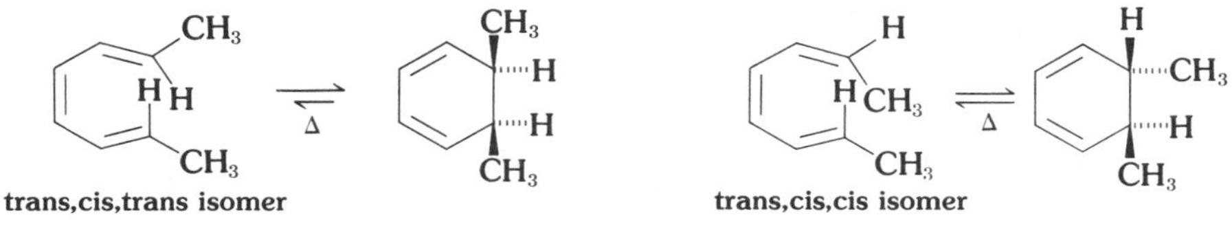

trans,cis,trans isomer

trans,cis,cis isomer

(16.48)

16·5 PERICYCLIC REACTIONS

A parallel configurational relationship is observed in the photochemical cyclization of substituted butadienes:

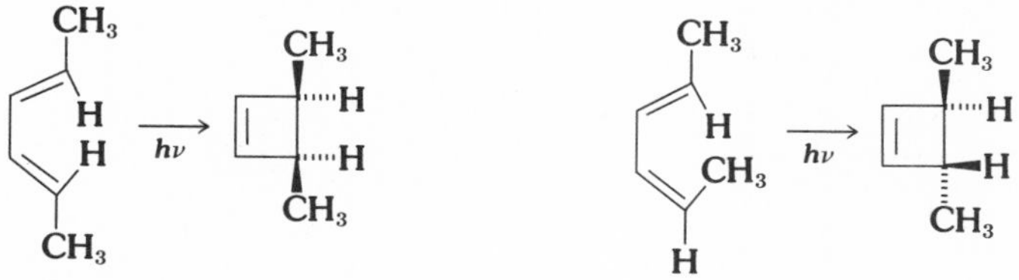

(16.49)

The stereospecificity of the Cope rearrangement, a [3,3] sigmatropic shift, is illustrated in Figure 16.9.

A third interesting feature of pericyclic reactions is that thermal and photochemical reactions often have a diametric yet complementary relationship. For example, cycloaddition reactions of the [2+2] kind almost never occur thermally but are common under photochemical conditions:

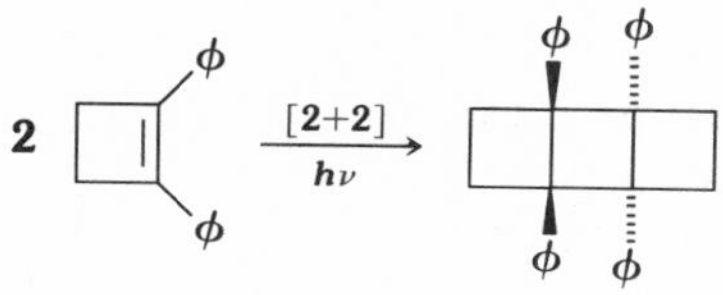

(16.50)

The reactions of bicycloheptadiene further illustrate this remarkable specificity. Despite the angle strain in the tetracyclic product, the photochemical transformation of the diene is an intramolecular [2+2] cycloaddition. In contrast, heat produces no cycloaddition, but instead results in fragmentation by a [4+2] cycloreversion process (a reverse Diels-Alder reaction):

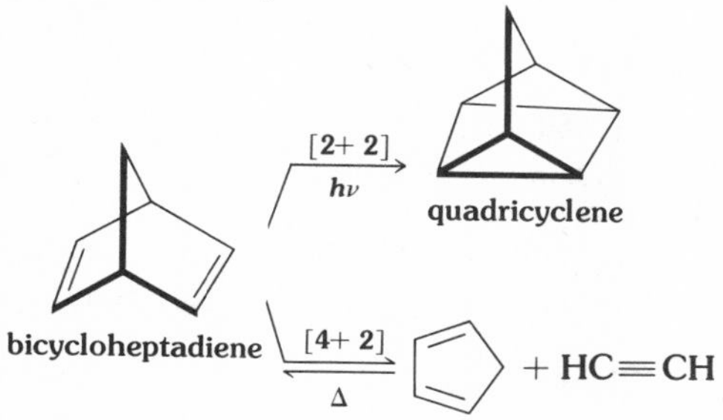

(16.51)

FIGURE 16·9 *Stereospecificity of the Cope rearrangement*

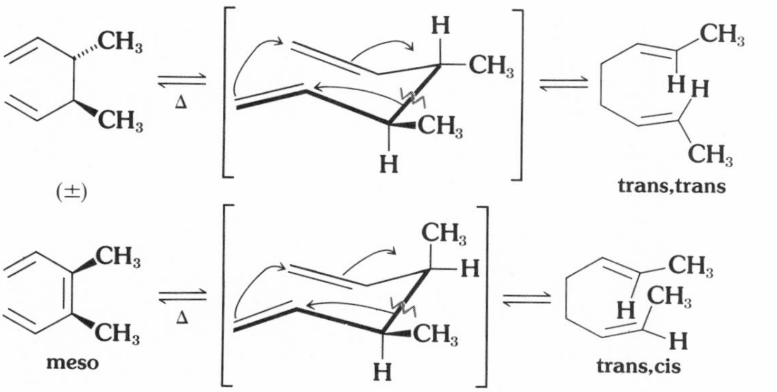

16·5 PERICYCLIC REACTIONS

Another manifestation of this dichotomy is seen in the reversed stereospecificity that sometimes occurs with a change from thermal to photochemical reaction conditions. For example, the electrocyclic transformation of *trans,cis,trans*-2,4,6-octatriene yields *cis*-5,6-dimethyl-1,3-cyclohexadiene under thermal conditions (equation 16.48), but gives only the corresponding trans isomer on photolysis:

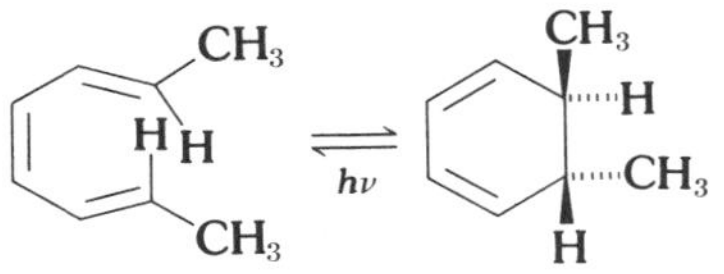

$$\text{(structures shown)} \quad \xrightleftharpoons{h\nu} \quad \text{(structures shown)} \tag{16.52}$$

STEREOCHEMICAL NOMENCLATURE FOR PERICYCLIC REACTIONS

The stereochemical course of electrocyclic reactions is not amenable to description by established terms, such as "retention" or "inversion." Consequently, a new nomenclature has been devised, as illustrated for the triene ⇌ cyclohexadiene system in Figure 16.10. For a σ bond to form between the end carbon atoms of the conjugated π-electron chain, these atoms must rotate approximately 90°. The rotations of the two end groups can be in the same direction, termed *conrotatory*, or the groups can rotate in opposite directions, termed *disrotatory*.† The thermal electrocyclic transformations shown in equations 16.48 are both disrotatory, whereas the photochemical case (equation 16.52) is conrotatory.

PROBLEM 16·10
a Classify the sigmatropic reactions in equations 16.49 as conrotatory or disrotatory.
b Thermal ring opening of *cis*-3,4-dimethylcyclobutene proceeds cleanly to give *trans,cis*-2,4-hexadiene. Is this reaction disrotatory or conrotatory?

Cycloaddition and sigmatropic reactions share one characteristic: a π-electron system (in the case of cycloadditions we have two distinct

†The prefix *dis*- means away from or opposed, as in *disconnect* or *disagree*. The prefix *con*- as used here means with or in common, as in *concur* or *congruence*.

FIGURE 16·10 *Conrotatory and disrotatory electrocyclic motion*

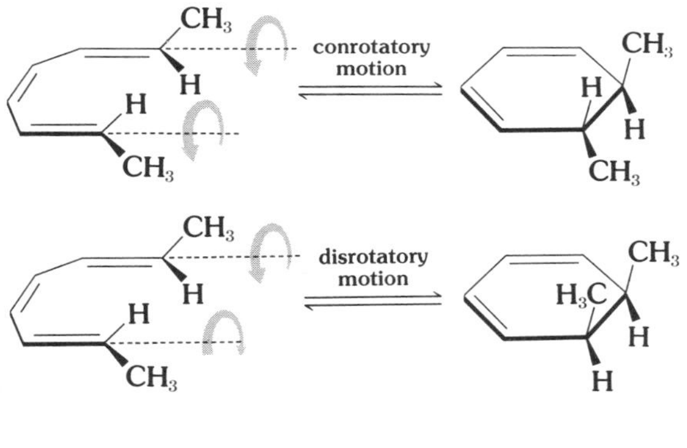

16·5 PERICYCLIC REACTIONS

π-electron systems) undergoes σ-bond formation and/or cleavage at its ends. In most reactions of this kind these two σ bonds are located on the same side of the plane defined by the atoms of the π system (this is approximately the common nodal plane of the π orbitals). We call such a relationship of bonds *suprafacial* (see Figure 16.11). The alternative orientation, with the newly formed or broken σ bonds on opposite sides of the nodal plane, is termed *antarafacial*.

The Diels-Alder reaction is an example of a doubly suprafacial cycloaddition and can be so designated by the introduction of a subscript s (or a) after the number indicating the π electrons in each component:

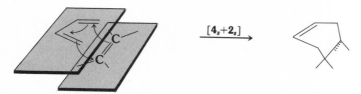

The [1,5] sigmatropic shift shown in equation 16.46 is also suprafacial:

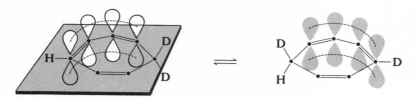

FIGURE 16·11 *Suprafacial bond orientation in pericyclic reactions*

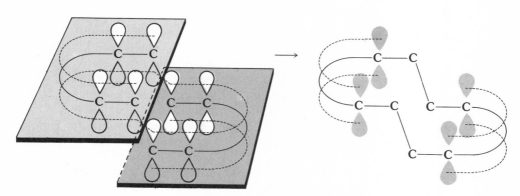

a suprafacial-suprafacial cycloaddition

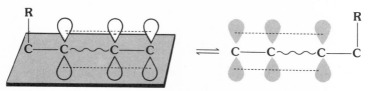

a suprafacial sigmatropic shift

Antarafacial reactions are relatively rare because most π-electron systems must be substantially twisted for concerted σ-bond formation to take place from opposite sides of the nodal plane. Only rather long conjugated systems, such as the 14-π-electron reactant in equation 16.47, can achieve this twisting without appreciable loss of conjugation. Reaction 16.47 is a $[14_a+2_s]$ cycloaddition.

16·6 THEORETICAL MODELS FOR PERICYCLIC REACTIONS

The many complexities and subtleties of pericyclic reactions present a formidable challenge to the principles and theories of modern organic chemistry. We can reduce this challenge to two closely related questions:

1 How can a seemingly unrelated array of stereospecific concerted reactions, involving various combinations of different functional groups, be consolidated within a single theoretical framework?

2 How can we reach a degree of understanding of these reactions that will permit us to make reliable predictions regarding the course of untested pericyclic transformations?

Since pericyclic reactions all involve a cyclic reorganization of electrons and bonds within the reacting complex, the answers to these questions should be found in a study of molecular-orbital interactions. Indeed, the primary factor which determines the mechanism or path followed by any chemical reaction is the necessity of maintaining maximum bonding throughout the reaction.

THE WOODWARD-HOFFMANN CONTRIBUTION

The most widely acclaimed and significant theoretical development in organic chemistry during the past decade is generally acknowledged to be the recognition and enunciation, by R. B. Woodward and Roald Hoffmann (Harvard University), of molecular-orbital symmetry control in concerted reactions. In 1965 Woodward and Hoffmann published a series of papers which, in essence, demonstrated that concerted reactions occur readily when there is congruence between the orbital-symmetry characteristics of the reactants and products.† Reactions lacking such congruence proceed poorly or not at all. *Congruence of orbital symmetry* means that the bonding character of all occupied molecular orbitals is preserved at all stages throughout the reaction. The greater the degree of bonding in the transition state of a concerted reaction, the lower the potential energy of that transition state, and the greater the rate of the reaction. The importance of orbital-symmetry (or phase)

---

†In a review article a few years later, Woodward and Hoffmann asserted: ". . . by utilizing the most simple but fundamental concepts of molecular orbital theory we have in the past three years been able to rationalize and predict the stereochemical course of virtually every concerted reaction."

16·6 THEORETICAL MODELS FOR PERICYCLIC REACTIONS

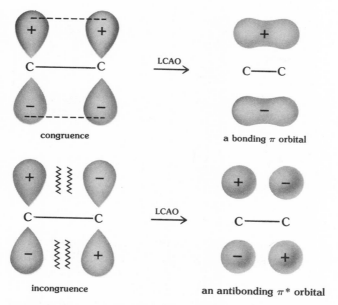

FIGURE 16·12 *Orbital congruence in LCAO bonding*

congruence is widely recognized in simple molecular-orbital theory. Recall from Chapter 4, for example, that adjacent p atomic orbitals may interact in a bonding (congruent) or an antibonding (incongruent) fashion, as shown in Figure 16.12. Woodward and Hoffmann extended this fundamental concept to systems undergoing concerted reactions. In order to do this rigorously, it is necessary to follow all the changes in orbital interactions and energies that take place as the reactants are transformed into products. Although a well-defined and relatively straightforward procedure exists for preparing such molecular-orbital correlation diagrams, it requires familiarity with certain "rules of the game" that are beyond our present scope.

THE FRONTIER-ORBITAL METHOD: CYCLOADDITION REACTIONS

Instead of the correlation-diagram approach, we shall treat pericyclic transformations by a simpler and less rigorous model, derived from the *frontier-orbital method* devised by K. Fukui of Japan. As an illustration of this approach, let us compare the Diels-Alder $[4_s+2_s]$ cycloaddition of 1,3-butadiene and ethylene with the $[2_s+2_s]$ cycloaddition of two ethylene molecules. If we ignore the phase relationships of the molecular orbitals, it might appear that, aside from the angle strain in the four-membered ring, both reactions are equally probable.†

---

†In fact, the $[2s+2s]$ transition state seems to allow a more precisely matched overlap of the p-orbital systems, since the terminal atoms of each reactant are about the same distance apart.

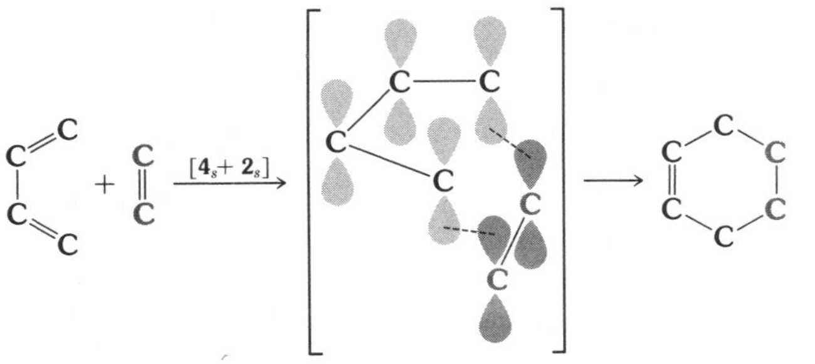

16·6 THEORETICAL MODELS FOR PERICYCLIC REACTIONS

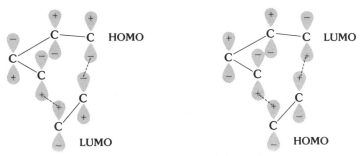

(16.55)

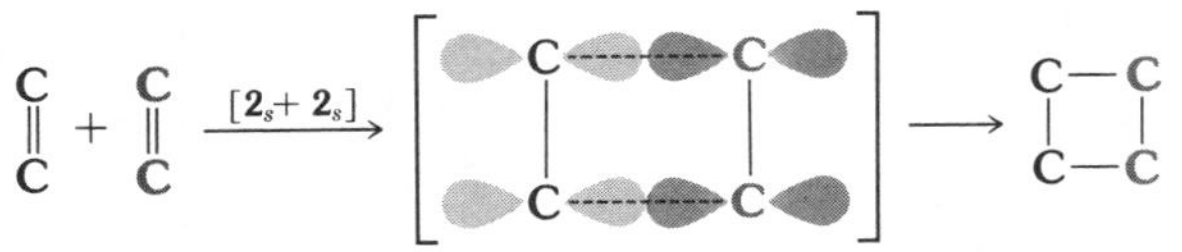

(16.56)

In the frontier-orbital method we make the reasonable assumption that in order to form bonds between the reactants, electrons will flow from the *highest-energy occupied molecular orbital (HOMO)* of one component to the *lowest-energy unoccupied molecular orbital (LUMO)* of the other component. Two such combinations are possible, and the first step in determining their phase congruence is to draw the HOMO and LUMO for each reactant. These are illustrated for the Diels-Alder reaction in Figure 16.13. In this case both possible supra-supra HOMO LUMO interactions are phase congruent; therefore, this $[4_s+2_s]$ thermal cycloaddition reaction should proceed through a bonding transition state. Such reactions are termed *symmetry allowed*:

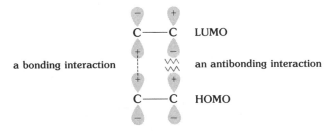

HOMO-LUMO interactions in a $[4_s+2_s]$ thermal cycloaddition

In contrast, the $[2_s+2_s]$ thermal cycloaddition does not have phase congruence and is said to be *symmetry forbidden*:

HOMO-LUMO interactions in a $[2_s+2_s]$ thermal cycloaddition

16·6 THEORETICAL MODELS FOR PERICYCLIC REACTIONS

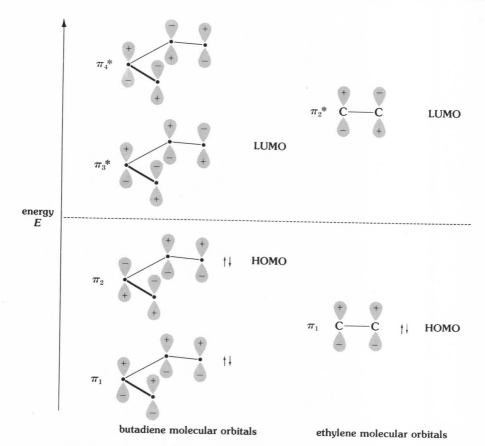

FIGURE 16·13 *Molecular orbitals for the Diels-Alder reaction*

If one of the bonding π electrons in a carbon-carbon double bond is promoted to the empty π_2^* orbital by photoexcitation (see Figure 9.4), the HOMO of this *excited state* has a phase symmetry which is congruent with the LUMO of an unexcited double bond. Thus, whereas $[2_s+2_s]$ thermal cycloaddition is forbidden, the $[2_s+2_s]$ photocycloaddition is symmetry allowed:†

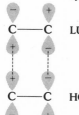

LUMO (ground state)

HOMO (excited state)

†These terms are useful as indicators of transition-state orbital congruence, but it is important to remember that other factors also influence the overall kinetic and thermodynamic characteristics of chemical reactions. For example, reasonable nonconcerted pathways may enable apparently "forbidden" reactions to take place. Furthermore, a symmetry-allowed reaction will usually have an activation energy of more than 15 kcal/mole due to other factors, such as rehybridization of atoms, steric hindrance, and changes in bond length.

564 16·6 THEORETICAL MODELS FOR PERICYCLIC REACTIONS

We can apply the frontier-orbital method to more complex reactions, such as the [6+2], [8+2], and [14+2] cycloadditions described in equations 16.43, 16.44, and 16.46, without going to the trouble of drawing all the π molecular orbitals in each unsaturated reactant. The only thing we really need to know is the orbital phase relationship at the terminal atoms of the HOMO and LUMO, and this is fortunately rather easy to determine. In the case of a stable molecule containing one double bond (such as ethylene), we know that the HOMO (π_1) has parallel phases at the ends of the π chain, whereas the LUMO has antiparallel phases. Each time another conjugated double bond (and its two π electrons) is introduced into the unsaturated system, this relationship reverses. These phase relationships are summarized in Figure 16.14.

PROBLEM 16·11
a Use the frontier-orbital method to explain the failure of the thermal $[6_s+2_s]$ cycloaddition reaction proposed in equation 16.44 and the success of the similar $[8_s+2_s]$ cycloaddition in equation 16.43.
b Suggest possible symmetry-allowed products from the following reaction:

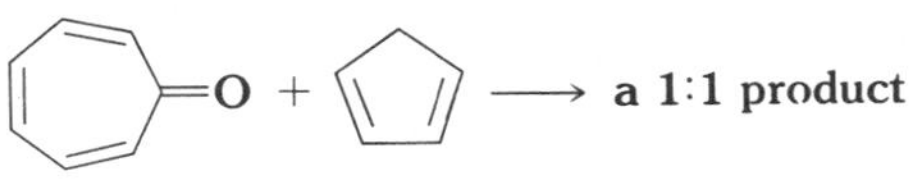

<chemical structure: cycloheptatriene with =O + cyclopentadiene ⟶ a 1:1 product>

c Explain the antarafacial course of reaction 16.47.

ELECTROCYCLIC REACTIONS

We can apply the frontier-orbital method to electrocyclic reactions such as 16.48 and 16.49 by viewing the ring-opening transformation as a cycloaddition of the breaking σ bond to the remaining π-electron system. For example, the cyclohexadiene ⇌ hexatriene interconversion in Figure 16.15 may be viewed as a σ-bond–diene interaction. In the thermal (ground-state) reaction, π_2 is the HOMO of the diene, and this can interact with the LUMO of the σ bond (σ^*) only if the latter cleaves and rotates in a *disrotatory* fashion:

(16.57)

Similarly, the LUMO of the diene (π_3^*) and the HOMO of the σ bond (σ) are also related in a disrotatory fashion:

(16.58)

16·6 THEORETICAL MODELS FOR PERICYCLIC REACTIONS

C------C HOMO for systems having 2, 6, 10, ..., $(4n + 2)$ π electrons
         LUMO for systems having 4, 8, 12, ..., $(4n)$ π electrons

a π orbital with parallel phases at the ends

C------C HOMO for systems having 4, 8, 12, ..., $(4n)$ π electrons
         LUMO for systems having 2, 6, 10, ..., $(4n + 2)$ π electrons

a π orbital with antiparallel phases at the ends

FIGURE 16·14 *A summary of HOMO and LUMO orbitals*

If, however, the diene moiety is electronically excited by the absorption of light, the HOMO of the excited diene becomes π_3^* and the LUMO is π_4^*. These frontier orbitals then interact with the σ orbitals in a *conrotatory* fashion:

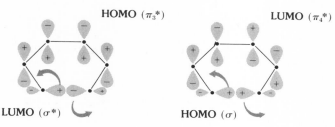

FIGURE 16·15 *Molecular orbitals for a six-electron electrocyclic reaction*

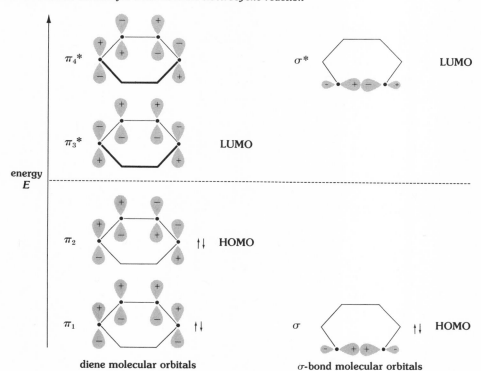

566   16·6 THEORETICAL MODELS FOR PERICYCLIC REACTIONS

Our theoretical model therefore clearly and unequivocally predicts that thermal cyclohexadiene ⇌ hexatriene electrocyclic transformations will be disrotatory, whereas the photochemical interconversions will be conrotatory.

> **PROBLEM 16·12**
> *a* Use the frontier-orbital method to explain the stereospecificity of reactions 16.49.
> *b* Suggest a reason for the dramatic difference in thermal reactivity between the stereoisomers in equation 16.45.

As a shortcut, we can make effective predictions regarding the stereospecificity of electrocyclic reactions by simply examining the HOMO of the open-chain polyene reactant. For example, the butadiene ⇌ cyclobutene transformation in Figure 16.16 is predicted to be conrotatory if it is thermal and disrotatory if it is photochemical.

> **PROBLEM 16·13**
> The following stereoisomerization has been observed. Explain the absence of any other stereoisomers among the products.

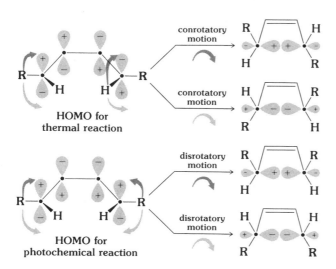

SIGMATROPIC REACTIONS

A simple method of analyzing sigmatropic rearrangements is first to imagine that the migrating group has undergone a homolytic cleavage from the π-orbital system, and then to ascertain whether this group could

FIGURE 16·16 Orbital congruence in a four-electron electrocyclic reaction

16·6 THEORETICAL MODELS FOR PERICYCLIC REACTIONS 567

maintain a bonding interaction with both its point of origin and the proposed rearrangement terminus in the HOMO of the π radical. For example, a sigmatropic hydrogen shift in 1,3-pentadiene would, by this approach, involve an interaction of the 1_s orbital of a hydrogen atom with the HOMO of the pentadienyl radical.† From the model in Figure 16.17 we predict that a [1,5] suprafacial shift is symmetry allowed, but a [1,3] shift would have to be antarafacial if a bonding transition state were to be maintained. Such a rearrangement would be structurally prohibited; indeed, reaction 16.42 shows an exclusive [1,5] suprafacial deuterium shift.

A similar analysis of the Cope rearrangement in Figure 16.9 leads us to an interaction between a pair of allyl radicals:

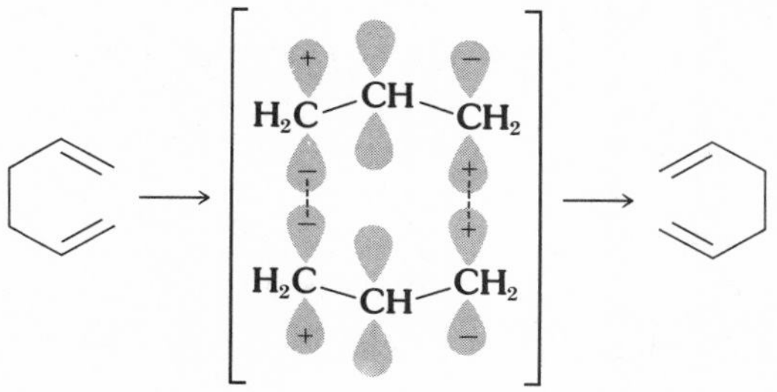

(16.59)

† In some π molecular orbitals the phase at a given carbon atom may be zero. Such carbon atoms are located on a nodal plane for that molecular orbital. The p-orbital component is left unmarked for these atoms.

FIGURE 16·17 *Orbital congruence in a sigmatropic hydrogen rearrangement*

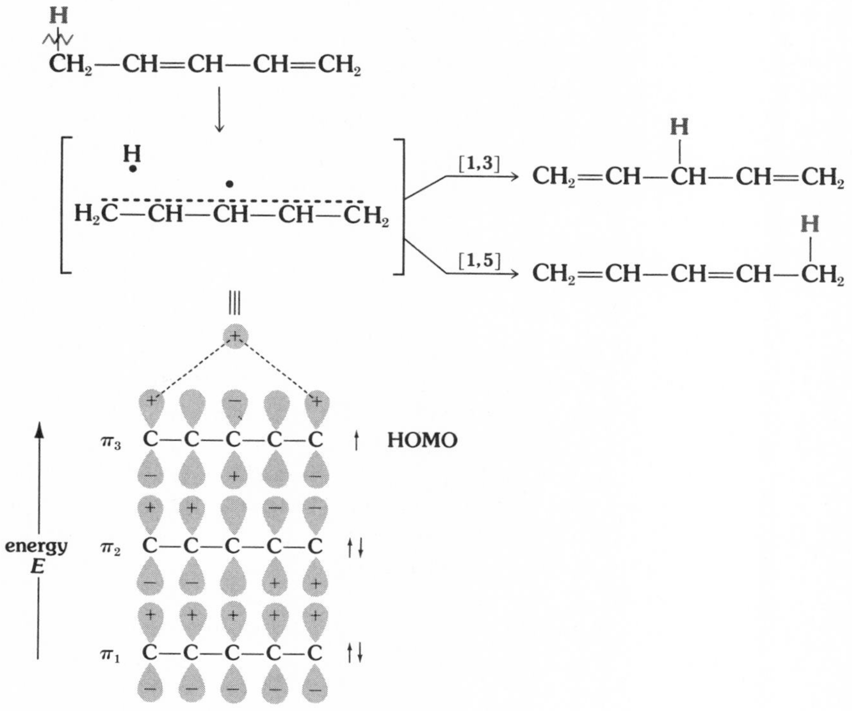

568   16·6 THEORETICAL MODELS FOR PERICYCLIC REACTIONS

The doubly suprafacial [3,3] sigmatropic rearrangement observed here is symmetry allowed and may proceed by either a chairlike or a boatlike six-membered cyclic transition state. Experimental evidence suggests that the chairlike arrangement shown in Figure 16.9 is normally preferred.

PROBLEM 16·14
Orbitals π_1, π_2, and π_3 of the pentadienyl radical are described in Figure 16.17. Draw π_4^* and π_5^*.

PROBLEM 16·15
a Show how the stereospecificity of the Cope rearrangement in Figure 16.9 requires a chairlike transition state.
b What kind of transition state is required for the following reaction?

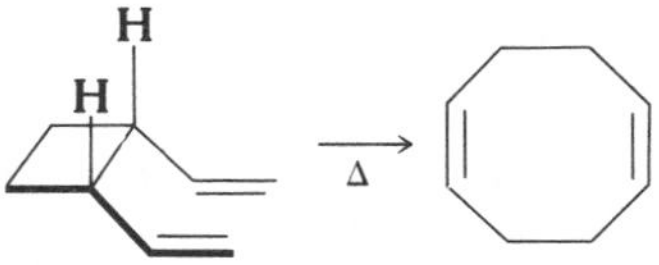

c The triene below is relatively stable despite the fact that its rearrangement to toluene should be highly exothermic. Suggest a reason for this stability.

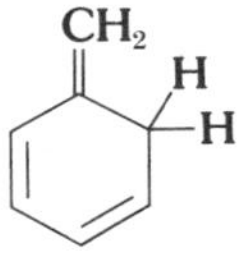

A SIMPLE EMPIRICAL RULE FOR PERICYCLIC REACTIONS

We can summarize the predictions of theoretical models by two easily remembered rules based on the number of delocalized electrons in the transition state of a pericyclic reaction. To do so we must first describe the reaction by a cyclic array of *electron-pair shifts* (indicated by curved arrows). For example,

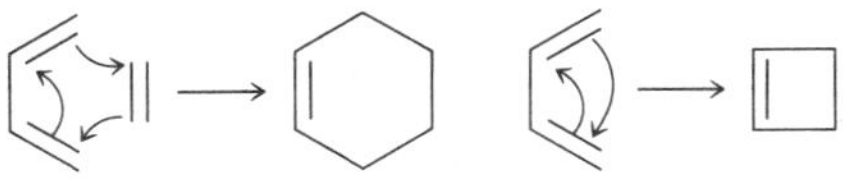

These reactions can then be classified according to whether an even or odd number of electron pairs (arrows) are involved in the transformation. An odd number of delocalized electron pairs corresponds to $4n + 2$ electrons (2, 6, 10, . . .) and may be termed *aromatic*. An even number of delocalized electron pairs corresponds to $4n$ electrons (4, 8, 12, . . .), and this we regard as an *antiaromatic* system. The rules for predicting allowed thermal (ground-state) and photochemical (excited state) pericyclic reactions are then as summarized in Table 16.5.

Since thermal cycloaddition and sigmatropic reactions involving relatively short and inflexible π-electron systems are usually restricted to suprafacial pathways, these rules permit us to make quick predictions

16·7 PERICYCLIC REACTIONS OF VITAMIN D

TABLE 16·5 *Rules for predicting allowed pericyclic reactions*

type of transition state	stereochemistry
thermal	
aromatic (odd number of arrows)	suprafacial-disrotatory
antiaromatic (even number of arrows)	antarafacial-conrotatory
photochemical	
aromatic (odd number of arrows)	antarafacial-conrotatory
antiaromatic (even number of arrows)	suprafacial-disrotatory

of the allowed or forbidden nature of reactions, as well as their preferred stereochemistry.

PROBLEM 16·16

a Predict the stereochemistry of the alkene formed in the following reaction:

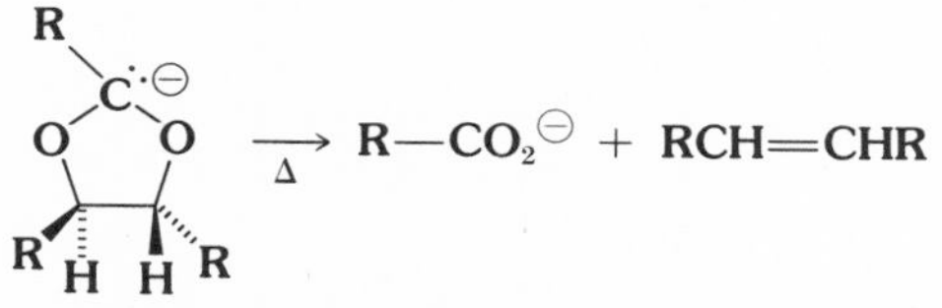

$$\text{(cyclic carbonate)} \xrightarrow{\Delta} R\text{—}CO_2^{\ominus} + RCH{=}CHR$$

b Explain the following transformation in terms of five successive pericyclic reactions, of which the first does not involve the quinone:

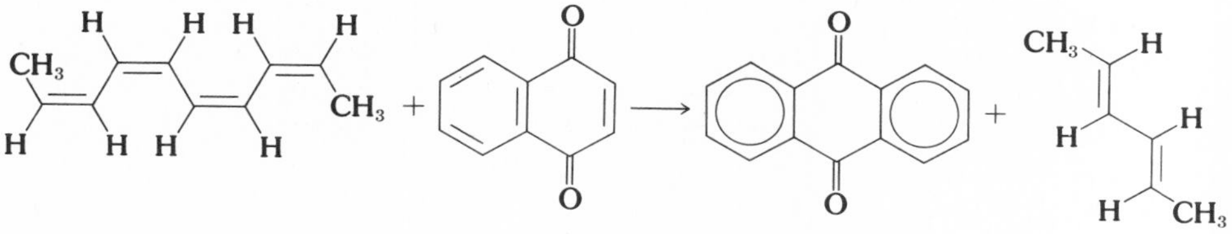

(triene + naphthoquinone → anthraquinone + diene)

16·7 PERICYCLIC REACTIONS OF VITAMIN D

Rickets, a disease of early childhood characterized by defective bone growth, is prevented by a fat-soluble substance that has come to be known as *vitamin D*. Substantial quantities of vitamin D are found in fish-liver oils, but its scarcity in most other foods can result in dietary deficiencies of the vitamin. Exposure to sunlight has long been known to be beneficial in preventing and treating rickets, a fact which led in 1924 to the discovery that a vitamin D deficiency in the diet could be remedied by irradiating the foodstuffs rather than the patient. Patents covering this method of increasing the antirachitic properties of foods yielded about $14 million in royalties to the Wisconsin Alumni Foundation over a period extending to 1945 (the initial research had been carried out by Steenbock at the Wisconsin Agricultural Experiment Station).

The importance of vitamin D to the health of human beings and livestock sparked a great deal of interest in this substance. Studies were

16·7 PERICYCLIC REACTIONS OF VITAMIN D

initiated in laboratories throughout the world to ascertain the nature of vitamin D. Two closely related compounds having antirachitic activity were isolated and proved to have very similar structural formulas (see Equation 16.60). Vitamin D_2 has an unsaturated side chain, whereas vitamin D_3 has a smaller, saturated side chain:

$$R = CH_3CHCH=CHCH(CH_3)CH(CH_3)_2 \qquad R = CH_3CHCH_2CH_2CH_2CH(CH_3)_2$$
$$\text{vitamin } D_2 \qquad\qquad\qquad\qquad \text{vitamin } D_3$$

Since many of the substances isolated in these studies were heat and light sensitive, progress was slow. However, it is now well established that a conjugated, doubly unsaturated sterol such as ergosterol undergoes a photochemical electrocyclic ring-opening reaction, giving an unstable intermediate, previtamin D, which is converted to vitamin D on mild heating:

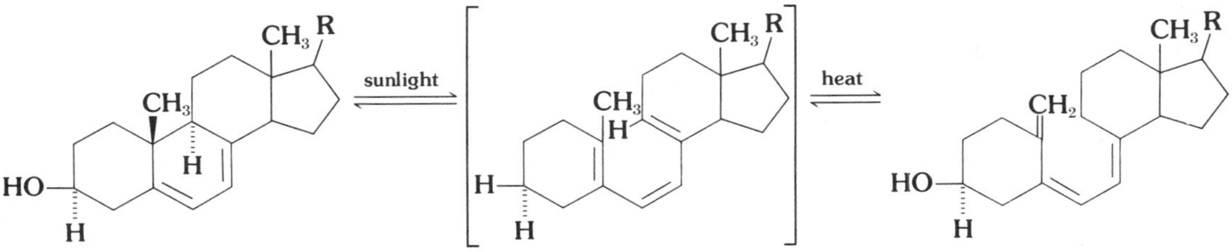

ergosterol ⇌(sunlight) [previtamin D] ⇌(heat) (16.60)

Figure 16.18 shows a number of other isomeric reaction products obtained under related photochemical and thermal reaction conditions. The unraveling of these interconversions, as well as several cis-trans double-bond isomerizations that are not shown here, must certainly be hailed as an outstanding achievement. Attempts to explain the complex interrelationships of these compounds met with frustration until the early 1960s, but it is now recognized that these transformations constitute a beautiful illustration of orbital symmetry control of pericyclic reactions.

PROBLEM 16·17

The novel $C_{10}H_{10}$ hydrocarbon bullvalene can undergo [3,3] sigmatropic rearrangements, the products of which are the same as the starting material (such rearrangements are called degenerate Cope rearrangements).

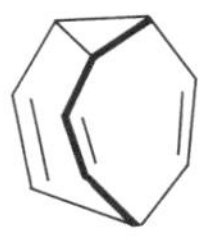

bullvalene

    a Write an equation illustrating this type of rearrangement in the bullvalene system.
    b The pmr spectrum of bullvalene at $-25°C$ shows two distinct signals at $\delta = 2.1$ and 5.7 (area ratio 2:3). At $60°C$ the prm spectrum shows only a single sharp signal, at $\delta = 4.2$. Explain the reason for this change.

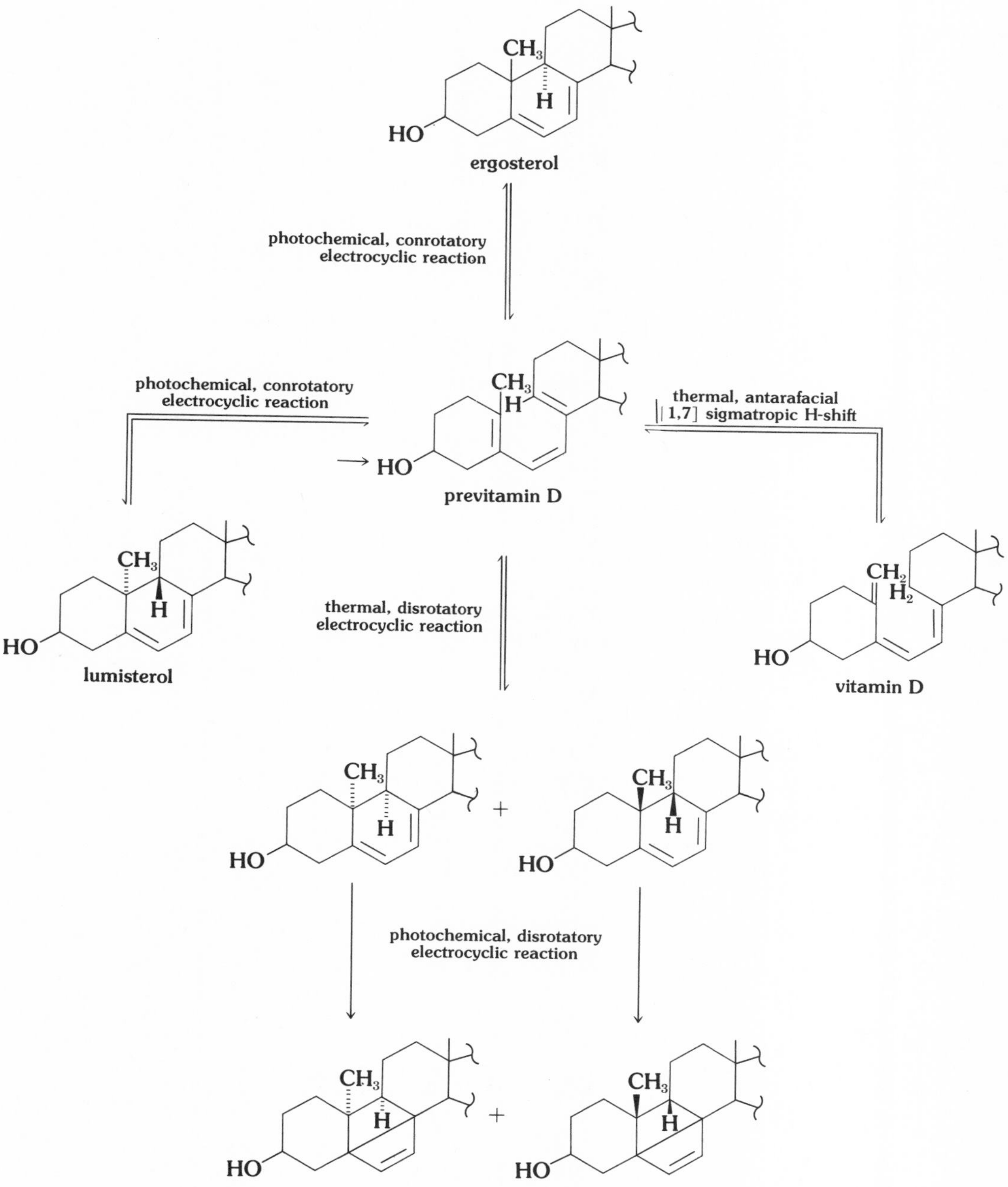

16·7 PERICYCLIC REACTIONS OF VITAMIN D

FIGURE 16·18 *Pericyclic transformations in the vitamin D series*

REFERENCES

W. A. Pryor, *Introduction to Free Radical Chemistry*, Prentice-Hall, Inc., Englewood Cliffs, N.J., 1966.

W. A. Pryor, *Radical Reactions*, McGraw-Hill Book Company, New York, 1966.

C. Walling, *Free Radicals in Solution*, John Wiley & Sons, Inc., New York, 1957.

D. C. Nonhebel and J. C. Walton, *Free Radical Chemistry*, Cambridge University Press, New York, 1974.

M. Bersohn and J. C. Baird, *An Introduction to Electron Paramagnetic Resonance*, W. A. Benjamin, Inc., New York, 1966.

E. S. Huyser, *Free-radical Chain Reactions*, Interscience, John Wiley & Sons, Inc., New York, 1970.

R. B. Woodward and R. Hoffmann, *The Conservation of Orbital Symmetry*, Verlag Chemie–Academic Press, New York, 1970.

K. Fukui, Recognition of Stereochemical Pathways by Orbital Interaction, *Accounts Chem. Res.*, 4:57 (1971).

R. G. Pearson, Symmetry Rules for Chemical Reactions, *Accounts Chem. Res.*, 4:152 (1971).

H. E. Zimmerman, The Möbius-Hückel Concept in Organic Chemistry, *Accounts Chem. Res.*, 4:272 (1971).

R. Lehr and A. Marchand, *Orbital Symmetry*, Academic Press, New York, 1972.

M. Orchin and H. Jaffe, *The Importance of Antibonding Orbitals*, Houghton Mifflin and Company, Boston, 1967.

H. Simmons and J. Bunnett, Orbital Symmetry Papers, American Chemical Society, Washington, D.C., 1974.

C. H. DePuy and O. L. Chapman, *Molecular Reactions and Photochemistry*, Prentice-Hall, Inc., Englewood Cliffs, N.J., 1972.

R. G. Pearson, *Symmetry Rules for Chemical Reactions*, Interscience, John Wiley & Sons, Inc., New York, 1976.

EXERCISES

16·1 Write structures for the species formed initially from the following reactions:

a $Br_2 \xrightarrow{h\nu}$

b $CH_3CH_2-O-N=O \xrightarrow{h\nu}$

c $CH_3\overset{O}{\underset{\|}{C}}-O-O-\overset{O}{\underset{\|}{C}}CH_3 \xrightarrow{\Delta}$

d $(CH_3)_3C-N=N-C(CH_3)_3 \xrightarrow{\Delta}$

e $\phi C(CH_3)_2-O-Cl \xrightarrow{h\nu}$

f $CH_3CH_2OOH + Fe^{2+} \longrightarrow$

g $\phi-SH + CH_3\dot{C}HCH_3 \longrightarrow$

h $\phi-CH=CH_2 + Cl\cdot \longrightarrow$

16·2 Write structural formulas for the products expected from these reactions:

a $(CH_3)_2C=CH_2 + HBr \xrightarrow{\text{radical scavengers}}$

b $(CH_3)_2C=CH_2 + HBr \xrightarrow{\text{peroxides}}$

c $(CH_3)_2CHCH_2\overset{\overset{O}{\|}}{C}ONa \xrightarrow[-e^\ominus]{\text{anode}}$

d 2 [phenol] $\xrightarrow{Fe(CN)_6^{3\ominus}}$

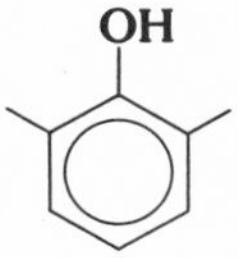

16·3 The following bond-dissociation energies for toluene and benzene can be added to the data given in Table 2.5 for the dissociation energies of C—H bonds (at 25°C):

$\phi CH_2{-}H$ 85 kcal/mole
$\phi{-}H$ 112 kcal/mole

a Given the following reaction,

$R-CH_2-H + Br\cdot \longrightarrow R-CH_2\cdot + HBr$

account for the fact that Br· reacts about 10^4 times faster with toluene ($R = \phi$) than with ethane ($R = CH_3$).

b Explain the exclusive formation of 1-bromo-1-phenylethane from the free-radical bromination of ethylbenzene:

$\phi-CH_2CH_3 + Br_2 \xrightarrow{h\nu} \phi-\underset{Br}{CHCH_3} + HBr$

c With Table 2.5 as a reference, arrange the following types of hydrogens in order of ease of abstraction by Br·:

benzylic   primary   tertiary   methyl   secondary   phenyl

16·4 Section 15.2 describes the effects of an alkyl side chain on electrophilic aromatic substitution. The benzene ring in turn influences the rate and position of free-radical substitution on the side chain (see Exercise 16.3). Show the structures of the chief organic products expected from each of the following reactions (if any).

a ethylbenzene + $Br_2 \xrightarrow{Fe}$

b ethylbenzene + $Br_2 \xrightarrow{h\nu}$

c [naphthalene] + $Br_2 \xrightarrow{h\nu}$

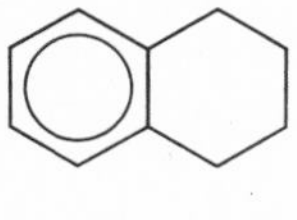

d ethylbenzene + $KMnO_4 \xrightarrow{\Delta}$

e benzene + $KMnO_4 \xrightarrow{\Delta}$

f ethane + $KMnO_4 \xrightarrow{\Delta}$

g [tetralin] + $KMnO_4 \xrightarrow{\Delta}$

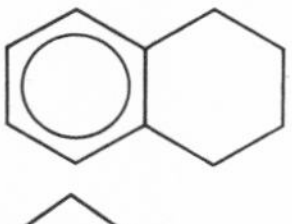

h [cyclopentane] + $KMnO_4 \xrightarrow{\Delta}$

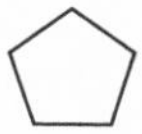

574 EXERCISES

16·5 Show all the steps in the free-radical bromination of ethylbenzene. Label each of the steps in this chain reaction as initiation, propagation, or termination.

16·6 Write a mechanism for the second step of the reaction in equation 16.28.

16·7 Explain the course of the following reactions:

a) $\phi\text{-}\cdots\text{-}O\text{-}N=O \xrightarrow{h\nu}$ an isomer $\xrightarrow{H_3O^{\oplus}}$ $\phi\text{-}CO\text{-}\cdots\text{-}OH$

b) pyrrolidine-NHCl with CH$_3$ $\xrightarrow{h\nu}$ protonated pyrrolidinium Cl$^{\ominus}$

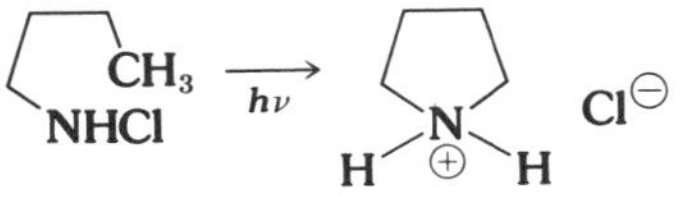

16·8 Polystyrene has a polymer chain of regularly alternating CH$_2$ and CHϕ units. Suggest a reason for this "head-to-tail" free-radical polymerization of styrene:

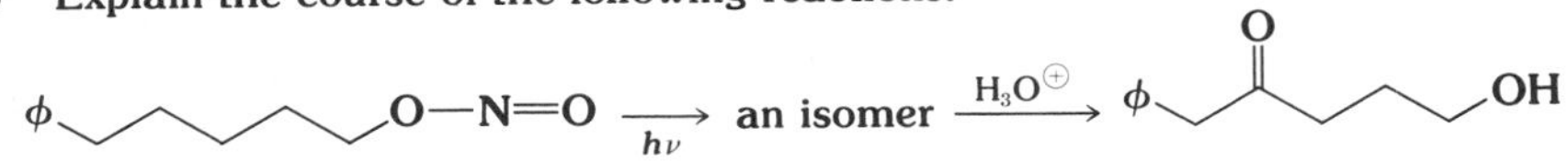

$\sim$CH$_2$CHCH$_2$CHCH$_2$CHCH$_2$CH$\sim$ (each CH bearing ϕ)
polystyrene

16·9 Show the product expected from each of the following reactions:

a) cis-5,6-dimethyl-1,3-cyclohexadiene $\xrightarrow{\Delta}$

b) cis-5,6-dimethyl-1,3-cyclohexadiene $\xrightarrow{h\nu}$

c) cyclooctatriene with D labels $\xrightarrow{h\nu}$

d) cyclooctatriene with D labels $\xrightarrow{\Delta}$

e) cyclooctatetraene $\xrightarrow{\Delta}$

f) cyclooctatetraene $\xrightarrow{h\nu}$

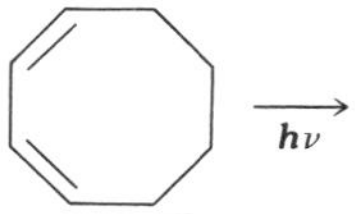

16·10 How could the following interconversions be accomplished?

a) (2Z,4Z)-2,4-hexadiene → (2E,4Z)-2,4-hexadiene

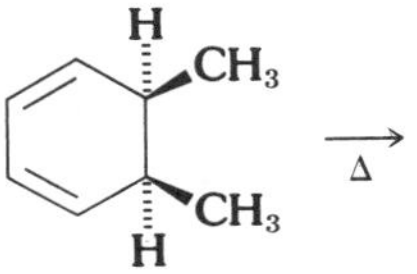

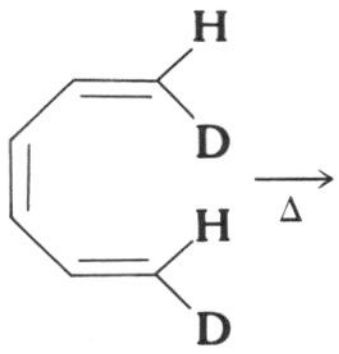

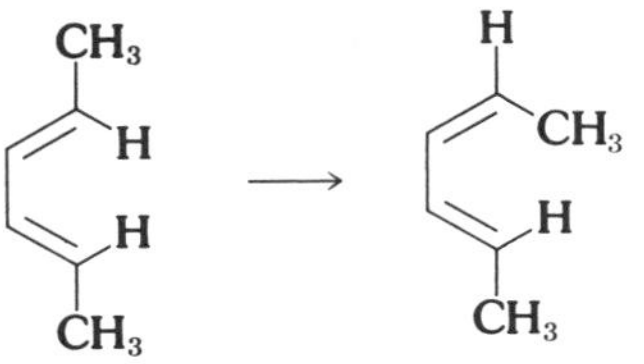

b) cis-5,6-dimethyl-1,3-cyclohexadiene → trans-5,6-dimethyl-1,3-cyclohexadiene

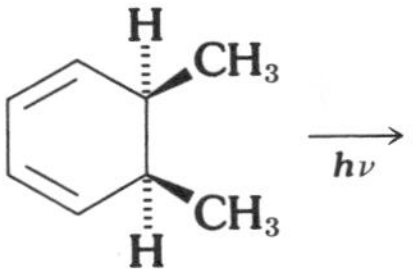

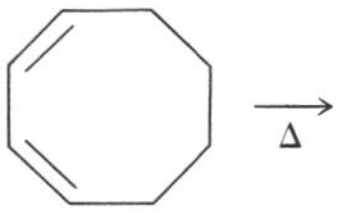

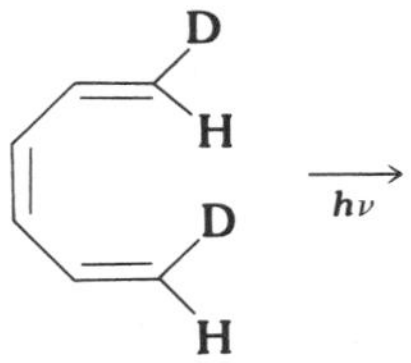

c) cis-dimethylcyclobutane → trans-dimethylcyclobutane

16·11 Explain the course of the following reactions:

a) [structure] →(140°, Δ) [structure] →(220°, Δ) [structure]

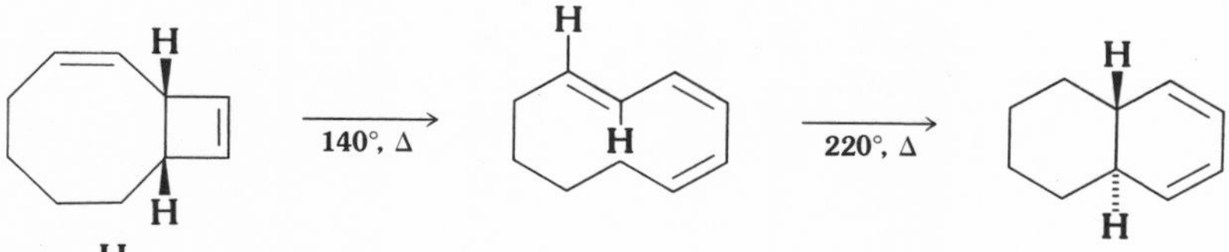

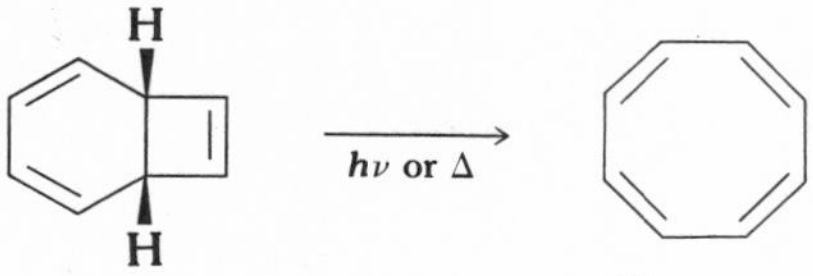

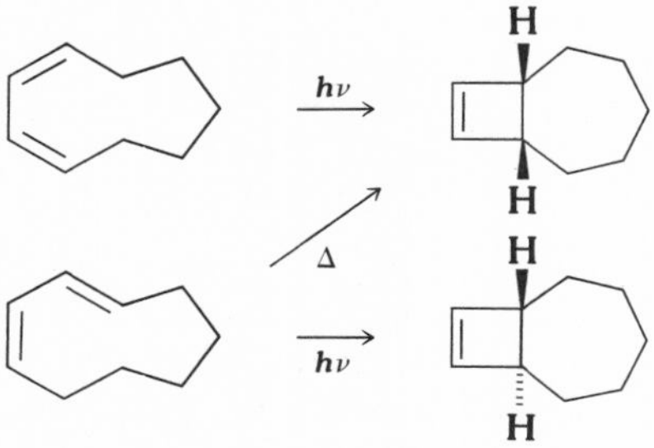

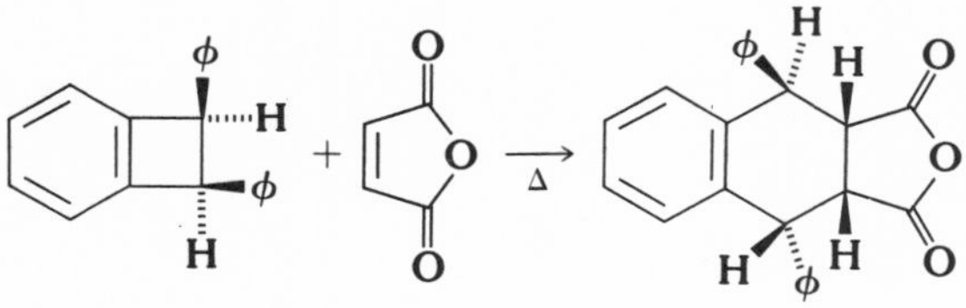

b) [structure] →(hν or Δ) [structure]

c) [structure] →(hν) [structure]; [structure] →(Δ) / →(hν) [structure]

d) [structure] + [maleic anhydride] →(Δ) [structure]

e) HOCH(H)−CH=CH−CH₂−CH=CH₂ →(Δ) CH₂=CH−CH₂−CH₂−CH₂−CHO

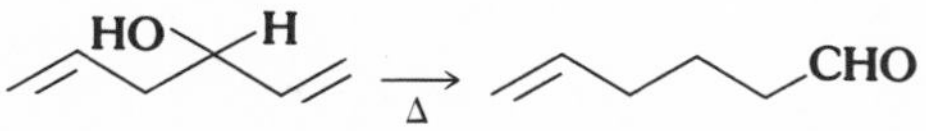

f) *meso*-3,4-dimethyl-1,5-hexadiene →(Δ) *trans, trans*-2,6-octadiene

(±)-3,4-dimethyl-1,5-hexadiene →(Δ) *cis,trans*-2,6-octadiene

16·12 1,3-Cyclopentadiene undergoes a thermal $[4_s+2_s]$ dimerization to give the endo stereoisomer as the major product (see Section 15.3):

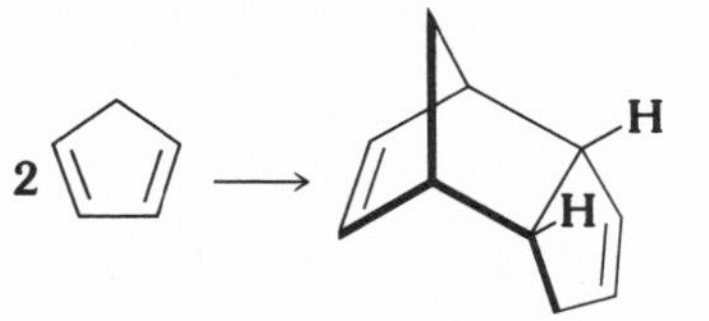

2 [cyclopentadiene] → [endo dicyclopentadiene]

endo isomer

Using the frontier-orbital method, examine this reaction and suggest a reason for the exclusive endo pathway.

16·13 Explain the highly stereospecific nature of the following rearrangement:

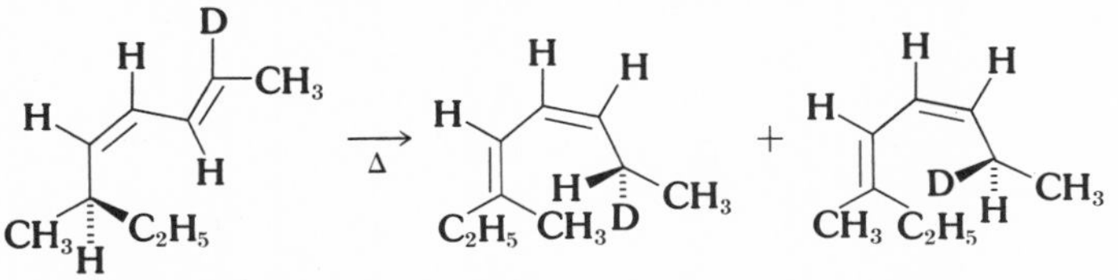

[structure, R, cis, trans] →(Δ) [structure, S, cis, trans] + [structure, R, cis, cis]

16·14 The relief of strain and the gain in resonance energy that takes place in the isomerization of Dewar benzene to benzene is of such a magnitude that the very existence of Dewar benzene is surprising. Suggest an explanation for the fact that this isomer of benzene can actually be made, isolated, and stored.

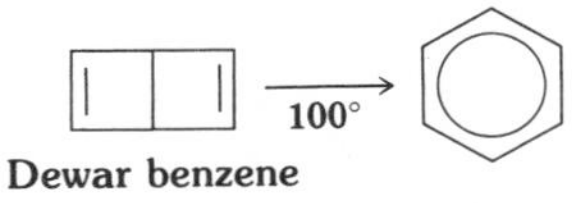

$$\text{Dewar benzene} \xrightarrow{100°} \text{benzene} \qquad \Delta H \approx -60 \text{ kcal/mole}$$

16·15 The structural constraints of the four-membered ring make it likely that the following reaction is *not* concerted. Suggest a mechanism.

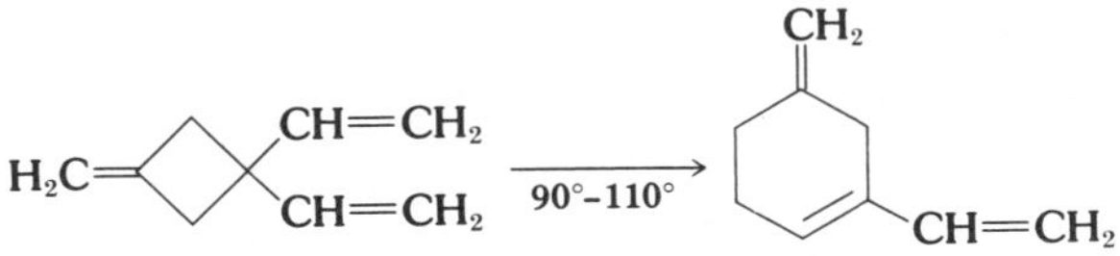

$$H_2C=\underset{\text{(four-membered ring)}}{\bigg\langle}\genfrac{}{}{0pt}{}{CH=CH_2}{CH=CH_2} \xrightarrow{90°-110°} \text{(cyclohexene with }=CH_2\text{ and }-CH=CH_2\text{ substituents)}$$

PROTEINS, PEPTIDES, AND AMINO ACIDS

PROTEINS, from the Greek *proteios*, meaning first, are a class of organic compounds which are present in and vital to every living cell. In the form of skin, hair, callus, cartilage, muscles, tendons, and ligaments, proteins hold together, protect, and provide structure to the body. In the form of enzymes, hormones, antibodies, and globulins, they catalyze, regulate, and protect the body chemistry. In the form of hemoglobin, myoglobin, and various lipoproteins, they effect the transport of oxygen and other substances within the body.

Most of us think of proteins as highly beneficial substances. Protein is a necessary part of the diet of all animals. Indeed, human beings can become seriously ill if they do not eat enough protein.† An example is the disease *kwashiorkor*, found in its severest form in parts of Africa and South America. Protein antibodies and antibiotics also help us to fight disease, and we warm and protect our bodies with clothing and shoes which are often protein in nature (wool, silk, and leather).

The deadly nature of protein toxins and venoms is perhaps less well known. The toxic protein secreted by the *Clostridium botulinum* bacillus is, in fact, the most powerful poison yet discovered. One microgram (10^{-6} g) of this substance, which acts by inhibiting the synthesis of acetylcholine (see Section 14.6) will kill roughly 1000 kg (over a ton) of mice. Most snake venoms are proteins, as is ricin, the poisonous principle of the castor bean (*Ricinus sanguineus* L. Euphorbiacae).

---

†Not all proteins have equal food value. Although peanuts have more protein than fish or eggs, the nutritional value of that protein is less than a third that of these other foods.

17·1 THE COMPOSITION OF PROTEINS

Despite the variety of their physiological roles and the differences in their physical appearance—silk is a flexible fiber, horn a tough rigid solid, and the enzyme pepsin is a crystalline solid—proteins are sufficiently similar in their molecular structure to justify our treatment of them as a single chemical family. Indeed, the proteins provide an excellent example of the extent to which variations on a simple structural theme can lead to an enormous diversity in properties and behavior.

17·1 THE COMPOSITION OF PROTEINS

The large molecules of proteins (molecular weights usually range from 10,000 to 5,000,000) are composed largely of carbon, hydrogen, nitrogen, oxygen, and smaller amounts of sulfur (traces of other elements may also be present). When proteins are treated with boiling acid or base solutions, they hydrolyze to small fragments which have been identified as α-amino carboxylic acids. In the course of many such degradative investigations applied to many different proteins, chemists have isolated from 20 to 30 distinct amino acids, most of which are listed in Table 17.1. Some of these naturally occurring α-amino acids have additional amino or carboxylic acid functions and are termed *basic* or *acidic amino acids*, respectively.

From x-ray diffraction studies of proteins it is clear that the amino acid units are bound together by a repeating sequence of amide bonds, called *peptide linkages*:

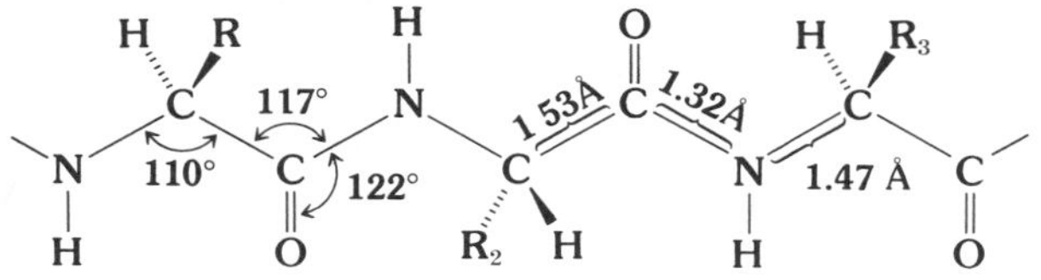

Proteins are therefore natural polymers consisting of long polyamide chains (a polypeptide) to which are attached various side chains or functions characteristic of each amino acid in the chain. The number of amino acid units comprising a given protein is usually very large; hence the possible combinations of 20 or more different amino acids is enormous. If, for example, we wanted to construct all the 10-unit peptide chains (decapeptides) that could possibly be formed by various combinations of 20 different amino acids, simple probability theory tells us there will be 20^{10} different combinations. Even if we reduce the variety of amino acids to four, the number of possible decapeptides will still exceed a million (10^6)!

PROBLEM 17·1

a The peptide hormone insulin (beef pancreas) contains 3.2% sulfur in the form of cystine (the cysteine dimer). What is the lowest molecular weight this hormone can have?

b The molecular weight of beef insulin is about 6000. How many cystine units are present in a molecule of insulin?

17·1 THE COMPOSITION OF PROTEINS

TABLE 17·1 *Naturally occurring amino acids*

name†	abbreviation	formula
glycine	Gly	$H-\underset{NH_2}{\overset{H}{\underset{|}{C}}}-COOH$
alanine	Ala	$CH_3-\underset{NH_2}{\overset{H}{\underset{|}{C}}}-COOH$
valine^e	Val	$\underset{CH_3}{\overset{CH_3}{\diagdown}}CH-\underset{NH_2}{\overset{H}{\underset{|}{C}}}-COOH$
leucine^e	Leu	$\underset{CH_3}{\overset{CH_3}{\diagdown}}CH-CH_2-\underset{NH_2}{\overset{H}{\underset{|}{C}}}-COOH$
isoleucine^e	Ileu	$CH_3-CH_2-\underset{CH_3}{\overset{}{\underset{|}{CH}}}-\underset{NH_2}{\overset{H}{\underset{|}{C}}}-COOH$
phenylalanine^e	Phe	$C_6H_5-CH_2-\underset{NH_2}{\overset{H}{\underset{|}{C}}}-COOH$
proline	Pro	$\begin{array}{c} H_2C-\overset{H}{\underset{|}{C}}-CO_2H \\ | \quad\quad | \\ H_2C \quad\quad N \\ \diagdown\;\;\diagup \;\;\diagdown \\ CH_2 \quad H \end{array}$
serine	Ser	$HO-CH_2-\underset{NH_2}{\overset{H}{\underset{|}{C}}}-COOH$
threonine^e	Thr	$CH_3-\underset{OH}{\overset{}{\underset{|}{CH}}}-\underset{NH_2}{\overset{H}{\underset{|}{C}}}-COOH$
tyrosine	Tyr	$HO-C_6H_4-CH_2-\underset{NH_2}{\overset{H}{\underset{|}{C}}}-COOH$
cysteine	CySH	$HS-CH_2-\underset{NH_2}{\overset{H}{\underset{|}{C}}}-COOH$

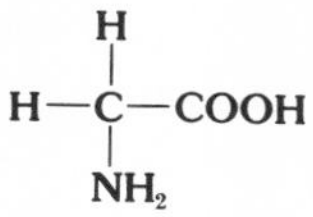

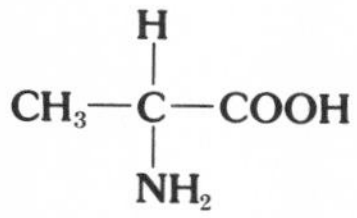

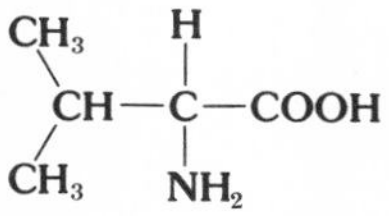

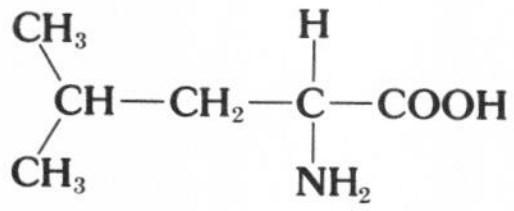

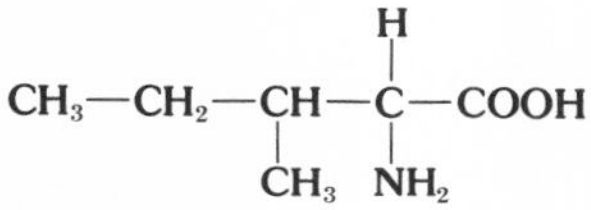

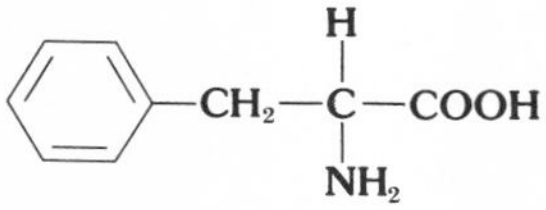

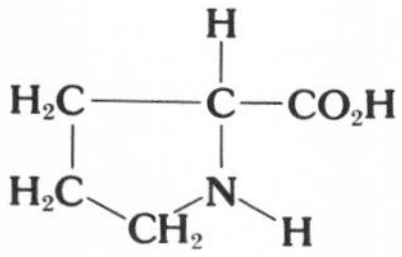

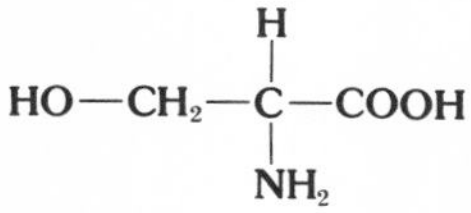

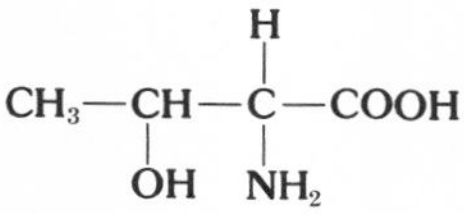

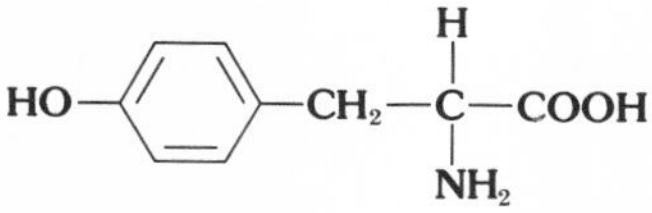

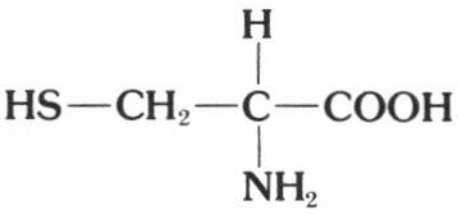

(continued)

17·1 THE COMPOSITION OF PROTEINS

TABLE 17·1 *Naturally occuring amino acids (continued)*

name†	abbreviation	formula
cystine	CySSCy	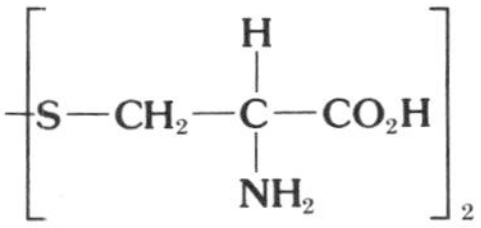 $\left[-S-CH_2-\underset{NH_2}{\overset{H}{C}}-CO_2H \right]_2$
methionine[e]	Met	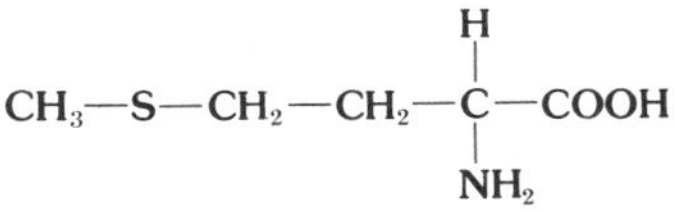 $CH_3-S-CH_2-CH_2-\underset{NH_2}{\overset{H}{C}}-COOH$
aspartic acid	Asp	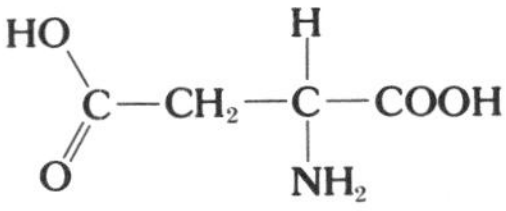 $\underset{O}{\overset{HO}{C}}-CH_2-\underset{NH_2}{\overset{H}{C}}-COOH$
glutamic acid	Glu	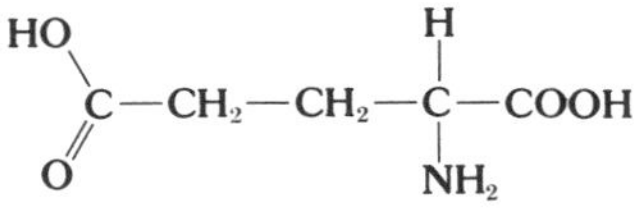 $\underset{O}{\overset{HO}{C}}-CH_2-CH_2-\underset{NH_2}{\overset{H}{C}}-COOH$
asparagine	Asp NH$_2$	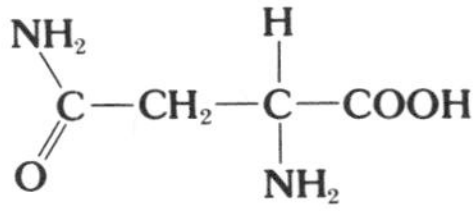 $\underset{O}{\overset{NH_2}{C}}-CH_2-\underset{NH_2}{\overset{H}{C}}-COOH$
glutamine	Glu NH$_2$	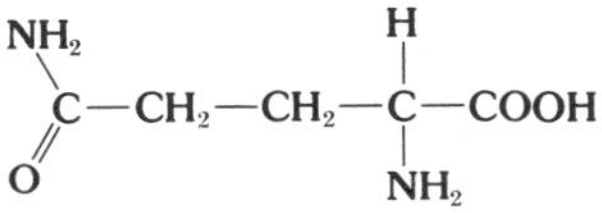 $\underset{O}{\overset{NH_2}{C}}-CH_2-CH_2-\underset{NH_2}{\overset{H}{C}}-COOH$
tryptophan[e]	Try	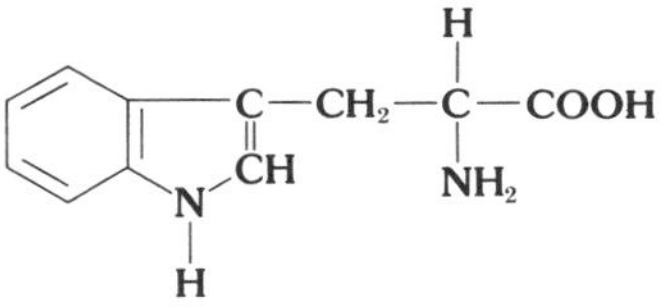 indole-$CH_2-\underset{NH_2}{\overset{H}{C}}-COOH$
lysine[e]	Lys	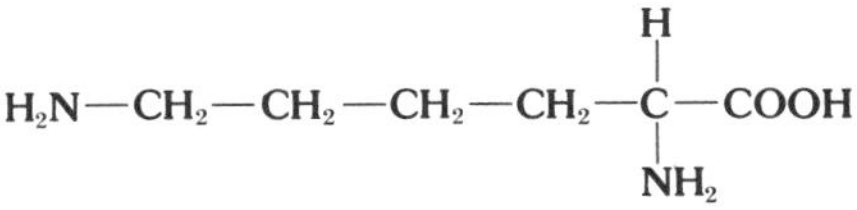 $H_2N-CH_2-CH_2-CH_2-CH_2-\underset{NH_2}{\overset{H}{C}}-COOH$
arginine	Arg	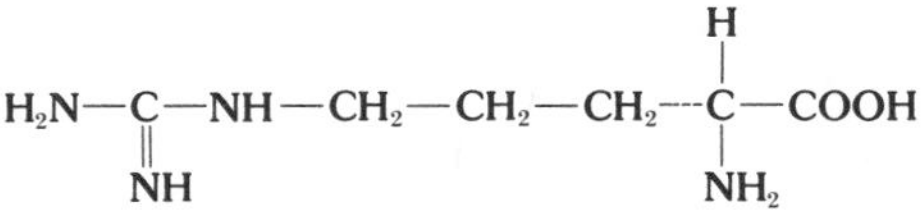 $H_2N-\underset{NH}{\overset{\|}{C}}-NH-CH_2-CH_2-CH_2-\underset{NH_2}{\overset{H}{C}}-COOH$
histidine	His	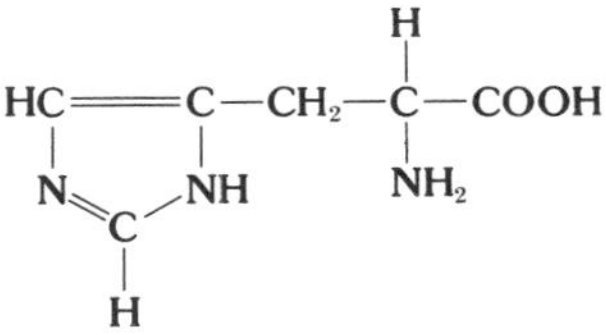 imidazole-$CH_2-\underset{NH_2}{\overset{H}{C}}-COOH$

†The superscript *e* refers to *essential amino acids*. These cannot be synthesized by the human metabolism and must therefore be present in the diet (usually bound in the form of protein) in order to maintain a proper nitrogen balance in the system.

17·2 CONFIGURATION AND PROPERTIES OF α AMINO ACIDS

Recognition of their key role in protein structure stimulated extensive investigations of both natural and synthetic α-amino acids, disclosing many characteristic and sometimes unexpected properties of these difunctional compounds.

1 All the natural amino acids save glycine have an asymmetric α carbon unit, and these are found in most cases to have an S configuration (sometimes called an L configuration in older texts). Recent studies, however, have identified significant amounts of natural R amino acids, particularly in invertebrate organisms.† The R amino acids are usually not incorporated in protein, but occur in free form or as simple amides or esters.

$$H_2N-\overset{CO_2H}{\underset{CH_3}{\overset{|}{C}}}-H \equiv \underset{H}{\overset{H_2N}{C}}\overset{CO_2H}{\underset{CH_3}{}} \qquad \underset{H_2N}{\overset{H}{C}}\overset{CO_2H}{\underset{CH_3}{}} \equiv H-\overset{CO_2H}{\underset{CH_3}{\overset{|}{C}}}-NH_2$$

S-alanine   R-alanine

2 The amino acids are nonvolatile crystalline solids, melting with decomposition at fairly high temperatures. In contrast, amines and carboxylic acids of equivalent molecular weights tend to be liquids or low-melting solids.

3 Amino acids are insoluble in nonpolar solvents such as benzene and hexane, but show appreciable solubility in water (aspartic acid, cystine, and tyrosine are exceptions). These aqueous solutions behave as though the solute has a very high dielectric constant.

4 The acidity and basicity constants of amino acids (they are amphoteric) are surprisingly low. Alanine, for example, has constants of $K_a = 1.9 \times 10^{-10}$ and $K_b = 5.1 \times 10^{-12}$, whereas most simple aliphatic carboxylic acids exhibit K_a values around 10^{-5} and the corresponding amines have K_b values of about 10^{-4}.

Many of the above properties can be explained as a consequence of inner-salt, or zwitterion, formation (see Section 13.1):

$$H_2\ddot{N}-\underset{H}{\overset{R}{\underset{|}{C}}}-\underset{\ddot{O}-H}{\overset{\ddot{O}}{\overset{\|}{C}}} \rightleftharpoons H_3\overset{\oplus}{N}-\underset{H}{\overset{R}{\underset{|}{C}}}-\underset{\ddot{O}{:}^{\ominus}}{\overset{\ddot{O}}{\overset{\|}{C}}}$$

neutral form   zwitterion form

Recall from Chapter 3 that acid-base equilibria favor formation of the weakest acid and weakest base. It is this factor that, in the present case, provides the driving force for zwitterion formation, inasmuch as ammonium ions ($R-NH_3^{\oplus}$) are weaker Brønsted acids than carboxylic acids, and carboxylate anions are weaker bases than amines.

†Recent analyses of carbonaceous meteors (such as the Murray and Murchison meteorites) have identified up to 17 different amino acids, 11 of which have not yet been observed in terrestrial protein. Many of these amino acids were found as nearly equal amounts of R and S enantiomers.

If strong acids or bases are added to aqueous solutions of amino acids, the corresponding conjugate acid or conjugate base is formed. Amino acids thus serve as buffers; that is, they hold the pH of their solutions fairly constant.

$$\overset{\oplus}{H_3}NCHRCO_2H \underset{H^\oplus}{\overset{-H^\oplus}{\rightleftharpoons}} \overset{\oplus}{H_3}NCHRCO_2^\ominus \underset{H^\oplus}{\overset{-H^\oplus}{\rightleftharpoons}} H_2NCHRCO_2^\ominus \qquad (17.1)$$

conjugate acid, predominates at pH < 2

predominates at pH 5 to 7

conjugate base, predominates at pH > 11

The pH at which positively charged amino acid ions are present in equal concentration to the corresponding negatively charged ions ($[\overset{\oplus}{H_3}NCHRCO_2H] = [H_2NCHRCO_2^\ominus]$) is called the *isoelectric point*. If an amino acid solution at this pH is placed in an electric field, it will exhibit no net migration of solute to either electrode. Most simple amino acids have isoelectric points in the range 5.0 to 6.5. Acidic amino acids such as aspartic acid and glutamic acid have isoelectric points of about 3.0, while the isoelectric point of basic amino acids such as lysine and arginine is 10.0 or higher.

PROBLEM 17·2
a Write equilibrium-constant expressions for K_a and K_b of glycine solutions in water.
b Would you expect an amino acid to be more soluble or less soluble in water at its isoelectric pH than at higher or lower values?

PROBLEM 17·3
Why should the isoelectric point for aspartic acid (2.8) be so much lower than that for leucine (6.0)?

17·3 THE PRIMARY STRUCTURE OF PROTEINS

Once the amino acid composition of a protein has been determined by hydrolysis, the next major challenge is determining its *primary structure*, the exact sequence of all the amino acids in the peptide chain.† One of the earliest approaches to this problem involved partial hydrolysis of a protein to a mixture of small fragments containing two, three, or more amino acids (enzymatic hydrolysis is often used for this purpose). Isolation and identification of these peptide fragments revealed part of the amino acid sequence of the original protein, and by matching a sufficient number of overlapping peptides the full sequence could be established.

More recent methods for accomplishing selective cleavage of terminal amino acids in proteins have proved especially useful in determining amino acid sequences near the ends of protein chains. The enzyme carboxypeptidase, for example, hydrolyzes peptide bonds adjacent to a free carboxyl group as shown in equation 17.2.

†If we consider the amino acid composition as being analogous to a molecular formula, the establishment of the correct sequence corresponds to the writing of a structural formula.

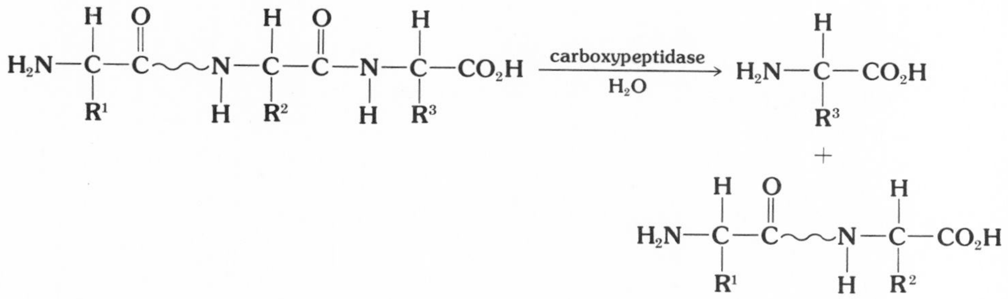

17·3 THE PRIMARY STRUCTURE OF PROTEINS 583

$$\text{H}_2\text{N}-\underset{\underset{R^1}{|}}{\overset{\overset{H}{|}}{C}}-\overset{\overset{O}{\|}}{C}\sim\sim\text{N}-\underset{\underset{R^2}{|}}{\overset{\overset{H}{|}}{C}}-\overset{\overset{O}{\|}}{C}-\text{N}-\underset{\underset{R^3}{|}}{\overset{\overset{H}{|}}{C}}-\text{CO}_2\text{H} \xrightarrow[\text{H}_2\text{O}]{\text{carboxypeptidase}} \text{H}_2\text{N}-\underset{\underset{R^3}{|}}{\overset{\overset{H}{|}}{C}}-\text{CO}_2\text{H} \quad (17.2)$$

$$+$$

$$\text{H}_2\text{N}-\underset{\underset{R^1}{|}}{\overset{\overset{H}{|}}{C}}-\overset{\overset{O}{\|}}{C}\sim\sim\text{N}-\underset{\underset{R^2}{|}}{\overset{\overset{H}{|}}{C}}-\text{CO}_2\text{H}$$

The *Edman degradation*, illustrated in Figure 17.1, is specific for free α-amino functions. Repetition of these methods discloses the order of amino acid units at each end of the peptide chain, and in the case of small to moderate-size peptides the complete sequence can be determined.†

PROBLEM 17·4
Write a mechanism for the Edman degradation. Why does the Edman reaction not cleave the entire polypeptide chain into thiazolinone fragments?

As an illustration of the sequencing procedure described above, let us consider a heptapeptide which on complete hydrolysis yields three glycines, two alanines, a valine, and a leucine. Suppose we find that the N-terminal amino acid (that having the free α-amino group) is one of the alanines, while the C-terminal residue (that amino acid having the free carboxyl group) is a glycine. We can then write the following partial formula for the heptapeptide:††

Ala·[Gly₂AlaValLeu]·Gly

†Machine automation of amino acid sequencing by terminal-group analysis has been successfully applied through ten or more cycles.

††By convention, peptides and proteins are always written so that the N-terminal amino acid is at the left and the C-terminal amino acid is at the right. In ambiguous cases an arrow is used, pointing to the nitrogen of the peptide bond.

FIGURE 17·1 *The Edman degradation*

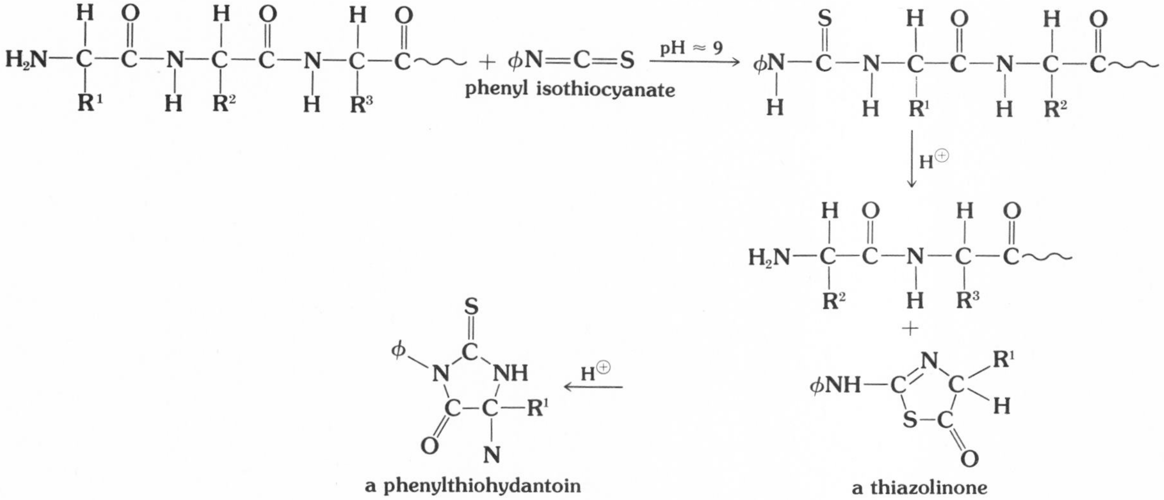

The sequence of five amino acids in the center of the peptide chain is at this point unknown; however, were we to find that partial hydrolysis of the heptapeptide gives the tripeptide Val·Gly·Leu and the dipeptides Gly·Gly and Leu·Ala (among others), we could then deduce the primary structure precisely:

Ala·Val·Gly·Leu·Ala·Gly·Gly

17·4 PEPTIDES

Peptides are arbitrarily defined as proteinlike substances having molecular weights less than 10,000. Depending on the number of amino acid units making up the peptide chains or rings, they may be classified as *dipeptides, tripeptides,* and so on up to *polypeptides;* however, the distinction between a polypeptide and a protein is not clear cut. Small peptides may be named as derivatives of the C-terminal amino acid. Thus the widely distributed tripeptide glutathione is properly named γ-S-glutamyl-S-cysteinylglycine. (Peptide bonding at the γ-carboxyl group of glutamic acid is not common in proteins.)

$$\overset{\oplus}{H_3}NCHCH_2CH_2CONHCHCONHCH_2CO_2H \equiv \gamma Glu \cdot Cy(SH) \cdot Gly$$
$$\underset{CO_2^{\ominus}}{|} \qquad \underset{CH_2SH}{|}$$

glutathione

NATURALLY OCCURRING PEPTIDES

Peptides have been extensively studied, partly because they are simple, easily acquired models for protein behavior, and were in fact instrumental in the determination of protein structure. However, they are also important natural products in their own right. Bradykinin, for example, is a hypotensive and smooth-muscle-stimulating principle formed by cleavage of plasma proteins.

Arg·Pro·Pro·Phe·Ser·Pro·Phe·Arg
      bradykinin

The posterior pituitary gland secretes several peptide hormones, the most important of which are the cyclic disulfides oxytocin and vasopressin.

```
   N-terminal                              N-terminal
     unit                                    unit
    Cy—Tyr—Ileu                             Cy—Tyr—Phe
   /        \                              /        \
  S          \                            S          \
  |           \                           |           \
  S            \                          S            \
   \            \                          \            \
   Cy—Asp(NH₂)—Glu(NH₂)                   Cy—Asp(NH₂)—Glu(NH₂)
   |                                       |
   Pro—Leu—Gly(NH₂)                        Pro—Arg—Gly(NH₂)
         C-terminal                              C-terminal
            unit                                    unit
        oxytocin                               vasopressin
```

17·4 PEPTIDES

The former stimulates the muscles of the uterus, while the latter exerts a pressor action (increases the blood pressure).

Cleavage of the disulfide bond in cystine is readily achieved by treatment with mild reducing agents (Chapter 8) or by oxidation to the sulfonic acid state through the action of performic acid:

$$R-S-S-R' \xrightarrow{HCO_3H} R-SO_3H + R'-SO_3H \tag{17.3}$$

Since the disulfide bonds in oxytocin and vasopressin are part of a ring, this cleavage does not result in a fragmentation of the peptide backbone. In the case of insulin, however, this oxidation leads to two peptide fragments of similar size. Consequently, we deduce that this hormone consists of two peptide chains held together by one or more disulfide units, as shown in Figure 17.2. The amino groups appended to some of the glutamic and aspartic acid residues indicate amide derivatives of the remaining carboxyl function.

Many microorganisms produce peptides which often have interesting physiological properties. The tyrocidines, for example, are a group of antibiotic cyclic peptides obtained from *Bacillus brevis*. They are unusual in that they contain some R amino acids. They also contain the uncommon amino acid ornithine (Orn), $H_2N(CH_2)_3CH(NH_3^{\oplus})CO_2^{\ominus}$.

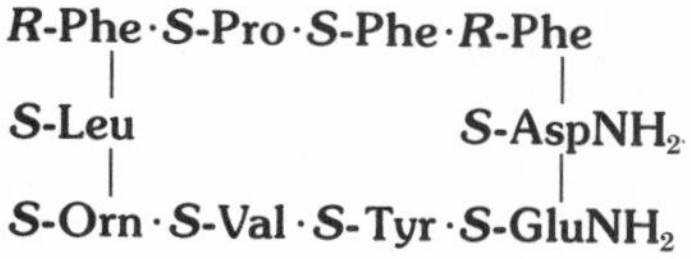

```
R-Phe·S-Pro·S-Phe·R-Phe
   |                |
 S-Leu           S-AspNH_2
   |                |
 S-Orn·S-Val·S-Tyr·S-GluNH_2
       tyrocidine A
```

Recent studies indicate that peptides may be responsible for some of the learning transferance originally attributed to DNA. The polypeptide scotophobin, which has been isolated from the brains of rats trained to avoid the dark, has been found to facilitate the acquisition of dark avoidance in both mice and goldfish.

Ser·Asp·Arg·GluNH_2·Glu·Gly·Lys·Ser·Ala·Gl
scotophobin
(this structure is uncertain)

FIGURE 17·2 *The structure of beef insulin*

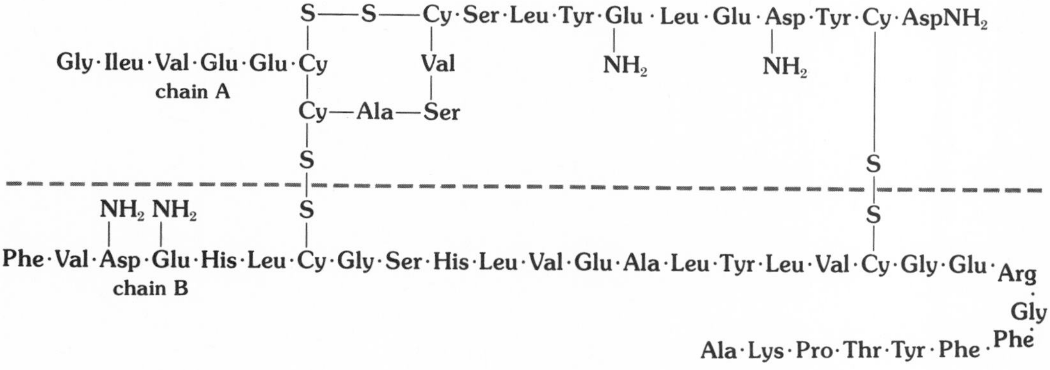

```
                S—S—Cy·Ser·Leu·Tyr·Glu·Leu·Glu·Asp·Tyr·Cy·AspNH_2
                |      |                  |         |         |
Gly·Ileu·Val·Glu·Glu·Cy    Val            NH_2      NH_2      |
     chain A          |                                       |
                      Cy—Ala—Ser                              |
                      |                                       |
                      S                                       S
- - - - - - - - - - - |- - - - - - - - - - - - - - - - - - - -|- - - - -
     NH_2 NH_2        |                                       |
      |   |           S                                       S
Phe·Val·Asp·Glu·His·Leu·Cy·Gly·Ser·His·Leu·Val·Glu·Ala·Leu·Tyr·Leu·Val·Cy·Gly·Glu·Arg
     chain B                                                              Gly
                              Ala·Lys·Pro·Thr·Tyr·Phe·Phe
```

17·4 PEPTIDES

PEPTIDE SYNTHESIS

The synthesis of peptides might at first appear to be simply a matter of amide formation, but the difunctional nature of the amino acid reactants has forced the development and use of highly selective protective groups. If, for example, we wish to prepare glycylalanine, some means of preventing the glycyl acylating reagent from attacking the amino group of another glycine molecule must be incorporated in the synthesis. The role of such protective groups is illustrated in Figure 17.3.

An amino protecting group must be introduced and removed under sufficiently mild conditions to avoid racemization of optically active amino acids or undesired reactions at other functional groups, and yet it must stand up to the conditions of the acylation step. Some useful protecting groups for this purpose are listed in Table 17.2.

Although it is possible to transform protected amino acids to powerful acylating agents such as acyl chlorides, experience has shown that it is generally better to use milder acylating species in order to avoid racemization. Derivatives such as p-nitrophenyl esters, acyl azides, and acyl imidazoles are thus preferred. The synthesis of glutathione in Figure 17.4 illustrates many of these methods.

FIGURE 17·3 *The role of protective groups in peptide synthesis*

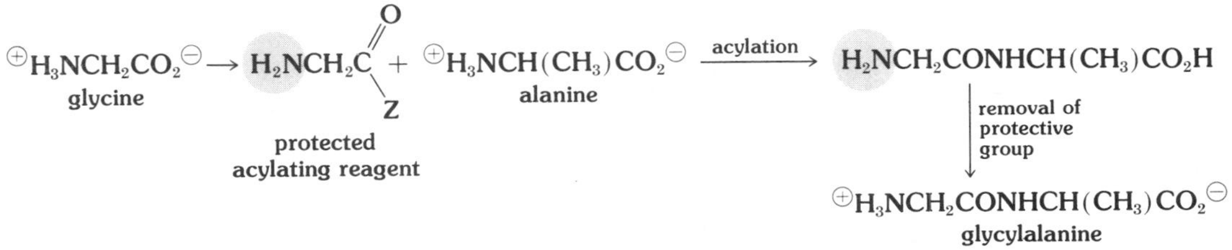

$$^{\oplus}H_3NCH_2CO_2^{\ominus} \rightarrow H_2NCH_2\overset{O}{\underset{Z}{C}} + {}^{\oplus}H_3NCH(CH_3)CO_2^{\ominus} \xrightarrow{\text{acylation}} H_2NCH_2CONHCH(CH_3)CO_2H$$

glycine   protected acylating reagent   alanine

↓ removal of protective group

$$^{\oplus}H_3NCH_2CONHCH(CH_3)CO_2^{\ominus}$$
glycylalanine

TABLE 17·2 *Protecting groups for amine functions*

protective group	introduction and removal
carbobenzoxy (benzylcarbamate)	$\phi CH_2OCOCl + H_2NCHRCO_2R' \xrightarrow{-HCl} \phi CH_2OCONHCHRCO_2R'$ removed by HBr in CH_3CO_2H or H_2 and Pd catalyst or by Na in NH_3
trityl (triphenylmethyl)	$\phi_3CCl + H_2NCHRCO_2R' \xrightarrow{-HCl} \phi_3CNHCHRCO_2R'$ removed by H_2 and Pd catalyst or aqueous acid
phthaloyl	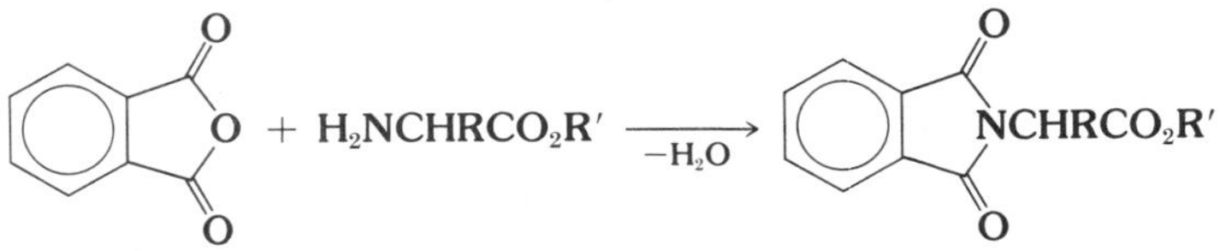 phthalic anhydride $+ H_2NCHRCO_2R' \xrightarrow{-H_2O}$ phthalimide-NCHRCO_2R' removed by hydrazine ⟶ phthalhydrazide
trifluoroacetyl	$(CF_3CO)_2O + H_2NCHRCO_2R' \longrightarrow CF_3CONHCHRCO_2R'$ removed by mild base hydrolysis

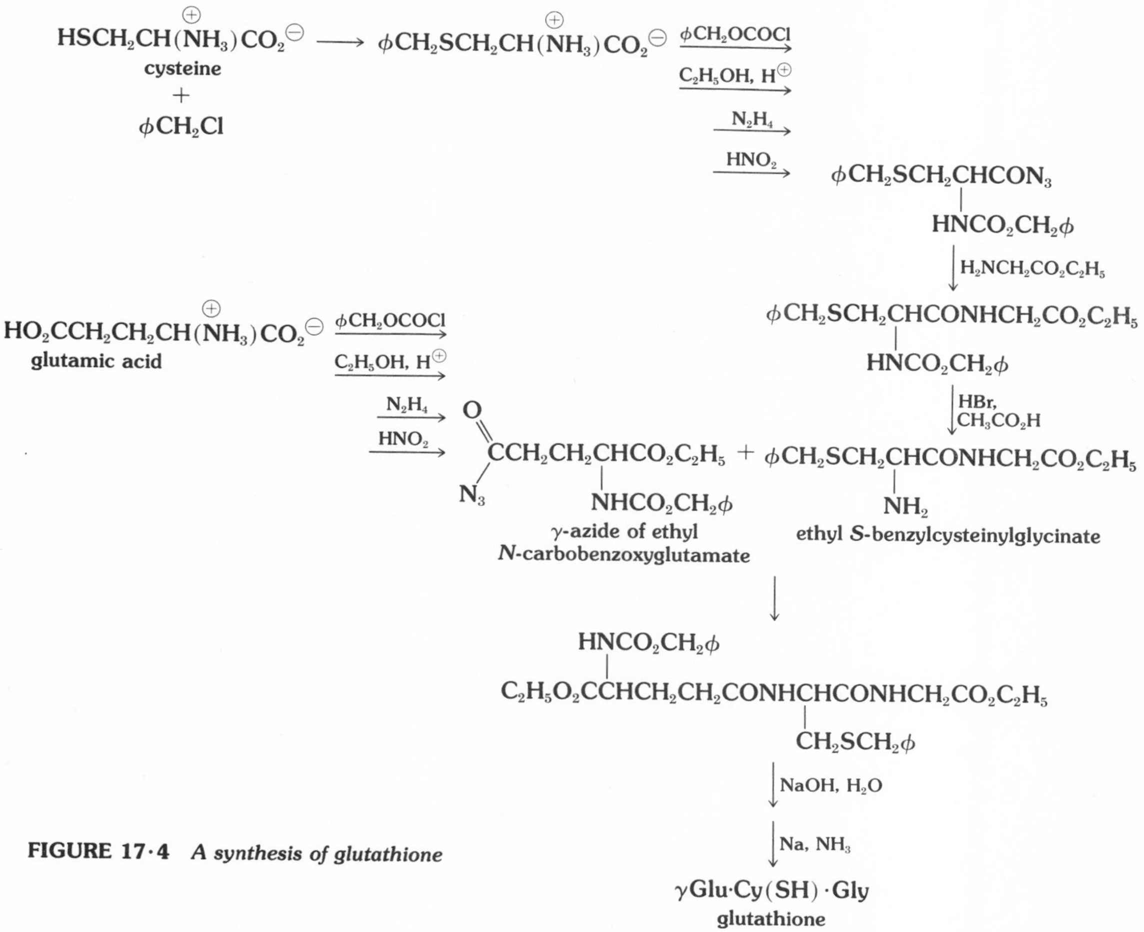

17·4 PEPTIDES

$$\text{HSCH}_2\text{CH}(\overset{\oplus}{\text{NH}_3})\text{CO}_2^{\ominus} \longrightarrow \phi\text{CH}_2\text{SCH}_2\text{CH}(\overset{\oplus}{\text{NH}_3})\text{CO}_2^{\ominus} \xrightarrow[\text{C}_2\text{H}_5\text{OH, H}^{\oplus}]{\phi\text{CH}_2\text{OCOCl}}$$

cysteine
+
$\phi\text{CH}_2\text{Cl}$

$$\xrightarrow{\text{N}_2\text{H}_4} \xrightarrow{\text{HNO}_2} \phi\text{CH}_2\text{SCH}_2\text{CHCON}_3$$
$$\qquad\qquad\qquad\qquad\quad |\!\!\!\text{HNCO}_2\text{CH}_2\phi$$

$\downarrow \text{H}_2\text{NCH}_2\text{CO}_2\text{C}_2\text{H}_5$

$$\text{HO}_2\text{CCH}_2\text{CH}_2\text{CH}(\overset{\oplus}{\text{NH}_3})\text{CO}_2^{\ominus} \xrightarrow[\text{C}_2\text{H}_5\text{OH, H}^{\oplus}]{\phi\text{CH}_2\text{OCOCl}}$$

glutamic acid

$$\phi\text{CH}_2\text{SCH}_2\text{CHCONHCH}_2\text{CO}_2\text{C}_2\text{H}_5$$
$$|\!\!\!\text{HNCO}_2\text{CH}_2\phi$$

$\downarrow \overset{\text{HBr,}}{\text{CH}_3\text{CO}_2\text{H}}$

$$\xrightarrow[\text{HNO}_2]{\text{N}_2\text{H}_4} \underset{\text{N}_3}{\overset{\text{O}}{\text{C}}}\text{CCH}_2\text{CH}_2\text{CHCO}_2\text{C}_2\text{H}_5 + \phi\text{CH}_2\text{SCH}_2\text{CHCONHCH}_2\text{CO}_2\text{C}_2\text{H}_5$$
$$\qquad\qquad\qquad\quad |\!\!\!\text{NHCO}_2\text{CH}_2\phi \qquad\qquad\qquad\quad |\!\!\!\text{NH}_2$$

γ-azide of ethyl
N-carbobenzoxyglutamate

ethyl S-benzylcysteinylglycinate

$\downarrow$

$$\overset{\text{HNCO}_2\text{CH}_2\phi}{|}$$
$$\text{C}_2\text{H}_5\text{O}_2\text{CCHCH}_2\text{CH}_2\text{CONHCHCONHCH}_2\text{CO}_2\text{C}_2\text{H}_5$$
$$\qquad\qquad\qquad\qquad\quad |\!\!\!\text{CH}_2\text{SCH}_2\phi$$

$\downarrow \text{NaOH, H}_2\text{O}$

$\downarrow \text{Na, NH}_3$

γGlu·Cy(SH)·Gly
glutathione

FIGURE 17·4 *A synthesis of glutathione*

PROBLEM 17·5

a In the above synthesis indicate which steps involve the introduction and removal of protecting groups and which steps result in peptide-bond formation.
b Why are acyl azides and acyl imidazoles better acylating agents than most amides? (*Hint:* The pK_a of HN_3 is 4.7.)
c Write an equation showing how a phthaloyl protective group can be removed from a primary amine function by treatment with hydrazine.

PROPERTIES OF SIMPLE PEPTIDES

The physical properties of small peptides are similar to those of amino acids such as glutamine. Thus a zwitterion structure predominates at neutral pH, and in the absence of acidic or basic side chains, the isolectric point of peptides is about 6.0. Peptides may be hydrolyzed by strong acids or bases, but in general they are relatively stable to moderate changes in temperature and pH.

PROBLEM 17·6

What characteristic physical properties would you expect the natural peptide bradykinin to exhibit?

17·5 SECONDARY AND TERTIARY STRUCTURES OF PROTEINS

In contrast to this similarity in physical properties, peptides often have unique and characteristic biological properties which may differ markedly from those of the component amino acids. The synthetic dipeptide ester S-Asp·S-PheOCH$_3$, for example, is over 100 times sweeter than table sugar (sucrose) despite the fact that S-Asp is tasteless and S-Phe is bitter. All other combinations of R- and S-Asp and Phe dipeptide esters have also proved to be bitter. Other more complex examples of physiologically active peptides are the hormones discussed above.

Proteins, in addition to having many different biological functions, exhibit a wide range of physical properties. On the basis of their solubility and relative stability, two broad categories are recognized:

Fibrous proteins As this name implies, these substances have an overall structure which is generally fibrelike. Fibrous proteins serve as the chief structural material of animal tissues. They consist largely of the *collagens* and *elastins* (the proteins of connective tissue, ligaments, and tendons), the *keratins* (found in skin, hair, feathers, horns, and hooves), the *myosins* (muscle protein), and *fibrin* (the protein formed in the clotting of blood.) As befits their important structural functions, these proteins are insoluble in water and are stable under moderate variations in temperature and pH.

Globular proteins Members of this class serve regulatory functions (enzymes, hormones, antibodies, etc.) and are much more sensitive to changes in temperature and pH than the fibrous proteins. They are usually soluble or form colloidal suspensions in water (the albumins) or aqueous electrolyte solutions (the globulins).

Many of these properties are best discussed in terms of the manner in which the polypeptide chains of the proteins are coiled and folded—their *secondary* and *tertiary structures*. Indeed, the biological function of many enzymes appears to be more a matter of macromolecular geometry than specific amino acid sequencing.

COILING AND FOLDING OF PEPTIDE CHAINS

Peptide (amide) bonds are planar as a consequence of resonance stabilization involving nitrogen electron delocalization:

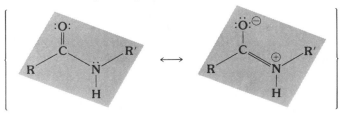

In a peptide chain two adjacent peptide planes are joined at the α carbon atom held in common and may assume a variety of twisted conformations, as shown in Figure 17.5. A fully extended peptide chain would,

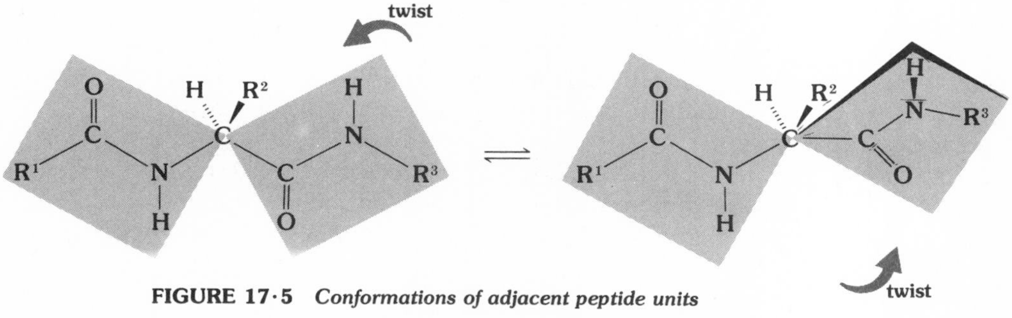

17·5 SECONDARY AND TERTIARY STRUCTURES OF PROTEINS 589

FIGURE 17·5 *Conformations of adjacent peptide units*

of course, have coplanar or parallel peptide units. The natural protein most closely approximating such a structure is silk fibroin. From x-ray diffraction measurements we find that silk consists of polypeptide chains stretched parallel to the fiber axis and hydrogen-bonded to neighboring

FIGURE 17·6 *The antiparallel β-pleated sheet found in silk fibroin*

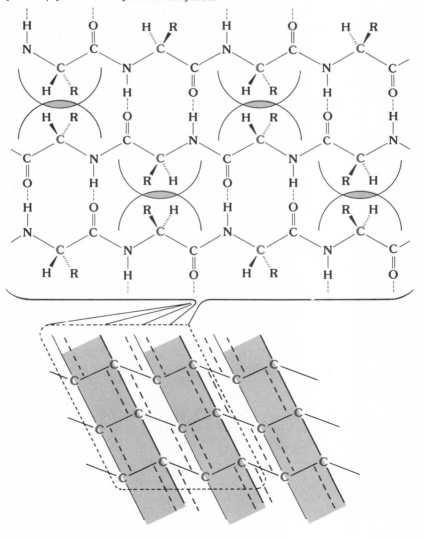

chains running in the opposite direction (antiparallel). As a consequence of nonbonded interactions between α substituents, the peptide functions twist slightly, allowing the peptide chains to contract. This results in the pleated sheet of antiparallel polypeptide chains shown in the lower portion of Figure 17.6.

(70 to 90%) of glycine, alanine, and serine in roughly a 4:2:1 ratio. Smaller amounts of other amino acids such as tyrosine, valine, arginine, aspartic acid, and glutamic acid are also present. Sequence studies on silk have shown that a characteristic hexapeptide grouping repeats for long distances in the peptide chains:

$$\sim[\text{Gly·Ser·Gly·Ala·Gly·Ala}]_n\sim$$

This feature, when incorporated into the antiparallel pleated-sheet structure of silk, results in the appearance of the serine and alanine side chains on one side of the pleated sheets. These sheets are then stacked with like-side chains nestling together, giving a highly ordered three-dimensional structure. Since there is no room in this lattice for larger side chains, such as those present in tyrosine and valine, these highly ordered, or crystalline, regions alternate with disordered or amorphous regions in the overall structure of silk.

Most fibrous proteins, such as the α-keratins that make up wool and hair, give x-ray diffraction patterns that differ significantly from those of silk. With the aid of careful measurements involving smaller peptides, Linus Pauling and R. B. Corey discovered that the peptide chains in the α-keratins adopt a *helical secondary structure* with 3.6 amino acid residues per turn (Figure 17.7). This *α-helix* arrangement appears to be a favored conformation of these polypeptides because it allows (and is stabilized by) intrachain hydrogen bonds between successive coils of the helix. A peptide chain constructed from the naturally occurring S amino acids can form either a right-handed or left-handed helix; however, the right-handed form is more stable and is the conformation adopted by the peptides of most α-keratins. The chiral nature of these helical conformations introduces an extra dextrorotatory factor into the optical activities of polypeptides. Optical-rotation measurements of proteins and peptides, in fact, can serve as a tool to disclose the approximate amount of α-helical coiling in the peptide chains.

The substituent groups that characterize specific amino acids extend outward from the spiral backbone of the α-helix, as shown in Figure 17.7). The stability of such helical conformations is therefore dependent on the size, shape, and charge of these substituents. Polyalanine, for example, has small uncharged side chains (CH_3), and in aqueous solution at pH = 7 it spontaneously assumes an α-helix form. Polyleucine, however, fails to coil in this fashion because of the steric hindrance introduced by the large butyl substituents. Since adjacent groups having like charges (positive or negative) will repel each other, it is not surprising to find that the relative stabilities of polylysine and polyglutamic acid conformations are pH dependent. At pH = 12 the primary amine functions in polylysine are neutral, and this polypeptide adopts an α-helix analogous to that favored by polyalanine. The mutual repulsion of the

17·5 SECONDARY AND TERTIARY STRUCTURES OF PROTEINS

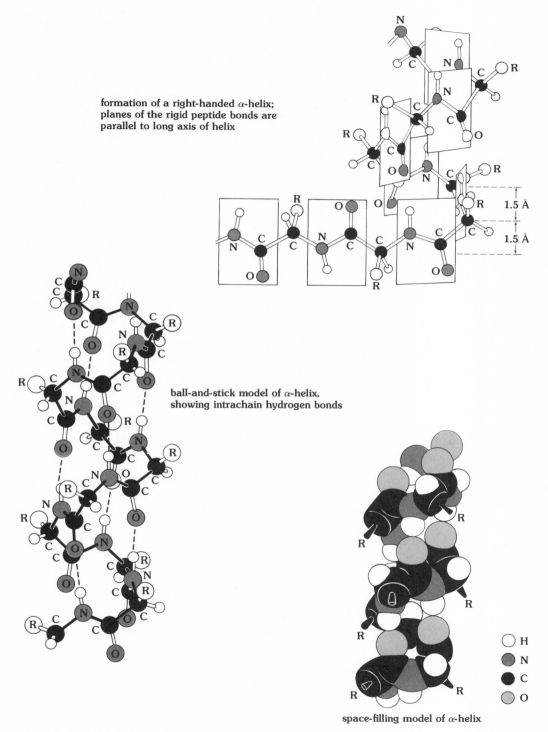

FIGURE 17·7 *The α-helix (after L. Pauling and R. B. Corey,* Proceedings of the International Wool Textile Research Conference, *B 249, 1955)*

592 17·5 SECONDARY AND TERTIARY STRUCTURES OF PROTEINS

carboxylate anions present in polyglutamic acid at pH greater than 7 disrupts the α-helix form and causes this peptide to exist as a randomly twisting chain, termed a *random coil*. At pH = 2 polyglutamic acid is an α-helix and polylysine a random coil (the side-chain ammonium ions repel each other). The amino acid proline introduces a unique perturbation. Twisting about the N—C_2 bond in proline is prohibited by the five-membered ring; consequently, proline interrupts the α-helix wherever it occurs in a peptide chain.

It is convenient to describe helical conformations of polypeptides by two numbers, n_m, where n is the number of amino acid units per turn of the helix and m is the number of atoms in the rings defined by closing the hydrogen bonds. According to this notation, the α-helix in Figure 17.7 is termed a 3.6_{13}-helix. Two other important helical conformations are the 3_{10}-helix and the 4.4_{16}-helix (π-helix), shown in Figure 17.8.

Helical polypeptide strands may themselves be coiled or woven together in what is referred to as the *tertiary structure* of proteins. Collagen, for example, is believed to have a fine structure consisting of three left-handed helices coiled together with a right-handed twist, so as to form a threefold superhelix called tropocollagen (see Figure 17.9). A high proportion of the unsubstituted amino acid glycine (about 33%) allows fairly close approach of the three strands, which are held together by hydrogen bonds. This interchain hydrogen bonding results in a strong and relatively rigid molecule, in contrast with the more flexible α-helix. Treatment of collagens with boiling water breaks apart the triple-stranded braids, converting it to the water-soluble protein gelatin. Since gelatins are more easily digested than most collagens, this process is important in the cooking of meat.

The keratins exhibit an even more complex tertiary structure, which we will not attempt to show here. In the α form helices are held parallel to each other by the disulfide linkages of cystine, an important component of the keratins. Soft keratins, such as skin and hair, have a rela-

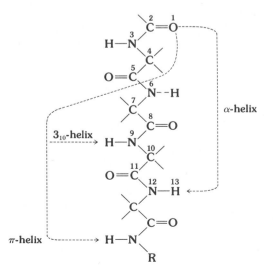

FIGURE 17·8 *Hydrogen bonding in various helices*

FIGURE 17·9 *A three-fold superhelix in collagen (after R. E. Dickerson and I. Geis,* The Structure and Action of Proteins, *Harper & Row, Publishers, New York, 1969, p. 41)*

tively low sulfur content and are more flexible and extensible than the high-sulfur hard keratins of nails, hooves, and horn. Hair and wool fibers are very extensible and when wet can be stretched to almost twice their normal length. In this process the hydrogen bonds of the α-helix are broken as it stretches, and the peptide chains ultimately adopt a β-sheet structure (β-keratin). The disulfide cross links provide a restoring force, which returns the fiber to its original state when the tension is removed. The permanent waving of hair involves an initial reductive cleavage of disulfide linkages with thioglycollate (equation 17.4), a curling and clamping of the hair in some desired fashion, and finally the formation of new disulfide bonds by a mild oxidation (equation 17.5):

$$\sim CH_2-S-S-CH_2\sim \; + \; 2HSCH_2CO_2^{\ominus}NH_4^{\oplus} \longrightarrow \sim CH_2SH \; + \; HSCH_2\sim \; + \; \begin{array}{c} SCH_2CO_2NH_4 \\ | \\ SCH_2CO_2NH_4 \end{array} \quad (17.4)$$

ammonium thioglycollate   ammonium dithioglycollate

$$\sim CH_2SH \quad HSCH_2\sim \; + \; \tfrac{1}{3}KBrO_3 \longrightarrow \sim CH_2-S-S-CH_2\sim \; + \; \tfrac{1}{3}KBr \; + \; H_2O \quad (17.5)$$

As we explore the nature of living tissues in greater detail, further structural variations in protein substances are recognized. Tropocollagen molecules, for example, pack together in an overlapping fashion to form collagen fibrils. These fibrils serve important connective and strengthening functions in living organisms. In tendons they are oriented parallel to the major axis of the tendon, while in the cornea they are arranged in laminated sheets.

DENATURATION

Experience has shown that a variety of conditions that do not disrupt peptide bonds will nevertheless destroy the biological activity of proteins. Such transformations are called *denaturations*, and the *denatured proteins* appear to have lost much of their characteristic secondary and tertiary structure, leaving a disorganized and randomly coiled peptide

TABLE 17·3 *Protein denaturation processes*

denaturing process	mechanism of operation
heat	hydrogen bonds disrupted by molecular vibration (coagulation of albumin in a frying egg)
ultraviolet radiation	action similar to heat (sunburn)
strong acids or bases	hydrogen bonds broken, salts formed (skin burns from mineral acids or alkali)
organic solvents (alcohol, acetone, etc.)	competition for hydrogen bonds (disinfectant action of 70% ethanol solutions)
urea solutions	competition for hydrogen bonds (laboratory precipitation of soluble proteins)
violent agitation	hydrogen bonds disrupted by shearing forces (beating egg white into a meringue)

chain. Since the α-helix strands and other organized segments of globular proteins are not as tightly grouped or oriented as in the fibrous proteins, globular proteins such as enzymes and antibodies are generally more susceptible to denaturation than their fibrous cousins. Denaturation is often followed by *coagulation* of the protein into an insoluble and biologically inactive solid mass. Examples of some denaturation processes are given in Table 17.3.

17·6 CONJUGATED PROTEINS

Up to this point we have been discussing *simple proteins*, defined as those yielding only α-amino acids and their derivatives on hydrolysis. Let us now consider a class of proteins known as *conjugated proteins*, which are characterized by nonprotein components or *prosthetic groups* (from the Greek *prosthesis*, an addition). Conjugated proteins include the following:

The *nucleoproteins;* the prosthetic group of these conjugated proteins is a nucleic acid (see Section 14.7).

The *lipoproteins;* the lipid prosthetic group in these proteins is variable, but usually contains glyceride, phosphatide, and cholesterol.

The *mucoproteins;* these substances have a carbohydrate prosthetic group, which almost always contains some hexosamine.

The *chromoproteins;* the prosthetic groups of this class are not easily classified except for their color (they are all pigments). Hemoglobin is an example.

The important conjugated protein hemoglobin serves as a vital oxygen-transfer agent in all higher animals. A person who weighs 70 kg (about 155 lb) contains about 910 g of hemoglobin distributed among some 5 billion red blood cells (erythrocytes). One liter of arterial blood at body temperature can dissolve and transport about 210 ml of oxygen, whereas the same fluid stripped of its hemoglobin will only carry 2 or 3 ml.

17·6 CONJUGATED PROTEINS

Thanks largely to the x-ray diffraction studies of M. Perutz and J. Kendrew at Cambridge University, we now have a fairly detailed picture of the structure of hemoglobin. The prosthetic group of this conjugated protein is called *heme* (or protoheme). As shown in Figure 17.10, four such groups are present in a hemoglobin molecule, each one wrapped or enfolded by a polypeptide chain consisting of either 141 amino acid residues (an α-chain) or 146 residues (β-chain). The almost spherical molecule formed by combining four of these moieties has a molecular weight of about 64,500.

The ferrous oxidation state of the iron atoms in hemoglobin is stabilized by hexacoordination, with four ligands provided by the porphin ring and two others by the imidazole side chains of suitably positioned histidine residues in the encircling polypeptide chain. An oxygen molecule can replace one of the imidazole units in the primary coordination sphere, giving rise to an oxyhemoglobin. Each of the four heme units in a molecule of hemoglobin can bind oxygen in this manner. Indeed, we find that the second, third, and fourth oxygen molecules are incorporated progressively more rapidly than the first, the last oxygen molecule being bound several hundred times faster than the first.† Arterial hemoglobin is thus almost completely oxygenated on leaving the lungs, and the circulatory system delivers this oxyhemoglobin to various body tissues, where the bound oxygen is released to be used or stored.

Metabolic oxidations generate acidic products such as carbonic and lactic acids, which dissolve in and lower the pH of surrounding fluids.

†Perutz has referred to this phenomenon by the Biblical phrase "To him who hath shall be given."

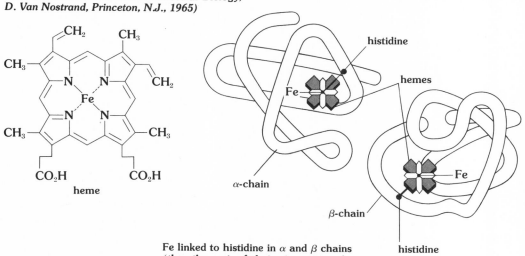

FIGURE 17·10 *The structure of hemoglobin (after R. F. Steiner, The Chemical Foundations of Molecular Biology, D. Van Nostrand, Princeton, N.J., 1965)*

Since hemoglobin binds oxygen less efficiently at low pH, the release of oxygen from the oxyhemoglobin is enhanced in this environment, freeing hemoglobin to assist in the transfer of CO_2 back to the lungs in the form of bicarbonate ions. The released oxygen can be stored in body tissues by the protein myoglobin, which contains one heme unit and is roughly one-quarter the size of hemoglobin. Myoglobin binds oxygen more strongly than hemoglobin, particularly at low pH. Consequently, the transfer of oxygen from the latter to the former is quite efficient. Diving mammals such as the porpoise and seal have a particularly high muscle content of myoglobin, enabling them to remain submerged for extended periods.

Since heme alone is unable to combine reversibly with oxygen, it is clear that the encapsulating polypeptide chains play a critical role in establishing the oxygen-transfer properties of hemoglobin. In fact, when heme is complexed with other proteins (as in the oxidases, catalases, and cytochromes), it exhibits very different characteristics.

The oxygen-transporting capability of hemoglobin can be destroyed by the action of several kinds of poisonous substances. Carbon monoxide, for example, replaces oxygen in the oxyhemoglobin complex, and being more tightly bound, it is difficult to remove. In the condition known as *methemoglobinemia* the iron is oxidized to the ferric state, in which form it is no longer able to carry oxygen. Infants are particularly susceptible to poisoning of this kind, which may be caused by high blood nitrite levels.

Many genetically transmitted abnormal hemoglobins have been identified. While not all of these are associated with significant physiological effects, in some cases a single change in one of the polypeptide chains may markedly influence the properties of this vital protein. One of the best-known examples of this relationship is the disease *sickle-cell anemia*, which results when one glutamic acid out of the 146 residues in the β-polypeptide chain is replaced by a valine unit. This change of a normally acidic side chain for a neutral group alters the solubility of the hemoglobin, which precipitates under certain conditions, thus altering the shape of the cell containing it. Oxygen transport is reduced and capillary clogging sometimes results, causing a crisis involving inflammation and fever. Recent studies have shown that urea and cyanate help to prevent or reverse the sickling process. The mechanism of their action is, however, not yet clear.

REFERENCES

K. D. Kopple, *Peptides and Amino Acids*, W. A. Benjamin, Inc., New York, 1966.

D. T. Elmore, *Peptides and Proteins*, Cambridge University Press, New York, 1969.

R. E. Dickerson and I. Geis, *The Structure and Action of Proteins*, Harper and Row, Publishers, New York, 1969.

H. D. Law, *The Organic Chemistry of Peptides*, John Wiley & Sons, Inc., New York, 1970.

A. L. Lehninger, *Biochemistry*, Worth Publishers, Inc., New York, 1970.

R. Barker, *Organic Chemistry of Biological Compounds*, Prentice-Hall, Inc., Englewood Cliffs, N.J., 1971.

M. Calvin and W. Pryor, *Organic Chemistry of Life*, part II: Macromolecular Architecture, W. H. Freeman and Company, San Francisco, 1973.

J. C. Kendrew, The Three-dimensional Structure of a Protein Molecule, *Scientific American,* December 1961, p. 96.

M. F. Perutz, The Hemoglobin Molecule, *Scientific American*, November 1964, p. 64.

EXERCISES

17·1 Name an amino acid in Table 17.1 that fits each description:

a Contains no chiral center
b Contains more than one chiral center
c An amide
d A dicarboxylic acid derivative
e Contains a phenolic group
f A thiol
g An alcohol
h Contains an indole ring

17·2 Show the product of each of the following acid-base reactions:

a $\overset{\oplus}{H_3N}CH_2\overset{O}{\underset{\|}{C}}O^{\ominus} \xrightarrow{HCl}$

b $\overset{\oplus}{H_3N}CH_2\overset{O}{\underset{\|}{C}}O^{\ominus} \xrightarrow{NaOH}$

c [pyrrolidine ring with $\overset{\oplus}{N}H_2$]—$CO_2^{\ominus}$ $\xrightarrow{HCl}$

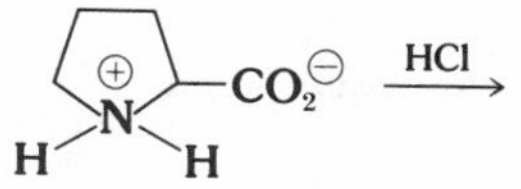

d [pyrrolidine ring with $\overset{\oplus}{N}H_2$]—$CO_2^{\ominus}$ $\xrightarrow{NaOH}$

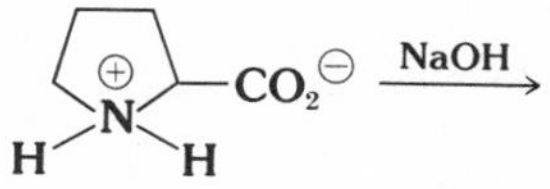

e $H_2N(CH_2)_4\underset{\underset{NH_3^{\oplus}}{|}}{CH}CO_2^{\ominus} \xrightarrow{HCl}$

f $HO_2C(CH_2)_2\underset{\underset{NH_3^{\oplus}}{|}}{CH}CO_2^{\ominus} \xrightarrow{NaOH}$

17·3 Account for the differences in the isoelectric points of the following pairs of amino acids:

a Phenylalanine 5.5 and aspartic acid 2.8
b Phenylalanine 5.5 and lysine 9.7
c Lysine 9.7 and tryptophan 5.9
d Aspartic acid 2.8 and asparagine 5.4

17·4 Using alanine as an example:

a Show the predominant ionic form of the amino acid in strong alkaline solution. What amino acid species are produced from this alkaline solution on addition of acid ($H^{\oplus}$)?

b Show the predominant ionic form of the amino acid in strong acidic solution. What amino acid species are produced from the acidic solution on addition of base ($OH^{\ominus}$)?

17·5 Repeat Exercise 17.4 for aspartic acid.

17·6 Repeat Exercise 17.4 for lysine.

17·7 *p*-Aminophenylactic acid exists almost entirely as the neutral compound shown below, whereas phenylalanine is predominately in the switterion form. Suggest a reason for this difference. *Hint:* Compare K_b values for aromatic and aliphatic amines.

p-aminophenylactic acid   phenylalanine   zwitterion

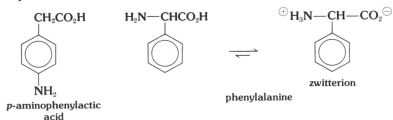

17·8 Draw structural formulas for each of the following peptides:

a Alanylglycine
b Glycylalanine
c Val·Phe·Leu
d Asp·Ser·Lys·Thr
e CySH·GluNH$_2$·Tyr·GlyCONH$_2$
f *S*-Asp·*S*-PheOCH$_3$

17·9 Draw structural formulas which show the stereochemistry of each compound:

a *S*-Ala·*S*-Phe
b *R*-Ala·*R*-Phe
c *R*-Ala·*S*-Ala·Gly
d *S*-Val·*S*-Pro

17·10 How many distinct tripeptides can be made from any combination of the amino acids *S*-glycine, *S*-alanine, and *S*-lysine? *Hint:* How many different combinations of the letters *g*, *a*, and *l* can be written?

17·11 Indicate the reagents and conditions necessary to accomplish the following synthesis of phenylalanine:

$$\phi CH_3 \longrightarrow \phi CH_2Cl \longrightarrow \longrightarrow \phi CH_2CH_2CO_2H \longrightarrow \phi CH_2\underset{\displaystyle Br}{\underset{|}{C}}HCO_2H \longrightarrow (\pm)\text{-phenylalanine}$$

17·12 Fill in the reagents and show mechanisms for each step in the following synthesis of alanine:

$$\begin{matrix} CH_3CO_2CH_2CH_3 \\ + \\ CH_3CH_2O_2CCO_2CH_2CH_3 \end{matrix} \longrightarrow CH_3CH_2O_2CCOCH_2CO_2CH_2CH_3 \longrightarrow \longrightarrow CH_3COCO_2H \longrightarrow (\pm)\text{-alanin}$$

Using the same synthesis method, outline a synthesis of (±)-leucine.

17·13 A chemist wishes to prepare the tripeptide *R*-Ala·*R*-Lys·*R*-Val, in which all the amino acids have the "unnatural" *R* configuration. His starting materials are the three racemic amino acids. To save time he decided to make the tripeptide from the racemic units and then carry out a single resolution, instead of conducting three separate resolutions on the amino acids before preparing the tripeptide. Discuss the merits and disadvantages of this idea.

17·14 Hypertensin, a decapeptide formed in the kidneys, produces an increase in blood pressure. The N-terminal amino acid of hypertensin, determined by Edman degradation, is aspartic acid. C-terminal analysis indicates leucine. Tripeptides formed on partial hydrolysis are Val·Tyr·Val, Asp·Arg·Val, His·Pro·Phe, and Phe·His·Leu. What is the primary structure of hypertensin?

17·15 A heptapeptide H is characterized by the following chemical behavior. What is its primary structure?

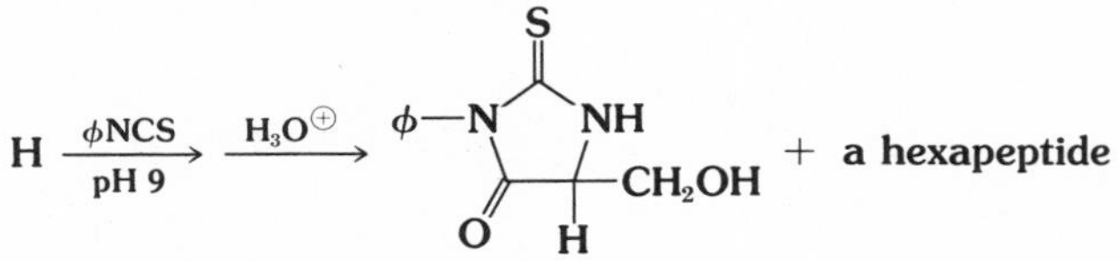

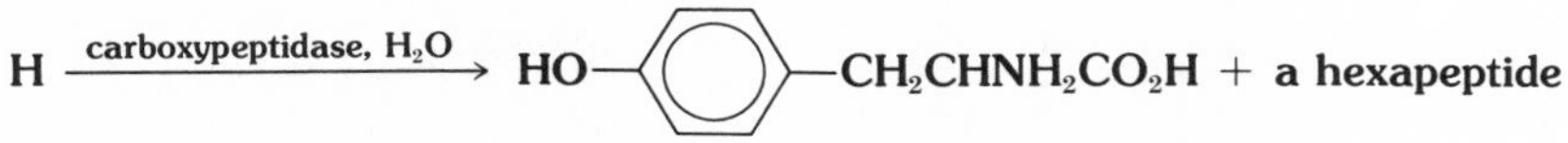

H $\xrightarrow[\text{pH 9}]{\phi\text{NCS}}$ $\xrightarrow{\text{H}_3\text{O}^{\oplus}}$ [phenylthiohydantoin of Ser with φ-N, NH, S, O, H, CH₂OH] + a hexapeptide

H $\xrightarrow{\text{carboxypeptidase, H}_2\text{O}}$ HO—⟨C₆H₄⟩—CH₂CHNH₂CO₂H + a hexapeptide

H $\xrightarrow[\Delta]{\text{3N, HCl}}$ 2 Gly, 1 Leu, 1 Phe, 1 Ser, 1 Tyr, 1 Pro

H $\xrightarrow{\text{1N, HCl}}$ Ser·Leu·Gly, Phe·Gly·Tyr, Pro·Phe·Gly, Leu·Gly·Pro, Gly·Pro·Phe

CARBOHYDRATES

CARBOHYDRATES are a widely distributed group of natural compounds which constitute one of our three major classes of food (the other two are fats and proteins). Sugars, starches, and cellulose are probably the best-known members of this group. Carbohydrate moieties are also found in a variety of complex substances such as glycosides, mucoproteins, and nucleic acids (Section 14.7). The name *carbohydrate* reflects the fact that ordinarily the ratio of hydrogen to oxygen in these compounds is 2:1, as indicated by the general formula $C_mH_{2n}O_n$ or $C_m(H_2O)_n$.

Low-molecular-weight carbohydrates are called sugars or *saccharides*. They are classified as *tetroses, pentoses,* or *hexoses* (*-ose* is the characteristic suffix used to denote a sugar), according to the number of carbon atoms present in their molecules. The aldehyde- or ketonelike behavior of most sugars leads to their further classification as *aldoses* or *ketoses*, depending on the location of the carbonyl function in the carbon skeleton.

Carbohydrates composed of chains of pentose or hexose units bonded together by ether linkages are referred to as *disaccharides* (two units), *trisaccharides* (three units), and in the case of large polymers as *polysaccharides*.

18·1 CONSTITUTION AND CONFIGURATION OF SIMPLE SUGARS

The most abundant and thoroughly studied sugar is the $C_6H_{12}O_6$ monosaccharide (+)-glucose, often called dextrose. Glucose occurs free in fruit juices and honey and can also be obtained by hydrolysis of starch and cellulose. The reactions outlined in Figure 18.1 served to establish the fundamental constitution of glucose as a linear *aldohexose*.

Since this structure has four chiral carbon units, we would expect to find $2^4 = 16$ stereoisomers having this constitution. One of these would be (+)-glucose. The problem of working out the correct configurations of all the aldohexoses, including glucose, appears formidable indeed, but the brilliant efforts of Emil Fischer during the 1890s culminated in this elucidation and subsequently (1902) led to a Nobel Prize.†

The configurations of the three important aldohexoses—glucose, mannose, and galactose—and the ketohexose fructose are shown here in the Fischer notation described in Figure 7.13. Fructose is the sweetest of these common sugars. We see that fructose is a constitutional isomer of the aldohexoses, whereas (+)-glucose and (+)-mannose are configurational epimers at C-2, and (+)-glucose and (+)-galactose are epimeric at C-4. Mirror-image configurations of these four isomers would represent the respective enantiomeric compounds—*L*-(−)-glucose, *L*-(−)-mannose, *L*-(−)-galactose, and *L*-(+)-fructose.††

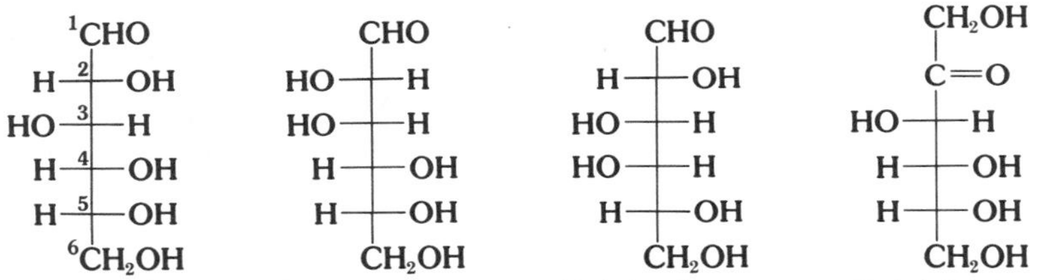

$$
\begin{array}{cccc}
^1CHO & CHO & CHO & CH_2OH \\
H-^2-OH & HO-H & H-OH & C=O \\
HO-^3-H & HO-H & HO-H & HO-H \\
H-^4-OH & H-OH & HO-H & H-OH \\
H-^5-OH & H-OH & H-OH & H-OH \\
^6CH_2OH & CH_2OH & CH_2OH & CH_2OH \\
D\text{-}(+)\text{-glucose} & D\text{-}(+)\text{-mannose} & D\text{-}(+)\text{-galactose} & D\text{-}(-)\text{-fructose}
\end{array}
$$

The symbols *D* and *L* refer to the configuration at C-5 in these sugars, relative to the reference compound (+)-*glyceraldehyde*, defined as a *D* configuration. Fischer had recognized that the 16 aldohexoses consisted of two enantiomeric families, a *D* series and an *L* series, and he used *D*-(+)-glucose as a reference compound to define the configurations of these sugars. However, because of some confusion caused by the three other chiral centers in the aldohexoses, the American chemist Rosanoff proposed that a reference compound having a single chiral center be chosen. Rosanoff suggested glyceraldehyde for this role, with any chiral compound derived from or transformed into (+)-glyceraldehyde classified

---

†Fischer's work in this area was also instrumental in confirming and winning general acceptance for the stereochemical theories of van't Hoff and LeBel.

††Although the chemical and physical properties of enantiomeric sugars are virtually identical, the chiral enzymes of living organisms can effectively discriminate between these isomers. Sugars in the *D* series, for example, are found to be fermented by yeast, whereas the *L* sugars are normally not fermented.

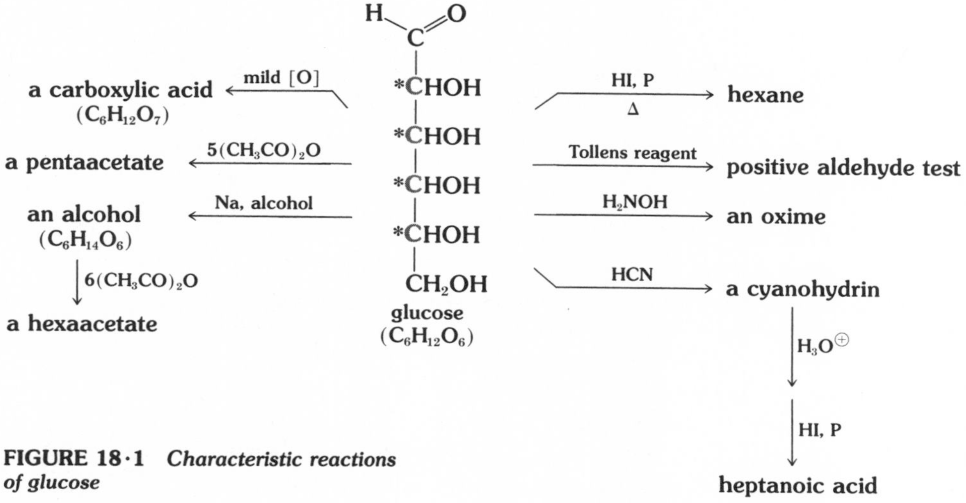

18·1 CONSTITUTION AND CONFIGURATION OF SIMPLE SUGARS

$$\text{a carboxylic acid} (C_6H_{12}O_7) \xleftarrow{\text{mild [O]}} \begin{array}{c} \text{H–C=O} \\ \text{*CHOH} \\ \text{*CHOH} \\ \text{*CHOH} \\ \text{*CHOH} \\ \text{CH}_2\text{OH} \\ \text{glucose} \\ (C_6H_{12}O_6) \end{array} \xrightarrow{\text{HI, P, }\Delta} \text{hexane}$$

- a pentaacetate ←5(CH$_3$CO)$_2$O—
- an alcohol (C$_6$H$_{14}$O$_6$) ←Na, alcohol—
- a hexaacetate ←6(CH$_3$CO)$_2$O—

- Tollens reagent → positive aldehyde test
- H$_2$NOH → an oxime
- HCN → a cyanohydrin →H$_3$O$^⊕$→ →HI, P→ heptanoic acid

FIGURE 18·1 *Characteristic reactions of glucose*

as D and those compounds experimentally related to (−)-glyceraldehyde designated L. Fischer's assignment of absolute configurations to the aldohexoses was arbitrary, but inspired, since his guess was later shown to be correct by an x-ray diffraction analysis of tartrate salts (the isomeric tartaric acids have been experimentally correlated with the sugars and with glyceraldehyde).

$$\begin{array}{cc} \text{H–C=O} & \text{H–C=O} \\ \text{H—OH} & \text{HO—H} \\ \text{CH}_2\text{OH} & \text{CH}_2\text{OH} \\ D\text{-(+)-glyceraldehyde} & L\text{-(−)-glyceraldehyde} \\ \text{(now called } R\text{)} & \text{(now called } S\text{)} \end{array}$$

Although the D and L configurational notation is useful for sugars, it is time consuming and sometimes ambiguous to make the necessary interconversions in other systems. Hence this notation has been supplanted by the convenient and unambiguous Cahn-Ingold-Prelog nomenclature described in Chapter 7. According to this newer system, the configuration at C-5 in (+)-glucose is R (rectus).

PROBLEM 18·1

Consider the following two reaction paths by which glyceraldehyde could be transformed into lactic acid. Would (+)-lactic acid have a D configuration or an L configuration?

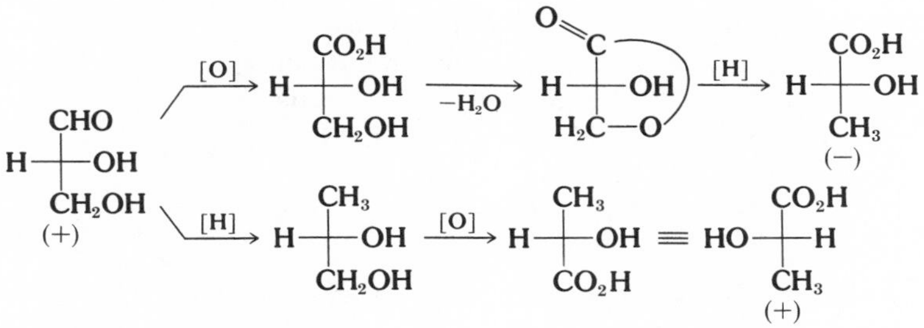

$$\begin{array}{c} \text{CHO} \\ \text{H—OH} \\ \text{CH}_2\text{OH} \\ (+) \end{array} \xrightarrow{[O]} \begin{array}{c} \text{CO}_2\text{H} \\ \text{H—OH} \\ \text{CH}_2\text{OH} \end{array} \xrightarrow{-H_2O} \begin{array}{c} \text{O=C} \\ \text{H—OH} \\ \text{H}_2\text{C—O} \end{array} \xrightarrow{[H]} \begin{array}{c} \text{CO}_2\text{H} \\ \text{H—OH} \\ \text{CH}_3 \\ (-) \end{array}$$

$$\xrightarrow{[H]} \begin{array}{c} \text{CH}_3 \\ \text{H—OH} \\ \text{CH}_2\text{OH} \end{array} \xrightarrow{[O]} \begin{array}{c} \text{CH}_3 \\ \text{H—OH} \\ \text{CO}_2\text{H} \end{array} \equiv \begin{array}{c} \text{CO}_2\text{H} \\ \text{HO—H} \\ \text{CH}_3 \\ (+) \end{array}$$

PROBLEM 18·2
Draw Fischer projections for all stereoisomers of fructose. Which are enantiomers? Which are diastereoisomers? Which pairs of isomers are epimeric at C-3?

18·2 CHEMICAL REACTIONS OF MONOSACCHARIDES

The high concentration of functional groups in sugar molecules gives rise to a variety of characteristic chemical reactions, some of which were outlined for glucose in Figure 18.1. Indeed, the ease with which these compounds undergo isomerization, oxidation, and dehydration, coupled with their very high water solubility and hygroscopic nature, makes them particularly difficult to work with. In the following discussion we shall examine many of the reactions which were instrumental in Fischer's investigation and configurational elucidation of the monosaccharides.

OSAZONE FORMATION

Nine years after his discovery that sulfite reduction of benzenediazonium salts gave phenylhydrazine, Emil Fischer announced that this reagent could serve as a powerful aid in the study of sugars. Specifically, treatment of an aldose with an excess of phenylhydrazine gave crystalline derivatives called *osazones*, which were much more easily handled than the sugars themselves (sugars tend to form syrups when not completely pure). In addition to providing an easy identification of different sugars through their characteristic melting points, the osazones also permitted certain configurational relationships to be seen:

$$
\begin{array}{c}
\text{CHO} \\
\text{H}\!-\!\text{OH} \\
\text{HO}\!-\!\text{H} \\
\text{H}\!-\!\text{OH} \\
\text{H}\!-\!\text{OH} \\
\text{CH}_2\text{OH} \\
(+)\text{-glucose}
\end{array}
\xrightarrow{3\phi\text{NHNH}_2}
\begin{array}{c}
\text{H}\!\diagdown\!\text{C}\!=\!\text{NNH}\phi \\
\text{C}\!=\!\text{NNH}\phi \\
\text{HO}\!-\!\text{H} \\
\text{H}\!-\!\text{OH} \\
\text{H}\!-\!\text{OH} \\
\text{CH}_2\text{OH} \\
\text{an osazone}
\end{array}
\xleftarrow{3\phi\text{NHNH}_2}
\begin{array}{c}
\text{CHO} \\
\text{HO}\!-\!\text{H} \\
\text{HO}\!-\!\text{H} \\
\text{H}\!-\!\text{OH} \\
\text{H}\!-\!\text{OH} \\
\text{CH}_2\text{OH} \\
(+)\text{-mannose}
\end{array}
\qquad (18.1)
$$

Since osazone formation destroys the local chirality at C-2 of an aldose, monosaccharides that are epimeric at C-2 (such as glucose and mannose) will give the same osazone. (−)-Fructose also gives an osazone identical with that obtained from (+)-glucose, thereby demonstrating that this ketohexose has the same configuration as the aldohexose at C-3, C-4, and C-5.

PROBLEM 18·3
Show the changes in oxidation number that take place during osazone formation.

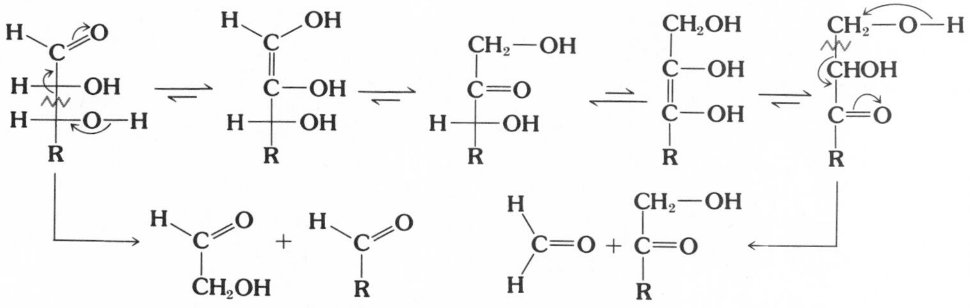

FIGURE 18·2 *Base-catalyzed isomerization and cleavage of a saccharide*

OXIDATION

Monosaccharides, both aldoses and ketoses, are readily oxidized by Fehling or Tollens reagent. Since the reagent is reduced in the process ($Cu^{+2} \longrightarrow Cu^{+1}$ for Fehling solution and $Ag^{+1} \longrightarrow Ag$ for Tollens reagent), these sugars are called *reducing sugars*. Although the oxidation of aldehydes is unexceptional (see Section 12.8), the facile reaction of ketoses under these mild oxidizing conditions may at first seem to be inconsistent with the relative resistance of simple ketones to such reagents. In fact, this behavior is characteristic of α-hydroxyketones that can exist as *enediol tautomers*:

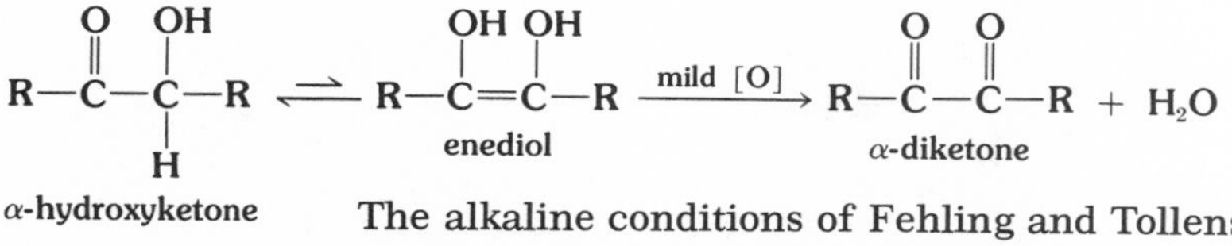

$$\underset{\alpha\text{-hydroxyketone}}{R-\underset{\underset{H}{|}}{\overset{\overset{O}{\|}}{C}}-\overset{\overset{OH}{|}}{C}-R} \rightleftharpoons \underset{\text{enediol}}{R-\overset{\overset{OH}{|}}{C}=\overset{\overset{OH}{|}}{C}-R} \xrightarrow{\text{mild [O]}} \underset{\alpha\text{-diketone}}{R-\overset{\overset{O}{\|}}{C}-\overset{\overset{O}{\|}}{C}-R} + H_2O \qquad (18.2)$$

The alkaline conditions of Fehling and Tollens reagents may also cause extensive isomerization of the carbonyl function and even chain cleavage, as shown in Figure 18.2.

Aldoses may be oxidized to monocarboxylic *aldonic acids* by treatment with hypobromous acid (bromine water) and to dicarboxylic *aldaric acids* by the action of nitric acid. Both these oxidations are illustrated in Figure 18.3. Stereochemical information can sometimes be obtained

FIGURE 18·3 *Aldose oxidations*

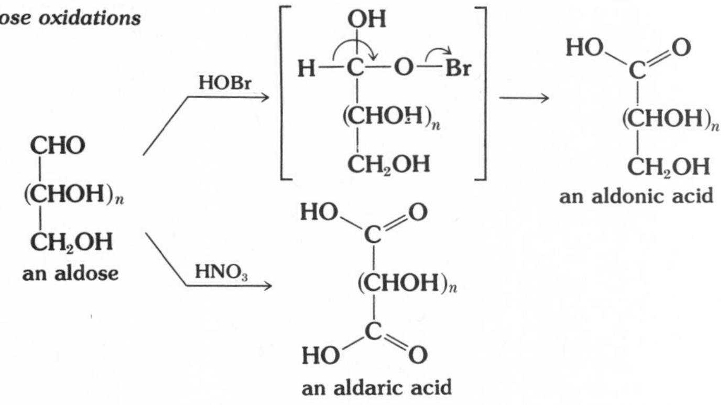

18·2 CHEMICAL REACTIONS OF MONOSACCHARIDES

from the latter transformation. For example, oxidation of (+)-galactose gives the optically inactive meso compound galactaric acid, whereas a similar oxidation of (+)-glucose yields an optically active aldaric acid:

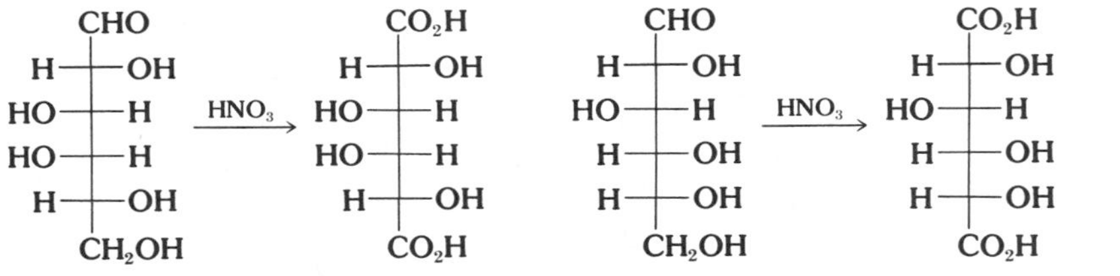

$$
\begin{array}{c}
\text{CHO} \\
\text{H}\!-\!\text{OH} \\
\text{HO}\!-\!\text{H} \\
\text{HO}\!-\!\text{H} \\
\text{H}\!-\!\text{OH} \\
\text{CH}_2\text{OH} \\
D\text{-}(+)\text{-galactose}
\end{array}
\xrightarrow{\text{HNO}_3}
\begin{array}{c}
\text{CO}_2\text{H} \\
\text{H}\!-\!\text{OH} \\
\text{HO}\!-\!\text{H} \\
\text{HO}\!-\!\text{H} \\
\text{H}\!-\!\text{OH} \\
\text{CO}_2\text{H} \\
\text{galactaric acid} \\
\text{(a meso compound)}
\end{array}
\qquad
\begin{array}{c}
\text{CHO} \\
\text{H}\!-\!\text{OH} \\
\text{HO}\!-\!\text{H} \\
\text{H}\!-\!\text{OH} \\
\text{H}\!-\!\text{OH} \\
\text{CH}_2\text{OH} \\
D\text{-}(+)\text{-glucose}
\end{array}
\xrightarrow{\text{HNO}_3}
\begin{array}{c}
\text{CO}_2\text{H} \\
\text{H}\!-\!\text{OH} \\
\text{HO}\!-\!\text{H} \\
\text{H}\!-\!\text{OH} \\
\text{H}\!-\!\text{OH} \\
\text{CO}_2\text{H} \\
D\text{-}(+)\text{-glucaric acid}
\end{array}
\qquad (18.3)
$$

The four chiral centers in galactaric acid must therefore be symmetrically arranged about a plane bisecting the C-3–C-4 bond.

SELECTIVE CHAIN SHORTENING AND LENGTHENING

Aldose chains may be selectively shortened from the aldehyde end by a suitable combination of mild oxidations. For example, the *Ruff degradation*, described in Figure 18.4, converts D-(+)-glucose to the aldopentose D-(−)-arabinose. A second application of this reaction converts D-(−)-arabinose to the tetrose D-(−)-erythrose.

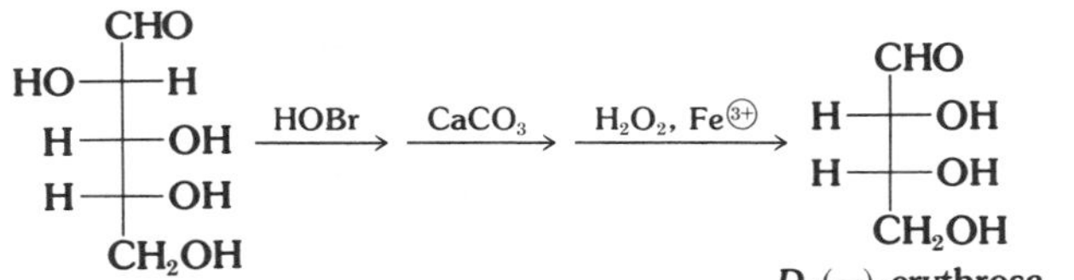

$$
\begin{array}{c}
\text{CHO} \\
\text{HO}\!-\!\text{H} \\
\text{H}\!-\!\text{OH} \\
\text{H}\!-\!\text{OH} \\
\text{CH}_2\text{OH} \\
D\text{-}(-)\text{-arabinose}
\end{array}
\xrightarrow{\text{HOBr}}
\xrightarrow{\text{CaCO}_3}
\xrightarrow{\text{H}_2\text{O}_2,\ \text{Fe}^{3+}}
\begin{array}{c}
\text{CHO} \\
\text{H}\!-\!\text{OH} \\
\text{H}\!-\!\text{OH} \\
\text{CH}_2\text{OH} \\
D\text{-}(-)\text{-erythrose,} \\
\text{a tetrose}
\end{array}
\qquad (18.4)
$$

Aldose chains may also be lengthened by one carbon atom through the intermediacy of a cyanohydrin derivative. This reaction sequence, shown in Figure 18.5, is known as the Kiliani-Fischer synthesis. Since a new chiral center (*C) is created at the cyanohydrin stage, this chain-lengthening process generates a pair of stereoisomers epimeric at this site. If we subject (+)-glyceraldehyde to the Kiliani-Fischer synthesis, two stereoisomeric tetroses, erythrose and threose, are formed.†

†The names of these tetroses give rise to the prefixes *erythro-* and *threo-*, used to designate the relative configuration of vicinal chiral centers (see Figure 15.8).

FIGURE 18·4 *The Ruff degradation*

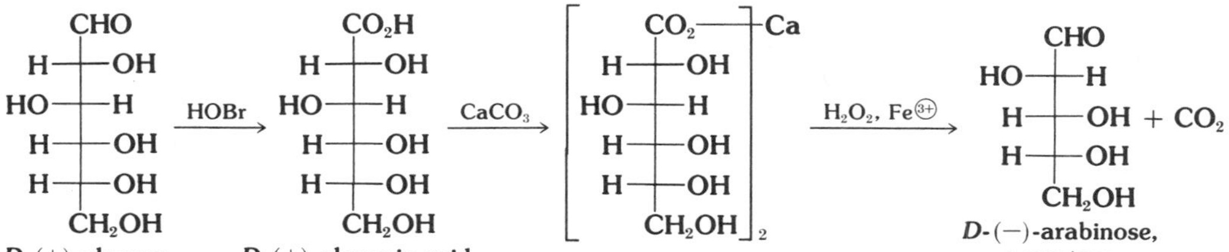

$$
\begin{array}{c}
\text{CHO} \\
\text{H}\!-\!\text{OH} \\
\text{HO}\!-\!\text{H} \\
\text{H}\!-\!\text{OH} \\
\text{H}\!-\!\text{OH} \\
\text{CH}_2\text{OH} \\
D\text{-}(+)\text{-glucose}
\end{array}
\xrightarrow{\text{HOBr}}
\begin{array}{c}
\text{CO}_2\text{H} \\
\text{H}\!-\!\text{OH} \\
\text{HO}\!-\!\text{H} \\
\text{H}\!-\!\text{OH} \\
\text{H}\!-\!\text{OH} \\
\text{CH}_2\text{OH} \\
D\text{-}(+)\text{-gluconic acid}
\end{array}
\xrightarrow{\text{CaCO}_3}
\left[
\begin{array}{c}
\text{CO}_2\!-\!\text{Ca} \\
\text{H}\!-\!\text{OH} \\
\text{HO}\!-\!\text{H} \\
\text{H}\!-\!\text{OH} \\
\text{H}\!-\!\text{OH} \\
\text{CH}_2\text{OH}
\end{array}
\right]_2
\xrightarrow{\text{H}_2\text{O}_2,\ \text{Fe}^{3+}}
\begin{array}{c}
\text{CHO} \\
\text{HO}\!-\!\text{H} \\
\text{H}\!-\!\text{OH} \\
\text{H}\!-\!\text{OH} \\
\text{CH}_2\text{OH} \\
D\text{-}(-)\text{-arabinose,} \\
\text{a pentose}
\end{array}
+ \text{CO}_2
$$

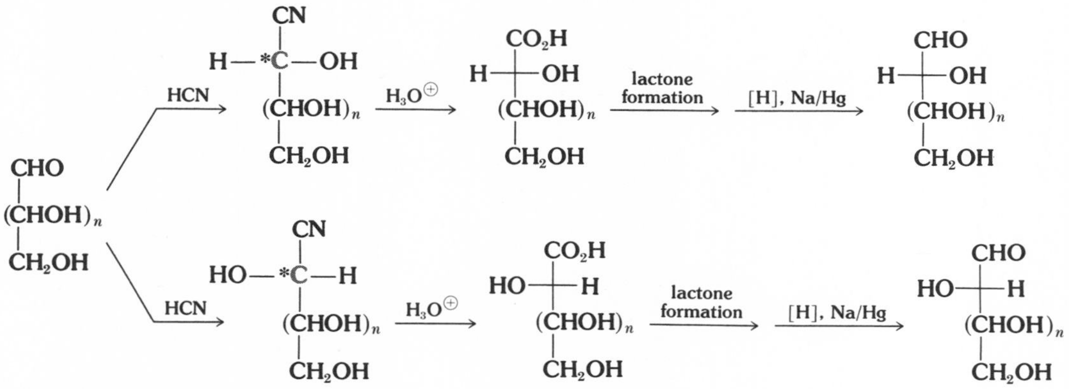

FIGURE 18·5 *The Kiliani-Fischer synthesis*

The configurational assignments for erythrose and threose rest firmly on their respective oxidations to meso- and (±)-tartaric acids when treated with nitric acid:

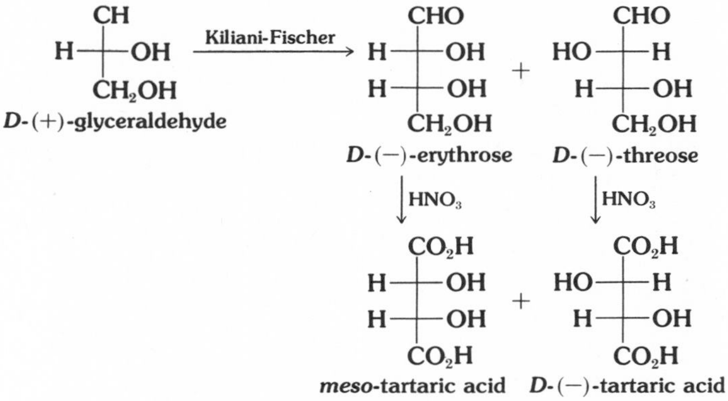

$$\text{D-(+)-glyceraldehyde} \xrightarrow{\text{Kiliani-Fischer}} \text{D-(−)-erythrose} + \text{D-(−)-threose} \xrightarrow{\text{HNO}_3} \text{meso-tartaric acid} + \text{D-(−)-tartaric acid} \tag{18.5}$$

18·3 THE CONFIGURATION OF (+)-GLUCOSE

We noted earlier that there are 16 stereoisomers (eight pairs of enantiomers) of the open-chain aldohexoses. When Emil Fischer began to study glucose in 1888, only a few monosaccharides were known. By using the reactions described in the last section, Fischer not only deduced the configuration of glucose, but was also able to synthesize and characterize all the other isomers. The following "proof of structure" illustrates the kinds of arguments Fischer used, although it does not retrace his complete reasoning.†

---

†Emil Fischer's extraordinary achievements in carbohydrate, protein, and porphyrin chemistry terminated in one of fate's striking ironies. His discovery of phenylhydrazine in 1875 provided him with a powerful tool which he used to good advantage in his work with sugars. Indeed, many chemists believe that the osazone reaction was the key that allowed Fischer to unlock the structure of the hexoses. From our present-day vantage point, however, it is clear that the symptoms of the fatal illness which ended Fischer's brilliant career were those of phenylhydrazine poisoning.

18·3 THE CONFIGURATION OF (+)-GLUCOSE

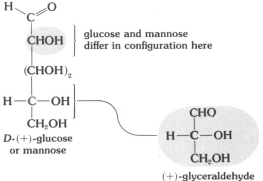

FIGURE 18·6 *Configurations of glucose and mannose*

1 Since (+)-glucose and (+)-mannose give the same osazone derivative (equation 18.1), these sugars must have identical configurations at all chiral centers except C-2. Fischer started with an assumption that the configuration at C-5 was the same as that of (+)-glyceraldehyde (see Figure 18.6).

2 The Ruff degradation of (+)-glucose and (+)-mannose gives in both cases the pentose (−)-arabinose, which must therefore incorporate the C-3, C-4, and C-5 configurations of glucose. As expected, the Kiliani-Fischer synthesis, when applied to (−)-arabinose, gives a mixture of (+)-glucose and (+)-mannose.

3 Oxidation of (−)-arabinose with nitric acid gives an optically active $C_5H_8O_7$ aldaric acid. This fact reduces the number of possible configurations of (−)-arabinose to two:

^{1}CHO	CHO	CHO	CHO
H—^{2}—OH	H——OH	HO——H	HO——H
H—^{3}—OH	HO——H	H——OH	HO——H
H—^{4}—OH	H——OH	H——OH	H——OH
^{5}CH$_2$OH	CH$_2$OH	CH$_2$OH	CH$_2$OH
I	II	III	IV
↓[O]	↓[O]	↓[O]	↓[O]
CO$_2$H	CO$_2$H	CO$_2$H	CO$_2$H
H——OH	H——OH	HO——H	HO——H
H——OH	HO——H	H——OH	HO——H
H——OH	H——OH	H——OH	H——OH
CO$_2$H	CO$_2$H	CO$_2$H	CO$_2$H
achiral (meso)	achiral (meso)	chiral	chiral

Only configurations III and IV (or their enantiomers) can give rise to an optically active aldaric acid. Since the configuration at C-4 (the original C-5 of glucose) has been arbitrarily fixed, the enantiomeric configurations are eliminated from consideration.

4 Oxidation of both (+)-glucose and (+)-mannose with nitric acid gives in each case an optically active $C_6H_{10}O_8$ aldaric acid. If we reflect

18·3 THE CONFIGURATION OF (+)-GLUCOSE

on the possible epimers formed from arabinose by a Kiliani-Fischer synthesis, we find that only configuration III is consistent with this fact, as illustrated by the following reaction equations:

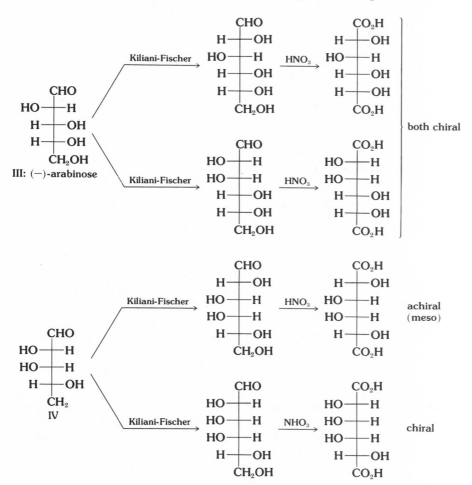

5   The configurations of (+)-glucose and (+)-mannose are, at this point in our argument, established as V and VI; the question remains. however, as to which is which.

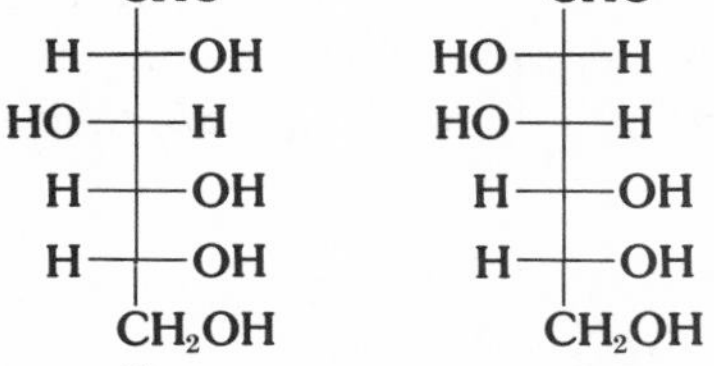

$$
\begin{array}{cc}
\text{CHO} & \text{CHO} \\
\text{H}\!-\!\text{OH} & \text{HO}\!-\!\text{H} \\
\text{HO}\!-\!\text{H} & \text{HO}\!-\!\text{H} \\
\text{H}\!-\!\text{OH} & \text{H}\!-\!\text{OH} \\
\text{H}\!-\!\text{OH} & \text{H}\!-\!\text{OH} \\
\text{CH}_2\text{OH} & \text{CH}_2\text{OH} \\
\text{V} & \text{VI}
\end{array}
$$

Fischer synthesized a new $C_6H_{12}O_6$ aldose, (+)-gulose, which he found gave the same aldaric acid as (+)-glucose. This means that (+)-glucose must have configuration V and (−)-gulose the inverted configuration VII:

CHO		CO₂H	CO₂H		CHO	
H—OH		H—OH	HO—H		HO—H	
HO—H	HNO₃→	HO—H	= HO—H	←HNO₃	HO—H	
H—OH		H—OH	H—OH		H—OH	
H—OH		H—OH	HO—H		HO—H	
CH₂OH		CO₂H	CO₂H		CH₂OH	
V: D-(+)-glucose					VII: L-(+)-gulose	

Configuration VI corresponds therefore to (+)-mannose.

Another way of viewing the argument used here is to consider the overall effect on glucose and mannose of a sequence of reactions which act to exchange the functional groups at the ends of the sugar carbon chains. Such an exchange leaves the structure of (+)-mannose unchanged but converts (+)-glucose to a different aldose, (+)-gulose:

$$\begin{array}{cccc}
\text{CHO} & \text{CH}_2\text{OH} & \text{CHO} & \text{CH}_2\text{OH} \\
\text{HO—H} & \text{HO—H} & \text{H—OH} & \text{H—OH} \\
\text{HO—H} & \text{HO—H} & \text{HO—H} & \text{HO—H} \\
\text{H—OH} & \text{H—OH} & \text{H—OH} & \text{H—OH} \\
\text{H—OH} & \text{H—OH} & \text{H—OH} & \text{H—OH} \\
\text{CH}_2\text{OH} & \text{CHO} & \text{CH}_2\text{OH} & \text{CHO} \\
\textit{D-(+)-mannose} & \textit{D-(+)-mannose} & \textit{D-(+)-glucose} & \textit{L-(+)-gulose}
\end{array}$$

(with terminal-group exchange between each pair)   (18.6)

PROBLEM 18·4

Suggest a method other than that used in step 4 of the preceding structure proof which could be used to distinguish between configurations III and IV for arabinose.

PROBLEM 18·5

The aldopentose (−)-ribose forms the same osazone as (−)-arabinose. Nitric acid oxidation of (−)-ribose gives an optically inactive $C_5H_8O_7$ aldaric acid. Write equations for these reactions showing the configurations of all reactants and products.

18·4 CYCLIC ANOMERS OF GLUCOSE

As Fischer and other chemists learned more about glucose, it slowly became clear that the pentahydroxy aldehyde structure, initially favored for this and other aldohexoses, could not account completely for its chemical behavior. One anomalous fact was that, although aldehydes generally give crystalline bisulfite addition products under mild reaction conditions, glucose does not:

$$R-\underset{H}{\overset{O}{\overset{\|}{C}}} + :\overset{\overset{\cdot\cdot}{O}:}{\underset{\overset{\|}{\underset{\cdot\cdot}{O}}-H}{S}}-\overset{\cdot\cdot}{\underset{\cdot\cdot}{O}}:^{\ominus}Na^{\oplus} \rightleftharpoons R-\underset{H}{\overset{H-O}{\overset{|}{C}}}-\overset{\overset{\cdot\cdot}{O}:}{\underset{\overset{\|}{\underset{\cdot\cdot}{O}}}{S}}-\overset{\cdot\cdot}{\underset{\cdot\cdot}{O}}:^{\ominus}Na^{\oplus} \quad (18.7)$$

Furthermore, the pentaacetate and pentamethyl ether derivatives of glucose are not oxidized by either Tollens reagent or Fehling solution

18·4 CYCLIC ANOMERS OF GLUCOSE

TABLE 18·1 *Some properties of (+)-glucose isomers*

D-(+)-glucose	melting point (°C)	specific rotation
α isomer	146	+113°
β isomer	150	+ 19°

and also do not give phenylhydrazone or bisulfite addition products:

$$C_6H_{12}O_6 \xrightarrow{5CH_3I,\ AgOH} \underset{\substack{\text{pentamethyl glucose}\\ \text{(no reaction with}\\ \text{Tollens reagent)}}}{C_{11}H_{22}O_6} \xrightarrow{H_3O^{\oplus}} \underset{\substack{\text{tetramethyl glucose}\\ \text{(reduces Tollens}\\ \text{reagent)}}}{C_{10}H_{20}O_6} \quad (18.8)$$

One of the ether groupings in the pentamethyl derivative is exceptionally reactive and undergoes hydrolysis in dilute acid to a tetramethyl derivative and methanol. This tetramethyl glucose derivative proved to reduce Tollens reagent as effectively as glucose itself.

In 1895 the French chemist Tanret isolated an isomeric form of (+)-glucose from hot concentrated solutions. He identified the new isomer as β-D-(+)-glucose, reserving the prefix α for the previously known isomer (see Table 18.1). Fresh aqueous solutions of either isomer slowly change in specific rotation, giving a final equilibrium value of +52.7°. We call this change in optical rotation *mutarotation*.

Although these facts are clearly inconsistent with the pentahydroxy aldehyde structure for glucose, the essential features of this formula are so firmly supported by experimental evidence that it cannot be completely abandoned. In order to accommodate the new facts, we must somehow diminish (but not eliminate) the aldehydic character of C-1 and also introduce a new chiral center into the molecule. On reflection, we see that these changes can be accomplished in one stroke by assuming that glucose exists in the cyclic hemiacetal form shown in Figure 18.7. We call the six-membered heterocyclic forms of sugars *pyranoses*,

FIGURE 18·7 *Pyranose structures for (+)-glucose*

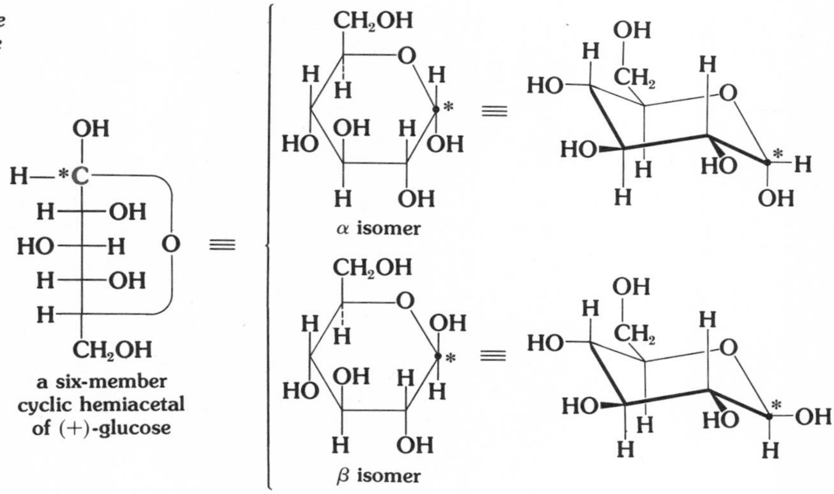

18·4 CYCLIC ANOMERS OF GLUCOSE

after the parent heterocycle pyran. The corresponding five-membered cyclic hemiacetals are referred to as *furanoses*, named after furan.

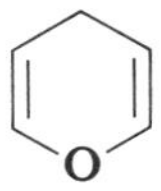

pyran    furan

The chemistry of glucose reflects, therefore, the existence of a one-sided but mobile equilibrium between open and cyclic structures, as outlined in Figure 18.8. Stereoisomeric carbohydrates which differ in configuration at C-1 are called *anomers*.

The reactions of glucose outlined in Figure 18.8 provide a striking illustration of an important relationship often encountered in organic chemistry: *A minor component of a mixture of molecular or ionic species in dynamic equilibrium may be responsible for most of the chemical transformations of this mixture.* In the present case, for example, the oxidation of glucose to gluconic acid, the formation of oxime or osazone derivatives from glucose, and the conversion of glucose to a methyl glucoside all involve the acyclic aldehyde form of the sugar despite its very low concentration at equilibrium (less than 0.01%). As the aldehyde is consumed by the chemical reaction, it is rapidly replaced by opening of the α- or β-pyranose anomers.

FIGURE 18·8 *Anomeric derivatives in the (+)-glucose family*

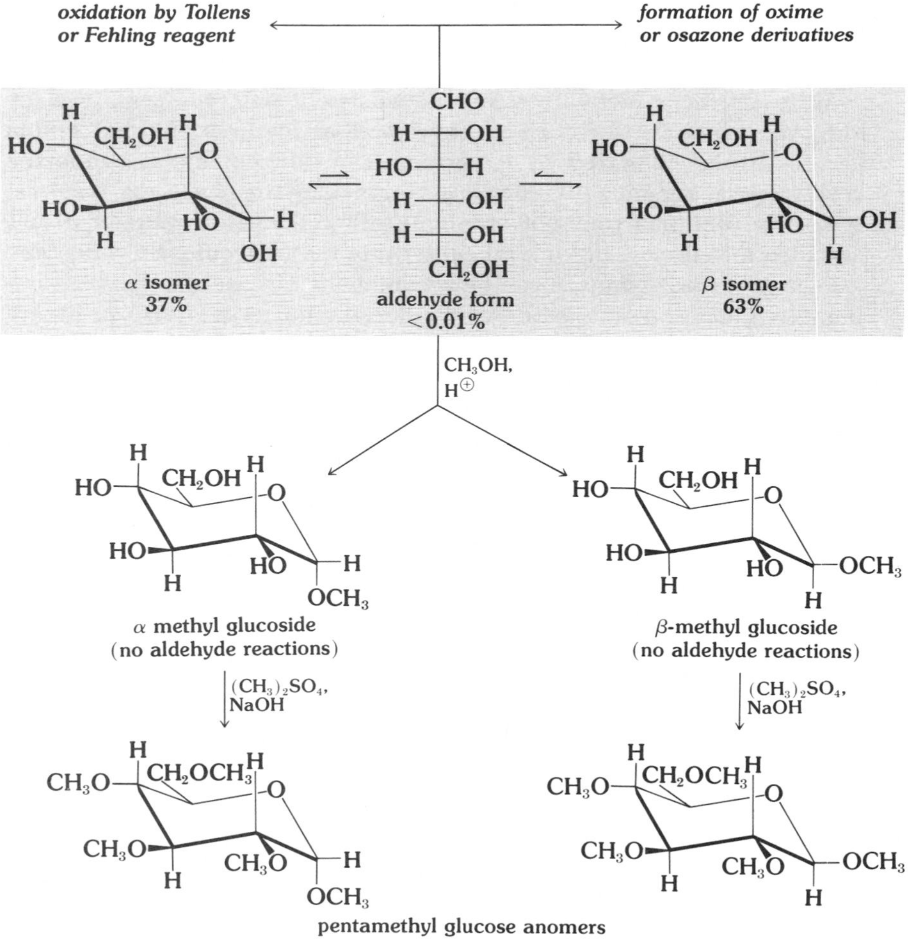

18·4 CYCLIC ANOMERS OF GLUCOSE 613

PROBLEM 18·6

Five- and six-membered rings are relatively strain free, and rings of both size are possible for the cyclic anomers of glucose.

a Write structures for the five-membered (furanose) forms of glucose.

b Write formulas for one anomer (either α or β) of the tetramethyl glucose formed in equation 18.8 from both the furanose and pyranose forms of glucose. Suggest a method for distinguishing these two forms of glucose. (Glucose has been shown to have the pyranose structure).

Glycosides are cyclic acetals, formed by the reaction of aldoses with alcohols, and may exist in α or β anomeric forms. Glycoside derivatives of glucose, called *glucosides*, may be prepared from simple alcohols such as methanol (see Figure 18.8) or from complex alcohols such as cholesterol. These compounds exist in a six-membered cyclic pyranoside structure.

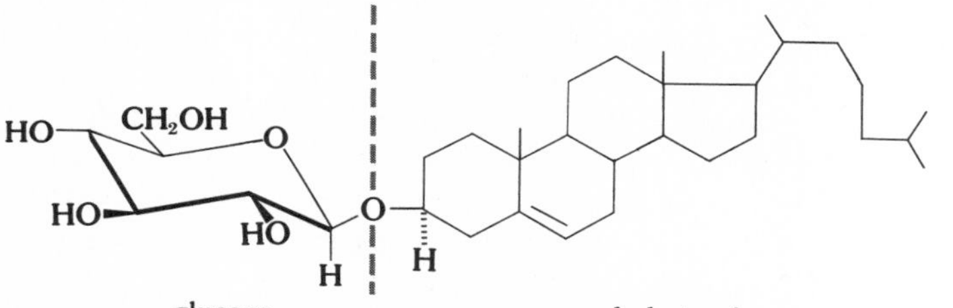

a β-glucoside of cholesterol

Since glycosides are acetals, they are relatively stable to base treatment, but they may be hydrolyzed to their sugar and alcohol (aglycone) components by the action of dilute mineral acids. Hydrolytic enzymes called *hydrolases* can also effect the hydrolysis of glycosides under very mild conditions and with high stereospecificity. The yeast enzyme maltase, for example, selectively hydrolyses α-glycosides, whereas the enzyme emulsin (from bitter almonds) acts only on β-glycosides. The result of both actions is the mutarotated sugar shown in Figure 18.9.

FIGURE 18·9 *Enzymatic hydrolysis of anomeric glycosides*

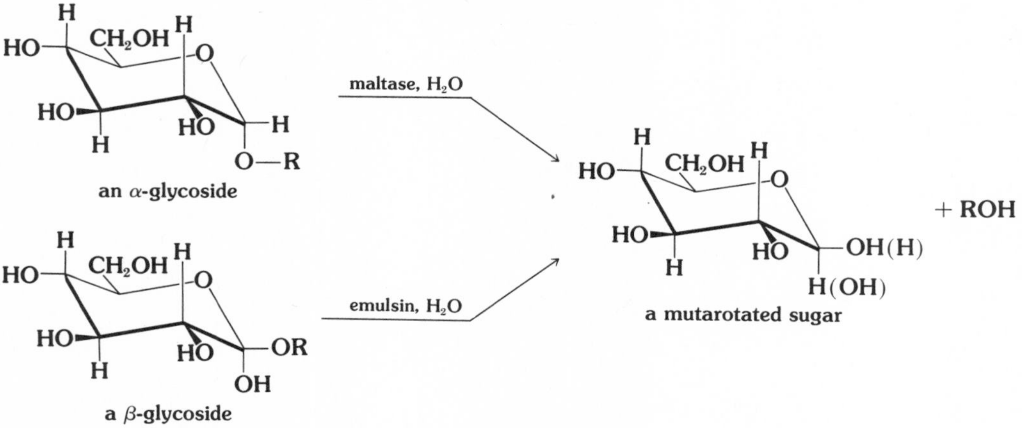

18·5 DISACCHARIDES AND POLYSACCHARIDES

Disaccharides are glycosides in which the aglycone or alcohol moiety is a second monosaccharide unit. On hydrolysis a disaccharide molecule yields two monosaccharide molecules. Since monosaccharides usually have five different hydroxyl groups, many different glycoside combinations of two sugars are possible. Investigations of naturally occurring disaccharides indicate that the C-4 hydroxyl function is most commonly involved in these glycosidic linkages.

(+)-MALTOSE

Partial hydrolysis of starch gives, along with (+)-glucose, the $C_{12}H_{22}O_{11}$ disaccharide (+)-maltose. Hydrolysis of maltose itself yields only (+)-glucose; hence we know that this disaccharide is composed of two glucose units. (+)-Maltose is a reducing sugar (it is oxidized by Tollens or Fehling reagent) and it reacts with phenylhydrazine to yield an osazone derivative. Oxidation of maltose by bromine water generates a monocarboxylic acid, $C_{11}H_{21}O_{10}$—CO_2H. The anomeric forms of (+)-maltose—α-(+)-maltose with a specific rotation of $+168°$ and β-(+)-maltose with a specific rotation of $+112°$—undergo mutarotation in aqueous solution, giving an equilibrium rotation of $+136°$. Taken together, these facts indicate that one of the glucose moieties in maltose still retains its reactive hemiacetal structure, as shown in Figure 18.10.

Before we can write a complete structure for (+)-maltose two questions remain to be answered:

1 What is the nature of the glycoside linkage in glucose unit I?
2 Which of the four hydroxyl groups in glucose unit II (the hemiacetal hydroxyl must be excluded) is involved in this linkage?

The first question is answered by the observation that (+)-maltose is rapidly hydrolyzed to glucose by the action of the enzyme maltase, but not by emulsin. These enzymes are known to effect selective hydrolysis of α- and β-glycosides, respectively; therefore (+)-maltose contains an α-glycoside linkage.

FIGURE 18·10 *A partial structure of (+)-maltose*

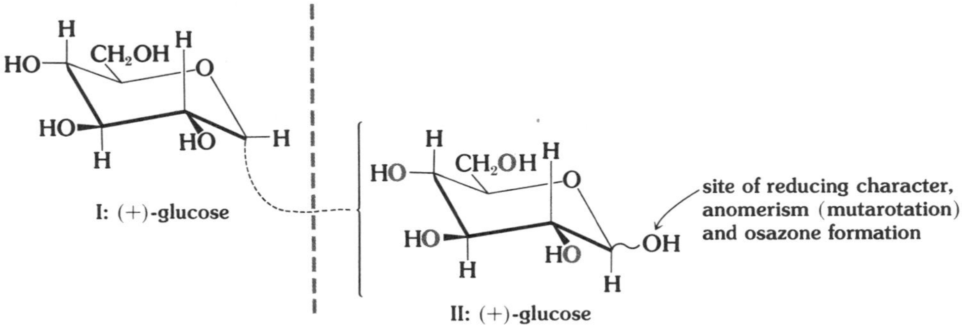

18·5 DISACCHARIDES AND POLYSACCHARIDES

The second question can be answered by marking all the free hydroxyl groups in maltose in such a way that they can be distinguished from those involved in the glycosidic linkage *after* the labeled maltose has been hydrolyzed to its glucose units. A methyl group makes an excellent label for this purpose, since the resulting methyl ethers are relatively stable to subsequent chemical reactions. Figure 18.11 shows the sequence of reactions that puts this strategy into practice. The first step is an oxidation of the aldehyde function in glucose unit II thus permitting the two glucose fragments to be easily distinguished at the end of the operation.

(+)-CELLOBIOSE

Acid-catalyzed hydrolysis of cellulose (for example, cotton fibers) yields (+)-glucose and a disaccharide fragment called *cellobiose*. (+)-Cellobiose resembles (+)-maltose in many respects. It has the molecular formula $C_{12}H_{22}O_{11}$, and it exists in α and β anomeric forms that undergo mutarotation. All the chemical transformations described for (+)-maltose proceed in an equivalent fashion for (+)-cellobiose, with the exception of the enzymatic hydrolysis. (+)-Cellobiose is hydrolyzed to two glucose moieties by the action of the enzyme emulsin, not by maltase. The

FIGURE 18·11 *Chemical structure proof for D-(+)-maltose*

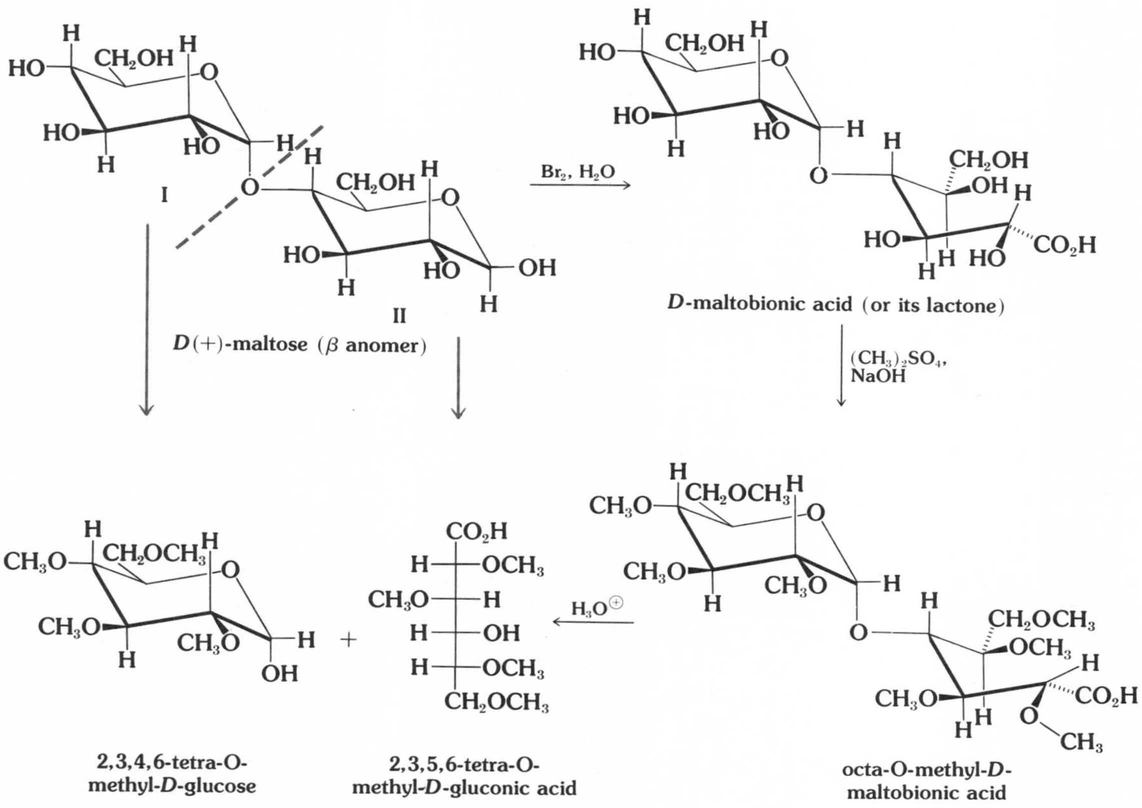

18·5 DISACCHARIDES AND POLYSACCHARIDES

linkage of glucose units in this disaccharide must therefore be a β-glycoside, rather than an α-glycoside, as in maltose:

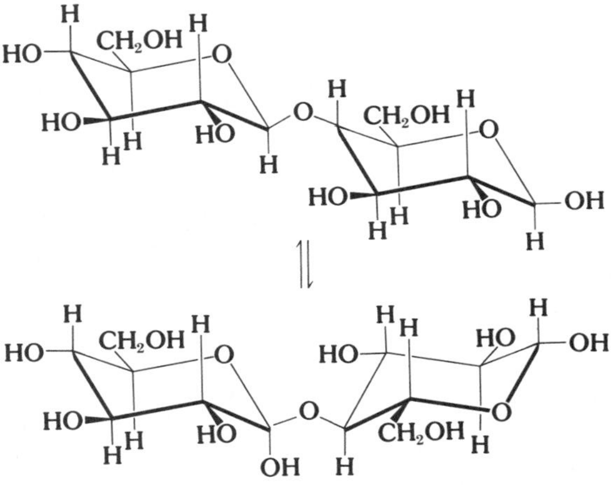

two conformers of (+)-cellobiose (β-anomer)

(18.9)

PROBLEM 18·7
a Draw a structural formula for the α anomer of (+)-cellobiose.
b Write equations for the sequence of oxidation, methylation, and hydrolysis as applied to (+)-cellobiose (see Figure 18.11).

(+)-LACTOSE

(+)-Lactose is a $C_{12}H_{22}O_{11}$ disaccharide found in the milk of mammals. Many adults, particularly those from countries where milk is not a dietary staple, have a metabolic intolerance for lactose. This intolerance is fairly widespread, a fact that severely restricts the use of dehydrated milk as a food supplement in many parts of the world. Cheese is often an acceptable food in these countries, since the lactose remains in the whey byproduct during the manufacturing process. Infants apparently have a digestive enzyme which acts to hydrolyze lactose, but this enzyme disappears on weaning.

Hydrolysis of (+)-lactose by dilute acid or the enzyme emulsin gives equal amounts of D-(+)-glucose and D-(+)-galactose. This disaccharide must therefore be a β-glycoside formed by the union of a molecule of D-(+)-glucose with one of D-(+)-galactose. Since lactose is a reducing sugar (it forms an osazone and exists in α and β anomeric forms), we are immediately faced with the question of which monosaccharide unit is the hemiacetal moiety and which is the acetal. To answer this question we need only effect a chemical transformation of the reactive aldehyde function (for example, an oxidation) and then ascertain which monosaccharide fragment has suffered the change. Figure 18.12 shows the application of this procedure to (+)-lactose. The location of the glycosidic linkage at the C-4 hydroxyl function on glucose can be demonstrated by the same oxidation, methylation, hydrolysis sequence we used to establish the structure of maltose.

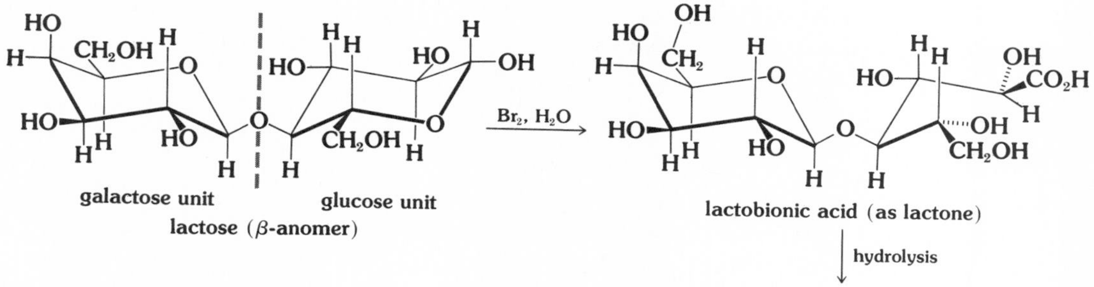

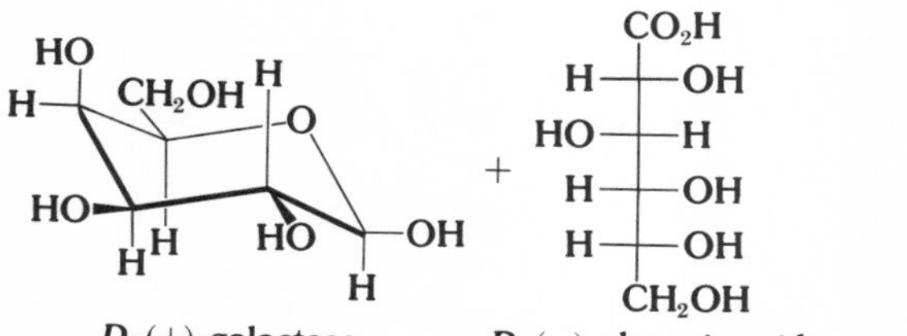

18·5 DISACCHARIDES AND POLYSACCHARIDES

FIGURE 18·12 *Characteristic reactions of (+)-lactose*

(+)-SUCROSE

The common table sugar we obtain from sugar cane and sugar beets is the $C_{12}H_{22}O_{11}$ disaccharide (+)-sucrose. Unlike the three disaccharides discussed earlier, (+)-sucrose is not a reducing sugar, does not exist in mutarotating anomeric forms, and it does not give an osazone derivative. These facts rule out the presence of a hemiacetal function in this disaccharide.

Acid- or enzyme-catalyzed hydrolysis of (+)-sucrose gives an equimolar mixture of D-(+)-glucose and D-(−)-fructose. Since the optical rotation of the sugar mixture changes from (+) to (−) as hydrolysis proceeds, the product mixture is often referred to as *invert sugar*:

$$\text{(+)-sucrose} \xrightarrow[\text{or maltase, H}_2\text{O}]{\text{H}_3\text{O}^\oplus} D\text{-(+)-glucose} + D\text{-(−)-fructose} \quad (18.10)$$
$$[\alpha]_D = +66.5° \qquad\qquad [\alpha]_D = +52° \qquad [\alpha]_D = -92°$$

At this point in our reasoning the evidence requires that sucrose be both a D-glucoside and a D-fructoside in order to avoid the presence of a reactive free carbonyl group. In other words, there must be a glycoside linkage between C-1 of glucose and C-2 of fructose. The configuration of this glycoside linkage is shown in Figure 18.13. The nature of the link-

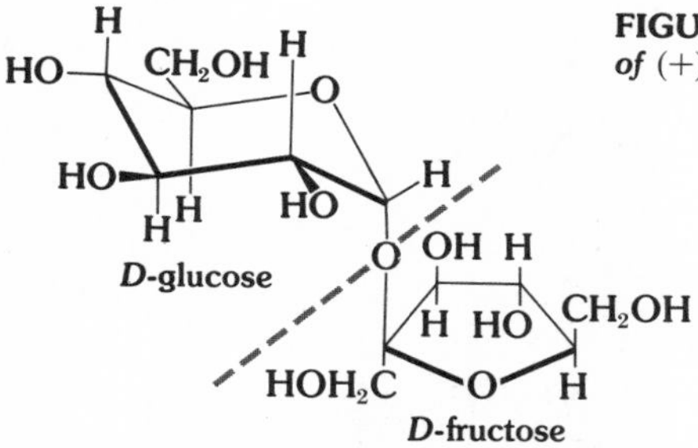

FIGURE 18·13 *The structure of (+)-sucrose*

α-D-glucopyranosyl β-D-(+)-fructofuranoside
or
β-D-fructofuranosyl α-D-(+)-glucopyranoside

age in each unit and the unusual furanose form of the fructose unit were established by a combination of chemical and physical measurements that we shall not go into here.

(+)-RAFFINOSE

There are many naturally occurring trisaccharides. One of these is (+)-raffinose, a $C_{18}H_{32}O_{16}$ nonreducing component of sugar beets, which on enzymatic hydrolysis gives D-galactose and sucrose. Methylation of raffinose, followed by hydrolysis yields methylated derivatives of D-fructose, D-glucose, and D-galactose, which indicate the nature of the glycosidic bonds joining these units together.

> **PROBLEM 18·8**
> Write equations showing the methylation and complete hydrolysis of (+)-raffinose.

18·6 POLYSACCHARIDES

Polysaccharides, also known as *glycans*, are polymeric substances composed of many hundreds or thousands of monosaccharide units bound together by glycoside linkages. Some of these compounds provide structural rigidity and strength to plants, whereas others serve as reserve food supplies.

CELLULOSE

More than half of the total organic carbon in the biosphere is cellulose. Cotton fibers are essentially pure cellulose, and the woody stems of bushes and trees are about half cellulose. Cellulose has the molecular formula $(C_6H_{10}O_5)_n$, where n is large (500 to 5,000), and a molecular

FIGURE 18·14 *Glycoside bonds in (+)-raffinose*

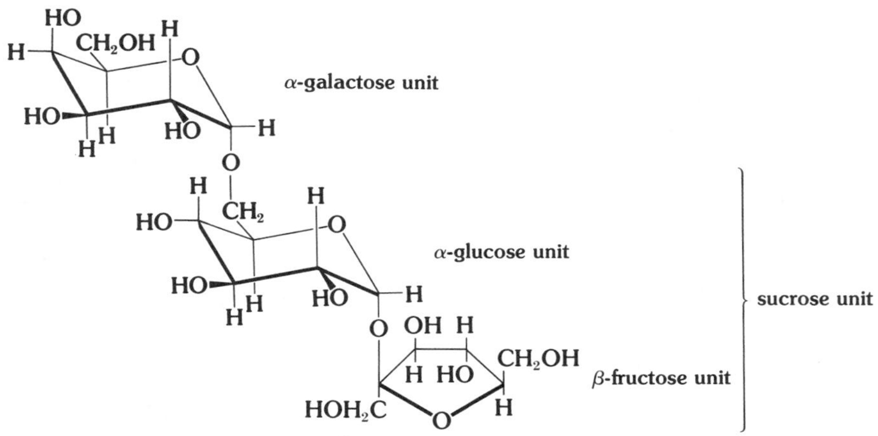

weight on the order of 100,000 to 1,000,000. Acid-catalyzed hydrolysis of cellulose yields D-(+)-glucose as the only monosaccharide, accompanied by some (+)-cellobiose in the case of incomplete hydrolysis. From these facts we can conclude that cellulose is a polymer of D-glucose units joined by β-glycoside linkages to C-4 of an adjoining unit:

cellulose

Some animals (the cow and termite, for example) have digestive enzymes capable of hydrolyzing cellulose into glucose; consequently, these organisms can feed directly on cellulose. Other animals, such as man, lack enzymes that will hydrolyze β-glucoside bonds, and they cannot use cellulose as a foodstuff.

In its role as a structural material, cellulose molecules are organized in bundles of parallel chains, cross linked by hydrogen bonds. These fibrils are in turn arranged in a regular, densely packed matrix, cemented together by other polymeric substances such as pectin or lignin.

The physical properties of cellulose (such as thermal stability, solubility in various solvents, or melting point) can be substantially modified by preparing ester or ether derivatives of the free hydroxyl groups in the polymer (three for each glucose unit in the chain). Some of these derivatives, such as cellulose nitrate, cellulose acetate, and methyl or ethyl ethers of cellulose, are useful plastics that can be spun into fibers, spread into films, and fashioned into a variety of other forms.

STARCH

Starch is a polymer of glucose, in which form plants accumulate and store food in roots, rhizomes, seeds, stems, tubers, and corms. Most animals depend on these plant starches for nourishment. In fact, all living organisms can be broadly classified according to whether they are self-feeding, *autotrophic,* or feed on other organisms, *heterotrophic.* Autotrophic organisms such as algae, diatoms, and various green plants employ the marvelous process of photosynthesis to produce their own carbohydrates from carbon dioxide and water:

$$n\text{CO}_2 + n\text{H}_2\text{O} \xrightarrow{\text{sunlight}} \underset{\text{a carbohydrate}}{(\text{CH}_2\text{O})_n} + n\text{O}_2 \qquad (18.11)$$

Although this reaction is strongly endothermic, the ability of these organisms to capture the energy in sunlight by its interaction with chlorophyll satisfies the stringent energy requirement.

Higher animals, most microorganisms, and nonphotosynthetic plants cannot consume carbon dioxide directly and require a more reduced form of carbon as a source of energy for their operation. The carbon "fuels" used by these heterotrophic organisms are usually carbohydrates,

18·6 POLYSACCHARIDES

fats and oils, and proteins. Monosaccharides such as glucose, or a polymer thereof, are potentially excellent energy sources, by what is essentially a reversal of the photosynthesis process:

$$C_6H_{12}O_6 + 6O_2 \longrightarrow 6CO_2 + 6H_2O \qquad \Delta G^* = -686 \text{ kcal/mole} \qquad (18.12)$$

Starch is found in plants in the form of microscopic granules, the size and shape of which are characteristic of the source. When starch is treated with hot water, the granules swell, and the resulting colloidal suspension can be separated into two components. The major fraction (about 80%) is an amorphous high-molecular-weight substance called *amylopectin*, and the minor component (about 20%) is a microcrystalline substance called *amylose*. However, different genetic strains of some plants, such as corn, may have starch with more amylose or more amylopectin. The characteristic blue color reaction of starch with iodine is due to the amylose fraction (amylopectin gives a red-purple color), which is able to incorporate up to 20% of its weight of iodine in an inclusion compound.

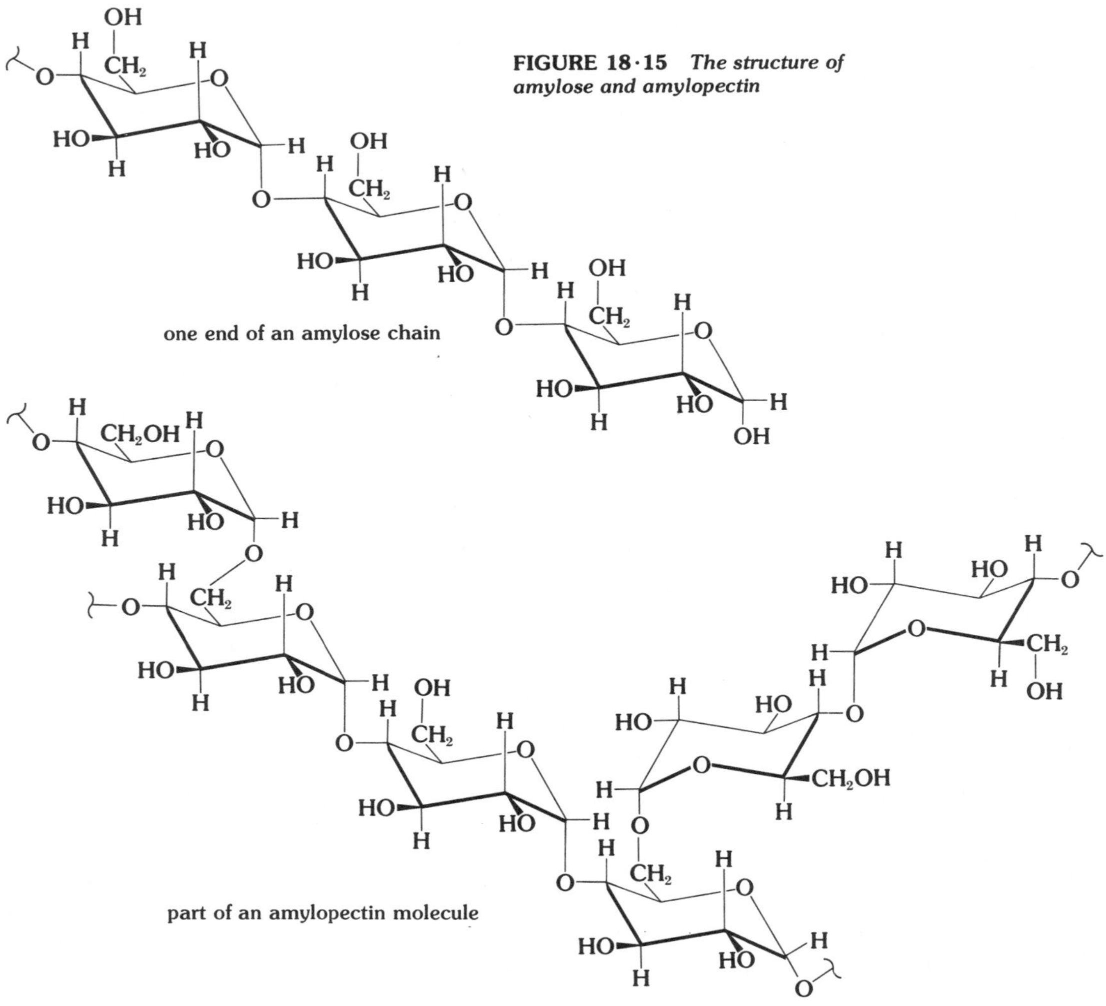

FIGURE 18·15 *The structure of amylose and amylopectin*

Acid-catalyzed hydrolysis of both amylopectin and amylose gives (+)-glucose, (+)-maltose, and an assortment of low-molecular-weight polysaccharides (corn syrup is such a mixture). Both starch components thus appear to be polymers of glucose, and their different properties are due in part to the degree of chain branching of the polymer. Physical and chemical measurements of amylose suggest it to be a mixture of unbranched chains of glucose units, varying in length from 1000 to 10,000 units. These glucose units are linked together by α-glucoside bonds and can be hydrolyzed by certain enzymes, such as maltase and amylase, which are specific for this type of glycoside.

Amylopectin appears to have a more highly branched structure, consisting of thousands of short glucose chains (about 20 units long) formed by 1,4-α-glucoside linkages, which in turn are joined in a random fashion to other chains or branches by 1,6-α-linkages.

PROBLEM 18·9
A useful method of probing the structures of polysaccharides is to carry out an end-group analysis; this is essentially a count of those pyranose units at the end of a chain in relation to the total monomer units. One way of doing this is to effect a total methylation of all free hydroxyl groups in the polysaccharides, followed by hydrolysis to a mixture of tri- and tetramethyl glucose derivatives.
a Show how a determination of the methyl glucose mixture allows us to arrive at an end-group count.
b What difference in end-group analysis would you expect for amylose versus amylopectin?

18·7 VITAMIN C

Ascorbic acid, or vitamin C, is an important biological reducing agent that appears to facilitate electron-transfer processes in living organisms:

$$\text{L-ascorbic acid} \rightleftharpoons \text{dehydro-L-ascorbic acid} + 2H^{\oplus} + 2e^{\ominus} \qquad (18.13)$$

Without ascorbic acid in their systems, most animals would die; however, nearly all are able to synthesize this vitamin from glucose by the sequential action of four liver enzymes, as shown in Figure 18.16. Unfortunately, the fourth enzyme, L-gulonolactone oxidase, is missing in man and a few other animals, such as guinea pigs, some monkeys, and birds. Consequently, these animals must consume ascorbic acid as part of their diet or suffer the chronic effects of its deficiency, which in extreme cases is the disease called scurvy.

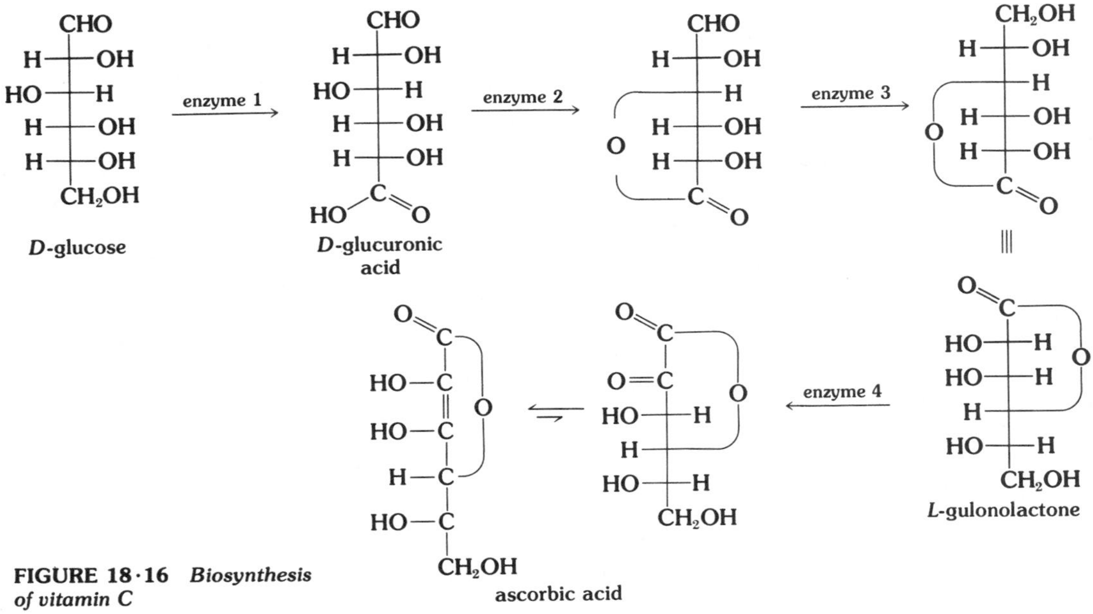

FIGURE 18·16 *Biosynthesis of vitamin C*

Some scientists, notably Irwin Stone, assert that the absence of this enzyme is due to a serious inborn error of carbohydrate metabolism. Apparently a mutation occurred on the gene that controls the synthesis of this enzyme, and the defective gene is now carried by members of the primate suborder Anthropoidea. Since the primitive monkeys belonging to the other primate suborder, Prosimii, do not have this genetic defect, the approximate time of the mutation appears to have been in the Paleocene epoch, about 60 million years ago. Because of the almost universal occurrence of ascorbate in both plants and animals, it is likely that the enzymes required for ascorbate synthesis evolved long before living organisms diverged into the plant and animal kingdoms.

The amount of ascorbic acid required to fully correct the hypoascorbemia of human beings is a matter of contention. Linus Pauling has calculated the optimal human intake to be 2 to 10 g daily, based on the caloric and ascorbate contents of fresh foods. Some physicians feel this is an unnecessarily large dose. A similar dispute also revolves around the alleged antiviral activity of ascorbic acid.

REFERENCES

E. A. Davidson, *Carbohydrate Chemistry*, Holt, Rinehart and Winston, Inc., New York, 1967.

R. J. McIlroy, *Introduction to Carbohydrate Chemistry*, Butterworth and Company Ltd., London, 1967.

R. D. Guthrie and J. Honeyman, *An Introduction to the Chemistry of Carbohydrates*, 3d ed., Clarendon Press, Oxford, 1968.

G. A. Aspinall, *Polysaccharides*, Pergamon Press, New York, 1970.

J. F. Stoddart, *Stereochemistry of Carbohydrates*, John Wiley & Sons, Inc., New York, 1971.

EXERCISES

18·1 Glucose is classified as an aldohexose. Classify each of the following saccharides:

```
         CHO              CH₂OH              CHO
a   H────OH         b      C=O         c  H────OH
        CH₂OH          H────OH            HO────H
                        CH₂OH             H────OH
                                          CH₂OH
```

18·2 Draw projection formulas for:
a The two glyceraldehyde stereoisomers
b All four aldotetroses
c The four D-aldopentoses
d The eight D-aldohexoses
e The four D-2-ketohexoses

18·3 Of the aldohexoses, only D-(+)-glucose, D-(+)-mannose, and D-(+)-galactose are found in abundance in nature. Draw Fischer projections for these sugars and their corresponding L-family enantiomers.

18·4 Explain the difference between:
a Racemization and mutarotation
b An anomer and an epimer
c An epimer and a diastereomer
d A D and an L sugar
e A pyranose and a furanose form of a sugar

18·5 Suggest a reason why D-glucose reacts with only 1 mole of CH_3OH to form an acetal.

18·6 Draw both possible chair conformations of β-D-glucopyranose. Which conformation is energetically most favorable? Why? How do the most favorable chair conformations of β-D-mannopyranose and β-D-glactopyranose compare energetically with β-D-glucopyranose?

18·7 Suggest a reason why β-D-glucose predominates (63%) over the α-D-glucose isomer (37%) at equilibrium (see Figure 18.7).

18·8 How many stereoisomers are possible for an aldohexose in the pyranose cyclic hemiacetal form?

18·9 Draw the structures for and name the compounds which are related to β-D-glucopyranose in the following way:
a The enantiomer
b The anomer
c An epimer
d A diastereomer (How many are possible?)

18·10 Indicate which carbon atoms in sucrose are acetal carbons (which would be carbonyl carbons in the open-chain forms of glucose and fructose?). Write a balanced equation for the hydrolysis of sucrose to glucose and fructose. How many moles of H_2O are required per mole of sucrose?

18·11 Fructose may exist in either a furanose or a pyranose form.
a Draw both the α and β anomers of D-fructofuranose.
b Draw both anomers of D-fructopyranose. Which hydroxyl is utilized in the formation of the pyranose cyclic hemiketal?

18·12 What two aldohexoses yield D-arabinose on Ruff degradation? What four aldohexoses yield D-erythrose as a result of two successive Ruff degradations?

18·13 D-(−)-ribose, an important aldopentose which occurs in the nucleic acid RNA, forms a furanose ring. Draw the two anomers of D-ribofuranose. *Hint:* See Problem 18.5 on page 610.

18·14 Amino sugars are a class of natural products in which one (or more) of the hydroxyl groups of a carbohydrate is replaced by an amino group.
a The most prevalent amino sugar is D-glucosamine, in which the 2OH of D-glucose has been replaced by NH_2. Draw both the open-chain and β-pyranose forms of D-glucosamine.
b The acetamide ($NHCOCH_3$) derivative of D-glucosamine is the monomeric unit in the polysaccharide *chitin,* the structural component of the exoskeletons of crustacea (lobsters, crabs, shrimp, etc.) and insects. Chitin is similar to cellulose, as it is composed of 1,4-glucosidic linkages of β-pyranose units. Draw a three-unit section of this aminopolysaccharide.

18·15 The reaction of D-glucose with sodium borohydride (or H_2/catalyst) yields the D-glucitol sorbitol ($C_6H_{14}O_6$). Sorbitol is found naturally in the berries of mountain ash (*Sorbus ancuparia*), and in high concentration (14%) in red seaweed (*Bostrychia scorpoides*). It is used extensively as a vehicle in prepared foods and cosmetics and as a sugar substitute for diabetics (about 60% as sweet as sucrose).
a What is the structure of sorbitol? Is sorbitol optically active?
b Would L-glucitol (made from L-glucose) be enantiomeric with D-glucitol?
c Write structures of all D-family aldohexoses which yield optically inactive hexols from $NaBH_4$ reduction.

18·16 According to the Cahn-Ingold-Prelog nomenclature system, D-(+)-glucose may also be called 2R,3S,4R,5R-2,3,4,5,6-pentahydroxyhexanal or 2R,3S,4R,5R-aldohexose. In this system what is the name of L-(+)-gulose (see Section 18.2)?

18·17 L-(−)-Sorbose is a ketohexose which is isomeric with fructose. It is prepared commercially by bacterial oxidation (*acetobacter suboxydans*) of sorbitol (Exercise 18.15 above) and is used in the manufacture of vitamin C. On treatment with excess phenylhydrazine, L-(−)-sorbose gives the same osazone as L-(+)-gulose. Write a Fischer projection for L-(−)-sorbose.

18·18 The disaccharide (+)-trehalose ($C_{12}H_{22}O_{11}$) is widely distributed in microorganisms, fungi, lichens, and insects. Like maltose and cellobiose, trehalose yields two equivalents of (+)-glucose on acid-catalyzed hydrolysis. However, it does not reduce Fehling solution, does not form an osazone, and does not mutarotate. Methylation of trehalose, followed by acid-catalyzed hydrolysis, yields two equivalents of 2,3,4,6-tetra-O-methylglucose.
a Suggest a structure for the (+)-trehalose molecule. What ambiguities (if any) exist in your answer?
b (+)-Trehalose has a large positive rotation, +178°. Since it is well known that α-glucosides have more positive rotations than β-glucosides, what configurational relationship in the glycoside linkage is likely?

18·19 How may the compounds in each pair easily be distinguished?
a Arabinose and ribose
b Arabinose and glucose
c Glucose and mannose
d Glucose and maltose
e Maltose and lactose
f Sucrose and cellobiose

THE STRATEGY OF ORGANIC SYNTHESIS

HISTORY and literature clearly picture the high value man has set on rare gems and precious metals; however, his esteem for certain organic substances is less widely recognized. Since ancient times, demand for the flavor and aroma of fine spices, the fragrance of exotic perfumes, and the bright colors of rare dyes has made these substances important items of commerce, often worth more than their weight in gold. Furthermore, a host of medicinally useful compounds have been discovered over the past century and a half—antibiotics, hormones, vitamins, and drugs which are generally acknowledged to have benefited mankind more than all the gold and diamonds so carefully hoarded. It is not surprising, therefore, that organic chemists have long sought to manufacture or synthesize these desirable substances from commonly available materials, just as the alchemists before them tried to transmute base metals into gold. Fortunately, organic chemists, unlike the alchemists, have been remarkably successful in their efforts. In this chapter we shall explore the general approach and some of the specific methods used by organic chemists in meeting the challenge of synthesis—that is, in answering the question, "How can it be made?".

Let us begin—as so often is the case—with a failure. In the spring of 1856 William H. Perkin, an eighteen-year-old student at the Royal College of Chemistry in London, attempted a synthesis of quinine in his home laboratory. Quinine, an important drug in the treatment of malaria, was available only from the bark of the South American tree *Cinchona officinolis,* and a decline in the native tree population had caused a large rise in the price of the drug. Very little was known then about the structure of quinine, other than its molecular formula

($C_{20}H_{24}N_2O_2$). Nevertheless, Perkin reasoned that the oxidation of a suitably constituted C_{10} amine, such as allyltoluidine ($C_{10}H_{13}N$), might generate quinine:

$$2 \underset{\substack{\text{allyltoluidine}\\(C_{10}H_{13}N)}}{\text{[2-methyl-4-(allylamino)benzene]}} + 3K_2Cr_2O_7 \xrightarrow{?} \underset{\text{quinine}}{C_{20}H_{24}N_2O_2} + H_2O \qquad (19.1)$$

This simple and direct synthesis approach failed completely, and from our vantage point over a century later, it is easy to see why. Thousands of isomers having the molecular formula of quinine are possible, but only one unique configuration of these 48 atoms constitutes a molecule of quinine. That the atoms of allyltoluidine should, in the course of one reaction, be selectively reorganized and combined in this specific manner is beyond all reasonable probability.

[structure of quinine]

Perkin's experiment was a failure only in the respect that it did not yield quinine, and his subsequent studies of aromatic-amine oxidations demonstrate the value of persistence. From the oxidation of a sample of impure aniline he obtained a purple dye he called *aniline purple* (also called *mauve*) which became the cornerstone of the synthetic dyestuff industry in Europe and made a fortune for its discoverer. The total synthesis of quinine was finally achieved in 1944 by R. B. Woodward and W. E. Doering of Harvard University.

The lessons of many failures and a few successes were quickly learned by organic chemists. The most important of these is that a haphazard or uninformed approach to a synthesis problem is almost surely doomed. Only the most careful and meticulous plans laid on a foundation of knowledge and facts will see one through to the desired compound. In some respects the synthetic chemist is like a master carpenter. Just as the carpenter builds complex structures by carefully shaping and combining wood and metal raw materials, the chemist manipulates his raw materials (selected from a group of several thousand commercially available compounds) by means of chemical reactions. He is thus able to synthesize substances having specific complex and/or novel molecular structures from simple starting materials.

The art and science of organic synthesis cannot be considered separately from the subjects of molecular structure and chemical reactivity, since these are the materials and tools of synthesis. One can no more aspire to synthesize new compounds without a good understanding of the structure and reactions of organic compounds than one could hope to create any object without knowing the properties of its components and the various ways of combining them. By the same token, advances in the methodology of organic synthesis have contributed substantially to our further understanding of chemical bonding and reaction mechanisms. Studies of structure and reactivity often require the use of new compounds, tailor-made to have specific properties. In this sense, then, organic synthesis bears a symbiotic relationship to structure and reactivity.

19·1 PLANNING A SYNTHESIS

In synthesizing any organic compound, the chemist is faced with three critical challenges:

1 Constructing the carbon framework or skeleton that characterizes the desired compound
2 Introducing and/or transforming functional groups in a way that will result in the functionality of the target compound
3 Exercising selective stereochemical control at all stages in which centers of stereoisomerism are formed or influenced

Recognition of these three concerns by no means implies that they represent discrete tasks, each to be attacked separately and solved in turn. A viable synthesis plan or strategy must correlate all operations so as to achieve an efficient and practical overall synthesis. The value of this conceptual approach is primarily one of organizing our thinking and factual recall. It is helpful, for example, to think in terms of the various common reactions that could be used to attach a two-carbon fragment to a substrate, so that when such a structural transformation is called for, the best tool can be chosen for the job.

TABLE 19·1 *Some common reactions for introducing a two-carbon fragment*

substrate	two-carbon fragment	reaction type	product
R—X	NaC≡CH	nucleophilic substitution (S_N2)	RC≡CH
RMgX	$\overset{O}{\overset{\diagup\;\diagdown}{CH_2—CH_2}}$	nucleophilic substitution (S_N2)	RCH_2CH_2OH
RMgX	CH_3CHO	Grignard addition	$RCH(OH)CH_3$
$R_2C=O$	C_2H_5MgX	Grignard addition	$R_2C(OH)C_2H_5$
ArCHO	CH_3CHO	aldol condensation	$ArCH=CHCHO$
$R_2C=O$	$CH_3CH=P\phi_3$	Wittig reaction	$R_2C=CHCH_3$
$RCOCH_2^{\ominus}$	C_2H_5X	enolate alkylation	$RCOCH_2CH_2CH_3$
ArH	C_2H_5X	Friedel-Crafts alkylation	$ArCH_2CH_3$
ArH	CH_3COX	Friedel-Crafts acylation	$ArCOCH_3$

19·1 PLANNING A SYNTHESIS

Some of the factors that must be considered in making such a choice are illustrated by the following examples.

EXAMPLE 1
How can 1-butanol be selectively converted to each of the three constitutionally isomeric monochloro derivatives of *n*-hexane?

Although the four-carbon chain of the starting material must clearly be extended by two carbon atoms for each isomer, the methods outlined in Table 19.1 will not be equally suited for use in each specific case.

1-Chlorohexane In 1-chlorohexane ($CH_3CH_2CH_2CH_2CH_2CH_2Cl$) the functional group is located at the end of the six-carbon chain. Consequently, we should try to select a reaction which will introduce this feature as a part of the carbon-skeleton extension. Since no simple methods exist for achieving the direct displacement of a hydroxyl group by an alkyl group, a reasonable first step would appear to be the replacement of this function by a more versatile chlorine or bromine atom.

$$CH_3(CH_2)_3OH + PBr_3 \xrightarrow{\Delta} CH_3(CH_2)_3Br \tag{19.2}$$

The resulting 1-halobutane can then be easily elaborated by acetylide displacement or Grignard addition to a two-carbon carbonyl or oxirane reactant:

$$CH_3(CH_2)_3Br + HC{\equiv}CNa \longrightarrow CH_3(CH_2)_3C{\equiv}CH + NaBr \tag{19.3}$$

$$CH_3(CH_2)_3Br \xrightarrow{Mg,\ ether} CH_3(CH_2)_3MgBr \xrightarrow{\overset{O}{\triangle}} \xrightarrow{H_2O} CH_3(CH_2)_5OH \tag{19.4}$$

The 1-hexanol formed in the latter reaction can be converted in one step to 1-chlorohexane. In contrast, a similar transformation of the acetylenic function in 1-hexyne requires several steps and is best accomplished via 1-hexanol itself.

$$CH_3(CH_2)_5OH + SOCl_2 \xrightarrow{\Delta} CH_3(CH_2)_5Cl + SO_2 + HCl \tag{19.5}$$

$$CH(CH_2)_3C{\equiv}CH \begin{array}{c} \xrightarrow{H_2,\ Pd/BaSO_4} CH_3(CH_2)_3CH{=}CH_2 \xrightarrow{B_2H_6} \xrightarrow{H_2O_2,\ NaOH} CH_3(CH_2)_5OH \\ \Big\uparrow NaBH_3 \\ \xrightarrow{R_2BH} CH_3(CH_2)_3CH{=}CHBR_2 \xrightarrow{H_2O_2,\ NaOH} CH_3(CH_2)_4CHO \end{array} \tag{19.6}$$

$[R = (CH_3)_2CHCH(CH_3)]$

The optimum overall route from 1-butanol to 1-chlorohexane is thus described by equations 19.2, 19.4, and 19.5.

2-Chlorohexane In devising an efficient synthesis of 2-chlorohexane ($CH_3CH_2CH_2CH_2CHClCH_3$) from 1-butanol, we again must keep in mind the ultimate location of the functional group. The ethylene oxide chain-lengthening method, so effective in the previous case, fails to control this factor. However, it can be satisfied by the addition of a Grignard reagent to acetaldehyde. Equations 19.2, 19.7, and 19.8 describe this synthesis.

19·1 PLANNING A SYNTHESIS

$$CH_3(CH_2)_3MgBr + CH_3CHO \xrightarrow{\text{ether}} \xrightarrow{H_2O} CH_3(CH_2)_3CH(OH)CH_3 \quad (19.7)$$

$$CH_3(CH_2)_3CH(OH)CH_3 + SOCl_2 \longrightarrow CH_3(CH_2)_3CHClCH_3 + SO_2 + HCl \quad (19.8)$$

Once again, 1-hexyne could also be used as an intermediate in this synthesis, but such an approach is less efficient:

$$CH_3(CH_2)_3C{\equiv}CH \begin{array}{c} \xrightarrow{H_2,\ Pd/BaSO_4} CH_3(CH_2)_3CH{=}CH_2 \xrightarrow{HCl} CH_3(CH_2)_3CHClCH_3 \\ \\ \xrightarrow{H_3O^{\oplus},\ HgSO_4} CH_3(CH_2)_3COCH_3 \xrightarrow{NaBH_4} CH_3(CH_2)_3CH(OH)CH_3 \end{array} \uparrow SOCl_2 \quad (19.9)$$

3-Chlorohexane The position of the functional group in 3-chlorohexane ($CH_3CH_2CH_2CHClCH_2CH_3$) favors a different approach from the previous two isomers. Since the functional carbon atom of 1-butanol becomes C-3 of the six-carbon chain formed by addition of a two-carbon atom fragment, we clearly ought to retain this reactive site in our chain-lengthening reaction. We can do this by oxidizing the starting material to butanal, extending the carbon chain by a Grignard addition, and then substituting a chlorine atom for the resulting secondary hydroxyl group:

$$CH_3(CH_2)_3OH + CrO_3 \xrightarrow[\text{pyridine}]{CH_2Cl_2,} CH_3(CH_2)_2\overset{\displaystyle O}{\underset{\displaystyle H}{\overset{\|}{C}}} \quad (19.10)$$

$$CH_3(CH_2)_2CHO + C_2H_5MgI \xrightarrow{\text{ether}} \xrightarrow{H_2O} CH_3(CH_2)_2CH(OH)CH_2CH_3 \quad (19.11)$$

$$CH_3(CH_2)_2CH(OH)C_2H_5 + SOCl_2 \longrightarrow CH_3(CH_2)_2CHClC_2H_5 + SO_2 + HCl \quad (19.12)$$

PROBLEM 19·1
Does the following reaction sequence offer a reasonable alternative to the synthesis of 3-chlorohexane shown above?

$$CH_3(CH_2)_2CHO + CH_3CH{=}P\phi_3 \longrightarrow CH_3(CH_2)_2CH{=}CHCH_3$$

$$CH_3(CH_2)_2CH{=}CHCH_3 + HCl \longrightarrow CH_3(CH_2)_2CHClCH_2CH_3$$

EXAMPLE 2
How can *meso-* and (±)-3,4-hexanediol be synthesized stereospecifically from starting materials having no more than two carbon atoms?

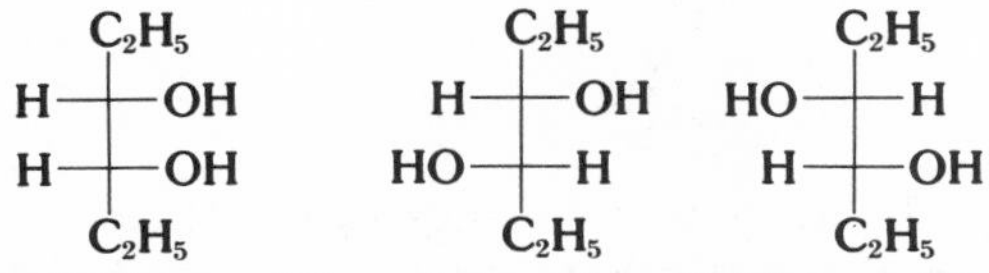

```
      C₂H₅              C₂H₅            C₂H₅
   H──┼──OH          H──┼──OH       HO──┼──H
   H──┼──OH         HO──┼──H         H──┼──OH
      C₂H₅              C₂H₅            C₂H₅
 meso-3,4-hexanediol    (+)- and (−)-3,4-hexanediol
```

Since most of the general methods for making vicinal glycols involve stereospecific hydroxylation or epoxidation of alkenes, as outlined in equations 19.13 and 19.14, this problem can be simplified to that of synthesizing *cis-* or *trans*-3-hexene.

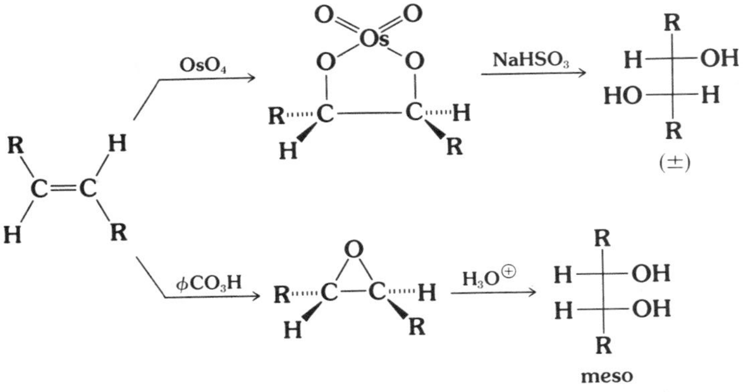

19·1 PLANNING A SYNTHESIS

$$\underset{HR}{\overset{RH}{C=C}} \quad \begin{array}{c} \xrightarrow{OsO_4} \\ \\ \xrightarrow{\phi CO_3H} \end{array} \quad \begin{array}{c} \text{osmate ester} \xrightarrow{NaHSO_3} \begin{array}{c} H-\!\!\!\!\!-OH \\ HO-\!\!\!\!\!-H \end{array} (\pm) \\ \\ \text{epoxide} \xrightarrow{H_3O^\oplus} \begin{array}{c} H-\!\!\!\!\!-OH \\ H-\!\!\!\!\!-OH \end{array} \text{meso} \end{array} \tag{19.13}$$

$$\underset{HR}{\overset{RH}{C=C}} \quad \begin{array}{c} \xrightarrow{OsO_4} \\ \\ \xrightarrow{\phi CO_3H} \end{array} \quad \begin{array}{c} \text{osmate ester} \xrightarrow{NaHSO_3} \begin{array}{c} H-\!\!\!\!\!-OH \\ H-\!\!\!\!\!-OH \end{array} \text{meso} \\ \\ \text{epoxide} \xrightarrow{H_3O^\oplus} \begin{array}{c} H-\!\!\!\!\!-OH \\ HO-\!\!\!\!\!-H \end{array} (\pm) \end{array} \tag{19.14}$$

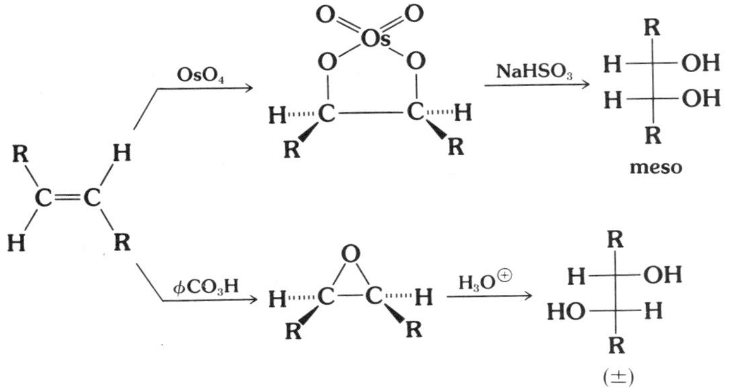

Chain lengthening by Grignard addition to ethylene oxide or acetaldehyde does not provide a simple and specific route to the required 3-hexenes. However, consecutive acetylide alkylation reactions, followed by reduction of the triple bond, provide an ideal means of achieving our goal:

$$HC\equiv CNa + C_2H_5Br \longrightarrow C_2H_5C\equiv CH \xrightarrow{NaNH_2} \xrightarrow{C_2H_5Br} C_2H_5C\equiv CC_2H_5 \tag{19.15}$$

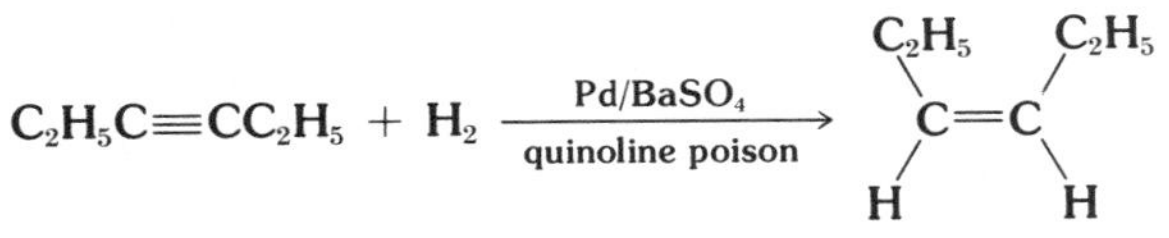

$$C_2H_5C\equiv CC_2H_5 + H_2 \xrightarrow[\text{quinoline poison}]{Pd/BaSO_4} \underset{HH}{\overset{C_2H_5C_2H_5}{C=C}} \tag{19.16}$$

$$C_2H_5C\equiv CC_2H_5 + Na + NH_3 \longrightarrow \underset{HC_2H_5}{\overset{C_2H_5H}{C=C}} \tag{19.17}$$

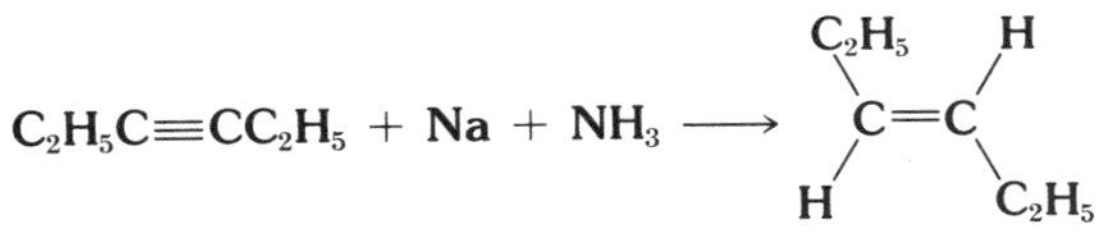

Either of the stereochemically distinct 3-hexenes in equations 19.16 and 19.17 can next be converted selectively to the isomeric 3,4-hexanediols by the two methods described in equation 19.13 (or 19.14). Alternatively, a single reaction (osmium tetroxide hydroxylation or epoxide opening) may operate stereospecifically on each of the isomeric 3-hexenes to give the two isomeric glycols. Thus several plausible solutions to this problem can be conceived.

19·2 SELECTIVE INTRODUCTION OF FUNCTIONAL GROUPS

EXAMPLE 3
How can toluene be converted to 3-phenylpropanoic acid?

Extension of the aromatic side chain of toluene by two carbon atoms can be accomplished in several ways which allow the simultaneous introduction of a functional group at the end of the chain. For example, selective functionalization of the methyl group (equation 19.18), followed by reaction of the derived Grignard reagent with ethylene oxide, could be used to insert the necessary two-carbon fragment. However, this would necessitate a subsequent oxidation of the resulting primary alcohol. A better approach (higher yield) is the malonic ester synthesis described in Chapter 14:

$$\phi CH_3 \xrightarrow[\Delta]{Cl_2(g)} \phi-CH_2-Cl + HCl \tag{19.18}$$

$$CH_2(CO_2C_2H_5) \xrightarrow{NaOR} NaCH(CO_2C_2H_5)_2 \xrightarrow[S_N2]{\phi CH_2Cl} \phi CH_2CH(CO_2C_2H_5)_2 \tag{19.19}$$

$$\phi CH_2CH(CO_2C_2H_5)_2 \xrightarrow[\Delta]{NaOH, H_2O} \phi CH_2CH(CO_2^{\ominus})_2 \xrightarrow[\Delta]{H_3O^{\oplus}} \phi CH_2CH_2CO_2H + CO_2 \tag{19.20}$$

This example illustrates an important general principle: It is sometimes advantageous to prepare intermediates having a *larger number of carbon atoms than the desired product*, with the understanding that the excess atoms will be removed at a later stage.

PROBLEM 19·2
Why do we not convert toluene to 3-phenylpropanoic acid by directly alkylating the enolate anion of ethyl acetate?

$$\phi CH_2Br + CH_2=C\begin{matrix}O^{\oplus}Na^{\ominus}\\ \\ OC_2H_5\end{matrix} \longrightarrow \phi CH_2CH_2CO_2C_2H_5 + NaBr$$

19·2 SELECTIVE INTRODUCTION AND MANIPULATION OF FUNCTIONAL GROUPS

Throughout the text we have discussed the characteristic chemical reactions of each common functional group. These reactions, which constitute a major part of the organic chemist's tools for synthesis, are tabulated in Table I (pages 670 to 679) according to the functional groups generated by their action.

As a rule, the most generally useful reactions are those which consistently and selectively operate on certain characteristic parts of molecules, usually one or more of the common functional groups. Application of a well-established body of such reactions to simple transformations of monofuctional compounds does not normally pose much of a problem, unless a functional group proves to be severely sterically hindered. The challenge of organic synthesis lies chiefly in devising effective multistep transformations of polyfunctional compounds and in the construction of complex carbon structures of specified configuration and functional-group positioning.

SELECTIVITY IN FUNCTIONAL-GROUP REACTIONS

Within a given reaction class there are often gradations of reactivity which permit selective transformations of polyfunctional substances. For example, reaction of 1,5-dibromo-3-methyl-*trans*-2-pentene with quaternary ammonium carboxylate salts proceeds exclusively at the allylic halide site:

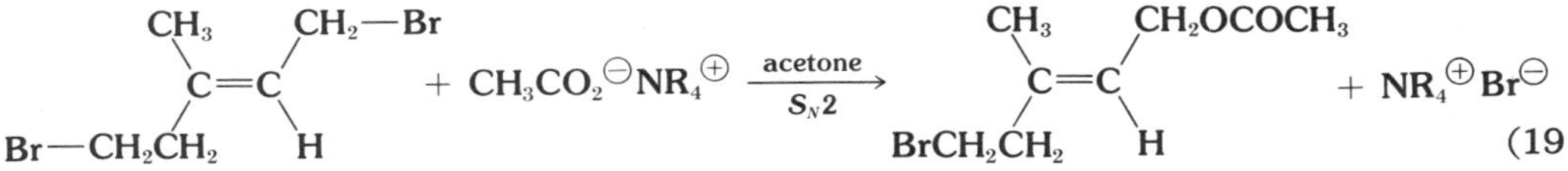

CH3
CH2—Br
C=C
Br—CH2CH2
H
+ CH3CO2⊖NR4⊕
acetone
SN2
CH3
CH2OCOCH3
C=C
BrCH2CH2
H
+ NR4⊕Br⊖

(19.21)

Reaction of 2,6-dichloro-1,2,3,4-tetrahydronaphthalene with a solution of potassium cyanide in aqueous dioxane gives, as the exclusive product, substitution of the aliphatic halogen atom:

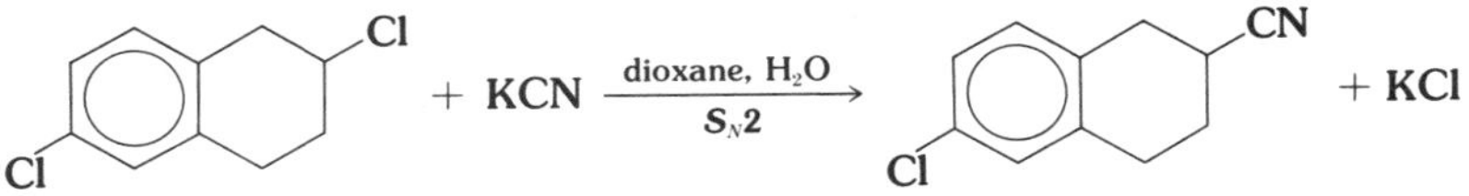

Cl
Cl
+ KCN
dioxane, H2O
SN2
CN
Cl
+ KCl

(19.22)

Fischer esterification of 2,6-dimethylterephthalic acid selectively transforms the 4-carboxyl function:

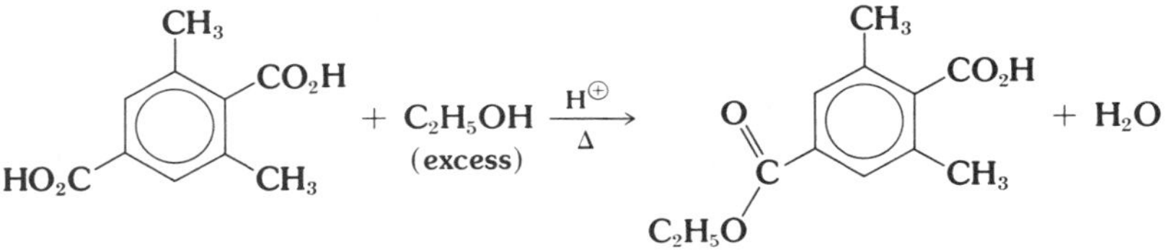

CH3
CO2H
HO2C
CH3
+ C2H5OH
(excess)
H⊕
Δ
CH3
CO2H
O
C
C2H5O
CH3
+ H2O

(19.23)

The difference in reactivity of the halogen functions in equations 19.21 and 19.22 is due to allylic *activation* of the C-1 bromide in the first case and a strengthening, or *deactivation*, of the aryl carbon-halogen bond in the second case (see Chapter 10). The selective esterification shown in reaction 19.23 is clearly a consequence of severe steric hindrance of the 1-carboxyl group.

ACTIVATING, DEACTIVATING, AND BLOCKING GROUPS

In cases where two functional groups have sufficiently similar reactivities to make discrimination by chemical reactions difficult, or where extreme reactivity or lack of reactivity interferes with a desired reaction, we can sometimes influence the outcome by using *activating groups*, *deactivating groups*, and/or *blocking groups*.

ACTIVATING GROUPS As we saw in Chapter 8, the hydroxyl function is a very poor leaving group in S_N2 reactions. However, conversion of this function to the corresponding methane (or *p*-toluene) sulfonate ester activates the oxygen function sufficiently to permit displacement reactions to proceed in good yield:

19·2 SELECTIVE INTRODUCTION OF FUNCTIONAL GROUPS

$$\underset{\underset{CH_3SO_2Cl}{+}}{CH_3\overset{OCH_2\phi}{\underset{|}{C}}HCH_2CH_2OH} \xrightarrow{pyridine} CH_3\overset{OCH_2\phi}{\underset{|}{C}}HCH_2CH_2OSO_2CH_3 \xrightarrow[acetone]{NaI,} CH_3\overset{OCH_2\phi}{\underset{|}{C}}HCH_2CH_2I \quad (19.24)$$

This transformation could not have been accomplished by direct treatment with hydrogen iodide or PI_3 because these reagents would have caused cleavage of the benzyl ether and subsequent substitution of the secondary hydroxyl function it was protecting.

Direct nitration of benzene, even under vigorous conditions, gives only poor yields of 1,3,5-trinitrobenzene, owing to the deactivating influence of the nitro substituents in the *m*-dinitrobenzene precursor. The modestly activating effect of the methyl group in toluene, however, shifts the balance of activating and deactivating forces so as to favor trinitration of this hydrocarbon. Subsequent oxidation and decarboxylation of TNT removes the activating methyl group and yields the desired trinitrobenzene.

$$\text{toluene} \xrightarrow[\Delta]{HNO_3, H_2SO_4} \text{TNT} \xrightarrow[\Delta]{KMnO_4} \text{(2,4,6-trinitrobenzoic acid)} \xrightarrow[\Delta]{NaOH} \text{1,3,5-trinitrobenzene} + NaHCO_3 \quad (19.25)$$

Carbon-carbon double bonds, in contrast to their facile reactivity to electrophilic reagents, are generally unreactive to nucleophiles such as amines and various carbanions. If, however, we activate this functional group by conjugation with a carbonyl or cyano group, as described in Chapter 15, the resulting combination undergoes rapid nucleophilic addition:

$$\text{piperidine} + CH_2=CH-\overset{O}{\underset{\|}{C}}-CH_3 \longrightarrow \text{N}-CH_2CH_2COCH_3 \quad (19.26)$$

In Chapter 12 we discussed alkylation reactions of the enolate conjugate bases of ketones. Efforts to alkylate acetone in this manner, however, give very poor results owing to self-condensation reactions of this simple ketone:

$$CH_3-\overset{O}{\underset{\|}{C}}-CH_3 \xrightarrow{NaH, THF} H_2C=\overset{O^{\ominus}Na^{\oplus}}{\underset{CH_3}{C}} \xrightarrow{condensation} \begin{array}{c} (CH_3)_2C(OH)CH_2COCH_3 \\ + \\ (CH_3)_2C=CHCOCH_3 \\ + \\ (CH_3)_2C=CHCOCH=C(CH_3)_2 \end{array} + \text{isophorone} \quad (19.27)$$

In practice, further activation of the α-hydrogen atoms in acetone by the presence of an additional carbonyl function (as in ethyl acetoacetate) permits rapid formation of a more stable conjugate base, which is less

19·2 SELECTIVE INTRODUCTION OF FUNCTIONAL GROUPS

susceptible to self-condensation and generally gives high yields of alkylation products. The activating ester function can then be removed by decarboxylation of the corresponding acid:

$$CH_3\overset{O}{\underset{\|}{C}}CH_2\overset{O}{\underset{\|}{C}}OC_2H_5 \xrightarrow{NaOC_2H_5} \underset{CH_3}{\overset{O^{\ominus}Na^{\oplus}}{\underset{|}{C}}}\!\!=\!\!\underset{CH}{\overset{}{C}}\!\!-\!\!CO_2C_2H_5 \xrightarrow{R-X} CH_3\overset{O}{\underset{\|}{C}}\underset{\underset{R}{|}}{CH}\overset{O}{\underset{\|}{C}}OC_2H_5$$

$$\Delta \Big\downarrow \begin{array}{l} NaOH, \\ H_2O \end{array}$$

$$CO_2 + CH_3-\overset{O}{\underset{\|}{C}}-CH_2-R \xleftarrow{H_3O^{\oplus}}{\Delta} CH_3\overset{O}{\underset{\|}{C}}\underset{\underset{R}{|}}{CH}\overset{O}{\underset{\|}{C}}\overset{}{\underset{O^{\ominus}Na^{\oplus}}{}}$$

(19.28)

PROTECTIVE OR BLOCKING GROUPS The highly basic and nucleophilic nature of amine functional groups makes them attractive targets for a variety of electrophilic reagents. Such functions must consequently be protected from unwanted attack during many reactions at other sites. Aniline, for example, suffers destruction on treatment with nitric acid and gives a tribromo derivative on mild bromination (see Chapter 11). However, partial deactivation of the amino group by an acyl substituent renders the substitution reactions of acetanilide and other similar amide derivatives much more tractable.

$$\underset{}{\text{PhNH}_2} + (CH_3CO)_2O \longrightarrow \underset{\text{acetanilide}}{\text{Ph-NHCOCH}_3} \xrightarrow[\Delta]{HNO_3} \underset{NO_2}{\text{Ph(NHCOCH}_3)} \xrightarrow[\Delta]{H_3O^{\oplus}} \underset{NO_2}{\text{Ph-NH}_2}$$

(19.29)

Amide and sulfonamide derivatives are the most commonly employed deactivating protective groups for amines (quaternary ammonium salts and tertiary amine oxides are also used on occasion). It is interesting to note, however, that the two acyl functions in phthalimide act to increase the acidity of the remaining nitrogen-bound hydrogen atom to a point where the conjugate base is easily formed and may be used as a strong nucleophile. In this respect the acyl substituents serve as activating groups, a role illustrated by the *Gabriel amine synthesis:*

$$\text{Phthalimide-N-H} \xrightarrow{KOH} \text{Phthalimide-N}^{\ominus}K^{\oplus} \xrightarrow{R-X} \text{Phthalimide-N-R}$$

$$\Delta \Big\downarrow \begin{array}{l} NaOH, \\ H_2O \end{array}$$

$$R-NH_2 + \underset{}{\text{benzene-1,2-}(CO_2^{\ominus})_2}$$

(19.30)

19·2 SELECTIVE INTRODUCTION OF FUNCTIONAL GROUPS

Ether or ester derivatives of hydroxyl groups may be used to protect this function from involvement in many kinds of substitution reactions, such as reaction 19.24, or from the action of oxidizing reagents, such as permanganate and dichromate:

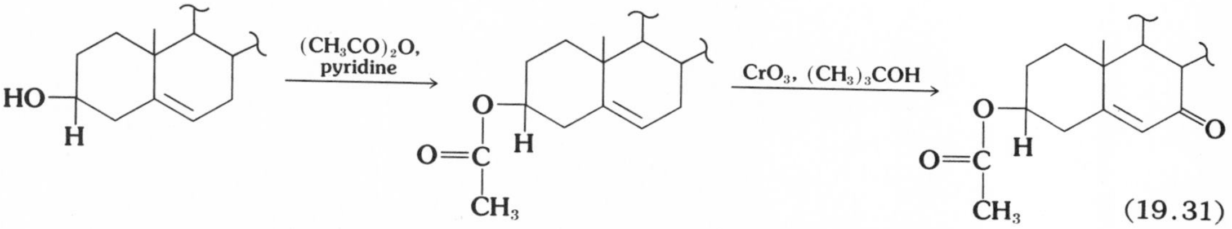

$$\text{(steroid-OH)} \xrightarrow{(CH_3CO)_2O,\ \text{pyridine}} \text{(steroid-OCOCH}_3) \xrightarrow{CrO_3,\ (CH_3)_3COH} \text{(oxidized steroid-OCOCH}_3) \quad (19.31)$$

The oxidation of cholesterol to cholestenone provides a good example of the use of protective groups in synthetic transformations. Direct oxidation with Cr^{+6} reagents does not provide an effective means of accomplishing this conversion, since the double bond and allylic methylene groups also suffer attack. To avoid this difficulty, the troublesome unsaturation can be removed by conversion to the corresponding dibromide. Then, after oxidation of the secondary hydroxyl group, the double bond may be regenerated by one of several methods.

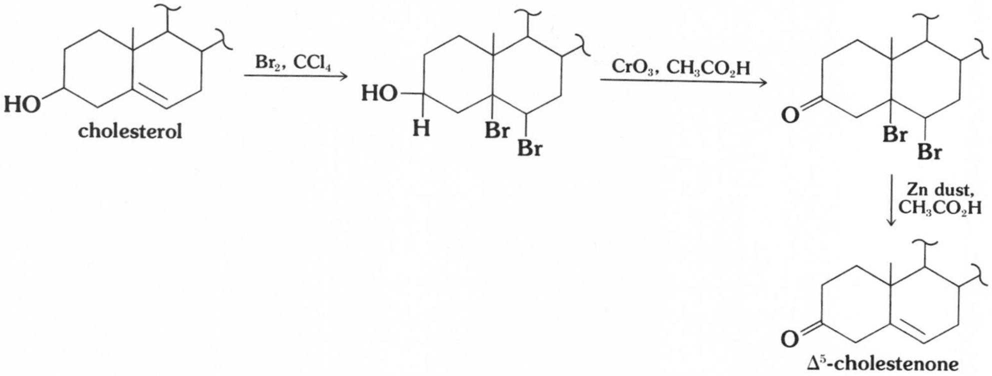

$$\text{cholesterol} \xrightarrow{Br_2,\ CCl_4} \text{(dibromide-OH)} \xrightarrow{CrO_3,\ CH_3CO_2H} \text{(dibromide ketone)} \xrightarrow{Zn\ \text{dust},\ CH_3CO_2H} \Delta^5\text{-cholestenone} \quad (19.32)$$

Ketal (or acetal) groups serve to protect ketones (and aldehydes) from attack by chemical reducing agents and organometallic reagents:

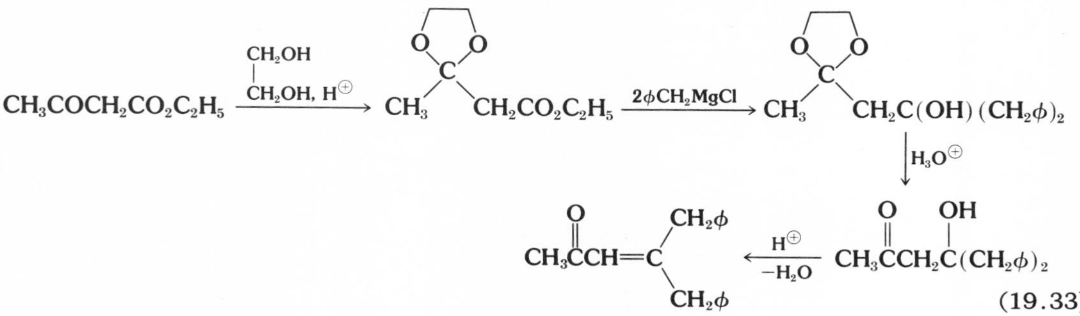

$$CH_3COCH_2CO_2C_2H_5 \xrightarrow[\text{CH}_2\text{OH},\ H^\oplus]{CH_2OH} CH_3\underset{O\diagdown\ \diagup O}{\overset{|}{C}}CH_2CO_2C_2H_5 \xrightarrow{2\phi CH_2MgCl} CH_3\underset{O\diagdown\ \diagup O}{\overset{|}{C}}CH_2C(OH)(CH_2\phi)_2$$

$$\downarrow H_3O^\oplus$$

$$CH_3\overset{O}{\overset{\|}{C}}CH=C\underset{CH_2\phi}{\overset{CH_2\phi}{\diagup}} \xleftarrow[-H_2O]{H^\oplus} CH_3\overset{O}{\overset{\|}{C}}CH_2\overset{OH}{\overset{|}{C}}(CH_2\phi)_2 \quad (19.33)$$

The nitro function often serves as a blocking group in electrophilic aromatic substitution reactions. It can be removed after fulfilling this purpose by reduction of the diazonium salt, prepared from the corresponding amine as shown in equation 19.34.

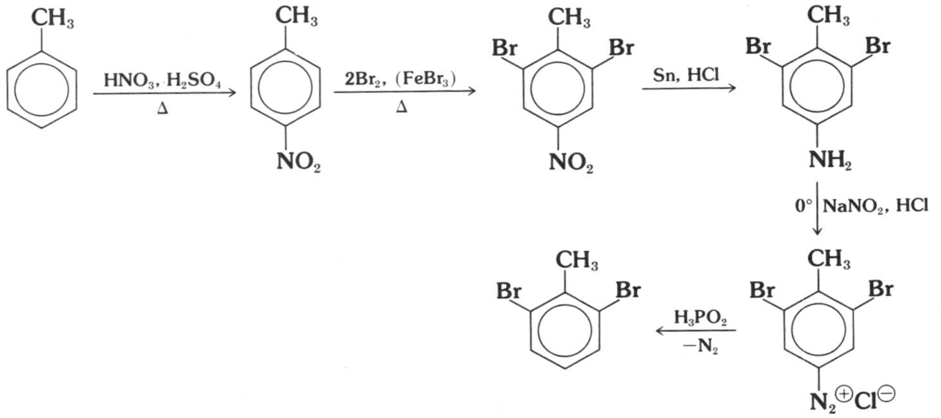

19·2 SELECTIVE INTRODUCTION OF FUNCTIONAL GROUPS

$$\text{toluene} \xrightarrow[\Delta]{HNO_3, H_2SO_4} \text{p-nitrotoluene} \xrightarrow[\Delta]{2Br_2, (FeBr_3)} \text{2,6-dibromo-4-nitrotoluene} \xrightarrow{Sn, HCl} \text{2,6-dibromo-4-aminotoluene} \xrightarrow[0°]{NaNO_2, HCl}$$

$$\text{2,6-dibromotoluene} \xleftarrow[-N_2]{H_3PO_2} \text{2,6-dibromo-4-diazoniumtoluene } N_2^{\oplus}Cl^{\ominus} \qquad (19.34)$$

Since electrophilic sulfonation reactions are reversible, the aryl sulfonic acid function is a potentially useful blocking group. One application of this technique is the following synthesis of o-nitroaniline. Direct nitration of acetanilide gives chiefly the para isomer.

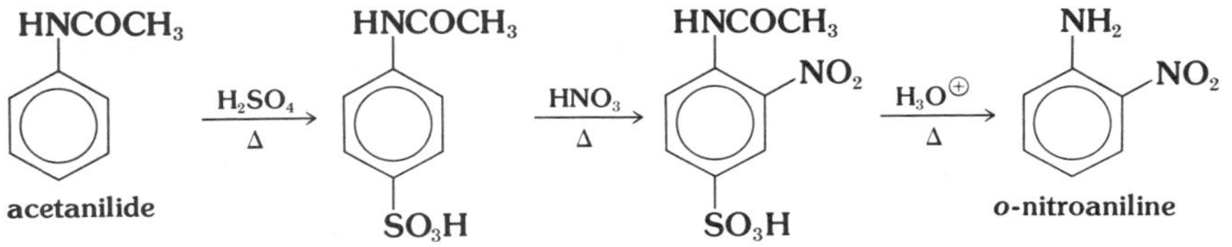

$$\text{acetanilide (HNCOCH}_3) \xrightarrow[\Delta]{H_2SO_4} \text{(HNCOCH}_3, \text{SO}_3\text{H)} \xrightarrow[\Delta]{HNO_3} \text{(HNCOCH}_3, NO_2, \text{SO}_3\text{H)} \xrightarrow[\Delta]{H_3O^{\oplus}} \text{o-nitroaniline (NH}_2, NO_2) \qquad (19.35)$$

PROBLEM 19·3
As noted in equations 19.34 and 19.35, it is sometimes necessary to remove a functional group completely—that is, to replace it with an appropriate number of hydrogen atoms. Indicate suitable procedures for removing the following functional groups. Note any restrictions your procedures may have.
 a The carbon-carbon double bond
 b The carbonyl group (aldehydes and ketones)
 c Halogen in alkyl or aryl halides (two steps)
 d Hydroxyl groups in alcohols (two steps)
 e Amino groups in primary amines (two or three steps)
 f The carboxyl group (three steps)

In summary, then, by activating, protecting, or blocking various sites and groups in a molecule the synthetic chemist is able to overcome some of the restrictions and limitations imposed by the fundamental properties of his starting materials. The following three characteristics are usually desirable for such operations:

1 The modifying (activating or protecting) group must be easily attached in high yield.
2 The modifying group must be resistant to the conditions of the reactions it is influencing.
3 After doing its job, the modifying group must be easily removed or transformed into some other desirable state.

19·3 CARBON-CARBON BOND-FORMING REACTIONS

Roughly 4000 to 5000 organic compounds are commercially available in moderate to large quantities and at reasonable prices (if price is not a factor, as many as 10,000 could probably be obtained). These compounds, which range in complexity from methane to chymotrypsin (an enzyme), constitute a reservoir of starting materials from which all other compounds—known and unknown—must be synthesized. Since the number of known organic compounds is near 3 million, and countless others can be imagined, the synthetic chemist must be adept at combining these starting materials in many different ways to meet the challenge of an almost infinite variety of molecular structures.

In the terminology of our earlier analogy, reactions which can be used to form carbon-carbon bonds may be likened to the nailing, screwing, and glueing operations of a carpenter. Since these reactions are critical to the strategy of organic synthesis, it behooves us to undertake a systematic cataloging of their characteristics. The most widely employed reactions of this kind are listed according to general mechanism in Table II on page 680. It is interesting to note in this connection that only a few fundamentally different kinds of reactions are used for making carbon-carbon bonds. The most common of these involves the interaction of a strong carbon nucleophile, such as an organometallic reagent or an enolate anion, with a weak carbon electrophile, such as a carbonyl group or an alkyl halide.

The reactions outlined in Table II can be used to produce a variety of structural transformations and elaborations on a wide range of different compounds. Table 19.2 shows some possible applications for a few frequently encountered synthesis operations.

PROBLEM 19·4
To which of the categories in Table II does each of the following reactions belong?

a. [reaction: α-tetralone + NaNH$_2$, ether; then CH$_3$CCl=CHCH$_2$Cl → α-alkylated tetralone with CH$_2$CH=C(Cl)CH$_3$ side chain]

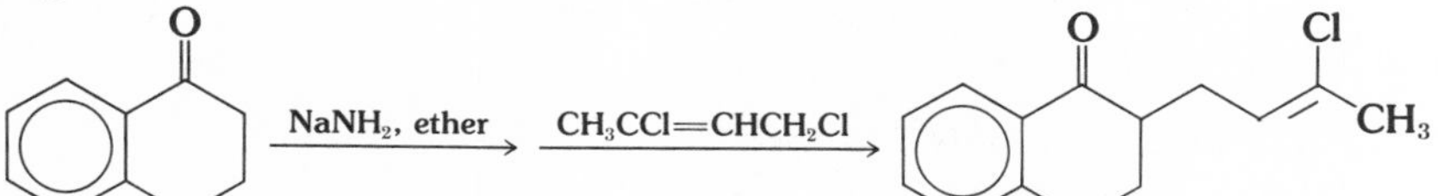

b. [reaction: cyclohexane bearing CH$_3$, CH$_2$COCH$_3$, CH$_2$CHO substituents, with R$_2$NH, benzene, Δ → bicyclic enone product with COCH$_3$ and CH$_3$ groups]

c. [reaction: cyclopentane with CH$_3$, CHO, and C(=CH$_2$)CH$_3$ substituents, with SnCl$_4$, benzene → bicyclic alcohol product with CH$_3$, OH, and =CH$_2$ groups]

(SnCl$_4$ is a Lewis acid)

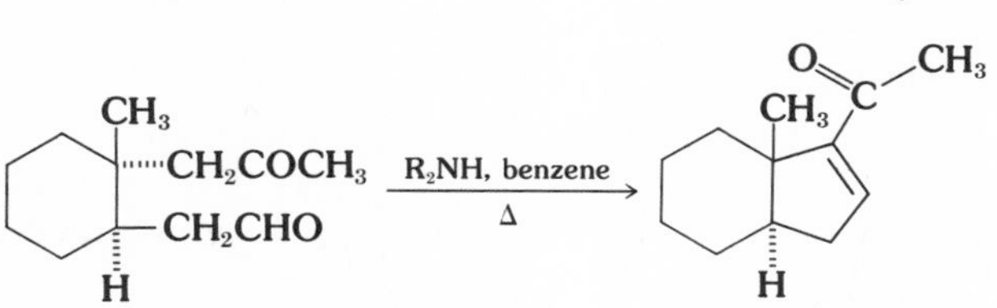

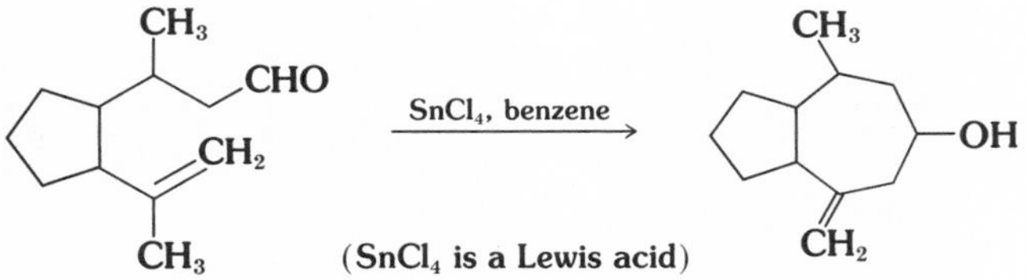

19·3 CARBON-CARBON BOND-FORMING REACTIONS

TABLE 19·2 *Important bond-forming operations*

synthesis operation	suitable reactions	substrates
increasing chain length: by one carbon atom	cyanohydrin formation	aldehydes
	organometallic addition	$RMgX + CH_2O$ or CO_2
	Wittig reaction	$\phi_3P=CH_2$ + aldehydes
	halomethane addition	alkenes
by two carbon atoms	See Table 19.1	
by three or more carbon atoms	terminal-alkyne alkylation	alkyl halides or tosylates
	organometallic addition	aldehydes
	Wittig reaction	aldehydes
	Kolbe electrolysis	carboxylic acids
	Cope rearrangement	dienes
introducing branching of a chain or ring	cyanohydrin formation	ketones
	organometallic addition	aldehydes and ketones
	alkylation	alkyl halides, ketones, esters, nitriles
	Michael reaction	aldehydes, ketones, esters, nitriles
	aldol condensation	aldehydes and ketones
	Claisen condensation	esters
	Wittig reaction	aldehydes and ketones
	Friedel-Crafts acylation or alkylation	alkenes, arenes
	Cope and Claisen rearrangements	dienes
	Carbonium-ion rearrangement	
carbon-ring formation: three-membered rings	internal alkylation	halo ketones or halo esters
	carbenoid addition	alkenes
four-membered rings	internal alkylation	halo ketones, halo esters, halo nitriles
	cycloaddition	alkenes

19·3 CARBON-CARBON BOND-FORMING REACTIONS

synthesis operation	suitable reactions	substrates
carbon-ring formation *(continued)*		
five-membered rings	internal alkylation	halo ketones, halo esters, halo nitriles
	Dieckmann and Thorpe condensations†	diesters or dinitriles
	aldol condensation	dicarbonyl compounds
	molecular rearrangement	larger or smaller rings
six-membered rings	internal alkylation	halo ketones, esters, nitriles
	Dieckmann and Thorpe condensations†	diesters and dinitriles
	aldol condensation	dicarbonyl compounds
	Diels-Alder reaction	diene and dienophile
	molecular rearrangement	larger or smaller rings
seven-membered and larger rings	Dieckmann and Thorpe condensations†	diesters and dinitriles
	acyloin condensation††	diesters
	molecular rearrangement	larger or smaller rings

†The Thorpe condensation is closely related to the Dieckmann condensation (Figure 14.12) but is considered superior for the synthesis of large rings:

$$\underset{\underset{CN}{|}}{\overset{\overset{CN}{|}}{(CH_2)_{14}}} \xrightarrow{R_2NLi} (CH_2)_{13}\overset{C=NH}{\underset{CHCN}{<}} \xrightarrow{NaOH, H_2O} (CH_2)_{13}\overset{C=O}{\underset{CHCO_2H}{<}} \xrightarrow[\Delta]{-CO_2} (CH_2)_{13}\overset{C=O}{\underset{CH_2}{<}}$$

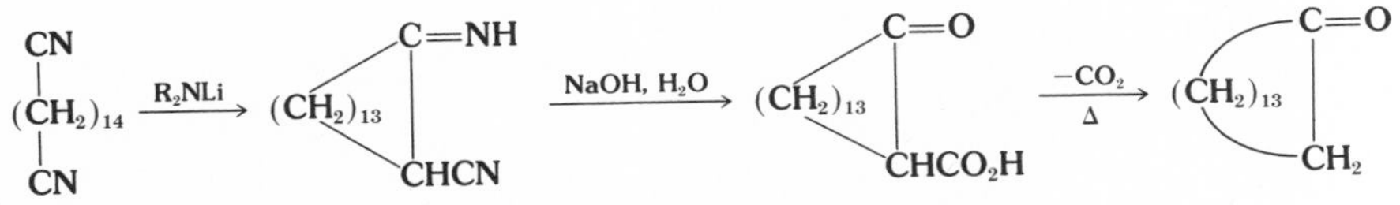

††The acyloin condensation is a reductive dimerization of esters occurring in the presence of sodium. It is particularly useful in the synthesis of large rings.

$$\underset{\underset{CO_2C_2H_5}{|}}{\overset{\overset{CO_2C_2H_5}{|}}{(CH_2)_8}} \xrightarrow[\text{(or NH}_3\text{, Na)}]{toluene, Na} (CH_2)_8\overset{C-ONa}{\underset{C-ONa}{\diagdown\diagup\|}} \xrightarrow{H_2O} \text{cyclodecanone with -OH and H}$$

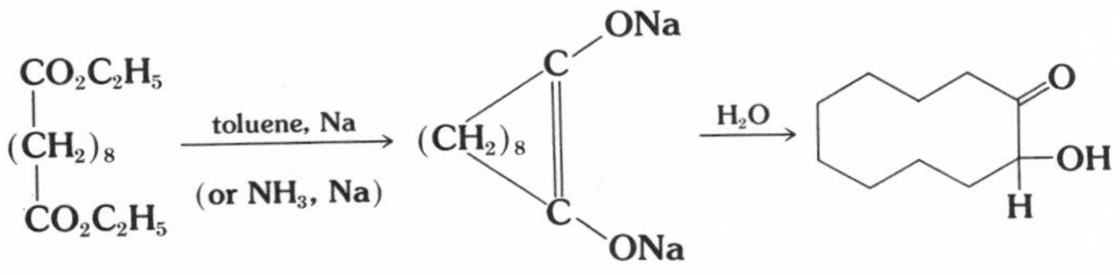

PROBLEM 19·5
Write equations giving a specific example for each of the following:
a Extending the length of a carbon chain by four carbon atoms by means of a Wittig reaction
b Introducing chain branching by a Michael reaction
c Introducing chain branching by a Claisen condensation
d Formation of a three-membered ring by an internal alkylation
e Formation of a seven-membered ring from a six-membered ring by a molecular rearrangement *Hint:* Try an amino-pinacol rearrangement

19·4 CONCEPTION AND DESIGN OF A SYNTHESIS

When faced with the task of synthesizing a complex target molecule, we must first analyze the problem in a systematic fashion. Several general principles have proved effective in simplifying such problems and in disclosing useful approaches.

SYNTHESIS STRATEGIES

IDENTIFY AND TAKE ADVANTAGE OF MOLECULAR SYMMETRY.
If a plane of symmetry passes through a molecule without bisecting any atoms, we can (at least, in principle) put together the final structure by combining the equivalent halves:

$$2(CH_3)_2CO \xrightarrow{Mg(Hg),\ ether} CH_3-\underset{\underset{CH_3}{|}}{\overset{\overset{O}{|}}{C}}\underset{}{\overset{Mg}{\diagdown\diagup}}\underset{\underset{CH_3}{|}}{\overset{\overset{O}{|}}{C}}-CH_3 \xrightarrow{H_2O} (CH_3)_2\underset{}{\overset{OH}{\underset{|}{C}}}\underset{}{\overset{OH}{\underset{|}{C}}}(CH_3)_2 \quad (19.36)$$

pinacol

In cases where a plane of symmetry passes through one or more carbon atoms, equivalent alkyl fragments may be attached to a suitably activated core unit.

$$2CH_3CH_2CH_2MgBr + CH_3CO_2C_2H_5 \xrightarrow{ether} \xrightarrow{H_2O} CH_3CH_2CH_2\underset{\underset{CH_3}{|}}{\overset{\overset{OH}{|}}{C}}CH_2CH_2CH_3 \quad (19.37)$$

4-methyl-4-heptanol

An approach of this kind is generally limited, however, to situations in which the plane of symmetry bisects only a few atoms or bonds.

TRY WORKING BACKWARD FROM THE TARGET MOLECULE.
By considering ways in which related compounds could be transformed into the target compound in one or two steps, we can generate a group of potential intermediates, some of which may suggest synthesis operations not obviously applicable to the original structure.

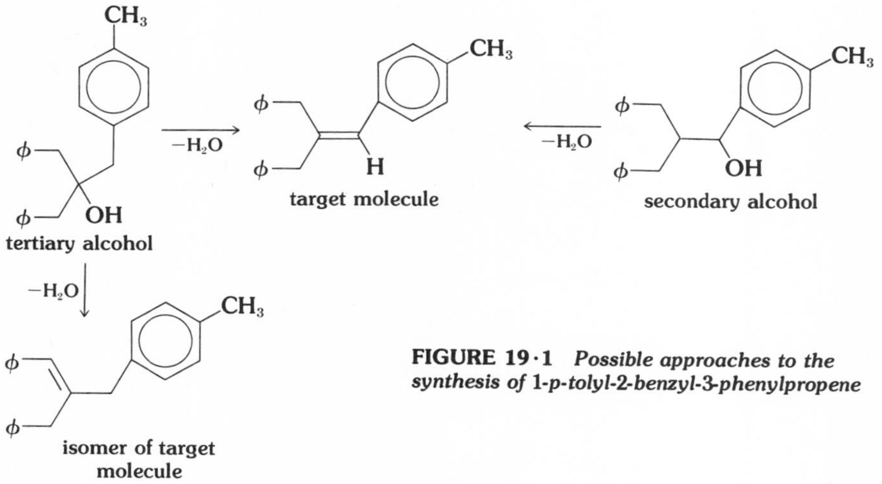

19·4 CONCEPTION AND DESIGN OF A SYNTHESIS

FIGURE 19·1 *Possible approaches to the synthesis of 1-p-tolyl-2-benzyl-3-phenylpropene*

Suppose, for example, we wish to prepare 1-*p*-tolyl-2-benzyl-3-phenylpropene from simple starting materials. We begin by considering methods for introducing the double bond, since this reactive function should be among the last features developed. One possibility (see Table I) is dehydration of the corresponding tertiary alcohol. However, as shown in Figure 19.1, this would undoubtedly produce an isomeric alkene in roughly double the yield of the target molecule. Dehydration of the corresponding secondary alcohol might work, but this would pose the danger of an acid-catalyzed double-bond shift. Further speculation about ways in which the desired double bond might be formed leads us eventually to two possibilities for a Wittig reaction:

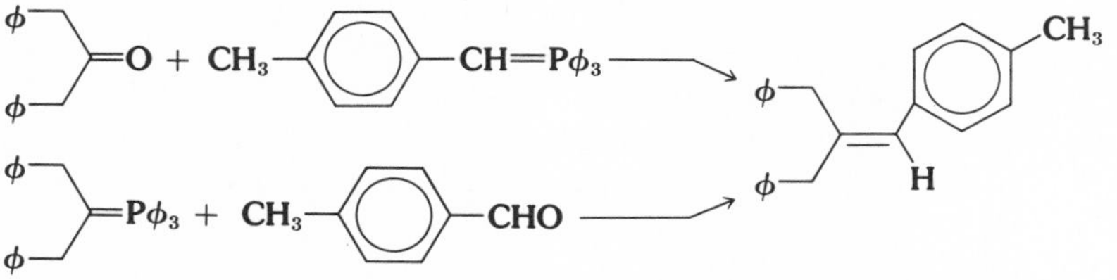

$$\phi_2C=O + CH_3\text{-}C_6H_4\text{-}CH=P\phi_3 \longrightarrow \text{product} \qquad (19.38)$$
$$\phi_2C=P\phi_3 + CH_3\text{-}C_6H_4\text{-}CHO \longrightarrow \text{product}$$

This approach is certainly specific and mild, and we can now transfer our attention to the dibenzyl ketone (or equivalent carbinol) fragment. The symmetry of this intermediate strongly suggests a preparation of the following sort:

$$2\phi CH_2MgBr + HCO_2C_2H_5 \xrightarrow{\text{ether}} \xrightarrow{H_2O} \phi CH_2\overset{OH}{\underset{|}{C}}HCH_2\phi \xrightarrow{CrO_3} \phi CH_2COCH_2\phi$$

$$\downarrow PBr_3$$

$$\phi CH_2CHBrCH_2\phi \xrightarrow{\phi_3P} \xrightarrow{C_4H_9Li} (\phi CH_2)_2C=P\phi_3 \qquad (19.39)$$

The remaining reactant is either *p*-methylbenzaldehyde or *p*-methylbenzylidene triphenylphosphorane, depending on the specific path being followed. In either case, its preparation poses no problem, as illustrated by reactions 19.40.

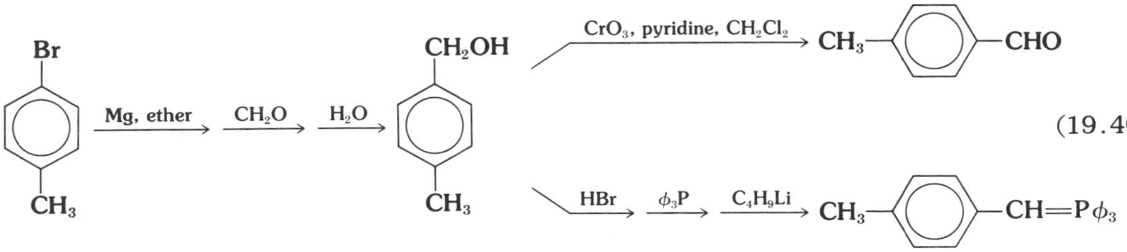

19·4 CONCEPTION AND DESIGN OF A SYNTHESIS

$$\underset{\underset{CH_3}{}}{Br-C_6H_4} \xrightarrow{Mg,\ ether} \xrightarrow{CH_2O} \xrightarrow{H_2O} \underset{\underset{CH_3}{}}{CH_2OH-C_6H_4} \begin{array}{c} \xrightarrow{CrO_3,\ pyridine,\ CH_2Cl_2} CH_3-C_6H_4-CHO \\ \\ \xrightarrow{HBr} \xrightarrow{\phi_3P} \xrightarrow{C_4H_9Li} CH_3-C_6H_4-CH=P\phi_3 \end{array} \quad (19.40)$$

PROBLEM 19·6

In Figure 19.1 the target molecule was β,β-dibenzyl-p-methylstyrene. Discuss the feasibility and relative merit of the following synthesis plan, with $(\phi CH_2)_2CHBr$ prepared as in equation 19.39:

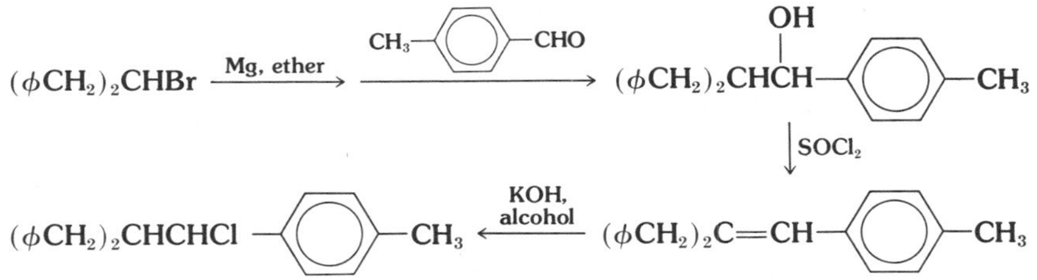

$$(\phi CH_2)_2CHBr \xrightarrow{Mg,\ ether} \xrightarrow{CH_3-C_6H_4-CHO} (\phi CH_2)_2CHCH(OH)-C_6H_4-CH_3$$

$$\downarrow SOCl_2$$

$$(\phi CH_2)_2CHCHCl-C_6H_4-CH_3 \xleftarrow{KOH,\ alcohol} (\phi CH_2)_2C=CH-C_6H_4-CH_3$$

RECOGNIZE KEY STRUCTURAL FEATURES (SYNTHONS).

We learn from our chemical experience to recognize within a target molecule certain units which offer particular advantage or promise for synthesis by virtue of their facile introduction and modification by well-established methods. Such subunits are called *synthons*. A given molecule may include many synthons, and some atoms of the molecule may be constituents of several overlapping synthons. Although a synthon may be almost as large as the target molecule itself or as small as a single hydrogen atom, the moderate-size synthons are usually more useful in synthesis design. Figure 19.2 illustrates the readily identifiable synthons in our previous target compound, 1-p-tolyl-2-benzyl-3-phenylpropene. Note that one of these is a single carbon. A variety of single-carbon

FIGURE 19·2 *Synthons of 1-p-tolyl-2-benzyl-3-phenylpropene*

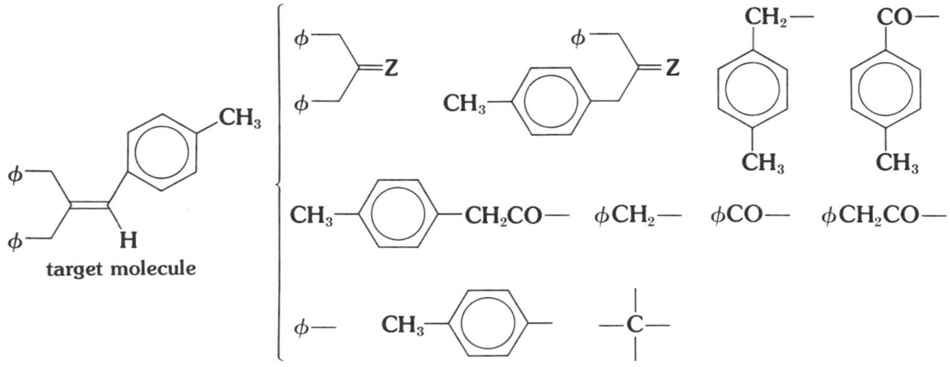

19·4 CONCEPTION AND DESIGN OF A SYNTHESIS

synthons are available, including CH_3I, CH_2O, $HCO_2C_2H_5$, and CO_2. In this case a highly oxidized derivative is desirable for the central carbon atom, because of the large number of alkyl groups to be attached to it.

PROBLEM 19·7

A major proportion of the reactions useful for making new carbon-carbon bonds involve the combination of carbon nucleophiles with carbon electrophiles. It is therefore useful to classify simple synthons according to whether they are nucleophilic or electrophilic in nature. Complete the following table by supplying specific examples of single-carbon synthons of both types, having a variety of oxidation states:

	specific examples	
synthon type†	electrophilic	nucleophilic
CH_3	————	————
ZCH_2	————	————
Z_2CH	————	————
Z_3C	————	————

†Z = substituents other than H or C

PAY PARTICULAR ATTENTION TO CHAIN BRANCHING AND RINGS.

Highly branched carbon atoms and carbon ring systems often present special difficulties in synthesis planning. By focusing our attention on these critical features, we can lessen the risk of designing ourselves into an unworkable "dead-end" position.

As a general rule, branched sites in complex molecules are a good place to begin dismantling the target structure into plausible synthons. *Quaternary carbon atoms*—carbon atoms bonded to four alkyl or aryl groups—pose a particular challenge, since many displacement reactions do not work well for generating such systems. Fortunately, Michael reactions and other conjugate addition reactions are very effective in this respect.

In the preparation of 9,10-dimethyl-*cis*-decalin shown in Figure 19.3, the problems presented by the bicyclic ring system and the two quater-

FIGURE 19·3 *Examples of conjugate addition reactions in synthesis*

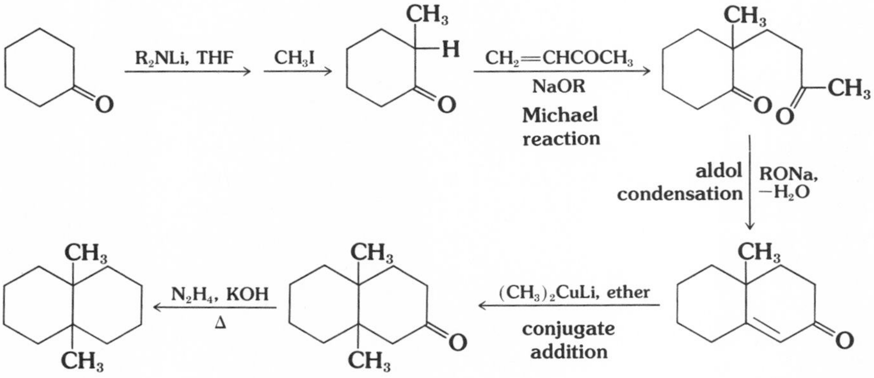

nary carbon atoms provide in their solutions the essence of the entire synthesis. The stereoselectivity of the final methyl-group addition is probable due to steric effects, which are easiest to visualize with the aid of molecular models.

A final point worth noting is that in certain cases selective cleavage of a previously introduced ring system can provide excellent positional and stereochemical control of side chains. This is particularly well illustrated by the following synthesis of all-*cis*-1,2,3,4-cyclopentanetetracarboxylic acid:

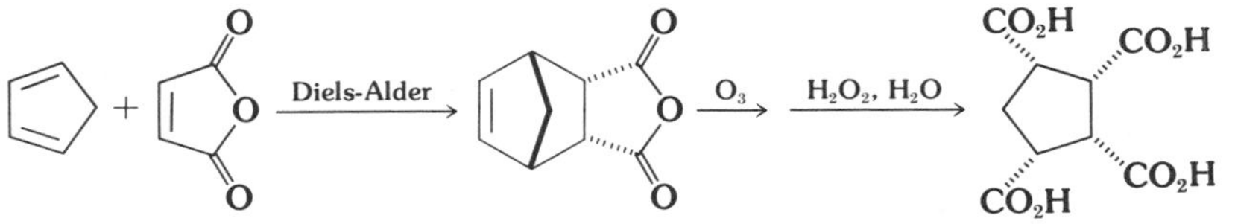

$$\text{(cyclopentadiene)} + \text{(maleic anhydride)} \xrightarrow{\text{Diels-Alder}} \text{(bicyclic anhydride)} \xrightarrow{O_3} \xrightarrow{H_2O_2, H_2O} \text{(all-cis tetracarboxylic acid)} \qquad (19.41)$$

A simpler or more efficient preparation would be difficult to conceive.

PROBLEM 19·8
Figure 19.3 shows a synthesis of 9,10-dimethyl-*cis*-decalin. Discuss the feasibility and relative merits of the following alternative plans:

[Scheme a: 2,3-dimethyl-1,4-benzoquinone + butadiene →Δ octalindione with two CH₃ groups at ring junction → reductions → 9,10-dimethyldecalin]

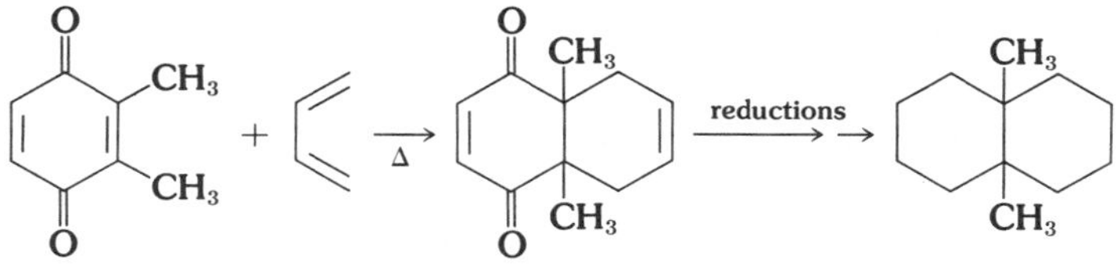

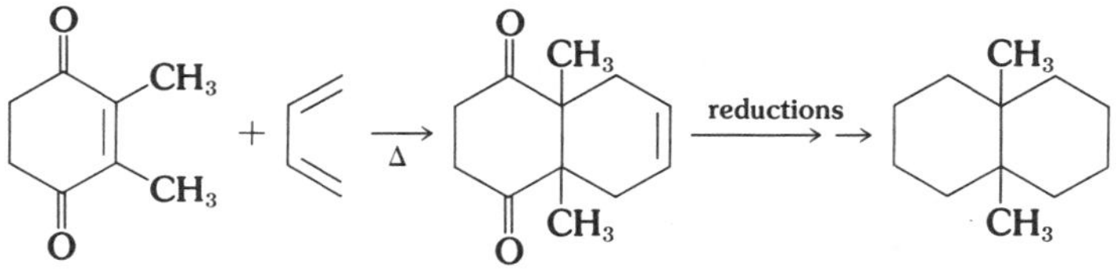

[Scheme b: 2,5-dimethyl-1,4-benzoquinone + butadiene →Δ dimethyl octalindione → reductions → dimethyldecalin]

[Scheme c: 2,3-dimethyl-2-cyclohexenone + butadiene →Δ dimethyl octalone → reductions → dimethyldecalin]

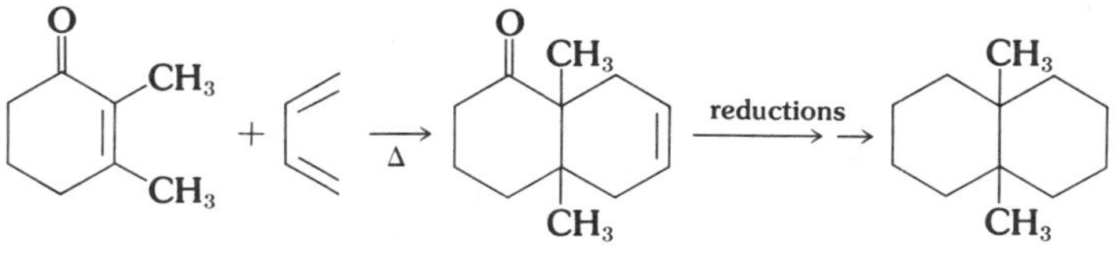

JUDGING A SYNTHESIS

In the most direct sense, it is a simple matter to judge a synthesis; we need only try it in the laboratory. If there are problems they will disclose themselves, and our speculations will be answered. However, since laboratory operations are expensive in terms of personnel time, consumable supplies, and equipment, it is best to anticipate and correct as many difficulties and weaknesses as possible before proceeding to the stage of physical experiment.

As a general rule, our goal is to convert the least expensive available starting materials into the desired product with a minimum of manipulation and in the highest possible overall yield. Of course, a balance of these conflicting objectives is often necessary. Thus a simple three-step

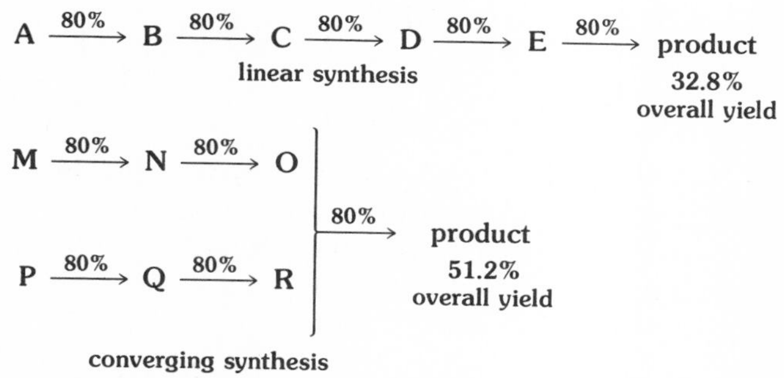

$$A \xrightarrow{80\%} B \xrightarrow{80\%} C \xrightarrow{80\%} D \xrightarrow{80\%} E \xrightarrow{80\%} \text{product}$$
linear synthesis   32.8% overall yield

$$M \xrightarrow{80\%} N \xrightarrow{80\%} O$$
$$P \xrightarrow{80\%} Q \xrightarrow{80\%} R$$
$$\Bigg\} \xrightarrow{80\%} \text{product}$$
51.2% overall yield
converging synthesis

FIGURE 19·4 *A comparison of synthesis yields*

procedure with moderate yield may be preferred over a high-yield five-step synthesis requiring some difficult operations. Since the overall yield is the product of the yields of the individual steps, it is best to keep the number of such steps low. Figure 19.4 shows a comparison of a *linear* five-step synthesis and a *converging* five-step synthesis. In both procedures the individual steps are all assumed to proceed in 80% yield. If each step in the linear synthesis were reduced to a 50% yield, the overall yield would drop to 3%!

It should be clear from our discussion here that just as a chain is no stronger than its weakest link, a synthesis is no better than its poorest step. Whenever possible, then, we should try to design alternative routes, or "detours," to all important intermediates. This provides some insurance against the failure of a critical step.

19·5 EXAMPLES OF SYNTHESIS PROBLEMS

SYNTHESIS 1

Suggest a sequence of reactions by means of which 2,6-dimethyl-4-heptanone can be synthesized from starting materials composed of no more than four carbon atoms.

$$(CH_3)_2CHCH_2\overset{O}{\underset{\|}{C}}CH_2CH(CH_3)_2$$

The clearcut symmetry about the carbonyl function in this molecule, together with the C_4 composition of the alkyl substituents, immediately suggests the following general approach:

$$2C_4H_9\text{—}Z + CY_n \longrightarrow C_4H_9CY_mC_4H_9 + 2YZ \tag{19.42}$$

Plausible combinations of functions Y and Z can be chosen from the ketone section of Table I; two such synthetic sequences are shown in the following equations:

$$(CH_3)_2CHCH_2Br + KCN \longrightarrow (CH_3)_2CHCH_2CN$$
$$(CH_3)_2CHCH_2Br \xrightarrow{\text{Mg, ether}} (CH_3)_2CHCH_2MgBr$$
$$\Bigg\} \xrightarrow{\text{ether}} \xrightarrow{H_2O} [(CH_3)_2CHCH_2]_2C\text{=}O \tag{19.43}$$

19·5 EXAMPLES OF SYNTHESIS PROBLEMS

$$(CH_3)_2CHCH_2MgBr \xrightarrow{CO_2} \xrightarrow{H_2O} (CH_3)_2CHCH_2CO_2H \xrightarrow{SOCl_2} (CH_3)_2CHCH_2COCl \quad (19.44)$$

$$[(CH_3)_2CHCH_2]_2Cd + (CH_3)_2CHCH_2COCl \longrightarrow [(CH_3)_2CHCH_2]_2C=O \quad (19.45)$$

In both these procedures the C_4 groups are added in a stepwise fashion, but a single-step combination of synthons would clearly be more efficient. Such a possibility can, in fact, be realized if we recognize that the corresponding secondary alcohol can be readily oxidized to the desired product and is thus a suitable alternative target:

$$2(CH_3)_2CHCH_2MgBr \xrightarrow{HCO_2C_2H_5} \xrightarrow{H_2O} [(CH_3)_2CHCH_2]_2CHOH \xrightarrow{CrO_3} [(CH_3)_2CHCH_2]_2C=O \quad (19.46)$$

SYNTHESIS 2
Suggest a synthesis for 2,7-dimethyl-4-octanone from starting materials having no more than four carbon atoms.

$$(CH_3)_2CHCH_2\overset{O}{\underset{\|}{C}}CH_2CH_2CH(CH_3)_2$$

Although the symmetry of the previous case is missing here, the same two isobutyl groups can be identified at the ends of the molecule. One possible solution to this problem would be to lengthen appropriate isobutyl reactants by one carbon atom, followed by application of a sequence analogous to reaction 19.43 or 19.45. However, such a synthesis would be rather lengthy. In searching for a more direct procedure, we can pursue the synthon approach by asking: "What symmetrically substituted *two-carbon synthon* can be readily transformed into —CH_2CO—?" One answer is the alkyne function:

$$(CH_3)_2CHCH_2Br + NaC\equiv CH \longrightarrow (CH_3)_2CHCH_2C\equiv CH$$
$$\downarrow NaNH_2$$
$$\downarrow (CH_3)_2CHCH_2Br$$
$$(CH_3)_2CHCH_2COCH_2CH_2CH(CH_3)_2 \xleftarrow{HgSO_4} H_3O^{\oplus} + (CH_3)_2CHCH_2C\equiv CCH_2CH(CH_3)_2 \quad (19.47)$$

SYNTHESIS 3
Suggest a synthesis for 3,3'-diaminodiphenylmethane starting with benzene, toluene, and simple aliphatic reactants.

The symmetry about the central carbon atom of this molecule is readily apparent, and this fact should again simplify our plans. Although several approaches to the diphenylmethane skeleton can be envisioned, a selection should be postponed until we have considered the nature and position of the functional groups.

Amino groups are sensitive to many reagents, and their nitro-group precursors are incompatible or unreactive with many carbon-carbon bond-forming reactions, such as Grignard additions and Friedel-Crafts reac-

19·5 EXAMPLES OF SYNTHESIS PROBLEMS

tions. It would therefore be prudent to avoid introducing these groups early in the synthesis. Assuming, then, that the functional groups will be inserted after the carbon skeleton is complete, the specific meta orientation of these groups presents a critical challenge. Direct nitration of diphenylmethane itself would give largely the 4,4'-dinitro derivative. Consequently, the central carbon atom will need to be modified so as to exert a meta-directing influence. A carbonyl function would be ideal for this purpose, as outlined in equation 19.48. Deoxygenation of the carbonyl group and reduction of the nitro groups can be accomplished simultaneously by catalytic reduction or the Clemmensen reduction:

$$\phi\text{-CO-}\phi \xrightarrow{\text{HNO}_3,\ \text{H}_2\text{SO}_4,\ \Delta} (\text{NO}_2)\phi\text{-CO-}\phi(\text{NO}_2) \xrightarrow{8\text{H}_2,\ \text{Pd/C}} (\text{NH}_2)\phi\text{-CH}_2\text{-}\phi(\text{NH}_2) \quad (19.48)$$

Diphenyl ketone (benzophenone) is easily prepared by any of several methods; for example:

$$\text{PhCH}_3 \xrightarrow{\text{KMnO}_4, \Delta} \text{PhCO}_2\text{H} \xrightarrow{\text{SOCl}_2} \text{PhCOCl} \xrightarrow{\text{C}_6\text{H}_6,\ \text{AlCl}_3,\ \Delta} \phi\text{-CO-}\phi$$

$$\text{C}_6\text{H}_6 \xrightarrow{\text{Br}_2,\ \text{Fe},\ \Delta} \phi\text{-Br} \xrightarrow{\text{Mg, ether}} \phi\text{MgBr} \xrightarrow{\text{HCO}_2\text{C}_2\text{H}_5} \xrightarrow{\text{H}_2\text{O}} \phi\text{CH}(\text{OH})\phi \xrightarrow{\text{CrO}_3} \phi\text{-CO-}\phi \quad (19.49)$$

SYNTHESIS 4

Suggest a synthesis for 2-ethyl-2-methyl-1,3-butanediol from starting materials having no more than two contiguous (directly bonded) carbon atoms.

$$\text{CH}_3\text{CH(OH)}-\underset{\underset{\text{CH}_2\text{OH}}{|}}{\overset{\overset{\text{CH}_3}{|}}{\text{C}}}-\text{C}_2\text{H}_5$$

The quaternary carbon atom in this molecule should be the center of our initial attention, since most of the carbon-carbon bonds to this site must be formed in our synthesis. If we start by assuming that the two alkyl synthons CH_3 and C_2H_5 might be introduced by appropriate alkylation reactions, we are left with a four-carbon synthon having functionality at C-1 and C-3: $\text{CH}_3\text{CH(OH)CCH}_2\text{OH}$. In a compound with a higher oxidation state, such as a β-keto ester, these functions could serve to activate C-2 for enolate alkylation:

$$\text{CH}_3\overset{\text{O}}{\overset{\|}{\text{C}}}\text{CH}_2\text{CO}_2\text{C}_2\text{H}_5 \xrightarrow{\text{NaOC}_2\text{H}_5} \xrightarrow{\text{C}_2\text{H}_5\text{Br}} \text{CH}_3\overset{\text{O}}{\overset{\|}{\text{C}}}\text{CH}(\text{C}_2\text{H}_5)\text{CO}_2\text{C}_2\text{H}_5 \xrightarrow{\text{NaOC}_2\text{H}_5} \xrightarrow{\text{CH}_3\text{I}} \text{CH}_3\overset{\text{OCH}_3}{\overset{\|}{\text{C}}}\text{C}(\text{C}_2\text{H}_5)\text{CO}_2\text{C}_2\text{H}_5 \quad (19.50)$$

Lithium aluminum hydride reduction of the substituted β-ketoester would then yield the desired product:

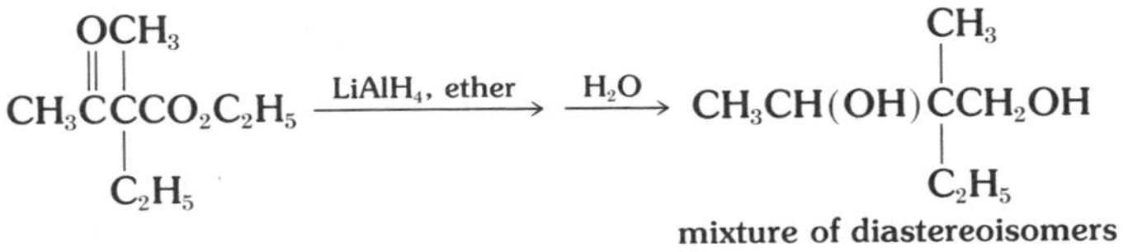
OCH3
CH3CCCO2C2H5
C2H5
LiAlH4, ether
H2O
CH3
CH3CH(OH)CCH2OH
C2H5
mixture of diastereoisomers

(19.51)

The ethyl acetoacetate used in equation 19.50 can, of course, be conveniently prepared by a Claisen condensation of ethyl acetate, thus meeting the stipulation of a starting compound having two contiguous carbon atoms:

$$2CH_3CO_2C_2H_5 \xrightarrow[\Delta]{NaOC_2H_5} \underset{\text{as the sodium salt}}{CH_3COCH_2CO_2C_2H_5} + C_2H_5OH \qquad (19.52)$$

SYNTHESIS 5
Suggest a synthesis of 1,4,6-trimethylnaphthalene from *para*-xylene and any other starting materials having no more than four carbon atoms.

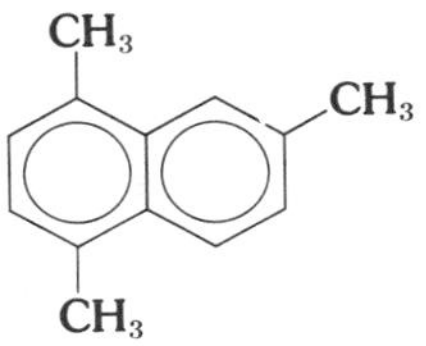
CH3
CH3
CH3

The grouping of atoms derived from p-xylene is clearly evident in the structure of the desired compound, and our first task must be to devise a method of attaching the second six-membered ring. Since both ends of a four-carbon chain must be bonded to the p-xylene, a terminally difunctional C_4 synthon is needed. Friedel-Crafts acylation, using succinic anhydride, appears to be ideal for this purpose:

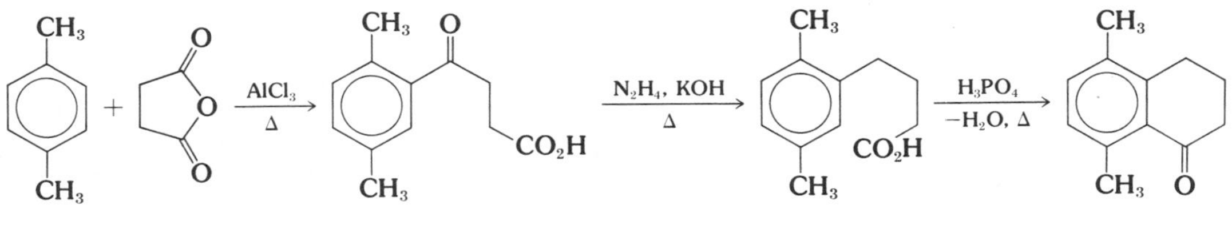
CH3
CH3
O
O
O
AlCl3
Δ
CH3 O
CO2H
CH3
N2H4, KOH
Δ
CH3
CO2H
CH3
H3PO4
−H2O, Δ
CH3
CH3 O

(19.53)

The remaining methyl group is easily inserted by enolate alkylation, and subsequent reduction, dehydration, and dehydrogenation generates the fully aromatic ring system.

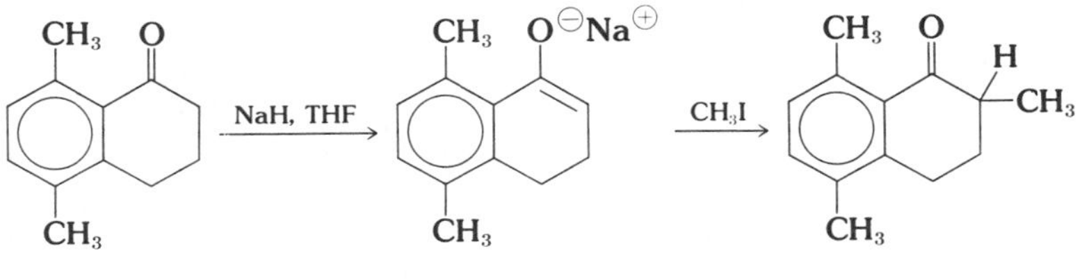
CH3 O
CH3
NaH, THF
CH3 O⊖Na⊕
CH3
CH3I
CH3 O
H
CH3
CH3

(19.54)

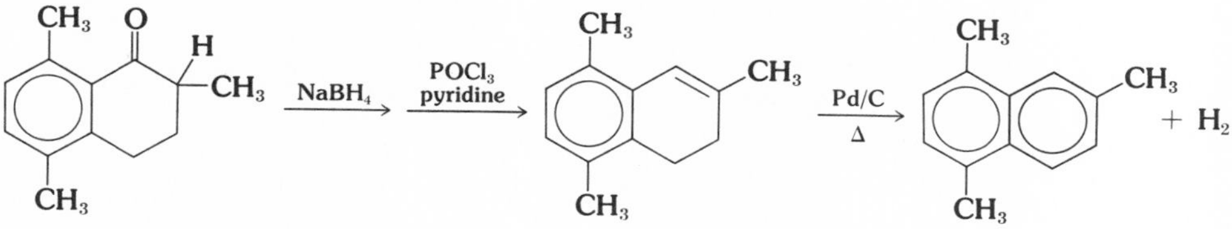

19·5 EXAMPLES OF SYNTHESIS PROBLEMS   649

$$\text{(tetrahydronaphthalenone with CH}_3\text{ groups)} \xrightarrow{\text{NaBH}_4} \xrightarrow[\text{pyridine}]{\text{POCl}_3} \text{(dihydronaphthalene)} \xrightarrow[\Delta]{\text{Pd/C}} \text{(1,4,6-trimethylnaphthalene)} + H_2 \quad (19.55)$$

PROBLEM 19·9
Why is the ketone function generated in the Friedel-Crafts acylation with succinic anhydride (equation 19.53) removed before the second six-membered ring is closed?

SYNTHESIS 6
Suggest a synthesis of 1,1-dimethylcyclopentane from the diketone 2,5-hexanedione and any other simple reagents.

The starting material, 2,5-hexanedione, is conveniently prepared by hydrolysis of 2,5-dimethylfuran:

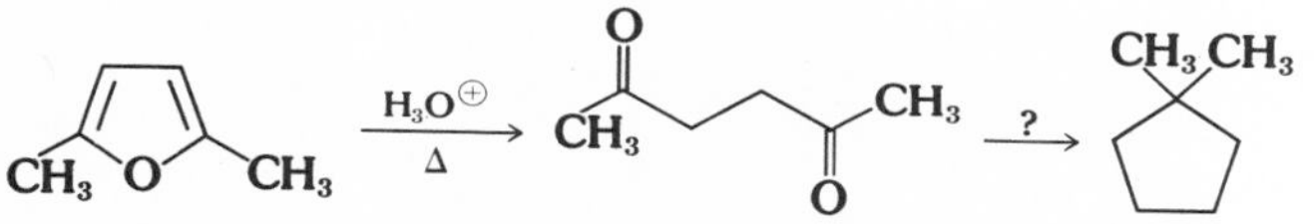

$$\text{2,5-dimethylfuran} \xrightarrow[\Delta]{H_3O^{\oplus}} CH_3\text{COCH}_2\text{CH}_2\text{COCH}_3 \xrightarrow{?} \text{1,1-dimethylcyclopentane} \quad (19.56)$$

Although the desired compound has only one more carbon atom than the starting material, considerable structural change will clearly be necessary. If we focus our attention on ring formation and branching, as suggested in the previous section, we can envision two general approaches to the final structure:

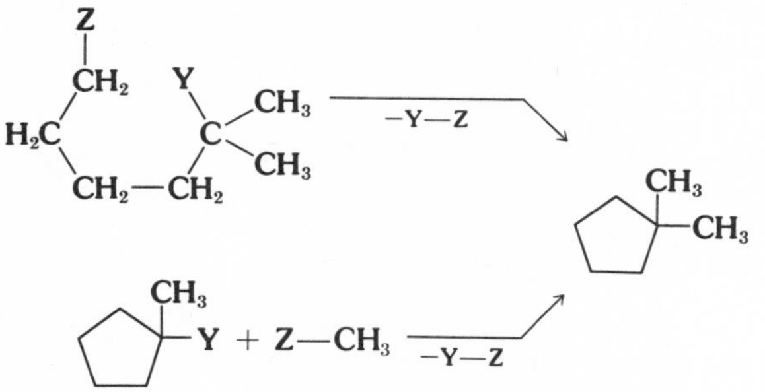

$$\begin{array}{c} \text{acyclic precursor with Y, Z, CH}_3, CH_3 \xrightarrow{-Y-Z} \\ \text{cyclopentane-Y} + Z-CH_3 \xrightarrow{-Y-Z} \end{array} \text{1,1-dimethylcyclopentane} \quad (19.57)$$

In the first approach introduction of the seventh carbon atom precedes cyclization to a five-membered ring, while in the second approach these operations are reversed.

At this point it is helpful to consider these operations as they might apply specifically to our starting material. Although the introduction of another methyl group at one of the carbonyl functions in 2,5-hexanedione poses no problem—indeed, it would be difficult to prevent the incorporation of a second methyl—cyclization of the resulting ketol would not be straightforward. However, cyclization of the starting diketone is easily effected by an intramolecular aldol condensation, as illustrated by reaction 19.58, and the methyl group can then be inserted by conjugate addition.

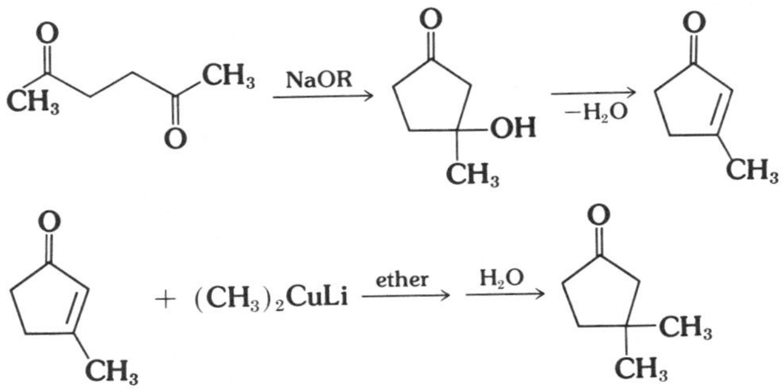

19·5 EXAMPLES OF SYNTHESIS PROBLEMS

$$CH_3\text{-CO-CH}_2\text{-CO-CH}_3 \xrightarrow{\text{NaOR}} \text{[3-methyl-3-hydroxycyclopentanone]} \xrightarrow{-H_2O} \text{[3-methylcyclopent-2-enone]}$$

$$\text{[3-methylcyclopent-2-enone]} + (CH_3)_2CuLi \xrightarrow{\text{ether}} \xrightarrow{H_2O} \text{[3,3-dimethylcyclopentanone]} \quad (19.58)$$

Finally, deoxygenation of the carbonyl group is easily accomplished by a Wolff-Kishner reduction (equation 12.43).

SYNTHESIS 7

Suggest a synthesis of the following spirobicyclic amine from acyclic starting materials having no more than three contiguous carbon atoms.

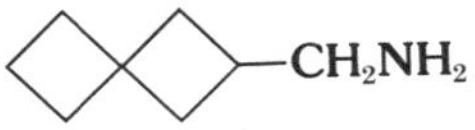

[spirobicyclic structure]—CH_2NH_2

The following guidelines result from an inspection of this structure:

1 The sensitive amine function should not be introduced until the last stages of the synthesis (consult Table I for methods).

2 The small strained rings require irreversible cyclization methods, such as enolate alkylations.

3 Because of the plane of symmetry in this molecule, formation of the quaternary carbon atom may be planned to coincide with ring formation:

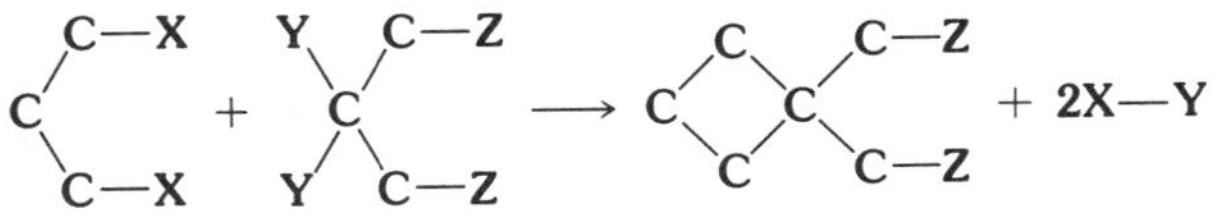

$$\begin{matrix} C-X \\ C \\ C-X \end{matrix} + \begin{matrix} Y & C-Z \\ C \\ Y & C-Z \end{matrix} \longrightarrow \begin{matrix} C & C-Z \\ C & C \\ C & C-Z \end{matrix} + 2X-Y \quad (19.59)$$

Equation 19.60 illustrates a specific application of this last suggestion, which generates a monocyclic intermediate incorporating six of the eight carbon atoms present in the final product:

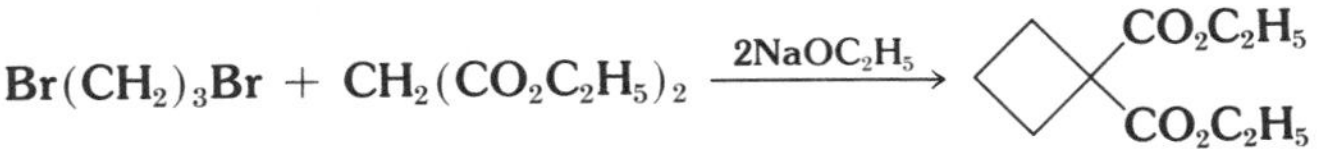

$$Br(CH_2)_3Br + CH_2(CO_2C_2H_5)_2 \xrightarrow{2NaOC_2H_5} \text{[cyclobutane-1,1-diyl]}\begin{matrix}CO_2C_2H_5\\CO_2C_2H_5\end{matrix} \quad (19.60)$$

At this point, two carbon atoms remain to be introduced and the second four-membered ring must be closed. If the ester functions in our intermediate could be transformed into primary bromides or sulfonate esters, we could then repeat the operation described in equation 19.60 to generate the second ring:

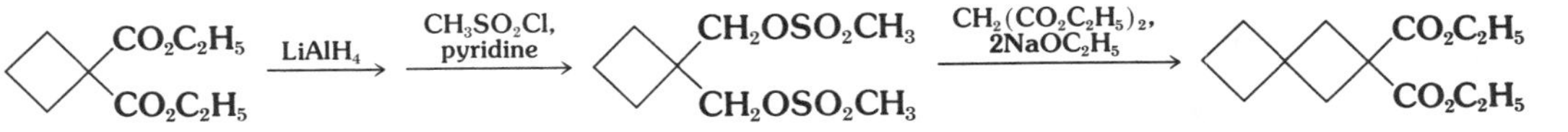

$$\text{[cyclobutane]}\begin{matrix}CO_2C_2H_5\\CO_2C_2H_5\end{matrix} \xrightarrow{LiAlH_4} \xrightarrow[\text{pyridine}]{CH_3SO_2Cl,} \text{[cyclobutane]}\begin{matrix}CH_2OSO_2CH_3\\CH_2OSO_2CH_3\end{matrix} \xrightarrow[2NaOC_2H_5]{CH_2(CO_2C_2H_5)_2,} \text{[spirobicyclic]}\begin{matrix}CO_2C_2H_5\\CO_2C_2H_5\end{matrix}$$

$$(19.61)$$

19·5 EXAMPLES OF SYNTHESIS PROBLEMS

The final stages of this synthesis consist of ester hydrolysis, decarboxylation, amide formation, and reduction:

$$\text{(structure with two CO}_2\text{C}_2\text{H}_5\text{)} \xrightarrow[\Delta]{\text{NaOH}} \xrightarrow[\Delta, -\text{CO}_2]{\text{H}_3\text{O}^\oplus} \text{(structure)-CO}_2\text{H} \xrightarrow[\text{LiAlH}_4]{\overset{\text{SOCl}_2}{\underset{\text{NH}_3}{\longrightarrow}}} \text{(structure)-CH}_2\text{-NH}_2 \quad (19.62)$$

SYNTHESIS 8

Suggest a synthesis of the following tricyclic ether from acyclic starting materials having no more than four contiguous carbon atoms.

At the outset it is clear that different synthesis methods will be required to construct the three rings of this molecule. Furthermore, the trans configuration of the heterocyclic ring fusion to the six-membered carbocycle poses a special problem which we must keep in mind while considering possible cyclization sequences.†

The advantages of approaching a synthesis problem from the product (that is, of working backwards) are particularly clear in this case, and we shall begin by assuming that the three-membered ring grouping should be one of the last structural features to be developed. Among the methods commonly used to form three-membered carbon rings, the Simmons-Smith reaction (see Table II) appears well suited for the last step in our synthesis:

$$\text{(bicyclic ether)} + \text{CH}_2\text{I}_2 + \text{Zn(Cu)} \xrightarrow{\text{ether}} \text{(tricyclic ether)} \quad (19.63)$$

The precursor used in this reaction contains a cyclohexene synthon, which at first glance might appear to be a potential Diels-Alder adduct:

$$\text{(diene)} + \text{(dihydrofuran)} \longrightarrow \text{(bicyclic ether)} \quad (19.64)$$

Such a reaction would, however, produce the cis fused stereoisomer, rather than the desired trans isomer. In order to circumvent this difficulty, we can postpone closing the heterocyclic ring until after the cyclohexene ring has been formed. This approach also permits some flexibility in the choice of dienophile substituents to activate this reactant:

$$\text{(diene)} + \text{(diester alkene)} \longrightarrow \text{(cyclohexene diester)} \xrightarrow[\text{H}_2\text{O}]{\text{LiAlH}_4} \text{(cyclohexene diol)} \quad (19.65)$$

†Both cis and trans configurations are possible here, in contrast to the three-membered ring fusion, which is constrained in a cis configuration.

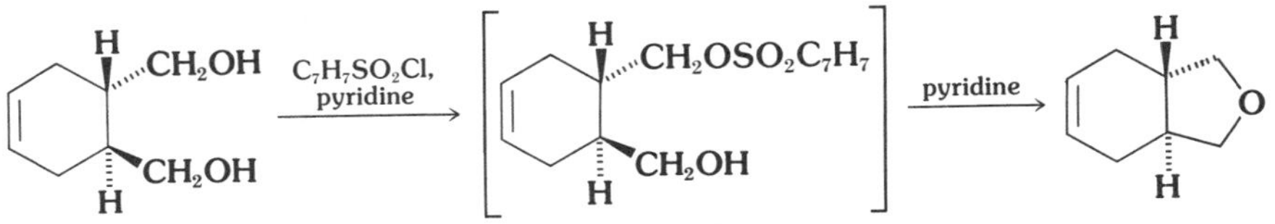

19·5 EXAMPLES OF SYNTHESIS PROBLEMS

$$\text{[scheme: diol} \xrightarrow{C_7H_7SO_2Cl, \text{ pyridine}} \text{mono-tosylate} \xrightarrow{\text{pyridine}} \text{bicyclic ether]} \tag{19.66}$$

SYNTHESIS 9
Suggest possible approaches to the synthesis of the tricyclic sesquiterpene copaene.†

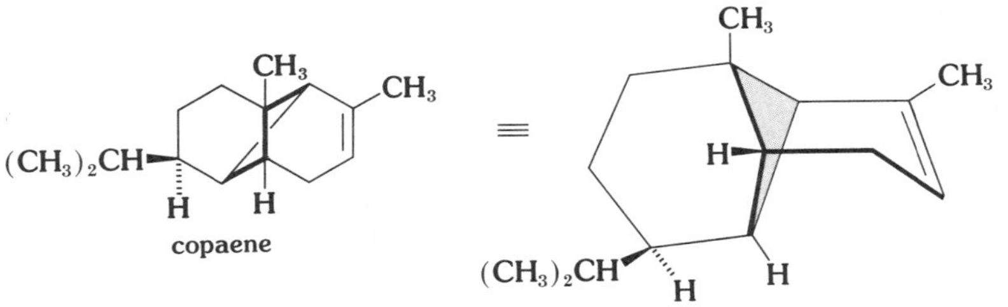

[structure of copaene with CH$_3$, CH$_3$, (CH$_3$)$_2$CH— substituents; equivalent three-dimensional representation shown]

This problem provides a taste of some of the thought and planning that go into a modern synthesis. The most demanding structural feature in this molecule is the double-bridged four-membered ring, which also contains the lone quaternary carbon atom. Two of the most common methods for preparing four-membered rings are *photocycloaddition reactions* of double bonds and *intramolecular enolate alkylations*. Let us consider each of these approaches in turn.

Several research groups have examined the internal cycloaddition reactions of 1,6-cyclodecadiene. However, the difficulties encountered in synthesizing properly substituted large-ring dienes, together with the problem of competing cycloaddition reaction paths have hindered the development of this approach.

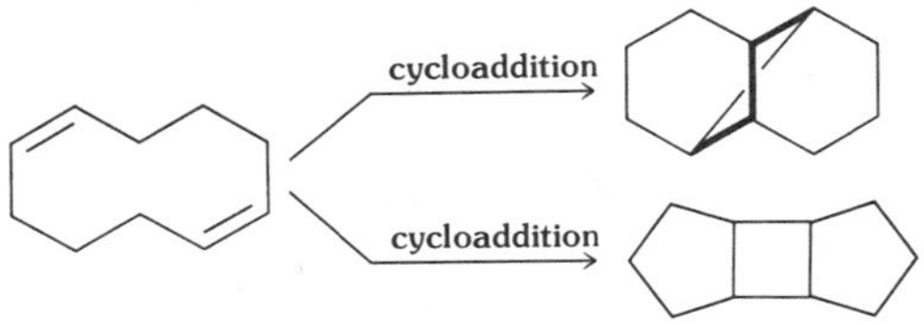

$$\text{[1,6-cyclodecadiene} \xrightarrow{\text{cycloaddition}} \text{two possible tricyclic products]} \tag{19.67}$$

The internal-alkylation approach to this bridged tricyclodecane ring system is derived from the fact that rupture of any of the bonds composing the four-membered ring gives a *cis*-decalin derivative. Conversely, a suitably substituted *cis*-decalin might be expected to cyclize to the desired ring system:

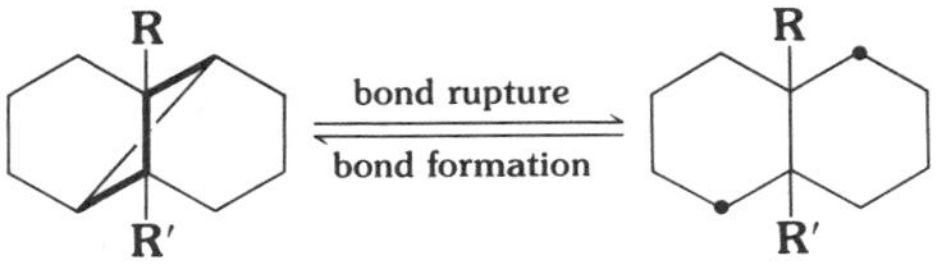

$$\text{[tricyclic with R, R'} \underset{\text{bond formation}}{\overset{\text{bond rupture}}{\rightleftharpoons}} \text{cis-decalin diradical with R, R']} \tag{19.68}$$

An application of intramolecular enolate alkylation to this transformation is shown in Figure 19.5.

---

†A total synthesis of copaene was reported in 1966 by Clayton Heathcock, professor at the University of California in Berkeley.

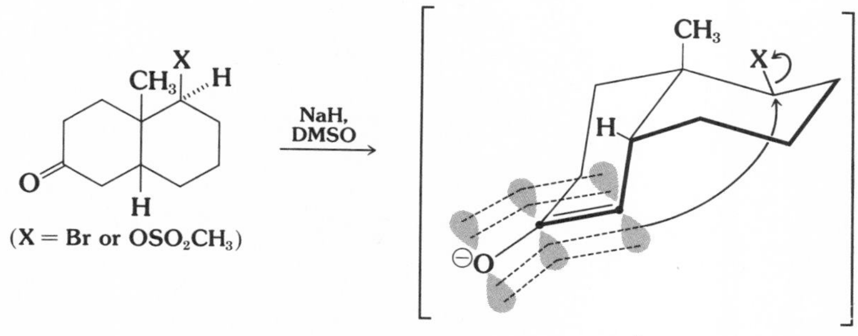

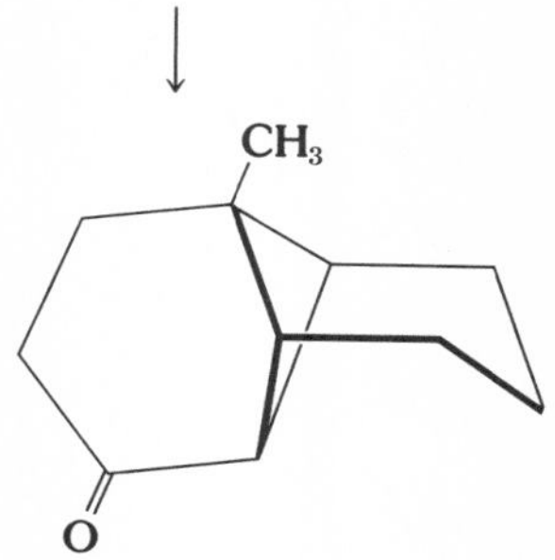

FIGURE 19·5 *Intramolecular enolate alkylation*

If we now dissect the copaene structure according to this internal-alkylation approach, the bicyclic species in Figure 19.6 can be considered as potential intermediates; the dots denote those carbon atoms that must be joined in the internal alkylation. Since all the alkyl substituents and the alkene function will probably not be present at the cyclization stage, each of these intermediates must be examined with the following questions in mind:

1 How can such a *cis*-decalin, having appropriately located carbonyl and leaving groups, be prepared?

2 After formation of the four-membered ring, how can the remaining alkyl groups be introduced?

3 How can the final introduction or modification of functional groups be achieved?

FIGURE 19·6 *Internal-alkylation approaches to the synthesis of copaene*

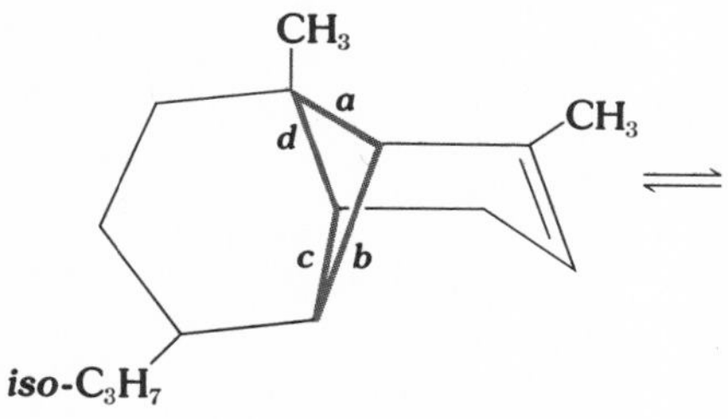

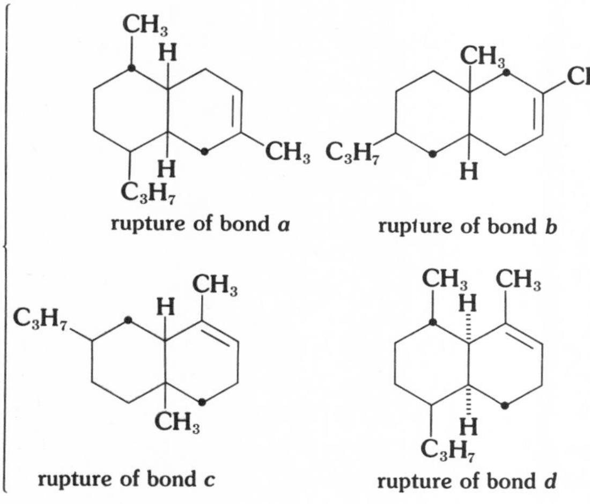

19·5 EXAMPLES OF SYNTHESIS PROBLEMS

For example, an intermediate related to the decalin resulting from cleavage of bond *a* might be prepared by a Diels-Alder reaction:

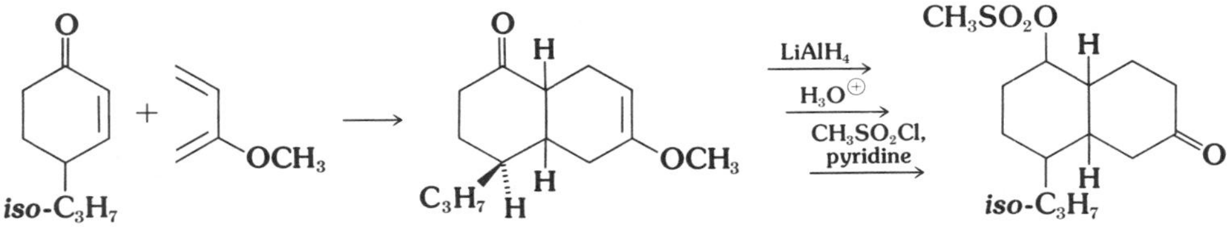

(19.69)

Cyclization of this intermediate would generate the desired ring system, but it would then be very difficult to insert the missing methyl substituent at the quaternary carbon atom (the other methyl group and the double bond should pose no problem):

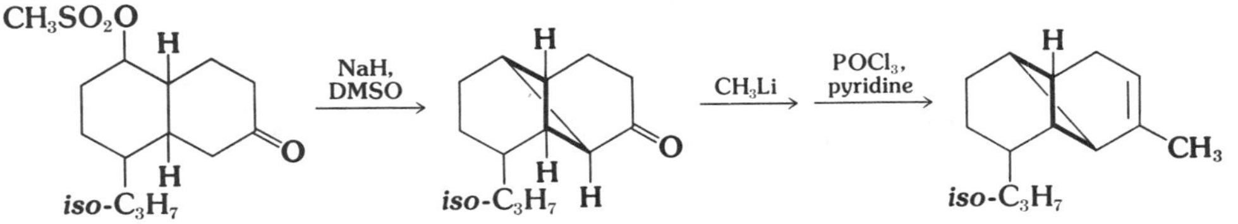

(19.70)

It would be fruitless to introduce the troublesome methyl group before the cyclization, since the resulting tertiary alcohol would undergo dehydration rather than substitution. Intermediates derived by cleavage of bond *d* also suffer this handicap.

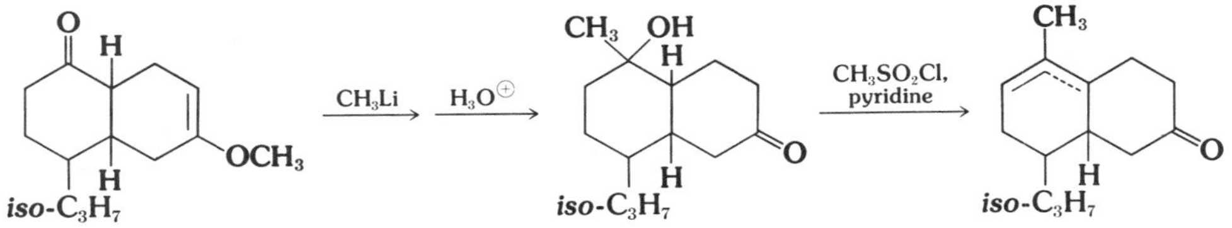

(19.71)

PROBLEM 19·10
a Write equations showing how the unsaturated ketone used in reaction 19.69 could be prepared from simple reactants.
b We have ignored several other potential problems in reactions 19.69 and 19.70. Can you see them?

If we examine intermediates related to cleavage of bond *b* or bond *c*, we find that the troublesome quaternary methyl group presents no exceptional difficulties, since it may be easily incorporated before cyclization:

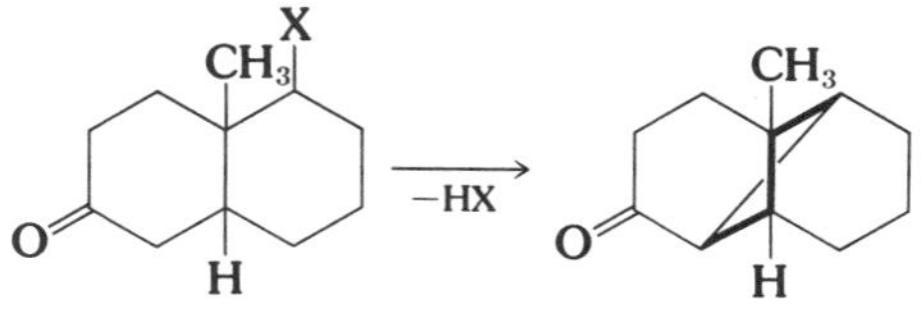

(19.72)

In this example the carbonyl function can serve as a "handle" for introducing the isopropyl group; however, we must be careful to provide a means for inserting the second methyl group and the double bond.

We might try to prepare an intermediate of this kind by a Diels-Alder reaction similar to that used earlier, but we would soon find that steric hindrance resulting from the methyl group interferes with this approach. A suitable alternative is fortunately available in that combination of Michael and aldol condensations known as the *Robinson annellation*:

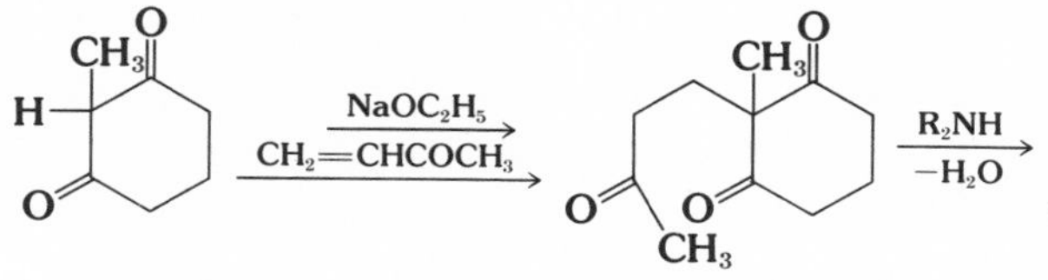

(19.73)

The intermediate diketone generated in this fashion can be modified to fit the requirements for formation of bond *b* by selective reduction of the saturated carbonyl group, catalytic reduction of the double bond (the hydrogen adds from the less hindered side), and E2 elimination of the corresponding tosylate. Cyclization at the keto tosylate stage should give the desired ring system, but we would be unable to introduce the remaining methyl group and the double bond. The mild base pyridine selectively effects elimination.

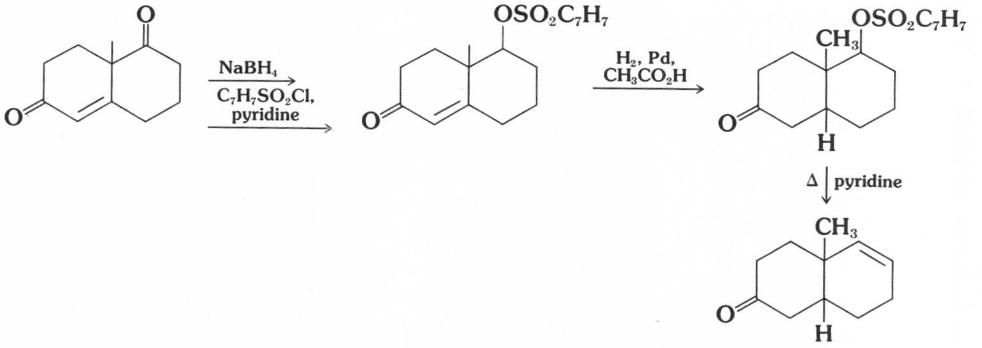

(19.74)

In the published synthesis, epoxidation of the double bond provided not only a leaving group at the desired carbon atom, but also a potential carbonyl function (by oxidation of the resulting secondary alcohol) at a site requiring the insertion of a methyl group and a double bond. Protection of the existing carbonyl group proved necessary during epoxidation in order to avoid Baeyer-Villiger rearrangement (see equation 15.86). Since the keto epoxide produced by hydrolysis of the protective group failed to cyclize when treated with base, the oxirane ring was opened by reaction with sodium benzylate (S_N2 displacement occurs at the least hindered carbon atom). The resulting secondary alcohol was then converted to a tosylate ester:

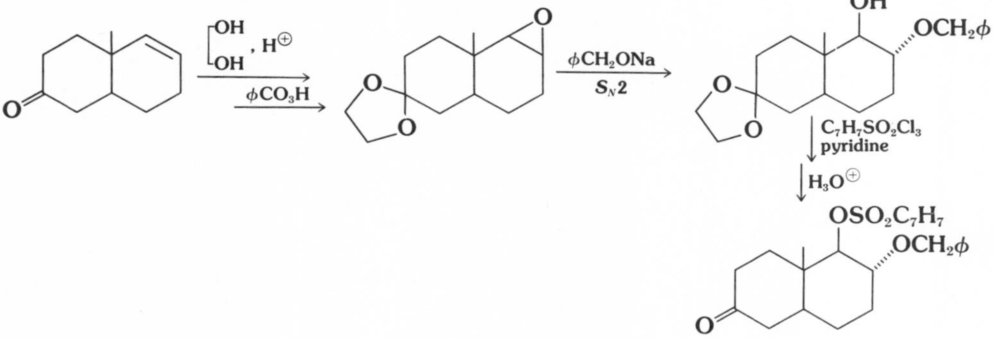

(19.75)

19·5 EXAMPLES OF SYNTHESIS PROBLEMS

This improvement in the leaving group permitted the desired cyclization, followed by acid-catalyzed removal of the benzyl protective group and addition of isopropyl lithium:

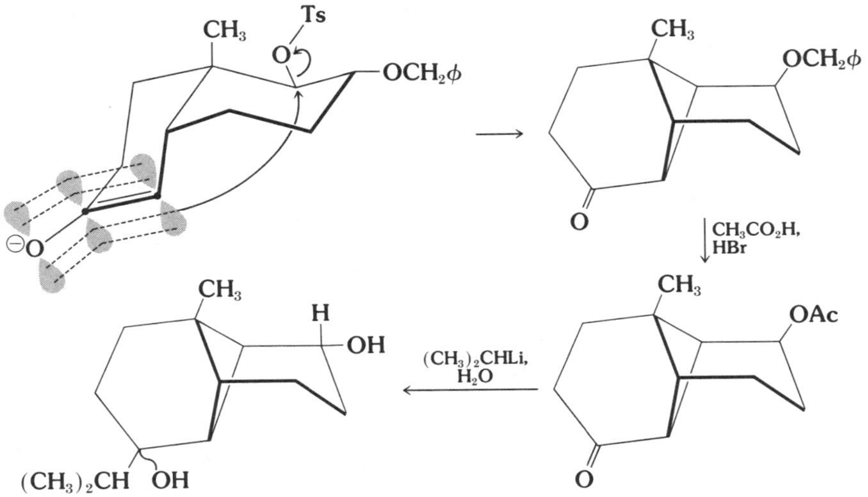

(19.76)

The remaining steps of the synthesis seem clear at this point. The tertiary alcohol must be dehydrated and the resulting alkene reduced; the secondary hydroxyl function must be oxidized to a ketone; and the final methyl group must be added to the carbonyl function, followed by dehydration of the resulting alcohol. In practice, however, several obstacles remained. Among these were the avoidance of acid-catalyzed rearrangement during dehydration and the control of stereochemistry during catalytic reduction (accomplished by the secondary hydroxyl group). These transformations were effected as shown in Figure 19.7.

FIGURE 19·7 *The synthesis of copaene*

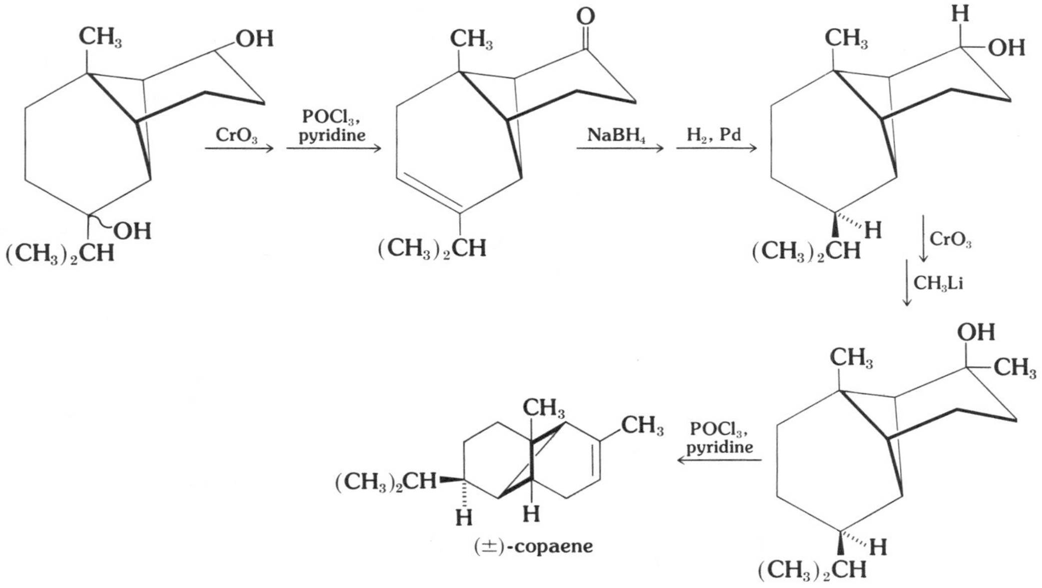

19·6 PROSPECTS IN SYNTHESIS

The fundamental questions that have stimulated the curiosity and challenged the creativity of chemists for the past 200 years remain unchanged. We identified these in the first chapter as:

What is it?
What does it do?
How can it be made?

Over the past three decades chemists have acquired some powerful tools with which they have obtained a great deal of information pertaining to the first two questions. For example, virtually every crystalline compound will relinquish the secret of its molecular configuration to a modern computer-controlled x-ray diffractometer. Furthermore, non-crystalline compounds, as well as solutions and other mixtures, can be scrutinized by a variety of sophisticated spectroscopic devices that are able to reveal the atom distribution and bonding in component molecules. Since answers to the second question must consist in part of an identification of all reaction products from characteristic transformations, the success of these new structure-elucidation methods has been far reaching indeed. These techniques, together with advances in efficient separation methods and the development of rapid-scanning devices for studying short-lived intermediates, have provided us with an understanding of chemical reactions scarcely dreamed of 30 years ago.

The facility and accuracy with which chemists now can answer the first two of our basic questions have improved their ability to resolve the third. Clearly, the better we understand how and why chemical reactions take place, the more effectively we can use them to achieve desired synthetic transformations. The fact remains, however, that despite this improvement, the technological leaps that have so transformed structure-elucidation and reaction-mechanism studies have been lacking in synthesis. To be sure, a host of new and useful reactions have been discovered in the past two or three decades; nevertheless, the procedures by which a synthesis expert analyzes a structure and develops a rational plan for its synthesis are largely unchanged.

The advantages of developing improved methods of solving synthesis problems have not gone unrecognized. Over the last ten years a number of chemists have tried to devise computer programs that would facilitate, and possibly duplicate, the complex creative thought processes by which individuals conceive of plausible synthesis pathways. Indeed, the experience and intelligence that enable the chemist to recall a multitude of facts about organic reactions, and from these to select a rational sequence of transformations to achieve desired changes, appear at first glance to be a poor match for the memory and computational speed of a modern computer. However, recent efforts to program computers to this end have not been particularly fruitful. In fact, a conference of the most active and creative researchers in this field concluded in 1972 that significant advances in computer-assisted synthesis would not appear for at least another decade. In the foreseeable future, then, it seems as though organic synthesis will continue to embody elements of creative speculation and art, and will display some of the finest achievements of human intellect.

In conclusion, it is useful to distinguish two different stages of a synthesis: the planning and the execution. This distinction was clearly made by W. S. Johnson of Stanford University in the report of the 1972 conference on computer-designed synthesis:

The present state of the art of synthesis is such that all well-planned synthetic schemes almost invariably fail to give the envisaged results at one, or more often, several stages. Thus a successful synthesis seldom follows and sometimes diverges dramatically from the original plan. It is this "fallibility phenomenon" which renders organic synthesis at least as much of a creative challenge at the execution stage as at the planning stage.

Therefore, even if the planning operations could be fully systematized, we will find that in practice chemistry is and will probably always be an experimental science.

REFERENCES

R. E. Ireland, *Organic Synthesis*, Prentice-Hall, Inc., Englewood Cliffs, N.J., 1969.

I. Fleming, *Selected Organic Syntheses*, John Wiley & Sons, Inc., New York, 1973.

N. Anand, J. S. Bindra, and S. Ranganathan, *Art in Organic Synthesis*, Holden-Day, Inc., San Francisco, 1970.

C. A. Buehler and D. E. Pearson, *Survey of Organic Synthesis*, John Wiley & Sons, Inc., New York, 1970.

W. E. Parham, *Syntheses and Reactions in Organic Chemistry*, John Wiley & Sons, Inc., New York, 1970.

R. S. Monson, *Advanced Organic Synthesis*, Academic Press, New York, 1971.

R. B. Woodward, *Perspectives in Organic Chemistry*, ed. by A. R. Todd, Interscience, John Wiley & Sons, Inc., New York, 1956.

S. E. Danishefsky and S. Danishefsky, *Progress in Total Synthesis*, vol. 1, Meredith Corporation, New York, 1971.

I. T. Harrison and S. Harrison, *Compendium of Organic Synthetic Methods*, Interscience, John Wiley & Sons, Inc., New York, 1971.

H. O. House, *Modern Synthetic Reactions*, 2d ed. W. A. Benjamin, Inc., New York, 1972.

J. S. Bindra and R. Bindra, *Creativity in Organic Synthesis*, vol. 1, Academic Press, New York, 1975.

M. Bersohn and A. Esach, Computers and Organic Synthesis, *Chem. Rev.*, 76:269 (1976).

S. Turner, *The Design of Organic Synthesis*, Elsevier Scientific Publishing Company, Amsterdam, 1976.

The following references are particularly useful for research:

Organic Synthesis, John Wiley & Sons, Inc., New York. Annual volumes and five collective volumes (volumes 1–49). This continuing series provides tested procedures for preparing organic intermediates.

Organic Reactions, John Wiley & Sons, Inc., New York. A continuing series, now at 22 volumes, providing critical reviews of important and useful reactions.

Reagents for Organic Synthesis, Fieser and Fieser–John Wiley & Sons, Inc., New York. Five volumes describing reagents of use to organic chemists. Suppliers, methods of preparation, physical constants, and examples of significant uses are given for each reagent.

Synthetic Methods of Organic Chemistry, Theilheimer–S. Karger Verlag, Basel. Annual compilations, beginning in 1946, of new methods for organic synthesis, arranged according to a system of bond formation and bond breaking.

EXERCISES

19·1 Consider the following simple five-carbon structure:

```
        C
        |
C—C—C—C
```

a Without indicating specific compounds, show all reasonable combinations of fragments consisting of four or fewer carbon atoms that could give this structure in one step. Indicate the sites of reactivity (for example, $C^\bullet + C^\bullet \longrightarrow C—C$).
b Write structures for the four constitutionally isomeric alcohols having the carbon skeleton above. How many of the general combinations you wrote in part *a* can be used to synthesize each of these alcohols? Be specific about the reactions you would use and try to conceive of as many different approaches as possible. *Note:* You should use only one carbon-carbon bond-forming reaction in each case, but you may modify functional groups as you see fit.

19·2 Each of the following molecules has in its structure and functional-group disposition the clue to a straightforward synthesis from less complex reagents. Write equations illustrating each synthesis.

a $C_2H_5COCH(CH_3)CO_2C_2H_5$

b cyclopentyl-C(OH)ϕ_2

c $(CH_3)_3C$-C$_6H_4$-COC(CH$_3$)$_3$

d bicyclic diester with H, CO$_2$CH$_3$ and H, CO$_2$CH$_3$ substituents

e $(\phi CH_2)_2 CHCO_2H$

f 2-isopropylidene thiacyclohexane with H and CH(CH$_3$)$_2$ on the exocyclic carbon

19·3 Some, and possibly all, of the following transformations require the use of activating or protecting (blocking) groups. Write equations showing these applications. In any case not requiring such modification, indicate the reagent that will effect the transformation.

a) 2-bromo-N-methylaniline → 2-bromo-4-nitro-N-methylaniline

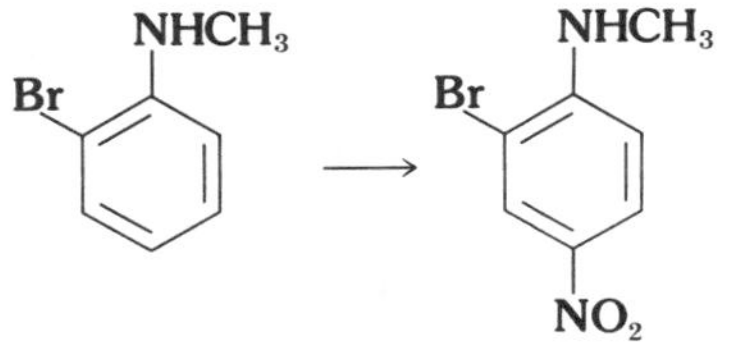

b) 2-methylcyclohexanone → 2,2-dimethylcyclohexanone

c) 2-methylcyclohexanone → cis-2,6-dimethylcyclohexanone (cis isomer)

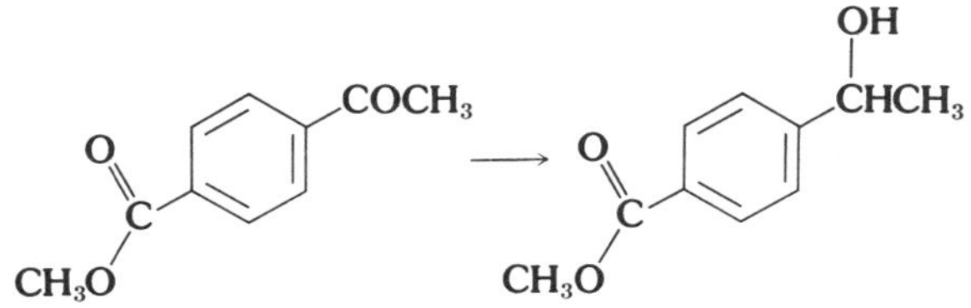

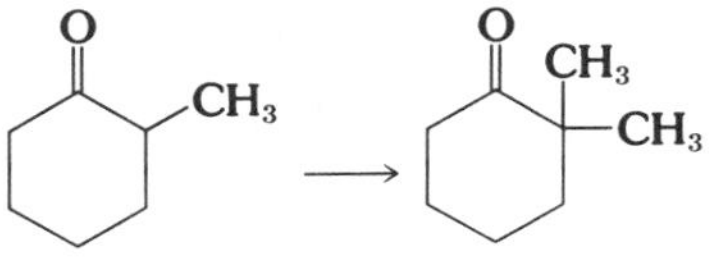

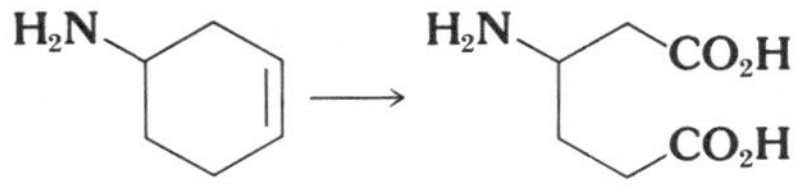

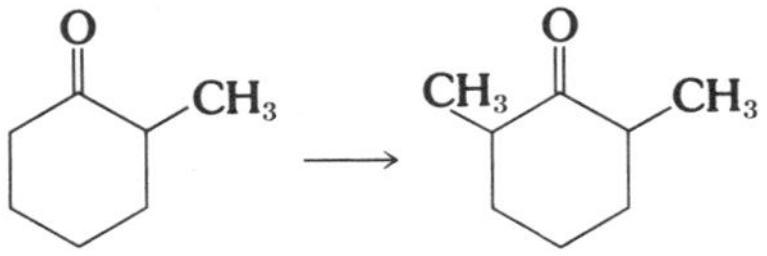

d) methyl 4-acetylbenzoate → 4-acetyl-α,α-diphenylbenzyl alcohol

e) methyl 4-acetylbenzoate → methyl 4-(1-hydroxyethyl)benzoate

f) 3-aminocyclohexene → 3-aminoglutaric acid (H$_2$N-CH(CO$_2$H)-CH$_2$-CO$_2$H)

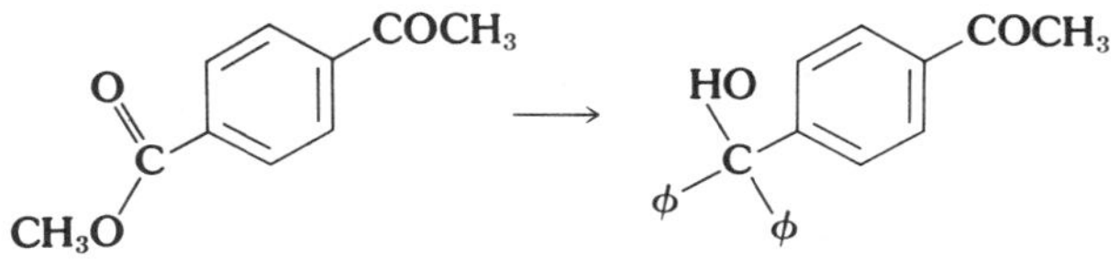

19·4 In the preparation of β-selinene, compound **3** was treated with methyl lithium, followed by hydrolysis, to yield **4**.
a What is the intermediate compound and how is it formed?
b At what point does epimerization occur? How? What is the driving force for epimerization?

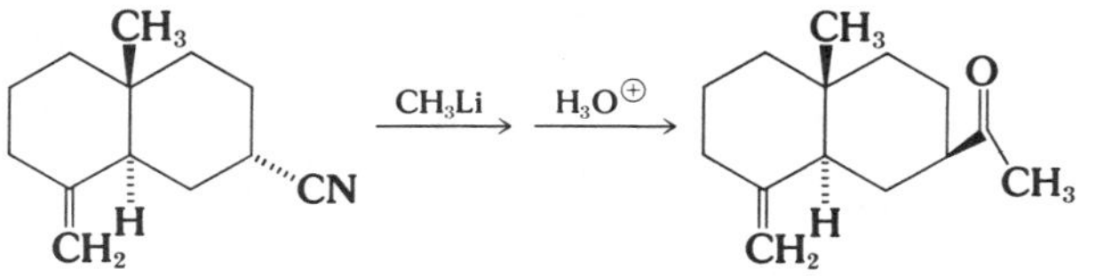

3 →(CH$_3$Li, H$_3$O$^{\oplus}$)→ **4**

19·5 In the total synthesis of cedrol, spiro compound **9** was used.
a What reagent may be used to promote the transformation to **9** from compound **8**? What type of reaction is this transformation?
b Why are the methyl and the carbomethoxy groups trans in compound **9**?

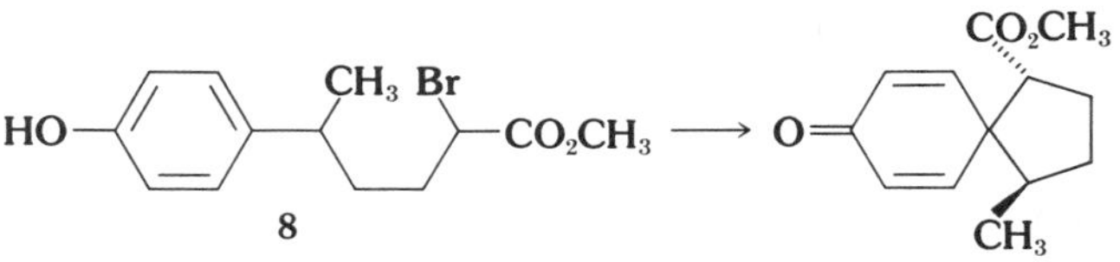

8 → **9**

19·6 A synthesis of α-onocerin involved the three transformations shown below.
a What reagent could be used to prepare compound 2 from 1?
b What type of reaction is the conversion of 2 to 3?
c Suggest a reason for the fact that only the isomer of 3 shown here was formed.

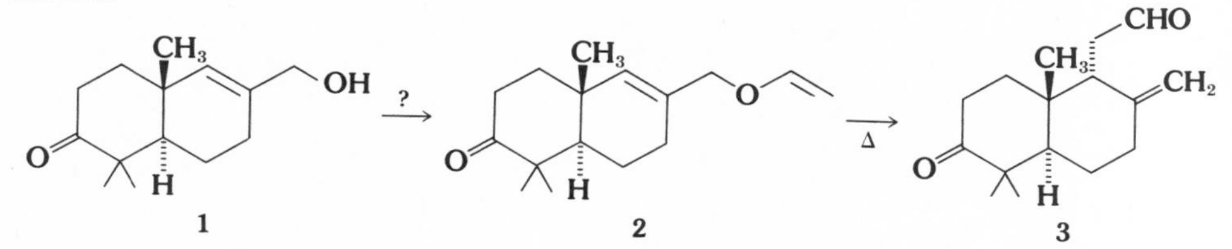

19·7 The first six steps of a total synthesis of the steroid cholesterol are shown below.
a Suggest a reason for the regio- and stereospecificity of the first transformation. What type reaction is this?
b Why was compound 2 treated with base? Show the mechanism of the transformation from 2 to 3.
c Is the stereochemistry of the hydride reduction leading to compound 4 important in determining the configuration of product 7? Explain your reasoning.
d Write a mechanism for the transformation of 4 to 5. Why was the hydroxyl at C-1 eliminated, but not the hydroxyl at C-4?
e What is the reagent of choice to convert 5 to 6?
f How can the acetyl group in compound 6 be removed to form 7?
g Suggest a reason why the methoxy compound 1 was used as the starting material, instead of the corresponding phenol.

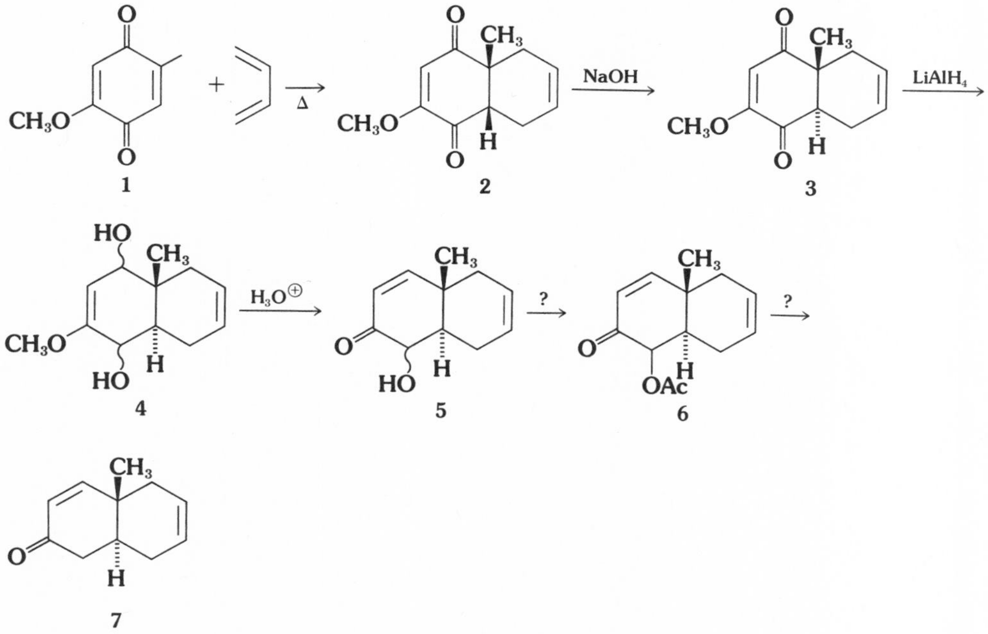

EXERCISES

19·8 In a synthesis of the male sex hormone androsterone, the side chain indicated in structure 11 was modified as shown:

a What is the structure of compound 12?
b Show the mechanism by which 12 is converted to 13.
c How does the reaction of 13 with water proceed? What happens to the acetamide group?
d Suggest an alternative route from compound 11 to 14.

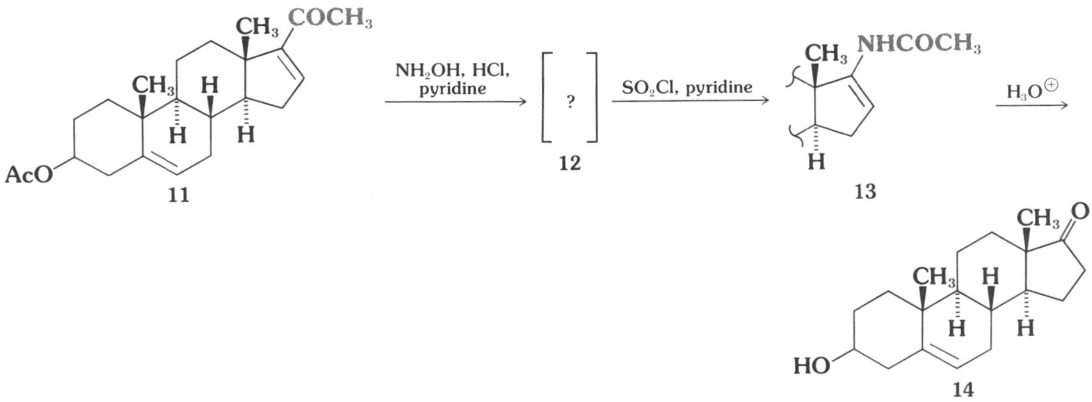

19·9 The reaction sequence shown below was a breakthrough in the synthesis of Aspidosperma alkaloids.

a Indicate the reagents which can be used to convert compound 15 to 16. Is the order of the conversions (ketone ⟶ ketal and ester ⟶ amide) important? Explain.
b Show the course of the reaction leading to 17. Would this conversion be successful if hydrolysis preceded reduction? Explain.
c What reagent could be used to prepare 18?
d Show the mechanism of the reaction leading from compound 18 to 19. Could potassium hydroxide be used as a base in this reaction instead of *t*-butoxide? Why was *t*-butoxide used?
f The conversion of 19 to 20 can be accomplished in three steps. What are they? Note that the amide carbonyl is removed, but the ketone remains intact in compound 20.

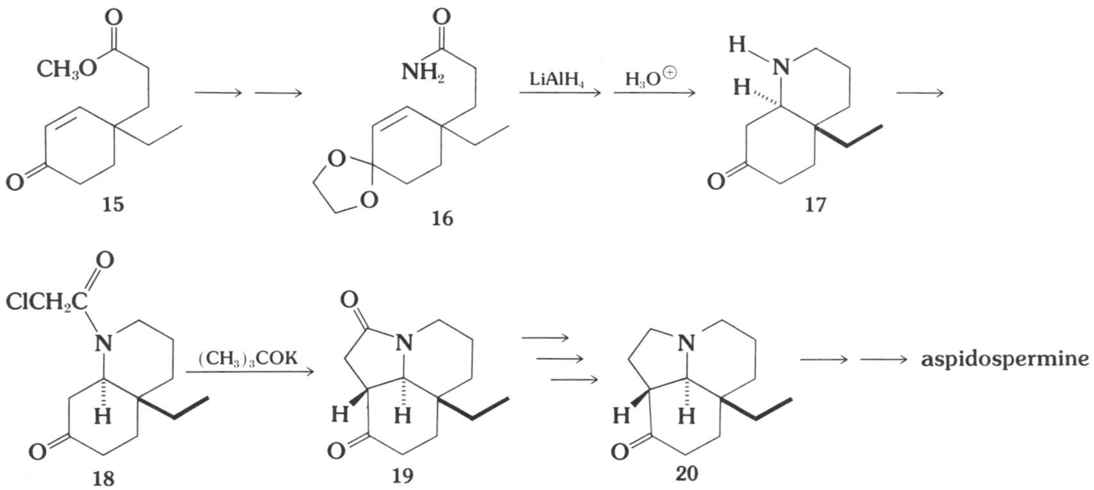

19·10 The last several steps of the synthesis of the steroid epiandrosterone are shown below.
a Show the mechanism of the reaction leading to compound 22.
b What intermediate is formed in the conversion of 22 to 23?
c The furfural group is removed by ozonolysis in the first of the steps from 23 to 24. Why was furfural added in the conversion of 21 to 22 when it is not wanted in the final product?
d Show the intermediates involved in the conversion of 23 to compound 24.
e The condensation leading from 24 to the final product is an important general reaction. What is its name? Show the pathway of this reaction.

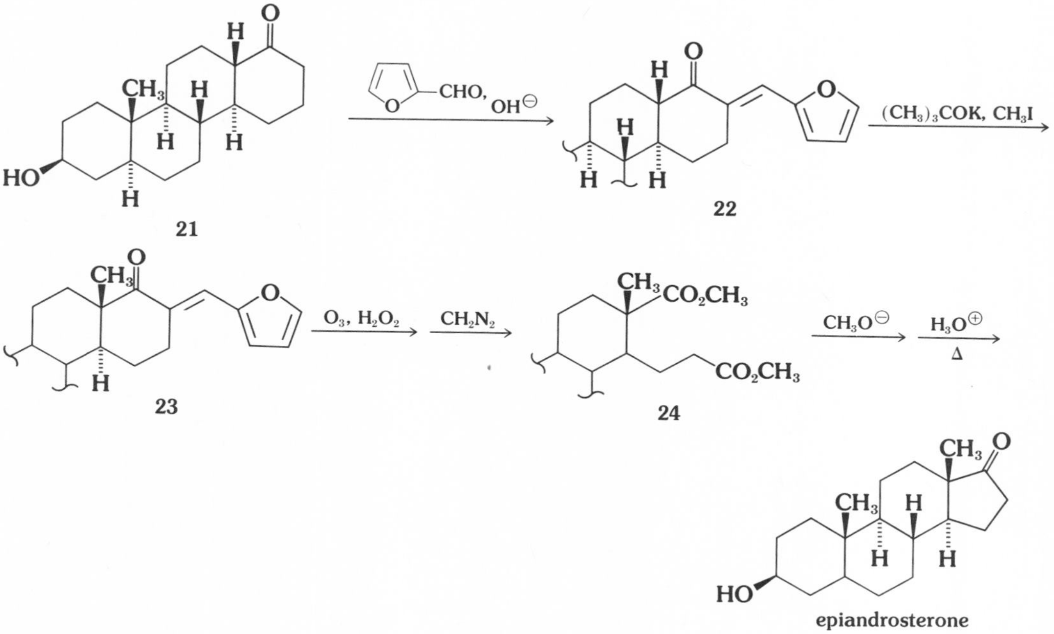

19·11 Shown below are the steps in a relatively short synthesis of α-caryophyllene alcohol.
a What reaction conditions are suitable for the formation of compound 28?
b Why is the attack by methyl lithium stereospecific to form 29?
c Write a mechanism for the conversion of 29 to the final product.

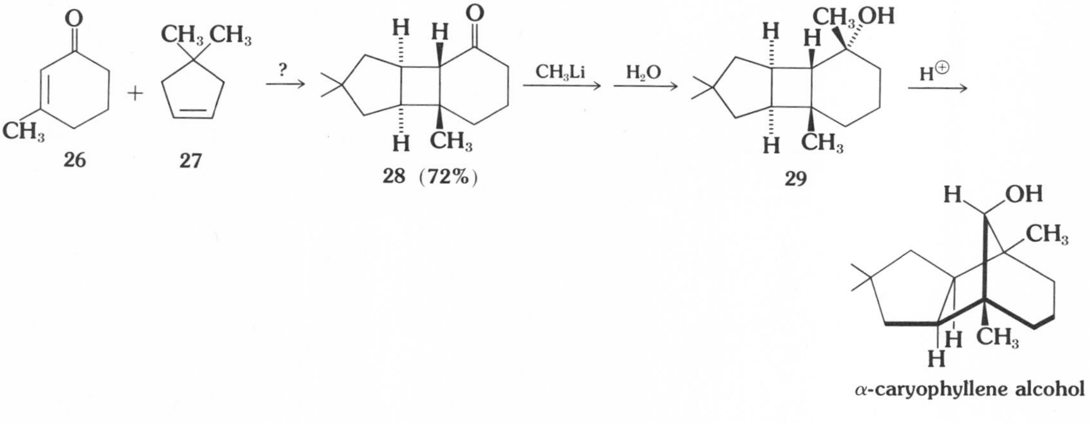

19·12 Several early steps in a total synthesis of longifolene are outlined below.

a Is the selective ketalization of 31 to form compound 32 reasonable? Explain. Why is it necessary to protect the saturated ketone as the ketal (32)?

b What reagent could be used to prepare 33 from 32? Suggest a reason why only one of the double bonds in diene 33 is oxidized by OsO$_4$.

c Would you expect *p*-toluenesulfonyl chloride to react specifically with diol 34 to form 35? Explain.

d Of what reaction type is the conversion of 35 to 36? Show the mechanism.

e Acid-catalyzed hydrolysis of ketal 36 causes rearrangement of the double bond. Suggest a reason for this rearrangement.

f What kind of reaction takes place to form the tricyclic system 38? Show the mechanism.

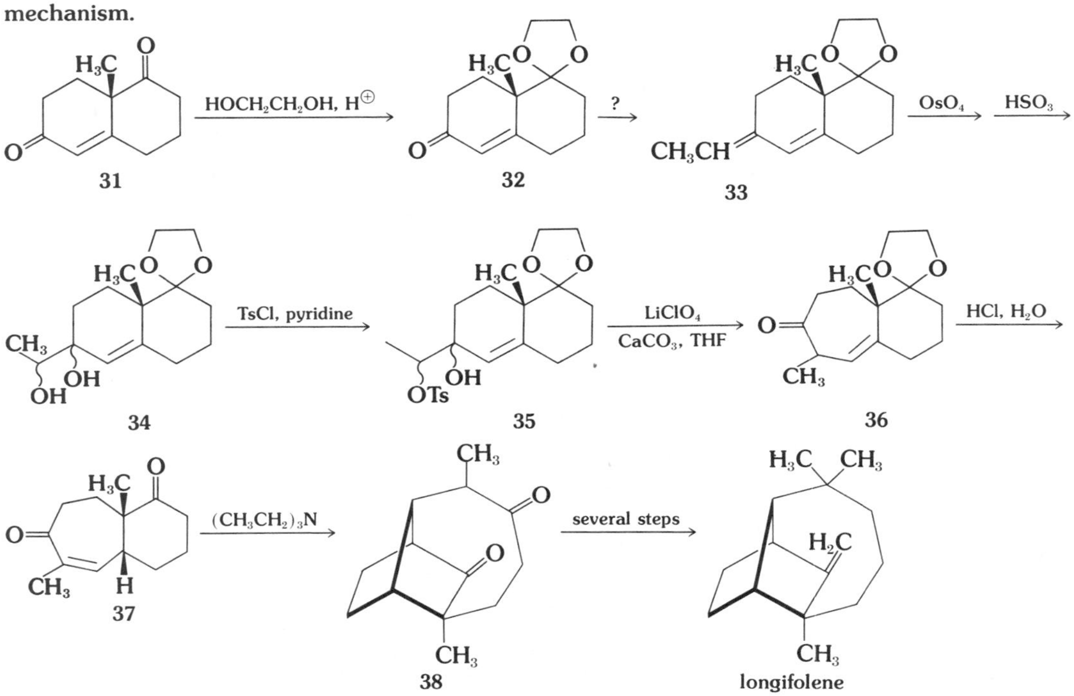

19·13 The reactions below represent a complete synthesis of β-eudesmol.

a Write a mechanism for the reaction leading to 2.

b Explain how sodium borohydride can reduce one of the double bonds of compound 2, but not the other. What product from the reduction of 2 contains the phenyl group?

c Suggest a reason for the stereoselective formation of alcohol 3.

d What would be the product of a direct sodium borohydride reduction of the starting ketone 1?

e What is the reagent of choice to convert alcohol 3 to bromide 4? Note that the reaction proceeds with retention of configuration.

f Write reactions for the four-step conversion of compound 4 to 5.

g What reagent would yield alcohol 6 from ester 5?

h A two-step procedure is required to go from compound 6 to 7. What are these steps?

i Is the configuration of the two new chiral centers in adduct 7 critical at this stage of the total synthesis? Explain.

j What is the purpose of treating ketone 8 with sodium methoxide?
k What reagent could be used to accomplish the transformation of ketone 9 to β-eudesmol?

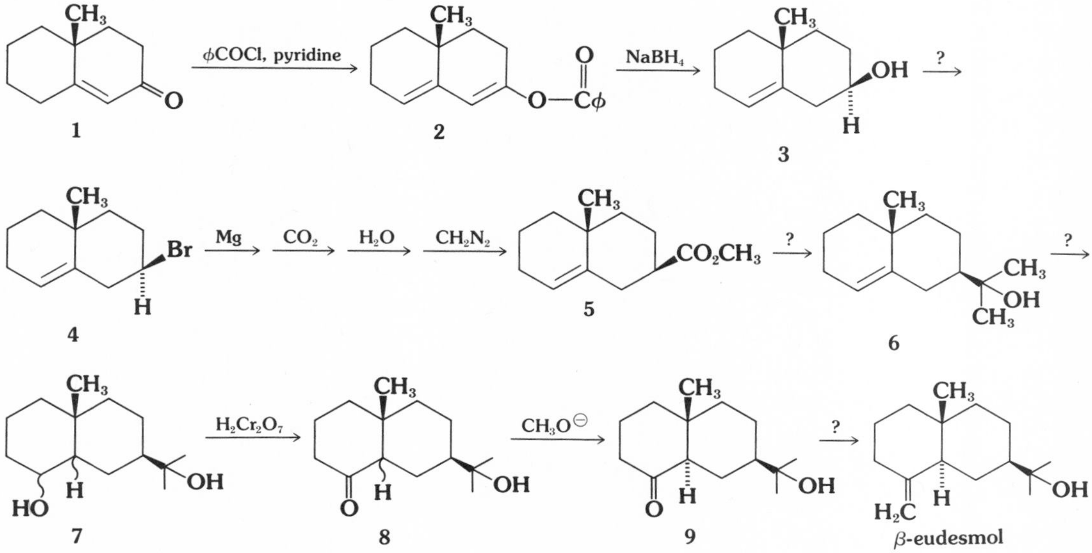

19·14 The following sequence shows the synthesis of propylure, a sex attractant of the pink bollworm moth (*Pectinophora gossypiella*). It is interesting to note that only the *trans*-5 double bond is active as a pheromone.
a Write equations for the preparation of compounds 11 and 12 from the corresponding alcohol and aldehyde, respectively.
b Why is it necessary to protect both the alcohol and aldehyde functions in the reaction leading to compound 13? What type of reaction is involved in the formation of 13?
c What is the nature of the reaction of alkyne 13 to form alkene 14? What conditions will lead stereoselectively to the corresponding cis alkene?
d The preparation of diene 15 was accomplished by a Wittig reaction. What was the reagent used?

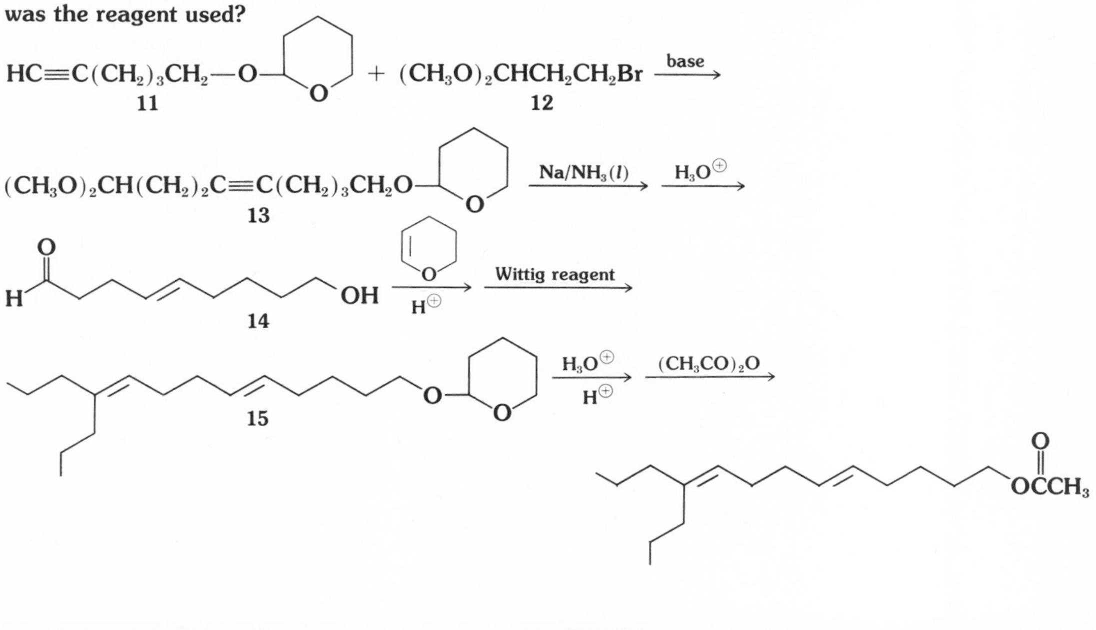

666 EXERCISES

19·15 Helminthosporal may be prepared by the synthesis outlined below.

a What is the purpose of introducing the formyl group into compound 18? What reaction did it facilitate? What reaction did it prevent? Write a mechanism for the reaction leading to compound 19.

b What type of reaction is employed in the conversion of compound 19 to 20? Write a mechanism for this reaction and suggest a reason why it proceeds in the direction it does (with loss of the formyl group).

c The transformation of dione 20 to the bicyclic compound 21 is an aldol-type condensation. Write a mechanism for this reaction and explain the direction the aldol condensation takes. What would be the most likely product if the condensation proceeded by another possible route?

d Show the pathway by which acetal 23 is formed from compound 22. Is the stereochemistry as expected?

e What would be the hydrolysis ($H_3O^\oplus$) product from the enolether function in compound 22? What is the net effect of the Wittig reaction ($21 \longrightarrow 22$) in the overall synthesis scheme?

f Compounds 22 and 23 each have aldehydes in a protected form. Suggest a reason why the acetal form was utilized in the remainder of the synthesis.

g Suggest an alternative route for the conversion of alkene 23 to dicarbonyl 24.

h Explain the direction the condensation takes to form compound 25.

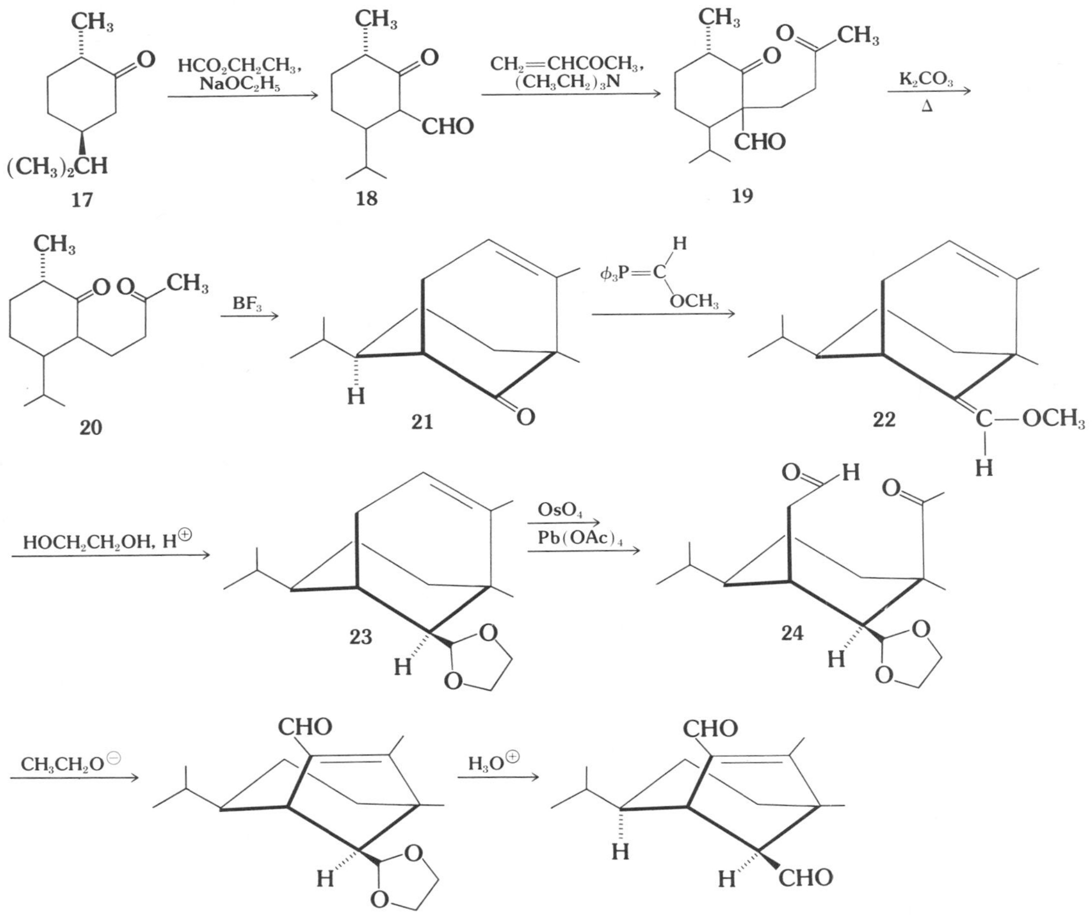

19·16 Given benzene, toluene, xylene (pure isomers), and any nonaromatic organic compounds having no more than three contiguous carbon atoms, as well as any inorganic reagents needed, write equations for all steps in the synthesis of each of the following products:

a [structure: methyl-substituted γ-butyrolactone with acetyl group]

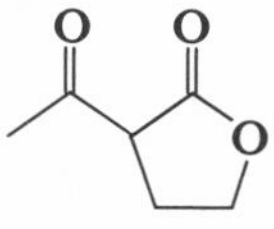

b [structure: 4-(methoxycarbonyl)cyclohexanone with CO$_2$CH$_3$]

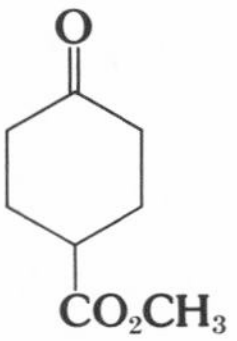

c C$_2$H$_5$CH(OH)CH(CH$_3$)CH$_2$CH=CH$_2$
   mixture of diastereoisomers

d (CH$_3$)$_2$C(OH)CH$_2$CH$_2$C(OH)(CH$_3$)$_2$

e (ϕCH$_2$CH$_2$)$_2$CHNH$_2$

f [structure: indan-1-one]

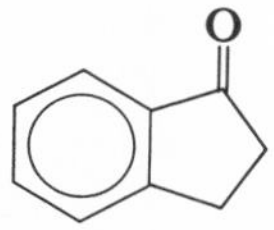

g [structure: indan-2-one]

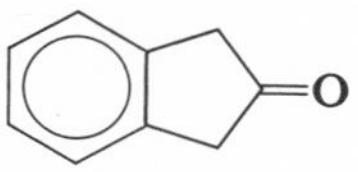

h (CH$_3$)$_3$CC(OH)(CH$_3$)$_2$

i [structure: 1,1,3,3-tetramethylcyclohexane]

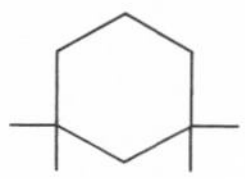

19·17
a Using reactants having no more than six carbon atoms, devise an alternative to the synthesis of 1,4,6-trimethylnaphthalene in equations 19.53 to 19.55.
b Using acyclic reactants having no more than six contiguous carbon atoms, devise a synthesis of 2-*t*-butylcyclopentanone.

TABLES

TABLE I   *Useful reactions for the selective introduction of functional groups*   670

TABLE II   *Commonly used carbon-carbon bond-forming reactions*   680

TABLE III   *Acidity constants and pK_a values*   685

TABLE I USEFUL REACTIONS FOR THE SELECTIVE INTRODUCTION OF FUNCTIONAL GROUPS

type of reaction and general equation	characteristic features	effect on carbon skeleton	chapter reference
alkyl halides			
direct halogenation of alkanes: $R-H + X_2 \xrightarrow{\Delta \text{ or } h\nu} R-X + HX$	poor selectivity, gives mixtures of products; $3°\,C-H > 2°\,C-H > 1°\,C-H$	none	2
addition of HX to alkenes (and alkynes): $R_2C=CR_2 + HX \longrightarrow H-\underset{R}{\underset{\|}{C}}-\underset{R}{\underset{\|}{C}}-X$	good regioselectivity (Markovnikov rule)	may cause rearrangement	5
addition of X_2 to alkenes (and alkynes): $R_2C=CR_2 + X_2 \longrightarrow X-\underset{R}{\underset{\|}{C}}-\underset{R}{\underset{\|}{C}}-X$	selective for vicinal dihalides; trans addition	usually none	5
hydroxyl substitution in alcohols: $ROH + HX$ (or PX_3 or SOX_2) $\longrightarrow R-X$	S_N2 or S_N1 mechanism, depending on R and conditions	may cause rearrangement	8
displacement of sulfonate esters: $ROSO_2R' + NaX \longrightarrow R-X + R'SO_2Na$	inversion of configuration (S_N2) for $R = 2°$ or $1°$	usually none	8
decarboxylative halogenation of acids (M = Ag, Hg, Tl, or Pb): $RCO_2M + X_2 \longrightarrow R-X + CO_2 + MX_n$	not stereospecific	loss of one carbon atom	14
aryl halides			
direct halogenation of aromatic systems: $Ar-H + X_2 \xrightarrow{Fe^{+3}} Ar-X + HX$	moderate to good selectivity, depending on substituents	none	9, 10, 15
substitution of aryl diazonium salts: $ArN_2^{\oplus} X^{\ominus} + Cu_2X_2 \longrightarrow Ar-X + N_2$	good selectivity	none	11

TABLE I

alkenes

dehydration of alcohols:

$RCH_2CH(OH)R + H_2SO_4(H_3PO_4) \xrightarrow{-H_2O, \Delta} RCH=CHR$

regioselective (Saytzeff rule); not usually stereospecific — may cause rearrangement — 8

dehydrohalogenation of alkyl halides:

$RCH_2CHXR + KOH \xrightarrow{alcohol} RCH=CHR + KX$

regioselective (Saytzeff rule); usually trans elimination — none — 3

dehalogenation of vicinal dihaloalkanes:

$RCHXCHXR + Zn \longrightarrow RCH=CHR + ZnX_2$

usually trans elimination — none — 3

reduction of alkynes:

$RC{\equiv}CR + H_2 \xrightarrow[\text{quinoline poison}]{Pd, BaSO_4}$ (cis alkene with R, R on same side)

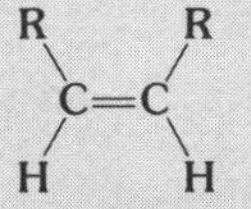

stereospecific to cis products — none — 6

$RC{\equiv}CR + 2Li \xrightarrow{NH_3(l)} $ (or Na) (trans alkene)

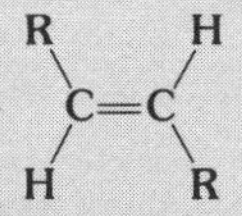

stereospecific to trans products — none — 6

pyrolysis of quaternary ammonium hydroxide salts and amine oxides:

$RCH_2CH_2\overset{\oplus}{N}(CH_3)_3 OH^{\ominus} \xrightarrow{\Delta} RCH=CH_2 + (CH_3)_3N + H_2O$

regioselective (Hofmann rule) — loss of alkyl groups on nitrogen — 11

$RCH_2CH_2\overset{\oplus}{N}(CH_3)_2\overset{\ominus}{O}\text{CH}_3 \xrightarrow{\Delta} RCH=CH_2 + (CH_3)_2NOH$

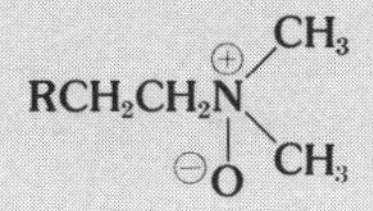

stereospecific cis elimination — loss of alkyl groups on nitrogen — 11

pyrolysis of esters:

$RCH_2CH_2OCOCH_3 \xrightarrow{\Delta} RCH=CH_2 + CH_3CO_2H$

stereospecific cis elimination — loss of carboxyl moiety — —

Wittig reaction:

$RCH=P\phi_3 + R'CHO \longrightarrow RCH=CHR'$

variable stereoselectivity — number of carbon atoms increases — 12

TABLE I USEFUL REACTIONS FOR THE SELECTIVE INTRODUCTION OF FUNCTIONAL GROUPS (continued)

type of reaction and general equation	characteristic features	effect on carbon skeleton	chapter reference
alkynes			
dehydrohalogenation of vicinal dihaloalkanes: $RCHXCHXR + 2NaNH_2 \longrightarrow RC\equiv CR + 2NaX$	requires a strong base; may give diene byproducts	none	6
acetylide displacement of alkyl halides: $RC\equiv CNa + R'X \longrightarrow RC\equiv CR' + NaX$	R' must be 1° or 2°; may give alkene byproducts	none	6
alcohols			
hydrolysis of alkyl halides: $R-X + H_2O + NaOH \longrightarrow ROH + NaX$	S_N2 or S_N1 mechanisms; may give alkene byproducts	S_N1 pathway; may undergo rearrangement	3
hydration of alkenes: $RCH=CHR + H_3O^{\oplus} \longrightarrow RCH_2CH(OH)R$	acid catalyzed; regioselective (Markovnikov rule)	may undergo rearrangement	5
hydroxylation of alkenes: $RCH=CHR + OsO_4 \longrightarrow \left[\begin{array}{c}\text{osmium}\\\text{complex}\end{array}\right] \xrightarrow{NaHSO_3} \begin{array}{c} OH\ \ OH \\ \|\ \ \ \| \\ RCH-CHR \end{array}$	selective for vicinal glycols; stereospecific cis addition; OsO_4 better than $KMnO_4$	none	5
reduction of carbonyl functions: $R_2C=O + LiAlH_4$ (or $NaBH_4$) $\longrightarrow R_2CHOH$ $RCO_2H + LiAlH_4 \longrightarrow RCH_2OH$	$LiAlH_4$ reduces all carbonyl groups; $NaBH_4$ does not reduce esters or acids	none	12
oxidation of alkylboranes: $RCH=CH_2 + B_2H_6 \longrightarrow RCH_2CH_2-B\begin{array}{c}R'\\ \\R'\end{array}$ $RCH_2CH_2-BR'_2 + H_2O_2 \xrightarrow{NaOH} RCH_2CH_2OH$	anti-Markovnikov orientation; cis addition retention of configuration in oxidation	none	5
ether cleavage: $ROR' + HX \longrightarrow ROH + R'X$ (X = Br or I)	limited usefulness; best for R' = CH_3, $(CH_3)_3C$, ϕCH_2	may cause rearrangement	8

TABLE I

Reaction			
ring opening of epoxides: $\underset{RHC-CHR}{\overset{O}{\triangle}} \xrightarrow{LiAlH_4} RCH_2CH(OH)R$ $\xrightarrow{H_3O^+} RCH(OH)CH(OH)R$	attack at $1°C > 2°C > 3°C$ specific for vicinal glycols	none may undergo rearrangement	12 8
Grignard addition to carbonyl and epoxide functions: $2RMgX + R'CO_2CH_3 \rightarrow R'R_2COH + CH_3OH$ $RMgX + \underset{\triangle}{\overset{O}{\triangle}} \rightarrow RCH_2CH_2OH$	$CH_2O \rightarrow 1°$ alcohol, $RCHO \rightarrow 2°$ alcohol, $R_2CO \rightarrow 3°$ alcohol, $RCO_2R' \rightarrow 3°$ alcohol	number of carbon atoms increases	12 14
phenols			
substitution of activated aryl halides: $O_2N-Ar-X + NaOH \xrightarrow{\Delta} O_2N-Ar-OH + NaX$	activated by ortho or para nitro groups and other electron-withdrawing functions	none	10
substitution of aryl diazonium salts: $Ar-N_2^+ X^- + H_3O^+ \longrightarrow ArOH + N_2$	good selectivity	none	11
alkali fusion of aryl sulfonate salts: $ArSO_3Na + 2NaOH \xrightarrow{\Delta} ArONa + Na_2SO_3$	severe reaction conditions	none	—
ethers			
Williamson synthesis: $R-X + R'ONa \longrightarrow ROR' + NaX$	R must be 1° or 2°; may give alkene byproducts	none	8
dehydration of alcohols: $2ROH + H_2SO_4 \xrightarrow{\Delta} ROR + H_2O$	may give alkene byproducts	may undergo rearrangement	8
addition of alcohols to alkenes: $(CH_3)_2C=CH_2 + ROH \xrightarrow{H^+} R-O-C(CH_3)_3$	regioselectivity (Markovnikov rule)	may undergo rearrangement	5

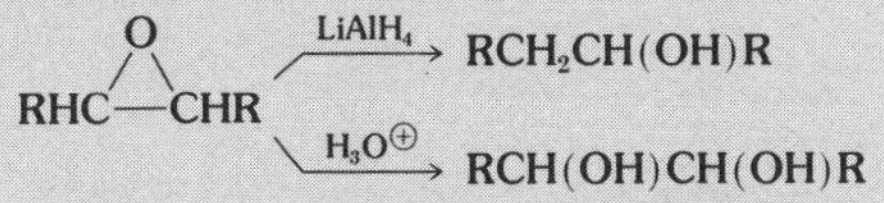

TABLE I USEFUL REACTIONS FOR THE SELECTIVE INTRODUCTION OF FUNCTIONAL GROUPS (continued)

type of reaction and general equation	characteristic features	effect on carbon skeleton	chapter reference
ethers			
peracid epoxidation: $R_2C{=}CR_2 + ArCO_3H \longrightarrow R_2C\overset{O}{\overset{\diagup\diagdown}{}}CR_2 + ArCO_2H$	cis addition	none	14
aldehydes			
oxidation of primary alcohols: $RCH_2OH + CrO_3 + \text{pyridine} \longrightarrow R{-}CHO$	overoxidation to acids may occur	none	8
reduction of acyl chlorides: $RCOCl + H_2 \xrightarrow{\text{Pd (poisoned)}} R{-}CHO$ $RCOCl + LiAlH(OC_4H_9)_3 \longrightarrow R{-}CHO$	overreduction to alcohols may occur	none	14
hydroboration and oxidation of terminal alkynes: $RC{\equiv}CH + [(CH_3)_2CHCH(CH_3)]_2BH \longrightarrow RCH{=}CHB(C_5H_{11})_2 \xrightarrow{NaOH, H_2O_2} RCH_2CHO$	good selectivity	none	6
Gatterman-Koch synthesis: $ArH + CO + HCl \xrightarrow{AlCl_3} ArCHO$	variable selectivity	gain of one carbon atom	14
oxidative cleavage of certain alkenes: $\begin{array}{c} RCH \\ \| \\ CHR \end{array} \xrightarrow{O_3} [\text{ozonide}] \xrightarrow{Zn, H_2O} 2RCHO$ $\xrightarrow{OsO_4} \xrightarrow{NaHSO_3} RCH(OH)CH(OH)R \xrightarrow{HIO_4} 2RCHO$	generally useful	fragmentation or ring cleavage	5
aldol condensation: $2RCH_2CHO \underset{}{\overset{NaOH}{\rightleftharpoons}} RCH_2CH(OH)CHRCHO$	α hydrogen atoms necessary	number of carbon atoms increases	12

ketones

oxidation of secondary alcohols: $R_2CHOH \xrightarrow{CH_3CO_2H}{CrO_3} R_2C{=}O$	generally very useful	none	8
hydration of alkynes (also hydroboration route): 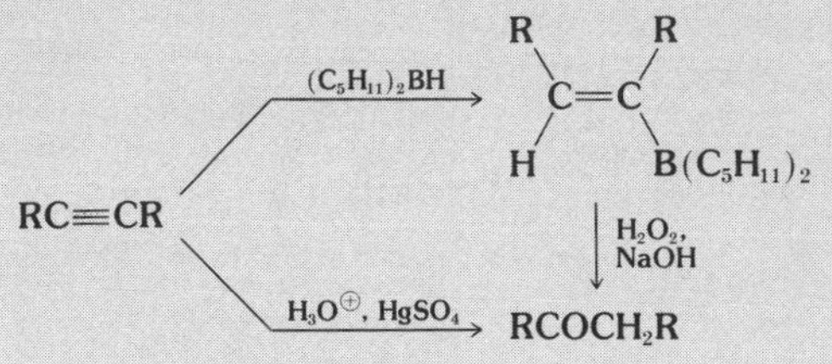 $RC{\equiv}CR \xrightarrow{(C_5H_{11})_2BH}$ $\underset{H}{\overset{R}{C}}{=}\underset{B(C_5H_{11})_2}{\overset{R}{C}} \xrightarrow{H_2O_2, NaOH} RCOCH_2R$ $\xrightarrow{H_3O^+, HgSO_4}$	regioselective (Markovnikov rule)	none	6
alkylation of ketone enolate anions: $R{-}X + H_2C{=}\underset{R'}{\overset{O^-}{C}} \longrightarrow RCH_2COR' + NaX$	R must be 1° or 2°; condensation byproducts	number of carbon atoms increases	11
acylation of organocadmium reagents: $R_2Cd + 2R'COCl \longrightarrow 2RCOR' + CdCl_2$	best for R = 1° and 2°	number of carbon atoms increases	14
Grignard addition to nitriles: $RMgX + R'CN \longrightarrow$ adduct $\xrightarrow{H_2O} RCOR'$	may give 3° alcohols as byproducts	number of carbon atoms increases	14
Friedel-Crafts acylation: $ArH + RCOCl \xrightarrow{AlCl_3} ArCOR$	variable selectivity; anhydrides may also be used	number of carbon atoms increases	14
pinacol rearrangement: $R_2C(OH)C(OH)R_2 \xrightarrow{H^+} R_3CCOR + H_2O$	limited to vicinal glycols	rearrangement	15
oxidative cleavage of certain alkenes (or corresponding glycols): $R_2C{=}CR_2 + O_3 \longrightarrow$ [ozonide] $\xrightarrow{Zn} 2R_2C{=}O$	generally useful	fragmentation or ring cleavage	5

TABLE I USEFUL REACTIONS FOR THE SELECTIVE INTRODUCTION OF FUNCTIONAL GROUPS (continued)

type of reaction and general equation	characteristic features	effect on carbon skeleton	chapter reference
carboxylic acids			
hydrolysis of acid derivatives: (Z = X, OR', or NR'$_2$) $$RCOZ + H_2O \xrightarrow[\Delta]{H^{\oplus} \text{ or } OH^{\ominus}} RCO_2H + ZH$$	acyl halides > anhydrides > esters > amides ≈ nitriles	none	14
oxidation of primary alcohols: $$RCH_2OH \xrightarrow{[O]} RCHO \xrightarrow{[O]} RCO_2H$$	aldehydes > alcohols	none	8, 11
carbonation of Grignard reagents (or alkyl lithiums): $$RMgX \xrightarrow{CO_2} RCO_2MgX \xrightarrow{H_3O^{\oplus}} RCO_2H$$	generally useful	number of carbon atoms increases	11
malonic ester alkylation and hydrolysis: $$RX + NaCH(CO_2C_2H_5)_2 \longrightarrow RCH(CO_2C_2H_5)_2$$ $$RCH(CO_2C_2H_5)_2 \xrightarrow[\Delta]{H_3O^{\oplus}} RCH_2CO_2H + CO_2$$	R must be 1° or 2°	number of carbon atoms increases	14
cyanohydrin formation and hydrolysis: 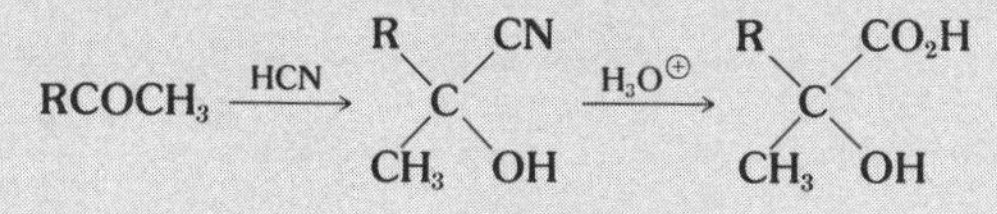 $$RCOCH_3 \xrightarrow{HCN} \underset{CH_3 \quad OH}{\underset{|\quad\;\;\;|}{R-C-CN}} \xrightarrow{H_3O^{\oplus}} \underset{CH_3 \quad OH}{\underset{|\quad\;\;\;|}{R-C-CO_2H}}$$	α hydroxy acids only; subject to steric hindrance	gain of one carbon atom	11
oxidative cleavage of alkenes (and alkynes): $$RCH=CHR \xrightarrow{[O]} 2RCO_2H$$	generally useful	fragmentation or ring cleavage	5, 6
side-chain oxidation of arenes: $$ArCH_2CH_2CH_3 \xrightarrow[\Delta]{KMnO_4} ArCO_2H$$	severe reaction conditions	number of carbon atoms may decrease	9
Baeyer-Villiger oxidation of ketones: $$RCOR + CF_3CO_3H \longrightarrow RCO_2R \xrightarrow{H_3O^{\oplus}} RCO_2H$$	variable selectivity (alkenes are epoxidized)	a carbon-carbon bond is cleaved	15
haloform oxidation of methyl ketones: $$RCOCH_3 + X_2 \xrightarrow{NaOH} RCO_2H + HCX_3$$	specific for this function	loss of one carbon atom	—

TABLE I

acyl halides

substitution reactions with inorganic acid halides:
$RCO_2H + PX_3$ (or SOX_2) $\longrightarrow$ RCOX   generally useful   none   14

anhydrides

dehydration of carboxylic acids:

$2RCO_2H + P_2O_5 \xrightarrow{\Delta}$ R—C(=O)—O—C(=O)—R   for symmetrical anhydrides   none   14

acylation of carboxylate salts:
$RCO_2Na + R'COCl \longrightarrow RCO_2COR' + NaCl$ generally useful   none   14

esters

Fischer esterification of acids:
$RCO_2H + R'OH \xrightleftharpoons{H^{\oplus}} RCO_2R' + H_2O$ retarded by steric hindrance   none   14

addition of acids to alkenes:
$RCO_2H + (CH_3)_2C{=}CH_2 \xrightarrow{H^{\oplus}} RCO_2C(CH_3)_3$ best for branched alkenes; regioselective (Markovnikov rule); may undergo rearrangement   none   14

displacement of alkyl halides by carboxylate:
$RCO_2Na + R'X \longrightarrow RCO_2R' + NaX$ best for R' = 1° ro 2°; S_N2 mechanism   none   14

diazomethane methylation:
$RCO_2H + CH_2N_2 \longrightarrow RCO_2CH_3 + N_2$ general for methyl esters   none   14

amides

acylation of amines:
$R{-}NH_2 + R'COX \longrightarrow RNHCOR'$ generally useful; anhydrides, esters, or ketenes may also be used   none   14

TABLE I USEFUL REACTIONS FOR THE SELECTIVE INTRODUCTION OF FUNCTIONAL GROUPS (continued)

type of reaction and general equation	characteristic features	effect on carbon skeleton	chapter reference
nitriles			
displacement reactions with cyanide ion: $R\text{—}X \text{ (or } ROSO_2R') \xrightarrow{NaCN} RCN + NaX$	best for $R' = 1°$ or $2°$; S_N2 mechanism	gain of one carbon atom	3
dehydration of primary amides: $RCONH_2 + P_2O_5 \xrightarrow{\Delta} R\text{—}CN + H_2O$	generally useful	none	14
Sandmeyer reaction of aryl diazonium salts: $ArN_2^{\oplus} X^{\ominus} + Cu_2(CN)_2 \longrightarrow Ar\text{—}CN + N_2$	generally useful for aryl nitriles	gain of one carbon atom	11
amines			
direct alkylation of ammonia and/or amines: $R\text{—}X + 2NH_3 \longrightarrow R\text{—}NH_2 + NH_4X$	usually poor; further alkylation occurs	none	11
Gabriel phthalimide synthesis: 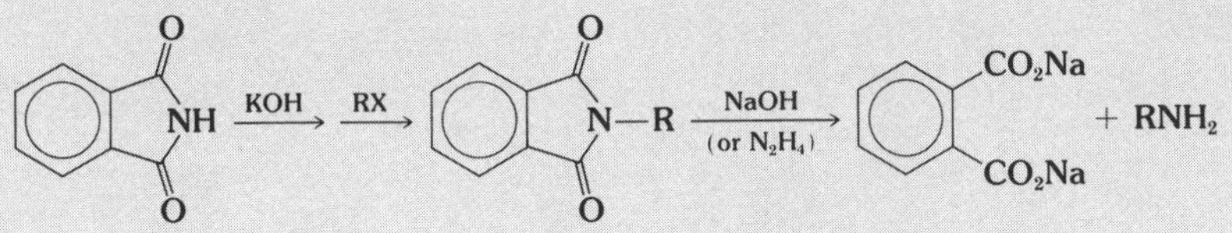 (phthalimide) $\xrightarrow{KOH}$ (K-phthalimide) $\xrightarrow{RX}$ (N-R phthalimide) $\xrightarrow[\text{(or } N_2H_4)]{NaOH}$ phthalate^{2-} + RNH_2	for primary amines only; R must be 1° or 2°	none	14
reductive alkylation of ammonia and primary amines: $R\text{—}NH_2 + R'CHO \longrightarrow [RN\text{=}CHR'] \xrightarrow{H_2, Pd} RNHCH_2R'$	useful for aldehydes and simple ketones	none	12
reduction of amides, nitriles, oximes, azides, and nitro functions:			
$RCONR'_2 + LiAlH_4 \longrightarrow RCH_2NR'_2$	for 1°, 2°, and 3° amines	none	14
$RCN + LiAlH_4 \longrightarrow RCH_2NH_2$	for 1° amines only	none	14
$R_2C\text{=}NOH + LiAlH_4 \longrightarrow R_2CHNH_2$	for 1° amines (limited)	none	—
$RX + NaN_3 \longrightarrow R\text{—}N_3 \xrightarrow{H_2, Pd} R\text{—}NH_2$	for 1° amines only	none	11
$Ar\text{—}NO_2 \xrightarrow{Sn, HCl} Ar\text{—}NH_2$	generally useful for 1° amines; $Fe + H_3O^{\oplus}$ or $Pd + H_2$ may also be used	none	11

Hofmann, Curtius, and Schmidt reactions:

$RCONH_2 + Br_2 + NaOH \longrightarrow R-NH_2 + NaHCO_3$ — for 1° amines only — loss of one carbon atom — 15

$RCON_3 + H_3O^{\oplus} \xrightarrow{\Delta} R-NH_2 + CO_2$ — for 1° amines only

quaternary ammonium salts

$R_3N + R'X \longrightarrow R_3NR'{}^{\oplus} X^{\ominus}$ — best for R' = 1° or 2° — no change in R or R' — 11

imines

$R-NH_2 + ArCHO \rightleftharpoons ArCH=N-R + H_2O$ — imines easily hydrolyzed — no change in R or Ar — 12

enamines

$R_2CHCOR' + R''_2NH \rightleftharpoons R_2C=C \begin{matrix} R' \\ NR''_2 \end{matrix} + H_2O$ — enamines easily hydrolyzed — no change in R groups — 12

amine oxides

$R_3N + H_2O_2 \text{ (or } R'CO_3H) \longrightarrow R_3\overset{\oplus}{N}-\overset{\ominus}{O}$ — general for 3° amines — none — 11

TABLE II COMMONLY USED CARBON-CARBON BOND-FORMING REACTIONS

type of reaction and general equation	characteristic features	chapter reference
reactions of strong carbon nucleophiles with electrophilic functions		
1,2 addition to carbonyl and nitrile functions		
Grignard and alkyl lithium reagents: $R-MgX$ (or Li) + $R'COR'' \longrightarrow RR'R''COM \xrightarrow{H_2O} RR'R''COH$ $R-MgX + R'C{\equiv}N \longrightarrow RR'C{=}NMgX \xrightarrow{H_2O} RCOR'$	very useful general reaction; gives 1° alcohols from formaldehyde, 2° alcohols from aldehydes, 3° alcohols from ketones, ketones from nitriles, and carboxylic acids from CO_2	12, 14
sodium acetylides: $RC{\equiv}CNa + R'COR'' \longrightarrow RC{\equiv}CC(OH)R'R''$	limited to terminal alkynes	6
aldol condensations and related reactions: $2RCH_2CHO \xrightarrow{OH^{\ominus}} RCH_2CH(OH)CHRCHO$ $CH_3CO_2C_2H_5 \xrightarrow[-78°]{R_2NLi} \left[CH_2{=}C{<}^{OLi}_{OC_2H_5}\right] \xrightarrow[-78°]{R'_2C{=}O} R'_2C(OH)CH_2CO_2C_2H_5$	intermolecular crossed condensations poor if components can serve as either donor or acceptor	12
1,2 addition followed by elimination or displacement		
Grignard and alkyl lithium reagents with esters: $2RMgX$ (or Li) + $R'CO_2R'' \longrightarrow R''OMgX + R_2R'COMgX \xrightarrow{H_2O} R_2R'COH$	good general synthesis of 3° alcohols	14
cyanohydrin formation ($CN^{\ominus}$ catalyst): $RR'CO + CN^{\ominus} \xrightarrow{HCN} RR'C{<}^{OH}_{CN} + CN^{\ominus}$	limited to R or R' = CH_3 or H and to certain cyclic ketones	12

TABLE II

organocadmium reagents and acyl halides:

$R_2Cd + 2R'COCl \longrightarrow 2R'COR + CdCl_2$

may give 3° alcohol as a byproduct; alkyl zinc reagents may also be used — 14

Claisen and Dieckmann condensations:

$2RCH_2CO_2C_2H_5 \xrightarrow{NaOC_2H_5} RCH_2COCHRCO_2C_2H_5 + C_2H_5OH$

retarded by further alkyl substitution of the α carbon atom; crossed condensations usually poor — 14

aldol condensation:

$RCOCH_3 + \phi CHO \xrightarrow{NaOR'} RCOCH=CH\phi + H_2O$

dehydration enhanced by conjugation — 12

Wittig reaction:

$RCH=P\phi_3 + R'_2CO \longrightarrow RCH=CR'_2 + \phi_3PO$

variable stereoselectivity; subject to steric hindrance — 12

1,4 addition to conjugated unsaturated systems

copper-containing organometallic reagents:

$RCH=CHCOR' + (CH_3)_2CuLi \longrightarrow CH_3(R)CHCH_2COR'$

subject to steric hindrance ($R'MgX + Cu_xX_2$ may also be used) — 15

Michael reaction:

$RCH_2NO_2 + 2CH_2=CHCN \xrightarrow{R'_2NH} \underset{NO_2}{RC(CH_2CH_2CN)_2}$

a very useful reaction; may be activated by $C=O$, $C\equiv N$, NO_2, SO_2, etc. — 15

displacement reactions

organometallic substitution of alkyl halides:

$RC\equiv CNa + R'X \longrightarrow RC\equiv CR' + NaX$

$RMgX + BrCH_2CH=CH_2 \longrightarrow RCH_2CH=CH_2$

R' must be 1°, 2°, allylic, or benzylic — 6

alkylation of enolate anion (M = Li, Na, or K):

$RCH=C\underset{R'}{\overset{O^{\ominus}M^{\oplus}}{\diagup}} + R''-X \longrightarrow RR'CHCOR' + MX$

$CH_2(CO_2C_2H_5)_2 + BrCH_2CH_2Br \xrightarrow{2NaOC_2H_5} \triangle(CO_2C_2H_5)_2$

$\phi CH_2CN + 2CH_2=CHCH_2Cl \xrightarrow{2NaOR} \phi\underset{CN}{\overset{|}{C}}(CH_2CH=CH_2)_2$

R'' must be 1°, 2°, allylic, or benzylic — 12, 14

TABLE II COMMONLY USED CARBON-CARBON BOND-FORMING REACTIONS (continued)

type of reaction and general equation	characteristic features	chapter reference
reactions of strong carbon electrophiles with nucleophilic functions		
Friedel-Crafts alkylation and acylation of aromatic systems		
acylation: $\phi\text{—H} + \text{RCOCl} \xrightarrow{\text{AlCl}_3} \phi\text{COR}$	deactivating substituents (such as NO_2) may prevent reaction	14
alkylation: $\phi\text{—H} + \text{R—X} \xrightarrow{\text{AlCl}_3} \phi\text{—R}$	anhydrides and H_3PO_4 or BF_3 may be used not as clean as acylation; R may undergo molecular rearrangement; polyalkylation may occur	9
acylation and alkylation of alkenes		
acylation: $(CH_3)_2C{=}CH_2 + \text{RCOCl} \xrightarrow{\text{AlCl}_3} (CH_3)_2CClCH_2COR$	β-chloro ketones may eliminate HCl	14
alkylation: $(CH_3)_2C{=}CH_2 + (CH_3)_3CH \xrightarrow{H_2SO_4} (CH_3)_3CCH_2CH(CH_3)_2$	technologically important	
radical coupling or addition reactions		
electrolytic oxidation of carboxylate salts (Kolbe reaction): $2RCO_2^{\ominus} \xrightarrow[-2e]{\text{anode}} [2RCO_2\cdot] \longrightarrow 2CO_2 + [2R\cdot] \longrightarrow R\text{—R}$	useful for large symmetrical molecules	16
pinacol reduction (M = Mg, Zn, or Al): $2R_2C{=}O + M \longrightarrow 2[R_2\overset{\ominus}{C}\text{—}\overset{\oplus}{O}M] \longrightarrow \underset{\text{HO OH}}{R_2C\text{—}CR_2}$	variable yields; often a side reaction in Clemmensen reduction	—
halomethane addition to alkenes: $RCH{=}CH_2 + HCX_3 \xrightarrow[\Delta]{\text{peroxides}} RCH_2CH_2\text{—}CX_3$	a radical-chain mechanism	5

Reaction	Notes	Ref.
polymerization ($Z = \phi$, Cl, —): $n\ ZCH=CH_2 \xrightarrow{\text{peroxides}} {\sim}CHZCH_2CHZCH_2{\sim}$	variable stereochemical control	5
carbenoid addition to alkenes		
carbene addition (X = H, Cl, Br): 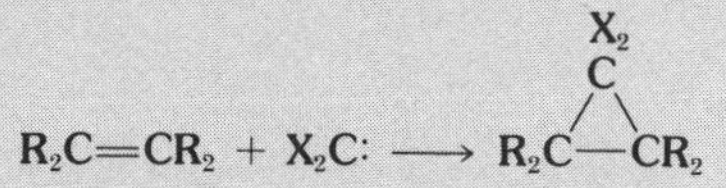 $R_2C=CR_2 + X_2C: \longrightarrow R_2C\!\!\begin{array}{c}X_2\\C\end{array}\!\!CR_2$	carbenes must be formed *in situ*	5
Simmons-Smith reaction: 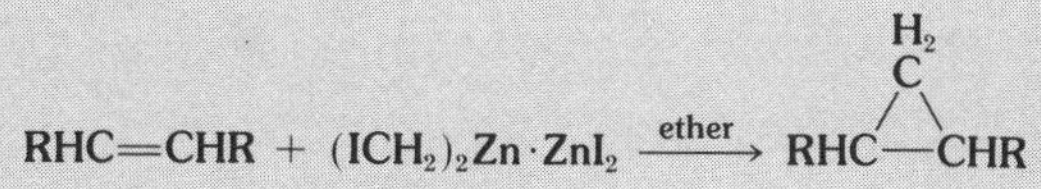 $RHC=CHR + (ICH_2)_2Zn \cdot ZnI_2 \xrightarrow{\text{ether}} RHC\!\!\begin{array}{c}H_2\\C\end{array}\!\!CHR$	generated from $CH_2I_2 + Zn(Cu)$; sensitive to steric hindrance	5
thermal and photochemical cycloaddition reactions		
Diels-Alder reaction (Z = electron-withdrawing groups) 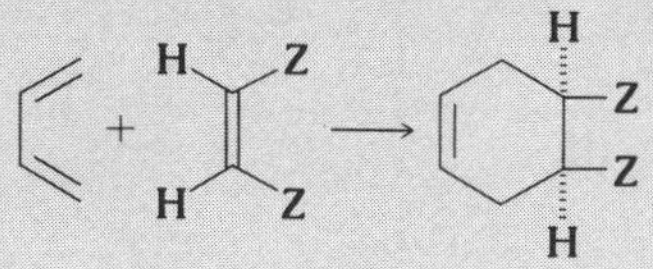	highly stereospecific	15, 16
photocycloaddition reactions 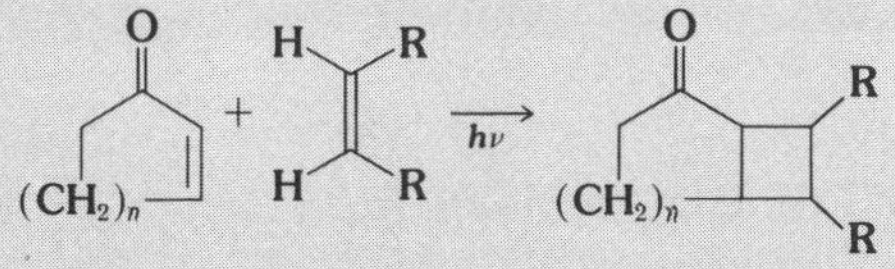	variable stereoselectivity and regioselectivity	16

TABLE II COMMONLY USED CARBON-CARBON BOND-FORMING REACTIONS (continued)

type of reaction and general equation	characteristic features	chapter reference
molecular rearrangement		
rearrangement to electron-deficient carbon atoms (carbonium ions)		
Wagner-Meerwein rearrangement: $\underset{\underset{Z}{\mid}}{R_3C}CHR \xrightarrow{-Z^{\ominus}} R_3C\overset{\oplus}{C}HR \longrightarrow R_2C\overset{\oplus}{C}HR_2 \longrightarrow$ alkenes, alcohols, etc.		15
pinacol rearrangement: $R_2C(OH)C(OH)R_2 \xrightarrow[-H_2O]{H^{\oplus}} R_2C(OH)\overset{\oplus}{C}R_2 \xrightarrow{-H^{\oplus}} RCOCR_3$		15
electrocyclic reactions		
Claisen rearrangement: 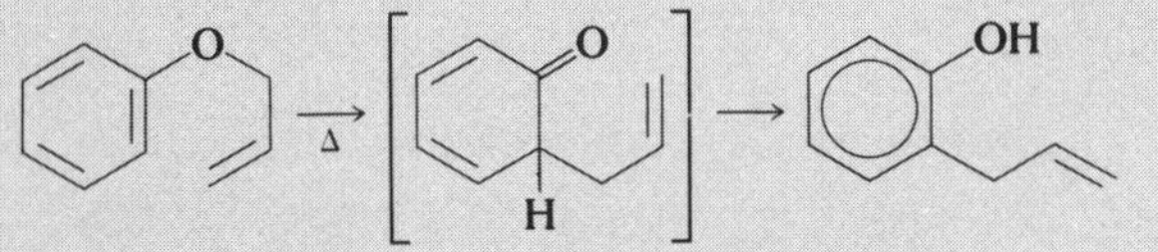		16
Cope rearrangement: 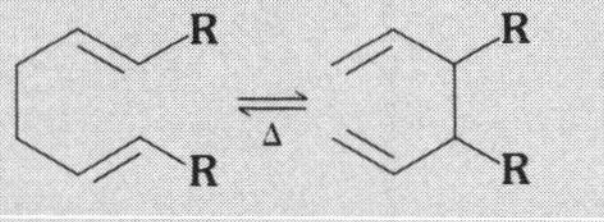		16

TABLE III ACIDITY CONSTANTS AND pK_a VALUES

acid	conjugate base	K_a	pK_a	acid	conjugate base	K_a	pK_a
$HClO_4$	$ClO_4^{\ominus}$	$\sim 10^{10}$	-10	$H_2NNH_3^{\oplus}$	H_2NNH_2	3.3×10^{-9}	8.48
$RC\equiv NH^{\oplus}$	$RC\equiv N$	$\sim 10^{10}$	-10	$(CH_3CO)_2CH_2$	$(CH_3CO)_2CH^{\ominus}$	$\sim 10^{-9}$	9.0
ϕSO_3H	$\phi SO_3^{\ominus}$	$\sim 10^7$	-7	$NH_4^{\oplus}$	NH_3	5.5×10^{-10}	9.25
HCl	$Cl^{\ominus}$	$\sim 10^7$	-7	$(CH_3)_3NH^{\oplus}$	$(CH_3)_3N$	1.67×10^{-10}	9.78
$H_3O^{\oplus}$	H_2O	55.0	-1.74	ϕOH	$\phi O^{\ominus}$	1.3×10^{-10}	9.89
CF_3CO_2H	$CF_3CO_2^{\ominus}$	1	0	$HCO_3^{\ominus}$	$CO_3^{2\ominus}$	5.6×10^{-11}	10.25
pyrrolium $\overset{\oplus}{N}H_2$	pyrrole NH	1	0	CH_3NO_2	$O_2NCH_2^{\ominus}$	6.2×10^{-11}	10.21
picric acid	picrate	0.4	0.4	$C_2H_5NH_3^{\oplus}$	$C_2H_5NH_2$	2.0×10^{-11}	10.70
CCl_3CO_2H	$CCl_3CO_2^{\ominus}$	1.7×10^{-1}	0.77	$CH_3COCH_2CO_2C_2H_5$	$CH_3COCHCO_2C_2H_5^{\ominus}$	$\sim 10^{-11}$	11.0
p-O_2N-C_6H_4-$NH_3^{\oplus}$	p-O_2N-C_6H_4-NH_2	0.1	1.0	H_2O_2	$HO_2^{\ominus}$	2.5×10^{-12}	11.60
FCH_2CO_2H	$FCH_2CO_2^{\ominus}$	2.2×10^{-3}	2.66	$CH_2(CO_2C_2H_5)_2$	$^{\ominus}CH(CO_2C_2H_5)_2$	3.16×10^{-14}	13.5
$ClCH_2CO_2H$	$ClCH_2CO_2^{\ominus}$	1.6×10^{-3}	2.80	$(H_2N)_3C^{\oplus}$	$(H_2N)_2C\!=\!NH$	2.0×10^{-14}	13.7
HF	$F^{\ominus}$	6.8×10^{-4}	3.17	H_2O	$OH^{\ominus}$	1.8×10^{-16}	15.7
HCO_2H	$HCO_2^{\ominus}$	1.7×10^{-4}	3.77	cyclopentadiene	cyclopentadienide$^{\ominus}$	$\sim 10^{-16}$	16.0
ϕCO_2H	$\phi CO_2^{\ominus}$	6.3×10^{-5}	4.20	C_2H_5OH	$C_2H_5O^{\ominus}$	$\sim 10^{-17}$	17.0
$\phi NH_3^{\oplus}$	ϕNH_2	2.5×10^{-5}	4.60	$\phi C\equiv CH$	$\phi C\equiv C^{\ominus}$	3.2×10^{-19}	18.5
CH_3CO_2H	$CH_3CO_2^{\ominus}$	1.7×10^{-5}	4.77	$(CH_3)_3COH$	$(CH_3)_3CO^{\ominus}$	$\sim 10^{-19}$	19.0
pyridinium $\overset{\oplus}{N}$-H	pyridine N	6.3×10^{-6}	5.2	CH_3COCH_3	$CH_3COCH_2^{\ominus}$	$\sim 10^{-20}$	20.0
H_2CO_3	$HCO_3^{\ominus}$	4.3×10^{-7}	6.37	$HC\equiv CH$	$HC\equiv C^{\ominus}$	$\sim 10^{-25}$	25.0
H_2S	$HS^{\ominus}$	$\sim 10^{-7}$	7.0	$CH_3CO_2C_2H_5$	$^{\ominus}CH_2CO_2C_2H_5$	$\sim 10^{-25}$	25.0
p-O_2N-C_6H_4-OH	p-O_2N-C_6H_4-$O^{\ominus}$	6.5×10^{-8}	7.19	CH_3CN	$^{\ominus}CH_2CN$	$\sim 10^{-25}$	25.0
ϕSH	$\phi S^{\ominus}$	$\sim 10^{-9}$	8.0	ϕNH_2	$\phi NH^{\ominus}$	$\sim 10^{-27}$	27.0
				$\phi_3 CH$	$\phi_3 C^{\ominus}$	$\sim 10^{-32}$	32.0
				CH_3SOCH_3	$CH_3SOCH_2^{\ominus}$	$\sim 10^{-33}$	33.0
				NH_3	$NH_2^{\ominus}$	$\sim 10^{-34}$	34.0
				ϕCH_3	$\phi CH_2^{\ominus}$	$\sim 10^{-35}$	35.0
				C_2H_4	$C_2H_3^{\ominus}$	$< 10^{-37}$	> 37.0
				ϕ-H	$\phi^{\ominus}$	$< 10^{-37}$	> 37.0
				C_2H_6	$C_2H_5^{\ominus}$	$< 10^{-40}$	> 40.0

ANSWERS TO TEXT PROBLEMS

CHAPTER 1

PROBLEM 1·1
Since 60.0% and 13.3% = 73.3%, we deduce there is 26.7% oxygen in the unknown compound. Thus the empirical formula is C_3H_8O:

carbon: $\dfrac{60.0}{12.01} \simeq 5.0 \qquad \simeq 3$

hydrogen: $\dfrac{13.3}{1.008} \simeq 13.3 \times \dfrac{1}{1.66} \simeq 8$

oxygen: $\dfrac{26.7}{15.99} \simeq 1.66 \qquad \simeq 1$

PROBLEM 1·2
Consider a proton exhibiting a resonance signal in the middle of the magnetic field range being scanned by a pmr spectrometer. If small local fields opposing the large applied field are generated at this proton, then the applied field must be increased in order to achieve a resonance condition. We say that such a proton is *shielded* from the applied field. Similarly, if small local fields reinforce the applied field, resonance will be observed at a lower applied field. Such protons are said to be *deshielded*.

PROBLEM 1·3
One way of proving the handshake problem is to reason from the first encounter. At this time two individuals have shaken hands an odd number of times (once each). All subsequent handshakes fall into one of three classes: (1) two odds shake and are converted to two evens, (2) two evens shake and are converted to two odds, (3) an odd and even shake and become an even and odd, respectively. Thus the even number of odd handshakes is never lost.
a If we think of the electron sharing that constitutes a covalent bond as an atomic handshake (a double bond would be two handshakes between the same atoms), we can classify the common light atoms according to the odd or even nature of their valency:

even-valenced elements: C(4), O(2), S(2)
odd-valenced elements: H(1), F, Cl, Br, I(1), N(3)

Stable compounds (those in which all valencies are satisfied) composed only of C, H, O, and S incorporate one kind of odd-valenced element, hydrogen. Consequently, the number of hydrogen atoms in such compounds must be even. The

688   ANSWERS TO TEXT PROBLEMS

introduction of other odd valency elements such as halogen or nitrogen would allow the number of hydrogen atoms to be odd or even as long as the total number of odd valence atoms remains even.

c   Stable formulas: $C_5H_8S_2$, $C_4H_7NO_2$, $C_{15}H_{25}O_2Cl_3$, and $C_{11}H_{19}N_2Cl$
    Unstable formulas: $C_3H_5N_2O$ and $C_{12}H_{20}NS$

PROBLEM 1·4

The formula for dimethyl sulfoxide is

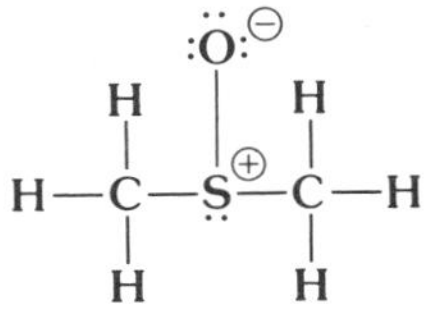

The following are isomer structures:

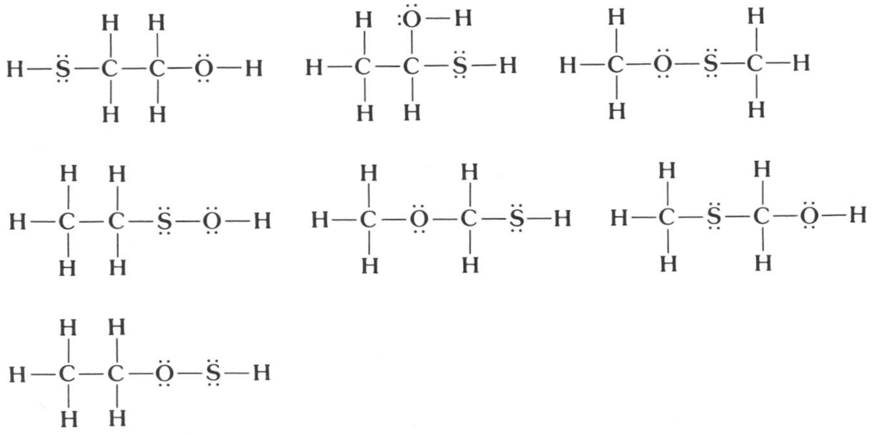

PROBLEM 1·5

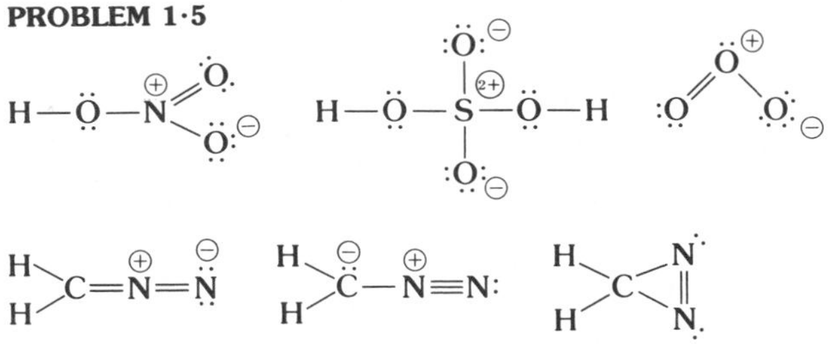

PROBLEM 1·6

a

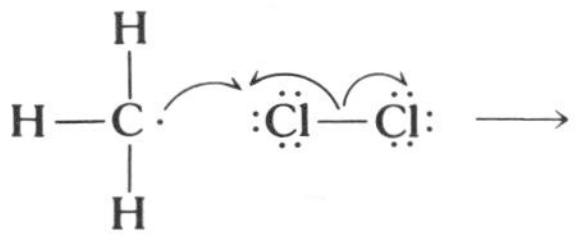

(sometimes abbreviated
$H_3C \cdot \frown Cl \frown Cl \longrightarrow$)

b

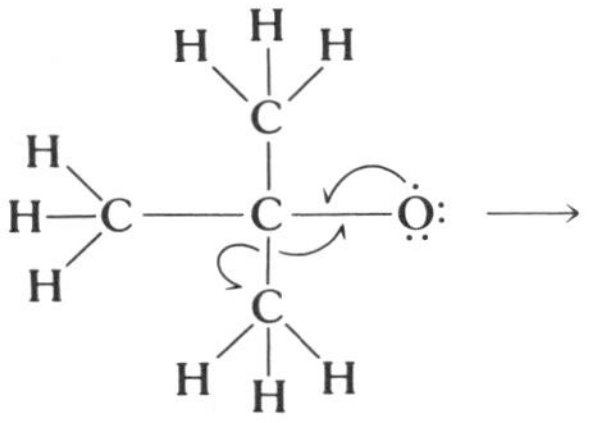

c

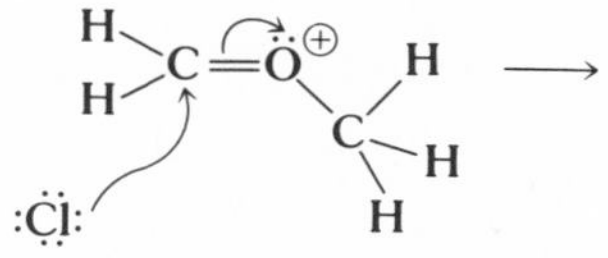

[structure showing H₂C=Ö⁺—H with CH₃ and :Cl:⁻ attacking]

d

[structure showing Br—CH₂—CH⁻—C(=O)—CH₂—H]

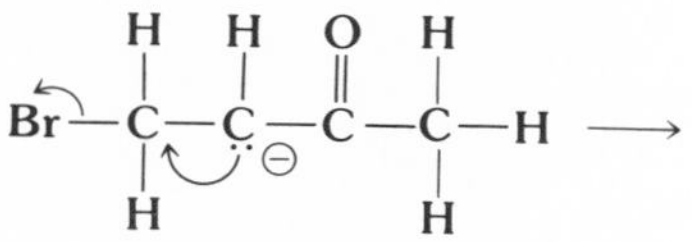

CHAPTER 2

PROBLEM 2·1
Structures a and e; b and g; c and h; and d and f.

PROBLEM 2·2
Staggered conformations:

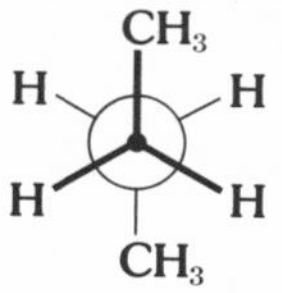

[Newman projection I: CH₃ top, H/H sides, CH₃ bottom between H's]

[Newman projection II (skew): CH₃ top-left, CH₃ top-right, H/H sides, H bottom]

I 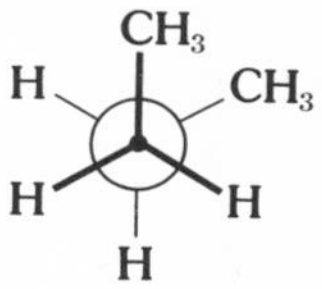   II

Eclipsed conformations: (called a *skew* conformation)

[Newman projection III: H₃C/CH₃ top eclipsed, H/H sides, H/H bottom]

[Newman projection IV: H₃C/H top eclipsed, H/H sides, H/CH₃ bottom]

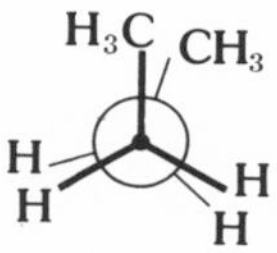

III 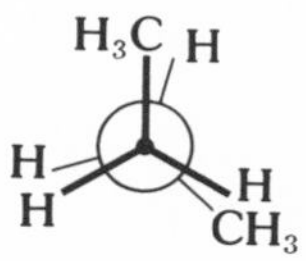   IV

Potential energy: I < II < IV < III

PROBLEM 2·3
Structures a and c; d and g are identical conformers. Structures a and b are different conformers of the same compound. Structures a and f; d and h are conformers of constitutional isomers.

PROBLEM 2·4
boiling points: $C_2H_5OH > CH_3OCH_3 > C_3H_8$ (78°, −24°, −45°, respectively)

water solubility: C_2H_5OH and CH_3OCH_3 more soluble than C_3H_8 (oxygen serves as an acceptor for hydrogen bonds)

PROBLEM 2·5

$2C_2H_2 + 5O_2 \longrightarrow 4CO_2 + 2H_2O$ or $C_2H_2 + \frac{5}{2}O_2 \longrightarrow 2CO_2 + H_2O$

$\Sigma_r = 2C-H(197.4) + C\equiv C(199.6) + \frac{5}{2}O=O(297.7) = 694.7$ kcal/mole

$\Sigma_p = 4C=O(768) + 2O-H(221.2) = 989.2$ kcal/mole

$\Delta H = 694.7 - 989.2 = -294.5$ kcal/mole (experimental $\Delta H = -300$)

PROBLEM 2·6
Begin by calculating the expected heat of reaction:

$\Sigma_r = 4 \times 98.7 + 2 \times 192.0 = 778.8$
$\Sigma_p = 3 \times 98.7 + 85.5 + 110.6 + 255.8 = 748.0$
$\Delta H = 778.8 - 748.0 = +30.8$ kcal/mole (strongly endothermic)

Even under the best of conditions, a great deal of thermal energy would have to be introduced in order to force this reaction to proceed.

PROBLEM 2·7

a The C—H bond energy given in Table 2.4 is an *average* bond energy. The remaining three C—H bonds in methyl will have bond-dissociation energies that, together with 102, will average to 98.7.

b $4 \times 98.7 = 394.8$ kcal/mole = total C—H bond energy in CH_4
$102 + 87 + 125 = 314$
$394.8 - 314 = 80.8$ kcal/mole = ΔH for remaining C—H

PROBLEM 2·8

Both n-pentane and neopentane have four C—C bonds; consequently, this contribution to the heat content of these isomers will be equal. The C—H bond contributions will be

neopentane: 12 primary C—H bonds = $12 \times 97 = 1164$ kcal/mole
n-pentane: 6 primary C—H bonds + 6 secondary C—H bonds = $582 + 570$
                = 1152 kcal/mole

The isomer with the greater total bond energy (neopentane) will have the lower potential energy and will therefore have a smaller heat of combustion:

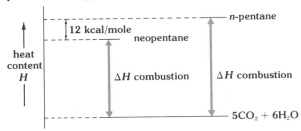

PROBLEM 2·9

An excess of chlorine over methane in the reactant mixture.

PROBLEM 2·10

a

	number of isomers		
	CH_3Cl	CH_2Cl_2	$CHCl_3$
square planar	1	2	1
pyramidal	1	2	1
tetrahedral	1	1	1

b Methane and its chlorine substitution products must have a tetrahedral configuration.

PROBLEM 2·11

From the bond energies in Table 2.4:

X = F $\Delta H = -165.3$ kcal/mole
X = Br $\Delta H = -10.7$ kcal/mole
X = I $\Delta H = +12.4$ kcal/mole

PROBLEM 2·12

$Cl\cdot + CH_4 \longrightarrow H—Cl + CH_3\cdot$ $\Delta H = 98.7 - 103.2 = -4.5$
$CH_3\cdot + Cl_2 \longrightarrow CH_3—Cl + Cl\cdot$ $\Delta H = 58.0 - 81.0 = -23.0$
$CH_4 + Cl_2 \longrightarrow HCl + CH_3Cl$ $\Delta H = -27.5$ kcal/mole

PROBLEM 2·13

$Cl\cdot + CH_4 \longrightarrow CH_3—Cl + H\cdot$ $\Delta H = +17.7$
$H\cdot + Cl_2 \longrightarrow H—Cl + Cl\cdot$ $\Delta H = -45.2$
$CH_4 + Cl_2 \longrightarrow HCl + CH_3Cl$ $\Delta H = -27.5$ kcal/mole

ANSWERS TO TEXT PROBLEMS   691

Since the first step of this chain mechanism is substantially endothermic, this mechanism is not as feasible as that given in Problem 2·12.

PROBLEM 2·14

isomer   expected pmr pattern

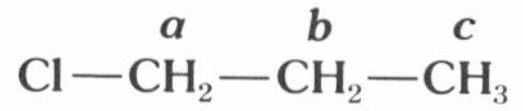

    a   b   c   three signals, area ratio 2:2:3 (a:b:c) 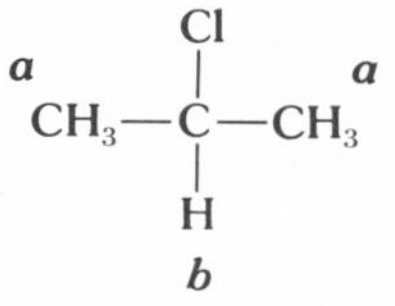
Cl—CH$_2$—CH$_2$—CH$_3$

$$\begin{array}{c} \text{Cl} \\ a \mid a \\ \text{CH}_3-\text{C}-\text{CH}_3 \\ \mid \\ \text{H} \\ b \end{array}$$ two signals, area ratio 6:1 (a:b)

PROBLEM 2·15

CH$_3$CH$_2$CHCl$_2$	CH$_3$CHClCH$_2$Cl	ClCH$_2$CH$_2$CH$_2$Cl	CH$_3$CCl$_2$CH$_3$
1,1-dichloropropane	1,2-dichloropropane	1,3-dichloropropane	2,2-dichloropropane
3 pmr signals at	3 pmr signals	2 pmr signals	1 pmr signal
(δ-1.0, 2.0, and 5.5)	(δ = 1.6, 3.7, and 4.1)		

PROBLEM 2·16

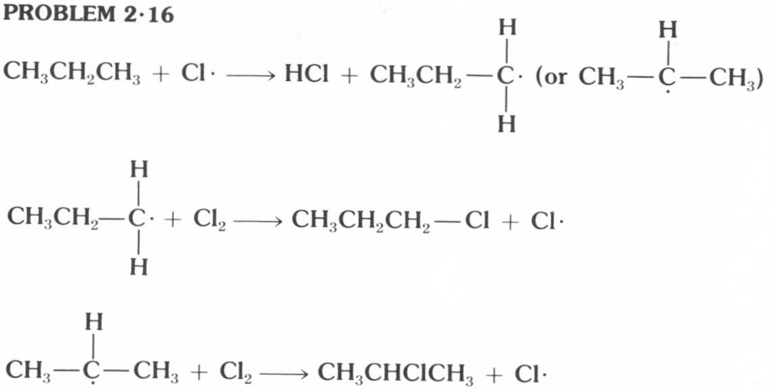

CH$_3$CH$_2$CH$_3$ + Cl· ⟶ HCl + CH$_3$CH$_2$—C·H$_2$ (or CH$_3$—C·H—CH$_3$)

CH$_3$CH$_2$—C·H$_2$ + Cl$_2$ ⟶ CH$_3$CH$_2$CH$_2$—Cl + Cl·

CH$_3$—C·H—CH$_3$ + Cl$_2$ ⟶ CH$_3$CHClCH$_3$ + Cl·

PROBLEM 2·17

a A C—H bond is broken and an H—Br bond is formed. All other bonds remain unchanged in proceeding from reactants to products. The relationship $\Sigma_r - \Sigma_p = \Delta H$ can therefore be simplified to

CH − HBr = 98.7 − 87.5 = +11.2 kcal/mole = ΔH

This reaction is endothermic.

b ΔH^* cannot be smaller than ΔH; consequently, the lowest possible ΔH^* is 11.2 kcal/mole.

PROBLEM 2·18

If a reaction has a positive ΔS^*, the transition state will be less highly ordered than the reactants. Since this is a thermodynamically favorable change, it should act to decrease the free energy of activation (ΔG^*). Such a decrease requires a negative sign for the entropy term in equation 2.21. A completely analogous argument can be made for reactions having a negative ΔS^*.

PROBLEM 2·19

Isobutane (CH$_3$)$_3$CH, has nine primary hydrogen atoms and one tertiary hydrogen atom. A straightforward statistical analysis would predict a 9:1 ratio of primary to tertiary monochlorination products. Since the observed ratio is only 2:1 (67:33), the tertiary product (2-chloro-2-methylpropane) must be favored by an energy factor of 4.5. Thus $9/1 \times 1/4.5 = 2/1$.

PROBLEM 2·20
The bonds being broken are C—H (98.7) and Cl—Cl (58.0) and the bonds being formed are C—Cl (81.0) and H—Cl (103.2):

$$\Delta H^{25°} = \Sigma_r - \Sigma_p = 156.7 - 184.2 = -27.5 \text{ kcal/mole}$$

Since the reaction is moderately exothermic, the energy factor is favorable. The mechanism suggested requires a three-body collision (a chlorine atom, a methane molecule and a chlorine molecule must all come together at the same time and in a highly ordered configuration). This is a very improbable event (ΔS^* will be negative and numerically large); consequently, the rate of such a reaction will be very small.

PROBLEM 2·21
The cyclopentane compound can exhibit cis-trans stereoisomerism. The cyclodecane compound can exist in nonidentical mirror-image configurations. This leads to a special kind of stereoisomerism which is discussed in Chapter 7.

PROBLEM 2·22
All are constitutional isomers except for the following stereoisomers: *b* and *e; d* and *g,* and *h* and *j*. The nomenclature is:

a methylcyclohexane
b *cis*-1,2-dimethylcyclopentane
c 1,1,3-trimethylcyclobutane
d *trans*-1,3-dimethylcyclopentane
e *trans*-1,2-dimethylcyclopentane
f *cis*-1-methyl-2-isopropylcyclopropane
g *cis*-1,3-dimethylcyclopentane
h *cis*-1-methyl-3-ethylcyclobutane
i *trans*-1-methyl-2-propylcyclopropane
j *trans*-1-methyl-3-ethylcyclobutane.

PROBLEM 2·23
Either of two methods may be used:

1 Determine the number of C—C bonds that must be broken in order to achieve a continuous chain of carbon atoms. This will be equal to the number of rings. Compound *b* is *pentacyclic:*

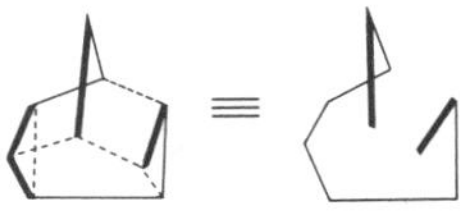

2 Every time a chain of carbon atoms joins to form a ring two hydrogen atoms must be lost; consequently, the carbon-to-hydrogen ratio is an indicator of the number of rings in a hydrocarbon. If we compare alkanes (C_nH_{2n+2}) with cyclo-alkanes (C_nH_{2n}), we see that a difference of two hydrogen atoms (for the same number of carbon atoms) indicates the presence of one ring. Thus compound *a* has the molecular formula $C_{14}H_{20}$, while an alkane with 14 carbon atoms should have 30 hydrogen atoms. The difference of 10 hydrogen atoms noted here must correspond to *five* rings.

ANSWERS TO TEXT PROBLEMS

PROBLEM 2·24

$C_8H_{16} + 12O_2 \longrightarrow 8CO_2 + 8H_2O$

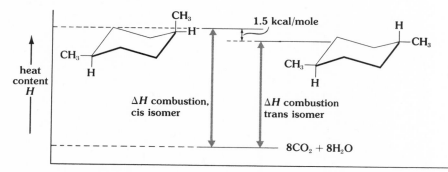

Since the cis isomer has an axial methyl group, it is less stable (has a higher heat content) than the trans isomer and therefore has a larger heat of combustion.

PROBLEM 2·25

In *trans*-decalin the two six-membered rings are fused to each other by equatorial bonds. Flipping either of the rings to another chair conformer is prohibited by the increased distance that would have to be bridged by the four-carbon chain.

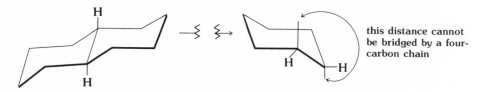

CHAPTER 3

PROBLEM 3·1

Methylene chloride has a greater density than water, whereas pentane is less dense. On mixing with water, the former will sink to the bottom of the container and the latter will float.

PROBLEM 3·2

Yes. The stretching frequency and bond energy of O—H is greater than C—H; C=O is greater than C=C or C—O.

PROBLEM 3·3

There are no C—H bonds; there may be C—F or C—Cl bonds. (The compound is CCl_4.)

PROBLEM 3·4

a
$CH_3-\underset{\underset{Br}{|}}{\overset{\overset{H}{|}}{C}}-CH_3$

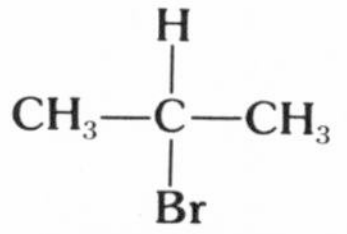

the methyl groups generate the $\delta = 1.71$ signal

b $ClCH_2CH_2CH_2Br$
  $\delta = 3.70 \quad \delta = 2.28 \quad \delta = 3.55$

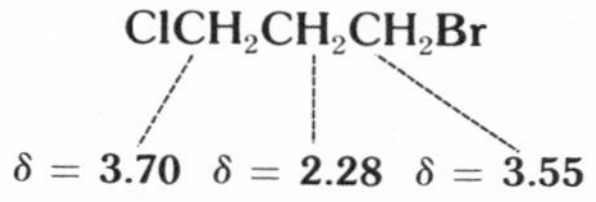

c

PROBLEM 3·5

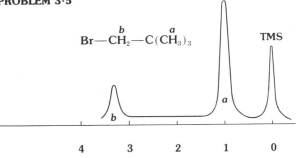

The intensity ratio of a to b is 9:2.

PROBLEM 3·6
The rate of reaction would be expected to decrease with increasing steric hindrance:

R = n-butyl > s-butyl > t-butyl

PROBLEM 3·7
The reaction of ethyl bromide with water (hydroxide ion) is proceeding by a different mechanism from the reaction involving t-butyl bromide.

PROBLEM 3·8

a reaction rate = $\dfrac{\text{moles}}{\text{liter} \times \text{sec}}$ = k_2 [moles/liter][moles/liter]

$k_2 = \dfrac{\text{liter}}{\text{moles} \times \text{sec}}$ = liters/mole/sec

b reaction rate = $\dfrac{\text{moles}}{\text{liter} \times \text{sec}}$ = k_1 [moles/liter]

$k_1 = \dfrac{1}{\text{sec}} = \text{sec}^{-1}$

c reaction rate = $k_2[C_3H_7Br][NaI]$
$1.1 \times 10^{-7} = k_2[0.01][0.01]$
$k_2 = 1.1 \times 10^{-3}$

Hence the rate of reaction is $1.1 \times 10^{-3}[0.2][0.1] = 2.2 \times 10^{-5}$ mole liter^{-1} sec^{-1}.

PROBLEM 3·9
If ethyl bromide and isopropyl bromide were reacted with the same nucleophile, the former would react at forty times the rate of the latter. In this case, however, the ethyl bromide is reacting with a nucleophile ($CH_3O^\ominus$) which is 52 times less reactive than the nucleophile ($SH^\ominus$) used in the isopropyl bromide reaction. The isopropyl bromide reaction is therefore expected to be $^{52}/_{40} = 1.3$ times as fast as the ethyl bromide reaction.

PROBLEM 3·10
A transition state is not an intermediate. The two dashed lines in the S_N2 transition state represent partial covalent bonds (approximately half a bond each); consequently, the total bond order for carbon remains 4. A pentavalent intermediate would have a carbon bond order of 5.

PROBLEM 3·11
Although the trigonal carbon of the carbonium-ion intermediate is flat, the molecular environment is not the same on both sides of this plane. Consequently, the rates of nucleophilic attack is not the same at each side and the product stereoisomer ratio will not be 50:50.

PROBLEM 3·12
The "back side" of the chlorine-bearing carbon atom is completely occupied by the carbon atoms of the two rings. An external nucleophile cannot approach in the manner required by an inversion transition state.

PROBLEM 3·13

a $(CH_3)_2CCH_2CH_3$ --- $\delta = 0.9$
    |
    OH $\delta = 1.7$
$\delta = 1.2$
$\delta = 5.0$

b $(CH_3)_3CCH_2OH$
    $\delta = 0.9$ $\delta = 5.0$
                $\delta = 3.5$

PROBLEM 3·14
a The carbonium-ion intermediate, $(CH_3)_2\overset{\oplus}{C}CH_2CH_3$, will react with the first nucleophile it encounters, in this case H_2O or C_2H_5OH. A mixture of products results.
b If the rearrangement step were not fast, the unrearranged primary carbonium ion would surely be trapped by reaction with the nucleophilic solvent, giving rise to $(CH_3)_3CCH_2OR$ as a product.

PROBLEM 3·15
If alkyl halides were in equilibrium with alkenes and hydrohalic acid, we would expect to find primary halides such as 1-chloropropane slowly isomerizing to isomeric halides such as 2-chloropropane, as a consequence of recombination (addition) of HX to the carbon-carbon double bond. This isomerization is seldom observed. Furthermore, such an elimination pathway would be relatively unaffected by variations in the base used to effect elimination, whereas the nature and concentration of the base is an important factor in many of these reactions.

PROBLEM 3·16
Such an explanation would correspond to the known reactivity order of CH groups toward halogen atoms (Chapter 2). However, this order reflects the relative stabilities of alkyl radicals, whereas the E2 elimination involves attack of a base on the CH group, and this is more likely to generate a carbanion intermediate than a radical. Carbanion stability increases in the order tertiary < secondary < primary. Steric hindrance and entropy would also favor attack of base at a primary CH group, but such a trend would not correspond to the Saytzeff rule.

PROBLEM 3·17

a $\underset{CH_3}{\overset{C_2H_5}{>}}C=CHCH_3$ $(C_2H_5)_2C=CH_2$
(cis and trans)

b [cyclopentene with CH$_3$ and CH$_3$ substituents]   [methylenecyclopentane with CH$_3$ substituent]

c [methylenecyclopentane with CH$_3$ substituent]

d [cyclohexane with =C(CH$_3$)$_2$]   [cyclohexane with =C(CH$_3$)(CH$_2$...)]

e [cyclohexene with CH$_3$]   [cyclohexene with CH$_3$]

PROBLEM 3·18
Cyclohexane conformations having equatorial chlorine are unable to assume an anti coplanar $E2$ transition state.

PROBLEM 3·19
a $NH_4^{\oplus}$, $H_3S^{\oplus}$, $C_2H_5OH_2^{\oplus}$, $(HO)_3\overset{\oplus}{S}{=}O$
b $HOSO_3^{\ominus}$, $SO_4^{\ominus}$, $NH_2^{\ominus}$, $CH_3^{\ominus}$

PROBLEM 3·20
a $A_1^{\ominus}$ will be a stronger base than $A_2^{\ominus}$ by a factor of 10^5.
b Water has a leveling effect on strong acids and bases. In other words, stronger acids than $H_3O^{\oplus}$ and stronger bases than $OH^{\ominus}$ cannot exist in water solution. The behavior of a strong acid (HA) and a strong base (B:$^{\ominus}$) in water is illustrated by the following equations:

$HA + H_2O \longrightarrow H_3O^{\oplus} + A^{\ominus}$
$B:^{\ominus} + H_2O \longrightarrow B{-}H + OH^{\ominus}$

PROBLEM 3·21
$(CH_3)_2C{=}CH_2$

PROBLEM 3·22
$(CH_3)_2C{=}CHCH_3$

PROBLEM 3·23
No. The carbonium-ion intermediate in an $E1$ elimination is approximately planar, so proton loss from either side is equally possible.

PROBLEM 3·24
$(CH_3)_3CCHClCH_3$ $(CH_3)_3CCH_2CH_2Cl$
   isomer A   isomer B

PROBLEM 3·25
1 $CH_3CH_2CH_3 + Cl_2 \xrightarrow{\Delta} CH_3CH_2CH_2Cl + CH_3CHClCH_3$
2 Separate the monochloro isomers by distillation
3 $C_3H_7Cl + Mg \xrightarrow{ether} C_3H_7MgCl$ (both isomers)
4 $CH_3CH_2CH_2MgCl + D_2O \longrightarrow CH_3CH_2CH_2D + Mg(OD)Cl$
5 $(CH_3)_2CHMgCl + D_2O \longrightarrow CH_3CHDCH_3 + Mg(OD)Cl$

PROBLEM 3·26
As an electrophilic substitution at oxygen.

CHAPTER 4

PROBLEM 4·1

a α-pinene

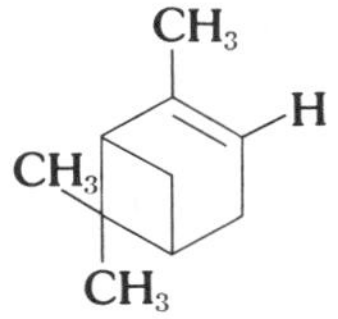

b β-pinene

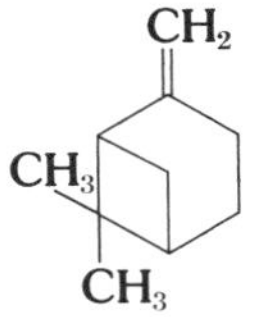

ANSWERS TO TEXT PROBLEMS

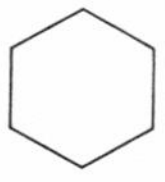

PROBLEM 4·2

a   CH₃   CH₃
     \\   /
      C=C
     /   \\
   CH₃   CH₃
     δ = 1.7

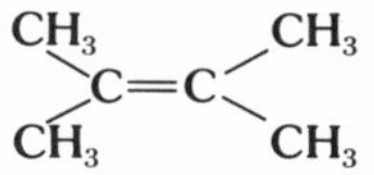

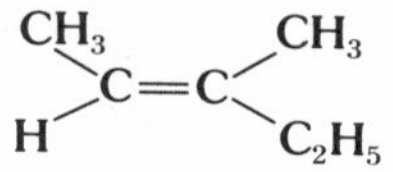

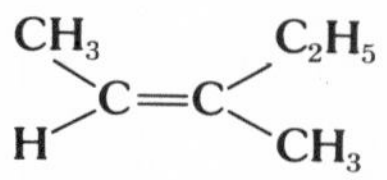

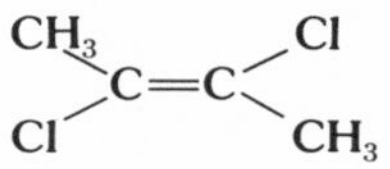

(cyclohexane)
     δ = 1.4

b   CH₃ CH₃
     \\ /
   H₂ — H₂
       /\\
      H H

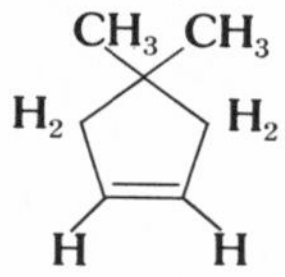

PROBLEM 4·3

Only two:

b   CH₃   CH₃   CH₃   C₂H₅
     \\   /   \\   /
      C=C   and   C=C
     /   \\   /   \\
    H   C₂H₅   H   CH₃

c   CH₃   Cl   CH₃   CH₃
     \\   /   \\   /
      C=C   and   C=C
     /   \\   /   \\
    Cl   CH₃   Cl   Cl

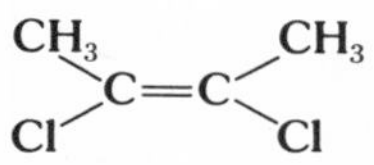

PROBLEM 4·4

a Two (cis and trans on the ring)
b None (no stereoisomers)
c Four (cis and trans on the ring and double bond)
d Four (cis and trans at both double bonds)

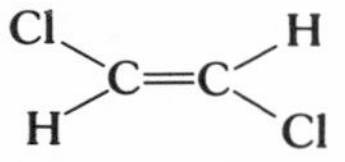

PROBLEM 4·5

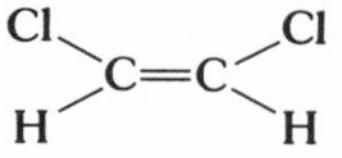

Cl   Cl   Cl   H
 \\   /   \\   /
  C=C   C=C
 /   \\   /   \\
H   H   H   Cl

cis isomer   trans isomer
(significant dipole   (zero dipole moment;
 moment)   bond dipoles cancel)

PROBLEM 4·6

In general we should construct hybrid bonding orbitals from the lowest-energy atomic orbitals available. This enables the bonding electrons to have the lowest possible energies. The 2s orbital is a lower-energy orbital than a 2p orbital.

PROBLEM 4·7

The carbon orbitals involved in the double bond can assume more *p* character (this decreases the angle between these orbitals), while at the same time the two orbitals used for bonding to the substituents have increased *s* character (and a larger orbital angle):

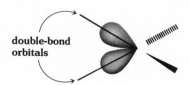

double-bond orbitals

If the *p* character is increased to 85% in these two orbitals, the *s* character drops to 15% and these become $sp^{5.65}$ hybrid orbitals. The remaining two orbitals must therefore be 35% *s* and 65% *p*, or $sp^{1.85}$ hybridized.

PROBLEM 4·8
Because the isomeric butenes have a common hydrogenation product, they are therefore thermodynamically linked to a single fixed reference point on the potential-energy scale. If the hydrogenation products are different, as in the case of ethene and 2-methylpropene, it is not possible to attribute differences in heats of hydrogenation specifically to alkene substitution effects.

PROBLEM 4·9

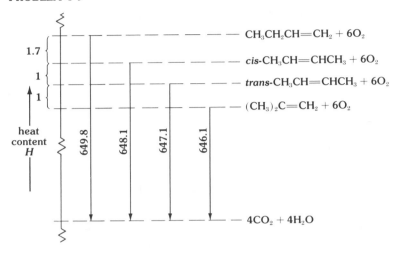

Heats of hydrogenation can be used in the same manner for all the isomers except 2-methylpropene. The hydrogenation product from this compound, isobutane, is different from the product obtained in the other cases.

CHAPTER 5

PROBLEM 5·1
If the impure cyclohexane is shaken or vigorously stirred with concentrated sulfuric acid, the cyclohexene is converted to cyclohexylsulfate, which dissolves in the sulfuric acid phase. The cyclohexane can then be washed with water, dried, and distilled.

PROBLEM 5·2
The two possible addition products are 2-chloro-2-methylpropane and 1-chloro-2-methylpropane. These compounds are easily distinguished by pmr spectroscopy: $(CH_3)_3CCl$ produces a single resonance signal at $\delta = 1.4$ and $(CH_3)_2CHCH_2Cl$ produces three resonance signals, at $\delta = 1.0$ (6H), $\delta = 1.7$ (1H), and $\delta = 3.6$ (2H).

PROBLEM 5·3
a An increase in $Cl^{\ominus}$ concentration increases the proportion of $(CH_3)_2CHCl$ in relation to $(CH_3)_2CHOCH_3$
b The CF_3 group acts to withdraw electrons from the double bond. Since hydrogen chloride is an electrophilic reagent, it will attack regions of high electron density faster than it does sites of lower electron density. Consequently, the rate of proton transfer to the electron-rich double bond of propene is greater

than to trifluoropropene. The electron-withdrawing properties of CF_3 destabilize adjacent carbonium-ion centers; therefore this group acts to reverse the Markovnikov rule. In the following reaction diagram $\Delta H_B^* > \Delta H_A^*$, and the rate at which product A is formed is greater than product B formation.

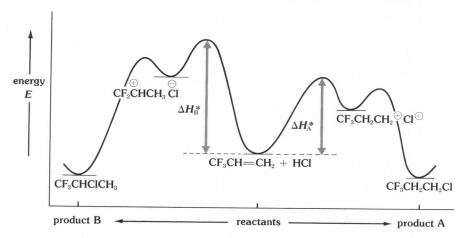

PROBLEM 5·4

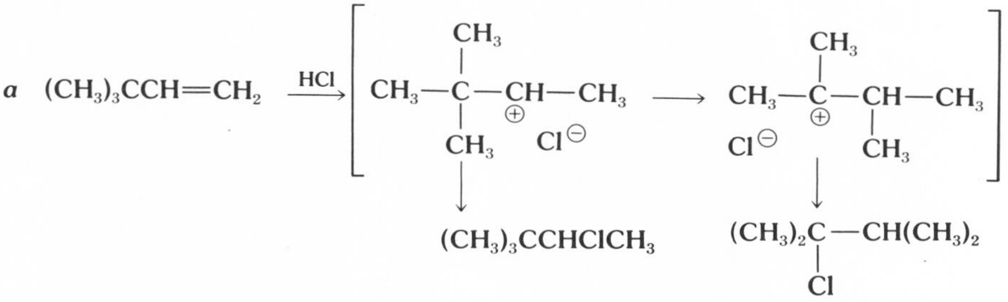

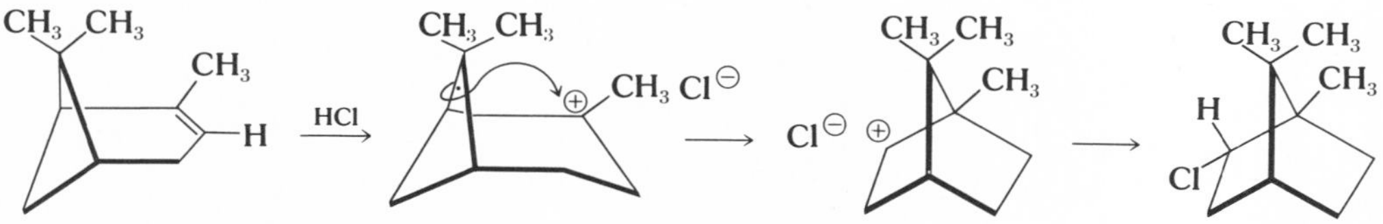

a $(CH_3)_3CCH=CH_2 \xrightarrow{HCl}$

[diagram showing:
$CH_3-\overset{CH_3}{\underset{CH_3}{C}}-\overset{\oplus}{CH}-CH_3 \quad Cl^\ominus \longrightarrow CH_3-\overset{CH_3}{\underset{\oplus}{C}}-\overset{}{\underset{CH_3}{CH}}-CH_3 \quad Cl^\ominus$]

↓ ↓

$(CH_3)_3CCHClCH_3 \qquad (CH_3)_2\underset{Cl}{C}-CH(CH_3)_2$

[Structural diagrams of α-pinene reacting with HCl via carbocation intermediates to give bornyl chloride]

α-pinene   bornyl chloride

PROBLEM 5·5

Stabilization of ions in the gas phase is poor (there is no solvation); therefore a mechanism involving carbonium-ion intermediates is unlikely to be operating. A radical-chain addition mechanism has one endothermic step:

$CH_2=CH_2 + Cl\cdot \longrightarrow Cl-CH_2-\dot{C}H_2$ $\qquad \Delta H = -17.8$ kcal/mole

$Cl-CH_2-\dot{C}H_2 + HCl \longrightarrow Cl-CH_2-CH_2-H + Cl\cdot$ $\qquad \Delta H = +6.2$ kcal/mole

A concerted (single-step) mechanism might be operating.

PROBLEM 5·6

By the addition of bromine to 1-hexene. Direct dibromation of hexane would give a mixture of 1,1-; 1,2-; 1,3-; 1,4-; 1,5-; 1,6-; 2,2-; 2,3-; 2,4-; 2,5-; 3,3-; and 3,4-dibromohexane isomers.

PROBLEM 5·7

a $CH_2=CH_2 + I_2 \longrightarrow I-CH_2-CH_2-I$

$\Sigma_r = 63.2 + 36.1 = 99.3$
$\Sigma_p = 2(51) = 102$
$\Delta H = 99.3 - 102.0 = -2.7$ kcal/mole

(The difference in bond energy between C—C and C=C is 63.2 kcal/mole.)
b Iodine forms a complex with the nucleophilic double bond in cyclohexene, hence the change in color. Complex formation is reversible. Slow addition of iodine to the double bond yields the colorless substance 1,2-diiodocyclohexane.

PROBLEM 5·8

$CH_3CH_2CH_2CH_2CH_2CH_2-B[CH(CH_3)CH(CH_3)_2]_2$

PROBLEM 5·9

a $CH_2=C(CH_3)CH_2CH_3 + H_3O^{\oplus} \longrightarrow (CH_3)_2\overset{OH}{\underset{|}{C}}CH_2CH_3$

b $CH_2=C(CH_3)CH_2CH_3 \xrightarrow{B_2H_6} \xrightarrow{H_2O_2,\ NaOH} HOCH_2-\overset{CH_3}{\underset{H}{\overset{|}{C}}}CH_2CH_3$

PROBLEM 5·10

[diagram of chair conformations with Cl and H, interconverting, then reacting with OH⁻ to give trans and cis chlorohydrins]

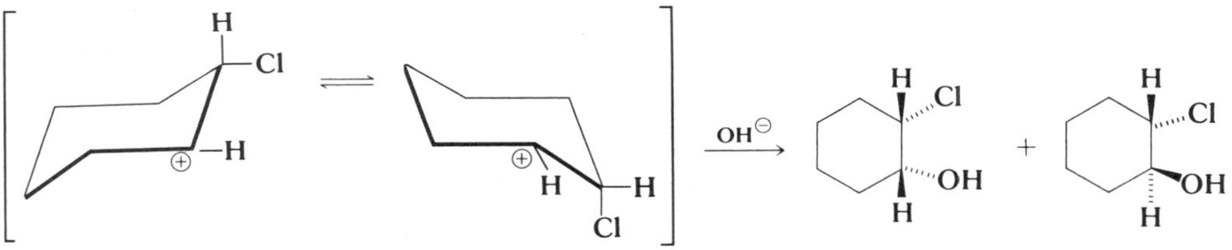

A mixture of cis and trans chloro alcohols is expected, as a result of hydroxide attack from both sides of the ion.

PROBLEM 5·11

The secondary carbonium ion formed by protonation of the alkene can rearrange to a more stable tertiary carbonium ion, whereas the cyclic bromonium ion formed by attack of electrophilic bromine on the alkene cannot rearrange to a more stable form.

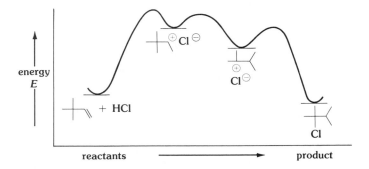

ANSWERS TO TEXT PROBLEMS 701

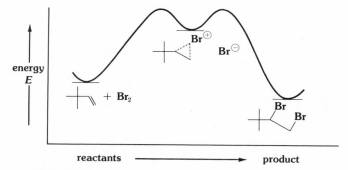

PROBLEM 5·12

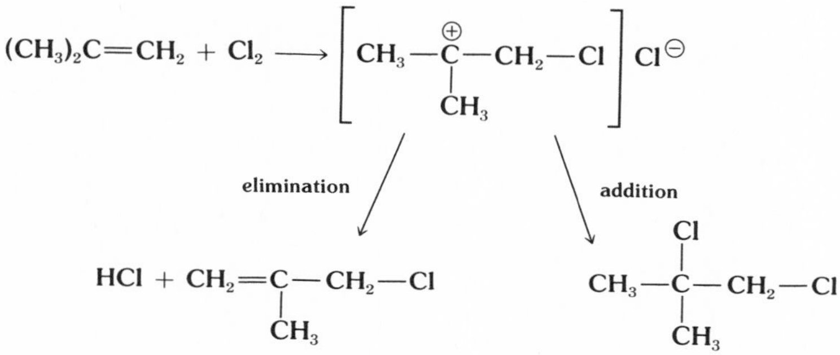

$$(CH_3)_2C=CH_2 + Cl_2 \longrightarrow \left[CH_3-\overset{\oplus}{\underset{CH_3}{C}}-CH_2-Cl \right] Cl^{\ominus}$$

elimination ↙   ↘ addition

$$HCl + CH_2=\underset{CH_3}{C}-CH_2-Cl \qquad CH_3-\underset{CH_3}{\overset{Cl}{C}}-CH_2-Cl$$

PROBLEM 5·13
Since electrons are shifting from a metal d orbital into a π^* orbital, the bonding character of the double bond is reduced and it becomes longer.

PROBLEM 5·14
$Cu^{\oplus}$ and Ni are isoelectronic; $Ag^{\oplus}$ and Pd are isoelectronic; $Hg^{2\oplus}$ and Pt are isoelectronic.

PROBLEM 5·15

$$CH_3-CH=CH_2 \xrightarrow{H_2SO_4} CH_3-\overset{OSO_3H}{\underset{|}{CH}}-CH_3$$

$\underline{-3\ -1\ -2}$ $\underline{-3\ \ 0\ \ -3}$
-6 no change -6

PROBLEM 5·16

$$CH_2=CHCH_2CH_2CH_3 \xrightarrow{O_3}$$
1-pentene

  ↗ Zn dust → $H_2C=O + O=CHCH_2CH_2CH_3$
  ↘ H_2O_2 → $CH_3CH_2CH_2CO_2H + CO_2$

$$CH_3CH=CHCH_2CH_3 \xrightarrow{O_3}$$
cis- and trans-2-pentene

  ↗ Zn dust → $CH_3CHO + CH_3CH_2CHO$
  ↘ H_2O_2 → $CH_3CO_2H + CH_3CH_2CO_2H$

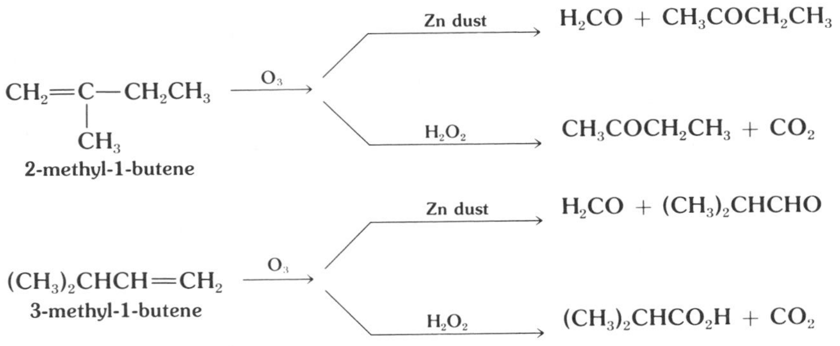

$$CH_2=C-CH_2CH_3 \atop \underset{CH_3}{|}$$
2-methyl-1-butene

$\xrightarrow{O_3}$

$\xrightarrow{\text{Zn dust}}$ $H_2CO + CH_3COCH_2CH_3$

$\xrightarrow{H_2O_2}$ $CH_3COCH_2CH_3 + CO_2$

$(CH_3)_2CHCH=CH_2$
3-methyl-1-butene

$\xrightarrow{O_3}$

$\xrightarrow{\text{Zn dust}}$ $H_2CO + (CH_3)_2CHCHO$

$\xrightarrow{H_2O_2}$ $(CH_3)_2CHCO_2H + CO_2$

PROBLEM 5·17

$CH_3CH_2CH=CHCH_2CH_3 \xrightarrow{O_3} \xrightarrow{H_2O_2} 2CH_3CH_2CO_2H$
cis and *trans*-3-hexene

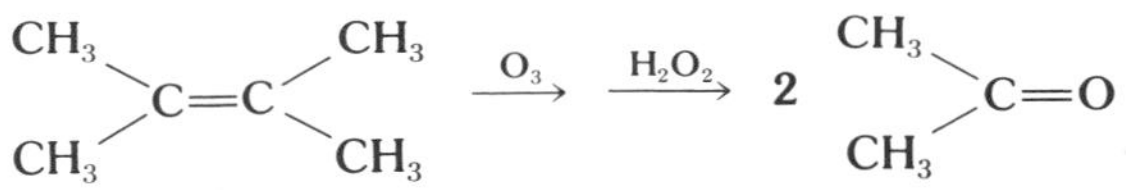

$$\underset{CH_3}{\overset{CH_3}{\diagdown}}C=C\underset{CH_3}{\overset{CH_3}{\diagup}} \xrightarrow{O_3} \xrightarrow{H_2O_2} 2 \underset{CH_3}{\overset{CH_3}{\diagdown}}C=O$$
2,3-dimethyl-2-butene

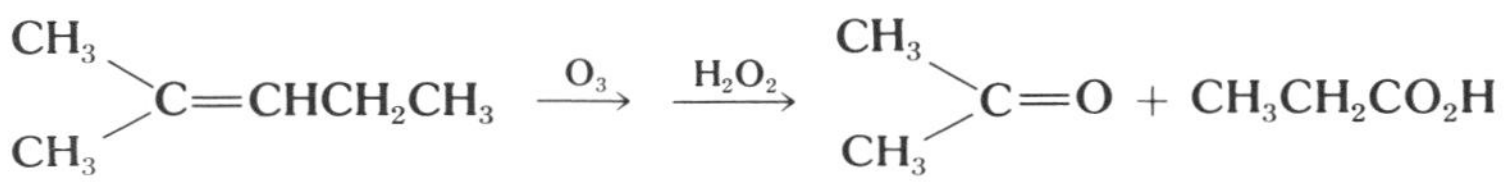

$$\underset{CH_3}{\overset{CH_3}{\diagdown}}C=CHCH_2CH_3 \xrightarrow{O_3} \xrightarrow{H_2O_2} \underset{CH_3}{\overset{CH_3}{\diagdown}}C=O + CH_3CH_2CO_2H$$
2-methyl-2-pentene

These alkenes can be distinguished by pmr spectroscopy.

PROBLEM 5·18

a A nucleophile cannot bond directly to the central oxygen atom, since this would require an expansion of the valence shell to accommodate 10 electrons. The electrophilic behavior of the terminal oxygen atoms toward a nucleophile (Nu) can be seen in the following equation:

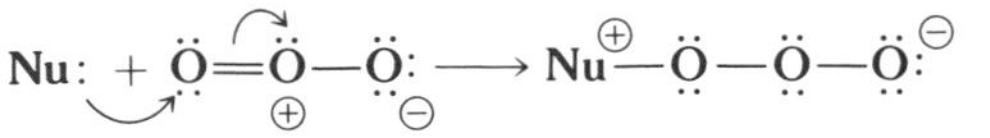

$$Nu: + \ddot{O}=\overset{\oplus}{\ddot{O}}-\ddot{\underset{\ominus}{O}}: \longrightarrow Nu\overset{\oplus}{-}\ddot{O}-\ddot{O}-\ddot{O}:^{\ominus}$$

b

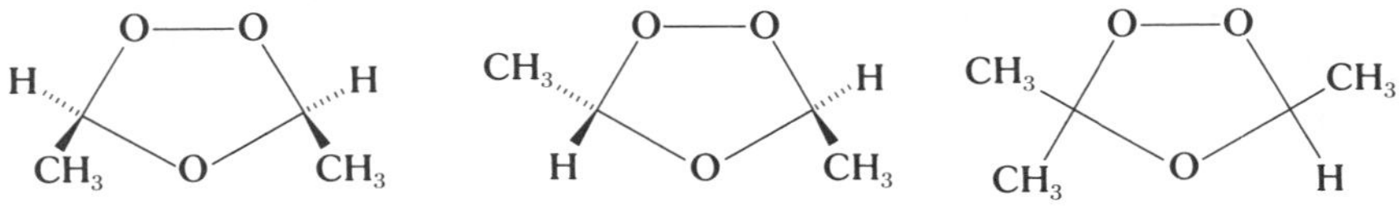

    cis   trans   no stereoisomers

PROBLEM 5·19

Bond-dissociation energy is defined as the energy ΔH of the reaction $R-H \longrightarrow R\cdot + \cdot H$. The varying bond dissociation energies of primary, secondary, and tertiary C—H bonds are a reflection of radical stability. Thus a tertiary C—H bond has a lower dissociation energy than secondary or primary bonds because a tertiary carbon radical is more stable than a secondary or primary radical.

ANSWERS TO TEXT PROBLEMS 703

PROBLEM 5·20

a $RCH=CH_2 + Cl_3C\cdot \longrightarrow R\dot{C}HCH_2CCl_3$ $\Delta H = -19.4$ kcal/mole

$R\dot{C}HCH_2CCl_3 + CCl_4 \longrightarrow RCHClCH_2CCl_3 + Cl_3C\cdot$ $\Delta H \approx 0$

$RCH=CH_2 + \cdot SCH_2CO_2H \longrightarrow R\dot{C}HCH_2SCH_2CO_2H$ $\Delta H = -1.8$

$R\dot{C}HCH_2SCH_2CO_2H + HSCH_2CO_2H \longrightarrow RCH_2CH_2SCH_2CO_2H + \cdot SCH_2CO_2H$ $\Delta H = -15.7$

b In each case one of the steps in the chain reaction is endothermic:

	X = Cl	I
$RCH=CH_2 + X\cdot \longrightarrow R\dot{C}HCH_2X$	$\Delta H = -17.8$	$+12.2$
$R\dot{C}HCH_2X + H-X \longrightarrow RCH_2CH_2X + X\cdot$	$\Delta H = +4.5$	-27.3

PROBLEM 5·21

a $F_2C=CFCl$ b $H_2C=O$ c $H_2C=C\underset{CH_3}{\overset{CO_2CH_3}{\diagdown}}$

PROBLEM 5·22

a $\sim CH_2-CH-CH_2-CH-CH_2-CH-CH_2-CH \sim$
         $|$ $|$ $|$ $|$
        CH_3 CH_3 CH_3 CH_3

b $\sim CH_2-CHCl-CH_2-CCl_2-CH_2-CHCl-CH_2-CCl_2 \sim$

PROBLEM 5·23

$CH_2=\overset{\oplus}{N}=\overset{\ominus}{N}: \xrightarrow{h\nu} N_2 + :CH_2$

$2H_2C: \longrightarrow CH_2=CH_2$

$:CH_2 + CH_2=CH_2 \longrightarrow$ cyclopropane

PROBLEM 5·24

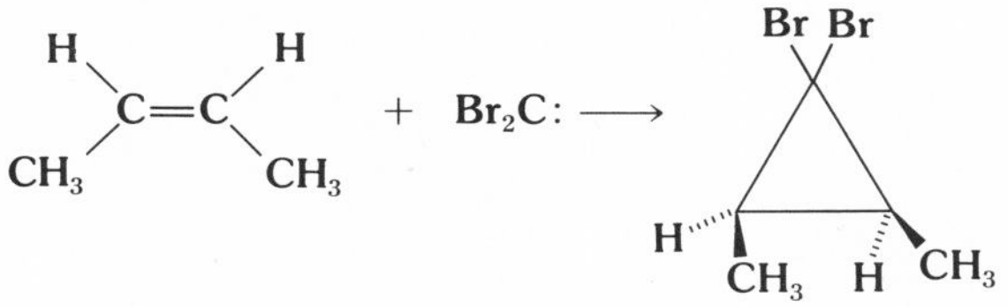

$\underset{CH_3}{\overset{H}{\diagdown}}C=C\underset{CH_3}{\overset{H}{\diagup}}$ + $Br_2C: \longrightarrow$ [cyclopropane with Br, Br on top carbon; H and CH$_3$ cis arrangement]

cis   cis

$\underset{H}{\overset{CH_3}{\diagdown}}C=C\underset{CH_3}{\overset{H}{\diagup}}$ + $Br_2C: \longrightarrow$ [cyclopropane with Br, Br on top carbon; CH$_3$ and H trans arrangement]

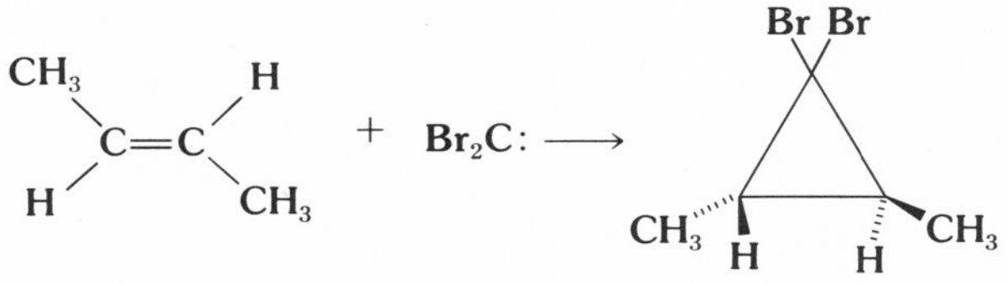

trans   trans

704 ANSWERS TO TEXT PROBLEMS

PROBLEM 5·25

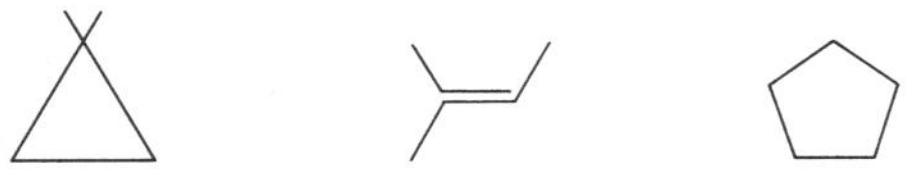

A: (triangle — methylcyclopropane/1,1-dimethylcyclopropane structure) B: (alkene) C: (cyclopentane)

A + Br$_2$ ⟶ (CH$_3$)$_2$CBrCH$_2$CH$_2$Br

B + Br$_2$ ⟶ (CH$_3$)$_2$CBrCHBrCH$_3$

B + O$_3$ ⟶ (ozonide) $\xrightarrow{H_2O_2}$ (CH$_3$)$_2$C=O + CH$_3$CO$_2$H

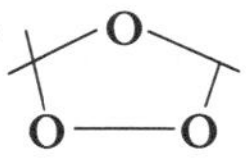

PROBLEM 5·26
Other conceivable ring openings:

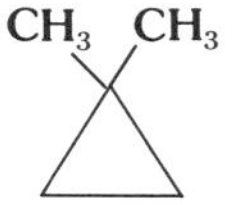

(CH$_3$)$_2$C-cyclopropane + HX ⟶ (CH$_3$)$_2$CHCH$_2$CH$_2$X
or
(CH$_3$)$_3$CCH$_2$X

If a stepwise reaction involving carbonium-ion intermediates is assumed, then both these modes of reaction would require the formation of a primary carbonium ion. This is a relatively unfavorable step.

CHAPTER 6

PROBLEM 6·1

a CF$_3$—C≡C—CH(C$_2$H$_5$)$_2$

b (cyclobutane with H substituents and two C≡CH groups)

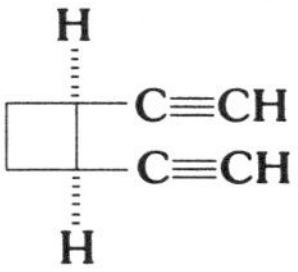

c trideca-1,11-dien-3,5,7,9-tetrayne

PROBLEM 6·2
a Since oxygen is divalent, it can be inserted into a C—H or C—C bond without changing the number of hydrogen atoms in the molecule. The presence of oxygen can therefore be ignored in making a comparison with the corresponding alkane. For example, the following compounds all have four carbon atoms and eight hydrogen atoms, and consequently have one ring or double bond:

(CH$_3$)$_2$C=CH$_2$ (CH$_3$)$_2$C(epoxide)CH$_2$ (CH$_3$)$_2$CHCHO

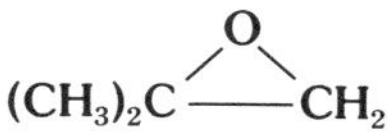

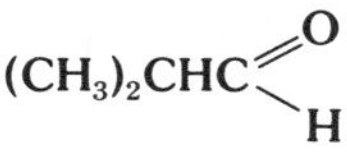

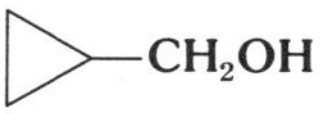

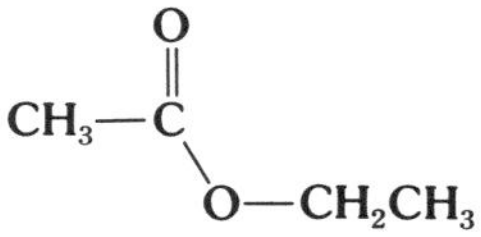

(cyclopropyl)—CH$_2$OH   CH$_3$—C(=O)—O—CH$_2$CH$_3$

(CH$_3$O)$_2$C=CH$_2$ (1,4-dioxane)   (1,3-dioxane-CH$_3$)

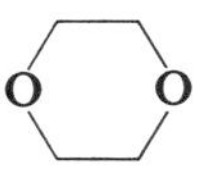

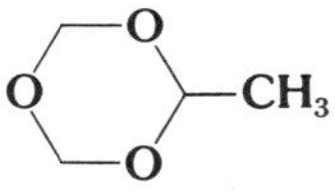

b C_6H_6 has four rings and/or double bonds, $C_5H_{10}O_3$ has one ring or double bond, and $C_{10}H_{16}SO$ has three rings or double bonds.
c The formula $C_7H_{10}O$ should be compared with the alkane composition C_7H_{16}:

$$\frac{16 - 10}{2} = 3 = \text{rings} + \text{double bonds}$$

Since catalytic hydrogenation adds one equivalent of hydrogen, the original compound has one double bond and two rings.
d The simplest expedient in calculating the ring–double-bond factor in a halogen-containing substance is to replace all halogen atoms by hydrogen and then carry out the calculation with this new formula.

PROBLEM 6·3

$C_2H_2 + \tfrac{5}{2}O_2 \longrightarrow 2CO_2 + H_2O \quad \Delta H = -300 \text{ kcal/mole}$
$\Delta H = \Sigma_{\substack{\text{reactant}\\\text{bonds}}} - \Sigma_{\substack{\text{product}\\\text{bonds}}} = -300$

$[2(98.7) + \tfrac{5}{2}(119.2) + C{\equiv}C] - [4(192) + 2(110.6)] = -300$
$\qquad\qquad [495.2 + C{\equiv}C] - [989.2] = -300$
$\qquad\qquad\qquad\qquad\qquad C{\equiv}C = 494 - 300$
$\qquad\qquad\qquad\qquad\qquad\qquad = 194 \text{ kcal/mole}$

PROBLEM 6·4
If the triple bond and the attached groups were to assume a bent configuration, such as

R
   \\
    C≡C
       \\
        R'   or   R\\C≡C\\R'

then stereoisomerism might be observed.

PROBLEM 6·5

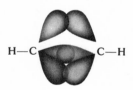

H—C   C—H

This model predicts that the triple bond should have a linear configuration and be shorter than the carbon–carbon double bond. The relative strength of the triple bond is not immediately apparent from the model, but as a rule, the shorter a bond is, the stronger it will be.

PROBLEM 6·6
a $HC{\equiv}CCH_2C{\equiv}CH$
b Eight σ bonds and four π bonds
c $CH_3{-}C{\equiv}C{-}C{\equiv}C{-}H$, signals at $\delta \approx 1.9$ and 3.0 with an area ratio of $3:1$, or $CH_2{=}C{=}C{=}C{=}CH_2$, a single signal at $\delta \approx 5.0$

PROBLEM 6·7

a $CH_3C{\equiv}CCH_3 + H_2 \xrightarrow{\text{Pd (poisoned)}} cis\text{-}CH_3CH{=}CHCH_3 \xrightarrow{Br_2} CH_3CHBrCHBrCH_3$

b $CH_3C{\equiv}CCH_3 + 2HBr \xrightarrow{Cu_2Br_2} CH_3CH_2CBr_2CH_3$

c $CH_3C{\equiv}CCH_3 + 2Br_2 \longrightarrow CH_3CBr_2CBr_2CH_3$

PROBLEM 6·8

The hydrogenation results tell us that compounds B and C must each contain two double bonds or a triple bond, whereas A has only an isolated double bond. Since all these compounds are isomers, the structure of A must incorporate a ring. The fact that B and C give a ketone on hydration indicates that they are the isomeric acetylenes 1-butyne and 2-butyne. The reaction of B with silver ion allows a choice to be made in this assignment:

$$CH_3CH_2C{\equiv}CH \xrightarrow{2H_2,\ Pd} CH_3CH_2CH_2CH_3 \xleftarrow{2H_2,\ Pd} CH_3C{\equiv}CCH_3$$

$$\text{B} \xrightarrow{2HCl} CH_3CH_2CCl_2CH_3 \xleftarrow{2HCl} \text{A}$$

$$\xrightarrow{H_3O^{\oplus},\ Hg^{2+}} CH_3CH_2\overset{O}{\underset{\|}{C}}CH_3 \xleftarrow{H_3O^{\oplus},\ Hg^{2+}}$$

$$CH_3CH_2C{\equiv}CH + Ag(NH_3)_2 \longrightarrow CH_3CH_2C{\equiv}C{-}Ag\ \text{(ppt.)}$$

The fact that A adds only one molar equivalent of HCl indicates that a three-membered ring is not present. The only reasonable assignment is that A is cyclobutene:

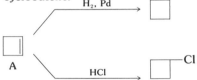

PROBLEM 6·9

They have prohibitively large activation energies. Since σ bonds are embedded within the molecular framework, they are less susceptible to attack by reactive species than are π bonds.

PROBLEM 6·10

$$RO\cdot + HBr \longrightarrow ROH + Br\cdot$$

$$\left.\begin{array}{l} C_4H_9C{\equiv}CH + Br\cdot \longrightarrow C_4H_9\dot{C}{=}CHBr \\ C_4H_9\dot{C}{=}CHBr + H{-}Br \longrightarrow C_4H_9CH{=}CHBr + Br\cdot \end{array}\right\} \text{chain reaction}$$

$$\left.\begin{array}{l} C_4H_9CH{=}CHBr + Br\cdot \longrightarrow C_4H_9CHBr{-}\dot{C}HBr \\ C_4H_9CHBr{-}\dot{C}HBr + H{-}Br \longrightarrow C_4H_9CHBr{-}CH_2Br + Br\cdot \end{array}\right\} \text{chain reaction}$$

Apparently a bromine atom stabilizes an adjacent radical more effectively than an alkyl substituent.

PROBLEM 6·11

a $C_4H_9C{\equiv}CH + H_3O^{\oplus} \xrightarrow{HgSO_4} C_4H_9COCH_3$

b $C_2H_5C{\equiv}CC_2H_5 + H_2 \xrightarrow{Pd\ (poisoned)} \underset{H}{\overset{C_2H_5}{\diagdown}}C{=}C\underset{H}{\overset{C_2H_5}{\diagup}}$

c $C_2H_5C{\equiv}CC_2H_5 + H_3O^{\oplus} \xrightarrow{HgSO_4} C_3H_7COC_2H_5$

d $C_4H_9C\equiv CH + R_2BH \longrightarrow C_4H_9CH=CHBR_2 \xrightarrow{\text{NaOH, H}_2\text{O}_2} C_5H_{11}C\overset{\displaystyle O}{\underset{\displaystyle H}{\diagup}}$
   $[R = (CH_3)_2CHCH(CH_3)]$

e $C_2H_5C\equiv CC_2H_5 + Na \xrightarrow{\text{NH}_3\,(l)}$ $\underset{C_2H_5}{\overset{H}{\diagdown}}C=C\underset{H}{\overset{C_2H_5}{\diagup}}$

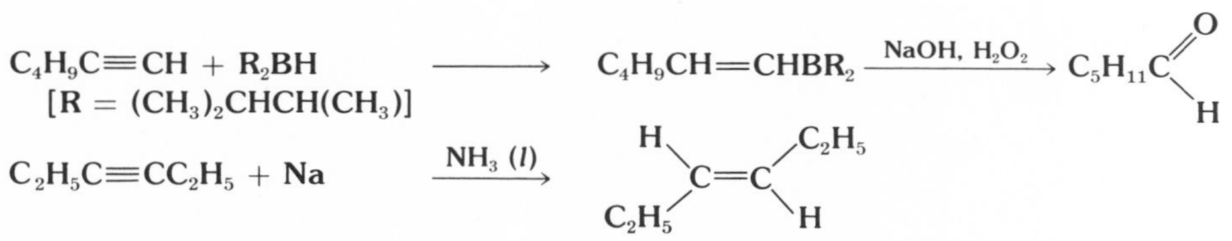

PROBLEM 6·12

From the mechanism depicted in Figure 6.7 we would anticipate a 3:1 ratio of 1,2,4 trisubstituted benzene isomer to the 1,3,5 isomer, and none of the 1,2,3 isomer.

PROBLEM 6·13

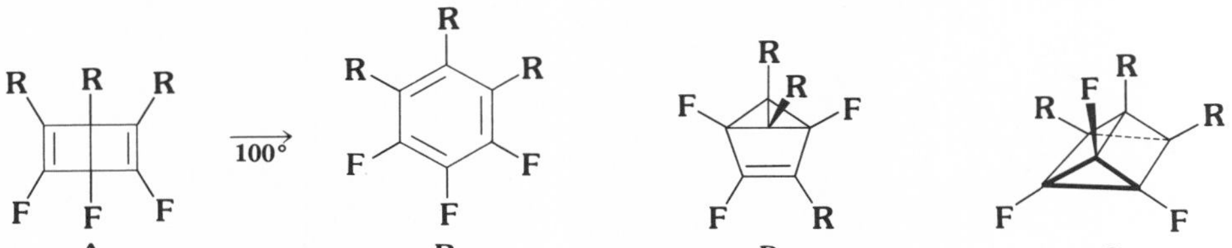

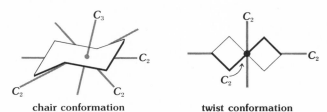

CHAPTER 7

PROBLEM 7·1

If the concentration of the sample in solution is changed, the observed rotation should show a proportional change. Thus, if the concentration of the sample is reduced to a quarter of its original value, the following changes in rotation should occur: $+40° \longrightarrow +10°$; $+220° \longrightarrow +55°$; $-140° \longrightarrow -35°$ (appears to be $+145°$); $-320° \longrightarrow -80°$ (appears to be $+100°$). The new rotations all give different polarimeter readings.

PROBLEM 7·2

C_3 ⋯ C_2 ⋯ C_2 ⋯ C_2

chair conformation

C_2 ⋯ C_2 ⋯ C_2

twist conformation

PROBLEM 7·3

a $\underset{H}{\overset{Cl}{\diagdown}}C\overset{\sigma}{=}C\underset{H}{\overset{Cl}{\diagup}}$

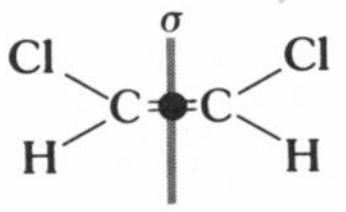

a plane of symmetry and a C_2 axis

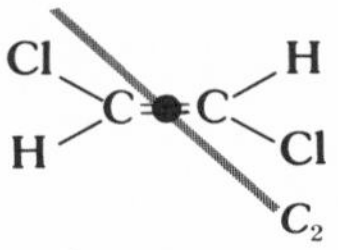

$\underset{H}{\overset{Cl}{\diagdown}}C=C\underset{Cl}{\overset{H}{\diagup}}$
C_2

a C_2 axis and a center of symmetry

b

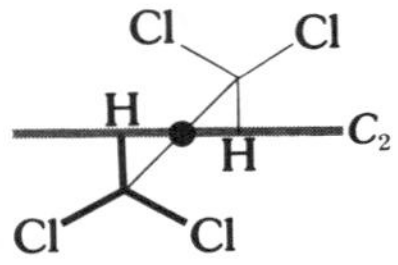

a center of symmetry
and a C_2 axis

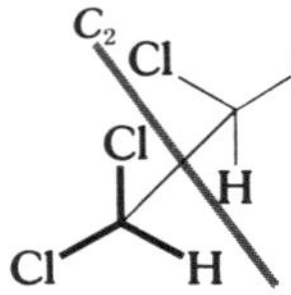

a C_2 axis
(no reflective symmetry)

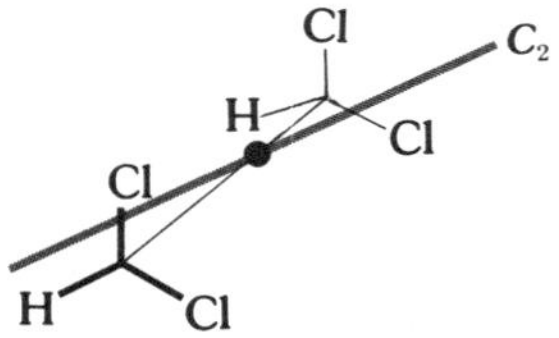

a plane of symmetry
and a C_2 axis

PROBLEM 7·4

Enantiomers:

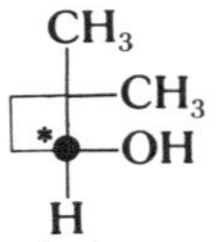

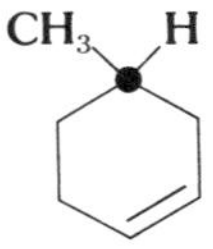

(±)   (±)   (±) $CH_3CH_2^*CH(OH)CH_3$

Constitutional isomers:

$CH_3CH_2CH(OH)CH_3$, $(CH_3)_2CHOCH_3$,   and $(CH_3)_2CHCH_2OH$

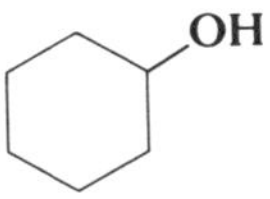

and 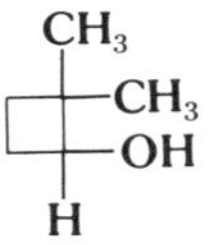 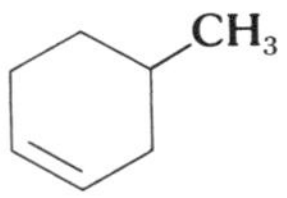   and

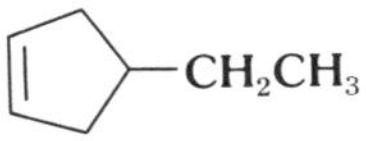

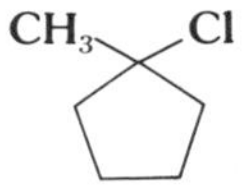

and

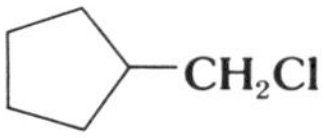

PROBLEM 7·5

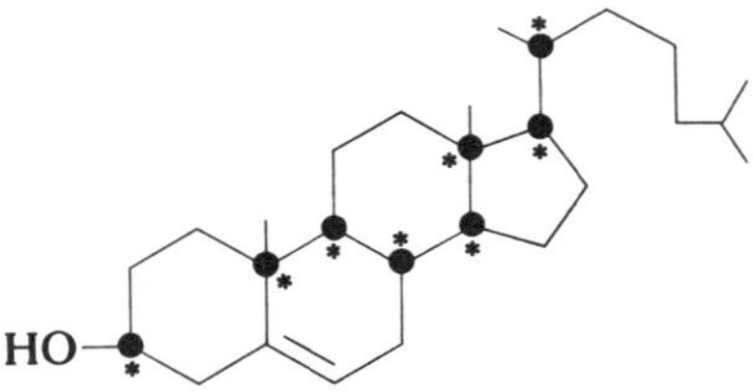

eight chiral carbon units

PROBLEM 7·6

a sp^2; sp,
b Two oriented at a right angle to each other.
c The terminal substituents lie in perpendicular planes.

PROBLEM 7·7

The allene grouping, $\begin{array}{c}R\\H\end{array}C=C=C\begin{array}{c}H\\R'\end{array}$ where $R' = HC\equiv C-C\equiv C$
and $R' = HO_2CCH_2CH=CH-CH=CH$.

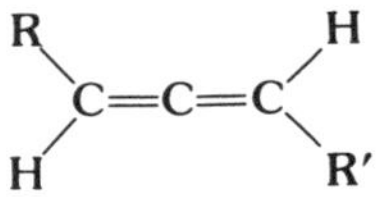

PROBLEM 7·8

The outside π orbitals in the triene are oriented at right angles to the central π orbital (the two central carbon atoms are sp-hybridized). The terminal carbon units are therefore coplanar, and the geometry of this molecule is that of an elongated alkene. If still another cumulated double bond is added, as in the tetraene, the terminal carbon units again become perpendicular to each other. Compounds belonging to the class

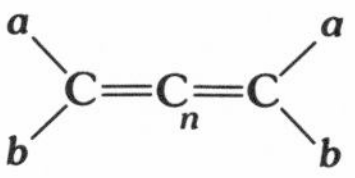

$$\underset{b}{\overset{a}{>}}C=\underset{n}{C=}C\underset{b}{\overset{a}{<}}$$

will be coplanar and may exhibit cis-trans isomerism if $n = 0,2,4,6. \ldots$ Compounds in which n is odd $(1,3,5, \ldots)$ will not be coplanar and will exhibit enantiomerism if a and b are different.

PROBLEM 7·9
Limonene dihydrochloride has a plane of symmetry and is therefore achiral. A substance composed of chiral molecules will show no optical activity if equal numbers of enantiomeric molecules are present in the sample.

PROBLEM 7·10

a

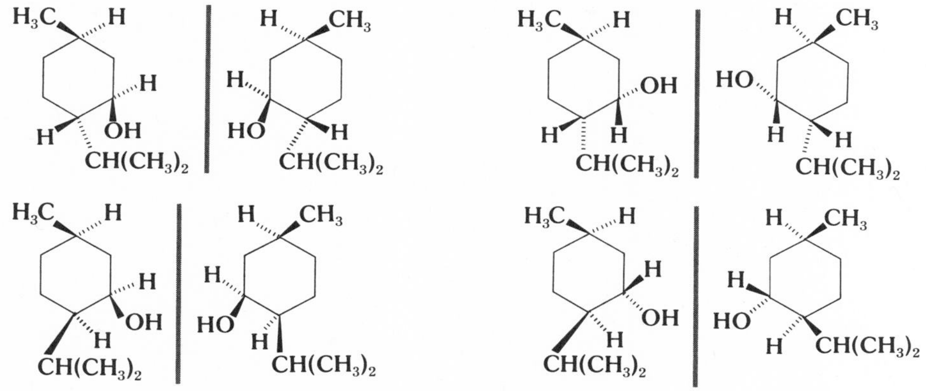

b

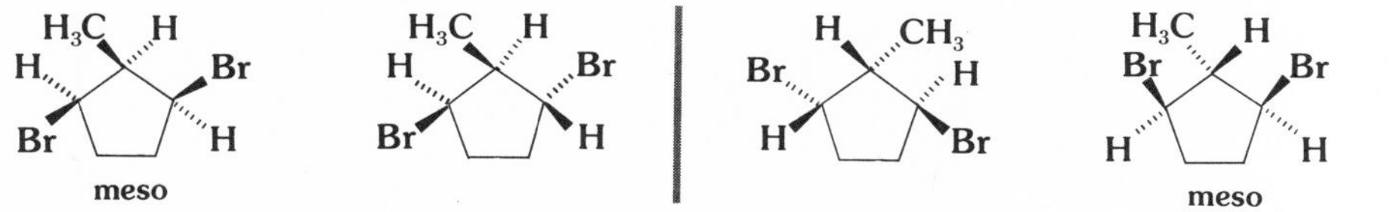

PROBLEM 7·11
a There are eight asymmetric carbon units in cholesterol, hence $2^8 = 256$ stereoisomers are possible.

b The allene is a chiral unit, and cis-trans isomerism about the double bond exists. Four stereoisomers are possible:

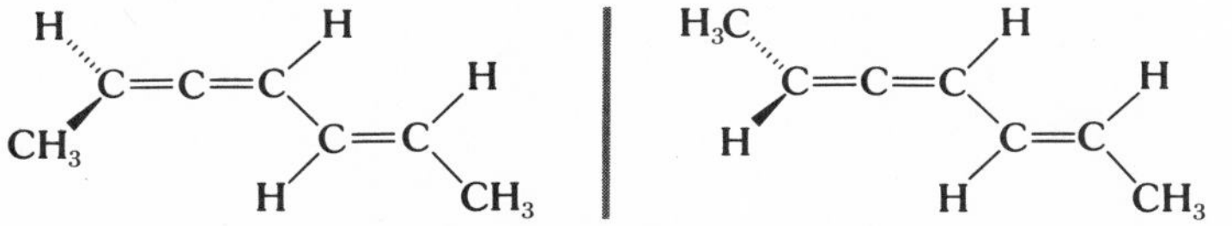

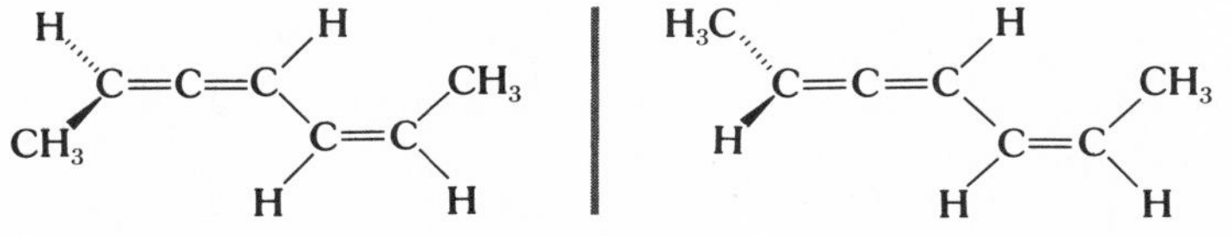

c A chiral carbon unit and a double bond are present. Four stereoisomers are possible:

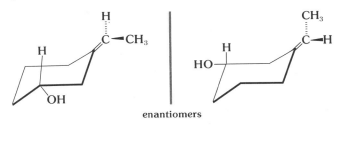

enantiomers

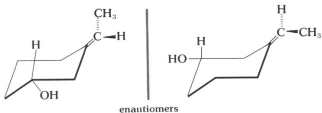

enantiomers

d Two chiral centers and a double bond are present. Two pairs of enantiomers exist.

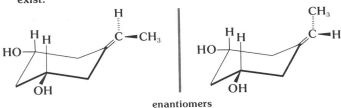

enantiomers

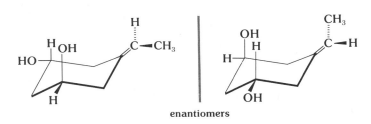

enantiomers

PROBLEM 7·12

a $(\pm)$-Y + $(-)$-Z $\longrightarrow$ $(+)$-Y$(-)$-Z + $(-)$-Y$(-)$-Z
   a pair of diastereoisomers

b The enantiomer of $(+)$-Y$(+)$-Z is $(-)$-Y$(-)$-Z; the enantiomer of $(-)$-Y$(+)$-Z is $(+)$-Y$(-)$-Z

c $(\pm)$-Y + $(\pm)$-Z $\longrightarrow$ $(+)$-Y$(+)$-Z + $(+)$-Y$(-)$-Z + $(-)$-Y$(+)$-Z + $(-)$-Y$(-)$-Z
             A   B   C   D

A and D are enantiomers, as are B and C. Therefore, although A can be separated from B and C, it cannot be separated from its enantiomer D because they have identical physical properties. Similarly, B can be separated from A and D, but not from C. Regeneration of the reactants from an A + D mixture (or B + C) will give equal amounts of $(+)$-Y and $(-)$-Y.

PROBLEM 7·13

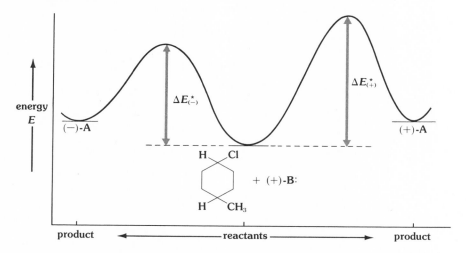

If an achiral base, such as sodium ethoxide, were used in this reaction, the transition states leading to (−)-A and (+)-A would be enantiomeric (not diastereomeric, as in the above case). Consequently, $\Delta E^*_{(-)} = \Delta E^*_{(+)}$, and the alkene enantiomers would be formed at equal rates.

PROBLEM 7·14
No.

PROBLEM 7·15

a No conclusions can be made about the stereochemistry of this reaction without further experimental evidence.

b Since this reaction does not take place at a center of chirality, retention of configuration is expected.

c Retention of configuration at all chiral centers except for that carbon bearing the hydroxyl function. The chirality is destroyed at that site.

PROBLEM 7·16

a

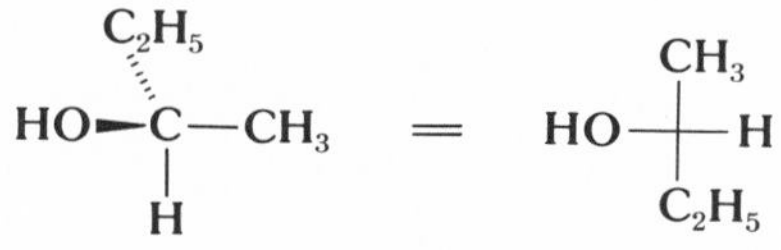

$$\begin{array}{c} C_2H_5 \\ HO-C-CH_3 \\ H \end{array} = \begin{array}{c} CH_3 \\ HO-\!\!\!\!-\!\!\!\!-H \\ C_2H_5 \end{array}$$

b

$$\begin{array}{c} R \quad CO_2H \\ H-\!\!\!\bullet\!\!\!-OH \\ HO-\!\!\!\bullet\!\!\!-H \\ R \quad CO_2H \end{array}$$

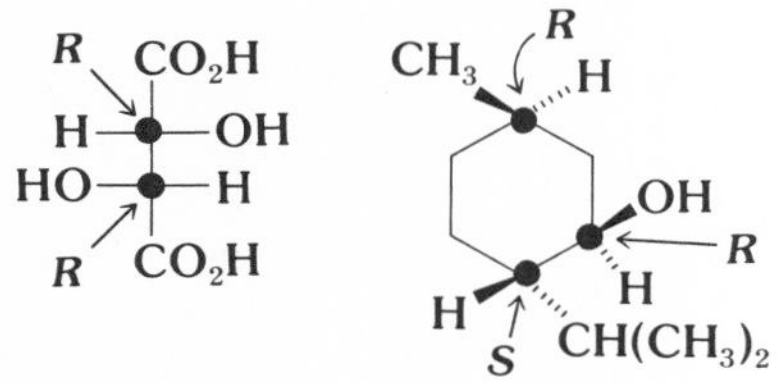

PROBLEM 7·17
The enantiomeric conformation is rapidly formed by rotation about the carbon-carbon bond:

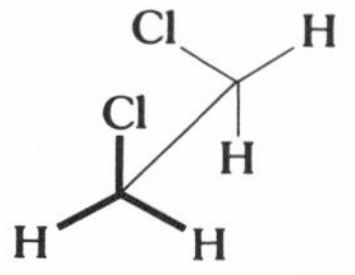

Since the enantiomeric conformations have equal free energies, they will be present in equal amounts.

712 ANSWERS TO TEXT PROBLEMS

PROBLEM 7·18

Stereospecific cis and trans addition of bromine to 1-butene give an identical product, (±)-1,2-dibromobutane.

PROBLEM 7·19

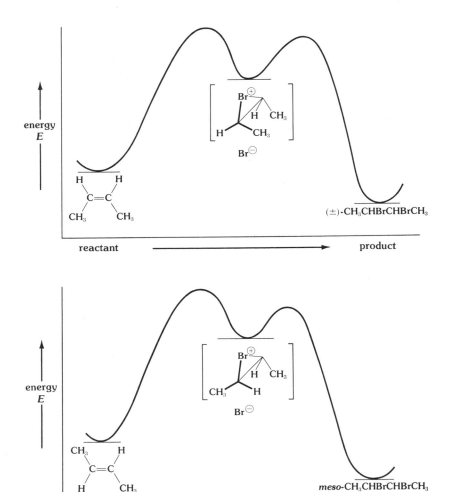

The (±)-CH$_3$CHBrCHBrCH$_3$ isomers can be resolved; the meso isomer cannot.

PROBLEM 7·20

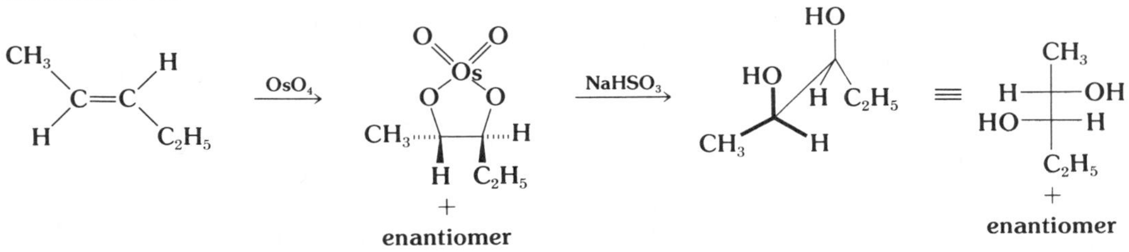

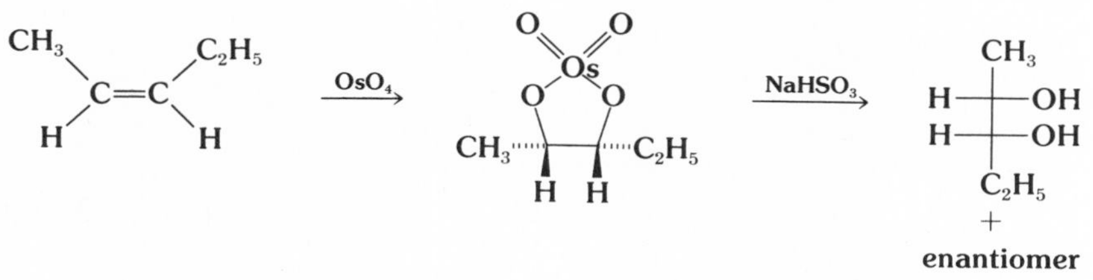

CHAPTER 8

PROBLEM 8·1

a	$CH_3CH_2CH_2CH_2CH_2OH$	1-pentanol, primary
	$CH_3(CH_2)_2{}^*CH(OH)CH_3$	2-pentanol, secondary
	$(C_2H_5)_2CHOH$	3-pentanol, secondary
	$(CH_3)_2CHCH_2CH_2OH$	3-methyl-1-butanol, primary
	$(CH_3)_2CH{}^*CH(OH)CH_3$	3-methyl-2-butanol, secondary
	$(CH_3)_2C(OH)CH_2CH_3$	2-methyl-2-butanol, tertiary
	$C_2H_5CH(CH_3)CH_2OH$	2-methyl-1-butanol, primary
	$(CH_3)_3CCH_2OH$	2,2-dimethyl-1-propanol, primary
b	$CH_3(CH_2)_3OCH_3$	1-methoxybutane
	$CH_3CH_2{}^*CH(CH_3)OCH_3$	2-methoxybutane
	$(CH_3)_2CHCH_2OCH_3$	2-methyl-1-methoxypropane
	$(CH_3)_3COCH_3$	2-methyl-2-methoxypropane
	$CH_3(CH_2)_2OC_2H_5$	1-ethoxypropane
	$(CH_3)_2CHOC_2H_5$	2-ethoxypropane

PROBLEM 8·2

a Hydrolysis of 3-chloropentane is preferred. Hydration or hydroboration of *cis*- or *trans*-2-pentene will give a mixture of 2-pentanol and 3-pentanol.

b $CH_3CH_2CH{=}CH_2 + H_3O^{\oplus} \longrightarrow CH_3CH_2CH(OH)CH_3$ (or from 2-butene)

$3CH_3CH_2CH{=}CH_2 + BH_3 \longrightarrow (C_4H_9)_3B \xrightarrow{H_2O_2,\ NaOH} 3CH_3CH_2CH_2CH_2OH$

$(CH_3)_2C{=}CH_2 + H_3O^{\oplus} \longrightarrow (CH_3)_3COH$

$3(CH_3)_2C{=}CH_2 + BH_3 \longrightarrow [(CH_3)_2CHCH_2]_3B \xrightarrow{H_2O_2,\ NaOH} 3(CH_3)_2CHCH_2OH$

PROBLEM 8·3

Intramolecular hydrogen bonding is proportionally more significant in the 1,2-diol than in the 1,4-diol. In other words, the intramolecular bonded species predominates in the 1,2-diol

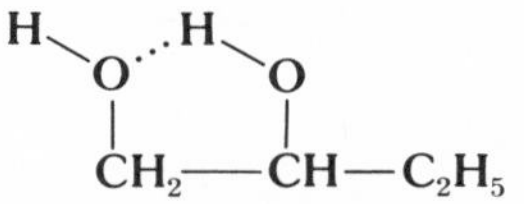

Properties that are related to intermolecular hydrogen bonding, such as elevation of boiling point and water solubility, are therefore less pronounced in the 1,2-diol than in the 1,4-diol.

PROBLEM 8·4

a An equal distribution of the two deuterons and one proton leads to a 66.7% concentration of CH_3OD.

b $CH_3OH + nD_2O \rightleftharpoons CH_3OD + HOD + (n - 1)D_2O$

If $2n/(2n + 1) = 0.9$, $n = 45$ (molar equivalents of D_2O).

PROBLEM 8·5
The greatest perturbing effect that the three methyl protons can exert on the adjacent hydroxyl proton occurs when all the methyl protons have the same spin (all parallel to or all antiparallel to the applied magnetic field). The resonance lines resulting from these interactions are the outermost parts of the quartet. Since the probability of such spin coincidence is low (see Figure 8.4), the intensity of these outer lines is also low (in relation to the inner lines, which result from more probable transition combinations).

PROBLEM 8·6

a
$$CH_3-\underset{\underset{Br}{|}}{\overset{\overset{H}{|}}{C}}-CH_3$$
The methyl protons appear as a doublet, and the tertiary-proton signal is split into seven lines.

$ClCH_2CH_2CH_2Br$ The central methylene group appears as a quintet, and the methylene groups bearing the halogen atoms each give rise to a triplet of lines.

$ClCH_2CCl_2CCl_2H$ Each of the proton signals in this compound is unsplit and they appear as singlets.

b The hydroxyl proton signal at $\delta = 5.2$ and the methyl proton signal at $\delta = 1.2$ are both split into triplets by the methylene protons. The CH_2 protons are split into a quartet (by CH_3), and this in turn is doubled by the OH splitting. A pair of overlapping quartets results. If a trace of acid or base is present, the rate of hydroxyl proton exchange increases to such a degree that its coupling with the methylene protons is no longer observed.

PROBLEM 8·7
1,2-Dimethoxyethane, $CH_3OCH_2CH_2OCH_3$, has two sets of constitutionally distinct groups of protons. Both sets of protons are on carbon atoms bonded to an oxygen atom; consequently, the chemical shifts will both be in the $\delta = 3.4$ to 4.0 region. Since the protons in the two sets are insulated from each other by the intervening oxygen atom, no splitting of the signals is expected, and two sharp singlets should be observed. The more intense signal for OCH_3 is expected at higher field than the less intense signal for OCH_2 because methyl groups generally appear at higher field than equivalently substituted methylene groups.

1,1-Dimethyoxyethane has three constitutionally different groups of protons, $CH_3CH(OCH_3)_2$. The most intense resonance in the pmr spectrum will be a singlet due to the six methoxy protons ($\delta \approx 3.4$). The three protons on the remaining methyl group should give rise to a resonance signal at $\delta \approx 1.2$ (a small downfield shift is due to the adjacent methoxy groups), which should be split into a doublet by the adjacent proton. The remaining proton appears as a quartet at $\delta = 4.5$ to 5.0. The coupling constants for both splittings will be equal and approximately 7 Hz.

PROBLEM 8·8
Rapid exchange of the hydroxylic protons results in equal average environments for all:

$C_2H_5OH + H_2O \rightleftharpoons C_2H_5OH + HOH$

PROBLEM 8·9
The second stage of reaction 8.5 is an acid-base equilibrium:

$$R-\overset{\overset{\oplus}{..}}{\underset{\underset{H}{|}}{O}}-Y + Z{:}^{\ominus} \rightleftharpoons R-\underset{\underset{H}{|}}{\overset{..}{O}}{:} + H-Z$$

acid   base   base   acid

The corresponding ether species is unable to undergo an equivalent transformation:

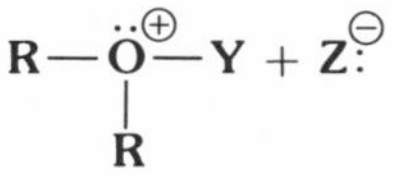

R—Ö⁺—Y + Z:⁻
 |
 R

PROBLEM 8·10

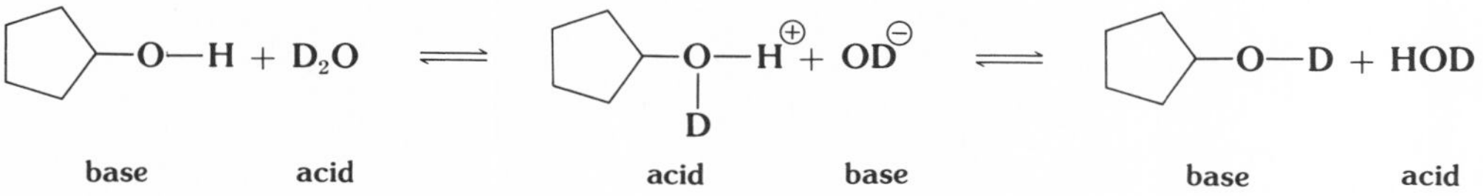

cyclopentyl—O—H + D$_2$O ⇌ cyclopentyl—O(D)—H⁺ + OD⁻ ⇌ cyclopentyl—O—D + HOD

base   acid   acid   base   base   acid

or

cyclopentyl—O—H + D$_2$O ⇌ cyclopentyl—O⁻ + D$_2$O⁺H ⇌ cyclopentyl—O—D + HOD

acid   base   base   acid   acid   base

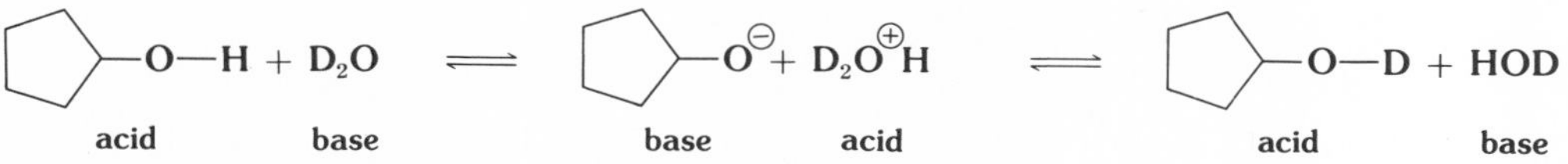

PROBLEM 8·11
Since the molecular weight of mangostin is 410, the 102-mg sample represents 0.25 millimoles of this substance. Assuming that methane behaves like an ideal gas, 0.75 millimoles of this gas are produced in the reaction of the mangostin sample with methyl Grignard reagent. Thus we conclude that mangostin has three hydroxyl groups.

PROBLEM 8·12

a cyclohexyl-O-C(=O)-C$_4$H$_9$ (with H on ring carbon)

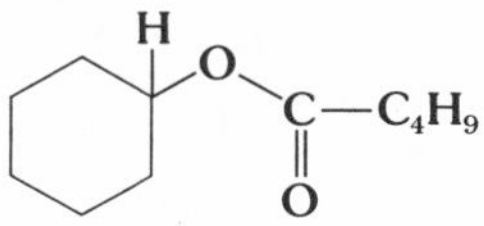

b (CH$_3$)$_2$CHCH$_2$O—S(=O)$_2$—CH$_3$

c CH$_3$—C(=O)—O— (attached to cholesterol skeleton)

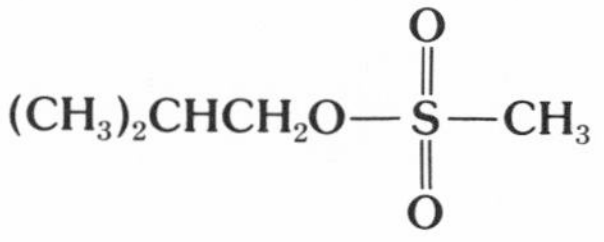

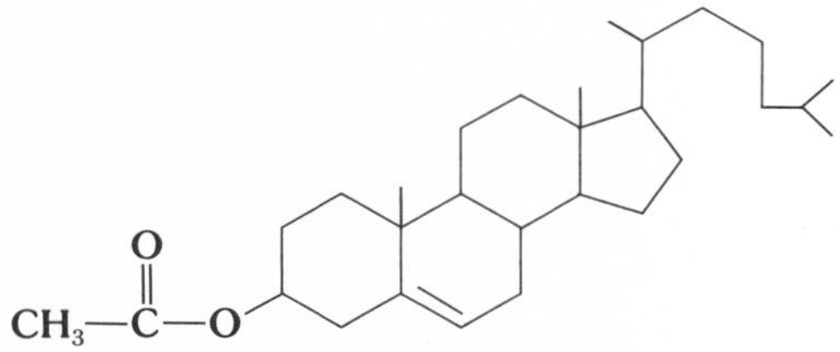

d cyclopentyl-CH$_2$-O-SO$_2$-C$_6$H$_4$-CH$_3$ (with H on ring carbon)

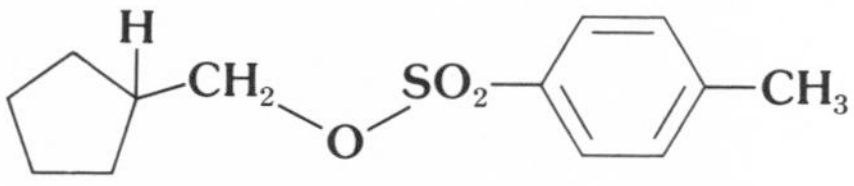

PROBLEM 8·13
a (CH$_3$)$_2$CH—O⁻Na⁺ + CH$_3$I ⟶ (CH$_3$)$_2$CHOCH$_3$
The reaction of methoxide base with an isopropyl halide would give some elimination products:

(CH$_3$)$_3$CCH$_2$O⁻Na⁺ + CH$_3$CH$_2$Br ⟶ (CH$_3$)$_3$CCH$_2$O—CH$_2$CH$_3$

The alternative combination of reagents would involve an S$_N$2 displacement reaction at a neopentyl carbon atom. Such a reaction would be sterically hindered.

b

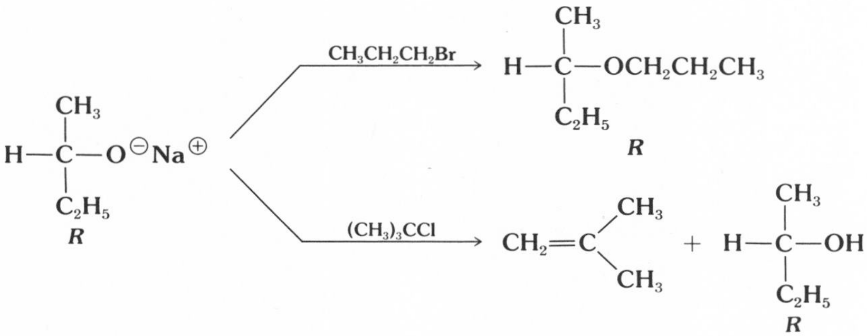

$$\begin{array}{c} CH_3 \\ | \\ H-C-O^-Na^+ \\ | \\ C_2H_5 \\ R \end{array}$$

—CH$_3$CH$_2$CH$_2$Br→ $\begin{array}{c} CH_3 \\ | \\ H-C-OCH_2CH_2CH_3 \\ | \\ C_2H_5 \\ R \end{array}$

—(CH$_3$)$_3$CCl→ CH$_2$=C(CH$_3$)(CH$_3$) + $\begin{array}{c} CH_3 \\ | \\ H-C-OH \\ | \\ C_2H_5 \\ R \end{array}$

ANSWERS TO TEXT PROBLEMS

PROBLEM 8·14

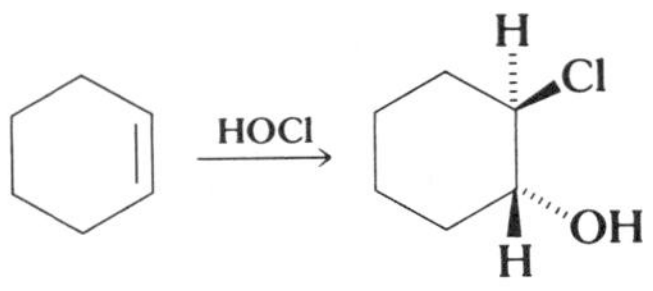

[cyclohexene + HOCl → trans-2-chlorocyclohexanol with H, Cl on one carbon and H, OH on the adjacent carbon, shown in chair form]

PROBLEM 8·15

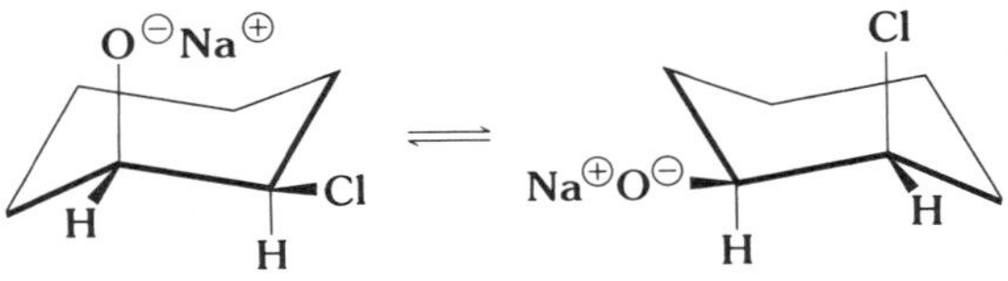

[chair conformation with O⁻Na⁺ axial up and Cl equatorial (with H's shown) ⇌ ring-flipped chair with Cl axial up and Na⁺O⁻ equatorial]

PROBLEM 8·16

[cis-2-butene + HOBr → bromohydrin (HO, H on one C; Br, CH₃ on other with specific stereochem) —NaOH→ epoxide (cis-2,3-dimethyloxirane)]

[trans-2-butene + HOBr → bromohydrin —NaOH→ trans epoxide]

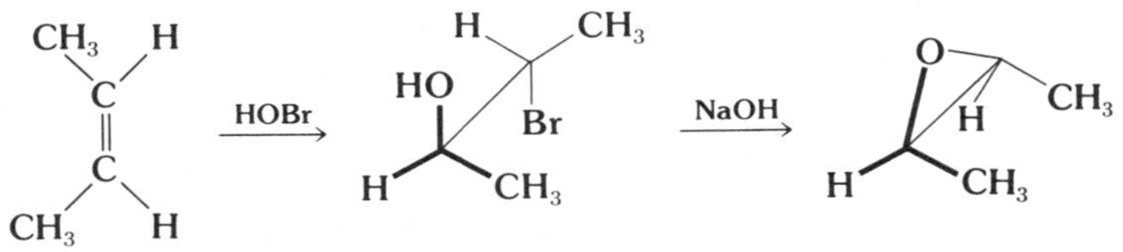

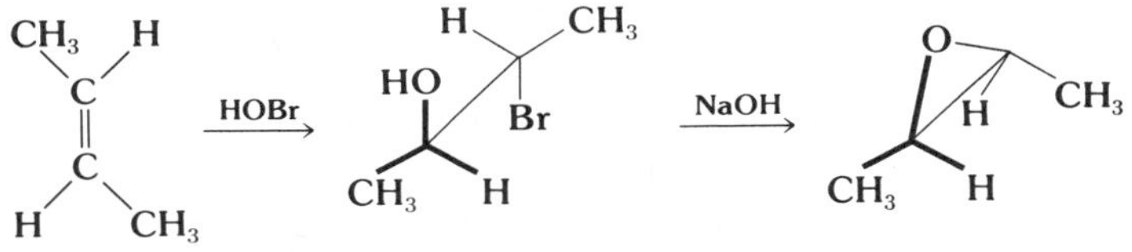

PROBLEM 8·17

a $CH_3CH_2CH_2OH$ $\xrightarrow{PBr_3}$ $CH_3CH_2CH_2Br$ $\xrightarrow{KCN}$ $CH_3CH_2CH_2CN$

$\xrightarrow{NaSH}$ $CH_3CH_2CH_2SH$

b Order of S_N2 reactivity:

$CH_3OSO_2CH_3 > (CH_3)_2CHOSO_2CH_3 > (CH_3)_3CCH_2OSO_2CH_3 \gg$

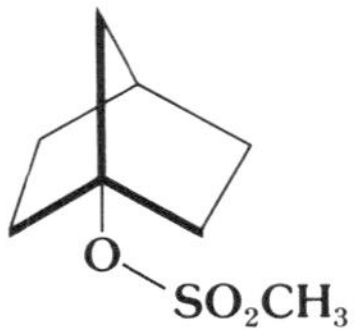

[norbornane-type cage compound with OSO₂CH₃ group at bridgehead]

The cage compound is completely unreactive, since it is impossible for a nucleophile to approach the functionalized carbon atom from the back side.

ANSWERS TO TEXT PROBLEMS 717

PROBLEM 8·18

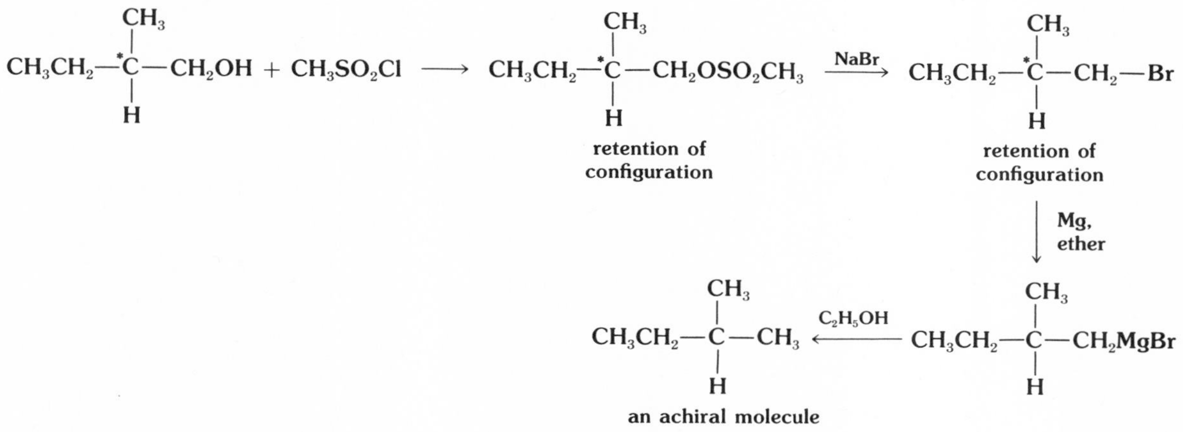

$$CH_3CH_2-\overset{*}{\underset{H}{\overset{CH_3}{C}}}-CH_2OH + CH_3SO_2Cl \longrightarrow CH_3CH_2-\overset{*}{\underset{H}{\overset{CH_3}{C}}}-CH_2OSO_2CH_3 \xrightarrow{NaBr} CH_3CH_2-\overset{*}{\underset{H}{\overset{CH_3}{C}}}-CH_2-Br$$

retention of configuration   retention of configuration

$$\downarrow Mg, ether$$

$$CH_3CH_2-\underset{H}{\overset{CH_3}{\underset{|}{C}}}-CH_3 \xleftarrow{C_2H_5OH} CH_3CH_2-\underset{H}{\overset{CH_3}{\underset{|}{C}}}-CH_2MgBr$$

an achiral molecule

PROBLEM 8·19

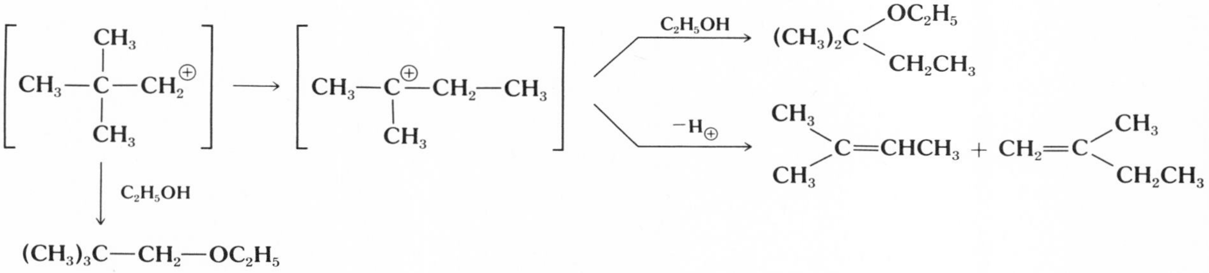

$$\left[CH_3-\underset{CH_3}{\overset{CH_3}{\underset{|}{C}}}-CH_2^{\oplus}\right] \longrightarrow \left[CH_3-\underset{CH_3}{\overset{\oplus}{\underset{|}{C}}}-CH_2-CH_3\right] \begin{array}{c} \xrightarrow{C_2H_5OH} (CH_3)_2C\overset{OC_2H_5}{\underset{CH_2CH_3}{\diagdown}} \\ \xrightarrow{-H^{\oplus}} \overset{CH_3}{\underset{CH_3}{\diagdown}}C=CHCH_3 + CH_2=C\overset{CH_3}{\underset{CH_2CH_3}{\diagdown}} \end{array}$$

$$\downarrow C_2H_5OH$$

$$(CH_3)_3C-CH_2-OC_2H_5$$

PROBLEM 8·20

An S_N2 displacement takes place at the less hindered group:

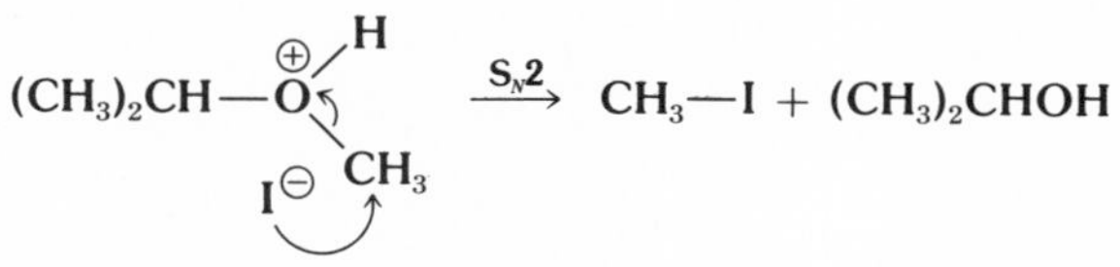

$$(CH_3)_2CH-\overset{\oplus}{O}\underset{\underset{CH_3}{\overset{|}{I^{\ominus}}}}{\overset{H}{\diagup}} \xrightarrow{S_N2} CH_3-I + (CH_3)_2CHOH$$

PROBLEM 8·21

$$(CH_3)_3CCH_2OH \xrightarrow{HBr} [(CH_3)_3CCH_2^{\oplus}] \longrightarrow CH_3-\underset{CH_3}{\overset{\oplus}{\underset{|}{C}}}-CH_2CH_3]Br^{\ominus} \xrightarrow{-H_2O} CH_3-\underset{CH_3}{\overset{Br}{\underset{|}{C}}}-CH_2CH_3$$

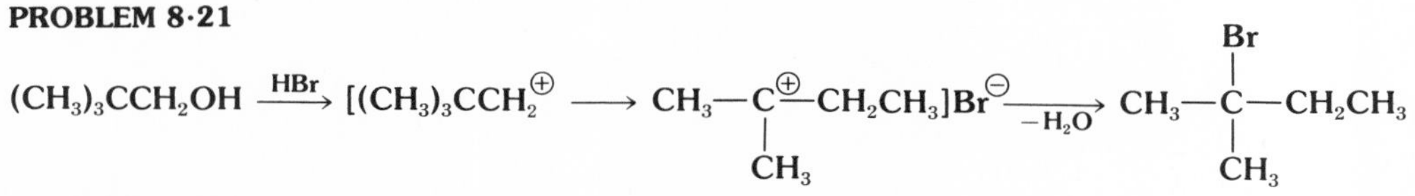

PROBLEM 8·22

$$\underset{}{\bigcirc}-O-C(CH_3)_3 \xrightarrow{H^{\oplus}} \left[\underset{}{\bigcirc}-\overset{H}{\underset{\oplus}{O}}-C(CH_3)_3\right]$$

$$\left[\underset{}{\bigcirc}-\overset{H}{\underset{\oplus}{O}}-C(CH_3)_3\right] \longrightarrow \underset{}{\bigcirc}-OH + \left[CH_3-\underset{CH_3}{\overset{CH_3}{\underset{|}{C^{\oplus}}}}\right] \xrightarrow{-H^{\oplus}} H_2C=C\overset{CH_3}{\underset{CH_3}{\diagdown}}$$

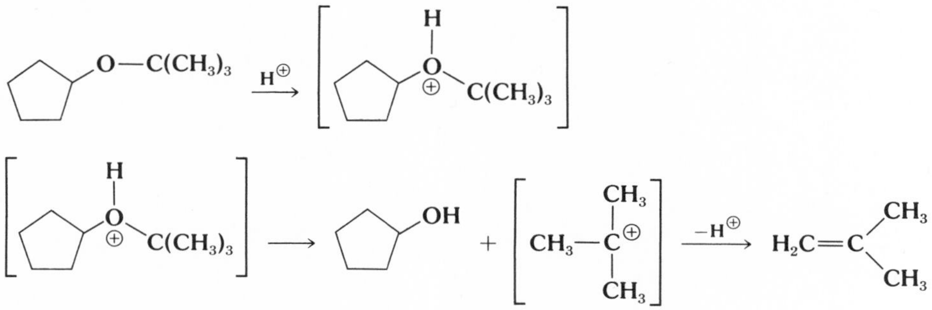

PROBLEM 8·23
The cis isomer is achiral, whereas the trans isomer is chiral. The trans isomer can be resolved under appropriate conditions. A less definitive method would involve measuring the boiling points of these isomers. The cis isomer will exhibit intramolecular hydrogen bonding; consequently, it will have a lower boiling point than the trans isomer.

PROBLEM 8·24

$$RCH_2\underset{base}{\overset{OH}{\underset{|}{C}}HR} + \underset{acid}{H_2SO_4} \rightleftharpoons RCH_2\underset{acid}{\overset{\overset{\oplus}{O}H_2}{\underset{|}{C}}HR} + \underset{base}{HSO_4^{\ominus}} \quad (1)$$

$$RCH_2\underset{acid}{\overset{\overset{\oplus}{O}H_2}{\underset{|}{C}}HR} \rightleftharpoons RCH_2-\underset{base}{\overset{\oplus}{C}{\underset{R}{\overset{H}{\diagdown}}}} + H_2O \quad (2)$$

$$RCH_2-\underset{acid}{\overset{\oplus}{C}{\underset{R}{\overset{H}{\diagdown}}}} + \underset{base}{H_2O} \rightleftharpoons \underset{base}{RCH=CHR} + \underset{acid}{H_3O^{\oplus}} \quad (3)$$

Step 2 is product determining in the alcohol dehydration mechanism.

PROBLEM 8·25
A molecular rearrangement is involved:

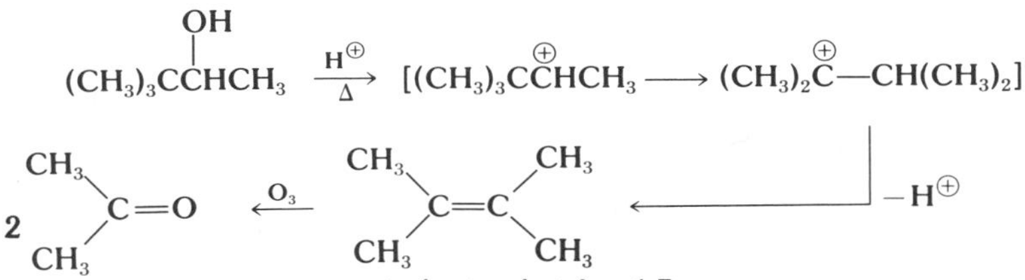

$$(CH_3)_3C\overset{OH}{\underset{|}{C}}HCH_3 \xrightarrow[\Delta]{H^{\oplus}} [(CH_3)_3C\overset{\oplus}{C}HCH_3 \longrightarrow (CH_3)_2\overset{\oplus}{C}-CH(CH_3)_2]$$

$$2\ \underset{CH_3}{\overset{CH_3}{\diagdown}}C=O \xleftarrow{O_3} \underset{CH_3}{\overset{CH_3}{\diagdown}}C=C\underset{CH_3}{\overset{CH_3}{\diagup}} \xleftarrow{-H^{\oplus}}$$

pmr: a single signal at $\delta = 1.7$

PROBLEM 8·26

a. cyclopentene + $H_2O \xrightarrow{H^{\oplus}}$ cyclopentanol $\xrightarrow{CrO_3, H_3O^{\oplus}}$ cyclopentanone

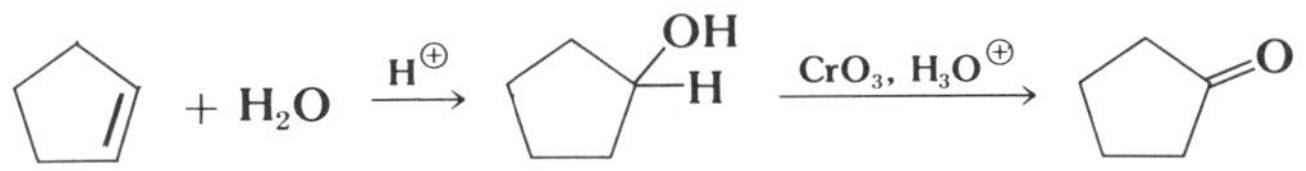

b. $(CH_3)_2CHCH_2OCH_3 \xrightarrow{HBr} CH_3Br + (CH_3)_2CHCH_2OH \xrightarrow{CrO_3} (CH_3)_2CHC\overset{O}{\underset{H}{\diagdown}}$

c. $(CH_3)_3COH \xrightarrow[\Delta]{H_2SO_4} H_2C=C\underset{CH_3}{\overset{CH_3}{\diagup}} \xrightarrow{HOBr} BrCH_2-\underset{CH_3}{\overset{OH}{\underset{|}{C}}}-CH_3$

d. (decalin derivative with OH and C=C) $\xrightarrow{H_2, Pt}$ (saturated decalin-ol) $\xrightarrow{CrO_3, H_3O^{\oplus}}$ (decalone)

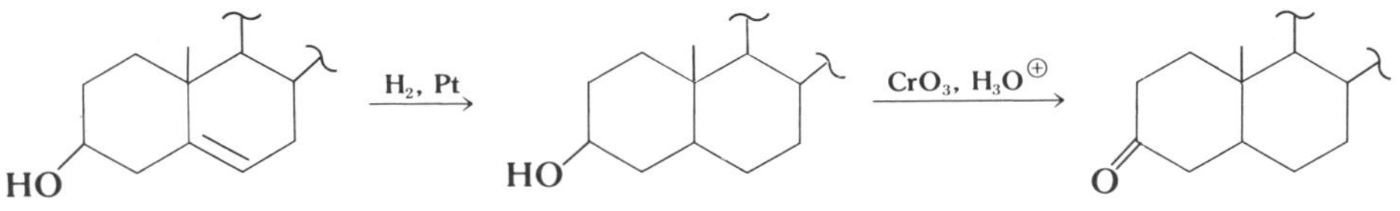

PROBLEM 8·27

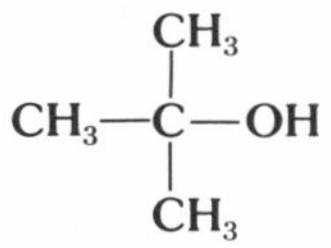

$$CH_3-\underset{\underset{CH_3}{|}}{\overset{\overset{CH_3}{|}}{C}}-OH \qquad (CH_3)_2CHCH_2OH \qquad (CH_3)_2CHOCH_3$$

A   B   C

1 $ROH + CH_3MgBr \longrightarrow ROMgBr + CH_4$
   $[R = (CH_3)_3C \text{ or } (CH_3)_2CHCH_2]$

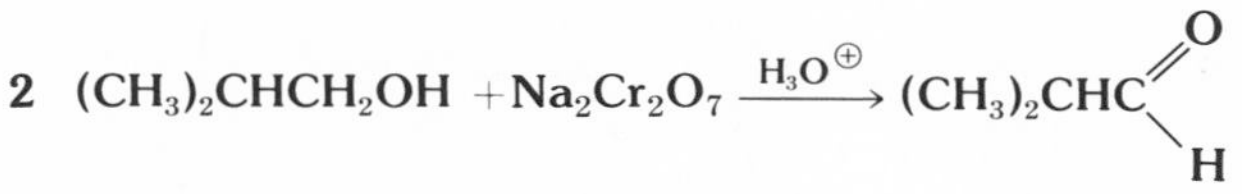

2 $(CH_3)_2CHCH_2OH + Na_2Cr_2O_7 \xrightarrow{H_3O^{\oplus}} (CH_3)_2CHC\underset{H}{\overset{O}{\diagdown\!\!\!/}}$

3 $(CH_3)_3COH \xrightarrow[\Delta]{H_3PO_4} CH_2=C(CH_3)_2$

4 $(CH_3)_2CHCH_2OH \xrightarrow[\Delta]{H_3PO_4}$

PROBLEM 8·28
Thiols and sulfides are easily prepared by S_N2 reactions of $HS^{\ominus}$ and $RS^{\ominus}$ with alkyl halides or sulfonate esters.

CHAPTER 9

PROBLEM 9·1
a Compare C_6H_6 with C_6H_{14}: $(14 - 6)/2 = 8/2 = 4$. Benzene must therefore have four rings and/or double bonds.
b Many possible structures can be written for C_6H_6. A few of these, arranged according to the number of rings, are shown below:

$CH_3-C\equiv C-C\equiv C-CH_3$ $H_2C=CH-C\equiv C-CH=CH_2$ $H_2C=C=C\underset{H}{\overset{}{\diagdown}}C=C=CH_2$
no rings

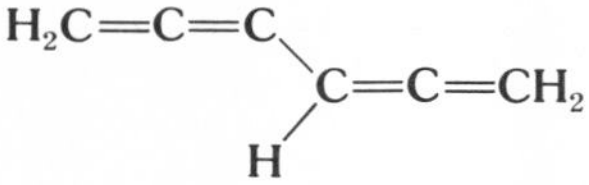

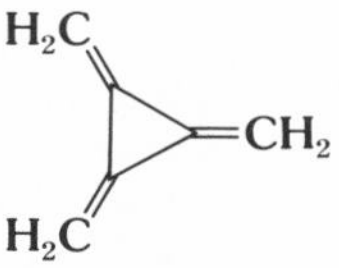

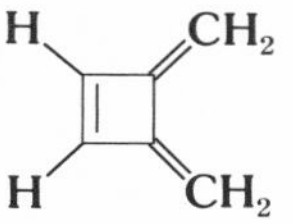

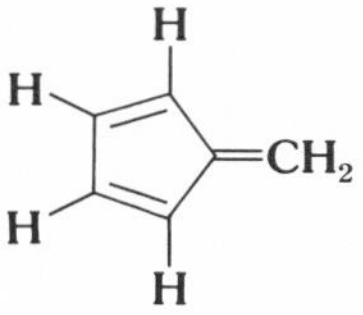

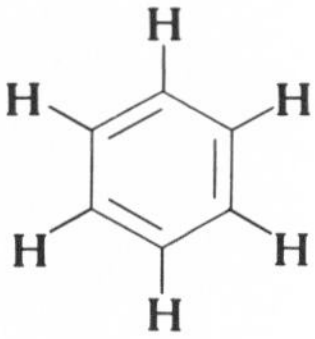

one ring

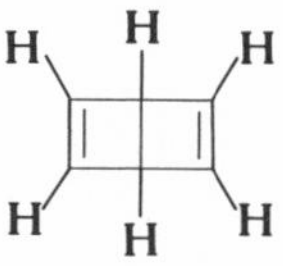

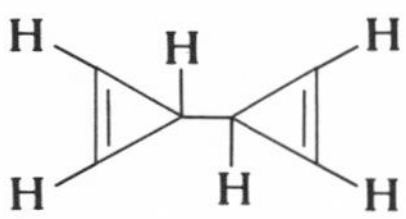

two rings

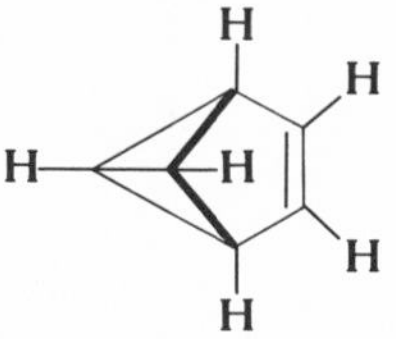

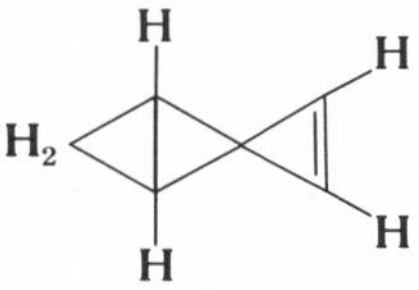

three rings 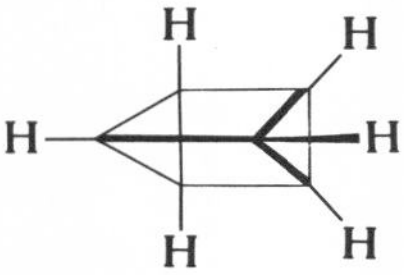   four rings

720 ANSWERS TO TEXT PROBLEMS

PROBLEM 9·2
The following would each yield a single monosubstituted derivative:

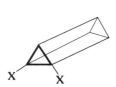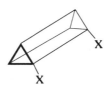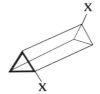

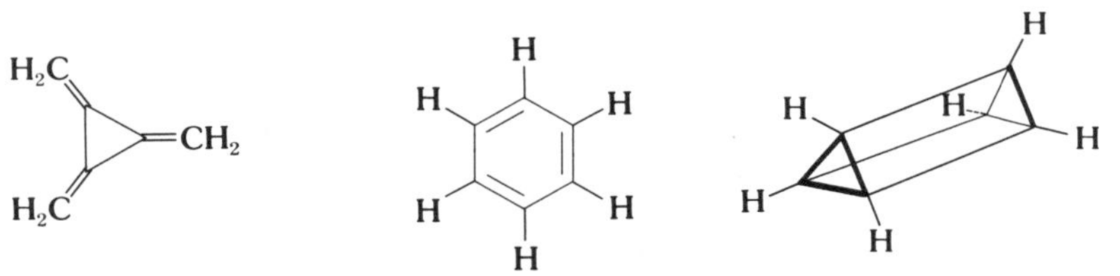

The disubstituted isomers of the cyclohexatriene are discussed in the next problem. The prismatic structure (prismane) would give three isomeric disubstituted derivatives:

The trimethylene cyclopropane would give four isomeric disubstituted derivatives:

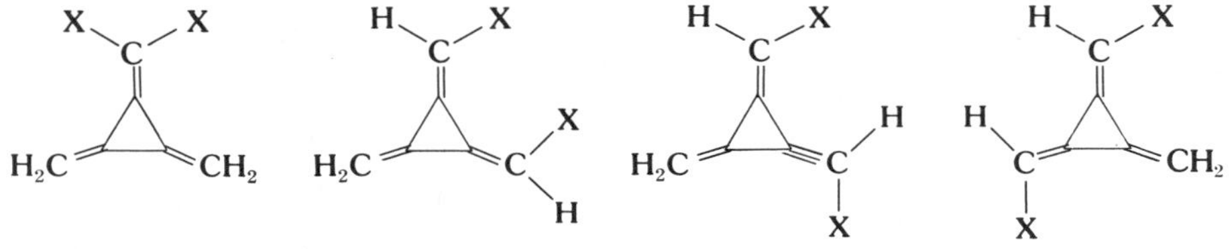

PROBLEM 9·3
Single cyclohexatriene isomers:

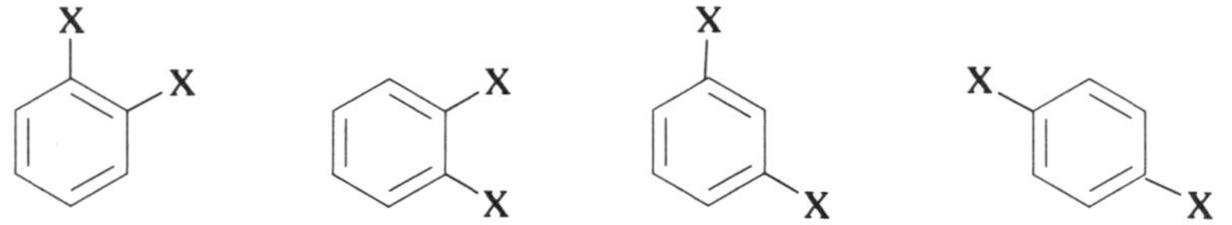

disubstituted

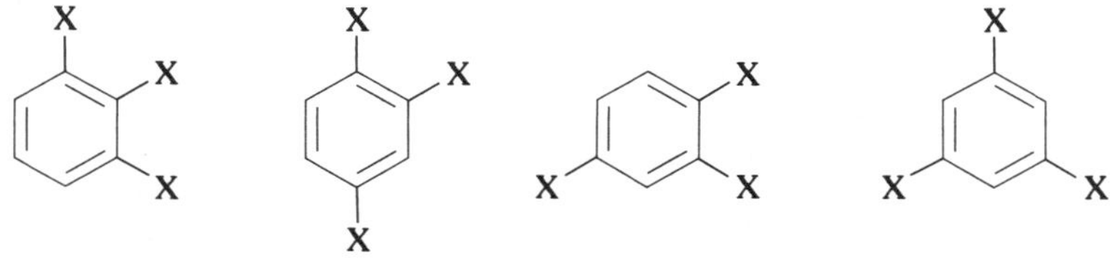

trisubstituted

Equilibrium mixture of isomers:

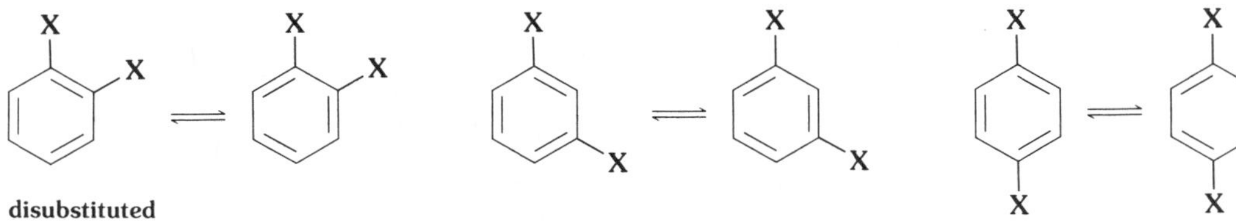

disubstituted

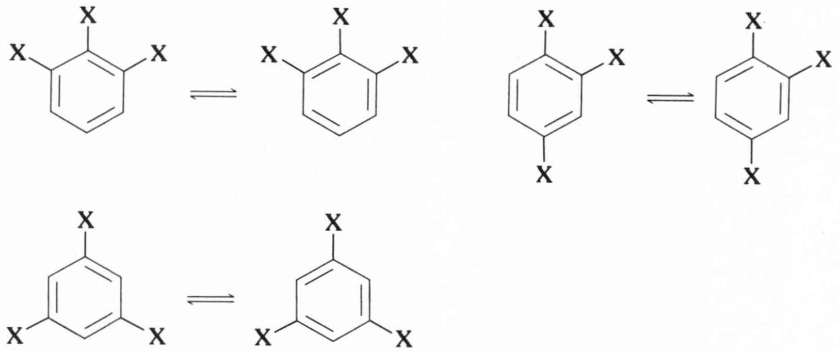

trisubstituted

PROBLEM 9·4

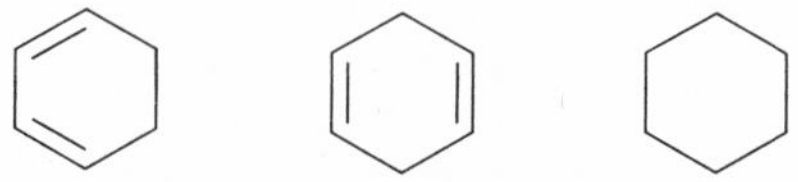

A B C

PROBLEM 9·5

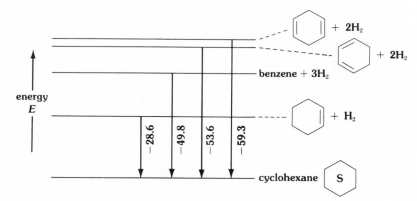

PROBLEM 9·6
Add the following equations:

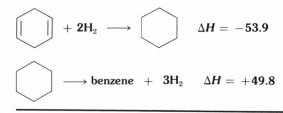

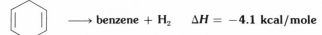

 $+ 2H_2 \longrightarrow$ $\Delta H = -53.9$

 $\longrightarrow$ benzene $+ 3H_2$ $\Delta H = +49.8$

 $\longrightarrow$ benzene $+ H_2$ $\Delta H = -4.1$ kcal/mole

PROBLEM 9·7

a These are cis-trans isomers: $\rightleftharpoons$
b Only electron pairs need be moved: $\longleftrightarrow$
c Only an electron pair need be moved: $\longleftrightarrow$
d These are tautomers: $\rightleftharpoons$

722 ANSWERS TO TEXT PROBLEMS

PROBLEM 9·8

a) Resonance structures of carbonate ion CO_3^{2-} showing three equivalent resonance forms.

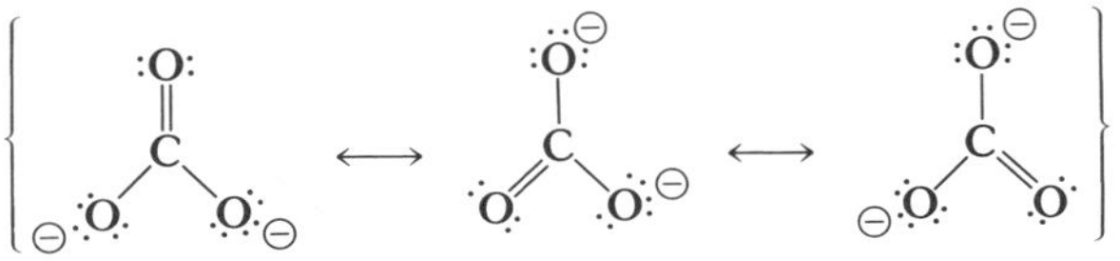

b) $\{:C=\ddot{O}: \longleftrightarrow :C\equiv\overset{+}{O}:^-\}$

PROBLEM 9·9

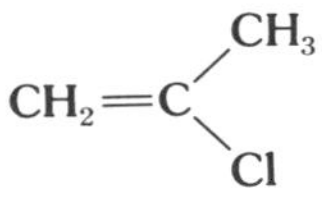

2-chloropropene: $CH_2=C(CH_3)Cl$

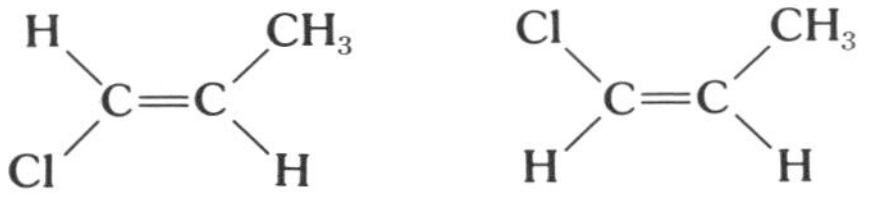

1-chloropropene (trans and cis forms shown)

$$CH_3CH=CH_2 + Cl\cdot \begin{cases} \xrightarrow{-HCl} [H_2\dot{C}-CH=CH_2] \xrightarrow{Cl_2} Cl-CH_2-CH=CH_2 + Cl\cdot \\ \xrightarrow{-HCl} [CH_3-\dot{C}=CH_2] \xrightarrow{Cl_2} CH_3-\underset{\underset{Cl}{|}}{C}=CH_2 + Cl \\ \xrightarrow{-HCl} [CH_3-CH=\dot{C}-H] \xrightarrow{Cl_2} CH_3CH=CHCl + Cl\cdot \\ \text{cis and trans} \end{cases}$$

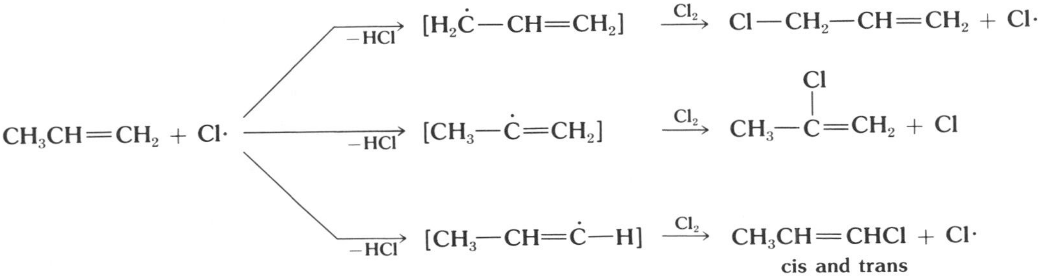

The allyl radical is more stable than the other intermediate radicals, owing chiefly to the delocalization of the unpaired electron:

$$\{\dot{C}H_2-CH=CH_2 \longleftrightarrow CH_2=CH-\dot{C}H_2\}$$

The hydrogen-abstraction step in the free-radical chlorination mechanism (the first step) thus favors formation of the allyl radical. Since this is also the product-determining step, allyl chloride is the major product. Note the C—H bond dissociation energies given in Table 2.5. These energies are inversely proportional to the relative stabilities of the radicals formed by homolysis.

PROBLEM 9·10

a) Structures derived from p-xylene with various chlorine substitutions on methyl groups and ring positions.

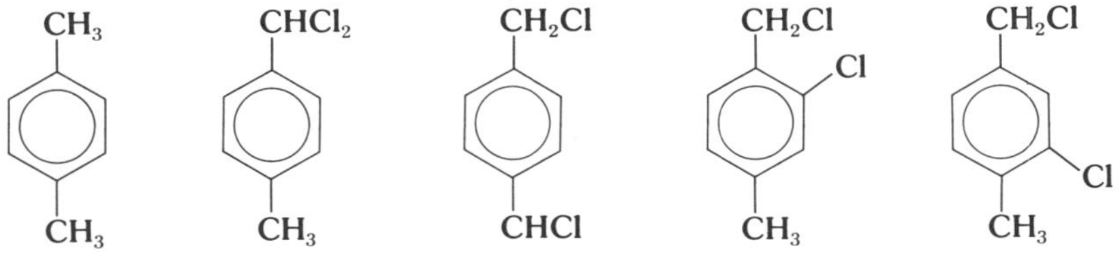

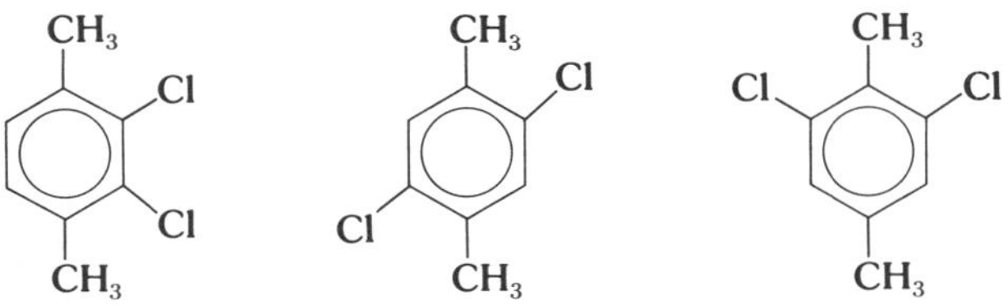

b *naphthalene* (structures of dichloronaphthalene isomers)

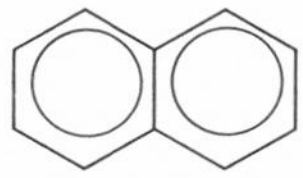

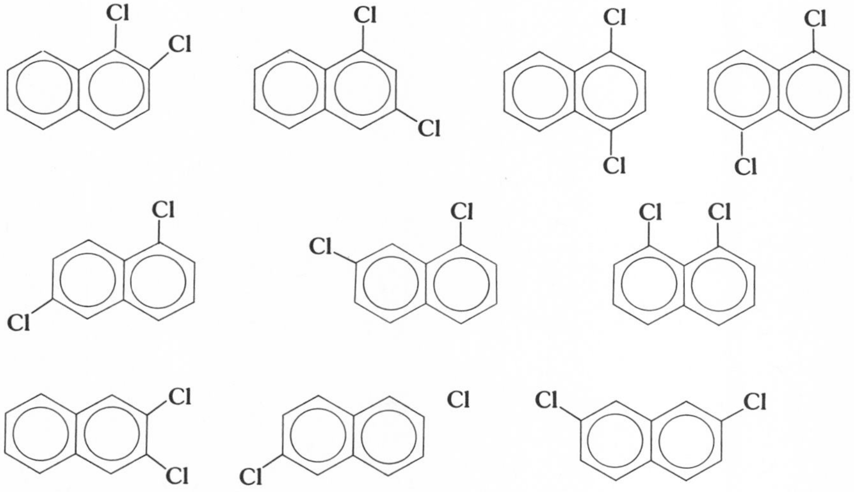

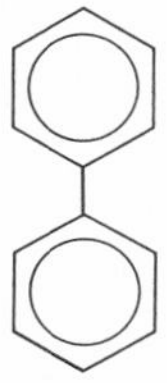

c *biphenyl* (structures of dichlorobiphenyl isomers)

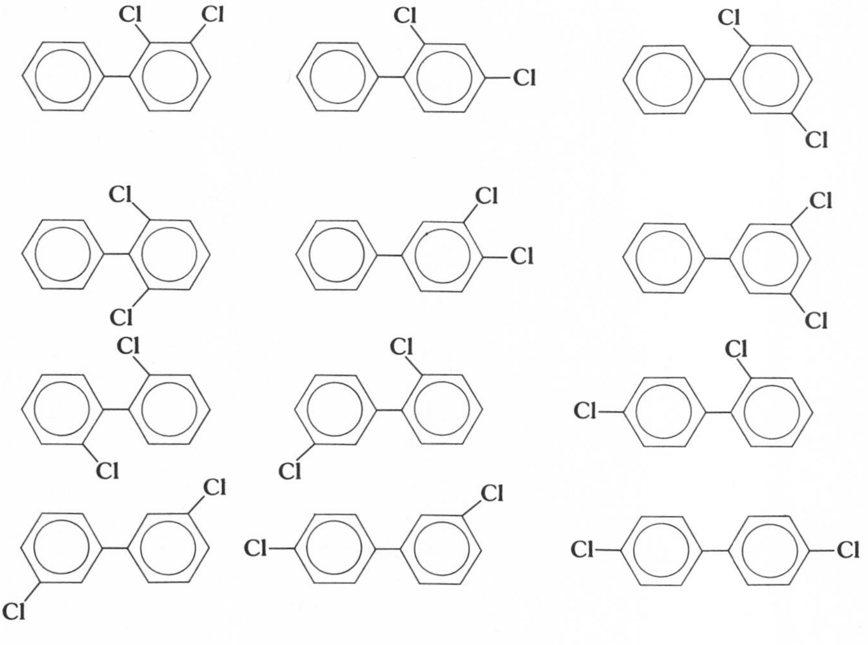

Note that

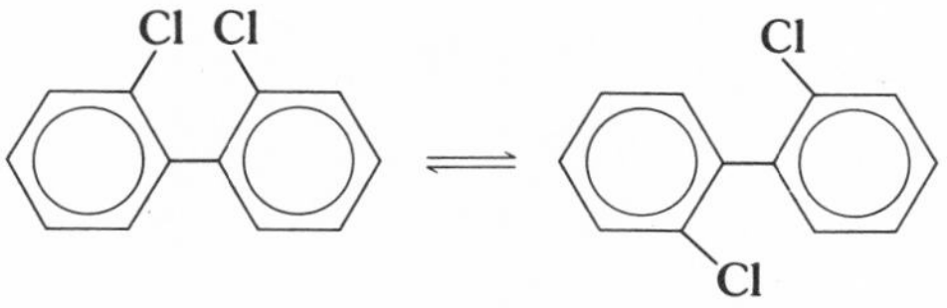

the two 2,2'-dichlorobiphenyl drawings are interconverted by rotation about the C—C bond joining the two aromatic rings. These are therefore conformers, not isomers.

PROBLEM 9·11

a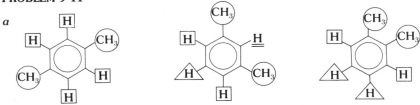

b If the aromatic protons in these isomers happened to have identical chemical shifts, all the isomeric xylenes would have very similar pmr spectra: a strong sharp peak at $\delta = 2.3$ and a smaller signal (two-thirds the area) at $\delta = 7.3$.

PROBLEM 9·12

a $\delta = 7.25$ { structure with $CH_3 \leftarrow \delta = 1.25$, $C-H \leftarrow \delta = 2.95$, CH_3 }

b structure with $H \leftarrow \delta = 6.78$, $CH_3 \leftarrow \delta = 2.25$

PROBLEM 9·13

$\delta = 7.17 \rightarrow$ indane structure with $H_2 \leftarrow \delta = 2.91$, $H_2 \leftarrow \delta = 2.04$, H_2

$\delta = 7.08 \rightarrow$ phenylcyclopropane structure with $\delta = 1.6$ to 2.0, H_2, H_2, $\delta = 0.5$ to 0.9

PROBLEM 9·14

The compound in which $R = H$ will racemize more rapidly than that in which $R = CH_3$ because steric compression will raise the activation energy of the latter reaction.

PROBLEM 9·15

Y is optically inactive because the flat coplanar configuration of its molecules is achiral. If compound X is re-formed from Y, the twisted molecules will again be chiral. The sample of X obtained by such a reaction will, however, be optically inactive, since equal amounts of the enantiomers will be formed.

PROBLEM 9·16

We see from the electronic formula of SO_3 that sulfur is electron deficient:

{ resonance structures of SO_3 } etc.

ANSWERS TO TEXT PROBLEMS   725

The sulfur atom may form covalent bonds with nucleophiles:

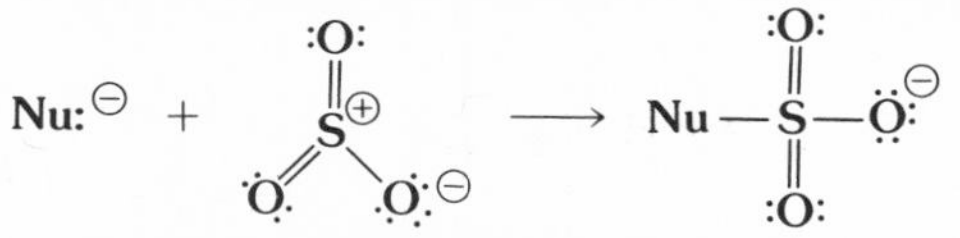

Nu:⊖ + O=S⊕(O⁻)(O⁻):O ⟶ Nu—S(=O)(:O⁻)—O:⊖

Part b of the following problem shows a specific example.

PROBLEM 9·17

a $HONO_2 + 2H_2SO_4 \rightleftharpoons NO_2^{\oplus} + 2HSO_4^{\ominus} + H_3O^{\oplus}$

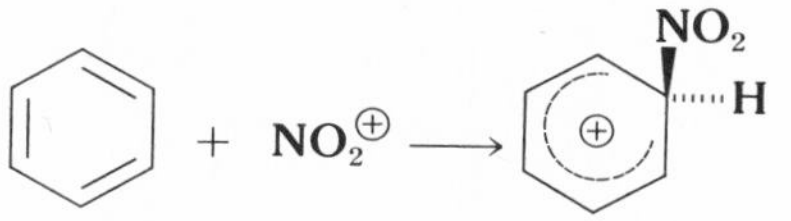

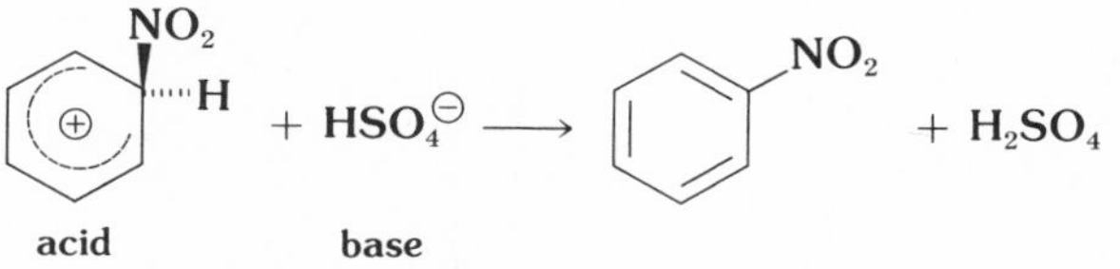

[benzene] + $NO_2^{\oplus}$ ⟶ [cyclohexadienyl cation with NO₂ and H]

[cation intermediate with NO₂, H] + $HSO_4^{\ominus}$ ⟶ [nitrobenzene] + H_2SO_4

  acid   base

b [p-xylene] + SO_3 —H₂SO₄→ [σ-complex with ⊖O₃S, H, CH₃, CH₃] ⇌—H₂SO₄— [σ-complex with HO₃S, H, CH₃, CH₃] + $HSO_4^{\ominus}$

                                                                                acid   base

[σ-complex HO₃S, H, CH₃, CH₃ acid] + $HSO_4^{\ominus}$ ⟶ [sulfonated xylene SO₃H, CH₃, CH₃] + H_2SO_4

                base

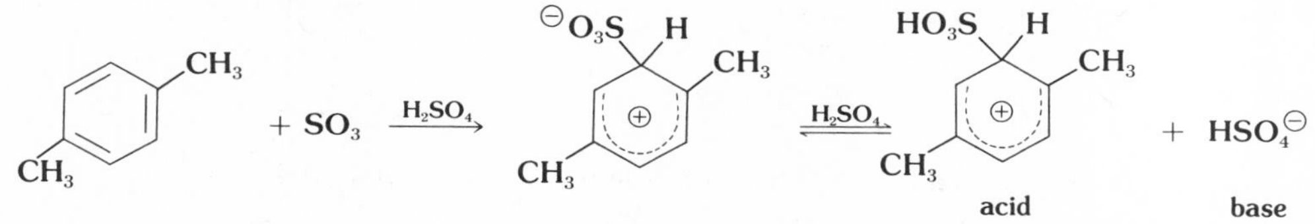

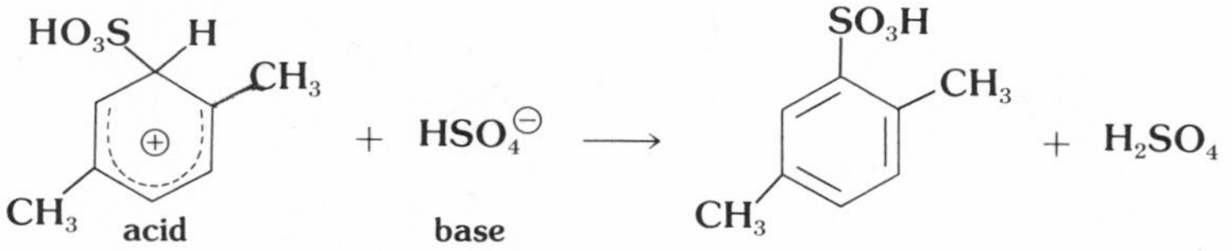

PROBLEM 9·18

a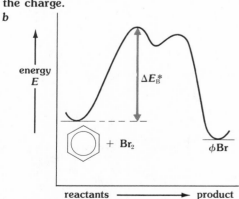

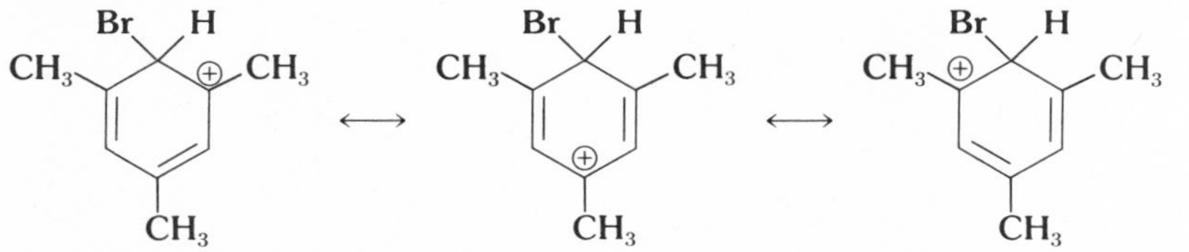

This intermediate is more stable. The methyl substituents assist in delocalizing the charge.

b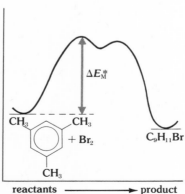

c Mesitylene is nitrated more rapidly than benzene.

PROBLEM 9·19

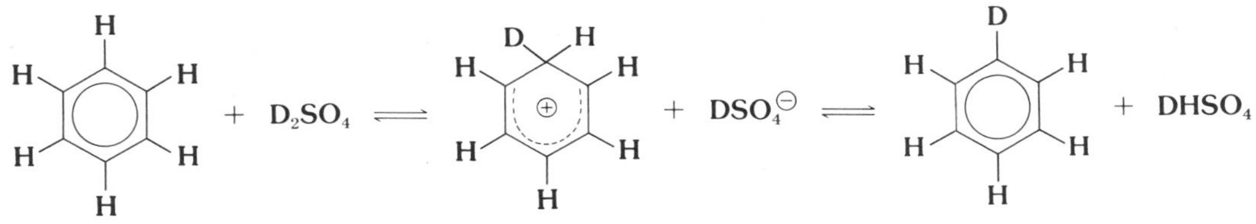

[Reaction: benzene + D$_2$SO$_4$ ⇌ deuterated arenium ion + DSO$_4^{\ominus}$ ⇌ deuterobenzene + DHSO$_4$]

PROBLEM 9·20

[Reaction: benzenesulfonic acid + H$_3$O$^{\oplus}$ ⇌ protonated arenium intermediate + H$_2$O ⇌ benzene + SO$_3$ + H$_3$O$^{\oplus}$]

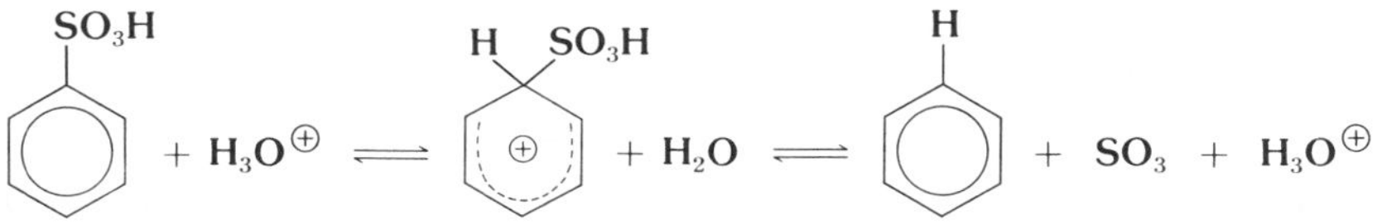

PROBLEM 9·21
Since the pyridinium ion is positively charged, it is less susceptible to attack by electrophilic reagents than is pyridine itself.

PROBLEM 9·22
No. The geometry about the double bond is twisted, and the overlap of p orbitals that gives rise to the π bond is consequently destroyed.

PROBLEM 9·23

a [Structures: thiazole; 2,5-dimethyl-1,3,4-oxadiazole; imidazole with N—H]

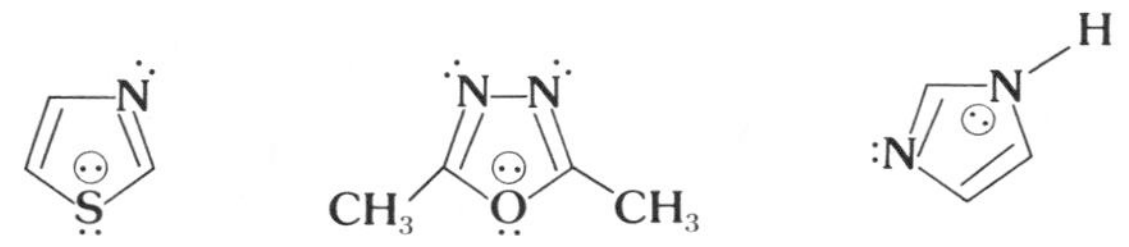

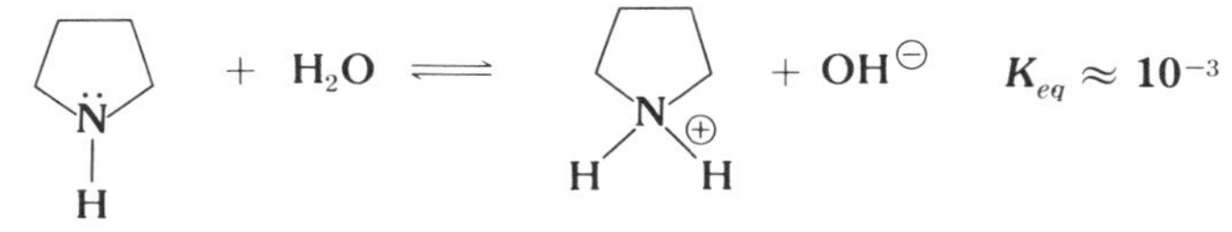

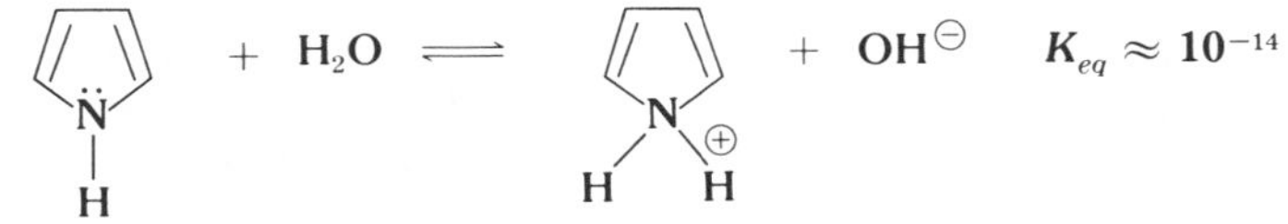

b
pyrrolidine + H$_2$O ⇌ pyrrolidinium + OH$^{\ominus}$ $K_{eq} \approx 10^{-3}$

pyrrole + H$_2$O ⇌ protonated pyrrole + OH$^{\ominus}$ $K_{eq} \approx 10^{-14}$

Pyrrolidine is a stronger base than pyrrole because the electron pair in pyrrole is delocalized in the aromatic system.

PROBLEM 9·24
The stability order predicted **solely** from the extent of charge delocalizations is

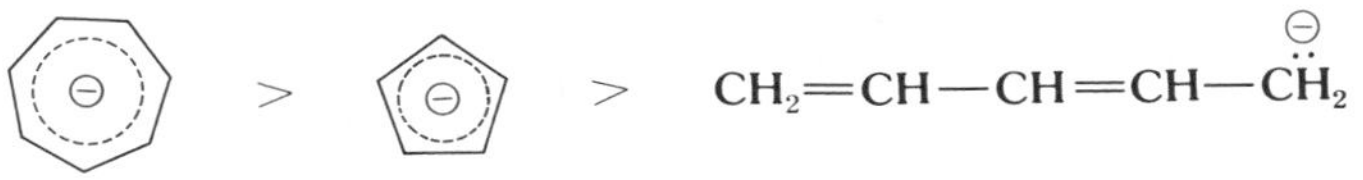

[cycloheptatrienyl anion] > [cyclopentadienyl anion] > CH$_2$=CH—CH=CH—$\overset{\ominus}{\text{C}}H_2$

PROBLEM 9·25

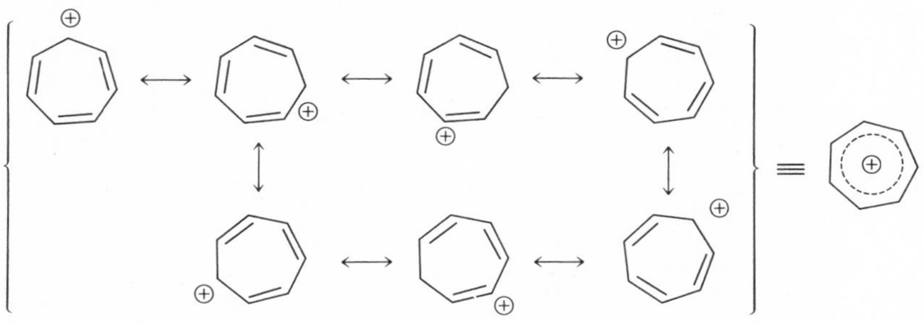

[structures showing resonance of cycloheptatrienyl cation]

PROBLEM 9·26

[structure of 1,6-methano[10]annulene with peripheral H atoms and CH₂ bridge]

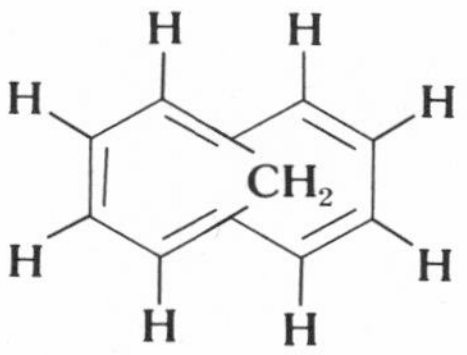

This is a ten-π-electron aromatic ring. The eight peripheral hydrogen atoms should be strongly deshielded ($\delta > 7$ ppm), whereas the two methylene protons on the single carbon bridge should be strongly shielded ($\delta < 0$ ppm). The observed values are: peripheral H atoms $\delta = 6.8$ to 7.5 and CH$_2$ group $\delta = -0.5$.

PROBLEM 9·27

In this case all the π electrons cannot be counted. One of the π bonds in the acetylenic function and the central π bond of the allene are oriented at 90° (no overlap) with the conjugated cyclic π system. We see, then, that this conjugated cyclic π system contains 14 electrons and should therefore be aromatic.

The aromaticity of this compound might manifest itself by showing exceptional stability relative to similar nonaromatic compounds. The pmr spectrum would be particularly informative. Eight protons occupy the deshielding region around the ring, whereas two protons lie in the inside shielding region.

CHAPTER 10

PROBLEM 10·1

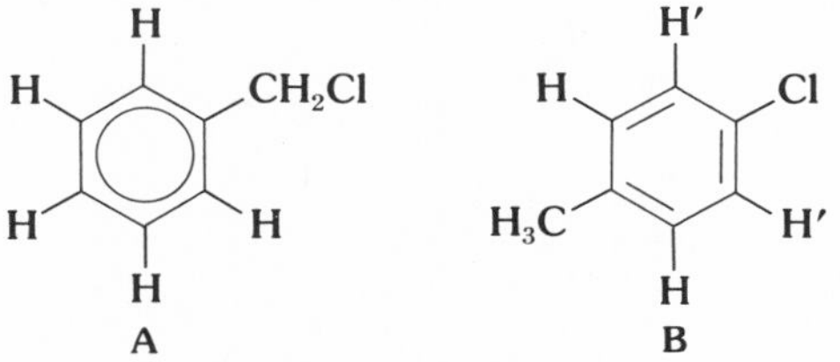

[structures of compound A (benzyl chloride) and compound B (4-chlorotoluene) with labeled H and H′ protons]

A   B

In compound A the aromatic protons all happen to have the same chemical shift. In compound B the aromatic protons ortho to the chlorine atom (H′) have a different chemical shift than the other two. These adjacent protons (H and H′) split each other.

728 ANSWERS TO TEXT PROBLEMS

PROBLEM 10·2

a Cyclohexyl chloride reacts with alcoholic silver nitrate (an S_N1 or S_N2 mechanism) to give a precipitate of AgCl. Chlorobenzene does not give this reaction.

b $I^{\ominus}$ + H₂C=C(CH₂)Cl ⟶ [hypothetical transition state with $\delta^{\ominus}$I---C---Cl$^{\delta\ominus}$] ⟶ I-CH=C(CH₂)H + Cl$^{\ominus}$

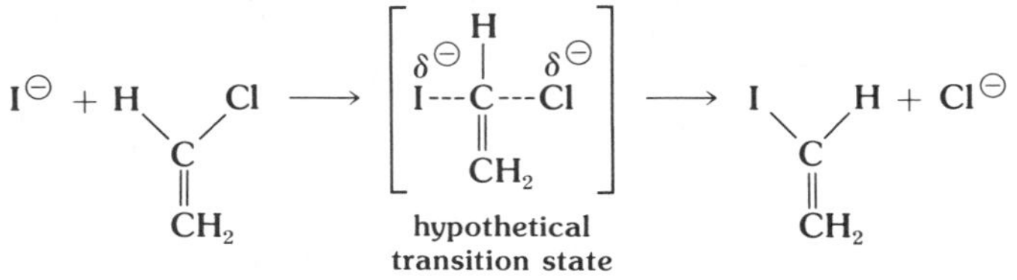

The hybridization of the carbon atom bearing the chlorine changes from sp^2 to sp in the transition state.

PROBLEM 10·3

The reaction described in Figure 10.3 should show second-order kinetics (first order in aryl halide and first order in nucleophile).

PROBLEM 10·4

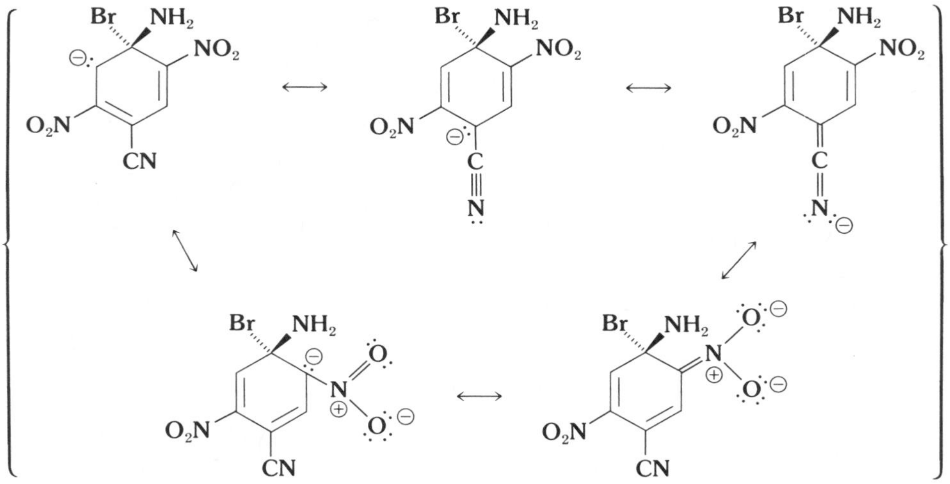

PROBLEM 10·5

a ortho

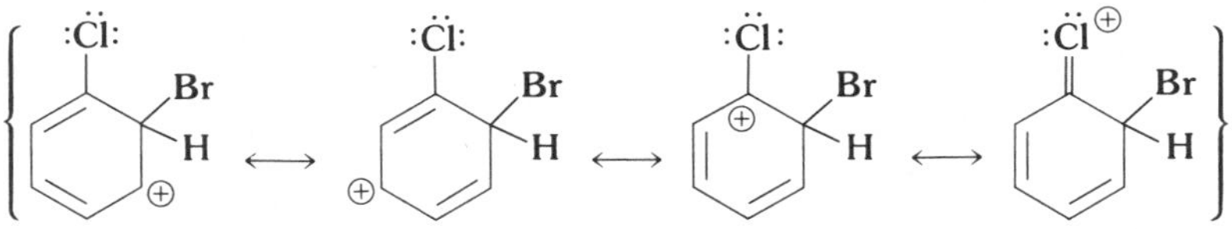

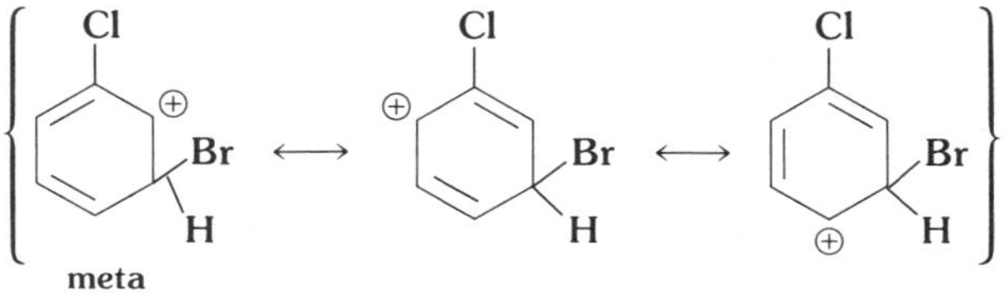

b meta

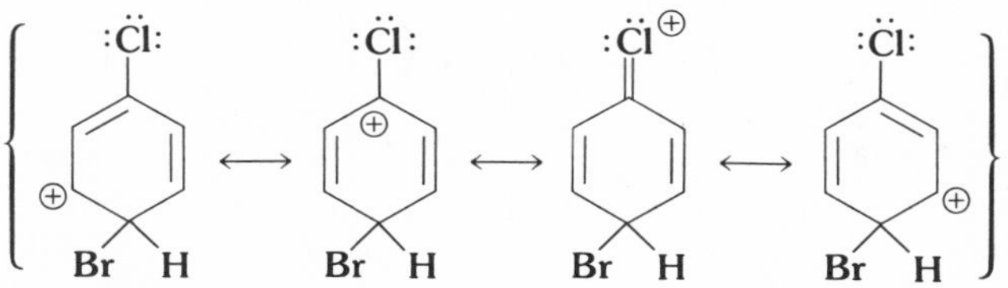

{:Cl: ⟷ :Cl: ⟷ :Cl:⁺ ⟷ :Cl:}
 ⊕   ⊕   ⊕
 Br H   Br H   Br H   Br H
para

b

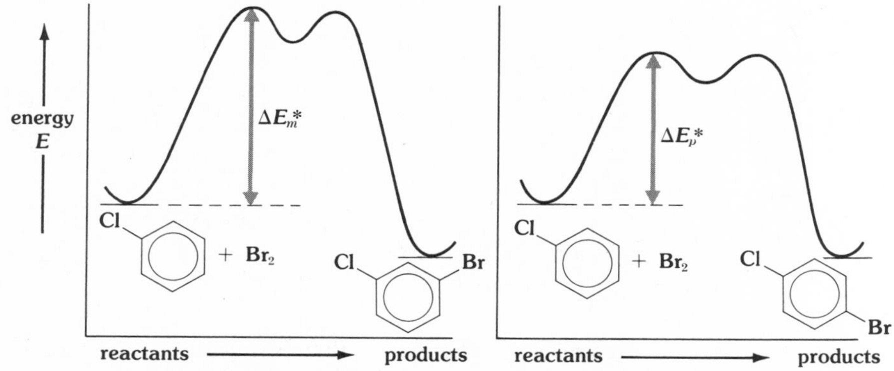

[Energy diagrams: meta substitution with ΔE_m^* (higher barrier); para substitution with ΔE_p^* (lower barrier). Reactants: chlorobenzene + Br$_2$; Products: chloro-bromobenzene isomers.]

PROBLEM 10·6

We can predict the course of these reactions by the following general rule: An electrophilic species (in these cases a proton or a carbonium ion) forms a covalent bond with the double bond so as to generate the most stable carbonium-ion intermediate. This intermediate then reacts with a nucleophile.

a $CH_2=CHBr + HBr \rightleftharpoons \left[CH_3-C\overset{H}{\underset{\overset{\oplus}{\ddot{B}r:}}{\diagup}} \longleftrightarrow CH_3-C\overset{H}{\underset{\ddot{B}r\oplus}{=}} \right] + Br^{\ominus} \longrightarrow CH_3-\underset{Br}{\overset{H}{\underset{|}{C}}}-Br$

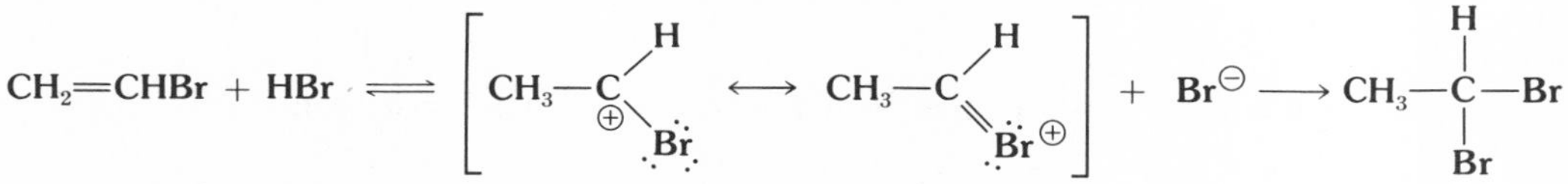

b $(CH_3)_3CCl + FeCl_3 \longrightarrow [(CH_3)_3C^{\oplus}]Cl^{\ominus} \xrightarrow{CH_2=CHCl}$

$\left[(CH_3)_3CCH_2-C\overset{H}{\underset{\ddot{Cl}:}{\diagup}}^{\oplus} \longleftrightarrow (CH_3)_3CCH_2-C\overset{H}{\underset{\overset{\ddot{Cl}:}{\oplus}}{=}} \right] \longrightarrow (CH_3)_3CCH_2CHCl_2$

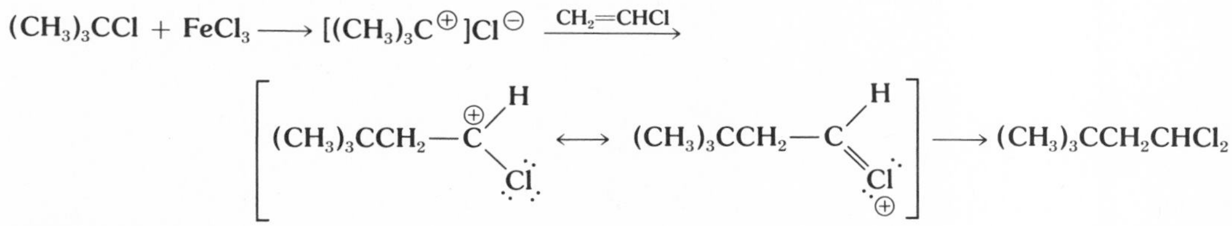

PROBLEM 10·7

a [Resonance/tautomer structures showing phloroglucinol (1,3,5-trihydroxybenzene) and its keto tautomers in equilibrium, culminating in 1,3,5-cyclohexanetrione.]

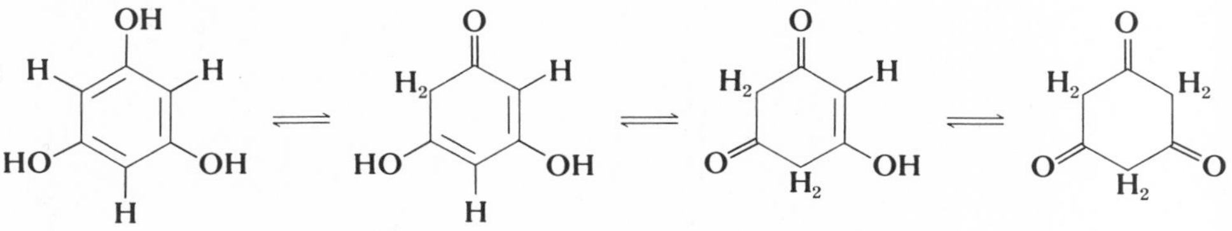

b [1,3-dihydroxybenzene (resorcinol) $\xrightarrow{H_2, Pt}$ 3-hydroxycyclohex-2-enol ⇌ enol form A ⇌ keto form B (1,3-cyclohexanedione)]

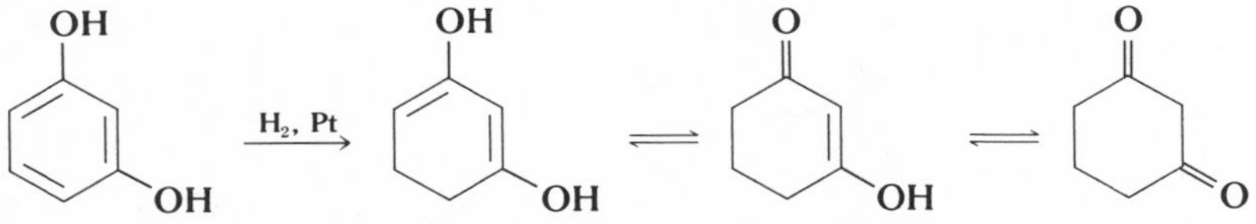

          A   B

The enol and keto tautomers A and B make up the stable dihydro product. Reduction stops here because A and B resist further hydrogen addition.

PROBLEM 10·8

a Acid-base equilibria always favor the weaker acid and the weaker base.

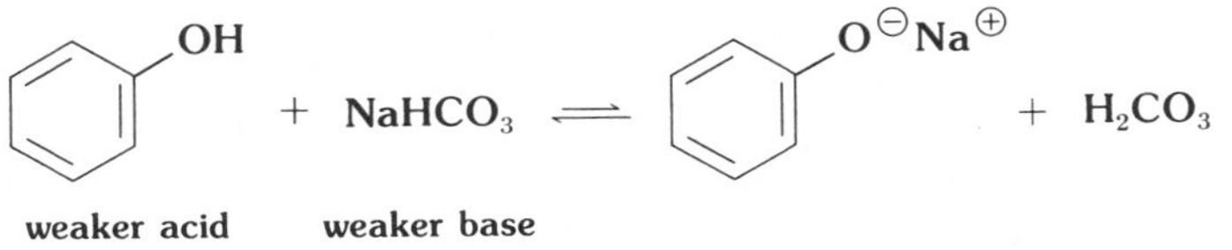

C₆H₅OH + NaHCO₃ ⇌ C₆H₅O⁻Na⁺ + H₂CO₃

weaker acid, weaker base

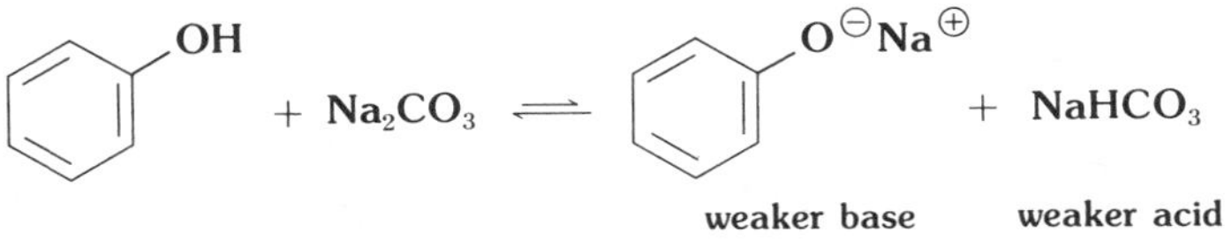

C₆H₅OH + Na₂CO₃ ⇌ C₆H₅O⁻Na⁺ + NaHCO₃

weaker base, weaker acid

b 2,4-dinitrophenol + NaHCO₃ (or Na₂CO₃) ⇌ 2,4-dinitrophenoxide Na⁺ + H₂CO₃ (or NaHCO₃)

weaker base, weaker acid

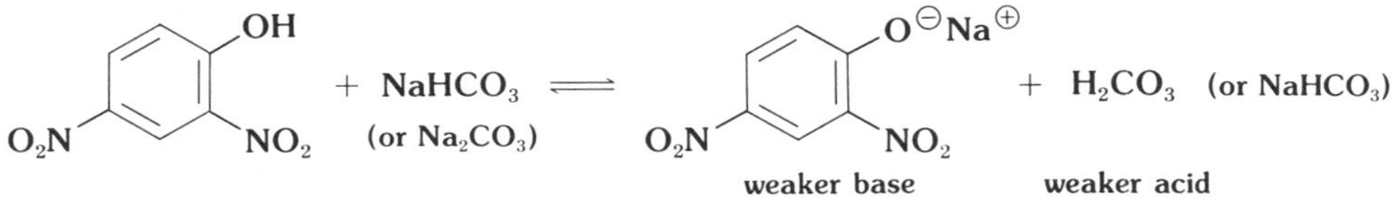

PROBLEM 10·9

The o-nitrophenol has intramolecular hydrogen bonding and is therefore less involved in intermolecular hydrogen bonding. As a result, the boiling point and water solubility of this isomer are reduced.

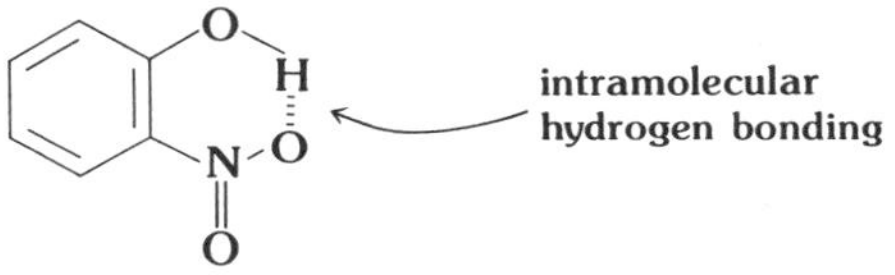

(structure showing intramolecular hydrogen bonding between OH and NO₂ groups on ortho positions)

PROBLEM 10·10

No. The nearest thing to a *meta*-quinone would have a three-membered ring:

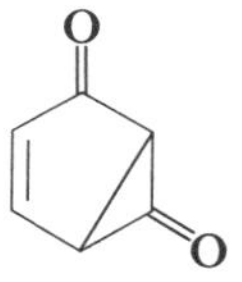

(structure of bicyclic compound with two C=O groups and a three-membered ring)

PROBLEM 10·11

Quinone I incorporates an unstable (antiaromatic) cyclobutadiene ring. After reduction this feature is no longer present.

PROBLEM 10·12

a 4-Br-C₆H₄-CH₂-CN

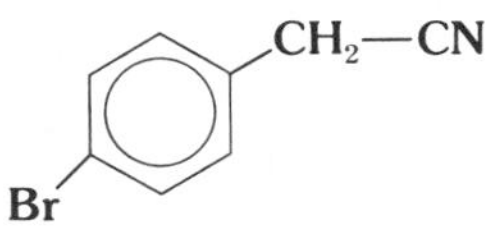

b (CH₃)(BrCH₂CH₂)C=C(H)(CH₂-OCOCH₃)

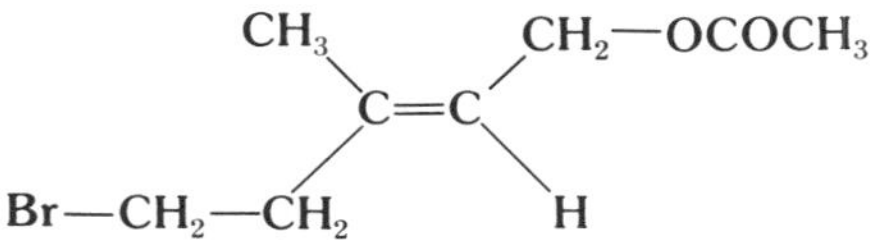

c No reaction

d (structure: 4-bromo-2,6-dimethyl aromatic with CH(CH₃)(I) and C(CH₃)₃ substituents) — No reaction; unlikely to be formed because of severe steric hindrance at the benzylic carbon atom

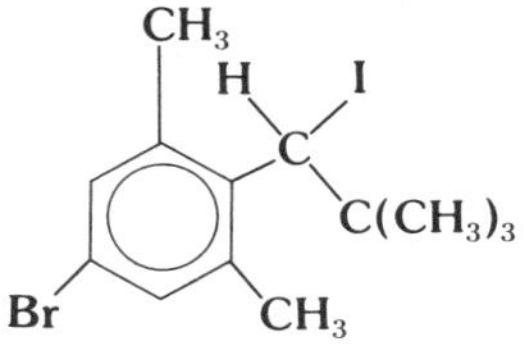

PROBLEM 10·13

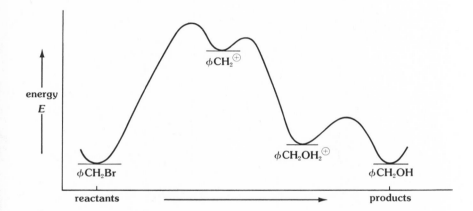

PROBLEM 10·14

$$\left\{ \phi_2\overset{\oplus}{C}-\underset{}{\bigcirc} \longleftrightarrow \phi_2C=\underset{\oplus}{\bigcirc} \longleftrightarrow \phi_2C=\underset{}{\bigcirc}\oplus \longleftrightarrow \phi_2C=\underset{\oplus}{\bigcirc} \right\}$$

Similar contributors can be written for the two other aromatic rings.

PROBLEM 10·15
Yes. A planar carbonium-ion intermediate is formed.

PROBLEM 10·16

a This is an S_N1 (solvolysis) reaction. Electron-donating substituents para to the benzylic position stabilize the carbonium ion intermediate, thus accelerating the rate of solvolysis:

$$\left\{ CH_3-\overset{..}{\underset{..}{O}}-\underset{}{\bigcirc}-\overset{\oplus}{C}\overset{H}{\underset{H}{\diagup}} \longleftrightarrow CH_3-\overset{..}{\underset{}{O}}\overset{\oplus}{=}\underset{}{\bigcirc}=CH_2 \longleftrightarrow \text{etc.} \right\}$$

Electron-withdrawing substituents act to destabilize the intermediate, thereby slowing the reaction:

$$\left\{ Z\leftarrow\underset{}{\bigcirc}-\overset{\oplus}{CH_2} \longleftrightarrow Z\leftarrow\underset{}{\bigcirc}\overset{\oplus}{=}CH_2 \longleftrightarrow Z\leftarrow\underset{(\text{poor})}{\bigcirc\oplus}=CH_2 \longleftrightarrow Z\leftarrow\underset{\oplus}{\bigcirc}=CH_2 \right\}$$

732 ANSWERS TO TEXT PROBLEMS

b

4-methoxybenzyl chloride + C₂H₅OH, ether → 4-methoxybenzyl ethyl ether + HCl

3-methoxybenzyl chloride + C₂H₅OH, ether → 3-methoxybenzyl ethyl ether + HCl

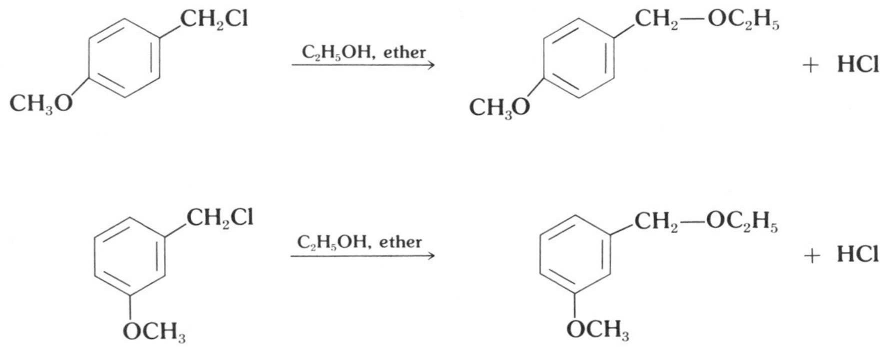

The *p*-methoxy isomer reacts faster than the meta isomer because the carbonium-ion intermediate in these solvolysis reactions is stabilized in the former case (see part *a*).

PROBLEM 10·17

a Combination of a benzyl cation with a nucleophile at the ortho or para positions would destroy the aromatic character of the ring. This transformation requires more energy (greater activation energy) than reaction at the benzylic carbon atom.

b Charge delocalization in the propargyl cation is less effective than that in the allyl cation because the relatively unstable allenic cation contributor is involved:

$$\{ HC{\equiv}C{-}\overset{\oplus}{C}H_2 \longleftrightarrow H{-}\overset{\oplus}{C}{=}C{=}CH_2 \}$$

Also, the ethynyl group is inductively electron withdrawing.

PROBLEM 10·18

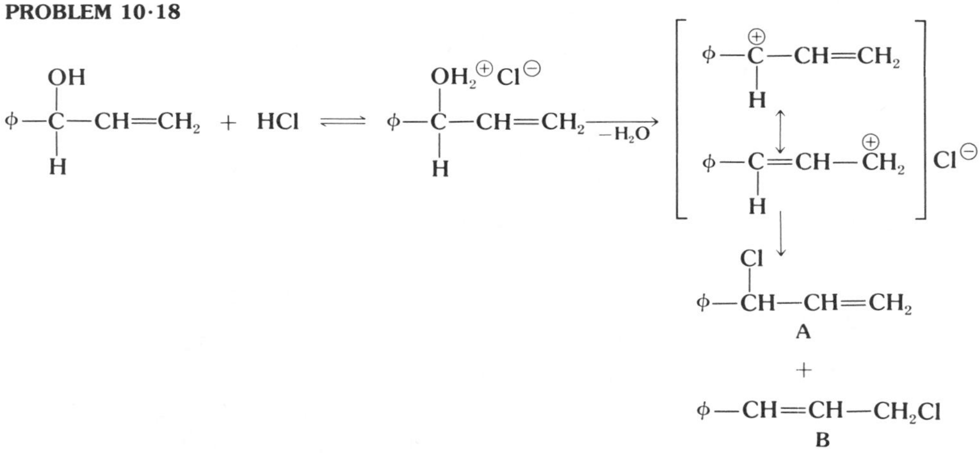

$$\phi-\underset{H}{\overset{OH}{\underset{|}{C}}}-CH=CH_2 + HCl \rightleftharpoons \phi-\underset{H}{\overset{\overset{\oplus}{O}H_2\ Cl^{\ominus}}{\underset{|}{C}}}-CH=CH_2 \xrightarrow{-H_2O} \begin{bmatrix} \phi-\overset{\oplus}{\underset{H}{C}}-CH=CH_2 \\ \updownarrow \\ \phi-\underset{H}{\overset{}{C}}=CH-\overset{\oplus}{C}H_2 \end{bmatrix} Cl^{\ominus}$$

$$\downarrow Cl$$

$$\phi-\underset{A}{CH}-CH=CH_2$$

+

$$\underset{B}{\phi-CH=CH-CH_2Cl}$$

Isomer B is more stable, since it incorporates a more substituted and conjugated double bond.

CHAPTER 11

PROBLEM 11·1

In coniine, piperidine; in nicotine, pyridine and pyrrolidine; in papaverine, isoquinoline; in serotonin, indole; in histamine, 1,3-diazole (or imidazole).

PROBLEM 11·2

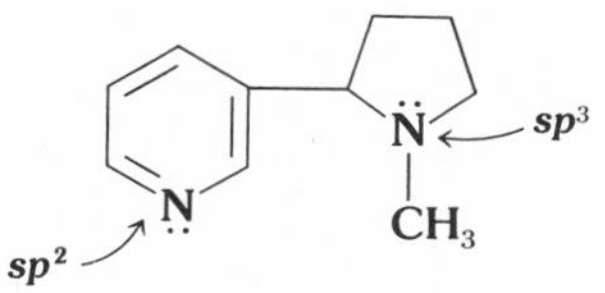

nicotine — pyrrolidine N: sp³; pyridine N: sp²
quinine — quinuclidine N: sp³; quinoline N: sp²
histamine — aliphatic NH₂: sp³; imidazole =N–: sp²; imidazole –NH–: sp² (aromatic)
thiamine — NH₂: sp³; pyrimidine N's: sp²
phenyl isocyanide — $\phi-\overset{\oplus}{N}\equiv \overset{\ominus}{C}:$, N: sp

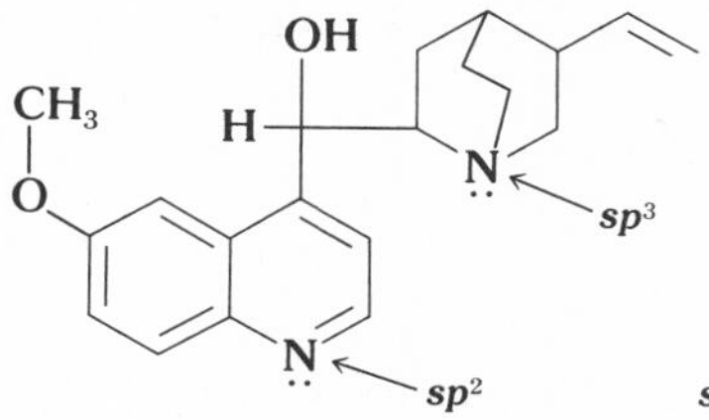

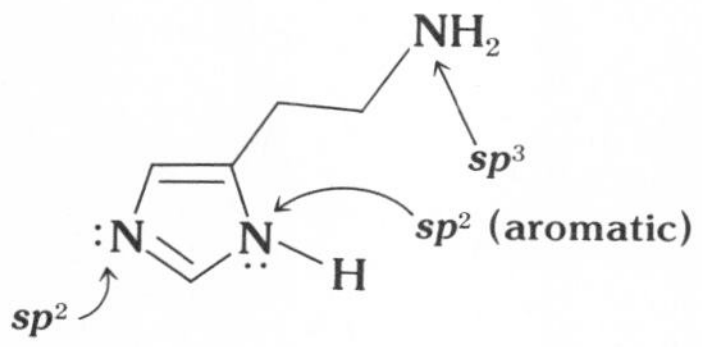

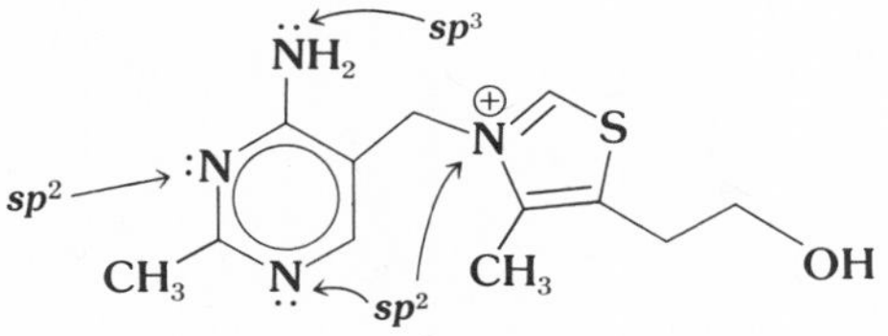

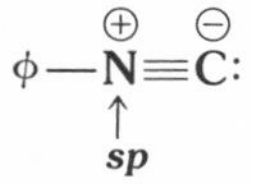

PROBLEM 11·3

a The planar transition state for amine inversion tends to have a trigonal configuration (bond angles 120°). Since the bond angles in this transition state are larger than the angles in the pyramidal form of the amine, the angle strain in the small aziridine ring is greater in the transition state than in the pyramidal form. The activation energy for inversion will therefore be larger for the aziridines than for acyclic amines.

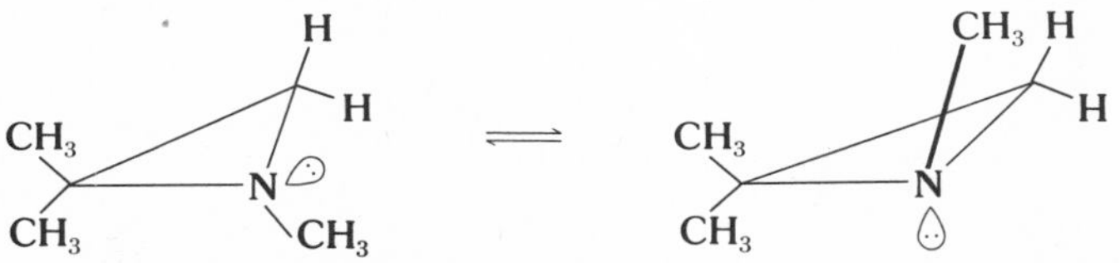

In the pyramidal configuration shown here the two methyl groups bonded to one of the aziridine carbon atoms are not constitutionally equivalent (one is cis to the N-methyl group and the other is trans). A similar nonequivalence exists for the hydrogen atoms of the ring methylene unit. We therefore expect to observe five discrete signals in the pmr spectrum, provided the inversion equilibrium shown above is slow compared with the pmr time scale. Three of these five signals will be due to the methyl groups (the N-methyl group will be found at a lower field than the others) and the other two come from the methylene hydrogen atoms (spin-spin splitting may be present).

If the inversion equilibrium is fast on the pmr time scale, the spectrometer will report an average of the proton chemical shifts. Thus the N-methyl signal will remain apart, the C-methyl groups will coalesce into one signal, and the methylene protons will similarly coalesce. This is observed at higher temperatures because the rate of inversion increases.

PROBLEM 11·4

If a solution of this mixture in benzene is shaken with 5% sodium hydroxide solution, the phenol will dissolve in the aqueous phase as its sodium salt,

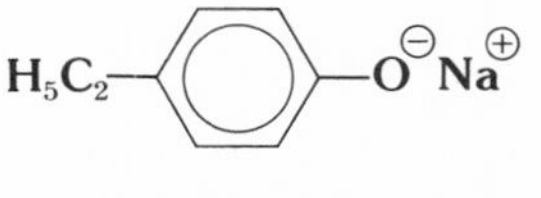

$H_5C_2-C_6H_4-\overset{\ominus}{O}\,\overset{\oplus}{Na}$

The remaining three components will be in the benzene phase, which can be separated from the aqueous phase by a separatory funnel.

Treatment of this benzene solution with 5% sulfuric acid solution (hydrochloric acid will serve equally well) converts the amine component to its water soluble "onium" salt:

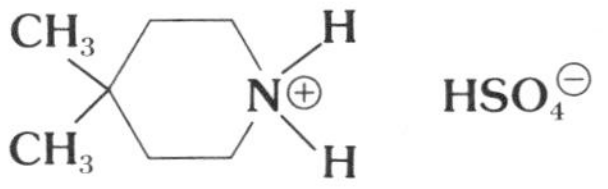

$$\underset{CH_3}{\overset{CH_3}{>}}\!\!\bigcirc\!\!\overset{H}{\underset{H}{N^{\oplus}}} \quad HSO_4^{\ominus}$$

Separation of the immiscible water and benzene phases leaves us with only two components in the benzene solution, the hydrocarbon and the ether. Since ethers are weakly basic, this mixture can be separated by treatment with concentrated sulfuric acid:

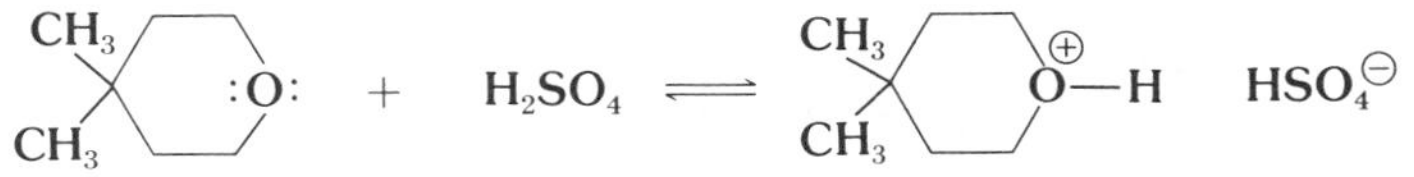

$$\underset{CH_3}{\overset{CH_3}{>}}\!\!\bigcirc\!\!:\!O\!: \;+\; H_2SO_4 \;\rightleftharpoons\; \underset{CH_3}{\overset{CH_3}{>}}\!\!\bigcirc\!\!\overset{\oplus}{O}\!-\!H \quad HSO_4^{\ominus}$$

Recovery of the neutral components from the salt solutions is achieved by a reversal of the salt-forming reactions:

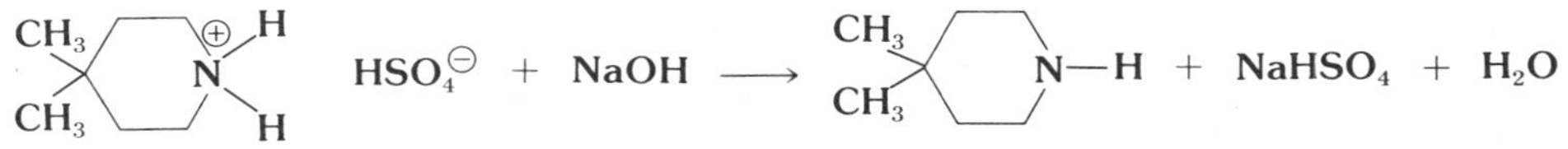

$$H_5C_2\!-\!\bigcirc\!-\!O^{\ominus}Na^{\oplus} \;+\; H_3O^{\oplus}Cl^{\ominus} \;\longrightarrow\; H_5C_2\!-\!\bigcirc\!-\!OH \;+\; NaCl \;+\; H_2O$$

$$\underset{CH_3}{\overset{CH_3}{>}}\!\!\bigcirc\!\!\overset{H}{\underset{H}{N^{\oplus}}} \; HSO_4^{\ominus} \;+\; NaOH \;\longrightarrow\; \underset{CH_3}{\overset{CH_3}{>}}\!\!\bigcirc\!\!N\!-\!H \;+\; NaHSO_4 \;+\; H_2O$$

$$\underset{CH_3}{\overset{CH_3}{>}}\!\!\bigcirc\!\!\overset{\oplus}{O}\!-\!H \; HSO_4^{\ominus} \;+\; H_2O \;\longrightarrow\; \underset{CH_3}{\overset{CH_3}{>}}\!\!\bigcirc\!\!O \;+\; H_3O^{\oplus}HSO_4^{\ominus}$$

PROBLEM 11·5

$K_a = 10^{-14}/K_b$

$C_2H_5NH_3^{\oplus}: K_a = 10^{-14}/5 \times 10^{-4} = 2 \times 10^{-11}$

$\phi NH_3^{\oplus}: K_a = 10^{-14}/4 \times 10^{-10} = 2.5 \times 10^{-5}$

$\bigcirc\!\!-\!NH_2: K_a = 10^{-14}/10^{-14} = 1$

PROBLEM 11·6

a

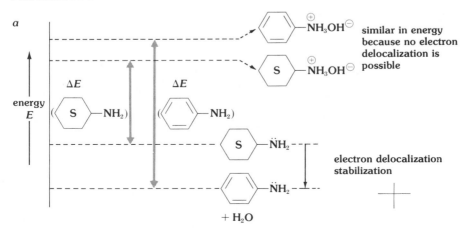

Aniline is a weaker base than cyclohexyl amine, primarily because p-π conjugation stabilizes the amine relative to its conjugate acid.

b Nitro substituents in the ortho or para positions of an aniline ring reduce the basicity of the amine function because they effect additional stabilization of the free amine through further delocalization of the nonbonding electron pair:

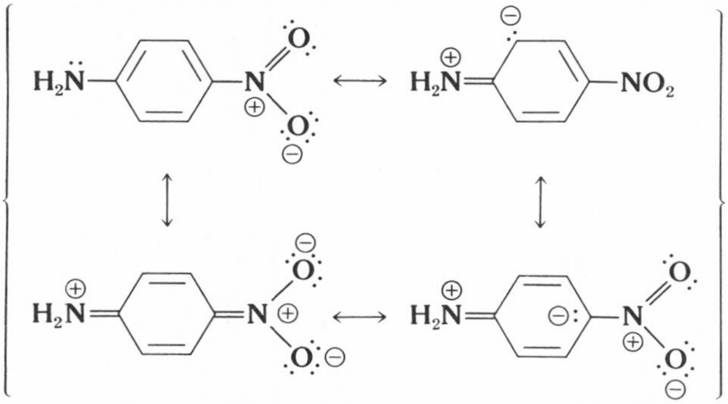

[Resonance structures showing delocalization of the amine lone pair onto the para-nitro group through the aromatic ring]

A meta nitro substituent cannot achieve such a delocalization and is therefore much less effective in influencing the basicity of the amine.

PROBLEM 11·7

ϕ—CH$_2$—$\ddot{\text{N}}$—CH$_2$—ϕ, from dialkylation.
    |
    ϕ

PROBLEM 11·8

$(C_2H_5)_2\ddot{N}H + HNO_2 \rightleftharpoons (C_2H_5)_2NH_2^{\oplus} NO_2^{\ominus}$

PROBLEM 11·9

N_2O_3 may be regarded as the anhydride of nitrous acid:

$O=N-OH + HO-N=O \longrightarrow O=N-O-N=O + H_2O$.

NOCl may be regarded as the acid chloride of nitrous acid.

PROBLEM 11·10

The nitroso function in an *N*-nitrosoamine behaves like a carbonyl function in reducing the basicity of the amine through electron-pair delocalization:

$R_2\ddot{N}-\ddot{N}=\ddot{O} \longleftrightarrow R_2\overset{\oplus}{N}=\ddot{N}-\ddot{O}:^{\ominus}$

PROBLEM 11·11

a Aniline is a primary amine. It will react with oxidizing agents either at nitrogen (as in equation 11.19) or at the ortho or para ring positions:

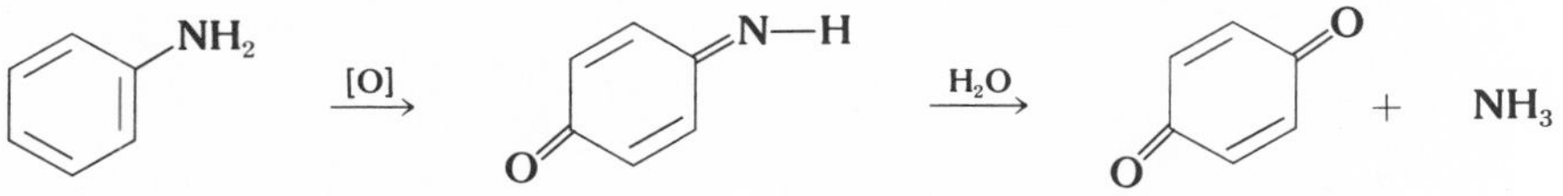

[Reaction scheme: aniline →[O]→ quinone imine →H₂O→ benzoquinone + NH₃]

Many other oxidation products are also possible.

b $(C_2H_5)_3\overset{\oplus}{N}-O-H \; X^{\ominus}$

PROBLEM 11·12

a We know that the thermodynamic stability of double bonds increases with increasing alkyl-group substitution (Section 4.4). Consequently, elimination reactions which proceed through transition states having extensive double-bond formation will favor those products with more highly substituted double bonds.

736 ANSWERS TO TEXT PROBLEMS

Conversely, reactions which favor the less substituted alkenes are likely to be characterized by a low degree of double-bond formation at the transition state.

two eliminations 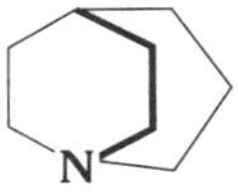 three eliminations 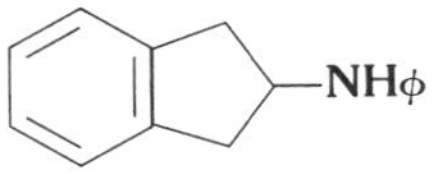 one elimination

PROBLEM 11·13

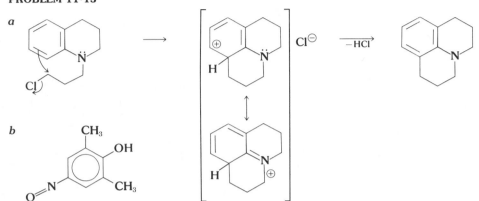

PROBLEM 11·14

a

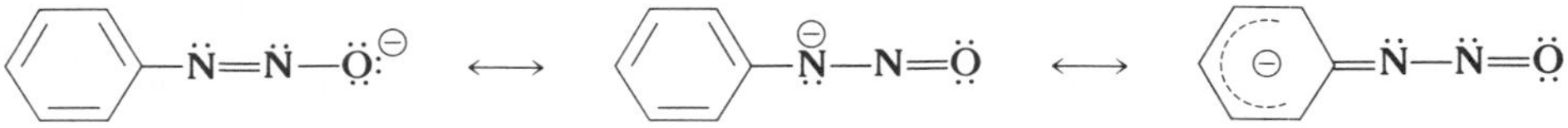

A para nitro substituent would provide additional charge delocalization and would therefore enhance the acidity of the corresponding diazoic acid.

b

PROBLEM 11·15

a

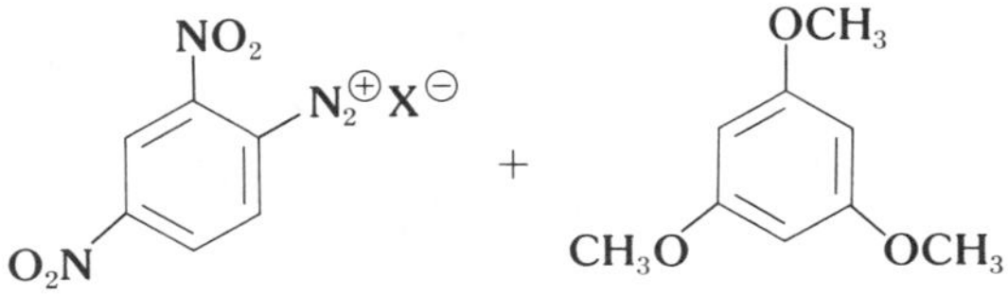

b

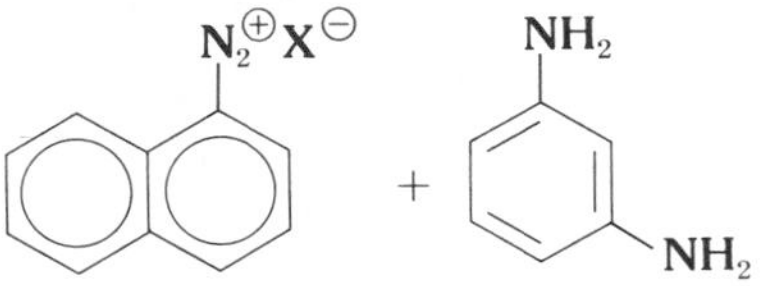

c

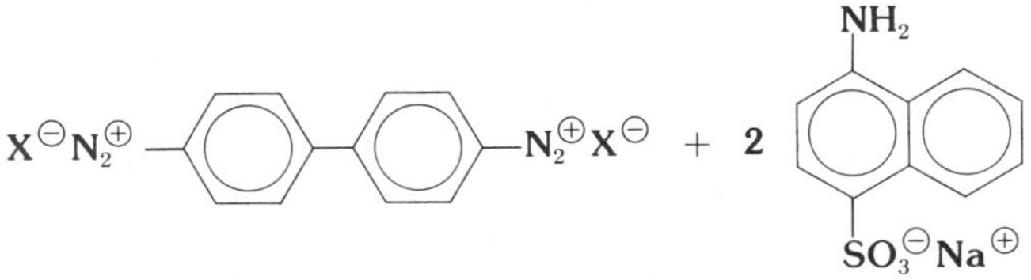

PROBLEM 11·16

a Direct iodination of benzene is very sluggish. An indirect method must be used:

$$\text{C}_6\text{H}_6 \xrightarrow{\text{HNO}_3, \text{H}_2\text{SO}_4} \text{C}_6\text{H}_5\text{NO}_2 \xrightarrow[\Delta]{\text{Fe, HCl}} \text{C}_6\text{H}_5\text{NH}_2 \xrightarrow[0°]{\text{NaNO}_2, \text{H}_2\text{SO}_4} \text{C}_6\text{H}_5\text{N}_2^{\oplus}\text{HSO}_4^{\ominus} \xrightarrow{\text{KI, H}_2\text{O}} \text{C}_6\text{H}_5\text{I}$$

b $(\text{CH}_3)_2\text{CHBr} + \text{KCN} \xrightarrow[\Delta]{\text{alcohol}} (\text{CH}_3)_2\text{CHC}\equiv\text{N} \xrightarrow{\text{LiAlH}_4, \text{ether}} (\text{CH}_3)_2\text{CHCH}_2\text{NH}_2$

c The cyano group is conveniently introduced onto an aromatic ring by a Sandmeyer reaction of a diazonium intermediate.

$$p\text{-BrC}_6\text{H}_4\text{NH}_2 \xrightarrow[0°]{\text{NaNO}_2, \text{H}_2\text{SO}_4} p\text{-BrC}_6\text{H}_4\text{N}_2^{\oplus}\text{HSO}_4^{\ominus} \xrightarrow{\text{Cu}_2(\text{CN})_2} p\text{-BrC}_6\text{H}_4\text{CN}$$

This approach would require the synthesis of p-bromoaniline from benzene. p-Bromoaniline can be prepared by the sequence in equation 11.33 or by a reverse sequence:

$$\text{C}_6\text{H}_6 \xrightarrow{\text{Br}_2, \text{Fe}^{3+}} \text{C}_6\text{H}_5\text{Br} \xrightarrow{\text{HNO}_3, \text{H}_2\text{SO}_4} o\text{-}/p\text{-BrC}_6\text{H}_4\text{NO}_2 \xrightarrow{\text{separation}} p\text{-BrC}_6\text{H}_4\text{NO}_2 \xrightarrow{\text{SnCl}_2, \text{HCl}} p\text{-BrC}_6\text{H}_4\text{NH}_2$$

d $\text{C}_6\text{H}_6 \xrightarrow{\text{part } a} \text{C}_6\text{H}_5\text{N}_2^{\oplus}\text{HSO}_4^{\ominus} \xrightarrow{\text{Cu}_2(\text{CN})_2} \text{C}_6\text{H}_5\text{CN} \xrightarrow{\text{LiAlH}_4, \text{ether}} \text{C}_6\text{H}_5\text{CH}_2\text{NH}_2$

e $\text{C}_6\text{H}_6 \xrightarrow{\text{part } a} \text{C}_6\text{H}_5\text{NH}_2 \xrightarrow{(\text{CH}_3\text{CO})_2\text{O}} \text{C}_6\text{H}_5\text{NHCOCH}_3 \xrightarrow[\Delta]{\text{HNO}_3} p\text{-O}_2\text{NC}_6\text{H}_4\text{NHCOCH}_3 + \text{ortho isomer}$

$\xrightarrow[\Delta]{\text{H}_3\text{O}^{\oplus}} p\text{-O}_2\text{NC}_6\text{H}_4\text{NH}_2 \xrightarrow[0°]{\text{NaNO}_2, \text{H}_2\text{SO}_4} p\text{-O}_2\text{NC}_6\text{H}_4\text{N}_2^{\oplus}\text{HSO}_4^{\ominus} \xrightarrow[\Delta]{\text{H}_3\text{O}^{\oplus}} p\text{-O}_2\text{NC}_6\text{H}_4\text{OH} \xrightarrow[\Delta]{\text{Fe, HCl}} p\text{-H}_2\text{NC}_6\text{H}_4\text{OH}$

f The most direct approach would be to dibrominate nitrobenzene and then reduce the nitro function:

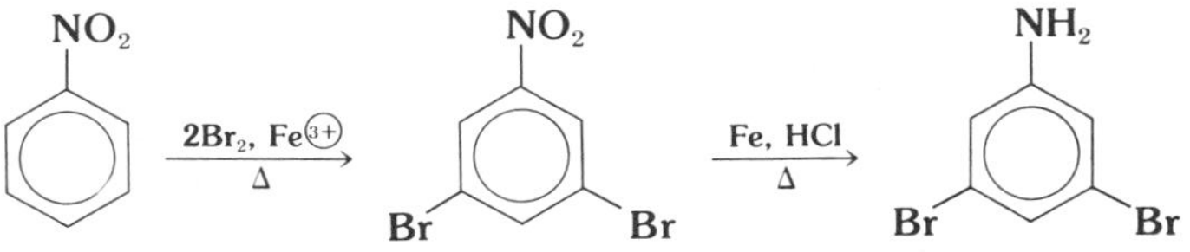

[NO₂-benzene] →(2Br₂, Fe³⁺, Δ) [3,5-dibromonitrobenzene] →(Fe, HCl, Δ) [3,5-dibromoaniline]

The problem here lies in the orientation of the second bromine atom. Since the first bromine will direct new substituents ortho and para to itself, the main dibromo product may be the 3,4-dibromonitrobenzene rather than the desired 3,5 dibromo isomer. We can circumvent this difficulty by using an amine activating group that will be removed in a later step:

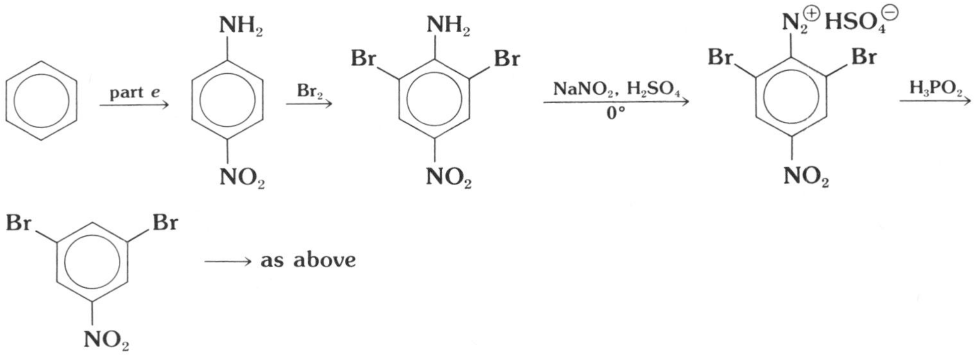

benzene →(part e) p-nitroaniline →(Br₂) 2,6-dibromo-4-nitroaniline →(NaNO₂, H₂SO₄, 0°) diazonium salt →(H₃PO₂)

3,5-dibromonitrobenzene ⟶ as above

CHAPTER 12

PROBLEM 12·1

If water enriched with an isotope of oxygen (such as ^{18}O or ^{17}O) is mixed with a carbonyl compound, the aldehyde or ketone incorporates the oxygen isotope:

$$R_2C=O + H_2^{18}O \rightleftharpoons R_2C(OH)(^{18}OH) \rightleftharpoons R_2C=^{18}O + H_2O$$

PROBLEM 12·2

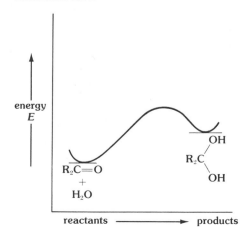
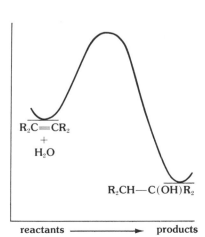

PROBLEM 12·3
If we take the simplest unsaturated ketone, 3-buten-2-one, as the reference compound, we begin with a basic λ_{max} of 215 nm for this chromophore. The examples given here clearly show that substituents on the double bond increase λ_{max} by about 10 nm each:

compound	number of substituents	λ_{max} (calculated)
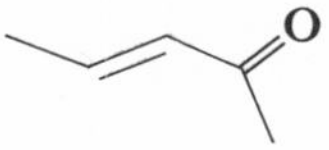 CH₃-CH=CH-C(=O)-CH₃	1	215 + 10 = 225 nm
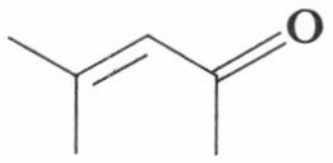 (CH₃)₂C=CH-C(=O)-CH₃ (with α-methyl)	2	215 + 20 = 235 nm
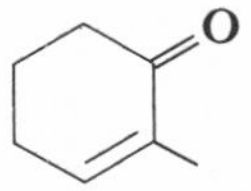 2-methylcyclohex-2-enone	2	215 + 20 = 235 nm

From examination of a much larger set of known enone compounds, these rules have been refined to give increments of +10 nm for α-alkyl substituents and +12 nm for β-alkyl substituents. Since aryl substituents extend the π conjugation of the chromophore, they usually exert a much stronger influence than alkyl groups.

PROBLEM 12·4
Unstrained saturated ketones ordinarily show a carbonyl stretching frequency in the 1705 to 1725 cm⁻¹ range. Aldehydes tend to have slightly higher frequencies. The introduction of a conjugated double bond lowers this stretching frequency by about 35 cm⁻¹. In contrast, ring strain (angle strain) in cyclic ketones increases the stretching frequency, as noted for cyclopentanone, cyclobutanone, and cyclopropanone.

PROBLEM 12·5
a The infrared measurements indicate a saturated ketone and the pmr signals indicate an ethyl group and a methyl ketone:

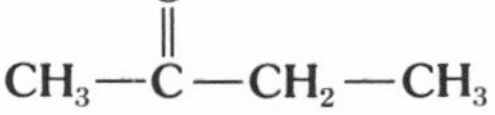

$$CH_3-\overset{O}{\underset{\|}{C}}-CH_2-CH_3$$

b The infrared spectrum indicates that there is no carbonyl or hydroxyl group. The pmr signals show two constitutionally distinct groups of protons having equal populations. The $\delta = 3.75$ group is next to the oxygen:

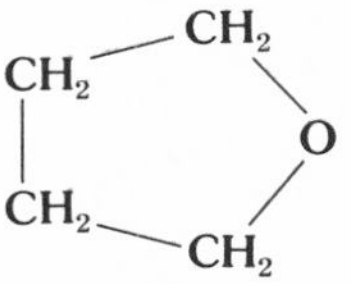

(tetrahydrofuran structure)

c The infrared spectrum indicates a saturated aldehyde. The pmr spectrum confirms an aldehyde and also shows an isopropyl group ($\delta = 1.2$ doublet):

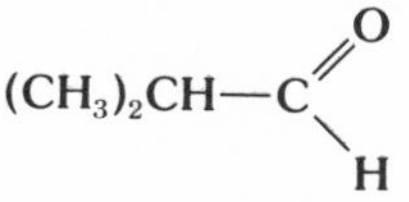

$$(CH_3)_2CH-C\overset{O}{\underset{H}{\diagdown}}$$

740 ANSWERS TO TEXT PROBLEMS

PROBLEM 12·6

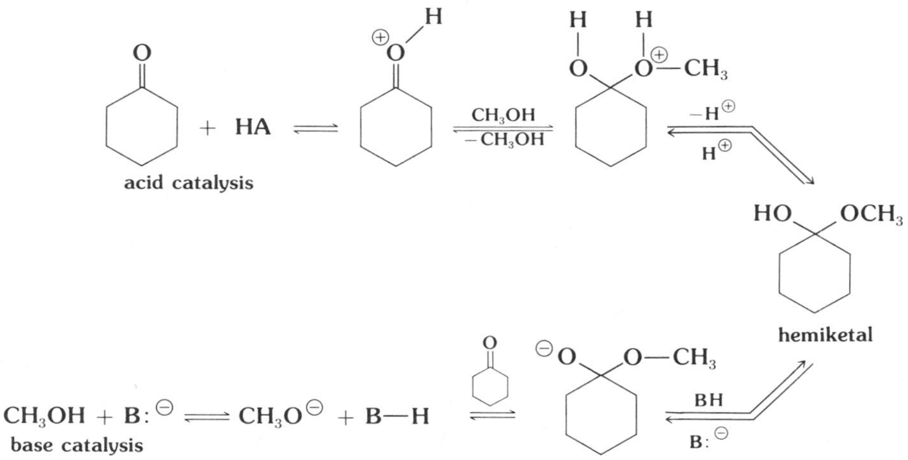

[Cyclohexanone] + HA ⇌ [protonated cyclohexanone, C=O⁺H] →(CH₃OH / −CH₃OH)← [cyclohexane ring with O–H and O⁺(H)–CH₃ groups at same carbon] ⇌(−H⁺ / H⁺) [hemiketal: cyclohexane ring with HO and OCH₃ groups]

acid catalysis

CH₃OH + B:⁻ ⇌ CH₃O⁻ + B—H , [cyclohexanone with ⁻O adjacent] ⇌ [cyclohexane ring with ⁻O and O—CH₃] ⇌(BH / B:⁻) [hemiketal]

base catalysis

PROBLEM 12·7

The carbonyl function is polar:

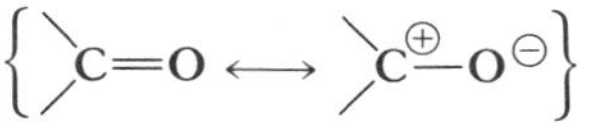

$$\{\,\mathrm{C{=}O} \longleftrightarrow \mathrm{C^{\oplus}{-}O^{\ominus}}\,\}$$

The presence of a positively charged or electrophilic group adjacent to a carbonyl function results in an unfavorable (high potential energy) juxtaposition of like charges.

PROBLEM 12·8

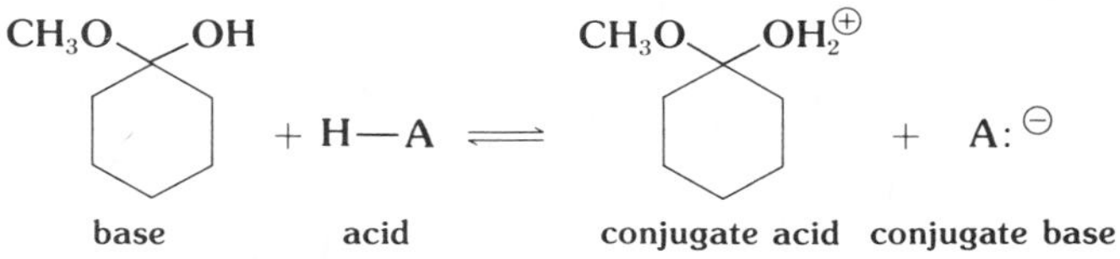

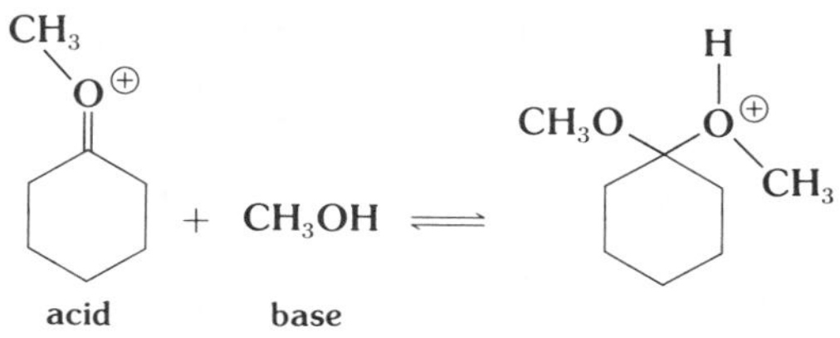

[Cyclohexane with CH₃O and OH substituents] + H—A ⇌ [Cyclohexane with CH₃O and OH₂⁺ substituents] + A:⁻

base   acid   conjugate acid   conjugate base

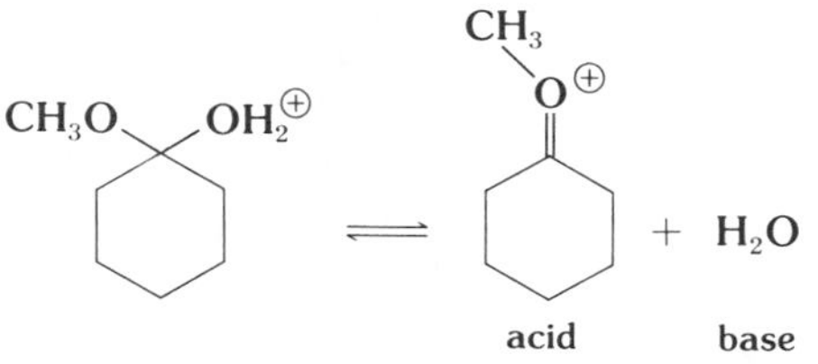

[Cyclohexane with CH₃O and OH₂⁺] ⇌ [Cyclohexanone-like with ⁺O–CH₃ (oxocarbenium)] + H₂O

acid   base

[Oxocarbenium with ⁺O–CH₃] + CH₃OH ⇌ [Cyclohexane with CH₃O and O⁺(H)–CH₃ substituents]

acid   base

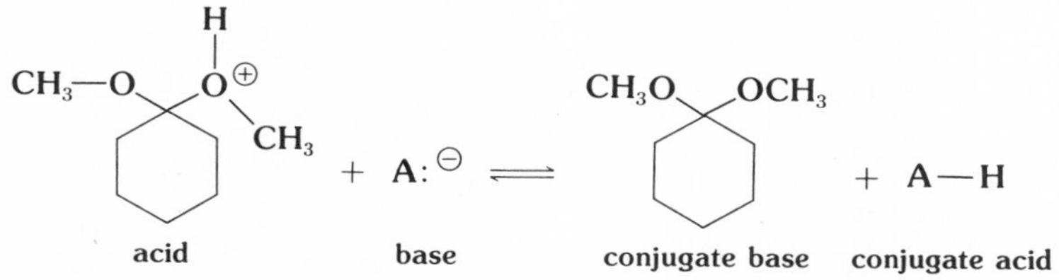

ANSWERS TO TEXT PROBLEMS 741

[Structural diagram showing acid-base equilibrium:]

CH₃—O, O⁺—H / CH₃ (on cyclohexane) + A:⁻ ⇌ CH₃O, OCH₃ (on cyclohexane) + A—H

acid   base   conjugate base   conjugate acid

PROBLEM 12·9

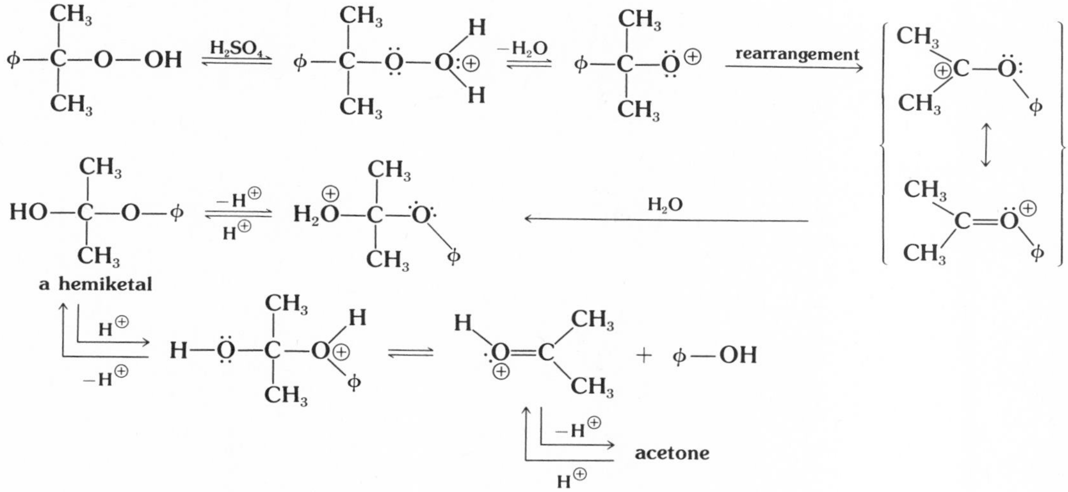

[Mechanism showing acid-catalyzed ketal hydrolysis starting from φ—C(CH₃)₂—O—OH with H₂SO₄, proceeding through protonation, loss of H₂O, rearrangement to give (CH₃)₂C—O⁺—φ resonance structures, then addition of H₂O to form hemiketal HO—C(CH₃)₂—O—φ, then proton transfer to give H—O—C(CH₃)₂—O⁺(H)—φ, loss of φ—OH to give protonated acetone (H)O=C⁺(CH₃)₂, and loss of H⁺ to give acetone.]

PROBLEM 12·10

a   CH₃—CH(OH)(OC₂H₅) + B: ⇌ CH₃—CH(O⁻)(OC₂H₅) ⇌ CH₃—CHO + C₂H₅O:⁻

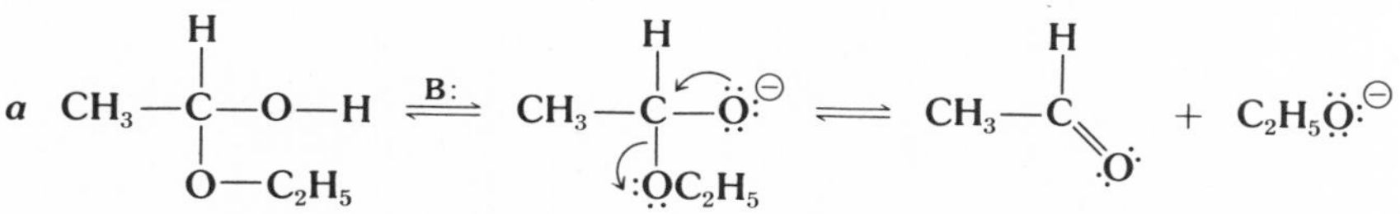

The corresponding acetal, CH₃CH(OC₂H₅)₂, does not have an acidic function, such as a hydroxyl group, and therefore does not react with a base.

b The third step in the acid-catalyzed conversion of a hemiketal to a ketal (see the answer to problem 12.8) is a bimolecular reaction between an organic cation and an alcohol. This becomes a unimolecular step in the cyclic ketal case:

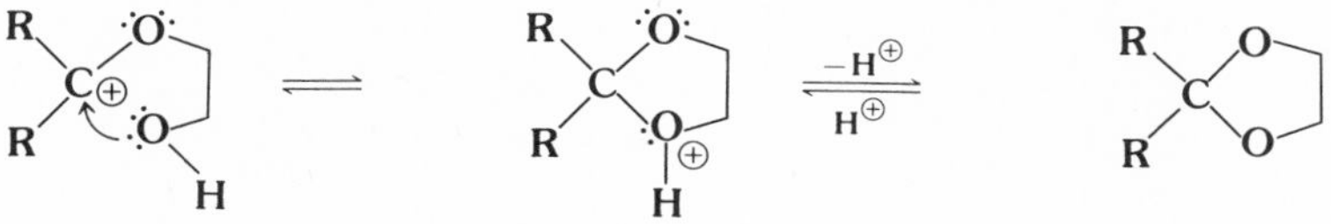

[Cyclic structure: R₂C⁺ with O on one side and O—H⁺ on other (five-membered ring) ⇌ R₂C in five-membered ring with O and O⁺—H ⇌(−H⁺/+H⁺) R₂C with cyclic —O—CH₂—CH₂—O—]

Consequently, cyclic ketals tend to form more rapidly than acyclic ketals.

c   R₂C=O + 2R'SH ⟶ R₂C(SR')₂ + H₂O
    ΔH(calc.) = 179 + 166 − (130 + 221.2) = −6.2 kcal/mole

The reaction is thus weakly exothermic.

PROBLEM 12·11

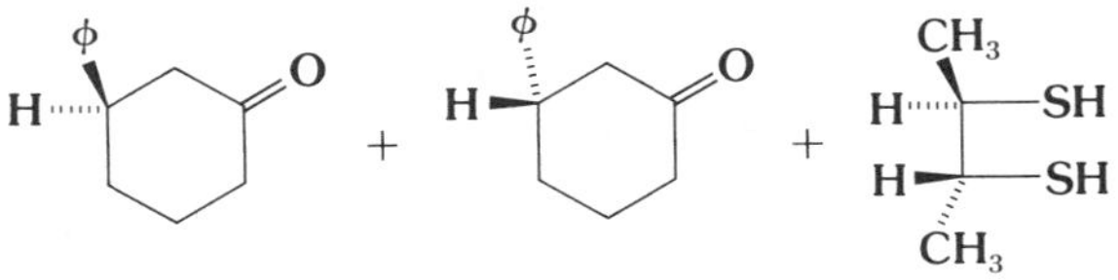

[Structures: (±)-3-phenylcyclohexanone + R,R-2,3-butanedithiol → mixture of diastereomeric thioketals]

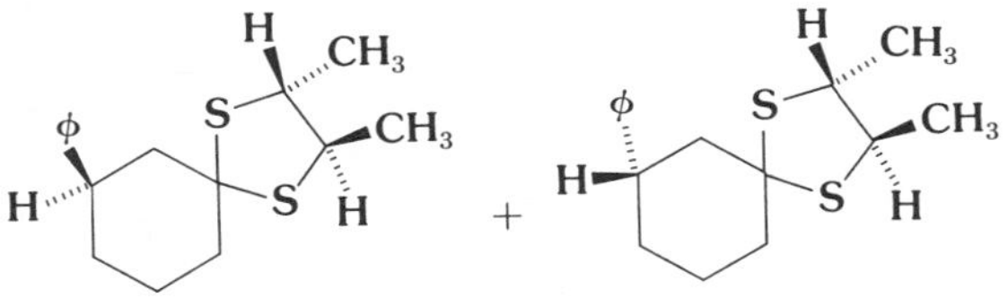

The diastereomeric thioketals have different physical properties and can therefore be separated by crystallization, distillation, or chromatography. Regeneration of the ketones then leads to optically pure (or partially pure) products. For example:

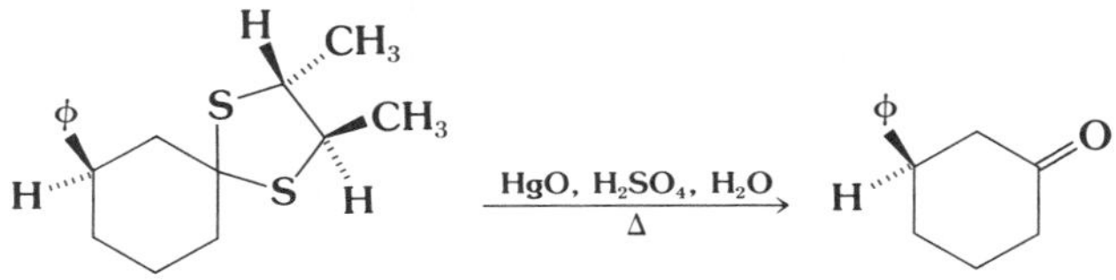

[Structure: R,R,R diastereomer] —HgO, H₂SO₄, H₂O, Δ→ [Structure: R enantiomer]

PROBLEM 12·12

The cyanohydrin of methone suffers nonbonded steric interactions which raise its potential energy. The equilibrium between the ketone and the cyanohydrin is therefore less favorable to the cyanohydrin than in the case of cyclohexanone.

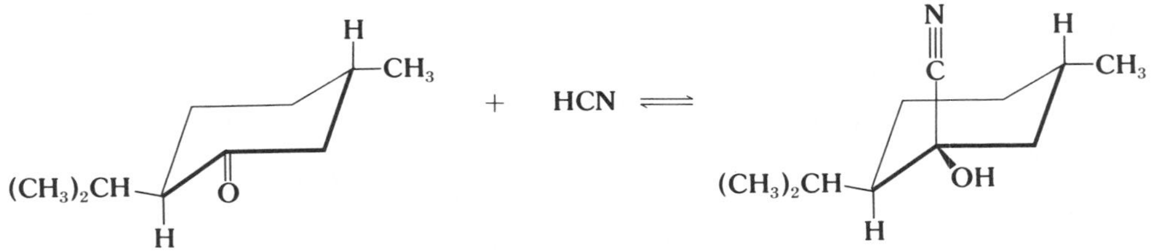

[Structure: methone] + HCN ⇌ [Structure: cyanohydrin]

PROBLEM 12·13

a Hydrazones can undergo a subsequent reaction with a second carbonyl containing molecule to give compounds called azines:

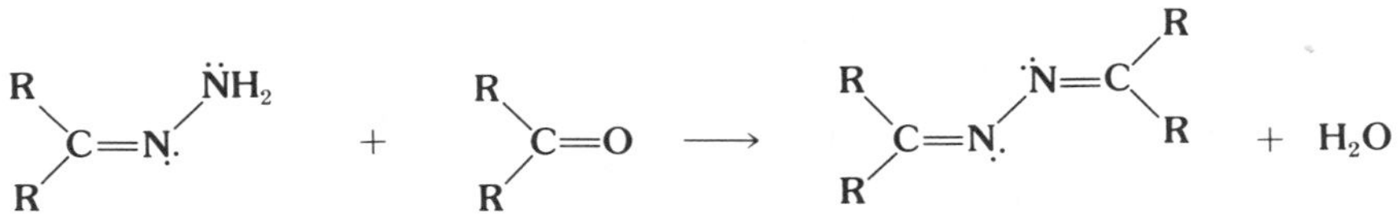

$$R_2C=N-NH_2 + R_2C=O \longrightarrow R_2C=N-N=CR_2 + H_2O$$

an azine

b The amino group adjacent to the carbonyl function is deactivated by delocalization of the nitrogen electron pair:

$$H_2\ddot{N}-\underset{\underset{\ddot{O}:}{\|}}{C}-\ddot{N}H-\ddot{N}H_2 \longleftrightarrow H_2\overset{\oplus}{N}=\underset{\underset{\ddot{O}:^{\ominus}}{|}}{C}-\ddot{N}H-\ddot{N}H_2$$

PROBLEM 12·14

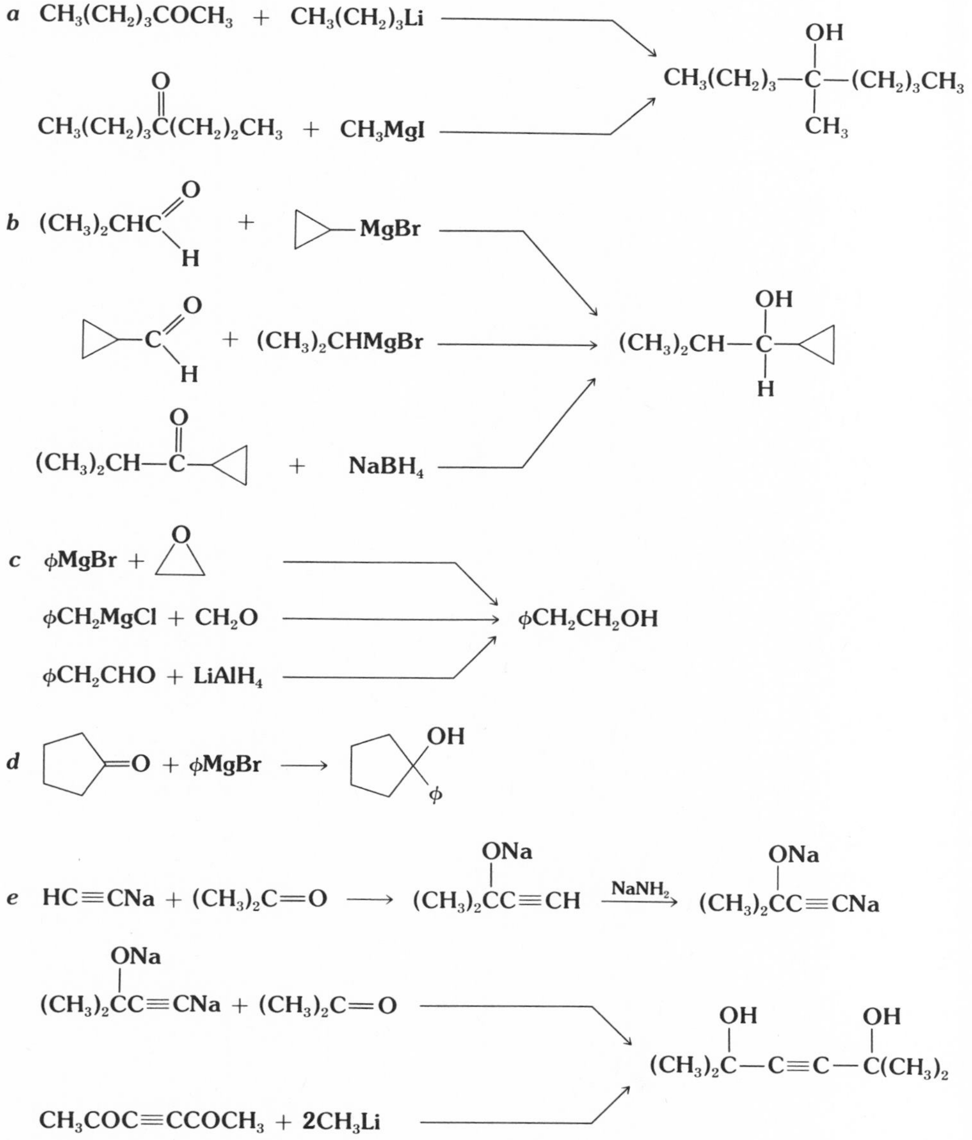

a $CH_3(CH_2)_3COCH_3 + CH_3(CH_2)_3Li \longrightarrow$

$CH_3(CH_2)_3\overset{O}{\overset{\|}{C}}(CH_2)_2CH_3 + CH_3MgI \longrightarrow$

$$CH_3(CH_2)_3-\underset{\underset{CH_3}{|}}{\overset{\overset{OH}{|}}{C}}-(CH_2)_3CH_3$$

b $(CH_3)_2CH\overset{O}{\overset{\|}{C}}{-}H +$ ▷—MgBr $\longrightarrow$

▷—$\overset{O}{\overset{\|}{C}}{-}H$ + $(CH_3)_2CHMgBr \longrightarrow (CH_3)_2CH-\underset{\underset{H}{|}}{\overset{\overset{OH}{|}}{C}}$—▷

$(CH_3)_2CH-\overset{O}{\overset{\|}{C}}$—▷ + $NaBH_4 \longrightarrow$

c $\phi MgBr +$ △O $\longrightarrow$

$\phi CH_2MgCl + CH_2O \longrightarrow \phi CH_2CH_2OH$

$\phi CH_2CHO + LiAlH_4 \longrightarrow$

d ⬠=O + $\phi MgBr \longrightarrow$ ⬠(OH)(ϕ)

e $HC \equiv CNa + (CH_3)_2C=O \longrightarrow (CH_3)_2\underset{\underset{}{|}}{\overset{\overset{ONa}{|}}{C}}C\equiv CH \xrightarrow{NaNH_2} (CH_3)_2\underset{\underset{}{|}}{\overset{\overset{ONa}{|}}{C}}C\equiv CNa$

$(CH_3)_2\underset{\underset{}{|}}{\overset{\overset{ONa}{|}}{C}}C\equiv CNa + (CH_3)_2C=O \longrightarrow$

$CH_3COC\equiv CCOCH_3 + 2CH_3Li \longrightarrow$

$(CH_3)_2\underset{\underset{}{|}}{\overset{\overset{OH}{|}}{C}}-C\equiv C-\underset{\underset{}{|}}{\overset{\overset{OH}{|}}{C}}(CH_3)_2$

PROBLEM 12·15

a $\phi_3P + CHCl_3 \xrightarrow{S_N 2} \phi_3\overset{\oplus}{P}—CHCl_2 \xrightarrow{(CH_3)_3COK} \phi_3P=CCl_2 + (CH_3)_3COH + KCl$
                              $Cl^{\ominus}$ [X]

b $\phi_3\overset{\oplus}{P}-CH_2(CH_2)_2CH_2Br \quad Br^{\ominus} \xrightarrow{\phi Li} \phi_3\overset{\oplus}{P}-\overset{\ominus}{\underset{H}{C}}-(CH_2)_2CH_2Br \longrightarrow \phi_3\overset{\oplus}{P}-\underset{H}{\square} \quad Br^{\ominus}$

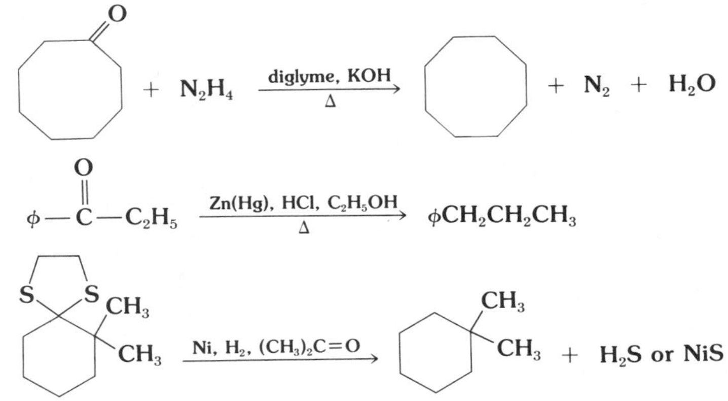

$\downarrow Li\phi$

$LiBr + \phi H + \phi_3P=\diamondsuit$ [Y]

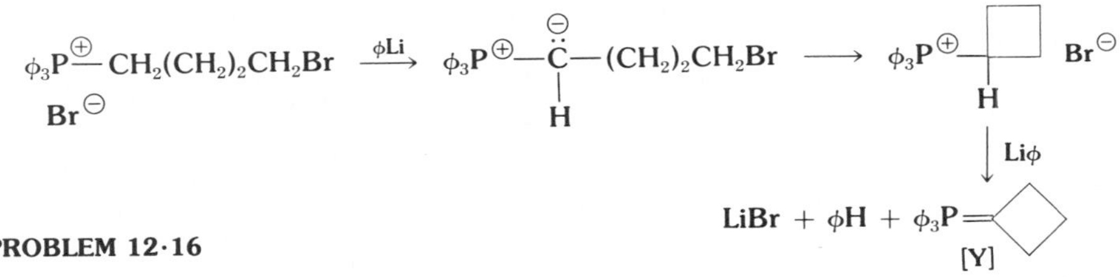

PROBLEM 12·16

a. cyclooctanone + N_2H_4 $\xrightarrow{\text{diglyme, KOH}, \Delta}$ cyclooctane + N_2 + H_2O

$\phi-\overset{O}{\overset{\|}{C}}-C_2H_5 \xrightarrow{Zn(Hg), HCl, C_2H_5OH, \Delta} \phi CH_2CH_2CH_3$

[dithiolane-spiro-cyclohexane with CH₃, CH₃] $\xrightarrow{Ni, H_2, (CH_3)_2C=O}$ [1,1-dimethylcyclohexane with CH₃, CH₃] + H_2S or NiS

b 3-methylcyclohexanone (with H indicated); 3-ethylcyclopentanone (with H indicated); 2-methyl-2-ethylcyclobutanone

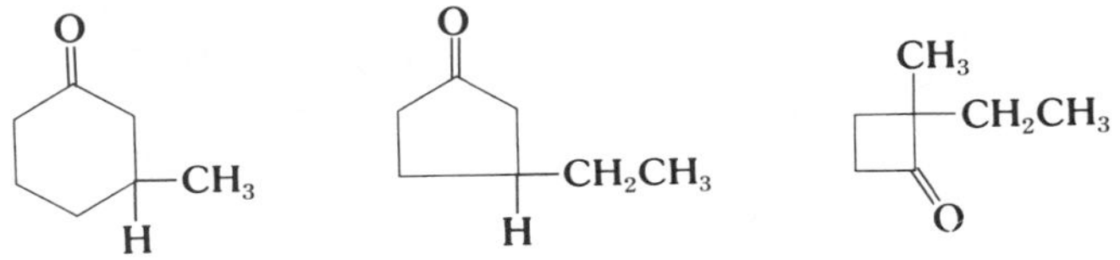

PROBLEM 12·17
A permanganate or chromate ester of a ketone hydrate cannot undergo the facile elimination reaction shown in equation 12.49. For example:

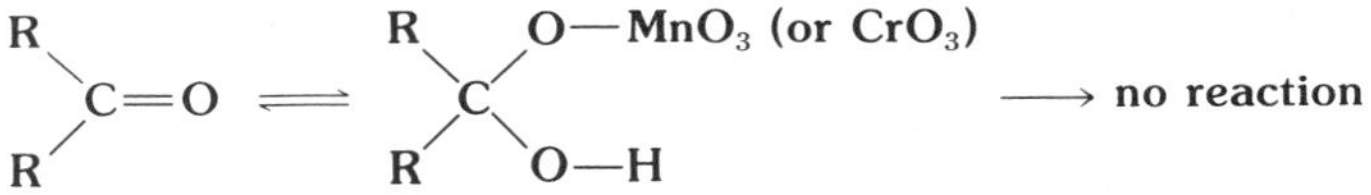

$\underset{R}{\overset{R}{>}}C=O \rightleftharpoons \underset{R}{\overset{R}{>}}C\underset{O-H}{\overset{O-MnO_3 \ (or\ CrO_3)}{<}} \longrightarrow$ no reaction

PROBLEM 12·18
This is a mixture of diastereoisomers. It is not a racemic modification and would probably (although not necessarily) be optically active.

PROBLEM 12·19

a

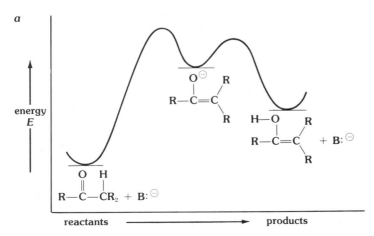

b Halogenation involves the attack of an electrophilic halogen molecule on a nucleophilic enol derivative. For a given halogen, the rate of reaction should therefore be proportional to the nucleophilicity of the enol intermediate. In base-catalyzed reaction systems the highly nucleophilic enolate anion will be present, whereas in acid-catalyzed systems the enol tautomer is the only nucleophile subject to attack by the halogen.

PROBLEM 12·20

The presence of a trace of $AlCl_3$ in the acetophenone-bromine reaction serves as a catalyst for enolization. If a large excess of this Lewis acid is introduced, it complexes the ketone transforming it into a meta-directing cationic substituent:

$$-\overset{O}{\underset{\|}{C}}-CH_3 \xrightleftharpoons{AlCl_3} \phi-\overset{\overset{\oplus}{O}-AlCl_3^{\ominus}}{\underset{\|}{C}}-CH_3 \xrightleftharpoons[HCl]{-HCl} \phi-\overset{O-AlCl_2}{\underset{|}{C}}=CH_2$$

$$Br_2 \xrightleftharpoons{AlCl_3} Br\,AlCl_3^{\ominus}\,Br^{\oplus}$$

$$\phi-\overset{\overset{\oplus}{O}-AlCl_3^{\ominus}}{\underset{\|}{C}}-CH_3 \xrightarrow{Br^{\oplus}} \left[\text{arenium ion intermediate} \right] \longrightarrow \text{3-bromoacetophenone}$$

PROBLEM 12·21

a That contributor having the negative charge located on the most electronegative atom will generally make the largest contribution to the resonance hybrid. In this case it will be the enolate anion (with the charge on oxygen).

b The conjugate bases of β-dicarbonyl compounds are exceptionally stable because of the extensive charge delocalization that is possible:

(three resonance structures of the pentane-2,4-dionate anion shown)

Note that the two contributors having the negative charge on oxygen are structurally and energetically equivalent. This leads to exceptional stabilization (see page 257).

PROBLEM 12·22

(reaction scheme: α-chloroallyl-substituted tetralone with α-H → NaH, THF; then CH_3I → α-methylated product)

b The conjugate base of 2,4-pentanedione serves as a general base in the $E2$ elimination of t-butyl bromide:

(acetylacetonate $Na^{\oplus}$) + $(CH_3)_3CBr \longrightarrow (CH_3)_2C=CH_2$ + (enol form of acetylacetone) + $NaBr$

PROBLEM 12·23

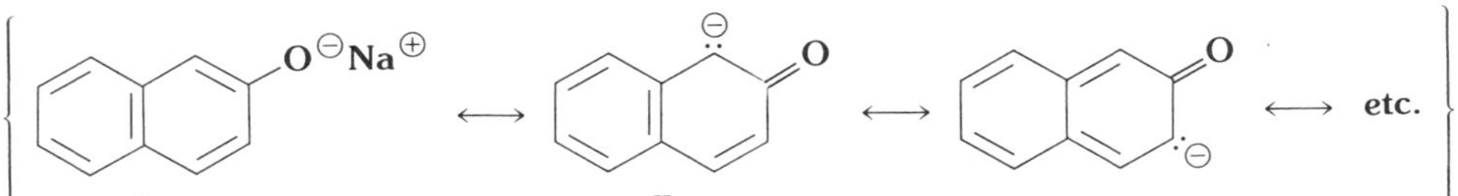

A resonance hybrid description of the conjugate base of 2-hydroxynaphthalene involves a number of contributing structures, the most important of which is that having the negative charge on oxygen (I above). Of the five contributors displaying the delocalization of charge in the naphthalene ring, II is the most significant because it retains the aromatic character of one benzene ring. The sites of electrophilic methylation correspond to nucleophilic sites in I and II. More precisely, the S_N2 transition states leading to methylation at oxygen and C-1 have lower energies than the transition states for methylation at other sites. The activation energies for reaction at these sites are therefore lower than the activation energies for other reactions.

PROBLEM 12·24

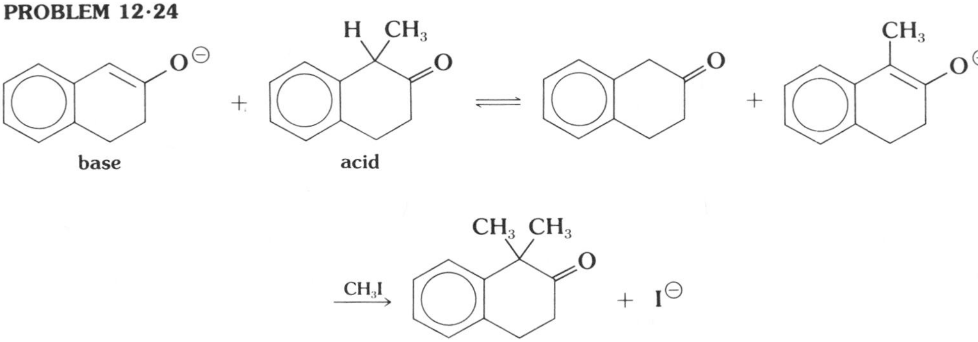

PROBLEM 12·25

a Reactions 12.70 and 12.71 describe the aldol condensation of aldehydes and ketones. In this reaction an alpha C—H bond and part of a carbonyl bond are broken, while C—C and O—H bonds are formed. The calculated heats of reaction are:

ΔH(aldehydes) = 90.5 + 98.7 − (110.6 + 82.6) = −4.0 kcal/mole
ΔH(ketones) = 93.5 + 98.7 − (110.6 + 82.6) = −1.0 kcal/mole

b $2\ CH_3CHO \longrightarrow CH_3-\underset{H}{\underset{|}{\overset{OH}{\overset{|}{C}}}}-O-CH=CH_2$

$\Delta H = (2 \times 90.5) + 98.7 - (110.6 + 85.5 + 63.2) = +20.4$ kcal/mole

2C=O π bonds + C—H − O—H + C—O + C=C π bond

This reaction is strongly endothermic and will have a higher activation energy than the aldol condensation.

PROBLEM 12·26

The presence of the carbonyl group enhances the acidity of the hydrogen atom

ANSWERS TO TEXT PROBLEMS 747

that must be removed in the dehydration. In some cases an enolate anion will be a discrete intermediate in these elimination reactions:

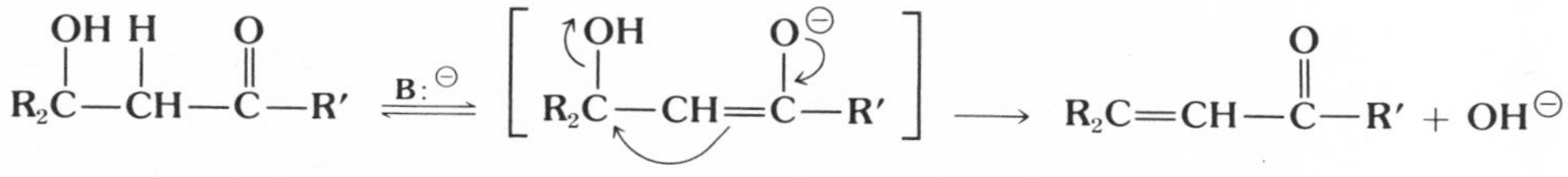

$$R_2C(OH)-CH(H)-C(=O)-R' \xrightarrow{B:^{\ominus}} \left[R_2C(OH)-CH=C(O^{\ominus})-R' \right] \longrightarrow R_2C=CH-C(=O)-R' + OH^{\ominus}$$

PROBLEM 12·27

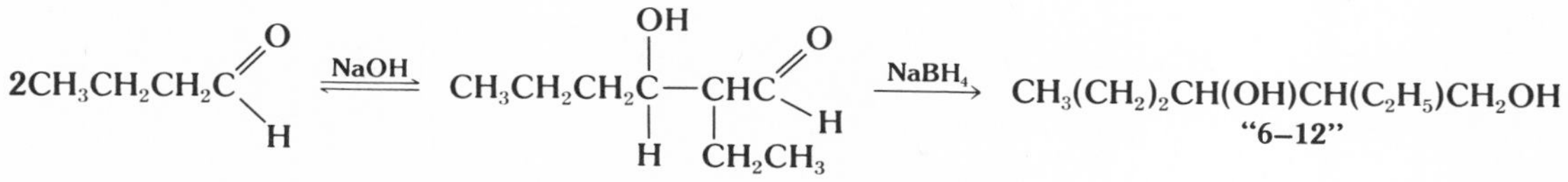

$$2CH_3CH_2CH_2CHO \xrightarrow{NaOH} CH_3CH_2CH_2C(OH)(H)-CH(CH_2CH_3)CHO \xrightarrow{NaBH_4} CH_3(CH_2)_2CH(OH)CH(C_2H_5)CH_2OH$$
"6–12"

PROBLEM 12·28

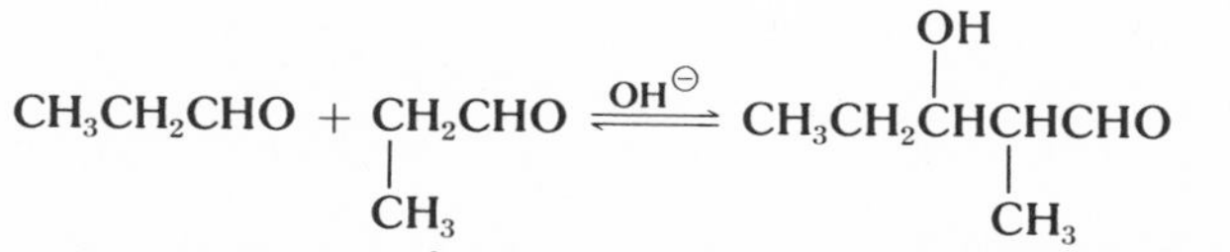

$$CH_3CH_2CHO + CH_2CHO(CH_3) \xrightleftharpoons{OH^{\ominus}} CH_3CH_2CH(OH)CH(CH_3)CHO$$
acceptor — donor — 3-hydroxy-2-methylpentanal

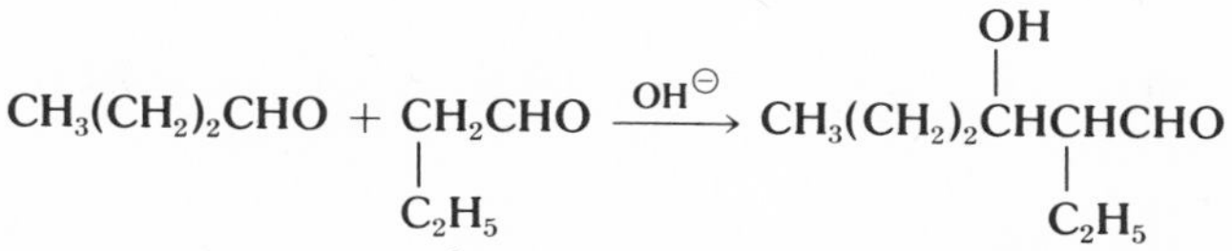

$$CH_3(CH_2)_2CHO + CH_2CHO(C_2H_5) \xrightarrow{OH^{\ominus}} CH_3(CH_2)_2CH(OH)CH(C_2H_5)CHO$$
acceptor — donor — 3-hydroxy-2-ethylhexanal

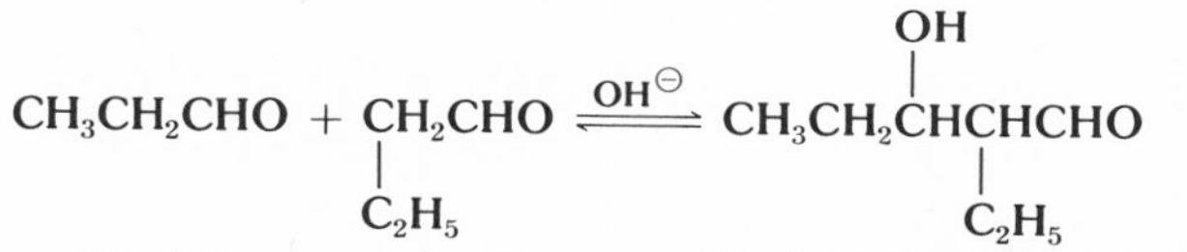

$$CH_3CH_2CHO + CH_2CHO(C_2H_5) \xrightleftharpoons{OH^{\ominus}} CH_3CH_2CH(OH)CH(C_2H_5)CHO$$
acceptor — donor — 3-hydroxy-2-ethylpentanal

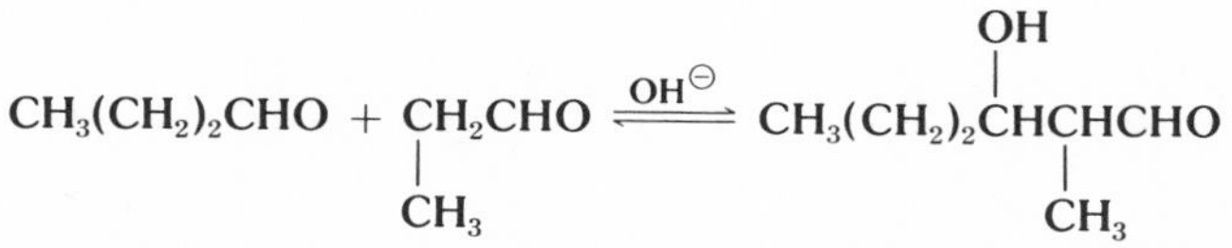

$$CH_3(CH_2)_2CHO + CH_2CHO(CH_3) \xrightleftharpoons{OH^{\ominus}} CH_3(CH_2)_2CH(OH)CH(CH_3)CHO$$
acceptor — donor — 3-hydroxy-2-methylhexanal

CHAPTER 13

PROBLEM 13·1

a For ϕ CH$_3$, -3; for ϕ CO$_2$H, $+3$

b [structure showing biotin-like ring with oxidation numbers: H$_2$C^{-1}, C^0, N–H, C^{+4}=O, S, C^0, C^0, N–H, H, H$_2$C^{-2}, H$_2$C^{-2}–C^{+3}O$_2$H]

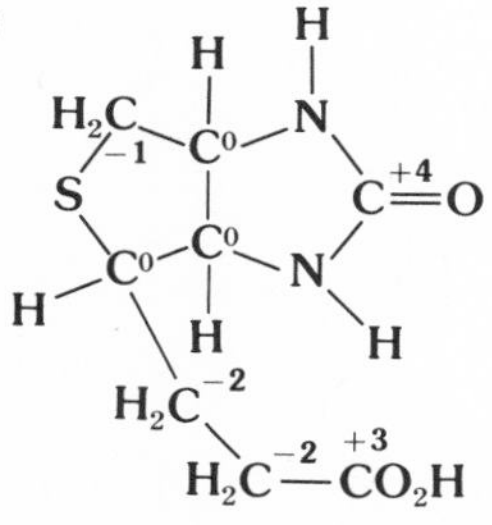

c

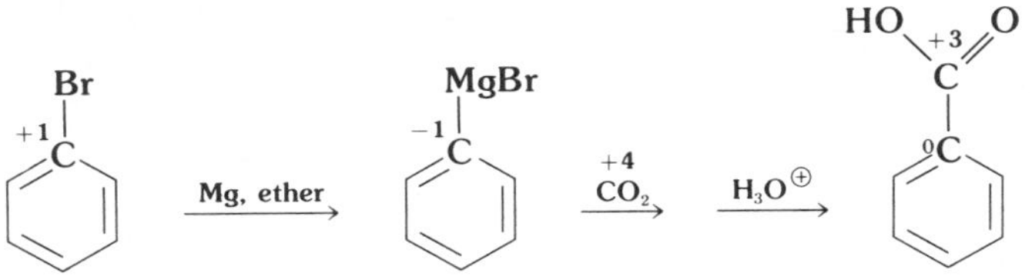

$\underset{\text{Mg, ether}}{\longrightarrow}$ ArCH(Br)(+1) → ArCH(MgBr)(−1) $\underset{\text{CO}_2}{\overset{+4}{\longrightarrow}}$ $\underset{\text{H}_3\text{O}^⊕}{\longrightarrow}$ ArC(+3)(O)OH

(benzyl bromide → benzylmagnesium bromide → phenylacetic acid via CO$_2$ then H$_3$O$^⊕$)

PROBLEM 13·2

a $CH_3CH_2CH=CH_2 \xrightarrow{B_2H_6} \xrightarrow{H_2O_2,\ NaOH} CH_3CH_2CH_2CH_2OH \xrightarrow{CrO_3,\ H_3O^⊕} CH_3CH_2CH_2CO_2H$

b $CH_3CH_2CH=CH_2 \xrightarrow{O_3} \xrightarrow{H_2O_2} CH_3CH_2CO_2H + CO_2$

c $CH_3CH_2CH=CH_2 \xrightarrow{HBr,\ peroxides} CH_3(CH_2)_3Br \xrightarrow{KCN} CH_3(CH_2)_3CN \xrightarrow{H_3O^⊕,\ \Delta} CH_3(CH_2)_3CO_2H$

d $CH_3CH_2CH=CH_2 \xrightarrow{HBr} CH_3CH_2CHBrCH_3 \xrightarrow{Mg,\ ether} \xrightarrow{CO_2} \xrightarrow{H_3O^⊕} CH_3CH_2CH(CH_3)CO_2H$

e C$_6$H$_6$ + Br$_2$ $\xrightarrow{Fe,\ \Delta}$ PhBr $\xrightarrow{Mg,\ ether}$ $\xrightarrow{CO_2}$ $\xrightarrow{H_3O^⊕}$ PhCO$_2$H

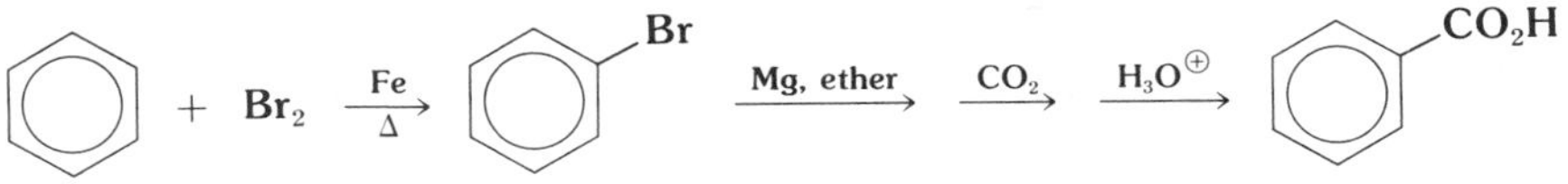

or

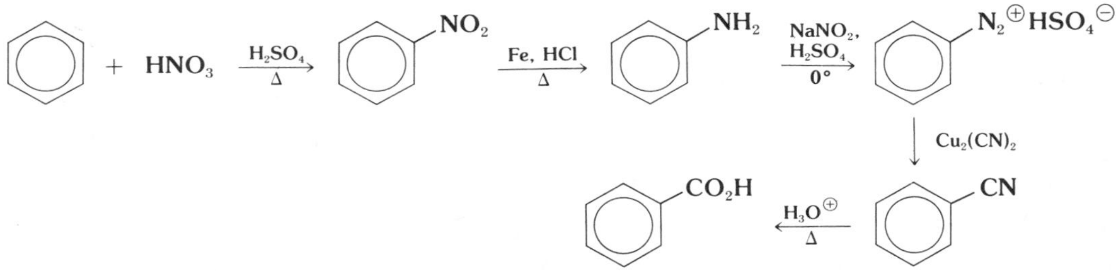

C$_6$H$_6$ + HNO$_3$ $\xrightarrow{H_2SO_4}$ PhNO$_2$ $\xrightarrow{Fe,\ HCl,\ \Delta}$ PhNH$_2$ $\xrightarrow{NaNO_2,\ H_2SO_4,\ 0°}$ PhN$_2^⊕$HSO$_4^⊖$ $\xrightarrow{Cu_2(CN)_2}$ PhCN $\xrightarrow{H_3O^⊕,\ \Delta}$ PhCO$_2$H

f PhNH$_2$ (as in e) + (CH$_3$CO)$_2$O → PhNHCOCH$_3$ $\xrightarrow{HNO_3,\ H_2SO_4}$ p-O$_2$N-C$_6$H$_4$-NHCOCH$_3$ + ortho isomer

$\xrightarrow{\Delta,\ H_3O^⊕}$ p-O$_2$N-C$_6$H$_4$-NH$_2$ $\xrightarrow{NaNO_2,\ H_2SO_4,\ 0°}$ p-O$_2$N-C$_6$H$_4$-N$_2^⊕$HSO$_4^⊖$ $\xrightarrow{Cu_2(CN)_2}$ p-O$_2$N-C$_6$H$_4$-CN $\xrightarrow{H_3O^⊕,\ \Delta}$ p-O$_2$N-C$_6$H$_4$-CO$_2$H

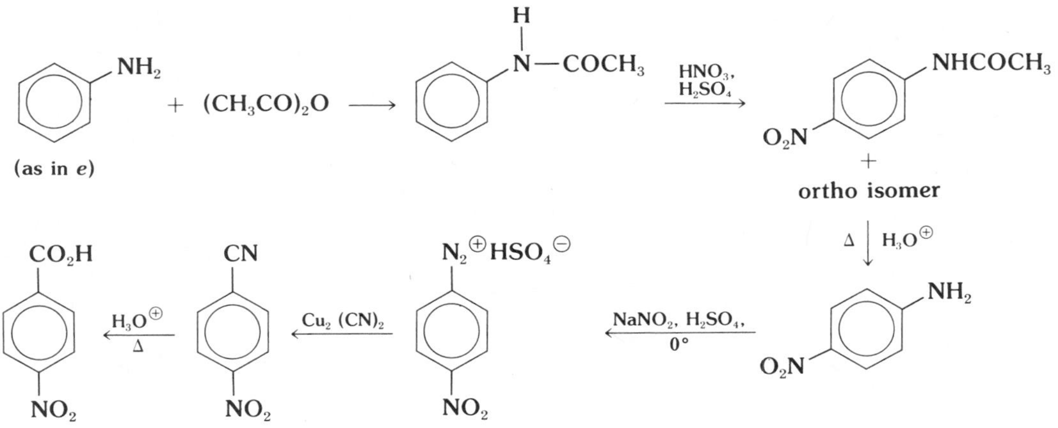

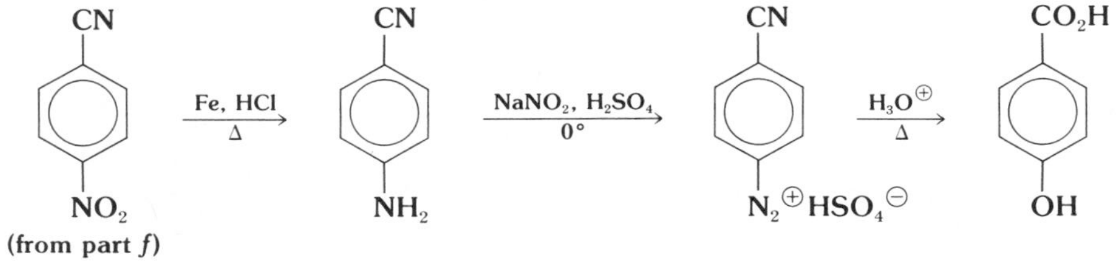

g p-O$_2$N-C$_6$H$_4$-CN (from part f) $\xrightarrow{Fe,\ HCl,\ \Delta}$ p-H$_2$N-C$_6$H$_4$-CN $\xrightarrow{NaNO_2,\ H_2SO_4,\ 0°}$ p-(HSO$_4^⊖$N$_2^⊕$)-C$_6$H$_4$-CN $\xrightarrow{H_3O^⊕,\ \Delta}$ p-HO-C$_6$H$_4$-CO$_2$H h [4-bromotoluene] $\xrightarrow{\text{Mg, ether}}$ $\xrightarrow{\text{CO}_2}$ $\xrightarrow{\text{H}_3\text{O}^{\oplus}}$ [4-methylbenzoic acid (CH$_3$ and CO$_2$H)] $\xrightarrow[\Delta]{\text{KMnO}_4}$ [terephthalic acid (CO$_2$H and CO$_2$H)]

PROBLEM 13·3

One factor is the inductive effect. The electronegative Z group withdraws electrons through the σ bond, making the carbonyl carbon atom more positive. This decreases the importance of the ionic contributor and increases the contribution of the double-bonded form to the hybrid. Another factor is the resonance effect. If the Z group has a nonbonding electron pair in its valence shell, it can delocalize the positive character of the carbonyl carbon atom by p-π conjugation:

$$R-C(=\ddot{O}:)(\ddot{Z}:) \longleftrightarrow R-C(-\ddot{O}:^{\ominus})(=\overset{\oplus}{Z}:) \longleftrightarrow R-C(-\ddot{O}:^{\ominus})(\overset{\oplus}{Z})$$

Such an interaction would increase the importance of the ionic contributor.

From our knowledge of aromatic substituent effects we can predict that the resonance effect of an amino substituent (Z = NR$_2$) will be much stronger than its inductive effect. Oxygen substituents (Z = OR) should show a less pronounced difference, still favoring the resonance effect, and halogen substituents (Z = Cl or Br) should have stronger inductive effects than resonance effects. The hybrid composition clearly affects the carbonyl stretching frequency of these derivatives, the higher frequencies being associated with carbonyl bonds that are harder to stretch (have greater double-bond character).

PROBLEM 13·4

At room temperature (25°) interconversion of the methyl groups by rotation about the C—N bond is slow ($\Delta E^* \approx 20$ kcal/mole).

$$\begin{array}{c} b \\ \text{CH}_3 \\ \diagdown \\ a\text{N}-\text{C} \\ \diagup \diagdown \\ \text{CH}_3 \text{H} \end{array} \rightleftharpoons \begin{array}{c} a \\ \text{CH}_3 \\ \diagdown \\ b\text{N}-\text{C} \\ \diagup \diagdown \\ \text{CH}_3 \text{H} \end{array}$$

This unusually high barrier to rotation about a "single" bond is another indication of the importance of nitrogen electron-pair delocalization in amides, a process which introduces double-bond character to the carbonyl carbon-nitrogen bond. This interconversion is more rapid at higher temperatures because a larger proportion of the sample molecules have sufficient kinetic energy to traverse the transition state. As a result, the pmr spectrometer "sees" a time average of the methyl-group environments.

PROBLEM 13·5

$$\text{CH}_3-\overset{\overset{\text{O}}{\|}}{\text{C}}-\text{O}-\text{CH}_2-\text{CH}_3$$

The infrared bands at 1740 and 1240 cm^{-1} indicate an ester. The pmr spectrum shows three sets of signals, the splitting patterns of which suggest the presence of

750   ANSWERS TO TEXT PROBLEMS

a methyl and an ethyl group. The large chemical shift of the methylene quartet ($\delta = 4.12$) indicates that it is bonded to an oxygen atom.

b $(CH_3)_2CH-O-\overset{\overset{\displaystyle O}{\|}}{C}-H$

The infrared bands at 1725 and 1240 cm^{-1} suggest an ester or a carboxylic acid, but the absence of hydroxyl absorption in the 2500 to 3300 cm^{-1} region seems to rule out an acid (C—H absorption at 2850 to 3000 cm^{-1} is expected in most organic substances). The 2725 cm^{-1} absorption is unusual and at first glance seems to point to an aldehyde. The pmr spectrum is very helpful in that it clearly requires the presence of an isopropoxy unit (the doublet and septet are very characteristic). The remaining one-proton signal at $\delta = 8.0$ is again suggestive of an aldehyde, a feature which is nicely met by the formate ester structure.

c $(CH_3)_2CHCO_2H$

The strong infrared signals in the 2500 to 3200, 1715, and 1230 cm^{-1} regions point to a carboxylic acid. The pmr signal at $\delta = 11.0$ clearly supports this assignment, especially since this proton rapidly exchanges with D_2O. The isopropyl group is again evident, but the smaller chemical shift of the tertiary proton ($\delta = 2.7$ here) indicates that the isopropyl moiety is bonded to carbon rather than oxygen.

PROBLEM 13·6

$(CH_3)_2CHCH_2CH_2OCOCH_3$
    isopentyl acetate

$CH_3(CH_2)_3OCO(CH_2)_2CH_3$
    butyl butyrate

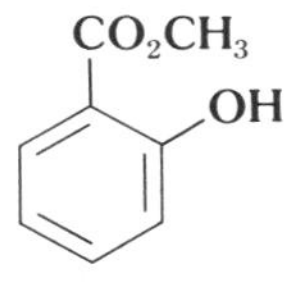

methyl salicylate

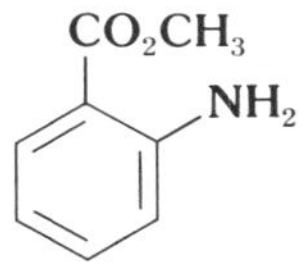

methyl anthranilate

$(CH_3)_2CHCH_2CH_2OCO(CH_2)_3CH_3$
    isopentyl valerate

PROBLEM 13·7

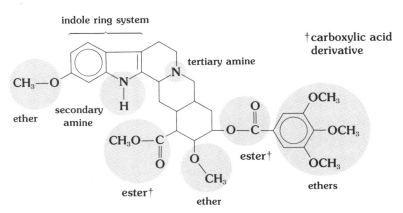

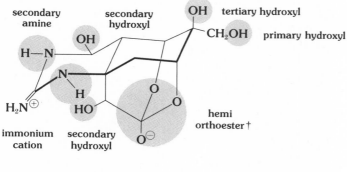

secondary amine | secondary hydroxyl | OH tertiary hydroxyl | CH₂OH primary hydroxyl | immonium cation | secondary hydroxyl | hemi orthoester † | ester † | secondary hydroxyl | a pyrrole | alkene | tertiary amine | hemiketal | ether

CHAPTER 14

PROBLEM 14·1

$$\phi CO_2H + H_2O \rightleftharpoons \phi CO_2^\ominus + H_3O^\oplus \qquad K_a = \frac{[\phi CO_2^\ominus][H_3O^\oplus]}{[\phi CO_2H]}$$

If the acid is half ionized, $[\phi CO_2H] = [\phi CO_2^\ominus]$. In this instance $K_a = [H_3O^\oplus]$ and $pH = pK_a$

PROBLEM 14·2

We can take advantage of the significant difference in the acidity of alcohols ($pK_a \approx 16$), phenols ($pK_a \approx 10$), and carboxylic acids ($pK_a \approx 5$). An ether solution of this mixture will, on extraction with sodium bicarbonate solution, lose the benzoic acid component to the aqueous-base phase. After separation of these immiscible phases the ether solution can be extracted again with the stronger base sodium hydroxide (5 to 10% solution). This extraction results in a separation of the phenol (aqueous layer) and the alcohol (ether layer).

PROBLEM 14·3

First, we recognize that 12 ml of 0.1 N solution of sodium hydroxide contains 1.2 milliequivalents (meq) of the base. Since our titration is to the point of complete neutralization, our 156-mg acid sample must also consist of 1.2 meq of acid. The equivalent weight of this acid is therefore $^{156}/_{1.2} = 130$.

PROBLEM 14·4

For reaction 14.10: $CH_2{=}CHCH_2CH_2CO_2H + ArSO_3H \rightleftharpoons$

For reaction 14.11:

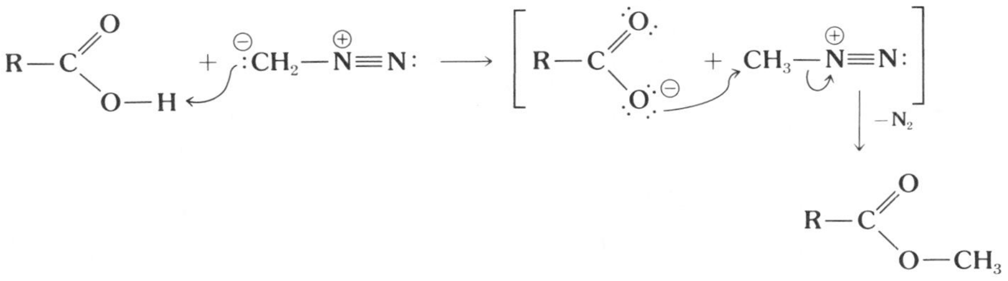

$$R-C\overset{O}{\underset{O-H}{\diagdown}} + {}^{\ominus}{:}CH_2-\overset{\oplus}{N}{\equiv}N{:} \longrightarrow \left[R-C\overset{{:}\overset{\oplus}{O}{:}}{\underset{{:}\overset{\ominus}{O}{:}}{\diagdown}} + CH_3-\overset{\oplus}{N}{\equiv}N{:} \right] \overset{-N_2}{\longrightarrow} R-C\overset{O}{\underset{O-CH_3}{\diagdown}}$$

PROBLEM 14·5

a $\quad K_{eq} = \dfrac{[C_3H_7CO_2C_2H_5][H_2O]}{[C_3H_7CO_2H][C_2H_5OH]}$

A doubling of the ethanol concentration will cause the concentration of butyric acid remaining at equilibrium to be reduced. The magnitude of this effect will depend on the size of the equilibrium constant. It should be clear from equation 14.12 that the mole fraction x of ester formed is equal to that of the water also produced. Therefore, if we assume equivalent amounts of acid and alcohol in the initial reaction mixture, we obtain

$K = \dfrac{x^2}{(1-x)^2} \quad$ and $\quad (K-1)x^2 - 2Kx + K = 0$

For $K = 2$ we find that $x = 0.586$ (58.6% of the acid is converted to ester at equilibrium). If we assume, however, that twice as much ethanol as acid is used in the reaction, then

$K_{eq} = \dfrac{x^2}{(2-x)(1-x)} \quad$ and $\quad (K-1)x^2 - 3Kx + 2K = 0$

In this case for $K = 2$ we then find that $x = 0.764$ (76.4% of the acid is converted to ester at equilibrium).

b Use an excess amount of ethanol and remove water as it is formed.

PROBLEM 14·6

The steric hindrance of the ortho methyl groups prevents addition reactions at the carbonyl function. This same steric hindrance, however, assists the elimination of water from a conjugate acid of mesitoic acid:

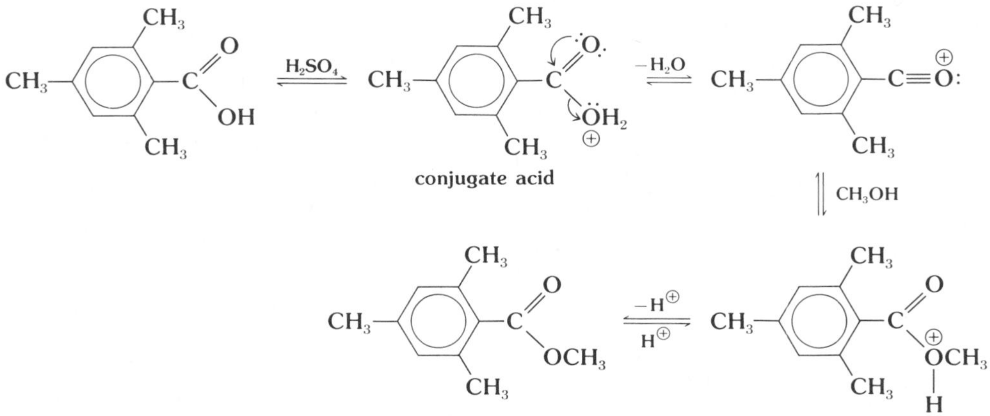

Note that another possible conjugate acid of mesitoic acid will also be formed reversibly, but it does not lead to any observable products:

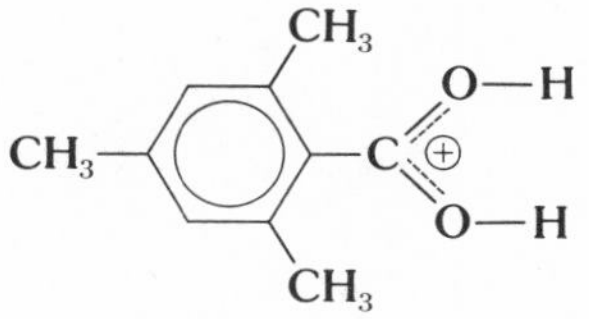

[Structure: 2,4,6-trimethylbenzoyl protonated diol cation]

PROBLEM 14·7

a [Mechanism showing 2-ethyl-2-carboxycyclohexanone with D on carboxyl O, losing CO₂ to give enol with OD, then D₂O giving 2-ethyl-2-deuterocyclohexanone with C₂H₅]

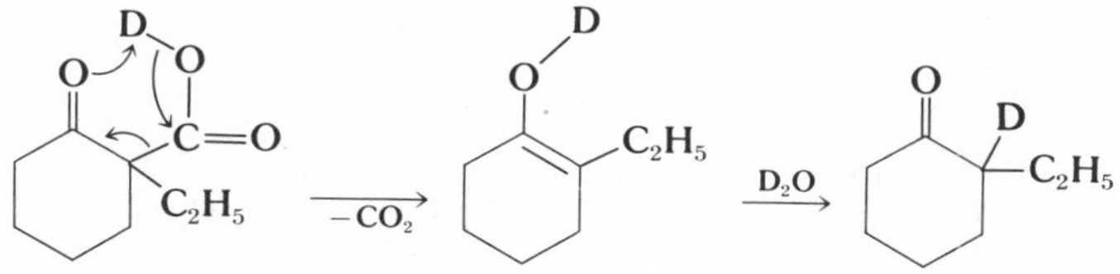

b The carbon-carbon double bond in the enol tautomer that would be produced by such a decarboxylation would be highly twisted by the geometrical constraints of the bicyclic ring system.

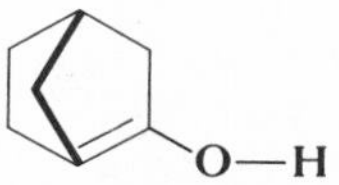

[Structure: bicyclic enol with O—H]

The activation energy for forming this unstable intermediate would be relatively high, and the reaction does not proceed.

PROBLEM 14·8

a ϕCOCl + O$_2$N—[benzene ring with NO$_2$]—NHNH$_2$ ⟶ ϕ—C(=O)—N(H)—NH—[benzene with NO$_2$, NO$_2$] + HCl

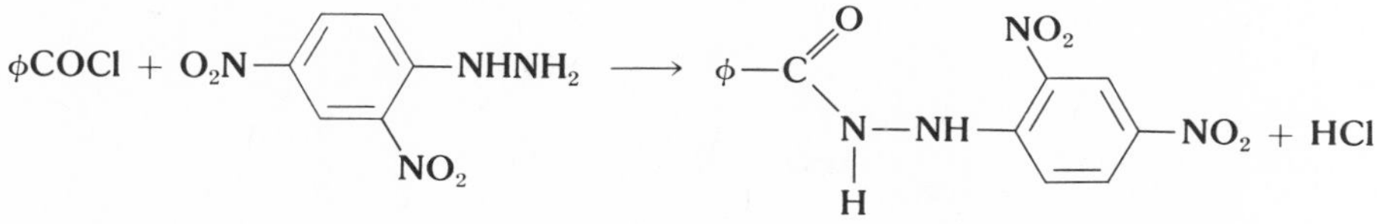

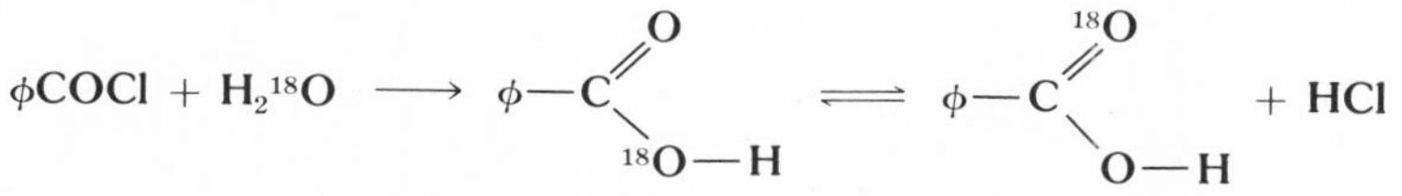

ϕCOCl + H$_2^{18}$O ⟶ ϕ—C(=O)—^{18}O—H ⇌ ϕ—C(=^{18}O)—O—H + HCl

b CH$_3$CO$_2$COCH$_3$ + O$_2$N—[benzene with NO$_2$]—NHNH$_2$ ⟶ CH$_3$—C(=O)—N(H)—NH—[benzene with NO$_2$, NO$_2$] + CH$_3$CO$_2$H

CH$_3$CO$_2$COCH$_3$ + H$_2^{18}$O ⟶ CH$_3$—C(=O)—^{18}O—H ⇌ CH$_3$—C(=^{18}O)—O—H

+

CH$_3$—C(=O)—O—H

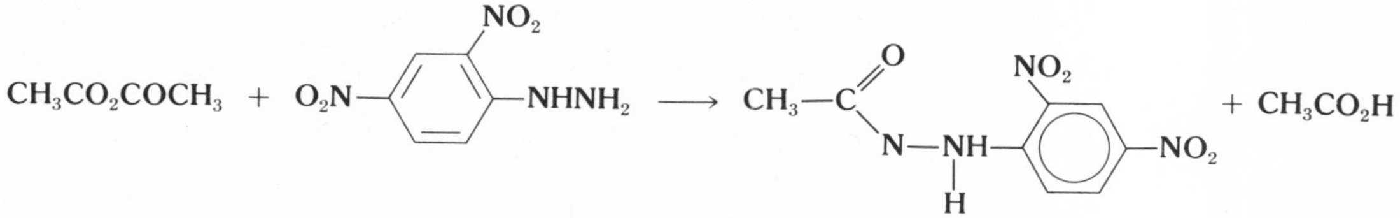

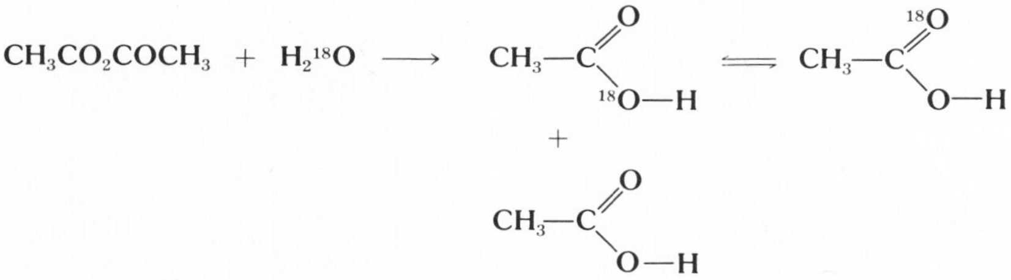

754 ANSWERS TO TEXT PROBLEMS

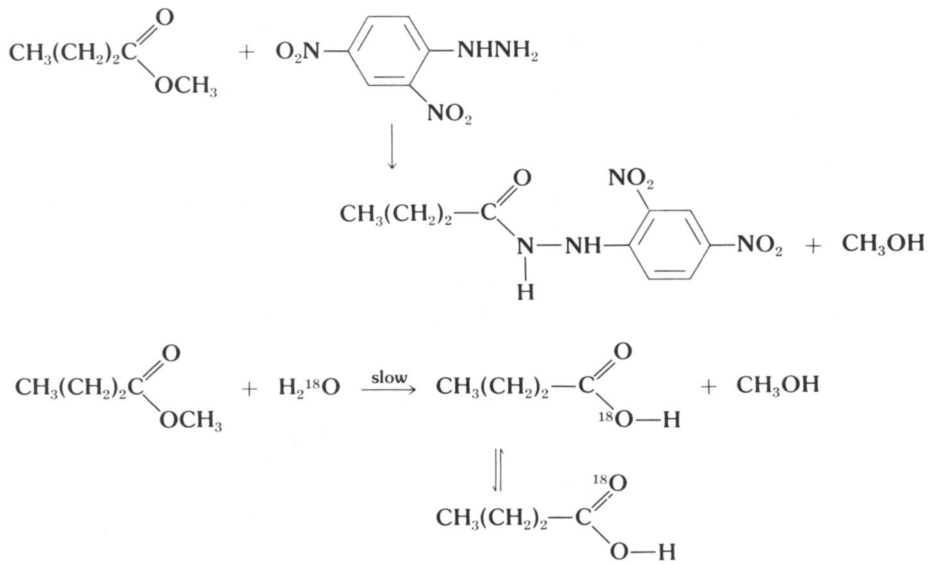

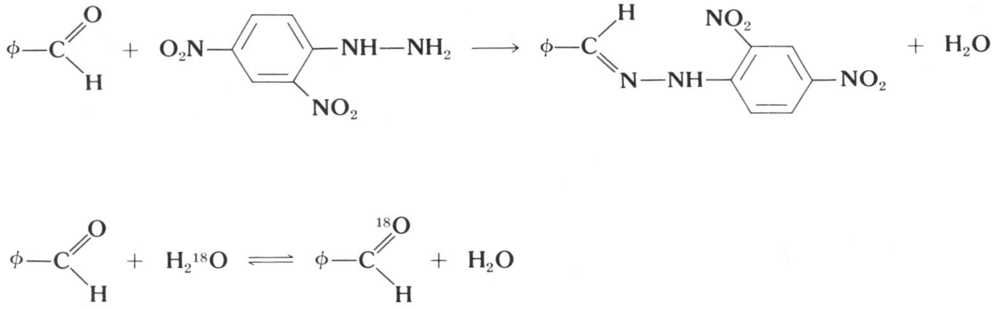

c $CH_3(CH_2)_2\underset{OCH_3}{\overset{O}{\overset{\|}{C}}}$ + $O_2N\text{-}\underset{NO_2}{\text{C}_6H_3}\text{-}NHNH_2$ $\longrightarrow$ $CH_3(CH_2)_2\overset{O}{\overset{\|}{C}}\text{-}\underset{H}{N}\text{-}NH\text{-}\underset{NO_2}{\text{C}_6H_3}\text{-}NO_2$ + CH_3OH

$CH_3(CH_2)_2\underset{OCH_3}{\overset{O}{\overset{\|}{C}}}$ + $H_2{}^{18}O$ $\xrightarrow{\text{slow}}$ $CH_3(CH_2)_2\overset{O}{\overset{\|}{C}}\text{-}{}^{18}O\text{-}H$ + CH_3OH

$\rightleftharpoons$ $CH_3(CH_2)_2\overset{{}^{18}O}{\overset{\|}{C}}\text{-}O\text{-}H$

d $\phi\text{-}\underset{H}{\overset{O}{\overset{\|}{C}}}$ + $O_2N\text{-}\underset{NO_2}{\text{C}_6H_3}\text{-}NH\text{-}NH_2$ $\longrightarrow$ $\phi\text{-}\underset{H}{\overset{}{C}}\text{=}N\text{-}NH\text{-}\underset{NO_2}{\text{C}_6H_3}\text{-}NO_2$ + H_2O

$\phi\text{-}\underset{H}{\overset{O}{\overset{\|}{C}}}$ + $H_2{}^{18}O$ $\rightleftharpoons$ $\phi\text{-}\underset{H}{\overset{{}^{18}O}{\overset{\|}{C}}}$ + H_2O

e CH_3COCH_3 + $O_2N\text{-}\underset{NO_2}{\text{C}_6H_3}\text{-}NH\text{-}NH_2$ $\longrightarrow$ $\underset{CH_3}{\overset{CH_3}{}}C\text{=}N\text{-}NH\text{-}\underset{NO_2}{\text{C}_6H_3}\text{-}NO_2$ + H_2O

$\underset{CH_3}{\overset{CH_3}{}}C\text{=}O$ + $H_2{}^{18}O$ $\rightleftharpoons$ $\underset{CH_3}{\overset{CH_3}{}}C\text{=}{}^{18}O$ + H_2O

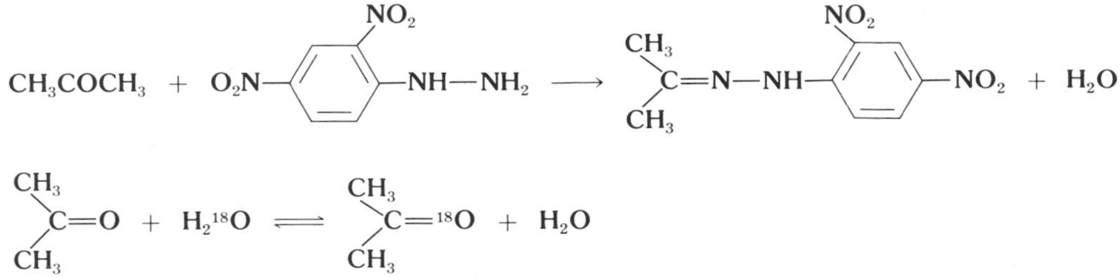

PROBLEM 14·9

a $RCOCl + H_2O \longrightarrow RCO_2H + HCl$
    $RCOCl + R'OH \longrightarrow RCO_2R' + HCl$
    $RCOCl + R'_2NH \longrightarrow RCONR'_2 + HCl$

b The amine function is tertiary; consequently, a stable amide derivative cannot be formed directly. Acylation of the tertiary amine gives an "-onium" salt, which is itself a good acylating agent:

$$\text{Ar}-\overset{\displaystyle O}{\underset{\displaystyle Cl}{C}} + (C_2H_5)_2NCH_2CH_2OH \rightleftharpoons \left[\text{Ar}-\overset{\displaystyle O}{C}\underset{\underset{\displaystyle Cl^{\ominus}\ \ ^{\oplus}C_2H_5}{|}}{\overset{\displaystyle C_2H_5}{\underset{\displaystyle }{\,}}}\text{N}-CH_2CH_2OH \right]$$

$$\Updownarrow$$

$$\text{Ar}-\overset{\displaystyle O}{\underset{\displaystyle O-CH_2-CH_2-\overset{\displaystyle C_2H_5}{\underset{\displaystyle C_2H_5}{N^{\oplus}}}-H\ Cl^{\ominus}}{C}} \longleftarrow \left[\text{Ar}-\overset{\displaystyle OH}{\underset{\displaystyle O}{C}}\overset{\displaystyle C_2H_5}{\underset{\displaystyle CH_2}{\overset{\displaystyle |}{N^{\oplus}-C_2H_5}}}\underset{\displaystyle CH_2}{\,}\ Cl^{\ominus} \right]$$

c In aqueous base, phenols ($pK_a \approx 10$) and hydrogen peroxide ($pK_a = 11.6$) are essentially entirely converted to their conjugate bases. The concentration of these negatively charged nucleophiles is therefore high compared with the conjugate base of the weaker acid water. Furthermore, the nucleophilicities of $HO_2^{\ominus}$ and $ArO^{\ominus}$ are greater than that of $OH^{\ominus}$ and much greater than that of neutral oxygen compounds such as water.

PROBLEM 14·10
Diazomethane has a nucleophilic carbon atom:

$$\left| H_2C=\overset{\oplus}{N}=\overset{\ominus}{\ddot{N}} \longleftrightarrow H_2\ddot{C}-\overset{\oplus}{N}\equiv N: \right|$$

$$R-\overset{\displaystyle O}{\underset{\displaystyle Cl}{C}} + H_2\overset{\ominus}{C}-\overset{\oplus}{N_2} \longrightarrow \left[R-\overset{\displaystyle O^{\ominus}}{\underset{\displaystyle Cl}{\overset{\displaystyle |}{C}}}-CH_2-\overset{\oplus}{N}\equiv N: \right] \longrightarrow R-\overset{\displaystyle O}{\underset{\displaystyle CH_2-\overset{\oplus}{N}\equiv N:}{C}}\ Cl^{\ominus}$$

If the ratio of reactants is 1:1, a displacement of N_2 by $Cl^{\ominus}$ can take place:

$$R-\overset{\displaystyle O}{\underset{\displaystyle Cl^{\ominus}\ CH_2-\overset{\oplus}{N}\equiv N:}{C}} \longrightarrow R-\overset{\displaystyle O}{\underset{\displaystyle CH_2-Cl}{C}} + N_2$$

If an excess of diazomethane is used, this reagent can serve as a base to neutralize the conjugate acid:

$$R-\overset{\displaystyle O}{\underset{\displaystyle CH_2-\overset{\oplus}{N}\equiv N:}{C}}\ Cl^{\ominus} + CH_2N_2 \longrightarrow R-\overset{\displaystyle O}{\underset{\displaystyle CHN_2}{C}} + CH_3-\overset{\oplus}{N_2}Cl^{\ominus} \longrightarrow CH_3-Cl + N_2$$

$$\text{base} \hspace{4cm} \text{conjugate acid}$$

PROBLEM 14·11

a

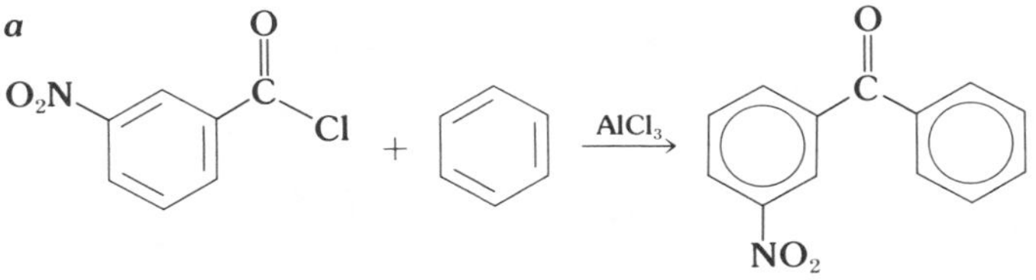

$$O_2N\text{-}C_6H_4\text{-}COCl + C_6H_6 \xrightarrow{AlCl_3} 3\text{-}O_2N\text{-}C_6H_4\text{-}CO\text{-}C_6H_5$$

Nitrobenzene is unreactive to Friedel-Crafts acylation.

b

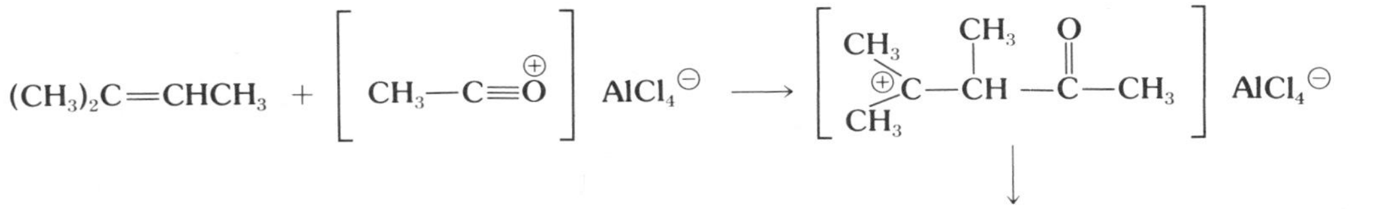

$$(CH_3)_2C=CHCH_3 + [CH_3-C\equiv\overset{\oplus}{O}]\ AlCl_4^{\ominus} \longrightarrow \left[\begin{array}{c}CH_3\ \ \ \ CH_3\ \ O\\ \overset{\oplus}{C}-CH-\overset{\|}{C}-CH_3\\ CH_3\end{array}\right] AlCl_4^{\ominus}$$

$\downarrow$

mixture of unsaturated ketones + HCl + AlCl$_3$
(see equation 14.37)

The acetylation of 1,1-diphenylethylene should proceed more rapidly than that of 2-methyl-2-butene, because the carbonium-ion intermediate formed in the first (rate-determining) step is further stabilized by resonance delocalization of the positive charge:

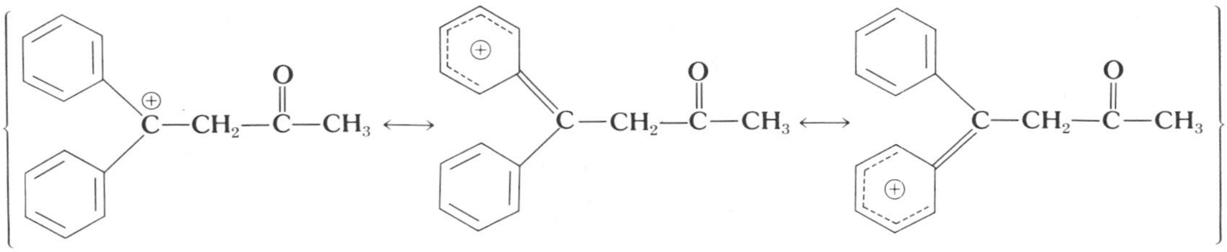

$$\left[(C_6H_5)_2\overset{\oplus}{C}-CH_2-\overset{O}{\overset{\|}{C}}-CH_3 \longleftrightarrow \cdots \longleftrightarrow \cdots \right]$$

c Acylation of anthracene by electrophilic attack at C-9 proceeds via a cationic intermediate which is relatively more stable than the corresponding intermediates resulting from attack at other sites (such as C-1 or C-2):

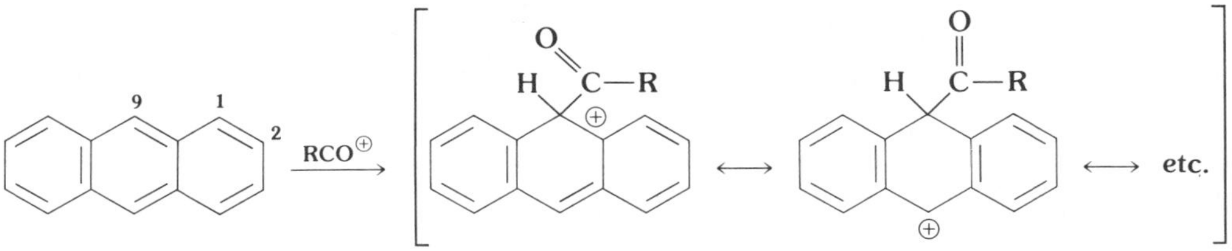

(anthracene) $\xrightarrow{RCO^{\oplus}}$ [resonance structures of C-9 acylated cation] $\longleftrightarrow$ etc.

Seven structures contribute to the resonance hybrid description of the C-9 intermediate (two are shown above; can you draw the other five?). All seven of these contributors have at least one intact benzene ring, in contrast to the C-1 and C-2 intermediates, which each have three (out of seven) contributors lacking any aromatic rings.

PROBLEM 14·12

The acetic anhydride may react with hydrogen chloride to give the very reactive acylating agent acetyl chloride:

$$(CH_3CO)_2O + HCl \rightleftharpoons CH_3COCl + CH_3CO_2H$$

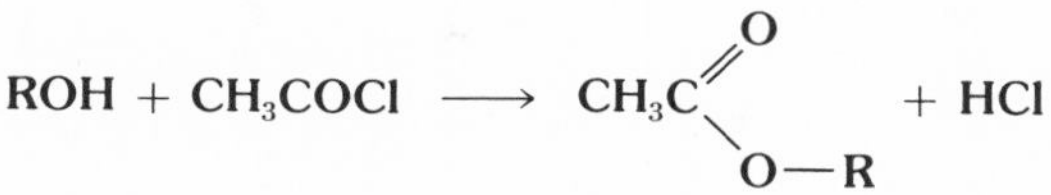

ANSWERS TO TEXT PROBLEMS 757

$$ROH + CH_3COCl \longrightarrow CH_3C(=O)O-R + HCl$$

PROBLEM 14·13

a The amide urea is less nucleophilic than hydroxyl amine. The nucleophilicity of the urea can be enhanced by converting it to its conjugate base:

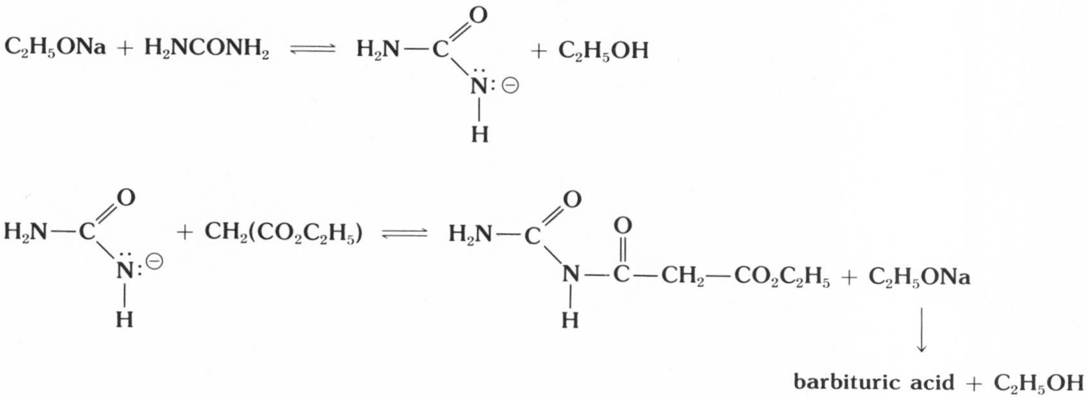

$$C_2H_5ONa + H_2NCONH_2 \rightleftharpoons H_2N-C(=O)-\overset{\ominus}{N}H + C_2H_5OH$$

$$H_2N-C(=O)-\overset{\ominus}{N}H + CH_2(CO_2C_2H_5)_2 \rightleftharpoons H_2N-C(=O)-N(H)-C(=O)-CH_2-CO_2C_2H_5 + C_2H_5ONa$$

$$\downarrow$$

barbituric acid + C_2H_5OH

b Yes: 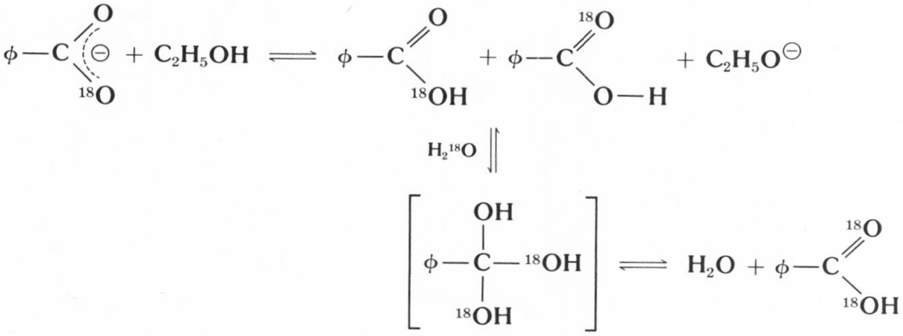 $\phi-C(\overset{\ominus}{O})(^{18}O) + C_2H_5OH \rightleftharpoons \phi-C(=O)(^{18}OH) + \phi-C(=^{18}O)(O-H) + C_2H_5O^{\ominus}$

$H_2{}^{18}O \updownarrow$

$$\left[\phi-C(OH)(^{18}OH)(^{18}OH) \right] \rightleftharpoons H_2O + \phi-C(=^{18}O)(^{18}OH)$$

c $R^1-C(=O)(OR^2) + HA \rightleftharpoons R^1-C(\overset{\oplus}{O}-H)(OR^2) + A^{\ominus} \xrightarrow{R^3OH} R^1-C(OH)(OR^2)(\overset{\oplus}{O}HR^3) \rightleftharpoons R^1-C(OH)(OR^3)(\overset{\oplus}{O}R^2/H)$

$\rightleftharpoons R^2OH + R^1-C(\overset{\oplus}{O}-H)(OR^3) \underset{H^{\oplus}}{\xrightarrow{-H^{\oplus}}} R^1-C(=O)(OR^3)$

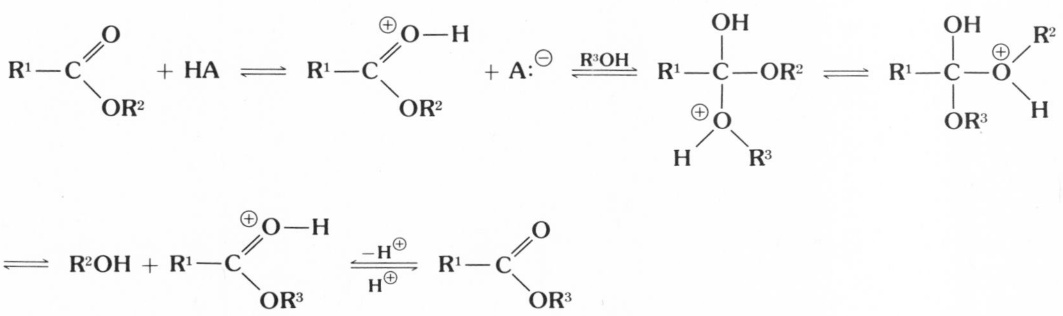

PROBLEM 14·14

a

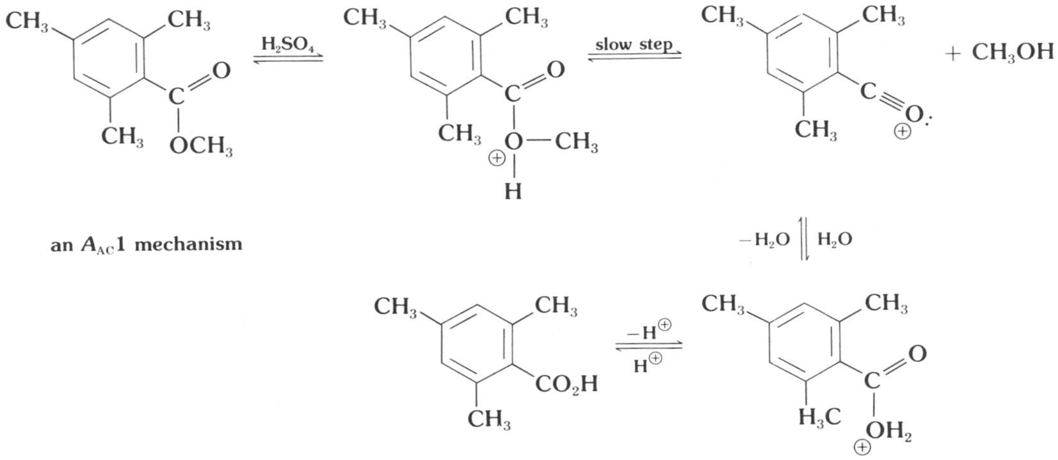

[Mechanism scheme showing methyl 2,4,6-trimethylbenzoate + H$_2$SO$_4$ ⇌ protonated ester (at OCH$_3$) → (slow step) acylium ion + CH$_3$OH; acylium ion + H$_2$O/−H$_2$O ⇌ protonated carboxylic acid ⇌ (−H$^+$/H$^+$) 2,4,6-trimethylbenzoic acid]

an $A_{AC}1$ mechanism

The loss of methanol from the conjugate acid shown above is favored by the relief of steric compression introduced by the ortho methyl groups (see problem 14.6). In an unhindered ester such as methyl benzoate, an isomeric conjugate acid, formed by protonation of the carbonyl oxygen atom, may be favored. Such intermediates normally react by a rate-determining addition of water, in contrast to the rate-determining loss of methanol in the $A_{AC}1$ mechanism. Hydrolysis by the latter $A_{AC}2$ mechanism would be very slow in ice water.

b

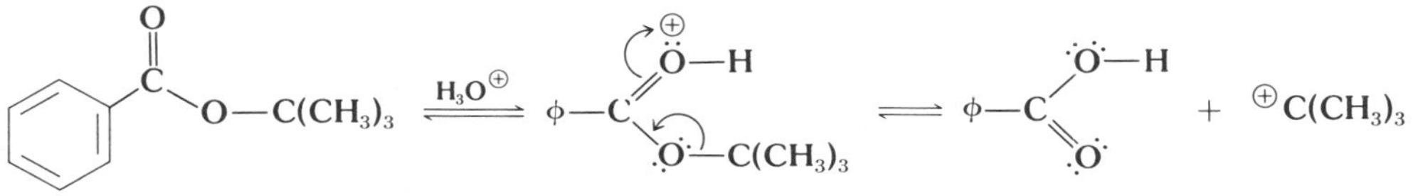

[Mechanism: PhC(O)O—C(CH$_3$)$_3$ + H$_3$O$^+$ ⇌ φ—C(OH)(Ö—C(CH$_3$)$_3$) ⇌ φ—C(OH)=Ö + $^+$C(CH$_3$)$_3$]

$(CH_3)_3C^\oplus$ + $2H_2O$ ⇌ $(CH_3)_3COH$ + $H_3O^\oplus$

PROBLEM 14·15

a

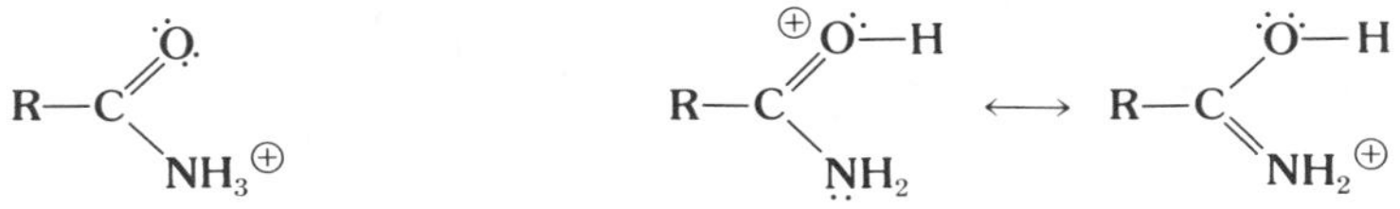

[Structures: R—C(=Ö)—NH$_3^\oplus$ (N-protonated conjugate acid) ; R—C(Ö—H)$^\oplus$—NH$_2$ ↔ R—C(Ö—H)=NH$_2^\oplus$ (O-protonated conjugate acid)]

Charge delocalization helps to stabilize the O-protonated species.

b Nylon 66 $\xrightarrow{\text{NaOH, H}_2\text{O}, \Delta}$ $^\ominus O_2C$—(CH$_2$)$_4$—CO$_2^\ominus$ + H$_2$N—(CH$_2$)$_6$—NH$_2$

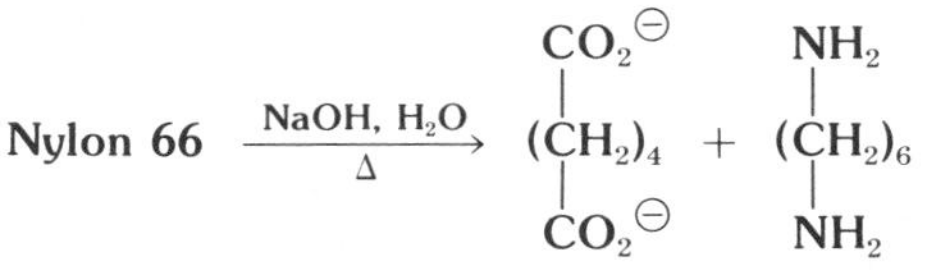

c The monocyclic lactam is stabilized to a much greater extent by nitrogen electron-pair delocalization. This is because of the unfavorable twisted nature of the double-bond contributor to the bicyclic compound (see problem 14.7*b*):

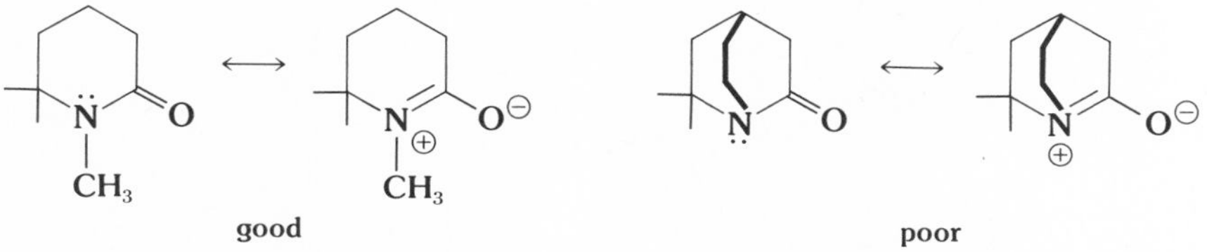

ANSWERS TO TEXT PROBLEMS 759

[Resonance structures: piperidinone with N–CH₃ (good) showing N: — C=O ↔ N⁺=C–O⁻; bicyclic bridgehead amide (poor) showing analogous resonance]

good   poor

The less stable lactone is hydrolyzed faster, since its carbonyl function is more reactive to addition reactions.

PROBLEM 14·16

$$R-C(=O)-O-R' + AlH_4^{\ominus} \xrightarrow{ether} R-\underset{H}{\overset{\ominus O---AlH_3}{\underset{|}{C}}}-OR' \longrightarrow R-C(=O)H + R'O^{\ominus}---AlH_3$$

$$R-C(=O)H + AlH_4^{\ominus} \xrightarrow{ether} R-CH_2-O^{\ominus}---AlH_3$$

PROBLEM 14·17

a $\phi-C(=O)OC_2H_5 + 2\,C_2H_5MgBr \xrightarrow{ether} \xrightarrow{H_3O^{\oplus}} \phi-\underset{OH}{\overset{|}{C}}(CH_2CH_3)_2 + C_2H_5OH$

b piperidine-N–C(=O)–CH₃ $\xrightarrow{LiAlH_4,\ ether}$ piperidine-N–CH₂CH₃ + H₂AlOLi

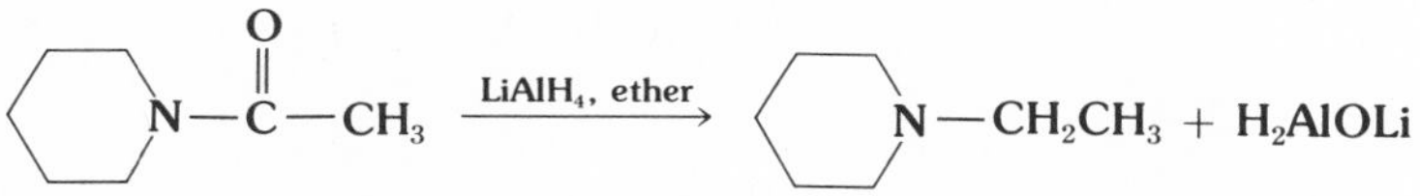

c $CH_3(CH_2)_{16}CO_2H \xrightarrow[\Delta]{SOCl_2} CH_3(CH_2)_{16}-C(=O)Cl \xrightarrow[-78°]{LiAl(OC_4H_9)_3H} CH_3(CH_2)_{16}-C(=O)H$

d norbornyl-C(=O)OH $\xrightarrow[\Delta]{SOCl_2}$ norbornyl-C(=O)Cl $\xrightarrow{(CH_3)_2Cd}$ norbornyl-C(=O)-CH₃ + CH₃CdCl

$\Delta \downarrow NH_3$ .... $H_2O \uparrow$

norbornyl-C(=O)NH₂ $\xrightarrow[\Delta]{P_2O_5}$ norbornyl-C≡N: $\xrightarrow{CH_3MgI}$ norbornyl-C(=N–MgI)–CH₃

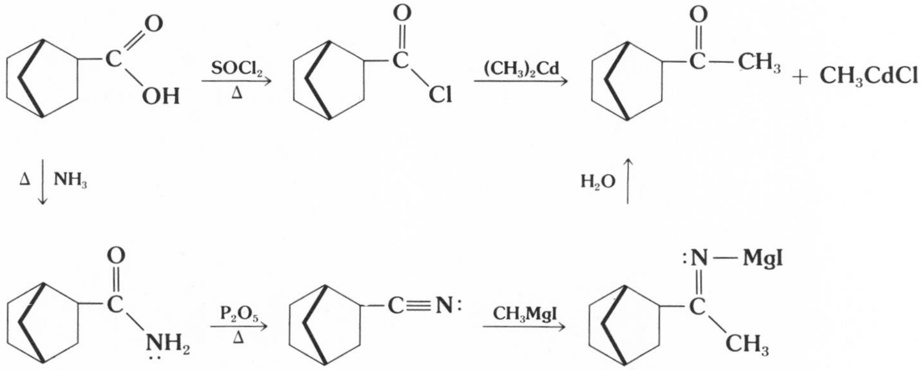

The conversion of a carboxylic acid derivative (such as RCOZ) to an aldehyde or ketone requires special techniques, since the metal hydride or organometallic reagents used for this purpose might continue to react with the carbonyl products to give alcohols:

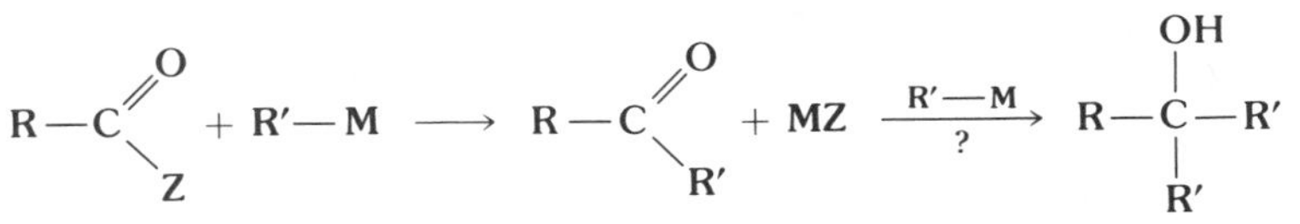

760 ANSWERS TO TEXT PROBLEMS

$$R-\underset{Z}{\overset{O}{C}} + R'-M \longrightarrow R-\underset{R'}{\overset{O}{C}} + MZ \xrightarrow[?]{R'-M} R-\underset{R'}{\overset{OH}{\underset{|}{C}}}-R'$$

(R' = H or alkyl
M = metal)

Two distinct strategies have been developed to avoid this problem:

1 A very reactive RCOZ derivative can be combined with low-reactivity metal reagents under sufficiently mild reaction conditions that subsequent addition does not take place. The reactions of acid halides with organocadmium reagents or with *tris t*-butoxyaluminum hydride at low temperature are examples of this approach.

2 The carboxylic acid derivative can be modified so that a carbonyl group is not formed in the initial step. If the carbonyllike functional group that is formed is less reactive than aldehydes or ketones, this intermediate may be isolated and converted to the aldehyde or ketone in a subsequent step. The reaction of nitriles with Grignard reagents is an example of this approach.

PROBLEM 14·18

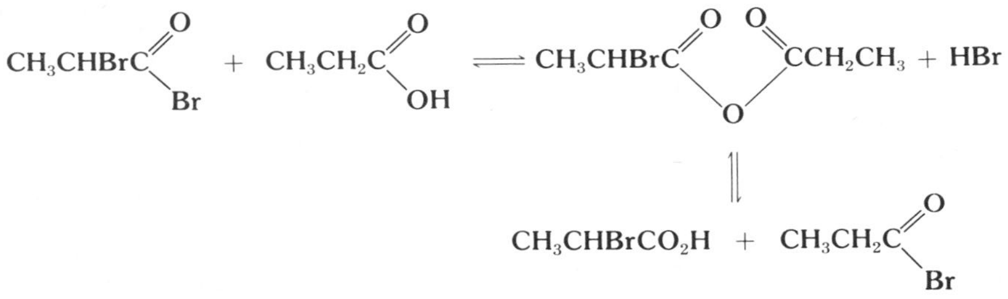

$$CH_3CHBrC\underset{Br}{\overset{O}{\diagdown}} + CH_3CH_2C\underset{OH}{\overset{O}{\diagdown}} \rightleftharpoons CH_3CHBrC\overset{O}{\diagdown}O\overset{O}{\diagup}CCH_2CH_3 + HBr$$

$$\Updownarrow$$

$$CH_3CHBrCO_2H + CH_3CH_2C\underset{Br}{\overset{O}{\diagdown}}$$

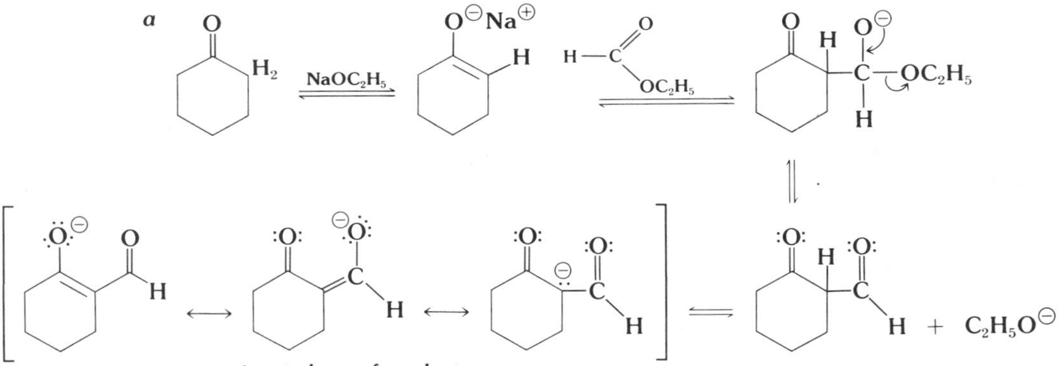

PROBLEM 14·19

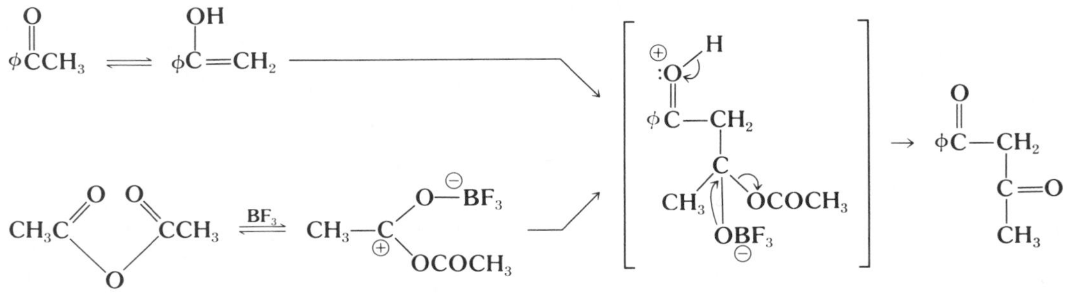

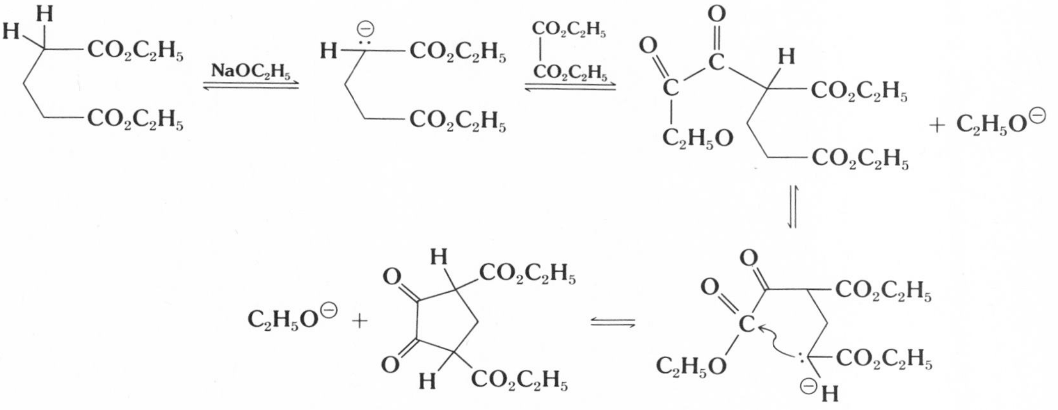

ANSWERS TO TEXT PROBLEMS

[Mechanism diagram showing Claisen condensation of diethyl glutarate with NaOC₂H₅, proceeding through enolate formation, intramolecular attack, and tetrahedral intermediate collapse to give the cyclic β-ketoester product plus C₂H₅O⁻]

b Consider the mechanism of Claisen condensation in this case:

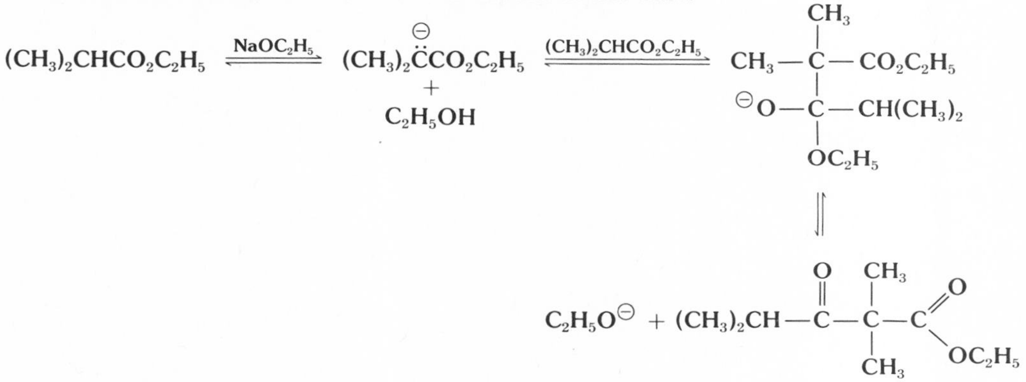

$$(CH_3)_2CHCO_2C_2H_5 \xrightleftharpoons[]{NaOC_2H_5} (CH_3)_2\ddot{C}CO_2C_2H_5 + C_2H_5OH \xrightarrow{(CH_3)_2CHCO_2C_2H_5}$$

$$CH_3-\underset{\underset{OC_2H_5}{\overset{\ominus O-C-CH(CH_3)_2}{|}}}{\overset{CH_3}{\underset{|}{C}}}-CO_2C_2H_5$$

$$\Updownarrow$$

$$C_2H_5O^\ominus + (CH_3)_2CH-\overset{O}{\underset{}{C}}-\underset{CH_3}{\overset{CH_3}{\underset{|}{C}}}-\overset{O}{\underset{OC_2H_5}{C}}$$

Since the β keto ester product formed here has no doubly activated α protons, it cannot be trapped or stabilized by conjugate-base formation. (This is the driving force in the Claisen condensation, as shown in Figure 14.12.) Steric hindrance by the geminal dimethyl grouping may also hinder carbon-carbon bond formation.

To improve this unfavorable condensation we can use a half-molar equivalent of a very strong base such as $\phi_3C:^\ominus Na^\oplus$ or $[(CH_3)_2CH]_2N:^\ominus Li^\oplus$. These bases effect complete conversion of the ester substrate to enolate intermediates. Condensation of the ester enolate with the remaining half of the ester substrate forms the Claisen-type product and liberates ethoxide ion. Since the ethoxide ion formed at the end of this sequence is a weaker base (is more stable) than the initiating base or the ester enolate, a significant driving force for the reaction results.

c $2C_2H_5CO_2C_2H_5 \xrightarrow{NaOC_2H_5} C_2H_5COCH(CH_3)CO_2C_2H_5$

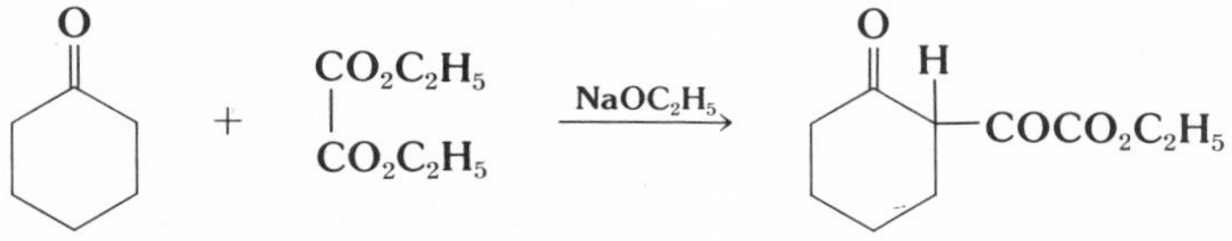

[Structure: cyclohexanone (donor) + diethyl oxalate CH(CO₂C₂H₅)₂ (acceptor) → NaOC₂H₅ → 2-(ethoxyoxalyl)cyclohexanone]

donor   acceptor

$\phi CO_2C_2H_5 + CH_3NO_2 \xrightarrow{NaOC_2H_5} \phi COCH_2NO_2$
acceptor   donor

PROBLEM 14·20

a The structural formula of 3-phenyl-2-benzyl propanoic acid shows that it can be viewed as an acetic acid moiety substituted with two benzyl groups:

$$\phi CH_2 - CHCO_2H$$
$$|$$
$$\phi CH_2$$

This, in fact, becomes a plausible synthesis pathway through the agency of malonic ester alkylation and decarboxylation:

$$CH_2(CO_2C_2H_5)_2 \xrightarrow{NaOC_2H_5} \overset{\oplus}{Na} : \overset{\ominus}{CH}(CO_2C_2H_5)_2 \xrightarrow{\phi CH_2Br} \phi CH_2CH(CO_2C_2H_5)_2$$

$$\downarrow NaOC_2H_5$$

$$(\phi CH_2)_2CHCO_2H \xleftarrow[\Delta]{H_3O^{\oplus}} \xleftarrow{NaOH, H_2O} (\phi CH_2)_2C(CO_2C_2H_5)_2 \xleftarrow{\phi CH_2Br} Na^{\oplus} \phi CH_2\overset{\ominus}{C}(CO_2C_2H_5)_2$$
$$+ $$
$$CO_2$$

b The structure of the target molecule has a five-membered carbocyclic ring, whereas the starting materials are acyclic. Clearly, one of the things we must do is devise a means of constructing this feature. Equation 14.94 provides an attractive means to this end, and the product of this Dieckmann condensation has the added benefit of a doubly activated α-carbon atom:

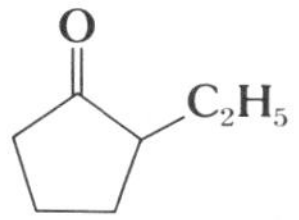

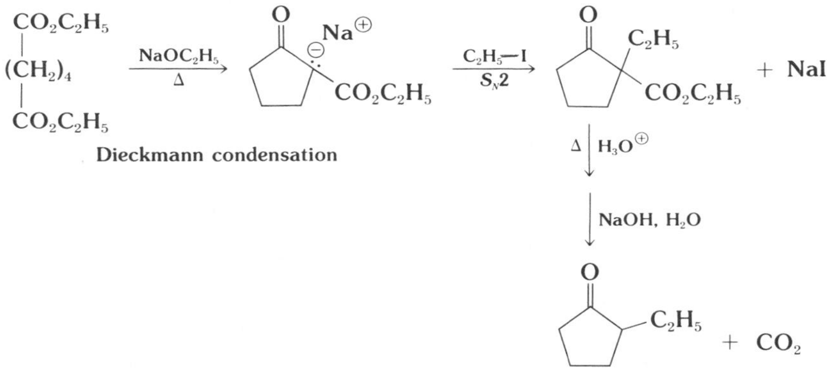

[Scheme: diethyl adipate → Dieckmann condensation → 2-carbethoxycyclopentanone sodium enolate → alkylation with C_2H_5–I (S_N2) → 2-carbethoxy-2-ethylcyclopentanone + NaI → $H_3O^{\oplus}$, Δ → NaOH, H_2O → 2-ethylcyclopentanone + CO_2]

c The target molecule in this case also has a five-membered carbocyclic ring, but the starting materials we are given do not permit us to construct it by a Dieckmann condensation. However, the 1,4-dibromobutane reactant suggests that ring formation might be achieved by a double alkylation (as in equation 14.98):

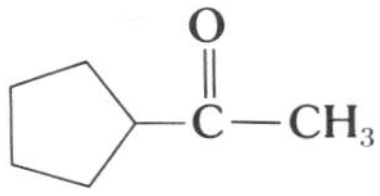

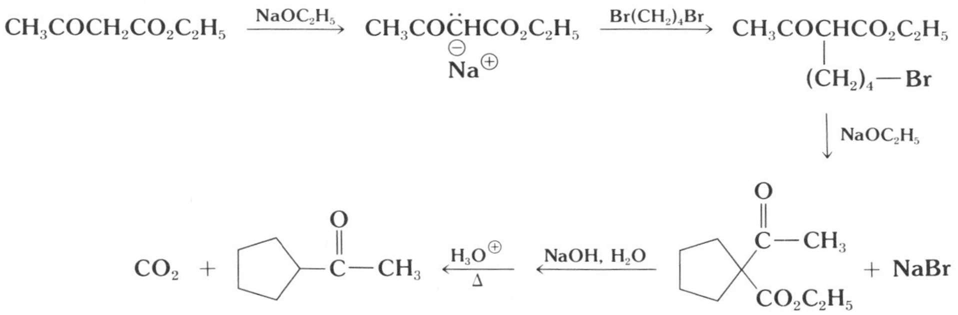

$$CH_3COCH_2CO_2C_2H_5 \xrightarrow{NaOC_2H_5} \underset{Na^{\oplus}}{CH_3CO\overset{\ominus}{C}HCO_2C_2H_5} \xrightarrow{Br(CH_2)_4Br} \underset{(CH_2)_4-Br}{CH_3COCHCO_2C_2H_5}$$

$$\downarrow NaOC_2H_5$$

[Scheme: cyclized intermediate 1-acetyl-1-carbethoxycyclopentane + NaBr → NaOH, H_2O → $H_3O^{\oplus}$, Δ → CO_2 + 1-acetylcyclopentane (cyclopentyl methyl ketone)]

ANSWERS TO TEXT PROBLEMS 763

PROBLEM 14·21
a First, since malonic ester is a relatively strong carbon acid ($pK_a = 13.3$), its conjugate base is readily formed by treatment with $NaOC_2H_5$ (pK_a of C_2H_5OH is about 16). A simple ester such as ethyl acetate is too weak an acid ($pK_a = 26$) to respond to this base in a similar fashion. Second, simple ester conjugate bases undergo self-condensation (Claisen condensation) so rapidly that they do not survive long enough to be alkylated in a separate step.
b Secondary amide conjugate bases are about 10^{10} times stronger (as bases) than ester enolates; consequently, the addition of an ester to an excess of such a base at low temperature gives rapid and quantitative conversion to the enolate conjugate base. Since the enolate base is not present with unreacted ester, no self-condensation occurs. During the alkylation step some condensation might take place, but it is retarded by the large *t*-butyl ester group.

PROBLEM 14·22
The first reaction is a reduction of a ketone to a secondary alcohol (as with $NaBH_4$ in chapter 12). The second reaction is a dehydration. The loss of water from β-hydroxyketones or esters is particularly facile. The third reaction is a hydrogenation of a double bond. In the laboratory this type of reaction is normally accomplished with hydrogen and a metal catalyst.

PROBLEM 14·23
a $CH_3\overset{*}{C}H_2CH_2\overset{*}{C}H_2CH_2\overset{*}{C}H=CH\overset{*}{C}H_2CH=\overset{*}{C}HCH_2\overset{*}{C}H_2CH_2\overset{*}{C}H_2CH_2\overset{*}{C}H_2CH_2\overset{*}{C}O_2H$
b The Claisen condensation uses base catalysts. Since the doubly activated methylene group in the acetoacetic ester product is more acidic than the methyl group, any subsequent reaction would probably take place at that site. The stability and hindrance of this conjugate base argue against any further reaction.

PROBLEM 14·24
a The conjugate bases of carboxylic acids are stabilized by charge delocalization:

$$R-C\overset{\ddot{O}\cdot}{\underset{\ddot{O}\colon^{\ominus}}{\diagdown}} \longleftrightarrow R-C\overset{\ddot{O}\colon^{\ominus}}{\underset{\ddot{O}\cdot}{\diagdown}}$$

The conjugate bases of peracids cannot be delocalized in an equivalent manner.
b Steric hindrance by the methyl group blocks approach of the peracid from that side of the double bond.

CHAPTER 15

PROBLEM 15·1
a

:NH_2—⟨ ⟩=O

This compound is a moderately strong base ($K_b \approx 10^{-5}$). It is rapidly acylated by acetic anhydride to

O=⟨ ⟩—N(H)COCH$_3$

and is reduced by NaBH$_4$ to

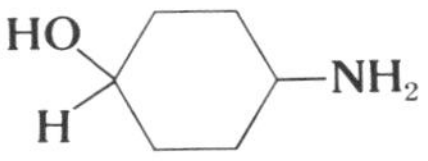

[structure: 4-aminocyclohexanol with HO, H on one carbon and NH$_2$ on the other]

Benzaldehyde reacts with both functional groups to produce

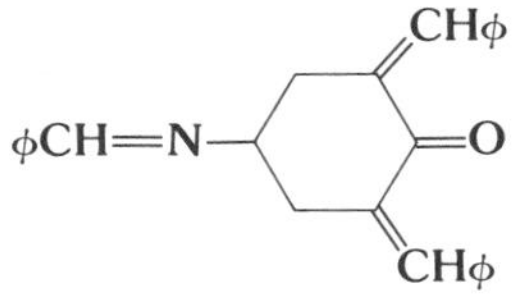

[structure: φCH=N— attached to cyclohexanone ring with =CHφ groups at both α-positions and C=O]

In contrast, amides are very weak bases and the carboxamide undergoes none of these reactions. Vigorous hydrolysis yields ammonia and a carboxylic acid:

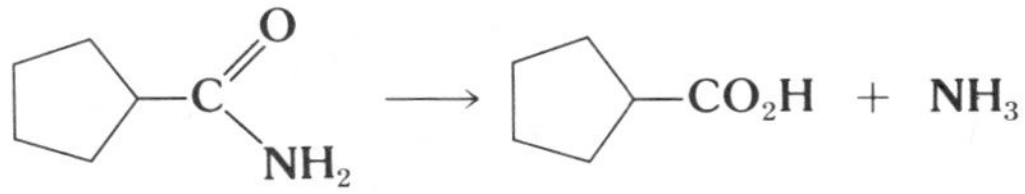

[structure: cyclopentane-C(=O)-NH$_2$ → cyclopentane-CO$_2$H + NH$_3$]

cyclopentane
carboxamide

b R—C(=O)—CH$_2$—CHR(OH)

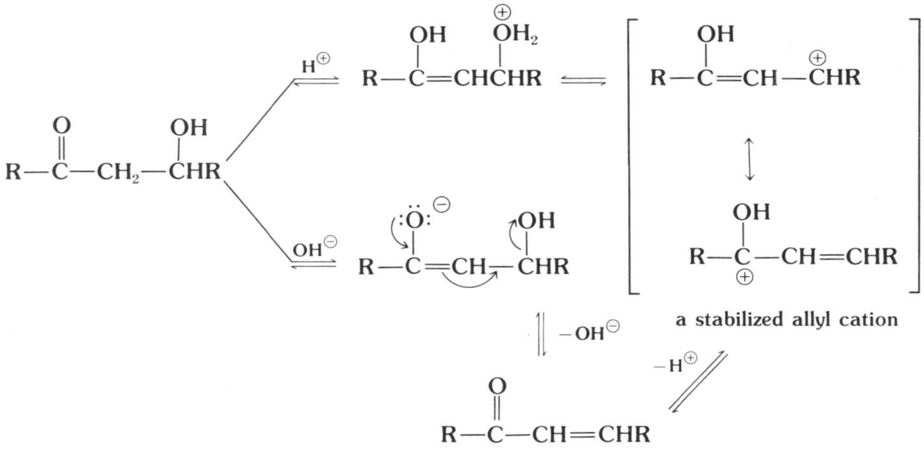

$\xrightarrow{H^+}$ R—C(OH)=CH—CHR(O$\overset{+}{H}_2$) ⇌ $\left[\begin{array}{c} \text{R—C(OH)=CH—}\overset{+}{\text{C}}\text{HR} \\ \updownarrow \\ \text{R—}\overset{+}{\text{C}}\text{(OH)—CH=CHR} \end{array}\right]$ a stabilized allyl cation

$\xrightarrow{OH^-}$ R—C(:Ö:$^-$)=CH—CHR(OH)

⇌ (−OH$^-$) ‖ $\parallel$ (−H$^+$)

R—C(=O)—CH=CHR

c [structure: cyclobutane with CH$_3$, O—C(=O)—CH$_3$, —C(=O)—CH$_3$, H groups] $\xrightarrow[\text{ester hydrolysis}]{\text{NaOH, H}_2\text{O}}$ [structure: cyclobutane with CH$_3$, OH, —C(=O)—CH$_3$, H] $\xrightarrow{^-OH}$ [open-chain: CH$_3$-C(=O)-CH$_2$-CH=C(CH$_3$)-O$^-$]

⇌

[cyclohexenone with CH$_3$] $\xleftarrow[-\text{H}_2\text{O}]{}$ [cyclohexanone with OH, CH$_3$] $\xleftarrow{\text{NaOH}}$ [open-chain diketone: O=C with CH$_3$ and chain to C(=O)—CH$_3$]

aldol condensation

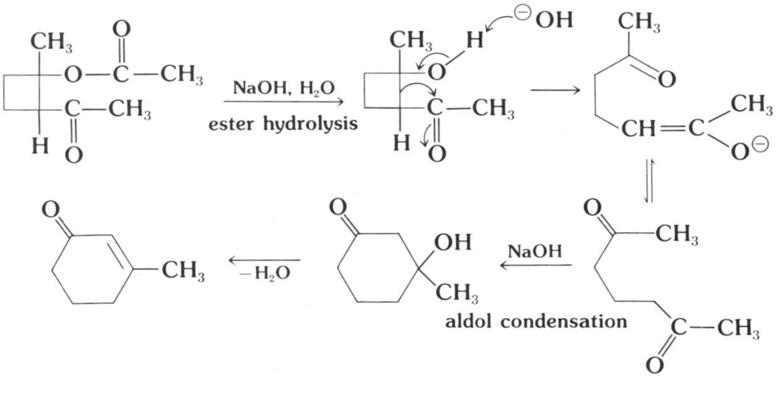

ANSWERS TO TEXT PROBLEMS 765

PROBLEM 15·2
a The trimethylammonium substituent is strongly electronegative and consequently deactivates the aromatic ring by inductive withdrawal of electron density. By introducing more and more methylene groups between the aromatic ring and the positive nitrogen atom, we reduce the inductive effect and eventually reach a point at which the methylene chain acts like an alkyl substituent.
b Halogen atoms are electronegative and show a strong electron-withdrawing inductive effect. In the cases shown here the halogen atoms are not bonded directly to the aromatic ring; electron donation by a resonance effect is therefore not possible. As we increase the number of benzylic halogen substituents or increase the electronegativity of the halogen (Cl < F), the inductive electron withdrawal should become stronger. This increases the meta directing influence of the side chain.

PROBLEM 15·3
a Since a nitro substituent deactivates an aromatic ring with respect to electrophilic substitution reactions, the initial product of a nitration reaction is ordinarily less reactive than the aromatic compound being nitrated. As a result, mononitro products predominate in most cases.
b The unusual behavior of durene under nitrating reaction conditions can be explained in part by the absence of strong resonance deactivation by the first nitro substituent. This conjugative deactivation requires that the NO_2 group be coplanar with the aromatic ring:

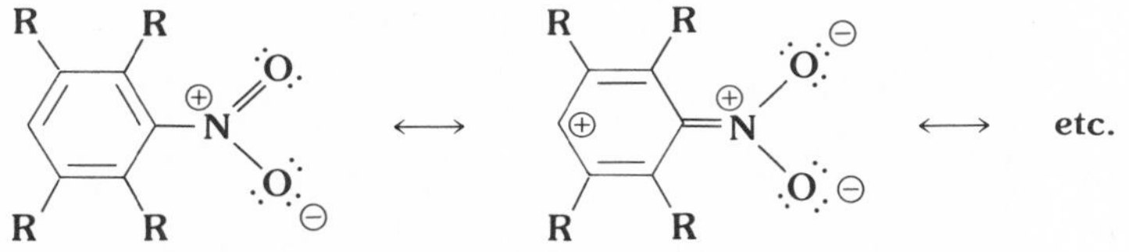

If the ortho positions to the nitro group are occupied by bulky substituents (R = CH_3), the necessary coplanarity is prevented, and the nitro function becomes only weakly deactivating (the inductive effect). Subsequent nitration para to the first nitro group is facilitated by the activating methyl groups.

PROBLEM 15·4
In all three of these cases we can account for the products by assuming that an initial electrophilic attack on the diene (or triene) proceeds so as to give the most stable cation intermediate:

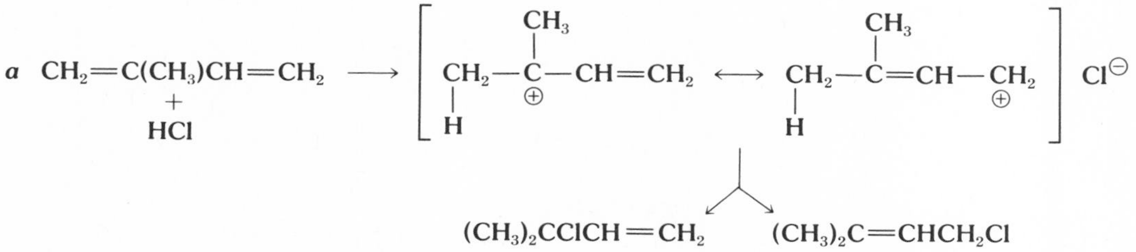

a $CH_2=C(CH_3)CH=CH_2$ + HCl $\longrightarrow$ [$CH_2-\overset{CH_3}{\underset{H}{C}}-CH=CH_2$ $\longleftrightarrow$ $CH_2-\overset{CH_3}{\underset{H}{C}}=CH-CH_2$] $Cl^{\ominus}$

$(CH_3)_2CClCH=CH_2$ $(CH_3)_2C=CHCH_2Cl$

b $CH_2=CHCH=CHCH=CH_2$
   $+$
   Br_2

$\longrightarrow$ $\begin{bmatrix} BrCH_2-\overset{\oplus}{C}HCH=CHCH=CH_2 \\ \updownarrow \\ BrCH_2CH=CH\overset{\oplus}{C}HCH=CH_2 \\ \updownarrow \\ BrCH_2CH=CHCH=CH\overset{\oplus}{C}H_2 \end{bmatrix}$ $Br^{\ominus}$

$\longrightarrow$ $BrCH_2CHBrCH=CHCH=CH_2$
$+$
$BrCH_2CH=CHCH=CHCH_2Br$

Reaction with nucleophiles is faster at the ends of the conjugated cation.

c $\phi CH=CHCH=CH_2$
   $+$
   Br_2

$\longrightarrow$ $\begin{bmatrix} \phi CH=CH\overset{\oplus}{C}HCH_2Br \\ \updownarrow \\ \overset{\oplus}{\phi C}HCH=CHCH_2Br \end{bmatrix}$ $Br^{\ominus}$ $\longrightarrow$ $\phi CH=CHCHBrCH_2Br$

This 1,2 addition product is the thermodynamically favored product because the remaining double bond is conjugated with the aromatic ring (this product is also kinetically favored).

PROBLEM 15·5

a

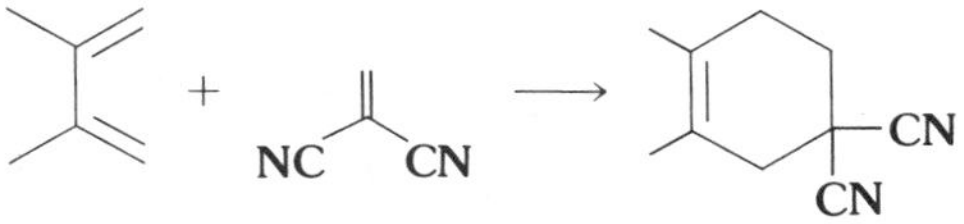

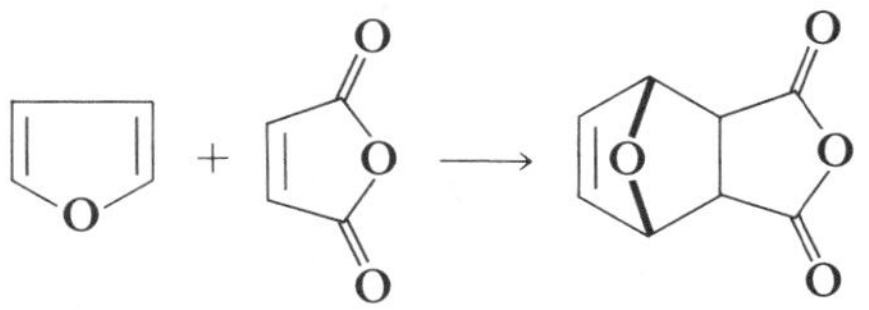

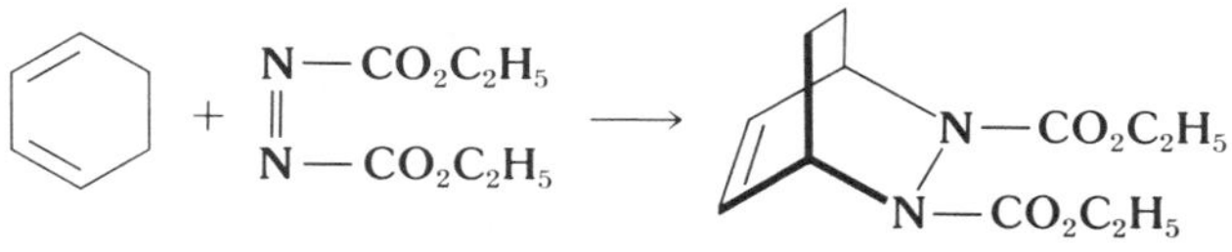

b X =    Y =

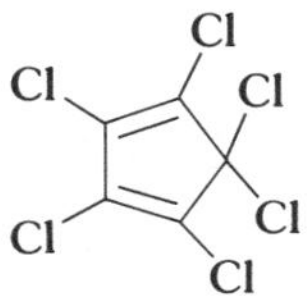

c [reaction: bicyclohexenyl + CH$_2$=CHNO$_2$ → octahydrophenanthrene-NO$_2$ adduct]

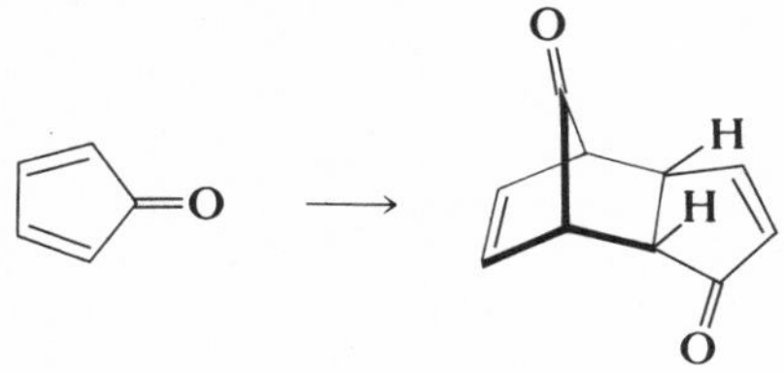

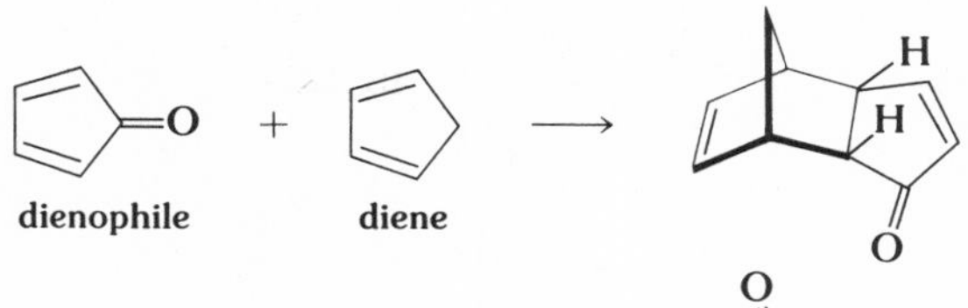

[reaction: 2-methyl-3-methylene-1,3-butadiene derivative + CH$_2$=CHCOCH$_3$ → cyclohexene product]

ϕCH=CHCH=CH—⟨C$_6$H$_4$⟩—CH=CHCH=CHϕ + 2HC≡CH

↓

[bis-phenyl product with two terphenyl rings]

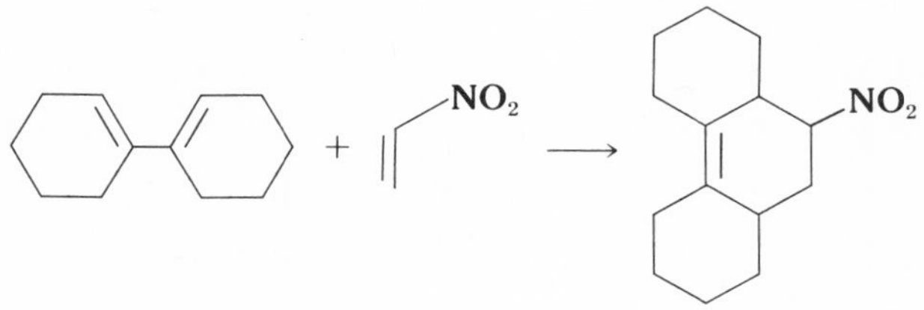

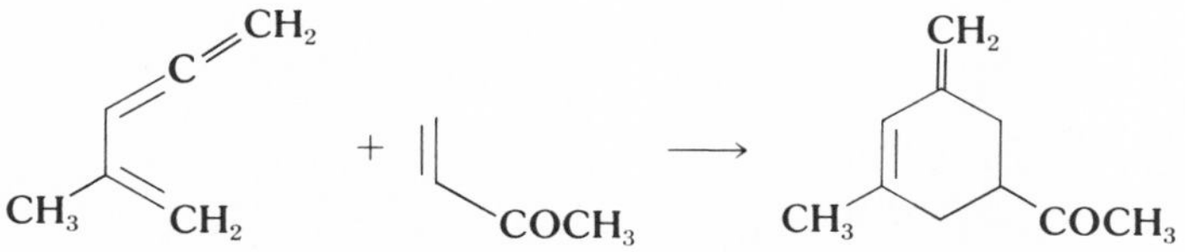

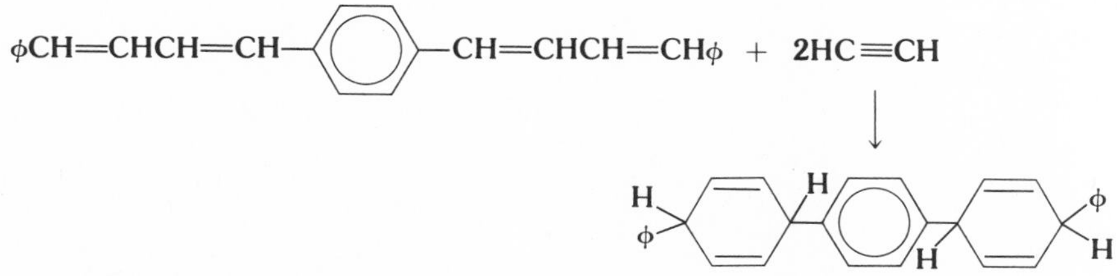

PROBLEM 15·6

a 2 [cyclopentadienone] → [endo dimer with two C=O]

[cyclopentadienone] + [cyclopentadiene] → [adduct]
 dienophile   diene

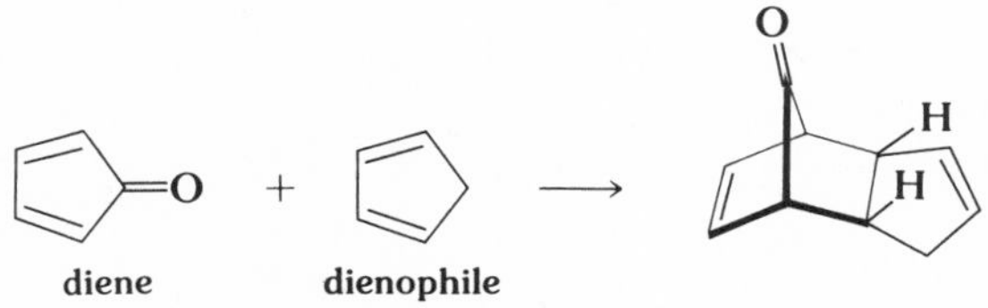

[cyclopentadienone] + [cyclopentadiene] → [adduct]
     diene   dienophile

b [3,4-dicyanofuran] + CF$_3$—C≡C—CF$_3$ ⇌ [oxabicyclic adduct with CN, CN, CF$_3$, CF$_3$] ⇌ [3,4-bis(trifluoromethyl)furan]

+

NC—C≡C—CN

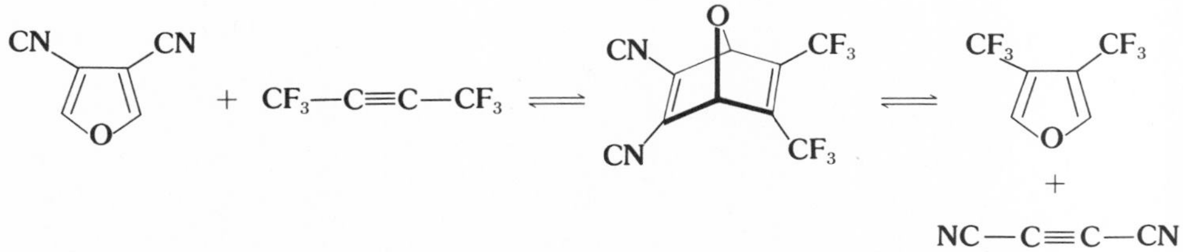

Apparently the reaction of the dicyanofuran with dicyanoacetylene is favored over its reaction with the hexafluoro-2-butyne dienophile. Similarly, the Diels-Alder reaction of the *bis*-trifluoromethylfuran with the latter dienophile is more favorable than its reaction with dicyanoacetylene. Since the Diels-Alder reactions are reversible, these favored adducts are the major products.

PROBLEM 15·7
Some conformations of *trans,trans,cis*-2,4,6-octatriene:

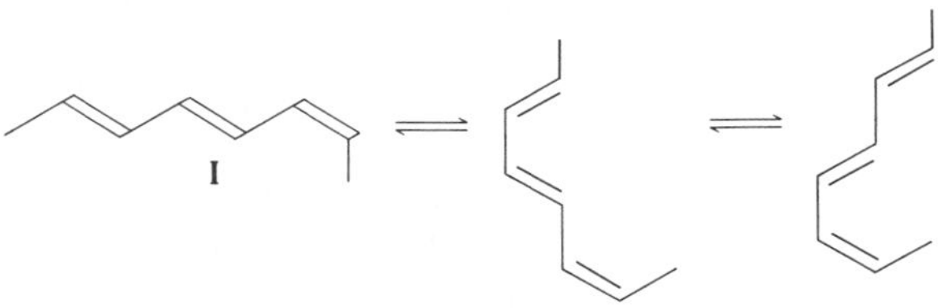

[Structures I, II, III shown in equilibrium]

The least sterically hindered cisoid diene is found in conformer II:

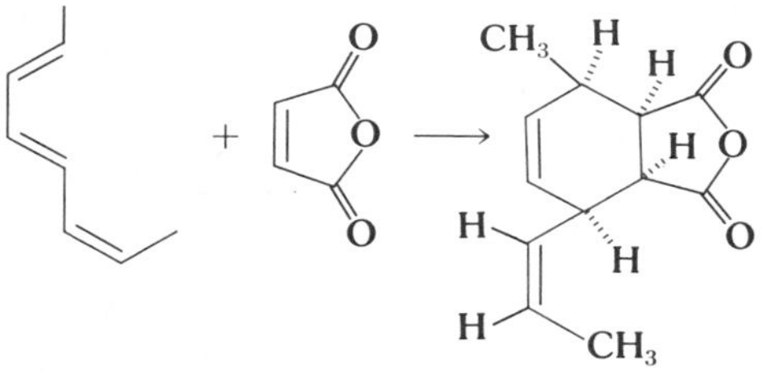

[Diels-Alder reaction of conformer II with maleic anhydride giving the cyclohexene adduct with CH₃, H stereochemistry shown]

PROBLEM 15·8

a $\phi SH + CH_2=CHCO_2C_2H_5 \xrightarrow{\text{1,4 addition}} \phi SCH_2CH_2CO_2C_2H_5$

b $N_2H_4 + (CH_3)_2C=CHCCH_3 \xrightarrow{\text{1,4 addition}}$ [hydrazine adduct] $\xrightarrow{-H_2O}$ [pyrazoline product]

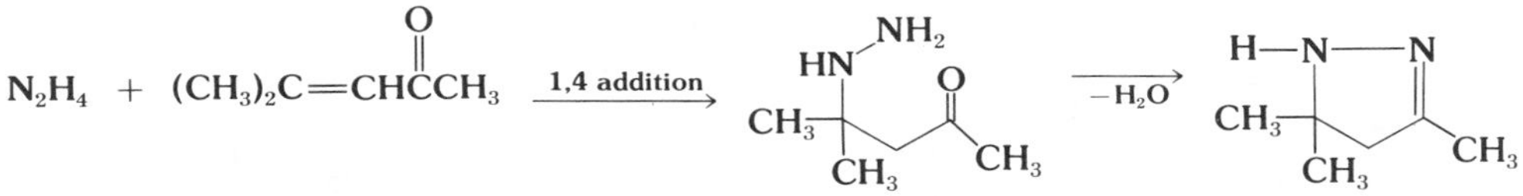

c $CH_2(CO_2C_2H_5)_2 + \phi CH=CHCO_2C_2H_5 \xrightarrow{NaOC_2H_5} \phi CHCH_2CO_2C_2H_5 \; | \; CH(CO_2C_2H_5)_2 \xrightarrow[\Delta]{NaOH} \xrightarrow{H_3O^\oplus} \phi CHCH_2CO_2H \; | \; CH(CO_2H)_2$

$\downarrow \Delta$

$\phi CH(CH_2CO_2H)_2 + CO_2$

PROBLEM 15·9

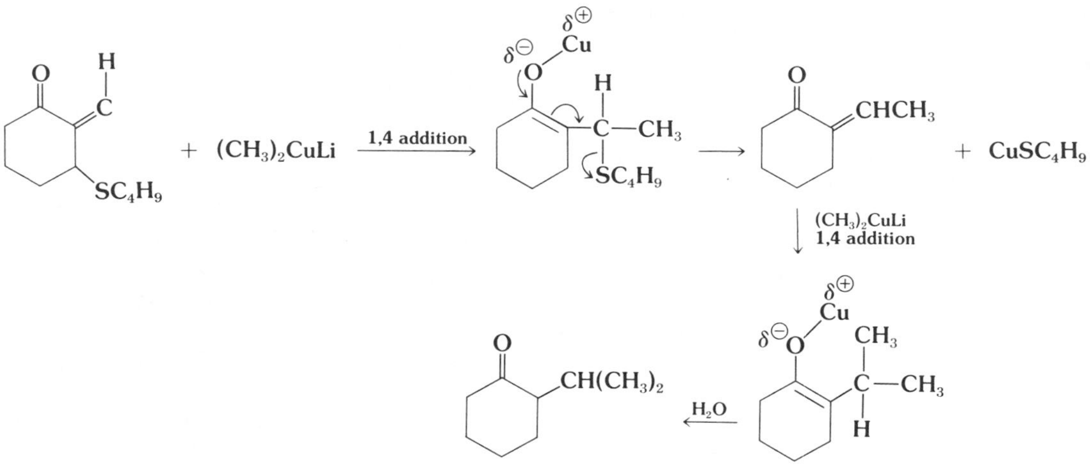

[Mechanism showing 2-(butylthio)cyclohexenone + (CH₃)₂CuLi → 1,4 addition via copper enolate intermediate → 2-ethylidenecyclohexanone + CuSC₄H₉ → second (CH₃)₂CuLi 1,4 addition → copper enolate → H₂O → 2-isopropylcyclohexanone]

ANSWERS TO TEXT PROBLEMS 769

PROBLEM 15·10
No. The intermediate bromonium ion is achiral (it has a plane of symmetry):

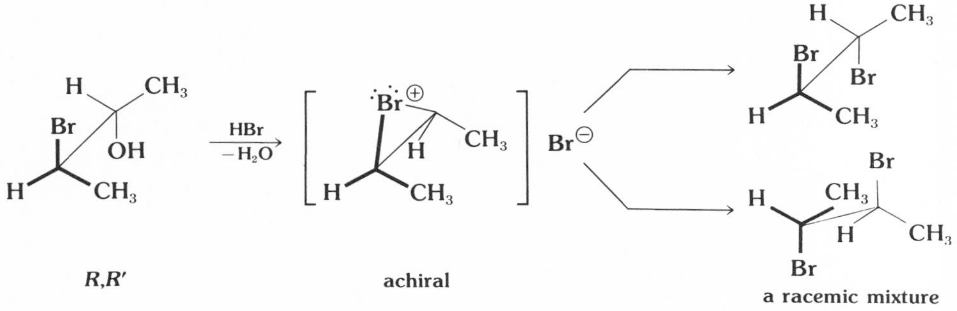

PROBLEM 15·11
Displacement of the sulfonate anion by an S_N2 mechanism would give the cis acetate ester exclusively. To account for the formation of the trans product and the apparent shift of the tritium label, we must assume a neighboring interaction by the methoxyl substituent:

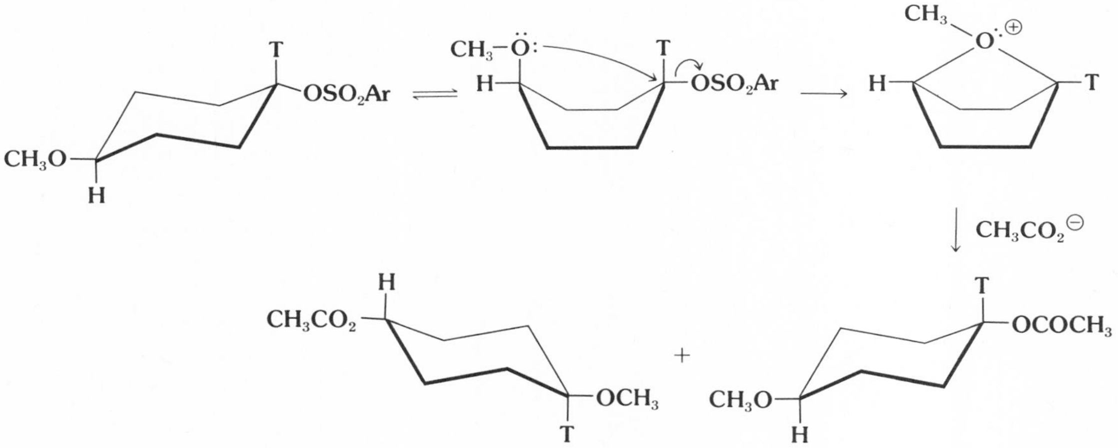

PROBLEM 15·12
A fast intramolecular acylation of an imidazole nitrogen atom gives an intermediate which is much more reactive than thio esters to hydrolysis:

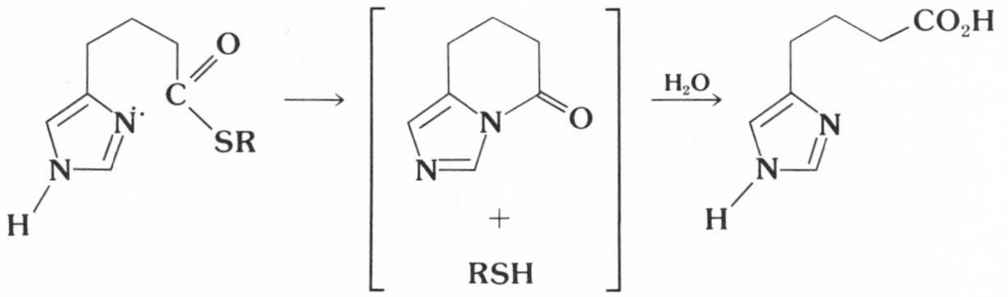

PROBLEM 15·13

a) [reaction scheme showing alkene with CH₃, CH₃, H, CH₂, H₂C-Cl groups cyclizing to form a cyclopropyl carbocation, then reacting with 2H₂O to give HO-C(CH₃)₂-cyclopropyl + H_3O^+]

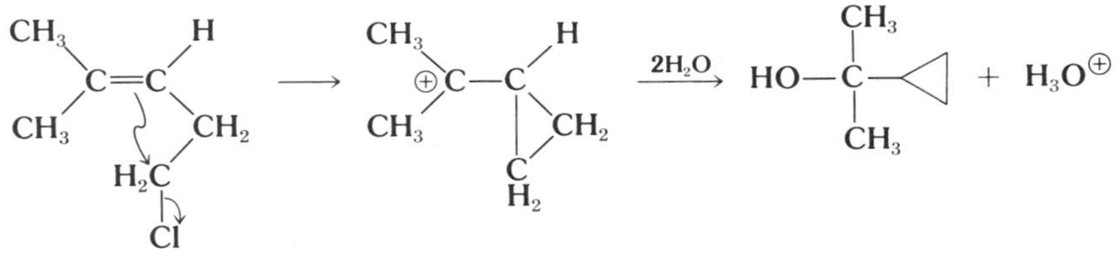

b) [alkyne in a ring with OSO₂Ar leaving group attacked by H₂O, forming enol with OH_2^+, then -H⁺ ketonization gives bicyclic ketone (decalone)]

[second alkyne substrate with OSO₂Ar giving a different ring-fused enol OH_2^+, then -H⁺ ketonization gives a different bicyclic ketone]

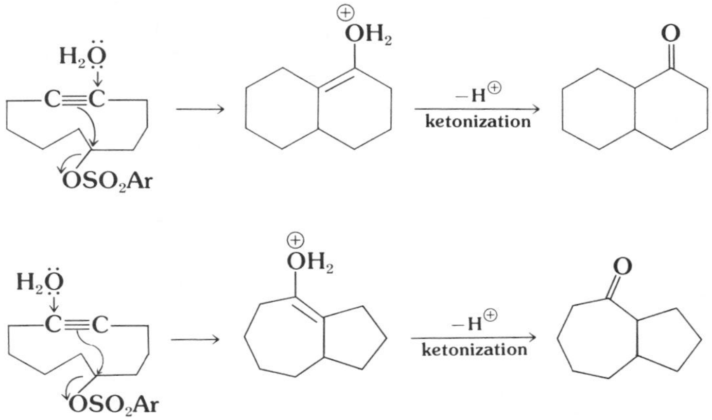

PROBLEM 15·14

[cyclodecene with Br⁺ adding and Br⁻ attacking transannularly to give two trans-bicyclic dibromides]

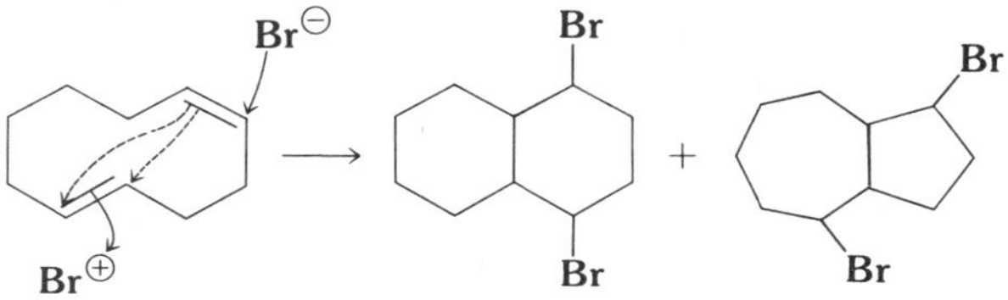

PROBLEM 15·15

a) Those sulfonate esters having double bonds at the Δ^5 position undergo solvolysis more rapidly than the saturated analog. Apparently the double bond provides anchimeric assistance in the ionization step:

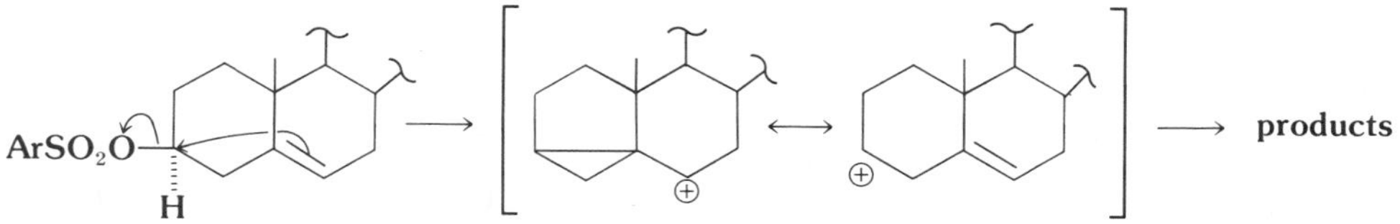

ArSO₂O—[decalin system with H]⟶ [bridged nonclassical cation ↔ classical cation] ⟶ products

Substituents or functional groups which further stabilize the positive charge in this intermediate provide additional rate enhancement. For example:

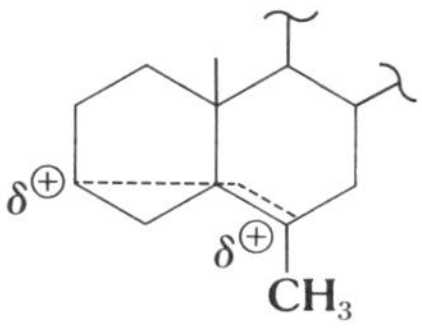

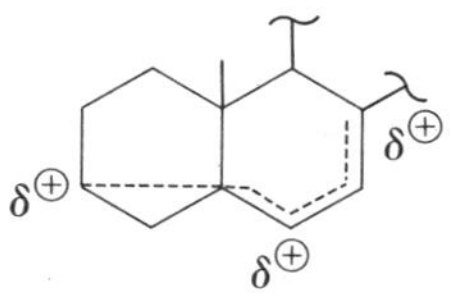

[two structures showing δ⁺ charge distribution, one with CH₃ substituent stabilizing the cation, another with extended conjugation showing δ⁺ at multiple positions] *b* Mustard gas and the nitrogen mustards are toxic because they can function as *bis*-alkylating agents to cross-link proteins and nucleic acids. The halogen atoms in this compound are both activated by the double bond (anchimeric assistance); consequently, it can also act as a *bis*-alkylating agent.

PROBLEM 15·16

a

$$\phi_2\underset{\underset{\text{OH}}{|}}{C}-\overset{\oplus}{C}H_2 \xrightarrow{\phi \text{ shift}} \phi-\underset{\underset{O}{\|}}{C}-CH_2-\phi$$

b Trifluoromethyl groups are strongly electron withdrawing and destabilize adjacent carbonium-ion intermediates. To effect a pinacol rearrangement in this case, an unfavorable ion must be formed:

$$CF_3-\underset{\underset{\phi}{|}}{\overset{\overset{OH}{|}}{C}}-\underset{\underset{\phi}{|}}{\overset{\overset{OH}{|}}{C}}-CF_3 \xrightarrow[-H_2O]{H^\oplus} \left[CF_3-\underset{\underset{\phi}{|}}{\overset{\overset{OH}{|}}{C}}-\underset{\underset{\phi}{|}}{\overset{\oplus}{C}}-CF_3 \right] \longrightarrow CF_3-\underset{\underset{\phi}{|}}{\overset{\overset{O}{\|}}{C}}-\underset{\underset{\phi}{|}}{\overset{\overset{\phi}{|}}{C}}-CF_3$$

(slow)

PROBLEM 15·17

c

[Mechanism showing ring expansion of a spiro cyclohexanone-cyclobutane system via protonation, ring expansion, and formation of spiro ketone]

PROBLEM 15·18
The relative stability of the phenonium-ion intermediates (see Figure 15.13) will be strongly affected by para (and ortho) substituents on the aromatic ring. The electron-donating methoxy group in the methoxy compound will stabilize the phenonium intermediate, making the mechanism in Figure 15.13 even more favorable. As a result, this threo isomer should give only racemic *threo*-acetate. The electron-withdrawing nitro group in the nitro compound destabilizes phenonium intermediates; consequently, the same mechanism fails to operate. The products can be explained by assuming competition between an S_N2 displacement by acetate and an ionization, leading chiefly to an $E1$ product.

PROBLEM 15·19

a [cyclopentanone] $=O + CF_3CO_3H \longrightarrow$ [δ-valerolactone] $+ CF_3CO_2H$

b [cyclohexanone] $+ HN_3 \xrightarrow{H^\oplus}$ [caprolactam] $+ N_2$

c

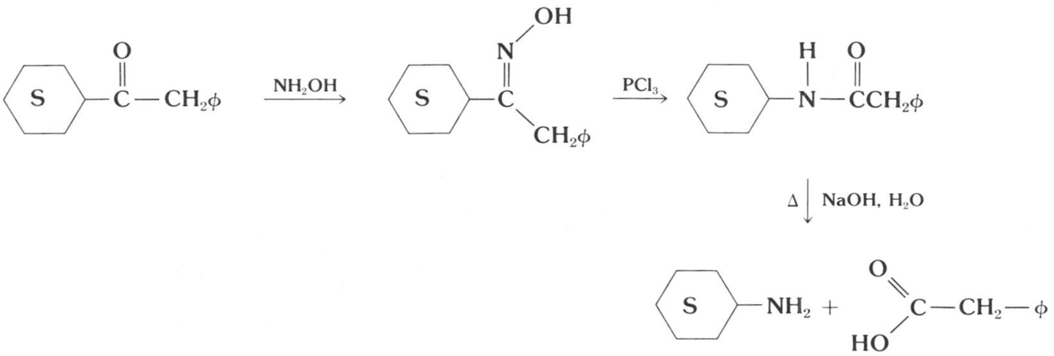

[Structure: S-ring-C(=O)-CH₂φ] —NH₂OH→ [Structure: S-ring-C(=NOH)-CH₂φ] —PCl₃→ [Structure: S-ring-N(H)-C(=O)CH₂φ]

Δ | NaOH, H₂O ↓

[S-ring-NH₂] + φ-CH₂-C(=O)-OH

CHAPTER 16

PROBLEM 16·1

a $\phi_3 C \cdot + \cdot \ddot{O}-\ddot{O} \cdot \longrightarrow (C_6H_5)_3C-\ddot{O}-\ddot{O}\cdot \xrightarrow{\phi C \cdot} \phi_3C-\ddot{O}:$
         a stable $\phi_3C-\ddot{O}:$
         diradical

$\phi_3C \cdot + I-I \longrightarrow \phi_3C-I + \cdot I$
(or I·)

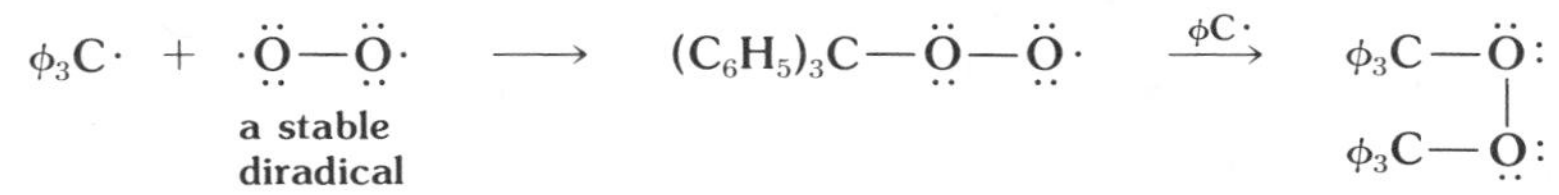

b Electronic (ultraviolet-visible) spectroscopy or pmr spectroscopy should be informative in distinguishing hexaphenylethane from the para-coupled dimer. Hexaphenylethane should have spectra that are similar to the spectra of benzene. The pmr spectrum, for example, should exhibit aromatic proton resonance exclusively. In contrast, the para-coupled dimer has olefinic and aliphatic protons, in addition to aromatic protons. Furthermore, the extended conjugation of the π-electron system in this dimer should give rise to strong absorption in the ultraviolet range at significantly longer wavelength (about 280 nm) than hexaphenylethane.

PROBLEM 16·2

If one of a pair of electrons occupying an atomic or molecular orbital is induced to change its spin state, then its partner must also change its spin in order to avoid violating the Pauli exclusion principle. The state resulting from such a double change will be indistinguishable from the initial state. Consequently, no esr signal will be observed.

PROBLEM 16·3

a Reaction 16.10 is an oxidation; $Fe(CN)_6^{3-}$ is the oxidizing agent. Reaction 16.11 is a reduction; Na is the reducing agent. Reaction 16.12 is an oxidation; Br_2 is the oxidizing agent.

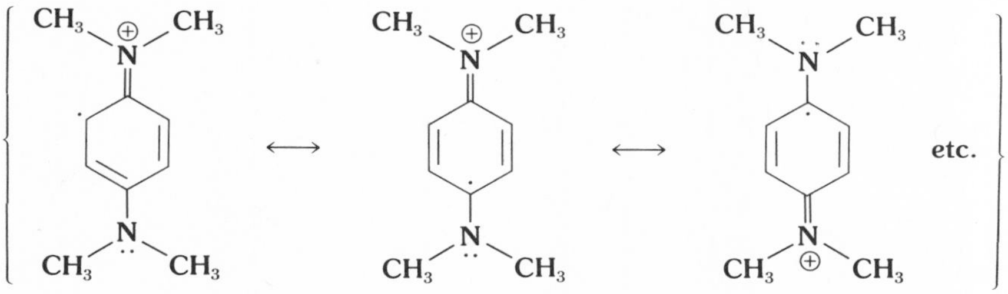

$$\left\{ \begin{array}{c} \text{[structure 1]} \end{array} \longleftrightarrow \begin{array}{c} \text{[structure 2]} \end{array} \longleftrightarrow \begin{array}{c} \text{[structure 3]} \end{array} \text{ etc.} \right\}$$

PROBLEM 16·4

a $\quad R\ddot{O}-\ddot{O}R \xrightarrow{\Delta} 2R\ddot{O}\cdot$

$R\ddot{O}\cdot + HCCl_3 \longrightarrow ROH + \cdot CCl_3$

$C_6H_{13}CH=CH_2 + \cdot CCl_3 \longrightarrow C_6H_{13}\dot{C}HCH_2CCl_3$ ⎫
⎬ chain reaction
$C_6H_{13}\dot{C}HCH_2CCl_3 + HCCl_3 \longrightarrow C_6H_{13}CH_2CH_2CCl_3 + \cdot CCl_3$ ⎭

The alternative addition of the trichloromethyl radical to C-2 of the alkene would yield a primary radical intermediate, which is less stable than the secondary radical intermediate shown above.

b $\quad R\ddot{O}\cdot + BrCCl_3 \longrightarrow ROBr + \cdot CCl_3$

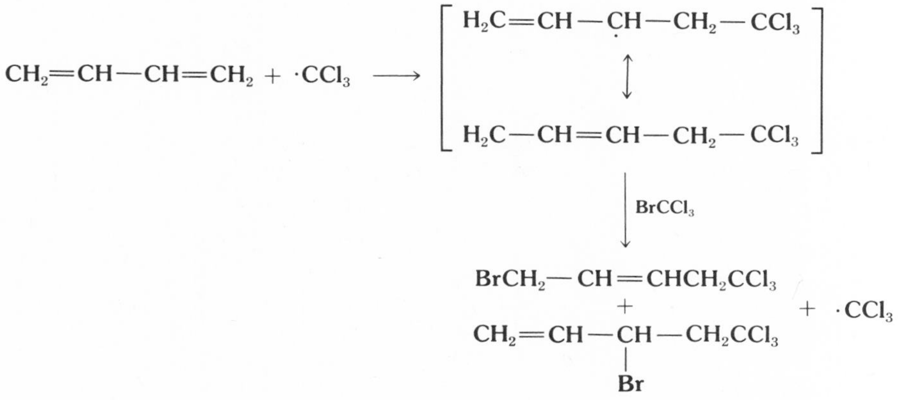

$$CH_2=CH-CH=CH_2 + \cdot CCl_3 \longrightarrow \left[\begin{array}{c} H_2C=CH-\dot{C}H-CH_2-CCl_3 \\ \updownarrow \\ H_2\dot{C}-CH=CH-CH_2-CCl_3 \end{array} \right]$$

$$\Big\downarrow BrCCl_3$$

$$\begin{array}{c} BrCH_2-CH=CHCH_2CCl_3 \\ + \\ CH_2=CH-CH-CH_2CCl_3 \\ | \\ Br \end{array} + \cdot CCl_3$$

PROBLEM 16·5

$\cdot \ddot{O}-\text{[2,6-di-t-butylphenyl]}-CH=\text{[2,6-di-t-butylcyclohexadienylidene]}=\ddot{O} + CH_3CH_2\cdot \longrightarrow CH_3CH_2-O-\text{[2,6-di-t-butylphenyl]}-CH-\text{[2,6-di-t-butylcyclohexadienylidene]}=O$

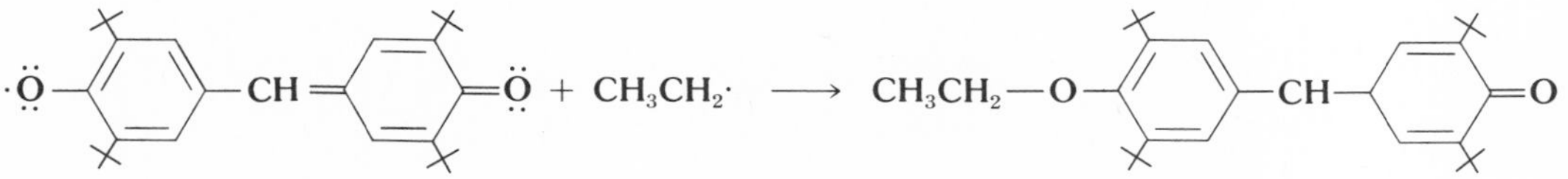

774 ANSWERS TO TEXT PROBLEMS

PROBLEM 16·6

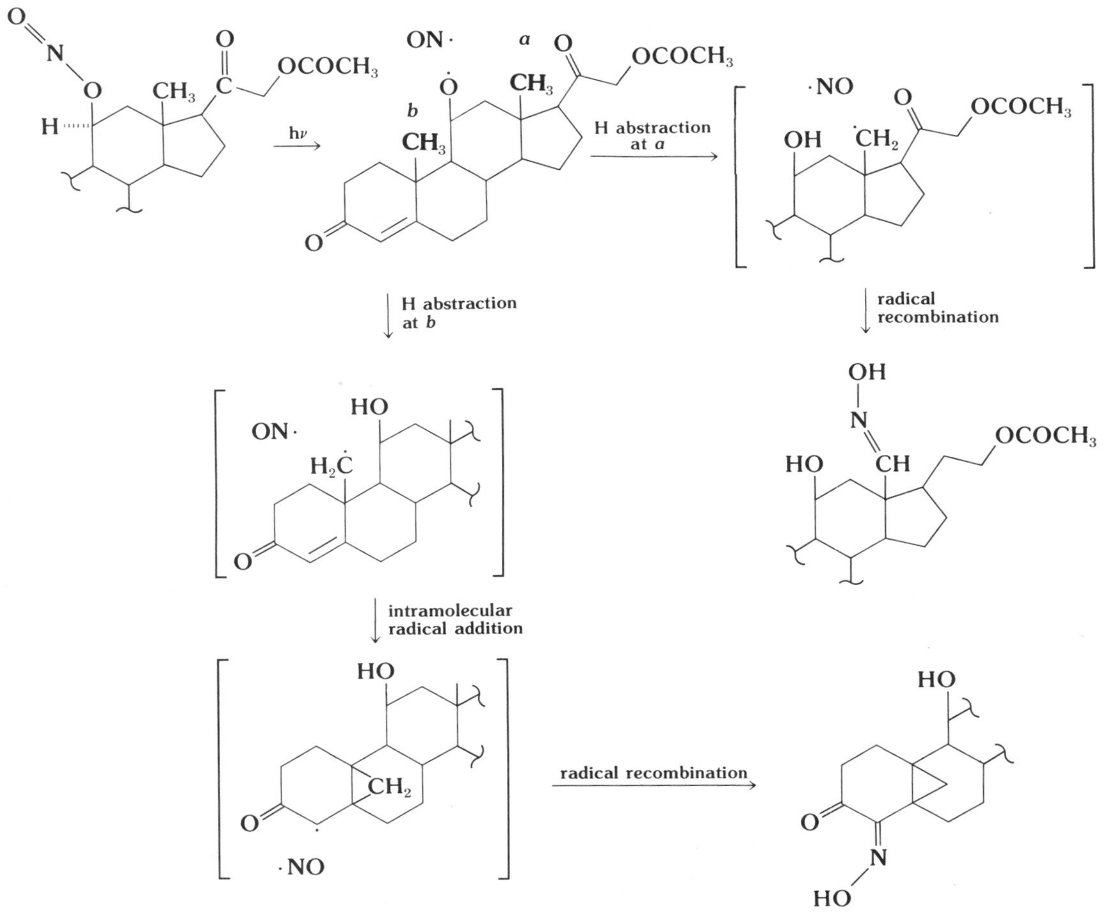

PROBLEM 16·7
Four π-bonding electrons are converted to four σ-bonding electrons in the cycloaddition reaction. This is usually an exothermic process.
a [8 + 2]
b [4 + 2]
c [4 + 2], [2 + 2]

PROBLEM 16·8
Comparison of these equations shows that reaction 16.35 (an example of an electrocyclic reaction) involves a change in the number of σ and π bonds in going from reactant to product. Reaction 16.37 (an example of a sigmatropic reaction) does not involve a change in the number of these bonds.

PROBLEM 16·9
This reaction is most likely not concerted. Diradical intermediates have been proposed; for example:

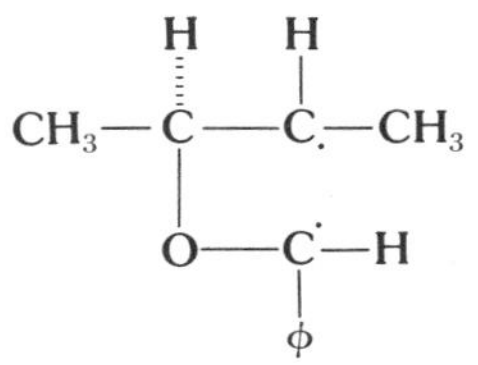

PROBLEM 16·10

a Both reactions are disrotatory.

b Conrotatory:

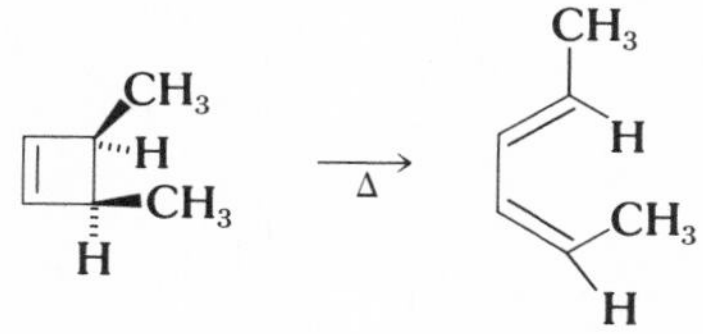

[structure: cis-3,4-dimethylcyclobutene with CH₃, H substituents → (2Z,4Z)-hexa-2,4-diene with CH₃, H groups, via Δ]

PROBLEM 16·11

a We may regard the cyclic polyenes in these reactions as coiled π-electron systems (the dashed line represents the σ bond holding the coiled arrangement in place). In the following shorthand notation for orbital phase relationships, the signs refer to the phases on one side of the molecular plane. The frontier orbitals for reactions 16.43 and 16.44 are:

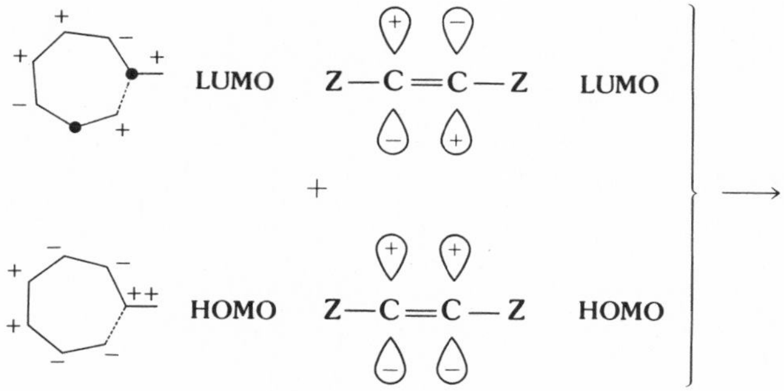

LUMO   Z—C=C—Z   LUMO

+

HOMO   Z—C=C—Z   HOMO

⟶ phase congruence for $[8_s + 2_s]$ cycloaddition

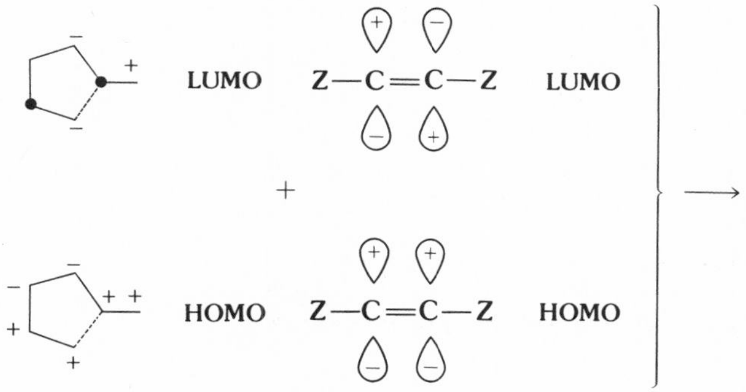

LUMO   Z—C=C—Z   LUMO

+

HOMO   Z—C=C—Z   HOMO

⟶ phase incongruence for $[6_s + 2_s]$ cycloaddition

b [structure: tropone] may serve as a mono-, di-, or triene.

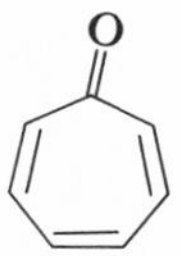

776 ANSWERS TO TEXT PROBLEMS

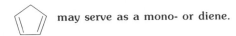 may serve as a mono- or diene.

Suprafacial cycloaddition reactions will be thermally allowed for the cases $[4_s+2_s]$ and $[6_s+4_s]$.

$[4_s+2_s]$ products:

exo + endo   exo + endo   + isomer   + isomer

$[6_s+4_s]$ products:

two stereoisomers

c If we regard the bicyclic reactant as a coiled heptaene, the notation used in part *a* gives us:

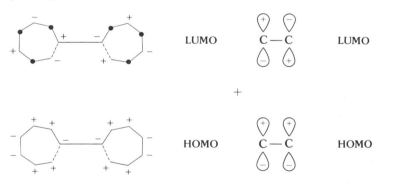

A $[14_s+2_s]$ cycloaddition is forbidden in a thermal mode, but the flexibility of the heptaene permits the allowed $[14_a+2_s]$ cycloaddition to take place. Smaller polyene systems are usually unable to assume the kind of configuration required for an antarafacial reaction.

PROBLEM 16·12
a The frontier orbitals for the photochemical reaction are

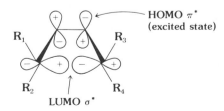

The disrotatory opening required by this analysis agrees with reaction 16.49.

ANSWERS TO TEXT PROBLEMS 777

b The frontier orbitals for the thermal reaction are

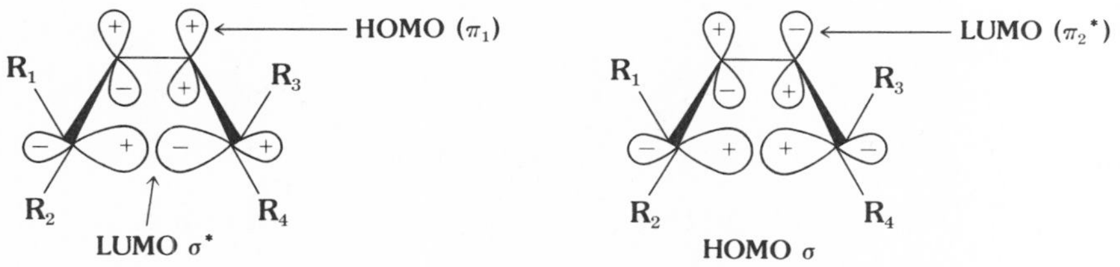

The conrotatory ring opening required by this analysis gives compound III only when $R^1 = R^4 = H$ (as in the case of I) and not when $R^1 = R^3 = H$ as in II:

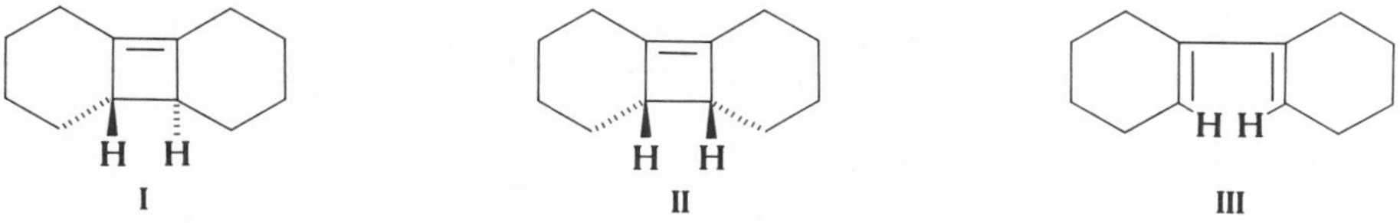

I   II   III

Conrotatory opening of II would necessarily give rise to an isomer of III with a trans double bond in one of the six-membered rings. Such a compound would be highly strained and capable of only momentary existence (see, for example, Section 4.5). The conversion of II to III is probably not a concerted reaction and may well proceed via a diradical intermediate.

PROBLEM 16·13
Following the discussion in problem 16.12, we expect that thermal butadiene $\rightleftharpoons$ cyclobutene interconversions will be conrotatory. Therefore:

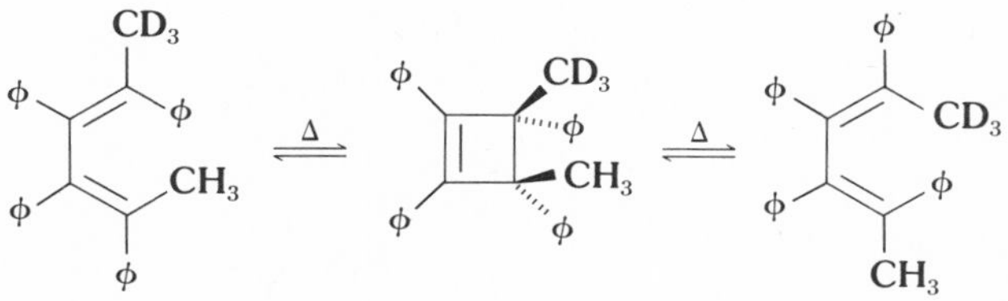

PROBLEM 16·14

$C-C-C-C-C \quad \pi_4^*$

$C-C-C-C-C \quad \pi_5^*$

PROBLEM 16·15
a Chair- and boatlike transition states will give different stereoisomeric dienes in this reaction:

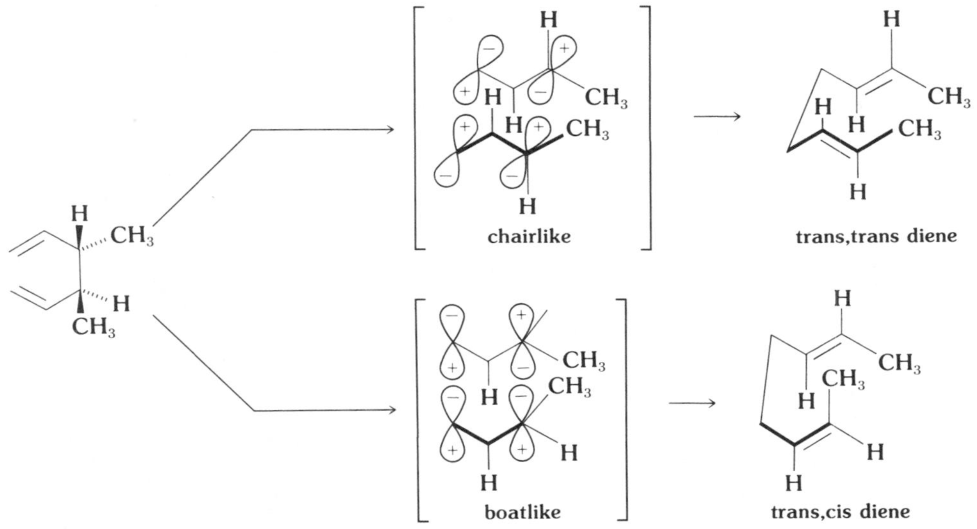

Similarly, the meso diene in Figure 16.9 will give the trans,cis diene via a chairlike transition state, but either the trans,trans or cis,cis diene via a boatlike transition state.

b A boatlike transition state is involved in this case. A trans double bond would be relatively unstable in the eight-membered ring, and this instability raises the energy of the chairlike transition state above that of the boatlike state.

c The carbon skeleton of this compound is flat and rigid; hence a concerted rearrangement to toluene can only occur in a suprafacial manner. The hydrogen shift required for this transformation would be either [1,3] or [1,7], neither of which is allowed. An allowed [1,5] sigmatropic hydrogen shift would simply regenerate the same triene. Acid-catalyzed rearrangement to toluene can proceed by the following mechanism:

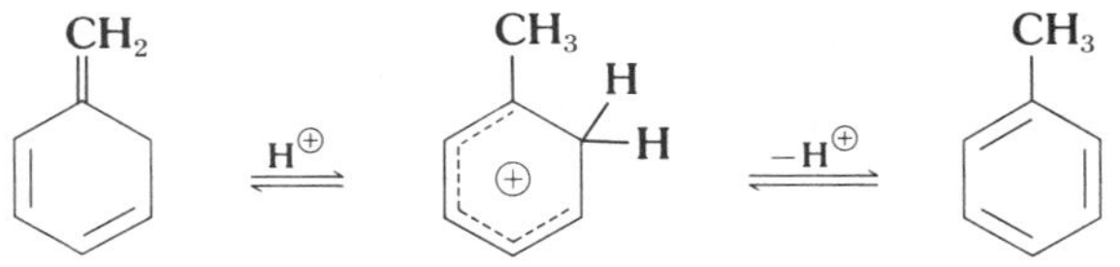

PROBLEM 16·16

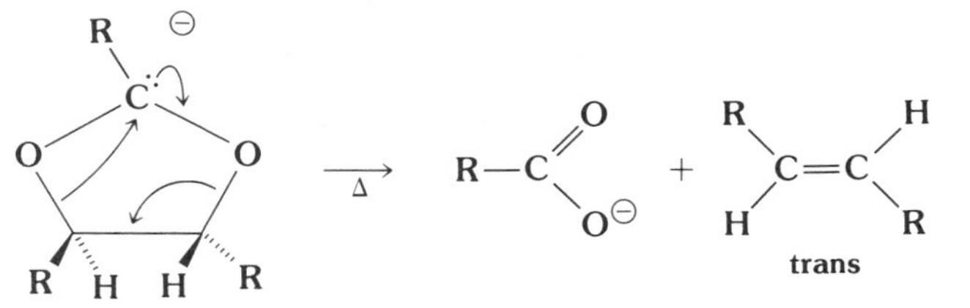

This reaction may be regarded as a cycloreversion of the [4 + 2] type (it has an aromatic transition state). The bonding changes will be suprafacial, consequently the trans alkene will be formed from the trans-disubstituted heterocyclic anion.

ANSWERS TO TEXT PROBLEMS 779

b One possible mechanism is

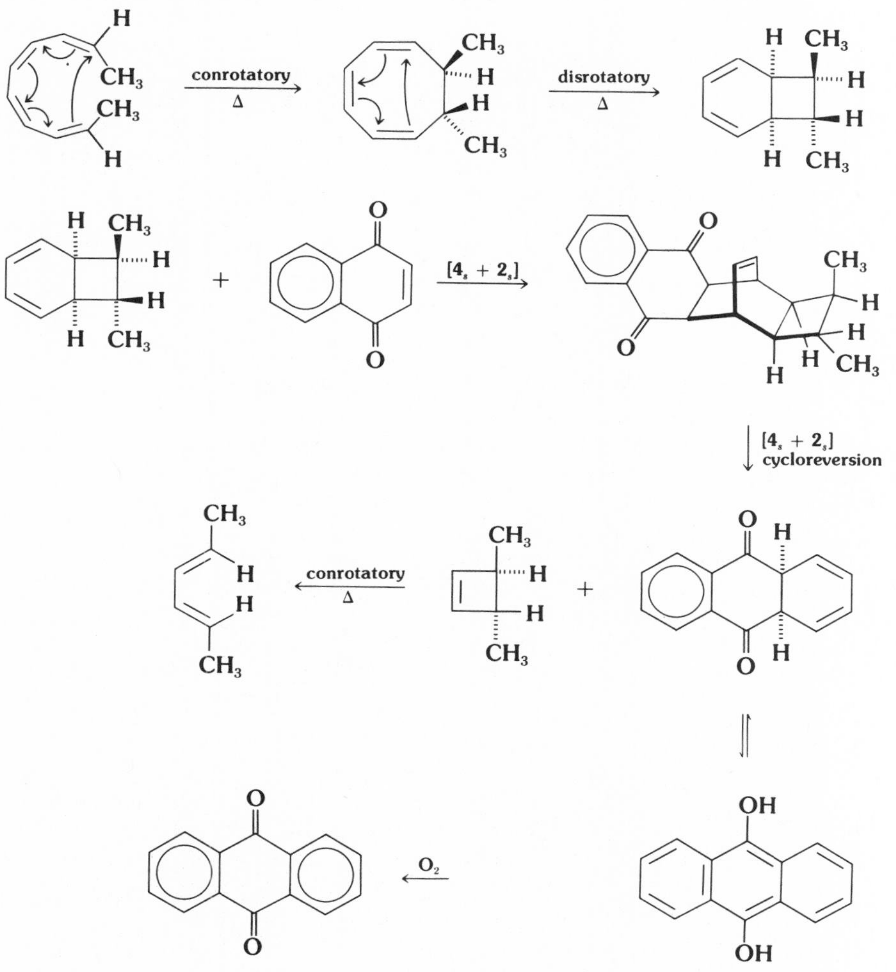

Another possibility is

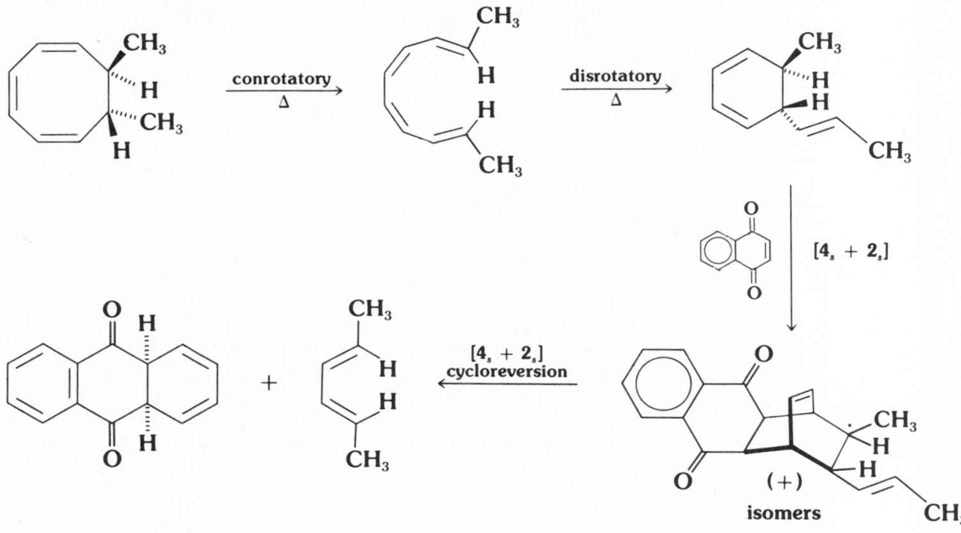

780   ANSWERS TO TEXT PROBLEMS

PROBLEM 16·17

a

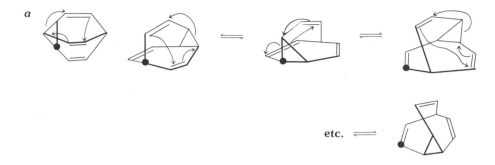

b Bullvalene has four constitutionally distinct sets of CH units: three making up the cyclopropane ring, three alkene units immediately adjacent to the cyclopropane moiety, three alkene units more remote from the cyclopropane, and one CH unit opposite the cyclopropane. As a result of successive Cope rearrangements, any CH units in bullvalene may occupy any of these four distinct sites (designated by the large dot in part a), and at 60° these rearrangements are taking place so rapidly that our pmr "camera" is seeing only a time average picture of all ten CH units. Here we encounter the remarkable case of a molecule for which no fixed bonds can be identified under conventional laboratory conditions of temperature (25°C) and time (minutes to hours). We call such compounds "fluxional." At −25°C the rate of the Cope rearrangements is slow enough to permit the pmr spectrometer to distinguish the olefinic from the aliphatic protons.

CHAPTER 17

PROBLEM 17·1

a The smallest amount of cystine that could be present in a molecule of insulin is one amino acid residue. Since cystine has two sulfur atoms, these would make a contribution of 64 atomic weight units to the molecular weight of insulin. The minimum molecular weight is therefore $(64 \times 100)/3.2 = 2000$.

b If the molecular weight of insulin is about 6000, the number of cystine units in each molecule must be proportionally larger than the minimum of 1 assumed in part a: $6000/2000 = 3$ cystine units/molecule.

PROBLEM 17·2

a $K_a = \dfrac{[H_2NCH_2CO_2^\ominus][H_3O^\oplus]}{[H_3N^\oplus CH_2CO_2^\ominus]}$

$K_b = \dfrac{[H_3N^\oplus CH_2CO_2H][OH^\ominus]}{[H_3N^\oplus CH_2CO_2^\ominus]}$

b Amino acids are, in general, fairly water soluble. However, this solubility varies with pH and is normally lowest at the isoelectric pH. This may be due to the fact that the proximate charges in the zwitterion are less efficiently solvated than the separated ions existing at high or low pH.

ANSWERS TO TEXT PROBLEMS   781

PROBLEM 17·3
Consider the following ion equilibria:

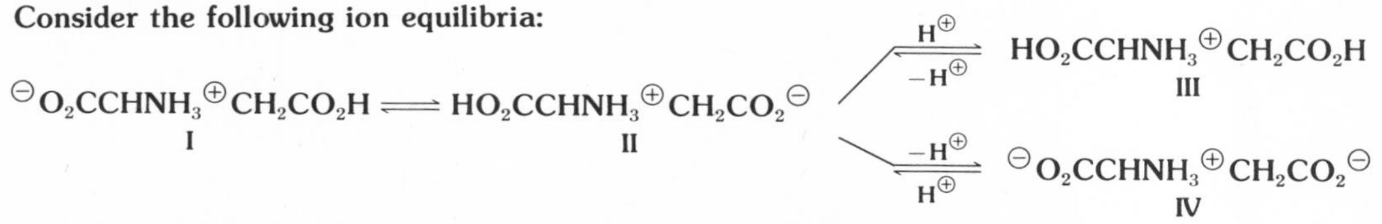

$$^{\ominus}O_2CCHNH_3^{\oplus}CH_2CO_2H \rightleftharpoons HO_2CCHNH_3^{\oplus}CH_2CO_2^{\ominus}$$
$$\text{I} \qquad\qquad\qquad \text{II}$$

$$\xrightarrow[-H^{\oplus}]{H^{\oplus}} HO_2CCHNH_3^{\oplus}CH_2CO_2H \quad \text{III}$$

$$\xrightarrow[H^{\oplus}]{-H^{\oplus}} {}^{\ominus}O_2CCHNH_3^{\oplus}CH_2CO_2^{\ominus} \quad \text{IV}$$

Ions I and II have a net electrical neutrality; III is a cation and IV is an anion. The second carboxyl function in aspartic acid is a source of an anionic species (IV) which is not possible in leucine or other monobasic amino acids. At neutral pH a significant concentration of IV will exist in aqueous solution (carboxylic groups are, after all, acidic). Consequently, it will be necessary to decrease the pH of such solutions if the formation of IV is to be suppressed to a point where anions and cations are present in equal concentrations (the isoelectric point).

PROBLEM 17·4

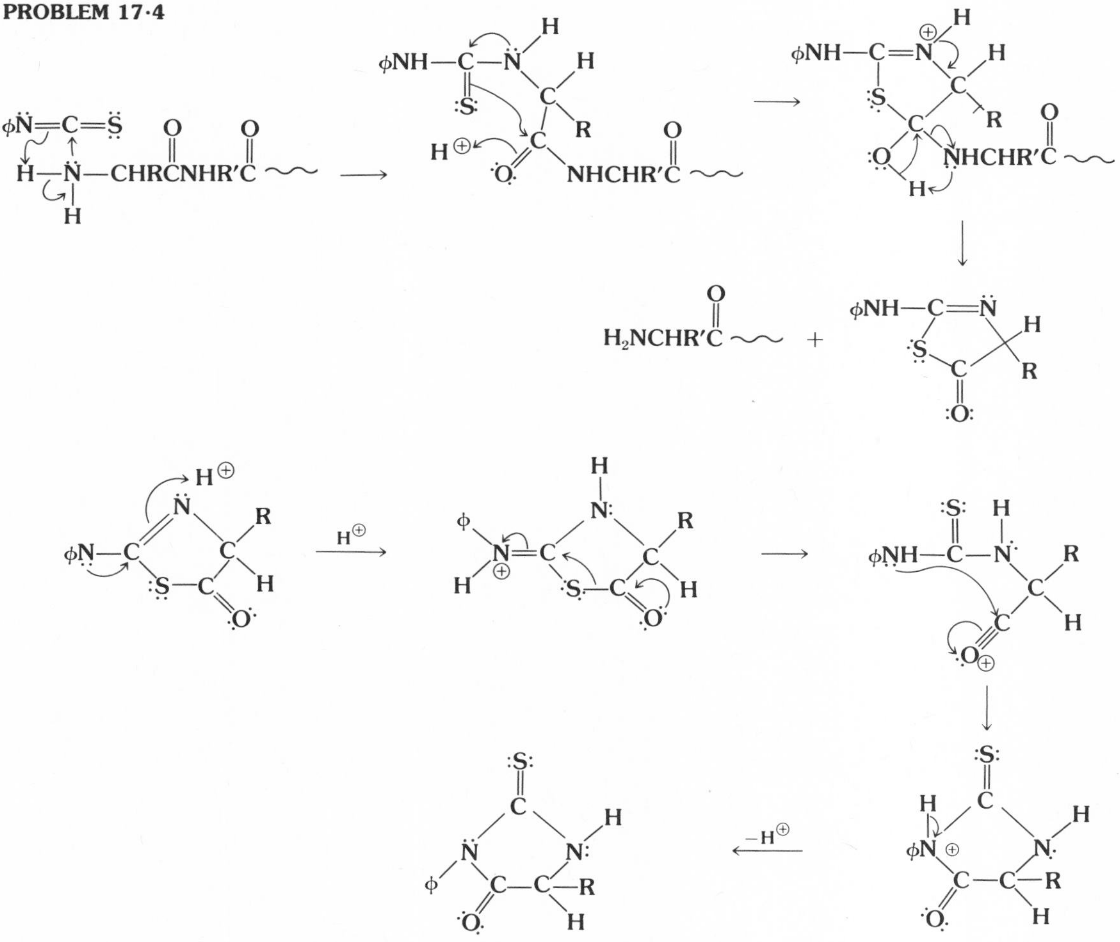

The first step of the Edman degradation involves substitution of the terminal amino group by the phenyl isothiocyanate. Once this has occurred, no further reaction takes place (amide nitrogen atoms are not reactive) until acid is added. Thus, the splitting off of the phenylthiohydantoin moiety and exposure of a new terminal amino group takes place after the isothiocyanate reagent has been removed.

782 ANSWERS TO TEXT PROBLEMS

PROBLEM 17·5

a

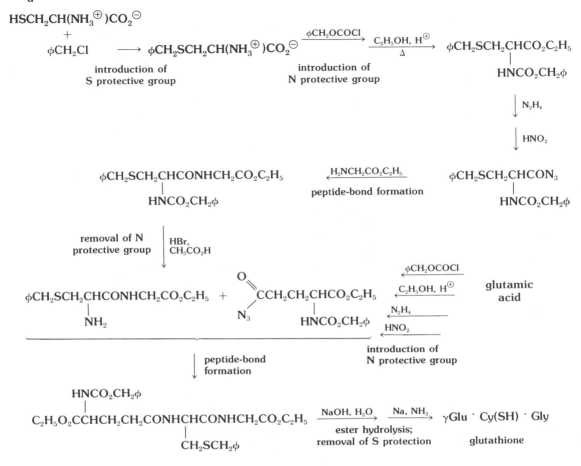

b In Section 14.5 we noted that the acylating ability of carboxylic acid derivatives (RCOZ) increased approximately in the same order as the acidity constants (K_a) of the Brønsted acids H—Z. Both of these trends seem to correlate with the stability (ease of formation) of $Z:^{\ominus}$. Amides are usually very poor acylating agents, just as ammonia and most amines are very weak Brønsted acids. Imidazole is an unusual amine in that it is more than 10^{20} times stronger as an acid than ammonia (it is also a fairly strong base). As noted in this problem, hydrazoic acid (HN_3) is also a relatively strong acid (even stronger than imidazole). Consequently, we should expect acyl imidazoles and acyl azides to be much better acylating agents than most amides.

ANSWERS TO TEXT PROBLEMS   783

PROBLEM 17·6
The terminal arginine units in bradykinin have guanidine substituents, and are therefore expected to be strongly basic (see Figure 11.5). The isoelectric point of arginine itself is about 10.8, so that for bradykinin should be similar. In fact, if we consider that the ratio of guanidine to free carboxyl groups is 2:1 in bradykinin and 1:1 in arginine, the former may well be the more basic. Bradykinin should be soluble in aqueous acid but relatively insoluble in organic solvents such as chloroform, ether, acetone, and hexane. Bradykinin may be optically active.

CHAPTER 18

PROBLEM 18·1

```
        I              II             III            IV
     CH₂OH          CH₂OH          CH₂OH          CH₂OH
      |              |              |              |
      C=O            C=O            C=O            C=O
HO——H           H——OH          HO——H           H——OH
 H——OH           H——OH          HO——H          HO——H
 H——HO           H——OH           H——OH          H——OH
     CH₂OH          CH₂OH          CH₂OH          CH₂OH
                         the D series


     CH₂OH          CH₂OH          CH₂OH          CH₂OH
      |              |              |              |
      C=O            C=O            C=O            C=O
 H——OH          HO——H            H——OH          HO——H
HO——H           HO——H            H——OH           H——OH
HO——H           HO——H           HO——H           HO——H
     CH₂OH          CH₂OH          CH₂OH          CH₂OH
                         the L series
```

Each compound in the D series is enantiomeric with the compound below it in the L series. All isomers that are not enantiomers are diastereomers. There are four pairs of epimers at C-3: D I and II, D III and IV, L I and II, and L III and IV.

PROBLEM 18·2
According to the experiment described in the first reaction sequence, the levorotatory lactic acid has a D configuration; it is related to D-(+)-glyceraldehyde. The second reaction sequence relates the enantiomeric dextrorotatory lactic acid to D-(+)-glyceraldehyde as well. Clearly, both enantiomers cannot have the same configuration, and this example discloses the ambiguity of the D,L notation.

784 ANSWERS TO TEXT PROBLEMS

PROBLEM 18·3

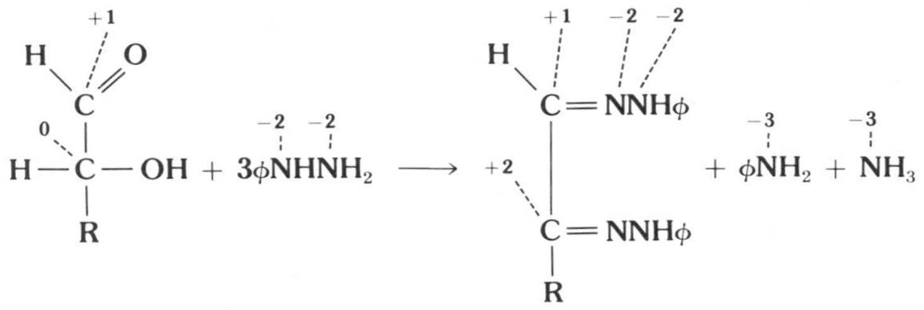

$$\underset{\overset{0}{H}}{\overset{+1}{H}}\!\!\diagdown\!\!\overset{\diagup\!\!\diagup\,O}{\underset{|}{C}}\,+\,3\phi NHNH_2 \longrightarrow \underset{\overset{+2}{\underset{|}{C}=NNH\phi}}{\overset{+1\;\;-2\;-2}{\overset{H}{\diagdown}\!\!\overset{|}{C}=NNH\phi}}\,+\,\overset{-3}{\phi NH_2}\,+\,\overset{-3}{NH_3}$$

(with $H-\underset{R}{\overset{|}{C}}-OH$ and R substituents)

Oxidation number changes are

carbon: $0 \longrightarrow +2$
nitrogen: $-2 \longrightarrow -3$; $-2 \longrightarrow -3$

PROBLEM 18·4

As noted in equation 18.4, D-(−)-arabinose undergoes the Ruff degradation to give D-(−)-erythrose, which in turn is oxidized by nitric acid to *meso*-tartaric acid (equation 18.5). If (−)-arabinose had configuration IV, it would have been converted in turn to (−)-threose and then to (−)-tartaric acid.

PROBLEM 18·5

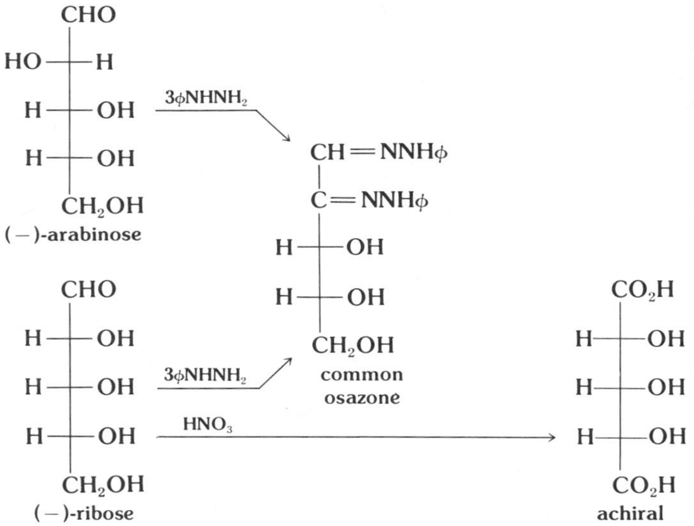

```
      CHO
  HO─┼─H
   H─┼─OH      3φNHNH₂
   H─┼─OH    ─────────↘
      CH₂OH              CH=NNHφ
  (−)-arabinose          │
                         C=NNHφ
      CHO             H─┼─OH
   H─┼─OH            H─┼─OH               CO₂H
   H─┼─OH  3φNHNH₂    CH₂OH            H─┼─OH
   H─┼─OH  ─────────↗  common          H─┼─OH
      CH₂OH   HNO₃     osazone         H─┼─OH
  (−)-ribose  ──────────────────────→     CO₂H
                                         achiral
```

PROBLEM 18·6

a   HOCH₂—CHOH   HOCH₂—CHOH
         ╱O╲   OH   ╱O╲   H
        │   │   │   │
        OH H   OH H
       H   H   H   OH
        H OH   H OH

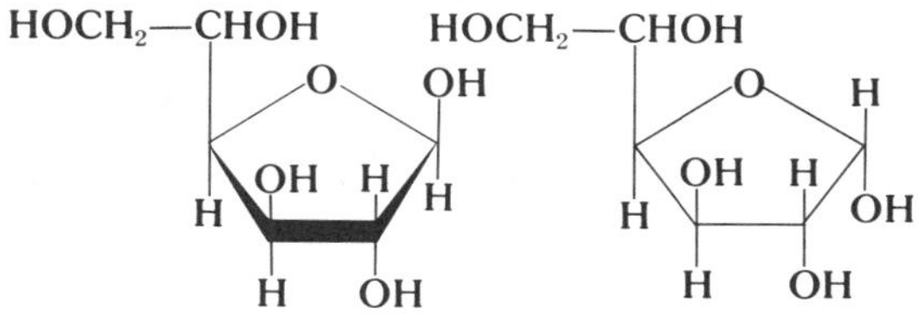

possible furanose anomers of (+)-glucose

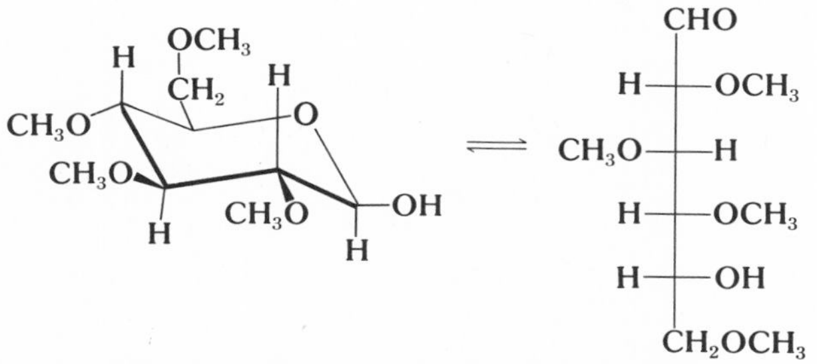

tetramethyl derivative of pyranose form of (+)-glucose

b

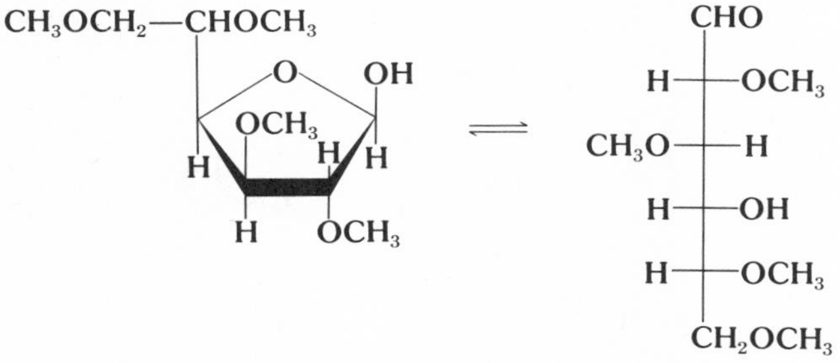

tetramethyl derivative of furanose form of (+)-glucose

c The open-chain (aldehyde) form of the tetramethyl glucose derivative has a free hydroxyl group at either C-4 or C-5, depending on the ring size of the cyclic anomer. Therefore determination of the ring size becomes a matter of finding out the location of the free hydroxyl group. In actual practice this was accomplished by oxidative cleavage of the hexose chain at the hydroxyl bearing carbon atom:

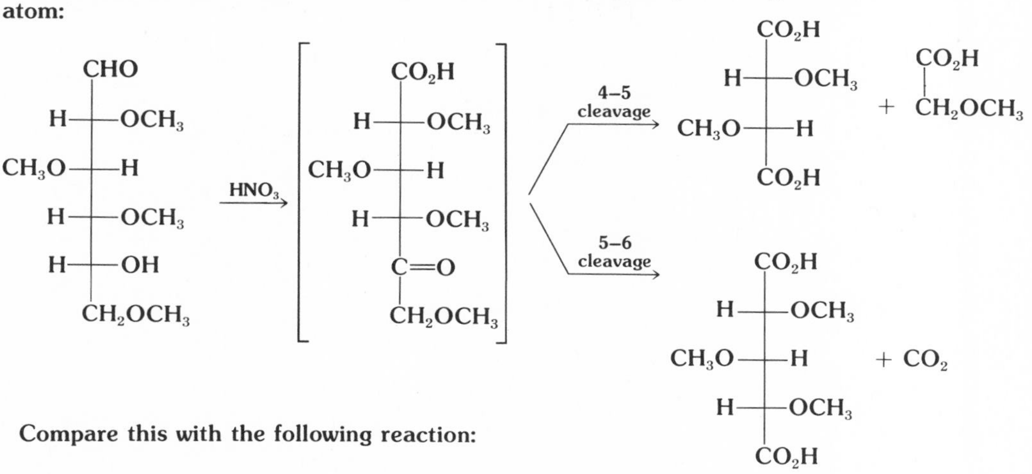

Compare this with the following reaction:

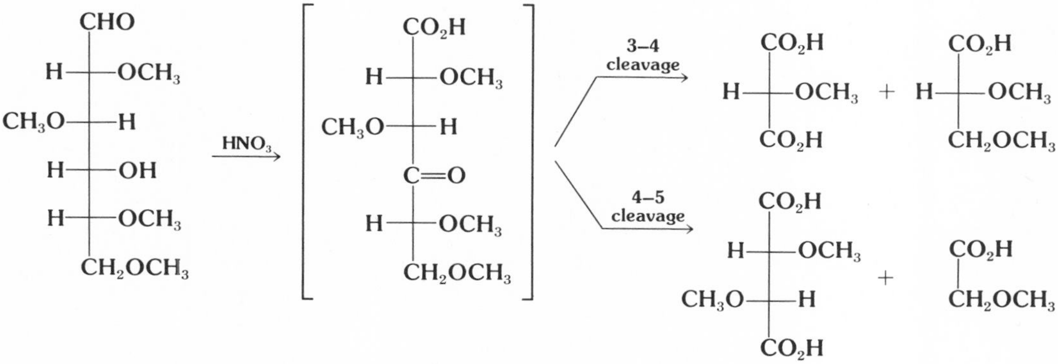

786 ANSWERS TO TEXT PROBLEMS

PROBLEM 18·7

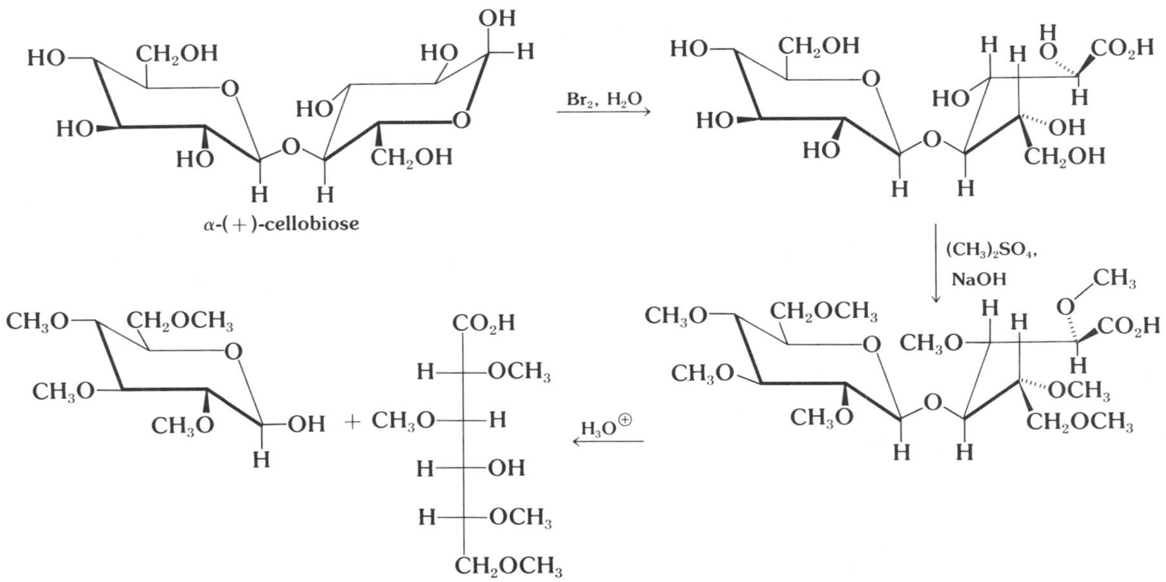

[Reaction scheme: α-(+)-cellobiose → (Br₂, H₂O) → carboxylic acid form → ((CH₃)₂SO₄, NaOH) → permethylated product → (H₃O⁺) → 2,3,4,6-tetra-O-methyl-D-glucose + open-chain 2,3,5,6-tetra-O-methyl-D-gluconic acid derivative]

PROBLEM 18·8

(+)-raffinose $\xrightarrow{(CH_3)_2SO_4,\ NaOH}$ permethylated raffinose $\xrightarrow{H_3O^\oplus}$ three methylated monosaccharide products

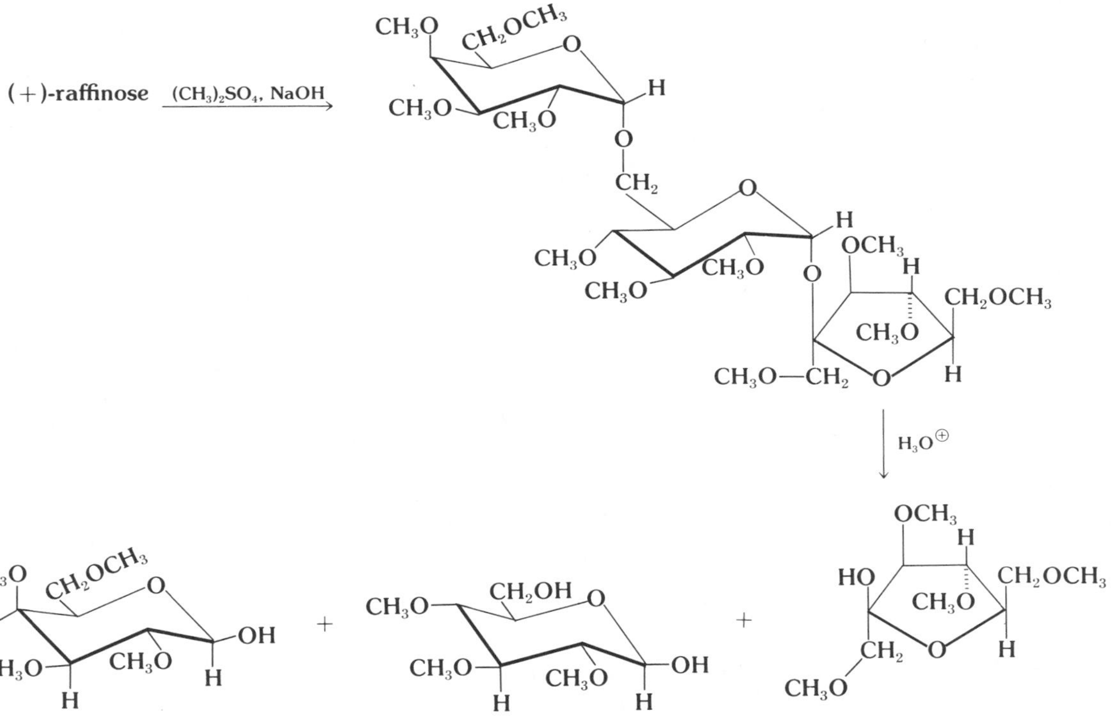

PROBLEM 18·9

Polysaccharides consisting of long unbranched chains of pyranose units will have many more interior units than end units. Each interior saccharide unit will have three free hydroxyl groups, whereas the terminal or end units have four hydroxyl functions. Trimethyl monosaccharide derivatives will therefore be much more abundant than tetramethyl derivatives in the hydrolysis products of the methylated polymer. For example, a sample of amylose, having an average chain length of 400 glucose units, will yield about 0.5% tetramethyl glucose compared with the trimethylated product.

A highly branched polymer such as amylopectin has a larger proportion of end units than the unbranched polymer. Consequently, the tetramethylated monosaccharide derivatives from such polymers are more abundant, accounting for 4 to 5% of the product mixture. As expected, the higher proportion of tetramethylated units from amylopectin is accompanied by a similar quantity of dimethylated units.

CHAPTER 19

PROBLEM 19·1
No. The reaction of 2-hexene with hydrogen chloride will not give a single product (will not be regiospecific). A mixture of 2- and 3-chlorohexanes will result.

PROBLEM 19·2
Conventional procedures for generating enolate anions from simple esters such as ethyl acetate usually result in Claisen condensations to β-ketoesters. See problem 14.21.

PROBLEM 19·3

a $\quad \mathrm{C}=\mathrm{C} \longrightarrow -\overset{H}{\underset{|}{C}}-\overset{H}{\underset{|}{C}}-$

Hydrogenation, with Pd, Pt, or Ni as a catalyst.

b $\quad \mathrm{C}=\mathrm{O} \longrightarrow \mathrm{CH}_2$

1 Wolff-Kishner reduction (N_2H_4, KOH, Δ); restricted to compounds which are stable to base.
2 Clemmensen reduction (Zn/Hg, HCl, CH_3CO_2H), restricted to compounds which are stable to acid.
3 Thioacetal formation ($HSCH_2CH_2SH$, BF_3) followed by desulfurization via a nickel catalyst (Raney nickel). This procedure may be limited by steric-hindrance effects.
c R—X ⟶ R—H
Grignard or lithium reagent formation (RMgX or RLi), followed by active hydrogen substitution (RMgX + H_2O ⟶ R—H + MgXOH). Other reactive functional groups (such as, C=O) are not tolerated.
d R—OH ⟶ R—H
1 Primary or secondary alcohols may be oxidized to aldehydes or ketones, followed by one of the procedures in part b.
2 A sulfonate ester derivative (R—OSO_2CH_3 or R—$OSO_2C_7H_7$) may be prepared, followed by $LiAlH_4$ displacement. Since the last step is essentially an S_N2 reaction involving a hydridelike nucleophile, it should be used only with primary or secondary alcohol derivatives.
3 A corresponding alkyl halide may be prepared, followed by the procedure of part c. Rearrangement may occur in the first step.

4 Dehydration of the alcohol to an alkene, followed by hydrogenation as in part a.

e R—NH$_2$ ⟶ R—H

1 Aromatic primary amines may be treated by the diazotization procedure shown in reaction 19.34.

2 Aliphatic amines may be exhaustively methylated (excess CH$_3$I), followed by Hofmann elimination and hydrogenation.

f R—CO$_2$H ⟶ R—CH$_3$

Reduction to a primary alcohol (LiAlH$_4$), tosylate formation (C$_7$H$_7$SO$_2$Cl + pyridine), and then hydride displacement (LiAlH$_4$).

PROBLEM 19·4

a *Reactions of strong carbon nucleophiles with electrophilic functions:* displacement reactions, alkylation of enolate anion

b *Reactions of strong carbon nucleophiles with electrophilic functions:* addition to carbonyl and nitrile functions, 1,2 addition followed by elimination or displacement, aldo condensation plus dehydration

c *Reactions of strong carbon electrophiles with nucleophilic functions:* acylation and alkylation of alkenes, alkene alkylation (intramolecular)

PROBLEM 19·5

a R—CHO + C$_3$H$_7$CH=Pϕ$_3$ ⟶ RCH=CHC$_3$H$_7$

or

$$R—CH=Pϕ_3 + \overset{O}{\underset{H}{\diagdown}}CCH_2CH_2CO_2CH_3 \longrightarrow RCH=CH(CH_2)_2CO_2CH_3$$

b CH$_3$CH=CHCO$_2$CH$_3$ + CH$_3$CH$_2$NO$_2$ $\xrightarrow{R_3N}$ CH$_3$CHCH$_2$CO$_2$CH$_3$
                                                                                                                        |
                                                                                              CH$_3$—CH—NO$_2$

c 2CH$_3$CH$_2$CO$_2$C$_2$H$_5$ $\xrightarrow[\Delta]{NaOC_2H_5}$ CH$_3$CH$_2$COCHCO$_2$C$_2$H$_5$
                                                                                                  |
                                                                                                 CH$_3$

d [cyclohexanone with —CH$_2$—Br substituent] $\xrightarrow{KOC(CH_3)_3}$ [bicyclic ketone]

e [cyclohexane with OH and —CH$_2$NH$_2$] $\xrightarrow[H_3O^{\oplus}]{NaNO_2}$ [cycloheptanone]   via   [cyclohexyl cation with migrating CH$_2$ and H—Ö:]

PROBLEM 19·6

This is a reasonable approach to the target molecule. Base-induced elimination of the chloride is preferable to acid-catalyzed dehydration because the latter conditions may cause isomerization of the double bond. The sequence of reactions shown in this problem has two more steps than the Wittig reaction path outlined in equations 19.38–19.40; the Wittig approach is therefore favored.

ANSWERS TO TEXT PROBLEMS 789

PROBLEM 19·7

synthon type	electrophilic	nucleophilic
CH_3	CH_3I, $(CH_3O)_2SO_2$	CH_3Li, CH_3MgX
ZCH_2	$H_2C=O$	ϕSO_2CH_2MgX
Z_2CH	$HCO_2C_2H_5$	$NaCH(NO_2)_2$
Z_3C	CO_2	$CN^\ominus$, $(\phi S)_3CLi$

PROBLEM 19·8

All these approaches use the Diels-Alder diene synthesis for preparing the cis-decalin ring system. Although this is generally a good idea, these specific cases will probably not proceed as shown.

a Since the Diels-Alder reaction is sensitive to steric hindrance, the quinone would react preferentially at the less substituted side:

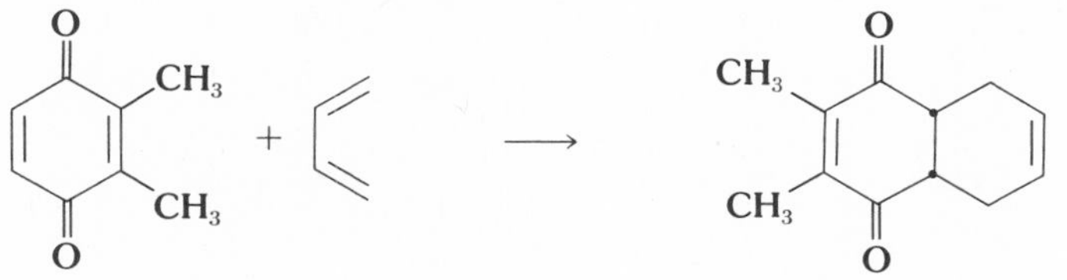

b The unsaturated diketone is unstable and will rapidly tautomerize to 2,3-dimethylhydroquinone, an aromatic compound.

c This proposal avoids the objections raised in parts *a* and *b*; however, in practice, this Diels-Alder reaction does not proceed because of steric hindrance.

PROBLEM 19·9

The aryl ketone formed in the initial Friedel-Crafts reaction acts to deactivate the benzene ring to further electrophilic attack. It is therefore necessary to remove this function before proceeding with the synthesis.

PROBLEM 19·10

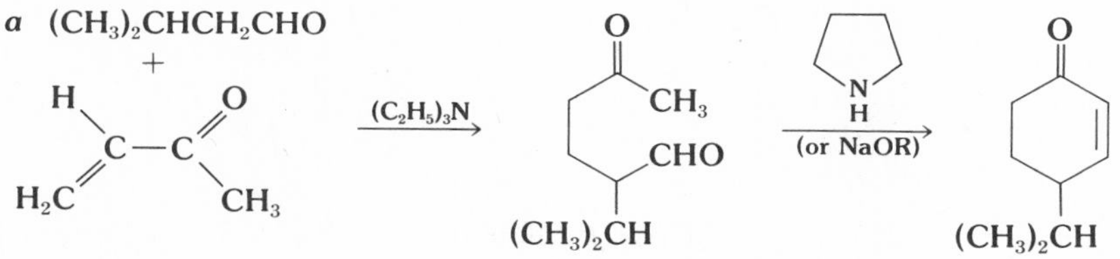

a $(CH_3)_2CHCH_2CHO$ +

Alternatively, a suitably substituted aromatic ring can be reduced:

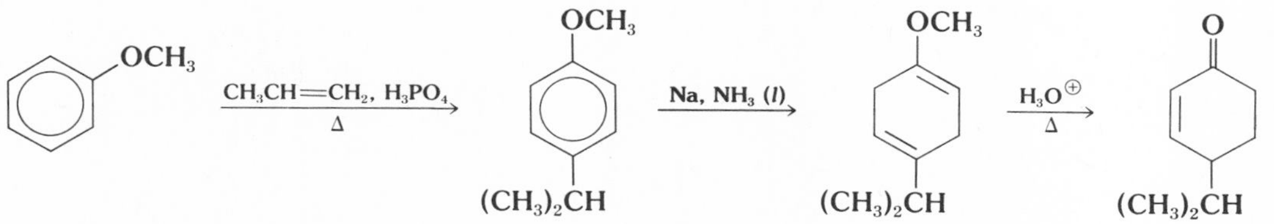

b In reaction 19.68 we have ignored an alternative orientation of the diene and dienophile:

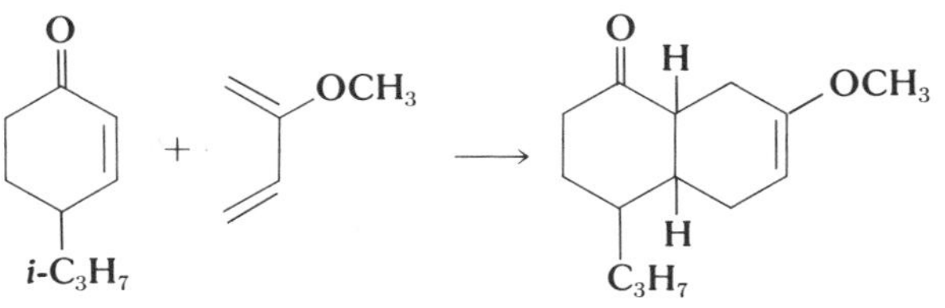

In practice, a mixture of products is obtained with the desired isomer (equation 19.68) predominating. The stereochemistry shown is based on steric hindrance and orbital-overlap effects.

In the first step of equation 19.69 several other elimination or fragmentation reactions might have occurred:

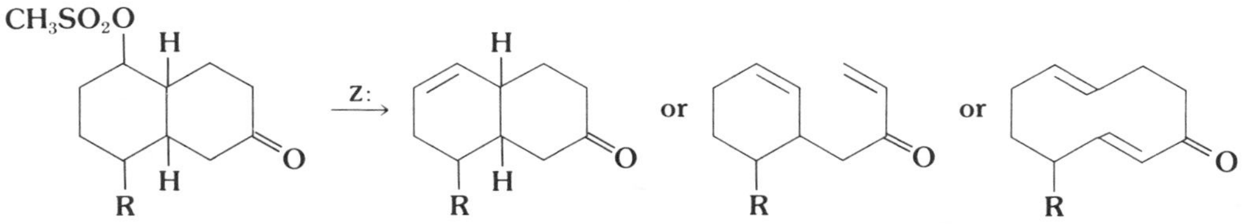

The dehydration of the tertiary alcohol formed in the reaction with methyl lithium might also give an isomeric alkene:

INDEX

abietic acid, 409
acetaldehyde, from acetylene, 164
  from ethanol, 362
  oxidation, 384
acetals, 371–373
acetate condensation reactions, 470–472
acetic acid, from acetaldehyde, 384
  acidity, 438
  infrared spectrum, 418
  ionization, 438–439
acetic anhydride, infrared spectrum, 418
acetoacetic acid, 409
acetone, dielectric constant, 85
  enol content, 386
  physical properties, 365–366
  preparation, from cumene hydroperoxide, 526–527
    from isopropyl alcohol, 362
    from molasses, 362
  reductive amination, 378
  ultraviolet absorption, 366
acetophenone, halogenation, 385
  reaction with Grignard reagent, 380
  self-condensation, 395–396
acetyl group, 363
acetylacetone, 468
acetylcholine, 458–459
acetylene (ethyne), 157, 166
  addition reactions, 162–165, 167
  from calcium carbide, 156
  dimerization, trimerization, and polymerization, 169–170
  ionization potential, 166
  salt formation, 161
  structure and bonding, 157–158
achirality, 184, 199
acid anhydrides, 413–414, 443–444, 451–452
  from acyl halide and carboxylic acid salt, 444
  carbonyl stretching frequency, 417
  from carboxylic acids, 443–444
  infrared spectra, 416–418
  from ketenes, 450
  mixed sulfonic and carboxylic, 460–461
  nomenclature, 413–414
  occurrence, 421, 431
  pmr spectra, 419–420

acid anhydrides (*cont.*)
  reactions, acylation, 451–452
    with alcohols, 221–222
    with amines, 337
    with lithium aluminum hydride, 463–464
    with organometallic compounds, 464–465
    with phenols, 305
acid halides (*see* acyl halides)
acidity, alcohols, 219–220
  aldehydes, 387–389
  amines, 323
  α-amino acids, 581
  carbon acids, 388–389
  crboxylic acids, 437–440, 476
  ethynyl hydrogen atoms, 160–162
  ketones, 388–389
  phenols, 302–304
  pK_a 437–439
  pyrrole, 277
  pyrrolidine, 277
acids, Brønsted-Lowry definition, 26, 94
  Lewis definition, 28
aconitic acid, 178
acrylic acid (propenoic acid), 409
acrylonitrile, from acetylene, 163
  copolymer with vinyl chloride, 147
  activating groups in aromatic substitution, 271, 272, 292–294
  in synthesis, 632–634
activation energy, addition to double bonds, 166
  addition to triple bonds, 166
  azo compound decomposition, 539–540
  peroxide homolysis, 539
  and reaction rates, 55–57
acyl chlorides (*see* acyl halides)
acyl group, reactivity with nucleophiles, 446
acyl halides, acylation, 444, 447–448
  carbonyl stretching frequency, 416
  from carboxylic acids and phosphorus chlorides, 443
  from carboxylic acids and thionyl chloride, 443
  formylation, 449
  in Friedel-Crafts acylation, 448–449
  infrared spectra, 416–417
  ketones from, 448, 464

acyl halides (*cont.*)
  nomenclature, 413–414
  phosgene, 449
  pmr spectra, 419
  reactions, with alcohols, 221–222, 447
    with alkenes, 448–449
    with amines, 447
    with aromatic compounds, 448–449
    with carboxylate salts, 444
    with complex metal hydrides, 463–464
    with diazomethane, 448
    with hydrogen peroxide, 448
    with organometallic compounds, 464–465, 477
    with phenols, 305, 448
  reduction to aldehydes, 463–464
acylating reagents, 446–455
  in peptide synthesis, 586
acylation, 445–455, 476
  with acyl halides, 447–448
  of alcohols, 447–448, 452–453
  with amides, 454–456
  of amines, 337–338, 344
  with anhydrides, 451–452, 460–461
  biological reactions, 457–462
  with esters, 452–454
  Friedel-Crafts, 448–449, 452
  mechanism, 446–447
acylium cation, 449
acyloin condensation, 639
adamantane, 65
addition reactions, 25
  of aldehydes and ketones, 369–384
  of alkenes, 127–129, 132–149
  of alkynes, 162–172
  of conjugated dienes, 495–497
  of cyclopropane, 149–150
  mechanism and stereochemistry, 129–131, 135–138, 369–370
  (*see also* electrophilic addition reactions, nucleophilic addition reactions)
adenine, 458
adenosine diphosphate (ADP), 461–462
adenosine monophosphate (AMP), 461–462
adenosine triphosphate (ATP), 461–462
  high-energy bonds in, 51
adipic acid, 410
ADP (*see* adenosine diphosphate)
adrenal hormones, 326–327, 361–362
adrenaline, 326
alanine, 579–580
alchemists, 625
alcohol (*see* ethanol)
alcohols, 205–239
  acidity, 219–220
  acylation, 447–448, 451–453
  alkoxides from, 222–223
  basicity of, 219–220
  boiling points, 208
  C—O stretching frequency, 209
  dehydration, 231–235
  dehydrogenation, 233–234
  determination of hydroxyl groups, 220
  electrophilic substitution, 221–224
  elimination reactions, 231–235
  esterification, 221–222, 441–443
  ethers from, 222–223
  functional group, 30

alcohols (*cont.*)
  hydrogen bonding, 208–209
  infrared spectra, 209–210
  nomenclature, 206–207
  nucleophilic substitution reactions, 225–230
  O—D stretching frequency, 110
  O—H stretching frequency, 209
  oxidation, 233–235, 412
  oxonium salt formation, 220
  pmr spectra, 209–216
  preparation, fermentation, 206
    Grignard synthesis, 380–381
    reduction of aldehydes and ketones, 378–380
    reduction of carboxylic acids, 445
    reduction of carboxylic acid derivatives, 463–464
  reactions, 238
    with acid anhydrides, 221–222, 451
    with acyl halides, 221–222, 447
    addition to aldehydes and ketones, 370–373
    addition to alkynes, 167
    addition to isobutylene, 221
    with alkali metals, 219
    with carbonium ions, 221
    with carboxylic acids, 221–222, 441–443
    with chromium VI compounds, 233–234
    with esters, 221–222, 452–453
    with halogens, 221
    with hydrogen halides, 229–230
    with organometallic reagents, 220
    with phosgene, 449
    with phosphorus trihalides, 227–228
    with sulfonyl chlorides, 221–222, 225–227
    with thionyl halides, 227–228
  reactivity of, 217–219
  solubility in acids, 220
  solubility in water, 208
  sources, 205–206
alcoholysis, 455
aldaric acids, 605, 608–610
aldehydes, 361
  acetals from, 371
  acidity, 388
  aldol condensation, 394–397
  alkylation of enolate anions, 392–393
  boiling points, 365–366
  Clemmensen reduction, 383
  conjugate acids and bases, 384–391
  derivatives, 375–376
  enolization, 385–391
  from formylation of aromatic rings, 449
  functional group, 363–364
  halogenation, 385–386
  hemiacetals from, 370–371
  hydration, 364–365, 370–374
  infrared spectra, 367–368
  nomenclature, 363–364
  oxidation, 383–384, 411
  from ozonolysis of alkenes, 142–143
  pmr spectra, 368
  preparation, from acyl halides, 463–464
    from alcohols, 233–235
    from alkenes, 142–143
    from alkynes, 164, 168–169
    from aromatic rings, 449
  reactions, addition of alcohol, 370–373
    addition of ammonia derivatives, 374–377

INDEX 793

aldehydes (cont.)
  reactions (cont.)
    addition of hydrogen cyanide, 373–374
    addition of water, 370–371
    with amalgamated zinc in acid, 383–384
    with chromic acid, 384
    with diborane, 379–380
    with halogens, 385–386
    with hydrazine, 383
    with metal hydrides, 378–379
    with organometallic compounds, 380–382
    with permanganate, 384
    with sodium hydroxide, 394–397
    with Tollens reagent, 384
    with ylides, 381–382
    reduction, 379–380
    reductive amination, 378
  solubility, 365–366
  sources, 361–362
  structure, 363–364
  test for, 384
  ultraviolet-visible spectra, 365–366
  Wittig reaction, 381–382
  Wolff-Kishner reduction, 383
aldol, 394
aldol condensation, 394–397
  acid-catalyzed, 397
  crossed, 396
  multiple condensations, 396–397
  in synthesis, 395–396
aldonic acids, 605
aldoses, 601
aldosterone, 549
alicyclic hydrocarbons, 35–36
  (see also cycloalkanes, cycloalkenes)
aliphatic hydrocarbons, 35
  (see also alkanes, alkenes, alkynes, dienes, cycloalkanes, cycloalkenes)
alkaloids, 310, 325–326, 425–426
alkanes, bond angles, 36
  bond energies, 47–51
  bond lengths, 36
  C—H stretching frequencies, 46
  chemical shift, 264
  combustion, 46–51
  condensed structural formulas, 36
  conformations, 39–40
  halogenation, 51–54
  homologous series, 36
  homologs, 36
  infrared spectra, 46
  molecular structure, 36–38
  nomenclature, 35–36, 40–43
  normal, 35–37
  physical properties, 43–46
  pmr spectra, 46
  reactivity, 46–54
alkenes, 107–152
  (see also cycloalkenes, dienes)
  anti-Markovnikov addition, 137, 145
  Baeyer test for, 141
  chemical shift, 264
  cis-trans isomerism, 112–114
  coordination with transition metals, 138–139
  Friedel-Crafts acylation, 448–449, 452
  functional group, 29
  halohydrin formation, 132–133

alkenes (cont.)
  heats of hydrogenation, 121–123
  from Hofmann elimination, 340–341
  hydration, 127–128
  hydrohalogenation, 127–131
  hydroxylation, 140–142
  infrared spectra, 110–111
  ionization potential, 166
  Markovnikov addition, 128–129
  mechanisms and stereochemistry of addition, 129–131, 135–138, 198
  mechanism of ozonolysis, 144
  nomenclature, 108–110
  oxidation, 140
  oxidative cleavage, 412
  ozonolysis, 142–143
  peroxide formation, 144–145
  pmr spectra, 110–112
  preparation, from amines, 340–342
    dehalogenation of vicinal dihalides, 108
    dehydration of alcohols, 108, 231–233
    dehydrohalogenation of alkyl halides, 108
    hydrogenation of alkynes, 108, 162
    reduction of alkynes, 167–168
    from substituted succinic acids, 445
    Wittig reaction, 381–382
  reactions, addition of Brønsted acids, 127–131, 136–137
    addition of carbenes, 148–149
    addition of carboxylic acids, 441
    addition of diborane, 133–134, 136–138
    addition of halogens, 115, 132
    addition of hydrogen, 115, 121–124
    addition of hypohalous acids, 133
    addition of Lewis acids, 132–134
    addition of radicals, 144–145
    addition of sulfuric acid, 115, 127
    addition of water, 128
    with dihalocarbenes, 149
    dimerization and polymerization, 145–148
    with ozone, 142–143
    with permanganate, 140–141
    with peroxy acids, 474–475
    with Simmons-Smith reagent, 149
  sources, 107
  stereoisomerism, 112–114
  structure, 110
  test for unsaturation, 132, 141
  transition-metal complexes, 138–139
  ultraviolet-visible spectra, 251
alkoxides, 222–223
alkyl azides, reduction, 350
alkyl boranes, formation from alkenes, 133–134
alkyl cyanides (see nitriles)
alkyl groups, nomenclature, 40–41
alkyl halides, carbonium-ion stability, 88–89
  conformations, 78
  dehydrohalogenation, 91–96
  elimination, 91–96
    E1 mechanism, 96
    E2 mechanism, 92–95
  Friedel-Crafts alkylation, 269
  functional group, 29
  hydrolysis, 80–81, 84–91, 315–316
  infrared spectra, 75–76
  nomenclature, 74
  physical properties, 74–75

alkyl halides (cont.)
  pmr spectra, 76–77
  preparation, from alcohols, 227–228
    from alkanes, 51–54
    from alkenes, 127–131
    from carboxylic acids, 445
  reactions, with acetylides, 161
    with alkoxides, 222–223
    with amines, 335–336
    with ammonia, 352
    with aromatic hydrocarbons, 269
    with cyanide ion, 79
    with cyanide salts, 456
    with halide ion, 79–84
    with hydroxide ion, 79, 91–95
    with metals, 97–99
    with phenols, 304
    with phosphines, 353
    with silver nitrite, 352
    with sodium azide, 350
    with sodium cyanide, 350
    with thiols, 79
    with water (hydrolysis), 80–81, 86–87, 90
    rearrangement in S_N1, 90–91
  substitution, nucleophilic, 79–91
    kinetics, 80–81, 86–87
    relative reactivities, 79–81, 83–84, 88–90
    S_N1 mechanism, 84–88
    S_N2 mechanism, 81–84
alkyl halosulfites, 228
alkyl isocyanates, 449
alkyl lithium compounds, acylation by carboxylic acid derivatives, 464
alkyl side chains, oxidation, 272–273
alkyl sufates, from alkenes, 127
alkylation, 477
  of acetylenes, 161
  of amines, 335–336
  of aromatic hydrocarbons, 151–156
  of enolate anions, 392
  of enolate derivates, 472–473
  Friedel-Crafts, 269
  of β-ketoesters, 468–470
alkynes, 155–172, 172
  acetylide salts from, 161
  acidity, 160–162
  bonding, 158–159
  chemical shift, 264
  functional group, 29
  hydration, 164–165
  hydrogenation, 162–163
  infrared spectra, 159–160
  ionization potential, 166
  nomenclature, 156–157
  nucleophilicity, 166–167
  oxidation, 169
  oxidative cleavage, 412
  ozonolysis, 169
  physical properties, 155
  pmr spectra, 159–160
  preparation, acetylene from calcium carbide, 156
    from alkyl halides and acetylides, 161
    from dihalides, 156
  reactions, 172–173
    addition of alcohols, 167
    addition of Brønsted acids, 163
    addition of diborane, 168–169

alkynes (cont.)
  reactions (cont.)
    addition of halogens, 163
    addition of hydrogen, 162–163
    addition of water, 164
    with alkali metals in ammonia, 167
    with carboxylic acids, 440–441
    dimerization, trimerization, and polymerization, 169–171
    with electrophilic reagents, 163
    with Grignard reagents, 161
    with nucleophilic reagents, 167–168
    with ozone, 169
    with permanganate, 169
    with strong base, 161
    with thiol bases, 167
  reactivity compared to alkenes, 166–167
  reduction, stereospecific, 167–168
  structure, 158–159
  ultraviolet-visible spectra, 251
allenes, chirality in, 185
allyl cations, 315, 514
allyl derivatives, S_N1 reactions, 315–316
allyl group, 109
allyl halides, 109
  S_N2 reactions, 312–313
almonds, bitter, 246
ambrettolide, 424
amides, acidity-basicity, 454–453
  from acylation of amines, 337, 452
  carbonyl stretching frequency, 416–417
  dehydration, 456
  Hofmann rearrangement, 526
  hydrogen bonding, 416
  hydrolysis, 454–456
  from ketenes, 451
  infrared spectra, 416–417
  nomenclature, 413–414
  occurrence, 424–429
  pmr spectra, 419
  protective group for amines, 344
  reactions, with bromine, 526
    with lithium aluminum hydride, 351, 463–464
    with nitrous acid, 455
    with phosphorus pentoxide, 456
    with thionyl chloride, 456
    with water, 454–455
  reduction, 351, 463–464
amine oxides, 340
  as leaving groups, 342
amines, 323–353
  acidities, 333
  N-acylation, 337–338, 344, 447, 451, 452
  N-alkylation, 335–337
  amides from, 337–338, 447, 451, 452
  aromatic-ring substitution, 343–344
  basicity, 333–335, 354
  boiling points, 330–331
  chirality, 328–329
  classification, 323
  Cope elimination, 342
  deamination, 338–339
  as derivatives of ammonia, 323
  electrophilic substitution at nitrogen, 335–340, 354
  elimination reactions, 340–341, 354
  exhaustive methylation, 340–341
  functional group, 30

amines (*cont.*)
    halogenation, 343–344
    Hinsberg test, 338
    Hofmann elimination, 341–342
    hydrogen bonding, 330
    infrared spectra, 330–331
    nitration of aryl amines, 343–344
    N-nitrosation, 338–339
    nitrous acid test for, 339
    nomenclature, 323–325
    N-oxidation, 339–340
    N—H stretching frequency, 330
    occurrence, 325–328
    pmr spectra, 330, 332
    preparation, from aldehydes and ketones, 378
        from alkyl azides, 350–351
        from alkyl halides, 335, 336, 350
        from amides, 351, 463
        from aryl halides, 336
        by Curtius rearrangement, 525–526
        by Gabriel phthalmide synthesis, 451
        by Hofmann reaction, 525–526
        from imines, 378
        from nitriles, 350–351, 463–464
        from nitro compounds, 350–352
        by reduction, 350–351, 463–464
        by Schmidt reaction, 525–526
    protection of amino group, 344
    quaternary ammonium salts, 336–337, 340–342
    reactions, with acid anhydrides, 336–337, 344, 451
        with acyl halides, 336–337, 447
        with alkyl halides, 335–338
        aryl amines with nitric acid, 343–344
        with aryl diazonium salts, 347
        with aryl halides, 336
        with carboxylic halides and anhydrides, 336–337
        with esters, 452
        with halogens, 343–344
        with hydrogen peroxide, 339–340
        with nitrous acid, 336–339
        resonance stabilization, 334–335
    solubilities, 330–331
    stereochemistry, 353
    sulfonation of aryl, 336–338
        with sulfonic acid halides and anhydrides, 336–337
α-amino acids, 578–582
    acidity, 581
    basicity, 581
    configuration, 581
    essential, 579–580
    isoelectric point, 582
    properties, 581–582
    from proteins, 578
    structure, 579–580
    zwitterion form, 581
2-amino-1-aryl propanes, 327
amino group, 324
aminobenzene (*see* aniline)
o-aminobenzoic acid, 409
α-aminocarboxylic acids (*see* α-amino acids)
1-amino-1-phenylethane, infrared spectrum, 331
    pmr spectrum, 332
p-aminotoluene, 324

ammonia, acylation, 445
    basicity, 333
    reaction, with aldehydes and ketones, 374
        with alkyl halides, 335, 351
ammonium salts, alkyl and aryl derivatives, 325
ammonolysis, 445
AMP (*see* adenosine monophosphate)
amphetamines, 334
amphoterism, 94
amylopectin, 620–621
amylose, 620–621
anchimeric assistance in substitution reactions, 508
angle strain, 61–62
anhydrides (*see* acid anhydrides)
aniline, 323
    basicity, 334–335
    bromination, 343
    nitration, 343
    physical properties, 331
    protection of amino group, 344
    resonance stabilization, 334–345
aniline purple (mauve), 626
anilinium sulfate, 325
anisole, 307
annulenes, 274–276, 278–280, 283
anomers, 611
antarafacial reactions, 559–560
anthracene, 260
anthranilic acid, 409
antiaromatic transition states, 568–569
antiaromaticity, 276
antibiotics, 300, 456, 471–472
antibonding orbitals, 116–117
anti conformation, 76, 93, 376
antihistamines, 327
anti-Markovnikov addition, to alkenes, 137, 145
    to alkynes, 168–167
    in radical reactions, 145
α-naphthylthiomrea (ANTU), 450
apocamphoyl peroxide, 550
D-(−)-arabinose, 606
arachidonic acid, 422
arenes (*see* aromatic hydrocarbons)
arginine, 580
aromatic compounds, 245, 283
    nomenclature, 259–260
    polysubstituted, 346–347
    (*see also* aromatic hydrocarbons, heterocyclic compounds, other specific families)
aromatic hydrocarbons, 35
    C—H stretching frequency, 262
    electronic absorption maxima, 261
    formylation, 449
    Friedel-Crafts acylation, 448, 452
    Friedel-Crafts alkylation, 269
    halogenation, 268
    Hückel rule, 276–282
    infrared spectra, 262
    nitration, 268
    nomenclature, 259–260
    oxidation, 272–273, 411–412
    pmr spectra, 262–264
    properties, 261–264
    reactions, with acid anhydrides, 452
        with acyl halides, 448
        with alkyl halides, 269
        with carbon monoxide and hydrogen chloride, 449

aromatic hydrocarbons (*cont.*)
  reactions (*cont.*)
    with halogens, 268
    with nitric and sulfuric acids, 456
    with permanganate, 272, 273, 412
    with sulfuric acid, 269
   sulfonation, 269
   ultraviolet-visible spectra, 261
   (*see also* benzene, naphthalene, other specific compounds)
aromatic substitution, mechanism for, 270–272
aromatic transition states, 568–569
aromaticity, of annulenes, 274–276
  criteria of, 273–274
  Hückel rule, 276–282
arrows, curved, 26
  resonance, 256
aryl amines (*see* amines)
aryl cyanides (*see* nitriles)
aryl diazonium salts (*see* diazonium salts)
aryl ethers, 305–307
2-arylethylamine derivatives, 327
aryl group, 289
aryl halides, 289–318
  compared with alkyl halides, 289–292
  elimination reactions, 295–296, 317
  electrophilic substitution of the aromatic ring, 297–299, 318
  formation of organometallic compounds, 296–297
  halogenation, 297
  mechanism for electrophilic substitution, 297–299
  mechanism for nucleophilic substitution, 293–294
  nomenclature, 260
  nucleophilic substitution of halide, 291–294, 317
  reactions, with halogens, 297
   with hydroxide ion, 295–296
   with metals, 296–297
   with nitric and sulfuric acids, 297
   with nucleophiles, 291–294
  sulfonation, 297
aryl hydrazines, 350
aryl lithium compounds, 296
aryne intermediates, 295–296
ascorbic acid
  dietary deficiency, 621
  structure, 622
asparagine, 580
aspartic acid, 580
aspirin, 451
  infrared spectrum, 13
asymmetric carbon, 199
atactic configuration, 193
atomic orbitals, 115–121
  hybridization, 117–118
  p, 115–116
  s, 115–116
  sp, 159
  sp^2, 119
  sp^3, 117–121
ATP (*see* adenosine triphosphate)
atropine, 425, 459
aureomycin, 73, 300
autonomic nervous system, 458–459
axial bonds, 64
axons (*see* synaptic transmission)
aza, 324
azetidine, 324

azine (*see* pyridine)
azide ion, 24
aziridine, 324
azo compounds, decomposition, 540
azo dyes, 348
azulene, aromaticity, 281

Baeyer test, 141
Baeyer-Villiger rearrangement, 526
barbituric acid, 452
barrelene, 341
bases, Brønsted-Lowry definition, 26, 94
  Lewis definition, 28
basicity, from acidity constants, 333–334
  alcohols, 219–220
  amines, 333–335, 354
  α-amino acids, 581
  ethers, 219–220
  nitrogen compounds,
  phosphines, 353
bathochromic shift, 261
batrachotoxin, 426
Beckmann rearrangement, 523–524
benadryl, 328
Benedict's reagent, 384
benzaldehyde, from bitter almonds, 246
  infrared spectrum, 367
  reduction, 379
  reductive alkylation, 378
  ultraviolet absorption, 366
benzaldoxime, 376
benzedrine, 327
benzene, acylation, 452
  alkylation, 269
  aromatic character, 273–274
  bond angles, 254
  bond lengths, 254
  configuration, 254
  conjugation in, 253–254
  electrophilic substitution reactions, 268–273
  Friedel-Crafts acylation, 452
  Friedel-Crafts alkylation, 269
  halogenation, 268
  heat of hydrogenation, 253
  hydrogenation, 247
  infrared spectrum, 262
  Kekulé structure, 247–248, 253–256
  mechanism for electrophilic substitution, 270–272
  molecular-orbital model, 254–256
  nitration, 268
  photochemical chlorination, 270
  pmr spectrum, 262–263
  resonance model, 256–258
  sulfonation, 269
  from trimerization of acetylene, 170
  ultraviolet spectrum, 254
  x-ray crystallography, 253–254
benzenediazonium chloride, 338
benzenesulfonyl chloride, 337
o-benzoquinone, 307
p-benzoquinone, 307–309
benzoyl group, 363
benzoyl peroxide, 447, 474
benzpyrene, 261
benzyl alcohol, from benzaldehyde, 380

benzyl cation, 313–314
benzyl derivatives, reactivity, 310–315, 318
benzyl group, 260
benzyl halides, reactions, 310–317
benzyl tosylates, 314–315
benzylic acid rearrangement, 532
benzyne, 296–297, 318
beriberi, 326
betaine, 382, 410
bicyclobutane, isomerization to butadiene, 554
bimolecular reaction, 81
  (*see also* elimination reaction, *E*2 mechanism, and substitution reactions, S_N2 mechanism)
biological acylations, 457
biosynthesis, 470–472
biotin, 410
biphenylene, 297
biphenyls, conformational enantiomerism, 265–267
  structure, 260
  ultraviolet spectra, 265–266
bisulfite addition to aldehydes, 610
Boltzman energy-distribution curve, 56
bombardier beetle, 309
bombykol, 243
bond angles, in acetylene, 158
  in carbonyl group, 364
  in carboxyl group, 407
  in cyclopropane, 61, 150
  in ethylene, 110
  linear, 23
  in methane, 36
  in nitrogen compounds, 328
  in peptides, 578
  tetrahedral, 22
  trigonal, 23
bond-dissociation energies, 50, 316
bond energies, 47–50, 364
bond lengths, in acetylene, 158
  in alkanes, 36
  in carbonyl group, 364
  in carboxyl group, 407
  in ethylene, 110
  in peptides, 578
bond polarity, nitrogen-oxygen bond in amine oxides, 339
bonds, axial, 64
  dashed, 22, 193
  equatorial, 64
  high-energy, 51
  pi (π), 118–120
  sigma (σ), 116–118
  tau (τ), 120–121
  wedge, 22, 193
  (*see also* covalent bonds, ionic bonds)
botulin, 459
bradykinim, 584
Bragg, W. H., 8
brain hormone, 430
bridged-cation intermediates, aziridinium ion, 513
  halonium ion, 508–511
  phenonium ion, 520–523
  sulfonium ion, 511
bromination (*see* halogenation)
bromine, electrophilic substitution, 268, 305, 385, 465
  trans addition to alkenes, 198
  (*see also* halogens)

bromobenzene, reaction with potassium amide, 295
2-bromobutyric acid, pmr spectrum, 419
1-bromopropane, β-elimination in, 91
3-bromopropene, 109
Brønsted acids, addition to alkenes, 127–131
  addition to alkynes, 163
Brønsted-Lowry acid and base definition, 26, 94
brucine, 426
bullvalene, 570
butanal, pmr spectrum, 369
butane, 331
  heat of combustion, 50
1-butanol, from molasses, 362
1-butene, 109
  addition of bromine, 132
  heat of hydrogenation, 122–123
2-butene, 109
  addition reactions, 128
  heat of hydrogenation, 122–123
  hydration, 186–187
cis- and *trans*-2-butenes, properties, 113
  stereochemistry of bromine addition, 198
2-butenoic acid, 409
butyl butyrate, 423
t-butyl chloride, hydrolysis, 96, 313
s-butylphenyl ketone, 385–386
1-butyne, 157
2-butyne, 157
butyraldehyde, 365–366
  pmr spectrum, 369

β-cadinene, 481
caffeine, 425
Cahn-Ingold-Prelog nomenclature rules, 193–196
camphor, 362
  enantiomers of, 188–189
cantharidin, 421, 431
capsaicin, 424
carbamate ester, 450
carbanions, 28
carbenes, addition to alkenes, 28, 148–149
carbohydrates, 601
  aldaric acids, 605–606
  aldonic acids, 605–606
  aldoses, 601–603, 612
  anomers, 611–612
  classification, 601
  configuration, 602–603
  cyanohydrin formation, 603, 606–607
  cyclic acetal formation, 610–613
  degradation by alkali, 605
  disaccharides, 614–618
  enediol tautomerism, 605
  energy from, 620
  esterification, 603, 619
  families, *D* and *L*, 602–603
  furanoses, 612
  glycosides, 612–613
  ketoses, 601–603
  Kiliani-Fischer synthesis, 606–607
  methylation, 611, 615, 621
  monosaccharides, 602–613
  mutarotation, 611
  osazone formation, 604–605
  oxidation, 605–606, 615–617

carbohydrates (*cont.*)
  polysaccharides, 618–621
  pyranoses, 611–613
  Ruff degradation, 606
  (*see also* specific compounds)
carbon, analysis for, 4
  oxidation states, 411–412
carbon acids, acidity, 388–390
carbon-carbon double bonds, 23
  analysis for, 142–144
  Baeyer test for, 141
  bond angles, 110
  conjugated, 110, 248
  cumulated, 110, 248
  infrared absorptions, 110–111
  isolated, 110
  LCAO models, 118–121
  length, 110
  pmr absorptions, 110–112
  rotational-energy barrier, 114
  thermodynamic stability, 122–124
  stereoisomer nomenclature, 112–113, 195–196
  stretching frequencies, 110, 160
  transition-metal complexes, 138–139
carbon-carbon single bonds, 20–21
  LCAO model, 117–118
  length, 110
  rotational barrier, 39–40
  stretching frequency, 160
carbon-carbon triple bonds, 21
  binding affinity for π electrons, 166–167
  bond length, 158
  infrared spectra, 159–160
  σ-π molecular-orbital model, 158–159
  pmr absorbtions, 159–160
  stretching frequency, 160
carbon dioxide, dipole moment, 23–24
  reaction with organometallic reagents, 381
carbon-hydrogen analysis, 4–5
carbon-nitrogen double bond, stereoisomerism, 376
carbon-oxygen double bond (*see* carbonyl group)
carbon tetrachloride, 52
  dipole moment, 23–24
carbonaceous meteors, 581
carbonium-ion intermediates, in addition reactions
  to alkenes, 127–131, 136–137
  in aromatic substitution, 271, 298
  in dehydration of alcohols, 231–232
  in $E1$ reactions, 96
  in solvolysis of sulfonate esters, 227
  in S_N1 reactions, 88–91
carbonium ions, 28
  delocalization of charge, 89, 310, 315, 514
  rearrangement, 517–523
  stability, 88–89, 291–292
carbonyl compounds, functional group, 30
  reactions, 364–365, 369–370
  spectroscopic properties, 368–369
  (*see also* aldehydes, ketones, carboxylic acids)
carbonyl group, 361
  infrared absorption, 367–368
  oxidation state of carbon, 413
  stretching frequencies, of aldehydes and ketones, 367–368
    of carboxylic acids and derivatives, 416
    in ketenes, 450
    of β-lactams, 456
  structure and bonding, 363–364

carboxyl group, 407
  oxidation state of carbonium, 413
  protection, 445
carboxylate ions, 438–439
carboxylation of organometallic reagents, 412
carboxylic acid derivatives, functional groups, 437
  nomenclature, 413
  relative reactivity, 445
carboxylic acids, 407, 432
  acid anhydrides from, 443
  acidity, 437–440, 476
  acyl halides from, 443
  acylation, 452
  alcohols from, 445
  alkyl halides from, 444
  biosynthesis, 470–472
  boiling points, 415–416
  carbonyl stretching frequencies, 416
  decarboxylation, 444
  essential fatty acids, 421
  esterification, 441–443
  in fats, 421
  functional derivatives, 407, 432
    (*see also* individual families)
  functional group, 30, 407
  halogenation, 465–466
  hydrogen bonding, 416
  infrared spectra, 416–418
  ionization, 438–439
  ketenes from, 450
  nomenclature, 407–411
  oxidative bis-decarboxylation, 444
  oxidative decarboxylation, 444
  peroxy acids from, 474
  in phospholipids, 422
  pmr spectra, 419–420
  preparation, carboxylation of organometallic reagents, 412
    from diethyl malonate, 468–469
    hydrolysis of acyl halides, 476
    hydrolysis of amides, 455
    hydrolysis of esters, 442, 453
    hydrolysis of nitriles, 412, 456
    from ketenes, 451
    oxidation of alcohols, 218, 411
    oxidation of aldehydes, 411
    oxidation of arenes, 411
    oxidative cleavage of alkenes and alkynes, 169, 412
  in prostaglandins, 422
  reactions, with acid anhydrides, 443, 452
    with alcohols, 221–222, 442–443
    with alkenes and alkynes, 440–441
    with amines, 336–337
    with diazomethane, 441
    with diborane, 445, 463
    with halogens, 465–466
    with hydrogen peroxide, 474
    with lithium aluminum hydride, 445
    with phosphorus pentoxide, 443
    with thionylchloride or phosphorus chlorides, 443
  reduction, 445, 463
  resonance stabilization of anion, 438–439
  salts, 438–440, 444, 445, 447
  solubilities, 416
carboxylic acid esters (*see* esters)
carnauba wax, 421

β-carotene, 252, 382
carvone, 361
caryophyllene, 107, 110
catalytic hydrogenation (see hydrogenation)
catechol, 300, 307
catnip, 424
cayenne pepper, 424
cedrol, 472, 474
(+)-cellobiose, reactions, 615
  structure, 615
cellulose, 146
  derivatives, 618–619
  structure, 618–619
cephalins, 422
cephalosporins, 428, 455
cetyl palmitate, 421
chain lengthening, examples in synthesis, 628–631, 638
chain reactions, addition to double bonds, 543–545
  chlorination of methane, 53
  polymerization, 544–545
charge delocalization (see resonance)
chelation, 390–391
chemical shift, 18
chemical stability, 88, 125
chiral reagents, 187, 189–190
chirality, 183–185, 199
  in allenes, 185
  in amine oxides, 336
  in amines, 329
  in biphenyls, 265–267
  configuration, 199
  in quaternary ammonium salts, 329
chitin, 624
chloral, 371
chloral hydrate, 371
chloramphenicol, 73
chlordane, 73
chlorination (see halogenation)
chlorobutyric acids, acidity, 438
cis- and trans-1-chloro-1-butene, 113
cis- and trans-2-chlorocyclohexanol, 224
  reaction with aqueous base, 224
1-chloro-3,3-dimethylcyclopentene, 109
chloroethene, 109
chloroform, 52
chloromethane, 51–52
chloromycetin, 73
1-chloro-1-phenylethane, 310–311
chlorophyll a, 429
chlorotetracycline, 73
cholesterol, 206
  acetylation, 451
  bromination, 135
  chiral centers, 185
cholic acid, 410
choline, 325, 458–459
cholinesterase, 458–459
chromate esters, 234–235
chromate oxidation, of alcohols, 234, 362, 411
  of aldehydes, 384, 411
  chromatography, 3
chromic acid oxidation (see chromate oxidation)
chromophore, 252
chromoproteins, 594
α-chymotrypsin, 455, 457
cinchona bark, 326
cineole, 205

cinnamaldehyde, 246, 361
cinnamon bark, 246, 361
cis-trans isomerism, alkenes, 112–114
  cycloalkanes, 60–61
  cycloakenes, 124
  polyenes in vision, 114
citral, 361
citric acid, 128, 409
civetone, 361
Claisen condensation, 466–467
Clemmensen reduction, 383
Clostridium botulinum, 577
cloves, 246
cocaine, 425
coenzyme A, 457–458, 470–471
cofactor, 457
colchicine, 310
collagen, 592
collision frequency, 54
combustion, calculation of heats of, 47–51
combustion analysis, 4–5
competitive reactions, kinetic vs equilibrium control, 377
  mechanism studies, 490
  in resolution, 190
  substituent effects in aromatic substitution, 490–491
composition of compounds, percentage, 5–6
concerted reactions, 554–555
condensed structural formulas, 36
configuration, 60, 68, 191–196
  absolute, 192
  alkylidene cycloalkanes, 185
  allenes, 185
  amino acids, 581
  Cahn-Ingold-Prelog nomenclature, 193–196
  cis-trans, 60
  trans-cyclooctene, 197–198
  D and L, 602–603
  hexahelicene, 267
  notation, 192–193
  and optical rotation, 191–192
  phosphines, 352
  quaternary ammonium salts, 329
  R and S, 193
  relative, 191
  seqcis-seqtrans, 195–196
  sequence rules for, 194–196
  spiroalkanes, 185
  sugars, 602–603
  twisted biphenyls, 265–266
  Z and E, 195–196
conformational analysis, 196–198
  angle strain, 63
  biphenyls, 265–267
  butane, 40
  cyclobutane, 67
  cycloheptane, 67
  cyclohexane derivatives, 63
  trans-cyclooctene, 197–198
  cyclopentane, 67
  decalin, 67
  1,3-diaxial interactions, 65
  cis-1,2-dichlorocyclohexane, 197
  1,4-dimethylcyclohexane, 66
  E2 elimination, 93–94
  eclipsing strain, 63
  ethane, 39

conformational analysis (*cont.*)
   factors in, 63
   D-(+)-glucose, 611–613
   monosubstituted cyclohexanes, 65
   pmr and, 64, 78
   pyranoses, 611
   tartaric acids, 196–197
   torsional strain, 63
conformational enantiomers, 162–163
conformational isomers (conformers), 63–65, 68, 76, 114, 196–198
conformations, 39, 68
   anti and gauche, 76
   boat, 63–65
   carbohydrates, 611–615, 619–621, 616
   conjugated dienes, 477
   cycloalkanes, 63–65
   cyclohexane, 63–65
   eclipsed, 39–63
   meso compounds, 196–197
   Newman projection, 39
   nonbonded repulsive interactions, 63
   proteins, 588–593
   sawhorse projection, 39
   staggered, 39
congo red, 349
congruence of orbital symmetry, 564–565
coniine, 326
conjugate-acid-base pairs, 94
conjugate acids as leaving groups, 229–230
conjugate addition reactions of organometallic reagents, 530–531
conjugated dienes (*see* dienes)
conjugated double bonds, 110
   heats of hydrogenation, 248–249
   molecular-orbital description, 249–250
   properties, 253
   thermodynamic stability, 248–252
   ultraviolet-visible spectra, 250–252
   (*see also* dienes, polyenes)
conjugated polyenes (*see* polyenes)
conjugated proteins, 594
conjugation, 283
conrotatory electrocyclic reactions, 558
constitutional isomers, 61, 64
coordinated alkynes, cycloaddition, 171
Cope elimination reaction, 342
Cope rearrangement, 553, 557
copolymerization, 147
coronene, 267
corticosterone acetate, 549
cortisone, 361–362
cotton, 618–619
coumarin, 424
coupling reactions, 348–349
covalent bonds, 20–25
cracking, 531
creatine, 440
   infrared spectrum, 14
cresols, preparation from aryl halides, 295
Criegie ozonolysis mechanism, 144
croconic acid, 309–310
crotonic acid, 409
crustecdysone, 430
crystal violet, 358
cumene hydroperoxide, acid-catalyzed rearrangement, 526–527

cumulated double bonds, 116
curare, 326, 426, 459
Curtius rearrangement, 526
curved arrows, 26
cyanocobalamin, 429
cyanohydrins, 373–374
cyclic anhydrides, 443
cyclic ketals, 373
cycloaddition reactions, 551–552, 558–560
cycloalkanes, 59–68
   cis-trans isomerism, 124
   heats of combustion, 62
   nomenclature, 59, 60
   physical properties, 59
   reactivity, 68
   stereoisomerism, 60–61
   strain in, 61–62
cycloalkenes, nomenclature, 109–110
   stereoisomers, 124
cycloalkynes ring size, 158
1,3-cyclobutadiene, 275–276
cyclobutane, 60–62
cyclobutanone, carbonyl stretching frequency, 368
2-cyclobutyl-3-methyl-1-butene, 109
cyclodecane, 62
cyclodecapentaenes, 280–281
cyclodecene, 124
cyclodecyne, 156
cycloheptane, 60, 62
cycloheptatriene, acidity, 278–279
cycloheptatrienyl cation, 279–280
cyclohexadiene, heat of hydrogenation, 253
cyclohexane, 60, 62, 63–65
   axial hydrogens, 64
   from benzene, 247
   boat conformation, 63–65
   chair conformations, 63–65
   combustion, 68
   conformations, 63–65
   equatorial hydrogens, 64
   pmr spectrum, 64
   strain in, 62–63
   symmetry elements, 182
   twist conformations, 63–65
cyclohexanol, 219, 331
cyclohexanone, addition of hydrogen cyanide, 373
   C-alkylation, 391
   carbonyl stretching frequency, 368
   condensation with benzaldehyde, 396
   enol content, 386
   enolate anion from, 390
   α-halogenation, 385
   hemiketal formation, 371
   ketal formation, 371
   keto-enol tautomerism, 300–301
   physical constants, 366
   reaction with sodium phenylacetylide, 381
   semicarbazone, 377–378
O-silylation, 391
   ultraviolet absorption, 366
   Wittig reaction, 382
cyclohexatriene, 247–248, 253–254
   (*see also* benzene)
cyclohexene, addition of dihalocarbene, 149
   addition of hypochlorous acid, 135
   complex with mercury, 139
   heat of hydrogenation, 253

INDEX 801

reaction with Simmons-Smith reagent, 149
cyclohexene oxide, 224
cyclohexenone, carbonyl stretching frequency, 368
  ultraviolet absorption, 366
cyclohexylamine, 331
cyclononane, 62
cyclononene, 124, 168
cyclooctane, 62
cyclooctatetraene, bromination, 275
  conformation, 275
  from cycloaddition of acetylene, 170
  heat of combustion, 274
  hydration, 275
  hydrobromination, 275
  hydrogenation, 274
cyclooctene, 124, 197–198
cyclopentadecaene, 162
cyclopentadiene, acidity, 278
cyclopentadienyl anion, aromaticity, 278–280
  neutralization, 278
  resonance stabilization, 279
cyclopentadienyl cation, reaction with hydroxylic solvents, 281
cyclopentane, 60
  heat of combustion, 62
  photochlorination, 68
  strain, 62
cyclopentanol, 379
cyclopentanone, 366
  alkylation, 392
  carbonyl stretching frequency, 366
  reduction, 379
cyclopentene, 109
  bromination, 137
  hydroboration, 132
cyclopentenone, 366
cyclopropane, 60–62
  bromination, 150
  strain, 61–62
  addition of sulfuric acid, 150
cyclopropane derivatives, from addition of carbenes to alkenes, 148
  from 1,3-dihalides, 98
cyclopropanone, carbonyl stretching frequency, 368
  hemiacetal from, 370
cyclopropyl alcohol (cyclopropanol), 207
cyclopropylcarbinyl cation, 523
cyclotetradecaheptaene, aromaticity, 281
cycloundecene, 124
p-cymene, infrared spectrum, 262
  pmr spectrum, 263
cysteine, 235, 580
cystine, 580
cytosine, tautomeric forms, 274

D configuration, 602–603
Dacron, 453
dashed bonds, 22, 193
DDT, 73
deactivating substituents in aromatic substitution, 490–491
deamination, 347
decarboxylation of β-ketoesters, 468–470
decarboxylation reactions, 477
  of carboxylic acids, 444–445
dehydration, of alcohols, 231–233
  in aldol condensation, 395

dehydrogenation, of alcohols, 233–235
  of cyclohexadiene, 254
delta (δ) chemical-shift scale, 18
denaturation, 594
dendrites (see synaptic transmission)
deshielded protons, 16–17
deshielding, 16, 17, 161, 263–264, 368
detergents, 423
deuterium, exchange, 209, 385, 466
  substitution reactions, 344–345
Dewar benzene, 445, 576
dextrorotatory compounds, 178
diacetyl peroxide, decomposition, 547
diacyl peroxides, 451
dialkyl sulfite, 228
1,2-diaminoethane, 331
diastereoisomers, 187–189, 199
  epimers, 351
  meso compounds, 188
  in resolution, 189–190
diazo coupling, 348–349
diazoic acid, 347
1,3-diazole (Imidazole), 325
diazomethane, methylene from, 92–93
  reaction with acyl chlorides, 448
  reaction with carboxylic acids, 440
diazonium salts, 354
  coupling reactions, 348–349
  as dibasic acids, 347
  mechanisms for substitution reactions, 345–347
  nucleophilic attack at nitrogen, 347–349
  preparation, 338–339
  reduction of, 350
  hydrobromination, 150
  hydrogenation, 150
  pmr signals, 150
  photochlorination, 68
  reactivity, 150
diborane, 477
  addition to alkenes, 133–134
  addition to alkynes, 168–169, 173
  reduction of aldehydes and ketones, 343–344
  reduction of carboxylic acids, 445, 463–464
1,2-dibromo-2-methylpropane, pmr spectrum, 77
1,2-dibromocyclopentane, 137
β-dicarbonyl compounds, 468–469
dicarboxylic acids, anhydrides from, 443
  esters, acyloin condensation, 639
    Dieckmann condensation, 468
  malonic esters in synthesis, 468–469
  nomenclature, 410
  oxidative bis-decarboxylation, 444–445
cis-1,2-dichlorocyclohexane, conformational racemization, 197
trans-1,4-dichlorocyclohexane, symmetry elements, 182–183
1,2-dichloroethane, infrared spectrum, 75
  pmr signals, 78
2,4-dichloro-3-ethyl-2-pentene, stereoisomers, 195–196
dichloromethane (methylene chloride), 51–52
1,1-dichloro-4-methylcylcodecane, 59
dichromate oxidation (see chromate oxidation)
Dieckmann condensation, 468
dielectric constant, 85
Diels-Alder reaction, 497–502
  cycloaddition classification, 552

Diels-Alder reaction (*cont.*)
  stereochemistry, 500–501
  substituent effects, 499–500
dienes, allenes, 185
  classification, 109–110
  conjugated, 110
    1,2-addition products, 496–497
    1,4-addition products, 496–497
    conformations, 497–499
    electrophilic addition to, 496–497
    heats of hydrogenation, 249
    thermodynamic stability, 248–250
    ultraviolet-visible spectra, 250–252
  cumulated, 110
  isolated, 110
diethyl malonate, 468
  in synthesis of carboxylic acids, 469
diffraction analysis, 8–9
dihalides, alkenes from, 98
  alkynes from, 156
  reactions, 98
dihalocarbenes, 148–149
m-dihydroxybenzene (resorcinol), 300
o-dihydroxybenzene (catechol), 300–307
p-dihydroxybenzene (*see* hydroquinone)
diisopropyl ether, infrared spectrum, 210
  pmr spectrum, 213
dimerization, of alkenes, 145–148
  of alkynes, 169–170
dimethyl ether, pmr spectrum, 17
dimethyl sulfate, reaction with phenols, 304
2,5-dimethyl-2,4-hexadiene, ultraviolet spectrum, 251
dimethylacetylene, 157
dimethylallylpyrophosphate, 473
N,N-dimethylaminoazobenzene, 349
3,3-dimethyl-1-butene, addition of halogen, 135
  reaction with diborane, 135
2,2′-dimethylbiphenyl, ultraviolet absorption, 266
3,3′-dimethylbiphenyl, ultraviolet absorption, 266
1,1-dimethylcyclohexane, 59
3,3-dimethylcyclohexene, hydroboration, 138
1,2-dimethylcyclopropane, 60
dimethylformamide, 85
N,N-dimethylformamide, pmr spectrum, 420
dimethylsulfoxide, dielectric constant, 85
  empirical formula, 5
  mass spectrum, 7
2,4-dinitrofluorobenzene, 336
dinitromethane, 469
dinitrophenol, 375
2,4-dinitrophenylhydrazine (DNP), 375–376
1,4-dioxane, 207, 208
  pmr spectrum, 215
1,2-dioxetane, 325
diphenyl-β-picrylhydrazyl, 537
dipole-dipole interactions, 45
dipole moments, 24
  of substituted benzenes, 307, 490–491
dipoles, bond, 23
  molecular, 24
directing effects in aromatic substitution, 491–494
diribonucleic acid (DNA), 461–462
  hydrogen bonds in, 45
disaccharides, 614
disparlure, 431
disrotatory electrocyclic reactions, 557–558
dissolving-metal reduction of alkynes, 167–168, 173
dissymmetry, 183, 199

distillation, 3–4
disulfide bond cleavage, 585
divinylacetylene from trimerization of aceytlene, 169
1,1-divinylcyclopropane, 109
DNA (*see* diribonucleic acid)
DNP (*see* dinitrophenolhydazine)
double bonds, 21
  (*see also* covalent bonds, carbon-carbon double bonds, carbonyl group)
drugs, 327–328
Dynel, 147

ΔE, 10
$E1$ (*see* elimination reactions), 96
$E2$ (*see* elimination reactions), 92–95
ecdysone, 430
eclipsing strain, 63
electrocyclic reactions, 553, 557–558
electromagnetic spectrum, 12
electron-donating groups as activating groups, 271, 272, 343–344, 490–491
electron-dot formulas, 20–21
electron-pair bonds, 20–25
electron solutions, 167
electron spin resonance (esr), 537–539
electron-transfer reactions, 541
electron-withdrawing groups, as activating groups, 292–294
  as deactivating groups, 490–491
electronic effects (*see* inductive effects, resonance effects)
electronic energy, 10
electronic excitation, 251
electronic spectroscopy (*see* ultraviolet-visible spectra)
electronegativity, 23–24
  sp-hybridized carbon atoms, 167
electrophiles, 28
electrophilic addition reactions, 127–138, 151, 495–497
  alkenes, 127–129, 132–134
  alkynes, 163–165
  conjugated dienes, 495–497
  mechanism and stereochemistry, 129–131, 135–138, 495–497
  (*see also* addition reactions)
electrophilic aromatic substitution, 268–273, 283, 490–495
  amines, 335–340, 343–344, 353
  aryl ethers, 305–307
  aryl halides, 297–299, 318
  benzene, 268–273
  competitive reactions, 490
  coupling of diazonium salts, 348–349
  Friedel-Crafts acylation, 448–449, 452
  inductive effect, 491–492
  mechanism, 270–272, 491–494
  orientation, 343–344
  phenols, 305–307, 318
  resonance effect, 493–495
  steric effects, 495
  substituent effects, 343, 344
electrophilic substitution reactions, of alcohols, 221–223, 238
  nitrosation of amines, 338–339

elimination reactions, 25, 239
  alkyl halides, 91–96
  alcohols, 231–233
  amines, 340–343
  aryl halides, 295–296, 317
  competition with nucleophilic substitution, 94–95
  vicinal dihalides, 98
  o-dihalobenzenes, 297
  $E1$ mechanism, 96
  $E2$ mechanism, 92–95
  $E2$ stereochemistry, 93
  α elimination, 28, 148–149
  β elimination, 91
  Saytzeff rule, 92
empirical formula, 5
emulsin, 613
enantiomers, 183–187, 199
  amines, 329
  biphenyls, 265–267
  configuration, 191–196
  Fischer projection formulas, 192
  interconversion, 197–198
  phosphines, 352–353
  physiological properties, 190
  racemization, 197–198
  resolution, 189–190, 199
endothermic reaction, 47
enediol tautomers, 487–488, 605
energy of activation (see activation energy)
energy diagram, 55–57
energy factors in reaction rates, 54
energy sources of pericyclic reactions, 554
enol content, carbonyl compounds, 386–387
enol intermediates, substitution at α carbon, 385–387
enol-keto tautomerism (see keto-enol tautomerism)
enolate anions, aldol condensation, 394–397
  alkylation reactions, 392–394
  carbon vs oxygen reactions, 391–392
  from carbonyl compounds, 390–391
  definition, 388
enolization of aldehydes and ketones, 385–391
enols, 164
  ferric chloride test for, 301
Enovid, 403
enthalpy (heat content), 47, 68
entropy, 58
entropy of activation, 58–59
enzyme catalysis, 457
enzymes, 457, 462–463
ephedrine, 324, 327–328
epimerization, 387
epimers, 387
epinephrine, 327, 458
epoxidation, 475
epoxides, 239
  from 1,2-halohydrins, 224
  from oxidation of alkenes, 230, 475
  reaction with organometallic compounds, 380–381
  ring-opening reactions, 230–231
equatorial bonds, 64
ergosterol, 570
ergot, 426
erythro diastereoisomers, 509, 606
erythromycin, 428
D-(−)-erythrose, 606–607
esr (see electron spin resonance)
essential amino acids, 579–580

essential fatty acids, 421–422
esterification, acid catalysis of, 441
  of alcohols, with anhydrides, 221–222, 451
    with acyl chlorides, 221–222, 447–448
    with carboxylic acids, 221–222, 441–442
    with esters, 221–222, 452–453
    with sulfonylchlorides, 225–226
  of carboxylic acids, 221–222, 441–442
  of dicarboxylic acids, 443
  mechanism, 441–442
  of phenols, 305
  of phosphorus acids, 461–462
  steric effects on rate, 441
  of sulfonic acids, 221–222, 459–460
esters, in biosynthesis, 470–472
  carbonyl stretching frequency, 417
  Claisen condensation, 466–468
  deuterium exchange, 466
  Dieckmann condensation, 468
  ester enolates, 466
  esterification mechanism, 441–442
  functional group, 413
  α-haloesters, reactions of, 466
  hydrolysis mechanism, 453
    $A_{AC}2$, 453
    $B_{AC}2$, 453
  infrared spectra, 416–417
  keto, 466–470
  nomenclature, 414–415
  occurrence, 421–422
  pmr spectra, 419–420
  preparation, from alkyl halides and carboxylate salts, 440–441
    from carboxylic acids and alcohols, 441–442
    from carboxylic acids and alkenes or alkynes, 441
    from carboxylic acids and diazomethane, 441
    from ketenes, 451
  racemization, 466
  reactions, with alcohols, 452
    with amines, 452
    with ester enolate anions, 466–468
    with lithium aluminum hydride, 463
    with organometallic compounds, 464–465
    with water, 453
  reactivity, 454
  reduction, 463
  saponification, 453
  transesterification, 453
estradiol, 300
ethanal (see acetaldehyde)
ethane, 118, 316
ethanol, 207
  catalytic air oxidation, 362
  dehydrogenation, 233
  diethyl ether from, 233
  ethene from, 233
  ionization, 438–439
  pmr spectrum, 17, 215
  physical properties, 208
  proton exchange in, 216
  reactions with sulfuric acid, 233
ethene (see ethylene)
ethers, 205–239, 237
  basicity, 219–220
  boiling points, 208
  cleavage, 305
  C—O stretching frequency, 209

ethers (cont.)
   functional group, 30
   hydrogen bonding, 208–209
   infrared absorptions, 209
   nomenclature, 206–207
   nucleophilic displacement reactions, 225–231
   preparation, diethyl ether from ethanol, 233
      from phenols, 304–305
      t-butyl ethers from alcohols, 221
      Williamson synthesis, 222–223
   reactions, 219, 229–230
   solubility in water, 208
   as solvents, 220
   spectroscopic properties, 209–211
ethoxy group, 206
p-ethoxyacetanilide, pmr spectrum, 420
ethyl acetate, infrared spectrum, 418
ethyl acetylene, 157
ethyl alcohol (see ethanol)
1-ethyl-2-methylcyclopentane, 59
trans-4-ethyl-2-methyl-4-hexen-2-ol, 207
ethylacetoacetate, 467–469
ethylene (ethene), 109–166
   catalytic oxidation, 230
   molecular orbitals, 118–121
   polymerization, 146–147
   structure, 110
ethylene imine, 324
ethylene oxide, (see oxitane)
ethyne (see acetylene)
ethynyl group, 156
ethynyl hydrogen, 160
eudesmol, 808m
eugenol, 246, 299–300
exothermic reaction, 47
extinction coefficient, 252
extraction, 4

farnesyl pyrophosphate, 473
fats, 421, 452
fatty acids, 407, 471
fatty esters, 452
Fehling reagent, 384, 605
ferric chloride test, 301
fibrous proteins, 588
fingerprint region (infrared), 13
fire ant, 358
fire extinguisher, 79
first-order reactions, 80
Fischer, Emil, 350, 602–607
Fischer projection, 192
flavoring agents, 423–424
fluoroacetic acid, 409, 426
folic acid, 460
formal charge, 24
formaldehyde, from methanol, 362
   reaction with ammonia, 375
formic acid, 85
formula, electron-dot, 20–21
   empirical, 5
   molecular, 5–6
   structural, 8–18
formyl chloride, 449
formyl group, 363
formylation of aromatic rings, 449
fragmentation reactions of radicals, 543

frankincense, 246
free energy of activation, 58–59
free radicals (see radicals)
freons, 73
Friedel-Crafts acylation, 448–449, 452
Friedel-Crafts alkylation, 269
   compared to acylation, 449
frontier-orbital model, 561–567
   cycloaddition reactions, 561–564
   Diels-Alder reaction, 562
   electrocyclic reactions, 564–566
   sigmatropic rearrangement, 566–567
D-(−)-fructose, osazone formation, 604
   structure, 602
   in sucrose, 617
Fukui, K., 561
fumaric acid, 410
functional-group interaction, hydroxyl and aromatic ring, 489
   (see also phenols)
   hydroxyl and carbonyl groups, 487–488
   hydroxyl and double bond, 489
   (see also enols)
   two hydroxyl groups, 487
functional groups, 29–30
   manipulation of, 631–636
furan, 325
   aromaticity, 276–277
furanoses, 612
furfuraldehyde semicarbazone, 377

g factor, 538
ΔG^* (free energy of activation), 58–59
Gabriel amine synthesis, 634
D-(+)-galactose, 602
   oxidation of, 605
gallic acid, 409
galvinoxyl, 537
ganglia, 458–459
gauche conformation, 76
gelatin, 592
geometric isomerism (see stereoisomerism)
geraniol, 472
   oxidative cleavage, 245
geranyl pyrophosphate, 473
globular proteins, 588
glucaric acid, from glucose, 605
glucosamine, 624
D-(+)-glucose, action of alkali on, 605
   in amylopectin structure, 620–621
   in amylose structure, 620–621
   α and β anomers, 611
   from arabinose, 608
   in cellobiose structure, 615–616
   from cellulose, 615, 619
   configuration, 607–610
   constitution, 602–603
   cyclic hemiacetal form, 610–612
   gluconic acid from, 606
   in lactose structure, 616–617
   in maltose structure, 614–615
   mutarotation, 611
   osazone formation, 604
   oxidation, 605
   pentaacetate, 603
   pentamethyl derivative, 611–612

D-(+)-glucose (cont.)
  Ruff degradation, 608
  from starch, 619–621
  stereoisomers, 602, 610–611
  sucrose structure, 617
glucosides, 612–613
glutamic acid, 580
glutamine, 580
glutaric acid, 410
  anhydride from, 443
glutathione, structure, 584
  synthesis, 587
glycans, 618
R-(+)-glyceraldehyde, 601, 607
glycerides, 421
glycerol, 205
glycine, 579
glycol cleavage, 487
glycosides, stability toward base, 613
Gomberg, Moses, 535–536
gossyplure, 431
grandisol, 431
Grignard reagents, from alkyl halides, 97
  from aryl halides, 296–297
  bis from dihalides, 297
  reactions, acylation by carboxylic acid
    derivatives, 464
    addition to aldehydes and ketones, 380–381
    with alkynes, 161
    with carbon dioxide, 381
    conjugate addition to enones, 506
    with oxirane, 381
Grignard, Victor, 97
griseofulvin, from polyketo acids, 472
  synthesis by Michael condensation, 505
guanidine, basicity, 335
L-gulonolactone oxidase, 621–622
L-(+)-gulose, 609–610
gutta-percha, 113

ΔH, (see heat of reaction)
ΔH^* (see heat of activation)
α-halogenated carboxylic acids and derivatives, 465–466
halogenation, of alcohols, 221
  of alkanes, 51–54
  addition to alkenes, 132–133
  addition to alkynes, 163
  of allyl derivatives, 316–317
  of aniline, 343
  of aryl halides, 297
  of benzene, 268
  of benzyl derivatives, 316–317
  of carboxylic acids and derivatives, 465–466
  of cyclooctatetraene, 275
  of ketones, 385
  of methane, 51–53
  of phenols, 305–306
  of propane, 54, 56–58
halogens, addition to alkenes, 132–133
  addition to alkynes, 163, 173
1,2-halohydrins, formation of epoxides, 224
halonium-ion intermediates, 508–510
Hammond postulate, 89–90
heat content H (enthalpy), 47, 68
heat of activation ΔH^*, 55–57, 164

heat of combustion, of isomeric butanes, 50
  of isomeric butenes, 123
  of cycloalkanes, 62
  of hydrocarbons, 47–51
heat of hydrogenation, of alkenes, 121–123
  for alkynes, 162
  of benzene, 253
  of cycloalkenes, 124
  of cyclohexadienes, 253
  of cyclooctatetraene, 274–275
  of pentadienes, 248–249
heat of reaction ΔH, 47
  in tautomerism, 164–165
helical conformations of proteins, 590–592
helium, molecular orbitals, 116–117
α-helix, 591
heme, 595
hemiacetals, 370–371
hemiketals, 370–371, 466–467
hemin, 429
hemlock (poison), 326
hemoglobin, 429, 595–596
heptane, 35–37
herbicides, 304
heroin, 451
heterocyclic amines, 324–325
heterocyclic compounds, aromaticity, 276–277
  nomenclature, 324–325
heterocyclic ethers, 207
heterolysis, 26–27
hexachlorophene, 427
hexahelicene, 267
hexamethylenetetramine, 375
hexane, 35–37
hexaphenylethane, attempted synthesis of, 535–536
hexoses, 601–602
4-n-hexylresorcinol, 301
high field, 16
Hinsberg amine test, 337–338
histamine, 327–328
histidine, 580
Hoffman, R., 560
Hofmann elimination, 341–342
Hofmann rearrangement, 525–526
Hofmann rule, 341
HOMO, 562
homologous series, 36
homologs, 36
homolysis, 26–27
hormones, adrenal, 327
  insect, 430–431
Hückel rule, 276–282, 283
  two-π-electron systems, 280
  six-π-electron systems, 276–280
  ten-π-electron systems, 280–281
  fourteen-π-electron systems, 281
  eighteen-π-electron systems, 282
hybrid atomic orbitals, 117
  (see also atomic orbitals)
hydration, aldehydes and ketones, 365, 370–371
  alkenes, 232–233
    compared to aldehydes and ketones, 365
  alkynes, 164–165
  2-butene, 186–187
hydrazines, 350
  reaction with aldehydes and ketones, 375
  in Wolff-Kishner reduction, 383
hydrazones, in Wolff-Kishner reduction, 375

hydrocarbons, 35
  (see also alkanes, alkenes, alkynes, aromatic hydrocarbons)
hydrogen, analysis for, 4
  axial and equatorial on cyclohexane, 64
  combustion, 48
  molecular orbitals, 116
hydrogen abstraction by radicals, 542
hydrogen bonding, 44–45, 237–238
  in alcohols, 208–209
  in amides, 416
  in amines, 330
  effect on boiling points, 44
  in carboxylic acids, 416
  in DNA, 45
  in ethers, 208–209
  and infrared absorption, 209–210
  in nucleic acids, 45
  in phenols, 301
  in water, 44
hydrogen cyanide, addition to aldehydes and ketones, 373–374
hydrogen peroxide, reaction with acyl chloride, 447
  reaction with amines, 339
hydrogenation, 121–122
  alkenes, 121–123
  alkynes, 162–163, 173
  benzene, 247
  cyclooctatetraene, 274
  heat of (see heat of hydrogenation)
hydrogenolysis, sulfides, 237
  thiols, 236
hydrohalogenation, alkenes, 127–129
  alkynes, 163
  aryl ethers, 305
  cyclooctatetraene, 275
hydrolases, 613
hydrolysis, alkenes, 128
  alkyl bromides, 80–81, 84–91
  alkyl chlorides, 87, 315–316
  amides, 455
  benzyl derivatives, 313–314
  t-butyl chloride, 96, 313
  esters, 453–454
  neopentyl bromide, 90
  nitriles, 412, 457
hydrophilic functions, 423
hydrophobic functions, 423
hydroquinone, equilibrium with quinone, 308
  oxidation, 307–309
  structure, 300
hydroxamic acid, 452
o-hydroxy benzoic acid (salicylic acid), 409, 451
β-hydroxydecanoic acid, 432
hydroxylamine, reaction with aldehydes and ketones, 375–376
hydroxyl group, 206
  analysis for, 220
  pmr signals, 210
  stretching frequency, 209
hydroxylation of alkenes, 140–142
hyperfine splitting, 538–539
hypoascorbemia, 621–622
hypohalites, photolysis of, 541
hypohalous acid, addition to alkenes, 133
hypophosphorous acid, reaction with aryl diazonium salts, 345–346

imidazole, 325
imides, nomenclature, 414
imido esters, 457
imines, 374
  reduction, 378
inductive effects, in aromatic substitution, 297–299, 491–492
  chloro substituents in acids, 439–440
  pmr shift by halogen substituents, 76–77
infrared absorption bands, characteristic values, 13
infrared spectra, acetic acid, 418
  acetic anhydride, 418
  alcohols, 209–210
  aldehydes and ketones, 367–368
  alkanes, 46
  alkenes, 110–111
  alkyl halides, 75–76
  alkynes, 159–160
  amines, 330–331
  1-amino-1-phenylethane, 331
  aromatic hydrocarbons, 262
  aspirin, 13
  benzaldehyde, 367
  benzene, 262
  carboxylic acids and derivatives, 416–418
  creatine, 14
  p-cymene, 262
  1,2-dichloroethane, 75
  diisopropyl ether, 210
  ethers, 209–210
  ethyl acetate, 418
  isopropyl alcohol, 210
  limonene, 111
  3-methyl-2-buten-1-ol, 210
  1-octene, 111
  phenols, 302
  progesterone, 367
  sulfides, 236
  thiols, 236
infrared spectroscopy, 12–14
  fingerprint region, 13
initiation of chain reactions, 53
insect hormones (see pheromones)
insect repellant "6-12," 395
insulin, 585
intermolecular forces, 44–45, 208
internal salts (zwitterions), 410, 581
inversion of configuration, 82
invert sugar, 617
ionic bonds, 19–20
ionic reactions, characteristics, 535
ionization (see acidity, basicity)
ionization potential, 19
  alkenes and alkynes, 166
isobutane, heat of combustion, 50
isobutylene, 109
  addition of alcohols, 221
  dimerization, 145–146
  heat of combustion, 123
  polymerization, 146
isocitric acid, 128
isocyanates, 449–450
  intermediates in rearrangement, 526
isoelectric point, 582
isolated double bonds, 110
  (see also carbon-carbon double bonds)
isoleucine, 579

isomenthone, isomerization, 387
isomerism, 8
  alkanes, 40
  alkenes, 112–114, 122–124
  alkyl halides, 76
  biphenyls, 265–267
  cis-trans (*see* stereoisomers)
  cycloalkanes, 60–62
  cycloalkenes, 124
  geometric (*see* stereoisomers)
  optical (*see* stereoisomers)
isomers, 8
  cis-trans, 112–114
    (*see also* stereoisomers)
  conformational (conformers), 63–65, 68, 76, 114, 196–198
  constitutional, 60–61, 64, 68
  diastereoisomers, 188
    (*see also* stereoisomers)
  enantiomers, 183–187
    (*see also* stereoisomers)
  epimers, 387
  geometric, 112–114
    (*see also* stereoisomers)
  optical (*see* stereoisomers)
  rotamers, 114
    (*see also* conformations)
  stereoisomers, 60–62, 112–114, 183–188, 387, 612
isopentyl acetate, 423
isopentenylpyrophosphate, 473
isopentylvalerate, 423
isophthalic acid, 410
isoprene, 472
isoprene rule, 472–474
isopropyl alcohol, 207
  infrared spectrum, 210
isoquinoline, 324
isotactic configuration, 193
IUPAC system of nomenclature, 40–41
  (*see also* individual compound families)

juglone, 308
juvabione, 430
juvenile hormone, 430

Kekulé, F. A., 247
Kekulé structure of benzene, 247–248
Kendrew, J., 595
α-keratin, 590, 592
β-keratin, 593
ketals, 371–373
ketene dimer, 450, 454
ketenes, preparation, 450
  reactions, 450
keto-enol tautomerism, 385–391
  acid and base catalysis, 387–388
  of cyclohexanone, 300–301
  halogenation, 385
  of phenols, 300–301
β-ketoesters, from Claisen condensation, 466–468
  reactions, 468–469

α-ketols, isomerization, 487, 605
  oxidation, 487–488, 605
  reduction, 487–488
β-ketols, dehydration, 488
  retroaldol cleavage, 488
ketones, 361–397
  acidity, 388–389
  aldol condensation, 394–397
  alkylation of enolate anions, 392–394
  Baeyer-Villiger rearrangement, 526
  boiling points, 366
  Claisen condensation, mixed, 467–468
  Clemmensen reduction, 383
  conjugate acids and bases, 387–389
  derivatives, 375
  enol intermediates, 385–391
  enolization, 385–387
  functional group, 364–365
  hydration, 365, 370–371
  infrared spectra, 367–368
  nomenclature, 363–364
  pmr spectra, 368–399
  photolysis, 541
  preparation, acetoacetate ester synthesis, 468–469
    Friedel-Crafts acylation, 448–449, 452
    by organocadmium compounds, 464–465
    from alcohols, 233–235
    from nitriles and Grignard reagents, 464–465
  racemization, 386–387
  reactions, 369–397
    addition of alcohols, 370–373
    with ammonia or amines, 374–376
    with diborane, 380
    addition of hydrogen cyanide, 373–374
    with hydrazoic acid, 524–525
    with lithium aluminum hydride, 379
    with organometallic compounds, 380–381
    with peracids, 526
    with sodium borohydride, 379
    addition of water, 365, 370–371
    with ylides, 382
  reduction, 379–380
  reductive amination, 378
  Schmidt reaction, 524–525
  solubility in water, 366
  sources, 361–362
  structure, 361
  ultraviolet-visible spectra, 366
  Wittig reaction, 382
  Wolff-Kishner reduction, 383
ketoses, 601
ketyl, 541
Kiliani-Fischer synthesis, 606, 607
kinetic order of reaction, 80
Kolbe electrolysis, 547–548
krebiozin, infrared spectrum, 14
kwashiorkor, 577

L configuration, 602–603
lactams, 413
lactic acid, 180, 409
  chiral carbon unit, 184
  enzyme selectivity, 190
  interconversion of enantiomers, 191
  optical isomers, 180

lactobacillic acid, 422
lactobionic acid, 617
lactones, 413
　from bifunctional molecules, 442
　occurrence in nature, 424
(+)-lactose, reactions, 616–617
　structure, 616–617
LCAO bonding model, 115–121
lead tetraacetate, cleavage of vicinal glycols, 487
leaving groups, in acylation of carboxylic acid derivatives, 446
　alkoxide ions, 225
　amine oxides, 342–343
　amines, 340
　conjugate acids, 229–230
　halide in S_N2 reactions, 84
　hydroxide ions, 225
　nitrogen, 344, 455
　phosphite ester intermediates, 227–228
　quaternary salts, 340–342
　sulfonate esters, 225–227
　sulfite ester intermediates, 227–228
lecithins, 422
lemon grass, 361
leucine, 190, 579
levorotatory compounds, 178
levulinic acid, from geraniol, 245
Lewis acid, 28
Lewis base, 28
ligands, 25–26
limonene, 107, 110, 472, 474
　chiral carbon unit, 184
　infrared spectrum, 111
　optical isomers, 180
　ozonolysis, 143
　pmr spectrum, 111
limonene dihydrochloride, 186
lindane, 73
linear bonds, 23
linen, 146
linoleic acid, 422
lipoproteins, 594
lithium in reduction of alkynes, 167
lithium aluminum hydride, 379
　reaction with t-butyl alcohol, 464
　reduction, aldehydes, 379–380
　　amides, 351, 463
　　carboxylic acids, 445
　　esters, 463
　　ketones, 379–380
　　nitriles, 351, 463
lithium tri-t-butoxyaluminum hydride, 464
low field, 16
LSD (see lysergic acid)
Lucite, 476, 545
lumisterol, 571
LUMO, 562
lysergic acid, diethyl amide of, 425
lysine, 580

ma huang, 327
macrolides, 428
maleic acid, 410
malic acid, 409
malonic acid, 410
malononitrile, 468

maltase, 613
D-maltobionic acid, 615
(+)-maltose, reactions, 614–615
　structure, 615
D-(+)-mannose, osazone formation, 604
　structure, 604
margarine, 421
marihuana, 8
Markovnikov's rule, 128–129, 163
mass spectrometer, 6
mass spectroscopy, molecular weights from, 6–8
　theory of, 6–8
mass spectrum, dimethyl sulfoxide, 7
mauve (aniline purple), 626
mecloqualone, 434
menthol, 205
menthone, 374, 387
mercuric ion, cyclohexene complex, 139
mescaline, 327
mesitoic acid, 443, 454
meso compounds, 188, 199
　conformations, 196–197
meta-directing groups, 491–493
metal chelates, 390
metal hydrides, 379, 463
methanal (formaldehyde), 363
methane, bond angles, 36
　bond lengths, 36
　Brode, Boord ball-and-stick model, 38
　chlorination, 51–53
　combustion, 49
　molecular models, 38
　molecular orbitals, 117
　structure, 36–37
　symmetry elements, 181
1,6-methanocyclodecapentaene, aromaticity, 281
methanol, 206–207
　catalytic air oxidation, 362
　dielectric constant, 85
　pmr spectrum, 211
　proton exhange and pmr spectrum, 216
　spin-spin splitting in pmr, 211–213
methemoglobinemia, 596
methionine, 580
methoxy group, 206
methyl anthranilate, 423
methyl chloride, 51–52
methyl glucosides, 612
methyl methacrylate, 476
　pmr spectrum, 420
methyl salicylate, 423
methylacetylene, 157
methylalcohol (see methanol)
methylamine, 323, 330
p-methylanisole, pmr spectrum, 263
methylation with diazomethane, 441
3-methyl-2-buten-1-ol, infrared spectrum, 210
2-methyl-2-butene, addition reactions, 128
　ozonolysis, 143
　in preparation of cis-alkenyl boranes, 168–169
3-methyl-1-chlorocyclopentanes, stereoisomers, 187–188
methyl p-chlorophenyl ketone, pmr spectrum, 369
cis- and trans-3-methyl-2-pentene, 113
4-methylcyclohexene, 109
cis-4-methylcyclohexanol, 207
5-methylcyctosine, tautomeric forms, 274
methylene, formation, 148

methylene chloride, 51–52
methylphenylketone (*see* acetophenone)
2-methylpropene, 109, 123
bis-methylsulfonylmethane, 469
mevalonic acid, 473
micelles, 423
Michael condensation, 504–505
mickey finn, 371
microns, 12
migraine, 426
models (*see* molecular models)
molar absorptivities, 252
molasses, fermentation, 362
molecular chirality, 183–185
  biphenyls, 266
molecular dipole, 24
molecular formula, determination, 2–6
molecular ion in mass spectroscopy, 6
molecular models, 38
  Brode, Boord ball and stick, 38
  Buchi, Brinkman, Dreiding, 38
  Corey, Pauling, Koltun, 38
molecular-orbital model, benzene, 254–256
  conjugated diene, 250
  six-π-electron heterocycle, 277
molecular orbitals, 21–23
  bonding and antibonding, 22
  LCAO model, 115–121
  pi (π), 118–120
  sigma (σ), 116–118
  tau (τ), 120–121
molecular rearrangement (*see* rearrangement)
molecular spectroscopy, 10–18
molecular structure, theoretical models, 19–25
molecular weight, mass spectrometry, 6–8
molozonide, 144
molting hormone, 430
monochromatic light, 177
monomer, 146
monosaccharides, chemical reactions, 604–607
  configuration, 604
morphine, 326, 451
morpholine, pmr spectrum, 332
mucoproteins, 594
multiple bonds, 21
  (*see also* covalent bonds, carbon-carbon double bonds, carbon-carbon triple bonds, carbonyl bonds)
muscone, 362
mustard gas, 511
mycomycin, 155
myoglobin, 596

NAD-NADH (*see* nicotinamide adenine dinucleotide)
naphthalene, aromaticity, 281
  structure, 260
naphthol blue-black B, 349
α-naphthylthiourea, 450
Natta, Giulio, 147
neopentyl bromide, hydrolysis, 90
  nucleophilic substitution, 80
nepetalactone, 424, 481
neurons (*see* synaptic transmission)
Newman projection, 39
niacin, 410
nicol prism, 177

nicotinamide adenine dinucleotide (NAD), 471
nicotine, 326
nicotinic acid, 410
ninhydrin, 371
nitration, aniline, 343
  aryl halides, 297–299
  benzene, 268
  phenols, 306
nitriles, carboxylic acids from, 412, 456
  functional group, 414
  hydrolysis, 412, 456
  imido esters from, 457
  nomenclature, 414
  preparation, from alkyl halides, 79
    from amides, 456
    from ketenes, 450
    from sulfonate esters, 456
  reactions, addition of alcohols, 457
    with Grignard reagents, 465
    with lithium aluminum hydride, 351, 463
    with water, 412, 456
  reduction, 351, 463
nitrite esters, photolysis, 451, 459
nitro compounds, reduction, 351
p-nitrobenzylchloride, 311
nitrogen atoms, hydridization states, 328–329
nitrogen compounds (*see* amines)
nitrogen mustards, 512–513
nitrosation of amines, 338–339
N-nitrosodiethylamine, 338
nitrous acid, 338
  reaction with amides, 455
  reaction with amines, 338–339
nmr (*see* nuclear-magnetic-resonance spectroscopy)
nomenclature, Cahn-Ingold-Prelog system, 193–196
  IUPAC system (*see* individual compound families)
nonclassical carbonium ions, 521–523
nonane, 35–37
nonbonded interactions, 63
norbornane, 65
norbornyl cation, pmr spectrum, 522
norbornyl tosylate, acetolysis, 522
norepinephrine, 327
novocaine, 447
nuclear-magnetic-resonance (nmr) spectroscopy, 15–18
  number of signals, 16–18
  signal frequency, 16
  signal intensity, 16
  signal splitting, 16, 211–214
  spectrometer, 15
  (*see also* proton-magnetic-resonance spectroscopy)
nuclear spin, 15
nucleic acids, hydrogen bonds, 45
  structure, 462
nucleophiles, 28
nucleophilic acyl substitution, acyl chlorides, 447–450
  amides, 454–456
  anhydrides, 451–452
  biological reactions, 457–459
  carboxylic acid derivatives, 446–456
  Claisen condensation, 466–468
  esters, 452–454
  Friedel-Crafts acylation, 448–449, 452
  phosgene, 449–450
nucleophilic addition, aldehydes and ketones, 369–382, 394–397

nucleophilic addition (*cont.*)
  aldol condensation, 394–397
  alkynes, 167–168, 173
  carbonyl compounds, 369–382, 446–456
  α,β-unsaturated carbonyl compounds, 503–506
  Wittig reactions, 382
nucleophilic addition reactions, 173
  (*see also* addition reactions)
nucleophilic aliphatic substitution, alcohols, 227–230
  alkyl halides, 79–91, 222–223, 291–294
  alkyl sulfonates, 225–227
  alkynes, 161
  ammonolysis of halides, 335–336, 350
  at carbon, 238
  ether derivatives, 229–231
  ether formation, 222–223
  ethers, 229–230
  α-halogenated acids, 466
  kinetics, 79–81
  neighboring-group participation, 507–515, 519–521
    anchimeric assistance, 508
    by aromatic rings, 519–521
    by carboxylate, 513
    by double bonds, 514–515
    by halogen, 508–510
    by nitrogen, 512–513
    by oxygen, 511
    stereochemistry, 507–509, 520
    by sulfur, 511
  reactions, alcohol derivatives, 225–230
    amines, 335–336, 350
    aryl diazonium salts, 344–346
  S_N1 mechanism, 84–88
  S_N2 mechanism, 81–84
nucleophilic reagents, order of reactivity, 83–84
  polarizability, 84
nucleophilicity, 83
  alkynes, 166–167
nucleoproteins, 594
nucleosides, tautomeric forms, 274
nucleotide, 462

octane, 35–37
1-octene, infrared spectrum, 111
oils, 421, 452
oleanolic acid, 409
olefins (*see* alkenes)
oleic acid, 409
  hydroxylation, 141
onium salts, 325, 336
opium, 326
optical activity, criterion for, 198, 199
optical isomers, 179–180
  properties, 180
  (*see also* enantiomers, stereoisomers)
orbitals, atomic, 115–121
  hybrid atomic, 117
  molecular, 115–121
  p, 116
  pi (π), 120
  s, 115–116
  sigma (σ), 116
  sp, 159

orbitals (*cont.*)
  sp^2, 119
  sp^3, 117
  tau (τ), 120
organic compounds, 1, 4
organic synthesis (*see* synthesis)
organoborane compounds, 133–134
organocadmium compounds, reaction with acyl halides, 464–465
organocuprous reagents, conjugate addition to enones, 506
organolithium compounds, formation from alkyl halides, 97–99
  reaction with aldehydes and ketones, 380–381
  reaction with carboxylate derivatives, 464
organomagnesium compounds, 99
  (*see also* Grignard reagents)
organomercury compounds, 99, 139
organometallic compounds, alchols from, 380–381, 464
  alkali metals, 167–168
  alkanes from, 98–99
  cadmium, 464
  carbon-metal bonds, 99, 138–139
  carboxylic acids from, 381, 412
  copper, 506
  cyclopropane derivatives from, 149
  ketones from, 464–465
  lithium, 97–99
  magnesium, 99
    (*see also* Grignard reagents)
  mercury, 99, 139
  preparation, from alkenes, 138–139
    from alkyl halides, 97–98
    from aryl halides, 296–297
    from dihalides, 97–98
    from methylene iodide, 149
  reactions, with alcohol, 98–99
    with acyl halides, 464–465
    with aldehydes, 380–381
    with alkenes, 149
    with alkynes, 14
    with ammonia, 98–99
    with carbon dioxide, 381, 412
    with esters, 0781m
    with ketones, 380–381, 506
    with nitriles, 464–465
    with oxirane, 381
    with water, 98–99
  transition metals with alkenes, 138–139
  zinc, 149
organozinc compounds, 149
orinase, 460
Orlon, 147
ornithine, 585
ortho esters, 457
ortho- and para-directing groups, 491–494
osazones, formation, 604
  structure, 604
osmium tetroxide, reaction with alkenes, 141
oxa, 324
oxalic acid, 410
oxane (tetrahydropyran), 207
oxaphosphetane, 382
oxetane, 207
  pmr spectrum, 214
oxidation, 239
  alcohols, 233–235, 362, 411

oxidation (cont.)
  aldehydes, 384, 411
  alkenes, 140
  alkyl side chains, 272–273
  alkynes, 169
  amines, 339
  arenes, 272–273
  catechol, 307
  hydroquinone, 307–308
  phenols, 318
  phosphines, 353
  saccharides, 605–606
  sulfides, 237
  thiols, 237
oxidation states, 140
  of carbon, 412–413
oxidative cleavage, 173
oxidative decarboxylation, 445
oximes, 375
  rearrangement, 523–524
oxirane (ethylene oxide), 207
  from 1,2 halohydrins, 224
  from oxidation of alkenes, 230, 475
  reaction with organometallic compounds, 381
  ring-opening reactions, 230–231, 381
oxocarbons, 309–310
oxolane (tetrahydrofuran), 207, 325
oxole (furan), 276–277, 325
oxonium salts, 220
oxyhemoglobin, 595
oxytocin, 584
ozonolysis, alkenes, 142–144,
  alkynes, 169
  mechanism, 144

papaverine, 326
para red, 349
paraffins (see alkanes)
paramagnetic resonance, 538
parasympathetic nervous system, 458
Pauling, Linus, α-helix, 590
  vitamin C requirements, 621–622
penicillins, 428, 455
1,3-pentadiene, resonance stabilization, 257, 279
pentadienes, heats of hydrogenation, 248–249
pentane, 35–37
  Buchi, Brinkman, Dreiding model, 38
2,4-pentanedione, 386–387, 389
2-pentene, 108
pentoses, 601
pepsin, 455
peptides, 584
  bonding, 578
  marking free amino groups, 336
  properties, 587
  synthesis, 586–587
peracids (see peroxyacids)
perbenzoic acid, 475
peresters, 475
perfumes, 423–424
pericyclic reactions, characteristics, 551
  empirical rule for predicting, 568–569
  energy requirement 554
  perplexing aspects, 555–558
  theoretical models, 560–567
    frontier-orbital, 561–567

pericyclic reactions (cont.)
  theoretical models (cont.)
    Woodward-Hoffman, 560–561
  types, 552
periodic acid, cleavage of vicinol glycols, 487
Perkin, William H. 625
permanent waving, 593
permanganate oxidation, aldehydes, 384
  alkenes, 140–141
  alkyl side chains, 272–273
  alkynes, 169
  arenes, 272–273
  frankincense, 246
  geraniol, 245
permanganate test, 141
peroxide-bond homolysis, 475, 539–540
peroxides, 237
  from alkenes, 144–145
peroxy acids and derivatives, 474–475
perphthalic acid, 474
Perutz, M., 595
phenacetin, 451
  pmr spectrum, 420
phenanthrene, 260
phenol coupling, 548
phenols, 299–318, 318
  acidity, 302–304
  acylation, 447, 451
  carbon vs oxygen alkylation of, 392–393
  dipole moment, 307
  distinguished from acids. 440
  electronic spectra, 304
  electrophilic substitution at the aromatic ring, 305–307
  esterification, 305
  ferric chloride test for, 301
  halogenation, 305–306
  hydrogen bonding, 301
  infrared spectra, 302
  keto-enol tautomerism, 300–301
  O—H bond polarity, 301
  O—H stretching frequency, 302
  occurrence, 299–300
  oxidation, 307–309
  pmr spectra, 302
  physical properties, 301–302
  polyhydric, 300
  preparation, from aryl halides, 295–296
    from aryl diazonium salts, 345
    from aryl ethers, 305
    from bromobenzene, 295
    from cumene hydroperoxide, 527
  reactions, with acid anhydrides, 305, 451
    with acyl halides, 305, 447
    with alkyl halides, 304
    with ferric chloride, 301
    with halogens, 306
    with nitric acid, 306
    with sulfonate esters, 304
    with sulfuric acid, 306
  reactivity of aryl C—O bond
  substitution at carbon, 305–307
  substitution at oxygen, 304–305
  Williamson ether synthesis, 304–305
phenoxide anions, 302–305
phenyl group, 260
phenylalanine, 579
phenylazide, 348

phenyldiazocyanide, 347
1-phenylethanol, 313–314
phenylhydrazine, 350
    reaction with aldehydes and ketones, 375
    reaction with sugars, 604
phenylisothiocyanate, 449
phenylpentazole, 348
phenylthiourea, 449
pheromones, 430–431
phloroglucinol, 300
phosgene, 449
    reactions with nucleophiles, 450
phosphate derivatives in living organisms, 461–463
phosphate esters, 415
phosphinate esters, 415
phosphines, 352–353
phosphinic acid, 415
    phosphinic acid derivatives, 415, 461
phosphite ester intermediates, 227–228
phosphite esters, 415
phosphocreatine, 462–463
phospholipids, 422
phosphonic acid, 415
phosphonium salt, 353
phosphoric amide, 415
phosphorus acid esters, 415, 461
phosphorus acids, 415
phosphorus oxychloride, reaction with phenols, 305
phosphorus pentachloride, reaction with carboxylic acids, 443
phosphorus pentoxide, dehydration of amides, 456
    reaction with carboxylic acids, 443
phosphorus tribromide, catalyst for halogenation of acids, 465–466
    reaction with alcohols, 228
phosphorus trichloride, reaction with alcohols, 227
    reaction with carboxylic acids, 443
phosphorus triiodide, reaction with alcohols, 228
phosphorylation, 461–462
pK_a, 437–438, 476
pmr (proton-magnetic-resonance) spectra, alcohols, 209–216
    aldehydes and ketones, 368
    alkanes, 46
    alkenes, 110–112
    alkyl halides, 76–77
    alkynes, 159–160
    amines, 332
    1-amino-1-phenylethane, 332
    aromatic hydrocarbons, 262–264
    2-bromobutyric acid, 419
    butanal, 368
    carboxylic acids and derivatives, 419
    cyclohexane, 64
    cyclopropanes, 150
    p-cymene, 263
    1,2-dibromo-2-methyl propane, 77
    diisopropyl ether, 213
    N,N-dimethylformamide, 420
    1,4 dioxane, 215
    ethanol, 17, 215
    ethers, 209–216
    p-ethoxyacetanilide, 420
    limonene, 111
    methanol, 211
    methyl p-chlorophenyl ketone, 368
    p-methylanisole, 263
    methylmethacrylate, 420

pmr (cont.)
    morpholine, 332
    phenols, 302
    oxetane, 214
    α- and β-pinene, 112
    2-propanethiol, 236
    sulfides, 236
    thiols, 236
    toluene, 262
    1,2,2-trichloropropane, 77
    2,4,4-trimethyl-2-(p-hydroxyphenyl)pentane, 302
    1,2,2-trimethylaziridine, 328
photochemical cycloaddition, 554, 557
photolytic bond homolysis, halogens, 541
    hypohalites, 541
    ketones, 541
    nitrite esters, 541
photosynthesis, 619
phthalic acid, 410
phthalimide, 451, 634
pi (π) bonds, (see bonds, molecular orbitals)
pi (π) electrons in double and triple bonds, 166–167
picric acid, 303
pinacol rearrangement, 517–518
α-pinene, reaction with diborane, 136
    reaction with hydrogen chloride, 131
α- and β-pinene, 107
    ozonolysis, 143
    pmr spectra, 112
piperidine, 324
piperine, 424
plane-polarized light, 177–178, 199
plasticizers, 305
β-pleated sheet in proteins, 589
Plexiglas, 476, 545
polar bonds, 23–24
polarimeter, 177–178, 199
polyenes, 109–110
    conjugated, 248–253
    cyclic conjugated, 274–276
    ultraviolet-visible spectra, 250–252
polyester, 453
polyethylene, 146–147
polyhydric phenols, 300
polyisobutylene, 146–147
polyketo acids, 472
polymerization, of alkenes, 145–148, 544–545
    of alkynes, 169–171
    and polymers, 146, 544–545
polysaccharides, 618
    end-group analysis, 621
polystyrene, 545
polyvinylchloride, 147, 545
porphin ring system, 429
potassium chloroplatinite, complexes with ethene, 138–139
previtamin D, 570–571
primary amines, N-acylation and sulfonation, 337
    Hinsberg test, 337
    nitrous acid test, 338
    (see also amines)
primary structure of proteins, 582–584
probability factor and reaction rates, 54, 57–59
procaine, 447, 459
progesterone, 8–9, 362
    infrared spectrum, 367
projection formulas, Newman, 39
    sawhorse, 39
proline, 579

propane, chlorination, 54, 56–58
2-propanethiol, pmr spectrum, 236
1-propanol, 207
2-propanol, 207, 210
propanone (see acetone)
propargyl group, 156
propargyl halides, reactions, 312–313
propene, 109
   addition reactions, 128
   C—H bond-dissociation energy, 316
   protonation, 130
   resonance forms, 258
propenoic acid (acrylic acid), 409
propiolic acid, 409
n-propyl alcohol, 207
n-propylamine, 331
propylene (see propene)
propyne, 157
propynoic acid (propiolic acid), 409
prostaglandin, 410, 422
   endoperoxide, 422
protective and blocking groups in synthesis, 634–636
protective groups, for amines, 344
   in peptide synthesis, 586
proteins, 577–597
   amino acid composition, 581–582
   amino acid sequence, 582–584
   classification and function, 581, 588
   conjugated, 594
   denaturation, 594
   in the diet, 577
   hydrolysis of amide bonds, 455
   hydrolytic cleavage, 578, 582
   marking free amino group, 0571m
   physiological function, 577–578
   primary structure, 582–584
   secondary and tertiary structure, 588–594
   terminal-group analysis, 582–584
proton-magnetic-resonance (pmr) spectroscopy, 16–18
   aromatic protons, 262–264
   chemical shifts, 18
      for hydrocarbons, 262–264
   delta (δ) scale, 18
   exchange of protons in alcohols, 209
   high field and low field, 16
   keto-enol composition, 387
   proton exchange, 215–216
   reference compound, 20
   shielding effects, 16–17
   signal multiplicity, 211–214
   spin-spin coupling, 214–216
   spin-spin interactions, 211–216
   (see also pmr spectra)
proton transfer, 26, 94
protons, deshielded, 16–17
   shielded, 16–17
purification methods, 2–3
purity, criteria of, 2–3
pyranoses, 611–612
pyrazine, aromaticity, 273
pyrethrin, 428
pyribenzamine, 328
pyridazine, aromaticity, 273
pyridine, 324
   aromaticity, 273–274
   reactivity relative to benzene, 274
pyridinium chloride, 325

pyridoxine, 326
pyrimidine, aromaticity, 273
pyrogallol, 300
pyrophosphates, 473
pyrrole, 324
   acidity, 277
   aromaticity, 276–277
   basicity, 277–278
pyrrolidine, 324
   acidity, 277
pyruvic acid, 410

quaternary ammonium bases, 325
quaternary ammonium salts, as leaving groups, 340–342
   preparation, 336
   resolution of, 329
queen bee substance, 435
quenching of radical reactions, 546
quinine, 326
   attempted synthesis, 625–626
   molecular formula, 5
   structure, 626
quinoline, 324
quinones, 307–309, 318
   equilibrium with hydroquinone, 308

R configuration, 193
racemic modifications, 180, 186–187, 199
   in resolution, 189–190
racemization, 187, 199
   conformational, 197–198
   enolizable ketones, 385–387
   norbornylmethanesulfonate in solvolysis, 227, 522
radical scavengers, 546
radicals, 28, 535
   detection, 537–538
   methods of generating, 539–542
   peresters and diacyl peroxides as initiators, 475, 539
   from peroxides, 237, 539
   quenching, 546
   reactions, abstraction, 542–543
      addition to alkenes, 144–145, 544–545
      chlorination of methane, 53
      coupling, 545–549
      fragmentation, 543
      polymerization, 544–545
      polymerization of methylmethacrylate, 475
      solvent-cage effects in, 547
   relative reactivity, 542
   stable radicals, 535–537
   stereochemistry of reactions, 550–551
radiomimetic compounds, 512
(+)-raffinose, 618
rate constant, 80–81
rate-determining step, 86
reaction mechanisms, 26–27
reaction rate, 80–81
reaction-rate theory, 54–59
reactive intermediates, 28–29
rearrangement, 25–26
   in alkyl halides, 90
   amino-pinacol, 519
   Baeyer-Villiger, 526

rearrangement (*cont.*)
  Beckmann, 523–524
  of carbonium ions, 861–871m
  of cyclooctatetraene, 275
  of cumene hydroperoxide, 527
  Curtius, 526
  of electron-deficient nitrogen, 523–526
  to electron-deficient oxygen, 526–527
  Hofmann, 526
  of neopentyl bromide, 90–91
  pinacol, 517–519
  Schmidt, 524–525
reducing sugars, 602
reduction, aldehydes and ketones, 379
  alkyl azides, 350
  alkynes, 167–168
  aromatic nitro compounds, 351
  aryl diazonium salts, 350
  carboxylic acids, 445
  carboxylic acid derivatives, 463
  enzymatic reduction, 471
  imines, 378
  nitriles, 463
  nitro compounds, 351
  (*see also* hydrogenation and specific compound families)
reductive alkylation reactions, 378
reductive deoxygenation, carbonyl compounds, 383
relative configurations, 191
reserpine, 425
resolution, amine oxides, 340
  of enantiomers, 189–190
  phosphines, 352
  quaternary ammonium salts, 328–329
resonance, 283
  allyl cation, 315
  allyl radicals, 316
  amines and their conjugate acids, 334–335
  benzene, 256–258
  benzyl cation, 314
  benzyl radicals, 316
  carbonyl group, 364
  carboxylate ion, 439
  cycloheptatrienyl cation, 279
  cyclopentadienyl anion, 279
  electrophilic substitution intermediates, 270, 298–299, 307, 493–494
  enolate anions, 388–389
  nitrate ion, 257
  nucleophilic substitution intermediates, 293–294
  oxocarbons, 310
  1,3-pentadiene, 257
  phenoxide anions, 303
  pyridine, 274
  rules, 256–257
  theory of, 256–258
  tropylium ion, 279
resonance energy, 257, 283
resonance hybrid, 256
resonance stabilization energy, 257
resorcinol, 300
retroaldol cleavage, 488
rhodizonic acid, 309
riboflavin, 326
ribonucleic acid (RNA), 461
D-(−)-ribose, 610
ricin, 577
rickets, 569

ring formation in synthesis, 638–639
ring strain, 61–62
rings, determination of number, 157–158
RNA (*see* ribonucleic acid)
Robinson annellation, 655
Rosanoff, M. A., 602
rotamers, 114
rotational barrier in conjugated dienes, 248
rotational energy, 10
rotational isomers, 196–198
rotational spectroscopy, 14
rubber, 113, 142, 146
Ruff degradation, of glucose, 606, 610
  of mannose, 610

S (entropy), 58
S configuration, 193
ΔS^* (entropy of activation), 58–59
$S_N 1$ reactions, alkyl halides, 84–88
  allyl derivatives, 315–316
  aryl diazonium salts, 344–346
  benzyl derivatives, 313–315
  mechanism and kinetics, 84–88
  rearrangement in, 90–91
  solvent effects, 85
  solvolysis of sulfonate esters, 226–227
  stereochemistry, 87–88
$S_N 2$ reactions, alkyl halides, 81–84
  alkylation of enolate anions, 392–393
  allyl halides, 312–313
  bynzyl halides, 310–312
  displacement of sulfonate esters, 225–227
  epoxide ring cleavage, 230–231
  formation of epoxides, 224
  hydrohalogenation of alcohols, 229
  mechanism and kinitics, 81–84
  nucleophilicity, 83–84
  propargyl halides, 312–313
  reactivity of leaving groups, 84
  stereochemistry, 82–83
  transition-state configuration, 82–83
$S_N 2$ *vs* E2 reactions, 94–95
saccharides, 601
saccharin, 461
salicylic acid, 409, 451
Sandmeyer reactions, 345–346
α-santonin, 481
saponification, 453
sarcosine, 409
sarin, 427, 459
saturated hydrocarbons (*see* alkanes, cycloalkanes)
sawhorse projection, 39
saxitoxin, 434
Saytzeff orientation rule, 92
  in alcohol dehydration, 232
  compared to Hofmann rule, 341–342
scavengers for radicals, 546
schizophrenia, 425
Schmidt reaction, 524–525
scotophobin, 585
second-order reactions, 80
secondary amines, Hinsberg test for, 337
  N-acylation and sulfonation, 337
  N-nitrosation, 338
  nitrous acid test for, 338
  (*see also* amines)

semicarbazide, 375
semicarbazones, 375
serine, 579
serotonin, 327, 425
shielded protons, 16
shielding regions, about a benzene ring, 262–263
   about a triple bond, 161
sickle-cell anemia, 596
sigma (σ) bonds, 116–118
sigmatropic reactions, 553, 566–567
silk, 146
silk fibroin, 589
silver-mirror test, 384
silver salts of carboxylic acids, 440
Simmons-Smith reagent, 149
soaps, 423, 453
sodium in reduction of alkynes, 167
sodium acetylides, alkylation, 161
   reaction with aldehydes and ketones, 381
sodium borohydride in reduction of aldehydes and ketones, 379
sodium ethoxide, reaction with benzyl halide, 310–311
sodium peroxide, reaction with anhydrides, 451
solvation, 85–86, 439
solvent-cage effects, 547
solvents for ultraviolet-visible spectroscopy, 250
solvolysis reactions, 313–314
   sulfonate esters, 226–227
sorbital, 624
sorbose, 624
spearmint, 361
specific rotation, 179, 199
spectroscopy, in structure identification, 6–8, 10–18
   time scale, 78
   (see also infrared, mass, nuclear-magnetic-resonance, proton-magnetic-resonance, ultraviolet-visible spectroscopy)
spectrum, electromagnetic, 12
spermaceti, 421
spin-spin interactions in pmr, 211–216
spin-spin splitting, 211–216, 238
squaric acid, 309
stability, chemical, 88, 125
   thermodynamic, 88, 125
stabilization energy, 257
starch, 619–621
sterculic acid, 107
stereochemistry, of Beckmann rearrangement, 524
   of bridged-ion reactions, 507–510, 520
   of carbonium-ion rearrangements, 507–510, 520, 522
   conformational analysis, 196, 198
   (see also conformational analysis)
   of Diels-Alder reaction, 500–502
   of electrocyclic reactions, 556–559
   electrophilic addition to alkenes, 135–138, 141
   of elimination reactions, 93–94
   of Hofmann reaction, 525
   hydrogenation of alkynes, 162, 167–168
   of neighboring-group effects, 507–510
   of pericyclic reactions, 556–559
   of radical reactions, 550–551
   of S_N1 reactions, 87–88
   of S_N2 reactions, 82–83
stereoisomers, absolute configurations, 192
   alkenes, 112–114
   alkylidenecycloalkanes, 185

stereoisomers (cont.)
   allenes, 185
   amines, 328–329
   biphenyls, 265–266
   chirality, 183–187
   cis-trans stereoisomers, 60–61, 112–114, 123–124
   configurational notation, 192–193
   cycloalkanes, 60–61
   cycloalkenes, 124
   D, L notation, 602–603
   determination of number, 187–188
   diastereoisomers, 187–189
   diazoic acid salts, 347
   enantiomers, 183–187, 199
   epimers, 387
   Fischer projection formulas, 192
   geometric, 60–61, 123–124
   glucose, 611
   hexahelicene, 267
   interconversion, 114, 197–198
   meso compounds, 188
   monosaccharides, 611
   optical isomers, 179–180
   oximes, 376, 524
   phosphines, 352
   polymers, 192–193
   R, S notation, 193–195, 611
   racemic modification, 180, 186–187
   racemization, 187
   relative configurations, 191
   resolution, 189–190, 199
   sequence rules, 193–196
   spiroalkanes, 185
stereoregular polymers, 193
steric hindrance, 80
   effect on rate of esterification, 442
   effect on rate of ester hydrolysis, 454
   effect on rate of S_N2 reactions, 83
steroid hormones, 362
stipitatic acid, 309
Stone, Irwin, 622
structural formula, 8
   condensed, 36
   determination of, 8–18
structure-composition relationship, 157–158, 172, 329–330
strychnine, 426
sublimation, 3
substitution nucleophilic bimolecular (see S_N2 reactions)
substitution nucleophilic unimolecular (see S_N1 reactions)
substitution reactions, 25–26
   (see also electrophilic substitution, nucleophilic substitution)
substrates, 457
succinic acid, 410
   oxidative bis-decarboxylation of, 445
succinimide, 414
(+)-sucrose, 617
sugars (see saccharides)
sulfanilamide, 460
sulfate ester, 415
sulfathiazole, 460
sulfides, 235–237, 239
sulfinate esters, 415
sulfinic acid, 415
sulfisoxazole, 460

sulfite ester intermediates, 227–228
sulfonamide, 337
sulfonate esters, 459
    nucleophilic reactions, 225–227
sulfonation, amines, 337
    aromatic rings, 460
    aryl halides, 297
    benzene, 269
sulfonyl halides, formation, 221–222
    reaction with alcohols, 221–222
    reaction with amines, 337
    reaction with arenes, 460
    reaction with silver salts, 460
    sulfonate esters from, 221–222
sulfonylation, 459–461
sulfur acids, 415
sulfuric acid, reaction with phenols, 306
suprafacial reactions, 559
symmetry in synthesis design, 640
symmetry-allowed reactions, 562–563
symmetry elements, 181–183, 199
symmetry-forbidden reactions, 562–563
symmetry operation, 199
sympathetic nervous system, 458–459
syn conformation, 376
synapse, 458–459
synaptic cleft, 458–459
synaptic transmission, 458–459
syndiotactic configuration, 193
synthetases, 470
synthesis, computer-assisted, 657
    examples of
        2-aminomethylspiro[3.3]heptane, 651
        copaene, 653
        3,3′-diaminodiphenylmethane, 646
        2,6-dimethyl-4-heptanone, 645
        2,7-dimethyl-4-octanone, 646
        1,1-dimethylcyclopentane, 649
        2-ethyl-2-methyl-1,3-butanediol, 647
        9-oxabicyclo[5.3.0^{3,5}]heptane, 651
        1,4,6-trimethylnaphthalene, 648
    planning of, 627–631, 640–645
    prospects in, 657–658
    strategy of, 625–667
    yields in, 645
synthons, 642

2,4,5-T, 427
tabasco, 424
tartaric acid, 192, 196–197, 409
tau (τ) bonds, 120–212
tautomerism, 164–165
    keto-enol of cyclohexanone, 300–301
    keto-enol of phenol, 300–301
tautomers, 165, 172
TCDD, 427–428
Teflon, 147
terephthalic acid, 410
terminal acetylenes, 160
termination of chain reactions, 53
terpenes, 472–474
α-terpiniol, 180, 184, 205
terramycin, 300
tertiary amines, N-acylation and sulfonation, 337
    Hinsberg test, 337
    nitrous acid test, 338

tertiary amines (cont.)
    (see also amines)
testosterone, 362
tetraalkylammonium hydroxides, 325
tetrachloromethane (carbon tetrachloride), 23–24
tetracycline, 300
tetrahedral bonding geometry, 22
tetrahydrocannabinol, 8–9
tetrahydrofuran, 207, 325
tetrahydropyran, 207
tetrahydrozoline, 434
3,3,6,6-tetramethyl-1,4-cyclohexadiene, 109
tetramethylsilane, use in pmr spectroscopy, 18
tetrazole, 325
tetrodotoxin, 426
tetroses, 601
thalidomide, 434
theobromine, 425
thermal decarboxylation, 444
thermodynamic stability, 88, 125
THF, 207, 325
thia, 324
thiamine, 326
1,3-thiazole, 325
thioacetals, 373
thioketals, 373
thiols, 235–237, 239
    addition to alkynes, 167
    infrared spectra, 236
    pmr spectra, 236
thionyl chloride, dehydration of amides, 456
    reaction with alcohols, 227–228
    reaction with carboxylic acids, 443
thiophene, 325
    aromaticity, 276–277
Thorpe condensation, 639
threo diastereoisomers, 508
threonine, 579
D-(−)-threose, 606–607
β-thujaplicin, 309
thymine, tautomeric forms, 274
thymol, 299
TMS, 18
tobacco, 326
Tollens reagent, oxidation of aldehydes, 384
    reaction with α-ketols, 487
    reaction with saccharides, 605
toluene, C—H bond-dissociation energy, 316
    pmr spectrum, 262
p-toluidine, 324
tonka bean, 424
toxicity, 426
tranquilizer, 425
transesterification, 453
transferase, 458
transition metals, complexes with alkenes, 139
    complexes with alkynes, 162, 171
transition state, 55–57
trehalose, 624
trialkylphosphine, 352
trialkylphosphites, 227
1,3,5-tribromobenzene, 346
trichloromethane, 52
2,4,5-trichlorophenol, 427
1,2,2-trichloropropane, pmr spectrum, 77
tricresylphosphate, 305
trigonal bonding geometry, 23
1,2,3-trihydroxybenzene, 300

1,3,5-trihydroxybenzene, 300
3,4,5-trihydroxybenzoic acid, 409
3,4,5-triiodonitrobenzene, 347
trimerization of alkynes, 169–170, 173
2,4,4-trimethyl-2-(p-hydroxyphenyl)pentane, pmr spectrum, 302
1,2,2-trimethylaziridine, 328
   pmr spectrum, 329
trimethylbenzylammonium hydroxide, 325
trimethylcyclopropenium cation, 280
trimethylene oxide (oxetane), 207, 214
2,4,6-trinitrophenol (picric acid), 303
triphenylmethane, 260
triphenylmethyl dimer, 536
bis-triphenylmethyl peroxide, 536
triphenylmethyl radical, esr spectrum, 538
   formation, 536
triphenylphosphine oxide, 353
triphenylphosphonium methylide, 353
triple bond (see alkynes, carbon-carbon triple bonds)
triton-B, 325
tropolones, 309
tropylium ion, 278–280
trypsin, 455
tryptophan, 580
tuberculostearic acid, 422
tubocurarine chloride, 326
twistane, 65
tyrocidines, 585
tyrosine, 579

ultraviolet-visible absorption, aldehydes, 366
   alkenes, 351
   alkynes, 251
   aromatic hydrocarbons, 261
   benzaldehyde, 366
   benzene, 254
   conjugation effect, 366
   cyclohexenone, 366
   2,2'-dimethylbiphenyl, 266
   3,3'-dimethylbiphenyl, 266
   ketones
   phenols, 304
   polyenes, 250–252
ultraviolet-visible spectra
   biphenyls, 265–266
   conjugated dienes, 250–252
2,5-dimethyl-2,4-hexadiene, 251
   frequency range, 12, 14
   solvents for, 250
unimolecular reaction, 85–87
unsaturated hydrocarbons, 121
   (see also alkenes, cycloalkenes, alkynes, arenes)
unsaturation, test for, 132, 141
urethane, 450
urushiols, 299–300

valine, 579
Van der Waals–London forces, 45
vanilla bean, 246, 361

vanillin, 299–300, 321
vasopressin, 584–585
vibrational energy, 10
vicinal dihalides dehalogenation, 98
   preparation, 132

vinyl acetate, from acetylene, 163
vinyl chloride, 109
   polymerization, 147
vinyl group, 109
vinyl halides, reactivity vs alkyl halides, 290–292
vinylacetylene, from dimerization of acetylene, 169
visible spectra (see ultraviolet-visible spectra)
vision, cis-trans isomerism in, 114
vital force, 1
vitamin A, 107, 110, 114, 252
vitamin B_1 (see thiamine)
vitamin B_2 (riboflavin), 326
vitamin B_6 (pyrridoxine), 326
vitamin B_{12}, 11, 429
   x-ray diffraction, 11
vitamin C (ascorbic acid), 621
vitamin D, pericyclic reactions, 570–571
   structure, 570
   in treatment of rickets, 569–570
vitamin K, 308

walnuts, 308
wavelength, 12
wavenumbers, 12
water, dielectric constant, 85
   dipole moment, 23–24
   hydrogen bonding, 44
waxes, 421
wedge bonds, 193
Williamson ether synthesis, 222–223, 304–305
Wittig reaction, 382
Woodward, R. B., 560, 626
wool, 146
Würster salt, 537

x-ray diffraction analysis, 8–9
   acetylenes, 158
   benzene, 253–254
   vitamin B_{12}, 11
xylene, 260

ylides, 353, 382

Ziegler catalyst, 147
Ziegler, Karl, 147
zwitterions, 410, 581

PERIODIC CHART OF THE ELEMENTS

IA	IIA	IIIB	IVB	VB	VIB	VIIB	VIIIB			IB	IIB	IIIA	IVA	VA	VIA	VIIA	VIIIA
1 H 1.00797																	2 He 4.0026
3 Li 6.939	4 Be 9.0122											5 B 10.811	6 C 12.01115	7 N 14.0067	8 O 15.9994	9 F 18.9984	10 Ne 20.183
11 Na 22.9898	12 Mg 24.312											13 Al 26.9815	14 Si 28.086	15 P 30.9738	16 S 32.064	17 Cl 35.453	18 Ar 39.948
19 K 39.102	20 Ca 40.08	21 Sc 44.956	22 Ti 47.90	23 V 50.942	24 Cr 51.942	25 Mn 54.9380	26 Fe 55.847	27 Co 58.9332	28 Ni 58.71	29 Cu 63.54	30 Zn 65.37	31 Ga 69.72	32 Ge 72.59	33 As 74.9216	34 Se 78.96	35 Br 79.909	36 Kr 83.80
37 Rb 85.47	38 Sr 87.62	39 Y 88.905	40 Zr 91.22	41 Nb 92.906	42 Mo 95.94	43 Tc (99)	44 Ru 101.07	45 Rh 102.905	46 Pd 106.4	47 Ag 107.870	48 Cd 112.40	49 In 114.82	50 Sn 118.69	51 Sb 121.75	52 Te 127.60	53 I 126.9044	54 Xe 131.30
55 Cs 132.905	56 Ba 137.34	57 La† 138.91	72 Hf 178.49	73 Ta 180.948	74 W 183.85	75 Re 186.2	76 Os 190.2	77 Ir 192.2	78 Pt 195.09	79 Au 196.987	80 Hg 200.59	81 Tl 204.37	82 Pb 207.19	83 Bi 208.980	84 Po (210)	85 At (210)	86 RN (222)
87 Fr (223)	88 Ra 226.05	89 Ac‡ (227)	104 (257)	105													

†*lanthanum series*

58 Ce 140.12	59 Pr 140.907	60 Nd 144.24	61 Pm (147)	62 Sm 150.35	63 Eu 151.96	64 Gd 157.25	65 Tb 158.924	66 Dy 162.50	67 Ho 164.930	68 Er 167.26	69 Tm 168.934	70 Yb 173.04	71 Lu 174.97

‡*actinium series*

90 Th 232.038	91 Pa (231)	92 U 238.03	93 Np (237)	94 Pu (242)	95 Am (243)	96 Cm (247)	97 Bk (249)	98 Cf (249)	99 Es (254)	100 Fm (253)	101 Md (256)	102 No (253)	103 Lr (257)

Atomic weights are based on $C^{12} = 12.0000$ and conform to the 1961 values.

PROTON CHEMICAL-SHIFT VALUES (CDCl$_3$ SOLUTION)

type of proton	chemical shift†	type of proton	chemical shift†
RCH$_3$	0.8–1.0	RCH$_2$CO$_2$R'	2.2–2.4
R$_2$CH$_2$	1.2–1.4	RCH$_2$COR'	2.2–2.5
R$_3$CH	1.4–1.7	RCH$_2$NR'$_2$	2.0–3.0
 R$_2$C(CH$_2$)(cyclopropyl)	0.2–0.8	RCH$_2$X (X = Cl, Br, I)	3.0–4.0
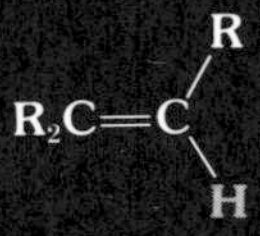 R$_2$C=CRH	4.5–6.5	RCH$_2$OR'	3.2–3.8
RC≡CH	2.0–2.8	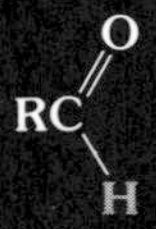 RCHO	9.0–10.0
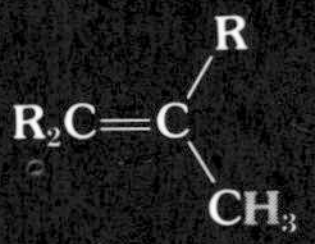 R$_2$C=C(R)(CH$_3$)	1.6–1.8	ROH	2.0–5.0
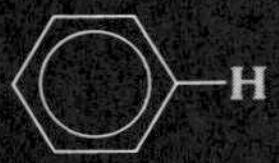		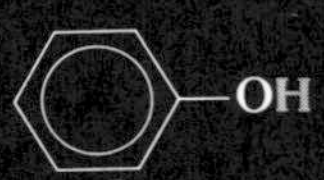 Ph–OH	4.0–7.0
Ph–H	7.3	RCO$_2$H	10.0–13.0
		RSH	1.0–3.0
Ph–CH$_2$R	2.2–2.4	R$_2$NH	1.0–4.0
		R$_3$NH$^{\oplus}$	6.0–9.0

†In parts per million from tetramethylsilane (δ values).

OTHER SOURCES OF SPECTROSCOPIC DATA

		page
TABLE 9·5	Electronic spectra of dienes	252
FIGURE 9·4	Electronic spectra of arenes	261
TABLE 10·2	Electronic spectra of phenol derivatives	304
TABLE 12·2	Electronic spectra of carbonyl compounds	366
TABLE 12·3	Infrared spectra of aldehydes and ketones	368
TABLE 13·5	Infrared spectra of carboxyl derivatives	417